BAYESIAN STATISTICS 2

BAYESIAN STATISTICS 2

Proceedings of the Second Valencia International Meeting

September 6/10, 1983

Edited by

J.M. BERNARDO
M.H. DEGROOT
D.V. LINDLEY
A.F.M. SMITH

1985

NORTH-HOLLAND – AMSTERDAM . NEW YORK . OXFORD

VALENCIA UNIVERSITY PRESS

ISBN: 0 444 87746 0

Jointly published by:

ELSEVIER SCIENCE PUBLISHERS B.V.
P.O. Box 1991
1000 BZ Amsterdam
The Netherlands

and

VALENCIA UNIVERSITY PRESS
Nave 2
46003 Valencia
Spain

Sole distributors for the U.S.A. and Canada:

ELSEVIER SCIENCE PUBLISHING COMPANY, INC.
52 Vanderbilt Avenue
New York, N.Y. 10017
U.S.A.

Typeset at the University of Valencia

Printed in The Netherlands

A MESSAGE FROM THE GOVERNOR OF THE STATE OF VALENCIA

Presidència
de la
Generalitat Valenciana

La Generalitat Valenciana te l'honor de patrocinar la serie de congressos internacionals sobre Métodes Bayesians que cada quatre anys tindran lloc a València.
Vull expressar el meu desig que aquest volum, que recull els resultats del Segon Congres, servesca per a potenciar l'interes per coneixer la Comunitat Valenciana, d'aquells que treballen amb metodes quantitatius moderns.

Joan Lerma

The Government of the State of Valencia is honoured to sponsor the series of International Meetings on Bayesian Statistics to be held in Valencia every four years.

I want to express my hope that these Proceedings of the Second Meeting will encourage all those who work in modern quantitative methods to pay regular visits to the State of Valencia.

Joan Lerma
Governor of the State of Valencia

PREFACE

The Second Valencia International Meeting on Bayesian Statistics took place from 6-10 September, 1983, at the Hotel Las Fuentes, Alcoceber, 100 kms north of Valencia, Spain, where the first such meeting had been held four years previously. As on that first occasion, the conference had a markedly international character, attracting over 130 participants from 24 different countries.

The Scientific Programme consisted of 28 invited papers by leading experts in the field, each of which was followed by a discussion, and 3 poster sessions of contributed papers. These Proceedings contain all but two of the invited papers, with associated discussion, together with a selection of 18 contributed papers, which were subject to a refereeing process. The concentration of expertise and the discussion format for invited papers combine to provide a lively, comprehensive overview of current research activity in Bayesian Statistics, encompassing a wide variety of topics ranging from foundational philosophy to practical case-studies.

Foundational issues are addressed by **Berger** and **Csiszar**: the former reexamines the likelihood principle in terms of its axiomatic justifications and relation to coherence; the latter considers a Bayesian justification for a version of the maximum entropy principle. Coherence is further considered in a paper by **Goldstein**, who examines the subtle conceptual problems which arise when temporal aspects of the consistency of probability judgements are taken into account, and also in a short contribution from **Seidenfeld** and **Schervish**.

Strategies for representing prior opinions in mathematical form are investigated by **Dickey** and **Ch.H. Chen**, who consider families of elliptical distributions for modelling subjective probabilities, and **Diaconis** and **Ylvisaker**, who consider approximating prior distributions by mixtures within conjugates families. Problems of ensuring the consistency of prior specifications when comparing alternative models are investigated by **Poirier** and **Sweeting**. Philosophical and technical issues relating to judgement, probability, evidence and measures of weights of evidence are discussed in papers by **Dempster** and **Good.**

The field of individual and group probability assessment receives considerable theoretical and practical attention in papers by **French, Lindley, Pratt** and **Schlaifer,** and **Press. French** provides a taxonomy and critical survey of previous work on group consensus problems; **Lindley** presents a technical solution to a version of the group problem; **Pratt** and **Schlaifer** and **Press** present case-studies on actual individual and group assessment behaviour. In addition, **DeGroot** and **Eriksson** contribute to the problem of assessing probability assessors.

Problems of modelling, with particular reference to robustness and diagnostics, are considered by **Pettit** and **Smith**, who concentrate on measures of outlyingness and influence in linear models, and **West**, who develops a Bayesian approach to generalized linear models. **Geisser** provides an up-to-date review of the predictive approach to inference and decision, which includes related ideas on diagnostics and model checking, and **Amaral** and **Dunsmore** consider measures of the information in a predictive distribution. **Berliner** considers another aspect of the robustness problem and **Spiegelhalter** provides an analytic treatment of the Cauchy sampling model. The mathematical and conceptual bases of certain models are examined in detail by **Consonni** and **Dawid** (in relation to structural invariance) and **Rubin** and **Sugden** (in relation to randomized experimentation and survey sampling). **Aitchison** emphasizes the need for new classes of models to deal with practical problems involving simplex sample spaces and **Albert** presents an approach to modelling association in contingency tables. **O'Hagan** and **Polasek** consider particular problems that arise with hierarchical models, and **Poli** presents a nonparametric approach to estimating a bivariate distribution function.

Specific decision problems are examined by **Bernardo** and **Bermudez** and by **Kokolakis** (who consider the choice of variables in classification problems), by **Ferrandiz** (who considers decisions regarding sharp null hypotheses) and by **Giovagnoli** and **Verdinelli** (who consider experimental design problems). **Winkler** discusses the effects of information loss in noisy and dependent processes on ensuing decisions.

Applications and the relevance of Bayesian ideas to other subject areas are explored by several authors: **W.C. Chen** *et al* present a Bayesian analysis of cancer survival curves; **Harrison** and **Ameen, Migon** and **Harrison** consider new methodology and case studies in the field of time series and **Jaynes** provides a time series case-study to illustrate the effects of highly informative priors in multiparameter situations: **Zellner** reviews Bayesian ideas in econometrics; **Kadane** and **Wasilkowski** explore links between Bayesian statistics and ϵ complexity theory in computer science; **Jewell** constructs Bayesian models for estimating the size of a population, and **Armero** considers Bayesian inference for a queuing problem. In addition, there is a short abstract describing **Mockus'** Bayesian approach to global optimization.

Finally, the important problems of implementing the Bayesian paradigm are addressed by **Stewart** and **Van Dijk** and **Kloek**, who explore different aspects of the application of Monte Carlo integration methods to Bayesian Statistics.

In our preface to the Proceedings of the First Valencia International Meeting, we noted the considerations that had led us to the

> "... concrete idea of a conference held in Europe, attempting to draw together, for the first time at a truly international level, a fairly exhaustive assembly of statisticians concerned - in something other than a purely hostile sense!- with Bayesian Statistics";

we added that

> "Perhaps we also had in mind that this might accelerate the process of fertilization and growth of Bayesian ideas...".

It is too soon for a proper evaluation of our success or failure in this latter enterprise. In the meantime, however, we note with pleasure the greatly increased number of participants in this Second Meeting, and the unique atmosphere of intellectual commitment combined with warm companionship which prevailed at Hotel Las Fuentes, and which is well captured in the chorus of the song written by Freeman and O'Hagan for the final conference dinner:

Mine eyes have seen the glory of the Reverend Thomas Bayes.
He is stamping out frequentists and their incoherent ways.
He has raised his mighty army at the Hotel Las Fuentes.

His troops are marching on.
Glory, glory, probability!
Glory, glory, subjectivity!
Glory, glory to infinity!
His troops are marching on.

(sung to the tune of *The Battle Hymn of the Republic*).

We are grateful for the financial and material support provided by: the Government of Spain; the Government of the State of Valencia; Valencia City Council; the University of Valencia; the US Office of Naval Research, and the British Council. In particular, we are delighted by the promise of future support contained in the personal message from the Governor of the State of Valencia (reprinted in these Proceedings immediately before this Preface, with a translation from the original Catalan).

In the process of preparing these Proceedings, and in many other aspects of the conference organization, we have received invaluable assistance from many members of the Departamento de Bioestadística and of the printing house of the University of Valencia. We especially wish to thank: **C. Armero, M.J. Bayarri, J. Bermúdez, Ch. Escandón, J. Ferrándiz, F. Mancebo, M. Rábena, L. Sáiz, Ll. Sanjuan** and **M. Sendra.**

J.M. Bernardo
M.H. DeGroot
D.V. Lindley
A.F.M. Smith

OCTOBER 1984

CONTENTS

PAGS.

I. INVITED PAPERS (with discussion)

II. CONTRIBUTED PAPERS

INVITED PAPERS

BAYESIAN STATISTICS 2, pp. 15-32
J.M. Bernardo, M.H. DeGroot, D.V. Lindley, A.F.M. Smith (Eds.)

Practical Bayesian Problems in Simplex Sample Spaces

J. AITCHISON
University of Hong Kong

SUMMARY

The simplex plays a major role in two important areas of statistical activity, as a parameter space in categorical data analysis and as a sample space in the analysis of compositional data. Thus in the Bayesian analysis of the first and in any form of analysis of the second a major problem is the specification of classes of distributions on the simplex. This paper reexamines the almost independent development of solutions in these two areas and suggests that each has now probably much to learn from the other.

Keywords: CATEGORICAL DATA; COMPLETE SUBCOMPOSITIONAL INDEPENDENCE; COMPOSITIONAL DATA; DIRICHLET DISTRIBUTIONS; LOGISTIC NORMAL DISTRIBUTIONS; NEUTRALITY; SIMPLEX.

1. INTRODUCTION

Most Bayesians, when subjected by their psychiatrists to the word stimulus *simplex*, would give the immediate response *parameter space* for categorical probabilities or some similar concept. Few would respond with *sample space* for compositional data. Yet there is a vast territory of statistical activity involving the simplex in ways other than as a parameter space and as yet relatively untouched by Bayesian hand or mind. For example, ten minutes spent turning the pages of geological journals will soon convince the reader that the variability of compositions of rocks, such as their major oxide percentages, or of sediments, such as their sand, silt, clay proportions, play a central role in geological discussion. As further evidence of the simplex as sample space are the ternary variation diagrams or two-dimensional simplices to be found every ten pages or so.

The purpose of this paper is to present a selected but challenging set of such practical problems, to advocate my own resolutions, to identify some open problems, and to invite comments and criticisms and, hopefully, alternative and better solutions. This Valencia meeting presents an excellent forum for the cross-fertilization of ideas on the simplex from both compositional data analysts and Bayesian categorical data analysts, hopefully to the advantage of both.

As a starting point for our discussion we present two illustrative compositional data sets in Tables 1 and 2 each consisting of five-part compositions of 25 samples of two types of rock, *bayesite* and *chayesite*. I apologize for the use of obviously contrived data but I have had to choose between referring the reader to real data sets too large to reproduce here and allowing him to see, handle and analyse smaller, presentable and typical data. The data sets presented are faithfully scaled-down versions of, and present as strong a challenge as, the real thing. The reader who is dissatisfied with this expository device can look for satisfaction with real data in some of the references cited.

Some questions can be immediately posed. Can we find a satisfactory way to describe the pattern of variability of bayesite? In particular, do we have a parametric class of distributions rich enough to model a wide range of compositional variability? If we are Bayesianly inclined what sensible prior distributions, preferably tractable, can we use to support the analysis? Do bayesite and chayesite differ significantly in composition? If so, can we devise a satisfactory means of classification on the basis of composition? If we are

TABLE 1. Compositions of 25 specimens of bayesite

Specimen no	Percentages A	B	C	D	E
B1	24.2	34.0	19.3	9.3	13.2
B2	26.1	49.0	6.6	6.9	11.4
B3	25.8	27.5	24.0	10.7	12.0
B4	27.3	49.6	5.8	7.8	9.5
B5	22.5	26.5	28.3	8.4	14.3
B6	24.4	28.4	24.7	10.9	11.6
B7	25.8	30.3	21.3	8.9	13.7
B8	27.6	32.9	16.8	8.6	14.1
B9	26.0	46.9	8.0	7.0	12.1
B10	23.5	52.1	6.6	7.0	10.8
B11	27.0	32.5	18.0	8.5	14.0
B12	24.6	40.1	14.5	9.4	11.4
B13	24.3	40.1	13.1	6.7	15.8
B14	22.4	22.4	32.9	9.4	12.9
B15	22.3	60.0	3.0	6.8	6.9
B16	26.5	37.1	14.7	8.8	12.9
B17	28.6	41.5	9.9	8.5	11.5
B18	29.5	41.0	9.7	8.5	11.3
B19	23.7	53.9	5.6	7.2	9.6
B20	24.5	24.1	19.2	11.5	10.7
B21	21.5	32.7	23.5	10.3	12.0
B22	23.4	30.9	23.4	10.3	12.0
B23	26.8	34.3	17.1	11.5	10.3
B24	27.0	36.2	14.7	8.2	13.9
B25	26.3	50.4	6.1	8.7	8.5

Bayesianly inclined what sensible prior distributions, preferably tractable, can we use to support the analysis? Do bayesite and chayesite differ significantly in composition? If so, can we devise a satisfactory means of classification on the basis of composition? If we are successful would it be possible to construct as effective a classification system on the basis of a subcomposition of the original composition? And so on?

TABLE 2 Compositions of 25 specimens of chayesite

Specimen no	Percentages A	B	C	D	E
C1	29.2	50.3	4.6	7.1	8.8
C2	25.9	39.9	13.4	7.9	12.9
C3	27.1	43.8	9.7	8.0	11.4
C4	25.7	49.9	6.1	5.5	12.8
C5	26.2	44.3	9.3	6.5	13.7
C6	29.7	38.7	11.7	9.1	10.8
C7	28.1	42.9	9.7	9.0	10.3
C8	21.8	33.2	23.2	8.7	13.1
C9	24.9	29.8	23.5	9.8	12.0
C10	25.0	24.2	28.7	9.1	13.0
C11	24.0	42.5	13.3	8.5	11.7
C12	24.9	45.9	10.1	8.4	10.7
C13	26.7	38.0	14.7	10.3	10.3
C14	25.6	41.2	13.0	9.9	10.3
C15	25.0	49.6	7.8	8.4	9.2
C16	24.0	36.5	18.5	10.3	10.7
C17	25.8	33.4	19.6	9.6	11.7
C18	28.5	41.3	10.2	8.6	11.4
C19	23.6	47.1	10.1	8.3	10.9
C20	27.1	40.2	12.5	8.8	11.4
C21	24.4	37.9	16.8	10.4	10.5
C22	21.9	25.8	30.4	10.3	11.6
C23	25.4	50.3	6.8	7.9	9.6
C24	21.9	33.5	21.8	6.9	15.9
C25	29.0	30.5	19.0	11.2	10.3

2. PARAMETRIC CLASSES OF DISTRIBUTIONS ON THE SIMPLEX

The natural sample space for recording a composition $x = (x_1,\ldots x_{d+1})$, of $d+1$ parts and satisfying the unit-sum constraint $x_1 + \ldots + x_{d+1} = 1$, is the d-dimensional simplex S^d in R^d:

$$S^d = \{(x_1,\ldots,x_d) : x_i > 0\,(i = 1,\ldots,d), x_1 + \ldots + x_d < 1\}, \tag{2.1}$$

sometimes more conveniently expressed symmetrically as a d-dimensional subspace of R^{d+1}:

$$S^d = \{(x_1,\ldots x_{d+1}) : x_i > 0\,(i = 1,\ldots,d+1), x_1 + \ldots + x_{d+1} = 1\}. \tag{2.2}$$

For simplicity of exposition we have defined S^d to be the strictly positive simplex so that we shall confine our attention to compositions without zero components and shall not here consider modifications required to the analysis to accommodate zeros.

Similarly the natural parameter space for a multinomial experiment with $d+1$ categories is the simplex:

$$S^d = \{(\theta_1,\ldots,\theta_{d+1}) : \theta_i > 0\,(i=1,\ldots,d+1), \theta_1 + \ldots + \theta_{d+1} = 1\}, \tag{2.3}$$

where $(\theta_1,\ldots,\theta_{d+1})$ denote the category probabilities.

In years to come the statistical historian will no doubt be in a much better position to trace the significant stages in the development of distributions over the simplex S^d. For our purposes here we may start by observing that up to about 1960 the only class considered appears to have been the Dirichlet class $D^d(\beta)$ with density function proportional to

$$x_1^{\beta_1 - 1} \quad \ldots \quad x_{d+1}^{\beta_{d+1}-1} \quad . \tag{2.4}$$

The early 1960s then saw the beginning of a new interest in the simplex from both Bayesian categorical data analysts and compositional data analysts with each developing apparently independently of the other.

In categorical data analysis the sequence of ideas has been roughly as follows:

(1) Meaningful hypotheses about categorical probabilities are associated with zero loglinear contrasts of $(\theta_1,\ldots,\theta_{d+1})$.

(2) With conjugate Dirichlet priors the posterior distributions of loglinear contrasts of $(\theta_1,\ldots,\theta_{d+1})$ are approximately normal (Lindley, 1964; Swe, 1964; Bloch and Watson, 1967).

(3) Because of its strong independence structure the Dirichlet class is not rich enough to provide an adequate range of prior distributions.

(4) A rich class of prior distributions can be specified directly in terms of normal loglinear contrasts (Leonard, 1973).

Thus Bayesians have been led to consideration of a reparametrisation ψ of θ through the multivariate logit or generalized logistic transformation:

$$\theta_i = \exp(\psi_i) / \sum_{j=1}^{d+1} \exp(\psi_j) \qquad (i = 1,\ldots,d+1) \tag{2.5}$$

and to specification of the prior by way of a $N^{d+1}(\mu,\Sigma)$ distribution on ψ. There is, of course, a small price to be paid for the loss of conjugacy but the transformation provides a flexible class of priors within which forms of exchangeability, quasi-independence and smoothness can be readily introduced by further modeling of μ and Σ, for example as in

Leonard (1973, 1975, 1977).

In compositional data analysis the sequence of developments has been rather different, along two main routes, (a) covariance analysis, (b) distribution theory.

(a1) The 'constant-sum' or 'closure' property, namely $x_1 + \ldots + x_{d+1} = 1$, leads to great difficulties in interpreting the covariance structure $\mathrm{cov}(x_i, x_j)$ $(i,j = 1,\ldots,d+1)$ of a composition (Chayes, 1960, 1962, 1971, 1983; Sarmanov and Vistelius, 1959).

(a2) Despite many attempts to overcome the difficulties by, for example, Mosimann (1962), Chayes and Kruskal (1966), Darroch (1969), Chayes (1971), Darroch and Ratcliff (1970, 1978) the problem remains essentially unresolved (Miesch, 1969; Kork, 1977) and there is among geologists a tendency to settle for detailed description of the difficulties rather than reexamine the basic structure of the problem.

(a3) A realization that a concept of covariance structure designed for R^d is not necessarily suited to S^d and that analysis of compositions is essentially concerned with the *relative* magnitudes of the parts (for example, the ratio x_i/x_j is preserved under the important operation of forming a subcomposition) leads to the introduction (Aitchison, 1981a) of a new covariance structure

$$\mathrm{cov}\left(\log \frac{x_i}{x_{d+1}}, \log \frac{x_j}{x_{d+1}}\right) \qquad (i,j = 1,\ldots,d), \tag{2.6}$$

free from the constant-sum difficulties of interpretation, or equivalently (Aitchison, 1983),

$$\mathrm{cov}\left(\log \frac{x_i}{g(x)}, \log \frac{x_j}{g(x)}\right) \qquad (i,j = 1,\ldots,d+1), \tag{2.7}$$

where $g(x) = (x_1 \ldots x_{d+1})^{1/(d+1)}$ is the geometric mean of the components of the composition.

(b1) Dissatisfaction with the Dirichlet class as a describer of compositional variability leads to a search for generalizations (Connor and Mosimann, 1969; Darroch and James, 1974; James, 1981) but with very limited success since the generalizations still possess overstrong independence properties (James, 1981).

(b2) A rich class of distributions on S^d can be induced from the $N^d(\mu,\Sigma)$ class on R^d through the generalized logistic transformation

$$x_i = \begin{cases} \exp(y_i)/\{\sum_{j=1}^{d} \exp(y_j) + 1\} & (i = 1,\ldots,d) \\ 1/\{\sum_{j=1}^{d} \exp(y_j) + 1\} & (i = d+1) \end{cases} \tag{2.8}$$

with inverse

$$y_i = \log(x_i/x_{d+1}) \qquad (i = 1,\ldots,d), \tag{2.9}$$

first considered in an unpublished report by Obenchain (1970), then in a specialized diagnostic setting by Aitchison and Begg (1976), and given an identity as the logistic-normal class $L^d(\mu,\Sigma)$ on S^d by Aitchison and Shen (1980).

The 'solutions' to the covariance analysis and distribution theory problems of compositional data analysis given in (a) and (b) clearly conform, and are also seen to be essentially the same as for the Bayesian approach to categorical data analysis. For both types of problem the class of logistic-normal distributions, for Bayesians over the simplex parameter space S^d and for compositionists over the simplex sample space S^d, has greatly enhanced the range of practical tools available for categorical and compositional data analysis. Naturally the practical problems in these two forms of data analysis differ substantially and so lead to quite different uses of the logistic-normal classes.

3. APPLICATIONS WITH SIMPLEX SAMPLE SPACES

The questions posed in §1 about the bayesite and chayesite compositions can now be readily answered in terms of the logistic-normal modeling of §2. We simply transform compositions $(x_1,\ldots x_{d+1})$ in S^d to logratio compositions

$$(y_1,\ldots,y_d) = \{\log(x_1/x_{d+1}),\ldots,\log(x_d/x_{d+1})\} \tag{3.1}$$

in R^d and the whole battery of standard multivariate normal theory is open to us, tests of multivariate normality (and hence of logistic-normality), linear modeling of the mean, discriminant analysis, conjugate normal-Wishart priors, multivariate-Student predictive densities. The asymmetry introduced into the logratio composition by the choice of a particular divisor is unimportant: the statistical procedures are invariant under the group of permutations of the components of the composition. A symmetric version based on the use of the geometric mean $g(x)$ as the common divisor, as in (2.7), is also available but at the cost of singular distributions and the use of pseudo-inverses.

Application to the bayesite and chayesite compositions with $d=4$ soon confirms that they are reasonably logistic-normal (Kolmogorov-Smirnov, Cramer-von Mises tests on univariate marginals, bivariate angles, d-dimensional radius), that their covariances are similar but their means significantly different (Hotelling's T^2 test). Thus if we wished to judge whether a new specimen with (A,B,C,D,E) composition (26.1,33.9,18.0,7.9,14.1) conforms to the bayesite pattern we might construct, following Aitchison and Dunsmore (1975, §11.4), a Student predictive density and hence compute the atypicality index, as the predictive probability of obtaining a new specimen more typical (of higher density) than the specimen currently under judgment. With the vague prior used by Aitchison and Dunsmore (1975, §11.2) the atypicality index for this specimen is 0.9990, thus casting some doubt on any claim that it is bayesite.

The fact that bayesite and chayesite compositions differ significantly in their logratio means suggests that it may be possible to construct a reasonable classification system for differentiating between the two types on the basis of their geochemical compositions. Rather than approach classification by modeling the conditional distributions of composition for given type we shall adopt the 'diagnostic paradigm' (Dawid, 1976) through the standard logistic discriminant model:

$$\log \frac{p(\text{bayesite}\,|\,x,\beta)}{p(\text{chayesite}\,|\,x,\beta)} = \beta_0 + \beta_1 y_1 + \ldots + \beta_d y_d \tag{3.2}$$

$$= \beta_0 + \beta_1 \log x_1 + \ldots + \beta_{d+1} \log x_{d+1}\,, \tag{3.3}$$

where $\beta_1 + \ldots + \beta_{d+1} = 0$. The advantage of the second form (3.3) is that it allows us to assess the effectiveness of any subcomposition such as

$$(x_1,\ldots,x_c)/(x_1+\ldots+x_c)$$

as a means of classification. The hypothesis that this subcomposition is as effective as the complete composition can be regarded as the parametric hypothesis that $\beta_{c+1}=\ldots=\beta_{d+1}=0$ and so can be tested by standard parametric hypothesis testing methods such as the Wilks (1938) asymptotic generalized likelihood ratio test.

In assigning classification probabilities to new cases we can adopt the Bayesian or predictive practice of integrating out the parameter by weighting with the posterior distribution, to obtain for example

$$\int_{R^{d+1}} p(\text{bayesite}\,|\,x,\beta)p(\beta\,|\,\text{data})d\beta \quad . \tag{3.4}$$

Adoption of a standard Bayesian normal approximation for the posterior allows reduction to a univariate integral which can be readily evaluated either by Hermitian integration or

TABLE 3. Distributions of predictive probabilities for classification as bayesite based on full composition and on the subcomposition formed from B, C, D, E.

	Full composition		**Subcomposition formed from B,C,D,E**	
Probability interval	**Bayesite**	**Chayesite**	**Bayesite**	**Chayesite**
0-0.05				
0.05-0.10		2		
0.10-0.15		12		
0.15-0.20		5		
0.20-0.25		1		
0.25-0.30				4
0.30-0.35		2	1	4
0.35-0.40		1	4	2
0.40-0.45	1			4
0.45-0.50	1	1	3	3
0.50-0.55			3	1
0.55-0.60	1		3	2
0.60-0.65	1		5	4
0.65-0.70			3	1
0.70-0.75	2		2	
0.75-0.80	2	1	1	
0.80-0.85	3			
0.85-0.90	10			
0.90-0.95	4			
0.95-1.00				

by replacing the logistic distribution function by a normal approximation (Lauder, 1978) to obtain a closed form. Table 3 shows the classification probabilities obtained by resubstitution of the 50 training set cases. Each of the five four-part subcompositions obtained by dropping out just one of the components was tested for effectiveness by the method described above and each was found to be very significantly poorer than the full composition. This loss of classification power can be seen in the predictive classification probabilities assigned on the basis of the subcomposition formed from B, C, D, E.

For further applications in compositional data analysis by the above and related techniques see Aitchison (1981b, 1983, 1984a,b), Aitchison and Bacon-Shone (1984), Aitchison and Lauder (1984), Aitchison and Li (1984), Aitchison and Shen (1984).

For Bayesians an area of application of particular interest may be the use of logistic-normal distributions to describe and analyse intra-person and inter-person variability in inferential trials in which subjects (such as clinicians) assign probabilities to a finite number of hypotheses (disease types) on the basis of information (diagnostic test results) presented for a sequence of cases (patients); see Aitchison (1981c) for a full discussion. A related problem is in the reconciliation of subjective probability assessments by Lindley, Tversky and Brown (1979) who in their use of normal log-odds models to describe subjective assessments implicitly use logistic-normal modeling. It may be that a fuller exploitation of the properties of the L^d class, such as the simple distributional properties of subcompositions, has still much to offer in this area of Bayesian investigation.

4. INDEPENDENCE IN THE SIMPLEX

In compositional data analysis the formulation and testing of hypotheses of independence in the simplex sample space play a central role (Aitchison, 1982). In the Bayesian analysis of categorical data the specification of prior distributions on the simplex parameter-space commonly involves the introduction of hyperparameters with certain independence properties, as for example in Leonard (1975). This raises the general question of whether Bayesians and compositionists have anything to learn from each other. I hope that we may find some answers in the discussion of this paper and see my role here as essentially a catalyst in the provision of a summary of some of the more important concepts of compositional independence since these will be less familiar to most readers.

The simplest approach to concepts of independence for compositions is to consider various forms of independence in relation to a single division $(x_1,\ldots,x_c|x_{c+1},\ldots,x_{d+1})$ and its corresponding *partition* (s_1,s_2,t), where s_1 and s_2 are the subcompositions formed from the left and right hand subvectors and t is the total of one of the subvectors, say $t = x_1 + \ldots + x_c$. Three useful forms of compositional independence can then be defined as follows:

(1) *Partition independence*: s_1,s_2,t independent.

(2) *Neutrality on the right:* (s_1,t) and s_2 independent.
Neutrality on the left: s_1 and (s_2,t) independent.
Neutrality: neutrality both on the right and on the left.

(3) *Subcompositional independence:* s_1 and s_2 independent.

Partition independence and subcompositional independence were introduced by Aitchison (1982). Neutrality on the right, introduced by Connor and Mosimann (1969) in relation to problems of biological growth, can be expressed equivalently as independence of the left-hand subvector $(x_1,\ldots,x_c)$ and the right hand subcomposition formed from

$(x_{c+1},\ldots,x_{d+1})$. Partition independence implies neutrality on the right (and equally its counterpart neutrality on the left) and neutrality of any form implies subcompositional independence. When we demand that these forms of independence hold for any partition and any permutation of the components of the composition then we have much more extreme forms of independence which we term *complete.*

Completeness is a strong form of independence and indeed provides characterizations of the Dirichlet class: any composition with complete partition independence or complete neutrality has necessarily a Dirichlet distribution. With such strong independence properties it comes as no surprise that the Dirichlet class fails to describe actual compositional variability and has largely been discarded by Bayesians as a reasonable prior distribution.

If any of independence properties (1)-(3) seem relevant to the specification of prior distributions for categorical data analysis then the compositional data analyst may be able to make a contribution because it is possible to support such independence properties as parametric hypotheses within wider classes of distributions on the simplex. We shall illustrate this in relation to neutrality properties in the next section.

One central form of independence in compositional data analysis is *complete subcompositional independence* which can be expressed as independence of every set of subcompositions formed from non-overlapping subvectors of the compositions. In Bayesian terms of constructing a prior distribution for categorical probabilities this would mean independent assignments to all pairs of odds ratios θ_i/θ_j, θ_k/θ_ℓ with i, j, k, ℓ different. Aitchison (1982) shows that complete subcompositional independence corresponds to a simple form of the logratio composition covariance matrix (2.6), namely

$$\Sigma = \operatorname{diag}(\lambda_1,\ldots,\lambda_d) + \lambda_{d+1} J_d, \tag{4.1}$$

where J_d is the $d \times d$ matrix with every element 1. An interesting challenge for Bayesians is then to find suitable forms of prior distribution for Σ to replace the customary unrestricted Wishart distribution. It is no easy matter because of the nasty nature of the $(\lambda_1,\ldots,\lambda_{d+1})$ space with its constraints

$$\lambda_i + \lambda_j > 0 \qquad (i \neq j = 1,\ldots,d+1). \tag{4.2}$$

5. SOME FURTHER DISTRIBUTIONS ON THE SIMPLEX

The logistic-normal class L^d clearly serves a useful purpose in both categorical and compositional data analysis. For some forms of compositional data analysis it has been necessary to invent other parametric classes and here we shall consider two which may be unfamiliar to Bayesians.

The logistic transformation (2.8) is by no means the only useful and simple form of logistic transformation from R^d to S^d. For example, the transformation

$$\begin{aligned} x_i &= \exp(y_i)/ \prod_{j=1}^{i} \{1+\exp(y_j)\} \qquad (i=1,\ldots,d),\\ x_i &= 1/ \prod_{j=1}^{d} \{1+\exp(y_j)\} \qquad (i=d+1), \end{aligned} \tag{5.1}$$

with inverse

$$y_i = \log\{x_i/(1-x_1\ldots-x_i)\} \qquad (i=1,\ldots,d), \tag{5.2}$$

could equally be used to induce a class of distributions $M^d(\mu,\Sigma)$ on S^d from the $N^d(\mu,\Sigma)$

class on R^d (Aitchison, 1981c, 1982). To distinguish between L^d and M^d we may use the terms *additive* and *multiplicative* logistic-normal classes because of the additive and multiplicative nature of the denominators in (2.8) and (5.1).

The importance of the M^d class in compositional data analysis is that it allows simple specification and testing of hypotheses of neutrality on the right for any c. The reason for this is that the subvector $(x_1,\ldots,x_c)$ and subcomposition formed from $(x_{c+1},\ldots,x_{d+1})$ are in one-to-one correspondence with $(y_1,\ldots,y_c)$ and $(y_{c+1},\ldots,y_d)$, respectively. Some rather obvious questions for Bayesians emerge. Because of its relation to a specific ordering of the components, may the M^d class have some relevance to the specification of prior distributions on category probability parameters when the categories are ordered? Are there any circumstances in prior specification where we might envisage the assignments to the category probabilities as taking place sequentially and if so is neutrality a useful concept of assignment? In the analysis of subjective performance in inferential tasks and in the reconciliation of probability assessments would testing between the separate classes L^d and M^d throw any light on the nature of the inference or assessment process?

As a tool of compositional data analysis the logistic-normal classes, being separate from the Dirichlet class, suffer from the limitation that extreme forms of independence such as complete neutrality, which characterize the Dirichlet class, cannot be tested within their framework. This can be remedied, and incidentally the objective of the Dirichlet generalizers realized, by the introduction of a more general class of distributions on S^d. The new $A^d(\beta,\Gamma)$ distribution can be most conveniently defined (Aitchison, 1984b) in terms of its logdensity function

$$c(\beta,\Gamma) + \sum_{i=1}^{d+1} (\beta_i-1)\log x_i - \sum_{i=1}^{d} \sum_{j=i+1}^{d+1} \gamma_{ij}(\log x_i - \log x_j)^2 \quad . \tag{5.3}$$

The D^d class corresponds to

$$\beta_i > 0\,(i=1,\ldots,d+1), \qquad \gamma_{ij} = 0\,(i<j=1,\ldots,d+1), \tag{5.4}$$

the L^d class corresponds to

$$\beta_1 + \ldots + \beta_{d+1} = 0, \tag{5.5}$$

and the generalization is bought at the expense of only one parameter more than for the L^d class and of some Hermitian integration in applications because of the absence of a closed form for $c(\beta,\Gamma)$. Some practical applications will be found in Aitchison (1984b) but the main purpose of introducing it here is to invite Bayesians to suggest suitable forms of prior distributions for $A^d(\beta,\Gamma)$ or some convenient reparametrisation of it.

I hope that I have set down enough to suggest that the exchange of ideas between categorical and compositional data analysis may bear some useful results. Interestingly it was that pioneer of categorical data analysis who in Pearson (1897) warned of the inherent difficulties underlying a too facile approach to compositional data analysis. More surprisingly the essence of a technique common to both forms of data analysis, the induction of new classes in awkward spaces by transformations from normal classes in R^d, lies even further back in history, in the invention of the lognormal class of distributions by McAlister (1879). Can our discussions in Valencia 1983 throw further light on these two fascinating and important areas of data analysis?

REFERENCES

AITCHISON, J. (1981a). A new approach to null correlations of proportions. *J. Math. Geol.* **13**, 175-189.

— (1981b). Distributions on the simplex for the analysis of neutrality. *Statistical Distributions in Scientific Work.* (C. Taillie, G.P. Patil and B. Baldessari, eds.), Vol. 4, pp. 147-156. Dordrecht, Holland: D. Reidel Publishing Company.

— (1981c). Some distribution theory related to the analysis of subjective performance in inferential tasks. *Statistical Distributions in Scientific Work* (C. Taillie, G.P. Patil and B. Baldessari, eds.), Vol. 5, pp. 363-385. Dordrecht, Holland: D. Reidel Publishing Company.

— (1982). The statistical analysis of compositional data. *J. Roy. Statist. Soc.* B**44**, 139-177. (with discussion).

— (1983). Principal component analysis of compositional data. *Biometrika* **70**, 57-65.

— (1984a). Reducing the dimensionality of compositional data sets. *J. Math. Geol.* **16**, in press.

— (1984b). A general class of distributions on the simplex. *J. Roy. Statist. Soc.* B. to appear.

— (1984c). The statistical analysis of geochemical compositions. *J. Math. Geol.* (to appear).

AITCHISON, J. and BACON-SHONE, J.H. (1984). A logconstrast approach to experiments with mixtures. *Biometrika*, to appear.

AITCHISON, J. and BEGG, C.B. (1976). Statistical diagnosis when the cases are not classified with certainty. *Biometrika* **63**, 1-12.

AITCHISON, J. and DUNSMORE, I.R. (1975). *Statistical Prediction Analysis.* Cambridge University Press.

AITCHISON, J. and LAUDER, I.J. (1984). Kernel density estimation for compositional data. Submitted to *Applied Statistics.*

AITCHISON, J. and LI, C.K.T. (1984). A new approach to classification from compositional data. Submitted to *J. Math. Geol.*

AITCHISON, J. and SHEN, S.M. (1980). Logistic-normal distributions: some properties and uses. *Biometrika* **67**, 261-272.

AITCHISON, J. and SHEN, S.M. (1984). Measurement error in compositional data. *J. Math. Geol.* **16**, in press.

BLOCH, D.A. and WATSON, G.S. (1967). A Bayesian study of the multinomial distribution. *Ann. Math. Statist.* **38**, 1423-1435.

CHAYES, F. (1960). On correlation between variables of constant sum. *J. Geophys. Res.* **65**, 4185-4193.

— (1962). Numerical correlation and petrographic variation. *J. Geol.* **70**, 440-452.

— (1971). *Ratio Correlation.* University of Chicago Press.

— (1983). Detecting nonrandom associations between proportions by tests of remaining-space variables. *J. Math. Geol.* **15**, 197-206.

CHAYES, F. and KRUSKAL, W. (1966). An approximate statistical test for correlations between proportions. *J. Geol.* **74**, 692-702.

CONNOR, R.J. and MOSIMANN, J.E. (1969). Concepts of independence for proportions with a generalization of the Dirichlet distribution. *J. Amer. Statist. Assoc.* **64**, 194-206.

DARROCH, J.N. (1969). Null correlations for proportions. *J. Math. Geol.* **1**, 467-483.

DARROCH, J.N. and JAMES, I.R. (1974). F-independence and null correlations of continuous, bounded-sum, positive variables. *J. Roy. Statist. Soc.* B **36**, 467-483.

DARROCH, J.N. and RATCLIFF, D. (1970). Null correlations for proportions II. *J. Math. Geol.* **2**, 307-312.

DARROCH, J.N. and RATCLIFF, D. (1978). No-association of proportions. *J. Math. Geol* **10**, 361-368.

DAWID, A.P. (1976). Properties of diagnostic data distributions. *Biometrics* **32**, 647-658.

JAMES, I.R. (1981). Distributions associated with neutrality properties for random proportions. *Statistical Distributions in Scientific Work* (C. Taillie, G.P. Patil and B. Baldessari, eds), pp. 125-136. Dordrecht, Holland: D. Reidel Publishing Company.

KORK, J.O. (1977). Examination of the Chayes-Kruskal procedure for testing correlations between proportions. *J. Math. Geol.* **9**, 543-562.

LAUDER, I.J. (1978). Computational problems in predictive diagnosis. *Compstat* 1978, 186-192.

LEONARD, T. (1973). A Bayesian method for histograms. *Biometrika* **60**, 297-308.

— (1975). Bayesian estimation methods for two-way contingency tables. *J. Roy. Statist. Soc.* B **37**, 23-37.

— (1977). Bayesian simultaneous estimation for several multinomial distributions. *Comm. Statist. A* **6**.

LINDLEY, D.V. (1964). The Bayesian analysis of contingency tables. *Ann. Math. Statist.* **35**, 1622-1643.

LINDLEY, D.V. TVERSKY, A. and BROWN, R.V. (1979). On the reconciliation of probability assessments. *J. Roy. Statist. Soc. A* **142**, 146-180 (with discussion).

McALISTER, D. (1879). The law of the geometric mean. *Proc. R. Soc.* **29**, 367.

MIESCH, A.T. (1969). The constant sum problem in geochemistry. *Computer Applications in the Earth Sciences* (D.F. Merriam, ed), pp. 161-167. New York: Plenum Press.

MOSIMANN, J.E. (1962). On the compound multinomial distribution, the multivariate β-distribution and correlations among proportions. *Biometrika,* **49**, 65-82.

OBENCHAIN, R.L. (1970). Simplex distributions generated by transformations. *Bell Laboratories Technical Report.*

PEARSON, K. (1897). Mathematical contributions to the theory of evolution. On a form of spurious correlations which may arise when indices are used in the measurement of organs. *Proc. Roy. Soc.* **60**, 489-498.

SARMANOV, O.V. and VISTELIUS, A.B. (1959). On the correlation of percentage values. *Doklady Akad. Nauk. SSSR.* **126**, 22-25.

SWE, C. (1964). The Bayesian analysis of contingency tables. *Ph.D. dissertation, University of Liverpool.*

WILKS, S.S. (1938). The large-sample distribution of the likelihood ratio for testing composite hypotheses. *Ann. Math. Statist.* **9**, 60-62.

DISCUSSION

A.M. SKENE (*University of Nottingham*):

In January 1982, Professor Aitchison presented a paper entitled "The Statistical analysis of compositional data" to the Royal Statistical Society. That paper was noteworthy, not only for the contribution it makes to our understanding of distributions on a simplex and for the insight it gives to the meaning of independence within such a space, but also for the fact that subsequent discussants could find very little to criticise. Today's paper bears heavily on the 1982 paper and the discussion which accompanied it. It is not so much an extension of the '82 paper as a representation of some of its principal results and open questions for our consideration. I make no criticism of the repetition. Such material deserves several airings. However, much of the previously reported discussion is still relevant and the warm reception given to the initial presentation of these ideas suggests that the scope for constructive comment on today's paper is small.

Despite the emphasis on sample spaces in the title, Professor Aitchison is presenting two sets of questions; one set relates to the analysis of compositional data, the other to the specification of a prior distribution for the parameter vector of a multinomial distribution. The specific problems described under the first heading are the specification of a prior for the covariance matrix when complete subcompositional independence is known to hold, and the specification of a prior for the β's and γ's of the generalisation of the Dirichlet family, $A^d(\beta,\Gamma)$. This family we note is a seven parameter family when the composition has just three components, with the number of parameters increasing rapidly as the dimension of the composition increases. I suspect that the specification of an informative prior for such a large number of "non-intuitive" parameters may prove beyond most practitioners and that a Bayesian analysis of compositional data will continue to make extensive use of logistic transforms and multivariate normal priors for some time to come. It is suggested that techniques of compositional data analysis may be used to investigate individuals' assessments of probability vectors, for example, in medical diagnosis. Given a reasonable set of alternative diagnoses, it is my experience that a clinician will frequently assign a probability of zero to at least one diagnosis. We need to be reminded of Professor John Anderson's discussion of the '82 paper which suggests that the methods advocated by Professor Aitchison can be sensitive to the treatment of these zeros.

Turning to the simplex as a parameter space, the various forms of compositional independence *can* be used by the Bayesian to good effect. A simple illustration is the following:

Jaundice in a patient may require surgical or medical treatment. One useful classification of the causes of jaundice identifies two forms of surgical jaundice (cancer vs non-cancer) and five medical conditions (viral hepatitis, chronic hepatitis, drug induced, alcoholic liver disease and 'rarer conditions'). Consider the problem of estimating the incidence rates for these 'diseases' and let S_1 and S_2 be two subcompositions corresponding to the surgical and medical conditions respectively. In specifying a prior for the incidence rates one could reasonably assume subcompositional independence between S_1 and S_2 and possibly neutrality on the left. However partition independence is probably not a valid assumption. A high proportion of medical cases may well be associated with a population where acute viral hepatitis is prevalent, i.e., knowledge of the proportion of medical cases is not independent of the proportions in S_2.

More seriously, the practical use of logistic transforms followed by multivariate normal priors is yet to be demonstrated. Appealing to the approximate normality of log linear constrasts is, I believe, suspect when cell frequencies are small. I would welcome guidelines in this respect and offer the suggestion that a detailed numerical investigation of exact and approximate Bayesian methods for small contingency tables may provide much needed insight.

J.D. BERMUDEZ (*Universidad de Valencia*):

I am a little puzzled with the adoption by Prof. Aitchison of the diagnostic paradigm rather than the sampling paradigm in the example of classification with compositional data developed in section 3.

The diagnostic paradigm has a very nice, intuitive appeal but it fails to provide a closed form solution, either using the logistic discriminant model or its most important alternative, the cumulative normal model (Aitchison and Lauder, 1979). This lack of solution in closed form implies the use of approximations, quite crude sometimes, such as that used by the author when he adopts a bayesian *asymptotic* normal approximation for the posterior, specially when in his bayesite-chayesite example there are 5 parameters and only 50 data in the training set. The use of this asymptotic approach leads to a result rather similar to the one provided by maximum likelihood; indeed, a look at Lauder's result (Lauder 1978, eq. 3.4), shows that the classification distributions obtained by this asymptotic method are maximum likelihood logistic probabilities a little shifted toward the uniform classification distribution.

Moreover, most of the paper is devoted to the defence of the logistic normal model for compositional data. Under the assumption of such a model the natural way to deal with the classification problem is through the sampling paradigm, because of the assumed normality of the log ratio compositions in each class. Thus, one would expect a better behaviour of the computationally simpler sampling paradigm solution than that corresponding to the diagnostic paradigm. The figure below shows the posterior log odds for each object in the training set given in tables 1 and 2. White squares represent bayesite elements while black squares are for those belonging to the chayesite class. The translation to probabilities of those log odds is given in the table. (The results presented here were obtained using the full composition).

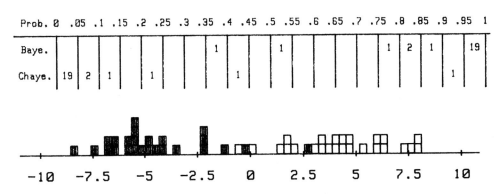

Prob. 0 .05 .1 .15 .2 .25 .3 .35 .4 .45 .5 .55 .6 .65 .7 .75 .8 .85 .9 .95 1

Baye.								1			1					1	2	1		19
Chaye.	19	2	1		1				1										1	

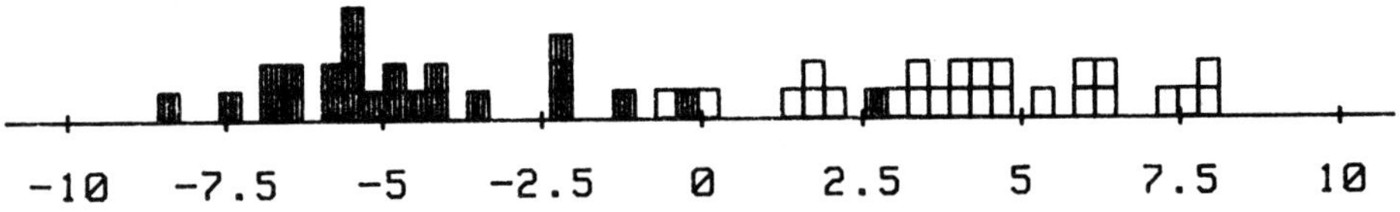

J.M. BERNARDO (*Universidad de Valencia*):

The author is to be congratulated for a brilliant summary of a number of apparently different problems which should benefit from cross-fertilization. Moreover, his emphasis on the logistic-normal classes of distributions is well justified in view of their apparent potential. Indeed, with multinomial data, those families seem to provide a feasible solution to the key problem in probabilistic classification of incorporating strong beliefs that similar facets should produce similar diagnostic distributions; I would like to see however a much more elaborated discussion of the technical consequences of further modelling μ and Σ, into the final distribution of the multinomial parameters.

I was specially interested in the brief discussion on probabilistic classification provided in Section 3; I was there less than convinced by the 'standard' normal approximation for the posterior distribution of β. Maybe the author could elaborate on his underlying assumptions and on the treatment given to the restriction $\Sigma\beta = 0$.

An alternative procedure for obtaining a probabilistic classification, described in Bernardo (1983), is to consider the situation as a decision problem where the action space is the class of logistic diagnostic distributions and the utility function is the logarithmic scoring rule. The results of using this technique on the data of Tables 1 and 2, in terms of the transformation $y_i = \log(x_i/x_1)$, $i = 2,3,4,5$ suggested by the author, is expressed in Table 4 in the format chosen by the author in his Table 3.

The discriminatory power of this alternative method is clearly, in this example, much higher than that of the method discussed in the paper. We note in passing the misleading probabilities we obtained with the only two misclassified items, namely B5, *Pr* (Bayesite) = 0.281 and C18, *Pr* (Chayesite) = 0.051. Are they outliers?

Probability interval	Author's method		Discussant's method	
	Bayesite	Chayesite	Bayesite	Chayesite
0-0.05				18
0.0.5-0.10		2		1
0.10-0.15		12		2
0.15-0.20		5		1
0.20-0.25		1		1
0.25-0.30			1	
0.30-0.35		2		
0.35-0.40		1		
0.40-0.45	1			1
0.45-0.50	1	1		
0.50-0.55				
0.55-0.60	1			
0.60-0.65	1		2	
0.65-0.70				
0.70-0.75	2			
0.75-0.80	2	1		
0.80-0.85	3			
0.85-0.90	10		2	
0.90-0.95	4		4	1
0.95-1.00			16	

TABLE 4. *Comparative classification performance of the methods discussed*

REPLY TO THE DISCUSSION

J. AITCHISON (*University of Hong Kong*):

First I should like to thank the three discussants for their kind remarks.

On Dr. Skene's discussion of the new $A^d(\beta,\Gamma)$ class I would like to point out that the parameter dimension is not so explosive with increasing d as he implies. For $d = 2$ this is 6 $(\beta_1,\beta_2,\beta_3,\gamma_{12},\gamma_{13},\gamma_{23})$, not 7 as he claims. More importantly the parameter dimension of $A^d(\beta,\Gamma)$ is $(d+1)(d+2)/2$, just one more, whatever the value of d, than the parameter dimension $d(d+3)/2$ of the corresponding d-dimensional normal class. Thus the problem of assigning priors may not be so much greater than that for multinormal parameters, provided some simple meaning can be ascribed to the extra parameter. I must admit that I have so far no progress to report on such an ascription.

On the question of handling zeros I would readily admit that the various methods of treatment that I have advocated in Aitchison (1982) can be sensitive. In the replacement of a zero component by $\epsilon > 0$ it is clearly always possible by choosing ϵ small enough to make the consequent $\log \epsilon$ dominate the whole analysis. Where the zeros are associated with rounding effects or denote traces then a sensitivity analysis is essential to gauge robustness to different choices of ϵ, but only over a sensible range of ϵ values. Where sensitivity is established there seems to me no alternative but to recognise the zeros as absolute or essential and to introduce conditional forms of model as indicated again in Aitchison (1982). When the zeros are confined to only a few of the parts this creates no great difficulty in handling but when essential zeros occur in many of the parts, as for example, in palaeoecological studies of foraminifera compositions at different levels of a core sample, I know of no satisfactory way of parametric modelling.

It should perhaps be reemphasized that there is not just a single form of zero problem, but several, so that ad hoc methods may have to be tailored to each individual situation. What is probably required at this stage of development is a set of case studies, to illustrate cases where there is robustness and cases where conditioning models or some as yet unadvocated alternative treatment is needed.

I was intrigued and delighted by Dr. Skene's use of simplex independence ideas in his partition analysis of the composition of probabilities assignable to 'surgical' and 'medical' forms of jaundice. It was the start of cross-fertilisation of ideas that I was hoping for.

I do not completely follow the arguments in Dr. Skene's scepticism in his final paragraph over the use of the L^d class for specification of priors on a multinomial probability vector. This choice of prior does not depend on any approximate normality of loglinear contrasts. My comment (2) in §2 was simply an attempt to identify a historical step, a view of how I thought logistic-modelling of priors had been arrived at within the Bayesian parameter space framework. What class we use for contingency probabilities is surely dependent on the ability of the class to describe our priors, and the L^d class is certainly more flexible than, for example, the Dirichlet class, and there is no difficulty in describing small cell probabilities in the prior. Moreover, it seems to me easier to express views about ratios of probabilities than about absolute probabilities. For some aspects of Bayesian analysis, for example in the combination of expert opinion as we see it in the Lindley (1984) paper at this conference, it seems that logistic-normal modelling is indeed for the moment the way to progress.

Bernardo and Bermudez have both directed their main attention to my example on classification of bayesite and chayesite. I should declare at this stage that these two compositional data sets are indeed simulated logistic-normal compositions so that Bermudez is correct in saying that it would have been more rational of me to have modelled through the sampling paradigm. This indeed was the approach to an earlier diagnostic problem used as an illustration by Aitchison and Shen (1980). My only reason for choosing the diagnostic paradigm here was to demonstrate how to handle compositions in the role of explanatory vectors, nothing more nothing less. Of course in a real practical situation there may be very compelling grounds (Dawid, 1976) for preferring the 'diagnostic paradigm' to the 'sampling paradigm' even when the compositions seem to be logistic-normally distributed.

Bernardo and Bermudez both appear to be claiming a better performance for their methods than the logistic discriminant method which I apply, on the grounds that they have 'greater discriminatory power'. They may be correct, but their use of the preponderance of diagnostic probabilities at the appropriate tails, as for example in Professor Bernardo's table, can be something of an illusion particularly in the resubstitution method of assessment. For example, with this criterion the estimative forms of either the sampling or the diagnostic paradigm will give even greater discriminatory power than obtained by either of the methods presented in the discussion. Yet we should be wary of the use of estimative methods because we know (Aitchison et al., 1977) that they can be less reliable at pinpointing the true diagnostic probabilities than predictive methods.

On a more general issue it seems to me that we have still a lot to learn about the overall performance as diagnostic probability assessors of the standard sampling and diagnostic paradigms even in circumstances where both are appropriate. In the comparison of the Bayesian predictive forms it may turn out that differences in the assessments are not so much due to suspect normal posterior approximations in the diagnostic paradigm as the different degrees of input of information fed to the system in the assignments of priors.

REFERENCES IN THE DISCUSSION

AITCHISON, J. (1982). The statistical analysis of compositional data (with discussion). *J. Roy. Statist. Soc.* B, **44**, 139-177.

AITCHISON, J. HABBEMA, J.D.F. and KAY, J.W. (1977). A critical comparison of two methods of statistical discrimination. *Appl. Statist.,* **26**, 15-25.

AITCHISON, J. and LAUDER, I.J. (1979). Statistical diagnosis from imprecise data. *Biometrika,* **66**, pp. 475-483.

AITCHISON, J. and SHEN, S.M. (1980). Logistic-normal distributions: some properties and uses. *Biometrika,* **67**, 261-272.

BERNARDO, J.M. (1983). Bayesian logistic diagnostic distributions. *Tech. Rep.:* Universidad de Valencia.

DAWID, A.P. (1976). Properties of diagnostic data distributions. *Biometrics*, **32**, 647-658.

LAUDER, I.J. (1978). Computational problems in predictive diagnosis. *Compstat 1978*, pp. 186-192.

LINDLEY, D.V. (1984). Reconciliation of discrete probability distributions. This volume.

BAYESIAN STATISTICS 2, pp. 33-66
J.M. Bernardo, M.H. DeGroot, D.V. Lindley, A.F.M. Smith (Eds.)

In Defense of the Likelihood Principle: Axiomatics and Coherency

JAMES O. BERGER
Department of Statistics (Purdue University)

SUMMARY

Belief in the Likelihood Principle was substantially advanced when A. Birnbaum showed it to be derivable from the apparently more natural Sufficiency and Conditionality Principles. This axiomatic development subsequently came under attack from a number of directions, among the most interesting being a criticism (by D.A.S. Fraser, G. Barnard and others) of the Sufficiency Principle for failure to take into account "structural" knowledge of the performed experiment. This criticism is addressed in this paper from two directions. First, a weak set of alternative axioms for the Likelihood Principle is developed. Second, ideas of coherency are employed to question the validity of knowingly violating the Likelihood Principle. In this development, arguments are presented for basing coherency on decision-theoretic concepts, rather than the more usual betting concepts. The basic conclusions of the paper also apply to other theories which can violate the Likelihood Principle, including many noninformative prior Bayesian theories.

Keywords: COHERENCY; COMPLETE CLASSES; CONDITIONALITY; INADMISSIBILITY; SUFFICIENCY; STRUCTURAL ANALYSIS.

1. INTRODUCTION

The Likelihood Principle (LP) received its major (non-Bayesian) impetus from ideas of R.A. Fisher and G. Barnard. (See Berger and Wolpert (1984) for references). It essentially states that any decision or inference in a statistical problem should involve the data and experiment only through the likelihood function of the unknowns given the observed data. The implications of the LP are farreaching, and the case for it is strong. The non-Bayesian case rests on various axiomatic developments of the LP from simpler believable principles, such as the Birnbaum (1962) development of the LP from the Conditionality and Sufficiency principles.

Arguments against the LP usually take one of four forms. First, are arguments concerning the "unintuitive" consequences of the LP, such as the consequent irrelevance of stopping rules in the final inference or decision and the incompatibility of the LP with significance testing and randomization analysis. Discussion and references concerning such matters can be found in Basu (1975) and Berger and Wolpert (1984), but in general it is hard to see how anyone believing in logic could reject the LP because of its consequences without rejecting at least one of the precepts upon which it is based. The three remaining arguments against the LP thus focus on either (i) the existence of the likelihood function, (ii) the validity of the Conditionality principle, or (iii) the validity of the Sufficiency principle.

As presented in Birnbaum (1962), the LP is dependent on the existence of a likelihood function (and in fact on the discreteness of the sample space). Thus, it can be argued that in nondiscrete nonparametric problems (and one is rarely completely sure of a parametric model) there may be no clearly defined likelihood function, and even when there is a likelihood function the nondiscreteness makes Birnbaum's arguments questionable. Basu (1975) answers this by arguing that, in reality, data is always discrete (because of limitations on observational accuracy) so that no difficulties arise. The importance of using continuous approximations to discreteness is well recognized, however, and validity of the LP for nondiscrete situations would, therefore, be comforting. A general version of the LP (called the Relative Likelihood Principle) was developed for essentially arbitrary situations in Berger and Wolpert (1984). It was shown to follow from conditionality and sufficiency, and also shown to have essentially the same consequences as the LP. Rejection of the LP on the grounds of nonexistence of likelihood functions can be circumvented, therefore.

The Conditionality Principle (CP) roughly states that, if an independent coin is flipped to decide between performing experiments E_1 and E_2, both pertaining to some unknown quantity θ of interest, then the *evidence about* θ obtained from the data should not depend in any way on the experiment not actually performed. First stated explicitly by Cox (1958), the CP seems completely obvious, but, rather startlingly, it is in sharp opposition to standard frequentist reasoning in statistics. The experiment not actually performed is another part of the sample space, and hence a frequentist would average over it in determining the performance of his "procedure". Numerous attempts to partially follow the CP and then go frequentist have been advocated, but do not seem to be ultimately justifiable (c.f. Berger and Wolpert (1984)). It is not, on the other hand, illogical for a frequentist to simply reject the CP, arguing (c.f. Neyman and Pearson (1933)) that the notion that one can obtain "evidence about θ from an experiment" is misguided; all one can do is evaluate how well a *procedure* that will be repeatedly used performs in the long run, and the procedure's performance will be an average over both E_1 and E_2. While logically viable, the artificiality of the position is clear. I doubt if many users of statistics would be willing to accept the point of view that one cannot obtain evidence about θ from an experiment. (Note that the CP and LP presuppose nothing about what this "evidence" is, or even that it is any single quantity).

The final criticism of the LP comes from ideas of Barnard (c.f. Barnard (1980, 1982), Barnard and Godambe (1982), and the discussion in Basu (1975)) and Fraser (c.f. Fraser (1963, 1968, 1972, 1979)) concerning the validity of the Sufficiency Principle. The criticism concerns the "sufficiency" of representing the experimental structure solely in terms of probability distributions on the sample space indexed by the unknown θ. This turns out to be a very difficult criticism to answer, and indeed it does not seem answerable in the same self-contained sense as the other criticisms. Although an attempt is made to deal with the issue axiomatically in Section 3, the attempt is something of a failure, one crucial axiom being suspect. Instead, therefore, we turn to Bayesian arguments of coherency and inadmissibility, in an attempt to indicate the difficulties in violating sufficiency. These arguments are, of course, more or less familiar, but because of the importance of the issue and the bearing that these arguments have on theories of inference such as Barnard's Pivotal Inference, Fraser's Structural Inference, and even various noninformative prior Bayesian theories, certain aspects of which may violate the LP, we will include a fair amount of detail and discussion. It should be stated at the outset that, in no sense, do we unequivocally resolve the controversy. The paper should be viewed more as an attempt to carefully state a Bayesian view of the issue.

Some Bayesians may wonder why there is any need to be concerned with the LP, except as a trivial consequence of the Bayesian position. There are essentially two reasons. The first is the purely pragmatic reason that promoting Bayesianism can often most effectively be done by first selling the LP, since this can be done without introducing the emotionally charged issue of prior distributions. The second reason is that the LP shows that Bayesians should be concerned with conditional (posterior) conclusions. This may seem to be a strange statement to most Bayesians, but it is certainly possible to be a Bayesian and not believe this. For instance, one could believe that only frequentist measures of procedure performance have validity, and yet, because of various rationality, coherency, or admissibility arguments, believe that the only reasonable procedures are Bayes procedures, and that the best method of choosing a procedure is through consideration of prior information and application of the Bayesian paradigm (c.f. the discussion by L. Brown in Berger (1984)). The posterior distribution will provide a convenient mathematical device for determining the best procedure, but (from this viewpoint) the overall frequentist (Bayes) performance of the procedure would be the relevant measure of accuracy. The LP directly attacks this view, arguing that thinking "conditional Bayes", not "frequentist Bayes", is important.

In Section 2 the needed notation will be given. Section 3 presents a weak set of principles which imply the LP, and discusses at which point the Barnard-Fraser criticism enters in. Section 4 outlines the scenario through which considerations of coherency and admissibility become relevant. In this section it is argued, as a side issue, that decision theoretic admissibility is a more valid evaluational tool than the more common (to Bayesians) betting coherency. Section 5 argues against violation of the LP because of these considerations, with particular attention given to the Barnard-Fraser criticism. Section gives some concluding remarks.

2. NOTATION

2.1. *The Experiment*

We will more or less follow Birnbaum's (1962) notation for reasons of familiarity. The first important issue is - what is an experiment? We will denote an experiment by

$$E = (X, \theta, \{P_\theta\}, (h, \omega)), \tag{2.1}$$

where X (a realization of which is the data and will be denoted x) is a random quantity taking values in a sample space $\mathbf{X}$ according to the probability distribution P_θ, the unknown aspects of this distribution being denoted by θ, an element of the parameter space Θ. (We will not overburden the description of E by including $\mathbf{X}$ or Θ in (2.1).) The unknown θ could include typical unknown parameters and also could index unknown functional features of the distribution. For instance, if $P_\theta(A) = G((A-\mu)/\sigma)$, where μ and σ are unknown and it is only known that $G \in \mathbf{S}$, some set of distributions, then θ would equal (μ,σ,G). We will, however, for simplicity restrict consideration to situations where a likelihood function exists, i.e. where $\{P_\theta, \theta \in \Theta\}$ is dominated by a measure ν, with respect to which we have densities

$$f(x|\theta) = \frac{dP_\theta}{d\nu} .$$

Also, in axiomatic discussions we will implicitly assume that $\mathbf{X}$ is discrete, so that $f(x|\theta)$ is well defined. (Again, as argued in Basu (1975), this assumption reflects ultimate reality; discussion of the philosophical validity of the LP in this setting is thus appropriate). The density, considered as a function of θ for fixed x, is of course the *likelihood function* for θ given that x is observed.

The final element of E in (2.1), the pair (h,ω), is included to allow consideration of the criticisms of Barnard and Fraser. Barnard, with his theory of Pivotal Inference (c.f. Barnard (1980, 1982) and Barnard and Sprott (1983)), and Fraser, with his theory of Structural Inference (c.f. Fraser (1968, 1972, 1979)), argue that it may sometimes be known how X, θ, and P_θ are related, and that this can be important information. Thus it may be known that

$$X = h(\theta,\omega)\,,$$

where ω is an unknown random quantity taking values in Ω according to a known distribution Q, and h is a known function from $\Theta \times \Omega \to \mathbf{X}$. (Often in Structural and Pivotal Inference, Q is known only to belong to some class $\mathcal{Q}$. For simplicity, we assume Q is known). This is actually more or less the "structural" formulation of the problem. The formulation in Pivotal Inference is based on "pivotals" $\omega = g(X,\theta)$ having known distributions. Typically g will be an appropriate inverse function of h, so the two approaches are very related. We will, for the most part, consider the structural formulation, although comments about differences for the pivotal model will be made. The structural model is sometimes called a functional model (c.f. Bunke (1975) and Dawid and Stone (1982)), but we will stick with Fraser's original term. The following example, from Fraser (1968) (and related to an example in Mauldon (1955)), illustrates the key issue.

Example 1. Suppose $X = (X_1, X_2)$, $\theta = (\sigma_1,\tau,\phi)$, and P_θ is bivariate normal with mean zero and covariance matrix

$$\Sigma = \begin{pmatrix} \sigma_1^2 & \tau\sigma_1 \\ \tau\sigma_1 & (\tau^2+\phi^2) \end{pmatrix}.$$

This could arise from either of the following two *structural* models:

(i) $\omega = (\omega_1,\omega_2)$ is bivariate normal, mean zero and identity covariance matrix, and

$$X = h(\theta,\omega) = (\sigma_1\omega_1,\quad \tau\omega_1+\phi\omega_2)\,; \tag{2.2}$$

(ii) ω is the same but

$$X = h^*(\theta,\omega) = (\tau'\omega_1+\phi'\omega_2,\ \sigma_2\omega_1)\,, \tag{2.3}$$

where $\sigma_2 = \sqrt{\tau^2+\phi^2}$, $\tau' = \sigma_1\tau/\sigma_2$, and $\phi' = \sigma_1\,\phi/\sigma_2$. In Barnard's setup one would write (2.2) and (2.3) as

$$\omega = (\omega_1,\omega_2) = (X_1/\sigma_1,\ (X_2-\tau X_1/\sigma_1)/\phi), \qquad (2.2)'$$

$$\omega = (\omega_1,\omega_2) = (X_2/\sigma_2,\ (X_1-\tau' X_2/\sigma_2)/\phi'), \qquad (2.3)'$$

and ω_1 and ω_2 would be the pivotals with known distribution upon which the inference would be based. In pursuing this example later in the paper, we will assume that independent observations $X^1,\ldots,X^n$ from the model are taken, giving the "sufficient" statistic $S = \sum_{i=1}^{n} (X^i)^t(X^i)$, which has a Wishart (n, Σ) distribution.

We do not presuppose that knowledge of (h,ω) is available, although our main concern will be the value of knowing (h,ω) if it is available. When it is available, we will say that we are in the "P-S situation" (for pivotal-structural).

2.2. *Evidence and Mixtures*

Of interest from an experiment is the "evidence" about θ obtained from knowledge of E and the data x. This will (following Birnbaum) be denoted $Ev(E,x)$, and could be any measure or collection of measures whatsoever (including frequency measures). The completely arbitrary nature of this concept should make it acceptable to most people. In Section 3 we will investigate principles that $Ev(E,x)$ should follow. (Dawid (1977) prefers to replace the concept of evidence by that of an inference pattern, and then talk about principles which inference patterns should follow. This might have some philosophical advantage, but we will stick to the more usual approach).

The final concept needed is that of a simple mixture of two experiments

$$E_1 = (X_1,\theta,\{P_\theta^1\}, (h_1,\omega_1)) \text{ and } E_2 = (X_2,\theta,\{P_\theta^2\}, (h_2,\omega_2)).$$

(It is important to note that, throughout the paper, θ will be assumed to be the same quantity in the two experiments). Let J be a random variable (independent of the X_i, ω_i, and any information about θ) taking on the values 1 and 2 with probabilities λ and $(1-\lambda)$, respectively. Then the *mixed experiment* E^λ is the experiment in which first J is observed, and then experiment E_J is performed. Thus the outcome of E^λ is the pair (J,X_J). We will often imagine, given E_1 and E_2, that we can perform E^λ, and hence can think of $Ev(E^\lambda, (j,x_j))$. Certain objections have been raised (c.f. Durbin (1970) and Kalbfleisch (1974)) concerning the treatment of E^λ as a "real" experiment, but it is clear that E^λ *could* be performed, and hence any proposed statistical methods should work well for it. Further discussion can be found in Birnbaum (1970), Dawid (1977), and Berger and Wolpert (1984).

3. PRINCIPLES OF EVIDENCE

This section is independent of the rest of the paper, and can be skipped by those less interested in axiomatic arguments. No attempt will be made to survey the wide array of principles that have been discussed and which lead to the LP. References include Birnbaum (1962, 1972), Basu (1975), Dawid (1977), Godambe (1979), Barnard and Godambe (1982), and Berger and Wolpert (1984). The key principles are the Weak Conditionality Principle below, and some version of the Sufficiency Principle. The weakest versions of sufficiency are Mathematical Equivalence (see Birnbaum (1972)) and the Distribution Principle (see Dawid (1977)) which essentially state that (h,ω) in (2.1) is irrelevant to $Ev(E,x)$. This section is an attempt to formulate weaker principles which lead to the LP. The effort is not as successful as had been initially hoped. We first list the principles.

Weak Conditionality Principle (WCP): $Ev(E^\lambda, (j,x_j)) = Ev(E_j,x_j)$. (Thus the evidence obtained from a simple mixture experiment is simply the evidence obtained from the experiment E_j actually performed).

Weak Ancillarity Principle (WAP): Suppose E_1 (as in (2.1)) consists of observing the random quantity $X_1 = (Y_1,Z)$, where Z has a known distribution independent of Y_1, ω_1, and θ, and E_2 consists of observing $X_2 = g(Y_1,Z)$, where g is a known function for which there exists a known function g^* such that $Y_1 = g^*(X_2)$. Then $Ev(E_1,x_1) = Ev(E_2,x_2)$. (The point here is that it seems clear that only Y_1 contains information about θ in E_1, and we can determine Y_1 *exactly* from X_2. But since X_2 depends only on Y_1 and the irrelevant Z, it can contain no more information about θ than Y_1. Note that passing from X_1 to X_2 and back to Y_1 via known functions, so the important structural elements of the experiments are preserved).

Weak Distribution Principle (WDP): Suppose E_i, $i=1,2$, consists of observing $X_i = (Y, Z_i)$, where Y is the same random variable and the Z_i are each 1 or 2 with probability $\frac{1}{2}$, independent of Y and θ (but not necessarily of the ω_i). Then $Ev(E_1, (y,j)) = Ev(E_2, (y,j))$ for $j = 1,2$.

It is the WDP that is potentially inconsistent with P-S analysis, in that the Z_i may contain structural information which is ignored because of the lack of dependence of the distribution of the Z_i on Y or θ. In some sense, the WDP is no more intuitively obvious than Birnbaum's principle of Mathematical Equivalence, but does appeal to a somewhat different intuition, namely the "frequentist" intuition behind sufficiency which states that if Y is a sufficient statistic (of (Y,Z_1)) for θ, then Z_1 could be replaced by any random variable with the same distribution without affecting any conclusions.

Likelihood Principle: If E_1 and E_2 are two experiments and there exist observations x_1' and x_2' in the respective experiments for which $f_1(x_1'|\theta) = cf_2(x_2'|\theta)$ for all θ (and some fixed constant c), then $Ev(E_1,x_1') = Ev(E_2,x_2')$.

Theorem 1. The LP is a consequence of the WCP, WAP and WDP.

Proof: Consider the mixed experiment E^λ, where $\lambda = 1/(1+c)$. By the WCP we have that

$$Ev(E^\lambda, (j,x_j')) = Ev(E_j,x_j') \,. \tag{2.4}$$

Next, define $Y = (J,X_J)$ (the outcome of E^λ) and

$$V_1 = \begin{cases} Y & \text{if } X_J \neq x_j' \text{ or } x_2' \\ 0 & \text{otherwise,} \end{cases} \tag{2.5}$$

and

$$Z_1 = \begin{cases} 1 & \text{if } X_j = x_1', \text{ or } V_1 \neq 0 \text{ and } Z = 1 \\ 2 & \text{if } X_j = x_2', \text{ or } V_1 \neq 0 \text{ and } Z = 2, \end{cases} \tag{2.6}$$

where Z is a random variable (independent of everything, probabilistically and structurally) taking values 1 and 2 with probability 1/2. Consider the experiment E^* with observation (Y,Z), and note that an application of the WAP (with $g(Y,Z) = Y$) shows that

$$Ev(E^\lambda,y) = Ev(E^*, (y,z)) \,. \tag{2.7}$$

Also, defining E^{**} as the experiment of observing $X^* = (V_1,Z_1)$ noting that (2.5) and (2.6) define $X^* = g(Y,Z)$, and observing that

$$g^*(X^*) \equiv \begin{cases} V_1 & \text{if } V_1 \neq 0 \\ (Z_1,x_{Z_1}') & \text{if } V_1 = 0 \end{cases} = Y,$$

the WAP can be applied to conclude that

$$Ev(E^*,(y,z)) = Ev(E^{**}, (v_1,z_1)) . \tag{2.8}$$

Note that

$$P_\theta(Z_1 = 1 \mid v_1) = \begin{cases} P(Z=1) = 1/2 & \text{if } v_1 \neq 0 \\ \dfrac{\lambda f_1(x_1'\mid\theta)}{\lambda f_1(x_1'\mid\theta) + (1-\lambda) f_2(x_2'\mid\theta)} = 1/2 & \text{if } v_1 = 0, \end{cases}$$

so that Z_1 is independent of both V_1 and θ. Application of the WDP to E^{**} shows that Z_1 could be replaced by any (structurally independent) random variable with the same distribution, and use of the WAP (as in (2.7)) then shows that $Ev(E^{**}, (v_1,z_1))$ does not depend on z_1. Combining this with (2.4), (2.7) and (2.8) yields the desired conclusion. | |

Since the WDP is in as much conflict with *P-S* analysis as Mathematical Equivalence, not much progress has been made. The point of developing alternative simple principles is the hope that they will spur those who question the LP into finding a clear counterexample to at least one of the principles. Unfortunately, "counterexamples" so far developed are either of the extermely involved variety (such as the Stopping Rule Paradox), or are of the form —here is an example of where 'My Method' clashes with Principle A— without a quantified demonstration of harm that would result in following Principle A. Of course, we don't believe valid counterexamples will be found. The reason is simply that repeated use of any method violating the LP seems likely to itself be demonstrably inferior. We turn now to this issue.

4. LONG RUN PERFORMANCE

The *practical* importance of considering the long run performance of statistical procedures or methods is certainly a matter open to debate, but one feature of long run performance seems clear: it cannot be right (philosophically) to recommend repeated use of a method if the method has "bad" long run properties. There have been two main approaches proposed for long run evaluations: decision theory and betting schemes. We will argue that the decision theoretic approach is the more satisfactory of the two (even for "inference" problems), although either approach strongly contraindicates violation of the LP. It is interesting that frequentist decision theory emerges (even for Bayesians) as an important testing ground for statistical theories.

4.1. *Decision Theoretic Evaluations*

The decision theoretic approach supposes that the result of the statistical investigation is to take an *action* $a \in \mathbf{A}$ (which could conceivably be the action to take a particular "inference"), the consequence of which, for given data x and when θ obtains, is the *loss* $L(x,a,\theta)$. It is also supposed that the statistical method being evaluated provides an action to take for each possible x, thus defining a statistical procedure $\delta(\cdot){:}\mathbf{X} \to \mathbf{A}$. (For the most part we will stick to nonrandomized procedures for simplicity). As usual in frequentist decision theory, we define the *frequentist risk* and the *Bayes risk* (with respect to a prior distribution π on Θ) as, respectively,

$$R(\theta,\delta) = E_\theta L(X,\delta(X),\theta), \text{ and } r(\pi,\delta) = E^\pi R(\theta,\delta) .$$

Of interest will be the following standard definitions. The procedure δ^1 is *strictly inadmissible* if there exists a δ^2 with $R(\theta,\delta^2) < R(\theta,\delta^1)$ for all θ, and is *extended inadmissible* if $R(\theta,\delta^2) < R(\theta,\delta^1)-\epsilon$ for all θ and some $\epsilon>0$. If $r(\pi,\delta^2) < \infty$ for all countably additive π, the above risk inequalities can be replaced by $r(\pi,\delta^2) < r(\pi,\delta^1)$ and $r(\pi,\delta^2) < r(\pi,\delta_1)-\epsilon$ for all countably additive π.

Following Hill (1974), and in a similar manner to many betting scenarios, we consider the following game.

Evaluation Game. Player 1 proposes use of δ^1 and Player 2 proposes δ^2. A master of ceremonies will choose a sequence $\theta = (\theta_1,\theta_2,\ldots)$, and for each θ_i the experiment E will be independently performed yielding an observation X_i (from the distribution P_{θ_i}, or equal to $h(\theta_i,\omega_i)$). Player j will use $\delta^j(x_i)$, paying to the other player his "loss" $L(x_i,\delta^j(x_i),\theta_i)$. After n plays, Player 2 will have won

$$S_n = \sum_{i=1}^{n} [L(x_i,\delta^1(x_i),\theta_i)-L(x_i,\delta^2(x_i),\theta_i)] .$$

Theorem 2. In the situation of the above game,

(a) If δ^1 is strictly inadmissible and $r(\pi,\delta^2) < \infty$ for all countably additive π, then

$$P_\pi(\lim_{n\to\infty} \inf \tfrac{1}{n} S_n > 0) = 1, \tag{3.1}$$

where P_π denotes the joint probability distribution of the X_i and θ_i.

(b) If δ^1 is extended inadmissible and the random variables $Z_i = [L(X_i,\delta^1(X_i),\theta_i) - L(X_i,\delta^2(X_i),\theta_i)]$ have uniformly bounded variances (i.e. $E_{\theta_i}[Z_i-E_{\theta_i}Z_i]^2 < K < \infty$ for all θ_i), then

$$P_\theta(\lim_{n\to\infty} \inf \tfrac{1}{n} S_n > \epsilon > 0) = 1, \tag{3.2}$$

for *any* sequence $\theta = (\theta_1,\theta_2,\ldots)$.

(c) If δ^1 is strictly inadmissible, Θ is closed, $R(\theta,\delta^1)$ and $R(\theta,\delta^2)$ are continuous in θ, and the moment condition in (b) holds, then (3.2) is valid for any *bounded* sequence θ (although ϵ could depend on the bound).

Comment 1. For bounded losses, the moment conditions in the theorem are all clearly satisfied. Even unbounded losses rarely cause a problem.

Comment 2. If (3.2) holds, then it also holds for P_θ replaced by P_π for any prior π, including finitely aaditive π. Also, Heath and Sudderth (1978) show that δ^1 is extended admissible only if it is Bayes with respect to some (possibly finitely additive) prior π.

Proof of Theorem 2. (a). If $r(\pi,\delta^1) < \infty$, then $E_\pi Z_i$ (E_π being expectation over the X_i and θ_i, and Z_i being as in part (b)) has finite expectation $\Delta_\pi = r(\pi,\delta^1)-r(\pi,\delta^2) > 0$. The result follows from the strong law of large numbers. If $r(\pi,\delta^1) = \infty$, the result follows by truncating the loss at a suitably large level.

(b) Define

$$\psi(\theta_i) = E_{\theta_i}(Z_i) = R(\theta_i,\delta^1)-R(\theta_i,\delta^2) > \epsilon.$$

By the strong law of large numbers,

$$\frac{1}{n} \sum_{i=1}^{n} [Z_i-\psi(\theta_i)] \to 0 \text{ almost surely,}$$

and the result follows easily. The proof of part (c) is similar. | |

The Evaluations Game seems to be a reasonably fair way of *testing* the performance a procedure. If δ^1 is certain to lose an arbitrarily large amount in comparison with δ^2, as occurs in the situations of Theorem 2, then δ^1 would seem to be theoretically inferior. (The word "theoretically" is inserted, because the practical difference in a realistic finite number of uses may be negligible). Extended inadmissibility seems very serious, in that δ^1 would always have a long run loss. Strict inadmissibility is less compelling, in that δ^1 is only guaranteed to lose against a countably additive prior π, or a bounded sequence θ (in the situation of part (c) of the theorem). But the fact that it will lose for *any* such π (even one of Player 1's choosing) or for *any* such θ, strikes us as sufficient reason to perceive δ^1 as not being fundamentally sound. Note that it is not necessary to *know* the bound on θ in case (c) to conclude that δ^1 loses. It is only necessary to know that there is some bound. (And, in reality, θ will be bounded; unbounded Θ are typically used only because one is not sure what the bound on θ should be). Again, it may be that δ^1 is justifiable as a good approximate rule, even if it is strictly or extended inamissible, but we would certainly hesitate to call any statistical method which led to δ^1 a *fundamentally* sound method.

Adopting a decision-theoretic viewpoint for evaluation can be criticized, especially for inference problems in which losses (if they exist at all) are vague or hard to formulate. This is not the place to argue the case for a decision-theoretic outlook, and indeed a justification of decision theory is not needed for our purpose here. Our goal is to judge the claim in P-S analysis (and other approaches) that the LP is invalid, because it ignores important features of the experiment. We will essentially try to argue that, in any decision problem, repeated violation of the LP will result in long run loss. Most statisticians would probably have qualms about trying to argue that, even if the LP should be followed in any decision problem, it need not be followed in inference problems. Essentially such an argument would be of the variety - "I know I'm right, but will not allow any quantifiable evaluation of my methods".

We will avoid the "unfair" possibility of taking an inference procedure and evaluating it with respect to a particular loss function. It is somewhat more fair to evaluate it with respect to a very wide range of loss functions (indeed, if a wide enough range of loss functions is allowed many "inadmissible" inference procedures become admissible, c.f. Brown (1973)), and strict inadmissibility for a wide range of reasonable losses should be a serious concern. More commonly, however, we will consider particular losses as given, and see where the following of P-S reasoning might lead us. Criticizing P-S reasoning (in particular, possible violation of the LP) in decision settings for which it was never intended is, of course, an uncertain undertaking, especially since it is not clear what P-S reasoning in decision contexts would be. Of relevance here is the following comment of Hill (1974):

> "But no matter what is meant by inference, if it is to be of any value, then somehow it must be used, or acted upon, and this does indeed lead back to the decision-theoretic framework. I suspect that for some 'inference' is used as a shield to discovery that their actions are incoherent".

As final comment, it should be mentioned that even Bayesians should be willing to submit their procedures (usually derived conditionally) to long run performance evaluations, especially when robustness is a serious concern or when improper prior distributions were used.

4.2. *Betting Evaluations*

Studying coherence in betting has a long tradition in statistics, especially Bayesian statistics. The typical scenario deals with evaluation of methods (usually inference methods) which produce, for each x, either a probability distribution for θ, say $q_x(\theta)$ (which could be a posterior distribution, a fiducial distribution, a structural distribution, etc.), or a system of confidence statements $\{C(x), \alpha(x)\}$ with the interpretation that θ is felt to be in $C(x)$ with probability $\alpha(x)$. For simplicity, we will restrict ourselves to the confidence statement framework; any $\{q_x(\theta)\}$ can be at least partially evaluated through confidence statements by choosing $\{C(x)\}$ and letting $\alpha(x)$ be the probability (with respect to q_x) that θ is in $C(x)$.

The assumption is then made (more on this later) that, since $\alpha(x)$ is thought to be the probability that θ is in $C(x)$, the proposer of $\{C(x), \alpha(x)\}$ should be willing to make both the bet that θ is in $C(x)$ at odds of $(1-\alpha(x))$ to $\alpha(x)$, and the bet that θ is not in $C(x)$ at odds of $\alpha(x)$ to $(1-\alpha(x))$. An evaluations game, as in Section 4.1, is then proposed, where the master of ceremonies again generates θ_i and X_i, Player 1 stands ready to accept bets on $\{C(x), \alpha(x)\}$, and Player 2 bets $s(x)$ at odds determined by $\alpha(x)$. Here, $s(x) = 0$ means no bet is offered; $s(x) > 0$ means that an amount $s(x)$ is bet that $\theta \notin C(x)$; and $s(x) < 0$ means that the amount $|s(x)|$ is bet that $\theta \notin C(x)$. (As discussed in Robinson (1979a), restricting $s(x)$ to satisfy $|s(x)| \leq 1$ is also sensible). The winnings of Player 2 at the ith play are

$$W_i = [I_{C(x_i)}(\theta_i) - \alpha(x_i)]\, s(x_i) ,$$

where $I_A(\theta)$ is 1 if $\theta \in A$ and 0 otherwise, and of interest is again the limiting behavior of $\frac{1}{n} \sum_{i=1}^{n} W_i$. If

$$P_\theta(\lim_{n\to\infty} \inf \frac{1}{n} \sum_{i=1}^{n} W_i > \epsilon > 0) = 1 ,$$

for all sequences $\theta = (\theta_1, \theta_2, \ldots)$, then $\{C(x), \alpha(x)\}$ is called *incoherent*, or alternatively $s(x)$ is said to be a *super relevant* betting strategy. If it is merely the case that for θ_i generated according to any countably additive π,

$$P_\pi(\lim_{n\to\infty} \inf \frac{1}{n} \sum_{i=1}^{n} W_i > 0) = 1 ,$$

then $\{C(x), \alpha(x)\}$ is *weakly incoherent* or $s(x)$ is *weakly relevant*. (These concepts can be found in this or related form in such works as Buehler (1959, 1976), Wallace (1959), Freedman and Purves (1969), Cornfield (1969), Pierce (1973), Bondar (1977), Heath and Sudderth (1978), Robinson (1979a, 1979b), and Lane and Sudderth (1983). Other general Bayesian works on coherency include Ramsey (1926), deFinetti (1937, 1974), Savage (1954) and Levi (1980)).

If $\{C(x), \alpha(x)\}$ is incoherent or weakly incoherent, then Player 1 will for sure lose money in the appropriate evaluations game, which certainly casts doubt on the validity of the probabilities $\alpha(x)$. A number of objections to the scenario can, and have, been raised, however, and careful examination of these objections is worthwhile.

Objection 1. Player 1 will have no incentive to bet unless he perceives the odds as slightly favorable. This turns out to be no problem if incoherence is present, since the odds can be adjusted by $\epsilon/2$ in Player 1's favor, and Player 2 will still win. If only weak incoherence is present, it is still often possible to adjust the odds by a function $g(x)$ so that Player 1 perceives that the game is in his favor, yet will lose in the long run, but this is not clearly always the case.

Objection 2. Weak incoherence has been deemed not very meaningful, since a sequence $\theta = (\theta_1, \theta_2, \ldots)$ could be chosen so that Player 1 is not a sure loser. However, the fact that Player 1 is a sure loser for any π (even one selected by himself) or any bounded θ (under certain reasonable assumptions) seems quite serious.

Objection 3. Of course, frequentists who quote a confidence level α for $\{C(x)\}$ remove themselves from the game, since they do not claim that α is the probability that θ is in $C(x)$, and hence would find the betting scenario totally irrelevant.

Objection 4. The game is unfair to Player 1, since Player 2 gets to choose when, how much, and which way to bet. Various proposals have been made to "even things up". The possibility mentioned in Objection 1 is one such, but doesn't change the conclusions much. A more radical possibility, suggested by Fraser (1977), is to allow Player 1 to decline bets. This can have a drastic effect, but strikes us as too radical, in that it gives Player 1 license to state completely silly $\alpha(x)$ for some x. It is after all $\{\alpha(x)\}$ that is being tested, and testing should be allowed for all x.

Objection 5. The most serious objection we perceive to the betting game is that $\{\alpha(x)\}$ is generally not selected for use in the game, but rather to communicate information about θ. It may be that there is no *better* choice of $\{\alpha(x)\}$ for communicating the desired information. Consider the following example, which can be found in Buehler (1971), and is essentially succesive modifications by Buehler and H. Rubin of an earlier example of D. Blackwell.

Example 2. Suppose $X = \theta + \omega$, where $P(\omega=1) = P(\omega=-1) = 1/2$, and $\theta \in \Theta = \{\text{integers}\}$. We are to evaluate the confidence we attach to the sets $C(x) = \{x+1\}$ (the point $(x+1)$), and a natural choice is $\alpha(x) = 1/2$ (since θ is either $x-1$ or $x+1$, and in the absence if fairly strong prior information about θ, either choice seems equally plausible). This choice can be beaten in the betting game, however, by betting that θ is not in $C(x)$ with probability $g(x)$, where $0 < g(x) < 1$ is an increasing function. (Allowing Player 2 to have a randomized betting strategy does not seem unreasonable). Indeed, the expected gain per bet of one unit, for any countably additive π on Θ, is $E^{\pi}[g(\theta+1) - g(\theta-1)] > 0$, so that $\alpha(x) = 1/2$ is weakly incoherent. (A continuous version of this example, mentioned in Robinson (1979a), has ω normal (0,1), $\Theta = \mathbb{R}^1$, $C(x) = (-\infty, x)$, and $\alpha(x) = 1/2$. Earlier examples of similar phenomena include the usual Student-t intervals, c.f. Stein (1961), Buehler and Fedderson (1963), and Brown (1967), and confidence intervals in the Behrens-Fisher problem, c.f. Fisher (1956). It should also be noted that Fisher originated the idea of looking at confidence, conditional on "relevant" subsets, which is the basis for many of the betting examples).

In this and other examples where $\{\alpha(x)\}$ loses in betting, one can ask the crucial question -Is there a better α that could be used? The question has no clear answer, because the purpose of α is not clearly defined. One possible justification for $\alpha(x) = 1/2$ in the above example is that it is the unique limiting probability of $C(x)$ for sequences of what could be called increasingly vague prior distributions (c.f. Stone (1970)). (A more formal Bayesian justification along these lines would be a robust Bayesian justification, to the effect that the class of possible priors is so large that the range of possible posterior probabilities for $(-\infty, x)$ will include ½ for all x). An alternative justification can be found by retreating to decision theory, attempting to quantify how well $\alpha(x)$ performs as an indicator of whether or not θ is in $C(x)$, and then seeing if there is any better α. For instance, using the quadratic scoring function of deFinetti (1962) (any proper scoring function is a possibility -see Good (1952), Savage (1971), Buehler (1971), and Lindley

(1982) for other scoring functions) as an indicator of how well $\alpha(x)$ performs, would mean considering the loss function

$$L(x,\alpha(x),\theta) = (I_{C(x)}(\theta)-\alpha(x))^2. \tag{3.3}$$

(For the moment, we are considering $\{C(x)\}$ as given, and worrying only about the choice of α. Note that, for any "posterior" distribution on θ, the optimal choice of $\alpha(x)$ for (3.3) is the posterior probability of $C(x)$, so (3.3) is a natural measure of the accuracy of α). One can then ask if there is a better α in terms of (3.3), employing usual decision-theoretic ideas. The answer in the case of Example 2 is - no. A standard limiting Bayes argument can be used to show that $\alpha(x) = 1/2$ is admissible for this loss, and hence no improvement (for all θ or all π) is possible. (The same cannot necessarily be said, however, if choice of $C(x)$ is brought into the picture. For instance, a reasonable overall loss for $\{C(x), \alpha(x)\}$ is

$$L(C(x),\alpha(x),\theta) = c_1(I_{C(x)}(\theta)-\alpha(x))^2 + c_2(1-I_{C(x)}(\theta)) + c_3\mu(C(x)),$$

where c_i are constants and μ is a measure of the size of $C(x)$. It can be shown in Example 2 that $\{C^*(x),\alpha^*(x)\}$, with $\alpha^*(x) \equiv ½$ and

$$C^*(x) = \begin{cases} \{x-1\} & \text{with probability } g(x) \\ \{x+1\} & \text{with probability } 1-g(x), \end{cases}$$

is a better procedure than the given $\{C(x),\alpha(x)\}$).

Decision-theoretic inadmissibility, with respect to losses such as (3.3), can be related to incoherency, and seems to be a criterion somewhere between weak incoherency and incoherency (c.f. Robinson (1979a)). This supports the feeling that it may be a more valid criterion than the betting criterion. This is not to say that the betting scenarios are not important. Buehler, in discussion of Fraser (1977), makes the important point that, at the very least, betting scenarios show when quantities such as $\alpha(x)$ "behave differently from ordinary probabilities". And as Hill (1974) says

> "...the desire for coherence...is not primarily because he fears being made a sure loser by an intelligent opponent who chooses a judicious sequence of gambles...but rather because he feels that incoherence is symptomatic of something basically unsound in his attitudes".

Nevertheless, Objection 5 often prevents betting incoherency from having a conclusive impact, and so decision-theoretic inadmissibility (with respect to an agreed upon criterion) is more often convincing.

Decision-theoretic methods of evaluating "inferences" such as $q_x(\theta)$ (i.e., distributions for θ given x) have also been proposed (c.f. Gatsonis (1981) and Eaton (1982)). For the most part, however, there has been little attention directed to these matters.

5. VIOLATION OF THE LP: INADMISSIBILITY AND INCOHERENCY

A violation of the LP will occur when there are two experiments E_1 and E_2, with $x_1' \in \mathbf{X}_1$ and $x_2' \in \mathbf{X}_2$ satisfying (for some positive constant c)

$$f_1(x_1'|\theta) = cf_2(x_2'|\theta) \text{ for all } \theta, \tag{5.1}$$

and for which different actions or conclusions would be recommended were x_1' or x_2' observed. Using the notation of Section 3, it is thus felt that

$$Ev(E_1,x_1') \neq Ev(E_2,x_2'). \tag{5.2}$$

Consider, in this situation, the mixed experiment $E^{1/2}$, in which $J = 1$ or 2 with probability $\frac{1}{2}$ each is observed (independent of elements of the E_i, both probabilistically and structurally), and experiment E_j is then performed. The Weak Conditionality Principle states that

$$Ev(E^{1/2}, (j,x_j)) = Ev(E_j,x_j),$$

which combined with (5.2) yields the conclusion

$$Ev(E^{1/2},(1,x_1')) \neq Ev(E^{1/2}, (2,x_2')). \tag{5.3}$$

Since x_1' and x_2' have proportional likelihood functions, behaving as in (5.3) violates sufficiency and cannot be Bayesian, and will be seen to entail inadmissibility and incoherency in a variety of situations. Note that the experiment $E^{1/2}$ preserves all structural features of E_1 and E_2, so the only possible objection to concluding (5.3) would be to the use of the WCP. It is our understanding, however, that Pivotal, Structural, and virtually all other approaches (except, of course "pure" frequentist theory) accept and extensively use the WCP.

As a final comment before proceeding, note that the formal setup implicitly involves discrete $\mathbf{X}$. As mentioned earlier, however, versions of the LP can be developed for continuous $\mathbf{X}$, the only essential difference being that (5.1) should be replaced by

$$f_1(x_1|\theta) = c(x_1) f_2(g(x_1)|\theta) \text{ for all } \theta \text{ and } x_1 \in B, \tag{5.4}$$

where $B \subset \mathbf{X}$ has $P_\theta^1(B) > 0$ for all θ, $c(x_1) > 0$ for $x_1 \in B$, and g: $\mathbf{X}_1 \to \mathbf{X}_2$ is one-to-one. (For a formulation of this without the assumption of densities, see Berger and Wolpert (1984).) All subsequent expressions should then be understood to hold with x_1' replaced by x_1 (in B) and x_2' replaced by $g(x_1)$. The set B is thus to be a set of positive measure for which x_1 and $x_2 = g(X_1)$ have proportional likelihood functions, and yet supposedly call for differing actions or conclusions. We will usually refrain from explicitely stating conditions or results in the continuous setting, but will nevertheless consider important continuous examples. We look first at the situation from a decision-theoretic viewpoint.

5.1. *Decision - Theoretic Evaluation*

Suppose $Ev(E^{1/2},(j,x_j))$ is decision-theoretic in nature, consisting of the action to be taken when (j,x_j) is observed, to be denoted $\delta((j,x_j)) = \delta_j(x_j)$, along with knowledge that the loss $L((j,x_j), \delta_j(x_j),\theta)$ is to be suffered. Note that this includes decision-theoretic inference, as mentioned in Section 4.2. We will only consider situation in which

$$L((1,x_1'),a,\theta) = L((2,x_2'), a, \theta) \text{ for all } a \text{ and } \theta, \tag{5.5}$$

so that any Bayes rule would be the same whether $(1,x_1')$ or $(2,x_2')$ is observed. (Of course, losses are usually of the form $L(a,\theta)$, with no dependence on the data. The possibility of data dependence is allowed to deal with losses like (3.3).) Because of (5.3), we are assuming that the proposed procedure $\delta^0((j,x_j)) = \delta_j^0(x_j)$ satisties

$$\delta_1^0(x_1') \neq \delta_2^0(x_2'), \tag{5.6}$$

and are out to establish that this is inadmissible. We first look at the most clearcut situation, that of convex loss.

5.1.1. *Convex Loss*

Suppose **A** is convex, and that L satisfies (5.5) and is strictly convex in a for all θ and the observations (j,x_j). Then the procedure

$$\delta^*((j,x_j)) = \begin{cases} \dfrac{c}{(c+1)}\,\delta_1^0(x_1') + \dfrac{1}{(c+1)}\,\delta_2^0(x_2') & \text{for } x_j = x_1' \text{ or } x_2' \\ \delta_j^0(x_j) & \text{otherwise,} \end{cases}$$

(where c is from (5.1)) satisfies (using (5.5) and strict convexity)

$$L((j,x_j'),\delta^*((j,x_j')),\theta) < \frac{c}{(c+1)}\,L((1,x_1'),\delta_1^0(x_1'),\theta) + \frac{1}{(c+1)}\,L((2,x_2'),\delta_2^0(x_2'),\theta). \qquad (5.7)$$

An easy calculation, using (5.1) and (5.5), then shows that (for the Experiment $E^{1/2}$)

$$R(\theta,\delta^0)\text{-}R(\theta,\delta^*) = \frac{(1+c)}{2c}\,f_1(x_1'|\theta)\Delta(\theta) > 0$$

where $\Delta(\theta)$ is the difference between the right and left hand sides of (5.7). The following lemma is immediate. (This is all, of course, a simple form of the Rao-Blackwell theorem).

Lemma 1. In the above situation, δ^0 is

(a) strictly inadmissible, providing x_1' has positive probability for all θ;

(b) extended inadmissible if, in addition to (a), $f_1(x_1'|\theta)$ and L are continuous in θ, and Θ is compact.

(The continuous analog of this lemma is also very easy).

Example 3. Suppose E_1 is binomial (n,θ) and E_2 is negative binomial (m,θ), where X_2 is the number of failures and m, the number of successes at which experimentation stops, is less than n. The densities are

$$f_1(x_1|\theta) = \binom{n}{x_1}\theta^{x_1}(1-\theta)^{n-x_1} \text{ and } f_2(x_2|\theta) = \binom{m+x_2-1}{x_2}\,\theta^m(1-\theta)^{x_2}$$

which are proportional when $x_1 = m \equiv x_1'$ and $x_2 = n-m \equiv x_2'$. Thus, in the mixed experiment $E^{1/2}$, the LP would call for the same action to be taken if either $(1,x_1')$ or $(2,x_2')$ were observed.

If now the goal is to estimate θ under quadratic loss $L = (\theta\text{-}a)^2$ (or any other strictly convex loss), and one uses different estimates of θ for $(1,x_1')$ and $(2,x_2')$, then Lemma 1(a) applies, and the behavior is strictly inadmissible. (Neither Pivotal nor Structural analysis would necessarily say that different actions should be taken in this problem, but Akaike (1982), in criticizing the LP, seems to say that different analyses are called for).

Example 1 (continued). Example 1 is an example where the same probability distribution can arise from different structural models. One component of Structural analysis is that of construction of "structural distributions" for any θ given the data, which can presumably be used, as are posterior or fiducial distributions, to make inferences or probability statements about θ. The structural densities, based on S, for $\theta = (\sigma_1,\tau,\phi)$ are given for the two models (2.2) and (2.3), respectively, by (see Fraser (1968))

$$\pi_1(\theta|s) = K_1(s)f(s|\sigma_1,\tau,\phi)\sigma_1^{-2}\phi^{-1}, \qquad (5.8)$$

$$\pi_2(\theta\,|\,s) = K_2(s)f(s\,|\,\sigma_1,\tau,\phi)(\tau^2+\phi^2)^{-1}\phi^{-1}. \tag{5.9}$$

(These correspond to the posterior distributions with respect to the right invariant Haar measures on the lower and upper triangular group decompositions of Σ).

We now consider the mixed experiment $E^{1/2}$, where E_1 and E_2 are the experiments of observing S (or really $X^1,\ldots,X^n$) from the models (2.2) and (2.3), respectively. Following structuralist theory (maintaining compatibility with the WCP), gives as the structural distribution for $E^{1/2}$

$$\pi(\theta\,|\,(j,s)) = \pi_j(\theta\,|\,s). \tag{5.10}$$

Note that the LP applies with $B = \mathbf{X}_1$ (see (5.4)), g chosen to be the identity map, and $c(x_1) = 1$. Thus, if different actions are to be taken for $(1,s)$ and $(2,s)$, as could well be called for if (5.10) is used, then strict inadmissibility can result. We consider two examples.

Case 1. Suppose it is desired to estimate Σ (which is equivalent to θ) under the strictly convex loss

$$L(\delta,\Sigma) = tr(\delta\Sigma^{-1}) - \log\det(\delta\Sigma^{-1}) - 2. \tag{5.11}$$

(The loss $L(\delta,\Sigma) = tr(\delta\Sigma^{-1} - I)^2$ would work similarly. The losses and following results are all well known, and are discussed in James and Stein (1961), Eaton (1970), Selliah (1964) and Takemura (1982)). It can be shown that the optimal estimator for this loss and the structural distribution (5.10) (treated as a posterior) is $\delta^0((j,s)) = \delta_j^0(s)$, where

$$\delta_1^0(s) = s_L \begin{pmatrix} \frac{1}{n+1} & 0 \\ 0 & \frac{1}{n-1} \end{pmatrix} s_L^t, \quad \delta_2^0(s) = s_U \begin{pmatrix} \frac{1}{n-1} & 0 \\ 0 & \frac{1}{n+1} \end{pmatrix} s_U^t \tag{5.12}$$

where $s = s_L s_L^t = s_U s_U^t$, s_L and s_U being lower and upper triangular, respectively. Also, $\delta_1^0(S)$ and $\delta_2^0(S)$ will differ with probability one, so the inequality (5.7) will hold with probability one. Thus δ^0 is strictly inadmissible. (For indications of how much improvement over δ^0 is possible, see Takemura (1982)).

Case 2. Suppose someone wants to know the "confidence" to be attached to a set $C \subset \mathbf{X}$, based on observation of (j,s) from $E^{1/2}$. Presumably the structuralist would assign confidence

$$\alpha((j,s)) = \int_C \pi_j(\theta\,|\,s)d\theta. \tag{5.13}$$

If now the "success" of such an inference is measured by an "inference loss" such as (3.3), which does satisfy (5.5) and is strictly convex in α, then strict inadmissibility results if C is such that $\alpha((1,S))$ and $\alpha((2,S))$ differ with nonzero probability (and there are many such C). Clearly, many other variations of this theme are possible.

5.1.2. *Complete Class Theorems*

Convex losses are, of course, rather special, and it would be nice to have general theorems concerning strict inadmissibility for other situations. We review below general theorems from statistical decision theory which can be of use in establishing strict inadmissibility. We use the common terminology that an *essentially complete* class of decision rules C is a class such that, if $\delta^0 \notin C$, then there exists a $\delta^* \in C$ such that

$$R(\theta,\delta^*) \leq R(\theta,\delta^0) \text{ for all } \theta, \tag{5.14}$$

while a *complete class* C is a class such that, in addition, $R(\theta,\delta^*) < R(\theta,\delta^0)$ for some θ. We are informal about technical conditions in the following.

Theorem 3. The class of all decision rules based on a sufficient statistic forms an essentially complete class (c.f. Ferguson (1967) or Berger (1980)).

Our interest in this result is, of course, that if (5.1) (or (5.4)) hold, then

$$T(J,X_J) = \begin{cases} (J,X_J) & \text{if } X_J \neq x_1' \text{ or } x_2' \text{ (or } x_1 \text{ or } g^{-1}(x_2) \notin B \text{ for the continuous case).} \\ X_1 & \text{otherwise} \end{cases}$$

is a sufficient statistic, and hence we can find a procedure based only on T (and hence satisfying the LP) which is as good as a procedure δ^0 satisfying (5.6). Unfortunately, we desire to show more: namely, that δ^0 is strictly inadmissible. There are usually two problems in doing this. First, it is necessary to show that the inequality in (5.14) is strict for some θ (i.e., that procedures based on a sufficient statistic form a complete class); and second, that the inequality can be extended to hold for all θ, or at least for all θ in the support of any possible prior π. Considering the last problem first, the following lemma is an easy consequence of complex analysis.

Lemma 2. Suppose Θ is a subset of $\mathbb{R}^m$, that (5.14) holds with strict inequality for some θ, and that both risk functions are analytic functions in each coordinate (as is frequently the case when dealing with exponential families and otherwise). Then $R(\theta,\delta^*) < R(\theta,\delta^0)$, except possibly for $\theta \in \Theta^*$, some set of discrete points having no limit point.

When the conditions of the lemma are satisfied, δ^0 is clearly strictly inadmissible for all nonatomic countably additive priors π. Usually it is possible to do even more: by slightly altering δ^* (using, say, a local averaging process), one can often get strict risk inequality for all θ.

The other concern mentioned above, that equality could hold in (5.14) for all θ, is more of a problem. We discuss below a few of the ways in which this could be attacked.

Lemma 3. If $R(\theta,\delta^*) = R(\theta,\delta)$ implies that $\delta^* = \delta$ (with probability 1 for all θ), then procedures based on a sufficient statistic form a complete class (and hence inequality will hold in (5.14) for at least one θ).

The condition in Lemma 3 can be verified for a number of situations (besides the obvious one of a convex loss). For instance, if the nonrandomized rules form a complete class (as in finite action problems with nonatomic densities -see, e.g., Dvoretsky, Wald, and Wolfowitz (1951), and in location parameter problems - see, e.g. Farrell (1964)), then this is easily seen to be satisfied. (If $R(\theta,\delta^*) = R(\theta,\delta)$, the randomized rule which chooses between δ^* and δ with probability 1/2 has the same risk, but by assumption can be improved upon). Another possible situation in which this can be verified is when $R(\theta,\delta)$ can be expressed as $E_\theta\, \psi(\delta(X))$ (thus $\psi(\delta(X))$ is an "unbiased estimator for risk" - c.f. Stein (1981)), ψ is a one-to-one operator, and $\{P_\theta\}$ is a complete family of distributions. For other situations in which the condition of Lemma 3 can be satisfied, see Brown, Cohen, and Strawderman (1980).

Lemma 4. The Bayes rules form a complete class if Θ is compact, $R(\theta,\delta)$ is continuous in θ for all δ, $L(\theta,\bullet)$ is lower semicontinuous, and **A** is a complete separable metric space and is compact or has a suitable compactification (c.f. Wald (1950) or Brown (1976)).

The Bayes rules also form a complete class in certain testing situations with compact null hypothesis (c.f. Brown, Cohen and Strawderman (1980)). The use of such a result is that it can sometimes be verified directly that δ^0 cannot be a Bayes rule. More generally, a complete class can sometimes be shown to consist of appropriate limits of Bayes rules, and it may be possible to show that δ^0 cannot be a limit of Bayes rules.

A number of examples of strict inadmissibility in, say, Example 1, could be developed using the results in this subsection. For instance, testing or finite action problems could be formulated, in which choosing the action according to the structural distribution (5.10) results in (effectively) a randomized and hence inadmissible rule, which could furthermore be shown to be strictly inadmissible via analyticity and monotonicity arguments. The point of this section, however, was more to convey the feeling that violation of the LP is in general, likely to result in some form of inadmissibility.

5.2. *Betting Evaluations*

First, it is easy to show that violation of the LP leads to, at least, weak incoherence.

Theorem 4. Consider the mixed experiment $E^{1/2}$ in the discrete case, and suppose a fixed set $C \subset \Theta$ is assigned "confidence" $\alpha((j,x_j'))$ when (j,x_j') is observed, where $\alpha((1,x_1')) \equiv \alpha_1 \neq \alpha_2 \equiv \alpha((2,x_2'))$. Consider the following strategy: define

$$L = \begin{cases} 1 & \text{if } \alpha_1 < \alpha_2 \\ 2 & \text{if } \alpha_2 < \alpha_1, \end{cases}$$

place no bet unless X_J equals x_1' or x_2', and then bet $c_J\alpha_J$ that $\theta \in C$ if $J = L$ and $c_J(1-\alpha_J)$ that $\theta \notin C$ if $J \neq L$, where $c_1 = 1$ and $c_2 = c$ (from (5.1)). The expected gain for this betting strategy (assuming odds corresponding to α_J are being given if (J,x_J') is observed) is $\frac{1}{2} f_1(x_1'|\theta)|\alpha_1-\alpha_2|$. Hence the probabilities assigned to C are weakly incoherent if $f_1(x_1'|\theta) > 0$ for all θ.

Proof. A straightforward calculation, using (5.1).

Comment 1. In the continuous case, the theorem also holds, the only changes needed being the replacement of α_j by $\alpha_j(x_j)$ (defined as the "confidence" in C if (j,x_j) is observed), betting only if $j = 1$ and $x_1 \in B$ or $j=2$ and $x_2 \in g(B)$ (see (5.4)), letting $L = L(x^*) = 1$ if $\alpha_1(x_1) < \alpha_2(g(x_1))$ and letting $L = 2$ otherwise (where x^* denotes occurrence of x_1 or $x_2 = g(x_1)$), and replacing c_j by $c_j(x_j)$, defined as 1 for $j = 1$ and as $c(g^{-1}(x_2))$ (see (5.4)) for $j = 2$. The expected gain is then

$$\int_B \tfrac{1}{2}\, |\alpha_1(x_1) - \alpha_2(g(x_1))|\, f_1(x_1|\theta)d\theta.$$

Comment 2. The above results only prove weak incoherence, although under suitable extra conditions they imply incoherence. To prove incoherence in general, results such as those in Heath and Sudderth (1978) and Lane and Sudderth (1983) can often be employed. Under minor technical conditions, these papers show that the quoted posterior (or structural distribution) $\pi(\theta|(j,x_j))$ for $E^{1/2}$ is incoherent unless it is the posterior for a (in general, finitely additive) prior on Θ. This seems unlikely to be the case if the LP has been violated. Situations like Example 1 require the finitely additive theorems, and it is not clear whether (5.10) is -or is not- the posterior for a finitely additive prior (we would guess not), but Example 3 requires only the countably additive theorems of Lane and Sudderth (1983) (which apply essentially whenever **X** or Θ is compact).

5.3. *The Stone Example*

As a final interesting example of these ideas, consider the well known example of Stone (1976) (essentially done earlier by Piesakoff (1950) in a less entertaining fashion), in which a soldier leaves a bar at 0, walking one block in each of a succession of randomly selected directions N, S, E, W, the only restriction being that he never immediately backtracks. The soldier trails a taut string, and at some point stops and buries a treasure. Let θ denote the path (a succession of the symbols N, S, E, and W) to this point. (Thus Θ is effectively the free group on two generators). The soldier then picks a random direction (by spinning a lady) walks one block in that direction (drawing up the string if necessary) and passes out. We observe his complete path X the next day, and have one guess as to where the treasure is. Letting ω denote a random variable that is N, S, E, or W with probability ¼ each, the above can be modeled "structurally" as $X = \theta\omega$, with the convention that, if the last two symbols in a string are opposite directions, they cancel.

Although, given x, the likelihood function for θ assigns value ¼ to each of xN, xS, xE and xW, and it would seem that this is the natural structural distribution for θ (obtained by taking θ to be $x\omega^{-1}$ with the probabilities associated wth ω), Fraser (in the discussion of Stone (1976)) argues that the correct structural distribution is

$$\pi_1(\theta\,|\,x) = \begin{cases} \frac{3}{4} & \text{if } x \text{ is an extension of } \theta \\ \frac{1}{12} & \text{if } \theta \text{ is an extension of } x \\ 0 & \text{otherwise.} \end{cases}$$

In arguing against the LP, Fraser, in the discussion of Hill (1981), constructs an alternative model for X and Θ as follows. Let x_0 denote a particular fixed path, suppose θ is as above, and let X be determined according to the probabilities $P(X=0\,|\,\theta=x_0) = 1$; $P(X=x_0\,|\,\theta=x_0z) = 1/4$ and $P(X=0\,|\,\theta=x_0z) = 3/4$ for z = N, S, E, and W; and $P(X=\theta\,|\,\theta) = 1/4$ and $P(X=0\,|\,\theta) = 3/4$ for other θ. (The soldier trails an *elastic* string, and after burying the treasure at the end of $\theta \neq x_0$ he passes out and has a 75 % chance of being snapped back to 0; the end of x_0, however, is very slippery, so if the soldier buries the treasure there and passes out he will be snapped back to 0 for sure. There also happens to be a "lady" who walks the streets within one block of the end of x_0 (her place of business), and if the soldiers passes out at x_0N, x_0S, x_0W, or x_0E and doesn't get snapped back to 0, the lady will take him back to x_0).

For this model, if the observation is $X = x_0$, the likelihood function again assigns value 1/4 to each of x_0N, x_0S, x_0E, and x_0W, and due to additional "symmetry" in the model, Fraser feels that this is an accurate representation of the probabilities of each θ: thus it appears that the recommended structural distribution (when x_0 is observed) is

$$\pi_2(\theta\,|\,x_0) = 1/4 \qquad \text{for } \theta = x_0\text{N}, x_0\text{S}, x_0\text{E}, \text{and } x_0\text{W}.$$

Consider now the mixed experiment $E^{1/4}$. The choice $\lambda = 1/4$ is more convenient than $\lambda = 1/2$. (For instance, suppose that on 1/4 of the nights the soldier leaves the bar at 0 with a lady and follow's Stone's scenario, and the other 3/4 of the nights he leaves alone and follows the alternate scenario. Upon finding the soldier in the morning we know which scenario eventualized, because if he is unaccompanied when he passes out he bumps his head as he falls). By the WCP, the structural distribution for θ in $E^{1/4}$ should satisfy

$$\pi(\theta \mid (j,x_0)) = \pi_j(\theta \mid x_0). \tag{5.15}$$

Needless to say, if someone were to repeatedly act in accordance with (5.15), he would be operating in an inferior fashion. From a betting viewpoint this is easy to establish. Let us instead, however, consider a decision-theoretic scenario.

A forgetful professor follows a permanently placed string to his office each day. The path happens to be x_0. Every so often, he notices a string running parallel to his and, if the strings run together for the entire journey, the professor is intrigued and mentions the curiosity to a soldier he sometimes finds at the end of x_0. The soldier, in such cases, tells the professor the situation and offers to let him look in one direction for the treasure (which is worth one unit). The soldier requires, however, that the professor pay him for this privilege, the cost being .7 if the soldier had not bumped his head and .2 if he had bumped his head (in which case he is confused and sells out cheaply). In the first case ($J=1$) the professor following (5.15) is 75 % "sure" that the treasure can be found by backtracking one block, and in the second case ($J=2$) feels that any direction is equally likely. He, hence, accepts the offer of the soldier in either case. For definiteness, let us suppose that, when $J=2$, the professor randomly chooses one of the three directions that extend the path x_0.

Let $m(x_0)$ denote the probability with which the above event happens, and let p_N, p_S, p_E, and p_W denote the probabilities with which the soldier buries the treasure at x_0N, x_0S, x_0E, and x_0W, given that $X = x_0$. One of the directions is the "backtrack" direction, say N (recall x_0 is fixed), so the professor's long run gain will be

$$m(x_0)\{1/4\,[p_N - .7] + 3/4\,[1/3\,(p_S - .2) + 1/3(p_E - .2) + 1/3(p_W - .2)]\} = m(x_0)(-.075).$$

Thus he will lose money if $m(x_0) > 0$. Of course, all the above probability calculations are based on assuming the *existence* of a true (countably additive) probability distribution π describing the soldier's path θ (but not on knowledge of this distribution).

It is of interest to see how a Bayesian would approach the problem. He would think about the generation of θ, perhaps deciding that the soldier has a probability p_n of having a path θ of length n, and assigning equal probability to all paths of length n, there being $N_n = 4 \cdot 3^{n-1}$ such paths. Thus the prior probability of a particular path of length n would be $\pi(\theta) = p_n/N_n$. Now if $X = x_0$ is observed, x_0 being of length m, the posterior probability that θ is x_0 "backtracked" can be calculated to be

$$9 \cdot p_{m-1}/(3\,p_{m+1} + 9\,p_{m-1}),$$

while the posterior probability that θ is any particular "extension" of x_0 is

$$p_{m+1}/(3\,p_{m+1} + 9\,p_{m-1}).$$

If p_{m+1} is thought to be approximately equal to p_{m-1}, then the posterior distribution is essentially π_1. The Bayesian, of course, feels this is reasonable for *either* of the two models when x_0 is observed (since the likelihood functions are the same), and will play the soldier's game but will always backtrack. (This Bayesian analysis is essentially that in Dickey's discussion of Stone (1976)).

6. CONCLUDING REMARKS

1. What is being criticized about Pivotal and Structural analysis is, at most, a very small part of the two theories. The major part of both theories is the reduction of the

original data and model to simpler entities which preserve all available information. The reductions are especially valuable in the very common situations where the structural model is known with considerable confidence, but the distribution of the error component, ω, is uncertain. All theories of inference, including Bayesian, can take advantage of the simplifications that result from such "necessary" reductions.

Pivotal and Structural analysis come into possible conflict with the LP only at the terminal stage of analysis. We are actually somewhat unclear as to when terminal Pivotal analysis conflicts with the LP; we have seen statements to the effect that it can, but not explicit examples. (This explains the emphasis on examples from the structural theory in this paper).

Structural theory can conflict with the LP at the terminal stage of analysis in two ways: first, when significance testing or frequentist confidence intervals are developed, and second, when structural distributions are created. We have concentrated on the problems arising with conflicting structural distributions, but again must emphasize that terminal analysis with structural distributions is only a very small part of Structural analysis. (Indeed, in Fraser (1979), very little emphasis is placed on structural distributions). Of course, the more classical frequentist type of terminal analysis can also conflict with the LP, but probably not too seriously. Indeed the degree to which Pivotal and Structural analysis violate the LP is, on the whole very small, and not really worth making an issue of, except that the theories purport to establish the lack of validity of the LP.

Incidentally, the "disproof" of the LP (by Structural analysis at least) seems to simply be the fact that the recommended terminal analysis, especially that based on structural distributions, can conflict with the LP. To us, however, the justification for terminal analysis with structural distributions is not on very firm footing (compared with the reduction analyses of structural theory), since it involves, at some point, a fiducial type inversion. Besides the examples in this paper, which point out questionable properties of such terminal analysis, there are the examples with non-amenable groups. Indeed, if in the situation of Example 1 the model is $X^t = A\omega^t$, where A is known only to be a nonsingular matrix and of interest is $\Sigma = AA^t$, then the structural analysis (c.f. Fraser (1973)) gives as a structural distribution the posterior distribution with respect to the (right and left) Haar measure on the full linear group, which is non-amenable. Use of this distribution is well known to result in extended inadmissibility and incoherence in a wide number of situations (such as in the estimation problems considered in Example 1 (continued)). Likewise, in the Stone example (where Θ is again non-amenable), the "natural" structural distribution for θ seems to be the one giving probability ¼ to each of the paths compatible with x, and is fairly clearly bad, and while Fraser gives a justification for the alternate structural distribution π_1, he does not clearly expose the error in the derivation of the "natural" structural distribution.

Of course, all "objective" statistical theories have trouble dealing with problems involving non-amenable groups (c.f. Bondar (1981)), and Structural analysis often fares better than most: it frequently bases the analysis on amenable subgroups of a non-amenable group, such as in the earlier version of Example 1. (Structural analysis also justifies use of the right invariant Haar measure as a noninformative prior, a justification also obtained from classical invariance theory, but lacking in many purely Bayesian theories).

2. While this paper concentrated on Pivotal and Structural theories because of their clearly voiced opposition to the full LP, there are, of course, many other theories of

inference which also violate the LP and are susceptible to the same inadmissibility and incoherency criticism. Among these are fiducial theory (c.f. Wilkinson (1977)), plausibility inference (c.f. Barndorff-Nielsen (1976)), and many noninformative prior Bayesian theories in which the noninformative prior depends on E, so that one may, in a situation like Example 3, end up using two different noninformative priors to process proportional likelihood functions. Examples of inadmissibility and incoherency are easy to construct for any such situation.

Of course, one "escape" available to any theory conflicting with the LP is to reject the Weak Conditionality Principle, for then, when faced with the mixed experiment $E^{\wedge}$, one could change the component experiment analyses. Rejecting what to many is the only "obvious" principle in statistics is not a very appealing escape, however.

3. Those who remain unconvinced by the inadmissibility and incoherency arguments, and stand fast in their objection to the LP because it ignores the structural model, must still take notice of certain related principles, such as the Stopping Rule principle of Barnard (1949) and the Censoring principal of Pratt (in Birnbaum (1962)). (See also Pratt (1965), Basu (1975), and Berger and Wolpert (1984) for very general versions). These principals have a crucial impact on statistics and can be derived using only the Weak Conditionality Principle and a structural version of sufficiency which preserves the structural information.
4. The LP is non-operational, in the sense that it does not say how the likelihood function is to be used. Cogent arguments can be given (c.f. Basu (1975) and Berger and Wolpert (1984)) that only Bayesian utilization of the likelihood function really makes sense. However, Bayesian analysis has to be concerned with sensitivity to the prior input, and, for a variety of practical and theoretical reasons, a sensible analysis from this viewpoint may formally violate the LP (usually, by having some aspects of the prior depend on E or the data). A theoretical justification for possible violation of the LP could be given in terms of Good's Type II Rationality (c.f. Good (1976)), but the practical necessities are fairly clear. In the same way, practical considerations may lead to a mild degree of inadmissibility or incoherency. (For extensive discussion of these issues and references see Berger and Wolpert (1984) and Berger (1984)). The LP (and admissibility and coherence) should be considered ideals, to which one should strive to adhere, rather than absolute prescriptions.

ACKNOWLEDGMENTS

I would like to thank Leon Gleser, Ker-Chau Li, and Herman Rubin for valuable discussions on these matters. This research was supported by the National Science Foundation under Grants MCS-8101670 and DMS-8401996.

REFERENCES

AKAIKE, H. (1982). On the fallacy of the likelihood principle. *Statistics and Prob. Letters,* **1,** 75-78.

BARNARD, G.A. (1949). Statistical inference. *J. Roy. Statist. Soc.* B, **11,** 115-139 (with discussion)

— (1980). Pivotal inference and the Bayesian controversy. *Bayesian Statistics* (J.M. Bernardo, M.H. DeGroot, D.V. Lindley and A.F.M. Smith, eds.) Valencia: University Press, 293-318 (with discussion).

— (1982). A coherent view of statistical inference. To be published in the proceedings of the Symposium on Statistical Inference and Applications. University of Waterloo, August, 1981.

BARNARD, G.A. and GODAMBE, V.P. (1982) Allan Birnbaum (memorial article). *Ann. Statist.,* **10,** 1033-1039.

BARNARD, G.A., JENKINS, G.M., and WINSTEN, C.B. (1962). Likelihood inference and time series. *J. Roy. Statist. Soc.* A, **125,** 321-372.

BARNARD, G.A. and SPROTT, D.A. (1983). The generalised problem of the Nile: robust confidence sets for parametric functions. *Ann. Statist.* **11,** 104-113.

BARNDORFF-NIELSEN, O. (1976). Plausibility inference. *J. Roy. Statist. Soc. B,* **38,** 103-131. (with discussion).

BASU, D. (1975). Statistical information and likelihood. *Sankhyā,* A, **37** 1-71 (with discussion).

BERGER, J. (1980). *Statistical Decision Theory: Foundations, Concepts and Methods.* New York: Springer-Verlag.

— (1984). The robust Bayesian viewpoint. *Robustness in Bayesian Statistics* (J. Kadane, ed.) Amsterdam: North-Holland.

BERGER, J. and WOLPERT, R. (1984). *The Likelihood Principle: A Review and Generalizations.* To appear in the Monograph Series of the Institute of Mathematical Statistics.

BIRNBAUM, A. (1962). On the foundations of statistical inference. *J. Amer. Statist. Assoc.,* **57,** 269-306 (with discussion).

— (1969). Concepts of statistical evidence. *Philosophy, Science and Method: Essays in honor of Ernest Nagel.* (S. Morgenbesser, P. Suppes, and M. White, eds.). New York: St. Martin's Press.

— (1970). On Durbin's modified principle of conditionality. *J. Amer. Statist. Assoc.,* **65,** 402-403.

— (1972). More on concepts of statistical evidence. *J. Amer. Statist. Assoc.,* **67,** 858-861.

BONDAR, J.V. (1977). On a conditional confidence principle. *Ann. Statist.,* **5,** 881-891.

BONDAR, J.V. and MILNES, P. (1981). Amenability: a survey for statistical applications of Hunt-Stein and related conditions on groups. *Z. Wahrscheinlich, verw. Gebiete,* **57,** 103-128.

BROWN, L.D. (1967). The conditional level of Student's t-Test. *Ann. Math. Statist.,* **38,** 1068-1071.

— (1973). Estimation with incompletely specified loss functions. *J. Amer. Statist. Assoc.,* **70,** 417-427.

— (1976). *Notes on Statistical Decision Theory.* (Unpublished lecture notes, Ithaca).

BROWN, L.D., COHEN, A., and STRAWDERMAN, W.E. (1980). Complete classes for sequential tests of hypothesis. *Ann. Statist.,* **8,** 377-398.

BUEHLER, R.J. (1959). Some validity criteria for statistical inference. *Ann. Math. Statist.,* **30,** 845-863.

— (1971). Measuring information and uncertainty. In *Foundations of Statistical Inference* (V.P. Godambe and D.A. Sprott, eds.). Toronto: Holt, Rinehart and Winston Publishing Co.

— (1976). Coherent preferences. *Ann. Statist.,* **4,** 1051-1064.

BUEHLER, R.J. and FEDDERSON, A.P. (1963). Note on a conditional property of Student's t. *Ann. Math. Statist.,* **34,** 1098-1100.

BUNKE, H. (1975). Statistical inference: Fiducial and structural vs. likelihood. *Math. Operationforsh. U. Statist.,* **6,** 667-676.

CORNFIELD, J. (1969). The Bayesian outlook and its application. *Biometrics,* **25,** 617-657 (with discussion).

COX, D.R. (1958). Some problems connected with statistical inference. *Ann. Math. Statist.,* **29,** 356-372.

DAWID, A.P. (1977). Conformity of inference patterns. *Recent Developments in Statistics* (J. R. Barra, et. al., eds.). Amsterdam: North Holland Publishing Co.

DAWID , A.P. and STONE, M. (1982). The functional-model basis of fiducial inference. *Ann. Statist.* **10**, 1054-1073.

DAWID, A.P., STONE, M. and ZIDEK, J.V. (1973). Marginalization paradoxes in Bayesian and structural inference. *J. Roy. Statist. Soc.* B, **35**, 189-233.

DEFINETTI, B. (1937). Foresight: its logical laws, its subjective sources. *Studies in Subjective Probability* (H.E. Kyburg and H.E. Smokler, eds., 1962). New York: Wiley.

— (1962). Does it make sense to speak of 'Good probability appraisers'? *The Scientist Speculates* (I.J. Good, ed.). New York: Basic Books.

— (1974). *Theory of Probability*. Volumes 1 and 2. New York: Wiley.

DICKEY, J.M. (1976). Approximate posterior distributions. *J. Amer. Statist. Assoc.*, **71**, 680-689.

DURBIN, J. (1970). On Birnbaum's theorem on the relation between sufficiency, conditionality, and likelihood. *J. Amer. Statist. Assoc.*, **65**, 395-398.

EATON, M.L. (1970). Some problems in covariance estimation. *Tech. Rep.*, **49**, Stanford University.

— (1982). A method for evaluating improper prior distributions. *Statistical Decision Theory and Related Topics III* (S. S. Gupta and J. Berger, eds.). New York: Academic Press.

FARRELL, R.H. (1964). Estimtors of a location parameter in the absolutely continuous case. *Ann. Math. Statist.*, **35**, 949-998.

FERGUSON, T.S. (1967). *Mathematical Statistics: A Decision-Theoretic Approach*. New York: Academic Press.

FISHER, R.A. (1956). *Statistical Methods and Scientific Inference.* Edinburgh: Oliver and Boyd.

FRASER, D.A.S. (1956). On the sufficiency and likelihood principles. *J. Amer. Statist. Assoc.*, **58**, 641-647.

— (1968). *The Structure of Inference.* New York: Wiley.

— (1972). Bayes, likelihood, or structural. *Ann. Math. Statist.*, **43**, 777-790.

— (1973). Inference and redundant parameters. *Multivariate Analysis-III* (P.R. Krisnaiah, ed.). New York: Academic Press.

— (1977). Confidence, posterior probability, and the Buehler example. *Ann. Statist.*, **5**, 892-898.

— (1979). *Inference and Linear Models.* New York: McGraw Hill.

FREEDMAN, D.A. and PURVES, R.A. (1969). Bayes methods for bookies. *Ann. Math. Statist.*, **40**, 1177-1186.

GATSONIS, C.A. (1981). *Estimation of the posterior density of a location parameter.* Ph.D. Thesis, Cornell University, Ithaca.

GODAMBE, V.P. (1979). On Birnbaum's mathematically equivalent experiments. *J. Roy. Statist. Soc.* B, **41**, 107-110.

GOOD, I.J. (1952). Rational decisions. *J. Roy. Statist. Soc. B,* **14**, 107-114.

— (1976). The Bayesian influence, or how to sweep subjectivism under the carpet. *Foundations of Probability Theory, Statistical Theories of Science,* Vol. II (W.L. Harper and C.A. Hooker, eds.). Dordrecht: Reidel.

HEATH, D. and SUDDERTH, W. (1978). On finitely additive priors, coherence, and extended admissibility. *Ann. Statist.*, **6**, 333-345.

HILL, B. (1974). On coherence, inadmissibility, and inference about many parameters in the theory of least squares. *Studies in Bayesian Econometrics and Statistics* (S. Fienberg and A. Zellner, eds.). Amsterdam: North Holland Publishing Co.

— (1981). On some statistical paradoxes and non-conglomerability. *Bayesian Statistics* (J.M. Bernardo, M.H. DeGroot, D.V. Lindley and A.F.M. Smith, eds.). Valencia: University Press, 39-66 (with discussion).

JAMES, W. and STEIN, C. (1961). Estimation with quadratic loss. *Fourth Berkeley Symposium Math. Statist. and Prob.* Berkeley: University of California Press.

KALBFLEISCH, J.D. (1974). Sufficiency and conditionality. *Proceedings of the Conference on Foundational Questions in Statistical Inference* (O. Barndorff-Nielsen, P. Blaesild and. G. Schou, ed.). Department of Theoretical Statistics, University of Aarhus.

KIEFER, J. (1977). Conditional confidence statements and confidence estimators *J. Amer. Statist. Assoc.,* **72,** 789-827, (with discussion).

LANE, D.A. and SUDDERTH, W.D. (1983). Coherent and continuous inference. *Ann. Statist.,* **11,** 114-120.

LECAM, L. (1955). An extension of Wald's theory of statistical decision functions. *Ann. Math. Statist.* **26,** 69-81.

LEVI, I. (1980). *The Enterprise of Knowledge.* Cambridge: MIT Press.

LINDLEY, D.V. (1982). Scoring rules and the inevitability of probability. *Int. Statist. Rev.,* **50,** 1-26.

MAULDON, J.G. (1955). Pivotal quantities for Wishart's and related distributions and a paradox in fiducial theory. *J. Roy. Statist. Soc.,* B, **17,** 79-85.

NEYMAN, J. and PEARSON, E.S. (1933). On the problem of the most efficient tests of statistical hypotheses. *Phil. Trans. Roy. Soc.* A, **231,** 289-337.

PEISAKOFF, M. (1950). *Transformation Parameters.* Ph.D. Thesis, Princeton University.

PIERCE, D.A. (1973). On some difficulties with a frequency theory of inference. *Ann. Statist.,* **1,** 241-250.

PRATT, J.W. (1965). Bayesian interpretation of standard inference statements *J. Roy. Statist Soc.* B, **27,** 169-203, (with discussion).

RAMSEY, F.P. (1926). *Truth and Probability in the Foundations of Mathematics.* Patterson, New Jersey: Littlefield, Adams, and Co., 1960.

ROBINSON, G.K. (1979a). Conditional properties of statistical procedures. *Ann. Statist.,* **7,** 742-755.

— (1979b). Conditional properties of statistical procedures for location and scale parameters. *Ann. Statist.,* **7,** 756-771.

SAVAGE, L.J. (1954). *The Foundations of Statistics.* New York: Wiley.

— (1971). Elicitation of personal probabilities and expectations. *J. Amer. Statist. Assoc.,* **66,** 783-801.

SELLIAH, J. (1964). *Estimation and testing problems in a Wishart distribution.* Ph.D. Thesis, Dept. of Statistics, Standford University.

STEIN, C. (1961). Estimation of many parameters. *Inst. Math. Statist. Wald Lectures,* Unpublished.

— (1981). Estimation of the mean of a multivariate normal distribution. *Ann. Statist.,* **9,** 1135-1151.

STONE, M. (1970). Necessary and sufficient conditions for convergence in probability to invariant posterior distributions. *Ann. Math. Statist.,* **41,** 1349-1353.

— (1976). Strong inconsistency from uniform priors, *J. Amer. Statist. Assoc.,* **71,** 114-125, (with discussion).

TAKEMURA, A. (1982). An orthogonally minimax estimator of the covariance matrix of a multivariate normal population. *Tech. Rep.* Stanford University.

WALD, A. (1950), *Statistical Decision Functions.* New York: Wiley.

WALLACE, D.L. (1959). Conditional confidence level properties. *Ann. Math. Statist.,* **30,** 864-876.

WILKINSON, G.N. (1977). On resolving the controversy in statistical inference. *J. Roy. Statist. Soc.* B, **39,** 119-171, (with discussion).

DISCUSSION

G.A. BARNARD (*Retired*)

Professor Berger is to be congratulated on the most through examination of the axiomatics of the likelihood principle that I have seen. He set out, I believe, to prove the general validity of the principle, using arguments other than those based on coherency, and frankly admits his lack of success. He is to be congratulated too, I believe, on this lack, since it indicates the rigour of his reasoning. In spite of some impressions I may have given to the contrary, I have never myself believed in the general applicability of the likelihood principle. My formulations and "advocacy" of the principle have been intended to call attention to the importance of the concept of likelihood -- something which badly needed doing, I think, and which, even now, is neglected in the textbook literature. When one is confronting a set of data reasonably judged consistent with a model involving unknown parameters, the first step should consist of an attempt to visualise the shape of the likelihood function. I do not know of any elementary text which tells students this (reckoning Jim Kalbfleish's book as not quite elementary), and yet the concept of likelihood is surely well within the reach of anyone likely to be concerned with making statistical inferences.

Those of us who had the pleasure of being at the first Valencia Conference will remember that toward the end we had a discussion on "where do we go from here"? the general conclusion of which was, that we should move towards the consideration of applications. And there have been a number of conferences since then which have gone in that direction. But when statisticians *as such* get together --as ditinct from medical, economic, or other specific kinds of statistician-- we are bound to spend some time on general issues, so long as we don't stray too far from applications. I would therefore like to convey my views on the likelihood principle via a "quasi-concrete" instance --not really a practical example, but one whose relationship to practice is reasonably close.

Suppose we are interested in the kind of problem which gave rise originally to the term "regression" --the correlation between stature of father and stature of son. Suppose we begin by taking it as known that the distribution of stature in the population is approximately normal, and has remained nearly invariant from generation to generation over a long period. This latter assumption has ceased to hold in advanced countries over the past half century; but for the present we shall ignore this just as, for simplicity, we have made the normality assumptions in spite of the fact that persons of negative height have not, so far, been observed. Then we can imagine that we have available a vast mass of data on stature which enables us to determine the mean and standard deviation of the distribution with negligible error. Then the data from a modern father-son pair can be expressed in the form (x,y), where x represents the deviation of father's height from the mean, in S.D. units, and y similarly represents that of the (eldest) son. Our model will imply that (x,y) are jointly normal, with joint density

$$(1/2\pi(1-\theta^2)^{1/2}) \exp -(x^2-2\theta xy+y^2)/2(1-\theta^2), \tag{1}$$

symmetric in x and y. This density has been used as an example of non-uniqueness of ancillaries, since x and y are each standard normal, whatever the value of Θ, and so either can be taken as ancillary, but both together cannot.

But the symmetry of (1) between x and y fails to express the fact that fathers necessarily precede sons. As is well known, Galton introduced the term "regression" -- more fully, regression towards the mean-- because he observed that sons of tall fathers tend to be less tall than their fathers, and conversely. The fact that fathers precede sons can be expressed by introducing e_1, e_2 to denote a pair of standard normal, independent variables, and expressing the model in the form

$$x = e_1 \tag{2}$$

$$y = \theta x + (1-\theta^2)^{1/2} e_2$$

or, equivalently,

$$e_1 = x \tag{3}$$

$$e_2 = (y-\theta x)/(1-\theta^2)^{1/2}$$

This model is in *pivotal form.* The "basic pivotals" --quantities in general which are functions of the observables and the parameters, whose distributions are taken as known, not involving the parameters--, are e_1, e_2, whose joint distribution is the circular standard normal. The general procedure of inference from a pivotal model is to transform by *known* 1-1 transformations the joint density of the pivotals in such a way that a maximal function of the basic pivotals is found which is constant on the parameter space. This is maximal ancillary (it is always essentially unique), and the inference problem is reduced by conditioning the distribution of the remaining transformed pivotals on the observed values of the maximal ancillary (Detailed examples are given in my paper printed in the proceedings of the first Valencia conference). In the present case, with just one observation pair, no transformation is necessary --the maximal ancillary is $e_1 = x$, and conditioning on this has no effect on the distribution of e_2, so that the reduced model is

$$e_2 = (y-\theta x)/(1-\theta^2)^{1/2} \quad \text{is } N(0,1) \tag{4}$$

By the familiar "confidence" argument we can deduce propositions about θ from (4), when the values of (x,y) are known. If we assert the proposition equivalent to $|e_2| < 1.96$, we shall be wrong with long run frequency 5 %; more directly, we can say, with given observations (x,y) substituted in the right hand side of (4), that unless an event of probability .05 or less has occurred, this expression must lie between -1.96 and +1.96.

Now consider two possible observation pairs, $(x,y) = (1,2)$ and $(x,y) = (2,1)$. We tabulate the values of e_2 as a function of θ:

	θ									
	−.9	−.7	−.5	−.3	−.1	+.1	+.3	+.5	+.7	+.9
$(2-\theta)/(1-\theta^2)^{1/2}$:	6.653	3.781	2.887	2.411	2.111	1.910	1.782	1.732	1.820	2.524
$(1-2\theta)/(1-\theta^2)^{1/2}$:	6.424	3.361	2.309	1.677	1.206	0.804	0.419	0.000	−0.560	−1.835

The inferences to be made from the two pivotals are quite different. From the second we find that $|e_2| < 1.96$ corresponds to θ falling between -0.399 and $+0.909$, and there is a $1-1$ relationship between possible values taken by e_2 and possible values for θ. From the first we find $|e_2| < 1.96$ corresponds to θ falling between 0.042 and 0.785, but the relationship between e_2 values and θ values is not $1-1$, and in particular, no θ makes e_2 less than 1.732. Whereas with the observation $(x,y) = (2,1)$ we can reasonably associate the improbability of e_2 exceeding 1.96 with the corresponding range of θ, and say, without qualification "unless an event of probability 0.05 or less has occurred, θ lies between 0.042 and 0.785", with the observation $(x,y) = (1,2)$ we know that an event of probability 0.042 *must* have occurred, since no value of θ makes e_2 less than 1.732; it is clearly illegitimate to associate the improbability of e_2 values with implausibility of θ values. Rather, the fact that (1,2) implies anyway that an improbable event has occurred means that this observation pair casts doubt on the correctness of the model. This corresponds to our intuitive interpretation, since (2,1) means that an exceptionally tall father has a less exceptionally tall son --nothing to be surprised at; but (1,2) means that a rather tall father has an exceptionally tall son-- the opposite of "regression", and, though not, of course, conclusive, a suspicious result, casting doubt on our assumption that stature is not trending upwards.

I could be argued that with the observation (2,1) the value 2 for e_1 itself is implausible --and if there were grounds for supposing that the distribution of parental statures might depart from normality in the direction of positive skewness, this should be scrutinised. The point I wish to make is, however, that the inferences to be drawn from (1,2) and from (2,1) *are* different, whereas the likelihood function is the same. That is, we have here a counter-example to the general applicability of the likelihood principle.

I would like to take the argument a little further. As I have pointed out elsewhere, the pivotal mode on inference includes the Bayesian. For example, we have only to add to e_1 and e_2 a further component of our basic pivotal:

$$e_3 = \theta$$

and assign to it whatever prior we judge appropriate --uniform over (-1, + 1), perhaps--and we shall then find that the maximal ancillary is now enlarged to the pair (x,y). Conditioning of this will now give the Bayesian posterior for θ in the usual way. This will, in particular, be symmetric as between (1,2) and (2,1) so that this additional assumption has the effect of masking the lack of symmetry between fathers and sons. How does this relate to arguments involving "coherence"?

The point has not, I think, been made sufficiently often, that "incoherent" procedures can be guaranteed to result in the possibility of a "Dutch book" only if the statistical model involved can be guaranteed correct. If we want to allow for the possibility that our mathematical model may not be right, then we may need to be cautious in "posting" our bets --to proceed "step by step", as Fisher was wont to emphasise. In the example we have used, the pivotal analysis procedure allows us to do this. Having set up the model (3) we can reduce it to (4), without introducing (5). *If* we have confidence in our model, we can then proceed to add the assumption (5) and we shall obtain the same final result as would have been obtained by assuming both (3) and (5) to start with. Nothing has been lost, and there has been a gain in our understanding of the "robustness" aspects of our model.

It has often been claimed that the fiducial argument involves incoherence because it involves integration over the sample space. Looked at from the pivotal point of view the integration involved is over the *pivotal* space, and this certainly does not necessarily involve incoherence. With the pivotal model (3) + (5), no Bayesian would object to integration over the conditional density of e_2, given (x,y), to obtain the posterior probability that Θ lies between, say, 0.2 and 0.8. But what about the pivotal model with (3) alone? Reduced to (4), we have to ask ourselve whether there is any reason to suppose that knowing (x,y), the distribution of e_2 is not $N(0,1)$. Clearly there is when $(x,y) = (1,2)$. But knowing $(x,y) = (2,1)$ do we have any reason for modifying the distribution of e_2 implied by the model? Clearly we would if we knew (5) to be true; but suppose we don't know this. What the fiducial argument does is to say that *ignorance* of Θ is to be interpreted, in suitable cases, as meaning that we may assign the same distribution to e_2, knowing $(x,y) =$ (2,1), as we did, not knowing (x,y) --and notice that the distribution is of e_2, *not* of Θ.

Robinson's examples show that the task of defining the phrase "in suitable cases" in this sentence is not going to be easy; but they do not demonstrate that it will be impossible.

So I do not think that those of Dr. Berger's arguments which are based on "coherence" will force us to accept the general applicability of the likelihood principle. On the basis of the model (3), reduced to (4), with the observation $(x,y) = (2,1)$, it rigorously follows that "unless an event of probability less than 0.05 has occurred in this experiment, Θ lies between 0.042 and 0.785" and this may be all that the scientist needs to know.

J.M. BERNARDO (*Universidad de Valencia*)

I have truly enjoyed this attactive discussion of the likelihood principle; the author makes clear indeed that *acting* on procedures which violate LP is definitely not sound. It is clear to me however that the compelling force of LP is *restricted* to analysis and/or decisions which are *conditional* to a particular set of data; indeed, any *evaluation* of a particular procedure must include integration over the sample space, thus formally violating the likelihood principle.

An important element in the evaluation of any inferential procedure is the *communication* of the inferential content of the data, conditional on the model assumed. The author's discussion of his Example 2 is particularly illuminating: no improvement is possible, as measured by a proper scoring rule, and yet the natural choice $\alpha(x) = 1/2$ can be beaten in the betting game; to me, this strongly suggests that scoring rules, rather than betting arguments, should be used to evaluate procedures designed to communicate information.

This is naturally related to the dependence of *reference* or noninformative distributions on the sampling rule, in violation of LP. For, to *evaluate* the importance of the prior information in a Bayesian analysis, it is necessary to obtain reference posterior distributions which act as an origin, and this evaluation violates LP. I will limit my comment to two specific examples mentioned in this paper:

(i) *Example 3.* In negative binomial sampling when sampling is continued until r successes are obtained, one is *assuming* that r successes *will* eventually appear, a different situation from that of ordinary sampling. This is duly reflected in the corresponding reference posterior distribution (Bernardo, 1979, p. 144) $p(\theta \mid n) \propto \mathrm{Be}(\theta \mid r, n-r+1/2)$ which is *not* proper if $r=0$. This was to be expected, for we are assuming that a success will appear and, in the absence of other information, we cannot make inferences otherwise. In ordinary sampling however the reference posterior turns out to be $p(\theta \mid r) \propto \mathrm{Be}(\theta \mid r+1/2, n-r+1/2)$ and we can make inferences even if $r=0$. I find this behaviour quite appealing.

(ii) *Flatland.* One does not need to use structural ideas to obtain the apparently appropriate solution suggested by Fraser and reproduced in Section 5.3 of the paper. Indeed, this may be obtained from a purely Bayesian point of view by maximizing the missing information (Bernardo, 1979, p. 141).

Finally, the author's concluding remarks in 6.4 seems to suggest that he reluctantly accepts violation of LP in evaluation procedures. I would like a more explicit statement of his view, specially in the particular evaluation procedure provided by reference distributions.

S. FRENCH (*Manchester University*)

At first sight the tactic of considering the long run performance of statistical procedures seems to lead to very persuasive arguments acceptable to both Bayesian or non-Bayesian alike. But are the arguments that acceptable to non-Bayesians? The doubting statistician is asked for a loss function; it is his choice what loss function is used. The argument then proceeds by showing that any dubious method can be beaten in terms of its long run average loss. Here 'dubious' means not satisfying the likelihood principle, but the argument is applicable in wider circumstances. However, why should the statistician's preferences about a sequence of losses be reflected by the long run average loss? To suggest that it should be is to insist that the loss function is an interval scale measurement of the statistician's preferences in circumstances of risk. In short, the loss function is a negative utility function. Now, if the statistician accepts that his preferences can be adequately represented by a utility function, he must also accept the Von Neumann-Morgenstern axioms (or some equivalent system). These axioms contain a *strong independence or cancellation principle* (see. e.g. Fishburn, 1982) which is quite akin to the likelihood principle. Indeed, statisticians who accept the underlying axioms of utility theory usually accept the underlying axioms of subjective probability as well: they are Bayesians. Thus it seems to me that this long run performance argument is really only truly persuasive to Bayesians and they are surely convinced of the validity of the likelihood principle anyway.

A.F.M. SMITH (*University of Nottingham*)

I think we should be careful not to accept a too literal rendering of the likelihood principle as committing us "to use only the data we've actually obtained and not to contemplate other data that we may have obtained but didn't". For example, I would not wish to rule out of court Good's device of imaginary observations (also referred to by Diaconis and Freedman as the "what if" principle), which is a valuable tool for assessing and checking inputs into Bayes' theorem. It seems to me that we need to draw a distinction between *procedures* (which should obey the likelihood principle) and *auxiliary inputs* (which may be arrived at in a variety of creative ways - some of which may violate the likelihood principle).

REPLY TO THE DISCUSSION

Dr. Barnard has provided a very thought provoking contribution. I should start, of course, by thanking him for his congratulations on my axiomatic failure: it is rare to receive praise for failing. Dr. Barnard's other introductory comment, on the crying need for the teaching of likelihood as one of the most important concepts in statistics, is also most welcome.

Dr. Barnard next presents an example, which certainly helps us to understand what pivotal analysis is all about. The example unfortunately strikes me as confirming what I have long suspected, that, despite the many attractive aspects of pivotal analysis, blind following of the paradigm can lead to inferior answers.

To discuss this example, let us begin with the assumption that e_1 and e_2 in (3) are *known* to be standard normal, so that only θ is unknown. If $x \neq y$, it is easy to check that the transformation between θ and e_2 cannot be one-to-one for one of the pairs (x,y) and (y,x). Although Dr. Barnard does not specifically recommend a "confidence set" for the non one-to-one case, presumably he would propose something quite different, even though the likelihood functions for the two pairs are the same. This provides a clear example where pivotal analysis would fail the "evalutions game". While thus providing a counterexample to pivotal analysis, from my viewpoint, it is incumbent to attempt to understand the example from Dr. Barnard's viewpoint. What follows is my understanding and reaction to this viewpoint.

Note, first, that it is a fact that the density of (1,2) is exactly equal to the density of (2,1) for any θ in (1), so that the two observations are equally supportive of the model in the absence of any other information. Dr. Barnard does point out, however, that in the "real world" problem there may be an upward trend in heights, in which case the observation of (1,2) would (in a Bayesian predictive sense) cast more doubt on the assumption (1) than would the observation (2,1). Different analyses might then, of course, be called for, but *not* analyses in violation of the LP, since a richer model is then being considered and the likelihoods for the richer model are relevant. To be more precise, if we introduce a quantity α to reflect a possible upward trend in heights, and work with an enriched model with density $f((x,y)\,|\,\theta,\alpha)$, there will indeed probably be a difference between $f((1,2\,|\,\theta,\alpha)$ and $f((2,1)\,|\,\theta,\alpha)$.

A related general point, made elsewhere by Dr. Barnard, is that, in the real world, one may have a pivotal model such as that given by (3), but be uncertain about the distributions of e_1 and e_2. While granting that this may typically be the case, it again in no way disproves the LP. The LP applies only when θ involves *all* unknown aspects of the model. Of course, when the unknowns are nonparametric, implementation of the LP will be hard, but pivotal analysis will run into difficultites also when e_1 and e_2 are known only to have some nonparametric family of distributions. Perhaps it is possible to argue that pivotal analysis can better deal with such nonparametric problems than can analysis (say, Bayesian) based on the LP. This would be interesting to see. Also, as Dr. Barnard discusses, pivotal analysis does have appeal in its step by step conditioning, allowing robustness at each step to be contemplated separately. None of this is relevant to the truth of the LP, however. It only refers to practical issues that may sometimes call for a "bending" of the LP (as will be discussed later).

I would think that Dr. Barnard would be somewhat disturbed, in his example, by the effect he observes of the introduction of prior information. To say that (1,2) and (2,1) convey drastically different information from a non-Bayesian pivotal viewpoint, yet convey *identical* information when *any* Bayesian input is introduced, strikes me as a violation of "continuity of information".

Dr. Barnard does not deal directly with the coherency issue, which is regretable. In particular, I would be interested in his views on the "inferior long run communication" aspect of violation of the LP, as opposed to merely betting incoherency.

His concluding statement, that it rigorously follows, upon observing (2,1), that "unless an event of probability less than .05 has occurred in this experiment, θ lies between .042 and .785" is correct in the conditional on x framework (which, however, should be specified in the statement). The value of such a statement to the scientist is brought into severe question, however, by this very example, for it is just as logically correct to make the statement, upon observing (1,2), that "unless an event of (conditional on x) probability .0409 has occurred (namely, $e_2 > 1.74$), θ lies *outside* the interval (.42,.57)". However, the data suggests strongly that just such a rare event has occurred, so what is the value of the statement? Dr. Barnard argues that observing (1,2) rules out use of the distribution of e_2 on the grounds of logical impossibility. This is just further reinforcement of the need to condition on all the data, however. Also, if this "ruling out" of use of e_2 is taken literally, a drastic discontinuity in inference is introduced at $x = y$: on one side of this line, inference can apparently be based on e_2 being $N(0,1)$, while on the other side the whole thing must apparently be rejected. Such discontinuity is very disturbing.

My comments concerning this example perhaps reflect a lack of understanding of the pivotal method, although this lack is related to the apparently allowable modification of pivotal analysis at any stage based on intuition. This relates to a point, raised at the beginning of Dr. Barnard's discussion, to the effect that it was time to move on to applications. The difficulty with this is that we can apparently take a simple problem such as that discussed here, arrive at completely different answers, and each maintain the other is wrong. Furthermore, Dr. Barnard is apparently unwilling to subject his answers to any type of validity test in repeated use, so what are we to do?

My own view is that we need to decide on certain things as basic, such as the LP and Bayesian criteria (since nothing else seems logically viable). *Then* we move on to examples and investigate how to best approximate this ideal. Here, pivotal analysis may have a lot to offer in terms of its isolation of robustness problems, and applications may help to determine this; but of what use is the study of applications if we can not agree what constitutes an acceptable answer?

Dr. Bernardo raises several deep and interesting issues. He first mentions that *evaluation of procedures* formally violates the LP. I prefer to think of the LP as simply not applying to such situations, or to design or to prediction problems. The distinction is, I think, important, in that many people seem to reject the LP because they feel it is meant to apply to *any* statistical problem, yet clearly does not. It is thus crucial to stress the limited domain of applicability of the LP, in order that it be taken seriously within this domain.

I am naturally pleased to have Dr. Bernardo express agreement on the issue of "scoring rule performance" versus "betting coherency". The former seems to be much more realistic and defensible.

We are less clearly in agreement on the issue of reference or noninformative priors, and their apparent violation of the LP. The issue here is, I think, partly one of orientation. The LP states simply that any information about θ, obtainable from E and X, is summarized in the likelihood function. Dr. Bernardo, however, refers not just to θ (or some interesting feature thereof), but instead to the evaluation of the "importance of the prior information in a Bayesian analysis". If one is to make sense of such a concept in absolute terms, it seems that a base prior is needed and that reference priors (or posteriors) are reasonable as a base. This does not in itself violate the LP, provided the reference posterior is used only for this measurement of "importance of prior information". Use of the reference posterior to draw *inferences about* θ, however, could violate the LP, and hence result in the inadequacies discussed in the paper. I am not completely clear as to whether or not Dr. Bernardo actually advocates such violations.

A separate issue would be the use to which this "importance of prior information" would be put. My initial feeling is that, as a concept, it will be less valuable in Bayesian analysis than will be attempts to measure the importance of the prior by simply taking a reasonable class of priors and observing the variation in inference or decision as the prior varies over the class. In other words, I am not confident that attempts to get at a "base prior" and consider variation from it are better than simple sensitivity studies.

As to the two examples mentioned by Dr. Bernardo, the fact remains that basing decisions or inferences on the reference posterior in Example 3 will result in long run loss in the "evaluation game", and the flatland example is easily handled by any sensible Bayesian analysis.

Dr. Bernardo concludes with a request for an explicit statement of my views on allowable violation of the LP, and its relation to reference posteriors. Stated succintly, my view is that violation of the LP can never be condoned theoretically. As a practical matter, however, the needed "auxilliary inputs" as Dr. Smith calls them (mainly the prior distributions to be considered) may depend on the data for a variety of reasons. Such reasons are extensively discussed in Berger and Wolpert (1984), and include those mentioned by Dr. Smith (with whom I am obviously in agreement). Indeed, from a practical perspective, I find noninformative prior Bayesian theory enormously succesful, with Dr. Bernardo's reference posterior being perhaps the best of the lot. This *practical* success of justification would not lead me to say, however, that it is *fundamentally* correct in Example 3 to use different posteriors to draw inferences about θ.

As a slight aside here, I feel it regretable that proposers of noninformative prior theories feel the need to argue that such theories are rigorously defensible. The various axiomatic rationality developments show, I think, that only proper prior Bayesian analysis is rigorously defensible. But one cannot quantify prior opinions exactly, so that the best that can be done is to strive to come close to this "ideal" answer. And, in many practical situations, use of noninformative priors will provide answers closer to the "ideal" than will hasty or unsophisticated use of proper priors. There is thus little need to defend noninformative priors, beyond saying simply that they are enormously successful practically. Indeed, theoretical defenses have tended to be counterproductive in that, when a counterexample is then found, some people feel free to reject the entire noninformative prior theory. This is not to say that efforts should not be directed towards better methods of obtaining reasonable noninformative priors. Such efforts are very worthwhile and Dr. Bernardo's approach to such is remarkably successful.

With regards to the comments of Dr. French, I am, in some sense, in agreement. Indeed, the point raised is reminiscent of the comment by I.J. Good, in the discussion of Birnbaum (1962), to the effect that derivation of the likelihood principle from conditionality and sufficiency is somewhat superfluous, since the only sensible justification for sufficiency is the Bayesian justification, which would yield the likelihood principle directly. My reply to all such observations is essentially to agree that (Bayesian) utility theory lies at the heart of the subject and that any other "justifications" are themselves only ultimately justifiable in these terms. This, however, misses the major point of these other "justifications", which is that it is usually far more effective to "sell" the likelihood principle to non-Bayesians by basing its justification on principles they already accept, rather than on the remote (to many) axiomatic utility developments. For whatever reasons, a very large number of non-Bayesians believe in sufficiency or believe in evaluating a procedure by looking at its long run average performance. Showing that these beliefs directly lead to the likelihood principle can have substantial impact. The utility axiomatic developments are easily ignorable by many as being too abstract, and can also, rightly or wrongly, be criticized at the axiomatic level (c.f., LeCam (1977)).

The specific "evaluations game" proposed in the paper can, I agree, be rejected as a method of evaluation. Yet it will seem fair and plausible to many who would not bother to consider the deeper axiomatic utility justification. Also, it is not the case that everyone who accepts utility theory also becomes a Bayesian. Indeed the opposite seems to be true, in some sense. Frequentists often turn to utility theory for justification of considering long run average behavior, while Bayesians often ignore long run average behavior by saying that good long run performance will follow automatically from good conditional performance. Thus the direct messages that (i) anyone considering long run average behavior had better take the LP seriously, and (ii) consideration of long run behavior can be important, even to Bayesians, strike me as important for the statistical profession, even if their fundamental logical basis is axiomatic expected utility theory.

One final point concerning this matter is that, via the evaluations game, one can explicitely demonstrate "something better". Saying that violation of the LP is wrong because of axiomatics does not have quite the impact of saying that violation of the LP would demonstrably lose in the evaluations game.

In conclusion, let me thank the discussants for their penetrating comments. The discussion has clarified a number of important issues in my mind.

REFERENCES IN THE DISCUSSION

BERNARDO, J.M. (1979). Reference posterior distribution for Bayesian inference. *J. Roy. Statist. Soc.* B, **41**, 113-142, (with discussion).

FISHBURN, P.C. (1982). The Foundations of Expected Utility. *Dordrecht, Holland: O. Reide.*

LECAM, L. (1977). A note on metastatistics or 'an essay toward stating a problem in the doctrine of chances'. *Synthese,* **36**, 133-160.

BAYESIAN STATISTICS 2, pp. 67-82
J.M. Bernardo, M.H. DeGroot, D.V. Lindley, A.F.M. Smith (Eds.)

The Choice of Variables in Probabilistic Classification

J.M. BERNARDO and J.D. BERMÚDEZ
Universidad de Valencia, Spain

SUMMARY

A conditional probability distribution over the possible categories, the diagnostic distribution, is the natural solution to the problem of classification with incomplete information. The choice of variables in probabilistic classification is analyzed here as a precise decision problem where the action space consists of interesting functions of the indicants, and the utility function is a proper scoring rule which describes the value of each particular diagnostic distribution conditional on each of the possible categories. In particular, the logarithmic utility function is shown to lead to minimizing the expected entropy, and the quadratic utility function is shown to lead to maximizing the expected squared norm of the resulting diagnostic distributions, as the appropriate criteria for the selection of the variables. The results are illustrated with a problem of probabilistic classification in Spanish politics.

Keywords: AUTOMATIC DIAGNOSIS; DISCRIMINANT ANALYSIS; ENTROPY; LOGISTIC REGRESSION; SELECTION OF VARIABLES; SCORING RULES.

1. INTRODUCTION

Let us consider a situation where it is desired to classify objects into one of a class of k exhaustive, mutually exclusive *categories*, $\{\delta_1,\ldots,\delta_k\}$, on the basis of the value of a multivariate *function* $z = f(x)$, $z = \{z_1,\ldots,z_m\}$, of a vector $x = \{x_1,\ldots,x_s\}$ of s possibly relevant *attributes*, given the information provided by a training data set $D = \{(x_{(j)},\delta_{(j)}), j=1,\ldots,n\}$ which consists of the vectors of attributes $x_{(j)}$ and corresponding categories $\delta_{(j)}$ of n objects classified with certainty. Since the information provided by z will not usually suffice to classify with certainty, we have to provide a probability distribution over the categories, the *diagnostic* distribution (Dawid, 1976) $\{p(\delta_i \mid z,D), i=1,\ldots,k\}$, conditional on the observed value $z = f(x)$ of the function used, and on the training data set D. Following Good and Card (1971) we shall term a *facet* each of the possible realizations of the vector $z = \{z_1,\ldots,z_m\}$ of observed *indicants* in the object to classify.

Let F be the class of those (multivariate) functions of the s attributes which the scientist considers worth studying. For instance, F may be a set of functions mapping the class of attributes into some of its parts, so that $z = f(x)$ would simply be a subset of x. The problem considered in this paper is that of the optimal choice of $f \epsilon F$, that is the choice of that multivariate function of the attributes which should be used for the probabilistic classification of new objects, not contained in the data set.

In Section 2 we analyze the choice of variables as a decision problem where the action space consists of that class of functions of the attributes in which the scientist is interested, and where the appropriate utility function turns out to be a proper scoring rule.

In Section 3 we study the decision criteria which emerge for the solution of the decision problem posed. In particular, we characterize both the conditions under which it is optimal to minimize the expected entropy of the resulting diagnostic distribution, and the conditions under which it is optimal to maximize its expected squared norm. This section concludes with a brief discussion of approximate search procedures.

Section 4 is devoted to the illustration of previous results within the context of a problem of probabilistic classification in Spanish politics. Finally, Section 5 contains further discussion and suggests areas for additional research.

2. THE DECISION PROBLEM

2.1. *Structure of the problem*

Let us consider in detail the formal structure of the decision problem posed by the choice of variables in probabilistic classification.

As illustrated in Figure 1, if the scientist chooses a function $f \epsilon F$ among those which he considers as possibly interesting and the object to classify has a vector of attributes x, he must produce a diagnostic probability distribution P, based on the data set D and the observed facet $z = f(x)$; as a result, he will obtain a consequence which should certainly depend on the true category δ of the object considered, but which may also depend on the chosen function f and on the observed facet z.

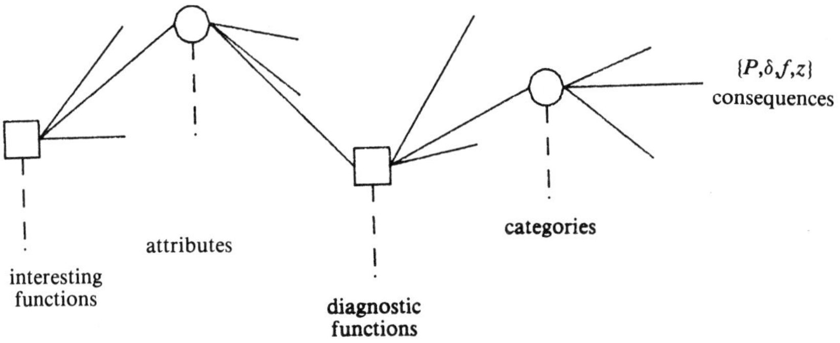

Figure 1. *Structure of the decision problem*

To solve the problem posed we must specify

(i) a (predictive) probability distribution $p(z \mid D)$ over the possible facets.

(ii) a (diagnostic) probability distribution $p(\delta \mid z, D)$ over the categories.

(iii) a utility function $u\{P, \delta, f, z\}$ over the consequences, which describes the goodness of the prediction P about the category of an object with facet $z = f(x)$, were that object found to belong to category δ.

We shall discuss these three elements in reverse order.

2.2. *The utility function*

A utility function of the type $u\{P,\delta\}$ is often named a *scoring rule*, for it describes the score which rewards the scientist if he produces a prediction P for the category of an object later found to belong to δ. Any sensible scoring rule must satisfy a consistency requirement. Indeed, the scientist's expected utility if he produces a prediction Q is

$$\sum_{i=1}^{k} u\{Q,\delta_i\}\, p(\delta_i)$$

where $P = \{p(\delta_i),\ i=1,\ldots,k\}$ describes his actual beliefs about δ. The utility function u should be chosen such that, as a function of Q, this expected utility is maximized if, and only if, $Q \equiv P$ for, otherwise, the scientist could maximize his expected reward by reporting a distribution Q different from P. A utility function which satisfies this requirement is called a *proper* scoring rule. The following are two well known examples of proper scoring rules.

(i) *logarithmic* (Good, 1950; Savage, 1971)
$u_1\{P,\delta\} = A \log p(\delta) + B(\delta)$

(ii) *quadratic* (Brier, 1950; De Finetti, 1965)
$u_2\{P,\delta\} = A\{2p(\delta) - \|P\|^2\} + B(\delta)$

$$\|P\|^2 = \sum_{i=1}^{k} p^2(\delta_i) .$$

For $k=2$ this reduces to $u_2\{P,\delta\} = A\{1-2(1-p(\delta))^2\} + B(\delta)$

In the context of our decision problem, making explicit the dependence on f and z of the utility functions and the dependence on z and D of the diagnostic distributions, the utility functions described above become

$$u_1\{P,\delta,f,z\} = A \log p(\delta \mid z,D) + B(\delta,f,z) \tag{1}$$

$$u_2\{P,\delta,f,z\} = A\{2p(\delta \mid z,D) - \|P\|^2\} + B(\delta,f,z) \tag{2}$$

where, in each case, A is an arbitrary *positive* constant and B an arbitrary *function* of δ, f and z.

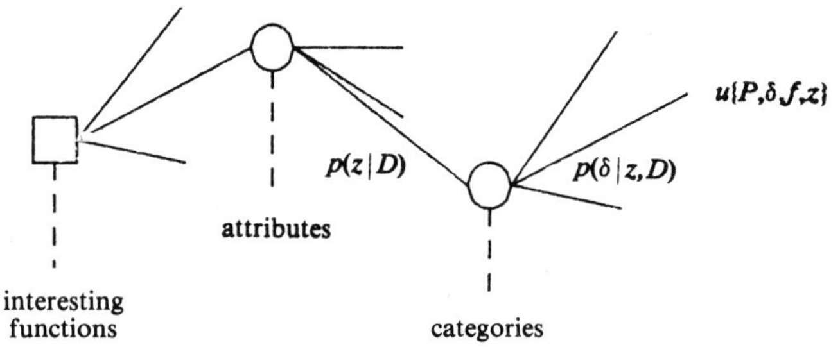

Figure 2. *The complete decision tree*

If u is a proper utility function, then the optimal diagnostic distribution P, i.e. that with the highest expected utility at the second decision node of Figure 1, necessarily coincides with the diagnostic distribution $p(\delta \mid z,D)$ associated with the possible categories. Thus, the decision problem described in Figure 1 is simply solved (see Figure 2) by maximizing in F the expected utility

$$u^*(f) = \int p(z|D) \sum_{i=1}^{k} p(\delta_i|z,D)\, u\{P,\delta_i,f,z\}\, dz \tag{3}$$

where $P = \{p(\delta_i|z,D)\,,\, i=1,\ldots,k\}$ and u is a *proper* utility function.

2.3. *The diagnostic distribution*

Our argument does *not* depend on the precise procedure used to derive the diagnostic distribution $\{p(\delta_i|z,D),\, i=1,\ldots,k\}$.

In a number of interesting situations the diagnostic distribution may be directly and subjectively assessed by the scientist. The usual Bayesian derivation of the diagnostic distribution requires the assumption of a conditional sampling model $\{p(z|\delta_i,\omega),\, i=1,\ldots,k\}$ for the facets and a prior distribution $p(\omega)$ for the parameters involved which are *not* easily specified. Indeed, no reasonably tractable *exact* solution seems to exist which covers the common situation where a *large* number of *both* discrete and continous *interrelated* indicants are to be used. However, work by Aitchison and Dunsmore (1975), Aitchison and Lauder (1979) and Aitchison and Shen (1980) provides some steps in this direction. Hand (1982) takes a semi-Bayesian approach by using kernel methods to estimate from the data the conditional sampling distribution $p(z|\delta_i)$.

Series expansion arguments lead one to expect that any diagnostic distribution could be well approximated by a *logistic* diagnostic distribution, i.e. a diagnostic distribution whose log-odds are a linear combination of the indicants $\{z_1,\ldots,z_m\}$. Indeed, from a Bayesian point of view, the two standard classical procedures to derive logistic diagnostic distributions, namely *linear discriminant analysis* (LDA) (Fisher, 1936; Wald, 1939; Anderson, 1958) and *maximum likelihood logistic regression* (Berkson, 1944; Cox, 1966; Anderson, 1972) may be seen as *large sample approximations* to a Bayesian logistic regression (Bernardo, 1983).

In Section 4, we shall illustrate our proposal for the selection of variables using linear discriminant analysis to derive the diagnostic distributions. As mentioned before, any other procedure would do, but LDA is probably the best known and certainly the quickest to compute. A natural extension of the geometrical argument originally proposed by Fisher (1936) which does *not* require any specific assumption about the sampling distribution of the facets, leads to

$$p(\delta_i|z,D) = \exp\{\beta_i\}\, p(\delta_k|z,D), \qquad i=1,\ldots,k-1$$

$$p(\delta_k|z,D) = 1/[1 + \sum_{i=1}^{k-1} \exp\{\beta_i\}]$$

where, for $i=1,\ldots,k-1$,

$$\beta_i = \lambda_{i0} + \sum_{j=1}^{m} \lambda_{ij}\, z_j$$

$$\lambda_{i0} = -\frac{1}{2}(\mathbf{m}_i - \mathbf{m}_k)'\, S^{-1}(\mathbf{m}_i + \mathbf{m}_k) + \log\{p(\delta_i)/p(\delta_k)\}$$

$$\lambda_i = (\lambda_{i1},\ldots,\lambda_{im})' = S^{-1}(\mathbf{m}_i - \mathbf{m}_k)$$

and where $\mathbf{m}_i$ is the vector of the mean values of the facets $z = \{z_1,\ldots,z_m\}$ of those objects in D which belongs to δ_i, S is their pooled covariance matrix and $p(\delta_i)$ is the prior probability of δ_i; there is ample evidence in the literature (see, e.g. Titterington *et. al.*, 1981, and references therein) which suggests that this provides sensible approximate diagnostic distributions with most types of data bank.

2.4. *The facets predictive distribution*

The precise derivation of the facets predictive distribution $p(z\,|\,D)$ and the subsequent evaluation of the integral in (3) seems intractable except, possibly, under multivariate normality assumptions on the facets sampling distribution.

However, with *prospective* data, the facets $\{z_{(j)}, j=1,\ldots,n\}$ of the n objects in the data set are a random sample from the facets population and, hence, they may be used to obtain a Monte Carlo approximation to the desired integral, namely

$$u^*(f) = \int p(z\,|\,D) \sum_{i=1}^{k} p(\delta_i\,|\,z,D)\, u\{P,\delta_i,f,z\}\, dz$$

$$\cong \frac{1}{n} \sum_{j=1}^{n} \sum_{i=1}^{k} p(\delta_i\,|\,z_{(j)},D)\, u\{P,\delta_i,f,z_{(j)}\} \tag{4}$$

With *retrospective* data, the facets $\{z_{ij}, j=1,\ldots,n_i\}$ of the, say, n_i objects in D which belong to category δ_i constitute a random sample from the facets of those objects in the population which belongs to δ_i. Since, moreover,

$$p(z\,|\,D) = \sum_{l=1}^{k} p(\delta_l)\, p(z\,|\,\delta_l,D)$$

the expected utility (3) may be approximated by

$$u^*(f) = \sum_{l=1}^{k} p(\delta_l)\, u_l^*(f)$$

$$u_l^*(f) \cong \frac{1}{n_l} \sum_{j=1}^{n_l} \sum_{i=1}^{k} p(\delta_i\,|\,z_{lj},D)\, u\{P,\delta_i,f,z_{lj}\} \tag{5}$$

In our experience, this kind of approximation produces in practice sensible results. It would be highly desirable however to see some research effort devoted to the production of more sophisticated numerical approximations to high-dimensional multivariate integrals such as (3).

3. SELECTION CRITERIA

3.1. *Logarithmic utility function*

An interesting characteristic of the logarithmic scoring rule lies in the fact that its value only depends on the diagnostic distribution through the probability which such distribution associates with the true category. A scoring rule which satisfies this requirement, i.e. a utility function u such that $u\{P,\delta,f,z\} = u_\delta\{p(\delta\,|\,z),\delta,f,z\}$ is called a *local* scoring rule. Note that, with this definition, the actual form of the utility function is allowed to depend on the true category.

It may be argued that preferences of a scientist faced with *pure* inference problems should be described by a *local* utility function. Indeed, such an *idealized* situation may be described as that in which only the *truth* matters. It is known (Savage, 1971; Bernardo, 1979) that with more than two categories the logarithmic function (1) is the *only* proper and local scoring rule; if $k=2$ the locality condition is vacous and the logarithmic is only one of the many proper scoring rules.

Substituting (1) into (3), we find that if the scientist's preferences are local, then the best function of the attributes is that which maximizes, in the class F of available functions, the expected utility.

$$u_1^*(f) = \int p(z|D) \sum_{i=1}^{k} p(\delta_i|z,D)\{A \log p(\delta_i|z,D) + B(\delta_i,f,z)\}\, dz$$

$$= -A \int p(z|D)\, H\{P\} dz + B(f) \tag{6}$$

where

$$H\{P\} = - \sum_{i=1}^{k} p(\delta_i|z,D) \log p(\delta_i|z,D)$$

is the *entropy* of the diagnostic distribution and

$$B(f) = \int p(z|D) \sum_{i=1}^{k} p(\delta_i|z,D)\, B(\delta_i,f,z) dz$$

is the expected value of $B(\delta_i,f,z)$.

We have thus proved a rather intuitive result:

In probabilistic classification with a logarithmic utility function and constant costs, one should select those variables which minimize the expected entropy of the resulting diagnostic distribution.

We now consider the meaning of the constants A and $B(f)$ which appear in (6), which may easily be deduced by considering two extreme situations. Indeed, for an ideal function of the attributes which produced a *perfect* diagnostic distribution, with no uncertainty left, the first element of (6) should vanish and the expected utility would be $B(f)$. On the other hand, for a *useless* function of the indicants which always produced a *uniform* diagnostic distribution, the expected utility would be $-A\log k + B(f)$, where k is the number of categories. Thus, if $C(f)$ is the *expected unitary cost* of using f, and V is the expected value of a perfect prediction, we must have

$$V - C(f) = B(f)$$
$$0 - C(f) = -A\log k + B(f)$$

and, hence, $A = V/\log k$. Substituting into (6), the expected utility becomes

$$u_1^*(f) = V\left\{1 - \frac{1}{\log k} \int p(z|D)\, H\{P\}\, dz\right\} - C(f), \tag{7}$$

which is the function to maximize in order to obtain the best available function of the attributes.

Note that (7) is of the standard additive form $V g_1(f) - C(f)$, where the *terminal* utility $g_1(f)$ —which is independent on the particular base in which the logarithms are computed— ranges from the value 0 for useless functions to the value 1 for a function which produces perfect predictions.

The exact computation of the integral in (7) would require some assumptions on the sampling distribution of the facets and may well prove to be pretty intractable. However, using the Monte Carlo methods described in section 2.4, it may be easily approximated. Indeed, with prospective data,

$$- \int p(z|D)\, H\{P\}\, dz \cong \frac{1}{n} \sum_{j=1}^{n} \sum_{i=1}^{k} p(\delta_i|z_{(j)},D) \log p(\delta_i|z_{(j)},D)$$

where $\{z_{(j)}, j=1,...,n\}$ are the n facets contained in the data set. Similarly, with retrospective data,

$$-\int p(z|D)\, H\{P\}\, dz = \sum_{l=1}^{k} p(\delta_l)\, I_l$$

$$I_l \cong \frac{1}{n_l} \sum_{j=1}^{n_l} \sum_{i=1}^{k} p(\delta_i|z_{lj},D) \log p(\delta_i|z_{lj},D)$$

where $\{z_{lj}, j=1,...,n_l\}$ are the n_l facets in the data set which correspond to category δ_l.

3.2. *Quadratic utility function*

With more than two possible categories, there are important differences between the behavior of local and non-local utility functions. For instance, if $\delta=\delta_1$ and $k=4$, the diagnostic distributions $P_1=\{0.4,0.6,0,0\}$ and $P_2=\{0.4,0.2,0.2,0.2\}$ would both have the same logarithmic utility, for they both give the same probability —namely 0.4— to the true category; however, the quadratic utility of P_1 would be much smaller than that of P_2 due to the better spread of 'wrong' probabilities in the second case. As far as this may be regarded as an interesting property, for it penalizes the misleading tendency of a distribution towards false categories, the quadratic utility function, always proper and always bounded, recommends itself.

Substituting (2) into (3), we find that if the scientist's preferences are described by a quadratic scoring rule, then the best function of the attributes is that which maximizes in F the expected utility

$$u_2^*(f) = \int p(z|D) \sum_{i=1}^{k} p(\delta_i|z,D) \left(A\{2p(\delta_i|z,D)-\|P\|^2\}+B(\delta_i,f,z)\right) dz$$

$$= A\int p(z|D)\, \|P\|^2 dz + B(f) \qquad (8)$$

where

$$\|P\|^2 = \sum_{i=1}^{k} p^2(\delta_i|z,D)$$

is the squared *norm* of the diagnostic distribution P and, again, $B(f)$ is the expected value of $B(\delta_i,f,z)$. Thus,

In probabilistic classification with a quadratic utility function and constant costs, one should select those variables which maximize the expected squared norm of the resulting diagnostic distribution.

A similar analysis to that performed with the logarithmic utility function enables us to identify the meaning of the constants A and $B(f)$ which appear in (8). Indeed, the expected utility of a perfect diagnostic distribution is $A + B(f)$. While that of a useless uniform distribution is $A/k + B(f)$. Thus, if V is the expected value of a perfect prediction and C(f) is the expected unitary cost of using f,

$$V - C(f) = A + B(f)$$
$$0 - C(f) = A/k + B(f)$$

so that $A = kV/(k\text{-}1)$ and

$$u_2^*(f) = V\frac{1}{k-1}\{k\int p(z|D)\|P\|^2 dz - 1\} - C(f) \qquad (9)$$

which, like (7), is of the form $Vg_2(f) - C(f)$ where $g_2(f)$ takes values in [0,1].

As mentioned in Section 2.4, the integral in (9) may be approximated by Monte Carlo methods using the facets contained in the training data set. With prospective data this yields

$$\int p(z|D)\, \|P\|^2 dz \cong \frac{1}{n} \sum_{j=1}^{n} \sum_{i=1}^{k} p^2(\delta_i | z_{(j)}, D)$$

where $\{z_{(j)},\ j=1,\ldots,n\}$ are the n facets contained in the data set. Similarly, with retrospective data

$$\int p(z|D)\, \|P\|^2 dz = \sum_{l=1}^{k} p(\delta_l) J_l$$

$$J_l \cong \frac{1}{n_l} \sum_{j=1}^{n_l} \sum_{i=1}^{k} p^2(\delta_i | z_{lj}, D)$$

where, again, $\{z_{lj}, j=1,\ldots,n_l\}$ are the facets in the data set which correspond to category δ_l.

3.3 *Search procedures*

To determine the best available function of the attributes it is necessary to compute the expected utility of all elements in the class F of interesting functions and to select that with the highest value. In practice, if F is a very large set this may imply a forbidding computational burden and, hence, approximate search procedures must be used.

We conclude this section by discussing in some detail the particular case where F consists of the class of functions mapping the set of attributes into the set of its parts, so that the problem is to determine the best subset of the attributes.

One may often assume that the cost of measuring a set of attributes is roughly proportional to the number of attributes measured; thus, if f_m denotes a function mapping the class of attributes into a subset of m elements, we would have $C(f_m) = c_0 + mc$. Since we have established that the expected utility of using f_m is of the form $Vg(f_m) - C(f_m)$, where $g(\cdot)$ is a terminal utility function defined on $[0,1]$, we immediately find that

(i) to compare subsets of attributes of the *same size* one has only to compare their g values.

(ii) to check whether it is worth *adding* a new attribute and using f'_{m+1} one has to check whether $g(f'_{m+1}) - g(f_m) > c/V$.

(iii) to check whether it is worth deleting an attribute and using f''_{m-1} one has to check whether $g(f_m) - g(f''_{m-1}) < c/V$.

It follows that, with *linear* costs, only the ratio c/V of the expected cost of a new attribute to the expected value of a perfect classification matters; this is mathematically equivalent to stopping the search procedure whenever the expected increase in utility falls below a preassigned level $\Delta = c/V$.

In our experience, an efficient approximate search procedure consists of a combination of the usual forward and backward methods, by which

Step 1 The scientist selects a set of attributes which he believes to be as informative and as mutually independent as possible.

Step 2 The algorithm determines that attribute whose deletion carries the minimum decrease in expected utility and deletes such an attribute if this is less than Δ. This step is repeated until no further deletion is possible.

Step 3 The algorithm determines that attribute whose addition provides the maximum increase in expected utility and adds such an attribute if this is larger than Δ. This step is repeated until no further addition is possible; then search is stopped.

In the next Section we provide an example of this search method in a problem of probabilistic classification in politics, using both the logarithmic and the quadratic utility functions.

4. EXAMPLE

In April 1983 the authors' Department agreed to analyze for the Spanish Socialist Party the results of an opinion poll on how people intended to vote in the *País Valenciano* —one of the 17 'states' in the new Spanish 'federal' system— in the local elections which took place on May 8th, 1983. A random sample of 3000 individuals over 18 years of age were interviewed and the following items were recorded —whenever possible— using previously agreed numerical codes:

x_1 = county
x_2 = sex
x_3 = age
x_4 = level of education
x_5 = political party voted for in past (1979) local elections
x_6 = political party voted for in past (1982) general elections
δ = voting intention in the 1983 local elections.

For the voting intention question, only the three parties with electoral possibilities in the *País Valenciano* were taken into account; thus, the class of categories used was

$$\{\delta_1 = \text{socialist},\ \delta_2 = \text{conservative},\ \delta_3 = \text{communist},\ \delta_4 = \text{other}\}.$$

As expected, nearly all recorded interviews contained x_1 to x_4 but a proportion of people refused to answer the overly political questions. Thus, only 70 % of the records contained x_5 and x_6 and only 61 % were complete. We were naturally interested in determining whether the attributes x_1 to x_6 really contained useful information with respect to voting intention; indeed, this information could be used not only to investigate the undecided vote, thus producing accurate predictions, but also to adapt accordingly the political campaign of the party.

In conversations with the politicians involved, we agreed that, on a [0,1] scale, a minimum increase in expected utility of 0.002 would be necessary before an effort to take this information into account in the political campaign made sense. They also agreed that x_6 (vote in 28th October 1982 general elections) was likely to be the most informative attribute.

The stepwise search procedure described in Section 3.3 was performed with the 1843 complete interviews and $\Delta = 0.002$, for both the logarithmic and the quadratic utility functions. Linear discriminant analysis, as described in Section 2.3 was used to compute the diagnostic distributions, and $\{x_6\}$ was used as the starting set. For both utility functions, the algorithm added successively the other political attribute (x_5; vote in the 1979 local elections), level of education (x_4), sex (x_2), age (x_3) and then stopped. The expected terminal utilities actually obtained are printed in Table 1.

Attributes	Logarithmic utilities	Quadratic utilities
x_6	0.2950	0.2939
x_6,x_5	0.3016	0.2969
x_6,x_5,x_4	0.3043	0.2997
x_6,x_5,x_4,x_2	0.3069	0.3021
x_6,x_5,x_4,x_2,x_3	0.3092	0.3043

TABLE 1. *Expected terminal utilities*

The decreasing marginal utility of adding new attributes is pretty obvious from a quick glance at the table. Moreover we found that, although the political attributes are much more informative when known, sociological attributes such as level of education, sex and age —in that order— also provide useful information. On the other hand, the fact that the county does not appear in the optimal subset proved a surprising homogeneity of the vote throughout the *País Valenciano*; this finding, later confirmed by the electoral results, was explained by some politicians by arguing that people actually took these elections as a kind of re-run of the recently held general elections where the socialists won by absolute majority, and did not pay much attention to really local politics.

Computing the expected terminal utilities which correspond to the $2^6 = 64$ subsets of the attributes, we later verified that for each size, none yielded higher values than those provided by the subsets progressively chosen by the algorithm.

To give a flavour of the actual implications of the results obtained we conclude this section by quoting a couple of the diagnostic distributions produced using the optimal subset of attributes: A university educated woman of 32 who voted socialist in the two past elections has probabilities 0.76, 0.08 and 0.09 of voting socialist, conservative or communist respectively; similarly, for a man of 75 with only primary education, who previously voted for the extinct 'center' party which governed Spain between the death of the Fascist dictator and October 1982, those probabilities were respectively, 0.56, 0.10 and 0.08. The results of the 1983 elections in the País Valenciano were: socialists 55 %; conservatives 31.4 %, communists 6.7 %, with the remaining 6.9 % divided among small parties which did not obtain any seat in the Local Parliament.

5. FINAL REMARKS

We have presented a unified, decision-theoretical approach to the problem of the selection of variables in probabilistic classification. The results contain as particular cases selection criteria often advocated on an *ad hoc* basis such as minimizing the entropy or maximizing the squared norm; from our point of view, this demonstrates, once again, that decision theory provides a sensible, systematic methodology to deal with *any* real problem.

In our example, the logarithmic and the quadratic utility functions give very similar results, but in other problems this might not be the case. Naturally, any other *proper* utility function could have been used within the framework presented; it would *not* be sensible however, to use an improper utility function such as the —ubiquitous and rather uninformative— error rate.

The choice of the procedure by which the diagnostic distributions are computed may well affect the final result. This was not the case however in our example, where both maximum-likelihood and Bayesian logistic regression gave rather similar results to those produced by linear discriminant analysis.

The precise derivation of the facets predictive distribution and the corresponding derivation of better approximations to the expected utilities are indeed possible under different assumptions on the sampling distribution of the facets. One of us (J.D. Bermúdez) is presently working in this direction.

REFERENCES

AITCHISON, J. and DUNSMORE, I.R. (1975). *Statistical Prediction Analysis.* Cambridge: University Press.

AITCHISON, J. and LAUDER, I.J. (1979). Statistical diagnosis for imprecise data. *Biometrika,* **66,** 475-483.

AITCHISON, J. and SHEN, S.M. (1980). Logistic-normal distributions: some properties and uses. *Biometrika,* **67**, 261-272.

ANDERSON, J.A. (1972). Separate sample logistic discrimination. *Biometrika,* **59**, 19-35.

ANDERSON, T.W. (1958). *An Introduction to Multivariate Statistical Analysis.* New York: Wiley.

BERKSON, J. (1944). Applications of the logistic function to bioassay. *J. Amer. Statist. Assoc.,* **39,** 357-365.

BERNARDO, J.M. (1979). Expected information as expected utility. *Ann. Statist.,* **7**, 686-690.

— (1983). Bayesian logistic diagnostic distributions. *Tech. Rep.* **8/83,** Universidad de Valencia.

BRIER, G.W. (1950). Verification of forecasts expressed in terms of probabilities. *Month Weather Rev.,* **78**, 1-3.

COX, D.R. (1966). Some procedures associated with the logistic qualitative response curve. *Research papers in Statistics: Festschrift for J. Neyman* (David, F.N. ed.) 55-71. New York: Wiley.

DAWID, A.P. (1976). Properties of diagnostic data distributions. *Biometrika,* **32**, 87-123.

DE FINETTI, B. (1965). Methods for discriminating levels of partial knowledge concerning a test item. *British J. Math. Statist. Psychol.,* **18**, 87-123.

FISHER, R.A. (1936). The use of multiple measurements in taxonomic problems. *Ann. Eugen,* **7**, 179-188.

GOOD, I.J. (1950). *Probability and the weighing of evidence.* London: Charles Griffin.

GOOD, I.J. and CARD, W.I. (1971). The diagnostic process with special reference to errors. *Meth. Inform. Med.,* **10**, 176-188.

HAND, D.J. (1982). *Kernel Discriminant Analysis.* Letchworth, UK: Research Studies Press.

SAVAGE, L.J. (1971). Elicitation of personal probabilities and expectations. *J. Amer. Statist. Assoc.* **66**, 783-801.

TITTERINGTON, D.M. *et al.* (1981). Comparison of discrimination techniques applied to a complex data set of head injured patients. *J. Roy. Statist. Soc.* A, **144**, 145-175. (With discussion).

WALD, A. (1939). Note on discriminant functions. *Biometrika*, **31**, 218-220.

DISCUSSION

I.R. DUNSMORE (*University of Sheffield*)

Towards the end of a conference it is sometimes difficult to maintain the momentum. The speaker has been magnificently successful in keeping the adrenalin flowing.

Of course we might have expected an enthusiastic delivery from Professor Bernardo and to hear such an illuminating talk. I thought initially that this paper was the panacea for decision theory models set within the framework of Raiffa and Schlaifer (1961), and that it would do 'all things for all men'. Theoretically perhaps it might, but practically there are many difficulties. However with judicious choice of utility structure, sensible if crude approximations in various places and simplifying restrictions on the possible functions $f \in F$, the authors have produced a most useful and workmanlike paper.

Perhaps the most important feature is the use of the diagnostic paradigm as discussed by Dawid (1976) and the incorporation of logistic distributions to approximate the diagnostic distribution $\{p(\delta_i | z,D),\ i=1,2,\ldots,k\}$. Bernardo's (1983) paper on this latter aspect is eagerly awaited. The diagnostic paradigm has much in its favour as compared to the alternative sampling paradigm in which we need to specify $\{p(z | \delta_i,\omega),\ i=1,2,\ldots,k\}$ and $p(\omega)$. The authors state that their argument does not depend on the precise procedure used to derive the diagnostic distribution. Whilst this might be so, one feature which needs some investigation is whether the diagnostic approximation ignores some possible information on the nature of the x values induced in some way by the particular choice of the functions f in F.

The approximations used for the expected utilities at the pre-posterior stage of analysis are somewhat crude, and their validity should be investigated. It is important to note that with retrospective data a prior $\{p(\delta_i),\ i=1,2,\ldots,k\}$ needs to be specified without reference to the training set D. Examples of the relative influence of this prior would be useful.

I did not quite understand in Section 3 the two results concerning purely inferential probabilistic classification in which one should select those variables which minimize the expected entropy or maximize the expected norm of the resulting diagnostic distribution. These seem to ignore the function $B(f)$ completely; or does purely inferential probabilistic classification mean that $B(\delta, f,z)$ is independent of f in the utility structure?

Finally there are some comments on the example. It would have been interesting to have been a fly on the wall at the conversations with the politicians at which it was agreed to take $\Delta = c/V = 0.002$. This critical factor seems to me to be a somewhat nebulous quantity to assess. An investigation into the effect of slight (or even large) variations in Δ would be illuminating.

Secondly, although the original 3000 individuals may have been a random sample, it seems unlikely that the 1843 complete interviews which were used to approximate the distributions will still be a random sample (prospective or retrospective). Can one simply ignore the missing data?

L. PICCINATO (*Università di Roma*)

The approach based on scoring rules has already been brilliantly applied to the problem of experimental design as is well known from previous work by J.M. Bernardo himself. The same can be said now of the present application. On the other hand, this problem too can be basically seen as a problem in preposterior analysis, the choice of action f (e.g. the subset of attributes one will observe) being essentially equivalent to the choice of the design of an experiment.

I would like to comment on two particular points, where some extensions seem possible and (I think) useful. For sake of simplicity, let us come back to the "political" example.

We learned that, with 5 or less attributes, attribute x_1 (county) is the least important from the point of view of predicting votes. It must be remarked however that this result is reached through an integration over the space of the possible objects to be classified. Then the relative irrelevance of county, which holds in an unconditional treatment of the data, could no longer be true for instance, for specified values of age (attribute x_3) for, let me suppose, the role of county could be more important for older people. In other words, if predictions are to be referred to some subclass of the involved population, appropriate conditioning must be introduced, restricting the integration. This is mathematically trivial, of course, but such explicitly stated extension would allow more flexibility and would prevent misunderstanding, that is referring results obtained *in general* to *special* categories.

As a second particular remark, let me point out first of all that in the example some answers, referring to more delicate attributes, are missing. I suppose that some sensible procedure has been used to overcome such difficulty, which was presumably seen as an undesirable departure from ideal conditions. Indeed, the authors provided us with a *general* method for an optimal design; nonetheless it would be interesting to deal *explicitly* even with a more specific problem of this kind: how much is worthwhile to insert questions having a given probability of not being answered? The problem can be made rather complex: even the refusal to answer could be seen as information, in some way connected with the fact to be predicted. I mentioned this problem as an example of a general attitude I would suggest: when a systematic theory of some situation is available, as in our case, it is time to deal with more complicated problems than before or, better, to improve the realism of the approach. In a sense, this is just what usually occurs with the Bayesian theory of experimental planning, contrasted with the classical one.

P. GARCIA-CARRASCO (*Universidad Complutense de Madrid*)

I would like to rise some technical points which may modify the range of applications of the results contained in this paper.

(i) The general form of the logarithmic and quadratic utility functions should also explicitly allow the dependence on f, and possibly on z, of the arbitrary constant A; this may well prove to be important in a realistic description of actual classification problems. As a consequence, the expected utilities (7) and (9) should generally depend on $V = V(f)$, the expected value of producing a perfect prediction using f.

(ii) The linear hypothesis on the expected costs certainly simplifies the mathematics but may be ill-suited to important applications. However, without this, or a similar assumption, the maximization of (7) and (9) does not reduce to minimizing the expected entropy or maximizing the expected norm.

(iii) The search algorithm proposed in 3.3 seems arbitrary. Is there any reason to prefer this to the standard one-in one-out procedure?

I.J. GOOD (*Virginia Polytechnic Institute*)

In a diagnostic tree, before one gets close to the end of the tree, it might be adequate to use a minimax-entropy procedure, as suggested by Good (1970, pp. 197-198). That is, one might minimize the expected entropy $-\Sigma p_i \log p_i$ when selecting a facet (where p_i is the current estimated probability of the ith diagnostic category), because the "rougher" the set of probabilities the closer one is to a unique diagnosis. But when estimating the probabilities, subject to various constraints, it often makes sense to maximize the entropy. Thus one might alternate between maximizing and minimizing. For further explanation see the cited reference. The use of "proper scores", other than the logarithmic one, could also be used in a similar minimaxing or maximinning procedure.

W. POLASEK (*Universität Wien*)

During the conference we have talked about what is an expert, what is "information". Now I want to add the question: What is diagnosis in statistics?

REPLY TO THE DISCUSSION

We are grateful to all discussants for their comments; in the following paragraphs we shall try to address them by topics in roughly the order in which they have been presented.

(i) Dr. Dunsmore is surely right when he points out the dependence of the result on the choice of F. Indeed, this is our action space and no theory will take us beyond the alternatives considered in F, but we do not think this depends on the precise procedure used to derive the diagnostic distribution.

(ii) We are indeed aware of the crudeness of the approximations used for the expected utilities, but we believe that Monte Carlo techniques are the best alternative to the high dimensional integrations involved. Certainly, those integrations could be carried out analytically with some restrictive assumptions such as multinomial sampling distributions for the facets, but those assumptions are rarely warranted in practice.

(iii) The expected utilities to maximize certainly include $B(f)$ or, equivalently, $C(f)$; thus, as Dr. Garcia-Carrasco remarks, the results on minimizing expected entropy or maximizing expected norm should be understood within functions f of the same cost. In scientific work however one is frequently interested in the predictive value of the attributes, irrespective of their observation cost; this situation is well described by pretending that $B(f)$ is constant and hence irrelevant.

(iv) The choice of Δ is certainly tricky. The value $\Delta = 0.002$ precisely means that the expected value of perfect information is judged to be 500 times larger than the expected cost of adding a new attribute into the study, but admittedly, this evaluation requires some effort. The argument followed in our example to arrive at a particular number was in terms of the proportion of the campaign budget to be used in the investigation of its optimal use.

(v) Both Dr. Dunsmore and Dr. Piccinato raise the problem of missing data, a general problem which certainly pervades most applied statistics. Both discussants are certainly right in questioning whether the subset of complete answers may generally be treated, as we did in our example, as a random subsample. We believe however that the assumption was roughly appropriate in our case, as partially shown by the close relation between our predictions and the final results; see Bernardo (1984) for more details on this point. A specific procedure to deal with missing data in probabilistic classification, which uses the type of conditioning mentioned by Dr. Piccinato, is currently being developed and will be presented elsewhere.

(vi) We certainly agree with Dr. Garcia-Carrasco on the possible dependence of A on f and even on z; we believe however that the simplifying assumption $A(f,z) = A$, which is essentially equivalent to the additive decomposition of the utility into terminal utility minus cost, may often be safely made; this was apparently the case in our example; she is also right in pointing out that the hypothesis of linear costs is frequently not realistic. We do not claim otherwise, but merely analyze its consequences.

(vii) The only way to guarantee that we obtain the optimal design is to compute the expected utility of all possible combinations of attributes. Since this is often a forbidding task, some simplification is necessary; we believe that our proposal is somehow in between the total calculation and the simplified one-in one-out procedure mentioned by Dr. Garcia-Carrasco; it takes more computer time but makes it more likely that we obtain the overall optimal solution.

(viii) The alternative procedure proposed by Dr. Good is intriguing. In this paper, we are specifically interested in the choice of variables and, as a consequence of our utility assumptions, minimizing the expected entropy appears to be an interesting particular case. Whether or not it might be appropriate to maximize the expected entropy in the totally different decision problem where the action space consists of possible estimates of the probabilities, would depend on the appropriateness of the underlying utility assumptions which would be necessary for this result to hold. We would certainly like to see an elaboration of this idea.

(ix) Dr. Polasek poses a very general question. We have used *diagnosis* in the very restricted sense of choice among alternative explanations, by extension of the standard use of this word in medicine; thus, by a diagnostic distribution we simply mean a probability distribution over a set of mutually incompatible alternatives. Naturally, the main point is that diagnosis, or classification, should always be couched in *probabilistic* terms.

REFERENCES IN THE DISCUSSION

BERNARDO, J.M. (1984). Monitoring the 1982 Spanish socialist victory: a Bayesian analysis. *J. Amer. Statist. Assoc.* **79**, 510-515.

GOOD, I.J. (1970). Some statistical methods in machine intelligence research, *Math. Biosc.* **6**, 185-208.

RAIFFA, H. and SCHLAIFER, R. (1961). *Applied Statistical Decision Theory*. Cambridge, Mass. The MIT Press.

BAYESIAN STATISTICS 2, pp. 83-98
J.M. Bernardo, M.H. DeGroot, D.V. Lindley, A.F.M. Smith (Eds.)

An Extended Maximum Entropy Principle and a Bayesian Justification

I. CSISZÁR
Mathematical Institute of the Hungarian Academy of Sciences
Budapest, Hungary

SUMMARY

Let Q be a probability measure and Π be a convex set of probability measures on a measurable space. The maximum entropy (or minimum discrimination information) principle calls for updating prior belief Q to conform with evidence Π by taking that $P \epsilon \Pi$ which minimizes I-divergence from Q. We show that while such a $P \epsilon \Pi$ need not exist, many of its desirable properties hold for an always uniquely defined P^* possibly not in Π. In particular, this P^* appears in a conditional limit theorem, which suggests a Bayesian justification for adopting P^* as an updating of the prior belief Q even in the case when $P \notin \Pi$.

Keywords: CONDITIONAL LIMIT THEOREM; EXPONENTIAL FAMILY; I-PROJECTION; KULLBACK-LEIBLER INFORMATION; MAXIMUM ENTROPY; UPDATING PRIOR BELIEFS.

1. INTRODUCTION

The principle of "maximum entropy" or "minimum discrimination information" is a valuable tool for assigning prior distributions, formulating statistical hypotheses, updating subjective probabilities, and also for estimation and hypothesis testing. It dates back to Boltzmann and was developed and extended by Gibbs, Jaynes, Kullback, and many others. Let me not enter here history and various application areas, let alone philosophical issues involved. For this, the reader is referred, e.g., to the conference proceedings "The Maximum Entropy Formalism" (MIT Press, Cambridge, Mass., 1979), particularly Jaynes' exposition therein, and —for statistical applications— to Kullback (1959), Gokhale and Kullback (1978). The aim of this paper is to suggest that some new mathematical results to be published in Csiszár (1984) extend the scope of the principle and can also be used to provide a kind of Bayesian justification for the extended (as well as the original) principle in a general context.

The organization of the paper is as follows. After summarizing below the basic terminology, Section II is devoted to a review of the maximum entropy principle, in the discrete case for simplicity. The new mathematical results are stated in Section III, and their implications for the maximum entropy principle are discussed in Section IV.

Throughout this paper, by a *probability distribution* (abbreviated as *PD*) on a set S we mean a (σ-additive) measure P on (a given σ-field F_s of subsets of) S such that $P(S) = 1$. If S is a finite set, we take for F_s the σ-field of all subsets of S and a *PD* on S will be identified with an $|S|$ -dimensional vector P with components $P(a)$, $a \epsilon S$, where $P(a)$ stands for the P-

measure of the one-point set $\{a\}$. The *expectation* (integral) of a function f on S with respect to a *PD* P will be denoted by $E_P f$. Let us emphasize that the formal terms "probability distribution" and "expectation" do not necessarily refer to "true" probabilities, physical, subjective or whatever. Indeed, a very important kind of *PD*'s will be the *empirical distributions* of samples, i.e., of sequences $\mathbf{x} = (x_1,\ldots,x_n) \in S^n$, defined by

$$P_{\mathbf{x}}(A) = \frac{1}{n} \sum_{i=1}^{n} 1_A(x_i) \qquad (A \in F_s) \tag{1.1}$$

where 1_A denotes the indicator function of the set A. The expectation with respect to $P_{\mathbf{x}}$ of any function f on S is its sample mean, i.e.,

$$E_{P_{\mathbf{x}}} f = \frac{1}{n} \sum_{i=1}^{n} f(x_i). \tag{1.2}$$

If S is a finite set then $P_{\mathbf{x}}$ is the vector of relative frequencies of the elements of S in $\mathbf{x}$, i.e., $nP_{\mathbf{x}}(a)$ is the number of occurrences of a in $\mathbf{x}$, for each $a \in S$.

For two *PD*'s P and Q on the same set S, the Kullback-Leibler (1951) information for discrimination $D(P\|Q)$ is defined as the P-expectation of the log-likelihood ratio $\log \dfrac{dP}{dQ}$, if P is absolutely continuous with respect to Q, and $D(P\|Q) = \infty$ else:

$$D(P\|Q) = \begin{cases} E_P\left(\log \dfrac{dP}{dQ}\right) & \text{if } P \ll Q \\ \infty & \text{if } P \not\ll Q. \end{cases} \tag{1.3}$$

This quantity plays a fundamental role in information theory, c.f., e.g., Csiszár and Körner (1981). For statistical purposes, $D(P\|Q)$ is a natural information-theoretic measure of (directed) divergence between P and Q, cf. Kullback (1959). In particular, $D(P\|Q) \geq 0$, with equality iff $P = Q$. More precisely,

$$2D(P\|Q) \geq |P-Q|^2 \tag{1.4}$$

where

$$|P-Q| = 2 \sup_{A \in F_s} (P(A) - Q(A)) \tag{1.5}$$

is the so-called variation distance of P and Q (for a proof of (1.4) cf. Csiszár (1967)).

In the sequel, $D(P\|Q)$ will be called informational divergence or simply *divergence*. If for a *PD* Q and a set of *PD*'s Π a unique $P^* \in \Pi$ with the property

$$D(P^*\|Q) = \min_{P \in \Pi} D(P\|Q) \tag{1.6}$$

exists, it is called —as in Csiszár (1975)— the I-projection of Q onto Π. By Theorem 2.1 of Csiszár (1975), the I-projection of Q onto Π always exists if Π is convex and closed in variation distance and $\inf_{P \in \Pi} D(P\|Q) < \infty$.

2. THE MAXIMUM ENTROPY PRINCIPLE

Let us first recall the simple Bayesian argument leading to the maximum entropy principle in a specific context. For simplicity, attention will be restricted throughout this section, except for the last example, to *PD*'s on a finite set S.

Suppose that in a sample $\mathbf{x} = (x_1,\ldots,x_n) \in S^n$ obtained by n drawings with replacement from S, the means of k given functions $f_1,\ldots,f_k$ were

$$\frac{1}{n} \sum_{i=1}^{n} f_j(x_i) = \gamma_j, \qquad j=1,\ldots,k. \tag{2.1}$$

Assuming uniform prior distribution on S^n, as the formulation suggests, we are interested in the posterior distribution given (2.1) (which is a *PD* on S^n) and, in particular, in its marginal on S (the posterior distribution of the first sample element, say). The following derivation dates back to Boltzmann (1877).

Let $N(P)$ denote the number of possible samples $\mathbf{x} \in S^n$ which have the same empirical distribution $P_{\mathbf{x}} = P$, i.e., of those samples in which each $a \in S$ occurs $nP(a)$ times. Then

$$N(P) = \frac{n!}{\prod_{a \in S} (nP(a))!} = \delta(P,n)\exp\{-n \sum_{a \in S} P(a)\log P(a)\} \tag{2.2}$$

where

$$(n+1)^{-|S|} \leq \delta(P,n) \leq 1 \qquad \text{for every } n \text{ and } P. \tag{2.3}$$

The loose bounds (2.3) fully suffice for our purposes and they have a very simple proof not needing even Stirling's formula, cf. Csiszár and Körner (1981), p. 30.

Let P^* be the *PD* which maximizes the entropy function

$$H(P) = - \sum_{a \in S} P(a) \log P(a) \tag{2.4}$$

as P ranges over the set of *PD*'s

$$\Pi = \{P{:}E_P f_j = \gamma_j, \qquad j=1,\ldots,k\}. \tag{2.5}$$

As the number of possible empirical distributions P grows only polynomially with n, it follows from (2.2), (2.3) and (1.2) that for large n the posterior distribution given (2.1) is practically concentrated on those $\mathbf{x} \in S^n$ whose empirical distribution is very close to P^*. Since this posterior distribution is symmetric in the n coordinates, this implies that its (either) marginal on S is also arbitrarily close to P^* if n is sufficiently large.

This argument extends with a minor modification also to the case when any prior of form Q^n (instead of the uniform one) is put on S^n, i.e., when the sample is obtained by independent drawings from a given *PD* Q on S. In this case, the prior probability of an $\mathbf{x} = (x_1,\ldots,x_n) \in S^n$ of empirical distribution $P_{\mathbf{x}} = P$ is

$$Q^n(\mathbf{x}) = \prod_{i=1}^{n} Q(x_i) = \prod_{a \in S} Q(a)^{nP(a)}$$

and (2.2) gives

$$Q^n(\{\mathbf{x}{:}\ P_{\mathbf{x}} = P\}) = N(P) \prod_{a \in S} Q(a)^{nP(a)} = \tag{2.6}$$

$$= \delta(P,n) \exp\{-n \sum_{a \in S} P(a) \log \frac{P(a)}{Q(a)} \}.$$

Now let P^* denote the *I*-projection of Q onto the set of *PD*'s (2.5), i.e., the *PD* minimizing the divergence

$$D(P\|Q) = \sum_{a \in S} P(a) \log \frac{P(a)}{Q(a)}$$

for $P\epsilon\Pi$. Then it follows from (2.6) that for large n the posterior distribution given (2.1) is sharply concentrated on those $\mathbf{x}\epsilon S^n$ whose empirical distribution is very close to P^*. Then also the marginal on S of this posterior will be arbitrarily close to P^*. Let us remark that the I-projection P^* has the form

$$P^*(a) = cQ(a) \exp \sum_{j=1}^{k} \vartheta_j f_j(a) \qquad (a\epsilon S) \tag{2.7}$$

for some $\vartheta = (\vartheta_1,\ldots,\vartheta_k)$ and $c>0$, provided that there exists some $P\epsilon\Pi$ such that $P(a)>0$ iff $Q(a)>0$.

As entropy maximization and divergence minimization are conceptually the same techniques, in this paper the term *maximum entropy principle* will refer to both, in the following sense: When a *PD* Q previously assigned to S has to be "updated" on the basis of information which specifies a set Π of feasible *PD*'s, the principle calls for assigning as updated *PD* the I-projection P^* of Q onto Π. In the sequel we assume, for simplicity, that Π is of form (2.5).

In the above formulation the interpretation of the previously assigned and updated *PD*'s was left unspecified deliberately. For adherents of subjective probabilities they usually represent subjective beliefs prior to and after the information specifying Π has been taken into account. Updating subjective probabilities has recently been discussed by Diaconis and Zabell (1982) in the case

$$\Pi = \{P\colon P(A_j) = \gamma_j, \qquad j=1,\ldots,k\};$$

they give extensive references also to the philosophical literature of the subject. For other statisticians, Q may be an original estimate of the unknown "true distribution" or an original null-hypothesis, and P^* a new estimate or modified null-hypothesis adjusted to the new information, c.f. Kullback (1959), Good (1963). In particular, when Q is the uniform distribution on S, its I-projection onto Π equals the *PD* maximizing $H(P)$ for $P\epsilon\Pi$. This P^* is interpreted as the subjective probability distribution representing the state of knowledge when the expectations of some given functions are known and no other information is available, cf. Jaynes (1957), (1979), or, alternatively, as a natural null-hypothesis to be tested, cf. Good (1963). The last viewpoint is particularly useful for the analysis of contingency tables, cf. also Gokhale and Kullback (1978). Since this is a Bayesian conference, in the sequel I shall use the subjective probability interpretation.

The previous simple Bayesian reasoning provides an operational justification of the maximum entropy principle in certain contexts, namely when (i) the interpretation of the *PD* Q initially assigned to S makes it reasonable to assign its n'th Cartesian power to S^n as prior distribution and (ii) the information specifying (2.5) as the set of feasible *PD*'s on S consists in the knwoledge of sample averages (2.1) for large sample size n.

An axiomatic approach leading to the "maximum entropy" rule of updating probabilities in a general context has been developed by Shore and Johnson (1980). The point I wish to make here is that the above operational (Bayesian) justification may also be relevant in quite a wide context, if only because it shows what "maximum entropy" updating is equivalent to.

Suppose, e.g., that in a large population, A_1 the joint distribution of certain characteristics $y_1,\ldots,y_\ell$ of individuals is of interest, and the available information is that the population means of the first k characteristics ($k\leq\ell$) are $\gamma_1,\ldots,\gamma_k$, say. Now, if in a population A_o the joint distribution of these characteristics is known to be Q, this Q may be chosen as representing initial beliefs, whose updating by the maximum entropy principle results in a *PD* of form

$$P^*(y_1,\dots,y_\ell) = cQ(y_1,\dots,y_\ell)\exp\sum_{j=1}^{k}\vartheta_j y_j, \tag{2.8}$$

cf. (2.7). By what has been said above, this same P^* would be obtained, loosely speaking, by Bayes' theorem if A_1 were a subpopulation of A_o. In statistical mechanics exactly such an assumption is often made: a particle system A_1 with known energy, say, is inbedded into a hypothetical population A_o by postulating that all conceivable configurations are equi-probable a priori; then one takes the posterior distribution given that a configuration with the known energy has been realized. This is almost the same argument as the original one of Boltzmann (1877) who accepted the *PD* P^* maximizing $H(P)$ subject to the energy constraint on the ground that the vast majority of configurations with this energy have empirical distribution very close to P^*. (This combinatorial fact was established at the beginning of this section).

The following line of reasoning may also deserve attention. Suppose that a person whose initial beliefs were represented by a *PD* Q on S, is given the information that the expectations of the functions $f_1,\dots,f_k$ for the "true distribution" are $\gamma_1,\dots,\gamma_k$. He interprets this information so that an experiment consisting on n trials (for large n) would result in a sample satisfying (2.1) with very good approximation. Then it is natural for him to update his prior beliefs by conditioning on the event (2.1). Of course, this can be done only if a prior distribution is assigned to S^n, rather than just to S. A person with firm initial belief that Q was the true physical distribution, would naturally assign Q^n as his prior. Then, as we have seen, the marginal on S of the posterior distribution would be (with as good approximation as desired as $n\to\infty$) just the I-projection of Q onto Π.

In my opinion, the last paragraph makes out a good case for the maximum entropy principle. Still, one may object that subjective prior beliefs about the result on n consecutive drawings are seldom represented by a *PD* of form Q^n. Rather, the general representation of such prior beliefs is (by De Finetti's theorem) a mixture of *PD*'s of this form. This means, informally, that one has independent drawings from an unknown Q, and the prior beliefs about the true Q are represented by a *PD* W defined on the set of all *PS*'s on S. Let us now see, what the previous argument gives in this general case. Suppose first that W has finite support giving weights $w_1,\dots,w_m$ to *PD*'s $Q_1,\dots,Q_m$ on S ($w_\ell > 0$, $\ell=1,\dots,m$; $\Sigma w_\ell = 1$). Then the prior probability of the set $\{\mathbf{x} : P_\mathbf{x} = P\}$ will be, from (2.6)

$$\sum_{\ell=1}^{m} w_\ell Q_\ell^n(\{\mathbf{x}: P_\mathbf{x} = P\}) = \delta(P,n)\sum_{\ell=1}^{m} w_\ell \exp\{-nD(P\|Q_\ell)\}.$$

Let P_ℓ denote the *PD* minimizing $D(P\|Q_\ell)$ for $P\epsilon\Pi$ (defined by (2.5)) and suppose that there is an index ℓ_o such that $D(P_{\ell_o}\|Q_{\ell_o}) < D(P_\ell\|Q_\ell)$ for $\ell\neq\ell_o$. Then the last equation implies that, for large n, the posterior distribution given (2.1) is sharply concentrated on the set of those $\mathbf{x}\epsilon S^n$ whose empirical distribution is very close to P_{ℓ_o}. Hence, as before, the marginal on S of this posterior distribution is also very close to P_{ℓ_o}. It is also seen that the *PD* representing posterior belief about the true Q gives almost full weight to Q_{ℓ_o}. By a continuity argument, one can get rid of the assumption that W (the prior belief about the true Q) has finite support. The result is (informally) stated below as a proposition, for I have never seen it stated before, even though presumably it has been known.

Proposition. Let W be a *PD* defined on the set of all *PD*'s on S and suppose that there is a unique pair of *PD*'s (P^*, Q^*) minimizing $D(P\|Q)$ for $P\epsilon\Pi$ and $Q\epsilon\Psi$ where Π is defined by (2.5) and Ψ is the support of W. Consider n independent drawings from an unknown distribution on S about which the prior beliefs are represented by W. Then the posterior distribution given that a sample satisfying (2.1) was drawn, has S-marginal arbitrarily close to P^*, and the posterior beliefs about the true Q are sharply concentrated in an arbitrarily small neighbourhood of Q^*, both when n is sufficiently large.

Let us emphasize that the exact distribution W representing the prior beliefs is irrelevant for this result, rather, a vague specification of the prior beliefs by Ψ (the support of W) is sufficient.

This proposition suggests the following *generalization* of the maximum entropy principle: When vaguely specified initial beliefs represented by a set Ψ of *PD*'s on S are to be updated on the basis of information specifying a set Π of feasible *PD*'s, and there exists a unique pair (P^*,Q^*) minimizing $D(P\|Q)$ for $P\epsilon\Pi$, $Q\epsilon\Psi$, represent the updated beliefs by P^*.

An interesting point is that while the initial beliefs were vaguely specified by a set Ψ of *PD*'s, the suggested updating results in a single *PD* P^*, provided that the uniqueness hypothesis is true, which appears to be the typical case. Let us remark that the need for minimizing $D(P\|Q)$ for $P\epsilon\Pi$, $Q\epsilon\Psi$ (for certain convex sets Π and Ψ of *PD*'s, and more generally, of arbitrary finite measures) occurs in a variety of problems of statistics and information theory, cf. Csiszár and Tusnády (1984), where also the convergence of the natural iterative minimization procedure is proved in a general context.

So far, attention has been restricted to finite S. Of course, the maximum entropy principle is used also in the non-discrete case. There is one conceptual difficulty, not to be entered here, concerning the suitable representation of the initial state of knowledge "complete ignorance". Another problem, for which a solution will be offered in Section IV, is the possibility of non-existence of I-projection. Let us conclude this section by an example showing that this is not an artificial problem, rather, it does occur in very simple real situations.

Example. Let S be the real line and Q be the standard normal distribution. Consider the single given function $f(x) = x^3$, and let Π be the set of all those *PD*'s P for which $E_P f$, i.e., the third moment of P, exists and equals 1 (or is at least 1). One easily sees by small modification of the tail of the normal distribution that to any $\epsilon > 0$ there exist *PD*'s $P\epsilon\Pi$ such that $D(P\|Q) < \epsilon$, while, of course, no $P\epsilon\Pi$ satisfies $D(P\|Q) = 0$. Hence the I-projection of Q onto Π does not exist.

3. STATEMENT OF MATHEMATICAL RESULTS

In this section we state the main results of Csiszár (1984). Their relevance for the maximum entropy principle and its Bayesian justification will be discussed in the next section.

Theorem 1. Let Q be a *PD* and Π be a convex set of *PD*'s on an arbitrary set S such that

$$D(\Pi\|Q) \overset{\text{def}}{=} \inf_{P\epsilon\Pi} D(P\|Q) < \infty . \tag{3.1}$$

Then there exists a unique *PD* P^*, not necessarily in Π, such that

$$D(P\|Q) \geq D(P\|P^*) + D(\Pi\|Q) \quad \text{for every } P\epsilon\Pi. \tag{3.2}$$

In particular, for any sequence of *PD*'s $P_n\epsilon\Pi$

$$D(P_n\|Q) \to D(\Pi\|Q) \quad \text{implies} \quad D(P_n\|P^*) \to 0. \tag{3.3}$$

Remark. If the *I*-projection of Q onto Π exists, the P^* of (3.2) equals this, and (3.2) reduces to (2.14) of Csiszár (1975). The (always existing) P^* of Theorem 1 will be called the *generalized I-projection* of Q onto Π. Notice that if the ordinary *I*-projection does not exist then $P^* \notin \Pi$. In most cases of interest $D(P^* \| Q) = D(\Pi \| Q)$, but not always; for a counterexample cf. Csiszár (1984).

In the sequel, $X_1, X_2 \ldots$ will denote a sequence of independent random variables, each distributed over S according to Q.

Theorem 2. Let P_n and $\tilde{P}^{(n)}$ denote the conditional distribution of X_1 and the conditional joint distribution of $X_1, \ldots, X_n$, respectively, when the empirical distribution of the random sample $(X_1, \ldots, X_n)$ is known to belong to a given convex set Π of *PD*'s on S. Then, under slight regularity conditions on Π, we have

$$\lim_{n \to \infty} D(P_n \| P^*) = 0 \tag{3.4}$$

and

$$\lim_{n \to \infty} \frac{1}{n} D(\tilde{P}^{(n)} \| P^{*n}) = 0 \tag{3.5}$$

where P^* is the generalized *I*-projection of Q onto Π and P^{*n} is its n'th Cartesian power.

Regularity conditions on Π sufficient for Theorem 2 are formulated in Csiszár (1984); measurability problems are also dealt with there. Notice that (3.4) follows from (3.5) by the easily checked identity

$$D(\tilde{P}^{(n)} \| P^{*n}) = D(\tilde{P}^{(n)} \| P_n^n) + nD(P_n \| P^*) .$$

(3.4) means that the conditional distribution P_n of X_1 converges to P^* in a sense stronger than in variation distance, cf. (1.4).

Theorem 2 is closely related to the so-called Sanov property, i.e., that the probability that the empirical distribution of the sample $(X_1, \ldots, X_n)$ belongs to Π, is of exponential order of magnitude $\exp\{-nD(\Pi \| Q)\}$ (under suitable regularity conditions; the original result of Sanov (1957) has been considerably extended, most significantly by Groeneboom, Oosterhoff and Ruymgaart (1979)).

As a general framework including most situations encountered in applications, consider any measurable mapping Ψ of S into a separable Banach space V, and let C be a convex subset of V. Then the set of *PD*'s defined by

$$\Pi = \{P: E_p \Psi \in C\} \tag{3.6}$$

satisfies the regularity conditions needed for Theorem 2, provided that some interior point v_o of C has the property

$$Q(\{s: \vartheta(\Psi(s)) < \vartheta(v_o) + \epsilon\}) > 0 \quad \text{for each} \quad \vartheta \in V' \text{ and } \quad \epsilon > 0. \tag{3.7}$$

Here V' denotes the dual of V, i.e., the set of all continuous linear functionals on V. The importance of the choice (3.6) lies in the fact that the empirical distribution of the sample $(X_1, \ldots, X_n)$ belongs to this Π iff

$$\frac{1}{n} \sum_{i=1}^{n} \Psi(X_i) \in C. \tag{3.8}$$

Further, the generalized I-projection P^* of Q onto a set of PD's of form (3.6) can be shown, as stated in the next Theorem, to belong to the exponential family of PD's $P \ll Q$ defined by

$$\frac{dP}{dQ} = c(\vartheta)\exp\vartheta(\psi(\cdot)) \qquad \text{for some} \qquad \vartheta \in V' . \tag{3.9}$$

Theorem 3. Let ψ be a measurable mapping of S into a separable Banach space V and let C be a convex subset of V such that some interior point ν_o of C has the property (3.7). Then the conditional distribution P_n of X_1 and the conditional joint distribution $\tilde{P}^{(n)}$ of $X_1,\ldots,X_n$ under the condition (3.8) satisfy (3.4), (3.5) where P^* is the generalized I-projection of Q onto the set of PD's (3.6). This P^* is an element of the exponential family of PD's defined by (3.9) namely the one which corresponds to the $\vartheta \in V'$ maximizing

$$\inf_{\nu \in C} \vartheta(\nu) - \log E_Q\{\exp \vartheta(\psi(\cdot))\} .$$

This theorem is proved in Csiszár (1984) for the more general case when V is an arbitrary locally convex topological vector space, but then an additional condition is needed on the image of Q under ψ, which is automatically fulfilled now. Condition (3.7) is equivalent to the condition there that ν_o should be in the convex closure of the support of the ψ-image of Q. Actually, for most applications it suffices to take for V the k-dimensional Euclidean space and for C the set of points $(y_1,\ldots,y_k)$ with $\alpha_j \le y_j \le \beta_j$, $j=1,\ldots,k$; in this case linear functionals $\vartheta \in V'$ can be identified with vectors $\vartheta = (\vartheta_1,\ldots,\vartheta_k)$ by

$$\vartheta(y_1,\ldots,y_k) = \sum_{j=1}^{k} \vartheta_j y_j .$$

4. IMPLICATIONS OF THE RESULTS

Consider the same problem which lead us in Section 2 to a Bayesian justification of the maximum entropy principle. Using the notations of Section 3, for a sequence of random variables $X_1,X_2,\ldots$ independently distributed on (an arbitrary set) S with common distribution Q, we are interested in posterior distributions when the means of k given functions in the random sample $(X_1,\ldots,X_n)$ are $\gamma_1,\ldots,\gamma_k$ cf. (2.1). Since in the non-discrete case the latter event usually has prior probability 0, we shall instead condition on

$$\alpha_j \le \frac{1}{n} \sum_{i=1}^{k} f_j(X_i) \le \beta_j, \quad j=1,\ldots,k, \tag{4.1}$$

where $\alpha_j < \gamma_j < \beta_j$, $j=1,\ldots,k$. Let us suppose that

$$Q(\{s\colon \sum_{j=1}^{k} \vartheta_j (f_j(s) - \gamma_j) < \epsilon\}) > 0 \quad \text{for every } \vartheta = (\vartheta_1,\ldots,\vartheta_k) \text{ and } \epsilon > 0. \tag{4.2}$$

Then Theorem 3 can be applied with the choice of V and C indicated at the end of Section 3, and with $\psi = (f_1,\ldots,f_k)$, for the assumption (4.2) means that $\nu_o = (\gamma_1,\ldots,\gamma_k)$ satisfies the hypothesis (3.7). It follows for the posterior distribution P_n of X_1 and the posterior joint distribution $\tilde{P}^{(n)}$ of $X_1,\ldots,X_n$ given (4.1) that

$$\lim_{n\to\infty} D(P_n \| P^*) = 0, \quad \lim_{n\to\infty} \frac{1}{n} D(\tilde{P}^{(n)} \| P^{*n}) = 0, \tag{4.3}$$

where P^* is the generalized I-projection of Q onto

$$\Pi = \{P: \alpha_j \leq E_P f_j \leq \beta_j, \qquad j=1,\ldots,k\}. \tag{4.4}$$

Also by Theorem 3, this P^* is determined by

$$\frac{dP^*}{dQ} = c \exp \sum_{j=1}^{k} \vartheta_j^* f_j \tag{4.5}$$

where $\vartheta^* = (\vartheta_1^*,\ldots,\vartheta_k^*)$ maximizes the expression

$$\inf_{\alpha_j \leq y_j \leq \beta_j} \sum_{j=1}^{k} \vartheta_j y_j - \log E_Q\{\exp \sum_{j=1}^{k} \vartheta_j f_j\}.$$

Let us recall that the first limit relation in (4.3) implies convergence in variation distance of the posterior distribution P_n of X_1 given (4.1) to P^*, cf. (1.5). For previous convergence results of that kind, under regularity conditions involving the existence of the I-projection of Q onto Π, we refer to van Campenhout and Cover (1981) and the references there. Actually, $D(P_n \| P^*) \to 0$ is a stronger form of convergence $P_n \to P^*$ than variation distance, e.g., it implies (cf. Lemma 3.1 of Csiszár (1975)) that

$$E_{P_n} f \to E_{P^*} f \text{ if } E_{P^*}\{\exp tf\} < \infty \quad \text{for} \quad |t| < \epsilon; \tag{4.6}$$

$P_n \to P^*$ merely in variation distance would imply $E_{P_n} f \to E_{P^*} f$ for bounded f only.

If the intervals $[\alpha_j, \beta_j]$ shrink to γ_j for each j, the generalized I-projections onto the correspondingly shrinking sets of PD's (4.4) will converge to a PD P_o^* still of form (4.5), now with ϑ^* maximizing

$$\sum_{j=1}^{k} \vartheta_j \gamma_j - \log E_Q \{\exp \sum_{j=1}^{k} \vartheta_j f_j\}.$$

This P_o^* is easily seen to be the generalized I-projection of Q onto

$$\Pi_o = \{P: E_P f_j = \gamma_j, \quad j=1,\ldots,k\}. \tag{4.7}$$

These results and the more general ones in Theorems 2 and 3 suggest the following *extended maximum entropy principle:* If initial beliefs represented by a PD Q are to be updated on the basis of information suggesting that the true distribution belongs to a specified convex set Π of PD's (such as in (4.4) or (4.7)), the updated beliefs should be represented by the generalized I-projection of Q onto Π.

It may not seem sensible to represent updated beliefs by a PD which possibly does not belong to the set of PD's Π suggested by the information available. Still, as a matter of philosophy, one may argue that information directly about expectations can hardly ever be available. Rather, such information is usually just an idealization of experimental evidence involving mean values for large samples. Even if it were known with certainty that the true physical distribution belongs to Π, a resolute subjectivist might find it OK to entertain such beliefs which are represented by a PD $P \notin \Pi$.

Actually, the above extended maximun entropy principle is not merely sensible, it is "right" at least to the same extent as is the Bayesian justification of the original maximum entropy principle discussed in Section 2, i.e., to the extent as it is right to identify "proper updating" of Q with the PD to which the posterior distribution P_n of X_1 given (4.1) converges as $n \to \infty$. Of course, P_n itself always belongs to the set (4.4) of PD's for every n. Incidentally, one sees from (4.6) that P^* can be outside Π only if

$$E_{P^*}\{\exp tf_j\} = \infty \quad \text{for every} \quad t > 0 \quad \text{or for every} \quad t < 0.$$

for some of the functions f_j appearing in (4.4).

It may be instructive to return to the example at the end of Section 2. In that example $P^* = Q$ = standard normal distribution, which is outside Π and, indeed, for $f(x) = x^3$ we have

$$E_{P^*}\{\exp tf\} = \infty \qquad \text{for every} \qquad t \neq 0.$$

The extended maximum entropy principle suggests to let the updated *PD* be Q itself, i.e., to disregard completely the information that the third moment of the true distribution is 1 (or ≥ 1). Since *PD*'s "arbitrarily close to Q" with such third moment exist, this, indeed, appears to be the right thing to do.

So far only the implications of the first limit relation in (4.3) were considered. Let us now turn to the implications of the second one. Notice first that it implies for any fixed m that

$$D(\tilde{P}_n^{(m)} \| P^{*m}) \to 0 \qquad \text{as} \quad n \to \infty \tag{4.9}$$

where $\tilde{P}_n^{(m)}$ denotes the posterior distribution of $X_1,\ldots,X_m$ given (4.1). In particular, $X_1,\ldots,X_m$ are asymptotically independent (as $n \to \infty$) under the posterior distribution, for any fixed m. Interestingly, the posterior joint distribution $\tilde{P}^{(n)}$ of the whole sample $(X_1,\ldots,X_n)$ under the condition (4.1) is (in a weaker sense) also close to independence. This is the content of the following lemma.

Lemma. For each n, let A_n be a subset of S^n such that $P^{*n}(A_n) < \alpha^n$ for some $0 < \alpha < 1$. Then the second limit relation in (4.3) implies that

$$\lim_{n \to \infty} \tilde{P}^{(n)}(A_n) = 0.$$

Proof. Using a basic property of informational divergence, we have

$$D(\tilde{P}^{(n)} \| P^{*n}) \geq n\,\tilde{P}^{(n)}(A_n) \log \frac{1}{\alpha} - h(\tilde{P}^{(n)}(A_n)) \log \frac{\tilde{P}^{(n)}(A_n)}{P^{*n}(A_n)} + \tilde{P}^{(n)}(A_n^c) \log \frac{\tilde{P}^{(n)}(A_n^c)}{P^{*n}(A_n^c)} \geq$$

where $A_n^c = S^n \setminus A_n$ and $h(t) = -t\log t - (1-t)\log(1-t)$.

As the function $h(t)$ is bounded in $0 \leq t \leq 1$, this proves the assertion.

The result just proved means that whatever probabilistic statement holds with probability tending to 1 exponentially fast for n independent random variables distributed according to P^*, the same statement holds with probability tending to 1 also under the posterior joint distribution $\tilde{P}^{(n)}$ (take for A_n the subset of S^n where the statement does not hold).

REFERENCES

BOLTZMANN, L. (1877). Beziehung zwischen dem zweiten Hauptsatze der mechanischen Wärmetheorie und der Wahrscheinlichkeitsrechnung respektive den Sätzen über das Wärmegleichgeiwicht. *Wien. Ber.* **76**, 373-435.

VAN CAMPENHOUT, J.M. and COVER, T.M. (1981). Maximum entropy and conditional probability. *IEEE Trans. Inform. Theory,* **IT-27**, 483-489.

CSISZÁR, I. (1967). Information-type measures of difference of probability distributions and indirect observations. *Studia Sci. Math. Hungar.*, **2**, 299-318.

— (1975). *I*-divergence geometry of probability distributions and minimization problems. *Ann. Prob.* **3**, 148-158.

— (1984). Sanov property, generalized *I*-projection and a conditional limit theorem. *Ann. Prob.*, (to appear).

CSISZÁR, I. and KÖRNER, J. (1981). *Information Theory: Coding Theorems for Discrete Memoryless Systems.* New York: Academic Press.

CSISZÁR, I. and TUSNÁDY, G. (1984). Information geometry and alternating minimization procedures. *Statistics and Decisions.* (to appear).

DIACONIS, P. and ZABELL, S.L. (1982). Updating subjective probability. *J. Amer. Statist. Assoc.* **77**, 831-834.

GOKHALE, D.V. and KULLBACK, S. (1978). *The Information in Contingency Tables.* New York: Marcel Dekker.

GOOD, I.J. (1963). Maximum entropy for hypothesis formulation, especially for multidimensional contingency tables. *Ann. Math. Statist.* **34**, 911-934.

GROENEBOOM, P. OOSTERHOFF, J. and RUYMGAART, F.H. (1979). Large deviation theorems for empirical probability measures. *Ann. Prob.* **7**, 553-586.

JAYNES, E.T. (1957). Information theory and statistical mechanics. *Phys. Rev.* **106**, 620-630.

— (1979). Where do we stand on maximum entropy? *The Maximum Entropy Formalism.* (Levine, R.O. and Tribus, M. eds.). Cambridge: MIT Press, 15-118.

KULLBACK, S. (1959). *Information Theory and Statistics.* New York: Wiley.

KULLBACK, S. and LEIBLER, R.A. (1951). On information and sufficiency. *Ann. Math. Statist.* **22**, 79-86.

SANOV, I.N. (1957). On the probability of large deviations of random variables (in Russian). *Math. Sbornik* **42**, 11-44.

SHORE, J.E. and JOHNSON, R.W. (1980). Axiomatic derivation of the principle of maximum entropy and the principle of minimum cross entropy. *IEEE Trans. Inform. Theory,* **IT-26**, 26-37.

DISCUSSION

G.A. BARNARD (*Colchester, UK*)

I must avow incompetence to discuss Dr. Csiszár's paper with the depth it deserves. We are here faced with choosing one from a family of distributions when certain moments of the true distribution are known. Such problems arise in statistical mechanics, when the energy, temperature, etc. of a statistical system are known, and it is required to "estimate" the distribution over a set of possible states. Maximal entropy principles have played a significant role in the analysis of such problems, but my impression, for what it is worth, is that such principles do not really provide a sound basis for deducing empirical facts. The history of changes in statistical mechanics, from pre-quantum theories to the unexpected distinction between bosons and fermions, to go no further, inclines me to suppose that we have here further instances of the phenomenon in the XIX century, when it was thought by some that Newton's third law of motion acquired a tautological character, in the form Force = Mass × Acceleration, this then being taken as the definition of Force. But when the modifications needed when the velocities involved were comparable with the velocity of light were considered, the true empirical basis of the law was more clearly understood.

That connections are to be found between Bayesian lines of reasoning and maximal entropy, or minimal information principles, I do not doubt. I believe it was Jean Ville who first drew attention to the fact that the Bayesian assumption of uniform prior probability over a finite set of alternatives could be expressed in terms of minimal information; and the development of posterior probabilities can be expressed in terms of growth of information. Then we had Kullback's impressive work on information theory in statistics.

The elegant inequalities arising in information theory may have more extended applications in statistical inference theory. For example, it may be that such results will enable us to demonstrate the conservative nature of the generalised Behrens-Fisher test procedure.

I hope discussions with Dr. Csiszár and others at this conference will help me, and others whose background resembles mine, to better understand their approach.

E.T. JAYNES (*Washington University, St. Louis*)

The question of the proper choice of Q is also of current importance in maximum-entropy image reconstruction and spectrum analysis. Incidentally, we feel uncomfortable at calling Q a "prior opinion" because that implies rather more than we wish to say. Mathematically, Q is just the basic "measure" of the sample space, or hypothesis space, on which our probabilities are defined. Its introduction is necessary to preserve invariance of the theory under a change of variables.

In each new application the question arises: what basic measure should we choose? In statistical mechanics the problem was solved long ago by Liouville's theorem on the invariance of phase volume under the equations of motion, which means that if we choose uniform density over phase volume the content of the theory will be independent of time. In fact, a current difficulty is that this problem was solved so long ago that its solution is taken for granted, and most students are unaware that the problem ever existed.

But prospective users of maximum entropy in new applications need to be aware of it, because the results depend on our choice of measure. In the applications to date which have reached the stage of final numerical evaluation by computer, a measure uniform on S^n has been used —sometimes by default because a user who is unaware of the problem is led to do this automatically (just as an antiBayesian who does not believe that he is using prior probabilities at all— and particularly denounces the use of a uniform prior— is thereby led automatically to the results that would have followed from use of a uniform prior).

The succesful maximum entropy results that have been achieved, noted in my presentation, indicate that this choice of measure (or what is the same thing, the choice of the variables over which the measure is uniform) is a good one in most cases. However, in image reconstruction there are hints that one may be able to do still better by a measure that already anticipates some positive correlation between adjacent picture elements, or "pixels". But a De Finetti type exchangeable mixture goes too far in the opposite direction, because it implies that correlations persist undiminished between pixels far apart.

An intermediate choice, that encourages positive correlation between nearby pixels but not between distant ones, allows the kind of "clustering" that often occurs in real images and real data. One possibility is a two-dimensional generalization of the sampling distribution of an autoregressive (AR) model, in which one can introduce AR coefficients up to some moderate order such as 4 or 5, and the Model automatically gives exponentially attenuating correlations beyond that distance.

The problem is currently under study, but only at the analytical theory stage, not yet the computer evaluation stage. We hope to have more definitive things to say about it in another year or two.

Finally, we note that there is an alternative interpretation of the maximum-entropy procedure itself, suggested by Dr. John Skilling of Cambridge University, that does not seem ruled out by any arguments or evidence yet known. Instead of regarding an entropy factor as part of a prior probability, one may think of it as part of a utility function; on this viewpoint we take the maximum entropy solution, not because it is more likely to be true than others, but because we prefer it as being more useful.

Since in decision theory the final decision depends only on the product of the prior and utility functions, not on the functions separately, many different interpretations can lead to the same actual solution. Although my own predilections still favor the prior probability viewpoint, it is possible that Skilling's utility viewpoint may prove in the future to have cogent advantages. For example, it may give us a rationale for using the present maximum entropy solution even when our prior information suggests a measure highly nonuniform on S^n.

T. SEIDENFELD (*Washington University, St. Louis*)

Let $P(\cdot)$ be an initial distribution. Suppose $Q(\cdot)$ is to be a revision of P subject to a set of constraints $C = \{c_i: \ldots\}$. Then the entropy principle enjoins us to select that Q (if one exists) which satisfies C and minimizes the shift from P according to the cross-entropy semi-metric

$$\int q(x) ln[q(x)/p(x)]dx.$$

Is the entropy principle consistent with the Bayesian postulates of *coherence* and *conditionalization*? (I understand conditionalization to require that, with "old" probability P and hypothesizing new evidence **e**, the "new" probability Q is $P(\cdot\,|\,e)$, i.e. Q is the "old" conditional probability). Williams (1980) argues that conditionalization is a special case of the entropy principle. Rosenkrantz (1977, p. 57) posits the same conclusion. The excellent presentations by Shore and Johnson (1980) and (1981) yield numerous results which suggest conditionalization is special case of a shift from "prior" to "posterior" according to the entropy rule.

I remain unconvinced of a simple agreement between Bayesian and entropy principles. In what follows I note three restrictions on the minimum entropy rule that serve to undercut the thesis that the entropy principle is a generalization of Bayesian theory.

1. Unlike the rule of conditionalization, the entropy principle does not apply to marginal distributions (on pain of inconsistent results). That is, with respect to a parameter of interest μ, even though the constraints C involve μ alone, the revised marginal distribution $Q(\mu)$ is *not* the result of a minimum shift from the initial marginal distribution $P(\mu)$. Instead, $Q(\mu)$ must be obtained by integrating-out over the full, joint distribution Q which is obtained by a minimum shift (subject to C) from the full, joint distribution P. (For a simple counter-example when this restriction is ignored, see Seidenfeld (1979, pp. 430-32)).

Of course, with conditionalization all that is required is the prior $P(\mu)$ and the likelihood $P(e\,|\,\mu)$, regardless the number of "nuisance" parameters there might be in the agent's joint distribution. The question, then, is how large must the algebra be to apply the entropy principle without running afoul of the difference between minimizing a shift in a marginal (average) and marginalizing a minimum shift? This problem is exacerbated by the next issue.

2. Conditionalization deals with "updating" P in the light of new evidence e, *consistent* with the background assumptions already apart of P. When it comes to duplicating basic properties of conditionalization, e.g. the posterior is invariant over the order with which evidence is accepted, advocates of the entropy rule suppose that the "shifts" result from addition of new, consistent constraints. For example, Shore and Johnson (1981, p. 478) report a property of the entropy principle, corresponding to path-invariance over successive shifts, that depends upon accumulating (consistent) constraints.

Unfortunately, translating instances of conditionalizing into the the language of minimum cross-entropy shifts does *not* preserve consistency of the new constraint set. Following Williams (1980), accepting new evidence: $X = x_0$ is represented by adding the constraint: $E[I_{x_0}] = 1$, where I_{x_0} is the indicator variable for the event $X = x_0$. By assumption, $X = x_0$ is new evidence; hence, $P(X = x_0) = k \neq 1$ for the initial distribution P. Suppose P, itself, is the result of a minimum cross-entropy shift subject to a constraint set C'. Then, since P satisfies $E[I_{x_0}] = k$, (Shore and Johnson, [1981, p. 474]) it is easy to verify that P remains the minimum shift for the constraint set $C'' = C' \cup \{E[I_{x_0}] = k\}$. However, against the constraint set C'', the new constraint $E[I_{x_0}] = 1$ is not a consistent addition. Of course, this formulation applies to every case of conditionalization. Because cross-entropy is merely a semi-metric (the triangle inequality does not hold generally), path-invariance does not obtain for the general case of shifts over constraint sets that are mutually inconsistent. Without new restrictions on the language of "constraints", I do not see how to capture the all important distinction between shifts that results from accepting new (consistent) evidence and shifts that are wholesale revisions in what is taken as evidence.

To put the problem simply, one cannot treat "constraints" like other propositions in the algebra over which probability is defined. To do so leads to the morass of iterated subjective probabilities (see De Finetti, (1972, p. 189)). What sense is a Bayesian to make of a "constraint"?

3. In the case of continuous random variables, the advantage of cross-entropy over entropy is that answers are invariant over equivalent random variables when the former is used. However, cross-entropy requires an initial distribution, P, and if rational agents are to agree on a specific Q once to constraint set C is identified, then additional regulations are needed to fix P.

Some use Jeffreys' (1961, §3.10) invariance theory to fill the gap (see Rosenkrantz (1977)). But Jeffreys' program is not consistent with Bayesian postulates. Inconsistencies are generated with simple, one parameter inference of the following form (see Seidenfeld (1979, p. 422) for an example)).

Let θ be the parameter of interest and observations of random variables X and Y constitute the data to be acquired. Suppose: (i) $P(X, Y|\theta) = P(X|\theta) \cdot P(Y|\theta)$ (independence) (ii) there is no common sufficient statistic for the combined data (x,y) and (iii) each distribution, $P(X|\theta)$ and $P(Y|\theta)$, admits a prior for θ by the invariance theory. Then, depending upon which observation is used to generate the initial distribution (in accord with invariance theory), there are two choices for the composite posterior $Q(\mu)$, given the combined data (x,y). The problem duplicates familiar fiducial inconsistencies with compound data lacking a common sufficient statistic but where smoothly invertible pivotals exist with each datum, separately.

How, then, can the entropy principle be a generalization of Bayesian theory if it depends upon Jeffreys' invariance rule for selecting a prior?

W. POLASEK (*Tech. Univ. Wien*)

I would be interested to learn how the extended maximum entropy principle can be related to Akaike's (1978) view of Bayesian procedures.

REPLY TO THE DISCUSSION

I thank all the discussants for their comments and also all those with whom I had fruitful discussions during this highly interesting meeting, for which Professor Bernardo is to be warmly congratulated.

I fully agree with Professor Barnard that connections between Bayesian lines of reasoning and the maximum entropy principle have been pointed out for quite some time. My aim was to demonstrate under substantially more general conditions than before that "maximum entropy updating" of probability assignments is, indeed, a Bayesian updating in an asymptotic sense, if the new information can be interpreted in terms of frequencies in a large (real or hypothetical) i.i.d. sample. Moreover, a solution is offered to the problem of what to do when the required extremal feasible distribution does not exist, as in the example at the end of Section II. In this paper the maximum entropy principle is interpreted strictly as a rule for updating probabilities, i.e., in addition to the new evidence we always rely also on a prior probability assignment. While this prior is most often taken as the uniform distribution, this is not claimed be justified by the maximum entropy principle; thus the danger of tautology hinted at by Professor Barnard is avoided.

Professor Jaynes' comments on his successes with suitably chosen non-independent priors in his applied work with the maximum entropy principle were very interesting. While a kind of non-independence was considered also in the present paper, I hope to return to this subject in greater generality on another occasion.

As Professor Seidenfeld remarked, maximum entropy updating is simply identical with Bayesian updating if the new evidence is that the probability of a particular event A must be 1. When the new evidence is not of this form, e.g., it consists in constraints on the expectations of certain $f_1, f_2, \ldots, f_k$, then Bayesian updating is not directly applicable. As Professor Seidenfeld puts it, "one cannot treat constraints like other propositions in the algebra over which probability is defined". On the other hand, such constraints may be readily identified with events (defined in terms of relative frequencies) in the n'th Cartesian power of the given space of events, in an asymptotic sense as $n \to \infty$. It is thus natural to use Bayesian updating in the product space, and project the obtained posterior onto the original space. By Theorem 2, this gives (in the limit as $n \to \infty$) the same result as the extended maximum entropy principle. Of course, when conditioning on a sample mean for a fixed sample size n, we get only an approximation to the "maximum entropy" solution. If the sample variance is also included in the conditioning, what we get will be an approximation to the maximum entropy solution based also on the corresponding second moment. As the prior probability assignment may also have been chosen by a process as above, we can naturally accomodate into our model the case when the prior probability assignment and the new evidence have the same degree of reliability, a situation recently investigated (in a special case) by Campbell (1983). The lack of "path invariance" for inconsistent constraints does not seem a drawback to me. Professor Seidenfeld claims that for constraints involving a marginal alone of some joint distribution, the "maximum entropy" updating of this marginal might differ from the corresponding marginal of the updated joint distribution. This, however, cannot happen in our model, as one sees from eq. (2.8). Of course, the maximum entropy principle does not give a rule for selecting the prior. The Bayesian is free to use the prior he feels most appropriate, disregarding any "rules" not consistent with Bayesian postulates.

Professor Polasek asks how the extended maximum entropy principle can be related to Akaike's view of Bayesian procedures. The answer is that I do not see any close relationship, mainly because the underlying problems are very different. Akaike considers statistical estimation problems, adopting the view that an unknown distribution (rather than a parameter) is to be estimated, and he uses $D(P\|Q)$ as a loss function measuring the loss when the unknown true distribution P is estimated by Q. The maximum entropy principle, as understood in this paper, relates to problems of a different kind, not within the scope of standard statistical decision theory, namely to updating priors to conform with evidence typically consisting in moment constraints. Although in both cases $D(P\|Q)$ appears as a measure of "distance" which should be minimized, a formal difference is that in Akaike's model minimization is performed with respect to the second variable while the maximum entropy principle calls for minimization with respect to the first one.

Finally, I have to call attention to two papers I became aware of after having presented this paper. Topsoe (1979) was first to systematically use generalized I-projection (under the name "relative center of attraction") and, in particular, to establish the important inequality (3.2). Jupp and Mardia (1983) proved a one-dimensional analogue of Theorem 3. The problem solved by our main result Theorem 2 was not considered in these papers.

REFERENCES IN THE DISCUSSION

AKAIKE, H. (1978). A new look at the Bayes procedure. *Biometrika* **65**, 53-59

CAMPBELL, L.L. (1983). Minimum information divergence estimation with inaccurate side information. *Tech. Rep.* **17**, Kingston, Ont, Canada: Queen's University.

DE FINETTI, B. (1972). *Probability, Induction and Statistics.* New York: Wiley.

JEFFREYS, H. (1961). *Theory of Probability.* 3rd. edition. Oxford: University Press.

JUPP, P.E. and MARDIA, K.V. (1983). A note on the maximum entropy principle. *Scand. J. Statist.* **10**, 45-47.

ROSENKRANTZ, R.D. (1977). *Inference, Method and Decision.* Dordrecht, Holland: D. Reidel.

SEIDENFELD, T. (1979). Why I am not an objective bayesian. *Theory and Decision,* **11**, 413-440.

SHORE, J.E. and JOHNSON, R.W. (1980). Axiomatic derivation of the principle of maximum entropy and the principle of minimum cross-entropy. *IEEE Trans. Information Theory,* **IT-26**, 26-37.

— (1981). Properties of cross-entropy minimization. *IEEE Trans. Information Theory,* **IT-27**, 472-482.

TOPSOE, F. (1979). Information theoretical optimization techniques. *Kybernetika* **15**, 8-27.

WILLIAMS, P.M. (1980). Bayesian conditionalization and the principle of minimum information. *Brit. J. Phil. Sci.* **31**, 131-144.

BAYESIAN STATISTICS 2, pp. 99-118
J.M. Bernardo, M.H. DeGroot, D.V. Lindley, A.F.M. Smith (Eds.)

Probability Forecasting, Stochastic Dominance, and the Lorenz Curve

DeGROOT, M.H.
Carnegie-Mellon University, Pittsburgh

and

ERIKSSON, E.A.
Royal Institute of Technology, Stockholm

SUMMARY

In this paper we consider the comparison of forecasters whose predictions are presented as their subjective probabilities of various future events. The concept of a forecaster being well-calibrated is reviewed, and the relationship that one well-calibrated forecaster is more refined than another is extensively discussed. The central purpose of the paper is to describe several other possible relationships between well-calibrated forecasters, each of which is equivalent to the relationship "more refined than".

In particular, it is shown that this relationship is equivalent to the relationship between distributions defined by the concept of second-degree stochastic dominance and to the relationship between distributions induced by their Lorenz curves. Other equivalent relationships are expressed in terms of the existence of certain types of stochastic transformations. The connection between these concepts and the concepts of majorization and Schur-convexity is described.

Keywords: CALIBRATION; COMPARISON OF FORECASTERS; MAJORIZATION; PREDICTION; REFINEMENT; SCHUR-CONVEXITY; SCORING RULES; STOCHASTIC TRANSFORMATION; SUBJECTIVE PROBABILITY.

1. INTRODUCTION

Consider a forecaster who at the beginning of each period n in a sequential process ($n=1, 2,...$) must specify his subjective probability that some particular event A_n will occur during that period. It is assumed that when the forecaster specifies his probability of A_n, he is aware of the values of various variables which he believes may be relevant to the occurrence of A_n. The knowledge of which of the previous events $A_1,...,A_{n-1}$ actually occurred will usually be part of the forecaster's information at the beginning of period n, but such knowledge is not essential to our model.

It is desired to evaluate and compare different forecasters on the basis of their subjective probabilities of the events $A_1,A_2,...,A_n$ and the subsequent observation of exactly which of these events occurred. It is assumed that these observations of the forecaster's performance are available over a large number of periods; i.e., it is assumed that n is large. Thus, the basic setting of this paper is the same as that presented in DeGroot and Fienberg (1982, 1983).

The forecaster might be either a meteorologist who at the beginning of each day must specify his probability that it will rain during that day at a particular location or an economist who at the beginning of each weekly or quarterly period must specify his probability that a particular interest rate or stock market average will rise during the period. In another context, the forecaster might be a medical diagnostician who must specify his probability that a patient has a particular disorder on the basis of the data available to him, and who subsequently learns the patient's true condition.

As in DeGroot and Fienberg (1982, 1983), we shall present the discussion in this paper in the context of a weather forecaster who day after day must specify his subjective probability x that there will be at least a certain amount of rain at some given location during a specified time interval of the day. We refer to the occurrence of this well-specified event simply as "rain". The probability x specified by the forecaster on any particular day is called his prediction for that day. Again, as in DeGroot and Fienberg (1982, 1983), we make the assumption that x is restricted to a given finite set of values $0 = x_0 < x_1 < \ldots < x_k = 1$, and we let $\mathbf{X}$ denote the set $\{x_0, x_1, \ldots, x_k\}$.

The overall performance of a forecaster is characterized by two functions defined on the set $\mathbf{X}$: (1) a probability function, or p.f., $\nu(x)$ which gives the probability or relative frequency with which the forecaster makes each possible prediction x, and (2) a function $\varrho(x)$ which gives the conditional probability or conditional relative frequency of rain, given that the forecaster's prediction is x. The quantities $\nu(x)$ and $\varrho(x)$ have the following three possible interpretations:

1. They may be regarded as limiting or theoretical values that would be obtained over an infinite sequence of days. This interpretation is most appropriate if daily meteorological conditions have an approximately stationary distribution over the days on which predictions are made and the forecaster's predictions are generated from a mathematical model or from simulations.

2. They may be the actual observed relative frequencies for the finite sequence of n days over which the forecaster's performance has been observed. This interpretation is also most appropriate when n is large and meteorological conditions have an approximately stationary distribution over the n days.

3. The value $\nu(x)$ may be regarded as an observer's subjective probability that the forecaster will make the prediction x on a specified or randomly chosen day in the future, and the value $\varrho(x)$ may be regarded as the observer's subjective conditional probability that it will rain on that day given that the forecaster's prediction is x. These subjective probabilities will be based on the observer's information about what the meteorological conditions will be on the chosen day and his beliefs about how the forecaster will perform under these conditions. In turn, these beliefs will be based in part on information available to the observer about the method by which the forecaster makes his predictions and on the forecaster's performance over the n days for which data are available. This interpretation is appropriate and useful even when meteorological conditions on the chosen day may be quite different from conditions on these n days.

A forecaster is said to be *well-calibrated* if $\varrho(x) = x$ for all $x \epsilon \mathbf{X}$ such that $\nu(x) \neq 0$. In other words, a forecaster is well-calibrated if and only if, under whichever of the three interpretations just presented is relevant, the forecaster's predictions can be accepted at face value in the sense that the conditional probability of rain is x given that the forecaster's prediction is x. A wide variety of theoretical and empirical reasons why forecasters will tend to be well-calibrated have been presented in DeGroot and Fienberg (1982, 1983), Dawid (1982a,b), Lad (1982), Murphy and Winkler (1977), Pratt (1962), and Kadane and Lichtenstein (1982), and will not be repeated here. Throughout the remainder of this paper we shall restrict our attention to the comparison and evaluation of well-calibrated forecasters.

In Section 2, the relationship of one well-calibrated forecaster being more refined than another is reviewed and new interpretations of this relationship are presented in terms of mean-preserving spreads and the existence of certain stochastic transformations. In Section 3, this relationship is shown to be equivalent to other relationships involving second-degree stochastic dominance and expectations of convex functions. In Section 4, its equivalence to relationships between Lorenz curves is described. Many of the relationships that are introduced in these sections have been extensively studied in economics and probability theory, and are well known to be equivalent to each other under various conditions. The central purpose of these sections is to show that they are also equivalent to the relationship of one well-calibrated forecaster being at least as refined as another, and to bring all these relationships together in a fixed framework under a unified set of conditions. Finally, in Section 5 we demonstrate the relationship between Schur-convex measures of the quality of forecasters and strictly proper scoring rules. Throughout the paper, the connection between this work and the study of inequality in the distribution of income in a population is explained and references to the economics literature are given.

2. INTERPRETING REFINEMENT

Consider two well-calibrated forecasters A and B whose predictions are characterized by the p.f.'s ν_A and ν_B. In DeGroot and Fienberg (1982, 1983), a concept of refinement was introduced which induced a partial ordering on the class of all well-calibrated forecasters. This concept is defined as follows:

A *stochastic transformation* $h(y|x)$ is a non-negative function defined on $\mathbf{X} \times \mathbf{X}$ such that

$$\sum_{y\epsilon\mathbf{X}} h(y|x) = 1 \qquad \text{for every } x\epsilon\mathbf{X}. \tag{2.1}$$

A well-calibrated forecaster A is defined to be *at least as refined* as another well-calibrated forecaster B if there exists a stochastic transformation $h(y|x)$ such that

$$\sum_{x\epsilon\mathbf{X}} h(y|x)\nu_A(x) = \nu_B(y) \qquad \text{for } y\epsilon\mathbf{X}, \tag{2.2}$$

$$\sum_{x\epsilon\mathbf{X}} h(y|x)x\nu_A(x) = y\nu_B(y) \qquad \text{for } y\epsilon\mathbf{X}. \tag{2.3}$$

In words, forecaster A is at least as refined as forecaster B is we can artificially generate, from A's prediction and an auxiliary randomization, a well-calibrated forecaster with the same p.f. ν_B as B. The following argument indicates why (2.3) ensures that the prediction generated from this process will be well-calibrated: Let X denote A's prediction and let Y denote the prediction generated from this process. Then $h(y|x) = Pr(Y=y|X=x)$ for any x such that $\nu(x) \neq 0$, and

$$Pr(\text{Rain}, Y=y) = \sum_{x\epsilon\mathbf{X}} Pr(\text{Rain}|Y=y,X=x)h(y|x)\nu_A(x). \tag{2.4}$$

However, the event "rain" must be conditionally independent of Y given X, since Y is generated from X by an auxiliary randomization. Hence,

$$Pr(\text{Rain}|Y=y,X=x) = Pr(\text{Rain}|X=x) = x \tag{2.5}$$

since A is well-calibrated, and it follows from (2.4) that

$$Pr(\text{Rain}, Y=y) = \sum_{x \in \mathbf{X}} x\, h(y|x) \nu_A(x). \tag{2.6}$$

Furthermore, it follows from (2.2) that

$$\begin{aligned} Pr(\text{Rain}, Y=y) &= Pr(\text{Rain} \mid Y=y) \nu_B(y) \\ &= y \nu_B(y) \end{aligned} \tag{2.7}$$

if and only if the prediction Y is well-calibrated. Thus, it is found from (2.6) and (2.7) that (2.3) will hold if and only if the prediction Y is well-calibrated.

If A is at least as refined as B and the p.f.'s ν_A and ν_B are not identically equal, then it is said that A is *more refined* than B. The partial ordering induced by this relationship is very strong. If A is more refined than B, then an observer who must choose between learning either the prediction of A or the prediction of B on a given day, at equal cost, should always prefer to learn the prediction of A, regardless of the decision problem in which the observer might use this information.

It is well-known and easily verified that for every well-calibrated forecaster, the mean of the distribution represented by the p.f. ν must have the same value μ. This constant μ is simply the overall probability of rain. Thus, the problem of comparing or ordering well-calibrated forecasters is equivalent to the problem of comparing or ordering all distributions on the set $\mathbf{X}$ with a given mean μ. This problem has been widely treated in economics in the context of comparing different possible income distributions in a population. Thus, if we regard ν as representing the distribution of income in some population, then we can consider comparing different distributions for which the mean income or, equivalently, the total income of the population remains fixed. The usual basis for comparing two such income distributions is whether one distribution represents greater inequality in the population than another. Some references which will lead to the vast literature on this subject are Sen (1973), Rothschild and Stiglitz (1970, 1973), Atkinson (1970), Dasgupta, Sen and Starrett (1973), Kolm (1976), Fei and Fields (1978), Eichhorn and Gehrig (1980), and Arnold (1981).

There is another interpretation of the relations (2.2) and (2.3) that has proven to be helpful in this income context. Suppose that X and Y are discrete random variables, each of which takes values in $\mathbf{X}$, and suppose that the joint p.f. of X and Y is defined as follows, for $x \in \mathbf{X}$ and $y \in \mathbf{X}$:

$$\Pr(X=x, Y=y) = \nu_A(x)\, h(y|x), \tag{2.8}$$

where $h(y|x)$ is a stochastic transformation satisfying (2.2) and (2.3). Then it follows from (2.8) that the marginal p.f. of X is ν_A, from (2.2) that the marginal p.f. of Y is ν_B, and from (2.3) that $E(X|Y=y)=y$ for $y \in X$ or, more simply, that $E(X|Y)=Y$.

This interpretation leads us to the following result. We shall denote the relationship that forecaster A is at least as refined as forecaster B by the symbols $A \geq B$, and the relationship that A is more refined than B by the symbols $A > B$ using, with an abuse of notation, the same symbol, than for arithmetic unequalities. To avoid confusion, it should be emphasized that our preference relation is the opposite of that common in economics. There, an income distribution with relatively little variability over the population or an investment with relatively little uncertainty is usually preferable, whereas a well-calibrated forecaster who makes relatively extreme predictions will be preferable.

Theorem 1. The relationship $A \geq B$ is satisfied if and only if there exist discrete random variables X and Y such that the marginal p.f. of X is ν_A, the marginal p.f. of Y is ν_B, and $E(X|Y) = Y$.

Proof. Suppose first that $A \geq B$. If we define the joint distribution of X and Y as in (2.8), then the requirements of the theorem will be satisfied. Conversely, suppose that there exist random variables X and Y satisfying the conditions of the theorem, and define $h(y|x) = Pr(Y=y|X=x)$. [The conditional p.f. $h(\cdot|x)$ may be defined arbitrarily for any $x \epsilon \mathbf{X}$ such that $Pr(X=x)=0$.] Then $h(y|x)$ will be a stochastic transformation satisfying (2.2) and (2.3). Hence, $A \geq B$. ■

Theorem 1 is closely related to the work of Strassen (1965) and to the idea of *mean-preserving spreads* or *transfers* in income distributions, as discussed by Rothschild and Stiglitz (1970, 1973), and which has roots in the early work of Pigou (1912) and Dalton (1920). This relation can be seen in the following simple corollary.

Corollary 1. The relationship $A \geq B$ is satisfied if and only if there exists a stochastic transformation $\eta(x|y)$ such that

$$\sum_{y \epsilon \mathbf{X}} \eta(x|y)\, \nu_B(y) = \nu_A(x) \qquad \text{for } x \epsilon \mathbf{X}\,, \tag{2.9}$$

$$\sum_{x \epsilon \mathbf{X}} x\, \eta(x|y) = y \qquad \text{for } y \epsilon \mathbf{X}\,. \tag{2.10}$$

Proof. By definition, $A \geq B$ if and only if there exists a stochastic transformation $h(y|x)$ satisfying (2.2) and (2.3). Define

$$\eta(x|y) = \frac{h(y|x)\nu_A(x)}{\nu_B(y)} \qquad \text{for } x,y \epsilon \mathbf{X}\,, \tag{2.11}$$

where $\eta(\cdot|y)$ may be defined to be an arbitrary conditional p.f. if $\nu_B(y)=0$. Then $\eta(x|y) \geq 0$ for $x,y \epsilon \mathbf{X}$ and $\sum_{x \epsilon \mathbf{X}} \eta(x|y) = 1$ for $y \epsilon \mathbf{X}$ because of (2.2). Furthermore, (2.9) follows from (2.1) and (2.10) follows from (2.3). ■

In the context of investment prospects, a standard interpretation of the relation $A \geq B$ is that X can be written as Y plus noise, i.e., $X = Y + Z$, where $E(Z|Y) = 0$. Hence, X is considered to be more risky than Y.

In the language of income distributions, Corollary 1 states that $A \geq B$ if and only if we can generate the income distribution ν_A from the income distribution ν_B in the following way: Consider all the people who have a particular income y under the distribution ν_B. Reassign all of these people to new income levels, and let $\eta(x|y)$ denote the proportion of them who are assigned to have new income x. This new assignment is to be carried out so that the total income of these people remains unchanged or, equivalently, so that their average income at their new levels has the same value y as before. Thus, when the people at income level y are spread over new levels, this reassignment must be a mean-preserving spread. This requirement is expressed by (2.10).

Since the definition of the relationship $A \geq B$ involves a stochastic transformation of the form $h(y|x)$, and Corollary 1 involves one of the form $\eta(x|y)$, we can combine these features to obtain the following result.

Corollary 2. The relationship $A \geq B$ is satisfied if and only if there exists a non-negative function p defined on the product set $\mathbf{X} \times \mathbf{X}$ such that

$$\sum_{x \epsilon \mathbf{X}} p\,(x,y)\nu_A(x) = 1 \qquad \text{for } y \epsilon \mathbf{X}, \tag{2.12}$$

$$\sum_{y \in \mathbf{X}} p(x,y)\nu_B(y) = 1 \qquad \text{for } x \in \mathbf{X}, \tag{2.13}$$

$$\sum_{x \in \mathbf{X}} p(x,y)\, x\nu_A(x) = y \qquad \text{for } y \in \mathbf{X}, \tag{2.14}$$

Proof. We know that $A \geq B$ if and only if there is a stochastic transformation $h(y|x)$ satisfying (2.2) and (2.3), and a stochastic transformation $\eta(x|y)$ satisfying (2.9) and (2.10). Accordingly, we may define $p(x,y)$ either by

$$p(x,y) = \frac{h(y|x)}{\nu_B(y)} \tag{2.15}$$

if $\nu_B(y) > 0$, or by

$$p(x,y) = \frac{\eta(x|y)}{\nu_A(x)} \tag{2.16}$$

if $\nu_A(x) > 0$. It follows from (2.11) that these definitions are equivalent when $\nu_A(x)\nu_B(y) > 0$, and it can easily be verified that (2.12)-(2.14) hold. ■

3. STOCHASTIC DOMINANCE

The theory of the comparison of forecasters is closely related to the theory of the comparison of statistical experiments as developed by Blackwell (1951, 1953), and from that development we can obtain further characterizations of the relationship $A \geq B$. For any well-calibrated forecaster, let F denote the distribution function (d.f.) corresponding to the p.f. ν; i.e., let

$$F(t) = \sum_{\{x: x \in \mathbf{X}, x \leq t\}} \nu(x) \qquad \text{for } 0 \leq t \leq 1 \tag{3.1}$$

In this section we shall again consider two arbitrary well-calibrated forecasters A and B, and we shall let F_A and F_B denote their d.f.'s.

Theorem 2. The relationship $A \geq B$ is satisfied if and only if

$$\int_0^s F_A(t)dt \geq \int_0^s F_B(t)dt \qquad \text{for all } 0 \leq s \leq 1. \tag{3.2}$$

Proof. For forecaster A, define

$$\begin{aligned} f_A(x|1) &= Pr(\text{Prediction } x \,|\, \text{Rain}) \\ &= x\nu_A(x)/\mu \quad \text{for } x \in \mathbf{X}, \end{aligned} \tag{3.3}$$

$$\begin{aligned} f_A(x|0) &= Pr(\text{Prediction } x \,|\, \text{No Rain}) \\ &= (1-x)\nu_A(x)/(1-\mu) \quad \text{for } x \in \mathbf{X}, \end{aligned} \tag{3.4}$$

and define $f_B(x|1)$ and $f_B(x|0)$ similarly. Then learning the prediction x of forecaster A on a specified day is equivalent to observing the value x in a statistical experiment in which the unknown p.f. of the observation must be either $f_A(x|1)$ or $f_A(x|0)$. It was shown in DeGroot and Fienberg (1982) that $A \geq B$ if and only if the experiment of learning the prediction of A is sufficient, in the sense of Blackwell, for the experiment of learning the prediction of B.

We shall now show that A is sufficient for B in this sense if and only if (3.2) is satisfied. To establish this result, we shall use a slight variation of Theorem 12.4.1 of

Blackwell and Girshick (1954), and to make it easier for the reader to relate our work to theirs, we shall use their notation. [Our theorem could also be obtained by a modification of the results of Bradt and Karlin (1956).]

Let

$$\alpha_A(x) = \mu f_A(x|1) + (1-\mu)f_A(x|0) \quad \text{for } x\epsilon\mathbf{X}, \tag{3.5}$$

$$S_A(t) = \{x{:}\mu f_A(x|1) \leq t\alpha_A(x)\} \quad \text{for } 0 \leq t \leq 1, \tag{3.6}$$

$$F_A(t) = \sum_{x\epsilon S_A(t)} \alpha_A(x) \quad \text{for } 0 \leq t \leq 1, \tag{3.7}$$

$$C_A(s) = \int_0^s F_A(t)dt \quad \text{for } 0 \leq s \leq 1, \tag{3.8}$$

and define similar quantities for forecaster B. Then it follows from Blackwell and Girshick (1954) that A is sufficient for B if and only if $C_A(s) \geq C_B(s)$ for all $0 \leq s \leq 1$. However, for a well-calibrated forecaster A, it also follows from (3.3) and (3.4) that $\alpha_A(x) = \nu_A(x)$ and that $S_A(t) = \{x{:}x \leq t\}$. Hence, $F_A(t)$ as defined by (3.7) is actually the d.f. defined by (3.1). Thus, the condition that $C_A(s) \geq C_B(s)$ for all $0 \leq s \leq 1$ is precisely the condition (3.2). The theorem now follows. ■

The relationship (3.2) between the d.f.'s F_A and F_B is known as *second-degree stochastic dominance*. A survey of the theory of this concept and its application in various fields is given by Fishburn and Vickson (1978). The next theorem, which goes back to Hardy, Littlewood, and Pólya (1919, 1934), now follows immediately because the inequality (3.9) in the theorem is well known to be equivalent to (3.2). A general proof is given by Fishburn and Vickson (1978), and in various settings in economics by Hanoch and Levy (1969), Hadar and Russell (1971), and many of the authors already cited in this paper.

Theorem 3. The relationship $A \geq B$ is satisfied if and only if, for every continuous, convex function g defined on the unit interval [0,1],

$$\sum_{x\epsilon\mathbf{X}} g(x)\nu_A(x) \geq \sum_{x\epsilon\mathbf{X}} g(x)\nu_B(x). \tag{3.9}$$

In applications in economics, it is natural to consider a result similar to Theorem 3 for concave, increasing functions. The interpretation for risky investments is that $A \geq B$ if and only if the expected utility with respect to ν_B is at least as large as the expected utility with respect to ν_A for all increasing concave utility functions. In the context of income distributions, the distribution ν_B is preferred to ν_A according to all social welfare functions formed by adding identical increasing, concave utility functions. The following result relates these notions to Theorem 3.

Theorem 4. Suppose that ν_A and ν_B are arbitrary p.f.'s concentrated on the interval [0,1]. Then (3.9) holds for all continuous convex functions defined on [0,1] if and only if the distributions represented by ν_A and ν_B have the same mean and (3.9) holds for all non-increasing, continuous convex functions defined on [0,1].

The proof will not be presented here. The next theorem was initially established by DeGroot and Fienberg (1982). A direct proof based on Theorem 2 can now be given.

Theorem 5. The relationship $A \geq B$ is satisfied if and only if

$$\sum_{i=0}^{j-1} (x_j - x_i)[\nu_A(x_i) - \nu_B(x_i)] \geq 0 \quad \text{for } j = 1,\ldots,k-1. \tag{3.10}$$

Proof. Both F_A and F_B are the d.f.'s of discrete distributions concentrated on the $k+1$ points $x_0, x_1, \ldots, x_k$. Hence, the integrals in (3.2) are piecewise linear functions of s with vertices at $0=x_0,\ldots,x_k=1$. It follows that the relation (3.2) will be satisfied for all $0\leq s\leq 1$ if and only if it is satisfied at each of the points $s=x_j$, for $j=1,\ldots,k$. Furthermore, it can be shown that for $j=1,\ldots,k$,

$$\int_0^{x_j} F_A(t)dt = \sum_{i=0}^{j-1} (x_j-x_i)\nu_A(x_i), \tag{3.11}$$

with the analogous relation for B. Thus, by Theorem 2, $A\geq B$ if and only if (3.10) is satisfied for $j=1,\ldots,k$. Finally, since the distributions represented by the d.f.'s F_A and F_B both have the same mean μ, there must be equality in (3.2) for $s=x_k$ or, equivalently, in (3.10) for $j=k$. Thus, the value $j=k$ can be and has been omitted from (3.10). ■

4. THE LORENZ CURVE

The Lorenz curve was originally proposed by Lorenz (1905) as a graphical way of indicating inequality in the distribution of income in a given population. A general definition due to Gastwirth (1971) is as follows: Suppose that F is the d.f. of an arbitrary non-negative random variable and, for $0<u<1$, define

$$F^{-1}(u) = \inf\{t: F(t)\geq u\}. \tag{4.1}$$

In contemporary statistics, F^{-1} is called the *quantile function* corresponding to the d.f. F. Suppose also that the distribution with d.f. F has a finite mean

$$\mu = \int_0^\infty x dF(x) = \int_0^1 F^{-1}(u)du. \tag{4.2}$$

Then the Lorenz curve $L(v)$ corresponding to the d.f. F is given by

$$L(v) = \frac{1}{\mu}\int_0^v F^{-1}(u)du \quad \text{for } 0\leq v\leq 1. \tag{4.3}$$

If F represents the distribution of income in a population, then for $0\leq v\leq 1$, $L(v)$ is the proportion of the total income of the entire population that is received by the poorest proportion v of the population. From this interpretation it is clear that $L(v)\leq v$ for $0\leq v\leq 1$, and that $L(v)=v$ for all $0\leq v\leq 1$ if and only if everyone in the population receives the same income μ; i.e., the distribution is concentrated on the single value μ.

For any d.f. F, the Lorenz curve $L(v)$ is a convex, nondecreasing function on the interval $0\leq v\leq 1$ such that $L(0)=0$ and $L(1)=1$. When F is the d.f. of a discrete distribution concentrated on just a finite number of points, as is true of all the d.f.'s we are considering in this paper, then $L(v)$ is also piecewise linear.

The Lorenz curve has been found to be a useful concept in various areas of reliability theory and statistics. Some references in these areas are Gastwirth (1972), Gail and Gastwirth (1978a, b) and Chandra and Singpurwalla (1981).

Now consider again two well-calibrated forecasters A and B, and let L_A and L_B denote the Lorenz curves corresponding to their d.f.'s F_A and F_B.

Theorem 6. The relationship $A\geq B$ is satisfied if and only if

$$L_A(v) \leq L_B(v) \text{ for all } 0\leq v\leq 1. \tag{4.4}$$

Proof. It was shown by Hardy, Littlewood, and Pólya (1929, 1934) that since the distributions represented by the d.f.'s F_A and F_B both have the same mean μ, then a necessary and sufficient condition that (4.4) hold is that (3.9) hold for every continuous convex function g. Thus, the theorem follows from Theorem 3. ■

Theorem 6 could have been established in several different ways since there are many results in the references that we have cited showing the equivalence of (4.4) to the various conditions that have themselves been shown earlier in this paper to be equivalent to the condition that $A \geq B$. Also, Atkinson (1970) demonstrated the equivalence of (4.4) and (3.2) when the distributions represented by the d.f.'s F_A and F_B are absolutely continuous and have the same mean. This result can be extended to the discrete distributions with which we are concerned here.

We say that the distribution represented by the d.f. F_A *majorizes* the distribution represented by the d.f. F_B if (3.2) holds. Marshall and Olkin (1979) have presented a comprehensive survey of the theory of majorization and they describe a long list of equivalent conditions, many of which are similar to the conditions that we have presented. However, their development is presented in the context of comparing two finite populations of values or two vectors of incomes rather than two probability distributions, so their conditions tend to have a different appearance from ours. For example, the existence of a function p satisfying conditions (2.12)-(2.14) in Corollary 2 corresponds to the well-known result of Hardy, Littlewood, and Pólya (1929, 1934) characterizing majorization for vectors in terms of the existence of an appropriate doubly stochastic matrix [see, e.g., Marshall and Olkin (1979), Chap 2]. In the terminology of Eichhorn and Gehrig (1980), our approach is "statistical" and the approach comparing two finite populations of values is "mechanistic".

The mechanistic approach is suitable for introducing the important concept of Schur-convexity. Let R_+^n denote the set of n-dimensional vectors with non-negative components. Then a real-valued function ϕ defined on R_+^n is said to be *Schur-convex* if $\phi(Px) \leq \phi(x)$ for all doubly stochastic $n \times n$ matrices P and all vectors $x \epsilon R_+^n$. An equivalent condition is that ϕ be a symmetric function of the n components of the vectors in R_+^n and the Pigou-Dalton condition that $\phi(x_1, s-x_1, x_3, \ldots, x_n)$ is increasing in x_1 over the interval $\frac{1}{2}s \leq x_1 \leq s$ be satisfied. Suppose that for any vector $x = (x_1, \ldots, x_n) \epsilon R_+^n$, the component x_j represents the income of person j in a population of n individuals, and that $\phi(x)$ represents a measure of the concentration of income when the income of each individual is specified by the vector x. Then the Pigou-Dalton condition states that concentration increases if income is transferred from person 2 to the wealthier person 1, and other incomes are held constant.

The Lorenz curve can easily be defined for a finite population of values and, hence, it may serve as a bridge between the statistical and the mechanistic approaches. The following theorem illustrates a standard result along these lines.

Theorem 7. Suppose that $x \epsilon R_+^n$ and $y \epsilon R_+^n$ satisfy the conditions $x_1 \leq x_2 \leq \ldots \leq x_n$, $y_1 \leq y_2 \leq \ldots \leq y_n$, and $\Sigma_{i=1}^n x_i = \Sigma_{i=1}^n y_i$.

Then

$$\sum_{i=1}^{k} x_i \leq \sum_{i=1}^{k} y_i \text{ for } k = 1, \ldots, n-1 \tag{4.5}$$

if and only if $\phi(x) \geq \phi(y)$ for all Schur-convex functions ϕ.

Proof. See, e.g., Hardy, Littlewood, and Pólya (1929, 1934), Dasgupta, Sen, and Starrett (1973), or Marshall and Olkin (1977). ■

Under the conditions of Theorem 7, it is said that x *majorizes* y if (4.5) holds. In words, Theorem 7 states that x majorizes y or, equivalently, that the Lorenz curve of the vector x is below that of y if and only if $\phi(x) \geq \phi(y)$ for all Schur-convex functions ϕ. Since a function of the form $\phi(x) = \Sigma_{i=1}^n g(x_i)$ is Schur-convex if g is convex, Theorem 7 is closely related to Theorems 3 and 4. These theorems can easily be reinterpreted under the mechanistic approach, in which case they imply that x majorizes y if and only if $\phi(x) \geq \phi(y)$ for certain special types of Schur-convex functions.

5. SCORING RULES

We now return to the statistical approach. Since the relationship $A \geq B$ induces only a partial ordering of the class of well-calibrated forecasters, a numerical measure of quality is often assigned to the forecasters in order to obtain a total ordering of this class. Thus, we wish to assign a measure of quality $m(\nu)$ to the p.f. ν of every possible well-calibrated forecaster. As we know, any p.f. ν that represents a discrete distribution concentrated on the points $x_0, x_1, \ldots, x_k$ with mean μ can be conceived of as the p.f. of the predictions of some well-calibrated forecaster. The values $m(\nu)$ should be assigned in such a way that the better the forecaster, the higher his measure of quality will be. It is natural to interpret this requirement to mean that if $A \geq B$, the $m(\nu_A) \geq m(\nu_B)$, with strict inequality unless the p.f.'s ν_A and ν_B are identical.

When the p.f.'s ν_A and ν_B represent possible distributions of income in a population, and $m(\nu)$ is to be regarded as a measure of the inequality of the distribution with p.f. ν, then the requirement just given is the natural extension under the statistical approach of the Pigou-Dalton condition stated in Section 4 for the mechanistic approach. Similarly, the concept of Schur-convexity can be directly extended to the statistical approach, and it can be seen that our requirement on the measure of quality m is precisely the requirement that m be strictly Schur-convex. Two simple examples of strictly Schur-convex measures of quality m are the variance and the Gini coefficient, which for any distribution is equal to twice the area in the unit square between the 45°-line and the Lorenz curve of the distribution.

For the evaluation of forecasters, it is useful to develop appropriate measures of quality from more basic considerations involving the concept of a scoring rule. Suppose that if a forecaster's prediction on a given day is x and rain occurs, then he receives a score $s_1(x)$, whereas if rain does not occur then he receives a score $s_2(x)$. If the forecaster's actual subjective probability of rain on a particular day is a and he announces the prediction x, then his expected score on that day will be

$$a\, s_1(x) + (1-a)s_2(x). \tag{5.1}$$

A scoring rule is said to be *strictly proper* if, for each fixed value of a in the unit interval $[0,1]$, (5.1) is maximized only at the value $x = a$. For a discussion of strictly proper scoring rules, with examples, see Staël von Holstein (1970) or DeGroot and Fienberg (1983).

It follows from Theorem 4 of DeGroot and Fienberg (1983) that if a forecaster is well-calibrated, then his overall or expected score will be

$$m(\nu) = \sum_{x \in \mathbf{X}} g(x)\nu(x), \tag{5.2}$$

where

$$g(x) = x\, s_1(x) + (1-x)s_2(x). \tag{5.3}$$

Furthermore, if the scoring rule based on s_1 and s_2 is strictly proper, then the function g defined on the interval $[0,1]$ by (5.3) will be strictly convex.

We conclude with the following result, which now follows directly from Theorem 3.

Theorem 8. Consider a strictly proper scoring rule based on the functions s_1 and s_2, and suppose that a measure of quality m is defined by (5.2) and (5.3). Then m is strictly Schur-convex.

ACKNOWLEDGMENT

This research was supported in part by the National Science Foundation under grant SES-8207295, the Swedish Council for Planning and Coordination of Research, the American Scandinavian Foundation, and the Stiftelsen Sixten Gemzéus.

REFERENCES

ARNOLD, B.C. (1981). Transformations which attenuate income inequality. *Tech. Rep.*, **76**, Dept. of Statistics, University of California, Riverside.

ATKINSON, A.B. (1970). On the measurement of inequality. *J. Econ. Theory,* **2**, 244-263.

BLACKWELL, D. (1951). Comparison of experiments. *Proc. Second Berkeley Symp. Math. Statist. Probability*, Berkeley: University of California Press, 93-102.

— (1953). Equivalent comparison of experiments. *Ann. Math. Statist,* **24**, 265-272.

BLACKWELL, D. and GIRSHICK, M.A. (1954). *Theory of Games and Statistical Decisions.* New York: Wiley.

BRADT, R.N. and KARLIN, S. (1956). On the design and comparison of certain dichotomous experiments. *Ann. Math. Statist.* **27**, 390-409.

CHANDRA, M. and SINGPURWALLA, N.D. (1981). Relationships between some notions which are common to reliability theory and economics. *Math. Operations Res.* **6**, 113-121.

DALTON, H. (1920). The measurement of the inequality of income. *Econ. J.* **30**, 348-361.

DASGUPTA, P., SEN, A., and STARRETT, D. (1973). Notes on the measurement of inequality. *J. Econ. Theory,* **6**, 180-187.

DAWID, A.P. (1982a). The well-calibrated Bayesian. *J. Amer. Statist. Assoc.* **77**, 605-610.

— (1982b). Objective probability forecasts. *Research Report N° 14*, Dept. of Statistical Science, University College London.

DEGROOT, M.H. and FIENBERG, S.E. (1982). Assessing probability assessors: calibration and refinement. *Statistical Decision Theory and Related Topics III, Vol. 1* (ed. by S.S. Gupta and J.O. Berger), New York: Academic Press, 291-314.

— (1983). The comparison and evaluation of forecasters. *The Statistician* **32**, 12-22.

EICHHORN, W., and GEHRIG, W. (1980). Measurement of inequality in economics. Discussion Paper N° 141. Institut für Wirtschaftstheorie und Operations Research, Universität Karlsruhe.

FEI, G.S. and FIELDS, J.C.H. (1978). On inequality comparisons. *Econometrica* **46**, 303-316.

FISHBURN, P.C. and VICKSON, R.G. (1978). Theoretical foundations of stochastic dominance. *Stochastic Dominance* (ed. by G.A. Whitmore and M.C. Findlay), Lexington, Massachusetts: Lexington Books, 39-113.

GAIL, M.H. and GASTWIRTH, J.L. (1978a). A scale-free goodness-of-fit test for the exponential distribution based on the Lorenz curve. *J. Amer. Statist. Soc.* **73**, 787-793.

— (1978b). A scale-free goodness-of-fit test for the exponential distribution based on the Gini statistic. *J. Roy. Statist. Soc., B,* **40**, 350-357.

GASTWIRTH, J.L. (1971). A general definition of the Lorenz curve. *Econometrica* **39**, 1037-1039.

— (1972). The estimation of the Lorenz curve and Gini index. *Rev. Econ. Statist.* **54**, 306-316.

HADAR, J. and RUSSELL, W.R. (1971). Stochastic dominance and diversification. *J. Econ. Theory* **3**, 288-305.

HANOCH, G. and LEVY, H. (1969). The efficiency analysis of choices involving risk. *Rev. Econ. Studies* **36**, 335-346.

HARDY, G.H., LITTLEWOOD, J.E. and PÓLYA, G. (1929). Some simple inequalities satisfied by convex functions. *Messenger Math.* **58**, 145-152.

— (1934). *Inequalities.* Cambridge: Cambridge University Press.

KADANE, J.B. and LICHTENSTEIN, S. (1982). A subjectivist view of calibration. *Tech. Rep. N°* **233**, Dept. of Statistics, Carnegie-Mellon University.

KOLM, S.-Ch. (1976). Unequal inequalities. I and II. *J. Econ. Theory* **12**, 416-442, and **13**, 82-111.

LAD, F. (1982). The calibration question. Unpublished report, Dept. of Economics, University of Utah.

LORENZ, M.O. (1905). Methods of measuring concentration of wealth. *J. Amer. Statist. Assoc.* **9**, 209-219.

MARSHALL, A.W. and OLKIN, I. (1979). *Inequalities: Theory of Majorization and Its Applications,* New York: Academic Press.

MURPHY, A.H. and WINKLER, R.L. (1977). Reliability of subjective probability forecasts of precipitation and temperature. *Appl. Statist.* **26**, 41-47.

PIGOU, A.C. (1912). *Wealth and Welfare.* New York: Macmillan.

PRATT, J.W. (1962). Must subjective probabilities be realized as relative frequencies? Unpublished seminar paper, Harvard University Grad. School of Bus. Administration.

ROTHSCHILD, M. and STIGLITZ, J.E. (1970). Increasing risk: I. A definition. *J. Econ. Theory* **2**, 225-243.

— (1973). Some further results on the measurement of inequality. *J. Econ Theory* **6**, 188-204.

SEN, A. (1973). *On Economic Inequality*, London and New York: Oxford University Press.

STAËL VON HOLSTEIN, C.A.S. (1970). *Assessment and Evaluation of Subjective Probability Distributions,* Stockholm: Economic Research Institute, Stockholm School of Economics.

STRASSEN, V. (1965). The existence of probability measures with given marginals. *Ann. Math. Statist.* **36**, 423-439.

DISCUSSION

P.K. GOEL (*Ohio State University*)

I am not much of an expert on the topic of probability forecasting. Thus my comments are those of an 'observer' a la DeGroot and Eriksson.

To start with, I must thank the authors for preparing an excellent compendium of results linking 'Stochastic Dominance' notions in multivariate analysis to Bayesian Calibration and Refinement. I do agree with DeGroot's comment during the discussion that this paper seems to be the final word on refinement and its interpretation.

However, I do have some reservations about the three suggested interpretations of $\nu(x)$ and $\varrho(x)$. The first and the second interpretations invoke the relative frequency definition of probability (infinite or finite sequence). It assumes an "Approximate Stationary Distribution" for meteorological conditions over a *large* number of days. I don't believe that this is a reasonable assumption. For the third interpretation, an observer and his subjective probabilities have been invoked in order to get rid of the undesirable stationarity assumption. However, it confounds the forecaster and the observer, and refinement and calibration cannot be attributed to the forecaster only.

In Section 2, the authors state that the problem of comparing or ordering well-calibrated forecasters is *equivalent* to the problem of ordering all distributions on the set **X** with a given mean μ. However, $\Sigma x\nu(x) = \mu$ doesn't imply that the forecaster is well-calibrated. Thus the two problems are not strictly equivalent.

Finally, the paper (not the presentation) ends with a theorem whose relevance to this context is not immediately apparent. Does it help in the construction of the score functions? I don't think so. Do the authors want to leave the impression that every Schur-convex function $m(\nu)$ can be thought of as an overall expected score? I hope that the authors will address this issue so that the theorem is not left hanging in the air.

D.J. SPIEGELHALTER (*MRC Biostatistics Unit, Cambridge*)

I find this paper very useful in clarifying the concept of "refinement" as applied to probabilistic forecasters; the relationship to work on income distribution is particularly pleasing. My comments concern the practical interpretation of the concept in an area such as medical forecasting, in which the predictive probabilities coming from a computer system are to be compared with those coming from clinicians. (See, for example, Spiegelhalter and Knill-Jones, 1984).

Suppose for illustration, there are four discrete levels at which the probability of "peptic ulcer" are given and both the computer and the doctors are well-calibrated. Then it is straightforward to show that the doctors are more refined than the computer if they make more judgements in both the lowest category and in the highest category (which in fact is the case for the system we are currently developing). When more categories are used the usual way of comparing two "forecasters" in this area is by 'Receiver Operating Characteristic' (ROC) analysis, in which it is assumed that some threshold θ is selected and every patient receiving probability $> \theta$ is allocated to the peptic ulcer (PU) class. Using the notation of the paper, a choice of $\theta = x_j$ is simply shown to lead to the error rates shown in Table 1. Plotting the "True positive" rate against the "False positive" rate for various choices of θ provides the ROC curve.

		Decision	
		Do not allocate to PU	*Allocate to PU*
Truth	*No Peptic Ulcer*	*'True negative rate'* $= \sum_{i=0}^{j=1} (1-x_i)\nu(x_i)/(1-\mu)$	*'False positive rate'* $= \sum_{i=j}^{k} (1-x_i)\nu(x_i)/(1-\mu)$
	Peptic Ulcer	*'False negative rate'* $= \sum_{i=0}^{j-1} x_i\nu(x_i)/\mu$	*'True positive rate'* $= \sum_{i=j}^{k} x_i\nu(x_i)/\mu$

TABLE 1 *Error rates using a probability threshold $\theta = x_j$ to allocate to PU*

If we consider θ as an unknown "state of nature", and a loss function is applied to the errors in Table 1, then A 'dominates' B in the decision theoretic sense if the expected loss in using A as a forecaster is less than or equal to that in using B as a forecaster for all choices of θ, and less than that of B for at least one choice of θ. (This contrasts with the usual decision-theory construction since the choice of decision rule is actually a choice of the sampling distribution). Suppose the loss incurred by a false positive is P and that by a false negative is N, then the expected loss when $\theta = x_j$ and we use a forecaster with predictions $\nu(\mathbf{x})$ is

$$R(\theta = x_j, \nu(\mathbf{x})) = P \sum_{i=j}^{k} (1 - x_i)\nu(x_i) + N \sum_{i=0}^{j-1} x_i \nu(x_i)$$

$$= P(1-\mu) - \sum_{i=0}^{j-1} \{P - (P+N)x_i\}\nu(x_i) \quad .$$

By comparing the above equation with expression (3.10), it is clear that if $P = \theta = x_j$ and $N = 1 - \theta = 1 - x_j$, A dominating B in the decision-theoretic sense given above is precisely equivalent to second-degree stochastic dominance. In fact, this is quite a plausible loss function, since the penalty for a false-positive error increases as the threshold increases. However, this interpretation does illustrate the stringency of the "refinement" criterion, and I would be interested to know of any suggestions for relaxing the requirements and hence making it more likely to be able to 'order' forecasters in practice.

Just as the risk-function analysis provides a partial ordering of forecasters, placing a prior distribution on the threshold θ and evaluating the Bayes risk will provide a total ordering which under certain assumptions is analogous to that provided by the scoring rules described in Section 5. For example, suppose $P = N$ in the loss function and a prior $p(\theta = x_j) = x_j^2 - x_{j-1}^2$; $j = 1,\ldots,k$ were assumed, then the Bayes risk is exactly the expected Brier score.

I would be interested to know the thoughts of the authors concerning a common way of achieving total ordering by comparing the areas under the ROC curve, which corresponds to the probability that a random diseased individual is given a higher probability of disease than a random patient without the disease (Hanley and McNeil, 1982).

R.L. WINKLER (*Indiana University*)

This paper describes some useful relationships between the comparison of forecasters generating probability forecasts and some concepts encountered frequently in the economics literature. This drawing together of concepts and results from different streams of research is important in that it can make researcher aware of relevant work from other fields and it may provide new insight by showing different ways to look at a problem.

The comparison in the paper is of well-calibrated forecasters making forecasts for the same situations. But the restriction to well-calibrated forecasters can be relaxed. If $\nu[\varrho(x)]$ is used instead of $\nu(x)$ when comparisons are made, any forecasters who are not well-calibrated have their forecasts modified so that they are calibrated before the forecasters are compared. Thus, for example, suppose that $\mu = 0.50$. For Forecaster A, $\nu_A(0.90) = 0.60$, $\nu_A(0.10) = 0.40$, $\varrho_A(0.90) = 0.70$, and $\varrho_A(0.10) = 0.20$. Forecaster B is well-calibrated with $\nu_B(0.60) = 0.50$ and $\nu_B(0.40) = 0.50$. Forecaster A gives more extreme probabilities than B but is not well-calibrated. If A^* represents A adjusted for miscalibration. then A^* is well-calibrated with $\nu_{A^*}(0.70) = 0.60$ and $\nu_{A^*}(0.20) = 0.40$. Here

A tends to be poorly calibrated in the sense of giving probabilities that are too extreme, and we should not take A's probabilities as our own. When we adjust for this miscalibration, we see that we get a well-calibrated "forecaster" who is more refined than B. In general, the label x that a forecaster assigns to each of a set of events judged equally likely by the forecaster is irrelevant as long as we know $\varrho(x)$ and can therefore work with calibrated probabilities.

A more difficult question concerns the comparison of forecasters in different situations. If the manager of a television station in Chicago is considering two forecasters, one from Pittsburgh and one from San Francisco, to handle the weather on the station's local news programs, where ν and ϱ are known for both forecasters at their current locations, how might the forecasters be compared? How can weather forecasters at different locations, in different seasons, and so on, be compared? Forecasting situations are not necessarily similar in terms of "difficulty".

A more basic question concerns the notion of calibration. If ϱ represents an observer's view of the forecaster's calibration, how is the observer's initial state of information taken into account? After seeing the forecast, the observer's probability could differ from the stated forecast because of the incorporation of the observer's own information, not just because of miscalibration on the part of the forecaster. Is there an implicit assumption that the forecaster has taken into account the same information possessed by the observer, perhaps stated formally by saying that the forecaster alone is as refined as the combination of the forecaster and the observer?

S. FRENCH (*University of Manchester*)

This is the first time that I have encountered the concept of refinement and, I confess, it worries me. The question that is being addressed is: when is one expert better than another. This immediately begs the question: better for whom? It seems to me that this latter question must in the final analysis be answered as better for the decision makers who ultimately use the expert's forecast. The authors suggest that, continuing the weather forecasting example, they are concerned with comparing forecasters for the television company which is to employ one of them; but surely that company wants to identify the forecaster whose views will be of use to the largest audience, and thus it is still the needs of the decision makers which should determine the better forecaster. Now, if the needs of the decision makers are the ultimate concern, the calibration of the experts is an irrelevancy. The most poorly calibrated expert, the chap who *always* states probabilities of 0 or 1 and is wrong, provides information to the decision maker which, once reversed, cannot be bettered. As shown in Morris (1974) and Lindley et al (1979), what concerns the decision maker are the conditional distributions $f_A(x|1)$, $f_A(x|0)$, $f_B(x|1)$ and $f_B(x|0)$ (see eqns. 3.3 and 3.4 in the paper) - or, more correctly, the likelihood ratios derived from these. Hence, I find the concentration in this paper on well-calibrated experts a distraction.

Since the conference I have read Blackwell's theory of sufficient experiments and its development into this concept of refinement. Roughly, I think the intention is to find sufficient conditions on the conditional distributions $f_A(x|1)$, $f_A(x|0)$, $f_B(x|1)$ and $f_B(x|0)$ such that expert A should always be preferred to expert B *whatever* the decision maker's prior and *whatever* his loss function. Since Blackwell's theory is undoubtedly correct and since the translation into the concept of refinement is apparently straightforward - particularly when done in the context of poorly calibrated experts (see DeGroot and Fienberg, 1983) - it would seem that this partial ordering of experts takes into account the needs of the decision makers whatever they may be. But is this so? It is based upon the assumption that $f_A(x|1)$, etc., can be defined independently of a decision maker and his particular problem. However, if we admit that people's opinions about events are not independent (French, 1980, 1983), these distributions will be conditional on the particular knowledge base of each decision maker. Thus the ordering is not independent of the decision maker's prior beliefs and much of its value is lost.

M. GOLDSTEIN (*Hull University*)

The authors cite a number of papers which provide "a wide variety of theoretical and empirical reasons why forecasters will tend to be well-calibrated". One of the papers cited is Lad (1982) and my understanding was that this paper was specifically intended to be an argument against the concept of calibration as it is used in the present paper.

T. SEIDENFELD (*Washington University, St. Louis*)

I would like to make two remarks about this interesting paper.

1. There is a real difference between an announced forecast for event E and the forecaster's personal probability of E. This difference is all important to understanding the deficiency with calibration (sense 2) when it is an improper score. Likewise, in the DeGroot-Eriksson subjectivist sense (3) of calibration, one should take care to distinguish between the forecaster's probability of E and his public forecast. If the forecaster B takes an interest in the decisions of the subjectivist A who judges him (B) calibrated (in sense 3), then B has reason to manipulate A's deliberations by way of the announced forecasts. For B, the forecast serves as his "vote" in A's decision. Gibbard (1973) and Zeckhauser (1973) provide a battery of unsettling results about the inevitability of manipulable voting systems. How should A react to this?

2. Calibration (sense 3) is a post-forecast consideration. But the partial ordering of forecasts by refinement is a pre-forecast relation. Why restrict comparisons by refinement to calibrated forecasts? Also, isn't it useful to consider calibration (sense 3) with respect to a group of forecasts (not just for forecasts taken singly)--understood on the analogy with sufficient statistics?

J.M. BERNARDO (*Universidad de Valencia*)

The authors are to be congratulated for their elegant connexion between the concept of stochastic dominance and the problem of evaluating forecasters. That said, I wonder what are the limitations which the authors perceive in the use of proper scoring rules to evaluate forecasters which have led them to consider an alternative approach which only lead to a partial ordering. Indeed, to agree on a particular scoring rule might be a practical problem but this should not stop us in using an appropriate utility measure just as the practical problems of specifying a prior should not stop us in using a Bayesian analysis.

I also wonder whether the relationship of the results in this paper with scoring rules may be extended beyond the results given in Section 5. In particular, it seems rather likely that if a forecaster is better than another for *any* proper scoring rule he is also better in one of the senses described in the paper, and the other way round.

REPLY TO THE DISCUSSION

We are grateful to all the discussants for their interesting and stimulating comments. We shall try to respond to all of the points that they have raised.

Professor Goel is correct in stating that one shortcoming of the frequency approach to calibration is that this approach is most appropriate for a stationary sequence of events. Furthermore, as Professor Goel, Professor Winkler, and Dr. French point out, under the subjective Bayesian approach mentioned in our paper, characteristics of a forecaster are defined relative to a given observer. Thus forecaster A is well-calibrated, or more refined than forecaster B, *in the opinion of the observer.* Another observer, or the same observer at a different time with different information, might not regard A as well-calibrated or as more refined than B.

The advantage of this subjective Bayesian approach to forecasting, calibration, and refinement, is that, in the language of Professor Winkler's example, it *does* allow the manager of a television station in Chicago to compare job candidates from Pittsburgh and San Francisco. The functions ν and ϱ for each forecaster represent the manager's opinion about how a candidate will perform in Chicago, based on whatever information the manager has about the forecaster. Thus, the manager may be fully aware of a forecaster's record in San Francisco and he will construct the functions ν and ϱ based on the extent to which, and the way in which, he feels that record would be relevant to the forecaster's performance in Chicago.

In particular, an observer would typically regard a forecaster as being well-calibrated, $\varrho(x) \equiv x$, only if he felt that the forecaster had all the relevant information that the observer had, and more. As suggested by Professor Winkler, this condition is essentially equivalent to the condition that in the opinion of the observer, the forecaster alone is *sufficient* for the combination of forecaster and observer, in the sense of DeGroot and Fienberg (1983).

As Professor Goel states, the mere fact that the mean of a forecaster's probability distribution ν is μ does not imply that the forecaster is well-calibrated. However, every distribution on $\mathbf{X}$ with mean μ *can be regarded* as the distribution of some (possibly hypothetical) well-calibrated forecaster. Hence, as we state in the paper, the problem of comparing or ordering all possible well-calibrated forecasters is equivalent to the problem of comparing or ordering all distributions on $\mathbf{X}$ with mean μ.

Professor Goel asks about the import of the final theorem in the paper. As we argue in the paper, if the class of all possible well-calibrated forecasters is to be ordered by some measure of quality, then that measure, regarded as a real-valued function on the space of all distributions on $\mathbf{X}$ with mean μ, should be strictly Schur-convex. The final theorem in the paper states that any measure of quality derived from a strictly proper scoring rule will have this desirable feature. As Professor Bernardo suggests, the choice of an appropriate Schur-convex function or an appropriate scoring rule in a given situation is analogous to the choice of a specific utility function or loss function in a decision problem.

Dr. Spiegelhalter is quite right in stressing that the relevance of our work in a field such as medical forecasting lies mainly in its relationship to the decision-making that follows from the predictions. It should be emphasized, however, that if one well-calibrated forecaster A is more refined than another one B, then a decision maker will attain a smaller expected loss in *any* decision problem that he faces by basing his decision on A's prediction rather than B's. This statement is not in conflict with Dr. French's observation that the comparison between A and B will, in general, depend on the nature of the decision problem that the observer faces and on the observer's own state of information. The strength and applicability of our results lie in the fact that if the observer regards both A and B to be well-calibrated and A to be more refined than B, then he will prefer to learn A's prediction rather than B's, *regardless* of his decision problem and *regardless* of any further detailed specification of his own prior distribution.

Thus, in Dr. Spiegelhalter's example, A's prediction will yield a smaller expected loss than B's for every choice of the threshold θ and all possible losses P and N resulting from the two types of wrong decisions. A general method of comparing forecasters based on their performance in two-decision problems is described by Schervish (1983).

Our refinement criterion for comparisons is very stringent and yields only a partial ordering of well-calibrated forecasters. A middle ground along the lines that Dr. Spiegelhalter mentions, between this partial ordering and the total ordering obtained from using a specific value of (θ,P,N), or a specific prior distribution for this value, would be to consider the "more complete" partial ordering obtained by demanding the usual Bayesian decision-theoretic domination for just a "reasonable" set of prior distributions for this value. Results for comparing statistical experiments along these lines are described in Goel and DeGroot (1979).

We are interested to read Dr. Spiegelhalter's example in which predictions can take only four values and a well-calibrated doctor is more refined than a well-calibrated computer if he makes more predictions in both the lowest and highest categories. (The proof of this result is not completely straightforward). A related result is presented in Corollary 2 of DeGroot and Fienberg (1984), which states in effect that for an arbitrary number of levels, the doctor will be more refined than the computer if the computer makes more predictions in every category *except* the lowest and highest.

Dr. Spiegelhalter requests our thoughts on the common way of comparing diagnostic systems in terms of the areas under their ROC curves. It is our understanding that this method corresponds to comparing the systems in terms of the following decision problem: One subject with the disease and one subject without the disease are to be selected at random and independently. The diagnostic systems are then ordered with respect to their overall probabilities of correctly classifying these two subjects, given the knowledge that there is exactly one of each type. This method of comparison is no more and no less reasonable than this particular decision problem.

Professor Winkler, Professor Seidenfeld, and Dr. French question the restriction of our paper to well-calibrated forecasters. As Professor Winkler suggests, one way to evaluate a forecaster who is not well-calibrated is simply to "recalibrate" his predictions and then use the methods described in our paper for well-calibrated forecasters. In fact, as Professor Winkler indicates, the forecaster's prediction x need not be a probability or even a number --it could be a descriptive phrase, for example-- as long as we know the function $\varrho(x)$ that translates the prediction into a well-calibrated probability. Dr. French is essentially heading toward the same point when he emphasized that it is the information that a forecaster's prediction conveys to the observer that is important, rather than the actual numerical value of the prediction. We quite agree. This approach for comparing and evaluating forecasters who are not necessarily well-calibrated is thoroughly studied in DeGroot and Fienberg (1982, 1983). The central purpose of the present paper, however, was to explore the beautiful connections between the concept of refinement and the concepts of stochastic dominance and the Lorenz curve that are present when we restrict ourselves to well-calibrated forecasters.

Dr. Goldstein mentions our reference to Lad (1982). In that paper, Lad points out that a forecaster being well-calibrated over *all* events to which he assigns probabilities is simply tantamount to his being coherent in these assignments. Hence, if the forecaster is not well-calibrated for a specific sequence of events, he might consider himself "unlucky".

We agree with Professor Seidenfeld that the observer should be aware that the forecaster may be trying to "manipulate" his opinion and may not be reporting "honest" subjective probabilities. The theory of strictly proper scoring rules is an attempt to encourage the forecaster to be honest, but an important short-coming of this theory is that it is implicitly based on very special assumptions about the forecaster's utility function. Schachter (1982) has made an interesting attempt to overcome this shortcoming. In principle, however, this type of difficulty is handled by the subjective Bayesian approach whereby the functions ν and ϱ represent the observer's opinion about the forecaster's behavior.

As Professor Seidenfeld suggests, it is important to consider forecasters who specify not just the probability of an event, but their entire predictive probability distribution of a random variable or random vector. Such problems have been widely considered in the forecasting literature from the viewpoint of scoring rules [see, e.g., Winkler (1983)]. The notions of calibration, refinement, and sufficiency, and their basic properties for this situation, are described in DeGroot and Fienberg (1982, 1984).

It is true, as Professor Bernardo and Dr. Spiegelhalter indicate, that the use of any particular scoring rule will induce a total ordering of forecasters, whereas our methods yield only a partial ordering. However, to repeat the statement that we made earlier in this response, if two well-calibrated forecasters are ordered by our approach then they will have this same ordering for *all* strictly proper scoring rules.

The converse is also true, as suggested by Professor Bernardo. Theorem 3 of our paper states that one well-calibrated forecaster A will be at least as refined as another one B if the expectation of every continuous convex function g on $[0,1]$ is at least as large under ν_A as under ν_B. Savage (1971) has shown that every strictly convex, differentiable g can be represented as in (5.3) of our paper for some functions s_1 and s_2 arising from a strictly proper scoring rule. Together, these properties are sufficient to show that if A achieves at least as high a score as B for every strictly proper scoring rule, then A is at least as refined as B.

REFERENCES IN THE DISCUSSION

DEGROOT, M.H. and FIENBERG, S.E. (1983). The comparison and evaluation of forecasters. *The Statistician* **32**, 12-22.

— (1984). Comparing probability forecasters: basic binary concepts and multivariate extensions. *Tech. Report,* **306**, Dept. of Statistics, Carnegie-Mellon University.

FRENCH, S. (1980). Updating of belief in the light of someone else's opinion. *J. Roy. Statist. Soc. A* **143**, 43-48.

— (1983). Calibration, refinement and the expert problem. *Notes in Decision Theory,* **144**, Dept. of Decision Theory, University of Manchester.

GIBBARD, A. (1973). Manipulation of voting schemes: A general result, *Econometrica,* **41**, 587-601.

GOEL, P.K. and DEGROOT, M.H. (1979). Comparison of experiments and information measures. *Ann. Statist.* **7**, 1066-1077.

HANLEY, J.A. and MCNEIL, B.J. (1982). The meaning and use of the area under a ROC curve. *Radiology,* **143**, 29-36.

LAND, F. (1982). The calibration question. *Unpublished report.* Dept. of Economics, University of Utah.

LINDLEY, D.V., TVERSKY, A., and BROWN, R.V. (1979). On the reconciliation of probability assessments. *J. Roy. Statist. Soc. A,* **142**, 146-180.

MORRIS, P.A. (1974). Decision analysis: expert use. *Management Science,* **23**, 679-693.

SAVAGE, L.J. (1971). Elicitation of personal probabilities and expectations. *J. Amer. Statist. Assoc.* **66**, 783-801.

SCHACHTER, R. (1982). An incentive approach to eliciting probabilities. *Tech. Rep.* **82-9.** Dept. of Industrial Engineering and Operations Research, University of California, Berkeley.

SCHERVISH, M.J. (1983). A general method for comparing probability assessors. *Tech. Rep.* **275**, Dept. of Statistics, Carnegie-Mellon University.

SPIEGELHALTER, D.J. and KNILL-JONES, R.P. (1984). Statistical and knowledge-based approaches to clinical decision-support systems, with an application in gastroenterology. *J. Roy. Statist. Soc.* **147**, 35-77 (with discussion).

WINKLER, R.L. (1984). On "good probability appraisers". *Bayesian Inference and Decision Techniques with applications.* (P.V. Goel and A. Zellner, eds.). Amsterdam: North-Holland (in press).

ZECKHAUSER, R. (1973). Voting systems, honest preferences and Pareto optimality, *American Political Science Review*, **67**, 934-946.

BAYESIAN STATISTICS 2, pp. 119-132
J.M. Bernardo, M.H. DeGroot, D.V. Lindley, A.F.M. Smith (Eds.)

Probability, Evidence, and Judgment

A.P. DEMPSTER
Harvard University

SUMMARY

The paper presents a summary, without technical details or examples, of some major issues facing the realistic practice of Bayesian statistics. I argue that formal Bayesian models are technological extensions of human capabilities, rather than internally wired human characteristics, whence the prescriptive rules for constructing such machines are a worthy field of study, and the nature of such machines should be regarded as subject to evolution and improvement. I stress that applications may result in probability assessments which are sometimes soft and sometimes hard. Bayesian analysis may be supplemented with sensitivity analysis to reflect softness, and belief function technology is suggested as a worthy extension of Bayesian statistics which can formally incorporate softness and has a formal mechanism for combining multiple sources of evidence. I stress that the social context of Bayesian statistics involves multiple actors, with the consequence that in professional practice probability assessment requires explicitly described elements of evidence and judgment.

Keywords: BELIEF FUNCTIONS; CONSTRUCTIVE ANALYSIS; EVIDENCE; HARD AND SOFT; JUDGMENT; NORMATIVE ANALYSIS; PRESCRIPTIVE ANALYSIS; PROBABILITY; PROBABILITIES; SENSITIVITY ANALYSIS; SUBJECTIVE EXPECTED UTILITY; SURPRISE.

1. INTRODUCTION

The technical feasibility of Bayesian statistics is rapidly improving as research and development activity exploits opportunities created by the computing revolution. My paper offers an analysis and critique of Bayesian statistics, not from an internal aspect of flourishing technical development, but rather from an external aspect of the real world analyses and problem-solving tasks which the technology is designed to aid. I wish to raise for discussion the question: are the current norms and prescriptions of Bayesian statistics adequate to the tasks?

From an external standpoint, the central function of Bayesian statistics is the provision of probabilities to quantify prospective uncertainties given a current state of knowledge. The uncertainties refer to questions of fact about natural and social phenomena and about the effect of human decisions on these phenomena. The external motivation can be purely scientific, but in statistical practice there are usually decision- or policy-analytic components.

How broadly should Bayesian statistics be defined? I believe that the statistics profession has been hindered by the orthodoxy of academic mathematical statistics over the past 50 years which has largely removed evaluation of prospective uncertainties from the domain of statistical science. Thus, although statistics is the dominant source of useful probabilistic technologies, statisticians are often perceived as narrowly focused, and new professions such as "decision analysis" or "risk analysis" are created to fill the void. The Bayesian movement should avoid a similar trap. Bayesian statistics should therefore experiment with a variety of paradigms. In particular, I hope that Bayesian statistics can assimilate extended technologies such as the diagnosis/robustness methods of Box (1980) and the belief function techniques of Shafer (1976, 1981, 1982, 1983). The ultimate standard should be professional consensus based on *post hoc* assessments of effectiveness for problem-solving.

A recurrent theme in the field of probability assessment is the question of balance between judgment and evidence. *Evidence* means knowledge either in the form of specific facts (e.g., statistical data) or arguments from theories or hypotheses which represent and structure broad areas of generally accepted facts. *Judgment* consists of commitments of belief in the relevance of evidence to the problem at hand. One caricature of Bayesian statistics maintains that the likelihood function of the data is based on objective evidence while the prior distribution requires only judgment. In fact, both evidence and judgment are involved in both parts of the formal structure, as they are in every probability assessment. Likelihood functions often rest on frequency-based sampling models, or similar applied probabilistic chance models, but the end use of Bayesian inference depends critically on *judgments*, first of the quality of the past evidence supporting parametric model assumptions, and second of the relevance of assumed exchangeability among sampled and unsampled units for elementary events. Likewise, perceptive consumers of Bayesian statistics always evaluate the evidential bases of prior distributions as well as the judgmental bases, and one can scarcely doubt that the practical impact of a formal Bayesian analysis is and should be sensitive to the consumer's assessment of quality of *evidence* underlying the prior distribution.

My purpose is to explore various themes related to *probability, evidence,* and *judgment* from the standpoint of professional practice and problem-solving. If Bayesian statistics can dispassionately recognize its weaknesses as well as its strengths, then perhaps it can grow and assimilate related technologies. These technologies may not conform strictly to the normative axioms of SEU (subjective expected utility) but they may be friendly amendments in the sense of addressing a common goal of improved assessment of uncertainty.

2. THE CONSTRUCTIVE/PRESCRIPTIVE ATTITUDE

Two important innovators have been stressing a similar fundamental attitude to the wide sense Bayesian enterprise.

Glenn Shafer's original writing on belief functions stressed that probability requires evidence (Shafer, 1976), but more recently he has been emphasizing in Shafer (1981, 1982, 1983) that probabilities are *constructed* from evidence. Now Shafer and Tversky (1983) refer to an active thought experiment which explicitly compares the current evidence with a scale of canonical examples. Shafer (1982) differentiates constructive probability from the frequentist, necessary, or personalist types described by Savage (1954), since these familiar types appear to assume that probabilities exist before they are constructed. Similarly, he queries the content of elicitation methods of obtaining personal probabilities:

> "What meaning and what persuasiveness do the answers have once it is admitted that there really are no predetermined probabilities in the back of our minds?"
> (Shafer, 1983)

Howard Raiffa has been developing a similar theme, proposing that we need a concept of *prescriptive* approaches to decision analysis to put alongside the more widely recognized normative and descriptive approaches. It is important to pay attention to how people actually think and behave in the real world (descriptive analysis), and it is important to propose and argue for specific general rules which formal assessments ought to obey (normative analysis), but there is much in the way professionals actually implement formal analyses (prescriptive analysis) which is not captured by the terms descriptive and normative, as these terms are currently used. In remarks prepared to open a recent conference he wrote:

> "In limited domains the SEU model may be used as a prescriptive tool in order to guide behavior, but this conscious effort involves a reflective thought process that is far more complex than the bare bones of the SEU model seems to indicate. Real people, in real situations, don't naturally act coherently, and one usually cannot discover via revealed behavior their latent probability distributions and utility functions. Rather, the way the SEU model is put to prescriptive use turns the model upside down. We don't start by assuming that the decision maker can, in an unaided fashion, compare any two alternatives but rather we test whether he can compare a few simple hypothetical consequences. Already in this limited domain he might exhibit intransitivities among the few consequences that he is willing initially to compare, but he then must be willing to reflect upon these inconsistencies and modify his preferences so that they line up transitively. In an iterative fashion he must be willing, in a particular instance, to act quite unnaturally: to deliberately police his choices of hypothetical simple situations, one by one, in order to conform to the desideratum of consistency. Gradually, if he is successful, a probability distribution over states and a utility function over consequences will emerge. These will literally have to be constructed bit by bit, and it is pure myth that latently these probabilities and utilities existed deep down and that the analyst merely has to cut away the fat in order to display the pre-existing structure. Next a leap of faith is required: the decision maker must be willing to use his probability and utility functions that he has laboriously constructed to calculate SEU's that will guide his selection of real-world alternatives".
>
> (Bell, Raiffa, and Tversky, 1983)

In correspondence quoted by Raiffa, Shafer pointed out that philosophers have often used the terms normative and prescriptive interchangeably, so that prescriptive may have too much the connotation of what one ought to do on high authority with "no alternative to be considered", but Shafer did write that prescriptive suggests actively helping real people, as when a doctor prescribes. Raiffa agreed that "constructive captures an essence of what we mean by prescriptive" but worried that constructive "may seem to some to be too systematic, too linear in conception, too structured". Raiffa indicated that no uniqueness conception is implied by prescriptive. Some may wish to approach construction "holistically (with perhaps a checklist of concerns)" while others may wish "to think in a decompose-recompose manner using the SEU framework". And, "there are a whole myriad of other frameworks".

With this legislative history in place, I believe that we have witnessed the emergence of a useful technical concept which can be labeled interchangeably by the terms constructive or prescriptive. The following remarks are intended to support this emergence and to suggest additional attitudes which may help to round out the concept.

I perceive a very basic distinction between constructive or prescriptive analysis and current approaches to descriptive studies of actual human probabilistic reasoning or decision-making. The implicit assumption of the latter is that the formal structures of prescriptive analysis are somehow embedded in the human psyche. It is as though one attempted to observe and understand the activity of a human cycling along a road, while regarding the bicycle simply as part of the human being, instead of viewing the bicycle as a deliberately constructed tool designed for a purpose. A formal probability or decision model, like a bicycle, exists abstractly apart from any user. Such tools evolve in large and small ways. Their intended purposes and modes of use must be learned. And the learning process results in many types and degrees of skill among users. Descriptive analyses which ignore and hence aggregate over technological complements of human behavior may confuse and miss essential aspects of probability assessments.

An important task is to describe and systematize the procedures which probability assessors or decision analysts use in prescriptive practice to match the analogs of bicycles to real needs. In the long paragraph quoted above, Raiffa refers to an iterative process whereby a decision analyst might approach a self-consistent SEU model, but Raiffa makes no explicit reference to how evidence or judgment enter the process. As noted above, Shafer (1981, 1982, 1983) and Shafer and Tversky (1983) emphasize judgmental comparison of evidence with a scale of canonical examples in a consciously designed thought experiment. In thinking about probability, I see a critical distinction between the *meaning* of a numerical probability, and the *source* of a numerical probability, and I see the catalogue of canonical examples as primarily an aid to communicating *meaning*-because many more of us are comfortable with odds quoted in familiar games of chance than we are with odds quoted in less controlled circumstances. The idea of a designed thought experiment is an attractive way to label the source mechanism, but still the set of hooks from human judgmental capabilities to external evidence is left undefined and unanalyzed. I believe that major opportunities exist for clarification and progress here, inviting both proposals for how one ought to relate evidence to uncertainty judgments, and examples and characterizations of how well-reputed and highly trained professionals actually do it. In effect, normative and descriptive attitudes need to be directed at the constructive/prescriptive problem.

Raiffa's distinction between normative and prescriptive was neatly captured by Jevons (1877), specifically in relation to the theory of probability:

> 'Nothing is more requisite than to distinguish carefully between the truth of a theory and the truthful application of the theory to actual circumstances''

It appears to me that the 20th century emphasis on axioms governing uncertainty and choice have left the normative theory pretty much in the form it had 100 year ago. Even the recent Dempster-Shafer extension to belief functions has roots 200 to 300 years ago in the work of Bernoulli and Lambert (Shafer, 1978). Such early giants of probability theory made their major gains while operating in a prescriptive mode on problems of physical and social analysis. We should follow that lead.

3. WHO IS THE MODELER/ANALYST/DECISIONMAKER?

The standard normative Bayesian theory postulates a single actor who is assumed to possess a well-defined formalization of the factual structure of observables, choices, and outcomes, of probabilistic uncertainty, and of numerical utilities. The idealized person was called 'you' by Good (1950), suggesting that the reader is being instructed on how to carry out his or her own analysis. Constructive/prescriptive analysis by professionals should, however, reflect the fact that in most situations several or many actors are involved. If the standard theory is to be applied, then the real world identification of the actor needs to be explained and justified. Alternatively, and perhaps preferably, the theory should be broadened to define roles for several actors.

Sometimes the actors may be individuals, or social groups represented by institutions whose utilities range from partially congruent to directly opposed. This line of thought leads to consideration of multiattribute utilities or to game theory. My concern is more with probability assessment, and with the situation of a Bayesian statistician functioning along with substantive experts in the service of a client whose goals are accepted as well-defined and legitimate.

One simple scenario involves only the statistician and client. Much of the bias of professional statisticians against Bayesian statistics stems from the belief that statisticians should not invoke technology whose evidential basis is missing or too soft to be communicated to the client. I agree with the bias, but not with the solution of replacing Bayesian analysis by something entirely different which succeeds only in hiding the problem. A preferable solution is to work to bring out and harden the evidentiary and judgmental bases of prior distributions and likelihoods, in a form which can be successfully communicated to the client. An alternative is to back off to a theory such as belief function theory which is weaker in the sense of making fewer demands for evidence and correspondingly providing more limited assessments of uncertainty. I suspect that both strategies have their place, depending on the problem.

The following anecdote illustrates my point that the client-consultant relationship may require Bayesians to pay more attention to the evidentiary basis of probability assessments. About a year ago I wrote to a set of colleagues asking them to rank a list of potential candidates for a position. Most responses included evidence in the form of thumbnail sketches of strengths and weaknesses. One prominent Bayesian gave a ranking, and even a numerical score to each candidate. But the evidence was missing, except in the case of one candidate unknown to the reviewer, who was given a score of 4 out of 10 with the explanation that in the reviewer's judgment we would not have put a less qualified candidate on our list while the reviewer would surely have been familiar with a more qualified candidate. I felt cheated, but it was my own fault for not specifically requesting the evidence. We all need to acquire the habit. Then the need for serious study of the mental processes relating evidence to probability will appear both natural and obvious.

A second scenario involves not only the client and statistician, but also multiple experts. For example, on a medical question there may be evidence from epidemiological studies, from population surveys, from animal studies, from clinical trials, and from clinical practice. What technologies, if any, should be invoked to combine different sources of evidence?

In principle, it is more desirable to pool the evidence and construct one probability assessment than it is to construct separate probability assessments and then combine, because the joint modeling effort almost certainly requires difficult judgments of independence or partial dependence which only make sense at the disaggregated level of empirical evidence. Indeed, the major challenge may be to construct a small world of factual knowledge which incorporates all the formal evidence. Log linear models should prove useful tools here, due to their flexible range of dependence parameters.

In a different direction, it should be mentioned that a central feature of belief function theory is a means for combining multiple sources of information which are judged independent. Extensions to partial dependence models are needed.

Whatever the normative limitations adopted, and whatever the range of models contemplated, Bayesian statistics needs formal prescriptive devices for combining multiple sources of evidence.

4. HARD AND SOFT PROBABILITIES

Consumers of probability assessments are familiar with the notion that some numerical probabilities are harder, or more certain, and others are softer, or more uncertain. This hard-soft dimension is rather vague, in the sense that no well-established single technology for quantification exists. The standard normative Bayesian theory recognizes only probability, not shades of probability, but there is a back-door prescriptive approach through sensitivity analysis which varies the formal assumptions and examines the corresponding variation of the resulting probability assessments. Extensions of Bayesian theory, such as belief function theory, or other theories of upper and lower probability, have an explicit technical mechanism for representing the hard-soft dimension, but these tools have not been widely tested in practice.

The canonical example of a hard probability is one derived from *evidence*, consisting of a relative frequency with a large denominator, and *judgment*, to the effect that the specific prospective application is not recognizable as selected on the basis of an observable characteristic associated with different evidence possessing acceptable hardness. Softening may come about for many reasons: the denominator may be small (sample size problems); the survey may not be of good quality, or may only be a vaguely recalled recollection of experience (data problems); or there may be several data sources of varying degrees of size and quality which are judged relevant to the event. The last is illustrated by John Venn's (1888) hypothetical question concerning which life tables apply to a consumptive Englishman living in Madeira.

All the factors of the preceding paragraph appear in generalized form in the context of more complex statistical designs and more complex sampling and stochastic models. Design type and quality are critical to prescriptive acceptability of Bayesian inferences. Bias is a fundamental statistical concept apart from any technical probability-based definitions of the term. Bias enters a design through weaknesses in protocols which specify measurement techniques and unit selection techniques, and through weakness in the execution of protocols. Finally, even flawlessly designed and executed protocols may yield results whose application to prospective new circumstances is biased in the sense that the new circumstances are judged sufficiently different that the effects of the differences must be quantitatively assessed.

Formal analysis of quantitative evidence depends on assumptions whose degrees of hardness must be communicated to the statistician's client in a convincing fashion. Assumptions are of many kinds. Parametric statistical models are often justified from informal hearsay or recalled empirical evidence which may deserve discounting. Similarly, scientific causal models range from well-established to highly speculative. Attempts to justify very small probabilities are generally based on trees of empirical evidence glued together with independence assumptions, or on other extrapolations beyond the reach of direct sampling experience. Any formal model depends on an implicit small world hypothesis that many obvious variables can be ignored. Assumptions which depend on forecasts, especially forecasts of human or social activities, may often be judged soft.

Sensitivity analysis of Bayesian inference may be expensive due to repeated attempts at model-construction, the computation of posterior probabilities and expectations under many Bayesian scenarios, and finally interpretation of the resulting variation in computed posterior probabilities and expectations. The technology is well worth pursuing, since without it Bayesian methodology has no way to convey the hard-soft dimension, and may therefore be judged unacceptable by clients. The awkwardness is that the Bayesian paradigm does not *per se* recognize the end result of a judgment-based range of posterior assessments.

By contrast, belief function theory has explicit mechanisms both for introducing softness into a source of evidence and for representing final softness. The mechanism for introducing softness is called *discounting*, and is defined and illustrated in many places by Shafer (e.g., Shafer, 1976, 1982). Softness in final uncertainty is achieved by replacing a single numerical probability by a pair of numbers called probability (a lower probability, or a kind of irreducible belief) and plausibility (or upper probability, or one minus the degree of irreducible belief in the complement). The theory needs much more prescriptive development and testing before its costs and benefits can be assessed.

5. THE EFFECTS OF UNUSUAL EVENTS

Major changes of perceptions of uncertainty often come in the form of events recognized as surprising or unexpected. Improbable events always happen, but some are singled out as surprising because of their potential or actual effects on human concerns, whether unexpected cures caused by an experimental drug, or unexpected disasters created by a nuclear generating plant failure. Laplace (1820) attempted to explain the phenomenon in terms of some outcomes being *extraordinary*, and some not, whence the occurrence of extraordinary events with small probability merit special attention. I tend to regard such interpretations as a valid use of subjective probability, as in ordinary significance testing. Bayesians tend to reject such retrospective or postdictive (Dempster, 1964) interpretations of probability, and attempt to incorporate the changes in probability assessment associated with unusual events within the normative Bayesian mechanism.

The essential point, however, is that the occurrence of unusual events often prompts a reexamination of the prescriptive foundations of an analysis. New variables are often introduced, or models are changed so that old variables treated as relatively unimportant are more influential in probability assessments. Old sources of evidence are examined more carefully, and new evidence may be sought. It seems to me implausible that such changes could all be characterized as an application of Bayes's theorem.

In any case, the relevant question is whether or not analyses constructed after an unusual event is recognized are biased. Is there an overreaction to the event? In pure science, one can protect against such bias by requiring that whole investigations be replicated, and make no pronouncements until the results are in. Applications to operational questions often are not permitted such replications, and decisions about effectiveness or risk must be made in real time.

My own view is that reconstructions prompted by unusual observations are important and necessary, and that all the mechanisms of prescriptive analysis may be employed and not just Bayesian rules, but that debiasing should be attempted by including in the model allowances for a wide range of extraordinary events which might have but did not occur, in addition to those which did occur.

6. ACKNOWLEDGMENTS

The issues and formulations of this paper were influenced by the discussion at two conferences which I attended in the summer of 1983: (1) Conference on Decision Making:

Descriptive, Normative and Prescriptive Interactions, 75th Anniversary Colloquium Series, Harvard Business School, June 16-18, 1983, organized by David E. Bell, Howard Raiffa, and Amos Tversky, (2) Workshop on Dealing with Uncertainty in Risk Analysis, Arthur D. Little, Inc., August 4, 5, 1983, organized by Joseph Fiksel, Louis A. Cox, Jr., and Helen D. Ohja.

REFERENCES

BELL, D.E., RAIFFA, H., and TVERSKY, A. (1983). Normative, descriptive, and prescriptive interactions. *Tech. Rep.* Harvard Business School.

BOX, G.E.P. (1980). Sampling and Bayes' inference in scientific modelling and robustness. *J. Roy. Statist. Soc. A* **143**, 383-404. (with discussion).

DEMPSTER, A.P. (1964). On the difficulties inherent in Fisher's fiducial argument. *J. Amer. Statist. Assoc.* **59**, 56-66.

GOOD, I.J. (1950). *Probability and the Weighing of Evidence.* London: Griffin.

JEVONS, W.S. (1877). *The Principles of Science: A Treatise on Logic and Scientific Method.* (second edition). New York: Macmillan. New York: Dover (1958).

LAPLACE, P.S. (1820). *Essai Philosophique sur les Probabilités.* English translation: *A Philosophical Essay on Probabilities.* New York: Dover, 1951.

SAVAGE, L.J. (1954). *The Foundations of Statistics.* New York: Wiley.

SHAFER, G. (1976). *A Mathematical Theory of Evidence.* Princeton: University Press.

— (1978). Non-additive probabilities in the work of Bernoulli and Lambert. *Arch. History Exact Sciences,* **19**, 309-370.

— (1981). Constructive probability. *Synthese* **48**, 1-60.

— (1982). Belief functions and parametric models. *J. Roy. Statist. Soc. B,* **44**, 322-352. (with discussion).

— (1983). Constructive decision theory. *Tech. Rep.* Harvard Business School.

SHAFER, G. and TVERSKY, A. (1983). Weighing evidence: the design and comparison of probability thought experiments. Research Paper: *Tech. Rep.* Harvard Business School.

VENN, J. (1888). *The Logic of Chance* (third edition). New York: Macmillan.

DISCUSSION

H. RUBIN (*Purdue University*)

It seems to me that Professor Dempster is trying to justify what people have been doing through ignorance of probability and statistics.

In combining evidence, as in any other situation, we must get the *client's* whole picture. The *client*, and not the *statistician*, must make the assumptions. The statistician may point out that certain assumptions are unimportant, and has the obligation to warn the client of consequences of his assumptions of which he is unaware. Asking stupid questions can only get stupid answers. I find it difficult to envision a situation where I would ask for a ranking, although I might ask for evaluations.

Probability is *not* relative frequency, and I personally have needed "harder" probabilities than can be obtained from any sample size which I can envision is physically possible. The probabilistic assumptions made in model specification are frequently "harder" than any sample will permit, and rightly so.

Professor Dempster seems clearly unhappy with many of the things a rational Bayesian approach forces him to accept, and consequently is searching for a way for mathematical results to help him out. Having been brought up in a more anti-Bayesian framework, I am also unhappy about the situation; however, if the truth disagrees with my philosophy, I must change my philosophy. Since mathematics is true (unless it is inconsistent, in which case there is so much to change that we must investigate everything anew, including the bases for Professor Dempster's arguments), we must accept the mathematics. We may look for robustness arguments to enable us to deal with difficult situations, but we must avoid selectively accepting those theorems which we like and rejecting those we do not like as a basis for behavior.

T. SEIDENFELD (*Washington University, St. Louis*)

Each of our speakers, one for roughly thirty years (Good, 1952), and the other for roughly sixteen (Dempster, 1967) has expressed sympathy for a liberalization of strict Bayesian theory. Their common motive, I believe, is to find a defensible representation for what, loosely put, is a state of ignorance. I allude to the view that a belief state may correspond to a set (or interval) of probability functions (or values). The roots of this intervalism differ, I suspect, for our two speakers.

My understanding is this. For Prof. Good the indeterminacy arises because we introspect imperfectly and this, too, is rational. So, I guess, we must see our own opinions (and values?) through uncertain eyes. For Prof. Dempster, I think the problem arises when trying to make sense of that old enigma, fiducial inference, where we would like to express the "ignorance" we profess about the unknown parameter. Thus, the intervalism is used to represent uncertainty, not due to introspective opacity, but uncertainty that is present when the *statistical* evidence fails to determine a precise "prior". On this approach it is incompletely described by the intervalism. Thus, e.g., we read Shafer (in his reconstruction of Dempster's position, 1982, p. 350), trying to avoid identifying his belief functions with Good's lower probabilities. However, the question I address to Prof. Dempster does not depend upon how the probability intervals arise or where the numbers come from.

In allowing a liberalization of the strict Bayesian theory for belief, there is a concomitant liberalization of decision theory. The change in postulates for decisions forces a separation from the "revealed preference" theory that is the occasional partner to strict Bayesianism. Let me illustrate.

For convenience, let us suppose that a belief-state is represented by a convex set of probability functions, D. The simplest form of dominance between acts is realized by the following decision rule, expressed in the language of choice functions.

Let X be a set of feasible options and $C(X)$ the choice set of "winners" from X. For each probability $d \in D$, denote expectation under d by $E_d[\bullet]$.

Rule$_1$: $x \in C(X)$ just in case there is no option $y \in X$ which, for each $d \in D$, satisfies

$$E_d[y] > E_d[x] \, .$$

Rule$_1$ is a binary choice rule. It identifies the "winners" as those feasible options which, in a pairwise choice over X, are expectation *un*dominated with respect to the set D.

Our concerns with dominance take us beyond binary comparisons. Consider a decision problem with these three feasible options:

x yields an outcome worth 1 utile if E occurs, 0 utiles otherwise.

y yields an outcome worth 0 utiles if E occurs, 1 utile otherwise.

z is the "constant act" worth. 0.4 utiles.

The choice problem is graphed in figure 1.

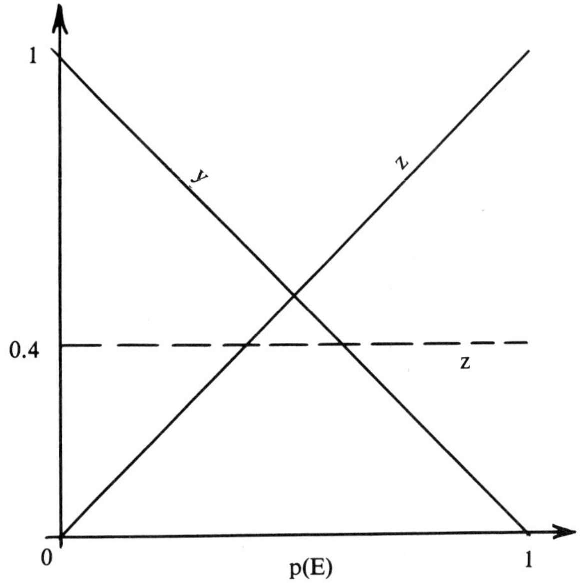

Act z fails to maximize expected utility over each probability $d \in D$. However, $rule_1$ -- binary comparisons-- does not detect this.

Savage (1954, pp. 123-124) notes this phenomenon with his example of choice over a set of "priors". Of course, Savage is thinking of the (convex) set of priors corresponding to a group decision. There is the familiar duality between individual and group choice. For instance, the well known theorem of Blackwell and Girshick (1954, p. 118) yields Arrow's Impossibility result by this duality --reinterpret the states as the individuals' preference profiles.

In an early paper, Good (1952, p. 114) offered a stricter requirement for admissibility than $rule_1$. Recently, I Levi (1980, p. 96) (also (1974)) has focused attention on it.

$Rule_2$: $x \in C(X)$ just in case there is some $d \in D$ which, for all $y \in X$, satisfies

$$E_d[x] \geq E_d[y].$$

If we use $rule_2$ (or even $rule_1$) for choice we do poorly in the economists' game of revealed preference. Not even $rule_1$ supports Independence of Irrelevant Alternatives. (IIA). Let us adopt A.K. Sen's taxonomy for choice functions (1977)[1] (using the duality of individual and social choice problems). Understand IIA to be the conjunction of Sen's properties α and β (hence, not Arrow's principle I).

Property α: (consistency over "contractions") If $S \subset T$ are feasible sets and $x \in C(T)$ then $x \in C(S)$, provided that $x \in S$.

Property β: (consistency over "expansions") If $S \subset T$ are feasible sets with $x \in C(S)$ and $y \in C(S)$, then $x \in C(T)$ if $y \in C(T)$.

[1] I thank I. Levi for bringing this important article to my attention.

Given a choice function $C(\bullet)$, define the (weak preference) relation R_C over elements x, y of a feasible set X by:

$$x R_C y \text{ if and only if for some } S \subset X, x \in C(S) \text{ and } y \in S.$$

Given a (complete and reflexive) relation R over pairs of feasible options, define the choice function generated by R:

$$\hat{C}(X,R) = \{x: x \in X \text{ and for all } y \in X, xRy\}.$$

Last, call a choice function *normal* if its weak-preference relation R_C regenerates itself through $\hat{C}$, i.e., if

$$C(X) = \hat{C}(X,R_C).$$

Sen shows (1977, p. 65) that a choice function is normal and generates an ordering if and only if it satisfies α and β.

That *rule*$_2$ violates property β is evident from the simple example (above). Using $X = \{x,y,z\}$ from figure 1, we see that $C(\{x,z\}) = \{x,z\}$ and similarly $C(\{y,z\}) = \{y,z\}$. But $C(\{x,y,z\}) = \{x,y\}$. For *rule*$_1$, consider the augmented feasible set $X' = \{w,x,y,z\}$ with w the mixed act: x with probability .5 and y with probability .5. Then $C(X) = X$ while $C(X') = \{w,x,y\}$.

Clearly, both *rule*$_1$ and *rule*$_2$ satisfy property α. But only rule$_1$ satisfies

Property γ: For each class $\mathbf{M}$ of feasible sets, if $x \in C(X)$ for each $X \in \mathbf{M}$, then $x \in C(\cup \mathbf{M})$.

To see *rule*$_2$ lacks γ, repeat the example used to show it fails β. That *rule*$_1$ satisfies γ is straightforward. Hence, by Sen's Proposition 8 (1977, p. 64)--a choice function is normal iff it satisfies α and γ--*rule*$_1$ is, whereas *rule*$_2$ is not normal.

Last, though not normal, *rule*$_2$ is "path independent", i.e. it has the property that $C(X \cup Y) = C(C(X) \cup C(Y))$. This follows from Sen's Proposition 19 (1977, p. 69)-- a choice function is path independent iff it satisfies α and

Property ϵ: If $X \subset Y$, then $C(Y)$ is not a proper subset of $C(X)$. Clearly, *rule*$_2$ has property ϵ.

Non-normality of *rule*$_2$ (despite its path independence) gives up all pretense to "revealed preference". It is not only that indifference is intransitive (recall both rules fail β while satisfying α), but with rule$_2$ the choice function cannot be recovered from examination of choices over feasible sets containing fixed pairs of options.

Rule$_2$ strikes me as the correct liberalization of the strict Bayesian policy to maximize expected utility. Thus, I find myself committed to abandoning the doctrine of "revealed preference". A philosophical platitude is germane. One person's *modus ponens* is another's *modus tollens.* Which way does logic's arrow point for our speaker?

Aside: Levi notes that, in effect, *rule*$_2$ can be weakened to *rule*$_1$ by having preferences go indeterminate over monotonic transformations of a given utility. This wipes out the difference between "second best" and "second worst" (see Levi, 1980, pp. 176-177) without abandoning *rule*$_2$. That is, *rule*$_1$ can be seen as the upshot of using *rule*$_2$ with ordinal (not cardinal) utility.

REPLY TO THE DISCUSSION

Professor Rubin's brief opening paragraphs apparently refer to my remarks on the sources of numerical probabilities for use in a specifical practical situation. His final paragraph comments on my acceptance, or lack thereof, of "a rational Bayesian framework". I will take these in reverse order.

Like him, I was brought up in a more or less anti-Bayesian environment, but I do not believe that any current unhappiness I might have with Bayes overlaps with his unhappiness. That is, I think that the arguments made 20 to 30 years ago by Savage and others, criticizing Wald's decision analysis from a Bayesian standpoint, were sufficiently compelling to dispel nagging doubts, and I am content with the mathematical precision, elegance, and coherence of Bayesian theory. My problem is that I am attracted also by the mathematical precision, elegance, and coherence of the Shafer-Dempster theory. He is wrong in suggesting that I seek salvation in "mathematical results". Since my paper contains no mathematics, I might have expected the opposite criticism. Mathematics *per se* is neutral on the issue of selecting a theory for practical use. In fact, neither mathematical truth nor philosophical truth, however finely honed these concepts may be in abstract modern academic discourse, seem to me to come close to resolving the choice of a theory. The only way forward that I see is to proceed case by case to build a catalogue of examples where instances of one theory or another have been constructed and appear to meet a modicum of scientific acceptance after careful critical analyses.

I agree with Herman that probability is *not* relative frequency. In many situations which command scientific approval, however, a relative frequency is agreed to be the source of a usable numerical probability. It is evident that formal Bayesian analyses require hard probabilities in the sense of mathematically precise specifications, and I agree that such precision is not attainable in fact, and perhaps not even in principle, from empirical frequencies. In a specific application, therefore, scientific evaluation requires bringing forth and assessing evidence which might be judged exchangeable with a canonical source such as a game of chance or a long run frequency. He wishes to put the responsibility for such judgments on his clients, which is desirable for clients sufficiently educated about the requirements for effective and acceptable use of methodology and techniques. I think it likely that better results will more often be obtained by a statistician and client working as a team.

Professor Seidenfeld suggests a distinction between Jack Good's lower probabilities and those of Dempster or Shafer, namely, that the former are drawn from imperfect introspection while the latter are intended to be based on evidence of one sort or another. This may very well be the case. More generally, Jack has good reason to be proud of many of his imperfect introspections whether or not they lead to lower probabilities. I believe, however, that Shafer and I make another distinction in our writing, namely that the mathematics of belief functions is more circumscribed than that of lower probabilities in a way which permits widespread use of the so-called Dempster rule of combination.

Professor Seidenfeld also asks me a question about Rule 1 vs. Rule 2 in relation to liberalizing Bayes. My response is that I would not formulate the question as he does. The Bayesian decision analyst's process is straightforward, to compute posterior expected utility for contemplated choices and select a maximizer. Using lower probabilities or belief functions, the extension is to compute lower posterior expected utilities and use them as guides which are unfortunately less decisive than precise numerical expectations. Interest focuses not on the definition and study of rules, since the problem is perceived as misformulated or nonsolvable in such terms. Instead, effort should be directed at formulating prescriptive methodology for constructing acceptable probabilities and utilities in specific real world circumstances.

To illustrate, consider his simple framing of a decision problem with choices x, y, and z. Suppose D is specified to be $\{d: .3 > d > .7\}$. Then Rule 1 says that x, y, and z are all admissible since none is dominated across the interval. Rule 2 eliminates z from consideration because there is no $d \in D$ for which z is preferred to both x and y. The lower expectations E_* for x, y, and z are respectively .3, .3, and .4. These should be interpreted as the amount of posterior expected utility the user can count on given his weak state of knowledge. The technology has not solved the problem of choice, but at least it appears that Rule 2 receives little support.

REFERENCES IN THE DISCUSSION

BLACKWELL, D. and GIRSHICK. M.A. (1954). *Theory of Games and Statistical Decisions.* New York: Wiley.

DEMPSTER, A.P. (1967). Upper and Lower Probabilities Induced by Multivalued Mapping. *Ann. Math. Statist.*, **38**, 325-339.

GOOD, I.J. (1952). Rational Decisions. *J. Roy. Statist. Soc. B,* **14**, 107-114.

LEVI, I. (1974). Indeterminate Probabilities. *J. Phil.*, **71**, 391-418.

— (1980). The Enterprise of Knowledge. Cambridge: The MIT Press.

SAVAGE, L.J. (1954). *Foundations of Statistics.* New York: Wiley.

SEN, A.K. (1977). Social Choice Theory: A Re-examination. *Econometrica,* **45**, 53-89.

SHAFER, G. (1982). Lindley's Paradox. *Amer. Statist. Assoc.*, **77**, 325-351 (with discussion).

BAYESIAN STATISTICS 2, pp. 133-156
J.M. Bernardo, M.H. DeGroot, D.V. Lindley, A.F.M. Smith (Eds.)

Quantifying Prior Opinion

PERSI DIACONIS
Stanford University

DONALD YLVISAKER
U.C.L.A.

SUMMARY

We investigate the approximation of a general prior by the more tractable mixture of conjugate priors. We suggest a new definition of conjugate prior for exponential families and offer a definition of conjugate prior for location families. A practical example, involving Bernoulli variables, is treated in detail.

Keywords: CONJUGATE PRIORS; EXPONENTIAL FAMILIES; MIXTURES.

1. INTRODUCTION

Bruno de Finneti has often emphasized the difference between the Bayesian standpoint and Bayesian techniques:

> "Bayesian Techniques, if considered as merely formal devices are no more trustworthy than any other tool (or *ad hoc* method) of the plentiful arsenal of 'objectivist statistics'."

In other words, there is more to Bayesian statistics than slapping down a convenient prior and computing Bayes rules. Let us illustrate these concerns through a simple example. Consider taking a specific penny and spinning it on its edge 50 times on a table. After observing the first 50 spins we are to predict the proportion of spins in a new series of spins and give an indication of how sure we are of our answer.

Any coherent Bayesian treatment of this problem can be interpreted as follows: Let S_n be the number of heads in the first n tosses. A parameter $p \in [0,1]$ can be introduced so that the law of S_n given p is binomial; further, there is a prior distribution (here taken to have a density) $f(p)$ on $[0,1]$, such that

$$P(S_n = k) = \binom{n}{k} \int_0^1 p^k (1-p)^{n-k} f(p) dp \,.$$

After observing $S_n = k$, the predictions will be based on the posterior

$$f(p \mid S_n = k) = \frac{p^k(1-p)^{n-k} f(p)}{\int p^k (1-p)^{n-k} f(p) dp} \,.$$

At issue here is the prior $f(p)$. Let us distinguish three categories of Bayesians (certainly a crude distinction in light of Good's (1971) 46,656 lower bound on the possible types of Bayesians).

1. *Classical Bayesians* (Like Bayes, Laplace and Gauss) took $f(p) \equiv 1$. A so called flat prior.

2. *Modern Parametric Bayesians* (Raiffa, Lindley, Mosteller) took $f(p)$ as a beta density $\beta(a,b;p) = \dfrac{\Gamma(a+b)}{\Gamma(a)\Gamma(b)} p^{a-1}(1-p)^{b-1}$. They note that this family contains a wide variety of distributional shapes, including the uniform prior $(a=b=1)$. With a beta prior, the posterior becomes especially simple $f(p \mid S_n = k) = \beta(a+k, b+n-k;p)$.

3. *Subjective Bayesians* (Ramsey, de Finetti, Savage). Take the prior as a quantification of what is known about the coin and spinning process.

As an example of this third approach, consider the way that Diaconis quantified his prior for the coin spinning experiment: "To begin with, there is a big difference between spinning a coin on a table and tossing it in the air. While tossing often leads to about an even proportion of heads and tails (indeed one can sort of prove this from the physics involved) spinning often leads to proportions like 1/3 or 2/3. Some basis for this opinion can be reported: I remember reading a story in the New York Times about a high-school teacher who had his class spin a penny 5000 times. The result was 80 % tails. When I was a graduate student, Arthur Dempster spun a coin on edge 50 times with a similar, skew result. It is a well known proposition around certain pool rooms that some coins have very strong regular biases when spun on edge (1964D pennies favor tails). The reasons for the bias are not hard to infer. The shape of the edge will be a strong determining factor - indeed, magicians have coins that are slightly shaved; the eye cannot detect the shaving, but the spun coin *always* comes up heads".

With this experience as a base, a bimodal prior seemed appropriate-spun coins tend to be biased, but not alway to heads. No beta prior is bimodal of course. A simple class of bimodal priors is given by mixtures of symmetric beta densities. Figure 1 shows the density

$$\{\beta(10,20;p) + \beta(20,10;p)\}/2.$$

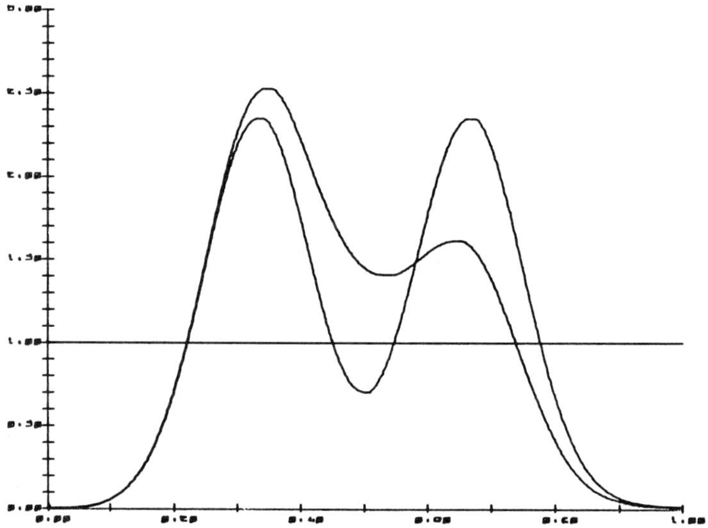

FIGURE 1. *Three prior distributions on [0,1]*

$\mathrm{I} = \beta(1,1)$

$\mathrm{II} = .5\beta(10,20) + .5\beta(20,10)$

$\mathrm{III} = .5\beta(10,20) + .2\beta(15,15) + .3\beta(20,10)$

On reflection, it was decided that tails had come up more often than heads in the past; further some coins seemed likely to be symmetric. A final approximation to the prior was taken as

$$.5\beta(10,20;p) + .2\beta(15,15;p) + .3\beta(20,10;p)\,.$$

All of these priors are of the form

$$f(p) = \sum_{i=1}^{n} w_i\beta(a_i,b_i;p)$$

for weights w_i and parameters a_i,b_i. Notice that the posterior of such a mixture of beta densities is again a mixture of beta densities

$$f(p\,|\,S_n=k) = \sum_{i=1}^{n} w_i'\beta(a_i+k_1\,b_i+n-k;p)\,.$$

Here the weights w_i' depend on n and k in a simple way.

$$w_i' = c\,w_i \int p^k(1-p)^{n-k}\beta(a_i,b_i;p)dp, \qquad \text{with } c \text{ chosen so } \Sigma\, w_i' = 1.$$

The mixture prior can be thought of as a weighted combination of "beta populations", the w_i measuring the prior degree of belief that the actual coin was chosen from the ith population. The posterior weights w_i' are proportional to the product of w_i and the relative likelihood of observing k successes in n trials in the ith population.

The penny was actually spun. After 10 spins there were 3 heads and 7 tails. Figure 2 shows the posterior distributions corresponding to the 3 priors of Figure 1. Note that the 3

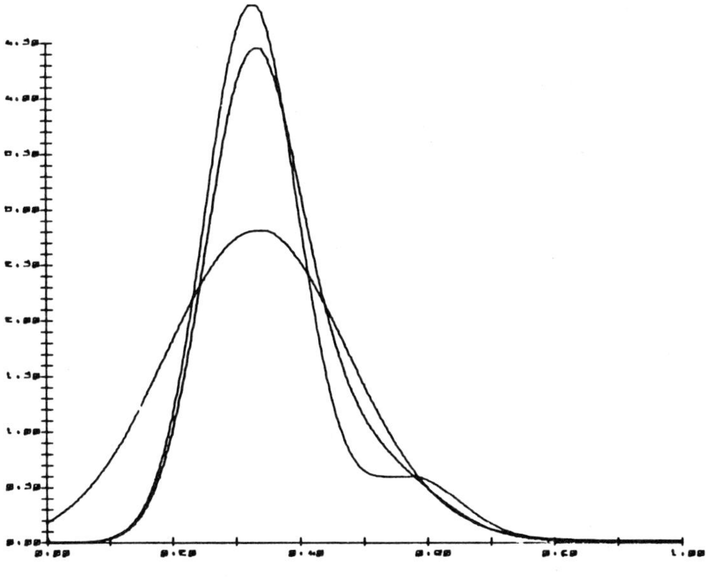

FIGURE 2. *The three posteriors after 3 heads in 10 trials*

I = $\beta(4,8)$

II = $.84\beta(13,27) + .11\beta(23,17)$

III = $.77\beta(13,27) + .16\beta(18,22) + .07\beta(23,17)$

modes agree (and point prediction from the 3 priors would be close) but the spreads are different, so that the variability assigned to predictions depend on the prior. After 50 spins there were 14 heads and 36 tails. The 3 priors are shown in Figure 3. They seem fairly close for any practical purpose.

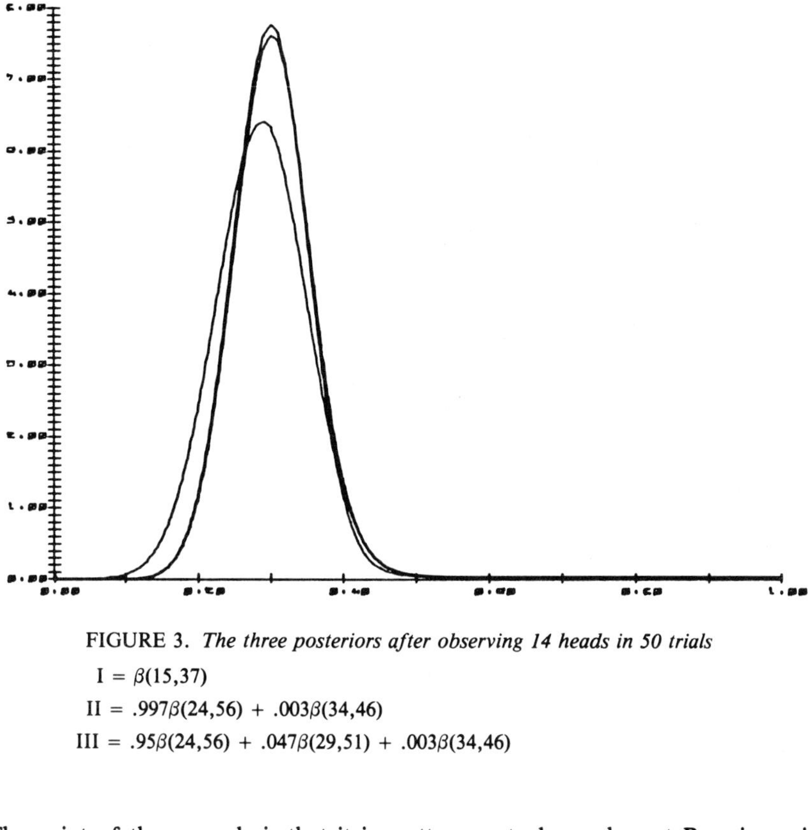

FIGURE 3. *The three posteriors after observing 14 heads in 50 trials*

I = $\beta(15,37)$

II = $.997\beta(24,56) + .003\beta(34,46)$

III = $.95\beta(24,56) + .047\beta(29,51) + .003\beta(34,46)$

The point of the example is that it is pretty easy to be an honest Bayesian using mixtures of conjugate priors. The computations for updating are straightforward. Of course, for "large samples" such careful quantification will not be important (at least in low-dimensional problems). However, for small or moderate samples, the prior matters.

It is natural to consider if we have gone far enough in considering mixtures of beta densities. Can any density be well approximated now, or are some opinions still ruled out. It is easy to see that any prior (density or not) can be well approximated by a mixture of beta densities. The reason is simple: by choosing a_i and b_i large, the densities $\beta(a_i,b_i;p)$ is close to a point mass at $a_i/(a_i+b)$ and mixtures of point masses are clearly dense in the weak star topology.

A more quantitative argument follows from

Theorem 1 Let $f(p)$ be a continuous density on [0,1]. Then, there exist $\{w_i,a_i,b_i\}_{i=0}^{n}$ such that

$$\max_p \left| f(p) - \sum_{i=0}^{n} w_i\beta(a_i,b_i;p) \right| \leq \tfrac{7}{4}\,\omega_f(1/\sqrt{n})$$

where the modulus of continuity is defined as

$$\omega_f(t) = \max_{|x-y|<t} |f(x)-f(y)| .$$

Proof. Consider the modified Bernstein polynomial

$$\sum_{i=0}^{n} w_i \beta(i+1, n-i+1; p), \quad w_i = \int_{i/n+1}^{i+1/n+1} f(x)dx .$$

(Note: The more usual Bernstein polynomial is

$$\Sigma f(\tfrac{i}{n}) \binom{n}{i} x^i(1-x)^{n-i} = \frac{1}{n+1} \sum_{i=0}^{n} f(\tfrac{i}{n})\beta(i+1, n-i+1, n-i+1; p)$$

this has weights $\frac{f(i/n)}{n+1} = w_i$.)

The usual non-probabilistic proof of the Wierstrass approximation theorem, using Bernstein polynomials, as given in chapter 2 of Lorentz (1966), goes through with the weights as given to yield the stated result. □

Remarks. For differentiable functions, $\omega_f(1/\sqrt{n}) \doteq c/\sqrt{n}$ for a constant c, so the approximation is of order $1/\sqrt{n}$. This is known to be the best possible rate of approximations by Bernstein polynomials as consideration of the density $f(x) = 4|x-\frac{1}{2}|$ shows. It is known that the best degree n polynomial approximation to a continuous function if of order $\omega_f(\frac{1}{n})$. (Jackson theorem). Indeed, it is possible to characterize the functions which can be well approximated by Bernstein polynomials, this work can be found by looking in the book by Lorentz or recent years of the Journal of Approximation Theory under the heading of saturation classes of Bernstein polynomials.

The purpose of the proof is *not* to suggest direct use of the Bernstein polynomial approximation, rather just to show that approximation is possible. In the example, 2 terms were chosen. In other examples, a small number of terms may be chosen so that the moments or a few quantities match exactly.

The purpose of the present paper is to indicate that the techniques used in the example apply fairly generally. A version of it holds for mixtures of conjugate priors in multivariate exponential families, and more generally yet. The structure of this paper is as follows: in section 2 we review the work of Diaconis and Ylvisaker (1979) on conjugate priors for exponential families. The main result here is that the standard families of priors can be characterized by a simple property of posterior linearity. In section 3 we indicate how the proposed definition of conjugate priors carries over to non-exponential families. This overlaps with work of Goldstein (1975). The final section discusses some definitions of approximation suitable for Bayesian inference. Here the results are less complete and there are many open research problems. Some of these problems are solved by Dalal and Hall (1983) who also work with mixtures.

2. CONJUGATE PRIORS FOR EXPONENTIAL FAMILIES

Conjugate priors are widely used for the usual exponential families of parametric statistics (normal, binomial, Poisson, gamma, etc.). We will suggest mixtures of such priors to approximate any prior. We begin by pointing out that the usual definitions are essentially vacuous.

Most often, a conjugate family is defined either as a family of priors that is closed under sampling or as a family of priors which is proportional to the likelihood. Consider the beta priors for coin tossing and observe that for any continuous non-negative function h the family

$$c\,p^{a}(1-p)^{b}h(p)dp$$

is closed under sampling and has a density (with respect to the carrier $h(p)dp$) which is proportional to the likelihood. Since h is an essentially arbitrary function, it would seem that *any* prior on $[0,1]$ is a conjugate prior. The usual family has $h(p) = 1$. In Diaconis and Ylvisaker (1979) we asked what additional properties of a prior give the families usually called conjugate priors. It turns out that such priors can be characterized by a condition of posterior linearity. For the binomial distribution this becomes

$$E\{p\,|\,S_n = k\} = ak+b \qquad k=0,1,2,\ldots,n\,.$$

A result like this holds for any of the standard families and actually characterizes the prior. Three of the main results will now be stated. We begin by stating the results for exponential families in their natural parametrizations. Following this we describe how the results transform to the more usual parametrizations. The results are equivalent. Any exponential family can be written in terms of the natural parametrization, and this allows a unified treatment.

Start with a fixed σ-finite measure μ on the Borel sets of $\mathbb{R}^d$--*the carrier measure*. Let $\mathcal{X}$ be the interior of the convex hull of the support of μ. Assume $\mathcal{X}$ is non-empty. Define $M(\theta) = \log \int e^{\theta\cdot x}\mu(dx)$ and let $\Theta = \{\theta\colon M(\theta) < \infty\}$. As usual, Holder's inequality shows that Θ is a convex set. It is called the *natural parameter space*. Throughout we assume that Θ is non-empty and open. The *exponential family* $\{P_\theta\}$ *of probabilities through* μ is defined as

$$dP_\theta = e^{x.\theta - M(\theta)}\,\mu(dx) \quad \theta\in\Theta\,.$$

As usual, the expectations under P can be determined by differentiating M:

$$E_\theta(X) = M'(\theta)\,.$$

Define a family $\{\tilde{\Pi}_{n_0,x_0}\}$ of measures on Θ by

$$\tilde{\Pi}_{n_0,x_0}(d\theta) = e^{n_0x_0\cdot\theta - n_0M(\theta)}\,d\theta, \quad n_0\in\mathbb{R}, \quad x_0\in\mathbb{R}^d.$$

If $\tilde{\Pi}_{n_0,x_0}$ can be normalized to a probability Π_{n_0,x_0} on Θ, it will be termed a *distribution conjugate to the exponential family* $\{P_\theta\}$ *of 2.1.* The next theorem determines for which (n_0,x_0) normalization is possible:

Theorem 2. a) If $\Theta = \mathbb{R}^d$, $\tilde{\Pi}_{n_0,x'_s}(\Theta) < \infty$ if and only if $n_0 > 0$, $x_0\in\mathcal{X}$.

b) If $\Theta \neq \mathbb{R}^d$ and $n_0 > 0$, $\tilde{\Pi}_{n_0,x_0}(\Theta) < \infty$ if and only if $x_0\in\mathcal{X}$.

The next theorem unifies many standard Bayesian calculations. It shows that x_0 is the prior mean of the parameter $E_\theta(X)$. It has been part of the folklore for years. A rigorous proof of the 1-dimensional case appears in Jewel (1974a,b).

Theorem 3. If θ has the distribution Π_{n_0,x_0} for $n_0>0$ and $x_0\in\mathcal{X}$ then

$$E(E_\theta(X)) = x_0\,.$$

Remark 1. The result gives posterior linearity for a sample $X_1,X_2,\ldots,X_n$ of size n from P_θ. Indeed, if Π_{n_0,x_0} is the prior for θ, the posterior density is $\Pi_{n_0+n,\ \frac{n_0x_0+n\bar{x}}{n_0+n}}$ with $\bar{x}$ the mean of the sample. Theorem 3 yields

$$E\{E_\theta(X)\,|\,X_1...X_n\} = \frac{n_0x_0+n\bar{x}}{n_0+n}.$$

Thus the posterior expectation of the mean parameter is a convex combination of the prior expectation of the mean parameter and $\bar{x}$. The weights are proportional to n_0 and the sample size n--in this sense n_0 may be thought of as a prior sample size. Novick and Hall (1965) have considered negative values of n_0 which yield improper "ignorance priors".

Remark 2. The argument for theorem 3 is integration by parts: consider the one-dimensional case. Then, as usual, $E_\theta(X) = M'(\theta)$ and

$$\int M'(\theta)e^{n_0x_0\theta-n_0M(\theta)}\,d\theta = \frac{e^{n_0x_0\theta-n_0M(\theta)}}{n_0}\Bigg|_{\underline{\theta}}^{\overline{\theta}} + x_0\int e^{n_0x_0\theta-n_0M(\theta)}\,d\theta\,.$$

The boundary terms vanish because Θ is open and the right side is x_0 times the correct norming constant. In higher dimensions, the argument is more complicated.

The next theorem gives a converse to Theorem 3 which characterizes conjugate priors.

Theorem 4. Let X be a sample of size one from P_θ and suppose the support of μ contains an open interval in $\mathbb{R}^d$. If θ has a prior distribution τ which is not concentrated at a single point and if

$$E\{E_\theta(X)\,|\,X\} = aX+b$$

for some constant a and vector b, then $a \neq 0$, τ is absolutely continuous with respect to Lebesgue measure, and

$$\tau(d\theta) = ce^{a^{-1}b.\theta-a^{-1}(1-a)M(\theta)}d\theta\,.$$

Remarks. Versions of Theorem 4 appropriate for discrete data are given in Diaconis and Ylvisaker (1979). The known results handle all the usual families, but they are still annoyingly incomplete.

Thus far we have assumed that the exponential family was given in its natural parametrization. Often, standard families are parametrized in other terms such as the parametrization involving p for the binomial. We now show how to transform the prior on Θ into a prior on any given parameter space to preserve linearity.

Let $\psi:\Theta\rightarrow\mathbb{R}^d$ be a diffeomorphism with range Θ_ψ and inverse ψ^{-1}. This transforms Lebesgue measure via multiplication by a Jacobian ψ'. The image of the prior Π_{n_0,x_0} becomes

$$e^{n_0x_0\psi^{-1}(t)-n_0M(\psi^{-1}(t))}\frac{1}{\psi'(\psi^{-1}(t))}\,dt \qquad \text{for } t\in\Theta_\psi\,.$$

If t is taken as parameter, so the family becomes

$$e^{x.\psi^{-1}(t)-M(\psi^{-1}(t))}\,\mu(dx) \quad t\in\Theta_\psi\,.$$

Then by standard properties of conditional expectation, we have as before

$$E\{E_t(X)\,|\,X_1...X_n\} = \frac{n_0x_0+n\bar{x}}{n_0+n}.$$

The conjugate family in terms of t is still closed under sampling and has posterior proportional to likelihood. The Jacobian factor simply specifies a choice of carrier measures which gives the additional property of posterior linearity. It is instructive to carry out the calculations for the standard families and see the standard conjugates, in their usual forms emerge at the end. Morris DeGroot has observed that the Jacobian factors always seem to merge in a nice way with the norming constant $M(\psi^{-1}(t))$. We do not know a theorem that makes this precise.

The final result of this section is the analog of Theorem 1 for a d-dimensional exponential family. The result shows that any prior (with or without a density) can be well approximated by a finite mixture of conjugate priors. The notation and assumptions are the same as in theorems 2,3, and 4.

Theorem 5. Let Θ be the natural parameter space of a d-dimensional exponential family. For any probability π on Θ, and any $\epsilon > 0$ there are weights w_i, and (n_i,x_i), $n_i > 0$, $x_i \epsilon \mathcal{X}$ such that if

$$\tilde{\pi}(d\theta) = \sum_{i=1}^{N} w_i\, c(n_i,x_i)\, e^{n_i x_i \cdot \theta - n_i M(\theta)}\, d\theta\ .$$

Then

$$d(\pi,\tilde{\pi}) < \epsilon$$

where d is the Prohorav metric.

Proof. It is well known, and easy to argue directly, that finite mixtures of point masses are weak star dense. A proof in a general setting is in Theorem 12.11 of Choquet (1969). Hence we must only show that any point mass can be weak star approximated by a prior of the form

$$c(n_0,x_0)e^{n_0 x_0 \theta - n_0 M(\theta)}\, d\theta\ . \tag{2.1}$$

To see this, differentiate (2.1), the maximum occurs at the unique value of θ satisfying

$$\nabla M(\theta) = x_0$$

It is straightforward to show that as n_0 tends to infinity, the prior (2.1) concentrates at this θ. Finally, for any $\theta_0 \epsilon \Theta$, $M'(\theta_0) = E_{\theta_0}(X) \epsilon \mathcal{X}$, so for any θ_0, there is an $x_0 \epsilon \mathcal{X}$ such that $M'(\theta_0) = x_0$. This completes the argument. □

Remarks. One can also emulate the proof using Bernstein polynomials. This has been carried out to yield an approximation with error term in unpublished thesis work of Mark Jacobson at Stanford University. Related results are in Lorentz (1953), Dubins (1983), Dalal (1978), and Dalal and Hall (1980, 1983). Again, the theorem above is just meant to indicate that mixtures are capable of approximating any prior.

3. LOCATION PARAMETERS

In Section 2 we suggest using linear posterior expectation as a definition of conjugate priors. This offers the possibility of moving away from the exponential family setting. The present section carries out this program for location parameter problems. The main characterization result is theorem 5 which extends a theorem proved by Goldstein (1975). Conjugate priors are suggested only as a convenient building block: We can show, in certain circumstances, that any prior can be well approximated by a mixture of conjugate priors.

We begin by giving a class of priors with linear posterior expectation. Let X be a d-dimensional random vector with distribution function F, and let θ be a random vector independent of X with prior distribution F^{*n} (n-fold convolution). For the location problem, $F(x\text{-}\theta)$, the observed variable is $Z = X+\theta$. In what follows, we do not want to assume that the prior has a mean. We thus define

$$E(\theta|Z) = g(Z) \text{ if and only if } E(\theta_i^+|Z)\text{-}E(\theta_i^-|Z) = g_i(Z)\ a.s.\ 1 \le i \le d$$

where the subscript denotes the ith coordinate. Conditional expectation in this sense is still linear and of course agrees with the usual notion when means are finite c.f. Strauch (1965). With this definition we have, provided the expectations exist,

$$E(\theta|Z) = \frac{n}{n+1} Z. \tag{3.1}$$

To see this, let $X_1,\ldots,X_n$ be independent of X and each other so that $S_n = X_1 + \ldots + X_n$ has the same distribution as θ. Then

$$\frac{n+1}{n} E\{S_n|S_n+X\} = E\{S_n|S_n+X\} + \frac{1}{n} E\{S_n|S_n+X\}$$

$$= E\{S_n|S_n+X\} + E\{X|S_n+X\} = E\{S_n+X|S_n+X\} = S_n+X.$$

For example, if X and θ are independent Cauchy variables, $E(\theta|X+\theta)$ exists and equals $\frac{X+\theta}{2}$. The class of priors can be widened by allowing inclusion of a known location parameter for the prior. With this in mind, we define a conjugate prior for the location parameter problem through F as any prior on $\mathbb{R}^d$ such that

$$E(\theta|Z) = aZ+b \quad \text{for real } a \text{ and } b \in \mathbb{R}^d. \tag{3.2}$$

For location problems, posterior linearity only holds for samples of size 1. For larger samples, $\bar{X}$ is not a sufficient statistic unless the distribution of X is normal, see Ferguson (1954) or DeGroot and Goel (1981) for more on this point.

It will surface in the proof of theorem 6 that (3.2) can hold only if $0 \le a \le 1$. The following lemma determines what happens at the boundary cases.

Lemma 1. Let X and θ be independent random variables and suppose $E(|X|)<\infty$. Then $E(\theta|X+\theta) = b$ if and only if θ is a.s. constant and $E(\theta|X+\theta) = (X+\theta)+b$ if and only if X is a.s. constant.

Proof. The lemma follows from a result in Doob (1953, p. 314). To bring things into that framework, let $Y = X+\theta$. Without loss of generality $b = 0$. Under the first assumption $0=E(\theta|X+\theta) = E(\theta+X|X+0) - E(X|X+\theta)$, so $E(X|Y) = Y$. Of course, $E(Y|X) = X$. This says that X and Y form a 2-term martingale in either order, and this is just what Doob shows is impossible when $E|X| < \infty$. The argument under the second assumption is similar. For related theorems see Girshick and Savage (1951, p. 1653) or Gilat (1971). □

Assume now that (3.2) holds for a location parameter problem and consider the question of uniqueness of the underlying prior distribution. To state theorem 6, let $X = (X_1,\ldots,X_d)'$ have distribution functions F not concentrated at a point and write $\lambda_{2n} = \int \{x_1^{2n} + \ldots x_n^{2n}\} dF$.

Theorem 6. Let X and θ be independent d-dimensional random vectors with neither X nor θ a.s. constant. Assume $E|X_i| < \infty$ for $i = 1,\ldots,d$ and that either

a) the characteristic function of X has no zeros or

b) $\sum_{n=1}^{\infty} \lambda_{2n}^{-1/2n} = \infty$.

If

$$E(\theta \mid X+\theta) = a(X+\theta)+b\,. \tag{3.3}$$

Then $0 < a < 1$ and the distribution of θ is uniquely determined.

Proof: We begin with $d = 1$ and show first that $E|\theta| < \infty$. Now $a^2 \neq a$ from lemma 1 and, by translating X if necessary, one can take $b = 0$. From (3.3) and the linearity of expectation,

$$E(\theta \mid X+\theta) = a(X+\theta) = aE(X+\theta \mid X+\theta) = aE(X \mid X+\theta)+aE(\theta \mid X+\theta)\,.$$

Hence

$$E(\theta \mid X+\theta) = \frac{a}{1-a}\,E(X \mid X+\theta)\,. \tag{3.4}$$

Use this in (3.3) to find

$$\theta = \frac{1}{1-a}\,E(X \mid X+\theta)-X\,. \tag{3.5}$$

By assumption, the right side of (3.5) is absolutely integrable so $E|\theta| < \infty$

When X and θ have finite expectations, it follows from lemma 1.1.1 of Kagan, Linnik and Rao (1973) that (3.3) holds if and only if the characteristic functions of X and θ satisfy

$$(1-a)\phi_\theta'(t)\phi_X(t) = a\phi_X'(t)\phi_\theta(t) \text{ for all } t\,. \tag{3.6}$$

Since $\phi_X(t)$ does not vanish in some interval I about 0, (3.6) gives for $t \in I$,

$$\phi_\theta(t) = \phi_X(t)^{\frac{a}{1-a}}\,. \tag{3.7}$$

Observe first that for (3.7) to hold with neither X nor ϕ constant, it must be that $a/(1-a) > 0$ and so $0 < a < 1$. Now if ϕ_X never vanishes, ϕ_θ is determined by (3.7). On the other hand, if the distribution of X satisfies b and so is determined by its moments, the corresponding moments of θ can be computed from (3.7) and satisfy the same determinedness condition. The proof of part a is complete.

Now suppose that X satisfies (b). We first argue that θ has moments of all order. For the rest of the argument, suppose we are given a pair of random variables θ_1, X_1 satisfying $E(\theta_1 \mid \theta_1+X_1) = a(\theta_1+X_1)$. For fixed $0 < a < 1$; we show that the existence of moments for X_1, $\lambda_{2m}^{(1)} = EX_1^{2m}$, implies the existence of moments for θ_1, $\mu_{2m}^{(1)} = E\theta_1^{2m}$. To see this let (θ_2,X_2) be an independent copy of (θ_1,X_1) and set $\theta = \theta_2-\theta_1$, $X = X_2-X_1$. It follows that

$$E(\theta \mid \theta+X) = a(\theta+X)\ ,$$

and X are symmetric with

i) $E(\theta_1-E\theta_1)^{2m} \leq E(\theta_2-\theta_1)^{2m} = \mu_{2m}$

ii) $\lambda_{2m} = EX^{2m} = \sum_{j=0}^{2m} \binom{2m}{j}(-1)^j \lambda_j^{(1)}\lambda_{2m-j}^{(1)} \leq 2^{2m}\lambda_{2m}^{(1)}\,.$

We shall show below that there is a C for which $\mu_{2m} \leq C^{2m}\lambda_{2m}$ for all m and then from i) and ii) we will have

$$E(\theta_1-E\theta_1)^{2m} \leq \mu_{2m} \leq C^{2m}\lambda_{2m} \leq (2C)^{2m}\lambda_{2m}^{(1)}\,.$$

From a knowledge of X_1 and its moments, we have $E(\theta_1) = \frac{a}{1-a}\,E(X_1)$ so it will generally be seen that if X_1 satisfies (b), so does $\bar{\theta} = (\theta_1-E\theta_1)$ and then, so does θ_1. To show that there is a C with $\mu_{2m} \leq C^{2m}\lambda_{2m}$ first observe that $E(\theta(\theta+X)^{2m-1}) = aE(\theta+X)^{2m}$. Thus

$$\sum_{j=1,\ j\text{ odd}}^{2m-1} \binom{2m-1}{j}\mu_{j+1}\lambda_{2m-1-j} = a \sum_{j=0,\ j\text{ even}}^{2m} \binom{2m}{j}\mu_j\lambda_{2m-j}$$

so that

$$0 = \sum_{j=0,\ j\text{ even}}^{2m} \left[\frac{a\binom{2m}{j}-\binom{2m-1}{j-1}}{\binom{2m}{j}}\right]\binom{2m}{j}\mu_j\lambda_{2m-j}\,.$$

We write this in the form

$$(*) \qquad \sum_{0\leq r\leq ma} \left(a-\frac{r}{m}\right)\binom{2m}{2r}\mu_{2r}\lambda_{2m-2r} = \sum_{ma\leq r\leq m} \left(\frac{r}{m}-a\right)\binom{2m}{2r}\mu_{2r}\lambda_{2m-2r}\,.$$

Determine the smallest r_0 so that $\frac{r_0}{r_0+1} > a$ and a C at least as large as $\left(\frac{2}{1-a}\right)^{\{1/(1-a)\}}$ so that

$$(**) \qquad \mu_{2r} \leq C^{2r}\lambda_{2r} \quad \text{for} \quad r = 1,2,\ldots,r_0.$$

For $m = r_0+1$, note that $ma = (r_0+1)a < r_o$ so that in the left hand side of (*) all $r \leq r_0$. Then the left hand side is

$$\leq \sum_{0\leq r\leq ma} \binom{2m}{2r} C^{2r}\lambda_{2r}\lambda_{2m-2r} \leq C^{2ma}2^{2m}\lambda_{2m}$$

while the right hand side of (*) exceeds $(1-a)\mu_{2m}$. Hence

$\mu_{2m} \leq \frac{1}{1-a}C^{2ma}2^{2m}\lambda_{2m} \leq C^{2m}\lambda_{2m}$ provided $C^{2m} \geq \frac{1}{1-a}C^{2ma}2^{2m}$. This requires that $C^{2m(1-a)}$

$\geq \frac{1}{1-a}2^{2m}$ or $C \geq \frac{2^{\{1/(1-a)\}}}{(1-a)^{\{/(2m(1-a))\}}}$. But in fact $C \geq \left(\frac{2}{1-a}\right)^{\{1/(1-a)\}} \geq \frac{2^{\{1/(1-a)\}}}{(1-a)^{\{1/(2m(1-a))\}}}$.

Therefore (**) is extended to $r_0 + 1$ with the same C as before. Performing the same induction step again requires only that $ma = (r_1+1)a$ *be* $< r_1$ and this is guaranteed inasmuch as $\frac{r_1}{r_1+1} > \frac{r_0}{r_0+1} > a$.

For $d>1$, the finiteness of $E|\theta_i|$ follows readily by the arguments used earlier. Take $b = 0$ again without real loss. Now let $\varrho \in \mathbb{R}^d$ and find from (3.3) that

$$E(\varrho\bullet\theta\,|\,X+\theta) = a\varrho\bullet(X+\theta) = E(\varrho\bullet\theta\,|\,\varrho\bullet(X+\theta))\,.$$

If the characteristic function of X has no zeros then the same is true of the characteristic function of $\varrho\bullet X$. Hence the one-dimensional version of the theorem implies that the distribution of $\varrho\bullet\theta$ is uniquely determined and so therefore is the distribution of θ. If the distribution of X satisfies b), the inequality $|\varrho\bullet X|^{2n} \leq m(X_1^{2n}+\ldots+X_d^{2n})$ with m depending on ϱ and d but not on n or X, implies that the distribution of $\varrho\bullet X$ is determined by its moments. The proof of the theorem is completed by another application of the one-dimensional version. □

Remark 1. Here is an example of independent random variables X and θ having finite means, different distributions and satisfying (3.3) with $a = \frac{1}{2}$, $b = 0$. The example makes it clear that some hypothesis on the distribution of X are required in Theorem 5 in order to guarantee uniqueness. Now (3.3) holds with $a = \frac{1}{2}$ and $b = 0$ if and only if $\phi_X'\phi_\theta = \phi_\theta'\phi_X$ as at (3.6). Let X have density

$$\frac{3\cdot 4^3}{\pi}\left(\frac{\sin\frac{x}{4}}{x}\right)^4 \qquad -\infty < x < \infty .$$

Such an X has a finite mean and a real characteristic function which is continuously differentiable and vanishes outside $(-1,1)$. Using Theorem 4.32 of Lucas (1970) we see that the function ϕ_θ which equals ϕ_X on $(-1,1)$ and has period 2 is the characteristic function of an arithmetic distribution. By construction $\phi_X'\phi_\theta = \phi_\theta'\phi_X$, since the two sides are equal in $(-1,1)$ and both sides vanish outside this interval.

The construction can be varied to give X and θ having different distributions but moments of all orders: Let $\phi(t)$ be the well known "tent" characteristic function supported on $(-1,1)$. Convolving $\phi(t)$ with itself leads to smoother and smoother functions with compact support--the example above is based on $\phi*\phi$. The function

$$\psi(t) = \mathop{*}_{n=1}^{\infty} \phi(2^n t)$$

may be shown to be an infinitely differentiable characteristic function with support on $[-1,1]$. If it is used to define X and its periodic continuation is used to define θ, we have an example with moments or all orders where $E\{\theta\,|\,X+\theta\} = \{X+\theta\}/2$ but the law of θ is not unique.

Remark 2. In theorem 5 we have used Carlemans sufficient condition for a distribution to be determined by its moments. We wondered if this could be replaced by the weaker condition that X was determined by its moments. Here is one result in that direction.

Proposition 1 If X is determined by its moments and θ is independent of X and satisfies

$$E(\theta\,|\,X+\theta) = \frac{1}{n}(X+\theta) \text{ for any fixed integer } n \geq 2,$$

then the distribution of θ is uniquely determined.

Proof. From (3.7), $\phi_X(t) = \phi_\theta(t)^{n-1}$ in a neighborhood of 0. This implies the moments of θ. We want to show that θ is determined by its moments. If not, then, by a fundamental theorem of moment theory, see Theorem B of Landau (1980), for any real t, there is a probability ψ with all the same moments as X, and so the same distribution as X, but ψ^{*n} has an atom at the point nt. Since t is arbitrary, there is a probability with the same moments as X but having an atom at any specified place. This contradiction proves the proposition. □

If it were true that being determined by moments was inherited by convolutions, general rational values of a could be handled; for from $\phi_X^n = \phi_\theta^m$ in a neigborhood of zero, and X determined would follow ϕ_X^n determined and then ϕ_θ determined by the above argument. Cristian Berg (1983) has provided a probability μ, which is determined by its moments but such that $\mu*\mu$ is not determined! For more on these matters, see Devinatz (1959).

Remark 3. We note the connection of the present section to a result in Martingale theory. If $\{X_i\}_{i=1}^\infty$ are i.i.d. random variables with $E|X_i| < \infty$, then the argument used in the introduction to this section shows that S_n/n is backward martingale. Here $S_n = \Sigma_{i=1}^n X_i$ and the martingale property is

$$E\{\frac{S_n}{n} \mid S_{n+1},S_{n+2},\ldots\} = \frac{S_{n+1}}{n+1}, \qquad n = 1,2,\ldots \tag{3.8}$$

using obvious generalizations of the argument in Theorem 6 we can show that if $\{X_i\}$ are independent random variables such that (3.8) holds and $E|X_1| < \infty$, then all the X_i have finite first moments. If ϕ_i is the characteristic function of X_i, (3.8) holds if and only if for all $k \geq 2$,

$$\phi_k' \prod_{j=1}^{k-1} \phi_j(t) = \frac{\phi_k(t)}{k-1} \left(\prod_{j=1}^{k-1} \phi_j(t)\right)' = \phi_{k-1}'(t) \prod_{j=1}^{k} \phi_j(t) \qquad \text{for all } t.$$

If $\phi_1(t) \neq 0$ then all the $\phi_i = \phi_1$. Using variants of the construction in remark 2 we can construct an infinite sequence of independent random variables, each with a different distribution, which satisfy (3.8).

Remark 4. For a distribution function F, let S_F be the set of real numbers a which can occur in (3.2). From (3.1), $n/(n+1) \in S_F$ for all $n = 1,2,\ldots$. If F is uniform on [0,1], these are the only numbers in S_F. If F is infinitely divisible with finite mean then $\phi_X{}^{a/1-a}$ is a characteristic function for every $a \in (0,1)$ so $S_F = (0,1)$. Conversely, if Φ has a mean, $\phi_F(t) \neq 0$ and $S_F = (0,1)$, Theorem 6 implies F is infinitely divisible.

Remark 5. When can any prior be approximated by mixtures of conjugate priors in a location parameters setting? If X is infinitely divisible then the conjugate priors have characteristic functions of the form $e^{-ith}\phi_X^{\alpha}$. As $\alpha \to 0$, this approaches a point mass at μ. We conclude that approximation is possible when X is infinitely divisible. It may be that the converse of this holds. If X is uniform on [0,1] then the only conjugate priors are translates of convolutions of uniforms; it is easy to show that finite mixtures of these are not weak star dense in all probabilities on the line.

Remark 6. It is possible to develop some theory of the above sort for scale parameters. As an example, we note the following. If X has a beta density with parameters a and b, and θ has a beta density with parameters $a+b$ and c, then

$$E\{\theta \mid X\theta\} = \frac{cX\theta}{b+c} + \frac{b}{b+c}\,.$$

4. SOME RESEARCH PROBLEMS

The material discussed above suggests many questions, both technical and philosophical. These are discussed here under the following headings: What are we doing: when are two priors close?; Stability; extensions to non-linear regression; matrices; and connections with de Finettis' Theorem.

4.1. *What are we doing: When are two priors close?*

The approximation theorems (1 and 6) involve a topology on the space of all priors. It is both practically relevant and philosophically natural to link the topology to the actual problem in hand. Thus to say when two priors are close entails specifying a use for the priors. Here are some specific suggestions drawn from work of Stein (1965).

Consider a decision problem specified by a family of probabilities $\{P_\theta\}_{\theta\in\Theta}$ and loss function L. Suppose that π_t represents a true prior and π_a an approximation. An observation x yields two posteriors π_t^x and π_a^x. Each of these will result in certain decision ("Bayes rules") $\delta_t(x)$ and $\delta_a(x)$. Here δ_t minimizes the risk $R(\pi_t,\delta) = E_t\{L(\theta,\delta(x))\}$. The difference in risk, if δ_a is used instead of δ_t can serve as a measure of separation between the two priors:

$$S(\pi_t,\pi_a) = R(\pi_t,\delta_a) - R(\pi_t,\delta_t) .$$

Remark. Of course the separation S is not a metric. Stein (1965) shows that it is not even symmetric. Some further discussion and interpretation is in Diaconis and Stein (1983). We have verified that when the statistical problem is estimation of a binomial parameter θ based on a sample of size n, with squared error as loss, then any prior can be approximated, in the sense of the separation S, by a mixture of beta priors. This is an easy case. Other loss functions, and unbounded parameter spaces certainly merit careful study. Jim Berger has shown us arguments that suggest that for estimating a normal mean with squared error as loss a Cauchy prior for the mean cannot be approximated in the sense indicated.

Stein has suggested an intermediate notion of separation which does not depend on the loss function: Let $\varrho(\cdot,\cdot)$ be a metric between probabilities. Consider

$$\varrho^*(\pi_t,\pi_a) = E_t\{\varrho(\pi_t^x,\pi_a^x)\}$$

where the expectation is taken with x given its true marginal distribution. Stein has sketched an argument to show that if $L(\cdot,\cdot)$ is smooth in its second argument, and ϱ is taken as the Hellinger distance:

$$\varrho(P_1,P_0) = \int (p_1^{1/2} - p_0^{1/2})\lambda$$

where p_i is the density of P_i with respect to the dominating measure λ, then

$$S(\pi_t,\pi_a) \leq c\,\varrho^*(\pi_t,\pi_a)$$

holds for some constant c, depending on L and perhaps π_t, but not on π_a. Thus, approximation in ϱ^* entails approximation in separation for a wide variety of problems. Again, it seems worthwhile to have examples and some "honest" theorems. It seems likely that if ϱ is taken as any metric metrizing the weak star topology, any prior can be approximated by mixtures of conjugate priors. Related issues are discussed by Kadane and Chuang (1978).

The language of "approximate" and "true" priors open a philosophical can of worms. Recent works by Shafer (1981), Jeffrey (1982) and Diaconis and Zabell (1982) emphasized the *constructive* nature of forming a prior - none of us have a true prior, sitting inside, waiting to be "elicited". Clearly, as we think about things different possibilities and refinements will leap to mind. The very act of thinking provides valuable "data" so that the true prior is always unknown, as is the "true position" of an electron.

One way around these difficulties is the conceptually difficult task of thinking "will it matter if I refine my prior further". Often, in well travelled problems, the fine details of a prior will not matter much. This is likely to be the case in low dimensional problems with reasonable sized samples.

A rather different argument against very careful specifications of a prior follows from work of Jeffrey (1982) and Diaconis and Zabell (1982). They argue that we often up date a prior to a posterior by methods different from Bayes theorem. This will tend to be true in "exploratory data analysis" situations where it is practically impossible to quantify a prior over a sufficiently high dimensional space, and our reaction to data is likely to be "I forgot all about that possibility".

4.2. *Stability*

It would be useful to have some more quantitative measures of the effect of small changes in prior specification on final decisions. One approach is to use the separation $S(\pi_t, \bullet)$ as a basis of influence function calculations. Ramsey and Novick (1980) have suggested similar things be done simultaneously for prior, likelihood and loss function.

4.3 *Extensions to Non-Linear Regression*

Consider now a general family of probabilities $\{P_\theta\}_{\theta\epsilon\Theta}$ and a parameter $\psi(\theta)$. When does

$$E\{\psi(\theta)\,|\,X\}$$

characterie a prior distribution on Θ? In section 2 we consider the special case of exponential families and linear regression. For exponential families in their natural parametrization and $\psi(\theta)=\theta$, a characterization result can be shown to hold. Thus, for a normal location problem, with scale 1, if the prior is proportional to $\theta^2\, e^{-\theta^2/2}$, $E\{\theta\,|\,X\} = \{X^3+6X\}/\{2X^2+4\}$, and the prior is uniquely characterized by this relation.

However, we cannot settle the uniqueness problem in any generality: For example, if P_θ is a normal location problem $\psi(\theta)$ is a polynomial in θ, then for a normal prior,

$$E\{\psi(\theta)\,|\,X\}$$

is a polynomial in X. We do not know if this characterizes normal priors.

4.4. *Matrices*

In exponential families and location problems, we can ask about priors with posterior linearity in terms of matrices. We here discuss the location problem in d-dimensions. When can one find independent θ and X with

$$E(\theta\,|\,\theta+X) = A(\theta+X)+b \tag{4.1}$$

for some $d\times d$ A and $b \in \mathbb{R}^d$? Assume $E|X| < \infty$ with $EX = E\theta = b = 0$. From page 11 of Kagan, Linnick and Rao (1973), (4.1) holds if and only if

$$\phi_X(s)(I-A)\nabla\phi_\theta(s) = \phi_\theta(s)A\nabla\phi_X(s)\,.$$

The matrix A may as well be taken non-singular. All possible A's occur when X is normal, see Jewel (1982) and the references cited there. If A is non singular and $A \neq aI$, then perhaps normality is the only case.

4.5. *Connections with de Finetti's Theorem*

It is possible to give parameter-free versions of some of the characterization results of section two. An elegant classical version is W.E. Johnson's theorem as discussed by Zabell (1982). Imagine a process $\{X_i\}$ taking $k \geq 3$ values. Suppose that $P\{X_{n+1}=i\,|\,X_1,\ldots,X_n\}$ only depends on the number of times i ocurred among $X_1,X_2,\ldots,X_n$. This is necessary and sufficient condition for the law of X_i to be a Dirichlet mixture of multinomials.

This can be generalized to characterize the exchangeable sequences which are conjugate prior mixtures of specified exponential families. It would take us too far afield to try to develop the modern theory of partial exchangeability here. A survey, in the language of Bayesian statistics, can be found in Diaconis and Freedman (1983). In general terms, researchers in this field have found additional notions of "symmetry" which imply that a sequence is a mixture of standard parametric families. We propose that a further condition can be given in terms of how one would predict X_{n+1} given $X_1,X_2,\ldots,X_n$, that will result in characterizations of the mixing measure. We will content ourselves with a single example.

Theorem 7. Let X_i $1 \le i < \infty$ take values in $\mathbb{R}_+$ and satisfy,

$$P\{(X_1,\ldots,X_n)\epsilon A\} = P\{(X_1,\ldots,X_n)\ \epsilon A+x\}$$

for all n and all Borel $A \subset \mathbb{R}_+^n$ with $x \epsilon \mathbb{R}^n$ satisfying $\Sigma\, x_i = 0$ and $A+x \epsilon \mathbb{R}_+^n$. Then X_i are a scale mixture of exponentials. If in addition,

$$E\{X_2|X_1\} = aX_1+b$$

then the mixing measure is a gamma distribution.

Sketch of Proof. It is easily verified that the symmetry condition implies X_i is exchangeable. As usual, the condition still applies when the law of X is conditioned on the tail field T. By deFinetti's theorem, the process conditioned on the tail field is i.i.d.. But this entails for all positive a_1 and a_2

$$P\{X_1>a_1+a_2|T\} = P\{X_1 > a_1,X_2>a_2|T\} = P\{X_1>a_1|T\}P\{X_1>a_2|T\}\,.$$

So X_i are exponential.

Because of exchangeability, the second conditions gives

$$E\{X_n|X_1\} = aX_1+b \quad \text{for any} \quad n \ge 2.$$

By the strong law for exchangeable variables the average of $X_2,X_3,\ldots,X_n$ converge to the mean $E_\theta(X_1)$, so we have

$$E\{E_\theta(X_1)|X_1\} = aX_1+b$$

and this was shown to characterize gamma priors in section two.

REFERENCES

BERG, C. (1983). On the Presevation of Determinacy Under Convolution. *Proc. Amer. Math. Soc.* (to appear).

CHOQUET, G. (1969). *Lectures on Analysis, Vol. I.* Benjamin, Reading, MA.

DALAL, S. (1978). A Note on the Adequacy of Mixtures of Dirichlet Processes. *Sakhyā, A,* **40**, 185-191.

DALAL, S. and HALL, J.H. (1980). On Approximating Parametric Bayes Models by Non Parametric Bayes Models. *Ann. Statist.,* **8**, 664-672.

DALAL, S. and HALL, W.J. (1983). Approximating Priors by Mixtures of Natural Conjugate Priors. *J. Roy. Statist. Soc. B,* **45**, 278-286.

DEVINATZ, A. (1959). On a Theorem of Levy-Raikov. *Ann. Math. Statist.,* **30**, 583-586.

DIACONIS, P. and FREEDMAN, D. (1983). Partial Exchangeability and Sufficiency. *Sakhyā.* (to appear).

DIACONIS, P. and STEIN, C. (1983). *Lectures in Decision Theory.* To appear.

DIACONIS, P. and YLVISAKER, D. (1979). Conjugate Priors for Exponential Families. *Ann. Statist.* **7**, 269-281.

DIACONIS, P. and ZABELL, S. (1982). Updating Subjective Probability. *J. Amer. Statist. Assoc.,* 822-830.

DOOB, J. (1953). *Stochastic Processes.* New York: Wiley.

DUBINS, L. (1983). Bernstein-Like Polynomial Approximation in Higher Dimensions. *Zeit. Wahr.* (to appear).

FERGUSON, T. (1955). On the Existence of Linear Regression in Linear Structural Relations. *Univ. Calif. Publications in Statistics,* **2**, 143-166.

GILAT, D. (1971). Some Conditions Under Which Two Random Variables are Equal Almost Surely and a Simple Proof of a Theorem of Chung and Fuchs. *Ann. Math. Statist.,* **42**, 1647-1665.

GIRSHICK, M.A. and SAVAGE, L.J. (1951). Bayes and Minimax Estimates for quadratic loss functions. *Proceeding of the Second Berkeley Symposium,* 53-74.

GOLDSTEIN, M. (1975). Uniqueness Relations for Linear Posterior Expectations. *J. Roy. Statist. Soc. B,* **37**, 402-405.

GOOD, I.J. (1971). 46,656 Varieties of Bayesians. Letter in *Amer. Statist.,* **25**, 62-63.

JEFFREY, R. (1982). *The Logic of Decision.* Chicago University of Chicago Press.

JEWEL, W.S. (1974). Credible Means are Exact Bayesian for Exponential Families. *Astin Bull.* **8**, 77-90.

— (1975). Regularity Conditions for Exact Credibility. *Astin Bull.,* **8**, 336-341.

— (1982). Enriched Multinormal Priors Revisited. *Tech. Report,* University of California, Berkeley.

KADANE, J. and CHUANG, D. (1978). Stable Decision Problems. *Ann. Statist.,* **6**, 1095-1110.

KAGAN, A.M., LINNICK, Yu.V., and RAO, C.R. (1973). *Characterization Problems in Mathematical Statistics.* New York: Wiley.

LANDAU, H. (1980). The Classical Moment Problem: Hilbertian Proofs. *J. Funct. Anal.,* **38**, 255-272.

LORENTZ, G.G. (1966). *Bernstein Polynomials.* University of Toronto Press, Toronto.

LORENTZ, H.C. (1980). *Approximation of Functions.* New York: Holt, Rinehart and Winston.

MORGAN, R.I. (1970). A Class of Conjugate Prior Distributions. *Tech. Rep.* Dep. of Statistics, University of Missouri, Columbia, MO.

NOVICK, M.R. and HALL, W.J. (1965). A Bayesian Indifference Procedure. *J. Amer. Statist. Assoc.* **60**, 1104-1117.

RAMSEY, J.O. and NOVICK, M.R. (1980). PLU Robust Bayesian Decision Theory. *J. Amer. Statist. Assoc.,* **75**, 901-907.

SHAFER, G. (1981). Constructive Probability. *Synthese,* **48**, 1-60.

STRAUCH, R.E. (1965). Conditional Expectations of Random Variables Without Expectations. *Ann. Math. Statist.,* **36**, 1556-1559.

STEIN, C. (1965). Approximation of Improper Prior Measures by Prior Probability Measures. *Bernoulli, Bayes, Laplace,* (J. Neyman, L. LeCam eds.), New York: Springer-Verlag.

ZABELL, S. (1982). W.E. Johnson's "Sufficientness" Postulate. *Ann. Statist.,* **10**, 1091-1099.

ZELLNER, A. (1980). On Bayesian Regression Analysis with *g*-Prior Distributions. *Tech. Rep.,* Graduate School Business, University of Chicago.

DISCUSSION

F.J. GIRON (*Universidad de Málaga, Spain*)

This paper deals with two main issues. The first one is an extension of the notion of conjugate families outside the exponential family, following the suggestion of Goel and DeGroot (1980) that "... perhaps one should use linear posterior expectation as the defining property of a conjugate distribution'' in view of a previous result of the authors in his 1979 paper.

The other important result is theorem 5, namely, that *any* prior can be approximated, in the sense of the weak star topology, by a finite mixture of conjugate distributions.

As stated in the introduction, the authors carry through the first point for the case of a location parameter, opening the way for further generalizations to scale parameters. They succeed in their attempt, for which I congratulate them, and develop a fine theory for the location parameter problem.

However, I want to make two minor points. Actually, the first one cannot really be considered an objection at all. In fact, in their setting for the location parameter problem, posterior linearity only holds for samples of size 1. Of course, outside the regular exponential family, sufficient statistics are not to be expected, so that posterior linearity can only hold for the case X is normally distributed, as proven in Goel and DeGroot (*loc. cit.*).

On the other hand, their definition of conjugate prior for the location parameter problem through F, in terms of linear posterior expectation, broadens the concept of conjugate distributions.

The second point is of a more fundamental nature. Consider the following examples, for which we follow the author's notation.

Example 1. Let X be uniformly distributed in [0,1], θ uniformly distributed in [−1,2], independently of X and let $Z = X + \theta$, so that $Z|\theta$ is uniform in $[\theta, \theta + 1]$,

Then the likelihood function $1(\theta|z)$ is

$$l(\theta|z) = I_{[z-1,z]}(\theta),$$

and the posterior distribution of θ given $Z=z$ is uniform in the interval [max $(z-1, -1)$, min $(z, 2)$], so that the posterior expectation

$$E[\theta|z] = \tfrac{1}{2}\{\max(z-1, -1) + \min(z, 2)\}.$$

Example 2. Suppose X follows an exponential distribution ($\lambda = 1$) and is independently distributed in θ, which is distributed as a right-truncated exponential with density

$$p(\theta) = \begin{cases} \mu e^{-\mu(\theta_o-\theta)} & \text{if } \theta \le \theta_o \\ 0 & \text{otherwise.} \end{cases}$$

If $Z = X + \theta$, then the conditional density of the observable Z given θ

$$f(z|\theta) = \begin{cases} e^{-(z-\theta)} & \text{if } z \ge \theta \\ 0 & \text{otherwise.} \end{cases}$$

Therefore, the posterior distribution of θ given $Z=z$ is

$$p(\theta|z) = \begin{cases} (\mu+1)\, e^{-(\mu+1)[\min(z,\theta_o)-\theta]} & \text{if } \theta \le \min(z,\theta_o) \\ 0 & \text{otherwise,} \end{cases}$$

and the posterior expectation is

$$E[\theta/z] = \min(z,\theta_o) - \frac{1}{1+\mu}.$$

In the first example, the posterior distribution is always uniform so that, according to the classical definition, the family of uniform priors is a conjugate family. However, the posterior expectation is *not* linear in the observation but *piece-wise linear.* Also note that in this case the boundary values of a (0 and 1) are admissible, and this does not contradict lemma 1 in section 3 of the paper.

The second example exhibits the same behaviour, so that the right-truncated exponential family is conjugate and the posterior expectation is piece-wise linear and continuous.

As a suggestion, piece-wise posterior linearity might serve as a basis for a definition of a conjugate family when the sample space depends on the parameter.

Turning now to the second important result, namely, theorem 5, I had found myself that for a few special families, such as normal, beta, gamma, etc., the above mentioned result was true using the same technique of approximating the constants, or degenerate, distributions by a suitable member of the conjugate family. Yet, I lacked a general proof of the result, which the authors provide in the paper.

This result is also extended in section 3 to the case of a location parameter problem when the underlying distribution is infinitely divisible.

Incidentally, note that if piece-wise linearity is considered, as in examples 1 and 2, then the family of all finite mixtures of uniform priors and the family of finite mixtures of right-truncated, two parameter, exponentials are both weak star dense in the set of all probability measures on the real line.

This theoretical result is also important from a practical view-point as follows from considerations take from the introductory example and from the following result which complements theorem 5.

With a slight change of notation, theorem 5 can be rewritten in the following form

Theorem 5. Let $\mathbf{b}$ be a conjugate family for the likelihood $f(x|\theta)$. Then for any prior Q on θ and $\epsilon > 0$, there exists weights w_i and $P_i \in \mathbf{b}$ such that, if $R = \sum_{i=1}^{N} w_i P_i$, $d_p(Q, R) < \epsilon$.

Furthermore if we denote by Q_x, R_x, P_{ix} the posterior distributions of Q, R, P_i, respectively, then there exist $\delta > 0$ such that

$$d_p(Q_x, R_x) < \delta,$$

where

$$R_x(d\theta) = \sum_{i=1}^{N} w_i' P_{ix}(d\theta) \qquad ; P_{ix} \in \mathbf{b},$$

$$w_i' = \frac{w_i f(x|P_i)}{\sum_{i=1}^{N} w_i f(x|P_i)},$$

and $f(x|P_i) = \int_\theta f(x|\theta)\, P_i(d\theta)$ is the predictive density of x given P_i.

D.V. LINDLEY (*Somerset, UK*)

I would like to use this interesting paper as an excuse to raise the question of what modern mathematics has to say for us statisticians. Diaconis mentions a result in the folklore. I look at this; it does not seem familiar but easily yields to integration by parts. Later we are told that this is correct "up to rigour". But what is rigour? Is integration by parts not rigorous? At one point in my hand-waving proof a limit has to be shown to be zero. Diaconis has to show a set is open. Is the difference material? I have recently read Morris

Kline's "Loss of Certainty" and find that he argues that mathematicians delude themselves in thinking their methods are totally rigorous, and that mathematics should relate more to the real world. Is it true that modern mathematics is just "papering up the cracks" or has it something original to say to us? I have heard John Hammersley express similar doubts concerning one branch, functional analysis. Perhaps these remarks of mine are simply caused by a generation gap and I am too wedded to the mathematics of 1940's Cambridge to understand 1980's Californian. But it would be good to have examples of original contributions to statistics that owe their ideas to later mathematics than mine. (One example, given to me after the discussion, was Stein's brilliant demonstration of the inadmissibility of the sample mean. This result arises from a desire to provide a rigorous proof of an "obvious" result, subsequently shown to be false, but nowhere appears to use mathematics that was not available to Neyman and Fisher.)

Take the case of the exponential family considered in this paper. The original idea was almost simultaneously produced by Pitman, Darmois and Koopmans. Pitman's paper was beautiful and unsophisticated. Koopmans' was heavy, more rigorous and indigestible. Darmois' I cannot remember, but knowing something of French mathematics of the period, it was probably both elegant and rigorous (by the standards of the day). Conjugate families are due to Wetherill and to Raiffa & Schlaifer. Does modern mathematics add substantially to these ideas? The converse type of result, that linearity implies conjugacy, is important because of the restriction it places on the usefulness of the linear tool. (Goel mentioned the importance of this, but does it affect the highly succesful GLIM programme?)

Techniques are not as important as concepts and originality. I am for a bit of "hand waving": it rarely lets one down and in the hands of brilliant people is almost always reliable. But I ask the question in a neutral spirit: has modern mathematics a substantial contribution to make to statistical science? The question is surely an important one. Forgive me, Persi, for using your paper as an excuse for raising the question. You are perhaps the best person to answer it.

J.M. BERNARDO (*Universidad de Valencia*)

The idea that most prior opinions may be well approximated by mixtures of conjugate priors has been tacitily accepted for a long time; I remember, for instance, that the issue was discussed in one of the London University Friday Statistical Seminars back in 1975. However, it is always dangerous in mathematics to take for granted 'obvious' results, ... which too often are later shown not to be true; thus, the precise statements given in this paper should be most welcome.

On the other hand, the distance between a theorem of existence and a constructive procedure to find the solution may be very large. We now *know* that opinions *may* be well described by some mixture of conjugate priors; I would like the author to comment on the procedures, as general as possible, which he suggests for *actually* expressing opinions in terms of such mixtures.

J. BERGER (*Purdue University*)

My comments are based on a (possibly faulty) memory of the very stimulating talk and ensuing lively discussion. Concerning the plethora of interesting results presented, I have a question, a picky comment, and a more serious comment.

The question concerns the analysis of the coin spinning experiment, in which Professor Diaconis used a mixture of beta priors for the probability p, that a spun coin would land heads, apparently as an illustration of a situation in which a mixture prior was needed. The motivation for this was not completely clear to me, however. Presumably, the probability of a head would be related to some physical difference, d (in, say, weights), between the two sides of the coin. For a completely unknown coin, the prior distribution of p should then have a similar shape to the distribution (among coins) of d, and I wonder why a mixture model for this would seem necessary? I am not trying to argue that single conjugate priors are sufficient - far from it; because of an inherent distrust of conjugate priors, for robustness reasons, I would like to see this as another example where conjugate priors are clearly inappropriate.

Professor Diaconis also discussed using finite mixtures of conjugate priors to approximate an arbitrary prior, and presented some very nice theorems to the effect that this could be done to any desired degree of accuracy. This possibility is being raised in many Bayesian quarters these days, due to the calculational ease of working with finite mixtures of conjugate priors. There is a very serious issue concerning such an approximation, however, namely the issue of whether this good approximation to the prior ensures that the posterior will also be well approximated. I think the answer, in general, is no.

To see this, consider the simple situation where $X \sim N(\theta,1)$, and my "true" prior, π_T, is Cauchy (0,1). Let

$$\pi_A = \sum_{i=1}^{m} \lambda_i \pi_i, \qquad \pi_i \text{ being } N(\mu_i, A_i),$$

be an approximating finite mixture of conjugate normal priors. Now it is easy to show that the posterior corresponding to π_T can be drastically different than that corresponding to π_A for specific observations x. Indeed, as $x \to \infty$, $\pi_T(\theta|x)$ becomes approximately equal to a $N(x,1)$ distribution while $\pi_A(\theta|x)$ becomes approximately equal to a $N(\mu(x),\varrho)$ distribution, where

$$\varrho = \frac{A^*}{1+A^*}, \qquad \mu(x) = x - \frac{1}{1+A^*}(x-\mu^*),$$

$$A^* = \max_i \{A_i\}, \qquad \text{and } \mu^* = \max_{i:A_i=A^*} \{\mu_i\}.$$

These distributions clearly differ drastically for large x.

Perhaps even more telling is that the approximation can be bad, from a posterior viewpoint, "on the average". For instance, if one were trying to estimate θ under squared error loss, the posterior mean would be used; call it $\delta_T(x)$ or $\delta_A(x)$ for π_T or π_A, respectively. Also let m_T denote the marginal distribution of X, and note that, as $x \to \infty$, a simple Taylor's expansion shows that

$$m_T(x) = \frac{1}{\pi(1+x^2)}(1+o(1)).$$

Now the "average" performance of δ_T is

$$\int\int (\theta-\delta_T(x))^2 \pi_T(\theta|x)\, m_T(x)\, d\theta\, dx < 1$$

(since δ_T must have smaller Bayes risk than the minimax rule), while the average performance of δ_A (with respect to the true prior) is

$$\int\int (\theta-\delta_A(x))^2 \pi_T(\theta|x)\, m_T(x)\, d\theta\, dx$$

$$\geq \int_K^{\infty} \int_{-\infty}^{\infty} (\theta-\delta_A(x))^2 \pi_T(\theta|x)\, m_T(x)\, d\theta\, dx$$

$$\cong \int_K^\infty \int_{-\infty}^\infty (\theta - \mu(x))^2 \frac{1}{\sqrt{2\pi}} e^{-1/2(\theta-x)^2} d\theta \frac{1}{\pi(1+x^2)} dx \quad \text{(for large } K\text{)}$$

$$\geq \int_K^\infty \frac{(x-\mu^*)^2}{(1+A^*)^2} \cdot \frac{1}{\pi(1+x^2)} dx = \infty .$$

This is just another illustration that the "tail" of the prior can be crucial to Bayesian robustness, and there is simply no way to appropriately approximate a polynomial tail (such as that of π_T) by a finite mixture of exponential tails (such as in π_A).

The final, nitpicky, comment I had concerned the "definition" of conjugate priors given by Professor Diaconis, essentially that given in Diaconis and Ylvisaker (1979). Although I appreciate the elegant characterization that this definition allows, I object to the fact that, if X has a p-variate normal distribution with unknown mean vector θ and known covariance matrix Σ, then a p-variate normal prior for θ (with mean vector μ and covariance matrix A not commuting with Σ) does not qualify under the definition as a conjugate prior. This is such an important situation and so "conjugate" in all other senses, that the definition of Diaconis and Ylvisaker should probably be called something else.

A.F.M. SMITH (*University of Nottingham*)

Would Diaconis like to comment further on the alternative approach to approximating arbitrary priors by mixtures of natural conjugate priors that was put forward recently by Dalal and Hall (1983)?

REPLY TO THE DISCUSSION

To Professor Bernardo:

I do not believe there are general rules for expressing prior belief. Each real problem has to be thought about on its own merits. Of course we have examples of previous analyses, familiar models, and notions of symmetry. We can try to break our problem into pieces such that each piece is similar to a familiar problem, where risks and benefits of particular assumptions are well understood. Mixtures are useful when there is a partition and a believable prior specification given each piece of the partition. Not all problems are amenable to such a decomposition, but I think there are enough examples to justify the present analysis.

To Professor Berger:

The point of the coin spinning example is that the coin is not "completely unknown". It was selected from a population displaying strong biases in both directions (presumably because of physical differences in the way the edges of coins are finished). A bimodal prior is forced by experience, reflecting (roughly) the results of dozens of coins spun hundreds of times each, together with a bit of physics.

I am delighted with your examples. For me, they emphasize the need for careful statements and honest proofs. They *also* show how "tails can wag dogs" and encourage me to find a metric for the weak star topology which is not so affected by what happens off at infinity. I've put some comments toward this end in the body of the paper.

On your comment about normal priors: as indicated in the section on 'matrices' I believe the only time we can have matrix scaling is for normal location problems; here, as elsewhere, the normal is special and I'd prefer to have a unified theory of exponential families and call the priors "augmented conjugate" as Jewel does.

To Professor Giron

Our paper tries to characterize the usual notion of conjugate priors and shows how, sometimes, any prior can be well approximated by mixtures. This works reasonably well for exponential families and less generally for location (and other families). Your examples seem like just the direction to head in: find tractable, understandable classes of priors and show that we can approximate any prior.

To Professor Lindley:

Why do I try to prove theorems instead of working in the older tradition of "+..."? Three reasons are correctness, communication, and aesthetics. Let me elaborate: On correctness - given my choice, I'd rather be "brilliant" than careful. Given my limitations, I simply can't see another route to "getting it right" than trying it out in real examples and trying to do the mathematics carefully. There are certainly great examples of the "+..." school - you don't even have to be Bayesian (e.g., Fisher or Cox) but there are so many more examples where the magic insight is missing and all one is left with is useless heuristics.

On communication - one of my delights at the Valencia meetings was meeting many young statisticians who learned modern mathematical statistics. When push come to shove, we would constantly revert back to standard mathematical usage to clarify discussions. This can be contrasted with trying to figure out an article written in "+..." language: there we find "theorem" with often no assumptions or clear conclusions; I find it frustrating to try to figure out what someone else has in mind without hints, and feel that someone should do their homework before publishing. One of the achievements of modern mathematical statistics is a common universal language, free of the mysteries of "inference".

On aesthetics - I find modern mathematics beautiful. The best work in mathematical statistics has the same appeal. So, if you like, I prove theorems because it makes me happy.

Professor Lindley asks for contributions of modern mathematics to statistical science. Many constructions that implicitly involve infinite dimensions are nowadays correctly and routinely used after very rocky beginnings. For example, robustness with its notion of influence curve (a function space derivative), the whole cannon of techniques from weak convergence and invariance principles as used in connection with Kaplan-Meier and Cox modelling.

Turning to Bayesian statistics we have the complete class theorems which have helped bridge the gap between decision theoretic and Bayesian methods - these are impossible without some topology and functional analysis. The work of Savage and de Finetti (and many of the rest of us) on de Finetti's theorem rests solidly on functional analysis - indeed the Hewitt-Savage paper is often cited by functional analysts as a first "Choquet theorem".

Finally, let us consider what is perhaps the finest synthesis of the Bayesian and decision theoretic schools: Stein's results and the whole sea of related work. It's true, that nowadays there are reasonably simple variants of Stein's result. *But* the result was born, bred, nurtured and raised in the heart of complete class theorems and invariance theory. The many far reaching extensions by Berger, Brown, Efron, Morris and a host of others use tools from every area of mathematics - from potential theory, through difussion through differential equations. One of the profound results of this work is an intimate connection between all of these areas, so tools developed one place can be used in any other. Thanks to this work many other simultaneous estimation problems (e.g., many Poisson variables) can be handled and thought about as they arise, without having to consult a few Guru's.

One virtue of mathematical language is its power to unify. Take Professor Lindley's question about putting the assumption "θ is open" in place of "the bounday terms vanish", in Theorem 3. The openness of θ also serves in Theorems 2 and 4; without it, we have a spate of extra conditions, one for each theorem; with it a modicum of order. There is also a practical reason: I believe openness is easier to check in several dimensions, while understanding the analog if integration by parts, Stoke's theorem, is hard.

As Professor Lindley points out, brilliant work can be done without modern mathematics. Further, the use of mathematical machinery does not guarantee correctness. To take the case suggested by Professor Lindley, consider the Koopman-Pitman-Darmois theorem: roughly, if a statistical problem admits finite dimensional sufficient statistics, then it is an exponential family. To make this precise, one needs a precise notion of "statistic" because there are continuous 1-1 functions from $\mathbb{R}^n$ into $\mathbb{R}$, some smoothness assumptions are needed: but how much smoothness? The earlier writers didn't worry about this in too detailed a way, so that it is difficult to determine if statistics involving absolute values are admissible. Later writers, like Dynkin and Brown did worry about it, and used substantial mathematical tools. Yet major theorems in their papers are simply wrong. The matter seems to be usefully settled, see Hipp (1974) which contains the relevant references.

To Professor Smith:

After our manuscript was finished, the very relevant paper by Dalal and Hall (1983) appeared. This contains a more general framework and some of the examples (Bernstein-like approximations) are developed in great detail. Additionally, they treat some of the problems suggested in section 4. Dalal has used mixtures of priors in other ways over the past 10 years (see their bibliography).

It would be interesting to trace the history of the use of mixtures. At the conference, George Barnard said he had used mixtures of priors in the 1940's, and several other groups seem to have discussed the possibility. Most of the results presented here were developed in 1974-75; they were written up because of the systematic prodding of Jay Kadane.

REFERENCES IN THE DISCUSSION

GOEL, P.K. and DEGROOT, M.H. (1980). Only normal distribution have linear posterior expectations in linear regression. *J. Amer. Statist. Assoc.* **75**, 895-900.

HIPP, C. (1974). Sufficient statistics and exponential families, *Ann. Statist.* **2**, 1283-1292.

BAYESIAN STATISTICS 2, pp. 157-182
J.M. Bernardo, M.H. DeGroot, D.V. Lindley, A.F.M. Smith (Eds.)

Direct Subjective-Probability Modelling Using Ellipsoidal Distributions

JAMES M. DICKEY and CHONG-HONG CHEN
State University of New York at Albany

SUMMARY

Interactive elicitation and fitting methods previously proposed for assesing Bayesian prior distributions in normal linear sampling are here extended to assessment of general subjective-probability models involving ellipsoidal distributions and concomitant variables without explicit sampling models. Ellipsoidal (sometimes called "elliptical") distributions are related to their marginal and conditional distributions in simple ways that allow assessment through univariate quantiles. Dependence on concomitant variables is modeled through assessable functional forms in the ellipsoidal location and scale parameters. New families of tractable spherical and ellipsoidal distributions are given based on Carlson's two-way multiple hypergeometric function. These generalize the multivariate normal and multivariate-t distributions. Assessment methods are developed in detail for scale mixtures of multivariate normal distributions.

Keywords: ASSESSMENT; ELICITATION; ELLIPSOIDAL DISTRIBUTIONS; ELLIPTICAL DISTRIBUTIONS; EXPERT OPINION; MULTIPLE HYPERGEOMETRIC FUNCTIONS; SUBJECTIVE PROBABILITY.

1. INTRODUCTION

It was asserted in the first Valencia conference (Dickey 1980) that subjective-probability modelling of expert opinion is more widely applicable than Bayes-theorem inference methods. Bayes-theorem methods require the expert assessment of "prior" probabilities, that is, prior to modeled statistical data. Often, however, important decisions must be based *primarily* on beliefs of experts, for example when proper sample data are not available, or not feasible to obtain. Commonly, the available data refer to something rather different than the context of the decision problem at hand. Or, there are times when it cannot honestly be said that the data are sampled according to a recognized formal statistical model.

In the end, it must be admitted that a *responsible* decision maker (same person here as the expert) will prefer to use his or her actual posterior opinion to some formal Bayes-theorem-generated posterior distribution, *especially* when the final opinion disagrees with the Bayes thing. Of course, the opinion must be formed in light of all available data and all economically obtainable data, and in light of feasible Bayes-theorem calculations and other probability calculations. It is difficult to include in a formal likelihood function all relevant information imparted by a body of data and its reporting process (Dawid and Dickey, 1977). It is difficult, posterior to data, to assess a prior distribution by pretending to consult ones state of mind prior to experiencing the report of the data. And then, what is to be done with new ideas and realizations that occur to the decision maker during the time interval since he has seen the report? In short, one does the best one can without shackling ones good sense by little-world theories of prior procedure optimality. (For further discussion of the incompleteness of Bayes-theorem inference, see Goldstein, 1981, 1983a,b.)

To the present authors, it seems good sense to express expert opinions in the form of subjective-probability models, assessed interactively and assessed *posterior* to all relevant data. Interactive posterior assessment means that the expert can look informally at Bayes-theorem and other probability calculations before and during the assessment process. What is expressed is informed personal opinion concerning unknown matters of fact. This is articulated and expressed by constructing and reporting a probability model, which can then be used to give subjective probabilities of events, point and interval predictions (jointly and marginally), dependence of opinion on concomitant and factor variables (e.g. using subjective response surfaces), and conditional opinions given new information, such as updating to account for further statistical data without expert reconsideration.

Two objections raised recently concern the value of a subjective-probability model and its assessment. First, does such a model lend undeserved authority and respectability to mere opinion? This is similar to the objection to teaching courses in statistics that they impart tools useful for telling lies; whereas, another attitude to have is that such courses can help make students less susceptible to being lied to. Yes, tools can be misused, but that is not the intention. Since real decisions are nearly always based eventually on opinions, it seems reasonable to express opinion with precision and in detail. Of course, in line with modern scientific dicta, experiments should be fully and objectively reported along with posterior opinion, and not irretrievably confounded with opinion.

A second objection has been expressed, "What is to prevent the expert from fooling himself?" Ultimately, of course, the answer is, "No guarantee". The intention, however, is that the interactive assessment procedure will expose the person to coherence checks and will stimulate detailed reasoning, explicit statement of assumptions, consideration of implications, and inquiry concerning justifications. Participation in the assessment process makes it harder to achieve self-deception, and perhaps deception of others if details of the assessment process are made public.

Objective accuracy of opinions, is, of course, another matter. Opinions known to be right are called knowledge; they are held by gods, and not men. However, the replacement of tacit assumptions by expressive subjective-probability models, which can be examined closely, promises to improve expert predictions and hence decisions.

Unfortunately, there is not much alternative methodology to compete with subjective-probability modelling, particularly for a single expert decision maker. The Brunswik lens model in psychology has led to methods using subjective response surfaces (Hammond 1980, Cook 1980), but without the benefit of subjective probability to provide uncertainty intervals accompanying the response-surface predictions. "Belief functions" put forth by Shafer (1976) have had an attraction for frequentists because of their promise to minimize the amount of subjective assessment. The calculations involved seem rather arbitrary. (For an important unanswered question, see Good 1982.) Lindley, Tversky and Brown (1979) report important work, close in spirit to the present paper, and concerned with general models. See also Lindley (1981, 1982, 1983).

Dickey (1980) suggested that the methods of Kadane *et al* (1980), for assessing Bayesian prior distributions in normal linear sampling, can be extended to more general subjective-probability models using ellipsoidal (or "elliptical") distributions, namely location-scale transforms of spherical (or "isotropic") distributions. (Spherical distributions have spherical or rotational symmetry.) Since a Bayesian multivariate-t predictive distribution was to be assessed first, in effect, it was suggested that the methods can be extended to other ellipsoidal families in place of the multivariate-t without explicit reference to an imbedded sampling model. The purpose of the present paper is to develop this idea.

Rotational symmetry in two dimensions means that the joint uncertainty concerning x_1, x_2 is such that the same uncertainty applies literally to $\tilde{x}_1 = x_1 \cos\theta + x_2 \sin\theta$ and $\tilde{x}_2 = -x_1 \sin\theta + x_2 \cos\theta$. In particular, x_1 has the same marginal subjective distribution as $\tilde{x}_1$ (a relation denoted by $x_1 \sim \tilde{x}_1$); for example, for $\tilde{x}_1 = \sqrt{2}(x_1+x_2)/2$. In n dimensions, each $x_i \sim n^{1/2}\bar{x}$. Hill (1969) argues that spherical distributions are relevant to subjective uncertainty. By change of variable, such arguments apply to ellipsoidal distributions.

It may be that ellipsoidal and spherical distributions are also useful as objective sampling models. Kariya and Eaton (1977), Jensen and Good (1981), and Tyler (1982) take such views. The emphasis in the present paper is on the modelling of subjective uncertainty, but methods of subjective assessment can often be easily converted to methods of fitting to sample data. For example, one might replace here the elicited subjective quantiles by sample quantiles.

Three levels of modelling will be involved. The first two concern the assessment of an ellipsoidal distribution. First, one must determine an underlying standard spherical distribution, of which the ellipsoidal distribution is a location-scale transform. This distribution can be accessed through the shape of the univariate margins of the ellipsoidal distribution. Secondly, the location vector and scale matrix of the ellipsoidal distribution are assessed through the location and scale parameters of univariate marginal and conditional distributions of the ellipsoidal distribution. Thirdly, functional dependence of the ellipsoidal location and scale parameters on additional concomitant and factor variables can be assessed.

Properties of spherical distributions relevant for their assessment will be developed here in Section 2, expanding on aspects of Kelker (1968, 1970). Tractable new families of spherical distributions for which assessment is feasible are given in the Appendix. Ellipsoidal distributions are developed in Section 3, and assessment methods are outlined, based on the univariate marginal and conditional distributions, in Section 4. Detailed methods are given for transforms of scale mixtures of normal (or similarly tractable) spherical distributions in Section 5.

2. SPHERICAL DISTRIBUTIONS

A *spherical* probability distribution is one that is invariant under rotations and reflections. Define the random vector $\mathbf{x} = (x_1,\ldots,x_n)'$ to be *spherical* if for every (nonrandom) matrix A satisfying $A'A = I$,

$$\mathbf{x} \sim A\mathbf{x} \quad . \tag{2.1}$$

This property is denoted by

$$\mathbf{x} \sim S(\mathbf{0}, I) \quad . \tag{2.2}$$

Equivalently, if $\mathbf{x}$ has a density, the contours of the density are zero-centered spheres. The probability measure for $\mathbf{x}$ decomposes into a mixture of relative-surface-area measures: $\mathbf{x} \sim S(\mathbf{0}, I)$ if and only if for the distribution P of $\mathbf{x}$,

$$P = \int P_r \, dP(r) \tag{2.3}$$

where P_r is supported on the sphere surface $x_1^2+\ldots+x_n^2=r^2$, over which P_r is uniform, $0\leq r<\infty$. In polar coordinates, P_r is a version of the conditional distribution of the angle variables given the radius r. These variables are then independent of r. The following theorem (expanding on a result of Kariya and Eaton 1977) will develop these conditional distributions in another form.

Define the usual Dirichlet family of distributions, $\mathbf{u}\sim$ Dirichlet $(\mathbf{b})$, having the density in any n-1 coordinates of the probability vector $\mathbf{u}=(u_1,\ldots,u_n)$, where each $u_i\geq 0$ and $u_+\equiv u_1+\ldots+u_n=1$,

$$p(\mathbf{u}) = B(\mathbf{b})^{-1}\,\Pi_1^n\, u_i^{b_i-1} \quad , \tag{2.4}$$

where $B(\mathbf{b}) = (\Pi_1^n\,\Gamma(b_i))/\Gamma(b_+)$, where $b_+\equiv b_1+\ldots+b_n$. The general moment is $E\,\Pi_1^n u_i^{c_i} =$ $= B(\mathbf{b}+\mathbf{c})/B(\mathbf{b})$. For the familiar ($v\sim$Beta$(b_1,b_2)$, $(v,1-v)\sim$ Dirichlet (b_1,b_2).

Lemma 2.1. If the random vector $\mathbf{x}=(x_1,\ldots,x_n)'$ is uniformly distributed on the hyperball $\mathbf{x}'\mathbf{x} \leq R^2$, then the random quantity $r=(\mathbf{x}'\mathbf{x})^{1/2}$ and the random vector $\mathbf{u}=(x_1^2,\ldots,x_n^2)'/r^2$ are independently distributed according to

$$r/R \sim \mathrm{Beta}(n,1) \quad , \tag{2.5}$$

$$\mathbf{u} \sim \mathrm{Dirichlet}(\tfrac{1}{2}\,\mathbf{1}_n) \quad , \tag{2.6}$$

where $\mathbf{1}_n=(1,\ldots,1)'$. Note that then $r^2/R^2 \sim \mathrm{Beta}(\frac{1}{2}\,n,\,1)$ and $(x_1^2+\ldots,x_k^2)/r^2=u_1+\ldots+u_k\sim$ $\mathrm{Beta}(\frac{1}{2}\,k,\,\frac{1}{2}(n-k))$, $k=1,\ldots,n$. □

Proof. Change variable from $\mathbf{x}$ to $(x_1,\ldots,x_{n-1},r)$, and then to $(u_1,\ldots,u_{n-1},r)$.

Denote a coordinatewise product of vectors by $\delta\times\mathbf{a} = (\delta_1 a_1,\ldots,\delta_n a_n)$, and a coordinatewise square root by $\mathbf{u}^{1/2}=(u_1^{1/2},\ldots,u_n^{1/2})'$.

Theorem 2.2. The random vector $\mathbf{x}$ has a spherical distribution if and only if it can be expressed in terms of the independent random quantities r,δ,u,

$$\mathbf{x} = r\cdot(\delta\times\mathbf{u}^{1/2}) \quad , \tag{2.7}$$

where $\mathbf{u}\sim$ Dirichlet $(\frac{1}{2}\mathbf{1}_n)$, and each

$$\delta_i = \begin{cases} -1 \text{ with prob. } \frac{1}{2} \\ \;\;\,1 \text{ with prob. } \frac{1}{2} \quad , \end{cases} \tag{2.8}$$

independently for $i=1,\ldots,n$. □

Proof. By (2.3) and the independence of $\mathbf{u}$ and r, (2.6) holds for all spherical distributions. The conditional distribution of $\mathbf{x}/r=\delta\times\mathbf{u}^{1/2}$ given r is obtained from (2.6) and (2.8). This is independent of r.

Corollary 2.3. The moments of spherical $\mathbf{x}$, if they exist, can be expressed in terms of a one-dimensional integral. For even integers $2s_i\geq 0$, $i=1,\ldots,n$,

$$E\,\Pi_1^n\,x_i^{2s_i} \tag{2.9}$$

$$= \lambda_{n,2s_+}B(\tfrac{1}{2}\mathbf{1}_n+\mathbf{s})/B(\tfrac{1}{2}\mathbf{1}_n)$$

where $s_+ \equiv s_1+\ldots+s_n$ and for any value s,

$$\lambda_{n,2s} = E(r^{2s}) \quad . \tag{2.10}$$

Whereas, this moment vanishes in the case of any odd integer power $2s_i \geq 0$. For example, the mean is $\mathbf{0}$ and

$$\mathrm{Var}(\mathbf{x}) = \lambda_{1,2} I_n \quad , \tag{2.11}$$

where $\lambda_{1,2} = E x_1^2 = \lambda_{n,2}/n$. Note that the moment (2.9) exists if $\lambda_{n,2s_+} \propto \lambda_{1,2s_+}$ is finite. □

If spherical $\mathbf{x}$ has a density it takes the form

$$g(\mathbf{x}) = \gamma_n(r^2) \quad , \tag{2.12}$$

for some function γ_n. The density of the radius itself will then be

$$p(r) = A_n r^{n-1} \gamma_n(r^2) \quad , \tag{2.13}$$

where $A_n r^{n-1}$ denotes the surface volume of the sphere $(\mathbf{x}'\mathbf{x})^{1/2} = r$, in which the surface for $r=1$ is

$$A_n = 2B(\tfrac{1}{2}1_n) = 2\pi^{n/2}/\Gamma(\tfrac{1}{2}n) \quad . \tag{2.14}$$

The corresponding relation for the density of r^2 is $q(r^2) = \frac{1}{2}A_n(r^2)^{n/2-1}\gamma_n(r^2)$. Both $q(r^2)$ and $p(r)$ (2.13) will be called *radial* distributions.

Much use will be made of this connection between the radial density of r or r^2 and the spherical density of $\mathbf{x}$. Note that, conversely, (2.13) yields the density of $\mathbf{x}$ from the density of r or r^2. Hence, spherical distributions can be characterized and studied in terms of their corresponding radial distributions.

2.1. *Spherical margins*

The characteristic function of a spherical distribution takes a special symmetric form. The random vector $\mathbf{x}$ is spherical if and only if its characteristic function satisfies

$$C_x(\mathbf{t}) = \psi(\mathbf{t}'\mathbf{t}) \tag{2.15}$$

for some function ψ of a real variable $\mathbf{t}'\mathbf{t}$.

A marginal distribution of a spherical distribution is also spherical, because of the form of its characteristic function, as follows. The joint form (2.15) has a simple relation to the characteristic function of the marginal distribution. For $\mathbf{x} = (\mathbf{x}_1', \mathbf{x}_2')'$, the subvector $\mathbf{x}_1$ has characteristic function,

$$C_{x_1}(\mathbf{t}_1) = \psi(\mathbf{t}_1'\mathbf{t}_1) \quad , \tag{2.16}$$

for the same function ψ as (2.15).

This does not mean, however, that the *density* function of $\mathbf{x}_1$ has the same form as the density function of $\mathbf{x}$, that is, marginal densities are not necessarily proportional to joint densities, as functions of the respective squared radii, apparently contrary to a statement by Muirhead (1982, p. 34). Indeed, if they are of the same form in such a way, then the distributions must be normal (Kelker 1968, Theorem 2.4). (See also Eaton 1981 and Kelker 1971.) The characteristic function and the density of a spherical distribution are related by a simple one-dimensional invertible Hankel transform (Lord 1954). However, the transform explicitly involves the dimensionality of the distribution as a parameter, and thus the density of $\mathbf{x}$ is not the same function of $\mathbf{x}'\mathbf{x}$ as the density of $\mathbf{x}_1$ is of $\mathbf{x}_1'\mathbf{x}_1$. Nor are they proportional.

Since the marginal distribution is itself spherical it can be developed in terms of its radial distribution, the distribution of r_1^2 for which $r^2 = r_1^2 + r_2^2$, where $r^2 = \mathbf{x}'\mathbf{x}$ and $r_i^2 = \mathbf{x}_i'\mathbf{x}_i$, $i=1,2$. The following formula relating the marginal density to the joint density appeared inconspicuously in Kelker (1968, p.16) and seems to have been ignored in much of the literature. (See, for example, Das Gupta *et al*, 1973, p.262, where the marginal density is left as a high-dimensional integral. However, see Eaton, 1981, for a sophisticated treatment.) The proof is immediate from the same change of variable as (2.7) and (2.13), but for n_2-dimensional $\mathbf{x}_2$ and r_2.

Theorem 2.4 If $\mathbf{x} = (\mathbf{x}_1', \mathbf{x}_2')'$ has a spherical distribution with joint density $g(\mathbf{x}) = \gamma_n(\mathbf{x}'\mathbf{x})$, the marginal distribution of $\mathbf{x}_1$ has a spherical density expressible as a one-dimensional integral,

$$g_1(\mathbf{x}_1) = \gamma_{n_1}(\mathbf{x}_1'\mathbf{x}_1) \quad , \tag{2.17}$$

where

$$\gamma_{n_1}(u_1) = \tfrac{1}{2} A_{n_2} \int_0^\infty u_2^{n_2/2-1} \gamma_n(u_1 + u_2) du_2 \quad . \tag{2.18}$$

In terms of the radial distributions $q_i(r_i^2)$, we have

$$q_1(u_1) = B(\tfrac{1}{2} n_1, \tfrac{1}{2} n_2)^{-1} \, u_1^{(n_1)/2-1} \int_0^\infty u_2^{(n_2)/2-1} (u_1 + u_2)^{-n/2+1} q(u_1 + u_2) du_2 \quad . \tag{2.19}$$

Note that $B(\tfrac{1}{2} n_1, \tfrac{1}{2} n_2)^{-1} = \Gamma(\tfrac{1}{2} n)/[\Gamma(\tfrac{1}{2} n_1)\Gamma(\tfrac{1}{2} n_2)] = \tfrac{1}{2} A_{n_1} \quad A_{n_2}/A_n$.□

The moments of the marginal distribution can be calculated, by Corollary 2.3, from its radial density (2.19). The moments of r_1 and r are related by

$$E(r_1^{2s}) = E(r^{2s}) \, . \, \frac{\Gamma(\tfrac{1}{2} n_1 + s)/\Gamma(\tfrac{1}{2} n_1)}{\Gamma(\tfrac{1}{2} n + s)/\Gamma(\tfrac{1}{2} n)} \quad , \tag{2.20}$$

by Corollary 2.3.

For an example illustrating Theorem 2.4, consider n-dimensional $\mathbf{x}$ uniform on the centered hyperball with radius R (context of Lemma 2.1). Then γ_n is constant and γ_{n_1} is proportional to $(R^2 - r_1^2)^{n_2}$. That is, by (2.13), $r_1^2/R^2 \sim \text{Beta}(\tfrac{1}{2} n_1, 1 + \tfrac{1}{2} n_2)$. Thus, the subvector $\mathbf{x}_1$ is not uniformly distributed.

Kelker (1970) expresses a spherical density of dimension n as proportional to a simple derivative of a marginal density of dimension $n-2$,

$$\gamma_n(u) = -\pi^{-1}\gamma_{n-2}'(u), \qquad a.e. u > 0 \quad . \tag{2.21}$$

In principle, this could be used to obtain odd-dimensional marginal distributions from an assessed univariate margin.

The following straightforward result will be very important for the theory of ellipsoidal distributions and their assessment. Indeed, ellipsoidal distributions are, themselves, the distributions of multivariate affine linear forms, the univariate margins of which are featured in the following theorem.

Theorem 2.5. All the coordinate marginal distributions of a given spherical distribution are identical; that is, $\mathbf{x} \sim S(\mathbf{0}, I)$ implies

$$x_i \sim x_1 \quad , \tag{2.22}$$

$i = 1, \ldots, n$. This is the same as the distribution of any homogeneous linear form whose coefficient vector has unit length. Furthermore, any affine linear form has a distribution which is a location-scale version of this same standard marginal distribution. To whit,

$$a + \mathbf{b}'\mathbf{x} \sim a + (\mathbf{b}'\mathbf{b})^{1/2} x_1 \quad . \; \square \tag{2.23}$$

We shall refer to the distribution of x_1 in (2.22), (2.23) as the *standard univariate margin*. We have seen how its density and moments are related to those of the vector $\mathbf{x}$.

2.2. *Spherical conditionals*

The conditional distributions of subvectors of a spherical distribution are also spherical. That is, they are defined by their radial distributions.

Theorem 2.6. If $\mathbf{x} = (\mathbf{x}_1', \mathbf{x}_2')'$ has a spherical distribution with joint density $\gamma_n(\mathbf{x}'\mathbf{x})$, the conditional distribution of $\mathbf{x}_2$ given $\mathbf{x}_1$ is spherical with density

$$g_{2\cdot 1}(\mathbf{x}_2 | \mathbf{x}_1) = \tilde{\gamma}_{n_2}(\mathbf{x}_2'\mathbf{x}_2; \mathbf{x}_1'\mathbf{x}_1)$$
$$= \gamma_n(\mathbf{x}_1'\mathbf{x}_1 + \mathbf{x}_2'\mathbf{x}_2)/\gamma_{n_1}(\mathbf{x}_1'\mathbf{x}_1) \quad . \tag{2.24}$$

This depends on both $\mathbf{x}_1$ and $\mathbf{x}_2$ only through r_1, r_2. With respect to the radial distributions, for the squared radii $u_i = r_i^2$, we have

$$q_{2\cdot 1}(u_2 | u_1) = B(\tfrac{1}{2} n_1, \tfrac{1}{2} n_2)^{-1}$$
$$u^{(n_2)/2-1} (u_1 + u_2)^{-n/2+1} q(u_1 + u_2)$$
$$/[u_1^{-(n_1)/2+1} q_1(u_1)] \quad . \quad \square \tag{2.25}$$

To continue with the example from the context of Lemma 2.1, suppose again that $\mathbf{x}$ is uniformly distributed on the centered hyperball of radius R, that is, $r^2/R^2 \sim \text{Beta}(\tfrac{1}{2} n, 1)$. Then (2.25) becomes $r_2^2/(R^2 - r_1^2) | r_1^2 \sim \text{Beta}(\tfrac{1}{2} n_2, 1)$, namely, $\mathbf{x}_2$ is conditionally uniform on the centered hyperball of radius $(R^2 - r_1^2)^{1/2}$. This contrasts with the nonuniform marginal distribution of $\mathbf{x}_1$.

3. ELLIPSOIDAL DISTRIBUTIONS

An ellipsoidal distribution can be generated as a location-scale transformation (affine transformation) of a spherical distribution. Write

$$\mathbf{y} \sim S(\mathbf{a}, U) \quad , \tag{3.1}$$

whenever

$$\mathbf{y} = \mathbf{a} + B\mathbf{x} \quad , \quad U = BB' \quad , \tag{3.2}$$

for some spherically distributed $\mathbf{x}$. Then, of course, $\mathbf{x} \sim S(\mathbf{0}, I)$. Since for such $\mathbf{x}$ and any k, $k\mathbf{x}$ is also spherical, then also $k\mathbf{x} \sim S(\mathbf{0}, I)$. Hence, if $\mathbf{y}$ satisfies (3.1), then also $\mathbf{y} \sim S(\mathbf{a}, k^2 U)$ for every $k \neq 0$.

Note that the distribution of $\mathbf{y}$ depends on B only through U and that the dimensionalities of $\mathbf{x}$ and $\mathbf{y}$ are not restricted. Indeed, the matrix U can be singular, and then $\mathbf{y}$ is confined to a linear manifold with probability one. Note that if U is non-singular, (3.1) holds if and only if

$$U^{-1/2}(\mathbf{y} + \mathbf{a}) \sim S(\mathbf{0}, I) \quad . \tag{3.3}$$

If, in addition, $\mathbf{x}$ and $\mathbf{y}$ in (3.2) have the same dimensionality, and $\mathbf{x}$ has a density $g(\mathbf{x}) = \gamma_n(\mathbf{x}'\mathbf{x})$, then $\mathbf{y}$ has the density

$$f(\mathbf{y}) = (\det U)^{-1/2} \gamma_n[U^{-1}((\mathbf{y} - \mathbf{a}))] \quad , \tag{3.4}$$

where we have used the double-parenthesis notation for a quadratic form, $M((\mathbf{z})) = = \mathbf{z}'M\mathbf{z}$.

The notation for the spherical property, $S(\mathbf{0}, I)$, does not identify a particular distribution, and the notation (3.1) does not identify a particular ellipsoidal distribution. However, we can specify a particular "*standard*" spherical distribution by mentioning a particular characteristic function, say $C_x(\mathbf{t}) = \psi(\mathbf{t}'\mathbf{t})$. Then write

$$\mathbf{x} \sim S_\psi(\mathbf{0}, I) \quad . \tag{3.5}$$

Denote the particular distribution satisfying (3.2) in terms of (3.5) by

$$\mathbf{y} \sim S_\psi(\mathbf{a}, U) \quad . \tag{3.6}$$

Then for parameters $\mathbf{a}$, U, this identifies a particular location-scale family.

The characteristic function of $\mathbf{y}$ (3.6) is

$$C_y(\mathbf{s}) = \exp(i\,\mathbf{a}'\,\mathbf{s})\psi(\mathbf{s}'U\mathbf{s}) \quad . \tag{3.7}$$

This form for the characteristic function provides a more succinct definition of the ellipsoidal property. Define $\mathbf{y}$ to have an *ellipsoidal* distribution if its characteristic function satisfies (3.7) for some $\mathbf{a}$, U, ψ.

Note that by (2.15), (2.16), the dimensionality of $\mathbf{x}$ is ambiguous in the representation (3.2) of $\mathbf{y} \sim S_\psi(\mathbf{a}, U)$. Given ψ, $\mathbf{a}$, U, any $\mathbf{x}$ and B for which $\mathbf{x} \sim S_\psi(\mathbf{0}, I)$ and $BB' = U$ will satisfy (3.2). For a specific nonsingular U, $\mathbf{x}$ can be taken to have the same dimensionality as $\mathbf{y}$. More generally, the dimensionality of $\mathbf{x}$ can be taken to equal the rank of U.

The ellipsoidal moments are, of course, simple functions of the spherical moments.

Corollary 3.1. For $\mathbf{y} \sim S_\psi(\mathbf{a}, U)$,

$$E\mathbf{y} = \mathbf{a} \tag{3.8}$$

$$\operatorname{Var}\mathbf{y} = \lambda_{1,2}\, U \tag{3.9}$$

(if they exist), where $\lambda_{1,2}$ is given by (2.10), (2.11) and (2.13) or (2.19).

More generally than the representation (3.2) with (3.5), an affine transformation of any ellipsoidal distribution is an ellipsoidal distribution. If $y \sim S_\psi(\mathbf{a}, U)$ and $\mathbf{z} = \mathbf{c} + D\mathbf{y}$, then by (3.7), $\mathbf{z} \sim (\mathbf{c} + D\mathbf{a}) + DU^{1/2}\mathbf{x}$ and

$$\mathbf{z} \sim S_\psi(\mathbf{c} + D\mathbf{a}, DUD') \tag{3.10}$$

In particular, the marginal distributions of an ellipsoidal distribution are ellipsoidal. To exhibit this, partition the dimensionality n of $\mathbf{y}$, $n = n_1 + n_2$, and conformably write

$$\mathbf{y} = \begin{pmatrix} \mathbf{y}_1 \\ \mathbf{y}_2 \end{pmatrix}, \quad \mathbf{a} = \begin{pmatrix} \mathbf{a}_1 \\ \mathbf{a}_2 \end{pmatrix}, \quad U = \begin{pmatrix} U_{11} & U_{12} \\ U_{21} & U_{22} \end{pmatrix} \tag{3.11}$$

Theorem 3.2. If $\mathbf{y} \sim S_\psi(\mathbf{a}, \mathbf{U})$, then

$$\mathbf{y}_1 \sim S_\psi(\mathbf{a}_1, U_{11}) \quad . \tag{3.12}$$

(Note that this generalizes the familiar marginal parameters in the usual joint normal theory.) □

Proof. Write $\mathbf{y}_1 = (I, 0)'\mathbf{y}$ and use (3.10).

Corollary 3.3. If $\mathbf{y}$ has the joint density (3.4), then the density for the marginal distribution (3.12) is

$$f_1(\mathbf{y}_1) = (\det U_{11})^{-1/2}\, \gamma_{n_1}[U_{11}^{-1}((\mathbf{y}_1 - \mathbf{a}_1))] \tag{3.13}$$

where γ_{n_1} satisfies (2.18) as a transform of γ_n.

An ellipsoidal distribution also has ellipsoidal conditional distributions. To see this, partition the quadratic form,

$$U^{-1}((\mathbf{y}-\mathbf{a})) \tag{3.14}$$

$$= U_{11}^{-1}((\mathbf{y}_1-\mathbf{a}_1)) + U_{22\cdot 1}^{-1}((\mathbf{y}_2-\mathbf{a}_{2\cdot 1}(\mathbf{y}_1))) \quad ,$$

where

$$\mathbf{a}_{2\cdot 1}(\mathbf{y}_1) = \mathbf{a}_2 + U_{21}U_{11}^{-1}(\mathbf{y}_1-\mathbf{a}_1) \tag{3.15}$$

$$U_{22\cdot 1} = U_{22} - U_{21}\,U_{11}^{-1}\,U_{12} \quad . \tag{3.16}$$

Theorem 3.3. If $\mathbf{y} \sim S_\psi(\mathbf{a}, U)$ and $\mathbf{y}$ has the joint density (3.4), then

$$\mathbf{y}_2 | \mathbf{y}_1 \sim S_{\tilde{\psi}}(\mathbf{a}_{2\cdot 1}(\mathbf{y}_1), U_{22\cdot 1}) \quad , \tag{3.17}$$

with density

$$f_{2\cdot 1}((\mathbf{y}_2|\mathbf{y}_1) = (\det U_{22\cdot 1})^{-1/2} \tag{3.18}$$

$$\tilde{\gamma}_{n_2}[U_{22\cdot 1}^{-1}((\mathbf{y}_2-\mathbf{a}_{2\cdot 1}(\mathbf{y}_1))); U_{11}^{-1}((y_1-\mathbf{a}_1))] \quad ,$$

where

$$\tilde{\gamma}_{n_2}(r_2^2; r_1^2) = \gamma_n(r_1^2 + r_2^2)/\gamma_{n_1}(r_1^2) \quad , \tag{3.19}$$

where γ_{n_1} and γ_n satisfy (2.18). □

Note that the standard conditional distribution and hence its characteristic function $\tilde{\psi}$ depend on the quadratic form $U_{11}^{-1}((\mathbf{y}_1-\mathbf{a}_1))$. The conditional location and scale parameters in (3.17) are the same as the familiar linear regressions and partial variance structure in the joint normal case. In this general case, however, as noted, the conditional distribution also depends, through it density $\tilde{\gamma}_{n_2}$, on the quadratic form in the condition variable, $U_{11}^{-1}((\mathbf{y}_1-\mathbf{a}_1))$. For example, the joint multivariate-t has a conditional distribution where, in effect, the quadratic form in the condition variable enters as a factor into the conditional scale. Thus $\mathbf{y}_2$ given $\mathbf{y}_1$ is multivariate-t with location $\mathbf{a}_{2\cdot 1}(\mathbf{y}_1)$ and scale proportional to the product $U_{11}^{-1}(\mathbf{y}_1-\mathbf{a}_1))U_{22\cdot 1}$.

4. ASSESSMENT OF ELLIPSOIDAL DISTRIBUTIONS

The following theorem may give what is, perhaps, the most important single result for the problem of how to assess an ellipsoidal distribution. The result derives immediately from Theorem 2.5 and the distribution (3.10).

Theorem 4.1. Any univariate affine form in an ellipsoidally distributed vector has a location-scale version of the standard univariate marginal distribution of the underlying spherical distribution: if $\mathbf{y} \sim S_\psi(\mathbf{a}, U)$ and $z = c + \mathbf{d}'\mathbf{y}$, then

$$z \sim (x + \mathbf{d}'\mathbf{a}) + (\mathbf{d}'\mathbf{U}\mathbf{d})^{1/2}x_1 \quad , \tag{4.1}$$

where $x_1 \sim S_\psi(0,1)$. This applies, in particular, to any of the univariate coordinate distributions of the ellipsoidal distribution,

$$y_i \sim a_i + (u_{ii})^{1/2}x_1 \quad , \tag{4.2}$$

where $U = (u_{ij})$.

The first problem in assessing an ellipsoidal distribution is to assess an underlying spherical distribution. Is it normal, and hence has independent coordinate distributions? Is it a scale mixture of normals (discussed in Section 5)? Is it a member of some other particular parameterized family (such as given here in the Appendix)? If so, what are its parameter values?

The location-scale form of the ellipsoidal coordinate random quantities y_i (4.2) provides a window into the univariate marginal distribution of the underlying spherical distribution. Once this standard marginal distribution is determined, the results of Section 2.1 can be used to obtain the underlying joint spherical distribution. For example, the one-dimensional integral equation (2.18) can be solved for the multivariate spherical density. Or if the parametric form (A.22), (A.24) is recognized and fitted to the standard univariate margin, a radial F distribution (A.16) is implied, which immediately gives the joint spherical density by (2.13).

Note, again, that the underlying joint spherical distribution of vector $\mathbf{x}$ is fully determined by its marginal univariate coordinate distribution, say the distribution of x_1. Of course, the common scale ambiguity, $S(\mathbf{0}, k^2 I) \sim S(\mathbf{0}, I)$, which appears between x_1 and its coefficient in (4.1), (4.2), can be removed by an arbitrary convention, as in the defined unit variance in a "standard" normal distribution.

Our problem of how to assess an ellipsoidal distribution then reduces to the two questions: (a) how to determine the standard marginal distribution x_1 from the location-scale-transformed margins of the ellipsoidal distribution y_i (4.2); and (b) given the underlying spherical distribution, how to assess the location and scale parameters a, U of the joint ellipsoidal distribution. The next two subsections address these questions, respectively.

4.1. *Determining the Standard Marginal Distribution*

The methods developed by Kadane *et al* (1980) for eliciting a univariate-t distribution extend rather directly to more general univariate location-scale distributions. Their procedure involves eliciting specific subjective quantiles z_q, where

$$q = P\{z \le z_q\} \quad . \tag{4.3}$$

For various univariate marginal and conditional distributions, one proceeds either by eliciting a variable's values z_q corresponding to specified cumulative probabilities q, or by eliciting the subjective cumulative probabilities q corresponding to specified values z_q. Probability-bisection methods and interval methods are especially convenient for this task (Garthwaite and Dickey 1983).

In a location-scale relation such as (4.2), ratios of quantile differences are the same for the y_i distribution as for the x_1 distribution and do not depend upon the location and scale parameters,

$$\begin{aligned}&(y_{iq_1}-y_{iq_2})/(y_{iq_3}-y_{iq_4})\\ &= (x_{1q_1}-x_{1q_2})/(x_{1q_3}-x_{1q_4}) \quad .\end{aligned} \tag{4.4}$$

This identity enables one to assess the shape of the standard spherical margin from elicited quantiles of a coordinate variable of the ellipsoidal distribution. Kadane *et al* (1980) determine in this way the degrees-of-freedom parameter for a multivariate-t model.

A special answer to the question of assessing the standard marginal distribution is developed in a later section here for underlying models that take the form of mixtures of known simple models. In these models only the mixing weights need be determined. Also then, the joint spherical distribution will be simply related to its margin.

4.2. *Determining the Location and Scale Parameters of the Joint Ellipsoidal Distribution*

The methods of Kadane *et al* (1980) successfully extend, again, for the problem of assessing ellipsoidal location and scale parameters. However, the discussion there of this problem is phrased in a language for situations when the model is a Bayesian prior average of imbedded normal sampling models, and that discussion is also unnecessarily entwined with the problem of assessing a dependence of these parameters on concomitant and factor variables. Space here will not allow a fully detailed account of the general methods, but a descriptive outline will be given.

Note first that the location and scale parameters of the marginal and conditional distributions of an ellipsoidal distribution, (3.12) and (3.17), are known simple functions of the joint parameters. The idea is that elicited quantiles of selected univariate marginal and conditional distributions can be used to solve for the joint parameters. However, the "standard" forms of the univariate marginal and conditional distributions must be determined first to carry this out. The preceding subsection addressed this initial task for the standard margin, which also will yield the standard joint spherical and hence the standard forms of the univariate conditional distributions.

The joint location parameter $\mathbf{a}$ is obviously best assessed directly through its coordinates a_i, which are the location parameters of the univariate margins y_i. Since by definition, spherical distributions are centered, we have

$$\text{Median}(y_i) = a_i \quad . \tag{4.5}$$

Similarly, the diagonal elements of the joint scale U are the scale parameters of the univariate margins, which can be determined as squared ratios of the two kinds of quantile differences,

$$u_{ii} = [(y_{iq_1} - y_{iq_2})/(x_{1q_1} - x_{1q_2})]^2 \quad . \tag{4.6}$$

How, then, to assess the off-diagonal elements of U? There are two approaches, the marginal method and the conditioning method. In the marginal method, scale parameters are assessed (similarly to (4.6)) for linear combinations of the coordinate quantities. For example, by (4.1),

$$y_i - y_j \sim S_\psi(a_i - a_j,\, u_{ii} - 2u_{ij} + u_{jj}) \quad , \tag{4.7}$$

and so an assessed scale for $y_i - y_j$ can be used to solve for u_{ij}, $i \neq j$. For the conditioning method, note that a univariate conditional location parameter $a_{i\cdot(1,\ldots,i-1)}(y_1,\ldots,y_{i-1})$ (3.15) is a linear function of the off-diagonal vector.

$$U_{i,(1,\ldots,i-1)} = (u_{i1}, u_{i2}, \ldots, u_{i,i-1}) \quad . \tag{4.8}$$

Systems of linear equations can thus be solved for (4.8) based on assessed conditional locations, as in Kadane *et al* (1980).

A complication that arises in both these methods is that a naive assessment procedure could produce scale matrices U which are not positive definite (or not positive semidefinite). This difficulty is surmounted by constructing the matrix U iteratively by succesive one-dimensional bordering, checking at each step that the new matrix will be positive definite, assuming that the preceding lower dimensional matrix is positive definite. This requires only that at the ith step $U_{ii\cdot(1,\ldots,i-1)} > 0$ (3.16), $i = 2,\ldots,n$. (We assume that U will be nonsingular, and hence that $\mathbf{x}$ and $\mathbf{y}$ have the same dimensionality.)

4.3. *Dependence of Location and Scale Parameters on Concomitant and Factor Variables*

Further-developed models can be considered that allow for dependence of location and scale parameters on additional concomitant and factor variables. Subjective point-predictions and corresponding uncertainties of an expert can thus be modeled in their dependence on various explanatory variables. Such a location function is termed a *subjective response surface*, and the scale function provides an associated nonconstant spread of predictive uncertainty. For example, a model of this kind is provided by the multivariate-t Bayesian predictive distribution based on a normal-linear-regression sampling model.

Kadane *et al* (1980) provide methods for assessing such a model which, again, will extend directly to general ellipsoidal distributions. Dickey (1980) treats the extended problem of fitting a subjective response surface by generalized least squares and considers the choice of weights for such fitting. Dickey, Dawid and Kadane (1983) consider the general multivariate-t and matrix-t cases. The basic idea is that the location vector for **y** can be elicited directly as subjective medians, and then a response surface function can be fitted to these elicited medians. The scale-matrix function is first assessed as a general positive-definite matrix, by the methods of Section 4.2, and then projected into a functional form quadratic in the concomitant variables. Alternatively, the functional form of the scale matrix can itself be built up by strategic choice of contemplated concomitant variable values (Dickey, Dawid and Kadane 1983). Garthwaite (1983) has a further method based on linearly constrained choices of concomitant variable values minimizing predictive uncertainty under the constraints.

5. SCALE MIXTURES

We develop in detail a particular tractable class of models, scale mixtures of normal (or similarly tractable) spherical distributions, and location-scale transforms of such distributions. These are simply related to their marginal and conditional distributions, and hence can yield simple assessment algorithms.

By a scale mixture constructed from a given spherical distribution, $S_{\psi*}(\mathbf{0}, I)$, is meant the marginal distribution of the vector **x** in a joint distribution of the random pair **x**, w, where the scaler w has an arbitrary distribution and **x** has the conditional distribution given w,

$$\mathbf{x} \mid w \sim S_{\psi*}(\mathbf{0}, wI) \quad . \tag{5.1}$$

Denote such a distribution for **x** by

$$\mathbf{x} \sim S_{\psi*}(\mathbf{0}, wI)*w \quad . \tag{5.2}$$

The distribution of w is called the *mixing* distribution, and the conditional distribution (5.1) is the *mixand*. In such a mixture, **x** has the cumulative probability function, $P(\mathbf{x}) = \int_0^\infty P(\mathbf{x} \mid w)dP(w)$, and the density (if it exists),

$$\gamma(\mathbf{x}'\mathbf{x}) = \int_0^\infty \gamma^*(\mathbf{x}'\mathbf{x}/w)w^{-n/2}\, dP(w) \quad . \tag{5.3}$$

For example, if w has a finite support set, $w_1 < w_2 < \ldots < w_\ell$, then the Stieljes integral in (5.3) becomes a summation,

$\Sigma_{j=1}^{\ell}\ \gamma^*(\mathbf{x}'\mathbf{x}/w_j)w_j^{-n/a}\ q_j$, where $q_j = P\ \{w = w_j\}$. But if w is continously distributed with probability density $p\ (w)$, then we have the usual Lebesgue integral, $\int_0^\infty \gamma^*(\mathbf{x}'\mathbf{x}/w)w^{-n/2}$

$p(w)dw$. (Ultimately, of course, any spherical distribution is a scale mixture of distributions supported and uniform on centered sphere surfaces.)

Obviously, a scale mixture constructed from a spherical distribution is spherical. The conditional and marginal distributions of the mixture are simply related to the conditionals and marginals of its mixand distributions, as follows.

Theorem 5.1. If $\mathbf{x}=(\mathbf{x}_1',\mathbf{x}_2')'$ has the scale-mixed spherical distribution (5.2), (5.3), then marginally $\mathbf{x}_1$ has the lower-dimensional scale mixture,

$$\mathbf{x}_1 \sim S_{\psi *}(\mathbf{0},\, wI)*w \quad , \tag{5.4}$$

with density

$$\gamma_{n_1}(\mathbf{x}_1'\mathbf{x}_1) = \int_0^\infty \overset{*}{\gamma}_{n_1}(\mathbf{x}_1'\mathbf{x}_1/w)w^{-n_1/2}\,dP(w) \quad , \tag{5.5}$$

in which the mixand density $\overset{*}{\gamma}_{n_1}$ is the (spherical) marginal form (2.18) from the joint mixand $\overset{*}{\gamma}_n$. The variable w in the marginal representation (5.4) has the same mixing distribution as in the representation (5.2) of $\mathbf{x}$. Also, the conditional distribution of $\mathbf{x}_2$ given $\mathbf{x}_1$ is a scale mixture,

$$\mathbf{x}_2|\mathbf{x}_1 \sim S_{\tilde{\psi}_w^*}(\mathbf{0},\, wI)*(w|\mathbf{x}_1) \quad , \tag{5.6}$$

with density

$$\tilde{\gamma}_{n_2}(\mathbf{x}_2'\mathbf{x}_2;\mathbf{x}_1'\mathbf{x}_1) = \int_0^\infty \tilde{\overset{*}{\gamma}}_{n_2}(\mathbf{x}_2'\mathbf{x}_2/w;\mathbf{x}_1'\mathbf{x}_1/w) \tag{5.7}$$

$$w^{-n_2/2}\,dP(w|\mathbf{x}_1) \quad ,$$

in which the mixand density $\tilde{\overset{*}{\gamma}}_{n_2}$ is the (spherical) conditional form (2.24) from the joint mixand $\overset{*}{\gamma}_n$. The probability mass or density of the mixing conditional distribution of w given $\mathbf{x}_1$ satisfies Bayes' formula,

$$p\,(w|\mathbf{x}_1) = \overset{*}{\gamma}_{n_1}(\mathbf{x}_1'\mathbf{x}_1\,/w)^{-n_1/2}\,p\,(w)/\gamma_{n_1}(\mathbf{x}_1'\mathbf{x}_1), \tag{5.8}$$

where $p\,(w)$ is the probability mass or density of w in (5.2) and (5.4). □

Three examples will be given in this section, all having normal mixands. Scale mixtures of spherical normal distributions are partially motivated by Kelker's (1970, Theorem 10) finding that such mixtures are the only spherical distributions which can be coherently extended in dimension indefinitely, that is, can be considered as marginal distributions of higher dimensional spherical distributions without an upper bound on the higher dimensionality. Of course, a spherical normal distribution is simply related to its marginal and conditional distributions. Hence, by Theorem 5.1, this will also be true of a scale mixture of normals. We shall use only the marginal distributions in the method to be given.

Example 1. Finite Mixture of Spherical Normal Distribution

If

$$\mathbf{x}|w \sim N(\mathbf{0},\, wI) \tag{5.9}$$

and the common variance has a finite number of possible values, $w=w_j$ with probability p_j for $j=1,\ldots,\ell$, then the subvector $\mathbf{x}_1|w \sim N(\mathbf{0},\, wI)$ for this same distribution of w.

Example 2. Continuous Mixture

The "conjugate prior" form can serve as a continuous mixing distribution for the same mixand (5.9),

$$w \sim s^2 \cdot (\nu/\chi^2_\nu) \quad . \tag{5.10}$$

Then $\mathbf{x}$ is multivariate-t, and so is $\mathbf{x}_1$,

$$\mathbf{x}_1 \sim N(0,s^2I)/(\chi^2_\nu/\nu)^{1/2} \quad . \tag{5.11}$$

Example 3. Finite Mixture of Multivariate-t Distributions

A hierarchical distribution for w, itself, can be constructed as a finite mixture of continuous distributions (5.10),

$$w\,|\,(\tilde{h}=h) \sim s_h^2(\nu_h/\chi^2_{\nu_h}) \quad , \tag{5.12}$$

where $\tilde{h}=h$ with probability π_h for $h=1,\ldots,H$. Then the resulting continuous distribution for w yields a finite mixture of multivariate-t distributions,

$$\mathbf{x}_1\,|\,(\tilde{h}=h) \sim N(\mathbf{0},\, s_h^2I)/(\chi^2_{\nu_h}/\nu_h)^{1/2} \quad , \tag{5.13}$$

where, again, $\tilde{h}=h$ with probability π_h, $h=1,\ldots,H$. This is still, of course, a continuous mixture of normals.

The operation of scale mixing and location-scale transformation commute, as follows.

Theorem 5.2. An ellipsoidal distribution obtained by a location-scale transform on a scale-mixed spherical distribution is a scale mixture of the same location-scale transformation on the mixand distributions. Furthermore, the two mixing distributions are the same. Symbolically, if $\mathbf{x}$ satisfies (5.2) and $\mathbf{y}=\mathbf{a}+B\mathbf{x}$, $U=BB'$, then

$$\mathbf{y} \sim S_{\psi *}(\mathbf{a},\, wU)*w \quad , \tag{5.14}$$

for the same random quantity w. □

Theorem 5.1 on marginal and conditional distributions of spherical $\mathbf{x}$ has an obvious modification for ellipsoidal $\mathbf{y}$. One merely inserts the ellipsoidal marginal and conditional location and scale parameters for modified (5.4) and (5.6), following (3.12) and (3.17).

To assess a location-scale transform of a scale mixture of spherical distributions, one proceeds as for any ellipsoidal distribution, by first using a coordinate marginal distribution to identify the underlying spherical distribution which in the present case is a mixture. By Theorem 4.1, each coordinate distribution is a univariate location-scale transform of the same univariate mixture. (These scale parameters are confounded, as in (4.1), (4.2), but only because of the arbitrary scaler standardization k of the underlying spherical distribution $S_\psi\,(\mathbf{0},k^2I)$, which can be chosen for convenience.) We give details for a finite mixture of normals.

Example 1 (Continued). When the underlying spherical distribution of an ellipsoidal distribution is a finite mixture of spherical normals (5.9), it lends itself to a tractable assessment method. The method has claims to broader applicability, that is, to finite mixtures of tractable distributions other than normal.

Suppressing subscripts, denote by y any particular coordinate of the ellipsoidally distributed vector $\mathbf{y}$. (The underlying spherical distribution can be assessed by working with any univariate margin of the ellipsoidal distribution). We have the univariate mixture for the ellipsoidal coordinate,

$$y\,|\,w \sim N(a,\, uw) \quad , \tag{5.15}$$

where $w=w_j$ with probability p_j for $j=1,\ldots,\ell$, and a and u denote the respective entries in the location and scale parameters for vector **y** (4.2). The location is easily elicited as a median,

$$a = y_{.50} \quad , \tag{5.16}$$

or some average of lower and symmetrically chosen upper quantiles. Because of the arbitrary scalar standardization of spherical **x**, mentioned earlier, we can assume without loss of generality that for this particular coordinate of **y**,

$$u = 1 \quad . \tag{5.17}$$

(Of course, such an assumption can be made for only one of the coordinates of **y**.) Now, once the mixands have been set up by specifying a reasonable list of possible w values w_j, all one needs is a way to assess the mixing probabilities p_j, $j = 1,\ldots,\ell$.

Given the elicited quartiles $y_{.75}$, $y_{.25}$, consider mixtures over the following scale values,

$$w_j = 2^{j-4}(y_{.75} - y_{.25})/1.34898 \quad , \tag{5.18}$$

for $j = 1,\ldots,7$ ($\ell = 7$, say). The denominator in (5.18) is the difference between the upper and lower quartiles of the usual standard normal distribution. This has the virtue that the mixture will become a single normal distribution with quartiles $y_{.75}$, $y_{.25}$ (as elicited) in the case $p_4 = 1$.

Finally, one can assess the mixing probabilities in the following way. Elicited quantiles $y_{q_k} \in k = 1,\ldots,K$ are related to the probabilities p_j, $j = 1,\ldots,\ell$, by the system of linear equations,

$$q_k = \Sigma_{j=1}^{\ell} P\{y \leq y_{qk} | w_j\} p_j \quad . \tag{5.19}$$

This can be "solved" for the mixing probabilities by least squares, or similar methods. The coefficients in (5.19) are available from tables as the standard normal cumulative probabilities,

$$\Phi[(y_{q_k} - a)/w_j^{1/2}] \quad . \tag{5.20}$$

It is good practice, of course, to check and adjust the fitted shape by working with further coordinates, either using (4.4), or using (5.19) again after assessing some of their scales u_{ii} by (4.6).

APPENDIX. NEW FAMILIES OF SPHERICAL DISTRIBUTIONS

Parameterized families of spherical distributions are easily developed by working with parameterized families of radial distributions. Equation (2.13) relates spherical distributions to radial distributions. Theorems 2.4 and 2.6 can be used to derive the corresponding parameterized marginal and conditional families. We begin by reviewing generalizations of the usual Dirichlet distributions in which the densities are proportional to the integrands in integral representations of B.C. Carlson's (1977, 1971) hypergeometric functions R.

A.1. *Background Distribution Theory*

Following Dickey (1983), we define a generalization of the Dirichlet distributions (2.4): denote the new distribution for a random probability K-tuple $\mathbf{u} = (u_1,\ldots,u_K)'$,

$$\mathbf{u} \sim D_a(\mathbf{b}, \mathbf{z}) \quad , \tag{A.1}$$

with the parameters, K-tuple $\mathbf{b}$, K-tuple $\mathbf{z}$, and real number a. This has density,

$$p(\mathbf{u}) = B(b)^{-1}(\Pi u_i^{b_i-1})\,(u_1 z_1 + \ldots + u_K z_K)^a \,/R_a(\mathbf{b},\mathbf{z}) \quad , \tag{A.2}$$

where the denominator R is equal to the integral of the numerator over the probability simplex. The moments are proportional to obvious ratios of R functions.

The Dirichlet distribution is the special case $a=0$, or the special case $\mathbf{z}=c\mathbf{1}$ (equal coords). Also, in the special case $a=-b_+ \equiv -\Sigma b_i$, the denominator simplifies,

$$R_{-b_+}(\mathbf{b}, \mathbf{z}) = \Pi\, z_i^{-b_i} \quad , \tag{A.3}$$

and then $\mathbf{u}$ has L.J. Savage's distribution, a simple transformation of the Dirichlet (Dickey 1968).

More generally, define

$$\mathbf{u} \sim D_a(\mathbf{b}, Z, \beta) \quad , \tag{A.4}$$

if the probability K-tuple $\mathbf{u}$ has density

$$p(\mathbf{u}) = B(\mathbf{b})^{-1}\,(\Pi_1^K u_i^{b_i-1})\,R_a(\beta, (\mathbf{u}'Z)')$$

$$/R_a(\mathbf{b}, Z, \beta) \quad , \tag{A.5}$$

where again the denominator R is the complete integral of the numerator. The new parameters here are the L-tuple β and the K-by-L matrix Z,

$$Z = \begin{pmatrix} \mathbf{z}_{1*} \\ \vdots \\ \mathbf{z}_{K*} \end{pmatrix} = (\mathbf{z}_{*1}, \ldots, \mathbf{z}_{*L}) \tag{A.6}$$

The density (A.5) simplifies considerably when $a=-\beta_+$. For then, by (A.3),

$$p(\mathbf{u}) = B(\mathbf{b})^{-1}(\Pi_1^K u_i^{b_i-1})\,\Pi_1^L(\mathbf{u}'\mathbf{z}_{*j})^{-\beta_j}\,/R_{-\beta_+}(\mathbf{b}, Z, \beta) \quad . \tag{A.7}$$

Of course, this gives a simplified form for $R_{-\beta_+}$.

We shall use the univariate forms of D_a (A.1) and D_a (A.4) ($K=2$), which generalize the beta distributions, and simple transformations that generalize the Snedecor-F distributions.

A.2. *Example 1. Beta-Distributed Squared Radius*

We generalize the spherical distribution uniform on the centered hyperball, used for illustration in Section 2, by taking a general beta distribution for the squared radius $r^2=\mathbf{x}'\mathbf{x}$,

$$r^2/R^2 \sim \mathrm{Beta}(\tfrac{1}{2}\alpha, \tfrac{1}{2}\beta) \quad . \tag{A.8}$$

For the subvectors of partitioned $\mathbf{x}=(\mathbf{x}_1', \mathbf{x}_2')'$, consider the squared radii $r_i^2=\mathbf{x}_i'\mathbf{x}_i$, $i=1,2$, for which $r_1^2+r_2^2=r^2$. The dimensions of $\mathbf{x}$, $\mathbf{x}_1$, $\mathbf{x}_2$ are n, n_1, n_2, respectively.

Theorem A.1. The Beta radial distribution (A.8) implies the conditional distribution for r_2^2 given r_1^2,

$$\varrho^2 | r_1^2 \sim D_{(1/2)(\alpha-n)}(\tfrac{1}{2}n_2, \tfrac{1}{2}\beta; 1, r_1^2/R^2) \tag{A.9}$$

where $\varrho^2 = r_2^2/(R^2 - r_1^2)$; and the marginal distribution for r_1^2,

$$r_1^2/R^2 \sim D_{(1/2)(\alpha-n)}\left(\tfrac{1}{2}n_1, \tfrac{1}{2}(\beta+n_2); Z; \tfrac{1}{2}n_2, \tfrac{1}{2}\beta\right) \quad , \tag{A.10}$$

where $Z = ((1, 1)', (1, 0)')$. □

These conditional and marginal distributions simplify for interesting special cases of the parent radial distribution (A.8). (a) For $\alpha = n$ we have

$$\varrho^2 | r_1^2 \sim \text{Beta}(\tfrac{1}{2}n_2, \tfrac{1}{2}\beta) \quad , \tag{A.11}$$

$$r_1^2/R^2 \sim \text{Beta}(\tfrac{1}{2}n_1, \tfrac{1}{2}(\beta + n_2)) \quad . \tag{A.12}$$

(Our original illustration had $\alpha = n$, $\beta = 2$.) Note that this subfamily is closed under successive marginalization and conditionalization.

(b) For $\alpha = n_1 - \beta$, we have

$$\varrho^2 | r_1^2 \sim \tilde{\varrho}^2/[\tilde{\varrho}^2 + (R^2/r_1^2)(1 - \tilde{\varrho}^2)] \quad , \tag{A.13}$$

where $\tilde{\varrho}$ has the distribution (A.11), and

$$r_1^2/R^2 \sim \text{Beta}(\tfrac{1}{2}(n_1 - \beta), \tfrac{1}{2}(\beta + n_2)) \quad . \tag{A.14}$$

The following expansion for general α may be found useful in assessing or fitting a coordinate marginal distribution ($n_1 = 1$). It is derived by use of the binomial expansion in the integrand in the density of r_1^2/R^2.

Corollary A.2. Under the radial distribution (A.8), if $\frac{1}{2}(\alpha - n)$ is a nonnegative integer, then the marginal distribution is a probability mixture of beta distributions,

$$r_1^2/R^2 | h \sim \text{Beta}(\tfrac{1}{2}n_1 + h, \tfrac{1}{2}(\alpha + \beta - n_1) - h) \quad , \tag{A.15}$$

where the mixing parameter h has the Beta-binomial distribution with power parameters $\frac{1}{2}n_1$, $\frac{1}{2}n_2$, and sample-size parameter $\frac{1}{2}(\alpha - n)$. That is, $h | \pi \sim \text{Binomial}(\frac{1}{2}(\alpha - n), \pi)$, where $\pi \sim \text{Beta}(\frac{1}{2}n_1, \frac{1}{2}n_2)$. □

The reader may find it interesting to generalize the Beta radial distribution in n dimensions here (A.8) to a distribution D_a (A.1) or (A.4)

A.3. *Example 2. F-Distributed Squared Radius*

For a second, rather broad example, assume that r^2 has a Snedecor-F distribution,

$$r^2/S^2 \sim F_{\alpha,\beta} \tag{A.16}$$

This general model has important and interesting special cases.

(a) For $\alpha = n$, $\mathbf{x}$ is distributed according to a multivariate-t distribution with β degrees of freedom,

$$\mathbf{x} \sim N^{(n)}(\mathbf{0}, (S^2/n)I)/(\chi_\beta^2/\beta)^{1/2} \tag{A.17}$$

where the numerator and denominator random quantities in (A.17) are independently distributed. This implies that the distributions for $\mathbf{x}$ associated with the radial distributions (A.16) offer a new generalization of the multivariate-t family.

(b) For $\beta \to \infty$, the family (A.16) becomes

$$r^2/S^2 \sim \chi_\alpha^2/\alpha \tag{A.18}$$

These interesting spherical distributions generalize the multivariate-normal; for if $\alpha = n$ in addition to $\beta \to \infty$,

$$\mathbf{x} \sim N^{(n)}(\mathbf{0}, (S^2/n)I) \quad . \tag{A.19}$$

(c) For $\alpha \to \infty$, we have

$$r^2/S^2 \sim \beta/\chi^2_\beta \quad . \tag{A.20}$$

This new family for $\mathbf{x}$ appears to contain no well known distributions.

Space here does not allow detailed development of these spherical sub-families. They will form the topic of a later paper. We give only the general conditional and marginal distributions.

Theorem A.3 The Snedecor-F radial distribution (A.16) implies the following marginal and conditional distributions for r_1^2, r_2^2, expressed in terms of $v_i = (\alpha/\beta)r_i^2/S^2$, $i = 1,2$. The conditional density of v_2 given v_1 is equal to

$$N^{-1} v_2^{n_2/2-1}(v_1+v_2)^{(\alpha-n)/2}(v_1+v_2+1)^{-(\alpha+\beta)/2} \quad , \tag{A.21}$$

where $N = B(\frac{1}{2}n_2, \frac{1}{2}(\beta+n_1))(v_1+1)^{-(\alpha+\beta-n_2)/2}$

$R_{(\alpha-n)/2}(\frac{1}{2}n_2, \frac{1}{2}(\beta+n_1); v_1+1, v_1)$. The marginal density of v_1 is

$$M^{-1} v_1^{(1/2)n_1-1}(v_1+1)^{-(1/2)(\alpha+\beta-n_2)} R_{(\alpha-n)/2}(\tfrac{1}{2}n_2, \tfrac{1}{2}(\beta+n_1) \; ; \; v_1+1, v_1) \quad , \tag{A.22}$$

where $M = B(\frac{1}{2}\alpha, \frac{1}{2}\beta)B(\frac{1}{2}n_1, \frac{1}{2}n_2)/B(\frac{1}{2}n_2, \frac{1}{2}(\beta+n_1))$. □

Just as the usual F distribution is a simple transformation of a Beta distribution, the new distributions (A.21), (A.22) are transforms of D_a distributions, as follows. Hence, the distributions (A.21) and (A.22) can be regarded as generalizations of Snedecor's F.

Corollary A.4. The F distribution (A.16) implies the conditional distribution for r_2^2 given r_1^2,

$$\varrho^2 | r_1^2 \sim D_{(\alpha-n)/2}(\tfrac{1}{2}n_2, \tfrac{1}{2}(\beta+n_1); (\alpha/\beta)r_1^2/S^2+1, (\alpha/\beta)r_1^2/S^2) \quad , \tag{A.23}$$

where $\varrho^2 = r_2^2/(r_1^2+r_2^2+(\beta/\alpha)S^2) = v_2/(v_1+v_2+1)$, and the marginal distribution

$$w \sim D_{(1/2)(\alpha-n)}(\tfrac{1}{2}n_1, \tfrac{1}{2}\beta; Z; \tfrac{1}{2}n_2, \tfrac{1}{2}(\beta+n_1)) \quad , \tag{A.24}$$

where $w = r_1^2/(r^2+(\beta/\alpha)S^2) = v_1/(v_1+1)$ and

$$Z = ((1,1)', (1,0)') \quad . \quad □$$

The simple form of M in the density (A.22) implies a simple normalizing constant for (A.24)

$$R_{(\alpha-n)/2}(\tfrac{1}{2}n_1, \tfrac{1}{2}\beta; Z; \tfrac{1}{2}n_2, \tfrac{1}{2}(\beta+n_1)) =$$
$$B(\tfrac{1}{2}\alpha, -\tfrac{1}{2}(\alpha-n))/B(\tfrac{1}{2}(\alpha+\beta), -\tfrac{1}{2}(\alpha-n)).$$

This same result follows also by applying a known transformation (Carlson 1977, p. 101, eq(20)) to the function R from (A.2) appearing in the definition of R from (A.5).

ACKNOWLEDGEMENT

This research was sponsored by the U.S. National Science Foundation Grant MCS-8301335.

REFERENCES

CARLSON, B.C. (1971). Appell functions and multiple averages. *S.I.A.M. J. of Math. Anal.* **2**, 420-430.

— (1977). *Special Functions of Applied Mathematics.* New York: Academic Press.

COOK, R.L. (1980). Brunswick's lens model and the development of interactive judgment analysis. (K.R. Hammond and N.E. Wascoe, eds.). *New Directions for Methodology of Social and Behavioral Science: Realizations of Brunswick's Representative Design.* San Francisco: Jossey-Bass.

DAS GUPTA, S., OLKIN, I., SAVAGE, L.J., EATON, M.L., PERLMAN, M., SOBEL, M. (1973). Inequalities on the probability content of convex regions for elliptically contoured distributions. *Proceedings of the Sixth Berkeley Symposium on Mathematical Statistics and Probability* **2**. (L.M. LeCam, J. Neyman and E.L. Scott, eds.) Berkeley: University of California Press, 241-265.

DAWID, A.P. and DICKEY, J.M. (1977). Likelihood and Bayesian inference from selectively reported data. *J. Amer. Statist. Assoc.* **72**, 845-850.

DICKEY, J.M. (1968). Three multidimensional-integral identities with Bayesian applications. *Ann. Math. Statist.* **39**, 1615-27.

— (1980). Beliefs about beliefs, a theory of stochastic assessments of subjective probabilities. *Bayesian Statistics.* (Bernardo, J.M. *et al* eds.). Valencia: University Press.

— (1983). Multiple hypergeometric functions: Probabilistic interpretations and statistical usses. *J. Amer. Statist. Assoc.* **78**, 628-637.

DICKEY, J.M., DAWID, A.P. and KADANE, J.B. (1983). Subjective-Probability Assessment Methods for Multivariate-t, Matrix-t, and Hierarchical Models. *Bayesian Inference and Decision Techniques with Applications: Essays in Honor of Bruno de Finetti,* (P.K. Goel and A. Zellner, eds.). Amsterdam: North-Holland.

EATON, M.L. (1981). On the projections of isotropic distributions. *Ann. Statist.* **9**, 391-400.

GARTHWAITE, P.H. (1983). *Assessment of prior distributions for normal linear models.* Ph.D. thesis. Dept. of Statistics, University College of Wales, Aberystwyth, U.K.

GARTHWAITE, P.H. and DICKEY, J.M. (1983). The double and single bisection methods for subjective assessment in a location-scale family. *Tech. Rep.* **4**, Dept. of Statistics, University of Aberdeen, U.K.

GOLDSTEIN, M. (1981). Revising previsions: A geometric interpretation. *J. Roy. Statist. Soc. B,* **43**, 105-130 (with discussion).

— (1983a). Separating beliefs. *Bayesian Inference and Decision Techniques with Applications: Essays in Honor of Bruno de Finetti.* (K. Goel and A. Zellner, eds.). Amsterdam: North-Holland.

— (1983b). Temporal coherence. (In this volume).

GOOD, I.J. (1982). Comment on "Lindley's Paradox" by G. Shafer. *J. Amer. Statist. Assoc.* **77**, 342.

HAMMOND, K.R., MCCLELLAND, G.H. and MUMPOWER, J. (1980). *Human Judgment and Decision Making: Theories, Methods, and Procedures.* New York: Praeger.

HILL, B. (1969). Foundations for the theory of least squares. *J. Roy. Statist. Soc.* B, **31**, 89-97.

JENSEN, D.R. and GOOD, I.J. (1981). Invariant Distributions Associated with Matrix Laws under Structural Symmetry. *J. Roy. Statist. Soc.,* B, **43**, 327-332.

KADANE, J.B., DICKEY, J.M., WINKLER, R.L., SMITH, W.S. and PETERS, S.C. (1980). Interactive elicitation of opinion for a normal linear model. *J. Amer. Statist. Assoc.* **75**, 845-854.

KARIYA, T. and EATON, M.L. (1977). Robust test for spherical symmetry. *Ann. Statist.* **5**, 206-215.

KELKER, D. (1968). Distribution theory of spherical distributions and some characterization theorems. *Tech. Rep.* **RM-210**, DK-1. Dept. of Statistics and Probability, Michigan State University.

— (1970). Distribution theory of spherical distributions and a location-scale parameter generalization. *Sankhyā A*. **32**, 419-430.

— (1971). Infinite divisibility and variance mixtures of the normal distribution. *Ann. Math. Statist.* **42**, 802-808.

LINDLEY, D.V. (1981). Reconciling continuous probability assessments. Manuscript. Decision Science Consortium, Inc., 7700 Leesburg Pike, Falls Church, Va. 22043.

— (1982). The improvement of probability judgments. *J. Roy. Statist. Soc.* A, **145**, 117-146.

— (1983). Reconciling discrete distributions. (In this volume).

LINDLEY, D.V., TVERSKY, A. and BROWN, R.V. (1979). On the reconciliation of probability assessments. *J. Roy. Statist. Soc.* A, **142**, 146-180 (with discussion).

LORD, R.D. (1954). The use of the Hankel transform in statistics. I. General theory and examples. *Biometrika*, **41**, 44-45.

MUIRHEAD, R.J. (1982). *Aspects of Multivariate Statistical Theory*. New York: Wiley.

PRESS, S.J. (1969). On serial correlation. *Annal. of Mathematical Statistics,* **40**, 188-196.

SHAFER, G. (1976). *A Mathematical Theory of Evidence*. Princeton: University Press.

TYLER, D.E. (1982). Radial estimates and the test for sphericity. *Biometrika,* **69**, 429-436.

DISCUSSION

D.V. LINDLEY (*Somerset, UK*)

Bayesian statistics today has inherited the yoke of sampling-theory statistics and we predominately take a likelihood and a prior, multiply and normalize to get a posterior. But there is more to the global view of uncertainty expressed through probability than that and we can contemplate and handle new problems. Dickey & Chen have studied one of these: the assessments of probabilities for several quantities and have made significant progress.

Their ideas are excellent and embrace a wider variety of possibilities than have heretofore been available but I feel that there is one important point omitted; namely coherence. The paradigm is based on coherence and consequently coherence should be included in any assessment programme as an integral part. The best way to prevent an expert "fooling himself" is to test his coherence. Here is an example. Suppose two random quantities, x and y, are being considered. They are assessed as having a bivariate, normal distribution with zero means. Only the dispersion matrix remains. Suppose that this is studied through var(x), $E(y|x)$ and var($y|x$) and suppose these are judged independent. Then coherence says that it is *not* possible that, interchanging the roles of x and y, var(y), $E(x|y)$ and var($x|y$) be independent. Most people having made one judgment of independence would be just as happy with the other. Coherence rules this out.

Of course, with a topic like this, we need to do lots of experiments with subjects to see whether they can make the judgments required by Dickey & Chen's theory. My guess would be that they will have difficulties with moments. They are very hard to assess. Fractiles, as used by Pratt & Schlaifer in their conference paper, are likely to be much better.

L. SIMAR (*Facultés Universitaires Saint Louis-Bruxelles and C.O.R.E.-Louvain-la-Neuve*)

The paper of Professor Dickey is an important contribution to the analysis of new families of distribution functions, and I have been impressed by the mathematics which are used to prove some remarkable properties of this family.

The main attractive idea of the paper could be summarized as follows: specify a structure of opinion only through predictives and choose a member of a large class of distribution functions as possible candidate for the predictive. This family of d.f. should be tractable and able to represent a wide variety of opinion structures. The claim of Professor Dickey is that the family of Elliptical distributions is suitable for this purpose.

The idea of working in statistical analysis ony through the predictives is an old attractive one, since it is much more easy to work on observables than with parameters (of a sampling model) which are often artificial and sometimes useless. Even in the formal (classical) Bayesian approach, what is needed is a joint distribution on the observables and the parameters and in most cases, the assessment of the predictive allows us to construct the prior for a given sampling model. However the main advantage of working with a sampling model, even when it is not necessary for the imbedded decision problem, is that it allows a so called "objective treatment" of the information available on the process: we know what we learn in observing, we have some guide-lines within which we know what we are doing.

In subjective probability modeling, we work with the essential (the observable real world) but we no longer have any "guide-lines" and the danger of errors, of misuse,...may be important. This point is quoted by Professor Dickey ("Tools can be misused"). In subjective probability modeling, what kind of misuses does he have in mind and how to prevent them?

I was very impressed by the results of Dickey and Chen on Elliptical distributions. We have there a very large family of d.f. characterized by some remarkable properties, simple computations (univariate integrals) of moments, of marginals, of conditionals, etc.... plus a description of an easy way of assessment (through univariate quantiles). Moreover, this family has some properties similar to those of the usual Normal case. However I wonder if from a practical point of view the assessment and the manipulations are so easy. In particular, how to choose practically the standard Univariate Spherical (without reference to some imbedded sampling model or to some objective information on the probability structure)? Some parametric forms are suggested in the appendix but how to recognized them? Can we hope for the aid of some computer programs? Further no internal checking procedures are set out, so what are the controls available in this general family to ensure that the elicited predictive is meaningful?

As a statistician I like to learn when observing and to measure or to appreciate what I have learnt. Theoretically, this can be achieved by considering the distribution of $y = (y_1,y_2)$ where y_1 is the past and y_2 the future; the analysis would then be the comparison of the marginal y_2 with the conditional $y_2|y_1$. But in this general context is this comparison practically possible or would we content ourselves with the comparison of the "Normal-like" moments?

All the above comments show that I would like to see how those important results may be used in a real world problem. In particular, does Professor Dickey have any empirical work which has shown the superiority of using this general family in comparison with a more specific model (like Multivariate-t)?

Finally, I will ask Dickey and Chen, the following three points:

(i) Is there any relation between the characterization of the Elliptical distributions and the work of Kagan-Linnik and Rao (1973) on characterization problems in statistics. (like properties of c.f., inducing linear regressions etc.)?

(ii) The marginals of a Spherical are also Spherical but not necessarily of the same type, except in the Normal case. Are there similar results for the conditionals of Spherical (and/or) Elliptical distributions?

(iii) As mentioned in the introduction by Dickey and Chen, Elliptical families could also be used as sampling models for a formal Bayesian analysis (with robustness in mind). Is there any hope of tractable results in this framework (prior, posterior on the parameters (a, U) etc.)?

J. BERGER (*Purdue University*)

Before making a few comments on the fairly radical philosophical position that seems to be espoused in the paper, I would like to say a few words about why this work strikes me as extremely valuable from a more traditional Bayesian viewpoint. In dealing with statistical problems with a multivariate normal sampling distribution, the usual conjugate normal priors suffer a certain lack of robustness in automatic use. (For some discussion and references see Berger (1983).) Prior distributions with flatter tails seem to offer some definite advantages. Indeed elliptical distributions have been put to this use as priors (c.f., Berger (1980)), and the rich and "elicitable" classes proposed by Professors Dickey and Chen thus strike me as having great potential as simply prior distributions. A question of some interest, in this regard, is - does an elliptical prior together with a normal sampling distribution (or, more generally, an elliptical sampling distribution) yield an elliptical predictive distribution and/or an elliptical posterior distribution? And, if the predictive distribution is elliptic, can the elicitation methods discussed in the paper be of help in getting at the elliptic prior by first eliciting the predictive distribution?

I called the philosophical position in the paper radical, not because I disagree with the authors' contention that a prior - sampling distribution breakup of knowledge may be unattainable, but rather because I feel that, in most problems, it is wise to attempt such breakups. Intuitive combination of different sources of information into a grand whole is very risky business, and I think it is crucial to emphasize that probabilistic decomposition and elicitation of factors separately should be attempted. For instance, it does not greatly bother me that practical Bayesian analysis cannot really be done in the prior to posterior mode in that, for realistic problems, prior elicitation must typically go hand in hand with study of the data. I will ultimately be satisfied with an answer only if the answer emerges as the answer for any reasonable quantification of my prior opinions, when combined with the data through Bayes theorem. In other words, one can fight "contamination" of prior beliefs (by the data) by insisting on rigorous sensitivity studies, and I would feel much more comfortable with such an approach than with attempting to intuitively understand what the data and prior information say via some grand gestalt. I recognize that the authors by no means seek to preclude such standard Bayesian investigations, as part of their overall elicitation attempt, but I would question the relegation of this activity to the role of only a supporting actor in the play.

F.J. GIRÓN (*Universidad de Málaga*)

Professors Dickey and Chen are to be congratulated for the elegant theory of spherical and ellipsoidal distributions.

Although their definition is simple and straightforward, the implications are far reaching not only from a theoretical point of view but also for their practical use in helping elicit probability distributions.

As shown in the Appendix, not only well known multivariate distributions, which appear naturally in Bayesian analysis as prior or predictive distributions, but some apparently new distributions are brought to light by this theory.

Yet I want to remark, from the perspective of an expert trying to elicit a distribution, three nice properties enjoyed by the elliptical distributions, which makes life easier for the probability assessor. This alone, could serve as a stimulus for a theoretical treatment of the subject.

While reading the paper, one gets the impression that the authors have set up first the principles and then have sought for a sufficiently rich and tractable family that meets these requirements. The result is that they found what they very aptly call "ellipsoidal distributions".

These principles are:

1st. principle: Marginals, and more generally, linear combinations of coordinates of ellipsoidal distributions are ellipsoidal (Th. 3.2)

2nd. principle: Conditionals of ellipsoidal distributions are also ellipsoidal (Th. 3.3)

3rd. principle: Scale mixtures of ellipsoidal distributions are ellipsoidal (Th. 5.2)

Whilst the first two principles seem basic in order to characterize ellipsoidal distributions, the 3rd. one does not seem so relevant but seems important for modelling when assessing complex underlying distributions.

One important feature of the ellipsoidal distributions, with important implications in the theory of portfolio selection, is that they contain as a special case a subclass of the multivariate stable distributions considered by S.J. Press (1970, 1971). His definition is closely related to the first principle by allowing linear univariate combinations of the vector to be univariate stable distributions. (Th. 4.1).

As a by-product of the paper one obtains the result, new to me, that the conditional distributions of certain multivariate stable distributions are stable.

This result seems difficult to prove directly, as stable densities only have a closed and compact form for very few values of the characteristic exponent (= ½, 1 and 2).

While the authors succeed in the task of determining the location and scale parameters of the joint-ellipsoidal distribution, with the only objection that the assessment of the matrix U in order to be positive definite may be *order* dependent, more research is needed for the more difficult problem of determining the standard marginal distributions.

For instance, suppose we believe the distribution to be assessed is a finite mixture of normal distributions. The authors present a simple method for eliciting the weights of the *mixing* distributions but they do not provide a method for determining the number of the terms in the mixture.

Finally, I want again to congratulate the authors and look forward to seeing the papers which will follow on from this one.

REPLY TO THE DISCUSSION

Before replying to specific points made by the discussants, I should like to clarify certain general issues. It is asserted in the paper that analyses and decisions are made (and should be made) based directly on post-data expert opinion. Perhaps, the only disagreement with whether they *are* so made is that analyses without specific decisions in sight commonly show little honesty concerning their reliance on opinion. The language of the myth of inference without an inferrer pervades our culture. I maintain, however, that analyses can be done more honestly and fully by articulating the operative opinions in probability-model form.

There is a theoretical side and a practical side to the direct basis in post-data expert opinion. In elementary decision theory, a person is admonished by rational preference considerations to choose (post-data) an act to maximize subjective-expected utility (using post-data subjective probability, of course). It is only by distorting the meaning of the theory by choosing ''procedures'' (act-valued functions on the sample space) *prior* to the statistical data that one can justify the automatic application of a Bayes-formula posterior distribution. The inevitable questions are, ''Why choose before seeing the data?'', and ''Why later follow the prior choice?'' In recent talks, F.J. Anscombe has called this distortion ''a-priori-ism''. Another term for it might be ''procedurism'' (Dickey 1975).

Practical inadequacies of a prior-ism were brought out in the paper under discussion. In mention again merely the problems of (a) unanticipated data, and (b) new ideas. In practice then, to reply to a point by Professor Berger, I would promote Bayes' formula from an underemployed star player to a strong supporting actor, with the dual roles of: (i) data analysis (as a tool among tools); and (ii) a point of reference, articulation, and justification for expert probabilities.

Years ago, I would argue with my frequentist colleagues, as follows. Suppose you were waiting for the statistical data from an experiment, and to while away the time, you carefully assessed your prior distribution (and sampling model). After receiving the data you [or your robot] calculated your serious frequentist procedures; and just out of curiosity, you [or your robot] also calculated Bayes' formula; and then you saw an important disagreement between the two analyses; and checking, fussing, and honest rethinking did not clear it up. What to do? Which analysis to use then? Just how does your argument go for such a situation?

Now, to *all* my procedurist colleagues, suppose that you also directly assess your post-data uncertainty; carefully, thoughtfully, with feedback and much calculation of various points of reference; and now you find important disagreement with the previous two analyses; and it does not go away. Now which? Exactly how does one argue that one of the previous two analyses should be preferred to the direct analysis? Just how do you tell the responsible expert or decision maker to go against his own considered judgement? If this issue is not dodged, I feel that we may see a new paradigm in statistics.

Professor Simar and Professor Berger raised the question of the role of sampling models. I wonder, about this, too. My background and my experience prejudice me in favour of their use. But not always. Neither this paper nor my paper in the previous Valencia conference (Dickey 1980) relies in any way on any assumption of a sampling model as a mixand.

Further yet, what is the role of distributions, at all, versus previsions? I feel the pathbreaking work of Michael Goldstein in this conference and elsewhere cannot be ignored. I like working with distributions, but there is writing on the wall. Professor Simar also raised the question of the role of probability conditioning. Again, I vacillate, recalling elegant work by Goldstein on ''separation of belief'', in which conditional probability is only one component of the anticipated reaction to information. My work on probability assessments has used conditional probability as if it were the whole anticipated reaction.

Professor Simar asked for ways to help prevent misuse of direct post-data assessment. One possibility is to require the availability of a detailed record of the assessment process, as performed. The effect of requiring that expert opinion be expressed in such documented personal probability form should be an increase in the level of responsibility. Of course, all the usual objective standards and requirements of subsequent availability can be applied to any experimental and observational data and its analysis on which the expert bases his opinion.

Before turning to more technical matters in the discussion, I would like to call attention to the work of Professor Lindley on "probability reconciliation", reported in this conference and elsewhere. This has been very stimulating for me, as he too treats problems of probability assessment. The differences between our approaches are intriguing. In Dickey (1980) I assumed an unknown subjective-probability distribution, with stochastic elicitations of its quantiles, and I recommended corresponding generalized least squares fitting in the assessment method. Professor Lindley, in his work, assumes that a person's elicitations are measurements with error on the actual unknown quantity of direct interest and he then carries out Bayesian inference to produce a posterior distribution as the "reconciled" (assessed) distribution. The conception is one of experts as being measuring instruments.

I agree with everything Professor Lindley says in his discussion of our paper, especially his call for more experience using assessment methods. One should note, however, that we do not try to assess moments, but rather quantiles (fractiles), as he recommends.

Professor Simar asks how to choose the standard one-dimensional marginal distribution in our method. A major point of the paper is that one can directly elicit quantiles of the one-dimensional margin of the ellipsoidal distribution. The shape, as indicated by ratios of quantile differences, is just the same as the shape of the one-dimensional standard distribution. This shape, in practice, can be fully specified by the expert's elicited quantiles, in as much detail as desired. There are no constraints, in principle. Once a shape is specified, the Abel type integral equation can be solved for the radial distribution and corresponding joint distribution. Of course, the use of a parametric family can simplify this process by removing the need to solve the integral equation. Simar's points (i) and (ii) require more time and space to answer adequately than here available. His point (ii) is affirmatively answered by Theorems 2.6 and 3.3 in the paper.

I am grateful to Professor Giron, who reminded us that S.J. Press's (1972) multivariate stable distributions can be ellipsoidal. He also raised the question, not faced in our paper, of how many mixands to have in the model. Even worse, we have not yet adequately considered the logically prior question of what distributions to use as potential mixands.

Professor Berger brings up the use of general ellipsoidal distributions as Bayesian prior distributions. Dickey (1968, 1974) developed multivariate-t prior distributions for the normal-regression location vector with prior independence of the scale; and Leamer (1978) treated general ellipsoidal priors. Unfortunately, the convolution of an ellipsoidal with an ellipsoidal distribution, as in a Bayesian predictive distribution, is not necessarily ellipsoidal. Witness the multivariate Behrens-Fisher distributions (Dickey 1968), which have two matrix parameters.

Finally, I would like to thank Professor Diaconis who in the oral discussion recommended to us all M.L. Eaton's excellent paper (1981) on spherical and ellipsoidal margins. Rob Kass of Carnegie-Mellon University subsequently brought to my attention Cambanis, Huang and Simons (1981) and recent technical reports by T.W. Anderson.

Note added in proof: Bruce Hill has pointed out in private correspondence that he used versions of Theorems 2.2 and 2.4 in work apparently predating Kelker. (See Hill, (1969) Foundations for the theory of least squares. *J. Roy. Statits. Soc. B,* **31**, 89-97; see also Hill (1975) On coherence, inadmissibility and inference about many parameters in the theory of least squares. *Studies in Bayesian Econometrics and Statistics: In Honor of Leonard Savage* (S.E. Fienberg and A. Zellner eds.) Amsterdam: North-Holland.)

REFERENCES IN THE DISCUSSION

BERGER, J. (1980). A robust generalized Bayes estimator and confidence region for a multivariate normal mean. *Ann. Statist.* **8**, 716-761.

— (1983). The robust Bayesian viewpoint. *Robustness in Bayesian Statistics* (J. Kadane, ed.). Amsterdam: North-Holland.

CAMBANIS, S., HUANG, S. and SIMONS, G. (1981). On the Theory of elliptically contoured distributions. *J. Multivariate An.* **11**, 368-385.

DICKEY, J.M. (1968). Three multidimensional-integral identities with Bayesian applications. *Ann. Math. Statist.,* **39**, 1615-1627.

— (1974). Bayesian alternatives to the *F*-test and least squares estimate in the normal linear model. *Studies in Bayesian Econometrics and Statistics* (S.E. Fienberg and A. Zellner, eds.). Amsterdam: North-Holland, 515-554.

— (1975). Holes in the sample space. *Amer. Statist.,* **29**, 131-132.

— (1980). Beliefs about beliefs, a theory of stochastic assessments of subjective probabilities. *Bayesian Statistics.* (Bernardo, J.M. *et al.* eds.), Valencia: University Press.

EATON, M.L. (1981). On the projections of isotropic distributions. *Ann. Statist.* **9**, 391-400.

KAGAN, A.M., LINNIK, Y.V. and RAO, C.R. (1973). *Characterization Problems in Mathematical Statistics.* New York: Wiley.

LEAMER, E. (1978). *Specification Searches.* New York: Wiley.

PRESS, S.J. (1970). Multivariate stable distributions. Paper presented at the 2nd. World Conference of Econometric Society, Cambridge.

— (1971). Estimation in univariate and multivariate stable distributions. Unpublished mimeo.

— (1972). *Applied Multivariate Analysis.* New York: Holt, Rinehart, and Winston.

BAYESIAN STATISTICS 2, pp. 183-202
J.M. Bernardo, M.H. DeGroot, D.V. Lindley, A.F.M. Smith (Eds.)

Group Consensus Probability Distributions: A Critical Survey

SIMON FRENCH
University of Manchester

SUMMARY

The formation of a group consensus probability distribution from the beliefs of its members has attracted much attention in the literature. Several procedures have been proposed, some on pragmatic grounds, others justified axiomatically. More fundamentally, the feasibility and desirability of such procedures have been discussed and several impossibility theorems proved. This paper surveys these contributions with particular regard to the context: why is such a consensus distribution needed.

Keywords: CALIBRATION; CONSENSUS OF OPINION; EXTERNAL BAYESIANITY; GROUP CHOICE; HONESTY; INDEPENDENCE OF EXPERTS; MARGINALITY; SUBJECTIVE PROBABILITY.

0. INTRODUCTION

0.0 *The Problems*

Suppose that n individuals each have opinions about events in a particular σ-field. Make the further utopian assumption that they are all Bayesians and so encode their beliefs as subjective probabilities. How should these be aggregated or summarised in a single probability distribution? This question begs another: why is such a summarised opinion needed. There seem to be three, possibly four reasons.

The expert problem. The group may be a panel of experts who have been asked for their advice by a decision maker, who faces a real, predefined decision problem. The decision maker is, or can be taken to be, outside the group. It is he alone that has the task of assimilating and aggregating their opinions. The group itself may never meet, but interact individually with the decision maker.

The group decision problem. The group itself may be jointly responsible for a decision. They may wish that to the outside world their decision making exhibit all the coherence embodied in the Bayesian paradigm of, say, Ramsay (1931) or Savage (1972). Thus they would wish to combine their beliefs and preferences into a group consensus probability distribution and a group utility function.

The text-book problem. The group may simply be required to give their opinions for others to use at some time in the future in as yet undefined circumstances. Here there is no predefined decision problem, but many potential ones that are only vaguely perceived. An everyday example occurs when several meteorologists are involved in giving a single forecast. In a sense there are two types of problem hiding under the same heading here:

those in which the group is responsible for collating their views and those in which an intermediary has that task. However, in another sense the problems are the same, because, whoever is responsible for collating the opinions, it is the needs of the ultimate, possibly unidentified users of that information that should be the primary determinant of the method of collation.

As I shall argue, the feasibility (and desirability) of forming a group consensus distribution depends on which of these problems obtains: a point that has not always been appreciated in the literature.

0.1. *The Dimension of Discussion*

Discussions such as this need issues and principles to give them direction; we will use several.

Marginalisation. Suppose that interest is focussed on a sub-σ-field of that over which the group members gave their opinions. It seems intuitively appealing that the same consensus distribution be obtained whether (i) their opinions are first combined into a consensus distribution over the complete σ-field and then a marginal distribution taken or (ii) the group members each give their marginal distributions over the sub-σ-field and a consensus distribution formed from these. McConway (1981) has shown that this *Marginalisation Property (MP)* is equivalent to the *Weak Setwise Function Property (WSFP).* This demands that the consensus probability of an event depends only on the event itself and the probabilities assigned to it by each of the group members; there is no dependence on their probabilities for other events.

External Bayesianity. Suppose that some objective evidence becomes available and that everyone concerned agrees on the likelihood function derived from this evidence. Then it seems reasonable to demand that the following two procedures give rise to the same posterior consensus distribution. (i) Each group member updates his prior belief through Bayes' Theorem and then the consensus distribution is formed. (ii) The consensus distribution of their prior distributions is first formed and then this is updated through Bayes' Theorem. This property of *External Bayesianity* was so termed by Madansky (1964). It is also known as *Prior-to-Posterior Coherence* (Zidek, 1983).

Probabilities or Data? De Finetti (1974) has argued cogently that probability only has existence for the person whose beliefs are modelled. Thus the probabilities elicited from a group member are probabilities *for him*; but what are they for anyone else? One is at liberty to adopt by conscious decision another person's probability for ones own, but one is not obliged to. One may choose to treat it simply as relevant data. Thus consensus procedures might assimilate the group's probabilities either as probabilities or as data.

Calibration. To interpret another's subjective probabilities one needs to know not only what substantive knowledge he possesses, but also how well calibrated he is, i.e., how good he is at encoding his beliefs as probabilities. The literature on calibration is growing apace and, in fact, encompasses several issues (Dawid, 1982; DeGroot, 1980; Kahneman *et al*, 1982; Lindley, 1982; Morris 1974, 1977, 1983). Here we shall not need to be precise about what is meant; instead we shall use the term *calibration* loosely to evoke such ideas as anchoring and availability as well as the more specific concepts such as frequency calibration.

Honesty. However good a probability assessor an individual is, in interactions with others it may be to his advantage to deliberately misstate his probabilities. This possibility of dishonesty clearly has implications for any consensus procedure.

Correlations between Group Member's Opinions. Two people's opinions are seldom independent of each other. The existence of a common language means that they will share many perceptions of the world. If they are both trained experts with knowledge relevant to an issue, then their opinions are likely to be substantially correlated. Such dependence is not easy to quantify, but it is surely important in discussing consensus of opinion.

Relative Expertise. While one may argue on democratic grounds that in matters of preference all men are equal, the same does not follow in matters of opinion and knowledge. We are not all equally expert in all matters. Thus any sensible consensus procedure should acknowledge the relative expertise of the group members.

Single Event or a larger Field of Interest? Several discussions of consensus of opinion hve focussed on the case in which there is only one event of interest. This simplifying assumption has the advantage of highlighting some issues; but, of course, it clouds others. For instance, it is impossible to discuss marginalisation unless the field is somewhat larger.

0.2 *Overview of Paper*

In turn a section is devoted to each of the expert, group decision and text-book problems. In each section the relevant literature is surveyed and reflected upon in the light of the issues raised above. I make no claim to the comprehensiveness in these surveys, nor am I impartial. Earlier, more comprehensive, and perhaps more impartial surveys have been given by Winkler (1968, 1981), Bacharach (1973), Dalkey (1976), Hogarth (1977), Press (1978) and Zidek (1983). Here my intention is to raise issues for discussion and to sound warnings on what I believe to be rather naive approaches to the problems.

1. THE EXPERT PROBLEM

1.0. *Problem Statement*

A decision maker is concerned with a specific decision problem the consequences of which depends upon events in a σ-field, $\mathcal{A}$. He asks n experts for their advice and each states his beliefs as subjective probabilities, p_k ($k = 1,2,\ldots,n$). Here p_k is to be interpreted as a probability mass or density function according to the context. The decision maker wishes to define his own probability distribution, π, in the light of their advice. π should clearly be a function of the p_k and arguably several other factors; but what function and what other factors?

1.1. *The Linear Opinion Pool*

Probably the commonest suggestion for the form of π is:

$$\pi = \sum_{k=1}^{n} \alpha_k p_k; \qquad \alpha_k \geq 0 \quad \forall k, \qquad \sum_{k=1}^{n} \alpha_k = 1.$$

Its origins date back at least to Laplace (Bacharach, 1979). Stone (1961) termed it the *Linear Opinion Pool.*

McConway (1981) has shown that when $\mathcal{A}$ contains at least three non-trivial, proper events the use of the linear opinion pool is equivalent to assuming the *Strong Setwise Function Property*, viz:

$$\pi(A) = f(p_1(A), p_2(A), \ldots, p_n(A)) \qquad \forall A \in \mathcal{A}, \qquad \text{(SSFP)}$$

i.e. the decision maker's probability of A dependes only on the n numerical probabilities assigned to A by the experts. Genest (1982) terms SSFP the *Locality Property*, while

Bordley and Wolff (1981) call it the *Context-free Assumption*. McConway also shows that SSFP is equivalent to both WSFP and the Zero Probability Property, viz:

$$\pi(A) = f(A, p_1(A), p_2(A), \ldots, p_n(A)) \qquad \forall A \in \mathcal{A} \qquad \text{(WSFP)}$$

and

$$\forall A \in \mathcal{A}, \quad p_k(A) = 0 \quad (k = 1,2,\ldots,n) \Rightarrow \pi(A) = 0 \qquad \text{(ZPP)}$$

Remembering that MP and WSFP are equivalent, we see that the linear opinion pool possess MP. Moreover, it is the only opinion pool to do so if we also accept ZPP. Wagner (1982) proves a similar set of results to McConway, but limits his discussion to finite σ-fields.

The linear opinion pool possesses a number of appealing properties: MP; ZPP; the weight α_k clearly reflects in some way the relative expertise of the kth expert; and it is both easily understood and easy to calculate. Nonetheless, I would argue that it is not an acceptable solution to the expert problem.

MP is not so obviously a desirable property. Suppose that the experts have different fields of expertise - and to some extent this will always be so. Suppose, say, that one expert's specialisation is forecasting over a particular sub-σ-field of $\mathcal{A}$, but that he has little experience relevant to the rest of $\mathcal{A}$. Then his opinion should have high weight in the marginal consensus distribution over that sub-σ-field, but much lower weight elsewhere. MP denies this.

SSFP is a very strong condition. Dalkey has shown that any consensus distribution which obeys SSFP cannot obey all the laws of mathematical probability (McConway, 1981; Bordley and Wolff, 1981). In this case there are problems in relation to conditional events, which, in turn, means that the linear opinion pool does not possess EB.

Continuing this point, SSFP implies that the α_k are fixed whatever marginal distribution is taken. Yet any decision maker would surely wish to revise the α_k if he gained information about the relative expertise of the experts. In fact, if one abandons SSFP and allows the possibility of revising the α_k, then a sensible procedure for updating the weights can be found (Roberts, 1965; Raiffa, 1968). The analysis is to all intents and purposes the standard Bayesian analysis of mixtures of models (see, e.g., Smith and Spiegelhalter, 1980). Moreover, once revision is allowed the procedure possesses EB. Nevertheless, it still has several faults; faults that are common to a number of other opinion pools.

1.2. *The Logarithmic and other Opinion Pools*

Several authors have suggested the *Logarithmic Opinion Pool*:

$$\pi(A) = \prod_{k=1}^{n} (p_k(A))^{\alpha_k}; \quad \alpha_k \geq 0 \,\forall k.$$

(Weerahandi and Zidek, 1981; Zidek, 1983). Indeed, it has several points in its favour. For instance, it possess EB and it may be justified in certain cases by the natural conjugate approach of Winkler (1968). Against this it must be noted that it possesses ZPP to a dramatic degree: if any expert assigns $p_k(A) = 0$, then $\pi(A) = 0$. However, the most telling reasons to my mind against the use of the logarithmic opinion pool count against the linear opinion pool, with or without weight revision, and a whole family of other conceptually possible opinion pools as well.

First, they all introduce weights α_k, which clearly reflect the relative expertise of the experts, but which are not operationally defined. How should they be chosen? To say that one expert is twice as good as another is a figure of speech, not an arithmetic statement. The problem is further complicated by the likelihood of correlation between the expert's opinions. Several pragmatic solutions have been proposed (see, e.g., Winkler, 1968), but to my knowledge none avoid a certain arbitariness.

Second, the issue of calibration is ignored and the p_k implicitly interpreted as probabilities, instead of as data. For instance, ZPP is not as innocuous as might appear. It is known that experts are often overconfident in assigning high and low probabilities. They tend to assign too low a probability to circumstances which they consider very unlikely (Kahneman *et al,* 1982). Thus, even if they all assign $p_k(A) = 0$, the decision maker may quite reasonably wish to correct for their overconfidence by taking $\pi(A)$ to be small but positive. It might be suggested that the p_k should be transformed through a *calibration function* (Morris, 1974, 1977, 1983) and then an opinion pool taken; but even this is dubious because it treats the transformed p_k as probabilities of the decision maker and they are not: for him they are data. It is rather as if a statistician transformed a set of observations to normality and *then forgot the normal errors.*

Third, it is assumed that the experts are honest and this is a strong, essentially unverifiable assumption. One might try to buy the expert's honesty by means of a scoring rule. But what works with a single expert need not work with a group. Suppose that (i) $\mathcal{A}$ has a finite basis $A_1, A_2, \ldots, A_m$, (ii) each expert is rewarded by a concave, monetary scoring rule $S(r_k, A_j)$ if he states r_k and A_j occurs, and (iii) each expert has a linear utility for money. Then:

$$\sum_{j=1}^{m} p_k(A_j) \sum_{i=1}^{n} S(p_i, A_j) \leq \sum_{j=1}^{m} p_k(A_j).n.\, S\left(\sum_{i=1}^{n} (p_i/n), A_j\right), \quad \forall k,$$

since S is concave. Thus each expert would expect the group to be rewarded more in total if they all stated $\Sigma p_i/n$ instead of their own true beliefs. Hence, if a scoring rule is used and group interaction allowed (so that they can each discover $\Sigma p_i/n$), Pareto's principle would encourage dishonesty not honesty. Dalkey (1976) gives a similar argument to this (but uses it to argue a virtue not a vice).

Because of all these arguments, I believe that the solution to the expert problem requires an approach that is distinctly different to that underlying opinion pools.

1.3. *Bordley's Approach*

Bordley (1983) takes an interesting thought-provoking approach. He confines attention to a single event of interest, A. Define $o_k = p_k/(1-p_k)$ $(k = 1, 2, \ldots, n)$, the kth expert's odds on A, and consider the vector $\mathbf{o} = (o_1, o_2, \ldots, o_n)$. Assume that the decision maker has an intuitive ordering, R_A, defined by $\mathbf{o} R_A \mathbf{o}' \Leftrightarrow$ he feels at least as sure about the event A ocurring when he hears the expert's odds $\mathbf{o}$ as when he hears $\mathbf{o}'$. Bordley assumes that the ordering R_A obeys sufficient consistency conditions to give an additive conjoint structure over the set of odds vectors $\{\mathbf{o}\}$ (Krantz *et al,* 1971). The two important assumptions here are that R_A is a weak order, i.e., comparable and transitive, and that independence or, as Bordley terms it, *noninteraction* holds. To define this latter property, let $I \subset \{1, 2, \ldots, n\}$ and rewrite $\mathbf{o}$ as $(\mathbf{o}_I, \mathbf{o}_{\bar{I}})$, where $\mathbf{o}_I$ are the odds of the experts in I and $\mathbf{o}_{\bar{I}}$ those of the rest. Then R_A is noninteractive if $\forall I$:

$$(\mathbf{o}_I^1, \mathbf{o}_{\bar{I}}^0)\, R_A\, (\mathbf{o}_I^2, \mathbf{o}_{\bar{I}}^0) \text{ for some } \mathbf{o}_{\bar{I}}^0 \Rightarrow (\mathbf{o}_I^1, \mathbf{o}_{\bar{I}})\, R_A\, (\mathbf{o}_I^2, \mathbf{o}_{\bar{I}}) \text{ for all } \mathbf{o}_{\bar{I}}.$$

In other words, if some experts change their opinion the effect that this has on the decision maker is independent of the fixed opinions of the other experts. By assuming some further house-keeping axioms, EB, and that $\pi(A) + \pi(A^c) = 1$, Bordley deduces the form of $\pi(A)$, which is most conveniently given in log-odds form. Let $o_D = \pi(A)/(1-\pi(A))$. Then

$$\log(o_D/o_o) = \sum_{k=1}^{n} \alpha_k \log(o_k/o_o),$$

where α_k and o_o are constants that can be determined by the usual procedures for fitting an additive conjoint structure to a decision maker's subjective judgements (see, e.g., Keeney and Raiffa, 1976). Bordley suggests that o_o may be interpreted as the decision maker's prior odds on A. By taking antilogarithms, it can be seen that Bordley essentially deduces the logarithmic opinion pool, but without interpreting the p_k as probabilities; they are just indications of likelihood. Moreover, the assumption of noninteraction is akin to, but much weaker than assuming noncorralation of the experts' opinions.

There are at least two difficulties with Bordley's approach, however. First, it is not obvious how it might be generalised to the case where there is more than one event of interest. Second, although o_o can be interpreted as the decision maker's prior odds, it is derived from the additive conjoint measurement procedure *after* the decision maker has heard (or, rather, imagined he has heard) the experts' opinions. Thus the procedure is not truely Bayesian in spirit.

1.4. *Morris's Axiomatic Approach*

Morris (1983) has like Bordley adopted an axiomatic approach; and he too is aware of the problem of calibration. Moreover, his methods are applicable to general $\mathcal{a}$. Initially, he asks how the decision maker should learn form one expert. Suppose that the decision maker has a genuine prior opinion π_o and he learns the expert's opinion p. Then Morris assumes that the decision maker should update his belief to $\pi = \phi(p,\pi_o)$. Note that the arguments of ϕ are probability distributions, although it is not assumed that the expert's distribution is calibrated. First, Morris assumes that ϕ possess EB. Next, he assumes one of two equivalent axioms; equivalent, that is, in the presence of Morris' other assumptions. One of these is:

> If both the decision maker and a calibrated expert have uniform priors, the updated distribution should also be uniform.

This I do not find objectionable; but what is worrying is that I find the equivalent assumption unacceptable:

> If a calibrated expert gives p to be a uniform distribution, then $\pi = \pi_o$.

Suppose that the decision maker's prior is sharply peaked. Then surely, on learning that an expert can give nothing other than a random prediction, the decision maker's confidence will be reduced and he will "spread out" his belief more.

The problem, I think, lies in Morris's assumption that $\pi = \phi(p,\pi_o)$. This assumes that π depends only on the functions π_o and p, and not on the reasons that lie behind the beliefs that they model. Thus p is treated in the same way whether it is based solely on the opinions of the expert or almost entirely on "objective" data that the expert has analysed statistically.

1.5. *Bayesian Updating*

My own preference for a solution to the experts problem is to have the courage of our Bayesian convictions: admit that the experts'opinions are simply data for the decision maker and so use Bayes'Theorem to update his beliefs. This has the immediate advantage that we can use the likelihood function to allow for the possibilities of miscalibration, dishonesty, and nonindependence of the experts. However, there is one further point that we should remember. If the decision maker were one of the experts, then we should allow for the possible correlation of his opinions with those of the others. Simply because we have conceptually separated him from the group does not mean that we should forget this possibility. I find it easiest to do this by conditioning the likelihood on the prior and have justified doing so elsewhere (French, 1980, 1981, 1982). Another means of achieving the same end is to remember that *all* the decision maker's probabilities are conditioned on H, the entire knowledge and previous thoughts of the decision maker (Jeffreys, 1961). Taking π_o, the prior, as a proxy for the relevant portion of H leads to essentially the same methods.

The following is a very brief summary of French (1981). Suppose that there is a single event of interest, A. It will be convenient to work in log-odds. Let $\lambda_k = \log(p_k/(1-p_k))$ and $\ell_0 = \log(\pi_o/(1-\pi_o))$. Further let $\lambda = (\lambda_1,\lambda_2,\ldots,\lambda_n)^T$ and $\ell = (\ell_0,\lambda^T)^T$. Bayes' Theorem tells us that

$$\pi(A\,|\,\lambda) = \pi(\lambda\,|\,A,\ell_0).\pi_o/\{\pi(\lambda\,|\,A,\ell_0)\pi o + \pi(\lambda\,|\,A^c,\ell_0)(1-\pi_o)\}$$

where $\pi_o = \pi(A)$ is the decision maker's prior for A and I have used π as a generic symbol for all the decision maker's beliefs.

Now if we assume that over a number of similar assessments the decision maker believes his and the experts' opinions to be jointly normally distributed, viz:

$$\ell|A \sim N(\mathbf{m},V) \qquad \text{and} \qquad \ell|A^c \sim N(\bar{\mathbf{m}}\,,\,V)\,,$$

we can derive

$$\lambda\,|\,A,\ell_0 \sim N(\mu(\ell_0),W) \qquad \text{and} \qquad \lambda\,|\,A^c,\ell_0 \sim N(\bar{\mu}(\ell_0),W),$$

where

$$\mu(\ell_0) = \mathbf{m}_1 + B(\ell_0\text{-}m_o), \qquad \bar{\mu}(\ell_0) = \bar{\mathbf{m}}_1 + B(\ell_0-\bar{m}_o),$$

$\mathbf{m} = (m_o,\mathbf{m}_1^T)^T$, and W and B are matrices simply derived from V. On applying Bayes' Theorem and a little algebra, one obtains

$$\ell(A\,|\,\lambda) = \{\mu(\ell_0)-\bar{\mu}(\ell_0)\}W^{-1}\{\lambda-(\mu(\ell_0)+\bar{\mu}(\ell_0))/2\} + \ell_0,$$

where $\ell(A\,|\,\lambda)$ is the decision maker's posterior log-odds.

Writing o_{ko} to be the antilogarithm of the kth component of $(\mu(\ell_0)+\bar{\mu}(\ell_0))/2$, this prescription becomes:

$$\ell(A\,|\,\lambda)-\ell_0 = \sum_{k=1}^{n} \beta_k\log(o_k/o_{ko}),$$

where the β_k are functions of $\mathbf{m}$, $\bar{\mathbf{m}}$ and V. Clearly this is very similar to, but not identical with Bordley's prescription. In fact, a little manipulation can produce a form identical to Bordley's, but with $ln(o_o) \neq \ell_0$, which confirms my remark that o_o was not a truely a prior odds. Clearly, this Bayesian prescription is also related to, but not the same as the logarithmic opinion pool.

EB is a property that, in spirit at least, treats the group members opinions as probabilities, not data. Thus we would not expect this model to obey EB when the experts' posterior log-odds are simply substituted for λ. But then if the decision maker knows that the group have observed some new data, it would no longer be appropriate to use the model:

$$\lambda|A,\ell_0 \sim N((\mu(\ell_0),W) \text{ and } \qquad \lambda|A^c,\ell_0 \sim N(\bar{\mu}(\ell_0),W).$$

Suppose, first, that the decision maker knows that the experts have observed some data and agreed among themselves on the log-likelihood, L, but that he does not know the data themselves. Assuming normality purely for the sake of the argument, let the decision maker's probability for the unknown value of L be

$$L \sim N(\eta,w^2).$$

This implies that he should update his prior log-odds to $(\ell_0 + \eta)$.

The decision maker now observes the expert's posterior beliefs, $\gamma = \lambda + L\mathbf{l}$, where

$$\gamma | A, \ell_0 \sim N(\xi(\ell_0), U) \quad \text{and} \qquad \gamma | A^c, \ell_0 \sim N(\bar{\xi}(\ell_0), U).$$

and $\xi(\ell_0) = \mu(\ell_0) + \eta\mathbf{l}$, $\bar{\xi}(\ell_0) = \bar{\mu}(\ell_0) + \eta\mathbf{l}$, and $U = W + w^2 J$, where J is a matrix with every element unity. The analysis of this model would now continue exactly as before to give the decision maker's posterior log-odds:

$$\ell(A|\gamma) = \{\xi(\ell_0) - \bar{\xi}(\ell_0)\}U^{-1}\{\gamma - (\xi(\ell_0) + \bar{\xi}(\ell_0))/2\} + \ell_0.$$

Since the decision maker has observed $\gamma = \lambda + L\mathbf{l}$ and not λ and L separately, it is reasonable that $\ell(A|\gamma) \neq \ell(A|\lambda) + L$, as EB would demand.
If the decision maker does know the data that the experts have observed, then the model develops naturally, but in a different way.
Suppose that an observation x is made and that all concerned agree that it has likelihood ratio on A of L. Then, if the decision maker first assimilates the experts'prior opinion and then the observation, we have

$$\begin{aligned}\ell(A|\lambda, X) &= L + \ell(A|\lambda)\\ &= L + \{\mu(\ell_0) - \bar{\mu}(\ell_0)\}W^{-1}\{\lambda - (\mu(\ell_0) + \bar{\mu}(\ell_0))/2\} + \ell_0.\end{aligned}$$

Suppose now that the decision maker and experts each assimilate x before the decision maker listens to the expert's advice. Then we must remember that the normal model is translated:

$$\ell + L\mathbf{1} | A \sim N(\mathbf{m} + L\mathbf{1}, V) \qquad \text{and} \qquad \ell + L\mathbf{1} | A \sim N(\bar{\mathbf{m}} + L\mathbf{1}, V).$$

Noting that

$$\mu(\ell_0 + L) = \mathbf{m}_1 + L\mathbf{1} + B(\ell_0 + L - m_o - L) = \mu(\ell_0) + L\mathbf{1} \text{ and similarly}$$

$\bar{\mu}(\ell_0 + L) = \bar{\mu}(\ell_0) + L\mathbf{1}$, we find that

$$\begin{aligned}\ell(A|\lambda, X) &= \{\mu(\ell_0 + L) - \bar{\mu}(\ell_0 + L)\}W^{-1}\{\lambda + L\mathbf{1} - (\mu(\ell_0 + L) + \bar{\mu}(\ell_0 + L))/2\} + \ell_0 + L\\ &= \{\mu(\ell_0) - \bar{\mu}(\ell_0)\}W^{-1}\{\lambda - (\mu(\ell_0) + \bar{\mu}(\ell_0))/2\} + \ell_0 + L.\end{aligned}$$

Hence this model obeys a variation of EB, in which the known value of L is filtered out of the expert's posterior belief.

Lindley (1983) has begun the task of extending this model to the case of more than one even5 of interest. Since he is presenting his results at this conference, I shall not discuss them here, although I shall suggest two directions for extending his work.

Lindley considers the case when $\mathcal{A}$ has a finite basis $A_1, A_2, \ldots, A_n$. Thus the decision maker's and experts' beliefs are modelled by discrete distributions.

Does it make conceptual sense for the decision maker to put smoothness demands on his posterior distribution; and, if so, how should this be done? I suspect that it does make sense if one embeds the problem in an exchangeable sequence of similar situations. Moreover, I suspect that the work of Leonard (1973) on smoothing histograms can be used to derive the appropriate procedures.

Suppose that some experts only give marginal distributions over a sub-σ-field of $\mathcal{A}$. How can this information be incorporated? What would be a rather straightforward analysis if the model were linear in the probabilities is by no means so simple when the model is linear in the log-probabilities.

1.6. *Discussion*

It should be apparent that I belief that the methods of the last paragraphs provide the solution, conceptually at least, to the expert problem. The issues of calibration, honesty, and correlation between opinions are all treated in a natural, straightforward manner. Relative expertise is quantified in a meaningful manner, and, indeed, all the quantities in the model are meaningful to the decision maker. (French, 1981; Zidek, 1983). It should be noted that the methods do not possess MP, but I have already argued that MP is perhaps not so desirable.

When the experts' opinions are interpreted as data, EB does not seem a necessary property. What is necessary is that the methods should give sensible prescriptions in circumstances when it is known that the experts have updated their opinion. I would argue that the models above are sensible. Of course, an interesting research topic would be to consider the situation in which there is disagreement about the likelihood function and, moreover, the decision maker asks the experts for their opinions both about $\mathcal{A}$ and about the likelihood function.

Finally, it should be noted that in the above I have assumed that the decision maker believes his own probabilities to be well calibrated. This may be so either because of his arrogance, because it is impossible for him to believe otherwise (Dawid, 1982), or because he has taken careful steps to improve his calibration (DeGroot, 1980; Lindley, 1982).

2. THE GROUP DECISION PROBLEM

2.0. *Problem Statement*

Suppose that a group with n members is jointly responsible for a decision in which they must select a strategy from a set $\{s_i\}$. We shall assume that this set is finite, although we admit the possibility of introducing an infinite number of hypothetical scaling alternatives to measure the members individual beliefs and preferences. The outcome of any strategy is uncertain; it depends on events in a σ-field, $\mathcal{A}$. Each member encodes his beliefs and preferences in a subjective probability distribution, p_k, and utility function, u_k, respectively. Thus each member has a comparable and transitive ranking R_k of $\{s_i\}$:

$$s_{i1}\, R_k s_{i2} \;\Leftrightarrow\; E_k u_k(s_{i1}) \geq E_k u_k(s_{i2}),$$

where $E_k u_k(s_i)$ is the expected utility with respect to p_k of s_i.

Suppose that the group wishes that to the outside world its behaviour is as coherent and rational as that of a single Bayesian. Thus they wish to define a group ranking R_G of $\{s_i\}$ such that there are functions π and ν consistent with the ranking in the sense that:

$$s_{i1}\, R_G\, s_{i2} \;\Leftrightarrow\; E_G \nu(s_{i1}) \geq E_G \nu(s_{i2}),$$

where $E_G\,(s_i)$ is the expected value of ν with respect to π under s_i. Two points should be made here.

(i) Although it is natural to term π and ν the group consensus probability distribution and group utility function, their purpose is *not* to model group belief and preference. Their *sole* purpose, at least in the problem as stated, is to ensure that R_G embodies the qualitative behaviour demanded by Bayesians of rational decision making. They are purely mathematical devices.

(ii) π may be a function of all the members' probability distributions *and* utility functions. Thus $\pi = \pi(p_1,p_2,\ldots,p_n,u_1,u_2,\ldots,u_n)$ and, similarly, $\nu = \nu(p_1,p_2,\ldots p_n,u_1,u_2,\ldots u_n)$. Often in the literature it is assumed that π does not depend on the members' utility nor ν on their probabilities. But to do so is to structure the problem further by assuming, first, that, contary to (i) above, π and ν have meaning in terms of group belief and preference, and, second, that group belief should be independent of individual preferences and group preference independent of individual beliefs.

2.1. *Impossibility Theorems*

Arrows' clasic work on social choice and the vast literature that has followed it shows that, in general, there is no fair and democratic way of forming R_G from the rankings $R_1, R_2, \ldots, R_n$ (Arrow, 1951, Kelly, 1978). However, Arrow's Theorem assumes that there are no restrictions on the possible forms of the individual rankings. Here the R_k are limited by the assumption that they are derived from expected utilities. Does this restriction allow the fair and democratic formation of R_G? Apparently not; except in the very special circumstances that either the members share the same utilities, $u_1 = u_2 = \ldots = u_n$, or they share the same probabilities, $p_1 = p_2 = \ldots = p_n$. (Raiffa, 1968; Bacharach, 1975).

Apart from making Arrow's demands of Pareto's Principle (PP), Independence of Irrelevant Alternatives (IIA), and Non-dictatorship (ND) on the formation of R_G, it seems sensible to demand that EB holds as well. EB follows naturally from the wish that the group acts to the outside world as a coherent decision maker. This further restricts the possibility of forming R_G (Madansky, 1978; Genest, 1983). It may also be sensible to demand MP. The arguments against MP in the context of the expert problem rest upon the possibility of variations in the relative expertise of the group members over $\mathcal{a}$. But arguments based upon expertise run counter to the spirit of the literature of social choice. Although it does not claim that all men are equally expert, democracy seems to demand that all men, expert or not, have equal voting power. In other words, it is only their preferences, R_k, between the possible strategies $\{s_i\}$ that matter; why they hold these preferences is irrelevant. In any case, assuming MP is closely related to assuming IIA. Consider a problem in which the outcomes of all but one strategy depend only on a sub-σ-field of $\mathcal{a}$. Deleting that strategy should not change preferences between the remaining ones (IIA) and it certainly will not do so if MP holds. This remark is supported by the fact that in the special cases when the group decision problem does have a solution π is a linear opinion pool of the p_k. (Bacharach, 1975).

Need the issue of calibration concern us? Probably not. If each group member believes himself to be calibrated, then the ranking R_k represents his preferences, however poorly calibrated he actually is. Thus, if we accept the thesis that, whatever the reasons underlying it, R_k is the object of importance, then we should ignore calibration issues. For a similar reason we need not be concerned by the possibility of nonindependence between the group members' beliefs and preferences.

The possibility of dishonesty is another matter. There is a large literature in social choice that shows essentially that it is impossible to devise a voting system which discourages tactical voting (Patternaik, 1978). In our problem let us assume that each member states p_k and u_k. Then a theorem of Gibbard (1973) suggests that, if all the members of the group (i) know the others' beliefs and preferences and (ii) understand the mechanism whereby the group decision will be made, then either there is a dictator or at least one member of the group has an incentive to state probabilities and utilities that do not reflect his true feelings. (Actually the statement of Gibbard's theorem does not apply directly. He assumes that each member has only a finite number of possible statements, whereas there are an infinite number of possible p_k and u_k. However, this finiteness is not used in the proof, although the finiteness of $\{s_i\}$ is). Thus any solution to the group decision problem must always acknowledge the possibility of dishonesty. Furthermore, if honesty is to be bought by a scoring rule, the rewards of that rule must be comparable with the rewards in the actual decision problem; in which case the argument in section 1.2 suggests that the scoring rule might encourage dishonesty.

2.2. *Bayesian Dialogues*

The literature of social choice allows that the group members may hold very disparate views. Yet, in reality groups are often formed of likeminded people. Moreover, in discussions amongst themselves it has been observed that they tend to move towards a position of all sharing the same beliefs and preferences. (Dalkey, 1969). Is it possible to formalise these discussions into a rational process, which guarantees that consensus will be reached? In keeping with the spirit of this paper we will concentrate on consensus of belief.

DeGroot (1974) analysed a model in which each group member revises his opinion by forming a linear opinion pool of all the group's stated p_k including his own. Arguing that each member should be prepared to revise his revised opinion by the *same* linear opinion pool, DeGroot generates an infinite dialogue, which he shows will converge to the position where they all share the same probability, providing that certain mild conditions hold. This model has been extended in two ways. Griffiths (1980) considers the case where other opinion pools than the linear might be used and again proves convergence under mild conditions. Chatterjee and Senata (1976) and Wagner (1982) consider linear pools, but suggest that the group members may use different weights at different rounds of the dialogue. This seems natural since the reasons for revising opinions at the first and subsequent rounds are different. At the first round group members hear eachother's opinions about α. In subsequent rounds they hear how they each have assimilated the others' opinions.

However, the above ignores the issue of calibration. It is not only conceptually possible but also quite reasonable that each member should consider himself to be well calibrated but that the rest are poorly calibrated. p_k is a probability to the kth member, but data to the rest. Thus at each round of the dialogue each member of the group faces an expert problem. In French (1981) I have briefly examined a version of this. It appears that under reasonable conditions each member will settle upon an opinion, i.e., the procedure will converge, but that this opinion need not be shared by other members.

Lastly, mention should be made of the work of Aumann (1976) and Bacharach (1979). These authors assume that differences in opinion between the group members exist because, and only because, they have made different observations. Important though this work is, it seems to ignore the fact that people's opinions differ not only because of different experiences, but also because of different attitudes, abilities and expertise.

2.3. *Supra Decision Maker*

One possible way out of the dilemma thrown up by Arrow's Theorem is to imagine that a benevolent decision maker sits over the group. He starts ignorant of all matters concerned with the decision, i.e., he has a vague prior over α . He then learns about α through and only through the group's beliefs, $p_1, p_2, \ldots, p_n$. Thus to form his opinion the supra decision maker faces an expert problem in which his prior is vague. Since it is usually assumed that the supra decision maker is paternalistic, it is reasonable to expect him to correct for the possible miscalibration of the members. (Note the change from section 2.1 in the spirit of approach here). The supra decision maker forms his utility function by an altruistic combination of the member's utilities (Keeney and Raiffa, 1976, Chap. 10).

The unfortunate thing about this supra decision maker approach is that he does not exist. (If he does, the group decision problem truely does become an expert problem). Thus the group must construct the supra decision maker *by agreement amongst themselves* about his behaviour. What likelihood function should he use in assimilating their beliefs? How should he combine their utilities altruistically? In short, in an attempt to solve their actual decision problem they have replaced it with several others. Moreover, these other problems may well be as intractable as the first (French, 1981).

2.4. *Discussion*

The literature of group decision making is full of confusing contradictions. A large number of "reasonable and sensible" solutions seem to exist for provenly insoluble problems. This suggests the underlying problem is poorly stated. My own view is that it is fallacious to suggest that a group can act as an entity and *decide*. Decisions can only be made by individuals with a free will. A group should be seen as a social process, which translates a voting pattern into a course of action. There is no single group decision, but one for each group member. How should he vote? Thus I see the group decision problem as n simultaneous, but, in a sense, quite independent versions of the expert problem. (French, 1981, 1983).

3. THE TEXT-BOOK PROBLEM

3.0. *Problem Statement*

Suppose that the group have been asked for their opinion, but that no particular decision is in prospect. How should the group summarise their opinion? In essence, we are being asked to look at the expert problem not from the point of view of the decision maker, but from that of the group. The group know that at some point in the future one or more, possibly unidentified decision makers will need to take note of their advice.

3.1. *Approaches involving an Intermediary*

Suppose first that an intermediary is responsible for collating the group's diverse opinions. One possibility is for him to update his beliefs as in the expert problem, and then report these as a summary. Thus all the theory of section 1.5 is relevant. However, this assumes that his subjective probability is the best report of his synthesis of the group's opinions. For him it most surely is, for it is *his* subjective probability. But to the ultimate users it is only data. Thus we must ask whether observing a person's subjective probability is the best way of eliciting his views on a matter. Given all the known biases and calibration problems (Kahneman et al, 1982), this is a dubious assumption. It is also an assumption that we might have questioned earlier, since we have been assuming throughout our treatment of the expert problem that group members should encode their beliefs as probabilities. Suffice it to say that it does not in principle affect the theory of section 1.5 if any other means of reporting opinion is used.

Press (1979, 1980) suggests the method of qualitative controlled feedback. Without going into details, the intermediary not only summarises group opinion in some way, but also summarises the reasons why these opinions are held. As a solution to the text-book problem I have much sympathy with this approach, since, although a summary, it retains a lot of informtion. I would, however, quarrel with many of the details. For instance, how is the ultimate decision maker to judge the relative expertise behind the various opinions if anonymity is maintained.

3.2. *Approaches in which the Group is responsible for the Summary*

Suppose now that the whole of the group is responsible for summarising their opinions. Savage (1972, Chap. 8) suggests that the whole of statistical theory is directly or indirectly aimed at the solution of this version of the text-book problem. However, that is to enlarge the problem considerably from the one that I had in mind. Nonetheless, statistics does, some suggest, present a direction in which to look for a solution. Statistical decision theory is based upon a family of "conventional" loss functions, i.e., negative utility functions, that apply in many situations of inference. If all the group members agree to adopt one of these for their own, a decision problem with unanimous agreement on the utility function arises: precisely one of the cases in which a sensible solution to the group decision problem may exist. Alternatively Weerahandi and Zidek (1981) suggest that one might look to bargaining theory. However, it should be remembered that bargaining theory is based upon some notion of bargaining strength. Whether such notions should underpin a summary of opinion is questionable.

3.3. *Discussion*

I became aware of the text-book problem only a few months ago. It is present in the undercurrents of the literature on consensus, but only occasionally does it rise to the surface. Moreover, whereas I am fairly sure of my position in discussing the expert and group decision problems, I have a very open mind on the text-book problem. Perhaps I can close with one thought that seems relevant. Any statistician, Bayesian or not, would think hard before condensing a data set into less than a sufficient statistic. So what is a sufficient statistic of the group's opinions?

4. CONCLUDING REMARK

If I had to summarise this paper in just one sentence (which is more than it is worth), it would be: when is a probability a probability and when is it data? Until one has a clear answer to that in mind, a straightforward solution to a consensus problem is likely to be elusive.

5. ACKNOWLEDGEMENTS

There are many people to whom I am grateful: the conference organisers; the Statistics Department at Warwick, where a seminar on consensus was held which stimulated much of the above; and my colleagues at Manchester for many helpful discussions.

REFERENCES

ARROW, K.J. (1951). *Social Choice and Individual Values*. New York: Wiley.

AUMANN, R.J. (1976). Agreeing to Disagree. *Ann. Math. Statist.* **4**, 1236-1239.

BACHARACH, M. (1973). *Bayesian Dialogues*. Unpublished manuscript. Christ Church, Oxford.

— (1975). Group Decisions in the Face of Differences of Opinion. *Mgmt. Sci.* **22**, 182-191.

— (1979). Normal Bayesian Dialogues. *J. Amer. Statist. Assoc.* **74**, 837-846.

BORDLEY, R.F. (1982). A Multiplicative Formula for Aggregating Probability Assessments. *Mgmt. Sci.* **28**, 1137-1148.

BORDLEY, R.F. and WOLFF, R.W. (1981). On the Aggregation of Individual Probability Estimates. *Mgmt. Sci.* **27**, 959-964.

CHATTERJEE, S. and SENATA, E. (1977). *J. App. Prob.* **14**, 89-97.

DALKEY, N.C. (1969). The Delphi Method: An Experimental Study of Group Decisions. *RAND Memo RM-5888-PR.*

— (1976). Group Decision Analysis in M. Zeleny (ed). *Multiple Criteria Decision Making Kyoto 1975.* Springer Verlag Lecture Notes in Economics and Mathematical Systems. **123**.

DAWID, A.P. (1982). The Well-Calibrated Bayesian. *J. Amer. Statist. Assoc.* **77**, 605-612.

DE FINETTI, B. (1974). *Theory of Probability Vol. 1.* Chichester: John Wiley.

DEGROOT, M.H. (1974). Reaching a Consensus. *J. Amer. Statist. Assoc.* **69**, 118-121.

FRENCH, S. (1980). Updating of Belief in the Light of Someone Else's Opinion. *J. Roy Statist. Soc.* **A143**, 43-48.

— (1981). Consensus of Opinion. *Eur. J. Opl. Res.* **7**, 332-340.

— (1982). On the Axiomatisation of Subjective Probabilities. *Theory and Decision* **14**, 19-33.

— (1983). A Survey and Interpretation of Multi-Attribute Utility Theory. In S. French, R. Hartley, L.C. Thomas and D.J. White, (eds). *Multi-Objective Decision Making.* London: Academic Press.

GENEST. (1982). External Bayesianity: an Impossibility Theorem. University of British Columbia. *Institute of Applied Mathematics and Statistics Tech. Rep.* 82-8.

GIBBARD, A. (1973). Manipulation of Voting Schemes: A General Result. *Econometrics* **41**, 587-601.

GRIFFITHS, H.B. (1980). Examiner's Meetings and the Arithmetico-Geometric Mean. *Bull. Inst. Math. and It's Applic.* **16**, 247-251.

HOGARTH, R.M. (1977). Methods for Aggregating Opinions. In. H. Jungermann and G. de Zeeuw, Eds. *Decision Making and Change in Human Affairs.* Dordrecht, D. Reidel Publishing Co.

JEFFREYS, H. (1961). *Theory of Probability.* 3rd. Edn. Oxford: University Press.

KAHNEMANN, D., SLOVIC, P. and TVERSKY, A. Eds. (1982). *Judgment under Uncertainty: Heuristics and Biases.* Cambridge: University Press.

KEENEY, P.L. and RAIFFA, H. (1976). *Decisions with Multiple Objectives,* New York: John Wiley.

KELLY, J.S. (1978). *Arrow Impossibility Theorems.* New York: Academic Press.

KRANTZ, D.H., LUCE, R.D., SUPPES, P. and TVERSKY, A. (1971). *Foundations of Measurement. Vol. 1.* New York: Academic Press.

LEONARD, T. (1973). A Bayesian Method for Histograms. *Biometrika* **60**, 297-308.

LINDLEY, D.V. (1982). The Improvement of Probability Judgments. *J. Roy. Statist. Soc.* **A145**. 117-126.

— (1984). Reconciliation of Discrete Probability Distributions. This Volume.

MCCONWAY, K.J. (1981). Marginalisation and Linear Opinion Pools. *J. Amer. Statist. Assoc.* **76**, 410-414.

MADANSKY, A. (1964). Externally Bayesian Groups. *RAND Memo* RM-4141-PR

— (1978). *Externally Bayesian Groups.* Unpublished Manuscript - University of Chicago.

MORRIS, P.A. (1974). Decision Analysis Expert Use. *Mgmt. Sci* **20**, 1233-1241.

— (1977). Combining Expert Judgments: A Bayesian Approach. *Mgmt Sci* **23**, 679-693.

— (1983). An Axiomatic Approach to Expert Resolution. *Mgmt Sci* **29**, 24-32.

PATTERNAIK, P.K. (1978). *Strategy and Group Choice,* Amsterdam, North Holland Publishing Co.

PRESS. S.J. (1978). Qualitative Controlled Feedback for Forming Group Judgments and Making Decisions. *J. Amer. Statist. Soc.* **73**, 526-535.

— (1980). Bayesian Inference in Group Judgment Formulation and Decision Making. *Bayesian Statistics*, (J.M. Bernardo, M.H. DeGroot, D.V. Lindley and A.F.M. Smith, eds.) Valencia: University Press.

RAIFFA, H. (1968). *Decision Analysis*. Reading Mass, Addison Wesley.

RAMSAY, F.P. (1931). Truth and Probability in the *Foundation of Mathematics and other Logical Essays*. London: Kegan Paul.

ROBERTS, H.V. (1965). Probabilistic Prediction. *J. Amer. Statist. Assoc.* **60**, 50-62.

SAVAGE, L.J. (1972). *The Foundations of Statistics*, 2nd. Edn. New York: Dover.

SMITH, A.F.M. and SPIEGELHALTER, D.J. (1980). Bayes Factors and Choice Criteria for Linear Models. *J. Roy. Statist. Soc.* **1342**, 213-220.

STONE, M. (1961). The Linear Opinion Pool. *Ann. Math. Statist.* **32**, 1339-1342.

WAGNER, C. (1982). Allocation, Lehrer Models and the Consensus of Probabilities. *Theory and Decision* **14**, 207-220.

WEERAHANDI, S. and ZIDEK, J.V. (1981). Multi-Bayesian Statistical Decision Theory. *J. Roy. Statist. Soc.* A.**144**. 85-93.

WINKLER, R.L. (1968). The Consensus of Subjective Probability Distributions. *Mgmt. Sci.* **15**, B61-B75.

— (1981). Combining Probability Distributions from Dependent Information Sources. *Mgmt. Sci.* **27**, 479-488.

ZIDEK, J.V. (1983). *Multi-Bayesianity: (1) Consensus of Opinion*. Unpublished Manuscript - University of London.

DISCUSSION

M. GOLDSTEIN (*Hull University*):

I suggest that the "textbook" problem can be subdivided to yield a version called the "statistician's textbook". This arises when a large number of experts each make a large number of jugdements about a large number of quantities. In addition there is a large amount of data about similar quantities. The large number of occurrences of the term large is the defining feature of the statistician's textbook problem as it defines the level at which the statistician almost certainly will be called in to demonstrate his "expertise". I would be very grateful for any comments/references on this problem. (My interest here is not purely theoretical!)

K. McCONWAY (*The Open University*):

I would like to start by congratulating Dr. French on producing a clear and well-structured review of this somewhat confussed field. I find his taxonomy of the three types of consensus-finding problem very valuable, particularly the notion of textbook problems.

Firstly, on the expert problem, I would accept some of the criticism of the linear opinion pool on the ground that the marginalization principle (MP) is inappropriate. I did say in my 1981 paper on the subject that MP might well be unreasonable when the experts are experts in different fields and have disparate prior knowledge. Also, the external Bayesianity (EB) criterior has always seemed inappropriate to me in the expert problem, particularly when the decision maker knows what data have been observed. Intuitively, this must give the decision maker extra information about the probability assessments, which must be reflected somehow in the consensus distribution.

In Section 1.2 of the paper, French makes three criticisms of opinion pools in general. First, he complains that the weights involved are not operationally defined. This is true of course, but I am not convinced that the extra complication in the Bayesian updating approach is outweighed by the fact that the probabilities and likelihoods in it can be given operational definitions. I certainly cannot think clearly about, for instance $\pi(\lambda|A,\ell_0)$ in terms of called-off bets. Secondly, French says that opinion pools treat the expert's stated probabilities as probabilities for the decision maker, rather than as data. I am not sure exactly what he means by this. Surely opinion pools merely prescribe a particular way of analysing these data. Of course, there are many possible ways of criticizing the way these data are used, but I do not think anyone claims they are not data.

Thirdly, French points out that the expert's honesty is a problem. I agree, but I am not at all sure how to solve it. In principle, as French says, it can be allowed for in Bayesian updating, but it is not clear in practice how this should be done. For instance, if the experts know which problem the decision maker is going to use their probabilities in, this will presumably affect their stated probabilities in a complex way. More work is needed in the area of coping with such difficulties.

Turning now to group decision problems, French says that group probability distributions and utility functions, if used, are "purely mathematical devices". I am not sure what he means by "mathematical", but I would say the same applied to individual decision making. The point there is to have a coherent pattern of preferences and actions which are in accord with, say, Savage's axioms. This is equivalent to acting *as if* one had subjective probabilities and utilities, and then maximised expected utility. In practice, the probabilities and utilities are *constructed* in order to arrive at a coherent pattern of preferences and actions. It is really no more than a helpful heuristic device to say that probabilities and utilities represent beliefs and preferences. Probability does not exist, even for an individual - it has to be constructed.

After this excellent review paper, I am sure workers in this area will still disagree. But at least it should be much clearer exactly what we are disagreeing about.

A. O'HAGAN (*University of Warwick*):

I must begin by thanking Simon French for his clear and stimulating "critical survey". In particular, I liked the way he interjected his own "critical" views sparsely enough not to disrupt the general dispassionate tone of a "survey". I will not be so restrained. The following are my wholly personal views of French's three categories.

The Expert Problem: It seems to me that this problem is essentially solved by Dennis Lindley's paper in this session. That is not to say that Lindley's solution is the only solution, but that his approach is the only proper approach. specifically, the decision-maker must evaluate his personal probability distribution over $\mathcal{A}$ given all the available data. The data are his prior informtion plus the expert's reported probabilities, so in Lindley's notation he must evaluate $p(A_j|Q,H)$ for each j.

Now, he is free to evaluate these probabilities in any way he chooses, but Lindley's approach contains two further devices that he would typically use.

1. He will naturally (but not necessarily) use Bayes' theorem in the form by Lindley in his equation (1.1), i.e.

$$p(A_j|Q,H) \propto p(Q|A_j,H)\, p(A_j|H)$$

2. To evaluate the likelihood $p(Q|A_j,H)$ he will naturally (but not necessarily) construct some kind of model.

Of the many models that might be constructed, Lindley offers us one based on multivariate normal distributions. Further work on the expert problem should address the tasks of constructing new models and of exploring further the consequences of Lindley's model.

The Group Decision Problem: The expert problem is easily solved because there is an individual decision-maker, and the Bayesian paradigm tells us exactly how individuals should act. Can it be extended to tell us how groups should act, and if so, how?

Of course, a group is composed of individuals, and only "makes decisions" through the individual decisions of its members. For instance, in a committee meeting I am constantly making decisions —when to speak, what to say, how to say it— and if a consensus is reached it is because each of us has decided not to argue any further. If a vote is required, each of us decides how to vote. So if we wish to study how groups actually "make decisions" we need nothing more than the usual Bayesian paradigm for individual decision-making. Yet the literature on this problem concerns "consensus probabilities" which are apparently not meant to be the personal probabilities of any identified individual, or "consensus decisions" which are apparently not meant to be the decisions of identified individuals acting coherently. It seem to be attempting to define a genuinely alternative paradigm for group decision-making. French has surveyed this literature quite thoroughly, and whichever way I look at it I am reminded of the attempts of frequentists to construct a theory of statistics; it is full of adhockery, paradox and vaguely motivated principles. The solution is the same. Abandon it. I agree with French, in his section 2.4 - only individuals have probabilities, and only individuals make decisions.

The Textbook Problem. Like French, I am less sure about this rather loosely-defined problem, but if it is simply a reporting problem then it is clearly better to report all the data than one person's assessment based on those data. Any condensation loses information. French also makes the pertinent point that expert's probabilities may not always be the most useful expression of their expertise, a question which is also mentioned in Michael Goldstein's paper at this conference.

T. SEIDENFELD (*Washington University, St. Louis*):

1. In his very useful taxonomy of the issues bound up with arriving at a group consensus, Professor French's discussion of the "group decision" problem (his §2) is predicated on the supposition that a collection of Bayesian agents would want to adopt the same (Bayesian) norms for the group deliberation as are satisfied by each individual. That is, Professor French suppose a Bayesian model for the group decision--complete with a coherent group-probability, a group-utility, and group-preference according to group-expected utility. The task, then, is to devise rules for amalgamating the individual probabilities and utilities into those for the "group", while ensuring the amalgamation is fair and secure against manipulation and... .

I do not see why the group should be Bayesian even when the individuals are. For simplicity, suppose the group is a collection of Bayesians who share a common utility function over particular welfare allocations; thus, we admit full cardinal comparability across individuals. Imagine, however, the social options involve uncertainties about which particular welfare allocations will be achieved and that the individuals do *not* agree in their assessments of personal probability for these uncertainties. In effect, in this example we face the kind of decisions problem considered by Savage in his *Foundations* (1954, pp. 123-124). Savage argues that, on grounds of expected utility, the group-admissible options are those which, for some convex combination of individual preferences, maximize expected utility. But this rule, reasonable as it is, fails to satisfy (A.K. Sen's) independence of irrelevant alternatives —it violates Sen's property β. In fact, the rule generates a *non-normal* group preference— it violates Sen's property γ. Is not this simple example reason enough to question the supposition that the group decision ought to have Bayesian model?

2. In his comment on the group decision problem, Professor Good suggests we might dodge Arrow's impossibility theorem by impugning Arrow's postulate of independence of irrelevant alternatives [IIA] (not to be confused with Sen's version: the conjunction of properties α and β). The challenge is to make clear what part of IIA to reject. IIA is multifaceted. Besides requiring that choice from a feasible set S not depend upon supersets $S' \supset S$, it prohibits interpersonal comparisons of utility and makes a tacit assumption of *normality* of the choice rule. Some range of solutions to Arrow-type problems is afforded by permitting interpersonal comparison of utility (see Roberts 1980), though difficulties persist in more general problem setting (as noted by French in paragraph 1 of §2.1). But merely weakening the ordering requirements on social preferences is not sufficient to escape oligarchies and vetos. [For this discussion see Sen (1977, p. 61 and pp. 70-71).]

REPLY TO THE DISCUSSION

S. FRENCH (*University of Manchester*):

I am grateful to all the contributors to the discussion for their helpful and stimulating comments. I am particularly grateful that they have not pilloried me as much as they might; they are too kind.

Perhaps I should say here what I failed to say in my paper; my interest in these problems is purely conceptual. I have found that by thinking about the Bayesian approach to consensus problems I have learnt a lot about the Bayesian approach in general. The very fact that more than one person's beliefs, and hence subjective probabilities, are involved makes it essential that one thinks very carefully and precisely about each step of the argument. Thus one comes to appreciate some subleties in the application of Bayes' Theorem that one might not appreciate in its more straightforward application to data arising in other contexts, e.g. scientific experimentation.

Given this, Michael Goldstein will not be surprised to hear that I cannot help him practically with his "statistician's textbook". My only comment is probabily not helpful. In the paper I said in relation to the textbook problem that it is the needs of the ultimate users of the information that is paramount in determining the method of collation. These users will be decision makers and the responsibility for those decisions lies with them. Thus, the method of collation should not be such that it implicitly takes the decisions for them.

Given that my interest is conceptual, I may be forgiven for not being unduly concerned that some of the quantities in the Bayesian solution to the expert problem are not easy to identify in practice. Kevin McConway is right that it is very difficult (but not, I think, impossible) to construct $\pi(\lambda \mid A, \ell_0)$ by thinking in terms of called-off bets. Equally how one allows for possible dishonesty in practise is an open question. However, surely a conceptual framework should admit all possibilities. Opinion pools are suspect to my mind simply because they do not allow for all eventualities.

Why do I emphasise my view that the expert's probabilities are data and that opinion pools treat them as probabilities? Simply because I believe this to be true. Perhaps linear opinion pools can be likened to the problem of finding a linear estimate of an appropriate quantity. But then where in the literature is a discussion of the statistical properties of such estimates, their likelihood function, even an admission that an "error term" exists?

Turning to the group decision problem, both Kevin McConway and Teddy Seidenfeld pick up my statement that group probabilities and utilities are purely mathematical devices used to ensure consistency of choice. What do I mean by this? Perhaps I was being over-emphatic, but there does seem to be a point to be made. Savage's axioms do not explicitly mention a decision maker's beliefs about events and preferences between consequences. The axioms are framed in terms of consistency of choice between actions. From these it becomes clear that the decision maker acts as if he wishes to maximise a certain expectation. That this expectation can be interpreted in terms of subjective probabilities and utilities is in a sense fortuitous. It derives naturally from the context of a single decision maker. I was simply making the point that in the case of a group of decision makers a similar interpretation is not so natural.

Finally turning to the comments of Tony O'Hagan, there seems to be a remarkable degree of consensus between us.

REFERENCES IN THE DISCUSSION

ROBERTS, K.W.S. (1980). Interpersonal Comparability and Social Choice Theory, *Review of Economic Studies,* **47**, 421-439.

SAVAGE, L.J. (1954). *Foundations of Statistics.* New York: J. Wiley.

SEN, A.K. (1977). Social Choice Theory: A Re-examination, *Econometrica*, **45**, 53-89.

BAYESIAN STATISTICS 2, pp. 203-230
J.M. Bernardo, M.H. DeGroot, D.V. Lindley, A.F.M. Smith (Eds.)

On the Prediction of Observables: A Selective Update

S. GEISSER
University of Minnesota, USA

SUMMARY

A brief review of past work involving the predictive approach to inference and decision is presented. Some selected recent developments, including predictive and estimative influence functions, data consistency, model checking, and calculating the probability that a future fraction of observables lies in a given set, are featured.

Keywords: INFLUENCE; MODEL CHECKING; OBSERVABLES; PREDICTIVE DISTRIBUTIONS.

1. INTRODUCTION

During the last decade the predictive approach in statistical inference and decision has become more widely accepted as statisticians realize its pertinence and applicability to real problems. This is particularly true of Bayesians, because predictive distributions are a natural consequence of the Bayesian attitude, although, admittedly it has taken time for some Bayesians to adopt this view. Recently, Stigler (1982) claimed that the foundation of Bayes' original argument was to assume that if nothing were known about the observable event (number of successes in N trials) then each value could be presumed equally likely and equal to $(N+1)^{-1}$. In other words the focus was on the observables rather than on the unknown parameter (the postulated probability of success on any given trial). If Stigler's argument, based on his scrutiny of Bayes' Scholium, is valid, then Bayes himself is the first Bayesian predictivist. One, however, may be puzzled as to why he couched his inference in terms of the parameter and not the chance of success of the $(N+1)$-*st* observation. Actually a calculation of the latter appears in the appendix which is due to Price, who communicated Bayes' posthumous essay. The calculation, as noted in the lectures of K. Pearson (1979), is somewhat obscure in the count of previous successes (whether it was one or two successes) and may not be consistent with the prior distribution of the parameter that was induced or assumed. Laplace (1774) correctly made the calculation and this came to be known as his now notorious "law of succession". Is it then, that Bayes was not as much a predictivist as Stigler asserts? First, although the calculation does not appear in the essay it does appear in the appendix, so it may have been suggested by Bayes. Secondly, the answer can better be determined by the specific problem that Bayes addressed. A ball was rolled on a unit square flat table and the horizontal coordinate of the final resting place was then assumed to be uniformly distributed in the unit interval. This perfectly reasonable assumption follows from the construction of the problem. A second ball is then rolled N times and one is informed as to the number of times the second ball came to rest to the left of the first ball without the actual horizontal coordinate of the first being disclosed. The problem is to infer the horizontal coordinate of the first ball. This is certainly predictivistic or observabilistic inference in the broad sense in that it includes events that have occurred but whose values are unknown to the inferrer. A retrodictive inference is of the same nature as a predictive one for realizable situations. That Bayes himself did not specifically discuss the chance of a success on the toss of the $(N+1)st$ ball for this problem reflects the fact that a different problem requires a different solution and hence should in no way mar Bayes' predictivistic credentials. However, when we go beyond this problem, and Stigler claims that Bayes did with regard to putting the prior on observables rather than on the parameter, then Laplace (1774) clearly undertook the next step at the posterior end, and even made the more general calculation of the chance that the next r out of M trials were successes, see also Condorcet (1786). Of course, Laplace was responsible for greatly widening the scope and applicability of probability calculations.

K. Pearson (1907, 1920) states that the fundamental problem of statistics was predictive and indicates how Laplace's type of calculation may be applied to a quantal response model.

Jeffreys (1939) extends this further by adopting a finite model, i.e. out of the total number of $N + M$ possible trials he assumes that before any of them are made, the chance for the total number of successes is the same for every integral value from 0 to $N+M$. After observing N trials, this yields the same result for the chance that the next r out of M trials are successes as the previous uniform prior assumption on the parameter itself.

A heated controversy broke out between Jeffreys (1932, 1933, 1934) and Fisher (1933, 1934) on a prediction problem. It concerned the probability that the third observation was included in the interval determined by the first two - all being independently and identically distributed. The controversy revolved more about the meaning of particular probability statements than about predicting. For a review of this controversy, see Lane (1981). Most of the problems Jeffreys deals with are physical measurement problems where the same true value is being measured imperfectly. Here the object is to infer limits on the true value so there is usually no need to predict future observations. Nonetheless, he still derived the predictive distribution for the $(N+1)st$ observation given the first N are a random sample from a normal distribution.

Fisher (1956) devotes several pages of his book to prediction. He discusses the "Bayesian" calculation of obtaining r successes out of M future trials having previously observed s out of N and further demonstrates the use of his fiducial argument for obtaining continuous predictive distributions. He makes some very penetrating remarks about the probabilistic prediction of observables and the capability of their verification as contrasted with probability statements about hypothetical parameters, as well as the connection between them. His views appear to be generally consonant with a predictivistic approach. But, because of a certain lack of clarity in his phraseology some of his assertions regarding predictions are capable of being interpreted in more than one way or perhaps misinterpreted.

Although, de Finetti and Savage, as far as I can discern, did not directly contribute to the methodology of statistical prediction, they have clearly provided the major philosophical underpinning for the observabilistic or predictivistic view. In fact, next to Bayes' theorem, de Finetti's exchangeability theorem is second to none in its importance for the subjectivistic view. It is also useful in demonstrating that a good deal of parametric inference can be viewed as a special or limiting case of observabilistic inference. The distribution of unobserved but realizable values or particularly important functions of them, as modified by values observed, is clearly what statistical inference is really about.

It is also to be noted that, philosophy notwithstanding, most of the methodology produced by Bayesians whether of the parametric or predictive variety tends to be more relaxed and approximate in its construction than one would anticipate given a strict subjectivistic viewpoint. The methods tend to be more pragmatic than subjective and largely conform to the spirit of the Bayes/Non Bayes compromises of Good (1965). They also concord with the view (shocking to some) that a parameter is often an unobservable construct of an approximate model mainly devised to facilitate the prediction of future observations. Hence a prior distribution for the parameter can be, to a degree, a matter of convenience.

In the next section we sketch out some of those areas for which predictive methods have been devised during the last twenty years. Featured in the third section are some recent predictive developments in regard to influential observations, data consistency and model checking. The last section discusses some results in predicting the number of future observations out of a total that lie in a chosen set. The latter can be considered a continuation, of sorts, of the Bayes-Laplace-Pearson-Jeffreys-de Finetti calculations.

2. APPLICATIONS OF BAYESIAN PREDICTIVISM

Forecasting in time series is a very natural enterprise and consequently a well developed field as attested to by its voluminous literature. We mention here only a few of the more notable statistical books that stress time series prediction; Wold (1938), Wiener (1949), Yaglom (1962), Whittle (1963), and Box and Jenkins (1970). Most of the orientation is non-Bayesian, and a good deal of effort is devoted to estimation e.g. Anderson (1971). Wold (1959) veered away from parametric estimation to a predictivistic approach for econometric data - but this apparently had no appreciable effect on other non-Bayesian workers. From here on I will stress only areas other than time series, where statistical prediction was evidently not as natural to the developers of statistical methodology and the critical focus was on the estimation of parameters and the testing of hypotheses about them.

The use of predictive distributions in Bayesian classification and discrimination problems was developed and elaborated upon by Geisser (1964, 1966, 1968) and in multivariate normal linear regression by Geisser (1965). Aitchison and Sculthorpe (1964) discussed prediction from a decision perspective mainly with regard to tolerance and coverage problems, see also Guttman and Tiao (1964). Dunsmore (1968, 1969, 1974) used the predictive approach to problems of calibration, life testing, regulation and optimization and Guttman (1967) to goodness-of-fit problems. Roberts (1965) and Geisser (1971) pointed out various non-traditional predictive areas, where Bayesians were using predictive distributions, such as classification, discrimination, certain hypotheses testing problems, design, sample size determination, sample surveys, goodness-of-fit, without necessarily acknowledging the fact. Geisser (1971) also pointed out that problems of ranking, selection and comparison handled, even by Bayesians, from an estimative viewpoint were better executed predictively. In short, predictive distributions had been grossly underutilized by Bayesians!

Prediction in growth curve situations was developed by Geisser (1970), Lee and Geisser (1972, 1975), and Fearn (1975). Finally, Aitchison and Dunsmore (1975) published the first text in statistical prediction covering a substantial number of predictivistic topics. For a detailed review, see Geisser (1976). The predictive approach has also been utilized by Akaike (1978) and Geisser (1979) in an attempt to induce objective prior distributions for parameters.

On the non-Bayesian side, except for the previously alluded to comments by Fisher and the early frequentist work by Wilks (1942) on distribution-robust tolerance regions, little attention had been paid to the predictive approach. A summary of classical frequentist tolerance procedures is given by Guttman (1970).

A new predictive impetus in low structure paradigms (and to problems with varying structure) derives from the work of Geisser (1974, 1975a, 1975b, 1979, 1980a, 1980b, 1980c, 1981), Stone (1974a, 1974b, 1977), Butler and Rothman (1980), Wahba (1977) and de Waal et al (1981) who use sample reuse procedures to make predictions, select models, estimate densities, and modify classification techniques, among other things.

That in the softer social, biological, and engineering sciences prediction should always have been a crucial factor, if not the crucial one, is well known, c.f. Meehl (1954). Recently, Jaynes (1980), has convincingly argued for a predictivistic version of statistical mechanics that is directed towards what he regards as the critical question: "Given the partial information that we do, in fact, have, what are the best predictions we can make of observable phenomena?"

In the next two sections some recent developments in the use of predictive distributions are presented and elaborated upon in some detail.

3. INFLUENTIAL OBSERVATIONS, DATA CONSISTENCY AND MODEL CHECKING

3.1 *Influential Observations*

How observations effect the estimation of certain parameters has been the focus of much attention for the last ten years. Cook (1977, 1979), Cook and Weisberg (1980, 1982), Andrews and Pregibon (1978), Hoaglin and Welsch (1978), Belsley, Kuh and Welsch (1980) have considered the problem of detecting observations which are influential in the estimation of regression parameters. Johnson and Geisser (1981, 1982, 1983) have considered the problem of influential observations using a Bayesian approach and derived methods both for the estimation of parameters and the prediction of future observations.

In general the goal is to detect those observations that are most influential in regard to decision making and inference, either in the estimative or predictive mode or both. A formal estimative Bayesian approach to this problem starts with defining a loss function

$$L\,(d,\theta)$$

representing the loss incurred by making decision $d(y)$ based on observing $y=(y_1,\ldots,y_N)$ at known covariates $x=(x_1,\ldots,x_N)$ when θ is the true value. One then calculates how d_i^*, which minimizes

$$E_\theta L(d(y_{(i)}),\theta),$$

where the expectation is over the posterior distribution of θ based on $y_{(i)}$ which is y with y_i deleted, changes when y_i is included to d^* which minimizes

$$E_\theta L(d(y),\theta).$$

In the predictive mode, influence is measured in how d_i^* which minimizes

$$E_Z L(d(y_i),z)$$

changes to d^* which minimizes

$$E_Z L(d(y),z)$$

where the expectation is taken over the predictive distribution of the set of future values Z at known covariates W and $L(d,z)$ represents the predictive loss incurred in making decision d when y is observed and z is the set of future realized values.

In particular a sort of "canonical" inferential methodology, which also can be derived from the above decision approach, was developed to ascertain how an observation (or set of them) influences the posterior distribution of a set of parameters of interest or the predictive distribution of a future set of observables. The approach is to compare the posterior (predictive) distribution of the parameters (future observables) with and without the set of observations whose influence is to be determined. Indicators of the discrepancy between the two distribution functions such as the Jeffreys-Good-Turing-Kullback-Leibler information measures, are used, c.f. Kullback and Leibler (1951). In particular, for estimation purposes, one computes the posterior marginal distribution of the set of parameters of interest and applies the previous notions to them.

Although the general inferential Bayesian methodology set out by Johnson and Geisser (1981, 1982, 1983) can be used in any situation requiring it, the most important applications have been in regression analysis. In regression problems the work of Cook (1977, 1979) is especially noteworthy. He proposed a statistic as an indicator of the influence that an observation has with regard to the estimation of a set of regression parameters. And as the regression problem is still of the widest interest, but by no means the only important application, we shall give prominence to it in the discussion here.

Consider a normal linear regression situation where

$$\begin{aligned} Y &= X\beta + e, & e &\sim N(0,\sigma^2 I) \\ Y' &= (Y_1,..,Y_N), & e' &= (e_1,\ldots,e_N) \\ x_i' &= (x_{i1},\ldots,x_{ip}), & \beta' &= (\beta_1,\ldots,\beta_p) \end{aligned} \tag{3.1}$$

and

$$X = \begin{pmatrix} x_{11} & x_{12} \ldots & x_{1p} \\ \vdots & \vdots & \vdots \\ x_{N1} & x_{N2} \ldots & x_{Np} \end{pmatrix} = \begin{pmatrix} x_1' \\ \vdots \\ x_N' \end{pmatrix} \tag{3.2}$$

with assumed prior density for β and σ^2, say

$$g(\beta,\sigma^2) \tag{3.3}$$

The first step in assessing the influence of individual observations with regard to the estimation of β alone, say, is the computation of

$$p(\beta) = p(\beta\,|\,y,X) \propto \int L(\beta,\sigma^2\,|\,y,X)\, g(\beta,\sigma^2)\, d\sigma^2 \tag{3.4}$$

and

$$p_{(i)}(\beta) = p(\beta\,|\,y_{(i)},X_{(i)}) \propto \int L(\beta,\sigma^2\,|\,y_{(i)},X_{(i)})\, g(\beta,\sigma^2)\, d\sigma^2 \tag{3.5}$$

where y the observed value of Y is decomposed such that $y_{(i)}$ is y with y_i deleted and similarly $X_{(i)}$ is X with the ith row deleted and $L(\bullet)$ is the likelihood function. Next we compute one of the information measures, say,

$$I_i(\beta) = I(p_{(i)},p) = E[\ln p_{(i)}(\beta) - \ln p(\beta)] \tag{3.6}$$

where, by definition, the expectation is taken with respect to the first density. All of the observations y_i are then ordered according to $I_i(\beta)$, the larger this value the more influential is y_i. Of course one can include σ^2 as well and if this is the goal we can calculate the joint information measure

$$I_i(\beta,\sigma^2) = I_i(\sigma^2) + E\,[I_i(\beta\,|\,\sigma^2)] \tag{3.7}$$

where

$$I_i(\sigma^2) = E\,[ln\,p_{(i)}(\sigma^2) \text{-}\ ln\,p(\sigma^2)], \tag{3.8}$$

and

$p_{(i)}(\sigma^2)$ and $p(\sigma^2)$ refer to $p_{(i)}(\sigma^2\,|\,y_{(i)},x_{(i)})$ and $p(\sigma^2\,|\,y,X)$ respectively; similarly

$$I_i(\beta\,|\,\sigma^2) = E\,[ln\,p_{(i)}(\beta\,|\,\sigma^2) - ln\,p(\beta\,|\,\sigma^2)] \tag{3.9}$$

and $I_i(\beta\,|\,\sigma^2)$ in (3.1) is averaged over the density $p_{(i)}(\sigma^2)$. This partition often helps to pinpoint the sources of influence. Details of this approach with examples have been worked out for the multivariate general linear model by Johnson and Geisser (1981).

When the stress is on prediction as it often should be in regression problems, it is necessary to calculate the predictive distribution of Z, the $m\times 1$ future vector to be observed for a given W, an $m\times p$ matrix, i.e.

$$Z = W\beta + e^* \qquad e^* \sim N(0,\sigma^2 I) \tag{3.10}$$

with and without y_i. Consequently,

$$f_{(i)}(z) = f_{(i)}(z\,|\,W,y_{(i)},X_{(i)}) = \int f(z\,|\,W,\beta,\sigma^2)\,p_{(i)}(\beta,\sigma^2)\,d\beta d\sigma^2 \tag{3.11}$$

$$f(z) = f(z\,|\,W,y,X) = \int f(z\,|\,W,\beta,\sigma^2)\,p(\beta,\sigma^2)\,d\beta d\sigma^2 \tag{3.12}$$

One then calculates

$$I_i(Z) = E\,[ln\,f_{(i)}(Z) - ln\,f(Z)] \tag{3.13}$$

as the predictive influence function (PIF) which is used for the relative assessment of influence of the y_i's with regard to predicting a future set of values at W. This is useful only in so far as prediction at W is at issue. If W is unknown but can be assigned probabilities then this can be incorporated into the assessment. If this is not the case, it has been found useful to set $W=X$, i.e. to essentially ascertain the effect of predicting back on the original set of independent variables as indicative of an overall assessment. The details of this procedure are given by Johnson and Geisser (1982, 1983).

To demonstrate the calculation in the simplest fashion, we use the vague prior for β and σ^2,

$$g(\beta,\sigma^2) \propto \frac{1}{\sigma^2}\,. \tag{3.14}$$

Let x_i' be the ith row of X then define

$$\begin{gathered} v_i = x_i'(X'X)^{-1}x_i, \qquad (N-p)s^2 = (y-\hat{y})'(y-\hat{y}), \\ \hat{\beta} = (X'X)^{-1}X'y, \qquad \hat{y} = X\hat{\beta}, \qquad \hat{y}_i = x_i'\hat{\beta}, \\ t_i^2 = \frac{(\hat{y}_i - y_i)^2}{(N-p)s^2(1-v_i)}\,, \qquad D_i = \frac{v_i(N-p)}{(1-v_i)p}\,t_i^2\,. \end{gathered} \tag{3.15}$$

(The statistic D_i was defined by Cook (1977) as a measure of the influence of y_i on the estimation of the set of regression coefficients β.)

Using these results we can calculate the various measures of influence previously defined. First we obtain $2I_i(\beta,\sigma^2)$ which is the sum of the two expressions,

$$2I_i(\sigma^2) = C + (N-1-p)t_i^2(1-t_i^2)^{-1} + (N-p)\,ln(1-t_i^2) \tag{3.16}$$

$$2E\,[I_i(\beta\,|\,\sigma^2)] = K + (N-1-p)\,\frac{v_i t_i^2}{(1-v_i)(1-t_i^2)} + \frac{v_i}{1-v_i} + ln(1-v_i) \tag{3.17}$$

where C and K are constants independent of the deleted observation. Although an explicit expression for $I_i(\beta)$ is not obtainable the following approximation, based on a "best" scaled multivariate normal approximation to a multivariate student distribution, should be more than adequate

$$2\hat{I}_i(\beta) = \frac{(N-p-2)v_i}{1-v_i} t_i^2 + ln(1-v_i) + p\left[\frac{N-p-2}{N-p-3} + ln\frac{N-p-3}{N-p-2} - 1 - ln(1-t_i^2) - \frac{t_i^2(N-p-2)}{N-p-3}\right]$$

$$- \frac{v_i}{1-v_i}\left[\frac{(N-p-2)(t_i^2-1)}{(N-p-3)}\right]. \tag{3.18}$$

It is to be noted that the lead term of the above expression, which reflects the effect of fit times leverage, is proportional to the influence function D_i proposed by Cook (1977). The other components reflect the effect on the variation and volume of the posterior distribution of β.

For the predictive influence function a similar "best" multivariate normal approximation to a multivariate student distribution is utilized. This results in

$$2\hat{I}_i(Z) = \frac{(N-p-2)v_i t_i^2(N-p-4)}{2(1-v_i)(N-p-3)} + \left[\frac{v_i(N-p-2)}{2(1-v_i)(N-p-3)} - ln\ 1 + \frac{v_i}{2(1-v_i)}\right] \tag{3.19}$$

$$+ N\left[\frac{N-p-2}{n-p-3}(1-t_i^2) - \ln\frac{N-p-2}{N-p-3}(1-t_i^2) - 1\right].$$

Again, the first term is proportional to Cook's D_i and measures mainly the change in location of the center of the predictive distribution according to some metric (lack of fit) multiplied by a measure of leverage. The second component is essentially a measure of leverage while the third components reflects the effect on the observational error. From the predictive viewpoint, an influential observation will be distantly observed, exhibit lack of fit and significantly alter the volume of the predicting ellipsoid.

For a very large sample it is clear that Cook's statistic, D_i, is adequate for the influence measure with respect to β, since the first term in $\hat{I}_i(\beta)$ is of an order of magnitude larger than the subsequent terms. However, as the sample grows, the joint influence measure $I_i(\beta,\sigma^2)$ depends not only on D_i but a convex function of $(N-p)t_i^2$ as well. With regard to $\hat{I}_i(Z)$, the predictive influence function, the situation is the same with greater relative emphasis on the convex function of $(N-p)t_i^2$.

Moreover, for small and moderate sample sizes all of the influence measures can differ considerably from D_i in the relative influence assigned to the observation.

For the details of applying these notions to univariate and multivariate general linear models including the influence of subsets of size k, alternative information measures, and the use of conjugate prior distributions, see Johnson and Geisser (1981, 1982, 1983). Particular data sets are analyzed in these papers to demonstrate the methods as well as to pinpoint differences in influence as compared with the method of Cook.

Other paradigms are currently being examined such as growth curves, time series, classification and discrimination and censored data situations to ascertain the influence that particular observables have on the resulting inferences and decisions that are to be made.

3.2. *Data Consistency*

Once relatively influential observations are detected, the data analyst is concerned with the source of the observation's influence. Are there substantive grounds for deletion of the observation? Upon investigation it might be ascertained that the value was inaccurately recorded or some departure from the experimental protocol or other adverse condition occurred that would render the observation defective. When there is no reason to suspect an observation's defective nature, other possibilities in regression situations are that it was distantly observed or inconsistent with the rest of the data in terms of the adequacy of the model. One can determine whether it was distantly observed simply enough by the relative values of v_{ii}. To check for its consistency with the rest of the data, an influential observation can be subjected to a predictive significance test. Aitchison and Dunsmore (1975) refer to it as the atypicality index of a new observation and discuss its use in several applications. On the assumption that the observations $y_{(i)}$ are mutually consistent in regard to the model, a small value of

$$Pr[Z{:}f_{(i)}(Z|X_i,X_{(i)},y_{(i)}) \leq f_{(i)}(y_i|X_i,X_{(i)},y_{(i)})] \tag{3.20}$$

could cast doubt on whether y_i is consistent with $y_{(i)}$. Here the future value Z is random and $y_{(i)}$ is fixed as opposed to the sampling situation where both are assumed random. This calculation, of course, presumes that y_i was chosen before the value was actually observed. If in fact y_i was so chosen because it had maximum influence, then, presumably, one ought to condition on this fact, but this greatly complicates the exact calculation of the "significance" value.

It is also possible to use, as a diagnostic,

$$d_i = f_{(i)}(y_i|X_i,X_{(i)},y_{(i)}) \tag{3.21}$$

Geisser (1980d) to search for possible observations that are inconsistent with the rest. Although d_i is obviously affected by transformations on y_i and $y_{(i)}$, this will not be a serious matter, see the discussions by Box (1980) and Stigler (1980). When an observation appears to be inconsistent with the rest, the data analyst may have to decide whether to retain or delete the observation or perform analysis with and without the offending observation. Certainly if the observation's influence is minimal it is of little consequence as to which alternative is chosen.

Thus far only one model has been entertained and a search for single observations or small subsets that were inconsistent with that model were conducted. For the linear regression problem of section 3; using the vague prior (3.14), it is easy to show that (3.20) is equivalent to

$$Pr\left(\frac{t_i^2}{1-t_i^2} \leq F(1,N-p-1)\right), \tag{3.22}$$

if one neglects to condition on the fact that $Z=y_i$ was chosen to maximize one of the influence functions.

3.3 *Model Checking*

Bayesian analysis is certainly most effective when it is applied to deciding or inferring which single model of an exhaustive set of models is most appropriate for a given goal or the most appropriate mixture of the models. For example, if models M_1 and M_2 with accompanying parameter sets α_1 and α_2 and prior probabilities q_1 and q_2 ($q_1+q_2=1$), are being entertained for observed data set $Y=y$, then the calculation of the ratio of

$$q_1 f(y|M_1) \,/\, q_2 f(y|M_2),$$

where

$$f(y|M_i) = \int f(y|M_i,\alpha_i)\, p(\alpha_i)\, d\alpha_i,$$

is appropriate for the comparison of M_1 and M_2. If prediction of a future observation is at issue then the calculation of the mixture is inferentially relevant

$$f(y_{n+1}|y) = q_1 f(y_{n+1}|y,M_1) + q_2 f(y_{n+1}|y,M_2). \tag{3.23}$$

Quite often either the models entertained are not exhaustive or one model conceptually appears, for certain reasons, pre-eminent in its explanatory or predictive potential and no other alternative model is entertained until this model's adequacy is sufficiently doubted.

Box (1980a, 1980b) working more or less along these lines devised an elegant approach to criticism of an entertained model. Assume that a single model M generating Y, given parameter set α, is structured such that

$$p(y,\alpha|M) = f(y|M,\alpha)p(\alpha|M) \tag{3.24}$$

where $f(y|M,\alpha)$ is a joint probability function of Y conditional on the parameter set α, specified by M and $p(\alpha|M)$ is the prior probability function of α. The marginal probability function of Y given the model M is

$$p(y|M) = \int p(y,\alpha|M)\, d\alpha. \tag{3.25}$$

Then Box asserts, that by referring y to $p(y|M)$ as in

$$\gamma = Pr\{Y{:}p(Y|M) < p(y|M)\} \tag{3.26}$$

or using some predictive checking functions, say, $g(Y)$ and referring g to $p(g|M)$ and

$$\gamma = Pr\,[g(Y){:}\, p(g(Y)|M) < p(g(y)|M)], \tag{3.27}$$

that a small value of γ is indicative of a tentative inadequacy of the model. A number of useful applications of this marginal predictive significance test were also presented. There is no doubt that this is a highly useful procedure. However, one must realize that both marginal probability functions, $p(y|M)$ and $p(g|M)$ given by (3.24) and (3.25) depend in a certain sense on the sampling distribution of Y or $g(Y)$. This particularly implies that in cases of optional stopping or censoring where the likelihood remains unaltered, the sampling distribution of the observables can critically depend on the stopping rule or the censoring mechanism, neither of which are inherently a component of the model that requires checking. Hence this type of model criticism can be confounded with the stopping rule or the type of censoring, which may have little to do with criticism of the model. A particular case of this is examined for Bernoulli sampling by Geisser (1983). In this situation it is not clear as to what aspect of the model is called in to question other than the stopping rule.

Let $X_1,X_2,\ldots$ be a sequence of i.i.d. Bernoulli trials with probability of success θ, and uniform prior probability for θ. Then for a fixed number n of trials where y successes are observed, the predictive probability function of Y is easily calculated to be

$$Pr(y|M) = \frac{1}{n+1} \qquad y=0,1,\ldots,n. \tag{3.28}$$

i.e. uniform for all admissible values of y. Hence no test of the type

$$Pr[p(Y|M) < p(y|A)] = \gamma \tag{3.29}$$

is available. Apparently predictive model criticism fails here.

If the experiment were terminated as soon as y successes were attained and resulted in n trials being observed, the predictive probability function of the number of trials is

$$Pr(N=n\,|\,M) = \frac{y}{n(n+1)} \qquad n=y,y+1,\ldots \quad . \tag{3.30}$$

The fact that the probability function is monotonically decreasing in n indicates that the Box procedure is now available i.e. if the observed $N=n_0$ is large enough relative to y, the model may be called into question. In fact,

$$Pr\,[N\geq n_0\,|\,M] = \sum_{n=n_0}^{\infty} \frac{y}{n(n+1)} = \frac{y}{n_0} = \gamma \tag{3.31}$$

where $\gamma=\hat{\theta}$, the MLE of θ. This implies that predictive model criticism here succeeds only for small $\hat{\theta}$. Sampling until a fixed number of failures is attained results in criticism increasing with $\hat{\theta}$. In either case the only aspect of the model that can presumably be called into question, other than the uniform prior, is the stopping rule. But this is absurd.

The major difficulty here is the strong dependence of the marginal predictive distribution on the stopping rule itself given the model of i.i.d. Bernoulli trials with a uniform prior. So, in essence whatever is to be criticized is confounded with the stopping rule. Hence caution must be exercised when using the Box procedure in certain situations.

In this connection it is to be noted that the predictive probability functions denoted in (3.20) and (3.21) are not susceptible to this criticism -- they do not contradict the likelihood principle.

3.4. *Illustration*

The data in Table 1 graphed in Figure 1, from Aitchison and Dunsmore (1975, p. 182), is used to illustrate some of the techniques discussed in the previous parts of this section. Here a simple linear regression of y on x is fitted and the relevant PIF and estimative influence measures are tabled.

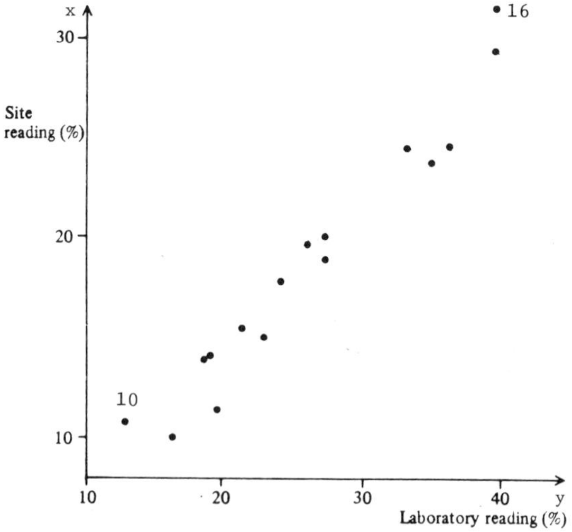

FIGURE 1 *Scatter diagram of laboratory and on-site measurements of 16 soil specimens*

TABLE 1

Water contents (percentages by weight) of 16 soil specimens determined by two methods and associated influence measure

Serial n° of specimen	*Laboratory* methods	*On-Site* method	$2\hat{I}_i(Z)$	$2I(\beta,\sigma^2)$	$2\hat{I}_i(\beta)$	$\frac{v_i}{1-v_i}$	t_i^2
1	35.3	23.7	.22	.31	.24	.11	.18
2	27.6	20.2	.05	.00	.01	.07	.01
3	36.2	24.5	.21	.31	.25	.13	.17
4	21.6	15.8	.04	.01	.03	.08	.03
5	39.8	29.2	.08	.04	.06	.30	.01
6	24.1	17.8	.04	.00	.02	.07	.03
7	16.1	10.1	.07	.03	.04	.21	.00
8	27.5	19.0	.05	.00	.01	.07	.01
9	33.1	24.3	.06	.01	.02	.12	.00
10	12.8	10.6	1.09	1.18	.82	.19	.31
11	23.1	15.2	.04	.01	.03	.09	.03
12	19.6	11.4	.15	.26	.23	.17	.13
13	26.1	19.7	.03	.02	.04	.07	.05
14	19.3	12.7	.06	.01	.02	.13	.00
15	18.8	12.6	.07	.01	.02	.13	.00
16	39.8	31.8	. .95	1.37	1.16	.48	.24

Specimens 10 and 16 are denoted on the graph of Fig. 1 because they are the most influential on all three measures. Specimen 10 is most influential from the point of view of prediction while specimen 16 is most influential for the estimation of β alone or β and σ^2 jointly. However specimen 16 is influential mainly because it is the most distantly observed, while specimen 10 is far less so. But with respect to being consistent with the rest of the data, specimen 10 is most deviant but it is well within the range of acceptability. In fact the predictive probability that the maximum of (3.22) exceeds .31/.69 must be reasonably large, since ignoring that condition yields.

$$Pr\,[F(1,13) \geq .45] \doteq .51. \tag{3.32}$$

For this simple regression situation, the graph itself reveals much the same information as the analysis of influential cases. It is clear that the model provides an adequate fit of the data[1] and the most influential cases are those that are most distantly observed but are consonant with the model. Such techniques however are much more useful in multiple and multivariate regression situations, where subsets of data are to be examined for their influence and graphical displays are unavailable.

[1] When I first studied this example I was quite surprised by the "goodness of fit" of the data as embodied in the result (3.32). I remarked on this fact to John Aitchison but he exhibited no surprise. The reason, he informed me, was that the data were ficticious and that one shouldn't manufacture bad data for a good book--or words to that effect.

4. THE PROBABILITY THAT A FUTURE FRACTION WILL LIE IN A GIVEN SET

A situation that often occurs in experimental work is where a sample of units is drawn from some "large" population of units and is measured with respect to some attribute or on a response to an administered agent. Before being measured the units are assumed indistinguishable with regard to the response (for the simplest paradigm) but they will inherently vary on the response, with the variation being natural and not an error of measurement (though measurement error may be an insignificant portion of the total variation). In this situation, inference and decision are often relevant for the response on a single future unit or several of them jointly or some special function of them depending on the goals of the investigation. For example, if an individual is about to take a new therapy whose response has been recorded for some experimental group of patients with whom this individual is assumed to be more or less indistinguishable with regard to the response, he would be interested only in the predictive distribution of a single future response (his own) or his chance of achieving some threshold value. A physician treating M such patients would be more concerned with the fraction of them that exceed the critical threshold. On the other hand, government health authorities who must deal with the possibility of an M which is a very large number and possibly not precisely known would be interested in the limiting fraction that exceeds the threshold or varying thresholds. The latter case is one situation where the limiting value of the function (which is a parameter, in the sense that it is potentially unobservable) may be of interest. Other situations may arise where the distribution for moderate or large M is rather complex or intractable. In such situations the distribution of the limiting value may serve as a convenient approximation. Sometimes an informative summary of sorts is wanted for the response but no particular fixed number of future values is of critical interest. In such a case one might want to focus on $M=1$ and $M\to\infty$. Even in this situation it would be clearly more informative to present a whole spectrum of values for M. Traditionally, most statistical analyses focus on the infinite case and in so doing make statements only about parameter values. Further, even when making a statement about a parameter, the parameter should be the limit of a sensible function of future observables. The predictivistic point of view is that a statistical model is introduced not because it is necessarily the "true" one —it surely isn't— but because it will serve as an adequate approximation. What is most often critical then is not the fictive parameters of the convenient and approximate formulation but the potential observables.

Consider the set of random variables $X=(X^{(N)};X_{(M)})$ where $X^{(N)}=(X_1,\ldots,X_N)$ are values that will be observed in the experiment and $X_{(M)}=(X_{N+1},\ldots,X_{N+M})$ are values to be predicted. Assume that the joint probability function of X is

$$f(x^{(N)};x_{(M)}\,|\,\alpha)=f(x_{(M)}\,|\,x^{(N)},\alpha)\,f(x^{(N)}\,|\,\alpha) \tag{4.1}$$

where α is a set of unknown parameters. For a given prior probability function $p(\alpha)$, we obtain the posterior density of α

$$p(\alpha\,|\,x^{(N)})\propto f(x^{(N)}\,|\,\alpha)\,p(\alpha) \tag{4.2}$$

and

$$f(x_{(M)}\,|\,x^{(N)})=\int f(x_{(M)}\,|\,x^{(N)},\alpha)\,p(\alpha\,|\,x^{(N)})\,d\alpha \tag{4.3}$$

is the predictive probability function of the future set $X_{(M)}$ given $X^{(N)}=x^{(n)}$.

Many problems in statistics are formulated such that the sequence X_i, $i=1,\ldots,N+M$ are independent and identically distributed. In such cases $X_{(M)}$ represents a set of exchangeable random variables so that each component, X_{N+i}, of $X_{(M)}$ has the same marginal distribution. Incidentally, because

$$f(x_{N+i}|x^{(N)}) = E_\alpha[f(X_{N+i}|\alpha)] \tag{4.4}$$

then from the point of view posterior squared error, $f(x_{n+i}|x^{(N)}$ is a "best" estimate of the common sampling distribution, Geisser (1971). Quite often one may be interested in a function (possibly vector valued) $g(X^{(M)})$ of the future values. Typically it may be the fraction of the observations that lie in some set I, say. Hence let

$$Y_i = \begin{cases} 1 & \text{if } X_{N+i} \in I \\ 0 & \text{otherwise} \end{cases} \tag{4.5}$$

and $R = Y_1 + \ldots + Y_M$, then $\bar{Y} = RM^{-1}$ represents the fraction of future values that lie in I. Further, set $\Theta(\alpha) = Pr(X_i \in I|\alpha)$, then via a simple conditioning argument

$$Pr[\bar{Y} = (r/M)] = \binom{M}{r} \int \theta^r(1-\theta)^{M-r} p(\alpha|x^{(N)})\, d\alpha. \tag{4.6}$$

When all the requirements of deFinetti's (1937) representation theorem are satisfied we also obtain

$$\lim_{M\to\infty} \bar{Y} = \Theta, \tag{4.7}$$

where Θ is a random variable with posterior probability function $p(\theta|x^{(N)})$ derivable from $p(\alpha|x^{(N)})$.

The first two moments are also easily obtained,

$$E(\bar{Y}) = \int_I f(x_{N+1}|x^{(N)})\, dx_{N+1} = q \tag{4.8}$$

$$\text{Var}(\bar{Y}) = q(1-q)\left[\frac{1-\varrho}{M} + \varrho\right] \tag{4.9}$$

where the common correlation coefficient is

$$\varrho = [Pr[Y_i = 1, Y_j = 1] - q^2]/q(1-q) \tag{4.10}$$

for $i \neq j$.

As an application, suppose we are dealing with a random sample from the translated exponential distribution where

$$F(x|\alpha,\gamma) = \begin{cases} 1-e^{-\alpha(x-\gamma)} & \alpha > 0, \\ & x > \gamma > -\infty \\ 0 & \text{otherwise.} \end{cases} \tag{4.11}$$

Assume that $X_1,\ldots,X_d$ are fully observed values while $X_{d+1},\ldots,X_N$ are censored at $x_{d+1},\ldots,x_N$ respectively. Let

$$\bar{x}_d = d^{-1}(x_1 + \ldots + x_d)$$
$$m_d = \min(x_1,\ldots,x_d) \tag{4.12}$$

and assume that $m = \min(x_1,\ldots,x_N) = m_d$ so that

$$m_d \leq \min(x_{d+1},\ldots,x_N). \tag{4.13}$$

This latter condition that there is no censored value less than the minimum of the fully observed values, is very often met in practice and greatly simplifies the likelihood function and the presentation of subsequent formulas, see Geisser (1982b). There is no inherent difficulty, however, in allowing for the contrary of the above case except that the formulas become more complex because of their piecewise nature.

Further, we assumne a conjugate prior density for α, γ to be

$$p(\gamma\,|\,\alpha) = N_0\alpha e^{\alpha N_0(\gamma - m_0)} \qquad \text{for } \gamma < m_0 \tag{4.14}$$

$$p(\theta) = [N_0(\bar{x}_0 - m_0)]^{d_0-1}\,\alpha^{d_0-2}\,e^{-\alpha N_0(\bar{x}_0 - m_0)}/\Gamma(d_0 - 1) \tag{4.15}$$

where $\alpha > 0$, $\bar{x}_0 > m_0$, and $1 < d_0 \leq N_0$ to insure that the distributions are proper.

From the above we can calculate

$$p(\gamma) = \frac{(d_0 - 1)(\bar{x}_0 - m_0)^{d_0-1}}{(\bar{x}_0 - \gamma)^{d_0}} \tag{4.16}$$

and

$$p(\alpha\,|\,\gamma) = [N_0(\bar{x}_0 - \gamma)]^{d_0}\alpha^{d_0-1}\,e^{-\alpha N_0(\bar{x}_0 - \gamma)}/\Gamma(d_0)\,. \tag{4.17}$$

Suppose we define the survival function as

$$\Theta = Pr\,[Z > z\,|\,\alpha,\gamma] = \begin{cases} e^{-\alpha(z-\gamma)} & \alpha > 0\,,\ z > \gamma > -\infty \\ 1 & \text{otherwise.} \end{cases}$$

then one can obtain the posterior distribution of Θ. For $0 < \theta < 1$

$$\theta^{N^*}\left(\frac{\bar{x}^* - m^*}{\bar{x}^* - z}\right)^{d^*-1} \qquad \text{for } z \leq m^* \tag{4.19}$$

$$Pr[\Theta \leq \theta\,|\,x^{(N)}] = \begin{cases} \theta^{N^*}\left(\dfrac{\bar{x}^* - m^*}{\bar{x}^* - z}\right)^{d^*-1} G\left(2N^*\left(\dfrac{\bar{x}^* - z}{m^* - z}\right)\log\theta\right) + 1 - G\left(2N^*\left(\dfrac{\bar{x}^* - m^*}{m^* - z}\right)\log\theta\right) & \text{for } m^* < z < \bar{x}^* \\[2ex] \theta^{N^*}\left(\dfrac{\bar{x}^* - m^*}{z - \bar{x}^*}\right)^{d^*-1} G\left(2N^*\left(\dfrac{z - \bar{x}^*}{m^* - z}\right)\log\theta\right) + 1 - G\left(2N^*\,\dfrac{\bar{x}^* - m^*}{m^* - z}\,\log\theta\right) & \text{for } z > \bar{x}^* \end{cases}$$

and

$$1 - G^*(-2N^*\log\theta) \qquad \text{for } z = \bar{x}^*$$

$$Pr\,[\Theta = 1\,|\,x^{(N)}] = \begin{cases} 1 - \left(\dfrac{z - m^*}{\bar{x}^* - z}\right)^{d^*-1} & z < m^* \\[2ex] 0 & z \geq m^* \end{cases} \tag{4.20}$$

where $G(u)$ represents the distribution function of a χ^2 variate with $2d^* - 2$ degrees of freedom and $G^*(u)$ a χ^2 variate with $2d^*$ degrees of freedom and

$$\begin{aligned} d^* &= d_0 + d \\ N^* &= N_0 + N \\ \bar{x}^* &= (N_0\bar{x}_0 + N\bar{x})/N^* \\ m^* &= \min(m_0, m). \end{aligned} \tag{4.21}$$

For the next observation X_{N+1} we can calculate the expectation of $Pr\,[X_{N+1} > z\,|\,\alpha,\gamma]$ w.r.t. $p(\alpha,\gamma\,|\,x^{(N)})$ which results in the predictive survival function

$$Pr\,[X_{N+1}>z] = \begin{cases} \dfrac{(N^*)^{d^*}(\bar{x}^*-m^*)^{d^*-1}}{(N^*+1)\,[z-m^*+N^*(\bar{x}^*-m^*)]^{d^*-1}} & z>m^* \\[2ex] 1-(N^*+1)^{-1}\left(\dfrac{\bar{x}-m^*}{\bar{x}^*-z}\right)^{d^*-1} & z\leq m^* \end{cases} \tag{4.22}$$

Using (4.6), we obtain, Geisser (1982b)

$$Pr(\bar{Y} = \tfrac{r}{M}|z) = \begin{cases} \left(\dfrac{\bar{x}^*-m^*}{\bar{x}^*-z}\right)^{d^*-1}\dbinom{N^*+r-1}{r}\Big/\dbinom{N^*+M}{M} & r<M, z<m^* \\[2ex] 1-\dfrac{M}{N^*+M}\left(\dfrac{\bar{x}^*-m^*}{\bar{x}^*-z}\right)^{d^*-1} & r=M,\ z<m^* \\[2ex] N^*\dbinom{M}{r}\displaystyle\sum_{j=0}^{M-r}\dbinom{M-r}{j}\dfrac{(-1)^j}{(N^*+r+j)}\left[1+\dfrac{(r+j)(z-m^*)}{N^*(\bar{x}^*-m^*)}\right]^{-(d^*-1)} \end{cases} \tag{4.23}$$

We further note, for the special limiting case of the conjugate prior on α and γ, namely the "noninformative" quasi prior which is

$$p(\alpha,\gamma)\propto\alpha^{-1}, \tag{4.24}$$

that the *'s are removed in (4.22) so that $d^* \to d$, $N^* \to N$, $\bar{x}^* \to \bar{x}$ and $m^* \to m$.

In cases where γ is known, by a simple translation, we effectively set $\gamma=0$ and use (4.17) as the prior density for α. In this case

$$Pr[\bar{Y}= \tfrac{r}{M}|z] = \binom{M}{r}(N^*\bar{x}^*)^{d^*}\sum_{j=0}^{M-r}\binom{M-r}{j}(-1)^j\,[N^*\bar{x}^*+z(r+j)]^{-d^*} \tag{4.25}$$

and similarly for the vague prior in this instance

$$p(\alpha)\propto\alpha^{-1}, \tag{4.26}$$

$d^*\to d$, $N^*\to N$ and $\bar{x}^*\to\bar{x}$, Geisser (1982a).

For large M, it is shown, Geisser (1982a) that the third expression of (4.23) and (4.25) can each be reasonably well approximated by replacing the Chi-squared distributions in (4.19) and its analogue for $\gamma=0$ by F distributions with appropriate degrees of freedom.

When there is no censoring and lack of knowledge as to the common sampling distribution of $X_1,\ldots,X_N$, the type of calculation given in (4.6) could be the basis for a reasonably robust Bayesian procedure. For example, if we assume a uniform prior distribution for Θ, in the Bayes-Laplace tradition, then we obtain

$$Pr\,[\bar{Y}= \frac{r}{M}\,|z] = \frac{\binom{r+s}{s}\binom{M+N-r-s}{N-s}}{\binom{M+N+1}{M}}\ , \tag{4.27}$$

where s is the number of X_1's, $i=1,\ldots,N$ that exceed z. It is to be recalled that the first use of (4.6) was the calculation of (4.27) made by Laplace (1774) where $p(\theta\,|x^{(N)})$ was based on a uniform prior for Θ as alluded to in the introduction.

This is much coarser method since it sacrifices the finer distinctions engendered by shifts in z and the distributional attributes of Θ.

Calculating (4.6) explicitly for many distributions is often difficult. For example, consider the simple normal case i.e. $X_1,\ldots,X_{N+M}$ are $N(\mu,1)$ without censoring. In this case

$$\Theta = Pr\,[Z>z\,|\,\mu] = 1-\Phi(z-\mu) \tag{4.28}$$

where $\Phi(\bullet)$ is the standard normal distribution function. In the simple situation where the quasi-prior for μ is uniform so that the posterior distribution for μ is $N(\bar{x},N^{-1})$, the calculation of (4.6)

$$\sqrt{\frac{N}{2\pi}}\binom{M}{r}\int[1-\Phi(z-\mu)^r\,\Phi^{M-r}(z-m)e^{-N(\mu-\bar{x})^2/2}\,d\mu \tag{4.29}$$

requires a series expansion and consequently an approximation. An alternative way of handling this problem is to calculate the joint predictive density of the set $X_{(M)}$,

$$f(x_{(M)}\,|\,x^{(N)}) = \int N^{1/2}\,\varphi(N^{1/2}(\bar{x}-\mu))\prod_{i=1}^{M}\varphi(x_{N+i}-\mu)\,d\mu \tag{4.30}$$

where $\varphi(\bullet)$ is the standard normal density. Hence the predictive distribution of the components of $X_{(M)}$ is exchangeable being a multivariate normal distribution with common mean $\bar{x}$, common variance $1+N^{-1}$ and common covariance N^{-1}. If we now require the probability that exactly r out M lie in the same interval i.e. $X_{N+i}>z$, this still remains a formidable calculation. The complexity increases when we permit σ^2 to be unknown and use the simple quasi-prior

$$p(\mu,\sigma^2)\propto 1/\sigma^2\,.$$

Now we obtain for the joint distribution of $X_{N+1},\ldots,X_{N+M}$ an exchangeable multivariate student distribution. This multivariate student distribution can be reasonably well-approximated by an exchangeable multivariate normal distribution with the same mean but slightly inflated variances and covariances. This has the advantage that an approximate solution for the more complex case depends on an exact solution for the simpler case.

In either case the probability that exactly r out of the next M observations will exceed z can be represented by

$$P_r = \binom{M}{r}Pr\,[X_{N+1}>z,\ldots,X_{N+r}>z,\,X_{N+r+1}\leq z,\ldots,X_{N+M}\leq z]$$

because the variables are exchangeable multivariate normal or student variables. Exact calculations for P_r require high speed computers but approximations or, perhaps, bounds similar to the type developed by Kounias (1968) and Hunter (1976) may be useful here.

4.1. *Examples*

In this section we present two examples illustrating the methods obtained for the exponential distribution

Example 1

A department store's past experience with a type of fluorescent light is given in days to failure as follows:

7	26	35	49	69	99	141
15	27	36	56	71	105	145
20	28	41	57	75	106	154
21	29	43	62	78	126	168
22	34	48	64	91	133	189

with 5 lights exceeding 196 days.

A new room in the store is being opened up and will require 10 lights. The manager wants to have an idea of the lifetime distribution of these 10 lights. Assuming an exponential survival distribution with the lower limit known to be 0 and using the quasi-prior of (4.26), we calculate

$$Pr\,[\bar{Y} \leq \frac{r}{10} \mid z]$$

for $r=0,1,\ldots,10$; $z=7, 14, 21, 28,\ldots,196$ (see table 2). This will provide all the necessary information concerning probability calculations on the lifetimes of the 10 lights.

Example 2:

The data set in the following table is reported in Pike (1966) and discussed by Kalbfleisch and Prentice (1980). The table gives the times from insult with a carcinogen to death for two differentially treated groups

TABLE 3

Days to Vaginal Cancer Mortality in Rats

Group 1	143,	164,	188,	188,	190,	192,	206,	209,	213,	216,
	220,	227,	230,	234,	246,	265,	304,	216*,	244*	
Group 2	142,	156,	163,	198,	205,	232,	232,	233,	233,	233
	233,	239,	240,	261,	280,	280,	296,	296,	323,	204*,
	344*									
Censored*										

Assuming a known lower level of 100 days, Kalbfleisch and Prentice use the exponential distribution to compare the groups and determine that there is no difference in survival between the two groups but indicated that the exponential fit was an inadequate description of the data. Pike (1966) found that the third power of the excess over 100 days was adequate to provide an exponential fit. Hence we assume $U=(X-100)^3$ is exponentially distributed with known minimum for U to be 0 and compute the predictive survival probability $Pr\,[X_{N+1}>x]$ using the quasi-prior of (4.24) for the two groups in Table 4 which is plotted in Figure 2. In Table 5 we present a brief comparison of the probabilities that the fraction of all future rats will survive beyond a varying threshold for the two groups. All of the computations indicate the survival superiority of group 2, assuming the adequacy of the model.

TABLE 2

Pr $[\bar{Y} \leq \frac{r}{10} | z]$ *for fluorescent light data by weeks* $w = z/7$

r \ w	1	2	3	4	5	6	7	8	9	10	11	12	13	14	15	16	17	18	19	20	21	22	23	24	25	26	27	28
0	.00	.00	.00	.00	.00	.00	.00	.00	.00	.00	.00	.01	.01	.01	.02	.03	.04	.05	.06	.07	.09	.11	.13	.15	.17	.20	.21	.24
1	.00	.00	.00	.00	.00	.00	.00	.00	.01	.02	.03	.04	.06	.08	.10	.13	.16	.20	.23	.27	.31	.35	.39	.43	.47	.50	.54	.57
2	.00	.00	.00	.00	.00	.01	.01	.03	.05	.07	.10	.14	.18	.23	.28	.33	.38	.44	.49	.54	.58	.63	.67	.70	.74	.77	.80	.82
3	.00	.00	.00	.00	.01	.03	.06	.09	.14	.20	.26	.32	.39	.45	.52	.58	.63	.68	.73	.77	.80	.83	.86	.88	.90	.92	.93	.94
4	.00	.00	.01	.02	.05	.10	.16	.23	.31	.40	.48	.55	.62	.68	.74	.78	.82	.86	.88	.91	.93	.94	.95	.96	.97	.98	.98	.98
5	.00	.01	.03	.08	.16	.25	.35	.45	.54	.63	.70	.76	.81	.85	.89	.91	.93	.95	.96	.97	.98	.98	.99	.99	.99	.99	1.00	1.00
6	.00	.04	.11	.22	.35	.47	.58	.68	.76	.82	.87	.90	.93	.95	.96	.97	.98	.99	.99	.99	1.00	1.00	1.00	1.00	1.00	1.00	1.00	1.00
7	.03	.14	.30	.46	.60	.72	.80	.86	.91	.94	.96	.97	.98	.99	.99	.99	1.00	1.00	1.00	1.00	1.00	1.00	1.00	1.00	1.00	1.00	1.00	1.00
8	.15	.39	.59	.74	.84	.90	.94	.96	.98	.99	.99	.99	1.00	1.00	1.00	1.00	1.00	1.00	1.00	1.00	1.00	1.00	1.00	1.00	1.00	1.00	1.00	1.00
9	.50	.75	.87	.93	.97	.98	.99	.99	1.00	1.00	1.00	1.00	1.00	1.00	1.00	1.00	1.00	1.00	1.00	1.00	1.00	1.00	1.00	1.00	1.00	1.00	1.00	1.00
10	1.00	1.00	1.00	1.00	1.00	1.00	1.00	1.00	1.00	1.00	1.00	1.00	1.00	1.00	1.00	1.00	1.00	1.00	1.00	1.00	1.00	1.00	1.00	1.00	1.00	1.00	1.00	1.00
Mode ($\bar{Y}$)	1.00	.90	.80	.80	.70	.70	.60	.60	.50	.50	.50	.40	.40	.40	.30	.30	.30	.30	.20	.20	.20	.20	.20	.10	.10	.10	.10	.10
Median ($\bar{Y}$)	.90	.90	.80	.80	.70	.70	.60	.60	.50	.50	.50	.40	.40	.40	.30	.30	.30	.30	.30	.20	.20	.20	.20	.20	.20	.10	.10	.10
Mean ($\bar{Y}$)	.93	.87	.81	.75	.70	.65	.61	.57	.53	.50	.46	.43	.40	.38	.35	.33	.31	.28	.27	.25	.23	.22	.20	.19	.18	.17	.15	.14

TABLE 4

Predictive probability that a random rat will survive x days when treated either as in group 1, $X_{1,N+1}$ or as in group 2, $X_{2,N+1}$.

x	150	160	170	180	190	200	210	220	230	240	250	260	270	280	290	300
$Pr[X_{1,N+1}>x]$	.95	.91	.86	.80	.72	.64	.55	.47	.38	.30	.23	.17	.12	.09	.06	.04
$Pr[X_{2,N+1}>x]$	.97	.95	.92	.89	.84	.78	.73	.67	.60	.53	.45	.39	.32	.26	.21	.16

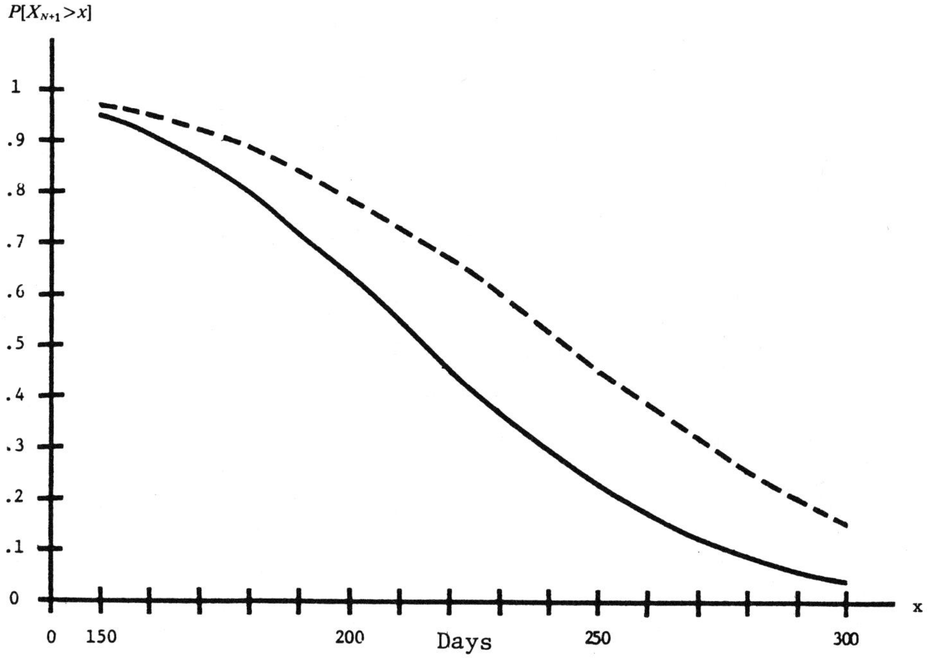

FIGURE 2

Comparison of survival probabilities for two groups of rats with differential carcinogenic insults

TABLE 5

Pr $[\Theta > \theta \mid x]$*: *Probability that at least a given fraction of all future rats survive beyond a specified threshold x for groups 1 and 2.*

x	200		225		250		275		300	
θ	1	2	1	2	1	2	1	2	1	2
.1	1.00	1.00	1.00	1.00	.97	1.00	.45	1.00	.05	.82
.2	1.00	1.00	1.00	1.00	.62	1.00	.07	.87	.00	.26
.3	1.00	1.00	.93	1.00	.20	.97	.01	.43	.00	.04
.4	1.00	1.00	.60	1.00	.03	.74	.00	.10	.00	.00
.5	.98	1.00	.19	.97	.00	.04	.00	.01	.00	.00
.6	.73	1.00	.02	.69	.00	.00	.00	.00	.00	.00
.7	.20	.97	.00	.15	.00	.00	.00	.00	.00	.00
.8	.01	.42	.00	.00	.00	.00	.00	.00	.00	.00
.9	.00	.00	.00	.00	.00	.00	.00	.00	.00	.00
$E(\Theta)$	.64	.78	.42	.63	.23	.45	.10	.29	.04	.16

* An entry of 1.00 in this table indicates that the value is between 1 and .995.
An entry of .00 indicates that the value is between 0 and .005.

ACKNOWLEDGEMENTS

I am indebted to Murray Clayton and Wesley Johnson for computational assistance. This work is supported in part by NIH Grant GM25271.

REFERENCES

AITCHISON, J. and DUNSMORE, I.R. (1975). *Statistical Prediction Analysis,* Cambridge: University Press.

AITCHISON, J. and SCULTHORPE, D. (1965). Some problems of statistical prediction. *Biometrika* **52**, 469-83.

AKAIKE, H. (1978). A new look at the Bayes procedure, *Biometrika,* **65**, 1, 53-59.

ANDERSON, T.W. (1971). *The Statistical Analysis of Time Series.* New York: Wiley.

ANDREWS, D. and PREGIBON, D. (1978). Finding outliers that matter. *J. Roy. Statist. Soc. B,* **40**, 85-93.

BELSLEY, D.A., KUH, E. and WELSCH, R.E. (1980). *Regression Diagnostics.* New York: Wiley.

BOX, G.E.P. (1980a). Sampling and Bayes' inference in scientific modelling and robustness. *J. Roy. Statist. Soc. A,* **143**, 383-430.

— (1980b). Sampling inference, Bayes' inference, and robustness in the advancement of learning. *Bayesian Statistics,* (J.M. Bernardo et al. eds.). Valencia: University Press, 366-370.

BOX, G.E.P. and JENKINS, M. (1970). *Time Series Analysis Forecasting and Control.* San Francisco: Holden-Day.

BUTLER, R. and ROTHMAN, E.D. (1975). Intervals based on reuse of the sample. *J. Amer. Statist. Assoc.* **75, 372,** 881-889.

CONDORCET, N.C. (1786). Réfexious sur le méthod de déterminer la probabilité des évènemens futurs, apres l'observation des évènemens passés. *Mémoirs de l'Académie Royale des Sciences,* 539-559.

COOK, R.D. (1977). Detection of influential observations in linear regression. *Technometrics,* **19,** 15-18.

— (1979). Influential observations in linear regression. *J. Amer. Statist. Assoc.* **74,** 169-174.

COOK, R.D. and WEISBERG, S. (1980). Finding influential cases in regression: a review. *Technometrics,* **22,** 495-508.

— (1982). *Residuals and Influence in Regression.* New York: Chapman and Hall.

DE FINETTI, B. (1937). La prèvision: ses lois logique, ses sources subjectives, *Annales de l'Institute Henri Poincaré,* **7,** 1-68.

DE WAAL, D. et al. (1981). Model selection, prediction and estimation for multivariate normal populations. *Journal of the South African Statistical Assoc.* **15,** 113-127.

DUNSMORE, I.R. (1968). A Bayesian approach to calibration. *J. Roy. Statist. Soc. B,* **30,** 396-405.

— (1969). Regulation and optimization. *J. Roy. Statist. Soc. B,* **31,** 160-70.

— (1974). The Bayesian predictive distribution in life testing models. *Technometrics,* **16,** 455-60.

FEARN, T. (1975). A Bayesian approach to growth curves. *Biometrika,* **62,** 1 89-100.

FISHER, R.A. (1933). The concepts of inverse probability and fiducial probability referring to unknown parameters. *Proc. of Roy. Soc. A,* **139,** 343-348.

— (1934). Probability likelihood and quantity of information in the logic of uncertain inference. *Proc. of Roy. Soc. A,* **146,** 1-8.

— (1956). *Statistical Methods and Scientific Inference.* Edinburgh: Oliver & Boyd.

GEISSER, S. (1964). Posterior odds for multivariate normal classification. *J. Roy. Statist. Soc. B.* **26,** 69-76.

— (1965). Bayesian estimation in multivariate analysis. *Ann. Math. Statist.* **36,** 150-159.

— (1966). Predictive discrimination. *Multivariate Analysis,* (P. Krishnaiah, ed.), New York: Academic Press, 149-163.

— (1970). Bayesian analysis of growth curves. *Sankhyā,* **A 32,** 53-64.

— (1971). The inferential use of predictive distributions. *Foundations of Statistical Inferences,* (Godambe and D. Sprott, eds.), New York: Rinehart and Winston, 456-469.

— (1974). A predictive approach to the random effect model. *Biometrika,* **61,** 101-107.

— (1975a). The predictive sample reuse method with applications. *J. Amer. Statist. Assoc.,* **70,** 320-328, 350.

— (1975b). A new approach to the fundamental problem of applied statistics. *Sankhyā,* B, **37,** 385-397.

— (1976). Review of statistical prediction analysis, by J. Aitchison and I.R. Dunsmore, *Bull. of the Amer. Math. Soc.* Sept., 683-688.

— (1979). Discussion on: Reference posterior distributions for Bayesian inference, by J.M. Bernardo. *J. Roy. Statist. Soc. B,* **41,** 136-137.

— (1980a). Sample reuse selection and allocation criteria. *Multivariate Analysis V,* (ed. P. Krishnaiah), Amsterdam: North-Holland, 387-398.

— (1980b). Predictive sample reuse techniques for censored data. *Bayesian Statistics* (J.M. Bernardo et al. eds.). Valencia, Spain: University Press. 430-468 (with discussion).

— (1980c). The estimation of distribution functions and the prediction of future observations. *Proc. Conf. Recent Developments in Statistical Methods and Applications.* Taiwan: Academia Sinica, 193-208.

— (1980d). Discussion: Sampling and Bayes' inference in scientific modelling and robustness, by G.E.P. Box. *J. Roy. Statist. Soc, A,* **143,** 416-417.

— (1981). Sample reuse procedures for predicting the unobserved portion of a partially observed vector. *Biometrika,* 243-250.

— (1982a). Aspects of the predictive and estimative approaches in the determination of probabilities. *Biometrics,* **38,** supplement, 75-93 (with discussion).

— (1982b). Predicting Pareto and exponential observables. *Tech. Report.* **408.** University of Minnesota.

— (1983). A remark on a model criticism technique. *Bull. Internat. Statist. Institute, 44th Session*, 925-927.

GEISSER. S. and DESU, M.M. (1968). Predictive zero-mean uniform discrimination. *Biometrika,* **55,** 519-24.

GEISSER, S. and EDDY, W.F. (1979). A predictive approach to model selection. *J. Amer. Statist. Assoc.,* **74,** 153-160. Corregendum, **75,** (1980), 765.

GOOD, I.J. (1965). *The estimation of probabilities: An essay on modern Bayesian methods.* Harvard: The MIT Press.

GUTTMAN, I. (1967). The use of the concept of a future observation in goodness-of-fit problems. *J. Roy. Statist. Soc. B,* **29,** 83-100.

— (1970). *Statistical tolerance regions: classical and Bayesian.* London: Griffin.

GUTTMAN, I. and TIAO, G.C. (1964). A Bayesian approach to some best population problems. *Annals of Math. Statist.* **35,** 825-35.

HOAGLIN, D.C. and WELSCH, R.E. (1978). The hat matrix in regression and ANOVA. *Amer. Statist.* **32,** 17-22: Corrigenda, **145.**

HUNTER, D. (1976). An upper bound for the probability of a union. *J. Appl. Prob.* **13,** 597-603.

JAYNES, E.T. (1980). The minimum entropy production principle. *Ann. Rev. Phys. Chemistry,* **31,** 579-601.

JEFFREYS, H. (1932). On the theory of errors and least squares. *Proc. Roy. Soc. London,* A, **138,** 48-55.

— (1933). Probability, statistics, and the theory of errors. *Proc. Roy. Soc. of London,* A, **140,** 523-535.

— (1934). Probability and scientific method. *Proc. Roy. Soc. of London,* A, **146,** 9-15.

— (1939). *Theory of Probability.* Oxford: University Press.

JOHNSON, W. and GEISSER, S. (1982). Assessing the predictive influence of observations. *Statistics and Probability Essays in Honor of C.R. Rao.* (Kallianpur, Krishnaiah, and Ghosh, eds.) Amsterdam: North-Holland, 343-358.

— (1983). A predictive view of the detection and characterization of influential observations in regression analysis. *J. Amer. Statist. Assoc.,* **78,** 137-144.

— (1981). Estimative influence functions for the multivariate general linear model. *Tech. Report.* **23,** University of California, Davis.

KALBFLEISCH, J.D. and PRENTICE, R.L. (1980). *The Statistical Analysis of Failure Time Data.* New York: Wiley.

KOUNIAS, E. (1968). Bounds for the probability of a union with applications. *Ann. Math. Statist.* **39,** 2154-2158.

KULLBACK, S. and LEIBLER, R.A. (1951). On information and sufficiency. *Ann. Math. Statist.* **22,** 79-86.

LAPLACE, P.S. (1774). Memoir sur la probabilité des causes par les évènements. *Memoirs de l'Académie Royale des Sciences,* **6,** 621-656.

LANE, D.A. (1981). Fisher, Jeffreys, and the nature of probability. *R.A. Fisher: An Appreciation.* (S. Fienberg and D. Hinckley, eds.) Berlin: Springer-Verlag, 148-160.

MEEHL, P.E. (1954). *Clinical versus Statistical prediction. A theoretical analysis and review of the Literature.* Minneapolis: University of Minnesota Press.

PEARSON, K. (1907). On the influence of past experience on futures expectation. *Philosophical Magazine,* **12,** 365-378.

— (1920). The fundamental problem of practical statistics. *Biometrika,* **13,** 1-16.

— (1979). *The History of Statistics in the 17th and 18th centuries.* (Pearson, E.S. eds.), London: Griffin.

PIKE, M.C. (1966). A suggested method of analysis of a certain class of experiments in carcinogenesis. *Biometrics,* **22,** 142-161.

ROBERTS, H.V. (1965). Probabilistic prediction. *J. Amer. Statist. Assoc.* **60,** 50-62.

STIGLER, S.M. (1980). Discussion on: Sampling and Bayes' inference in scientific modelling and robustness by G.E.P. Box. *J. Roy. Statist. Soc. A,* **193,** 416-417.

— (1982). Thomas Bayes' s Bayesian inference. *J. Roy. Statist. Soc. A,* **145,** 250-258.

STONE, M. (1974a). Cross-validatory choice and assessment of statistical predictions. *J. Roy. Statist. Soc. B,* **36,** 111-147 (with discussion).

— (1974b). Cross-validation and multinomial prediction, *Biometrika,* **61,** 509-515.

— (1977). An asymptotic equivalence of choice of model by cross-validation and Akaike's criterion, *J. Roy. Statist. Soc. B,* **39,** 44-47.

WAHBA, G. (1977). Optimal smoothing of density estimates. *Classification and Clustering,* New York: Academic Press, 423-458.

WHITTLE, P. (1963). *Prediction and Regulation by Linear Least Squares Methods.* Princeton, N.J.: VanNostrand.

WIENER, N. (1949). *The Extrapolation, Interpolation and Smoothing of Stationary Time Series with Engineering Applications,* Cambridge Mass: The MIT Press.

WILKS, S.S. (1942). Statistical prediction with special reference to the problem of tolerance limits. *Ann. Math. Statist.* **13,** 400-409.

WOLD, H. (1938). *A Study in the Analysis of Stationary Time Series,* Stockholm: Almquist and Wicksell.

— (1959). Ends and means in econometric model building, *Probability and Statistics: The Harold Cramer volume,* New York: Wiley, 355-434.

YAGLOM, A.M. (1962). *An Introduction to the Theory of Stationary Random Functions,* Englewood Cliffs, N.J.: Prentice-Hall.

DISCUSSION

I.R. DUNSMORE (*University of Sheffield, U.K.*)

It is always inspiring to hear or read about Professor Geisser proclaiming the predictivistic approach and rallying his troops. As he says, predictive distributions have been grossly underutilized by Bayesians and others. As the week has progressed it has been interesting to note that there have been more and more references to predictive distributions. Here Geisser gives a short review of predictivism and a selection of applications - but it should be noted that in his references the number of papers which actually use predictive distributions is disappointingly small. It would be difficult to comment on all the aspects mentioned in the review, so I will confine my attention to just two.

Johnson and Geisser's (1983) predictive approach to influential observations in Section 3.1 is an excellent illustration of the usefulness and intuitive satisfactoriness of the predictive methodology. Obviously from previous papers this week they have not been alone in working along these lines. My only criticism lies in the backtracking incurred by setting $W=X$, that is by looking at the influence of observations if you were going to predict again at all the sames values (including the ones you are omitting). This seems to me a somewhat dangerous exercise and provides an overall assessment whose meaning is somewhat mystical. It is surely better to provide a predictive influence function for any given W. In particular if predictions at single values are of importance, so that W is a $1\times p$ matrix, we may consider the effect of X_i on prediction at W for each i.

As an illustration consider the data in Section 3.4. Professor Geisser has in fact reversed the x and y values from the original data. The values for $2\hat{I}_i(Z)$ are very similar to those shown, except that observation 16 appears to be the most influential under the dubious measure based on $W=X$. If we consider prediction at single future values with

$$W = (1, w), \qquad X_i = (1, x_i)$$

and write

$$u = WS^{-1}W', \qquad \ell_i = x_i S^{-1}W',$$

then the approximate predictive influence measure is given by

$$2\hat{I}_i(Z) = \frac{(N-4)t_i^2}{(1+u)(1-v_i)} X_i S^{-1}W' WS^{-1}X_i$$

$$+ \frac{(N-4)\ell_i^2(1-t_i^2)}{(N-5)(1+u)(1-v_i)} - ln \left[1 + \frac{\ell_i^2}{(1+u)(1-v_i)}\right]$$

$$+ \frac{(N-4)}{(N-5)}(l-t_i^2) - ln \left[\frac{(N-4)}{(N-5)}(l-t_i^2)\right] - 1.$$

Figure 1 shows the values of $2\hat{I}_i(Z)$ for some of the more influential of the points. (All the other values of x_i have values of $2\hat{I}_i(Z)$ consistently lower than 0.01). We see that if we are predicting in the range w = 10-23, then observation 10 is the most influential, whereas for $w = 23-40$ observation 16 takes over this role. Notice how observations 1 and 3 have more influence than 10 if values of w in the range $30-40$ are envisaged. Notice also that these influential points tend to have the larger residual values.

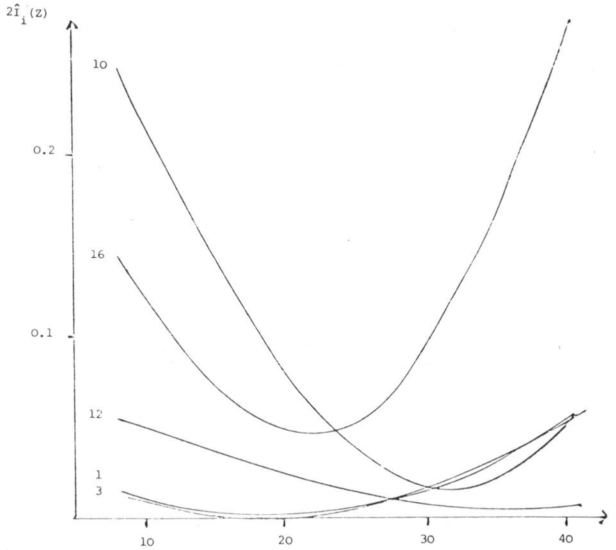

Figure 1- *Predictive influence measure $2\hat{I}_i(Z)$ plotted against w for some of the more influential of the points.*

Perhaps one of the most important points in the paper is made at the beginning of Section 4 where the distinction is emphasized between inferences and decisions for the responses of (i) a single future unit or a future group of M units, and (ii) a future group where M is very large and perhaps not even known precisely. Herein lies the rationale and motivation behind the predictive approach. Surely much more emphasis should be placed on the prediction of future (observable) values than on statements about (unobservable) parameter values.

Perhaps I can close with an appropriate analogy. In the United Kingdom over the past few years we have seen the formation of the SDP/Lib alliance. The Liberals have been around for a long time, working away steadily and becoming more popular once more although not always getting their due recognition. Perhaps here they are Lindley's Integrated Bayesians. The SDP is a relatively new party - there has been much publicity but it is taking some time to reach the threshold before they really take off. If I may be allowed to break the Conference rule, perhaps here they are Seymour's Distributional Predictivists. Working together as an alliance in the Bayesian field (as in the nature of the two papers presented) there should be no stopping them.

L. PICCINATO (*Universitá di Roma*)

As Professor Geisser reminded us, interest in the predictive approach in Statistics has been growing for many years and now its importance seems generally accepted among Bayesians. There remain difficulties for the other schools because of the lack of appropriate tools, but this does not concern us at least on this occasion.

I would like to discuss a point of a general nature: in my opinion it would be useful to distinguish among different attitudes even *within* a predictivistic framework. I would say that Geisser deals essentially with predictive *problems*, in the framework of a standard hypothetical approach. In particular he introduces among other things a *statistical model* and predictions are conditional on its validity or its approximate validity.

A much more radical approach is however emerging. Let me call it the *completely predictive approach*. In that approach, the potential observations are seen as a stochastic process, indexed by time and characterized by a well defined and *unique* probability measure, namely the prior predictive distribution. The predictive inference at a given time is simply the updating of the probability measure, through conditioning on the known segment of results. Slight generalizations are obvious but not worth discussing here.

Needless to say, there is no *logical* opposition between the completely predictive approach and the standard one. If the observations are exchangeable (that is, if they provide random samples in the usual sense), the de Finetti representation theorem assures that a statistical model is *implicit* in the prior predictive distribution; and vice versa, starting with the standard approach, one could always look at the marginal distributions of the observations. Such an approach, coming back to the initial position of de Finetti (before his contacts with Savage he considered only this kind of predictivism) has been recently firmly advocated by some Italian authors (see Cifarelli and Regazzini, 1982). Last year for instance a text book was published, intended for a first year course (Daboni and Wedlin, 1982) where just this approach was considered as the basic inferential scheme. Essentially similar arguments were also used, at least for special problems, by different authors, for instance by Lindley and Phillips (1976) presenting the Bernoulli process "from a higher viewpoint" as they say, quoting Felix Klein (but the critical issue, now, is: *higher* or *more fundamental* ?), and by Dawid (1982) dealing with the problem of building statistical models.

I am not claiming, now, the primacy of any specific approach. I would only point out that if the focus has to be on the observables, the method used by Geisser is a compromise: parameters are introduced, but they are soon integrated out. On the other hand the compact form of inference, in the completely predictive approach, could provide a tool which is too rigid in practice: for example because the methods corresponding to *intermediate structure* situations (in the sense of Geisser, 1980) do not seem any available any more. Nevertheless, I am generally unable to understand the real meaning of a prior probability assessment without evaluating the implied characteristics of the prior predictive distribution; that is, referring to the observables in the strictest sense.

Summing up, while welcoming the treatment of the unusual and very interesting predictive problems provided by Geisser, which might well serve as new chapters to be added to our textbooks, I think that the debate about predictivism will continue, involving among other things the issue I mentioned. One cannot ignore the fact that so many authors now seem to suggest that our textbooks have to be rewritten and not simply implemented. It can be predicted that some compromise will be reached, presumably reducing the now overwhelming role of the classical statistical models; in any case, further exploration of the links connecting the two approaches seems to be in order.

L.I. PETIT (*Goldsmiths' College, London*)

I share Dr. Dunsmore's concern with the choice of $W = X$. Cook and Weisberg (1982) have suggested that for each value of i the maximum of the predictive influence function over all possible choices of $\mathbf{W}$ be found. Can I ask what Professor Geisser thinks of this suggestion? In the context of designed experiments it is clear that we can reduce the possible influence of points with large values of v_i by replication, see Huber (1975) and Box and Draper (1975).

REPLY TO THE DISCUSSION

I respond to Professor Dunsmore's criticism of setting $W = X$ as a canonical measure in the predictive case. First, I shall reiterate that if one knows at what values of W predictions are to be made--there is no problem. Now Dunsmore suggests that we pretend that we are predicting one-at-a-time and not jointly, and examine $\hat{I}(Z)$ over a plausible range of values for a single W. When $p = 2$, this is easily accomplished as Dunsmore demonstrates for the simple linear regression example that I used. However for $p > 2$ plotting and interpreting $\hat{I}(Z)$ over a multidimensional region rather than an interval is a task not easy to execute. My view is that, as long as one is not sure of where or how many predictions are to be made, the canonical measure is a worthwhile first pass at the data to indicate cases which may bear further investigation. As noted by Dr. Pettit, Cook and Weisberg suggested that the Johnson-Geisser procedure be altered to compare maxima of the PIF's for all possible values of W (marginally). I have difficulty in finding this informative because; maxima may occur beyond that region for which the model is plausible; comparisons of PIF's lose their force when the observations are not at the same point or points because PIF's are interpretable as a relative measure rather than an absolute measure; there is no compelling reason to presume that predictions will be made marginally rather than jointly.

Professor Dunsmore correctly states that I reverse the x and y values from the original data. Because this was a calibration problem and I wanted to characterize the influence of observations with regard to this aspect I used inverse regression which can be shown to be a Bayesian solution to calibration when a student prior distribution is used, by completing the analysis initiated in Aitchison and Dunsmore (1975, p. 191).

Not only do I have no disagreement with Professor Piccinato's comments but I strongly support the Roman view which involves expunging Greek letters (in statistical analyses) whenever possible. My view as Professor Piccinato has discerned has also been to use a parametric formulation, when necessary, as a convenient modeling device in order to get on with the job of predicting observables. The introductory paragraph in section 4 of my paper here, as well as many of the references to my earlier papers therein, detail my conviction that observables are the proper foci of statistical analyses. A secondary use of parameters, as I have also mentioned here and previously, occurs when it is necessary to consider the limiting value of function of a large number of observables (in some situations it may be necessary and often it is convenient). Further, I have also expressed grave doubts about the ability of assessing prior distributions for entities whose real existence I often doubt unless expressed as a limit in the previous sense. I also strongly prefer assessing and modeling observables, as can be seen from the paper referenced by Professor Piccinato, Geisser (1980), in particular, see p. 377-378. Consequently, when applicable, the completely predictive approach is clearly the preferable alternative. But its implementation for many statistical paradigms appears more difficult than convenient ''compromise'' modeling methods.

I have held these views for about 20 years. At the Waterloo Symposium in 1970 I.J. Good (1971), in a brief discussion of my paper, drew attention to a lecture of de Finetti he had attended not long before, in which de Finetti expressed a completely predictive attitude regarding observables. Since de Finetti's views clearly antedated my own--it would be appropriate to style my approach, which in addition to the completely predictive program includes compromising when necessary and introducing predictive sample reuse (PSR) procedures, Geisser (1974. 1975). in non-stochastic frameworks, as the *Via Minne-Roma*. (This can also imply a watered down or diluted Roman approach since Minne, an American Indian word, refers to water as in Minnesota--the land of lakes.)

REFERENCES IN THE DISCUSSION

BOX, G.E.P. and DRAPER, N.R. (1975). Robust Designs. *Biometrika,* **62**, 347-352.

CIFARELLI. D.M. and REGAZZINI, E. (1982). Some considerations about mathematical statistics teaching methodology suggested by the concept of exchangeability. *Exchangeability in Probability and Statistics.* (Koch, G. and Spizzichino, F. eds.). Amsterdam: North-Holland. 185-206.

COOK, R.D. and WEISBERG, S. (1982). *Residuals and Influence in Regression.* New York: Chapman and Hall.

DABONI, L. and WEDLIN, A. (1982). *Statistica. Un'introduzione all'impostazione neo-bayesiana.* Torino: UTET.

DAWID, A.P. (1982). Intersubjective statistical models. *Exchangeability in Probability and Statistics.* (Koch, G., and Spizzichino, F. eds.), Amsterdam: North-Holland, 217-232.

GEISSER. S. (1980). A predictive primer. *Bayesian Analysis in Econometrics and Statistics.* (Zellner, A. ed.), Amsterdam, North-Holland, 363-382.

GOOD, I.J. (1971). Comments on The inferential use of predictive distributions, by S. Geisser. *Foundations of Statistical Inference.* (V. Godambe and D. Sprott, eds.), New York: Rinehart and Winston, 467.

HUBER, P.J. (1975). Robustness and Designs. *A Survey of Statistical Design and Linear Models.* (Srivastava, J.N. ed.). Amsterdam: North Holland.

LINDLEY, D.V. and PHILLIPS, L. (1976). Inference for a Bernoulli process (a Bayesian view). *Amer. Statist.* **30**, 112-9.

BAYESIAN STATISTICS 2, pp. 231-248
J.M. Bernardo, M.H. DeGroot, D.V. Lindley, A.F.M. Smith (Eds.)

Temporal Coherence

MICHAEL GOLDSTEIN

(*University of Hull*)

SUMMARY

We argue that the property of temporal coherence should be interpreted strictly as requiring coherence at each individual moment of time. The relationship between changes in your conditional previsions and changes in your beliefs for a single random quantity is derived under this interpretation of temporal coherence. The notion of a separation of beliefs is defined and is related to the general structure of the revision of belief. The above ideas are then combined to derive the relationship between changes in a collection of conditional previsions and changes in a collection of beliefs.

Keywords: COHERENCE; LINEAR PREDICTION; PREVISION; SEPARATION OF BELIEFS.

1. INTRODUCTION

You consult an expert as to the relevance of a particular piece of evidence E (as yet unobserved) in establishing the truth of a hypothesis H. He informs you that his assessment is that $P(H|E) = 0.8$. He also informs you that he intends to consider this question further, and that if you were to ask him the same question on the following day (still before observing E) then he would give you a difference answer say $\hat{P}(H|E)$. Clearly, he does not yet know the value of $\hat{P}(H|E)$, but as it is a well defined random quantity, he is prepared to express a prevision (expectation) for its value. In fact, he informs you, his prevision for $\hat{P}(H|E)$ is 0.7. What information does this statement contain, and how can you use it? How worthwhile would it be to revisit the expert tomorrow? If you choose not to revisit him, which conditional probability value will be relevant if you observe E? Suppose that you are eliciting this conditional judgement to assist you in constructing a decision tree, to select a course of action which must be chosen now. The justification for using the decision tree to guide your current actions is supposed to be that if you subsequently reach a particular point on the tree, for example, if you observe E, then your actual probabilities will change to the specified conditional probabilities. If you know now that this will not be the case, (e.g. you intend to revisit the expert and insert his revised values), then is there any alternative justification for using the decision tree? If not, how should you choose your current action? Have you asked the expert the right questions e.g. is his current prevision for his future conditional probability a useful quantity to elicit, or are there more important aspects of his revisions of belief that you should now be asking him to consider?

On a more fundamental level, is the expert being coherent in his judgement? (For example, he would be incoherent if he specified both that $P(H) = 0.8$ and also that this prevision for $\hat{P}(H)$ was 0.7 where $\hat{P}(H)$ is his revised value for $P(H)$ which he will assert tomorrow). If the expert is being coherent, then what information does he convey by his statement $P(H|E) = 0.8$? As his assessment will change before E is observed it clearly does not represent the value that he will state for $P(H)$ if he observes E. Does it represent his "best guess" in some average sense as to the value he will state?

Even this is not necessary. For example, suppose that he is sure that he will discover the occurrence or otherwise of H before he discovers the ocurrence of E. In this case $\hat{P}(H|E)$ will take value 1 if H does occur, or zero otherwise, so that presumably the best guess for the future value of $P(H|E)$ is not the current assessment of $P(H|E)$ but the current assessment of $P(H)$. (Note that this example also answers our question as to whether the expert can coherently express the beliefs we have assigned to him.)

The question as to what information is conveyed by the expression $P(H|E)$ is central to the Bayesian argument. Most developments of Bayesian theory proceed by defining conditional probabilities and deriving their properties in terms of called off penalties or bets, determined before the conditioning event is revealed. This definition is then taken to represent your actual beliefs about H, having seen E. This transition is not justified and is rarely even pointed out. However, it is clearly a false transition on both theoretical and practical grounds. As no coherence principles are used to justify the equivalence of conditional and a posteriori probabilities, this assumption is an arbitrary imposition on the subjective theory. As Bayesians rarely make a simple updating of actual prior probabilities to the corresponding conditional probabilities, this assumption misrepresents Bayesian practice. Thus Bayesian statements are often unclear (for example, what does it mean to say that your probability statements should be conditional on all of your information? are you envisaging a called off penalty or not?). The practical implication is that Bayesian theory does not appear to be very helpful in considering the kind of question that we have raised about the expert and his changing judgements.

Our intention is to make explicit the extent to which the subjective theory can answer such questions. In section 2 we define our approach to temporal coherence. In section 3 we consider the relationship between changes in your conditional previsions and changes in your beliefs for a single random quantity. Then, following some discussion of general principles, in the final section we consider the relationship between changes in a collection of conditional previsions and changes in your collection of beliefs. Finally, we provide an appendix in which we suggest that there would be fundamental difficulties in trying to axiomatize the relationship between conditional and posterior probabilities beyond the level which we shall specify in section 2.

2. TEMPORAL COHERENCE

For any random quantity X and any event H, the prevision of X, $P(X)$, and the conditional prevision of X given H, $P(X|H)$, are defined as follows. $P(X)$ is the value $\bar{x}$ you would choose if confronted with a penalty $K(X-\bar{x})^2 = L$. (K is a constant defining the units of loss.) $P(X|H)$ is the value $\bar{x}$ you would choose if the penalty L were replaced by the penalty HL, where we are using H as the indicator function for the event (i.e. $H = 1$ if H occurs, otherwise $H=0$), Coherence implies that $P(X + Y) = P(X) + P(Y)$ and $P(HX) = P(X|H)\,P(H)$. These assessments are all made at the same point in time. If you make assessments for a set of random quantities at different time points, then all that coherence implies is that your collection of previsions made at each individual time point should obey the coherence properties. We will say that your beliefs concerning these random quantities are *temporally coherent* at a particular moment if your current assessments are coherent and you also believe that at each future time point your new current assessments will be coherent.

The usual notion of coherence does not require that your beliefs will cohere (for example, you might be presented with a problem which is too complicated for you to logically unravel). Instead it is a statement of the necessary and sufficient conditions to avoid being a sure loser. The notion of temporal coherence is similar in that it is the necessary and sufficient condition for you to now be confident that at each future time point you will not be a sure loser. (In other words, you now accept that however you change your assessments for $P(X)$, $P(Y)$ and $P(X + Y)$ they will always obey the relation $P(X + Y) = P(X) + P(Y)$ and so forth.)

The subjective theory is a description of all those consequences which follow strictly from observing the requirement of coherence. Subject to the conditions of coherence you have complete freedom of choice in evaluating previsions. In the same way, subject to the conditions of temporal coherence, you have complete freedom of choice in changing your previsions. Thus, the subjective theory does not tell you what your future beliefs ought to be, but instead provides ways of describing your current attitudes to your future beliefs. Bayesian methodology namely updating prior to posterior beliefs by conditionalisation, has a useful and clearly defined role to play in this description.

That role is outlined in Goldstein (1984a) and will be briefly reviewed in section 4 below. Indeed the interplay between the specific information about your changes in belief provided by your conditional beliefs and your complete freedom in modifying your beliefs is central to the theory. The problem we raised in section 1 is how the theory should incorporate your further freedom, namely that you may also freely modify your conditional beliefs at any time. We might represent a Bayesian as an individual who at various time points offers up conditional probability models for Bayesian analysis, while in between these time points he uses various heuristics -from sensitivity analyses to peer criticism- in order to change these conditional probability models. If this Bayesian is actually you, i.e. a temporally coherent individual whose judgements conform to subjective theory, then that theory will describe the precise interpretation that you should now give for your future Bayesian lifestyle. Our analysis of the temporally coherent Bayesian is in two parts. Firstly, in section 3 we identify the constraints on your individual changes in conditional prevision. Then, having clarified in section 4 the role of conditional prevision within subjective theory, in section 5 we place your changes in conditional probability models into the context of your changes in belief. Finally, some readers might feel that with a little more ingenuity we could replace the probabilistic relationships that we derive between your beliefs at different times by logical relationships that tell you what you should believe at some future time. For their benefit we include an appendix where we point out some of the basic difficulties that such a program would involve.

3. CHANGES IN CONDITIONAL PREVISIONS

At this moment, you delare a prevision $P(X)$ for a random quantity X. At some well defined future time point T, you intend to declare a revised value for $P(X)$. Call this revised value $P_T(X)$. (T may be any precisely defined future time point, random or otherwise, e.g. "3 p.m. on Wednesday afternoon" or "when this experiment is completed'' or so forth.) At present $P_T(X)$ is a random quantity. If you are temporally coherent, then provided that X is bounded your previsions must satisfy.

$$P\,P_T(X) = P(X) \tag{1}$$

(A ''sure-loser'' argument for this relation is given in Goldstein (1983).)

Notice that we are not assuming some partition of the possible experiences that you might undergo between now and time T, and neither are we imposing any countable additivity requirement on $P(\cdot)$. However, we will throughout impose the requirement that X is bounded.

Equation (1) is the basic ''temporal'' property of prevision from which follows the geometric representation of your revision of belief as outlined in Goldstein (1981). As we are interested here in the interplay between your conditional previsions and your actual revisions of belief, we now develop an analogous relation to (1) for conditional prevision.

One of the most important features of conditional probability is that it ''behaves like'' probability, i.e. for a general event A with $P(A)>0$, the function $P(\cdot|A)$ is a nonnegative, finitely additive set function, giving value 1 for the certain event. In precisely the same way, conditional prevision behaves like prevision, and we will now complete the identification, by showing that conditional prevision does indeed obey the temporal property of prevision, i.e. relation (1). Thus, suppose that you now declare a value $P(X|A)$. At time T (still before the occurrence or otherwise of A is made known to you), you intend to declare a revised value $P_T(X|A)$. We must show that if you condition all your beliefs on A then you can still recover relation (1). We have the following result.

Theorem (1) If you are temporally coherent, then your previsions must satisfy

$$P[P_T(X|A)\,|A] = P(X|A) \tag{2}$$

(Thus, in particular, the geometric representation outlined in Goldstein (1981) also applies if you condition every belief on the same conditioning event.)

Proof $P[P_T(X|A)\,|A] = P[P_T(X|A)A]/P(A)$

(using (1)) $= P\,P_T[P_T(X|A)A)]/P(A)$

(as, at time T, $P_T(X|A)$ is a constant) $= P[P_T(X|A)\,P_T(A)]/P(A)$

(as you are temporally coherent) $= \left[P\,\dfrac{P_T(XA)}{P_T(A)}\;P_T(A)\right]/P(A)$

(using (1)) $= P(XA)/P(A)$

$= P(X|A)$ Q.E.D

NOTES

(i) The result (2) is a special case of a slightly more general result which may be proved similarly. Suppose that E is a refinement of B (*i.e.* $E = 1 \Rightarrow B = 1$), with the property that $P_T(E|B) = P(E|B)$ (with probability 1). For example, we could set $E = B$, giving the above theorem, or we might have $E = CB$, where C is an event which you currently judge to be independent of B and which you are confident that you will still judge independent of B at time T. Then your beliefs must satisfy

$$P[P_T(A|B)|E] = P(A|B)$$

(ii) In section (1), we observed that in general $P[P_T(X|A)] \neq P(X|A)$. From the above theorem, we see that the possible discrepancy occurs because of your latitude in specifying $P[P_T(X|A)\ |\bar{A}]$ where $\bar{A}$ is the complement of A. More specifically,

$$P[P_T(X|A)] - P(X|A) = P(\bar{A})\,[P(P_T(X|A)|\bar{A}) - P(X|A)]$$

This is intuitively reasonable. Whatever additional information or reflections occur to you by time T should have the effect of making $P_T(X|A)$ "more like X". This still makes $P(P_T(X|A)|A)$ "like" $P(X|A)$, but it tends to make $P(P_T(X|A)|\bar{A})$ more like $P(X|\bar{A})$ than like $P(X|A)$. In other words, large discrepancies between $P\,P_T(X|A)$ and $P(X|A)$ correspond to your belief that you will receive fresh information or insights into the value of X by time T.

The main consequence of theorem (1) is as follows. If you consider a finite partition $\Pi = (H_1,\ldots,H_k)$ (i.e. each H_i or 0 or 1 and $\Sigma H_i = 1$), and you assess all of the values $P(X|H_i)$, $i = 1,\ldots,k$, then informally you might consider that, in some sense that we must make precise, you are prepared to replace your current prevision $P(X)$ by the random prevision $\Sigma H_i P(X|H_i)$, which we write $P(X|\Pi)$. (Thus, if H_i occurs then $P(X|\Pi)$ will take the value of $P(X|H_i)$. We are following the notation of deFinetti (1974), section 4.7.) This is the way that conditional probabilities are treated within the usual Bayesian formulation which is why they are interchangeably referred to as conditional probabilities or posterior probabilities. We have argued against this interpretation in section (1). However, there is a rather more limited sense in which $P(X|\Pi)$ can be considered as a substitution for $P(X)$, and we will make this precise in the next section. For now, let us consider the corresponding temporal properties of $P(X|\Pi)$. At time T, you will replace each $P(X\ |H_i)$ by the corresponding value $P_T(X|H_i)$ so that $P(X|\Pi)$ will be replaced by $P_T(X|\Pi) = \Sigma H_i P_T(X|H_i)$.

What is the relationship between $P(X|\Pi)$ and $P_T(X|\Pi)$?

Returning for a moment to relation (1), one interpretation we could give is that the change in your prevision, i.e. $P_T(X) - P(X)$, cannot be "systematically predicted" from your current beliefs. We will now show that relation (2) implies a similar relationship between $P(X|\Pi)$ and $P_T(X|\Pi)$, namely that you cannot use $P(X|\Pi)$ to predict the change in your prevision i.e. $P_T(X|\Pi) - P(X|\Pi)$. We have the following theorem, which is the basic temporal property of conditional prevision which we shall use in constructing our general interpretation in the next subsection.

Theorem (2). The quantity $Q = P_T(X|\Pi) - P(X|\Pi)$ has the following properties

(i) $P(Q) = 0$

(ii) Q and $P(X|\Pi)$ are uncorrelated

(iii) $P(Q|\Pi)$ is identically zero.

(Notice that property (i) is a further generalisation of relation (1). Property (ii) says that we cannot form a linear prediction of Q from $P(X|\Pi)$. Property (iii) is a strengthening of property (ii) as it is equivalent to the restriction that the general regression function of Q on $P(X|\Pi)$, i.e. $f(x) = P(Q|P(X|\Pi) = x)$ is identically zero.)

Proof. We will prove property (iii), the other properties following similarly.

We must show that $P(Q|H_i) = 0$, $i = 1,\ldots,k$. We have

$$P(Q|H_i) = P(\Sigma(P_T(X|H_j) - P(X|H_j))H_j|\ H_i)$$

$$= \frac{1}{P(H_i)} P((P_T(X|H_i) - P(X|H_i))H_i)$$

$$= P(P_T(X|H_i) - P(X|H_i)|\ H_i)$$

$$= 0, \text{ by theorem 1} \qquad \text{Q.E.D.}$$

Thus, in this important, practical sense, you cannot predict your changes in conditional belief. Notice that having shown that $P\,(Q|\Pi)$ is identically zero, we may immediately deduce that Q cannot be predicted by any quantity $P\,(Y|\Pi)$ for any choice of random quantity Y. This will be relevant in the next subsection, when we move from considering your changes in beliefs about a single random quantity to the more interesting problem of considering your changes in beliefs about a collection of random quantities.

4. CHANGES IN COLLECTIONS OF PREVISIONS: GENERAL IDEAS

Subjectivist theory concerns the relationship between the beliefs that you express about each of a collection of random quantities. How the theory develops is governed by the organising principal which is initially applied to collect together your stated beliefs. In the usual Bayesian formulation, this organisation is into a joint probability distribution over every random quantity that is to play a role in the analysis. In a development based on prevision, the natural way to organise your stated beliefs is into an inner product space. This is described in deFinetti (1974) sections 2.8, 4.17, and is taken as the starting point for Goldstein (1981). Briefly, you begin with a collection of random quantities (X_i) of interest. (Some of the quantities may be functions of other of the quantities. For example, we would essentially return to the standard Bayesian formulation if we included in our list every function of every random quantity of interest.) You then form the affine vector space L of random quantities X. (Each X is represented by a vector and linear combinations by linear combinations.) Then your beliefs are expressed by considering L as an inner product space A with inner product and norm defined by

$$(X, Y) = P(X - P(X))(Y - P(Y)), \quad \|X\| = (X,X)^{1/2}$$

We restrict attention to quantities X with $P\,(X^2) < \infty$. To avoid the need to deal with equivalence classes, we adjust each quantity X in L by substracting the prior prevision $P(X)$.

One advantage of expressing your beliefs in terms of the inner product space is that you must explicitly decide which aspects of your uncertainty (i.e. which functions of which random quantities) you will assess, rather than being forced to pretend that you have assessed all aspects. Conversely, because you are not forced to specify all aspects of your uncertainties, you may more easily extend your analysis over a wide variety of such quantities. A perhaps more debateable advantage of this formulation is that unless you have spent half your life as a probabilist, you may find linear operator theory easier than measure theory. For example, the basic results in exchangeability when formulated in this way transform the problem of choosing appropriate limiting probability distributions into a simple problem of finding the closure of an appropriate inner product space (details forthcoming in Goldstein (1984b).

At the heart of the subjective theory is the representation of how your beliefs change over time. This is expressed by the change in your inner product over A. In general, we can represent this change in terms of a compact linear operator over A (details in Goldstein (1981)). However, we may gain certain general insights into your changes of belief without full specification of this linear operator, by the device of the separation of beliefs. Details are given in Goldstein (1984a), but the basic idea is as follows. You try to decompose the space A as a subspace of the orthogonal sum of two spaces A_1, A_2 with the property that not only are A_1, A_2 orthogonal but also the corresponding "belief" spaces $A_{1,T}$, $A_{2,T}$ are orthogonal. What we mean by a belief space is that we take each random quantity X in the original inner product space and replace it by the corresponding random quantity $P_T(X)$. This is again an inner product space with the same inner product (i.e. prior covariance).

The fundamental role of the belief space is briefly indicated in Goldstein (1981) and is made rather more explicit in Goldstein (1984a,b). For our purposes, the relevant result is that if $A \subseteq A_1 \oplus A_2$ and $A_T \subseteq A_{1,T} \oplus A_{2,T}$ (where $\oplus$ denotes orthogonal sum), then A_1, A_2 separate beliefs over A with respect to T, in the following sense. For any X, $Y \in A$, denoting X_1, X_2 as the orthogonal projections of X into A_1, A_2 respectively, we have

$$P(X,Y)_T = P(X_1, Y_1)_T + P(X_2, Y_2)_T \tag{3}$$

where $(.,.)_T$ denotes the value you will assign for the inner product at time T (and is, at the present, a random quantity).

In other words, in order to assess your beliefs about the information that you will acquire by time T (i.e. the change in the inner product over A), it is sufficient to separately assess the information you will acquire about A_1, and the information you will acquire about A_2. Your overall changes in belief will immediately follow by relation (3). In particular, if we intuitively measure information by the expected reduction in the squared norm, then from (3) this quantity may be decomposed as the sum

$$(X,X) - P(X,X)_T = (X_1,X_1) - P(X_1,X_1)_T + (X_2,X_2) - P(X_2,X_2)_T \tag{4}$$

In particular, if you find it straightforward to evaluate for each $X \in A$, the term $P(X_1,X_1)_T$, then you may unambiguously refer to the shrinkage in the inner product over A_1 as the information you expect to receive about A by consideration of A_1, while perhaps regarding the information you will gain from considering A_2 as a "bonus" which you will pick up at time T.

Much statistical practice, when tightly defined, does lead to such a situation. In particular, what is relevant for our purposes is that we can now make precise the discussion of the previous section as to the way in which we may consider $P(X|\Pi)$ as a replacement for $P(X)$. The result, again from Goldstein (1984a) is as follows. Take the original space A. Add each element H_i of the partition (remember each H_i is a random quantity namely the indicator function of the corresponding event) to form an enlarged inner-product space $\hat{A}$ (i.e. the space spanned by the original elements of A, plus the additional elements $H_1,\ldots,H_k$). Now let A_1 be the subspace $\hat{\Pi}$ spanned by the elements $H_1,\ldots,H_k$ and A_2 be $\hat{\Pi}^{\perp}$, the orthogonal complement of $\hat{\Pi}$ in $\hat{A}$. Let T be any time at which you will certainly know which H_i has occurred (e.g. T might be defined to be the time at which the outcome H_i is revealed to you). Now the result is that $A_1 = \hat{\Pi}$ and $A_2 = \hat{\Pi}^{\perp}$ separate beliefs over A with respect to T.

What this has to do with conditional prevision is that with A_1 so defined, the quantity X_1 i.e. the orthogonal projection of X into A_1 becomes $X_1 = P(X|\Pi)$. In other words, for each $X \in A$, we write

$$X = P(X|\Pi) + V,$$

where V is uncorrelated with $P(X|\Pi)$. Further

$$P_T(X) = P(X|\Pi) + U$$

where U is uncorrelated with $P(X|\Pi)$. Thus, although you cannot consider that your posterior prevision at time T will be equal to the conditional prevision corresponding to the observed outcome, you should at present consider this posterior prevision, as equal to the conditional prevision plus a random quantity which cannot be predicted by your specification over Π (i.e. $P(U|\Pi)$ is identically zero).

In terms of the comments after relation (4) above, you may unambiguously consider the gain in information (i.e. change in inner product over A) that you expect to obtain purely by observing which outcome H_i occurs and updating your beliefs strictly to the corresponding conditional beliefs. This leaves open to further specification the gain in information over A_2. Comparing the usual Bayesian formulation, what we would like to see changed is the logic used to explain why you should consider conditional probabilities and the lack of interest, at least within the theory, in picking up the "bonus" over A_2. Notice, in particular, that it is the gathering of all the quantities of interest into a well-defined structure (i.e. the inner product space A), that allows us to be precise about the status of our inferential statements.

Finally, we note the following general points. Firstly, there is no reason why this separation need be into two subspaces, and the same definitions and results apply for a separation into k subspaces $A_1,\ldots,A_k$, the requirement in this case being that the spaces $\{A_i\}$ are mutually orthogonal and that the derived spaces $\{A_{i,T}\}$ are also mutually orthogonal. (For example, the geometric development of Goldstein (1981) demonstrates how this reduction can be carried forward to the stage where each A_i is spanned by a single element, namely an eigenvector of a certain linear operator.) Secondly, there is nothing intrinsically "special" about the indicator functions of events, as prevision does not distinguish probabilities from the assessment of any other random quantities. As we are here focusing on conditional previsions, we have described results in terms of partitions into disjoint events. However, exactly the same results hold if instead of adding elements H_i you form the space $\hat{A}$ by adding any set of random quantities $U = (U_1,\ldots,U_r)$ to the space A. The result is the same as above namely that if you will certainly know the value of each member of U by time T, then U and $U^{\perp}$ separate beliefs over A with respect to T.

5. CHANGES IN COLLECTION OF CONDITIONAL PREVISIONS

You assess conditional previsions to resolve now some of the uncertainty that you would otherwise have to resolve in the future. This provides a partial guide to your choice of actions now. If you want a better guide, then you need to consider further your present attitude to your future uncertainties. For each $X \varepsilon A$, the portion of your uncertainty resolved by conditioning is $(P(X|\Pi), P(X|\Pi))$, which we denote by $C(X|\Pi)$.

Expanding the inner product gives

$$C(X|\Pi) = \sum_i P(H_i)\, P^2(X|H_i)$$

Before you had specified any conditional previsions, the only restriction for coherent specification of your prevision for each $(X,X)_T$ was that $P(X,X)_T \leq (X,X)$. When you specify your conditional previsions, the only further restriction is that the upper bound for each $P(X,X)_T$ is reduced to $(X,X) - C(X|\Pi)$.

If you assess $P(X,X)_T$ to be near this upper bound for each X, then this specification provides a reliable guide to current action. This may happen if you think that the evidence will be so overwhelming that the upper bound will be near zero. In such cases, you have an "automatic inference" which can be programmed into a "Bayesian machine" so that you need no longer concern yourself with the outcome. However, in most situations, you do not relinquish this control, and the argument of this paper is that your considerations as to the adequacy of the usual Bayesian analysis do not lie in some meta-theory but are expressible directly within the subjective theory itself. What you must assess is the quantities $P(X,X)_T = (X,X) - (P_T(X), P_T(X))$.

These are previsions for well defined random quantities, so that provided that you are willing to make previsions for all random quantities, then in principle you need only compare the quantities $(P_T(X), P_T(X))$ and $C(X|\Pi)$, to determine the extent to which you should allow the inference to be automated. We may try to construct a statistical methodology based on such comparisons, but we will better be able to discuss this when we have completed the theoretical argument concerning changes in conditional prevision.

We consider your changes in conditional prevision for the same reason that we consider conditional prevision itself, namely to help to resolve now some of your future uncertainty. One of the reasons why your posterior beliefs may not equal your current conditional beliefs is that your conditional beliefs may themselves change at some intermediate time. Indeed if your conditional previsions are considered sufficiently important to carry the main burden of the analysis, then such changes should be an important factor in this discrepancy.

We now complete our description of the relationship between your changes in conditional probability and your changes in belief. Thus, at the present time you have formed a collection C of random quantities of interest, and you are also considering a partition $\Pi = (H_1,\ldots,H_k)$, for which the outcome H_i will be revealed to you by time T. You assess, for each $X \varepsilon C$, the quantity $P(X|\Pi)$, viewing it as a substitute for $P(X)$ after H_i in the sense of section 4. However, you also intend that at time t, which will certainly occur before H_i is revealed to you, you will reconsider this problem, and in particular, you will assess for each H_i and X, a new value $P_t(X|H_i)$. Thus, as in section 3 you are now aware that at time point t you will replace each $P(X|\Pi)$ by the value $P_t(X|\Pi)$. Define the difference= as $Q(X|\Pi) = P_t(X|\Pi) - (X|\Pi)$. How should you analyse this, and how should you value the information due to this extra effort?

The answer is that your new considerations have induced a further separation of beliefs. This is as follows. With notation as in section 4, suppose you have formed the collection C into the inner product space A. You have formed the further subspace $\hat{\Pi}$ spanned by $H_1,\ldots,H_k$ and added this space to A giving $\hat{A}$ as before. In addition you have introduced, for each $X \varepsilon A$ a further set of random quantities, of form $P_t(X|H_i)$. As we have suggested in section (3), it is not particularly the quantities $P_t(X|H_i)$ which are of interest, but rather the quantities $H_i(X)$ defined by $H_i(X) = H_i P_t(X|H_i)$.

Thus, suppose that you have further enlarged $\hat{A}$ to A^*, by adding to A the space Π^* which is spanned by all of the elements H_i and all of the elements $H_i(X)$, over all i and all $X \varepsilon A$. Let $\Pi^{*\perp}$ be the orthogonal complement of Π^* in A^*. Finally, let $\hat{Q}$ be the subspace spanned by all the changes in your assessed values i.e. all quantities of the form $H_i(P_t(X|H_i) - P(X|H_i))$ over all i and $X \varepsilon A$. We have the following result.

Theorem 3 The spaces $\hat{\Pi}$, $\hat{Q}$ and $\Pi^{*\perp}$ separate beliefs over A with respect to T. In particular, for any $X, Y \varepsilon A$

(i) $(X,Y) = (P(X|\Pi), P(Y|\Pi)) + (Q(X|\Pi), Q(Y|\Pi)) + (R(X|\Pi), R(Y|\Pi))$

and

(ii) $P(X,Y)_T = P(R(X|\Pi), R(Y|\Pi))_T$,

where for any Z, $R(Z|\Pi) = Z - P(Z|\Pi) - Q(Z|\Pi)$, satisfies the relations

$$P(R(Z|\Pi)\,|H_i) = P(R(Z|\Pi)P_t(U|H_i)\,|H_i) = 0$$

for each i and each $U \varepsilon A$.

Proof. At time T you will certainly know the value of each element of Π^*. Thus, by the remarks at the end of section (4) the spaces Π^* and $\Pi^{*\perp}$ separate beliefs over A with respect to T. Further, it follows from theorem (2) that $\hat{Q}$ and $\hat{\Pi}$ are orthogonal spaces, so that $\Pi^* = \hat{Q} \oplus \hat{\Pi}$, As $\hat{Q}_T = \hat{Q}$, $\hat{\Pi}_T = \hat{\Pi}$ the spaces $\hat{Q}$, $\hat{\Pi}$ separate Π^* with respect to T. Thus, $\hat{\Pi}$, $\hat{Q}$, $\Pi^{*\perp}$ separate A with respect to T.

Let Z_1, Z_2, Z_3 be the projections of Z into $\hat{\Pi}$, $\hat{Q}$, $\Pi^{*\perp}$. As noted in section (4), $Z_1 = P(Z|\Pi)$. Now observe that for any element $H_j(P_t(Y|H_j)-P(Y|H_j))$ the inner product

$$
\begin{aligned}
&(Z-Q(Z|\Pi), H_j(P_t(Y|H_j)-P(Y|H_j)))\\
&\quad = P(ZH_jP_t(Y|H_j)) - P(ZH_j)\,P(Y|H_j)\\
&\quad - P(H_j(P_t(Z|H_j)-P(Z|H_j))\,(P_t(Y|H_j)-P(Y|H_j))\\
&\quad = P(H_j\,P_t(Y|H_j)\quad (Z-P_t(Z|H_j)) && \text{(using theorem (1))}\\
&\quad = P(P_t(P_t(Y|H_j)\,(ZH_j - H_jP_t(Z|H_j)) && \text{(using relation (1))}\\
&\quad = P\ P_t(Y|H_j)\,(P_t(ZH_j)-P_t(ZH_j))\\
&\quad = 0
\end{aligned}
$$

Thus, $Z-Q\ (Z|\Pi)$ is orthogonal to $\hat{Q}$ and so $Z_2 = Q(Z|\Pi)$. Relation (i) follows immediately, and property (ii) is an immediate consequence of the separation of beliefs, as the inner product over Π^* is zero at time T. Finally, the conditions on $R(Z|\Pi)$ in the theorem are restatements of the orthogonality of $\Pi^{*\perp}$ to $\hat{\Pi}$ and to $\hat{Q}$. Q.E.D.

Thus, although your changes for the values of individual conditional previsions appear to have no simple interpretation, as soon as we view these conditional previsions simply as coefficients in the random quantity $\Sigma H_iP(\bullet|H_i)$, then the effect of such changes is immediately apparent. Notice, in particular, the importance of taking prevision rather than probability as the starting point for the theory. Taking probability as a starting point, you would begin by specifying a joint conditional probability measure for all the quantities of interest. You would then consider the change, at time t, to a random joint conditional probability measure. You would then need to specify a prior probability measure over the set of joint conditional probability measures. Finally, you would need to specify your joint probability measure for the conditional probability measure chosen at time t, and your unconditional probability measure specified at time T. This seems ridiculous, particularly as we have shown in theorem (3) the basic simplicity of the relationships that we are seeking. This is a good illustration of the comments made in section (4) in particular concerning the way that you may extend your analysis over a wide variety of quantities precisely because you are not constrained to specify all aspects of your uncertainty about each quantity. We might argue that even if you prefer to work in terms of probability, you still need the theory of prevision to explain how the theory of probability works!.

How should you value your intention to revise your conditional beliefs? At the beginning of this section, we observed that the effect of your original conditional specification was to reduce the upper bound for each $P\ (X,X)_T$ from (X,X) to (X,X) --$C(X|\Pi)$. by introducing the quantities $H_1,\ldots,H_k$ into A. Your further considerations follow by introducing the quantities $H_i(X)$ into A. From theorem (3), the only effect of this is to reduce the upper bound for coherent specification of $P\ (X,X)_T$ by a further amount namely $(Q(X|\Pi), Q(X|\Pi))$, which we denote $C_t(X|\Pi)$. Expanding this expression and using theorem (1), we have

$$C_t(X|\Pi) = \sum_i \mathrm{var}(P_t(X|H_i)|H_i)\,P(H_i) \tag{5}$$

$C_t(X|\Pi)$ is your valuation for your decision to replace $P\ (X|\Pi)$ by $P_t\ (X|\Pi)$. Informally, you judge that it is important to revise your conditional previsions if you now have high uncertainty for the revision that you will make to the conditional prevision that you will eventually use. Assessing quantities such as (5) is an unfamiliar task. However, these uncertainties exist whether you quantify them or not. What we have demonstrated is that within subjective theory the quantification must be of form (5). The larger the values of the quantities $C_t(X|\Pi)$, the more reason you will have to postpone automating your inferences at least to time t. The difference between the quantities $P\,(X,X)_T$ and the new upper bound will reflect whether you should now plan to automate your inference at time t.

Of course, the question of whether and when you should prefer convenience to accuracy is for you to decide. All we have tried to do is to identify those quantities that you must assess in order to make this choice. However, notice that because our analysis is based on the assessment of well-defined random quantities it can be made the subject of a precise statistical treatment. For example, consider a "pure" Bayesian process, such as an expert system, which holds a collection of conditional judgements of some hypothetical expert, these judgements being made now before the individual case (i.e. patient to be diagnosed, house to be insured or whatever) has been presented. Suppose that the raw material of the programme involves not only judgements on the space A but also judgements on all the quantities $P_T\,(X)$, $H_t(X)$ etc., so that the formal structure is that of the augmented inner product space.

We would conduct a series of trial runs of the programme in conjunction with actual expert judgements on the same trials. (Presumably, we would need to do this anyway, to check the programme). In each case, the expert would declare his values for the quantities $H_t(X)$ and $P_T\,(X)$ and these along with the observed X values would then be fed back into the programme as data. With the assumption of exchangeability between trials, the inner product over the augmented space would automatically be updated eventually to converge to some limiting concensus value. The technical machinery that we require for this analysis is a formulation of exchangeability which is expressed in terms of prevision rather than probability. As in our comments following theorem (3), you don't want to specify a gigantic joint prior measure over all the possible ways that your beliefs may change.

We need a formal definition of exchangeability to cover exchangeability between inner product spaces (i.e. one space for each trial), deducible directly from the prevision statements that you actually make, along with corresponding representation theorem and a formal updating procedure. (Full details will be given in Goldstein (1984b)). The aim will be to achieve a concensus inner product over the augmented space, and then use our calculations above to identify decision points at which it would be more efficient to refer the decision out of the programme and back to the expert, and also to incorporate at the final stage (patient diagnosis, insurance valuation) a judgement as to the scope for the expert to improve on the automated analysis. This illustrates our general point that for an automated inferential system to be useful it should have a well defined relationship to an actual inference procedure.

We conclude with some general points. Firstly, we have been considering your changes in conditional probabilities because conditional probabilities play a basic role in Bayesian methods, and they provide a simple link between projections into subspaces as required for the general theory and more familiar intuitive conditioning arguments. However, as we remarked at the end of section (4), prevision makes no distinction between events (as represented by indicator functions) and any other random quantities. This is also the case here and the above results follow similarly for your assessment now of the change at time t of your projection into a general space U spanned by $(U_1,\ldots,U_r)$ where you will know the value of each U_i by time T, namely that the changes in your projections at time t induce a further separation of beliefs.

Secondly, though we have considered a single time point $t < T$, at which you will revise beliefs, you could more generally consider a series $t_1 < t_2 < \ldots < t_r < T$, of points at which you intend to revise your beliefs. The results are as above, i.e. you will write

$$X = P(X|\Pi) + (P_{t_1}(X|\Pi) - P(X|\Pi)) + \ldots + (P_{t_r}(X|\Pi) - P_{t_{r-1}}(X|\Pi)) + Z,$$

and this induces a separation of beliefs into $(r + 2)$ disjoint orthogonal systems, for which the inner product $(P_{t_i}(X|\Pi) - P_{t_{i-1}}(X|\Pi), P_{t_i}(X|\Pi) - P_{t_{i-1}}(X|\Pi))$ gives the value of the ith revision of your conditional previsions. The analysis proceeds iteratively, for example with each t_i corresponding to a decision node on a decision tree, perhaps deleting succesively from the list values t_i for which your path along the decision tree is unaffected. This continues until your immediate decision is clarified and a provisional series of nodes is identified at which the tree will be recalculated. This procedure is repeated when you reach the first node so identified. (Presumably, you make analogous calculations all the time. Simultaneously, your hand reaches for a cup of tea, you are talking to someone, you keep an eye on the clock - somewhere just below the level of your consciousness, your brain is constantly rescheduling the times at which your eyes must shift from one task to another).

Finally, to keep technicalities to a minimum we have been considering the effect of proposed revisions of belief on each X individually. However, the subjective theory concerns the changes in your whole collection of beliefs. In the geometric development of Goldstein (1981), this was summarized by the eigenstructure of the self-adjoint operator $T{:}A \Rightarrow A$, defined for each $X \varepsilon A$ by $TX = P_A P_T(X)$, where P_A is the projection operator into A. We have written P_T as $P_T(X) = P(X|\Pi) + Q(X|\Pi) + R(X|\Pi)$, where P, Q, R are projections onto mutually orthogonal subspaces. Thus the questions we have considered may be analysed over the whole space by comparing the eigenstructures of three operators $P_A P(\bullet|\Pi)$, $P_A Q(\bullet|\Pi)$ and $P_A R(\bullet|\Pi)$.

APPENDIX

Axiomatics and Temporal Coherence

Some attention has been paid to the circumstances under which conditionalisation carries the extra meaning of representing your posterior beliefs (Teller (1973) is one of the first attempts at partial justification. French (1982) is interesting for our purpose in explicitly considering the individual at different time points. In contrast, Good (1977) presents arguments as to why we should explicitly consider changes in previsions arising from pure thought.) The approaches that I have seen attempt to place further restrictions on the nature of the conditioning events and the ways in which you receive the information about them in order to achieve a partial identification of conditional and posterior beliefs. These restrictions do not seem to be intuitively obvious or logically compelling, and are almost tautological (i.e. you should update your beliefs using Bayes theorem if you think that you should and not otherwise). These treatments may make useful practical points, but there are fundamental difficulties in trying to identify logical relationships between conditional and posterior beliefs. The purpose of this paper is to discuss those relationships which do exist rather than those that don't exist. However, it is interesting to consider whether the subjective language could possibily imply this identification under any circumstances. This is really a digression from our main purpose, but it is valuable to explore the limitations of theory.

The subjective language is useful because it "coheres", this coherence being achieved through the logical force of the sure-loser argument. However, in circumstances where you would choose to be a sure loser, and in particular where you could simultaneously be a sure loser and a sure winner, coherence no longer follows.

To see how this may occur consider the following example. (We shall argue below that the example is not really so extreme as it appears but that there is a basic methodological problem involved). Thus, suppose that you explain to a skeptical friend that provided that a random quantity is well defined then you can always assign a prevision for its value. Suppose that your friend then asks you to specify your prevision for X where X is the number that he will write on a piece of paper in five minutes time. He tells you that he will determine the value of X as follows.

If you fail to declare a prevision, he will assign $X = 0$. If you assign $P(X) = 5$, he will assign $X = 2$. If you assign any other value for $P(X)$, then he will define X to be $P(X) + 10$. From the definition of prevision, you should presumably assign $P(X) = 5$, but clearly your further assessments will not cohere (e.g. you might also be asked to specify $P(2X)$ and you would specify $P(2X) = 4$). If you feel inclined to argue that your true prevision for X is 2 and that you only "pretend" to announce a prevision of 5, then tighten the example up by eliminating the role played by your friend and write down the value of X yourself by the above rules.

The point is this. The sure loser argument supposes a strictly "objective" status for the random quantity X, i.e. that, in some sense, the value of X is already written down on the piece of paper. Under these circumstances the sure loser argument is convincing, because it is based on comparing the penalties incurred by two different assignments of prevision for each possible value of the random quantities. However, as soon as the choice of prevision affects the outcome, as in the above example, this comparison is no longer meaningful.

This mutability of the X's is a feature of many substantitive problems. A teacher assesses previsions for pupil performance and the pupils' performance subsequently reflects the teacher's prevision, a businessman makes his prevision for the profitability of an option and tries to make true his prevision and so forth.

Pragmatically a reasonable response is to recognise that the effect of your specification on the outcome has two components. Firstly, there are the reasoning processes which underlay the specification and secondly, there is the specification itself. Consider the example of the teacher, above. If the teacher has taught the pupils already and is assessing previsions for their future performance under his further teaching, then presumably it is his preconceptions which affect their future performance and the previsions he makes are simply reflections of these preconceptions.

However, if the teacher is assessing previsions of their future performance for the purpose of giving these previsions to a different teacher who will be subsequently teaching them, then it is arguably the previsions themselves which are causing the effect on the later performance. Thus, it would seem reasonable to apply the theory in the former case but perhaps not in the latter. Our tentative conclusion is that we should be wary of applying the theory under circumstances in which the previsions themselves, as opposed to the reasoning process underlying the previsions, have an effect on the outcomes.

How does this relate to temporal coherence? Suppose that we have axiomatised temporal coherence in some non-tautological way. The coherence argument is that if you assign $P(W) = 1$ and $P(V) = 2$ and $P(W + V) = 4$ then, despite any reasons you might have for making these previsions the way that you did, you are incoherent and you should change some of your specifications. In the temporal context, you make some specifications at one time point while leaving others to be made at a further time point. If the temporal axiom is non-tautological (i.e. you are looking to the theory for guidance rather than simply to reflect whatever you choose to do anyway) then your specifications at the further time point are affected not just by your reasoning processes but also by the prevision statements that you have made. (In comparison, when you assigned previsions for W, V and $W + V$ you could consider that it was your reasoning processes that imposed coherence. However, if you imagine that the previsions for W and V are fixed, i.e. frozen at a previous time point, then it is the previsions themselves which affect the specification for $W + V$). In other words, a temporal coherence axiom would incorporate precisly the type of assessment problem to which the subjective theory does not apply. Thus, you would not be able to treat your future previsions as random quantities, for which you could now take expectations. (For example, the justification for decision trees would be lost). Looking at this more closely, I think that the specific point of conflict here arises from the different views we may hold as to whether prevision is operationally defined.

Each assessment of a prevision is operational in that it involves consideration of a well-defined penalty. However, the penalty itself is notional in that you certainly do not consider yourself bound to pay it (Certainly if you specify a continuous probability distribution, you will not feel bound to pay the uncountably infinite number of penalties implied by these assessments). More precisely, I think that you are supposed to consider your attitude to a penalty which arises quite separately from your intention to assess a prevision. In this sense, prevision is operationally defined not in quite the same way that length is operationally defined, but rather in the sense of emotional attributes to which it is somewhat closer. For example, we might make an operational definition of "forgiveness", by saying that you have this property if, when someone does you harm, you do not attempt revenge. Implicit in this definition is that the person doing you harm is not part of an experiment to determine whether you have the forgiveness property, in which case your reaction might be very different. If we accept a similar view for prevision, then this suggests that the reason that temporal coherence axioms put such a strain on the subjective theory is that they attempt to transform purely notional penalties considered at one time point in order to specify certain previsions into actual penalties which are supposed to be relevent retrospectively in considering previsions at a subsequent time point. This underlines our previous argument, namely that temporal coherence cannot be axiomatised into the subjective framework.

REFERENCES

DE FINETTI, B. (1974). *Theory of Probability Vol. 1.* London: Wiley.

FRENCH, S. (1982). On the Axiomatisation of Subjective Probabilities. *Theory and Decision,* **14**, 19-33.

GOLDSTEIN, M. (1981). Revising Previsions. A Geometric Interpretation. *J. Roy Statist. Soc. B,* **43**, 105-130 (with discussion).

— (1983). The Prevision of a Prevision. *J. Amer. Statist. Assoc.,* **78**, 817-819.

— (1984(a)). Separating Beliefs. (To appear in the volume *Bayesian Econometrics and Statistics in honour of Professor de Finetti.*)

— (1984(b)). Exchangeable Situations. (in submission).

GOOD, I.J. (1977). Dynamic Probability, Computer Chess and the Measurement of Knowledge. *Machine Intelligence,* **8**. (E.W. Elcock and D. Michie Eds.) London and New York: Ellis Harwood and Wiley.

TELLER, P. (1973). Conditionalization and Observation. *Synthese,* **26**, 218-258.

DISCUSSION

D.V. LINDLEY (*Somerset, U.K.*):

Goldstein raises a problem that I was not aware existed; but he is right, there is substantial difficulty in how probabilities, or previsions, change with time. He is to be congratulated on bringing it to our attention and discussing it so admirably. His solution is a "loose" one: my preference is for a "tight" one. The notation $P(X)$ for a prevision is inadequate because it fails to recognize the circumstance under which the prevision is made. Suppose you contemplate X and assign $P(X)$. You then learn the truth of an event A (for example, get some data). The prevision changes, yet Goldstein's notation does not reflect the incorporation of A, only the time change. I prefer Jeffreys' notation $P(X|H)$ where H indicates what is known when the prevision is assessed. Even this notation is inadequate and a better one is $P(X|\,Y;H)$ (Lindley, D.V., 1982). This is the prevision of X, conditional on Y, knowing H; where Y refers to something whose truth is uncertain and requires the associated bet to be called-off if Y is false. When made only H is known. (Notice $P(X|H)$ does not require a called-off bet.) Then Goldstein's problem can be solved if the axiom is made that

$$P(X|\,YZ;H) = P(X|\,Y;ZH)$$

so that experiencing Z does not change a prevision conditional on Z. This axiom enables us to talk about temporal coherence. It has the property that it does not refer to time explicitly. This may be an advantage since it is not time that matters but the change in H over time.

The example in the Appendix is important (and, I agree, not extreme). It can be studied by treating it as a decision problem in which the decision is the prevision to be announced. Some decisions lead to the announced value truly being a prevision (for example, proper scoring rules) but others, like this one, do not. A doctoral thesis by Ross Schachter at Berkeley discussed the important question of how a prevision can be elicited. A practical example concerns a geologist advising an oil company about the presence of oil at a site. Should he state his true prevision when he knows the value is going to affect the company's decision to drill and hence the testing of his statement?

SEIDENFELD, T. (*Washington University, St. Louis*)

I agree with Professor Goldstein that it is important to examine with care the temporal aspects of de Finetti's criterion of coherence for previsions. Even with the aid of Shimony's (1955) device of called-off bets —used to represent a conditional bet from the current point of view— there is no dynamic, inter-temporal import to coherence. Shimony's argument establishes an atemporal version of conditionalization, *not* a temporal version.

To the best of my knowledge, coherence alone does not dictate how previsions ought to change over time. Thus, I am intrigued by Professor Goldstein's parenthetical statement

(§3) that a "sure loser" argument establishes the central result:

$$P(X) = P(P_T(X)); \tag{1}$$

where $P(\cdot)$ is current prevision and $P_T(\cdot)$ is the prevision at some future time T. How does (1) follow from coherence without combining the proceeds of bets made at the current prevision of X, $P(X)$, with the proceeds of bets made at the future prevision of X, $P_T(X)$? Of course, to argue for (1) in this manner would be to add to coherence exactly what is to be proven. Is there some othe route to (1)?

The claim that coherence, by itself, places few restriction on "updating" over time (a special case of Goldstein's question) is the subject of a lively debate in philosophic journals. Recent exchanges between Kyburg (1980) and Levi (1978 and 1981) help to clarify some of the many subtleties lurking here. From a different perspective, Hammond (1976) addresses a related controversy: problems of sequential choice with changing tastes.

A. O'HAGAN (*University of Warwick, U.K.)*

Dennis Lindley complains that Michael Goldstein's probabilities are not conditional on "*H*", the usual symbol for the background knowledge or "history". This highlights a difference in understanding of probability. Goldstein's probabilites are defined only in terms of bets, and to him no other definition is sufficiently tight. To Lindley, and to me, probabilities represent degrees of belief. Admittedly such an interpretation is based on a less precise definition, but it is the only sense in which I can usefully emply probabilities in practice. Lindley argues that when Goldstein's expert changes his mind and reassesses his probabilities it is because of "*H*". He may remember some fact which hitherto he had forgotten, or see some unrecognised connection, or just reason in a different way. Whatever his justification for changing his assessment, if no new observations have been made, then that reason must have been implicit in "*H*". (The case of intervening observations would not need any temporal coherence argument.) The expert's various different assessments of the probability are due to the impossibility of ever comprehending simultaneously all the implicatios of "*H*".

It seems, then, that Lindley's expert's probabilities are not really conditioned on "*H*" either, but on various subsets of it. The subsets are loosely defined in terms of which aspects and consequences of "*H*" the expert has taken into account when assessing the required probability. Such probabilities can therefore be regarded as approximations to the ideal, but unattainable, probability conditioned on the whole of "*H*". As an interpretation of what really occurs when people assess and re-assess probabilities, I prefer a formulation like this, rather than Goldstein's abstract approach. However, we should not overlook another important feature of re-assessment. When a re-assessed probability value supplants the previous assessment it should be accompanied by a feeling that the new value is somehow a better approximation to the ideal value. According to my meagre understanding of Goldstein's paper, his Theorem 3(i) appears to address and quantify this feeling (albeit rather obliquely).

S. FRENCH (*Manchester University*):

Isn't the introduction of an "expert" into your argument a red-herring? Surely you mean De Finetti's "You"?

REPLY TO THE DISCUSSION

M. GOLDSTEIN (*University of Hull*)

Professor Lindley argues that we may sidestep the problems of temporal coherence by appropriate choice of notation and axioms. How can this be possible? Simply consider any situation in which you are about to receive new information, but you are not sure what form it will take or how you will respond to it (in short, any real life situation). When you come to revise your beliefs, then what you know may have changed in ways previously undreamt of. If you recognise the possibility of such changes then what is the relationship between your stated beliefs at the two different time points? This is not a question of axiomatics, but of analysis. My argument is that, however you choose to observe and assess your experiences, coherence at each individual time point imposes a natural probabilistic structure on your current beliefs about your future beliefs. This is the irreducible minimum structure which must apply to any revision of belief. Of course, you may sometimes impose more structure. In simple cases you may construct a plausibly exhaustive list of relevant outcomes and you may even be prepared to treat your "called off bet" probabilities as your actual posterior probabilities. In more complicated cases you may pretend that you can still do this, to set up a simple model as the starting point for analysing your beliefs. This is of considerable practical importance, but it is still only a special case of no more or less intrinsic importance than any other revision of beliefs.

Professor Lindley suggests that this special case can be made more general by imposing certain axioms. But previsions are not theoretical quantities. They are actual statements of your beliefs. Either axioms are logically compelling in which case they are unnecessary as your beliefs will, in any case, conform to them, or they are not compelling in which case they are irrelevant to your actual specifications. (Compare the example I discuss in the appendix. When the sure loser principle has logical force, it does not need to be an axiom. As soon as it fails to be compelling, your actual previsions will no longer respect it.) Thus, I am very sceptical of any attempt at temporal axiomatics. However, I must confess a particular aversion to any axiom based on metaphysical quantities such as an H denoting "what is known". From reading Lindley (1982), I gather that H is a list of all the quantities that you know, either through empirical or logical discovery. How could I construct such a list? Does it refer to things that I know, or things that I am sure that I know, or things that I happen to remember? Whatever it is, it is not an objective, operationally defined quantity, and so it has no place as a conditioning event, which is always well defined and always requires a called off bet. (Why ruin a precise piece of notation? You can always keep H by writing $P_H(X)$ or somesuch.) Perhaps it would be better to say that you are free to define your notation in any way that you choose. However if you blur the distinction between called off bets and posterior probabilities in your basic definitions, then your theory will never tell you anything about the rich relationship between the two concepts. This seems a high price to pay for an axiom that you will hardly ever follow, if only because your "H" will always be changing in unforeseen ways, so that you will in any case need the more general analysis I have given.

As to Dr. O'Hagan's comments, previsions do reflect your beliefs. However, your beliefs can be quantified in different ways, and I prefer to make clear the basis of my quantification. There is no particular distinction in my analysis between reassessing old experiences and having new experiences (not quite the same as "intervening observations"!). I am puzzled by an approach which regards all your specifications as "approximations to the ideal, but unobtainable, probability conditioned on the whole of *H*''. What could be more abstract than that? I prefer to simply analyse the statements you make on their own terms. I am grateful to Professor Seidenfeld for the interesting references. I hope that we will soon see some relevant material in the statistical literature! As to the question concerning the relation $P(X) = P(P_T(X))$, all I can suggest is that you should consult the referenced paper, Goldstein (1983) (which will have appeared by the time that these proceedings are published). In that paper, I give a sure loser argument, based on your bets at the present time, without making any combination of the bets at the two time points. Finally, in answer to Dr. French, yes the expert is a red-herring. However, my feeling was that if I began the paper by asserting that "you" intended to change your beliefs in the manner I attributed to the expert, then many readers of the paper would have immediately thought that "they" would never intend such changes. Hopefully, showing that the expert could intend such changes might encourage "you" to consider such possibilities as well.

REFERENCES IN THE DISCUSSION

GOLDSTEIN, M. (1983). The prevision of a prevision. *J. Amer. Statist. Assoc.* **78**, 817-819.

HAMMOND, P.J. (1976). Changing tastes and coherent dynamic choice. *Review of Economic Studies,* **43**, 159-176.

KYBURG, H.E. (1980). Conditionalization. *Journal of Philosophy*, **77**, 98-114.

LEVI, I. (1978). Confirmational conditionalization. *Journal of Philosophy,* **75**, 730-737.

— (1981). Direct inference and confirmational conditionalization. *Philosophy of Science,* **48**, 532-552.

LINDLEY, D.V. (1982). The Bayesian approach to statistics. *Some Recent Advances in Statistics.* (De Oliveira, J.T. and Epstein, B., eds.). London: Academic Press, 65-87.

SHIMONY, A. (1955). Coherence and the axioms of confirmation. *Journal of Symbolic Logic,* **20**, 8-20.

BAYESIAN STATISTICS 2, pp. 249-270
J.M. Bernardo, M.H. DeGroot, D.V. Lindley, A.F.M. Smith (Eds.)

Weight of Evidence: A Brief Survey

I. J. GOOD
Virginia Polytechnic Institute and State University

SUMMARY

A review is given of the concepts of Bayes factors and weights of evidence, including such aspects as terminology, uniqueness of the explicatum, history, how to make judgments, and the relationship to tail-area probabilites.

Keywords: BAYES FACTORS; BAYESIAN LIKELIHOOD; CORROBORATION; DECIBANS; TAIL-AREA PROBABILITIES; WEIGHT OF EVIDENCE.

1. INTRODUCTION

My purpose is to survey some of the work on weight of evidence because I think the topic is almost as important as that of probability itself. The survey will be far from complete.

"Evidence" and "information" are related but do not have identical meanings. You might be interested in *information* about Queen Anne but in *evidence* about whether she is dead. The expression "weight of evidence" is familiar in ordinary English and describes whether the evidence in favour or against some hypothesis is more or less strong. The Oxford English Dictionary quotes T.H. Huxley (1878, p. 100) as saying "The weight of evidence appears strongly in favour of the claims of Cavendish", but C.S. Peirce used the expression in the same year so I suspect it was familiar before that date. The expression "The weight of the evidence" is even the title of a mystery story by Stewart or Michael Innes (1944). Moreover, Themis, the Greek goddess of justice is usually represented as carrying a pair of scales, these being for weights of evidence on the two sides of an argument.

A jury might have to weigh the evidence for or against the guilt or innocence of an accused person to decide whether to recommend conviction; a detective has to weigh the evidence to decide whether to bring a case to law; and a doctor has to weigh the evidence when doing a differential diagnosis between two diseases for choosing an appropriate treatment. A statistician can be said to weigh evidence in discrimination problems, and also, if he is not a Neyman-Pearsonian, when he applies a significance test.

In all these examples it seems obvious that the weight of evidence ought to be expressible in terms of probabilities, although the appropriate action will usually or always depend on utilities as well. At least three books have both the words "probability" and "evidence" in their titles (Good, 1950; Ayer, 1972; Horwich, 1982), as well as Dempster's lecture at this conference, and this again shows the close relationship of the two topics.

I believe that the basic concepts of probability and of weight of evidence should be the same for all rational people and should not depend on whether you are a statistician. There should be a unity of rational thought applying, for example, to statistics, science, law, and politics. This assumption will set the tone of my survey. No concept is fundamental if only statisticians use it.

Suppose now that we have some hypothesis or theory, H, and some event, evidence, or experimental result called E. For example, H, might be the hypothesis that an accused person is guilty of some crime, and then the negation of H, denoted by $\overline{H}$, means that he is innocent. Of course H and $\overline{H}$ will seldom be simple statistical hypotheses in this legal example. A Bayesian, of whatever kind, assumes that it is meaningful to talk about such probabilities as $P(E|H\&G)$, $P(E|G)$, and so on, where G denotes background information such as that it is bad for the health to have guns fired at one. To economize in notation I shall usually take G for granted and omit it from the notation, so that the various probabilities will be denoted by $P(E|H)$ etc. These probabilities might be logical probabilities, known as "credibilities", or they might be subjective (meaning personal) or multisubjective (multipersonal); and they might be partially ordered, that is interval-valued, with upper and lower values, or they might have sharp (numerical) values. Although I believe partially-ordered probabilities to be more fundamental than sharp values, as I have said in about fifty publications (for example, Good, 1950, 1962a), I shall base my discussion on sharp values for the sake of simplicity. The discussion could be generalized to partially-ordered probabilities and weights of evidence but only with some loss of clarity. Any such generalization, if it is valid, should reduce to the "sharp" case when the intervals are of zero width. I would just like to remind you that inequality judgments of weights of evidence can be combined with those of probabilities, odds and other ratios of probabilities, expected utilities and ratios of them, etc., for improving a body of beliefs (for example, Good, 1962a, p. 322). I have not yet understood Shafer's theory of evidence, which is based on Dempster's previous work on interval-valued probabilities. Aitchison (1968) seems to me to have refuted the approach that Dempster took at this time, at least in some circumstances. At any rate, as I said, the present paper is based on sharp probabilities.

Other possible names for weight of evidence are "degree of corroboration" and "degree of confirmation", but the latter name was spoilt by Carnap because he made the mistake of calling a credibility a "degree of confirmation" thus leading philosophers into a quagmire of confusion into which some of them have sunk out of sight. This disaster shows the danger of bad terminology. Moreover, the expression "weight of evidence" is more flexible than the other two expressions because it allows such natural expressions as "the weight of evidence is against H". It would be linguistically unnatural to say "the degree of corroboration (or confirmation) is against H".

2. DERIVATION OF THE EXPLICATUM

I intend presently (meaning "soon") to discuss the history of the quantitative explication of weight of evidence, but it will be convenient first to mention a method of deriving its so-called explicatum from compelling desiderata. Let $W(H{:}E)$ denote the weight of evidence in favour of H provided by E, where the colon is read "provided by". If there is some background information G that is given all along, then we can extend the notation to $W(H{:}E|G)$. I mentioned that partly to show that we cannot replace the colon by a vertical stroke.

It is natural to assume that $W(H{:}E)$ is some function of $P(E|H)$ and of $P(E|\overline{H})$, say $f[P(E|H), P(E|\overline{H})]$. I cannot see how anything can be relevant to the weight of evidence other than the probability of the evidence given guilt and the probability given innocence, so the function f should be mathematically independent of $P(H)$, the initial probability of H. But $P(H|E)$, the final probability of H, should depend only on the weight of evidence and on the initial probability, say

$$P(H|E) = g[W(H{:}E), P(H)].$$

In other words we have the identity

$$P(H|E) = g\{f[P(E|H), P(E|\overline{H})], P(H)\}$$

On writing $P(H) = x$, $P(E) = y$, and $P(H|E) = z$ we have the identity:

$$z \equiv g\{f[\frac{zy}{x}, \frac{y(1-z)}{1-x}], x\}.$$

It can be deduced from this functional equation that f is a monotonic function of $P(E|H)/P(E|\overline{H})$ (Good, 1968, p. 141) and of course it should be an increasing rather than a decreasing function. If H and $\overline{H}$ are simple statistical hypotheses, and if E is one of the possible experimental outcomes occurring in the definitions of H and $\overline{H}$, then $P(E|H)/P(E|\overline{H})$ is a simple likelihood ratio, but this is a very special case. In general this ratio is regarded as meaningful only to a Bayesian. It could be called a ratio of *Bayesian likelihoods*.

We can think of weights of evidence like weights in the scales of the Goddess of Justice, positive weights in one scale and negative ones in the other. Therefore we would like weight of evidence to have the additive property

$$W[H{:}(E\&E')] = W(H{:}E) + W(H{:}E'|E), \tag{1}$$

provided that this does not force us to abandon the result already established that $W(H{:}E)$ is a function of $P(E|H)/P(E|\overline{H})$. We can in fact achieve (1), uniquely up to a constant factor, by taking

$$W(H{:}E) = \log \frac{P(E|H)}{P(E|\overline{H})}. \tag{2}$$

Now we can easily see, by four applications of the product axiom of the theory of probability, namely $P(A\&B) = P(A)P(B|A)$, that

$$\frac{P(E|H)}{P(E|\overline{H})} = \frac{O(H|E)}{O(H)} \tag{3}$$

where O denotes odds. The odds corresponding to a probability p are defined as $p/(1-p)$. Some numerical examples of the relationship between probability and odds are shown in Table 1. In ordinary betting terminology odds of 2 are called odds of 2 to 1 on, and odds of ½ are called odds of 2 to 1 against, while odds of 1 are called "evens".

TABLE 1. *Probability and Odds*

Probability	Odds
0	0
1/10	1/9
1/3	1/2
1/2	1
2/3	2
9/10	9
1	∞

The right side of equation (3) can be described in words as the ratio of the final odds of H to its initial odds, or the ratio of the posterior to the prior odds, or the factor by which the initial odds of H are multiplied to give the final odds. It is therefore natural to call it the *factor in favour of H provided by E* and this was the name given to it by A.M. Turing in a vital cryptanalytic application in WWII in 1941. He did not mention Bayes's theorem, with which it is of course closely related, because he always liked to work out everything for himself. When I said to him that the concept was essentially an application of Bayes's theorem he said "I suppose so". In current Bayesian literature it is usually called the *Bayes factor in favour of H provided by E*. Thus weight of evidence is equal to the logarithm of the Bayes factor. The Bayes factor and weight of evidence are Bayesian concepts because the probabilities $P(H)$, $P(H|E)$, $P(E|H)$, and $P(E|\bar{H})$ are all in general regarded as meaningless by anti-Bayesians.

The additive property (1) simplifies if E and F are both independent given H and given $\bar{H}$. This condition usually requires that both H and $\bar{H}$ should be simple statistical hypotheses, a point of which Herman Rubin reminded me privately after the lecture.

The formula (3) occurs in a paper by Wrinch and Jeffreys (1921, p. 387); and Jeffreys (1936), called weight of evidence "support", but in this book, Jeffreys (1939), he dropped this expression because he always assumed that $O(H) = 1$ so that there $W(H:E)$ reduced to the logarithm of the final odds. His motive in concentrating on this special case must have been to try to sell fixed rules of inference: his original aim was to arrive at rules defining impersonal credibilities though his judgments of these were inevitably personal to him (Good, 1962b, p. 556). Whether they will become highly multipersonal is an empirical matter.

The basic property of weight of evidence can be expressed in words thus:

"initial log-odds plus weight of evidence = final log-odds".

Incidentally Barnard (1949), who, independently of Turing and of Wald, invented sequential analysis, called log-odds "lods". Good (1950), following a suggestion of J.B.S. Haldane, called it "plausibility", but "log-odds" is short enough and is self-explanatory.

It is sometimes convenient to write $W(H/H':E)$, read "the weight of evidence in favour of H as compared with H', provided by E", as a shorthand notation for $W(H:E|H \vee H')$. Of course, if $H \vee H'$ is given then $\bar{H} = H'$.

The Fisher-Neyman factorability condition for sufficiency (Fisher, 1925, p. 713; Neyman, 1925) can be expressed in terms of weight of evidence. I'll express it in terms of a class **H** of hypotheses instead of in terms of parameters. Let $f(E)$ be some function of the evidence. If $W[H/H':f(E)] = W(H/H':E)$ for all pairs H,H' of hypotheses in the class **H**, then $f(E)$ is sufficient for the hypotheses. Here $f(E)$ need not be a scalar or vector; it might be a proposition. This is a Bayesian generalization of the concept of sufficiency because $W(H/H':E)$ is not always an acceptable concept for the non-Bayesian. It could be called Bayesian sufficiency or "efficaciousness" (Good, 1958). For legal purposes, $f(E)$ is a possible interpretation of what is meant by "the whole truth and nothing but the truth", when there are two or more hypotheses to be entertained. Of course approximate Bayesian sufficiency is all that can be demanded in a court of law. Anyone who swears to tell the whole truth has already committed perjury.

For applications of weight of evidence, apart from the many applications for Bayesian tests of standard statistical hypotheses, see Good (1983e, p. 161).

3. SOME HISTORY

In the draft of my talk I said that C.S. Peirce (1878), is an obscurely written paper, had failed to arrive at the correct definition of weight of evidence owing to a mistake (see

Good, 1981b). But I intend to amend this comment in the discussion, in my reply to Dr. Seidenfeld. Levi (1982) agrees that Peirce made a mistake although he thinks it was different from the one I thought he made. Levi points out that Peirce was anti-Bayesian and to some extent anticipated Neyman and Pearson.

The definition of weight of evidence as the logarithm of the Bayes factor was given independently of Good (1950) by Minsky and Selfridge (1961); and again independently Tribus (1969) used the term "evidence" for weight of evidence. Kemeny and Oppenheim (1952) used the expression "factual support for a hypothesis" (provided by evidence), and their desiderata led them to the formula

$$\frac{P(E|H)-P(E|\overline{H})}{P(E|H)+P(E|\overline{H})}.$$

This is an increasing function of $W(H:E)$, namely sinh $\{W(H:E)/2\}$.

The philosopher Karl Popper (1954) proposed desiderata for corroboration and he said (1959, p. 394) "I regard the doctrine that the *degree of corroboration or acceptibility cannot be a probability* as one of the most interesting findings of the philosophy of knowledge". This fundamental contribution to philosophy was taken for granted by a dozen British cryptanalysts eighteen years before Popper published his comment, the name used being "score" or "decibannage". Moreover we used the best explicatum, which is not mentioned by Popper although this explicatum satisfies his desiderata.

In 1970, at the Second World Congress of the Econometric Society, Harold Jeffreys said it had taken fifty years for his work with Dorothy Wrinch to be appreciated by the statistical community, and he predicted that it would be another fifty years before the philosophers were equally influenced (or words to that effect). Recently the slowness of professional philosophers to use the correct explicatum for degree of confimation (or corroboration), namely $W(H:E)$, has been exemplified by Horwich (1982, p. 53). He suggests two measures, $P(H|E)$ - $P(H)$ and $P(H|E)/P(H)$, of which the latter had been used by J.L. Mackie (1963). Although both these explicata satisfy the additive property (1), neither is a function of $W(H:E)$. To see that $P(H|E)-P(H)$ is inappropriate as a measure of degree of corroboration consider (i) a shift of probability of H from 1/2 to 3/4, (ii) a shift from 3/4 to 1, and (iii) a shift from .9 to 1.15. In each case $P(H|E)-P(H) =$ 1/4, but the degree of corroboration seems entirely different in the three cases, especially as case (iii) is impossible! A similar objection applies to Mackie's suggestion $P(H|E)/P(H)$ and to its logarithm. For further discussion of Horwich (1982) see the review by Good (1983b). That review contains some other applications of the concept of weight of evidence to philosophical problems such as that of induction.

The unit in terms of which weight of evidence is measured depends on the base of its logarithms. The original cryptanalytic application was an early example of sequential analysis. It was called Banburismus because it made use of stationery printed in the town of Banbury; so Turing proposed the name "ban" for the unit of weight of evidence when the base of the logarithm is 10. Another possible name, especially in a legal context, would be a "themis" partly because Themis was the goddess of justice, and partly because Themis is the name of the tenth satellite of Saturn. But "ban" is shorter, more convenient, and historically justified. Turing called one tenth of this a *deciban* by analogy with a *decibel* in acoustics, and we used the abbreviation *db*. Just as a decibel is about the smallest unit of difference of loudness that is perceptible to human hearing, the deciban is about the smallest unit of weight of evidence that is perceptible to human judgment. It corresponds to a Bayes factor of 5/4 because $\log_{10}5 = .70$ and $\log_{10}4 = .60$. A bit is 3.01 db.

When I arrived at Bletchley the work on Banburismus had been going for some weeks and the entries on the Banbury sheets were of the form 3.6 meaning 3.6 db. I proposed first that the decimal point should be dropped so that the entries would be in centibans, and better still that the unit could be changed to a half-deciban or *hdb* with little loss of accuracy. This very simple suggestion saved much writing and eyestrain, and probably decreased the number of arithmetical errors. It may have cut the time for Banburismus by a factor of 2 and time was of the essence. One of my colleagues, Alex Kendrick, suggested the name "bonnieban" for the *hdb*.

The concept of weight of evidence is formally related to the logit transformation, $x = \log[P/(1-P)]$, although here P is a *cdf*. I don't think it explains why the logit transformation is useful, but it might have suggested the transformation to Fisher because Jeffreys was one of his students. Jeffreys (1936) discussed the logarithm of the Bayes factor and Fisher & Yates (1938) suggested the logit transformation.

4. A SIMPLE EXAMPLE

As a simple example, suppose we are trying to discriminate between an unbiased die and a loaded one that gives a 6 one third of the time. Then each occurrence of a 6 provides a factor of $\frac{1/3}{1/6} = 2$, that is, 3 *db*, in favour of loadedness while each non-6 provides a factor of $\frac{2/3}{5/6} = \frac{4}{5}$, that is, 1 *db*, against loadedness. For example, if in twenty throws there are ten 6's and ten non-6's then the total weight of evidence in favour of loadedness is 20 *db*, or a Bayes factor of 100. The corresponding tail-area probability for getting ten or more 6's if the die is fair is about 1/1670. A Bayes factor is always smaller than the reciprocal of a tail-area probability (Good, 1950, p. 94), and in this example it is smaller by a factor of 16.7.

5. HOW TO MAKE JUDGMENTS

Even when H and $\overline{H}$ are simple statistical hypotheses, in which case a Bayes factor is equal to a likelihood ratio, the terminology of Bayes factors and weights of evidence has more intuitive appeal. This intuitive appeal persists in the general case when the weight of evidence is not the logarithm of a likelihood ratio. I conjecture that juries, detectives, doctors, and perhaps most educated citizens, will eventually express their judgments in these intuitive terms. In fact, in legal applications, it must be less difficult to judge $P(E|H)/P(E|\overline{H})$ or $O(H|E)/O(\overline{H})$, or its logarithm, than to judge $P(E|H)$ and $P(E|\overline{H})$ separately because these probabilities are usually exceedingly small, often less than 10^{-100}. Of course the official responsibility of juries is more to judge $P(H|E)$ if they think in terms of probabilities. They are supposed to exclude some kinds of evidence, such as previous convictions, but they probably do allow for these convictions when they know about them, judging by some experiences of Hugh Alexander (c. 1955) when he served on a British jury.

A problem that arises both in legal and medical applications is in deciding what is meant by the initial probability of H. For example, if the accused is regarded as a random person in the world, his initial probability of guilt in much smaller than if he is known to live in the town, or village, where the crime was committed. For this reason it might often be easier to judge $O(H|E)$ directly than to compute it as $O(H)F(H{:}E)$ where F denotes the Bayes factor. Perhaps the best judgmental technique is to split the evidence into pieces and to check your judgments for consistency. For example, you could make separate judgments of (i) $O(H|E\&E')$ and (ii) $O(H|E)F(H{:}E'|E)$, while realizing that these should be equal. Some people, after some training, might find it easier to work with the additive logarithmic form, instead of or in addition to the multiplicative form, that is, to use log-odds and weights of evidence (or log-factors) instead of odds and Bayes factors. It is convenient that factors of 2, 4, 5, 8, 10 and 20 correspond closely to weights of evidence of 3, 6, 7, 9, 10 and 13 decibans respectively. Themis should be grateful to Zeus for giving us just ten fingers.

I believe that this device of splitting the evidence into two or more groups corresponds to a psychologically natural manner of evaluating evidence. First some evidence E makes you suspicious, and you estimate the odds of H as somewhere near evens; then some more or less independent evidence arrives, perhaps in the form of a new witness, and this peps up the odds by a factor that you can judge separately. (Similarly an antibayesian, unaware that he is really a Bayesian, will choose null hypotheses of non-negligible prior probabilities, and then test them). It might help the judgment to recognize consciously that the chronological order of hearing evidence is not entirely relevant, and to imagine that it had arrived in some other order. In legal applications, one example of a convenient piece of evidence, that can be mentally separated from the rest, is the discovery of a strong motivation for the crime. An alibi is another example, in the opposite direction.

6. EXPECTATIONS AND MOMENTS OF BAYES FACTORS AND OF WEIGHTS OF EVIDENCE. ENTROPY

In 1941, or perhaps in 1940, Turing discovered a few simple properties of Bayes factors and weights of evidence. One curious result, which was independently noticed by Abraham Wald, was, in Turing's words "The expected factor in favour of a wrong hypothesis is 1". This fact can be better understood from its very simple proof: Suppose the possible outcomes of an experiment are E_1, E_2, E_3,... and that the hypothesis H is true. If E_i is an observed outcome the factor against H is

$$F(\overline{H}{:}E_i) = \frac{P(E_i|\overline{H})}{P(E_i|H)} .$$

Its expectation given the *true* hypothesis H is

$$\mathbf{E}[F(\overline{H}{:}E_i)|H] = \sum_i \frac{P(E_i|\overline{H})}{P(E_i|H)} \,.\, P(E_i|H)$$

$$= \sum_i P(E_i|\overline{H}) = 1. \tag{4}$$

This result seems surprising at first sight, and not just because of its simplicity. If $\overline{H}$ is false we expect the Bayes factor in its favour to be less than 1 in most experiments. The only way to get an expected value of 1 is if the distribution of the Bayes factor is skewed to the right, that is, when the factor against the truth exceeds 1 it can be large.

To exemplify (4), let's consider the example concerning a die that we considered before and suppose that the die is really a fair one. Then, on one throw of the die, there is a probability of 1/6 that the factor in favour of loadedness is $\frac{1/3}{1/6} = 2$ and a probability of 5/6 that the factor in favour of loadedness will be 4/5. Hence the expected factor in favour of loadedness when the die is unloaded is $1/6 \times 2 + 5/6 \times 4/5 = 1/3 + 2/3 = 1$. Thus Turing's theorem can be used as a check of the calculation of a Bayes factor.

It is disturbing that one can get a large factor against the truth. This point will emerge again later in this talk.

Let $f = F(H:E)$. Then the nth moment of f about the origin given H is equal to the $(n+1)$st moment of f given $\overline{H}$; that is,

$$\mathbf{E}(f^n | H) = \mathbf{E}(f^{n+1} | \overline{H}). \tag{5}$$

The case $n = 0$ is Turing's result, just discussed. It can be further proved that $E(f^\alpha | H)$ is an increasing function of α for $\alpha > 0$. Better results will be published elsewhere.

This follows from an algebraic inequality that might date back to Duhamel & Reynaud (1823, p. 155); see Hardy, Littlewood, and Pólya, (1934, p. 26). By letting $\alpha \to +0$ we find, as I shall show in a moment, that

$$\mathbf{E}[W(H:E) | H] \geq 0 \tag{6}$$

or in words, the expected weight of evidence in favour of the truth is non-negative, and vanishes only when $W(H:E) = 0$ for all E of positive probability. This is of special interest because weight of evidence is additive so its expected value is more meaningful than that of a Bayes factor. This inequality was pointed out to me by Turing in 1941, with a different proof. Regarded as a piece of algebra it is merely an elementary inequality. What makes it interesting is its interpretation in terms of human and therefore statistical inference. That is why I regard it as reasonable to attribute it to Turing although it was also applied to statistical mechanics by Gibbs (1902, p. 136).

The monotonic property of $\mathbf{E}(f^\alpha | H)$ can be written

$$\Sigma \frac{p_i^{\alpha+1}}{q_i^\alpha} \text{ increases with } \alpha \quad (\alpha \geq 0) \tag{7}$$

where $p_i = P(E_i | H)$, $q_i = P(E_i | \overline{H})$.
But the left side is 1 when $\alpha = 0$, so

$$\Sigma p_i \left(\frac{p_i}{q_i}\right)^\alpha \geq 1,$$

that is,

$$\Sigma p_i \exp(\alpha \log \frac{p_i}{q_i}) \geq 1.$$

Therefore

$$\Sigma p_i [1 + \alpha \log \frac{p_i}{q_i} + \ldots) \geq 1.$$

By taking α small we get

$$\Sigma p_i \log \frac{p_i}{q_i} \geq 0 \tag{8}$$

which states that $\mathbf{E}(\log f | H) \geq 0$. Thus (7) can be regarded as a generalization of (6) or (8). The fact that (8) is an algebraic theorem confirms that weight of evidence is correctly explicated, although I hope you are already convinced. See also Good (1983d).

One way to interpret (6) or (8) is that in expectation it pays to acquire new evidence, if arriving at the truth is your objective. An explicit proof in terms of decision theory, that it pays in your own expectation to acquire new information, without reference to weight of evidence, was given by Good (1967, 319-321); but see also Good (1974) (where it was shown that, in some one else's expectation, it does not necessarily pay you). The principle is related to what Carnap (1947) called "the principle of total evidence": Locks' recommendation to use all the available evidence when estimating a probability.

Turing's inequality can of course be interpreted in terms of discrimination between two multinomials, a familiar problem in cryptanalysis. If $p_1, p_2, \ldots, p_n$ are the category probabilities under the true hypothesis H, and are $q_1, q_2, \ldots, q_n$ under hypothesis $\overline{H}$, then the expression (8) is equal to the expected weight of evidence "per letter" in favour of H.

Sometimes one of the hypotheses is that of equiprobability, say that $q_1 = q_2 = \ldots = q_n = 1/n$. Then the expected weight of evidence becomes $\Sigma p_i \log(np_i)$ and this is equal to $\log n + \Sigma p_i \log p_i$. The expression $-\Sigma p_i \log p_i$ is usually called "entropy" because it is a form that entropy often takes in statistical mechanics (Boltzmann, 1964, p. 50; Gibbs, 1902, p. 129). That is because it is convenient for some purposes to divide phase space into equal volumes, in virtue of Liouville's theorem. (In thermodynamics, which is explained by statistical mechanics, the entropy has a different definition). The entropy expression $-\Sigma p_i \log p_i$ occurs prominently in Shannon's theory of communication, but his coding theorems can be somewhat better expressed in terms of expected weight of evidence in my opinion (Good & Toulmin, 1968).

Apart from its central position in human inference, one reason that expected weight of evidence is more fundamental than entropy is that it is applicable to continuous variables without ambiguity. This fact is related to its "splitative" property in the discrete case. That is, $\Sigma p_i \log(p_i/q_i)$ is unchanged if one of the categories is split into two categories in a random manner such as by spinning a coin. Among its names apart from "expected weight of evidence" are "relative entropy", "cross-entropy", "dinegentropy", and "discrimination information". (Should this one be "discrimination evidence"?)

The symmetrized expression $\Sigma(p_i - q_i)\log(p_i/q_i)$, called "divergence" by Kullback (1959), had been used also by Boltzmann (1964, p. 54) in statistical mechanics, Jeffreys (1946) in his theory of invariant priors, and by me for discrimination problems in cryptanalysis during World War II.

For deciding in advance whether Banburismus was likely to be successful, Turing estimated the expected weight of evidence. While doing so he discovered another theorem which I shall now state.

Suppose that the weight of evidence in favour of H, when H is true, has a normal distribution with mean μ and variance σ^2, and suppose our unit is the "natural ban". Then $\sigma^2 = 2\mu$. In other words

$$\mathrm{var}\{W(H:E)\,|\,H\} = 2\,\mathbf{E}\{W(H:E)\,|\,H\}. \tag{9}$$

Moreover

$$\mathbf{E}\{W(\overline{H}:E)\,|\,H\} = -\mathbf{E}\{W(H:E)\,|\,H\} = -\mu\,. \tag{10}$$

This result was later published by Peterson, Birdsall and Fox (1954) in connection with radar. The result is surprising so I shall give the proof.

Let x be an observed weight of evidence in natural bans. Since the weight of evidence tells us just as much as E does about the odd of H, we have

$$W(H:E) = W\{H:W(H:E)\} = W(H:x).$$

Therefore the ratio of the probability densities

$$\frac{P.D.(x\,|\,H)}{P.D.(x\,|\,\overline{H})} = e^x\,. \tag{11}$$

Assume that x (or rather the corresponding random variable) has the distribution $N(\mu,\sigma^2)$ so that the probability density of x, given H, is

$$\frac{1}{\sigma\sqrt{2\pi}}\exp -\left[\frac{(x-\mu)^2}{2\sigma^2}\right]\,.$$

Then by (11)

$$P.D.(x|\overline{H}) = \frac{1}{\sigma\sqrt{2\pi}} \exp\left[-\frac{(x-\mu)^2}{2\sigma^2} - x\right].$$

Therefore

$$\frac{1}{\sigma\sqrt{2\pi}} \int_{-\infty}^{\infty} \exp\left[-\frac{(x-\mu)^2}{2\sigma^2} - x\right] dx = 1$$

and from this it follows that $\sigma^2 = 2\mu$ and also that the distribution of x given $\overline{H}$ is $N(-\mu,\sigma^2)$.

If we use decibans the formula $\sigma^2 = 2\mu$ becomes converted to $\sigma = \sqrt{\mu 20 \log_{10} e} = \sqrt{8.686\mu} \cong 3\sqrt{\mu}$. Thus the standard deviation is much larger than one might have guessed, a fact that in the application to radar is disturbing. For example, if the expectation is 16 *db*, which corresponds to a Bayes factor of 40, there is a probability of 1/6 that the weight of evidence will exceed $16 + 3\sqrt{16} = 28$ *db*, corresponding to a factor of 600, and a probability of 1/6 that it will be less than 4 *db*, corresponding to a factor of only 2½. Also there is a chance of 1/740 that the Bayes factor against the truth will exceed 100.

For generalizations of this theorem of Turing's to the more realistic case where it is assumed that the weight of evidence is only approximately normally distributed near its mean see Good (1961), which dealt with false-alarm probabilities in signal detection. The results can then be even more disturbing than in the case of strict normality and I hope this fact is well known to the defence departments of all countries that are civilized enough to possess an atom bomb.

Good & Toulmin (1968, Appendix B) and Good (1983f) give other relationships between the moments and cumulants of weight of evidence for the general case. Such identities can be deduced from the elegant formal identity $\phi(t+i) = \overline{\phi}(t)$ where ϕ and $\overline{\phi}$ denote the characteristic functions of $W(H{:}E)$ given H and $\overline{H}$ respectively. For example, when the moments exist,

$$\overline{\mu}_s' = \sum_{\nu=0}^{\infty} (-1)^\nu \mu_{s+\nu}'/\nu! = e^{-\Sigma}\mu_s', \qquad \mu_s' = \sum_{\nu=0}^{\infty} \overline{\mu}_{s+\nu}'/\nu! = e^{\Sigma}\overline{\mu}_s',$$

where μ_s' and $\overline{\mu}_s'$ denote moments about 0, and where Σ, just here, denotes the suffix-raising operator. There are similar identities for the cumulants. The cases $s = 0$ and $s = 1$ are of special interest.

When Turing judged the value of Banburismus by estimating an expected weight of evidence he was in effect treating weight of evidence as if it were a utility. It may be regarded as a *quasi-utility*, that is, an additive substitute for utility expressed in terms of probabilities. If you recall Wald's theorem that a minimax procedure is one that can be regarded as using a least favourable prior, you are led to the idea of minimizing expected weight of evidence or maximizing entropy in the selection of a prior. (Compare Good, 1969; Bernardo, 1979). Although minimax procedures in statistical inference are controversial they have the advantage of having invariant properties. The idea of using maximum entropy for choosing a prior was suggested by Jaynes (1957), though without mentioning the minimax property. For a recent statement of my views on maximum entropy see Good (1983c).

In the design of an experiment the entire distribution of weight of evidence, and in particular its variance, is of interest, and not just its expectation. In this respect weight of evidence, like money, is not an exact substitute for utility.

Expected weight of evidence is basic to the non-Bayesian approach to significance testing of Kullback (1959).

For some relationships between expected weight of evidence and errors of the first and second kinds see Good (1980). Other properties of weight of evidence can be located through the indexes of Good (1983e).

7. TAIL-AREA PROBABILITIES

A Fisherian might try to interpret weight of evidence, in its ordinary English sense, in terms of tail-area probabilities in tests of significance. Suppose then that a client comes to a Fisherian with experimental results E and he wants to know how much evidence this provides against some null hypothesis H, or even whether H is supported if that is possible. The client does not want to reject H too readily for he considers it to be simpler than its rivals and so easier to work with. For example, if he did not have experimental results he would have "accepted" H in the sense of assuming that its observational implications were approximately correct. (Should this be the definition of a null hypothesis?) The situation occurs, for example, when other hypotheses involve additional parameters. This by the way explains why it is not always better to replace a significance test by an estimation procedure. This point was made, for example, by Arnold Zellner in discussion at the 21st SREB-NSF Meeting on Bayesian Inference in Econometrics in 1980 in response to someone who was trying to knock significance tests. For several of my own views concerning significance tests see Good (1981a).

Suppose you apply a significance test and get a tail-area probability or P-value of .0455. (It is irrelevant to most of my discussion whether this is a single or double tail.) How should you report this to your client? The answer is not as simple as it seems. You might report this result to the client in one of the following ways, depending on your philosophy and on the client's philosophy, and on the practical background of the problem:

(i) "The hypothesis is 20 to 1", as in the lines from War of the Worlds: "The chances of anything coming from Mars are a million to one. But still they come!" I hope it's not a million to one *on*!

(ii) "The odds against the hypothesis are about 20 to 1" (a familiar fallacy perpetrated by reputable scientists).

(iii) "The probability of getting so extreme an outcome is .0455 if the null hypothesis is true", where the meaning of "more extreme" needs to be stated. It can't mean that the probability density is small because the density can be made arbitrarily small, even where the mode originally occurred, by applying a suitable transformation to the independent variable. (Compare the usual attack against "Bayes's postulate" of a uniform distribution!).

(iv) "The probability of getting so extreme a result is less than .05 if the null hypothesis is true."

(v) "Reject H."

(vi) "Reject H because $P < .05$."

(vii) "I wouldn't reject H (as a good approximation) because H is *a priori* so probable." For example, suppose a coin gave 61 heads and 39 tails, H being the hypothesis that the coin is fair. (Here the double-tail-area, allowing for a continuity correction, is .036.)

(viii) "I wouldn't reject H because the cost to you of assuming $\overline{H}$ is too great."

(ix) "The result is not decisive: collect more data if it is practicable."

(x) "You should have consulted me in advance so that we could have decided on a rejection procedure in the Neyman-Pearson fashion."

(xi) Ask the client "At what threshold P-value would you reject H?" Or arrive at a rejection level by discussion with the client.

(xii) None of the above.

I'm going to consider an example where «None of the above» is appropriate because the null hypothesis should be clearly accepted, not rejected. Let's imagine that the following game is being played at a gambling casino. An urn is known to contain 100 black and white balls. You pay an entrance fee, and the game consists in extracting one ball at a time. You win a dollar whenever a black ball is extracted. After each gamble the ball is returned to the urn and the urn is well shuffled, so the sampling is with replacement. Assume that each ball has probability 1/100 of being selected. Suppose that the game is played N times and that there are r successes and $N - r$ failures. We formulate the null hypothesis that there are 50 balls of each colour.

We are dealing with a binomial sample, and the standard deviation of r, given the null hypothesis, is $\sqrt{[N\frac{1}{2}(1-\frac{1}{2})]} = \frac{1}{2}\sqrt{N}$. For convenience assume that N is a perfect square and that $r = \frac{1}{2}N + \sqrt{N}$. Thus the bulge is 2σ and the double tail-area probability P = .0455 so the result is «significant at the 5 % level». (I'm ignoring the continuity correction). I am now going to prove uncontroversially and without explicit Bayesianity that if N is large enough this outcome does not undermine the null hypothesis, in fact it supports it. This shows that it is incorrect to say that a null hypothesis can never be supported but can only be refuted, as one so often hears.

In this problem, the possible values of the binomial parameter p are 0, .01, .02,...,.99, 1.00, though the values 0 and 1 will have been ruled out if $r \neq 0$ or N.

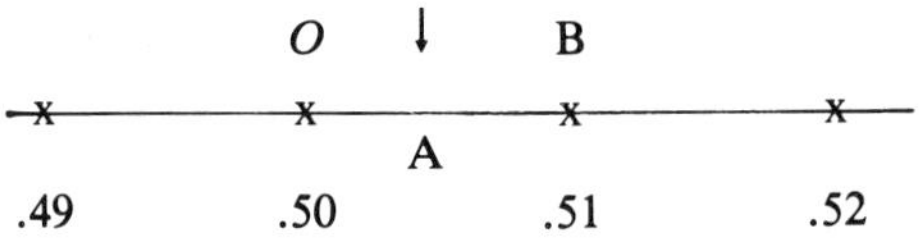

In this diagram the possible values of p are marked with crosses. The observed fraction r/N of successes is marked by an arrow at the point A. The null hypothesis corresponds to the point O and the closest possible value for p, to the right of $p = \frac{1}{2}$, is at B where p = .51.

The point A corresponds to a fraction $r/N = \frac{1}{2} + N^{-1/2}$. Thus, if N is large enough, the distance OA is much shorter than the distance AB. It is therefore obvious that if N is large enough our tail-area probability of .0455 supports the null hypothesis and the null hypothesis becomes more and more convincing as $N \to \infty$, corresponding to this fixed tail-area probability. A similar argument can be used even if the binomial parameter is continuous but it is not so clear-cut. It shows that a given P-value means less for large N. (Jeffreys, 1948, p. 222; 1961, p. 248; Hill, 1982; Good, 1983a). A possible palliative is to use *standardized tail-areas*. That is, if a small tail-area probability P occurs with sample size N we could say it is *equivalent to a tail-area probability of* $P\sqrt{100/N}$ *for a sample size of 100* if this is also small (Good, 1982b). The topic is closely related to the possibility of "sampling to a foregone conclusion" by using optional stopping when tail-area probabilities are used without any Bayesian underpinning. The earliest reference I know for this form of cheating is Greenwood (1938) and other references are given by Good (1982a).

Here is a Bayesian solution to the problem of the one hundred black and white balls in an urn. If there were only one rival H_p to the null hypothesis $H_{1/2}$, the Bayes factor against $H_{1/2}$ would be

$$F(H_p|H_{1/2}:r) = \frac{\binom{N}{r} p^r (1-p)^{N-r}}{\binom{N}{r} 2^{-N}} = (2p)^r (2-2p)^{N-r}$$

$$\begin{aligned} &= (1+q)^r(1-q)^{N-r} \qquad (\text{where } q = 2p-1) \\ &= \exp[r \log(1+q) + (N-r)\log(1-q)] \\ &= \exp[r(q - \tfrac{1}{2}q^2 + \ldots) - (N-r)(q + \tfrac{1}{2}q^2 + \ldots)] \\ &\approx \exp(2q\sqrt{N} - \tfrac{1}{2}q^2 N) \qquad (\text{when } r = \tfrac{1}{2}N + \sqrt{N}). \end{aligned}$$

If we wanted to compute an exact Bayes factor against $H_{1/2}$ we would need to take a weighted average of the Bayes factors corresponding to each p (or q) the weights forming a prior distribution $P(H_p)$. But we don't need to do this in the present case because we obtain the maximum weight of evidence against $H_{1/2}$ by ignoring all values of q except $2 \times .51 - 1 = .02$. Thus the factor against $H_{1/2}$ is *at most* $\exp(\frac{\sqrt{N}}{25} - \frac{N}{5000})$, corresponding to a weight of evidence of $\frac{\sqrt{N}}{25} - \frac{N}{5000}$ natural bans. Here is a small table:

TABLE 2. *Evidence in favour of* $H_{1/2}$ if $P = .0455$

N	Weight of evidence *in favour* of $H_{1/2}$	Bayes factor *in favour* of $H_{1/2}$
40,000	≥ 0	≥ 1
90,000	≥ 6 nat. bans	≥ 400
1,000,000	≥ 160 nat. bans	$\geq 3 \times 10^{69}$

Thus in this example it is possible to get a lot of evidence in favour of the null hypothesis under circumstances where a dogmatic use of the 5 % rejection level would be ludicrous.

The primary lesson to be learnt from this example is that tail-area probabilities need to be used cautiously. If you use tail-area probabilities, perhaps you should always make an honest effort to judge whether your use of them is in violent conflict with your judgment of the Bayes factor or weight of evidence against the null hypothesis. In human thought, weight of evidence is a more fundamental concept than a tail-area probability. There was no Greek goddess who rejected hypotheses at the 5 % level with one tail in each scale!

Berkson (1942) criticised the view that a small P-value is evidence against a null hypothesis. He admits that he used to adopt the usual view but argues, without mentioning Bayes or Jeffreys, that (i) a small P-value is not evidence against the null hypothesis unless an alternative can be suggested that would make this low value more probable; (ii) values of P in the range (.3, .7) can support the null hypothesis for large samples. There is an error on page 333 where he says that "small P's are more or less independent, in the weight of the evidence they afford, of the numbers in the sample". (See also page 332). Otherwise Berkson's paper was largely Bayesian although he didn't notice it. Everybody is to some extent a Bayesian especially when using common sense.

ACKNOWLEDGEMENT

This work was supported in part by an N.I.H. Grant number GM18770.

REFERENCES

AITCHISON, J. (1968). Comment on a paper by A.P. Dempster, *J. Roy. Statist. Soc. Ser. B* **30**, 234-236.

ALEXANDER, C.H.O'D. (c. 1955). Oral communication.

AYER, A.J. (1972). *Probability and Evidence*. Columbia University Press.

BARNARD, G.A. (1949). Statistical inference, *J. Roy. Statist. Soc. B* **11**, 115-149 (with discussion)

BERKSON, J. (1942). Tests of significance considered as evidence. *J. Amer. Statist. Assoc.* **37**, 325-335.

BERNARDO, J. (1979). Expected information as expected utility, *Ann. Statist.* **7**, 686-690.

BOLTZMANN, L. (1964). *Lectures on Gas Theory*, Berkeley: University of California Press. Trans. by Stephen G. Brush from the German, *Gastheorie* of 1896-1898.

CARNAP, R. (1947). On the application of inductive logic, *Philosophy and Phenomenological Research* **8**, 133-148.

DUHAMEL, J.M.C. & REYNAUD, A.A.L. (1823). *Problèmes et développements sur diverses parties des mathématiques*. Paris.

FISHER, R.A. (1925). Theory of statistical estimation, *Proceedings of the Cambridge Philosophical Society*, **22**, 700-725.

FISHER, R.A. & YATES, F. (1938). *Statistical Tables for Biological, Agricultural, and Medical Research*. Edinburgh: Oliver and Boyd.

GIBBS, J.W. (1902). *Elementary Principles in Statistical Mechanics*. London: Constable; Dover reprint, 1960.

GOOD, I.J. (1950). *Probability and the Weighing of Evidence*. London: Charles Griffin; New York: Hafners.

— (1958). Significance tests in parallel and in series, *J. Amer. Statist. Assoc.*, **53**, 799-813.

— (1961). Weight of evidence, causality, and false-alarm probabilities, *Information Theory, Fourth London Symposium* (1960). London: Butterworth.

— (1962a). Subjective probability as the measure of a non-measurable set, *Logic, Methodology, and Philosophy of Science* (Nagel, E., Suppes, P., Tarski, A., eds.) California: Stanford University Press. Reprinted in *Studies in Subjective Probability*, 2nd. edn. (H.E. Kyburg & H. E. Smokler, eds.; Huntington, N.Y.: R.E. Krieger), 133-146; and in Good (1983e), 73-82.

— (1962b). Review of Harold Jeffreys *Theory of Probability*, Third edn. London: Oxford University Press. *The Geophysical Journal of the Royal Astronomial Society*, **6**, 555-558. Also in *J. Roy. Statist. Soc. Ser. A*, **125**, 487-489.

— (1967). On the principle of total evidence, *British Journal for the Philosophy of Science* **17**, 319-321.

— (1968). Corroboration, explanation, evolving probability, simplicity, and a sharpened razor, *British Journal for the Philosophy of Science*, **19**, 123-143.

— (1969). What is the use of a distribution?, *Multivariate Analysis II* (ed. P.R. Krishnaiah); New York: Academic Press, 183-203.

— (1974). A little learning can be dangerous, *British Journal for the Philosophy of Science* **25**, 340-342.

— (1980). Another relationship between weight of evidence and errors of the first and second kinds, C67 in *Journal of Statist. Comput. & Simul.* **10**, 315-316.

— (1981a). Some logic and history of hypothesis testing, in *Philosophy in Economics* (ed. Joseph C. Pitt); Dordrecht: D. Reidel, 149-174. Also in Good (1983e), 127-148.

— (1981b). An error by Peirce concerning weight of evidence. *Journal of Statist. Comput. & Simul.* **13**, 155-157.

— (1982e). Comment on a paper by G. Shafer. *J. Amer. Statist. Assoc.* **77**, 342-344.

— (1982b). Standardized tail-area probabilities, C**140** in *Journal of Statist. Comput. & Simul.* **16**, 65-66.

— (1983a). The diminishing significance of a fixed *P*-value as the sample size increases: a discrete model, C**144** in *Journal of Statist. Comput. & Simul.* **16**, 312-314.

— (1983b). Review article of Paul Horwich, *Probability and Evidence*, Cambridge University Press, 1982; *British Journal for the Philosophy of Science,* **35**, 161-166.

— (1983c). Review of "The Maximum Entropy Formalism", Raphael D. Levine & Myron Tribus, eds. (1979). *J. Amer. Statist. Assoc.,* **78**, 987-989.

— (1983d). When are free observations of positive expected value?, C**161** in *Journal of Statist. Comput. & Simul.,* **17**, 313-315.

— (1983e). *Good Thinking: The Foundations of Probability and its Applications.* University of Minnesota Press.

— (1983f). Moments and cumulants of weights of evidence, C**162** in *J. Statist. Comput. & Simul.,* **17**, 315-319; **18**, 85.

GOOD, I.J. & TOULMIN, G.H. (1968). Coding theorems and weight of evidence, *J. Inst. Math. Applics.* **4**, 94-105.

GREENWOOD, J.A. (1938). An empirical investigation of some sampling problems, *Journal of Parapsychology* **2**, 222-230.

HARDY, G.H., LITTLEWOOD, J.L. & PÓLYA, G. (1934). *Inequalities.* Cambridge: University Press.

HILL, B.M. (1982). Comment on a paper by G. Shafer. *J. Amer. Statist. Assoc.* **77**, 344-347.

HORWICH, P. (1982). *Probability and Evidence.* Cambridge: University Press.

HUXLEY, T.H. (1878). *Physiography: an Introduction to the Study of Nature.* London: Macmillan; New York: D. Appleton. 2nd edn.

JAYNES, E.T. (1957). Information theory and statistical mechanics, *Physical Review* **106**, 620-630; **108**, 171-190.

JEFFREYS, H. (1936). Further significance test, *Proceedings of the Cambridge Philosophical Society,* **32**, 416-445.

JEFFREYS, H. (1939/1948/1961). *Theory of Probability.* Oxford: Clarendon Press.

— (1946). An invariant form for the prior probability in estimation problems, *Proceedings of the Royal Society, A.,* **186**, 453-461.

KEMENY, J.C. & OPPENHEIM, P. (1952). Degrees of factual support, *Philosophy of Science,* **19**, 307-324.

KULLBACK, S. (1959). *Information Theory and Statistics.* New York: Wiley.

LEVI, I. (1982). Private communication, December 9.

MACKIE, J.L. (1963). The paradox of confirmation. *British Journal for the Philosophy of Science* **13**, 265-277.

MINSKY, M. & SELFRIDGE, O.G. (1961). Learning in random nets, in *Information Theory: Fourth London Symposium* (Colin Cherry, ed.). London: Butterworhts, 335-347.

NEYMAN, J. (1925). Su un teorema concernente le cossidetti statistiche sufficienti, *Giorn. Inst. Ital. Attuari,* **6**, 320-334.

PEIRCE, C.S. (1878). The probability of induction, *Popular Science Monthly,* reprinted in *The World of Mathematics,* **2**, (ed. James R. Newman); New York: Simon and Schuster, 1956, 1341-1354.

PETERSON, W.W., BIRDSALL, T.G. & FOX, W.C. (1954). The theory of signal detectability, *Trans. Inst. Radio Engrs. PGIT*-**4**, 171-212.

POPPER, K. (1954). Degree of confirmation, *British J. Philosophy Sc.* **5**, 143-149.

— (1959). *The Logic of Scientific Discovery*. London: Hutchinson.

STEWART, J.I.M. ("Michael Innes") (1944). *The Weight of the Evidence* London: Gollancz; also Harmondsworth, Middlesex, England: Penguin Books, 1961.

TRIBUS, M. (1969). *Rational Descriptions, Decisions and Design*. New York: Pergamon Press.

WRINCH, D. & JEFFREYS, H. (1921). On certain fundamental principles of scientific inquiry, *Philosophical Magazine, Series 6*, **42**, 369-390.

H. RUBIN (*Purdue University*)

Possibly some of my difficulties with philosophy are due to my allergy to horseradish (see Professor Seidenfeld's comments). However, some philosophy is necessary.

The consideration of situations in which the state of nature is highly restricted is necessary to clarify thinking. However, one must resist the temptation, made 99.99 % of the time by users of statistics, to believe the model. There is no conceivable way that I can state my prior or posterior probabilities in the last example in the paper; all that can be said is that it is reasonable (or unreasonable) to *act* as if the results come from Bernoulli trials with probability .5. I completely agree with Professor Good that a "significant difference" is *not* the proper criterion here. If there was a relative frequency of 50.1 % in 10^6 trials, on this evidence I personally would "accept" the hypothesis; with 10^{12} trials I would reject it; and with 10^9 trials I would think hard about the matter.

The model given is, in practice, *never* correct. Thus we can only use the evidence to decide which actions or statements to make. If a hypothesis is broad enough, it can be true; if it is too specific, it must be false, but it may still be appropriate and reasonable to act as if it is true.

T. SEIDENFELD (*Washington University, St. Louis*)

As a philosopher interested in "foundations", I take delight in the opportunity to comment on the papers of our distinguished speakers. Let me preface these remarks, more in the form of questions, with an admission of my perception of the role of philosophy in a session titled "Probability and Evidence". To paraphrase Larisa in Pasternak's *Dr. Zhivago* (chapter 13, §16), philosophy is like horseradish. It is good if taken in small amounts in combination with other things. But it is not good in large amounts by itself. The risk with philosophy, as with horseradish, is the temptation to use ever stronger concentrations to maintain the sensation of that first taste. Soon you are serving up pure horseradish!

Professor Good's savory recipe calls for a dash of philosophy in the form of an explication of "weight of evidence". Explication, you recall, is the business (made into an industry thanks to Carnap) of making clear and precise an inexact concept (the *explicandum*) taken from everyday language. The explicandum has all the obscurity typical of presystematic talk. In explication, the vague explicandum is replaced by an *explicatum* which, though similar to the original notion, must be exact, fruitful and simple. *Explication* is Carnap's (1950) explicatum for the explicandum "philosophical analysis".

Carnap begins his *Logical Foundations* in the hope of providing an explication of "probability". In 600 pages that follow, he struggles to defend the thesis of probability as a logical relation. In so far as Carnap's attempt at explication is not successful, I think it fair to say he does not meet the requirement of *usefulness*. Carnap's effort with logical probability fails to yield productive conceptual tools for reconstructing, e.g. statistical inference. For one, he misses completely the important problem of the "reference class" for direct probability: how do we reconcile information from different statistical "populations" concerning some common "individual"?

I have a parallel concern with Good's explication. His account is exact and, no doubt, simple enough. But how does "weight of evidence" serve a useful purpose in solving problems of inference or decision? Let me argue, briefly, that two natural, candidate roles for an explication of *weight* are not fulfilled by Good's explicatum. Then it will be up to the author, himself, to point out what he intends for his creation.

J.M. Keynes, in chapter 6 of his *Treatise* (1921), raises the subject of weight of evidence along with the caveat that he remains uncertain how much importance to attach to the question. For Keynes, *weight of evidence* cannot be defined by probability as he sees weight monotonically increasing with increasing evidence. To use Keynes' metaphor (p. 77) *weight* measures the sum of favourable and unfavourable evidence whereas probability indicates the difference between these two. Keynes suspected that this hazy notion of *weight* plays a role in decisions separate from the role of probability. I do not think what Keynes had in mind requires a violation of expected utility theory. One interpretation of his query is to ask for a measure of weight of evidence that would help determine when a decision maker has adequate evidence for a (terminal) choice. That is, I propose we understand Keynes's problem with *weight* as his groping for a formulation of the stopping problem to which *weight* would offer the key to a solution.

In discussing the requirement of total evidence he writes,

> Bernoulli's second maxim, that we must take into account all the information we have, amounts to an injunction that we should be guided by the probability of that argument, amongst those of which we know the premisses, of which the evidential weight is greatest. But should not this be reenforced by a further maxim, that we ought to make the weight of our arguments as great as possible by getting all the information we can? It is difficult to see, however, to what point the strengthening of an argument's weight by increasing the evidence ought to be pushed. We may argue that, when our knowledge is slight but capable of increase, the course of action, which will, relative to such knowledge, probably produce the greatest amount of good, will often consist in the acquisition of more knowledge. But there clearly comes a point when it is no longer worth while to spend trouble, before acting, in the acquisition of further information, and there is no evident principle by which to determine *how far* we ought to carry our maxim of strengthening the weight of our argument. A little reflection will probably convince the reader that this is a very confusing problem. (pp. 76-77).

Some sixteen years ago, Good published a philosophical note (1967) in which he, like Keynes before him, connected the requirement of total evidence with the stopping problem. The upshot of that note is the result (also reported in Savage (1954, pp. 125-126)) that procrastination is best when observations are cost-free and not (almost surely) irrelevant. But, that finding as well as the general theory of optimal stopping is tangential to Good's concept of *weight*. Of course, with enough *weights* we recover the likelihood function. Hence, the *weights* are sufficient (though hardly a *reduction* of the data). Except in special cases, however, the stopping rule is not a function merely of the *weights*. Is there some reason to think Keynes was on the right track when he posited weight of evidence to solve optimal stopping? It seems to me current wisdom would label this a dead-end approach. Nor does Good's explicatum serve such a purpose.

A second role weight of evidence might conceivably play is in fixing belief. When is it reasonable to add a consistent belief on the basis of new evidence? An informal reply is: you are justified in coming to believe a proposition when the weight of the new evidence is strong enough in its favor. Unfortunately, it seems Good's explicatum does nothing to defend this intuition.

The point is simple. Total evidence requires we respect equivalences implied by all we know. If $\mathbf{h}_1$ and $\mathbf{h}_2$ are equivalent given all our evidence, then whatever epistemic stance we take toward the one we take toward the other. To believe the one is to believe the other. But *weight of evidence* (here, of a kind with relevant measures of "support") does not conform to the needed invariance.

For example, let $X_i = 0,1$ ($i = 1,2$) be two Bernoulli trials. Suppose the personal probability is symmetric and exchangeable and satisfies: $p(X_i = 0) = .5$ and $p(X_1 + X_2 = 0) = .05$. Hence, $p(X_1 + X_2 = 2) = .05$ and $p(X_1 + X_2 = 1) = .9$. Let $\mathbf{e}$ be the new evidence: $X_1 = 1$.

Let $\mathbf{h}_1$ be the hypothesis that $X_1 + X_2 = 2$ and $\mathbf{h}_2$ the hypothesis that $X_2 = 1$. Given $\mathbf{e}$, $\mathbf{h}_1$ and $\mathbf{h}_2$ are equivalent. But $\mathbf{e}$ has positive *weight* for $\mathbf{h}_1$ and negative *weight* for $\mathbf{h}_2$. If we use *weight* to account for our presystematic talk (weight measures reason for/against adding belief), then we have the incoherent conclusion that, given all we know, $\mathbf{e}$ is evidence for and against the same belief. It is an elementary and familiar exercise to show this phenomenon ubiquitous.

In a recent paper with D.Miller, Sir Karl Popper (1983) expresses concern over failure of "positive relevance" to respect such equivalences. Thus, I do not agree with Good (p. 8) when he speculates that *weight* satisfies Popper's desiderata for degree of corroboration or acceptability.

In short, my question to Professor Good is this one. What shall I do with *weight of evidence*?

REPLY TO THE DISCUSSION

Teddy Seidenfeld was kind enough to send a copy of his comments before the meeting. His main question was "What shall I do with weight of evidence". I think there must be some misunderstanding because my answer is so simple. My answer is that the weight of evidence provided by E should be added to the initial log-odds of the hypothesis to obtain the final log-odds. Or equivalently, the Bayes factor is multiplied by the initial odds to give the final odds. The final odds are then combined with utilities to make rational decisions. Weights of evidence and Bayes factors resemble likelihood, they have the merit of being independent of the initial probability of the hypothesis. Moreover the technical meaning of weight of evidence captures the ordinary linguistic meaning and that is my main thesis.

In the example, used by Seidenfeld to question the explication of weight of evidence, he had effectively the table of probabilities,

X_1 \ X_2	0	1	
0	.05	.45	.5
1	.45	.05	.5
	.5	.5	1

with $H_1 \colon X_1 = X_2 = 1$ $\qquad P(H_1) = .05,\ O(H_1) = 1/19$

$H_2 \colon X_2 = 1$ $\qquad P(H_2) = ½,\ O(H_2) = 1$

$E : X_1 = 1$ $\qquad$ (H_1 and H_2 are logically equivalent, given E).

Bayes factor provided by E		*Initial odds*	*Final odds*
H_1	19/9	1/19	1/9
H_2	1/9	1	1/9

The fact that the final odds of H_1 and H_2 are equal, given E, is consistent with the fact that H_1 and H_2 are equivalent given E. The evidence E, that $X_1 = 1$, supported H by increasing its odds to 1/9, and undermined H_2 by decreasing its odds to 1/9. Before E was known, H_1 and H_2 were not equivalent and their initial odds were not equal. The ocurrence of E has simply changed the situation. Seidenfeld seems to have confused $W(H{:}E)$ with $W(H{:}E|E)$. The latter expression is equal to zero. That is, once E is *given* it supplies no further weight of evidence. To imagine that it does is like trying to double the true weight of evidence. The error is prevented by noticing the distinction between the vertical stroke which means "given" and the colon which means "provided by".

Popper made a different mistake when he apparently equated corroboration with acceptability, and when I said that his remark about corroboration had been previously taken for granted I should have made it clear that I was referring only to his statement that degree of corroboration cannot be a probability. My definition of weight of evidence does essentially satisfy all the desiderata for corroboration laid down by Popper in the Appendix dealing with the topic in his *Logic of Scientific Discovery* (Popper, 1959, pp. 400-401). The meaning of "essentially" here is spelt out in Good (1960, p. 321); for example, I replace Popper's bounds of ± 1 on degree of corroboration, by $\pm\infty$. Perhaps Popper has since shifted his position.

The alleged proof by Popper & Miller (1983) of the impossibility of inductive probability is unconvincing, and I have written a note to *Nature* arguing this (Good, 1983h).

A special case of weight of evidence was used by Peirce (1878) although he did not express it in Bayesian terms; in fact, as Isaac Levi has pointed out, Peirce anticipated the Neyman-Pearson theory to some extent. Incidentally, when a Neyman-Pearsonian asserts a hypothesis H, he unwittingly provides a Bayes factor of $(1-\alpha)/\beta$ in favour of H; and, when he rejects H, he similarly provides a Bayes factor of $(1-\beta)/\alpha$ against H. (See Good, 1983g; Wald, 1947, p. 41). These results are based on the assumption that we know the values of α and β, and we know what recommendation is made by the Neyman-Pearsonian, and nothing else. We can achieve this state of ignorance by employing a Statistician's Stooge who, by definition, is shot if he tells us more than we ask him to.

I have referred to practical applications in my paper, such as to sequential analysis, an example of which was Banburismus. The Bayes factor is also used throughout Harold Jeffreys's book on probability, though he nearly always assumes that the initial odds are 1. Every Bayesian test of a hypothesis can be regarded as an application of the concept of weight of evidence. Perhaps the most important applications, like those of probability, are the semiquantitative ones in the process of rational thinking as an intelligence amplifier.

Keynes's definition of weights of arguments, in which he puts all the weights in one scale, whether they are positive or negative, is like interpreting weight of evidence as the weight of the documents on which they are printed. I think, if not horseradish, it is at least a crummy concept in comparison with the explicatum of weight of evidence that I support. Keynes himself said of his discussion (1921, p. 71) "... after much consideration I remain uncertain as to how much importance to attach to it. The magnitude of the probability of an argument ... depends upon a *balance* between what may be termed the favourable and the unfavourable evidence...". In other words he clearly recognizes that Themis is right to use both scales. It is a standard English expression that the weight of evidence favours such and such. Of course this refers to the *balance* of the evidence, *not* to the sum of all the pieces irrespective of their signs.

If you *must* have a quantitative interpretation of Keynes's "weight of arguments", just compute the weights of evidence in my sense for each "piece" of evidence and add their absolute values. This then is yet another application of my explicatum, to give a somewhat quantitative interpretation to the crummy one. But Keynes's discussion of this matter is purely qualitative.

Seidenfeld raises the question of whether weight of evidence can be used for deciding when to stop experimentation. My answer is that weight of evidence is only a quasiutility, as I stated in my paper. When you have a large enough weight of evidence, diminishing returns set in, where the meaning of "large enough" depends on the initial probability and on the utilities. Weight of evidence is a good quasiutility, and it is fine that the expected weight of evidence from an observation is nonnegative. But it cannot entirely replace expected utility as a stopping rule. When a judge's estimate of the odds that an accused person is guilty or innocent reaches a million-to-one on, the judge is apt to say "Finis" and bang down his gavel. This is because, in his implicit or explicit opinion, the expected gain in utility from seeking new evidence is not worth the expected time it would take to acquire.

I turn now to the public comments made by Herman Rubin. He stated that $P(E|H)$ is not well defined if H is a composite hypothesis. This is certainly true in non-Bayesian statistics but in "sharp" Bayesian statistics it is assumed to have a sharp value. For the sake of simplicity most of my exposition was based on the sharp Bayesian position. My aim was to discuss weight of evidence without going into the foundations of probability.

Dr. Rubin mentioned that, in my example of sampling with replacement from a bag of 100 black and white balls, if he obtained 50.1 % white drawings in a sample of a trillion, he would reject the model. So would I, but my example was based on a fixed P-value of 0.045. I deliberately selected not too small a P-value so that the model itself would not come under suspicion.

I agree further that precise models are seldom exact, but they are often useful on grounds of simplicity. Compare, for example, Good, 1950, p. 90; 1983e, p. 135.

REFERENCES IN THE DISCUSSION

CARNAP, R. (1950). *Logical Foundations of Probability*. Chicago: University Press.

GOOD, I.J. (1952). Rational decisions. *J. Roy. Statist. Soc. B.,* **14**, 107-114.

— (1955). Contribution to the discussion on the Symposium on Linear Programming. *J. Roy. Statist. Soc., B,* **17**, 194-196.

— (1960). Weight of evidence, corroboration, explanatory power, information, and the utility of experiments. *J. Roy. Statist. Soc., B,* **22**, 319-331; **30** (1968), 203.

— (1963). Maximum entropy for hypothesis formulation, especially for multidimensional contingency tables. *Ann. Math. Statist.* **34**, 911-934.

— (1967). On the principle of total evidence. *British Journal for the Philosophy of Science,* **17**, 319-321.

— (1980). Some history of the hierarchical Bayesian methodology. *Bayesian Statistics* (Bernardo, J.M., DeGroot, M.H., Lindley, D.V. and Smith, A.F.M. eds.) Valencia: University Press, 489-510 & 512-519 (with discussion).

— (1983g). A correction concerning my interpretation of Peirce, and the Bayesian interpretation of Neyman-Pearson 'hypothesis determination'", C**165**, *J. Statist. Comput. & Simul.* **18**, 71-74.

— (1983h). The inevitability of probabilistic induction. *Nature* **310**, 434.

KEYNES, J.M. (1921). *A Treatise on Probability*. London: Macmillan.

POPPER, K. and MILLER, D. (1983). A proof of the impossibility of inductive probability. *Nature* **302,** 687-688.

SAVAGE, L.J. (1954). *The Foundation of Statistics*. New York: Wiley.

SEIDENFELD, T. (1979). Why I am not an objective Bayesian: some reflections prompted by Rosenkrantz. *Theory and Decision 11,* 413-440.

WALD, A. (1947). *Sequential Analysis*. New York: Wiley.

BAYESIAN STATISTICS 2, pp. 271-298
J.M. Bernardo, M.H. DeGroot, D.V. Lindley, A.F.M. Smith (Eds.)

Normal Discount Bayesian Models

J.R.M. AMEEN and P.J. HARRISON
University of Warwick, U.K.

SUMMARY

Normal Discount Bayesian Models (N.D.B.M.'s) are introduced in order to overcome practical disadvantages associated with Dynamic Linear Models. A discount vector replaces the D.L.M. system variance matrix; introduces conceptual simplicity; removes much ambiguity; and provides a system transition model which is invariant to the measurement scale of independent and control variables. The necessary operational recurrence equations are given and a relationship with D.L.M.'s established. For Constant models, limiting results are derived and a correspondence obtained between the limiting forecast distributions of N.D.B.M.'s and Constant D.L.M.'s as well as between their limiting Forecast Functions and those of A.R.I.M.A. processes. A number of applications are discussed.

Keywords: BAYESIAN FORECASTING; DISCOUNT WEIGHTED ESTIMATION; DYNAMIC LINEAR MODELS; EXPONENTIALLY WEIGHTED REGRESSION; INTERVENTION; KALMAN FILTERING; MULTIPROCESS MODELS; NORMAL DISCOUNT BAYESIAN MODELS; NORMAL WEIGHTED BAYESIAN MODELS.

1. INTRODUCTION

1.1. *General*

A bright future for the application of Bayesian forecasting, Dynamic Linear Models and Kalman filtering was predicted by Harrison and Stevens (1971, 1975, 1976). All the facilities seemed to be provided. The approach uses mathematics and statistics as a language. Given a logically consistent description of the way in which a forecasting method should deal with a wide class of situations then a model can be constructed to satisfy the requirement. The models are structured. This facilitates meaning and provides a natural way of blending subjective information and data. The principle of superposition enables an easy way of building linear models from separate components. The structuring also leads to robust models in which the information contained in one model component is protected against a major disturbance affecting another component. Initially the models can operate with little or no data, estimate model parameters on-line; discriminate between rival models as in multi-process models class I; and model major changes using multi-process models class II. Further the models include all the well known A.R.I.M.A. models, regression models, and statistical linear Normal models as special cases. In fact one of the attractions of the approach is that it does not revolutionise, in the sense of wishing to overthrow existing methods. Instead it extends these methods by offering a reformulation which introduces many facilities which are not elegantly available in the original classical formulations. In view of the prospective and now largely realised developments in micro computing the future of Bayesian forecasting seemed assured.

It must be acknowledged that, in practice, the methods have not been adopted as quickly or as widely as anticipated. One reason is the lack of relevant literature and software. Another is that the Bayesian approach asks practitioners to use mathematics and statistics descriptively. This requires skill and practice. Unfortunately descriptive mathematics is often neglected in educational curriculums. Hence the attractions of black box methods. Mechanical methods are limited and further progress in modelling demands a more thoughtful approach. Where the need has been great Bayesian methods have proved very successful. But where the motivation has been weak many practitioners have not been prepared to devote sufficient effort. They have been content to use less powerful and flexible models.

The favourite applied Bayesian models are Normal Dynamic Linear Models (D.L.M.'s). Practitioners need to be well aquainted with parametric modelling and with state space stochastic difference equations. This requires modellers to be familiar with the Normal probability distribution representation of innovations. The specification of the associated system variance matrices has proved a major obstacle. Even experienced people find that they have little natural quantitative feel for the elements of these matrices.

The object of this paper is to introduce a class of Normal Discount Bayesian Models founded upon the discount concept. The innovation or system variance matrix is replaced by a set of discount factors. This has a number of advantages which make the models preferable to practitioners. In particular many modellers have a natural feel for discounting. Furthermore once the discount factors are chosen, established methods for on-line estimation of the observation variance may be applied. This is important since experience has also shown that few practitioners are able to properly assess this variance. They confuse it with the one step ahead forecast error variance. It is expected that the introduction of the discount concept will help considerably in furthering applications of Bayesian forecasting.

1.2 *The Discount Concept*

Two desirable properties of applied mathematical models are ease of application, and conceptual parsimony. Hence the attraction of discount factors in methods of sequential estimation. A previous paper, Ameen and Harrison (1984) defines the method of Discount Weighted Estimation (D.W.E.) which generalizes the estimation method of Exponentially Weighted Regression (E.W.R.) promoted by Brown (1963). In the simplest situation a single discount factor β describes the rate at which information is lost with time so that, if the current information is now worth I units, then its worth with respect to a period k steps ahead is β^k I units. However if a system has numerous characteristics then the discount factors associated with particular components may be required to take different values. The D.W.E. method provides a means of doing this but it is strictly a point estimation method.

Generally decision makers require information and forecasts as probability distributions or support functions and the major objective of this paper is to provide Bayesian Forecasting methods founded upon the discount concept. This concept has already been successfully applied in the I.C.I. forecasting package MULDO, Harrison (1965) and Harrison and Scott (1965), the I.C.I. Multivariate Hierarchical forecasting package, Harrison, Leonard and Gazzard (1977), and in other applications. However, the former is a particular method which does not easily generalise, Whittle (1965), and the other applications have been based upon Constant Dynamic Linear Models (D.L.M.'s) which have limiting forecast functions equivalent to those derived using E.W.R., Godolphin and Harrison (1975), Harrison and Akram (1983). The use of such models has involved practitioners specifying a system variance matrix **W** which has elements that are proportional to functions of a discount factor. They are thus indirect applications of discounting. Practical problems arise because of the non-uniqueness of **W** and because the lack of familiarity with such matrices causes application difficulties and leads practitioners to other methods. Non-uniqueness arises since there exists an uncountable number of time shift reparameterisations which have identical forecast distributions. Further, in the general application of D.L.M.'s the **W** matrix is not invariant to the scale on which the independent variables are measured and practitioners experience difficulty in applying even simple regression type D.L.M.'s.

This paper develops a class of Normal Discount Bayesian Models (N.D.B.M.'s) which eliminates the system variance matrix **W**. Instead a discount vector is introduced which associates possibly different discount factors with different model components. Such a discount factor converts its components posterior precision P_{t-1} at time, t-1, to a prior precision $P_t^* = \beta P_{t-1}$ for time t. The term precision is used in its Bayesian sense but may also be thought of as a Fisherian measure of information.

The use of discount factors overcomes the major disadvantages of the system variance **W**, since ambiguity is removed, the discount vector is invariant to the scale on which both the independent and dependent variables are measured, and the methods are easily applied. Because of the conceptual simplicity and ease of operation it is anticipated that the N.D.B.M. approach will find many applications in dynamic regression, forecasting, time series analysis, the detection of changes in process behaviour, quality control and in general statistical modelling where the observations are performed sequentially or ordered according to some index.

1.3. *Schemata*

Section 2 briefly describes the estimation method of Discount Weighted Estimation (D.W.E.). After defining Dynamic Linear Models (D.L.M.'s) a relationship between the two is established in terms of identical Forecast functions. An extensive class of Normal Weighted Bayesian Models (N.W.B.M.'s) is introduced in section 3. This class includes all Normal D.L.M.'s. However attention is centred upon a particular subset of models founded on the discount concept. This set of Normal Discount Bayesian Models (N.D.B.M.'s) is defined and all the necessary operations for forecasting and updating are given. The relatioship between N.D.B.M.'s and Exponentially Weighted Regression (E.W.R.), D.W.E. and D.L.M's is discussed. One method for the on-line estimation of the observation variance is given.

The class of Constant N.D.B.M.'s is examined in section 4. After defining similar models and a method of transforming from one model to any other similar model, limiting results are derived. These are not only of theoretical interest but are also of practical use since convergence is often fast. In studying Constant D.L.M.'s, both statisticians and control engineers have found difficulty in obtaining limiting results because of the troublesome Riccati equations. However, for N.D.B.M.'s and thus for the set of Constant D.L.M.'s which have equivalent Forecast functions, the results are obtained directly. In particular the limiting parametric pecision matrices are derived. This immediately gives the limiting value of the adaptive or gain vector and all the limiting parametric and forecast distributions. A limiting relationship between the observations and the one step ahead forecast errors is given for any Constant N.W.B.M. This enables a correspondence to be established between a subset of Constant N.W.B.M.'s and A.R.I.M.A. models.

Section 5 is concerned with the application of N.D.B.M.'s and N.W.B.M.'s. The construction of models is briefly discussed and then a Constant N.D.B.M. is applied to

Irish Turkey Poult sales. In routine operation an intervention facility would be provided. In order to illustrate one way in which discount factors can be used effectively to 'age' the past data with respect to some model components but not others, an artificial series was generated by the superposition of a stochastic linear trend; a stochastic harmonic of period 12; Normal white noise; and major disturbances. The analysis is given in 5.3. Two useful types of N.W.B.M. based upon N.D.B.M.'s are Modified N.D.B.M.'s, discussed in 5.4 and Extended N.D.B.M.'s, discussed in 5.7. Practical procedures for estimating the observation variance on-line are discussed in 5.5. In particular, slowly changing variances and variance power laws are discussed. A brief discussion of N.D.B.M.'s in relationship to the analysis of designed experiments is given in 5.6. An application to multi-process modelling is given using prescription data. This illustrates the automatic treatment of outliers and sharp trend changes. All the data sets are given in an Appendix.

1.4. *Related Work*

Much related work has been done in the fields of control engineering, statistics, econometrics and operational research. In discounting, the work of Brown (1963) is basic and other authors are Jones (1966), Morrison (1969), Gilchrist (1967), Godolphin and Harrison (1975), McKenzie (1976), Brown, Durbin and Evans (1975), Harrison and Akram (1983) and Durbin (1983). In Bayesian modelling and forecasting the books of DeGroot (1970), Box and Tiao (1973), Zellner (1971) and Aitchison and Dunsmore (1975) are to be mentioned as are the works of Lindley and Smith (1972), Smith and West (1983) and Smith (1977, 1979, 1981). Other work of note on dynamic and sequential modelling is that of Plackett (1950), Astrom (1970), Young (1974) and Whittle (1969). The filter algorithm of Kalman (1963) is widely used with dynamic models and, although there are major interpretive differences from the Bayes parametric models, state space models are related, Priestley (1980). Although this paper is written in Bayesian terms, the approach may be considered as a special case of Information Forecasting, Ameen and Harrison (1983a), in which the Likelihood and Support ideas of Fisher (1950), Edwards (1972), Barnard (1951) and Barnard, Jenkins and Winsten (1962) have been influential.

2. D.W.E. AND NORMAL D.L.M.'s

2.1. *Discount Weighted Estimation (D.W.E.)*

D.W.E. is the subject of the paper by Ameen and Harrison (1984). The triple $\{\mathbf{F}_t, \mathbf{G}, \mathbf{B}\}$ defines the D.W.E. method for each time period such that the local model for an outcome Y_t is

$$Y_t = \mathbf{F}_t \theta + \epsilon_t$$

where Θ is a parameter vector and the random variable ϵ_t has zero mean and is independent of the data set
$D_{t-1} = \{(y_{t-1}, \mathbf{F}_{t-1}),\ldots,(y_1,\mathbf{F}_1), D_0\}$. The Forecast Function is

$$F_t(k) = E\left[Y_{t+k} \mid D_t, \mathbf{F}_{t+k}\right] = \mathbf{F}_{t+k}\,\mathbf{G}^k\,\mathbf{m}_t$$

where, for a given discount matrix $\mathbf{B} = \text{diag}(\beta_1,\beta_2\ldots\beta_n)$, $0<\beta_i\leq 1$ $(i=1\ldots n)$, the estimate $\mathbf{m}_t$ of θ given D_t may be obtained from the recurrence relationships

$$\mathbf{m}_t = \mathbf{G}\,\mathbf{m}_{t-1} + \mathbf{A}_t\, e_t$$

$$\mathbf{R}_t = \mathbf{B}^{-1/2}\,\mathbf{G}\,\mathbf{Q}_{t-1}^{-1}\,\mathbf{G}'\,\mathbf{B}^{-1/2}$$

$$\mathbf{A}_t = \mathbf{R}_t \mathbf{F}_t'/(1+\mathbf{F}_t\mathbf{R}_t\mathbf{F}_t')$$

$$\mathbf{Q}_t^{-1} = (\mathbf{I}-\mathbf{A}_t \mathbf{F}_t)\,\mathbf{R}_t$$

$$e_t = y_t - \mathbf{F}_t \mathbf{G}\,\mathbf{m}_{t-1}$$

No matrix inversions are required using these relationships but the notation $\mathbf{Q}_t^{-1}$ arises from the alternative recurrences for $(\mathbf{A}_t, \mathbf{Q}_t)$ which although not as practically useful, are more easily derived as

$$\mathbf{Q}_t = \mathbf{B}^{1/2}(\mathbf{G}^{-1})'\mathbf{Q}_{t-1}\mathbf{G}^{-1}\mathbf{B}^{1/2} + \mathbf{F}_t\mathbf{F}_t',$$

$$\mathbf{A}_t = \mathbf{Q}_t^{-1}\mathbf{F}_t'$$

where $\mathbf{G}$ is generally of full rank. Further, comparison with Exponentially Weighted Regression (E.W.R.) recurrences shows that the D.W.E. $\{\mathbf{F}_t, \mathbf{G}, \beta\,\mathbf{I}\}$ method is identical to that of E.W.R.

A Constant D.W.E. method $\{\mathbf{F}, \mathbf{G}, \mathbf{B}\}$, in which $\mathbf{F}$, $\mathbf{G}$ and $\mathbf{B}$ are constant for all t, represents what is called the time series method. Let $\mathbf{G} = \text{diag}(\mathbf{G}_1,\ldots,\mathbf{G}_r)$ be in Jordan form with λ_i being the eigenvalue of $\mathbf{G}_i$ and write $\mathbf{I}_i$ as the identity matrix of the same rank as $\mathbf{G}_i$. Then if $\mathbf{B} = \text{diag}(\beta_1\,\mathbf{I}_1,\ldots,\beta_r\mathbf{I}_r)$, provided $0<\beta_i<|\lambda_i|^2$ for $i=1,\ldots,r$, $\lim_{t\to\infty} \mathbf{Q}_t = \mathbf{Q}$ and $\lim_{t\to\infty} \mathbf{A}_t = \mathbf{A} = \mathbf{Q}^{-1}\mathbf{F}'$ both exist.

It will be evident from section 3 that the D.W.E. method generalizes to triples $\{\mathbf{F}, \mathbf{G}, \mathbf{B}\}_t$ and to the case of observation vectors.

2.2. *Normal Dynamic Linear Models (N.D.L.M.'s)*

A D.L.M. is defined by the quadruples $\{\mathbf{F}, \mathbf{G}, V, \mathbf{W}\}_t$ for each index t, Harrison and Stevens (1976). For univariate observations the interpretation is

$$Y_t = \mathbf{F}_t\theta_t + v_t \qquad v_t \sim N[0; V_t]\,,$$

$$\theta_t = \mathbf{G}_t\theta_{t-1} + \mathbf{w}_t \qquad \mathbf{w}_t \sim N[\mathbf{0}; \mathbf{W}_t]$$

where, usually it is assumed that $\mathbf{F}_t$, $\mathbf{G}_t$, V_t and $\mathbf{W}_t$ are known; that, without loss in generality, v_t and $\mathbf{w}_t$ are independent random vectors; and that θ is an $n\times 1$ parameter vector. A D.L.M. for which $\{\mathbf{F}, \mathbf{G}\}$ is constant is defined as observable if and only if $[\mathbf{F}', (\mathbf{F}\,\mathbf{G})',\ldots,(\mathbf{F}\,\mathbf{G}^{n-1})']$ is of rank n. If V and $\mathbf{W}$ are also independent of time the model is called a Constant D.L.M.

Letting $D_t = \{y_t, D_{t-1}\}$, then, if initially $(\theta_0|D_0) \sim N[\mathbf{m}_0;\mathbf{C}_0]$ it follows that $(\theta_t|D_t) \sim N[\mathbf{m}_t;\mathbf{C}_t]$. Defining $\hat{y}_t = E[Y_t|D_{t-1}] = \mathbf{F}_t\mathbf{G}_t\mathbf{m}_{t-1}$, $\hat{Y}_t = \text{Var}[Y_t|D_{t-1}] = \mathbf{F}_t\mathbf{R}_t\mathbf{F}_t' + V_t$ where $\mathbf{R}_t = \text{Var}[\theta_t|D_{t-1}] = \mathbf{G}_t\mathbf{C}_{t-1}\mathbf{G}_t' + \mathbf{W}_t$, $\mathbf{A}_t = \mathbf{C}_t\mathbf{F}_t'V_t^{-1}$ and $e_t = y_t - \hat{y}_t$, the natural Bayesian recurrence relationships are

$$\mathbf{m}_t = \mathbf{G}_t\mathbf{m}_{t-1} + \mathbf{A}_t e_t$$

$$\mathbf{C}_t^{-1} = \mathbf{R}_t^{-1} + \mathbf{F}_t'V_t^{-1}\mathbf{F}_t.$$

The Forecast function $F_t(\bullet)$ is defined for $k>0$ by

$$F_t(k) = E[Y_{t+k}|D_t] = \mathbf{F}_{t+k}\prod_{i=1}^{k}\mathbf{G}_{t+i}\,\mathbf{m}_t\,.$$

However, in practice the variance recurrence is better expressed in the equivalent conditional Normal form

$$\mathbf{C}_t = (\mathbf{I} - \mathbf{A}_t\mathbf{F}_t)\mathbf{R}_t$$

with $\mathbf{A}_t = \mathbf{R}_t\,\mathbf{F}_t'\,\hat{Y}_t^{-1}$.

An alternative expression for $\mathbf{C}_t$ is $\mathbf{C}_t = \mathbf{R}_t - \mathbf{A}_t\,\hat{\mathbf{Y}}_t\,\mathbf{A}_t'$ which is a representation of the variance recurrence relation in the filter of Kalman (1963).

2.3. *A Relationship between D.W.E. and N.D.L.M.'s*

The recurrence relationships of 2.1 and 2.2 suggest a connection between estimation using D.W.E. and estimation using a D.L.M. Consider D.W.E. $\{\mathbf{F}, \mathbf{G}, \mathbf{B}\}_t$ with prior setting $(\mathbf{m}_0, \mathbf{Q}_0)$ where $\mathbf{Q}_0$ is a full rank precision-type matrix. Let $\mathbf{H}_t = \mathbf{B}_t^{-1/2}\,\mathbf{G}_t$. Define any D.L.M. $\{\mathbf{F}_t, \mathbf{G}_t, V, \mathbf{W}_t\}$ for which $\mathbf{W}_t = (\mathbf{H}_t\,\mathbf{Q}_{t-1}^{-1}\,\mathbf{H}_t' - \mathbf{G}_t\,\mathbf{Q}_{t-1}^{-1}\,\mathbf{G}_t')\,V$ as a Corresponding D.L.M.

2.3.1. *Theorem*

Let $(\theta_0\,|\,D_0) \sim N[\mathbf{m}_0;\ \mathbf{Q}_0^{-1}\,V]$. Then for D.W.E. $\{\mathbf{F}, \mathbf{G}, \mathbf{B}\}_t$, all Corresponding D.L.M.'s produce Forecast Functions identical to those obtained by D.W.E. Further $(\theta_t\,|\,D_t) \sim N[\mathbf{m}_t;\ \mathbf{C}_t]$ where $\mathbf{m}_t$ is the D.W.E. estimate and $\mathbf{C}_t = \mathbf{Q}_t^{-1}\,V$.

Proof: Using induction, suppose true for t-1. From the D.L.M. results in 2.2

(i) $\mathbf{R}_t = \mathbf{G}_t\,\mathbf{Q}_{t-1}^{-1}\,\mathbf{G}_t'\,V + \mathbf{W}_t = \mathbf{H}_t\,\mathbf{Q}_{t-1}^{-1}\,\mathbf{H}_t'\,V$ and since $\mathbf{C}_t^{-1} = \mathbf{R}_t^{-1} + \mathbf{F}_t'\,\mathbf{F}_t\,V^{-1}$

$$= [(\mathbf{H}_t^{-1})'\,\mathbf{Q}_{t-1}\,\mathbf{H}_t^{-1} + \mathbf{F}_t'\mathbf{F}_t]\ /\ V = \mathbf{Q}_t/V \text{ we have } \mathbf{C}_t = \mathbf{Q}_t^{-1}\,V.$$

(ii) $E[\theta_t\,|\,D_t] = \mathbf{G}_t\,\mathbf{m}_{t-1} + \mathbf{A}_t\,e_t = \mathbf{m}_t$ the D.W.E. estimate since

$$\mathbf{A}_t = \mathbf{C}_t\,\mathbf{F}_t'/V = \mathbf{Q}_t^{-1}\,\mathbf{F}_t',\ \text{the D.W.E. } \mathbf{A}_t.$$

(iii) the Forecast Function is $E[Y_{t+k}\,|\,D_t] = \mathbf{F}_{t+k}\prod_{i=1}^{k}\mathbf{G}_{t+i}\,\mathbf{m}_t$, as for D.W.E.

2.3.2. *Corollaries*

(i) For $t \geq 0$, the D.W.E. $\{\mathbf{F}_t, \mathbf{G}_t, \mathbf{I}\}$ gives a Forecast Function identical to that of the D.L.M. $\{\mathbf{F}_t, \mathbf{G}_t, V, \mathbf{0}\}$.

(ii) For the Constant D.W.E. $\{\mathbf{F}, \mathbf{G}, \mathbf{B}\}$, if all the eigenvalues of $\mathbf{H}$ lie outside the unit circle and, if the D.L.M. is observable, then as $t\to\infty$, $\text{Lim}\{\mathbf{Q}_t, \mathbf{A}_t, \mathbf{C}_t, \mathbf{R}_t\} = \{\mathbf{Q}, \mathbf{A}, \mathbf{C} = \mathbf{Q}^{-1}\,V, \mathbf{H}\,\mathbf{Q}^{-1}\,\mathbf{H}'\,V\}$ exists and $\text{Lim}\ \mathbf{W}_t = \mathbf{W} = (\mathbf{H}\,\mathbf{Q}^{-1}\,\mathbf{H}' - \mathbf{G}\,\mathbf{Q}^{-1}\,\mathbf{G}')\,V$ exists. This result is used in section 4 but it may be noted that the Corresponding D.L.M. as defined above convergences to a Constant D.L.M. $\{\mathbf{F}, \mathbf{G}, V, \mathbf{W}\}$.

2.3.3. *Comment*

In D.L.M. terms the above setting for $\mathbf{W}_t$ is unusual in its dependence upon $\mathbf{C}_t$, the uncertainty of the observer concerning θ_t given D_t. The concept, that the observers view of the future development of the process depends upon his current information is also adopted in Entropy Forecasting, Souza and Harrison (1977), in Smith (1979) and in Information Forecasting, Ameen and Harrison (1983).

3. NORMAL DISCOUNT BAYESIAN MODELS (N.D.B.M.'s)

3.1. *Normal Weighted Bayesian Models (N.W.B.M.'s)*

The class of N.W.B.M.'s is extensive containing the class of N.D.L.M.'s. A member N.W.B.M. is defined by a quadruple $\{\mathbf{F}, \mathbf{G}, \mathbf{V}, \mathbf{H}\}_t$ for each integer $t>0$. This states that, if the posterior parameter distribution is

$$(\theta_{t-1}|D_{t-1}) \sim N[\mathbf{m}_{t-1}; \mathbf{C}_{t-1}],$$

then the prior distribution is

$$(\theta_t|D_{t-1}) \sim N[\mathbf{G}_t\, \mathbf{m}_{t-1}; \mathbf{R}_t],$$

where $\mathbf{R}_t = \mathbf{H}_t\, \mathbf{C}_{t-1}\, \mathbf{H}_t'$. Further, the observation probability model is

$$(\mathbf{Y}_t|\theta_t) \sim N[\mathbf{F}_t\, \theta_t; \mathbf{V}_t].$$

No restrictions are imposed upon **F**, **G** or **H** but **V** and **C** are variance matrices. It follows that

$$(\theta_t|D_t) \sim N[\mathbf{m}_t; \mathbf{C}_t]$$

with the familiar recurrences most conveniently expressed as

$$\mathbf{m}_t = \mathbf{G}_t\, \mathbf{m}_{t-1} + \mathbf{A}_t\, \mathbf{e}_t;\ \mathbf{C}_t = (\mathbf{I} - \mathbf{A}_t\, \mathbf{F}_t)\, \mathbf{R}_t$$

$$\mathbf{A}_t = \mathbf{R}_t\, \mathbf{F}_t' \hat{\mathbf{Y}}_t^{-1};\ \mathbf{e}_t = \mathbf{y}_t - \mathbf{F}_t\, \mathbf{G}_t\, \mathbf{m}_{t-1};\ \hat{\mathbf{Y}}_t = \mathbf{F}_t\, \mathbf{R}_t\, \mathbf{F}_t' + \mathbf{V}_t.$$

The class of N.W.B.M.'s includes all linear and non-linear models for which the prior and posterior distributions are Normal. If $\mathbf{W}_t = \mathbf{H}_t\, \mathbf{C}_{t-1}\, \mathbf{H}_t' - \mathbf{G}_t\, \mathbf{C}_{t-1}\, \mathbf{G}_t'$ is positive semidefinite then the conditional distribution $(\theta_t|\theta_{t-1}) \sim N[\mathbf{G}_t\theta_{t-1}; \mathbf{W}_t]$ may be introduced to provide coherent lead time forecasts. Thus with setting $\mathbf{H}_t = (\mathbf{G}_t\, \mathbf{C}_{t-1}\, \mathbf{G}_t' + \mathbf{W}_t)^{1/2}\, \mathbf{C}_{t-1}^{-1/2}$ it is evident that any Normal D.L.M. $\{\mathbf{F}, \mathbf{G}, \mathbf{V}, \mathbf{W}\}$ is a N.W.B.M. In this paper the concern is mainly with a subset of models called N.D.B.M.'s and a subset of Modified N.D.B.M.'s.

3.2. *Normal Discount Bayesian Models (N.D.B.M.'s)*

It is evident that the discount concept is simply introduced to Bayesian forecasting by replacing the D.L.M. system equation by a discounted relationship between the posterior parameter precision for θ_{t-1} and the prior parameter precision for θ_t. An N.D.B.M. is defined by a quadruple $\{\mathbf{F}, \mathbf{G}, \mathbf{V}, \mathbf{B}\}_t$ and is a particular N.W.B.M. $\{\mathbf{F}, \mathbf{G}, \mathbf{V}, \mathbf{H}\}_t$ where always $\mathbf{H}_t = \mathbf{B}^{-1/2}\, \mathbf{G}_t = \mathbf{G}_t\, \mathbf{B}^{-1/2}$. Interest will be centred on models structured such that $\mathbf{G}_t = \text{diag}\{\mathbf{G}_1, \mathbf{G}_2 \ldots \mathbf{G}_r\}_t$ has a block diagonal form; the block $\mathbf{G}_i$ is of full rank n_i; and, if $\mathbf{I}_i$ is the identity matrix of dimension n_i, then $\mathbf{B}_t = \{\beta_1\, \mathbf{I}_1, \ldots, \beta_r\, \mathbf{I}_r\}_t$ where $0<\beta_i\leq 1$ for all $i=1\ldots r$. The N.D.B.M. $\{\mathbf{F}, \mathbf{G}, \mathbf{V}, \mathbf{B}\}_t$ states that if

$$(\theta_{t-1}|D_{t-1}) \sim N[\mathbf{m}_{t-1}; \mathbf{C}_{t-1}],$$

then

$$(\theta_t|D_{t-1}) \sim N[\mathbf{G}_t\, \mathbf{m}_{t-1}; \mathbf{R}_t]$$

where $\mathbf{R}_t = \mathbf{H}_t\, \mathbf{C}_{t-1}\, \mathbf{H}_t' = \mathbf{B}_t^{-1/2}\, \mathbf{G}_t\, \mathbf{C}_{t-1}\, \mathbf{G}_t' \mathbf{B}_t^{-1/2}$,

and that the observation probability model is

$$(\mathbf{Y}_t|\theta_t) \sim N[\mathbf{F}_t\, \theta_t; \mathbf{V}_t].$$

3.3. *Forecasting and Updating an N.D.B.M.*

3.3.1 The one-step-ahead forecast is $(\mathbf{Y}_t|D_{t-1}) \sim N[\hat{\mathbf{y}}_t;\hat{\mathbf{Y}}_t]$ where $\hat{\mathbf{y}}_t = \mathbf{F}_t\,\mathbf{G}_t\,\mathbf{m}_{t-1}$;

$$\hat{\mathbf{Y}}_t = \mathbf{F}_t\,\mathbf{R}_t\,\mathbf{F}_y' + \mathbf{V}_t\,.$$

3.3.2 On receiving the observation $\mathbf{y}_t$, the posterior parameter distribution is

$$(\theta_t|D_t) \sim N[\mathbf{m}_t;\,\mathbf{C}_t] \text{ where defining } \mathbf{e}_t = \mathbf{y}_t - \hat{\mathbf{y}}_t \text{ and } \mathbf{A}_t = \mathbf{R}_t\,\mathbf{F}_t'\hat{\mathbf{Y}}_t^{-1}$$

$$\mathbf{m}_t = \mathbf{G}_t\mathbf{m}_{t-1} + \mathbf{A}_t\,\mathbf{e}_t;\ \mathbf{C}_t = (\mathbf{I}-\mathbf{A}_t\,\mathbf{F}_t)\,\mathbf{R}_t.$$

3.3.3 The Forecast 'Function' $\mathbf{F}_t\,(\bullet)$ is defined for $k>0$ by

$$\mathbf{F}_t(k) = E[\mathbf{Y}_{t+k}|D_t] = \mathbf{F}_{t+k}\ \prod_{i=1}^{k}\ \mathbf{G}_{t+i}\ \ \mathbf{m}_t\,.$$

3.3.4 Following 2.3, the coherent joint forecast distribution may be derived using a D.L.M.

$$\mathbf{Y}_{t+k} = \mathbf{F}_{t+k}\,\theta_{t+k} + \mathbf{v}_{t+k} \qquad \mathbf{v}_{t+k} \sim N[0;\,\mathbf{V}_{t+k}]$$

$$\theta_{t+k} = \mathbf{G}_{t+k}\,\theta_{t+k-1} + \mathbf{w}_{t,k} \qquad \mathbf{w}_{t,k} \sim N[0;\,\mathbf{W}_{t,k}]\,.$$

Let $\mathbf{R}_{t,1} = \mathbf{H}_{t+1}\,\mathbf{C}_t\,\mathbf{H}_{t+1}'$. Writing $\mathbf{W}_{t,k} = \{w_{ij}\}$ and $\mathbf{R}_{t,k} = \{r_{ij}\}$ the recommended practical recursive relationships for $\mathbf{W}_{t,k}$ are

$$w_{ij} = (1-(\beta_i\beta_j)^{1/2})r_{ij}$$

$$\mathbf{R}_{t,k+1} = \mathbf{H}_{t+k+1}\,(\mathbf{I}-\mathbf{A}_{t,k}\,\mathbf{F}_{t+k})\,\mathbf{R}_{t,k}\,\mathbf{H}_{t+k+1}'$$

$$\mathbf{A}_{t,k} = \mathbf{R}_{t,k}\,\mathbf{F}_{t+k}'\,(\mathbf{V}_{t+k} + \mathbf{F}_{t+k}\,\mathbf{R}_{t,k}\,\mathbf{F}_{t+k}')^{-1}$$

For univariate series no matrix inversions are required.

3.3.5 $\mathbf{V}_t = (\mathbf{I}-\mathbf{F}_t\mathbf{A}_t)\hat{\mathbf{Y}}_t$ is true for all N.W.B.M.'s and is useful in the practical on-line estimation of $\mathbf{V}_t$ as in 5.5.

3.4. *Comments on N.D.B.M.'s*

(i) Given a common initial setting $(\theta_0|D_0)$, the N.D.B.M. $\{\mathbf{F}_t,\,\mathbf{G}_t,\,\mathbf{V}_t,\,\mathbf{I}\}$ and the D.L.M. $\{\mathbf{F}_t,\,\mathbf{G}_t,\,\mathbf{V}_t,\,\mathbf{0}\}$ have identical posterior parameter and joint forecast distributions.

(ii) The N.D.B.M. $\{\mathbf{F}_t,\,\mathbf{G}_t,\,V,\,\beta\,\mathbf{I}\}$ with initial setting $[\mathbf{m}_0;\mathbf{C}_0]$ has an identical forecast function to the D.W.E. $\{\mathbf{F}_t,\,\mathbf{G}_t,\,\beta\,\mathbf{I}\}$ with the initial setting $(\mathbf{m}_0;\,\mathbf{Q}_0 = \mathbf{C}_0^{-1}\,V)$ and this is an E.W.R. forecast function.

(iii) Given equivalent initial setting as in (ii), the N.D.B.M. $\{\mathbf{F}_t,\,\mathbf{G}_t,\,V,\,\mathbf{B}_t\}$ has an identical forecast function to that of the D.W.E. $\{\mathbf{F}_t,\,\mathbf{G}_t,\,\mathbf{B}_t\}$ and $\mathbf{C}_t = \mathbf{Q}_t^{-1}\,V$.

(iv) Given no missing observations, the N.D.B.M. procedure defined in the previous subsections is coherent since then $\mathbf{C}_{t,i} = \mathbf{C}_{t+i}$ and $\mathbf{W}_{t,i} = \mathbf{W}_{t-1,i+1} = \mathbf{W}_{t+i}$ for all t and $i>0$.

(v) A Constant N.D.B.M. is such that $\{\mathbf{F},\,\mathbf{G},\,V,\,\mathbf{B}\}_t$ is constant for each $t>0$. It is often referred to as the Time Series Model. The Constant N.D.B.M. $\{\mathbf{F},\,\mathbf{G},\,V,\,\mathbf{B}\}$ has a limiting forecast distribution equivalent to the limiting forecast distribution of the Constant D.L.M. $\{\mathbf{F},\,\mathbf{G},\,V,\,\mathbf{W}\}$ where $\mathbf{W} = (\mathbf{H}\,\mathbf{Q}^{-1}\,\mathbf{H}' - \mathbf{G}\,\mathbf{Q}^{-1}\,\mathbf{G}')V$. (see 2.3).

3.5. *Estimating the Observation Variance*

A number of approaches have been adopted for estimating the observation variances $\mathbf{V}_t$ and an effective practical procedure is given in 5.4. Here we mention the approach of Smith (1977) and West (1982) which is based upon an idea of DeGroot (1970). It is assumed that the variance $\mathbf{V} = \mathbf{I}/\phi$ where ϕ is an unknown constant. The probability model is

$$(\mathbf{Y}_t | \theta_t \phi) \sim N[\mathbf{F}_t \theta_t; \mathbf{I}/\phi]$$

$$(\theta_{t-1} | D_{t-1} \phi) \sim N[\mathbf{m}_{t-1}; \mathbf{C}_{t-1}/\phi]$$

$$(\phi | D_{t-1}) \sim \Gamma(\alpha_{t-1}/2, n_{t-1}/2)$$

with this Gamma p.d.f. having a kernel

$$\exp[(n_{t-1}/2 - 1)\log \phi - \alpha_{t-1}\phi/2] \, .$$

It follows that, because of Normality, the recurrence relationship for $\mathbf{m}$, $\mathbf{C}$, $\mathbf{R}$, $\hat{\mathbf{y}}_t$ and $\hat{\mathbf{Y}}_t$ are exactly as defined in 3.1 and 3.2 with the setting $\mathbf{V} = \mathbf{I}$. And

$$(\theta_t | D_{t-1} \phi) \sim N[\mathbf{G}\,\mathbf{m}_{t-1}; \mathbf{R}_t/\phi]$$

$$(\mathbf{Y}_t | D_{t-1} \phi) \sim N[\hat{\mathbf{y}}_t; \hat{\mathbf{Y}}_t/\phi]$$

$$(\theta_t | D_t \phi) \sim N[\mathbf{m}_t; \mathbf{C}_t/\phi]$$

$$(\phi | D_t) \sim \Gamma(\alpha_t/2; n_t/2)$$

where $n_t = n_{t-1} + 1$ and $\alpha_t = \alpha_{t-1} + \mathbf{e}_t' \hat{\mathbf{Y}}_t^{-1} \mathbf{e}_t$. As usual $\mathbf{e}_t = \mathbf{y}_t - \hat{\mathbf{y}}_t$. The joint distribution $(\mathbf{Y}_{t+1}, \phi | D_t)$ is readily obtained and $(\mathbf{Y}_{t+1} | D_t)$ is derived by integrating out ϕ. In the univariate case

$$(Y_{t+1} - \hat{y}_{t+1})/(\hat{Y}_{t+1}/\hat{\phi})^{1/2} \sim t,$$

Students t-distribution with n_t degrees of freedom. This method is operationally elegant and is properly Bayesian. It is not easy to retain the elegance when generalising to many cases where the correlation structure of $\mathbf{V}$ is unknown or where $\mathbf{V}_t$ is not a constant. Practitioners may prefer the robust variance estimation methods discussed in 5.5.

4. LIMITING RESULTS FOR CONSTANT N.D.B.M.'s

4.1. *Introduction*

There has been a continued interest in deriving limiting values for the parameter variance $\mathbf{C}_t$ and the adaptive vector $\mathbf{A}_t$ associated with observable Constant D.L.M.'s but the difficulty in solving Ricatti equations has restricted progress. However, for Constant N.D.B.M.'s $\{\mathbf{F}, \mathbf{G}, V, \mathbf{B}\}$ these values can be obtained directly. Hence the results also apply to the set of Constant D.L.M.'s $\{\mathbf{F}, \mathbf{G}, V, \mathbf{W}\}$ which have limiting forecast distributions equivalent to those of Constant N.D.B.M.'s. These results are relevant to practice since convergence is often fast and, in order to achieve conceptual and parametric parsimony, previous effort has been devoted to determining Constant D.L.M.'s which have limiting forecast functions equivalent to those obtained by the application of E.W.R., Harrison and Akram (1983) and Roberts and Harrison (1984).

Similar Models and the method of transforming from one similar model to another are defined. Limiting results for $\mathbf{C}$ and $\mathbf{A}$ are stated first for models similar to a model with a diagonal $\mathbf{G}$ structure and then for general Constant N.D.B.M.'s. The limiting relationship between the observations and the one step ahead prediction errors is obtained for N.W.B.M.'s. This leads to the establishment of a relationship between ARIMA models and Constant N.W.B.M.'s.

4.2. *Similar Models and Reparameterisation*

Let the matrices $\mathbf{G}$ and $\mathbf{G}_1$ be similar so that there exists a full rank matrix $\mathbf{L}$ such that $\mathbf{L}\ \mathbf{G}\ \mathbf{L}^{-1} = \mathbf{G}_1$. Then the two observable Constant N.W.B.M.'s $\{\mathbf{F}, \mathbf{G}, V, \mathbf{H}\}$ and $\{\mathbf{F}_1, \mathbf{G}_1, V, \mathbf{H}_1\}$ are defined as similar models if $\mathbf{H}_1 = \mathbf{L}\ \mathbf{H}\ \mathbf{L}^{-1}$. From observability it follows that $\mathbf{T} = [\mathbf{F}', (\mathbf{F}\ \mathbf{G})', \ldots, (\mathbf{F}\ \mathbf{G}^{n-1})']'$ and the corresponding $\mathbf{T}_1$ are both of full rank. Hence $\mathbf{L} = \mathbf{T}_1^{-1}\ \mathbf{T}$. If θ is the parameter vector of the first model then the reparameterisation $\psi = \mathbf{L}\,\theta$ produces the model $\{\mathbf{F}_1, \mathbf{G}_1, V, \mathbf{H}_1\}$.

4.3. *A Particular Limit Result*

Let $0 < \beta_i < |\lambda_i^2|$, $\mathbf{W} = \{w_{ij}\}$, $\mathbf{C} = \{c_{ij}\}$ and $\mathbf{A} = (A_1, A_2, \ldots, A_n)'$. If $\mathbf{F} = (1, 1, \ldots, 1)$, $\mathbf{G} = \mathrm{diag}(\lambda_1, \ldots, \lambda_n)$, $\mathbf{B} = \mathrm{diag}(\beta_1, \ldots, \beta_n)$ and if $u_i = \beta_i^{1/2}/\lambda_i$ $(i = 1 \ldots n)$ are distinct then

(i) $$c_{ij} = V(1 - u_i\, u_j) \prod_{k \neq j}^{n} \frac{1 - u_k u_i}{u_k - u_j} \prod_{h \neq i}^{n} \frac{1 - u_h u_i}{u_h - u_i}$$

(ii) $$A_i = (1 - u_i^2) \prod_{j \neq i}^{n} \frac{1 - u_i u_j}{1 - u_i/u_j}$$

(iii) $$w_{ij} = (1 - (\beta_i \beta_j)^{1/2})\, c_{ij}/(u_i u_j)$$

(iv) $$\hat{Y} = V / \prod_{i=1}^{n} u_i^2$$

Proof:

(i) $\mathbf{Q} = \{q_{ij}\}$; $q_{ij} = (1 - u_i u_j)^{-1}$; and $\mathbf{C} = \mathbf{Q}^{-1}\, V$ so

$$\mathbf{Q}^{-1} = \{p_{ij}\},\ c_{ij} = V p_{ij} \text{ as stated.}$$

(ii) By induction. Let $\mathbf{A}(n)$ be the solution for $\lambda_1, \ldots, \lambda_n$. For $n \geq 1$ and $i \leq n$

$$\sum_{j=1}^{n} q_{ij}\, A_j(n) = 1 \tag{1}$$

$$A_{n+1}(n+1) = (1 - \sum_{i=1}^{n} q_{n+1,j}\, A_j(n+1))/q_{n+1,n+1}$$

so substituting in (1) with $(n+1)$

$$\sum_{i=1}^{n} \frac{1 - u_j/u_{n+1}}{1 - u_j\, u_{n+1}}\, q_{ij}\, A_j(n+1) = 1. \tag{2}$$

Noting $A_1(1) = 1 - u_1^2$; the sets of equations (1) and (2) have unique solutions; and upon deriving $A_{n+1}(n+1)$ by symmetry it follows that

$$A_j(n+1) = \frac{1 - u_j\, u_{n+1}}{1 - u_j/u_{n+1}}\, A_j(n) \qquad \text{for } j = 1, \ldots, n.$$

(iii) easily derived from $\mathbf{W} = \mathbf{B}^{-1/2}\,\mathbf{G}\,\mathbf{C}\,\mathbf{G}'\,\mathbf{B}^{-1/2} - \mathbf{G}\,\mathbf{C}\,\mathbf{G}'$.

(iv) $V = \hat{Y} - \mathbf{F\,R\,F}' = (1-\mathbf{F\,A})\,\hat{Y} = (1 - \sum_{i=1}^{n} A_i)\,\hat{Y} = \hat{Y} \prod_{i=1}^{n} u_i^2.$

Comment

If $\beta_i = \beta$ for all i then (ii) reduces to the E.W.R. result of Dobbie (1963). The A_i's may take any values in the complex plane. The theorem is of practical interest mainly for periodic models with distinct complex eigenvalues lying inside or on the unit circle. For a real observation series a similar model with real **G** would be adopted and the corresponding limit values are easily derived. In the above, **W** is the system variance matrix in the Constant D.L.M. {**F**, **G**, *V*, **W**} which produces a limiting Forecast Function equivalent to that obtained from the N.D.B.M. {**F**,**G**, *V*,**B**}.

4.4. *A General Limit Result*

Any observable Constant N.D.B.M. {**F**, **G**, *V*, **B**} is similar to the canonical Constant N.W.B.M. $M_c = \{\mathbf{F}_c, \mathbf{G}_c, V, \mathbf{H}\}$ where, writing $\mathbf{B} = \text{diag}\{\beta_1,\ldots,\beta_n\}$, $\mathbf{F}_c = (1,0\ldots0)$ and $\mathbf{H} = \{h_{ij}\}$ with $h_{ii} = -h_{i,i+1} = \lambda_i/\beta_i^{1/2} = \mu_i$ for $i=1,\ldots,n$ and $h_{ij} = 0$ otherwise, $\mathbf{H}^{-1} = \{a_{ij}\}$ with $a_{ij} = 1/\mu_j$ for $j \geq i$ and $a_{ij} = 0$ otherwise. It follows that for model M_c, $\lim_{t\to\infty} \mathbf{C}_t = \mathbf{C}$ $= \mathbf{Q}^{-1} V$ exists where the Precision recurrence can be rearranged to give the Liapounov equation

$$\mathbf{H}'\,\mathbf{Q} = \mathbf{Q}\,\mathbf{H}^{-1} + \mathbf{H}'\,\mathbf{F}_c'\mathbf{F}_c.$$

This allows an easy sequential term by term evaluation of $\mathbf{Q} = \{q_{ij}\}$:-

$$q_{11} = \mu_1^2/(\mu_1^2 - 1);\; q_{12} = q_{11}/(\mu_1\mu_2 - 1)$$

$$q_{1,i} = \mu_1\mu_{i-1}\,q_{1,i-1}/(\mu_i\mu_1 - 1) \qquad \text{for } i>2$$

and $q_{i,k} = \mu_{i-1}\,q_{i-1,k}/\mu_i + S_{i,k}/\mu_k$

where $S_{i,k} = q_{i,k} + S_{i,k-1} = \sum_{j=1}^{k} q_{ij}$, and $q_{ij} = q_{ji}$. It follows from the results of section 3, that

(i) $\mathbf{C} = \mathbf{Q}^{-1}\,V$

(ii) $\mathbf{A} = \mathbf{Q}^{-1}\,\mathbf{F}_c'$, the first column of $\mathbf{Q}^{-1}$ with $A_1 = 1 - \prod_{i=1}^{n} \mu_i^{-2}$

and $A_n = (-1)^{n+1}(1-A_1) \prod_{i=1}^{n} (\mu_i\mu_n - 1)$

(iii) $\hat{Y} = V/(1-A_1) = V \prod_{i=1}^{n} \mu_i^2$

(iv) $\mathbf{W} = \mathbf{H\,C\,H}' - \mathbf{G}_c\,\mathbf{C}\,\mathbf{G}_c'$.

4.5. *A Limit Theorem*

Let {**F**, **G**, *V*, •} be a Constant N.W.B.M. which has a full rank **G** matrix with eigenvalues $\lambda_1\ldots\lambda_n$, or be any Constant Normal D.L.M. If the model is observable and $\lim_{t\to\infty} \mathbf{C}_t = \mathbf{C}$ exists as a proper variance matrix, then denoting the eigenvalues of $\mathbf{C\,R}^{-1}\,\mathbf{G}$, or equivalently $(\mathbf{I} - \mathbf{A\,F})\,\mathbf{G}$, by $\varrho_i \quad (i=1,\ldots,n)$,

$$\lim_{t\to\infty} \left[\prod_{i=1}^{n} (1-\lambda_i B)y_t - \prod_{i=1}^{n} (1-\varrho_i B)e_t \right] = 0.$$

Proof: Let $\mathbf{P}_1(B)$ and $\mathbf{P}_2(B)$ be row vectors and $P_3(B)$ and $P_4(B)$ scalars, all being polynomials of maximum order n-1, in the Backward Shift Operator B. Then since every square matrix satisfies its own characteristic equation

(i) $y_{t+1} = \mathbf{F}\,\mathbf{G}\,\mathbf{m}_t + e_{t+1}$ and $\mathbf{m}_t = \mathbf{G}\,\mathbf{m}_{t-1} + \mathbf{A}_t\,e_t$ give

$$\prod_{i=1}^{n} (1-\lambda_i B) y_{t+1} = \prod_{i=1}^{n} (1-\lambda_i B) e_{t+1} + P_1(B)\mathbf{A}_t\, e_t,$$

(ii) As $t \to \infty$, from Bayes Theorem, using prediction matrices,

$$\text{Lim}\,[\mathbf{C}^{-1}\mathbf{m}_t - \mathbf{R}^{-1}\mathbf{G}\,\mathbf{m}_{t-1} - \mathbf{F}'\,V^{-1}y_t] = 0$$

so

$$\text{Lim}\,[\mathbf{m}_t - \mathbf{C}\,\mathbf{R}^{-1}\,\mathbf{G}\,\mathbf{m}_{t-1} - \mathbf{A}y_t] = 0$$

(iii) Define $\mathbf{M}_t = \mathbf{C}\,\mathbf{R}^{-1}\,\mathbf{G}\,\mathbf{M}_{t-1} - \mathbf{A}\,y_t$ with $\mathbf{M}_0 = \mathbf{m}_0$

and $x_{t+1} = \mathbf{F}\,\mathbf{G}\,\mathbf{M}_t + e_{t+1}$, then

$$\prod_{i=1}^{n} (1-\varrho_i B) e_{t+1} = \prod_{i=1}^{n} (1-\varrho_i B) x_{t+1} + P_2(B)\mathbf{A}\,y_t$$

As $t \to \infty$ $\mathbf{C}_t \to \mathbf{C}$, $\mathbf{R}_t \to \mathbf{R}$, $\mathbf{A}_t \to \mathbf{A}$, $\mathbf{M}_t \to \mathbf{m}_t$ and $x_t \to y_t$

$$\text{Lim}\left[\prod_{i=1}^{n} (1-\varrho_i B) e_t - (1+BP_4(B))y_t\right] = 0$$

and from (i)

$$\text{Lim}\left[\prod_{i=1}^{n} (1-\lambda_i B) y_t - (1+BP_3(B))e_t\right] = 0$$

The order of the B polymonials correspond and, since these equations are true for all allowable (λ_i, ϱ_i) $i=1...n$, the coefficients of B^i can be equated to obtain the result.

Comment: This result includes the E.W.R. results of Godolphin and Harrison (1975), McKenzie (1976) and the D.W.E. result of Ameen and Harrison (1984). The Normal assumption is only used in (ii) and that result can be obtained from the limiting forms of the variance updating recurrence relationship arising from minimum variance unbiased linear estimation. Hence the results may be extended beyond the Normal models.

4.6. *Relationship with ARIMA Processes*

Let Y_t be a random time series generated according to an ARIMA process

$$\prod_{i=1}^{n} (1-\lambda_i B) Y_t = \prod_{i=1}^{n} (1-\varrho_i B) a_t$$

where $0<|\lambda_i| \leq 1$, $0<|\varrho_i|<1$, $(i=1...n)$ and where a_t is a weakly stationary white noise process with zero mean and Autocovariance Generating Function $\gamma(B) = \sigma^2$. The appropriate Box-Jenkins (1970) predictor replaces a_t by the one step ahead prediction error e_t' and it is well known that $\text{Lim}\; e_t' = a_t$. Applying the appropriate Dynamic Model $\{\mathbf{F}, \mathbf{G}, \mathbf{V}, \cdot\}$ to the realised series y_t, $\text{Lim}\left[\prod_{i=1}^{n} (1-\lambda_i B) y_t - \prod_{i=1}^{n} (1-\varrho_i B) e_t\right] = 0.$

Hence $\text{Lim}\,|e_t - e_t'| = 0$ and, with probability one, the limiting Box-Jenkins forecast function is equivalent to that of the Dynamic Model.

For an unbalanced ARIMA process

$$\prod_{i=1}^{p} (1-\lambda_i B) Y_t = \prod_{i=1}^{q} (1-\varrho_i B) a_t$$

let $n = \max(p,q)$. Then given any $\epsilon > 0$, if $n=p$ (or $n=q$) by taking p-q of the ϱ's (or $q-p$ of the λ's) appropriately close to zero, $\text{Lim}\,|e_t - e_t'| < \epsilon$ is assured. Thus all ARIMA processes can be modelled by Constant N.W.B.M.'s and the Constant N.D.B.M.'s model a subset.

5. N.D.B.M.'s IN PRACTICE

5.1. *Introduction*

The natural way of constructing a linear model is to build from a set of smaller linear models. This utilizes the familiar Principle of Superposition which states that any linear combination of linear models is a linear model. For statistical models the principle extends to Normal linear models based upon the fact that any multivariate Normal distribution can be decomposed into a set of component multivariate Normals. Hence in building a N.D.L.M. $\{\mathbf{F}_t, \mathbf{G}, \mathbf{V}_t, \mathbf{W}_t\}$ practitioners usually construct the model so that $\mathbf{G} = \text{diag}(\mathbf{G}_1, \mathbf{G}_2, \ldots, \mathbf{G}_r)$ where the block $\mathbf{G}_i$ is associated with a meaningful model component. This block diagonal construction tends to be the natural operational form. N.D.B.M.'s dispense with the awkward variance $\mathbf{W}$ and its role is taken by a discount matrix $\mathbf{B}_t = \text{diag}(\beta_1, \ldots, \beta_n)$ which generally associates a single discount factor with each block, so that if $\mathbf{G}_i$ is n_i square and $\mathbf{I}_i$ is the n_i square identity matrix $\mathbf{B}_t = \text{diag}(\beta_1 \mathbf{I}_1, \ldots, \beta_r \mathbf{I}_r\}$. Sometimes $\mathbf{G}$ will have r distinct eigenvalues $(\lambda_1, \ldots, \lambda_r)$ and $\mathbf{G}_i$ will be similar to the Jordan form for an eigenvalue λ_i of multiplicity n_i. But for real observations processes, where complex eigenvalues are concerned, it is usual to consider conjugate pairs in the same block. For example, for a pair of conjugate complex eigenvalues $(\lambda e^{iw}, \lambda e^{-iw})$ of multiplicity one, the form we adopt is $\mathbf{G} = \lambda \begin{pmatrix} \cos w & \sin w \\ -\sin w & \cos w \end{pmatrix}$. This could represent a damped sine wave of period $2\pi/w$ and would typically have a single associated discount factor $0 < \beta \leq \lambda^2$.

5.2. *An Application of a Constant N.D.B.M.*

The Turkey Poult data given in the Appendix shows the monthly sales of one day old chicks from hatcheries in Eire. There is clearly a growth in sales over the ten years and an annual seasonal pattern. Many events took place such as changes in feed prices and poult prices, attempts to promote Easter turkey consumption and turkey sausage meat in order to reduce the seasonality in sales and events arising from the continual competition with broilers and the meat industry. The desirability of incorporating information on these events either by modelling them or through intervention or multi-process models emphasizes the importance of flexible operational systems. However, for illustration, a simple trend seasonal Constant N.D.B.M. $(\mathbf{F}, \mathbf{G}, V, \mathbf{B})$ is applied to the logarithm of the data with $\mathbf{F} = (\mathbf{F}_1, \mathbf{F}_2, \ldots, \mathbf{F}_7)$, $\mathbf{G} = \text{diag}(\mathbf{G}_1, \ldots, \mathbf{G}_7)$, $\mathbf{B} = \text{diag}(\beta_1 \mathbf{I}_2, \beta_2 \mathbf{I}_{11})$, where $\mathbf{F}_i = (1,0)$, $i = 1 \ldots 6$, $\mathbf{F}_7 = 1$; $\mathbf{G}_1 = \begin{pmatrix} 1 & 1 \\ 0 & 1 \end{pmatrix}$ represents a linear growth trend component and the seasonal component comprises six harmonics with the block for the kth harmonic being $\mathbf{G}_{k+1} = \begin{pmatrix} \cos(kw) & \sin(kw) \\ -\sin(kw) & \cos(kw) \end{pmatrix}$ for $k = 1, \ldots, 5$ with $w = \pi/6$ and $\mathbf{G}_7 = -1$. β_1 is the trend discount factor and β_2 the seasonal discount factor. The one step ahead median point forecasts are shown in Fig. 1. The initial vague prior postulated a level lying in (1.5,

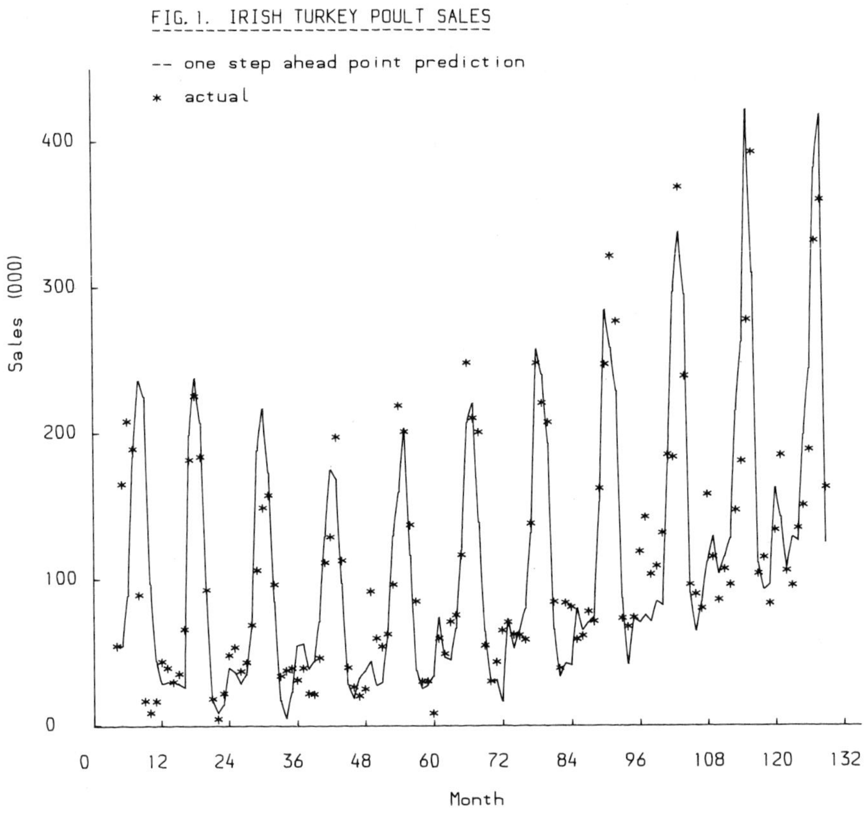

2×10^{17}) with 95 % confidence and log growth and log seasonality $N[0;100\,\mathbf{I}]$. Throughout $\beta_1 = .975$ and $\beta_2 = 0.925$. The variance V was estimated on line by a variance estimator (Section 5.4) and throughout, remained at about .04. Two examples of the application of D.W.E. (Ameen and Harrison, 1984) may be regarded as applications of Constant N.D.B.M.'s and that relating to the U.S. Air Passenger data allows comparisons with many other methods since it has been so widely analysed. However the turkey data is much more typical of the series met by forecasters. The above trend seasonal N.D.B.M. applied to the Airpassenger data attains a Mean Absolute Deviation of 2.3 % for the one-step-ahead error. But almost any forecast method appears good when applied to that data series. Hence its popularity in papers demonstrating stationary and differenced stationary models. It is evident that, like most practical series, the turkey data is far from stationary. One or two variables such as feed price and total meat supply might beneficially be introduced but they also need to be estimated. Many of the events which contribute large variation do so over a short period. Often such an event is anticipated but can only be subjectively described. This description might quote an expected time effect of the event and also importantly the additional uncertainty associated with it. Bayesian methods are ideal for introducing subjective information and blending it with data. A complete Bayesian system can show vast improvements in forecasts compared to the performance of classical models. This has been demonstrated in companies like Cadbury Ltd. However it is difficult to convince in academic papers since acceptable comparative standards have not yet been established.

5.3. *An Application of Intervention using Discount Factors*

Intervention involves changing a routine or existing probability model often by introducing subjective information. The object of structuring a model is to enable changes to be made to particular model components in a way that leaves other components largely unaffected. This gives the model robustness. Typically interventions are specified as transfer functions (Box and Tiao (1975)) but, in Bayesian Dynamic Models, as transfer probability distributions which not only introduce an expected effect but also extra uncertainty associated with the change. In the following example one useful way in which additional uncertainty can be specified through the discount factors is illustrated. For an assessment of the effectiveness of this intervention an artificial data series is used comprising 120 months observations. The generating model combined a linear trend dynamic model and a 12 period seasonal effect derived from a single first harmonic. The routine observation random error was distributed independently $N\,[0;\,400]$. However the following major impulses and events were imposed:

(i) 200 was substracted from the intermediate observation at $t=32$, in order to simulate a maverick or outlier;

(ii) immediately after $t=36$, the process 'deseasonalised level' was reduced by 270 to give a ramp or jump;

(iii) following $t=60$ the linear growth was reversed in sign from roughly 8 units per period to -8, giving a 'slope change';

(iv) following $t=85$, the linear growth was again reversed in sign and simultaneously the seasonal amplitude increased by 50 %;

(v) data points 113, 114 and 115 were eliminated to give a period of three missing observations.

The N.D.B.M. $\{\mathbf{F}, \mathbf{G}, V_t, \beta_t\}$ is applied with $\mathbf{F} = (1, 0, 1, 0)$;
$\mathbf{G} = \text{diag}\left[\begin{pmatrix} 1 & 1 \\ 0 & 1 \end{pmatrix} ; \begin{pmatrix} \cos w & \sin w \\ -\sin w & \cos w \end{pmatrix}\right]$ with $w=\pi/6$; and $\mathbf{B} = \text{diag}(\beta_1,\beta_2,\beta_3,\beta_4)_t$. Apart from intervention times, $\beta_1 = \beta_2 = 0.9$; $\beta_3 = \beta_4 = 0.95$; and $V_t = 400$. A vague initial prior $(\theta_0 | D_0) \sim N\,[(75, 9, 85, 1)'; 2000\,\mathbf{I}]$ was adopted. When the major changes (i) to (iv) are about to occur it is assumed that the type of forthcoming event is known but that there is *no* available information on the size or even the sign of the coming change. The maverick observation was communicated for $t=32$ and the updating procedure treats it as a missing observation since model components need to be protected from the inadequate description of any statistical description of outliers. Foreknowledge of the jump trend change at $t=37$ was signalled by $(\beta_1, \beta_2) = (0.9^{40}, 1)$ and for the trend growth change at $t=61$ by $(\beta_1,\beta_2) = (1, 0.9^{80})$. In each instance only the trend marginal distributions was updated since, in practice, it is desirable to protect information on model components from the effects of imprecise descriptions of sharp changes in other components. At these intervention times a Modified N.D.B.M. is applied as described in 5.4. The simultaneous sudden change in trend and seasonal at $t=86$ is signalled by setting $\mathbf{B}_{86} = (1, (0.9)^{80}, (0.95)^{30}; (0.95)^{30})$. The three missing observations were dealt with naturally by the posterior distributions of θ_t for $t=113$, 114 and 115 simply being the prior parameter distribution $(\theta_t | D_{112})$. Fig. 2 graphs the observations and one step ahead expectations in order to demonstrate the power of the intervention method. Since in the routine data generation $V = 400$, then without major disturbances the limiting Mean Absolute Deviation (M.A.D.) of the one step ahead forecast errors would be about 17.5. The performance in terms of the M.A.D. for each

year is tabulated omitting the outlier e_{32}, the jump e_{37} and the three missing errors e_{113}, e_{114}, e_{115}, indicated by *.

YEAR	1	2	3*	4*	5	6	7	8	9	10*
M.A.D.	39.5	20.3	20.5	20.3	15.0	17.2	17.4	20.1	21.5	15.8

The average run length of errors with the same sign was 2.1 and the first lag autocorrelation coefficient of the errors was -0.07. The routine discount factors and the intervention discount factors were not chosen to optimize a performance criterion but are rounded figures which were thought to be appropriate bearing in mind that it is usually preferable to err on the side of underestimating discount factors (Harrison, 1967).

5.4. *Modified N.D.B.M.'s*

In some cases of intervention like those in 5.3 and in non-linear modelling as applied by Migon and Harrison (1983), a class of N.W.B.M.'s called Modified N.D.B.M.'s is used. Consider $(\theta_{t-1}|D_{t-1}) \sim N\,[\mathbf{m}_{t-1};\ \mathbf{C}_{t-1}]$ where, corresponding to any diagonal block structure for $\mathbf{G} = \text{diag}\,\{\mathbf{G}_1, \mathbf{G}_2 \ldots \mathbf{G}_p\}$, at time t-1, the partitioned structure of $\mathbf{C}_{t-1}$ is $\{\mathbf{C}_{ij}\}$, of $\mathbf{R}_t$ is $\{\mathbf{R}_{ij}\}$, for $i,j = 1 \ldots p$, and for $\mathbf{B}_t$ is diag $\{\mathbf{b}_1, \ldots, \mathbf{b}_p\}$. Then a Modified N.D.B.M. $\{\mathbf{F}, \mathbf{G}, V, \mathbf{B}\}_t$ defines $(\theta_t|D_{t-1}) \sim N\,[\mathbf{G}\,\mathbf{m}_{t-1};\ \mathbf{R}_t]$ where $\mathbf{R}_{ii} = \mathbf{b}_i^{-1/2}\,\mathbf{G}_i\,\mathbf{C}_{ii}\,\mathbf{G}_{ii}'\,\mathbf{b}_i^{-1/2}$ $i = 1 \ldots p$, but

$$\mathbf{R}_{ij} = \mathbf{G}_i\,\mathbf{C}_{ij}\,\mathbf{G}_j' \quad \text{for } i \neq j.$$

FIG. 2: AN EXAMPLE OF INTERVENTION ON AN ARTIFICIAL DATA SET

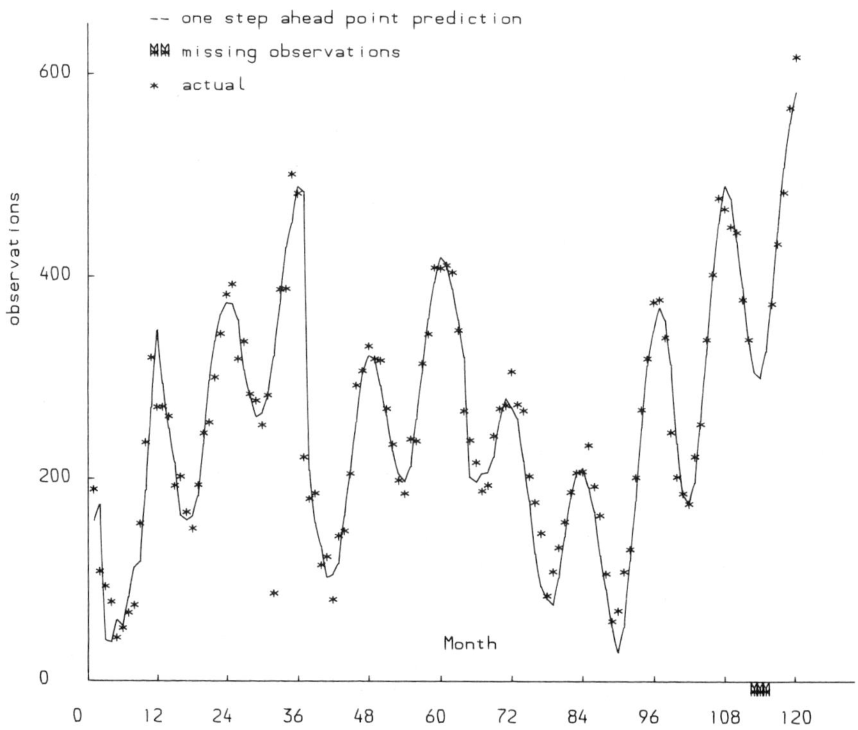

All other updating and forecast function calculations are as for the usual N.D.B.M. However, although often $\mathbf{b}_i = b_i \mathbf{I}_i$ this is not necessary for the Modified model.

The reason for introducing this model class is that there may be a requirement that the system evolution for the different blocks $\mathbf{G}_i$ be considered as independent of one another. The information from the observation series Y_t is distributed amongst the elements of θ and naturally produces parameter covariances. However this does not signify that that covariance should be developed over time by discounting or by any other means. For example, consider the application of a major disturbance on one or more of the parameters of a given block. This may be signalled, as in 5.3, by using a discount factor β^N where N is chosen to have the effect of ageing the past history relevant to that component by N periods. In general it may not be desirable to apply the term $\beta^{-N/2}$ to the covariances with other blocks since this is likely to have the effect of introducing the major disturbance to those blocks. The definition of Modified N.D.B.M.'s is widened. Consider $\mathbf{G} = [\mathbf{G}_{ij}]$ as any matrix written in a desired structural form. Let $\mathbf{G}\ \mathbf{C}_{t-1}\ \mathbf{G}' = \Sigma = [\Sigma_{ij}]$ in a corresponding partitioned form. Then define $\mathbf{R}_{ii} = \mathbf{b}_i^{-1/2}\,\Sigma_{ii}\,\mathbf{b}_i^{-1/2}$ and, for $i \neq j$, $\mathbf{R}_{ij} = \Sigma_{ij}$.

Of course where constraints on the parameter vector θ are expressed by the variance matrix it will be necessary to group the relevant subset of parameters.

5.5. *Estimating the Observation Variance*

In addition to the method given in 3.5, a number of approaches have been adopted for estimating the observation variance V, for example Smith and West (1983), Harrison and Stevens (1975), Cantarelis and Johnston (1983) and Leonard and Harrison (1977). A simple but effective robust procedure can be obtained using the relationship of 3.3.5,

$$\mathbf{V}_t = (\mathbf{I} - \mathbf{F}_t\,\mathbf{A}_t)\hat{\mathbf{Y}}_t\,.$$

For a univariate time series, defining $d_t^2 = (1-\mathbf{F}_t\mathbf{A}_t)e_t^2$, a natural estimate for an unknown constant V is $\hat{V}_t = X_t/n_t$, where $X_t = X_{t-1} + d_t^2$ and $n_t = n_{t-1} + 1$. Initially (X_0, n_0) may be chosen such that $X_0/n_0 = \hat{V}_0$ is a point estimate for V and n_0 is the accuracy expressed in terms of a number of degrees of freedom, or of equivalent observations. In the analysis of 3.5 it is seen that $\hat{V}_t = 1/E[\phi\,|\,D_t]$. Hence, if required, forecasts can be produced, as in 3.5, using a Student t-distribution with n_t d.o.f. In practice it may be wise to protect the estimate from outliers and major disturbances. Outlier prone distributions, O'Hagan (1979), can be introduced. However one simple effective practical method is to define $d_t^2 = (1-\mathbf{F}_t\,\mathbf{A}_t)\,\mathrm{Min}[e_t^2, k\,\hat{Y}_t]$ where k is generally chosen in the interval [4,6]. In those cases in which it is suspected that V varies slowly over time a discount factor β may be introduced so that

$$X_t = \beta X_{t-1} + d_t^2 \quad \text{and} \quad n_t = \beta n_{t-1} + 1.$$

This procedure is easily applied and experience with both pure time series and regression-type models is encouraging. The theory and an application in financial econometrics are given in Harrison and Johnston (1983). However, because of the skew distribution of d_t^2 it is wise to choose $0.95 < \beta \leq 1$. Further, if the initial prior of the parameter vector θ is vague, then it is recommended that variance learning commences at time $(n+1)$ where n is the vector dimension. In stock control, with positive observations, an empirical variance law $V_t = a\,\hat{y}_t^{2b}$ with $b = 0.75$ is often used. A theoretical model is given in Stevens (1974). An estimate $\hat{a}_t$ of a is then derived from $Z_t = \beta\,Z_{t-1} + d_t^2/\hat{y}_t^{1.5}$; $n_t = \beta\,n_{t-1}+1$; and $\hat{a}_t = Z_t/n_t$. Future estimates of V are $\hat{V}_{t+k} = \hat{a}_t(E[Y_{t+k}\,|\,D_t])^{1.5}$. If the power is also unknown and the variance law written $V_t = a\,\hat{y}_t^{2b}$, the Support for (a,b) may be defined as $S(a,b\,|\,D_t) = \beta\,S(a,b\,|\,D_{t-1}) + (d_t^2/(a\,\hat{y}_t^{2b}) - \log a - 2b\log \hat{y}_t)/2$. The maximally supported values may be used as in the I.C.I. forecasting package MELO, Harrison and Pearce (1972). With such a method it is recommended that $d_t^2 = (1-\mathbf{F}_t\mathbf{A}_t)\,\mathrm{Min}[e_t^2, 4\hat{Y}_t]$, that $\beta \geq 0.98$ and that strong prior support be applied.

5.6. *N.D.B.M.'s and the Analysis of Designed Experiments*

Whenever experiments are performed sequentially or are ordered by an index t, the assumption of model constancy should be questioned. For example in the production of Nylon Polymer, a batch of nylon salt is taken from a salt reservoir, charged to one of a bank of n autoclaves, cooked, nylon polymer extruded, chopped, dried and sampled. Using a constant linear model a statistical analysis of the Amine End Group quality characteristic on over a thousand batches concluded that there is no significant between autoclaves effect. However a dynamic linear model revealed a stochastic autoclave effect to be the most significant source of A.E.G. variation. Further, there were sharp changes in the effects following autoclave cleanouts. In this type of situation, where effects are approximately stationary, it is very likely that static linear models applied to large amounts of data will fail to identify major sources of variation.

The D.L.M. $\{\mathbf{F}, \mathbf{G}, \mathbf{V}, \mathbf{W}\}_t$ has been applied to designed experiments but the choice of $\mathbf{W}$, which is not invariant to the measurement scale of $\mathbf{F}$, has been troublesome. The application of N.D.B.M.'s $\{\mathbf{F}, \mathbf{G}, \mathbf{V}, \mathbf{B}\}_t$ removes most of the difficulty. For example consider sequentially replicated blocks of a 2^2 Factorial design where it is thought that the block and treatment effects vary slowly with time. The N.D.B.M. $\{\mathbf{F}, \mathbf{I}, \mathbf{V}, \mathbf{B}\}$ with

$$\mathbf{F} = \begin{pmatrix} 1 & 1 & 1 & 1 \\ 1 & 1 & -1 & -1 \\ 1 & -1 & 1 & -1 \\ 1 & -1 & -1 & 1 \end{pmatrix}$$

may provide an appropriate model with $\theta_t = (\theta_1,\theta_a,\theta_c,\theta_{ac})_t$ where θ_1 is the block, θ_a the effect of treatment A and θ_{ac} the interaction effect. The variances $\mathbf{V}$ and $\hat{\mathbf{Y}}$ may be estimated on-line.

5.7. *Extended N.D.B.M.'s*

In many applications an N.D.B.M. provides an adequate model but other applications may require a more general N.W.B.M. This is particularly the case when $\mathbf{G}$ is singular and when high frequencies and some types of stochastic transfer responses are to be modelled. The Extended N.D.B.M. is defined by the quintuplet $\{\mathbf{F}, \mathbf{G}, \mathbf{V}, \mathbf{B}, \mathbf{W}\}_t$ where, given $(\theta_{t-1}|D_{t-1}) \sim N[\mathbf{m}_{t-1}; \mathbf{C}_{t-1}]$, this defines

$$(Y_t|\theta_t) \sim N[\mathbf{F}_t\theta_t; \mathbf{V}_t]\,,$$

$$(\theta_t|D_{t-1}) \sim N[\mathbf{G}_t\mathbf{m}_{t-1}; \mathbf{R}_t]\,,$$

$$\text{and } \mathbf{R}_t = \mathbf{B}_t^{-1/2}\mathbf{G}_t\mathbf{C}_{t-1}\mathbf{G}_t'\mathbf{B}_t^{-1/2} + \mathbf{W}_t.$$

For example, in the sequential factorial design of 5.6, the block effects may be independent or exchangeable so that $(\theta_{1t}|D_{t-1}) \sim IN[\theta_{5t}; \sigma^2]$ where θ_{5t} is unknown and either static ($\beta_5 = 1$) or very slowly moving. An appropriate Extended N.D.B.M. might then be $\{\mathbf{F}, \mathbf{G}, V, \mathbf{B}, \mathbf{W}\}$ where writing $\mathbf{f}$ as the design matrix in 5.6, $\mathbf{F} = \begin{pmatrix} \mathbf{f} & \mathbf{0} \\ \mathbf{0}' & 0 \end{pmatrix}$,

$\mathbf{G} = \begin{pmatrix} 0 & (0\,0\,0\,1) \\ \mathbf{0} & \mathbf{I}_4 \end{pmatrix}$, $\beta_1 = \beta_5$ generally and $\mathbf{W}_t = \{w_{ij}\}$ with $w_{11} = \sigma^2$ as the only non zero element. Although this reintroduces a system type matrix $\mathbf{W}$, in most cases it will be sparse and easily specified.

Similarly, in modelling correlated observation errors such as those generated by a stationary second order autoregressive process $\nu_t = (1-\phi_1 B)(1-\phi_2 B)\delta_t$, models of the type

$$\left\{(1,0,\mathbf{0}_n);\ \left(\begin{array}{cc:c} \phi_1 & 1 & \mathbf{0} \\ 0 & \phi_2 & \mathbf{F}_{t+2} \\ \hdashline \multicolumn{2}{c:}{\mathbf{0}} & \mathbf{G} \end{array}\right);\ 0;\ \mathrm{diag}(1,1,\mathbf{B});\ \mathbf{W}\right\}$$

may be preferred. The only non zero element of $\mathbf{W}$ is $w_{22} = V$ and this is easily estimated on-line. For a particular extension of E.W.R. to Generalised E.W.R. which considers stationary observation errors see Harrison and Akram (1983).

5.8. *A Multiprocess Application*

Multiprocess models are described in Harrison and Stevens (1971, 1976). Class 2 models are constructed in order to deal automatically with various specified types of major change or disturbance. In this application to five years medical prescription data (Appendix) the routine model is the full linear growth and seasonal model used in the Turkey example with $\beta_1 = 0.95$, $\beta_2 = 0.975$. V is estimated as about 0.36. Just two major types of disturbance are considered, namely sharp trend changes and outliers or mavericks. The model transition probabilities are constant throughout so that, at time t, the prior

FIG. 3. PRESCRIPTION DATA. MULTIPROCESS MODELLING.

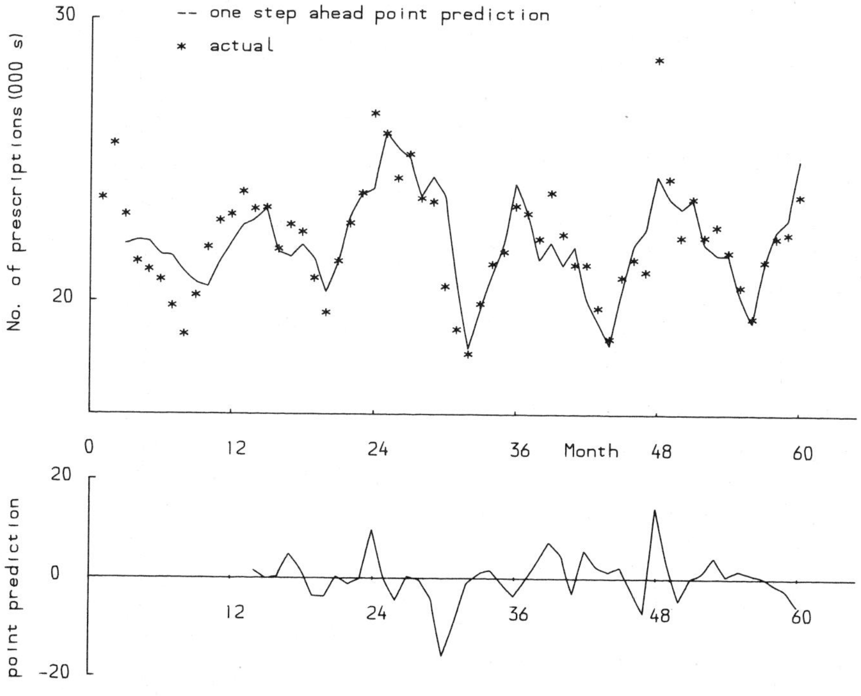

probability that there is an outlier is $\pi_2 = 0.05$ and that there is a sharp trend change is $\pi_3 = 0.025$. The procedure is then similar to that adopted by Harrison and Stevens.

For each of the three models, at time t-1, the posterior information is $\{(\mathbf{m}_i, \mathbf{C}_i); P_i\}_{t-1}$ ($i=1,2,3$). The probability that at time t-1 model i operated and that at time t model j operates is $\pi_j\ P_i$ and the corresponding prior parametric distribution $(\theta\,|\,M_i\ M_j\ D_{t-1}) \sim N[\hat{\theta}_{ij};\ \mathbf{R}_{ij}]$ is obtained in the usual way with $\mathbf{R}_{ij} = \mathbf{B}^{-1/2}\,\mathbf{G}\,\mathbf{C}_i\,\mathbf{G}'\,\mathbf{B}^{-1/2}$ for $j=1,2$ and $\mathbf{R}_{i,3} = \ell^*\,\mathbf{B}^{-1/2}\,\mathbf{G}\,\mathbf{C}_i\,\mathbf{G}'\,\mathbf{B}^{-1/2}\,\ell^{*\prime}$ where $\ell^* = \text{diag}(1., 1., \mathbf{0}')$. For control purposes the outlier variance V_{12} is set such that the one step ahead forecast variance $\hat{Y}_{13} = \hat{Y}_{12}$. The point prediction for Y_t at time t-1 which is plotted in Fig. 3. is derived using an inverted Normal Loss function, Lindley (1976), Smith, Harrison and Zeeman (1981) since quadratic loss functions are inappropriate for use with multimodal distributions. The posterior probabilities P_{ij}, that model i operated at time $t-1$ and model j at time t are obtained by $P_{ij} \propto L(y_t\,|\,M_i\ M_j\ D_{t-1})\ \pi_j\ P_i\ (i,j=1,2,3)$. The nine posterior parametric distributions $(\theta\,|\,M_i\,M_j\,D_t)$ are obtained as usual except that for mavericks ($j=2$) the posterior is taken as the prior mean. These two practical steps protect model components from what are necessarily inadequate descriptions of major changes concerning other model components. The usual Harrison-Stevens collapsing procedure is used to reduce the nine posterior distributions to three.

In the application the initial estimates were, level $\sim N\,[22;10]$, growth $\sim N\,[0;1/4]$ and seasonality $\sim N\ [\mathbf{0};25\mathbf{I}]$. After the first thirteen months the seasonal pattern is differentiated from the trend and, as can be seen in Fig. 3, the point predictions show an improved accuracy. At month 24 a minor positive jump in trend occurs and at month 30 a negative jump occurs as the prescription charge is raised. In the latter case the point prediction error is -15 % and is followed by an error of -5 % as the dilemma between responding to an outlier and a sharp trend change is resolved. For month 48, the point error is 14 % but the following errors are not large and the causal influenza epidemic is correctly treated as an outlier.

Naturally, in order to improve accuracy and to anticipate major trend changes, prescription charge should be introduced as an independent variable. Here we use the multiprocess approach simply to illustrate the automatic treatment of sharp changes in a readily appreciated context.

The use of N.W.B.M.'s in multiprocess modelling is the subject of a paper by Ameen and Harrison (1983b). That paper introduces an economical procedure based upon a Cumulative Sum (CUSUM) control test. The idea is that a preferred forecasting model is used until the Cusum controller signals a probable significant event. A multiprocess phase then operates until it is again judged that the single preferred model is satisfactory. The multiprocess operation differs from the one given in this paper. The prescription data is one of four illustrative examples.

APPENDIX

Monthly Sales of Turkey Poults in Eire (000's)

MONTH \ YEAR	72	73	74	75	76	77	78	79	80	81	82
JAN.	—	39.0	52.9	38.7	91.0	58.8	70.0	58.3	141.8	114.5	184.0
FEB.	—	29.2	36.4	21.1	59.0	48.1	61.3	60.6	102.5	85.2	105.0
MAR.	—	34.9	42.4	20.6	53.3	70.1	60.9	77.1	108.2	106.2	95.0
APR.	54.6	65.4	68.2	45.6	61.8	74.7	58.0	70.6	130.7	95.6	134.0
MAY.	165.0	181.5	105.8	110.9	95.9	116.1	137.6	161.5	184.5	146.5	150.0
JUN.	207.9	225.2	148.6	128.6	218.2	247.5	247.2	246.5	192.9	180.1	188.0
JUL.	189.1	183.3	156.9	196.7	200.4	209.6	219.7	320.4	367.7	276.8	331.0
AUG.	89.3	92.0	95.7	112.1	131.5	199.7	206.4	275.7	238.4	391.3	359.0
SEP.	16.6	17.4	33.0	39.0	84.0	53.9	83.5	92.5	95.4	102.9	162.0
OCT.	8.3	3.7	36.9	25.4	29.1	29.1	38.1	68.7	89.0	114.0	—
NOV.	16.0	21.1	38.5	19.3	29.3	42.7	83.0	73.1	79.3	82.6	—
DEC.	43.3	47.4	30.3	24.2	7.4	64.2	80.1	118.0	157.3	132.7	—

Artificial Data Series

MONTH \ YEAR	1	2	3	4	5	6	7	8	9	10	11
JAN.	189	270	392	221	318	411	273	233	377	449	—
FEB.	108	261	318	179	317	404	267	192	340	444	—
MAR.	93	192	335	185	269	347	202	163	246	377	—
APR.	77	201	283	114	234	267	176	106	202	338	—
MAY.	42	166	276	122	198	238	146	59	185	—	—
JUN.	52	150	253	80	185	216	84	69	175	—	—
JUL.	67	193	282	143	239	187	108	108	222	—	—
AUG.	75	244	86	148	237	193	132	130	254	373	—
SEP.	155	255	387	205	314	242	157	201	338	433	—
OCT.	236	300	388	292	343	269	187	268	402	483	—
NOV.	320	343	501	307	409	272	206	319	478	567	—
DEC.	270	382	482	331	408	306	207	375	467	617	—

Prescription Data (scaled numbers)

Year \ Month	JAN.	FEB.	MAR.	APR.	MAY.	JUN.	JUL.	AUG.	SEP.	OCT.	NOV.	DEC.
1	23.7	25.6	23.1	21.4	21.1	20.8	19.8	18.8	20.2	21.9	22.8	23.1
2	23.9	23.3	23.3	21.8	22.7	22.4	20.8	19.6	21.4	22.7	23.8	26.6
3	25.9	24.4	25.2	23.6	23.5	20.5	19.0	18.1	19.9	21.3	21.7	23.4
4	23.1	22.2	23.8	22.4	21.3	21.3	19.8	18.7	20.8	21.5	21.0	28.6
5	24.3	22.3	23.6	22.3	22.6	21.7	20.5	19.4	21.4	22.3	22.4	23.7

REFERENCES

AITCHISON, J. and DUNSMORE, I.R. (1975). *Statistical Prediction Analysis.* Cambridge: University Press.

AMEEN, J.R.M. and HARRISON, P.J. (1984). Discount weighted estimation. *J. of Forecasting,* **3**, 285-296.

— (1983a). Information forecasting. Warwick. *Res. Rep. N° **31**.*

— (1983b). Discount Bayesian Multiprocess Modelling with CUSUMS. *Proceedings of International Time Series Conference,* Nottingham, 1983, North-Holland.

ASTROM, K.J. (1970). *Introduction to Stochastic Control Theory.* New York: Academic Press, Inc.

BARNARD, G.A. (1951). The theory of information. *J. Roy. Statist. Soc.* B, **13**, 46-64.

BARNARD, G.A., JENKINS, G.M. and WINSTEN, C.B. (1962). Likelihood inference and time series. *J. Roy. Statist. Soc.* A. **125**, 321-372.

BOX, G.E.P. and JENKINS, G.M. (1970). *Time Series Analysis, Forecasting and Control.* San Francisco: Holden-Day.

BOX, G.E.P. and TIAO, G.C. (1973). *Bayesian Inference in Statistical Analysis.* Addison-Wesley.

— (1975). Intervention analysis with applications to economic and environmental problems. *J. Amer. Statist. Ass.,* **70**, 70-79.

BROWN, R.G. (1963). *Smoothing, Forecasting and Control.* San Francisco: Holden-Day.

BROWN, R.L., DURBIN, J. and EVANS, J.M. (1975). Techniques for testing the constancy of regression relationships. *J. Roy. Statist. Soc.* B, **37**, 149-192.

CANTARELIS, N. and JOHNSTON, F.R. (1983). On-line variance estimation for the steady state Bayesian forecasting model. *J. Time Series Analysis,* **3**, 225-234.

DEGROOT, M.H. (1970). *Optimal Statistical Decisions.* New York: McGraw-Hill.

DOBBIE, J.M. (1963). Forecasting periodic trends by Exponential smooltring, *Op. Res.* **11**, 908-918.

DURBIN, J. (1983). Extensions of the Brown and Holt-Winters forecasting systems and their relation to Box-Jenkins models. *International Conference held at Valencia,* (O.D. Anderson, ed.). North-Holland, 7-18.

EDWARDS, A.W.F. (1972). *Likelihood.* Cambridge: University Press.

FISHER, R.A. (1950). *Contributions to Mathematical Statistics.* New York: Wiley.

GILCHRIST, W.G. (1967). Methods of estimation involving discounting. *J. Roy. Statist. Soc.* B, **29**, 355-369.

GODOLPHIN, E.J. and HARRISON, P.J. (1975). Equivalence theorems for polynomial projecting predictors. *J. Roy. Statist. Soc.* B, **37**, 205-215.

HARRISON, P.J. (1965). Short-term sales forecasting. *J. Roy. Statist. Soc.* C, **15**, 102-139.

— (1967). Exponential smoothing and short-term sales forecasting. *Man. Sci.,* **13**, 821-842.

HARRISON, P.J. and AKRAM, M. (1983). Generalized exponentially weighted regression and parsimonious dynamic linear modelling. *Time Series Analysis,* **3**, (O.D. Anderson, ed.) North-Holland 19-42, (with discussion).

HARRISON, P.J., LEONARD, T. and GAZZARD, T.N. (1977). An application of multivariate hierarchical forecasting. Warwick *Res. Rep.,* **15**.

HARRISON, P.J. and PEARCE. (1972). The use of trend curves as an aid to market forecasting. *Ind. Mark. Manage.,* **2**, 149-170.

HARRISON, P.J. and SCOTT, F.A. (1965). A development system for use in short-term sales forecasting investigations. Warwick *Res. Rep.* N° **26**.

HARRISON, P.J. and STEVENS, C.F. (1971). A Bayesian approach to short term forecasting. *Oper. Res. Quart.,* **22**, 341-362.

— (1975). "Case studies". Warwick Res. *Rep. N°* **14**.

— (1976). Bayesian forecasting. *J. Roy. Statist. Soc.* B, **38**, 205-247.

HARRISON, P.J. and JOHNSTON, F.R. (1983). A regression method with non stationary parameters. Warwick *Res. Rep.* N° **35**. To be published in *J. of O.R.*

JONES, R.H. (1966). Exponential Smoothing for multivariate time series. *J. Roy. Statist. Soc.* B, **28**, 241-251.

KALMAN, R.E. (1963). New methods in Wiener filtering theory. In *Proceedings of the First Symposium on Engineering Application of Random Function Theory and Probability.* (J.L. Bogdanoff and F. Kozin, eds.). New York: Wiley.

LEONARD, T. and HARRISON, P.J. (1977). Bayesian updating for the steady state Kalman filter. Warwick *Res. Rep.* N° **15**.

LINDLEY, D.V. (1976). A class of utility functions. *Ann. Statist.,* **4**, 1-10.

LINDLEY, D.V. and SMITH, A.F.M. (1972). Bayes estimates for linear models. *J. Roy. Statist. Soc.* B, **34**, 1-41.

McKENZIE, E. (1976). An analysis of general exponential smoothing. *Oper. Res.,* **24**, 131-140.

MIGON, H.S. and HARRISON, P.J. (1983). Warwick Statistical Research Reports on Non-Linear modelling.

MORRISON, N. (1969). *Introduction to sequential smoothing and prediction.* McGraw Hill.

O'HAGAN, A. (1979). On outlier rejection phenomena in Bayes inference. *J. Roy. Statist. Soc.* B, **41**, 358-367.

PLACKETT, R.L. (1950). Some theorems on least squares. *Biometrika,* **37**, 149-157.

PRIESTLEY, M.B. (1980). State-dependent models: A general approach to non-linear time series analysis I. *Time-Series Analysis,* **1**, 47-71.

ROBERTS, S.A. and HARRISON, P.J. (1984). Parsimonious modelling and forecasting of seasonal time series. *Eur. J. Oper. Res.* **16**, 365-377

SMITH, A.F.M. and WEST, M. (1983). Monitoring renal transplants: An application of the multi-process Kalman filter. *Biometrics,* **39**, 867-878

SMITH, J.Q. (1977). *Problems in Bayesian Statistics Relating to Discontinuous Phenomena, Catastrophe Theory & Forecasting.* Ph.D. thesis, University of Warwick.

— (1979). A generalization of Bayesian steady forecasting model. *J. Roy. Statist. Soc.* B, **41**, 378-387.

— (1981). The multi parameter steady model. *J. Roy. Statist. Soc.* B, **43**, 256-260.

SMITH, J.Q., HARRISON, P.J. and ZEEMAN, E.C. (1981). The analysis of some discontinuous decision processes. *Eur. J. Oper. Res.,* **7**, 30-43.

SOUZA, R.C. (1981). A Bayesian entropy approach to forecasting: the multi state model. *Time Series Analysis* (Houston, Tex.) North Holland, 535-542.

SOUZA, R.C. and HARRISON, P.J. (1977). A Bayesian entropy approach to forecasting. Warwick *Res. Rep.* N° **19**.

STEVENS, C.F. (1974). On the variability of demand for families of items. *Oper. Res. Quart.*, **25**, 411-420.

WEST, M. (1982). *Aspects of recursive Bayesian estimation.* Ph.D. thesis, University of Nottingham.

WHITTLE, P. (1965). Recursive relations for predictors of non-stationary processes. *J. Roy. Statist. Soc.* B, **27**, 523-532.

— (1969). A view of stochastic control theory. *J. Roy. Statist. Soc.* A, **132**, 320-334.

YOUNG, P.C. (1974). Recursive approaches to time series analysis. *Bul. Inst. Maths. and Applic.* **10**, 209-224.

ZELLNER, A. (1971). *An introduction to Bayesian inference in economics.* Wiley.

DISCUSSION

U.E. MAKOV (*Chelsea College, London*)

The authors can be congratulated on an attractive paper which deals with the realistic situation of estimating or predicting state parameters in the case where the observation and state covariance matrices are unknown. They propose several methods which perform well and are simple to apply.

It should be emphasized that none of the methods discussed is truly (coherent) Bayesian. For if one believes that both observation and state parameters are described by a D.L.M., the only coherent Bayes solution is via the Kalman filter equations. Thus even if there are genuine difficulties in assessing the state covariance matrix **W**, an elegant introduction of a discount factor is clearly an ad-hoc intervention and should therefore be referred to not as 'Bayes' but as 'approximate' 'neo' or 'quasi'-Bayes procedure.

Suppose we replace the unknown **W** by an estimate $\hat{\mathbf{W}}$. If estimation or prediction are allowed to continue with time, a divergence is likely to occur which will amount to a total breakdown in the data processing method. In both N.D.B.M. and modified N.D.B.M. the authors adopted $\hat{\mathbf{W}} = 0$, i.e., ignoring **W** altogether. This can artificially increase the 'gain' A_t (via R_t) which is associated with the 'innovation' e_t, and thus can lead to a divergence. In order to prevent it, the authors reduce R_t by a factor β and thus attenuate the impact of ignoring W. Since β is not chosen in any optimal (Bayesian?) way, the authors should comment on the sensitivity of their methods to different values of β. Are there values for which the prediction procedures are less accurate or 'unstable'? Can the authors comment on the accuracy of their methods in long term prediction ($>$ 1)?

The paper starts with a brief account of the bright future of Bayesian forecasting as predicted in the 70's by Harrison and Stevens. In practice, however, the present is not as bright as anticipated and one reason given is the "lack of relevant...literature". In fact such very relevant literature, though from different contexts, exists in abundance in the engineering literature. It is unfortunate that Bayesian statisticians spend so much of their time attempting to devise methods which have been investigated and reported and are ready to be used. The idea of discounting the influence of past data on the estimate of the current state is not new. In Sorenson and Sacks (1971) a recursive 'fading memory filter' was suggested as a means for overcoming the destructive influence of model errors in Kalman filter. The method which assumed a known **W** introduced a discount (exponential) factor which, in effect, does exactly what β is meant to do.

Several other methods have been proposed to identify the unknown noise covariance matrices of the D.L.M. and in a different approach attention was focused on the direct estimation of the steady-state Kalman gain (A_t) without using covariance matrices. See Lee (1980) for some references.

It must be pointed out that some of the 'engineering' methods are quite cumbersome compared to the rather uncomplicated procedures suggested today, which were demonstrated to perform so elegantly.

M. MOUCHART (*Université Catholique de Louvain, Belgium*)

I am not sure I have correctly understood how far the authors refer to different models (i.e. to different data generating processes) or to different methods (or procedures). For instance, is, in section 2, the DWE *method* defined as a local *model*?

I suggest the authors exhibit a model for which their methods work nicely while other (reasonable) methods would fail, and to characterize the class of models for which their methods are either optimal or, at least, perform better than "reasonable" competitors. Also a detailed example would probably have been more informative than sketching a lot of possible applications. Furthermore, initial conditions are given scarce attention only. Does it mean that the proposed methods are applicable for large samples only? More generally, making the assumptions relative to the data generating process more explicit would undoubtedly help the reader to a better appreciation of the genuine contribution of this paper to the existing literature on Bayesian forecasting.

A.F.M. SMITH (*University of Nottingham*)

In those applications of multiprocess dynamic linear models with which I am most involved (mainly in medical monotoring -see, for example, Smith and West, 1983), emphasis is on calculating posterior probabilities of process states, rather than on obtaining "smoothed forecasts". Whereas the discount method looks like a very promising breakthrough for the latter problem, I am not clear that it would help much in the "monitoring" context.

REPLY TO THE DISCUSSION

We thank the discussants for their contributions. Much has been missed and misunderstood but, from a readers' viewpoint this is probably good since further discussion provokes clarification.

We regret that Dr. Makov did not attend the meeting and so missed the verbal presentation of the paper. The Bayesian approach to forecasting is quite different from most other approaches and Dr. Makov makes the surprising statement that none of the methods discussed is truly (coherent) Bayesian. This is particularly odd since section 3.3 is entirely devoted to defining a coherent forecast distribution for the main N.D.B.M. Hence the comments about the introduction of a discount factor as a clearly ad hoc intervention seem misplaced unless the ad hoc is applied generally to all mathematical models since they all are simply personal descriptions of reality.

Perhaps the biggest misunderstanding of Dr. Makov comes in his failure to grasp the idea of a discount matrix. He makes a rather snappy comment that it is 'unfortunate that Bayesian statisticians spend so much of their time attempting to devise methods which have been reported and are ready to be used'. This implies that Sorenson and Sacks (1971) first introduced the idea of discounting. It should be very clear that in forecasting this was introduced in the late 1950's by Brown and was widely applied in the early 1960's. The statistical paper of Harrison (1965) pointed out the great weakness of using a simple discount factor for different model components and it is stressed again in this paper. Dr. Makov obviously misses the point. What the referenced engineering literature has done is to use just one discount factor for all model components. This amounts simply to applying Brown's method of E.W.R. under the name of fading memory. Harrison introduced the idea of different discount factors for different model components since it is evident that the future relevance of current estimates can vary between model components.

For example in the I.C.I. short-term forecasting method MULDO which was developed in 1964, two major model components are identified as trend and seasonal, Harrison (1965), Harrison and Scott (1965). The trend is often a local description of a relatively fast changing 'deseasonalised' process level and current estimates are not regarded as particularly relevant for more than 12 months ahead. Hence this trend is often discounted at a rate of $\beta_1 = 0.9$ or less. The seasonal pattern is usually regarded as much more durable and relevant to future years and so is often discounted at a much slower rate of say $\beta_2 = 0.97$. The introduction of a discount *matrix* is one of the key points of this paper and we are amazed that this could be overlooked by Dr. Makov who continually refers to it as a scalar or a factor. Consequently his comments about the E.W.R. or fading memory single methods doing exactly what β is meant to do are attributable to a complete failure to grasp a fundamental point. Dr. Makov takes us up on our comment on the lack of relevant literature for practitioners. He says that such literature exists in abundance in engineering literature. We agree that a lot of papers relate to Kalman Filtering as opposed to Bayesian Forecasting. It must be remembered that the Kalman Filter is a dynamic version of the Gauss Markov theorem and bears the same relationship to Bayesian Forecasting as regression does to Bayesian statistics. The fact that we so often use Normal Models makes the link particularly important. The Kalman Filter simply provides recurrence relationships for the mean and variance of a parameter vector estimator. The Bayesian approach is a general method of estimation. For example, in this paper we provide joint distributions of the parameter vector and the observation variance even when the latter is not constant but follows some relationship or itself varies stochastically. The approach also generalizes to Non-Normal Distributions and to estimating such things as discount factors although we acknowledge that such extensions often require numerical methods. However, there is little doubt that forecasting practitioners find the bulk of the literature on both Kalman filtering and Bayesian forecasting unreadable and this is what we mean by lack of practical relevance.

Dr. Makov also appears to consider a D.L.M. to have a constant covariance matrix $\mathbf{W}$. As defined by Harrison and Stevens (1971 and 1976) this is certainly not true. It is totally undesirable to be imprisoned by such concepts as stationarity or derived stationarity. Although stationarity concepts have advanced a restricted theory they have hindered successful applications of dynamic statistical models. There is a further important point that is also not widely appreciated which is that even if it is a constant, the covariance matrix $\mathbf{W}$ is nearly always overparameterized with respect to the joint forecast of all future observations. In section 1.2 we refer to this as the non uniqueness of the $\mathbf{W}$ matrix. Usually there is an uncountable set of pairs $(V, \mathbf{W})$ which will produce exactly the same limiting joint forecasts. Examples are given in Harrison and Akram (1983).

Consider a simple model such as

$$\left\{ (1,0)\,;\begin{pmatrix} 1 & 1 \\ 0 & 1 \end{pmatrix};\, V;\, \mathbf{W} = \begin{pmatrix} W_1 + aW_2 & aW_2 \\ a\,W_2 & W_2 \end{pmatrix} \right\}$$

then for all a such that $(2a-1)^2 \leq 1+4W_1/W_2$, given corresponding initial priors, the joint forecasts and estimates are identical for all times. This point needs to be appreciated since it reveals a major obstacle in selecting **W** and shows the relationship between the choice of **W** and the meaning of the parameter vector. Dr. Makov quotes the reference of Lee (1980). Tucked away at the end of that short paper is a comment on uniqueness which hardly does justice to the major obstacle to estimating a **W** matrix even if one does assume it constant. The problem is simply dismissed as saying that, if there is non uniqueness, put extra restrictions of the Variance matrix! Tell that to a practitioner and he is going to get very upset. Hence it is important practically, that the discount approach overcomes this ambiguity. It is also practically important that the problem of **W** not being invariant to the measurement scale of **F** is also overcome.

The comment about replacing **W** by an estimated value leading to a divergence and breakdown of a data processing method can be true. It is always possible to point to extreme cases and in simple models Harrison (1967) investigated the effect of using the wrong models. However, for observable time series type models, the discount procedure will not do this if the restriction $0<\beta<|\lambda|^2$ is followed for each eigenvalue. In this case the information received in a finite series of observations about a state parameter is not less than that which is lost due to the system changes. The comment on accuracy and long term prediction is completely answered for Time Series cases by our limiting results in Section 4. These results are independent of the 'true' values $(V,\mathbf{W})$ and will be of much theoretical interest since it has always been difficult to derive limiting results for the general class of D.L.M. The values of V and **W** can be far from appropriate yet convergence of the joint forecast distributions of $(Y_{t+1},\ldots,Y_{t+m})$ is ensured. Furthermore, from 4.5, it is a simple task to assess the effects of a non optimal pair $(V,\mathbf{B})$ by expressing the one step ahead error

$$e_t = \prod_{i=1}^{n} [(1-\pi_i B)/(1-\varrho_i B)]\alpha_t$$

where the π_i's and α_t refer to the 'true model'. In the cases of independent variables comprising the $\mathbf{F}_t$ vector the question of ensuring an adequate flow of information is just the standard problem associated with design matrices: if you keep a variable at a constant value then naturally no information on its effect counters the time loss and so its associated variance increases, and possibly explodes.

Professor Mouchart asks whether the D.W.E. method is defined as a local model. D.W.E. is essentially a generalisation of E.W.R. and is a method of estimation. However, in order to estimate, it is necessary to have some model in mind and, in view of the discount approach, we talk about local models. He also suggests that we exhibit a model for which our methods work nicely but on which other reasonable methods fail. We would stress that we do not take the stance that a model describes the 'true' generating process of data but that a model expresses the way in which we wish to view the data. We are concerned with the practical performance on real data in real situations, not with optimality with respect to imprisoning theoretical assumptions. We view statistics and mathematics as a powerful language which is capable of expressing almost any views we have about a process, and of specifying how we wish our views on the process and its future to change in response to specific events. On this point the Bayesian approach is much misunderstood. It appears that knowledge and training in old concepts relating to time series and forecasting is preventing many people from even understanding the underlying concepts and principles of new approaches. For example, we are asked if, since initial conditions are given scarce attention, the methods only apply to large samples. This is again a surprising question since one of the major properties of Bayesian forecasting is its power to operate on little data or even no data at all if subjective information is available.

Professor Smith raises the question of monitoring where the main aim is the detection of changes. The basis of many operational forecasting systems is that of Management by Exception. Here a routine Bayesian procedure processes information and produces forecasts unless exceptional circumstances arise. The exceptions arise in two ways. First information received, by say Marketing Department, may lead to a 'feed forward' intervention. The second exception is when a control system detects that the forecast system appears inadequate, in that the forecast errors are unusually large. Then, often a feedback message will be produced indicating that well founded subjective information would be helpful. In addition automatic procedures such as multiprocess models come into effect in order to deal with major changes. Originally in the 1960's one of us used Cusum controllers, Harrison and Davies (1964), and later developed the Multiprocess models, Harrison and Stevens (1971). Recently we have combined the two as described in Ameen and Harrison (1983b) and as applied in section 5.8 of this paper. The models may all be based on discount factors or on the modified or extended models. We find this useful and effective in implementing systems.

REFERENCES IN THE DISCUSSION

HARRISON, P.J. and DAVIES, O.L. (1964). The use of Cumulative Sum (CUSUM). Techniques for the Control of Routine Forecasts of Product Demand. *J.OR.S.A.* **12**, 325-333.

LEE, T.T. (1980). "A direct approach to identify the noise covariances of the Kalman filtering", *IEEE Trans Automatic Control AC*-**25**, 841-842.

SORENSON, H.W. and SACKS, J.E. (1971). Recursive fading memory filtering, *Information Sciences* **3**, 101-119.

BAYESIAN STATISTICS 2, pp. 299-328
J.M. Bernardo, M.H. DeGroot, D.V. Lindley, A.F.M. Smith (Eds.)

Bayesian Analysis of Survival Curves for Cancer patients following treatment

W.C. CHEN[1], B.M. HILL, J.B. GREENHOUSE[2] and J.V. FAYOS[3], M.D.

University of Michigan

SUMMARY

A generalization of the mixture model of Berkson and Gage is formulated and used for the analysis of survival data from cancer. A subjective Bayesian approach is taken towards inference and decision-making. The methods are illustrated both with data concerning carcinoma of the endometrium, and with leukemia data previously analysed by D.R. Cox. The procedures developed are effective for very small samples, including cases where only a handful of patients have died, and a moderate number are still surviving.

Keywords: BAYESIAN ANALYSIS; CANCER SURVIVAL ANALYSIS; MIXTURE MODEL; WEIBULL DISTRIBUTION

1. INTRODUCTION

This article is concerned with the statistical analysis of survival times following treatment for cancer. Our work is based upon a model generalizing that of Boag (1949) and Berkson and Gage (1952). The approach to inference and decision making is the subjective Bayesian approach of de Finetti (1975) and L.J. Savage (1961, 1972).

Berkson and Gage proposed that a proportion C of patients treated for cancer might be effectively cured in the sense that their mortality experience would follow, approximately, the appropriate actuarial tables for the general population without cancer; while a proportion $1-C$ would be subject to additional forces of mortality. Letting $L(x)$ denote the proportion of patients surviving x time units after treatment, their model becomes the mixture

$$L(x) = CL_0(x) + (1-C)L_0(x)[V_\theta(x)], \qquad (1.1)$$

where $L_0(x)$ is the actuarially obtained proportion of people (without the particular cancer in question) surviving x time units, and Berkson and Gage took $V_\theta(x) = e^{-\theta x}$ to represent the additional force of mortality for uncured patients. They then showed that by least squares curve fitting of (1.1) to the life tables for cancer, that they were able to estimate C and θ, obtaining remarkably good fits in view of the simplicity of (1.1) and the complexity of the data.

(1) This article is dedicated to the memory of the late Wen-Chen Chen, who had been Assistant Professor of Statistics at Carnegie-Mellon University.

(2) Present address: Assistant, Professor Department of Statistics, Carnegie-Mellon University.

(3) Professor of Radiology at the University of Miami.

There are several questions and difficulties inherent in the Berkson-Gage model. First and foremost there is the intriguing question as to whether, and if so, why, the simple exponential form of (1.1) used by Berkson and Gage should apply to something so complex as cancer. We examine this by considering a Weibull generalization, replacing $V_\theta(x)$ in (1.1) by $e^{-(\theta x)^\alpha}$, and then analysing the likelihood function $L(p,\theta,\alpha)$ based upon both uncensored and censored data. For the two sets of data that we analyse, one of which concerns endometrial carcinoma, and the other of which is the leukemia data of Cox (1972), we find a sharp likelihood function which points towards $\alpha = 2$ rather than the $\alpha = 1$ of Berkson and Gage. On the other hand, the estimated survival curve that we obtain is for many practical purposes well approximated by that obtained with $\alpha = 1$.

An important feature of our analysis is that by use of the likelihood principle we are able to factor out certain proportionality constants that depend on the known distribution of survival times for the population of «normals», and thus greatly simplify the likelihood function. Another feature of our analysis is that it is extremely effective, even for very small samples, with only a handful of deaths. This is of extreme importance for making early decisions as to the efficacy of treatments, since inferior treatments can be rejected without the necessity of waiting to observe a substantial number of deaths. A related advantage is that we are able to deal with withdrawals in a simple and natural way, not requiring artificial assumptions, and avoid difficulties in the Berkson-Gage approach that arise from the choice of the grouping intervals for the life table. Finally, whereas the Berkson-Gage method of curve fitting is sensitive, in principle, to dependencies amongst the errors, changing variances, and some other problems associated with the fitting of monotonic functions, our methods of inference are based upon the likelihood function, and thus automatically allow for such considerations. (The difficulties in the treatment of such data from a regression point of view are spelled out in footnote 5 of Berkson and Gage on page 514. However, they argue for a least squares approach as opposed to a maximum-likelihood approach, on the grounds of simplicity. The Bayesian approach taken in our article rests on the evaluation of the likelihood function, and therefore automatically builds in precisely those assumptions that correspond to our model. It is also simple to implement using modern computer technology.)

In Section 2 we put forth and motivate our model. The likelihood function is obtained and simplified, both for a non-parametric model, and for Makeham's Law, a parametric model of special interest and importance. In Section 3 we formulate the decision problem from the subjective Bayesian point of view, and suggest certain quantities of importance in the comparison of treatments. Section 4 then suggests methods for analysing and utilising the likelihood function and posterior distribution to solve the decision problem, while Section 5 applies these methods to cancer data. Finally, in Section 6 we discuss the key aspects of our approach, and compare it with other methods currently being employed in the analysis of cancer survival data, and in Section 7 we discuss some special issues that have been raised in connection with our analysis.

2. STATISTICAL MODEL AND LIKELIHOOD FUNCTION

We consider patients who have been treated for a particular type and stage of cancer. For such a patient the data that we wish to analyse consists of his age when treatment occurred, and if he has died, possibly due to cancer, his age at death; if he is known to be still alive, or if he has died from some apparently unrelated cause (e.g., struck by lightning), or if his life status is presently unknown (e.g., because of withdrawal from study), then in place of age at death we take the last age at which he was known to be alive, which follows from the likelihood principle, as is discussed in Section 6. For convenience we shall refer to the first such group of patients as "dead", and the second, as "still surviving", although such terminology is not to be taken literally. Now let X_i be the random time from treatment to death for the ith patient classified as dead, and let Y_j be the random time from treatment until the last time known to be alive for the jth patient classified as still surviving. Our data then consists of pairs (t_i,x_i), $i=1,\ldots,m$, where (t_i,x_i) is the event that the ith dead patient was treated at age t_i, and died at age t_i+x_i, so that x_i is the observed value of the random variable X_i; and of pairs (s_j,y_j), $j=1,\ldots,n$, where (s_j,y_j) is the event that the jth patient still surviving was treated at age s_j, and was known to have survived at least to age s_j+y_j (but not to any greater age), so that $X_j \geq y_j$.

We now define our concept of a cure, which is essentially that of Boag (1949) and of Berkson and Gage (1952). By a "cure" we shall mean that the survival time of the patient will be just as though he had not had either the cancer or its treatment. Strictly speaking it seems unlikely that cures ever occur in such a strong sense, and that even though the tumor should be completely terminated the survival time of the patient would be shortened, if only because of the time interval during which the tumor was present. Nonetheless it seems appropriate to use such a definition because, on the one hand, clinical experience and statistical analysis (for example, that of Berkson and Gage) suggest that such shortening is relatively small; and, on the other hand, there would be very little data pertaining directly to the distribution of survival times for a more realistic definition of cure. See also Berkson and Gage (1952, p. 504, p. 508) on this point.

Given that a patient is cured in our sense, then the probability distribution for his survival time is that for the appropriate subset of the general population (same sex, race, etc.) not having the specific cancer in question. Suppose that this distribution is F, where if W is a random variable representing the random survival time of such a cured patient, we take $F(x) = Pr\{W \geq x\}$. We shall also use the notation

$$F(x;t) = Pr\{W \geq t+x \mid W \geq t\},$$

to represent the conditional probability of survival to age $t+x$, given survival to age t. Now a great deal is know about F from the vast actuarial experience accumulated, and it may either be estimated by smoothing appropriate life tables, as in Berkson and Gage; or, as is sometimes done here, assumed to have a specific parametric form, which is widely used in actuarial work. This parametric form, known as Makeham's Law, will be explored at the end of this section. However, most of our analysis does not depend upon the form of F, so that unless explicitly stated otherwise, we shall view F as an arbitrary distribution for a non-negative random variable.

Now suppose that a patient is not cured in our sense. In this case we assume that his survival time W has a distribution belonging to a parametric family defined by

$$G_\theta(x;t) = Pr\{W \geq t + x \mid W \geq t, \theta\},$$

where θ is an unknown parameter vector to be estimated from the data. However, it must be kept in mind that strictly speaking $G_\theta(x;t)$ only has meaning for a patient who contracted the cancer, was treated for it at age t, and for whom the treatment failed to cure the cancer. It is thus unlike $F(x;t)$, which in principle has meaning for the "normal" population throught life. For example, if t is the age at which treatment occurs, then we think of $G_\theta(x;t)$ as defining a distribution for survival beyond age t, and *both* θ and t serve as parameters for the distribution. Note also that a patient who has the cancer and is treated for it, but for whom the treatment is ineffective, may have a very different mortality experience from one who has the cancer but is not treated for it, since the treatment itself may be harmful, or alternatively may ameliorate it to some extent even though it does not cure the cancer. Because the intervention of the treatment at age t is implicit in $G_\theta(x;t)$, one cannot view the latter as representing merely the mortality experience of patients with the cancer in question.

With this understanding, we can now state the probability model for a patient who contracted the cancer at some earlier age and who received treatment at age t. If p is the probability that the treatment cures the cancer, then the overall model for survival to age $t + x$ is

$$Pr\{W \geq t + x \mid W \geq t, p, \theta\} = p\,F(x;t) + (1-p)G_\theta(x;t). \tag{2.1}$$

It is implicit in this model that neither p nor θ depend on the age at which treatment occurs.

Now let $V_\theta(x;t) = G_\theta(x;t)/F(x;t)$, where it is assumed throughout that $F(x;t)>0$. Following Berkson and Gage, $V_\theta(x;t)$ can be interpreted in terms of the additional force of mortality due to the specific cancer, and (2.1) can be written as

$$Pr\{W \geq t+x \mid W \geq t,p,\theta\} = F(x;t)[p+(1-p)V_\theta(x;t)]. \tag{2.2}$$

Because $F(x;t)$ cannot depend upon p or θ (and in fact is effectively known from actuarial data) this simplification has a very important consequence, namely, that for many purposes it is only necessary to specify a parametric model for $V_\theta(x;t)$, rather than for $G_\theta(x;t)$. This will become clear in our analysis of the likelihood function arising from (2.2). We note that (2.2), with $V_\theta(x;t) = e^{-\theta x}$, and $F(x;t) = L_0(x)$, is identical with the Berkson-Gage model (1.1). Typically we shall choose $V_\theta(x;t)$ as a function that does not depend upon t, say $V_\theta(x)$, with $0 \leq V_\theta(x) \leq 1$. The reasons for doing so will be discussed in Section 6. Although various parametric forms may prove interesting, the data analysis in this article is performed for $V_\omega(x) = e^{-(\theta x)^\alpha}$, i.e., the upper tail of a Weibull distribution with parameter $\omega = (\psi,\alpha)$, and $\theta>0$, $\alpha>0$. With the Weibull specification for $V_\omega(x)$ the probability distribution for survival of uncured patient is

$$G_\theta(x;t) = e^{-(\theta x)^\alpha}F(x;t) \tag{2.3}$$

This can be given the following interpretation. Given treatment at age t, survival to age $t + x$ for an uncured patient is probabilistically equivalent to survival to age $t + x$ for a cured patient together with an event $Z \geq x$, where Z is a random variable having the Weibull distribution $Pr\{Z \geq x\} = e^{-(\theta x)^\alpha}$ and Z is independent of all other random variables involved in the process for cured patients. Such an interpretation, of course, is possible with any $V_\theta(x)$ replacing the Weibull, provided only that $1 - V_\theta(x)$ is a probability distribution function. This in turn lends support to an interpretation of $V_\theta(x)$ as an additional force of mortality over and above that for cured patients, or alternatively, in terms of competing risks. Our interest in the Weibull form arises from the general importance of the Weibull distribution in reliability theory, and because with $\alpha = 1$ it reduces to the exponential model of Berkson and Gage, and generalizes their model in a natural way. One of the aims of this article is to determine whether the Berkson and Gage exponential model provides a satisfactory approximation within the general Weibull model, which will be discussed in Section 6. Finally we note that the Weibull form can be derived from the more general representation

$$V_\theta(x) = \exp\left[-(\theta)^\alpha\alpha \int_0^X s^{\alpha-1}e^{-\beta s}\,ds\right],$$

by setting $\beta = 0$. Although when $\beta > 0$, $V_\theta(+\infty) > 0$ so that the probabilistic interpretation of $V_\theta(x)$ is lost, we can still obtain a mathematical analogue of the hazard function, namely

$$-\frac{d}{dx}\ln[V_\theta(x)] = \theta^\alpha\alpha x^{\alpha-1}e^{-\beta x},$$

or $\theta^\alpha \alpha x^{\alpha-1}$ when $\beta = 0$.
The last of course is the hazard function for the Weibull distribution, and is increasing for $\alpha > 1$, decreasing for $\alpha < 1$. Although we will not pursue it here, we note that the general case with $\beta > 0$ yields a hazard function that is not monotonic, and thus offers even more flexibility. In our formulation it is not necessary that $1 - V_\theta(x)$ be a probability distribution function, and the additional parameter β may prove of some value.

We are now ready to obtain the likelihood function for the data (t_i, x_i), $i = 1,\ldots,m$, (s_j, y_j), $j = 1,\ldots,n$. We assume $G_\theta(x;t) = V_\theta(x)F(x;t)$, and define $g_\theta(x;t) = -\frac{d}{dx}[G_\theta(x;t)]$,

$v_\theta(x) = -\frac{d}{dx}[V_\theta(x)]$, $\quad f(x;t) = -\frac{d}{dx}[F(x;t)]$, so that

$$g_\theta(x;t) = V_\theta(x)f(x;t) + v_\theta(x)F(x;t). \tag{2.4}$$

If the survival times for the $m+n$ patients are independent random variables, then from (2.1), the likelihood function for the parameters p, θ, is

$$L(p,\theta) \propto L_1(p,\theta)L_2(p,\theta), \qquad \text{where}$$

$$L_1(p,\theta) \propto \prod_{i=1}^{m} [p\, f(x_i;t_i) + (1-p)g_\theta(x_i;t_i)], \qquad \text{and}$$

$$L_2(p,\theta) \propto \prod_{j=1}^{n} [p\, F(y_j;s_j) + (1-p)G_\theta(y_j;s_j)]. \tag{2.5}$$

For the population of "normals", i.e., those without the specific cancer, let $F(x)$ be the probability of survival to at least age x, so that $F(x;t) = F(t+x)/F(t)$, and $f(x;t) = f(t+x)/F(t)$, where $f(x) = -\frac{d}{dx}[F(x)]$.

Finally, let $\mu(x) = f(x)/F(x)$ be the hazard function corresponding to F. Since we are assuming that F is effectively known from actuarial tables, the quantity $[\prod_{i=1}^{m} F(x_i;t_i)]^{-1}$ is also known, and will serve merely as a proportionality constant in the likelihood function. If $g_\theta(x;t)$ as given by (2.4) is substituted into (2.5), and the resulting expression is multiplied by the above proportionality constant, we obtain

$$L_1(p,\theta) \propto \prod_{i=1}^{m} [p\mu(t_i+x_i) + (1-p)\{\mu(t_i+x_i)V_\theta(x_i) + v_\theta(x_i)\}]. \tag{2.6}$$

This is then the exact likelihood function arising from the m "dead" patients, i.e., the uncensored data. Similarly, factoring out the proportionality constant $\prod_{j=1}^{n} F(y_j;s_j)$ yields

$$L_2(p,\theta) \propto \prod_{j=1}^{n} [p + (1-p)V_\theta(y_j)], \tag{2.7}$$

and this is the exact likelihood function arising from the n "survivors", or censored data. The overall exact likelihood function is thus obtained by multiplying $L_1(p,\theta)$, as given by (2.6), and $L_2(p,\theta)$, as given by (2.7). We note that although only $V_\theta(\cdot)$ enters into (2.7), the values $\mu(t_i+x_i)$ are needed in order to evaluate (2.6). Since we have assumed that F is known, it follows that $\mu(\cdot)$ is also known, and in practice the required $\mu(\cdot)$ values can be obtained either from standard actuarial tables, or alternately can be estimated with great precision from the extensive data that is available.

If the $\mu(t_i+x_i)$ are sufficiently small, then $L_1(p,\theta)$ can be approximated by

$$\hat{L}_1(p,\theta) \propto (1-p)^m \prod_{i=1}^{m} v_\theta(x_i). \tag{2.8}$$

For the Weibull model, with $V_\omega(x) = e^{-(\theta x)^\alpha}$, the approximation (2.8) yields the overall approximate likelihood function

$$\hat{L}(p,\theta,\alpha) \propto \hat{L}_1(p,\theta,\alpha)L_2(p,\theta,\alpha)$$

$$\propto (1-p)^m (\theta^\alpha\alpha)^m [\prod_{i=1}^{m} x_i]^{\alpha-1} e^{-\sum_{i=1}^{m}(\theta x_i)^\alpha} \prod_{j=1}^{n} [p+(1-p)e^{-(\theta y_j)^\alpha}]. \tag{2.9}$$

If $\alpha = 1$, as in the Berkson-Gage model, (2.9) gives

$$\hat{L}(p,\theta) \propto (1-p)^m\theta^m e^{-\theta\sum_{i=1}^{m} x_i} \prod_{j=1}^{n} [p+(1-p)e^{-\theta y_j}]. \tag{2.10}$$

When the ages t_i at which treatment occurred for the "dead" patients are known, the exact $L_1(p,\theta)$ should ordinarily be used, instead of $\hat{L}_1(p,\theta)$. However, sometimes these ages will be unknown, and in this case it will be necessary to use either $\hat{L}_1(p,\theta)$, or arbitrarily chosen values $\mu(t_i+x_i)$ (perhaps a constant value) and use these to obtain $L_1(p,\theta)$. This will be discussed further in Section 6.

Now let us consider the use of Makeham's Law for survival times. This family of distributions has been widely used by actuaries (Miller 1949), (Jordan 1952), for mortality of persons beyond youth. Although $L_2(p,\theta)$ does not depend upon the choice of F, the exact $L_1(p,\theta)$ does, through the hazard function $\mu(\bullet)$; also, as we shall see in Section 3, it is necessary to specify F in order to evaluate some of the quantities useful in decision-making as to choice of treatment.

A non-negative random variable W is said to have the Makeham distribution, F, with parameters a, b, c, if, for $x \geq 0$,

$$Pr\{W \geq x\} = F(x) = \exp -[ax + (b/c)(e^{cx}-1)] \tag{2.11}$$

When $c<0$, (2.11) defines a probability distribution if $a>0$ and $a + b > 0$; when $c>0$, it is necessary that either $b = 0$ and $a > 0$, or that $b > 0$ and $a + b > 0$; and, interpreting $c = 0$ as yielding $b/c(e^{cx}-1) = bx$, when $c = 0$, it is necessary that $a+b>0$. The hazard function $\mu(x)$ for the distribution is

$$\mu(x) = \mathrm{a} + be^{cx}, \tag{2.12}$$

and we note that it is strictly increasing if $c > 0$, constant if $cb = 0$, and strictly decreasing if $cb < 0$. Also, $F(x;t) = \exp -[ax + (b'/c)(e^{cx}-1)]$, where $b' = be^{ct}$, so that conditional upon survival to age t, the survival time still has a Makeham distribution with only a change in the value of the parameter b. If $V_\omega(x) = e^{-(\theta x)^\alpha}$ and $G_\omega(x;t) = V_\omega(x)F(x;t)$, then the hazard function for the survival time of an uncured patient (with θ and t as parameters) is $\theta\alpha(\theta x)^{\alpha-1} + a+b'e^{cx}$. For the Berkson and Gage exponential model $\alpha = 1$, and so this hazard function is $a'+b'e^{cx}$, where $a' = a+\theta$. Hence, in this case the $G_\omega(x;t)$ distribution remains of Makeham form with only a change in the value of the parameter a. This may be contrasted with the proportional hazards model (Breslow 1975) where the hazard function changes by a multiplicative constant.

For the decision-theoretic analysis of Section 3 certain expectations of the distribution $G_\omega(x;t)$ are required. Although in general there are no closed form solutions available, we observe that if the hazard function of the distribution of the nonnegative random variable

W is of the form $\mu(x) = \theta\alpha(\theta x)^{\alpha-1} + a + be^{cx}$, then

$$Pr\{W \geq x\} = \exp[-\int_0^x \mu(s)ds], \quad \text{and}$$

$$E[W] = \int_0^\infty Pr\{W \geq x\}dx = \int_0^\infty \exp-[(\theta x)^\alpha + ax + b/ce^{cx}]dx. \tag{2.13}$$

One can express, using similar methods, the expectation of a monotonic function of W. When $\alpha = 1$, so that the distribution of W is Makeham with positive parameters a, b, c, then the above expectation can be simplified to

$$E(W) = K \int_{b/c}^\infty z^{-1-a/c} e^{-z} dz, \quad \text{where } K = b^{-1}(b/c)^{1+a/c}e^{b/c}. \tag{2.14}$$

The integral appearing in (2.14) is related to Pearson's form of the incomplete gamma function (Abramowitz and Stegun 1965, p. 262).

3. STATISTICAL INFERENCE AND DECISION ANALYSIS

We take a Bayesian approach to inference and decision theory. See Jeffreys (1961, Lindley (1965), DeGroot (1970), for introductions to Bayesian inference and decision theory, and Hill (1963, 1965, 1975) for applications.

We shall assume that the prior knowledge that we desire to incorporate in our analysis is represented by a prior density function $P(p,\theta)$, with respect to Lebesgue measure. The product of $P(p,\theta)$ and $L(p,\theta)$, normalized to have unit area, i.e.,

$$P''(p,\theta) = P(p,\theta)L(p,\theta) \left(\int_0^1 \int P(p,\theta)L(p,\theta)dp\, d\theta\right)^{-1},$$

is then the posterior density function of the parameters. In conjunction with a utility function, the posterior distribution can be used for all purposes of inference and decision making.

We shall now proceed to suggest and examine those quantities which we view as most important in measuring the effectiveness of a treatment. We assume that on the basis of data as described above, and in conjunction with some prior distribution for (p,θ), that a posterior distribution of these parameters has been determined. All probability statements are then conditional upon the data.

We imagine a new patient, having the cancer in question, and contemplating taking treatment at age t. Let W be the random quantity representing the age at which this patient will die, conditional upon treatment at age t. Then we are interested in evaluating, for him, the function

$$R(x;t) \equiv Pr\{W \geq t+x \mid \text{treatment at age } t, \text{ data}\}. \tag{3.1}$$

If this patient is regarded as one for whom the model applies, then from (2.2), with $V_\theta(x;t) = V_\theta(x)$,

$$Pr\{W \geq t+x \mid \text{treatment at age } t, p, \theta, \text{data}\} = F(x;t)[p+(1-p)V_\theta(x)],$$

so integration with respect to the posterior distribution of (p,θ) yields

$$R(x;t) = F(x;t)\, E[p + (1-p)V_\theta(x) \mid \text{data}]. \tag{3.2}$$

$R(x;t)$ can then be interpreted as the posterior probability of surviving to age $t+x$, conditional upon treatment at age t; while $F(x;t)$ is the posterior probability of surviving to age $t+x$, conditional upon survival to age t, for a person without the specific cancer (or alternatively, a «cured» patient); and $F(x;t)\, E[V_\theta(x;t)\,|\,\text{data}]$ is the corresponding posterior probability for a patient given the treatment but not cured. We note that this last posterior probability distribution may be quite different from that for a patient not given the treatment. Clearly $R(x;t) \leq F(x;t)$, and as x grows large, $R(x;t)/F(x;t)$ tends to $E(p\,|\,\text{data})$, which can be interpreted as the posterior probability of a cure.

We view the function $R(x;t)$, or equivalently, the posterior distribution of survival time, given that the treatment in question is performed at age t, as the primary aid to clinical decision making. We shall make available tables of this function for various forms and stages of cancer, treatments (including no treatment), functions $V_\theta(x)$, and prior distributions on (p,θ). If, for simplicity, we ignore side effects, recurrences of the tumor, and all other aspects of the problem except for age at death, W, then the utility of performing the treatment upon the patient can be expressed in the form

$$E\,[U(W)\,|\,\text{treatment at age } t\,],$$

where $U(w)$ is a utility for survival to age w and W has the distribution of (3.1). We shall not attempt to specify the form that such a utility function should take, either for the patient or the physician. If $U(w) \equiv w$, then of course we have simply the expected life time, given treatment at age t, namely,

$$R\,(t) \equiv E\,[W\,|\,\text{treatment at age } t] = \int_0^\infty R\,(x;t)\,dx. \tag{3.3}$$

Although we shall make use of $R\,(t)$ as well as $R\,(x;t)$, it is important to note that many individuals may tend to have strictly concave utility functions when age is large, just as with utility functions for money. Thus an extra yeaı of life may be viewed as more valuable at age 50 than at age 100. Although we shall not discuss this question further here, there could be substantial advantages gained by giving careful consideration to the form of the utility function.

Now suppose that we contemplate two different treatments, with associated functions $R^{(i)}(x;t)$, $i=1,2$. These functions of x may cross one another. For example, if p_i is the parameter for treatment i, then although, say, $E\,(p_1\,|\,\text{data}) > E\,(p_2\,|\,\text{data})$, so that for large x, $R^{(1)}(x;t) > R^{(2)}(x;t)$, it may well be that treatment 2 yields a larger expectation of life, conditional upon *not being cured*, and that $R^{(2)}(x;t) > R^{(1)}(x;t)$ for small values of x. This serves to emphasize the fact that inspection of the entire function $R\,(x;t)$ will ordinarily be necessary in order to make intelligent decisions, when no explicit utility function is incorporated into the analysis. (If clinical trials are contemplated then of course still other considerations become relevant, such as two armed bandits, but we do not examine this here).

We now consider the evaluation of (3.2) for the case $V_\theta(x;t) = e^{-(\theta x)^\alpha}$, and F a Makeham distribution. First suppose $\alpha = 1$. Then, assuming treatment at age t,

$$E\,[W\,|\,W \geq t, p, \theta] = p\,E\,[W_0\,|\,W_0 \geq t] + (1-p)E\,[W_\theta\,|\,W_\theta \geq t] \tag{3.4}$$

where W_0 is a random variable having the Makeham distribution with parameters a_0, b_0, c_0, and W_θ is a random variable having the Makeham distribution with parameters $a_0+\theta, b_0$, c_0. If $M\,(a, b, c)$ denotes the expectation of the Makeham distribution with parameters a, b, c, then $M\,(a, b, c)$ is given by (2.14). It follows from (2.2) and (3.4) that

$$E\,[W\,|\,W \geq t, p, \theta,] = M\,(a_0, b_0', c_0)p + M\,(a_0+\theta, b_0', c_0)(1-p),$$

where $b_0' = b_0 e^{c_0 t}$. Integration with respect to the posterior distribution of (p,θ) then yields

$$R\,(t) = M\,(a_0, b_0', c_0)\,E[p\,|\,\text{data}] + E\,[(1-p)M\,(a_0+\theta, b_0', c_0)\,|\,\text{data}]. \tag{3.5}$$

In the general case, where α need not be 1, the same type of analysis can be made, except that now W_ω has the distribution $G_\omega(x;t) = e^{-(\theta x)^\alpha} F(x;t)$.

Although in general the functions $R(x;t)$ and $R(t)$ will have to be evaluated by numerical analysis, when the posterior distribution of (p,θ) is sufficiently sharply concentrated about some point, say, $(\hat{p}, \hat{\theta})$, then as an approximation we may employ

$$\hat{R}(x;t) = F(x;t)\,[\hat{p} + (1-\hat{p})e^{-(\theta x)^\alpha}] \tag{3.6}$$

and

$$\hat{R}(t) = \int_0^\infty \hat{R}(x;t)\,dx. \tag{3.7}$$

From a conventional statistical point of view, with $(\hat{p},\hat{\theta})$ the maximum-likelihood estimate of (p,θ), (3.6) and (3.7) can be viewed as maximum-likelihood estimates of $R(x;t)$ and $R(t)$, respectively.

Finally, let $\hat{s}(x) = \hat{p} + (1-\hat{p})e^{(\theta x)^\alpha}$. It is this function that we use in our analysis of cancer data below as an indicator of survival time for patients given the treatment. For our data, $\hat{s}(x)$ provides a smooth approximation to the Kaplan-Meier survival curve. Note, however, the direct way in which age at treatment enters into (3.6) through $F(x;t)$, and for which there appears to be no analogue in other treatments of survival curves, except perhaps when age is treated as a covariate.

4. ANALYSIS OF THE LIKELIHOOD FUNCTION

In this section we study the likelihood functions $L(p,\theta,\alpha)$ and $L(\phi_1,\phi_2,\phi_3)$ where $\phi_1 = \ln[p/(1-p)]$, $\phi_2 = -\ln\theta$, and $\phi_3 = \alpha$. The methods used are appropriate for both exact and approximate likelihood functions. At the close of this section the relationship between the suggested analysis of the likelihood function and a complete Bayesian analysis is examined.

Define the functions

$$H_1(p) = \ln L(p,\hat{\theta}(p),\hat{\alpha}(p)), \qquad H_2(\theta) = \ln L(\hat{p}(\theta),\theta,\hat{\alpha}(\theta)),$$

$$H_3(\alpha) = \ln L(\hat{p}(\alpha),\hat{\theta}(\alpha),\alpha),$$

$$\tilde{H}_1(\phi_1) = H_1(e^{\phi_1}/(1+e^{\phi_1})), \qquad \tilde{H}_2(\phi_2) = H_2(e^{-\phi_2}),$$

$$\text{and} \quad \tilde{H}_3(\phi_3) = H_3(\phi_3),$$

where $(\hat{\theta}(p),\hat{\alpha}(p))$ maximizes $L(p,\theta,\alpha)$ for fixed p, and $(\hat{p}(\theta), \hat{\alpha}(\theta))$ maximizes $L(p,\theta,\alpha)$ for fixed θ, etc. The overall maximum-likelihood estimates $(\hat{p},\hat{\theta},\hat{\alpha})$ or $(\hat{\phi}_1,\hat{\phi}_2,\hat{\phi}_3)$ can then be obtained by maximizing any of the $H_i(\cdot)$ or $\tilde{H}_i(\cdot)$. We note that if $L(p,\theta,\alpha)$ were approximately trivariate normal centered at $(\hat{p},\hat{\theta},\hat{\alpha})$, then $H_1(p)$, $H_2(\theta)$ and $H_3(\alpha)$ would be approximately quadratic functions in the vicinity of $\hat{p}$,$\hat{\theta}$, and $\hat{\alpha}$, with maxima at $\hat{p}$, $\hat{\theta}$, and $\hat{\alpha}$, respectively. Of course the same statement could be made for $L(\phi_1,\phi_2,\phi_3)$ and the $\tilde{H}_i(\phi_i)$. From the known asymptotic behavior of likelihood functions (Lindley 1975, Ch. 7), (DeGroot 1970, Ch. 10), we can anticipate approximately such behavior. Of course this can and should be checked in each case for the realized likelihood functions. When such approximate normality holds, then the sample information

$$I_i \equiv -\frac{\partial^2 H_i(z)}{\partial^2(z)}, \quad i=1,2,3, \text{ or } \tilde{I}_i \text{ if } \tilde{H}_i \text{ is used in place of } H_i,$$

evaluated at its maximum, $\hat{z}_i$, say, can be interpreted (from a Bayesian point of view) as the reciprocal of the sampling variance associated with the corresponding variable. In other words, for a particular parameter, say ϕ_1, it is approximately as though a normal measurement $\hat{\phi}_1$, with known variance $\tilde{I}_1^{-1}$, had been made. The sample information matrices

$$J_1 = -\begin{bmatrix} \dfrac{\partial^2 \ln L}{\partial^2 p}, & \dfrac{\partial^2 \ln L}{\partial p \partial \theta}, & \dfrac{\partial^2 \ln L}{\partial p \partial \alpha} \\ & \dfrac{\partial^2 \ln L}{\partial^2 \theta} & \dfrac{\partial^2 \ln L}{\partial \theta \partial \alpha} \\ & & \dfrac{\partial^2 \ln L}{\partial^2 \alpha} \end{bmatrix}, \text{ and } J_2 = -\begin{bmatrix} \dfrac{\partial^2 \ln L}{\partial^2 \phi_1}, & \dfrac{\partial^2 \ln L}{\partial \phi_1 \partial \phi_2}, & \dfrac{\partial^2 \ln L}{\partial \phi_1 \partial \phi_3} \\ & \dfrac{\partial^2 \ln L}{\partial^2 \phi_2} & \dfrac{\partial^2 \ln L}{\partial \phi_2 \partial \phi_3} \\ & & \dfrac{\partial^2 \ln L}{\partial^2 \phi_3} \end{bmatrix}$$

evaluated at their respective maxima, have a similar interpretation in terms of the trivariate distributions. We note that a likelihood or Bayesian approach leads one to focus attention on such *sample* informations, as opposed to the more conventional Fisherian information, since the sample informations pertain to the realized likelihood functions, and do not average over all possible data sets, as does the Fisherian information. We also note that approximate normality is to be anticipated for $L(\phi_1,\phi_2,\phi_3)$ in a stronger sense than for $L(p,\theta,\alpha)$. This is true in part because the ϕ_1, $i=1,2$, can take on any real value, while p and θ are constrained to be such that $0 \leq p \leq 1$, and $\theta > 0$. Another reason for such an anticipation stems from the fact that if X has the beta distribution with parameters a and b, where a and b are moderately large, then $\ln[X/1-X]$ is approximately normally distributed (Lindley 1965, p. 147). Now from the approximate form of $L_1(p,\theta,\alpha)$ in (2.8), we anticipate a factor of the form $(1-p)^m$. Similarly from inspection of $L_2(p,\theta,\alpha)$ we see that when some y_j is nearly 0, then the factor $p + (1-p)e^{-(\theta y_j)^\alpha}$ is nearly 1, thus contributing negligible information. For large y_j, on the other hand, $p + (1-p)e^{-(\theta y_j)^\alpha}$ is approximately p, so that we anticipate a factor of the approximate form $p^{n_1}(1-p)^m$ in the likelihood function, where there are n_1 "large" y_j, i.e., individuals who have survived sufficiently long to be regarded nearly as cures. This suggests that the overall likelihood function has a factor nearly of the beta form, and therefore that in terms of $\ln[p/(1-p)]$ we should have approximate normality.

Finally, we would like to relate the above analysis of the likelihood function to a full Bayesian analysis. First, we observe that when sample sizes are moderately large, then it can be anticipated, due to the phenomenon of stable estimation (DeGroot 1970, p. 198) that the likelihood function will ordinarily be sharply concentrated in a neighborhood of its maximum, relative to the prior density. In this case, to a first approximation, the likelihood function will be nearly proportional to the posterior density, over a region having high posterior probability. For small samples, of course, this will not be the case, and in any event it is wise to check the approximation for the realized likelihood function, and some appropriate class of prior densities. Second, although for small sample sizes (or for uninformative data, generally), there are substantial gains possible by incorporating realistic forms of prior knowledge, many of the advantages of our approach stem from the model itself, and from the Bayesian formulation of the problem, as in the previous section. Since our use of the likelihood function and maximum-likelihood estimates can be justified from the classical point of view, as well as the Bayesian, in order to reach the widest audience it seems appropriate to focus upon those aspects of the analysis that are common. As we have already noted many advantages of the Bayesian approach, such as the simplification of the likelihood function, the use of sample information instead of Fisherian information, and the choice of $R(x;t)$ as the fundamental aid to decision making, are already incorporated into our analysis. Thirdly, the type of analysis of the likelihood function that we have suggested can be applied equally well to the posterior densities. $P(p,\theta,\alpha)L(p,\theta,\alpha)$ or $P(\phi_1,\phi_2,\phi_3)\ L(\phi_1,\phi_2,\phi_3)$, and the fact that we have implicity taken $P(\phi_1,\phi_2,\phi_3) \equiv 1$, should not deter others from using whatever prior distributions they choose. For clinical decision making we certainly plan to incorporate what we consider to be realistic forms of knowledge, and a utility function $U(\bullet)$. However, it would only complicate matters unnecessarily to do so here.

Finally we would like to add a word of caution with regard to the use of improper prior distributions. Suppose, for example, one contemplated choosing a prior density of the form

$$P(p,\theta,\alpha) \propto p^{a-1}(1-p)^{b-1}\theta^{c-1}e^{-d\theta}\alpha^{f-1}e^{-g\alpha},$$

I.e., with p having a beta distribution with parameters $a > 0$, $b > 0$, θ having a gamma distribution with parameters $c > 0$, $d > 0$, α having a gamma distribution with parameters $f > 0$, $g > 0$, and p, θ and α a priori independent. Such a family is reasonable rich, and is particularly convenient for the analysis. In this case the exact posterior density would be

$$p''(p,\theta,\alpha) \propto p^{a-1}(1-p)^{b-1}\,\theta^{c-1}\,e^{-\theta d}\,\alpha^{f-1}\,\theta^{-g\alpha}$$
$$\times \prod_{i=1}^{m} [p\mu(t_i+x_i) + (1-p)e^{-(\theta x_i)^\alpha}\{\mu(t_i+x_i)+\theta\alpha(\theta x_i)^{\alpha-1}\}]$$
$$\times \prod_{j=1}^{n} [p + (1-p)e^{-(\theta y_j)^\alpha}]. \tag{4.1}$$

A conventional improper prior density would take $a = b = c = d = f = g = 0$, in which case $P(p,\theta,\alpha) \propto [\alpha\theta p(1-p)]^{-1}$. However, inspection of (4.1) reveals that in this case the posterior density will remain improper, unless some $\mu(t_i+x_i) = 0$, and some $y_j = \infty$. This situation is closely analogous to that arising in Bayesian analysis of variance components (Hill 1965, p. 811) and suggests that improper prior distributions of the above form are not appropriate. Note that $P(p,\theta,\alpha) \propto [\alpha\theta p(1-p)]^{-1}$ implies

$$P(\phi_1,\phi_2) \propto 1,$$

so that in using $L(\phi_1,\phi_2,\phi_3)$ as an approximate posterior distribution, we should think of the approximation as appropriate only in a suitable neighborhood of $(\hat{\phi}_1,\hat{\phi}_2,\hat{\phi}_3)$. In effect this amounts to viewing $\ln[p/1-p]$ and $\ln[1/\theta]$ as uniform only over some huge but finite interval, rather than over the whole real line, as would be the case if $P(\phi_1,\phi_2)$ were identically constant. One might also take $\ln \alpha$ as independently uniform over some finite interval, a priori, but this is less compelling for α than for the scale parameter θ because of the special significance attached to $\alpha = 1$ and $\alpha = 2$.

5. ANALYSIS OF CANCER DATA

In this section, we apply some of the theory derived above to the analysis of two data sets. The first data set, consisting of three groups, A, B and B*, concerns cancer of the endometrium. From the years 1955 through 1974, 189 patients with carcinoma of the endometrium were treated with preoperative irradiation at the University of Michigan. All of the cases were Stage II, (disease confined to the uterus; good surgical risk).

The treatment consisted of giving external irradiation initially with a 250 Kv unit and in more recent years, with Co60 radiation (a more penetrating beam). In cases in which it was not possible to give external irradiation, mainly because of obesity, the patient had a radium implant. In either case, whether treated by radium or external irradiation, the patient had surgery 6 weeks after irradiation. After surgery the uterus was examined for depth of penetration of the tumor into the uterine wall. The patients were then divided into three groups according to degree of penetration. The groups and values of m and n are:

A = (1) No residual tumor present or carcinoma *in situ* ($m=1$, $n=91$).

B = (2) Tumor penetrating one half or more of the wall of the uterus ($m=11$, $n=18$).

B* = (3) Tumor penetrating the wall of the uterus but less than half of its thickness ($m=5$, $n=63$).

(Group A had only one death, and although it can be analysed by our methods, it is relatively uninformative to do so. We have analised all three of these groups, but present only B from the first data set.)

The second data set is from a clinical trial for the treatment of leukemia, and has been described and analysed previously by Gehan (1965) and Cox (1972). Group C is the group receiving 6-mercaptopurine for the maintenance of remission in acute leukemia ($m=9$, $n=12$).

The statistical methods described previously will now be applied to the two groups, B and C. The raw data consist of survival time in months after treatment for group B, and length of remission in weeks for group C. For group B, the ages at which treatment occurred are known, and so it is possible to obtain both the approximate and exact likelihood functions. For group C the ages are unknown, and so only the approximate likelihood function can be obtained. Note also that the intepretation of the underlying model is different for group C, because although p, θ, and α can retain the same intrinsic meaning as in our general model, $F(x;t)$ must now be interpreted in terms of the appearance of symptoms for the appropriate population without acute leukemia. There would seem to be little data upon which to base an estimate of $F(x;t)$. However, since the approximate likelihood function does not depend upon $F(x;t)$ in any way, it is still possible to make the analysis, and the estimated survival curve in fact turns out to be a smooth approximation to the Kaplan-Meier estimate. This is discussed further in Section 6 in connection with the interpretation of the approximate likelihood function.

Table I gives the ordered survival times for the two groups, and the ages, t_i, and values, $\mu(t_i+x_i)$, for group B. Tables IIA and IIB give the maximum likelihood estimates for (ϕ_1,ϕ_2,ϕ_3) and (p,θ,α), and standard errors and correlations, as derived from both the approximate and exact likelihood functions (using sample informations in place of Fisherian informations, as discussed in Section 4). Tables IIIA and IIIB give our estimated survival curve, $\hat{S}(x)$, based upon the Weibull model and the Berkson and Gage model, and for comparison the Kaplan-Meier product limit estimate (Kaplan and Meier, 1958), and the Life-Table estimate (Jordan, 1952, Ch. 14).

Figures A, and B display some of the functions of Table III for the two data sets, and Figures C-H display the univariate "marginal" log likelihood functions for the ϕ_i, $i=1,2,3$. The quadratic form of these functions supports our use of normal theory in approximating the various posterior distributions.
Contour plots of the bivariate likelihood functions, not shown here, also support use of the normal approximation.

It is interesting to note that use of the exact likelihood function, in place of the approximate likelihood function, makes little change in both the estimates $(\hat{\phi}_1,\hat{\phi}_2,\hat{\phi}_3)$ and $\hat{S}(x)$. As remarked earlier, in principle the exact likelihood function is preferable whenever the (x_i,t_i) are known and the $\mu(t_i+x_i)$ are known or can be estimated with enough precision. In practice, however, there will always be some question as to whether the population used for estimation of the function $\mu(\cdot)$ is appropriate for the data at hand. Considerations regarding robustness of Bayesian inference (Hill 1975, 1980; Berger 1982) will then sometimes favor use of the approximate likelihood function. At this time we are not in a position to make a definitive recommendation with regard to group B. The fact that $\hat{S}(x)$ based upon the approximate likelihood function gives a better approximation to the Kaplan-Meier survival curve than does $\hat{S}(x)$ based upon the exact likelihood function should not necessarily be viewed as support for use of the approximate likelihood function. The role of $F(x;t)$ in our model should be kept in mind, as well as the fact that the Kaplan-Meier curve implicity involves an averaging over the different ages of treatment represented in the data, while $\hat{S}(x)$ does not.

6. DISCUSSION

In this section we shall discuss the key aspects of the present approach, and relate it to some current methods of analysing cancer survival data.

The most important difference between our approach and others lies in the model. Ours is based upon the mixture concept of Berkson and Gage, and this seems to have been rejected by most workers in favor of parametric or non-parametric models not explicitly involving mixtures (Gehan 1969, Kaplan and Meier 1958). The reasons for such rejection are not apparent. Berkson and Gage have presented strong evidence that the simple mixture, with V_θ exponential, works surprisingly well for some cancer data (even with very large sample sizes), and this is supported by our own data analysis. Their observation (Berkson and Gage 1952, p. 503) that the survival curve for the specific cancer (plotted on semi-log paper) becomes parallel to that for the general population after a sufficient period of time, suggesting that, after the non-curables have died out, the remaining portion of the treated cancer patients behave just like the general population, seems to lend support to the mixture model. Of course, as we have shown, there is no need to restrict V_θ to be exponential, and within the more general Weibull family there is some real improvement in fit possible. For both data sets (despite the great difference between the leukemia data and the other data), $\hat{\alpha}$ is close to 2 for the approximate likelihood function. Since $\alpha = 2$ yields a linear hazard function, this value may be of special significance, just as $\alpha = 1$ is of special significance. On the other hand, with regard to the estimated survival curve, $\hat{S}(x)$, the effect of using $\hat{\alpha}$ in place of $\alpha = 1$ is not of great practical importance, so that in this sense our use of the Weibull model lends support to the exponential model of Berkson and Gage as a useful approximation.

The proportional hazards model (Breslow 1975) is a special case of our model with general $V_\theta(x;t)$. It is special both in not using the mixture formulation (although it could easily be incorporated into a mixture model), and in choosing a very special form for $V_\theta(x;t)$. Thus if

$$Pr\{W \geq t+x \mid W \geq t,\theta\} = Z(F(x;t),\theta)$$

gives the survival distribution for the proportional hazards model, where θ does not depend upon x or t, then such a model implies

$$\frac{\partial \ ln\ Z(F(x;t),\theta)}{\partial x} = K_\theta \frac{\partial \ ln\ F(x;t)}{\partial x}$$

Since the underlying model is that the force of mortality is greater for the cancer patients than for the general population, we must also have $K_\theta > 1$. In terms of our model for the non-cured patients, however, we have,

$$Z(F(x;t),\theta) = F(x;t)\; V_\theta(x;t),$$

so that the proportional hazards model is the special case of our general model in which $p = 0$ and $V_\theta(x;t) = [F(x;t)]^{K_\theta - 1}$, for $K_\theta > 1$. We feel that strong theoretical arguments and empirical evidence would be required to justify such a restriction. We remark that much of the simplification of the likelihood function by Breslow under the proportional hazards model is, as we have shown, still possible in the more general model.

Next let us consider our assumption $V_\theta(x;t) = V_\theta(x)$, so that the additional force of mortality does not depend upon the age at which treatment occurs. This can be motivated through the fact that the first-order effect of age has already been taken into account in $F(x;t)$. With this assumption our model separates out the factor $V_\theta(x)$, as representing the non-age dependent additional force of mortality due to the cancer and its (ineffective) treatment, from the age dependent ordinary force of mortality. This has some extremely important consequences, for example, $\hat{R}(x;t) = F(x;t)[\hat{p} + (1-\hat{p})e^{-(\theta x)^\alpha}] = F(x;t)\hat{s}(x)$, so that in using our methods a patient merely adjusts $\hat{s}(x)$, as obtained from the data, by multiplying by the known $F(x;t)$ (for his age t), to obtain the estimate of his probability of surviving x time units. Unless $F(x;t)$ is nearly 1 there will be a real adjustment due to $F(x;t)$, and $\hat{s}(x)$ by itself would be an overestimate of the patient's survival probability. This suggests that perhaps the Kaplan-Meier estimate, which for our data is well-approximated by $\hat{s}(x)$ based upon the approximate likelihood function, should also be adjusted for the age at treatment (for example, with age as a covariate), unless of course $F(x;t)$ is nearly 1. However, the precise role of age at treatment seems to be brought out much more clearly in our analysis than in that of other methods of analysis. We note also that under our model the asymptote of the Kaplan-Meier curve, for large x, can be interpreted as the probability of cure.

The last discussion of the role of age of treatment is also pertinent to understanding of the approximate likelihood function (2.9) that is obtained by setting $\mu(t_i+x_i) = 0$, $i = 1,\ldots,m$. In a crude sense this approximation is equivalent to the assumption that $F(x_i;t_i)$ is large compared to $G_\theta(x_i;t_i)$ for the m "dead" patients (or more accurately that the $v_\theta(x_i)/\mu(t_i+x_i)$ are large), so that these patients are very likely to have been uncured. This, however, does not imply that the mixture model has been abandoned since the factor $(1-p)^m$ for the uncensored data can only arise within the mixture model, and of course the mixture model is still being used for the censored data. When either ages at treatment are unknown (as with the leukemia data) or when it is plausible to take $F(x_i;t_i) \approx 1$, the approximate likelihood function can serve a very useful purpose. On the other hand we regard it as advantageous that when both the t_i and $\mu(t_i+x_i)$ are known, our model allows us to make appropriate modifications. The estimate $\hat{\phi}_j$ are continuous functions of the $\mu(t_i + x_i)$, so that when the latter are sufficiently small it makes little difference whether the exact or the approximate likelihood function is used. For large values $\mu(t_i + x_i)$ the following heuristic argument suggests the nature of the modification obtained by using the exact likelihood function in place of the approximate likelihood function. In effect a large $\mu(t_i+x_i)$ increases the chance that a particular "dead" patient was in fact cured, and hence tends to raise the estimate of p. Also, for such a patient, if x_i is large, then it influences the estimate of θ to a lesser extent than do the smaller x_j, so that the estimate of θ tends to increase by use of the exact likelihood function. Both of these phenomena are seen to occur in the data we analyse. Of course the above heuristic explanation is only meant to provide some insight, and precise understanding is to be obtained from careful study of the likelihood function.

The next major difference between our work and that of others lies in our use of the Bayesian approach, and consequent reliance upon careful analysis of the likelihood function. The reason that we were able to obtain a relatively simple analysis of this rather complicated model is because we were able to factor out of the likelihood function proportionality constants that depend only on $F(x;t)$. In non-Bayesian approaches the distribution of such proportionality constants must usually be allowed for, as for example in the analysis of data from a binomial distribution with fixed number of trials, as compared to analysis of the "same" data from a negative binomial distribution, with fixed number of successes (DeGroot 1970, p. 165). But even apart from running counter to the likelihood principle, such allowance for the distribution of proportionality constants would enormously complicate the analysis. Thus although in the present article the estimates we use are in fact maximum-likelihood estimates, we are not concerned with the very complicated sampling distributions of such estimates. Rather we examine the likelihood function itself, and use it to assess the intrinsic precision of our "estimates", particularly via the sample informations. As figures *C-H* suggest, the "marginal" likelihood functions are very nearly of Gaussian form, and for many purposes the overall posterior distribution can be regarded as approximately trivariate normal centered at $(\hat{\phi}_1, \hat{\phi}_2, \hat{\phi}_3)$. The approximate posterior distribution of (p,θ,α) can then be obtained either from that of (ϕ_1, ϕ_2, ϕ_3) or directly via $L(p,\theta,\alpha)$. Comparisons between different groups such as *A* and *B* can then be made using standard normal theory, for example, to obtain the posterior distribution of the difference between ϕ_1 for the two groups.

There is an interesting question regarding the simplification of the likelihood function for the censored patients that so far has been ignored. The Y_j are the random withdrawal times for the «survivors», and an event $Y_j = y_j$ means that the corresponding patient was last known to be alive at age $s_j + y_j$. In our evaluation of $L_2(p,\theta,\alpha)$ we implicitly replaced such data, $Y_j = y_j, j=1,\ldots n$, by the implied events $X_j \geq y_j, j=1,\ldots n$, where $X_j \geq y_j$ means that the corresponding patient lives at least y_j units after treatment. Such a replacement is fairly standard in the literature, but the subjective Bayesian approach sheds some light upon the assumption underlying this replacement. The equation that expresses such an assumption is

$$\begin{aligned} P(Y_j=y_j, j=1,\ldots,n \mid p,\theta,\alpha) &= P(Y_j=y_j, j=1,\ldots,n \mid X_j \geq y_j, j=1,\ldots,n,p,\theta,\alpha) \\ &\quad \times P(X_j \geq y_j, j=1,\ldots,n \mid p,\theta,\alpha) \\ &\propto P(X_j \geq y_j, j=1,\ldots,n \mid p,\theta,\alpha). \end{aligned}$$

Thus the underlying assumption can be described as asserting that conditional upon survival at least y_j time units after treatment, the precise value $Y_j = y_j$ of the "withdrawal" is totally uninformative about the parameters (p,θ,α), so that $P(Y_j = y_j, j=1,\ldots,n \mid X_j \geq y_j, j=1,\ldots,n, p,\theta,\alpha)$ is merely another proportionality constant. If, for example, the death occurs due to causes thought to be unrelated to the cancer, such as being struck by lightning or a meteorite, then this would be the case. This assumption can sometimes be justified as an approximation by the approach of (Hill 1975, 1980) towards robust Bayesian inference, but should not be made automatically, since it is easy to imagine and construct models under which the exact withdrawal times would be informative. Although for the data analysed in this article we believe the assumption to be appropriate, at least as a first approximation, we think that it would be well to keep such considerations in mind. Note that from a non-Bayesian point of view it may be difficult, or even impossible, to talk about a distribution for the Y_j, since such withdrawal times may be governed by a host of subjective factors.

The final difference between our analysis of cancer survival data and other methods lies in our use of the subjective Bayesian decision-theoretic formulation of the problem. This leads us to $R(x;t)$, $R(t)$, and more generally, the posterior expected utility $E[U(W)|\text{treatment at age } t, \text{data}]$, for a given treatment, as the measure of the effectiveness of the treatment. To the best of our knowledge such quantities do not appear explicitly in other formulations. It should be clear how the methods we have proposed, for example, the study of $\hat{R}(x;t)$, or $\hat{s}(x)$, can be modified to allow for various utility functions, prior distributions, and other sources of information, such as covariates, as well as more general models $V_\theta(x;t)$ for the additional force of mortality. This is a highly non-trivial advantage of the subjective Bayesian approach over other approaches which do not easily allow for modifications.

7. SOME ISSUES IN BAYESIAN ANALYSIS OF CANCER SURVIVAL CURVES

Since readers are presumably knowledgeable regarding the basics of Bayesian inference, I thought it might be interesting to address some of the more sophisticated and difficult aspects of the subject, using the survival model to illustrate and discuss these issues. The questions I would like to discuss, some of which have been raised by kind (and some by not so kind) readers concern:

1. Bayesian inference versus (subjective) Bayesian modelling.
2. What is meant by a cure to cancer?
3. Is the problem of assessing survival probabilities after treatment for cancer, in a serious and meaningful way, beyond the capability of even the subjective Bayesian approach? Is it just too difficult?
4. Does the analysis we have proposed in effect amount to using the general population as a control group, somewhat like a merely historical (hysterical?) control)
5. What about Bayesian robustness considerations? If there is reason to think that the patients are of a very special subpopulation, perhaps a high risk group even apart from their cancer, is it better to use $F(x;t)$ as in the article, to use some other $F(x;t)$ even though it is based upon very scant data, to use the analysis based upon the approximate likelihood function, or to punt?

To my way of thinking these questions are intimately related, and although my answers will appear as answers to specific questions, in fact the answer to each will bear upon the other.

1. There has been much more attention given to the problems of Bayesian inference and decision-making than to the much deeper problem of modelling from a subjective Bayesian viewpoint. The latter raises difficult questions concerning the role and meaning of parameters, and verifiability. (Some of these questions appear in a different guise in the statistical literature, for example, the controversy over prediction as opposed to parameter estimation). de Finetti, in his magnificent volumes on the theory of probability, and elsewhere, has of course discussed these questions with his usual vision and insight. Although I cannot go into these questions here, I would like to say that in my opinion the

mixture model of Boag, and of Berkson and Gage, does exemplify some of the important features of subjective Bayesian modelling. At the heart of the de Finetti and Savage theory is the idea that as much as possible of the available information in a problem be utilized in the analysis. In the cancer problem this information includes the fact that, at least for certain cancers, many people seem to return to their former health status and live out apparently ''normal'' lives. This is what Berkson and Gage tried to argue by showing that even very crude estimates of the hazard function reveal that after a period of time during which the treated cancer patients are subject to high mortality, the hazard function for the survivors becomes very close to that for the general population. It seems to me that to entirely ignore such information (as is commonly done) is disastrous.

2. When one tries to incorporate such information into the analysis of the data, an objection that is often raised is that there is no way to give an adequate definition of what is meant by a cure for cancer. I agree that there is no way at present to do this, and perhaps there never will be. Does this mean that one should ignore the observation of Berkson and Gage concerning the apparent resumption of normal lives? Does anyone imagine that there are adequate definitions of quarks or even electrons? Yet scientists do not ignore quarks now, nor electrons back when they were only dimly understood. Another way to answer this particular objection is to ask whether there is an adequate definition of what is meant by a cure for the common cold or for a sore throat. Next, it seems appropriate to observe that since our model includes $p=0$ as a special case, if it were really the case that for a particular cancer there was nothing remotely resembling a cure, then presumably the data should so indicate. As you can see from the article the data indicate nothing of the sort. Finally, one can, if so inclined, regard our mixture model without the cure interpretation and merely as a generalization of standard models, for example the simple Weibull model, although personally I think this loses much of the meaning of the model.

3. The analysis of cancer survival data does push the subjective Bayesian approach, both modelling and inference, close to its limits. This is meant here only in the sense of providing answers in which one can put faith. (The Kaplan-Meier and Cox methods, are for me, so preposterous in this sense that I should think that those who advocate such analyses must have some other aim in mind than providing meaningful answers to important questions). With regard to the subjective Bayesian approach the crunch comes when one chooses an $F(x;t)$, or equivalently a hazard function, when one has reason to regard the treated patients as being from some special subpopulation, perhaps a high risk group, and yet one has little or no direct data upon which to base the choice of the hazard function. There seem to be three meaningful ways to try to deal with this problem. First, one can simply modify the standard $F(x;t)$ according to ones best judgment, and use the exact likelihood function; second, one can use the approximate likelihood function as a sort of default analysis; third, one can put in covariates for each patient to represent what is known about both his own $F(x;t)$ and his potential for cure. The trouble with the first is that it may make the analysis too sensitive to prior opinion about $F(x;t)$ in a situation where there is no real opportunity to learn much about it from data. As discussed in Section 6, the approximate likelihood function is appropriate when the patient, if cured, is under relatively low risk, but by the same token this analysis sacrifices much when this is not the case (however, in my opinion this analysis is still vastly superior to the methods now in use). The third method, use of covariates, is in principle the most sensible thing to do, but it would introduce several new parameters, and may make serious analysis very difficult. Despite these problems, and they are very real problems, there is some hope for an effective analysis, which I will discuss when I turn to robustness considerations.

4. I regard the question of historical controls more in the nature of a misunderstanding than a problem. We do not use the actuarial $F(x;t)$ tables as a control, but rather as an adjustment to allow us to obtain better inference about survival probabilities for the treated patients. We can so use these tables both for a single treatment or, to compare two different treatments or stages of the cancer, with an adjustment for each. We could run into troubles of the type mentioned above if, for example, in a comparison of two treatments, we thought the appropriate $F(x;t)$ for one treated group was quite different than for the other, and one or both was not sufficiently well known. However, even in this case it is not a problem of historical controls, but rather a question of Bayesian robustness.

5. The question of robustness with regard to both prior distribution and likelihood function is, of course, always around to plague us. (The distinction between prior distribution and likelihood function is more blurred than usual here, since $F(x;t)$ can be regarded as part of the model. However, those familiar with random effects models will recognize that such blurring is not really all that uncommon, and that this really comes down to some of the deeper questions alluded to earlier regarding subjective Bayesian modelling). In a recent article, Berger (1982), reformulated the robustness question and made some valuable contributions. My own point of view differs slightly, and has been expressed in several articles, but especially in Hill (1980). When two different Bayesian analyses of the same data lead to radically different inferences and/or decisions, then all one can hope to do is to become aware of this situation (as for example in random effects models), if possible perform an experiment to resolve the difficulty, and if this is not possible, choose as best one can. In the cancer problem one can and should try several plausible $F(x;t)$ functions if one is not happy with the standard table, also try the approximate likelihood analysis, and determine whether any, and if so, which, of these analyses, lead to materially different conclusions. If one is fortunate, then only very (subjectively) unlikely $F(x;t)$ functions will so affect the conclusions, i.e., which of two treatments appears superior. If one is unlucky, then slight changes in $F(x;t)$, so slight that one has little basis for choice, will so affect the conclusions. This is not an enviable position to be in, but when it's there it's there, and unlike Unbayesians, we can at least learn to be aware of the situation.

TABLE I

Survival times for Groups B and C

Group	Events: Survival times x_i	Events: Age t_i	Events: Normal Force of Mortality $\mu(t_i+x_i)^1$	Censored: Survival times y_j
B	3**	59***	0.0007	*3,9,10,14,14,24,31,31,37,39, 52,70,75,76,97,100,103,108
	8	77	0.0041	
	11	63	0.0010	
	11	66	0.0014	
	11	69	0.0024	
	14	59	0.0010	
	16	56	0.0007	
	16	56	0.0007	
	19	64	0.0014	
	35	70	0.0024	
	36	80	0.0065	
C	6**	.	.	**6,9,10,11,17,19,20, 25,32,32,34,35
	6	.	.	
	6	.	.	
	7	.	.	
	10	.	.	
	13	.	.	
	16	.	.	
	22	.	.	
	23	.	.	

(1) *Vital Statistics of the United States, 1976,* Vol. II - Mortality, Part A, National Center for Health Statistics. Yearly basis.

* months ** weeks *** years

TABLE II A

Maximum Likelihood Estimates of $\phi_1\ \phi_2\ \phi_3$: Groups B and C

	Group B Cancer of the Endometrium: Exact MLE	SE	Approximate MLE	SE	Group C Cox Data: Approximate MLE	SE
$\hat{\phi}_1$	0.188	0.442	0.160	0.607	-0.235	0.577
$\hat{\phi}_2$	2.989	0.189	3.000	0.272	2.751	0.287
$\hat{\phi}_3$	1.793	0.577	1.783	0.569	2.051	0.590
$\text{Corr}(\hat{\phi}_1,\hat{\phi}_2)$	-0.195		-0.331		-0.301	
$\text{Corr}(\hat{\phi}_1,\hat{\phi}_3)$	0.009		0.0		0.224	
$\text{Corr}(\hat{\phi}_2,\hat{\phi}_3)$	0.221		0.178		0.518	

TABLE II B

Maximum Likelihood Estimates
of p, θ, α: Groups B and C

	Group B Cancer of Endometrium				Group C Cox Data	
	Exact		Approximate		Approximate	
	MLE	SE	MLE	SE	MLE	SE
$\hat{p}$	0.546	0.107	0.540	0.151	0.442	0.142
$\hat{\theta}$	0.051	0.009	0.050	0.014	0.064	0.018
$\hat{\alpha}$	1.793	0.578	1.783	0.569	2.051	0.590
Corr$(\hat{p},\hat{\theta})$	0.170		0.321		0.330	
Corr$(\hat{p},\hat{\alpha})$	0.012		0.0		0.225	
Corr$(\hat{\theta},\hat{\alpha})$	-0.214		-0.196		-0.529	

TABLE III A

Estimates of the survival function for Group B

	Exact Likelihood		Approximate Likelihood			
Times from treatment	Weibull $\hat{S}(x)$	Berkson-Gage $\hat{S}(x)$	Weibull $\hat{S}(x)$	Berkson-Gage $\hat{S}(x)$	Kaplan-Meier	Life -Table
0	1.000	1.000	1.000	1.000	1.000	1.000
3	0.985	0.939	0.985	0.937	0.966	1.000
8	0.918	0.854	0.919	0.850	0.930	1.000
11	0.865	0.812	0.867	0.807	0.814	1.000
12	0.846	0.799	0.849	0.794	0.814	0.818
14	0.809	0.775	0.812	0.770	0.775	0.818
16	0.773	0.754	0.776	0.748	0.689	0.818
19	0.722	0.724	0.726	0.718	0.646	0.818
24	0.654	0.684	0.656	0.678	0.646	0.655
35	0.572	0.622	0.571	0.616	0.592	0.655
36	0.569	0.617	0.567	0.612	0.538	0.606
48	0.549	0.579	0.544	0.574	0.538	0.546
60	0.546	0.557	0.540	0.553	0.538	0.546

TABLE III B

Estimates of the survival function for Group C

Time from treatment	Approximate Likelihood		Kaplan-Meier	Life-Table
	Weibull $\hat{S}(x)$	Berkson-Gage $\hat{S}(x)$		
0	1.000	1.000	1.000	1.000
6	0.927	0.860	0.857	1.000
7	0.903	0.839	0.807	1.000
10	0.816	0.778	0.752	1.000
12	0.754	0.740	0.752	0.737
13	0.734	0.722	0.690	0.737
16	0.638	0.670	0.627	0.737
22	0.517	0.576	0.538	0.737
23	0.503	0.561	0.448	0.737
24	0.492	0.549	0.448	0.456
36	0.444	0.407	0.448	0.456

Figure A: Estimates of the survival function for Group B

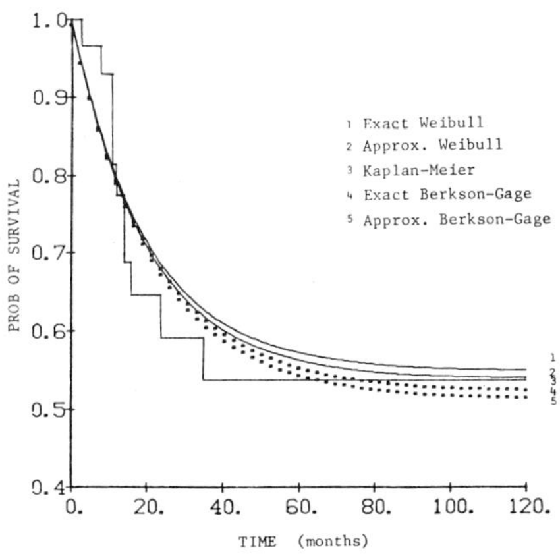

Figure B: Estimates of the survival function for Group C

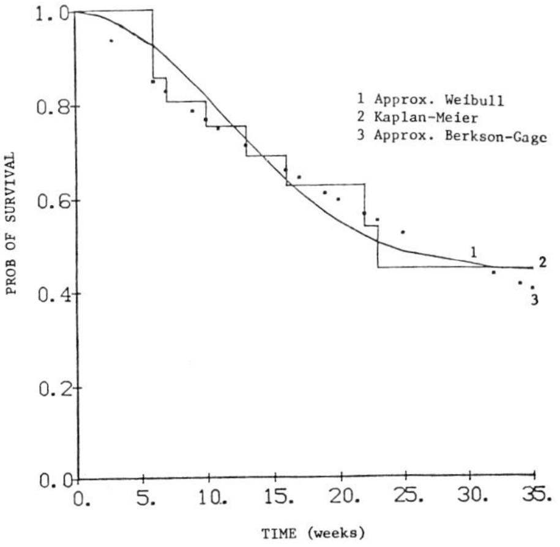

Figure C: $H_1(\phi_1)$ obtained from the exact log likelihood function

Group B

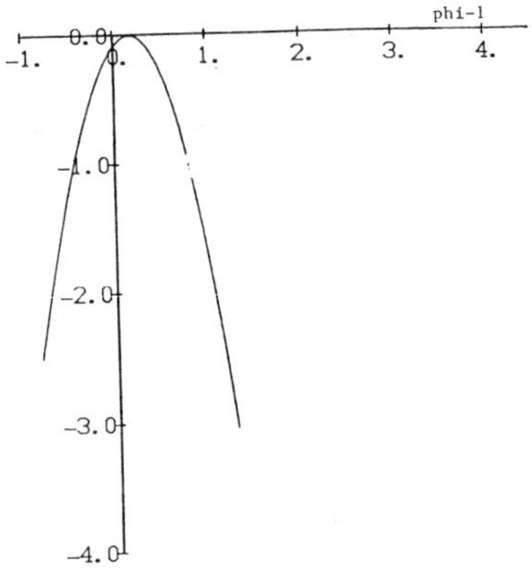

Figure D: $H_1(\phi_1)$ obtained from the approximate log likelihood function Groups B and C

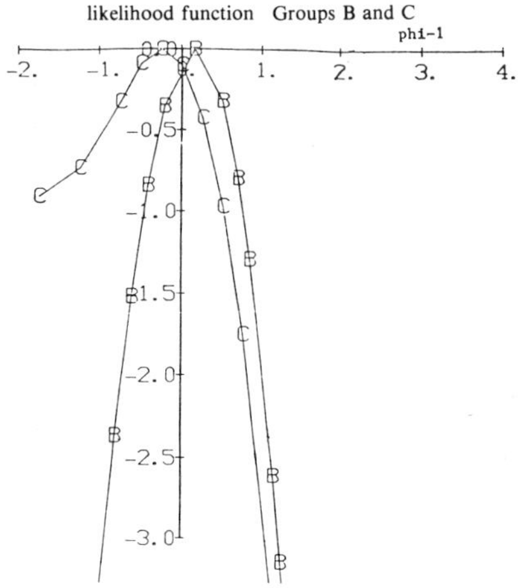

Figure E: $\tilde{H}_2(\phi_2)$ obtained from the exact log likelihood function

Group B

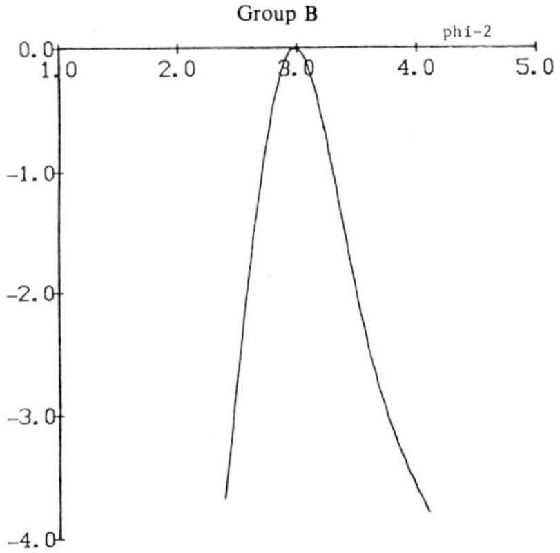

Figure F: $\tilde{H}_2(\phi_2)$ obtained from the approximate log likelihood function Groups B and C

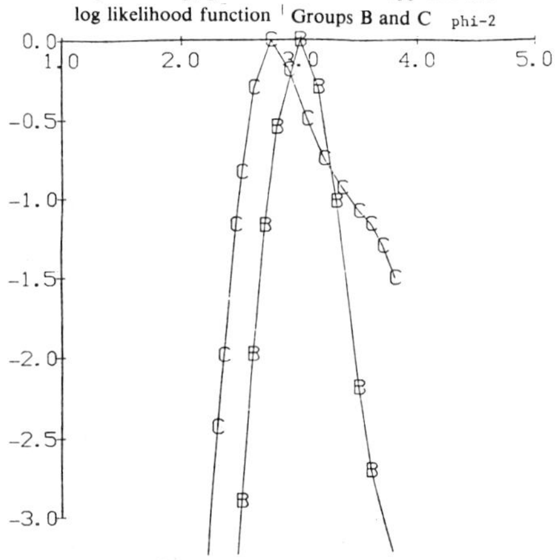

Figure G: $\tilde{H}_3(\phi_3)$ obtained from the exact log likelihood function

Group B

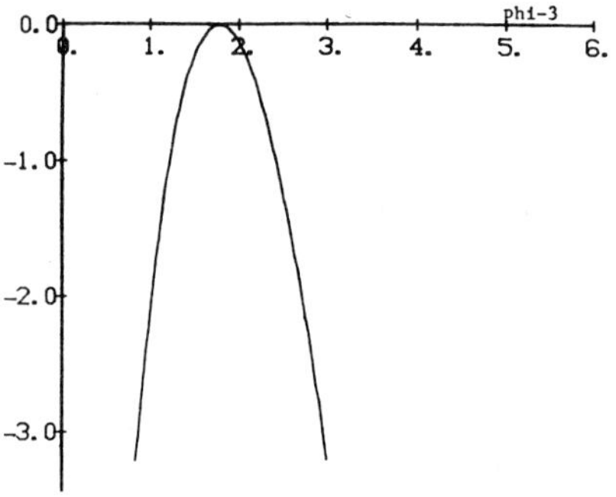

Figure H: $\tilde{H}_3(\phi_3)$ obtained from the approximate log likelihood function

Groups B and C

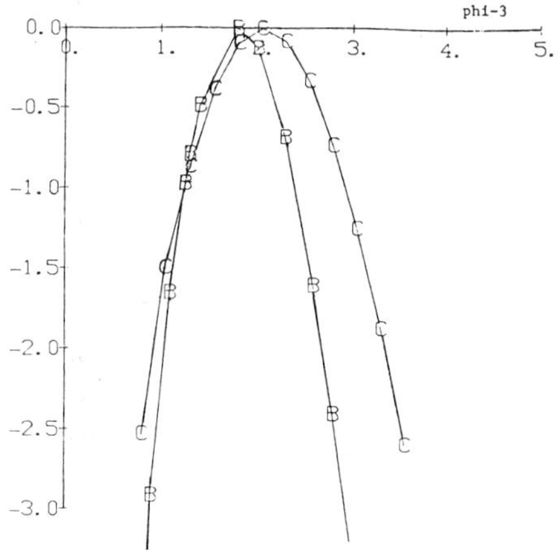

REFERENCES

BERGER, J.O. (1982). The Robust Bayesian Viewpoint, *Tech. Rep.* **82-9**, Department of Statistics, Purdue University, and in *Robustness in Bayesian Statistics,* (J. Kadane, ed.) North-Holland, Amsterdam (1984).

BERKSON, J. and GAGE, R.P. (1952). Survival Curve for Cancer Patients Following Treatment, *J. Amer. Statist. Assoc.*, **47**, 501-515.

BOAG, J.W. (1949). Maximum Likelihood Estimation of the Proportion of Patients Cured by Cancer Treatment, *J. Roy. Statist. Soc.*, **11**, 15-53.

BRESLOW, N.E. (1975). Analysis of Survival Data Under the Proportional Hazards Model, *International Statistical Review*, **43**, 45-57.

COX, D.R. (1972). Regression Models and Life-Tables, *J. Roy. Statist. Soc. B* **39**, 187-220.

DE FINETTI, B. (1975). *Theory of Probabilty*, **1** London: John Wiley.

DEGROOT, M.H. (1970). *Optimal Statistical Decisions*, New York: McGraw-Hill.

GEHAN, E.A. (1965). A generalized Wilcoxon Test for Comparing Arbitrarily Single-Censored Samples, *Biometrika* **52**, 203-224.

— (1969). Estimating Survival Functions from the life Table, *J. Chronic Disease,* **21**, 629-644.

HILL, B.M. (1963). The Three-Parameter Lognormal Distribution and Bayesian Analysis of a Point-Source Epidemic, *J. Amer. Statist. Assoc.*, **58**, 72-84.

— (1965). Inference about Variance Components in the One-Way Model, *J. Amer. Statist. Assoc.*, **60**, 806-825.

— (1975). A Simple General Approach to Inference about the Tail of a Distribution, *Ann. Statist.* **3**, 1163-1174.

— (1980). Robust Analysis of the Random Model and Weighted Least squares Regression *Evaluation of Econometric Models*, (J. Kmenta and J. Ramsey, eds.), New York: Academic Press, 197-217.

JEFFREYS, H. (1961). *Theory of Probability*, Third Edition. Oxford: University Press.

JORDAN, CH. W. Jr. (1952). *Society of Actuaries' Textbook on Life Contingencies,* The Society of Actuaries.

LINDLEY, D.V. (1965). *Probability and Statistics 2*, Cambridge: University Press.

KAPLAN, E.L. and MEIER, P. (1958). Nonparametric Estimation from Incomplete Observations, *J. Amer. Statist. Assoc.*, **53**, 457-481.

MILLER, M.D. (1949). *Elements of Graduation*, Monograph of the Actuarial Society of America and the American Institute of Actuaries.

SAVAGE, L.J. (1961). The Subjective Basis of Statistical Practice. *Tech. Rep.* University of Michigan.

— (1972). *The Foundations of Statistics*, Second Revised Edition. New York: Dover

DISCUSSION

A.M. SKENE (*University of Nottingham*)

This paper is to be welcomed as it provides a clear illustration of the need to adopt the Bayesian approach to applied problems in Statistics. It is important to be reminded that data is most often collected with a view to making certain decisions or perhaps predictions. This is no less true in studies of survival as Professor Hill and his co-authors ably demonstrate.

However, several features of this paper leave me puzzled and somewhat disappointed. The authors emphasize the need to specify relevant prior distributions and utility functions, yet they avoid doing so. The statistical community at large is aware that such components are an integral part of the Bayesian philosophy. It would seem that any practical illustration of the power of the Bayesian approach should demonstrate that such functions can be chosen after consultation and mature reflection. The paper comments on some of the difficulties in practice of distinguishing between likelihood and prior. However, the adoption of the mixture model must surely be based on some reasonably strong belief that such a a model is appropriate. It seems logical therefore to expect that some prior knowledge should also be available as to likely values of the mixing probability.

I was also disappointed to discover the readiness with which maximum likelihood estimation was adopted in lieu of a more comprehensive investigation of the posterior distribution, particularly when only three parameters are involved. This approach was justified by the authors with an appeal to the phenomenon of stable estimation and a cursory inspection of the likelihood in the region of the mode. However, an overly strong reliance on the asymptotic properties of posterior densities can lead to two forms of oversight. First, tail behaviour of the joint posterior may appear nearly normal, yet be sufficiently different from normal to make the tails of marginal distributions interesting. Secondly, and perhaps more important, is the fact that stable estimation implies that the posterior distributions is unimodal and thus the Bayesian statistician may be content with finding one mode. Failure to fully explore the likelihood surface can mean that interesting features of the data or possible deficiencies in the model can be missed.

My plea is that the authors match their strength of feeling as to the necessity of the Bayesian approach with a similar commitment to the numerical aspects of the analysis. Then a valuable case study in Bayesian statistics will emerge.

R.E. BARLOW (*University of California, Berkeley*)

I liked this paper. It is a good and useful paper. It is one of the very few papers concerning survival analysis written from a Bayesian point of view. A related Bayesian paper which perhaps should be referenced is Lindley (1979).

As the authors indicate, the "natural" parameterization of a life distribution model is the hazard function or "force of mortality". Relevant properties of this function are discussed in detail in Chapter 4 of Barlow and Proschan (1975). A property particularly apropos this paper is that mixtures of different exponential distributions have a decreasing force of mortality. Decreasing force of mortality is therefore an artifact of the Berkson-Gage mixtures model discussed in this paper when life distributions being "mixed" have approximately constant but different forces of mortality in the age interval of interest. If p is the probability that the treatment cures the cancer, then, in this model, the overall model for survival to age $t + x$ is

$$p\overline{F}(x \mid t) + (1-p)\overline{G}(x \mid t)$$

where $\overline{F}(x \mid t)$ is the cohort actuarial probability of survival to age $t + x$ given survival to age t and $\overline{G}(x \mid t)$ is the similar probability for members of the cohort with the cancer who have been treated. Even though p may be extremely small (or reflect cures unrelated to the treatment), the mixture life distribution may still exhibit decreasing force of mortality thus perhaps erronously leading to the belief that the treatment is helpful. This observation should perhaps be considered relative to model robustness considerations.

This paper gives us a methodology superior to the least squares fit used by Berkson-Gage.

J.C. NAYLOR (*Derbyshire College of H.E., UK*)

The results presented by Chen et al depend heavily on the use of a maximum likelihood type of approximation to the posterior density. Such an approximation is subject to severe risks of error both in the numerical methods used (numerical differentiation is difficult) and in assessing the validity of the statistical approximation used. This approximation involves the assumption that the posterior is sufficiently near normal for the behaviour near the mode to adequately describe both marginal densities and tail behaviour. Such assumptions are very difficult to check.

An alternative approach is to use numerical integration methods; this then involves no statistical approximations, only the use of generally more reliable numerical methods. For problems such as this (three parameters and some belief in "near normality") the methods of Naylor and Smith (1982) may be routinely applied.

J. BERGER (*Purdue University*)

It is always a pleasure to see a serious Bayesian effort to grapple with a hard real problem, and the authors have done an admirable job. Although one can always quibble over assumptions and details, the authors' approach is an excellent illustration of the need for a Bayesian to concentrate his "subjective" efforts on what he feels will be the most influential aspects of the situation. In this problem, the authors concluded that

(i) the possibility that some patients would be "cured" could have a significant impact and must be involved; and

(ii) partly to implement (i) and partly to incorporate possibly significant age dependence, the model should be constructed as a "baseline" survival function $F(x;t)$ multiplied by a treatment dependent correction factor.

Subsequent modeling was deemed less influential (or subsequent prior information more vague), so standard models and maximum likelihood analyses were sensibly employed for convenience. Of course, the likelihood principle was followed throughout, an important practical consideration.

There are two points I would appreciate clarified. The first concerns the classification of "deaths". Apparently death-by-lightning will be classified as a censored observation. Is the general idea that any medical death will be classified as an x_i, and any non-medical death as a y_j, in which case the baseline $F(x;\, t)$ would refer to medical risk? Or could one try to define only cancer deaths as x_i and non-cancer deaths as y_j, in which case the baseline $F(x;\, t)$ would refer to only cancer risks in the general population? Or is this meant to be left open, the idea being that one tries to classify as many clearly non-cancer deaths as y_j as is possible, subject to the availability of data from which to estimate $F(x;\, t)$ for the remaining death factors.

This is related to my second question, namely what is the meaning of $s(x)$? In the paper it is referred to as the "survival time", but its mathematical definition would seem to imply that it is the survival time conditioned on the fact that death did not occur by any method that would have been classified as a y_j.

As a final comment, the utility analysis and robustness philosophy espoused in the paper are very appealing. I would have loved to see an actual example of either with numbers. Have the authors tried such?

A.P. BASU (*University of Missouri, Columbia*)

Because of recent theories on development of cancerous cells, like two stage development of cancer, age is clearly a variable which need to be included as a covariate. Thus consideration of a model like Cox's regression model, where covariates are included, makes more sense in studying survival analysis. I would like to know if the authors have considered including covariates in their model.

REPLY TO THE DISCUSSION

(by B.M. Hill)

I would like to adress first the general points made by Skene, that there is a "need to specify relevant prior distributions and utility functions, yet they avoid doing so", and secondly, the general point, made by both, Dr. Skene and Dr. Naylor, concerning the use of numerical integration as opposed to the approximation we use for what I would call the "marginal likelihood function".

Certainly I am in complete agreement with Skene as to the ultimate need to make the analysis with several prior distributions, several utility functions, and I would add, with several likelihood functions, and several methods of approximation. I am not sure exactly what Skene would consider relevant choices, or if he and I could agree upon just what were the relevant choices, but obviously our intent is to put forth results which we believe are pertinent and appropriate for the analysis of survival data. This is stated quite clearly in the article. However, I think that Skene has ignored one important point. I think it is fairly obvious that if one works with a very bad statistical model then no matter how carefully one makes the numerical analysis, one is not going to overcome the inherent limitations of the model that is being used. Conversely, if one is using a sensible statistical model, then it is difficult (although not impossible) to foul things up too badly, even when making a non-Bayesian analysis of the data. For this reason I feel that the first step that must be made is to improve the statistical models that are now being used for the analysis of survival data, and this is the primary goal of our article. The improvements that I expect to come later, such as the incorporation of more realistic prior distributions and utility functions, are real and substantial, but are simply not of the same order of magnitude as those that come from using a reasonable model to begin with. This is the reason that we focus more on the model than on the detailed choice of prior distribution. I must also say that our experience has been that those in what I call the "statistical cancer industry" are very reluctant to drop the models that are in common use, no matter how silly such models are shown to be. In order to make a convincing case to the few who are capable of some measure of objectivity, it is important to separate out, so far as possible, two separate and different controversial issues; namely, the issue concerning the mixture model, and the issue concerning Bayesian statistics. I want to reach the widest possible audience of statisticians and medical people who are interested in the analysis of survival data, and capable of intelligent mature decision. I do not want to turn any off merely because they have misconceptions about the Bayesian theory of statistics. On the other hand, I am not about to make a non-Bayesian analysis of the problem, or put forth as my posterior distribution something that doesn't come close. I do regard the prior distribution we have used as relevant, and as leading to a posterior distribution that is appropriate for the problem, to a reasonable degree of approximation; and I do not think that one can overemphasize the fact that any Bayesian analysis of real data involves many approximations at many different levels, which is part of the art of the practice of Bayesian statistics. At a later date

I hope to incorporate more realistic prior information about p and the other parameters, but for now the primary focus must be on getting rid of the silly models currently in use, and I think this is best done by separating, so far as possible, the model issue from the Bayesian issue. (And perhaps I should mention that there is also a controversial medical issue as to the effect of using high energy radiation as done here in GROUP B, so that there is not going to be much of a consensus in the medical community as to an appropriate choice of prior distribution. If we had done as Skene suggests, then I am afraid that the three different issues would have been hopelessly confounded, and the key point concerning the model would have been obscured by these other issues).

I wish to thank Professor Barlow for his kind comments. I am in full agreement with his comments as to the effect of the mixture model in yielding an apparent decrease in the force of mortality. There is another aspect of this question which is perhaps also worthy of note. It is sometimes referred to as the "selection effect". Suppose we envision the population as composed of two types of units, namely, the "fit" and the "unfit". Here fitness is thought of as not relating to the cancer, but rather to the general health of the unit. Under the force of a disease such as cancer, it is not implausible that the less fit will be less likely to survive the onslaught of the cancer, and hence that the survivors after a substantial period of time are more likely to include the fitter members of the population. Hence one anticipates that such survivors will exhibit an apparently better hazard function than perhaps even the general population without cancer. I don't know of anyone naive enough to think that this implies that having cancer improves one's long term prospects (although I expect that such people exist even amongst professional statisticians). The correct interpretation, I believe, is that having survived cancer for a substantial period of time is an indication that one was fitter to begin with. This question is discussed in an article that is now in preparation.

The point raised by Naylor is an interesting one. It is true that our method for obtaining an approximate "marginal likelihood function" is most appropriate when the likelihood function is approximately normal, and sharp relative to the prior distribution, and that this can be difficult to check. I agree with him that it would be well worth while to make the numerical integration, and see what differences arise. There are several reasons why we did not do this. The first is the simple practical reason that after the death of Professor Chen, who had been doing the programming, and was an experienced numerical analyst, we did not have available anyone to undertake the numerical integration. The second reason is that my own experience with the approximate method is that when used with judgment it provides close agreement to the results of numerical integration, as for example is vividly demonstrated in my analysis of the three-parameter lognormal distribution referred to in the article. The third reason is the same reason that we did not go to the type of Bayesian analysis that Skene would like to have seen; namely, that our primary aim was to improve the type of model that is used for the analysis of survival data. If we had gone to a numerical integration, and if with a certain diffuse prior the results had been materially different from the results obtained by our approximation, then in addition to the other issues that are involved, there would be an issue concerning the choice of prior distribution and method of numerical analysis, and I am afraid that I don't think that the choice between our approximate method and numerical integration is quite so simple as Naylor implies. For one thing, the two methods are sensitive to quite different aspects of the prior distribution and the likelihood functions. If the results of the two analyses had been materially different, then I would have wanted to know why, and would have felt obligated to make the numerical integration for several different prior distributions, in an attempt to learn the implications of the various diffuse or proper prior distributions that seem plausible. Based upon our analysis, however, I do not think the results will be materially different, and it is my belief that the marginal posterior distributions that we give will turn out to be fairly robust to the precise specification of the (diffuse) prior distribution and to the method of numerical analysis.

Of course, in the final stages of the implementation of our model it will be appropriate to make many such sensitivity studies, but at the present time I think it would only raise some delicate issues about Bayesian robustness that would detract from the simple but vital point that the main improvement comes from using a sensible model as opposed to those now in common use. Furthermore, I fully expect that our method of approximation will generally provide quite close agreement to the results of a numerical integration with an appropriate diffuse prior, and as explained at the outset I think this is in fact the case for our data.

I am grateful for Professor Berger's perceptive comments, and for the opportunity to discuss the interesting and important points that he has raised. There are a variety of ways that the problem of classification of deaths can be dealt with, and although none are entirely satisfactory, I think the problem is manageable. The basic difficulty is that every death, or other form of withdrawal, is surrounded by its own set of circumstances, and these circumstances can contain subtantial information as to whether or not the person in question was cured. Thus the data which should be used to form a likelihood function is not merely of the form "death at age $t+x$", or "withdrawal from the study at age $s + y$", but rather such statements together with a description of the relevant circumstances under which they occurred. To begin with we should break the deaths up into three groups. The first group consists of those deaths where we are virtually certain that the death was in fact due to the cancer, as for example might be revealed by autopsy, and hence where the cancer was not cured (or else was cured and then reoccurred independently, which of course is highly improbable). For this group of deaths the mixture model is not appropriate, and the data are really simply that the patient was not cured together with a direct observation $t + x$ from the distribution of the uncured population. The second group consists of those deaths where the circumstances are such that we are virtually certain that the patient was in fact cured, and hence where again the mixture model is not appropriate, and where the only data of interest is the fact that the patient was cured, since in this case the age at death would be irrelevant for inference about the parameters of the uncured population. The third group consist of those deaths where the circumstances contribute little or no information as to cure, and it is really only in this situation that the mixture model, as used in the article, is appropriate. Needless to say there is ordinarily going to be some real doubt as to whether or not a patient was cured, and it would appear to be a conservative approach to focus primarily upon this third group, which is what we have done. Next, within this third group we distinguish a subgroup of cases where the circumstances surrounding the death are viewed as uninformative not only about cure status but also about all parameters of the model. For example, if an individual is walking to work on a bright sunny day and is struck dead on the spot by a meteorite, or a brick from a skyscraper, such a death would be viewed by most of us (but not, to my surprise, by all of us, as for example Professor Paul Meier, who in a similar example asked why the person was walking around during a thunderstorm) as totally uninformative regarding cure status and all the parameters of the model. A formal analysis, via the likelihood principle, much as in Section 6, would simply calculate the probability of such a death at age $t + x$, conditional upon survival to $t + x - e$, as not dependent upon cure or the parameters, and then letting e go to zero, would yield the result that the likelihood for such a death is proportional to the probability of survival to age $t + x$, and hence treat this death precisely as though a withdrawal had occurred at such time. This is why we treat such a death as a y. Of course such an evaluation is merely a subjective evaluation, and cannot be proven to be 'correct', but I suspect that anyone with any sense would make exactly this evaluation. Unfortunately, very few people understand the likelihood principle, and I believe this is part of the reason for confusion on this simple point.

Now let us consider the questions raised by Berger. Not all non-medical deaths could be treated as the censored data, since, for example, a death in an automobile accident, where the auto was driven at high speed by the person treated for cancer, might in fact be a form of suicide, and thus indicative of lack of cure, and provide substantial information about the parameters for the uncured group. Similarly, not all medical deaths would necessarily be regarded as the uncensored data, since, for example, death due to myocardial infarction would ordinarily be regarded as independent of whether or not cancer of the thyroid had been cured, and such a death might well be treated like a withdrawal. So what we do is, as Berger suggests, to use the individual circumstances surrounding each death or withdrawal to aid in classifying the death or withdrawal. For those that are put into group three, and for whom the mixture model is thus deemed appropriate, we try to decide whether or not a death should be treated as a censored observation, using the likelihood principle along with a subjective judgment as to whether the particular death in question contributes any information regarding the parameters over and above the information contained in survival to the age at which death occurred. What then are the implications regarding the interpretation of $F(x;t)$ and $s(x)$? Should we exclude, as best we can, those types of deaths that would be classified as censored, from our definition and assessment of $F(x;t)$? I don't think there is any simple single answer to this question. First of all one must decide whether the function $R(x;t)$ of Section 3 should include only some subset of the deaths, for example those thought possibly related to the cancer, or whether it should include all deaths, as we have defined it. Either approach could be taken, and this really depends upon the type of information we want to provide the patients and physicians. The primary difficulty with the first approach is that in obtaining appropriate actuarial tables to assess $F(x;t)$ when only a subset of deaths is considered. On the other hand, if we take the second approach, as in fact we have done, then Berger asks whether it is not then requisite to eliminate, from the $F(x;t)$ that we use, all deaths that would have been classified as censored. It might well be desirable to do so, but I don't think any real distortion is obtained by not doing so. First of all we must keep in mind, as the suicidal automobile driver example illustrates, that it is not only the general type of death that enters in to how the death is classified, but the individual circumstances surrounding the death. Thus I would not want to treat all automobile deaths as indicative of lack of cure, and in fact the great majority would probably be best treated as withdrawals. Secondly, there is a fundamental question as to the appropriate degree of detail to go into in regard to each death. If we were to record only the fact that death occurred at age such and such, then of course $F(x;t)$ should include all types of death, and none of the deaths would be treated as censored. This seems to me to be too crude an approach, and would throw away important information. At the other extreme would be the case where every known circumstance surrounding the death would become part of the data, each presumed cause of death would be given, and unfortunately the problem would become impossible to analyse. So we take an intermediate point of view, recording and using those particular circumstances that we regard as especially informative, and ignoring others. A little thought reveals that there is a delicate question as to the appropriate data upon which to condition for the analysis, underlying the question as to the appropriate choice of $F(x;t)$. I do agree with Berger, however, that to the extent that we can identify those general types of death that would almost always be treated as censored, that it would be preferable to eliminate these from our assessment of $F(x;t)$, provided of course that we can do so satisfactorily, and that it would make a material difference to do so. (In fact for the data that we analysed none of the deaths were treated as censored, and I see no reason not to use $F(x;t)$ for all deaths, as we have done. As for the question what would we have done if we had in fact obtained some deaths that we would have treated as censored, it would be well to recall Pratt's amusing example concerning the likelihood principle and censoring of data).

As to Berger's last question, I am afraid that I have as not yet made the type of utility and robustness analyses that I have espoused. My only defense is that in the seven years since this article was first written it has been subjected to such a barrage from those bastions of the statistical cancer establishment that are responsible for the absurd methods of analysis that have become commonplace, and who have little interest in or understanding of the likelihood principle, that for a long time I lost interest in the problem. This was compounded by the tragic death of Professor Chen who had been doing the programming for the analysis. At present I and some colleagues have returned to the problem.

In response to Professor Basu, I regard it as one of the advantages of our approach that age at which treatment occurs is an integral part of the model, rather than a covariate as in the Cox model. As far as genuine covariates are concerned, yes, it is quite easy to incorporate them into the model, and a program has been written to do so.

REFERENCES IN THE DISCUSSION

BARLOW, R.E. and PROSCHAN, F. (1975). *Statistical Theory of Reliability and Life-Testing.* New York: Holt.

COX, D.R. (1972). Regression model and life tables. *J. Roy. Statist. Soc. B* **34**, 197-220 (with discussion).

LINDLEY, D.V. (1979). Analysis of life tables with groupings and withdrawals. *Biometrics* **35**, 605-612.

NAYLOR, J.C. and SMITH, A.F.M. (1982). Applications of a method for the efficient calculation of posterior distribution. *Appl. Statist.* **31**, 219-225.

BAYESIAN STATISTICS 2, pp. 329-360
J.M. Bernardo, M.H. DeGroot, D.V. Lindley, A.F.M. Smith (Eds.)

Highly Informative Priors

E.T. JAYNES
(*Washington University, St. Louis*)

SUMMARY

After discussing the role of prior information in statistical inference, historically and in current problems, we analyze the problem of seasonal adjustment in economics. Litterman (1980) has shown how informative priors for autoregressive coefficients can improve economic forecasts. We find that in seasonal adjustment informative priors can have a much greater effect on our conclusions. In our model, even the dimensionality of the joint posterior distribution of the irregulars depends on prior information about the seasonal component; and some functions of the irregulars can be determined more accurately than in sampling theory

Keywords: BRUTE STACK; FOURIER SERIES; JEFFREYS PRIOR; LEAST SQUARES ESTIMATES; NECESSARY VIEW; NONINFORMATIVE PRIORS; PRIOR INFORMATION; SAMPLING THEORY METHODS; SEASONAL ADJUSTMENT; STATISTICAL MECANICS; SUBJECTIVITY; WEIGHTED AVERAGE.

1. INTRODUCTION

The statistical problems envisaged in our pedagogy are almost always ones in which we acquire new data D that give evidence concerning some hypotheses $H, H',\ldots$ (this includes parameter estimation, since H might be the statement that a parameter lies in a certain interval); and we make inferences about them solely from the data. Indeed, Fisher's maxim, "Let the data speak for themselves" seems to imply that it would be wrong —a violation of "scientific objectivity"— to allow ourselves to be influenced by other considerations such as prior knowledge about H.

Yet the very act of choosing a model (i.e. a sampling distribution conditional on H) is a means of expressing some kind of prior knowledge about the existence and nature of H, and its observable effects.

This was noted by John Tukey (1978), who observed that sampling theory is in the curious position of holding it decent to use judgment in deciding which parameters should be present in a model; but then indecent to use judgment in estimating their values. He saw the Bayesian method as something which "allows one to do the indecent thing while modestly concealed behind a formal apparatus".

Here we do this indecent thing, and note how it changes the problem of seasonal adjustment, an unusual problem in that all the parameters are nuisance parameters. But it is just for this reason that our formal apparatus enables us take into account things beyond the technical means, and even the concepts, of sampling theory. Jimmie Savage advocated noninformative priors on the grounds that they didn't make much difference; we advocate informative priors on the grounds that they do make a very important difference. As

Litterman (1980) found in a similar problem, taking into account cogent information not contained in the sampling distribution can improve the accuracy and reliability of our conclusions.

In the following two Sections we digress to comment on the strange history of the prior information issue, with its controversies still not entirely resolved, and to note some other new applications of informative priors. Our seasonal adjustment calculation starts in Section 4.

2. NECESSARIANS

There is a surprisingly wide range of philosophical views about the role of prior information. Since the 1930's a common view has been that it is just plain wrong to take prior information into account. As a student in the late 1940's, the writer was strongly indoctrinated with this view. Indeed, in most orthodox works the term "prior infomation" does not appear at all. Perhaps it was felt that since prior knowledge is hard to document, the user would be under the temptation to slip in prior opinions, in the guise of prior knowledge -- a terrible sin in the view of some (van Dantzig, 1957).

But then there was a seemingly violent swing to the opposite extreme view of the so-called "subjective Bayesians". Jimmie Savage (1954) proclaimed it his intention to incorporate prior opinions —not prior knowledge— into scientific inference (or at least into the reasoning of an idealized being called "the person", thought of as a normative model for scientific inference).

He rejected formal principles (symmetry, maximum entropy) by which prior probabilities might express, and be determined by, prior knowledge; and accused those of us who advocated such principles of holding "necessary" views of probability and of claiming to get something for nothing (Savage, 1981, p. 731). Indeed, he took it as a fundamental tenet (Savage, 1954, p. 3) that two "persons" with the same prior knowledge might assign different prior probabilities without either being unreasonable.

These philosophical differences leave a practical scientist, concerned with problems of the real world, with the uncomfortable feeling of being caught in the middle. From his viewpoint, it then appeared that there were two active camps, holding opposite extreme positions equally unreasonable and inapplicable to his problems of inference.

An earlier discussion (Jaynes, 1968) stressed the need to find a safe middle ground between these extremes, which recognizes the relevance of cogent prior knowledge and the need to take it into account (in seeming disagreement with Fisher); but also recognizes the claims of logical consistency (in seeming disagreement with Savage). That is, we took it as our fundamental "Desideratum of Consistency" that in two problems where we have the same relevant prior information, we should assign the same prior probabilities; and showed that this desideratum is already sufficient to determine priors in some cases. Savage (1981, p. 736) proceeded to dismiss this as "an unusual necessarian position".

It appears to us, however, that this position and desideratum are neither unusual nor "necessarian" as Savage defined that term. Zellner (1982) has also noted the contrast between Savage's definition of "necessary views" and the actual position of Jeffreys, whom he accused of holding them. Indeed, it would be hard to cite anyone, in the entire history of probability theory, who has ever held a "necessary" view (Keynes perhaps came closest).

Reliable judgments on these matters cannot, however, be made merely by examining a writer's philosophical remarks. Doubtless all of us have had the experience of writing some interpretive statement, while our minds were preoccupied with one context; and later seeing it in print and wishing that we had made a different choice of words, since the statement as published was misleading if taken in a different context.

Any author who has written extensively over many years can be made to appear to hold almost any position one wishes to impute to him, by a carefully selective quotation of his philosophical remarks, made in many different contexts. The comments of Kass (1982) on Jeffreys demonstrate this very nicely. Even Karl Popper (1959) can be made to seem a Jeffreys type Bayesian if one quotes only from his earlier chapters and not the later ones.

For this reason we think that the issue of being or not being a "philosophical necessarian" is slippery, probably undecidable, and therefore not very relevant. What is relevant and decidable is the actual content of one's calculations; whether he can be termed a "functional necessarian".

In the early 1960's I had tried, in correspondence and conversation with Jimmie Savage, to persuade him that Laplace and Jeffreys were not necessarians, and in fact not only their philosophical remarks taken as a whole —but far more importantly, their methods of calculation— show the opposite of the position that he called "necessary".

Their use of probability theory, far from supposing that probability measures the extent to which one proposition, out of logical necessity, confirms the truth of another, clearly denies this. For the numerical value of $p(A \mid B)$ does not depend only on A and B; it depends also on the sample space, or hypothesis space, in which A is embedded.

Even though we do not change the propositions A and B, when we change the set of alternatives against which A is being compared, we clearly —and rightly— change $p(A \mid B)$. This "anti-necessary" dependence is always present implicitly, and sometimes appears explicitly, in the calculations of Laplace and Jeffreys. It seems, then, that the only functional necessarians are the sampling theorists who use significance tests that make no reference to alternatives. But Jimmie's only reaction to my arguments was to include me in his list of necessarians! More specifically (Savage, 1981, p. 542), I became "a latter-day necessarian".

Jimmie Savage thus poses a curious problem to us: so much of what he said was absolutely correct, deeply insightful, and of timely importance to statistics, that his failure to appreciate the work of Jeffreys stands out as an inexplicable puzzle. Fifteen years before Savage, Jeffreys (1939) had not only enunciated the same Bayesian principles and anticipated Savage's generalities about prior probabilities; but he also constructed explicit priors and demonstrated their successful application to real problems with a thoroughness that Savage never approached. Yet in what must have been his final judgment, Savage (1981, p. 727) still clung to his original position of 1954; in his eyes it was Jeffreys' theory that was seriously incomplete, and he saw no cogency in even its "ostensible beginnings".

My own view —then and now— has been that Savage's theory is seriously incomplete for real applications, just because of its failure to deal seriously with prior knowledge. Jeffreys had at least made a good start on remedying this.

Recently (Jaynes, 1984) I made a strenuous effort to resolve this puzzle and reconcile our different positions. There seem to be two possible explanations. Firstly, it may be that Savage gave Jeffreys a cursory reading, through the eyes of one indoctrinated by the "orthodox" teaching of the time, before his own independent thinking on the subject had matured; and just never overcame a wrong first impression.

As a second possible explanation, we note that his "necessarian" charges were directed mainly at physicists. I now conjecture that the problems Savage had in mind were quite different from the ones physicists faced. Part of the blame for this is mine, for in 1963 I had ample opportunity to point out to him, in conversation, certain technical details (existence of a deeper hypothesis space and prior knowledge about it) that gave many of our problems —and also many of engineering and economics— more structure than those of the statistics textbooks; but failed to do so.

As a result, it may be that Jimmie Savage never needed to go beyond the "diffuse prior" mentality, because he never faced a problem in which there was specific prior information that needed to be taken into account. So it appeared to him that those of us who were seeking a formal apparatus by which one can construct informative priors, were "necessarians" doing things that seemed to him unnecessary.

3. NEW APPLICATIONS

Coming back to the present, it seems to us that if Bayesian theory is ever to lay any claims of full logical consistency, a high priority research problem must be the development of the formal apparatus that can realize the aforementioned desideratum by converting specific prior information into specific prior probability assignments, in a wider variety of problems. We face this need not only for logical reasons, but also for pragmatic ones. Today, there are important new applications where informative priors are not just window dressing; but required for the application to succeed at all.

To date we have made substantial progress in this direction, in the principles of Group Invariance, Maximum Entropy, Marginalization (Jaynes, 1981), and Coding Theory (Rissanen, 1983). As noted below, many current problems are now being solved routinely and successfully by these principles. But it is basically an open-ended program and much remains to be done.

These new applications are very recent; for the most part they have come into their own, in the sense of general recognition and acceptance, only since our last meeting here four years ago. Before then, our "necessarian" strivings did not have much effect on the treatment of real problems, because in the one problem (Statistical Mechanics) where a formal principle for determining priors was most needed, we had it already from J. Willard Gibbs -- only masquerading unrecognized under a different name. In most of the other problems then being considered it was only illogical and inefficient, not fatal, to ignore prior information.

Indeed, once a model has been set up, in the "classical" problems of inference studied in the past the data D were so much more informative than our prior information I about the values of the parameter that it would have made little difference whether we used I or not. Then from a pragmatic standpoint "orthodox" or sampling theory methods were satisfactory unless there were technical problems —nuisance parameters, lack of sufficient or ancillary statistics, a rectangular likelihood function, etc.— that sampling theory has not learned to deal with in a satisfactory way.

But from the standpoint of logic and principle, problems of inference are basically ill-posed if prior information is not considered. Even if our model is not freely chosen but imposed on us from above, the answer to the query: "What do you know about H after seeing the data D" depends "necessarily" on this: "What did you know about H before seeing D?"

All of us recognize this in our everyday inferences; in trying to guess whether it will rain today we take into account not only how the sky looks now, but also what the weather map showed yesterday. A medical diagnostician could be accused of malpractice if he failed to take into account the available information about a patient's medical history as well as his present symptoms. In many real situations, it would be foolhardly to "let the data speak for themselves".

Nevertheless, most current Bayesian practice tends to imitate sampling theory in that one incorporates little or no prior information beyond the choice of the model, and so seeks "noninformative" priors. The Bayesian formal apparatus can then be expected to out-perform sampling theory only when the latter faces some technical problem of the aforementioned kind. But the great potential advantage of Bayesian methods lies in exploiting this unused capability of taking prior information into account with informative priors.

Probably the most impressive example of the power of prior information is the Statistical Mechanics of J. Willard Gibbs (1902). For many years this was presented in a language and conceptual framework so different that most authors saw it as an unfinished, and only partly satisfactory, attempt to apply the laws of physics; and did not recognize it as a problem of inference at all. Only recently has Gibbs' work been seen in its simplicity and generality, as a method of inference in which prior information about multiplicity W converts a vague, ill-posed problem into a well-posed, accurately solvable one.

The Gibbs formalism is based on the maximization of entropy $S = \log W$ subject to the constraints of our data. This is, from our present standpoint, not an application of a law of physics, but simply locating the peak of a distribution that contains W (the "size" of that deeper microscopic hypothesis space that I failed to communicate to Jimmie Savage) as a factor. The accuracy of our predictions (sharpness of that peak) is due to the fact that W is an enormously rapidly varying function of the macroscopic quantities —and of course, that we know the laws of physics well enough to calculate it correctly.

Once this had been recognized, it was evident that the reasoning was equally applicable in other problems than thermodynamcs. The recent advances in the techniques for Spectrum Analysis (Burg, 1975, Childers, 1978; Currie, 1981, Jaynes, 1982), Image Reconstruction (Gull & Daniell, 1978, Frieden 1980, Gull & Skilling, 1980), the determination of crystallographic and biological macromolecular structure from X-ray scattering data (Bricogne, 1982; Wilkins, et al, 1983; Bryan, et al, 1983), and estimating mathematical functions from a few moments (Mead & Papanicoleau, 1984) have resulted from this recognition of the Maximum Entropy principle. In effect, it is a rule for constructing informative priors when we have partial prior information that restricts the possibilities significantly but not completely.

Also among Statisticians and Economists, several recent Bayesian works have recognized the importance of prior information and the growing need for general methods for constructing informative priors. Much thought and effort has gone into techniques for elicitation of such priors from subject-matter experts; see, for examaple, Kadane (1980), Winkler (1980) and references therein. Arnold Zellner's presentation at this meeting has an impressive survey of recent uses of informative priors in Econometrics.

As already noted, Litterman (1980) showed that economic forecasts using an autoregressive model can be improved by using informative priors that express common-sense judgments about the autoregressive coefficients (i.e. they surely fall off with increasing lag). We apply the same idea to seasonal adjustment, showing how similar common-sense judgments about the harmonic content of the seasonal component can improve our estimates of the irregular component. It appears that in the seasonal adjustment case the effect may be greater.

4. BAYESIAN SEASONAL ADJUSTMENT

We have a discrete time series y_t of length N; think of it as a monthly economic report over $N/12$ years:

$$y_t = s_t + e_t, \quad 1 \le t \le N \tag{1}$$

composed of a periodic seasonal component: $s_t = s_{t+12}$, and the part e_t, variously termed "irregular", "error", or "noise". The seasonal component is represented by a finite Fourier series, containing 12 parameters, $(A_0 \ldots A_6, B_1 \ldots B_5)$:

$$s_t = A_0 + \sum_{k=1}^{6} [A_k C(kt) + B_k S(kt)] \tag{2}$$

where $C(kt) \equiv \cos(2\pi kt/12)$, $S(kt) \equiv \sin(2\pi kt/12)$. Define, for uniform summation limits, $B_0 = B_6 = 0$. The inversions

$$A_0 = (1/N) \sum_{t=1}^{N} s_t \ , \tag{3a}$$

$$A_k = (2/N) \sum_t s_t C(kt) \qquad 1 \le k \le 5 \tag{3b}$$

$$A_6 = (1/N) \sum_t (-)^t s_t \tag{3c}$$

$$B_k = (2/N) \sum_t s_t S(kt) \qquad 0 \le k \le 6 \tag{3d}$$

are exact if N is a multiple of 12, as we suppose here.

There is still another parameter, the standard deviation σ of our prior density for the irregulars, $p(e_1 \ldots e_N | I)$, where I denotes the prior information. With no creative imagination, we simply follow custom by assigning the iid Gaussian prior:

$$e_t \sim N(0,\sigma), \qquad 1 \le t \le N. \tag{4}$$

The rationale by which this custom can be justified is a rather lengthy topic; we think it is far better justified than is usually realized. More comments about this are in Appendix A.

Like any other model, this one can be extended endlessly; in particular we could combine seasonal adjustment with detrending by adding to (1) terms linear, quadratic, etc. in t. We keep our model as simple as possible so as not to obscure the point to be made; and also to heed Arnold Zellner's wise advice about "sophisticatedly simple" models. Further reasons include the accuracy of our estimates and the need for diagnostic checks from limited data, discussed in Appendix C.

As noted in the Introduction, this problem is unusual in that all of these parameters are nuisance parameters. Our goal, just the opposite of the usual Bayesian goal, is to estimate the "noise" instead of the "signal".

The calculation may be organized as follows. Using the abbreviations:

I = prior information
$y = (y_1 \ldots y_N)$, data
$e = (e_1 \ldots e_N)$, irregulars
$s = (s_1 \ldots s_N)$, seasonal values
$A = (A_0 \ldots B_5)$, seasonal parameters,

we want the joint posterior density of $(e_1 \ldots e_N)$ conditional on the data and the prior information

$$p(e|yI) = \int\int p(e|\sigma AyI) p(\sigma A|yI)\, d\sigma dA \ . \tag{5}$$

But if y and A are given, then e is known; so σ is irrelevant in the first factor: $p(e|\sigma yAI) = p(e|yAI)$. Then σ integrates out of the second factor, leaving,

$$p(e|yI) = \int p(e|yAI)\, p(A|yI) dA \ . \tag{6}$$

Direct evaluation of (6) would be tedious because as a function of A, $p(e|yAI)$ is nonzero only on a complicated set of points. But we can avoid these intricate details by calculating first the N-fold Fourier transform of (6): using the notation $r.a = \sum_t r_t a_t$,

$$E(e^{ir.e}|yI) = e^{ir.y} E(e^{-ir.s}|yI) = e^{ir.y} f(r) . \tag{7}$$

So in general the calculation could proceed in three steps:

(A) Evaluate the joint posterior density $p(A|yI)$ of the seasonal parameters.

(B) Calculate, with s given by (2), the characteristic function

$$f(r) = \int e^{-ir.s} p(A|yI)\, dA . \tag{8}$$

(C) Invert $f(r)$, translating by the data y:

$$p(e|yI) = (2\pi)^{-N} \int f(r)\, e^{ir.(y-e)}\, d^N r . \tag{9}$$

Part (A) is the conventional Bayesian exercise. The joint likelihood of all the parameters is proportional to $p(y|A\, I)$:

$$L(A_0 \ldots B_5 ;\sigma) = \sigma^{-N} \exp(-Q_L/2) \tag{10}$$

where Q_L is the quadratic form

$$Q_L(A_0 \ldots B_5) = \sigma^{-2} \sum_{t=1}^{N} \{y_t - \sum_{k=0}^{6} [A_k C(kt) + B_k S(kt)]\}^2 \tag{11}$$

from which we find that the joint maximum likelihood estimates of $(A_0 \ldots B_5)$ are given by (3) with s_t replaced by y_t (this being exact if N is a multiple of 12 as supposed; otherwise new small terms of relative order N^{-1} would be present).

We assign independent Gaussian priors for our seasonal parameters:

$$A_k \sim N(a_k, \sigma_k), \qquad 0 \le k \le 6 \tag{12a}$$

$$B_k \sim N(b_k, \sigma_k), \qquad 1 \le k \le 5 \tag{12b}$$

and define for formal reasons, $b_0 = b_6 = 0$. Then their joint prior distribution has another quadratic form:

$$p(A_0 \ldots B_5|I) \propto \exp(-Q_P/2) \tag{13}$$

where

$$Q_P = \sum_{k=0}^{6} \sigma_k^{-2} [(A_k - a_k)^2 + (B_k - b_k)^2]. \tag{14}$$

In general the joint posterior density of the seasonal parameters will be

$$p(A|yI) \propto p(A|I) p(y|AI) = p(A|I) \int p(y|A\sigma I) p(\sigma|AI) d\sigma . \tag{15}$$

But if we assign independent priors to σ and A, we have $p(\sigma|AI) = p(\sigma|I)$, and our result is

$$p(A_0 \ldots B_5|yI) \propto \int p(\sigma|I) \sigma^{-N} \exp[-(Q_L + Q_P)/2] d\sigma . \tag{16}$$

If σ is supposed known in advance, then $p(\sigma|I)$ is a delta function concentrated on a single point and in (16) we need only keep the exponential term. If σ were initially completely unknown, then the Jeffreys prior $p(\sigma|I) = 1/\sigma$ would be appropriate, and (16) would be a multivariate t-distribution with the same quadratic forms. Realistic prior information is presumably intermediate between these extremes. The choice of $p(\sigma|I)$ is discussed further in Appendix B.

For present purposes (to illustrate an entirely different point: the effect of prior information about the seasonal component), even these extremes do not lead us to very different conclusions unless our data sharply contradict the informative prior. But in that case a diagnostic check would lead us to doubt the correctness of the model or the prior information. The way in which Bayesian theory automatically provides the needed diagnostic checks is explained in Appendix C. In the following we shall, therefore, suppose σ known.

Alternatively, one could say that we are calculating only $p(A|\sigma yI)$, and an integration of our final result with respect to any $p(\sigma|I)$ can still be performed.

This enables us to bypass steps (B) and (C) above; for it is obvious that $p(e|yI)$ must be a multivariate Gaussian, determined by its first and second moments. Evaluating the Fourier transforms (which amounts to calculating all moments) is not needed. The distribution (16) reduces to

$$p(A|\sigma yI) \propto \exp[-(Q_L + Q_P)/2] \,. \tag{17}$$

Expanding these merged quadratic forms we have, to within an irrelevant additive constant,

$$Q_L + Q_P = \sum_{k=0}^{6} M_k [(A_k - \hat{A}_k)^2 + (B_k - \hat{B}_k)^2] \tag{18}$$

in which, as before, $B_0 = B_6 = b_0 = b_6 = 0$; and in consequence $\hat{B}_0 = \hat{B}_6 = 0$ from (20d) below. The reciprocal variances (exact if N is a multiple of 6) are

$$M_k = \frac{N}{\sigma^2} + \frac{1}{\sigma_k^2} \,, \qquad k = 0,6 \tag{19a}$$

$$M_k = \frac{N}{2\sigma^2} + \frac{1}{\sigma_k^{\ 2}} \,, \qquad 1 \leq k \leq 5 \tag{19b}$$

and the mean values (the optimal estimates with a symmetric loss function):

$$\hat{A}_0 = M_0^{-1} \left[\frac{N}{\sigma^2} \cdot \frac{1}{N} \sum_t y_t + \frac{1}{\sigma_0^2} a_0 \right] , \tag{20a}$$

$$\hat{A}_k = M_k^{-1} \left[\frac{N}{2\sigma^2} \cdot \frac{2}{N} \sum_t y_t C(kt) + \frac{1}{\sigma_k^2} a_k \right] , \qquad 1 \leq k \leq 5 \tag{20b}$$

$$\hat{A}_6 = M_6^{-1} \left[\frac{N}{\sigma^2} \cdot \frac{1}{N} \sum_t (-)^t y_t + \frac{1}{\sigma_6^2} a_6 \right] , \tag{20c}$$

$$\hat{B}_k = M_k^{-1} \left[\frac{N}{2\sigma^2} \cdot \frac{2}{N} \sum_t y_t S(kt) + \frac{1}{\sigma_k^2} b_k \right] . \qquad 0 \leq k \leq 6 \tag{20d}$$

The Bayes estimates (20) are weighted averages of the prior estimates a_k, b_k, and the maximum likelihood estimates; a rather old result. In his *Essai Philosophique* (1814) Laplace discusses a similar problem where the prior distribution (14) is —as it could well be in our problem— actually the posterior distribution from a different set of data, and records his pleasure at finding this rule by calling it "une analogie remarquable de ce poids, avec ceux des corps comparés à leur centre commun de gravité". This is possibly the origin of the term "weighted average".

Thanks to what we shall term (in conformity with Stigler's Law of Eponymy; see Appendix A) the Gaussianity of $p(A|yI)$, we need now only find the posterior expectations and covariance matrix for the irregulars:

$$\hat{e}_t = E(e_t|\sigma yI) \tag{21}$$

$$R_{tr} = E(e_t e_r | \sigma y I) - \hat{e}_t \hat{e}_r \, . \tag{22}$$

We find for the former

$$\hat{e}_t = y_t - g_t - \sigma^{-2} \sum_{r=1}^{N} R_{tr} y_r \tag{23}$$

where

$$g_t \equiv \sum_{k=0}^{6} [a_k C(kt) + b_k S(kt)] / M_k \sigma_k^2 \tag{24}$$

is a kind of shrunken prior estimate of the seasonal component s_t, in which different harmonics are weighted according to their prior variances. (23) may also be written as $\hat{e}_t = y_t - \hat{s}_t$ where $\hat{s}_t$ is Laplace's "plus avantageux" weighted average estimate of s_t.

The covariance matrix R is found to be

$$R_{tr} = \sum_{k=0}^{6} M_k^{-1} \cos \frac{2\pi k(t-r)}{12} \tag{25}$$

which, like any covariance matrix, must be positive semidefinite; a direct proof of this which also determines the rank of R is given below.

The joint posterior distribution of the irregulars is therefore

$$p(e | yI) \propto \exp[-\tfrac{1}{2}(e-\hat{e})' R^{-1}(e-\hat{e})] \, . \tag{26}$$

This solution reveals a great deal of interesting (and to the writer unexpected) insight into the seasonal adjustment problem. To see the kind of results that are in (23) - (26) the next Sections examine the effect of different kinds of prior information (I_1, I_2, ...), starting with very simple special cases. Even the seemingly trivial cases are instructive.

How would sampling theory deal with this problem? One answer is given by Tukey *et al* (1980), They would also subtract from the data an estimate of the seasonal as in (23); but say "few would argue" that the proper way to estimate that seasonal is by the monthly averages, which are the least squares estimates:

$$(\hat{s}_t)_{ST} = (12/N) \sum_m y_{t+12m} \, . \tag{27}$$

This is doubtless the most obvious thing to do. In similar problems it is much used also (to reduce the amount of data to be analyzed) by geophysicists, who call it a "Brute Stack". But we seem to be among those few; the Bayesian estimates in (23) appear so totally different from (27) that it is not clear whether there is any case in which they would agree. We shall try to understand this difference, which arises entirely from prior information that brute stacking ignores.

5. SIMPLE PRIOR INFORMATION

Example 1 Let the prior information be: I_1 = "There is no oscillating seasonal component, but there may be a *DC* offset A_o." Although this seems too trivial to be worth analyzing, let us do it anyway. Mathematically, from the general solution (23), (25) we are to pass to the limit

$$a_k \rightarrow 0, \; b_k \rightarrow 0, \; \sigma_k \rightarrow 0, \qquad 1 \le k \le 6$$

Then $M_k^{-1} \rightarrow 0$, $1 \le k \le 6$ and (23), (25) reduce to

$$R_{tr}^{(1)} = M_0^{-1} = \frac{N}{\sigma^2} + \frac{1}{\sigma_0^2} \, , \qquad 1 \le t,r \le N \tag{28}$$

$$\hat{e}_t^{(1)} = y_t - \left(\frac{N}{\sigma^2} + \frac{1}{\sigma_0^2}\right)^{-1} \left(\frac{1}{\sigma_0^2} a_0 + \frac{N}{\sigma^2} \bar{y}\right) \tag{29}$$

with $\bar{y} = N^{-1} \Sigma y_t$, the sample mean. The solution corrects for the unknown offset A_0 by subtracting from the datum y_t our "best" weighted average (20a) estimate of A_0.

If now $\sigma_0 \to 0$ (we know in advance that $A_0 = a_0$) this reduces, as it should, to

$$\hat{e}_t = y_t - a_0 \tag{30}$$

and in the opposite limit $\sigma_0 \to \infty$ (we have no prior knowledge of A_0) it becomes

$$\hat{e}_t = y_t - \bar{y} \tag{31}$$

and we must "let the data speak for themselves", having nothing else to rely on.

Out of all the Bayesian results that are in Case 1, Eq. (31) appears to be the only one that could have been found also from sampling theory. Of course, the prior estimate a_0 and weighting factors of (29) can hardly emerge from a theory which does not admit prior distributions. But more important are the correlations of the different e_t in their joint posterior distribution expressed by the matrix R. Clearly, if e_1 and e_2 are negatively correlated in this distribution, then we can estimate $(e_1 + e_2)$ more accurately than $(e_1 - e_2)$, while the opposite is true if they are positively correlated, as is the case here.

If our goal is to estimate some function $f(e_1 \ldots e_N)$, and not just the e_t individually, these correlations in the posterior distribution can greatly affect both our estimate of f and its accuracy. Yet it appears to us that sampling theory, far from being able to take this into account, lacks even the conceptual basis and vocabulary in which one could *state* that such logical interconnections exist.

The reason for this is clear when we note from (25) that these correlations come entirely from the prior information; the posterior covariance matrix R can be known before one has the data y. Faced with this observation, a sampling theorist will tend first to question the cogency and trustworthiness of the information contained in R. For that reason, we have pointed out the phenomenon in a simple, intuitive case.

A mechanical application of the sampling theory result (27) would estimate the sum or difference of e_1, e_2 by

$$(\hat{e}_1 \pm \hat{e}_2)_{ST} = [y_1 - (\hat{s}_1)_{ST}] \pm [y_2 - (\hat{s}_2)_{ST}] \tag{32}$$

and since the sampling distributions of y_1 and y_2 are independent, would ascribe to both the sum and difference estimates the same accuracy, given by their sampling standard deviation of $(24\sigma^2/N)^{1/2}$. Of course, a sampling theorist would perceive at once that in this case it would be better to use all his data in the estimator (31), reducing the error. But if he tried to judge the errors in the estimates by the sampling variances of the estimators, he would find that

$$E[(y_1 + y_2)^2 | A_0] = 4A_0^2 + 2\sigma^2/N \tag{33a}$$

$$E[(y_1 - y_2)^2 | A_0] = 2\sigma^2/N \tag{33b}$$

and thus conclude that the sum and difference estimates have equal probable error $(2\sigma^2/N)^{1/2}$.

In contrast to this the Bayesian, looking instead at the posterior distributions, would find from (28), (31)

$$E[(e_1 + e_2)^2 | yI_1] = (\hat{e}_1 + \hat{e}_2)^2 + 4\sigma^2/N \tag{34a}$$

$$E((e_1 - e_2)^2 | yI_i] = (\hat{e}_1 - \hat{e}_2)^2 \tag{34b}$$

and conclude that the probable error in estimating (e_1+e_2) is larger by a factor $\sqrt{2}$ than indicated by (33); but that (e_1-e_2) can be estimated with perfect accuracy.

But this is obviously the case; for if we know that there is no oscillating seasonal component, then however poorly we may know the offset A_0, we evidently do know the difference

$$e_1-e_2 = (y_1-A_0)-(y_2-A_0)$$

exactly. Indeed, all differences (e_m-e_n) are known exactly.

Presumably, an alert sampling theorist will see this also, and will decide not to use sampling theory to estimate (e_1-e_2). Would he still use the sampling theory for (e_1+e_2)? Perhaps; if so, the Bayesian will agree with this estimate, but will say that (33a) is overoptimistic about its accuracy, by that factor $\sqrt{2}$.

But again the Bayesian result is obviously correct, for the sampling distribution variances $(2\sigma^2/N)$ in (33) are entirely irrelevant. In the sampling theorist's scenario we are to think of repeating all this many times with A_0 fixed; then our estimates would indeed vary according to (33). But the true values of the e_t would vary along with them by the same amount, so the accuracy of our estimates of them would have nothing to do with (33). It is determined, rather, by the accuracy of our estimate of A_0. In estimating (e_1-e_2), whatever error may be in it cancels out as noted. But in estimating (e_1+e_2) the identical error occurs twice; whatever errors we are making in our estimates of e_1 and e_2 separately are not independent, which would lead to a variance $(2\sigma^2/N)$ as in (33); they are perfectly correlated, leading instead to $(4\sigma^2/N)$ as in (34).

The point of this trivial example is to make it obvious that the prior information contained in R, far from lacking in cogency and trustworthiness, has restored both the agreement with deductive reasoning, and the recognition of the perfectly correlated errors, that a mechanical application of sampling theory rules would miss.

Of course, we do not suggest that a competent sampling theorist would fail to see these points in such a simple case; we think that common sense would force him to the Bayesian results. But he would be hard put to give a sampling theory justification for them, since that common sense is using prior information that the formalism of sampling theory does not recognize. In the more subtle cases to be considered next these effects of prior information are still present and just as cogent; but they are no longer obvious.

6. SIMPLE PRIOR INFORMATION

Example 2. Now consider the prior information: I_2 = "The seasonal component is purely sinusoidal of period 12, with no *DC* offset". We are to pass to the limit $a_0 \to 0$, $\sigma_0 \to 0$, and

$$a_k \to 0,\ b_k \to 0,\ \sigma_k \to 0, \qquad 2 \le k \le 6 \tag{35}$$

Now the covariance matrix (25) reduces to

$$R_{tr}^{(2)} = M_1^{-1} \cos \frac{2\pi(t-r)}{12} \tag{36}$$

and the estimate of irregulars to

$$\hat{e}_t^{(2)} = y_t - \hat{A}_1\, C(t) - \hat{B}_1\, S(t) \tag{37}$$

where $\hat{A}_1$, $\hat{B}_1$ are given by (20) and as in (2), $C(kt) = \cos(2\pi kt/12)$, $S(kt) = \sin(2\pi kt/12)$ are the k'th harmonic seasonal sinusoids. The solution now subtracts from the data our "best" weighted average estimate of the first harmonic seasonal part. Again, in the limits $\sigma_1 \to 0, \sigma_1 \to \infty$ this goes into what our common sense tells us it should.

But again, the interesting result is in what the covariance matrix (36) tells us. As we shall prove in the next Section, R is now of rank 2; so our solution tells us that there are $(N-2)$ algebraically independent functions of the irregulars:

$$f_m(e_1 \dots e_N), \qquad 1 \le m \le (N-2) \tag{38}$$

that can be estimated exactly. To the writer and several others, this result was at first glance very far from obvious indeed. It first appeared in our attempt to solve this case by the Fourier transform method (8), (9), in which puzzling divergent integrals appeared in what was thought to be a straightforward, highly convergent Gaussian calculation. Some study was required before we were convinced that it is, after all, correct. But once understood, this result also can be made to seem "obvious" by the following argument.

Factor the joint posterior distribution (26) into a two-point distribution and a conditional distribution:

$$p\,(e_1 \dots e_N | yI) = p\,(e_3 \dots e_N | e_1 e_2\, yI)\, p\,(e_1,e_2 | yI)\,. \tag{39}$$

Now if we know that there is only a first harmonic seasonal component, then the model equations (1), (2) reduce to

$$y_t = A_1\, C(t) + B_1\, S(t) + e_t \tag{40}$$

But if e_1, e_2 are given in additional to the data y_t, then we can solve (40) for the two unknowns A_1 and B_1. Then all the subsequent values $(e_3 \dots e_N)$ are also known. In other words, the possible vectors (e) compatible with our information do not vary over a manifold of dimension N; the posterior probability $p\,(e_1,e_2|yI)$ on a two-dimensional manifold already contains full information and the conditional probability in (39) is a product of delta-functions.

This is why the Fourier transforms diverged, when we tried to jump directly into the limit $\sigma_k = 0$ at the beginning of the calculation. The difficulty is avoided if we first work out the general solution in the safe territory where all $\sigma_k > 0$, and then approach various limits from it.

Given e_1 and e_2, the extrapolation to all t is just

$$e_t = y_t + (y_1 - e_1)\,\frac{S(t-2)}{S(1)} - (y_2 - e_2)\,\frac{S(t-1)}{S(1)}\,, \qquad 1 \le t \le N \tag{41}$$

for (41) has the required form of y_t plus a first harmonic and is obviously true for $t=1$, $t=2$; so it must be true for all t. Likewise, given any two values e_m, e_n the interpolation and/or extrapolation determining the others is

$$e_t = y_t + (y_m - e_m)\,\frac{S(t-n)}{S(n-m)} - (y_n - e_n)\,\frac{S(t-m)}{S(n-m)}\,, \qquad 1 \le t \le N \tag{42}$$

This is evidently most stable when $|S(n-m)| = 1$; i.e. when $(n-m)$ is 6 months, 18 months, etc. It fails when $(n-m)$ is a multiple of 12; for given e_n we know that $e_{n+12} = y_{n+12} - (y_n - e_n)$ and only one piece of information has been given.

If our prior information had been that there can be only two harmonics, for example $k = 1$, $k = 2$, present in the seasonal component, then this argument would work if we factor $p\,(e|yI)$ into $p\,(e_1,e_2,e_3,e_4|yI)$ and a probability conditional on $(e_1 \dots e_4)$. Then, given any four non-redundant values of the irregular, the others would be known exactly.

The joint posterior density of $(e_1 \ldots e_N)$ is non-zero only on a manifold of dimension 4; and so on.

We have here a kind of "anti-collinearity" phenomenon. Singularity of a covariance matrix R does not mean that some components of the vector e cannot be estimated from the data; it means just the opposite. That is, if R is of rank r, then the N-dimensional space S_N of its eigenvectors has a subspace S_r of dimension r, spanned by those eigenvectors of R with non-zero eigenvalues, in which all the posterior probability is concentrated. Vectors with non-zero projections into the complementary $(N-r)$-dimensional subspace S_{N-r} have zero posterior probability. But to keep them out of S_{N-r} requires $(N-r)$ algebraically independent conditions on $(e_1 \ldots e_N)$; hence $(N-r)$ independent functions of $(e_1 \ldots e_N)$ are determined exactly. Collinearity is not bad, but good!

There is a mathematical lesson to be learned from this example: from a casual glance at the posterior distribution (26) one would at first say that if R becomes singular, then the quadratic form

$$Q = (e-\hat{e})' R^{-1}(e-\hat{e})$$

blows up and (26) becomes meaningless. But the quantity of interest is not Q, but $\exp(-Q/2)$, which does not blow up. As an eigenvalue of R tends to zero, the "thickness" of the basic support set of the distribution (26) in the direction of the corresponding eigenvector goes smoothly and continously to zero. In the limit the support set lies on a manifold of smaller dimensionality. Instead of blowing up, we therefore have a mathematically well-behaved and useful solution; (26) is non-zero only when $(e-\hat{e})$ lies entirely in the subspace S_r.

It seemed worthwhile to stress this point, because some working on the lore of improper priors have become entangled in "paradoxes" much less subtle than this. In working with any kind of singular mathematics we recommend very strongly the procedure that theoretical physicists have learned, over many years, to follow: (I) start from safe territory where everything is finite, convergent, and well-behaved and there is no question about what is the correct solution; (II) approach the singular cases cautiously, as limits from this.

The limit of a sequence of "good" solutions may or may not be a "good" solution in itself. A mathematically well-behaved limit is one wherein certain quantities just become smaller and smaller and eventually disappear, leaving behind a simpler analytical expression. If the limit is not well-behaved in this way (but instead, for example, blows up or oscillates forever), then the limit cannot be interpreted as a valid solution to the problem, and any attempt to find a solution by jumping directly into that limit would have led to nonsense, *the cause of which cannot be seen by looking only at the limit.* We think that most of the recent paradoxing in statistics could have been averted by following this "cautious approach" policy.

7. EFFECT OF NEW PARAMETERS

We have seen the cases in which the prior information I_1 tells us that only the DC offset A_o is present, and I_2 that only the first harmonic component is present. Suppose now I_3 = "Both are present". How are our results changed?

There is a common "folk-theorem" in the statistical literature to the effect that adding more unknown parameters in a problem must lead to a deterioration in the accuracy with which we can estimate the old parameters; and so one should not do this unless the data clearly call for it. Carrying this further, several authors state that it is fundamentally impossible to estimate more than N parameters from N data points; and practically impossible to estimate more than a small fraction of that number.

Although we feel intuitively that there must be some truth in these folk-theorems, we have never seen a proof of either. But from our general solution we can learn something about their validity and unstated qualifications.

First, note that if we are estimating the seasonal parameters (A_0 ... B_5), we have from (18) that their posterior distributions are independent. Therefore our estimate of A_1 and its accuracy are the same whether A_0 is considered known, or whether it is also being estimated. The folk-theorem is simply wrong in this case.

There is a fairly general class of situations that includes seasonal adjustment and many other problems, in which these effects are easy to understand. Consider two different problems; in (a), we are estimating an "old" parameter β, with prior information I_a that specifies a model containing only β. In problem (b) we are still estimating β but there is a "new" parameter ϕ, also unknown. The class of situations considered is that in which model (a) corresponds to setting $\phi = 0$. Then the posterior distribution of β in the two problems are:

$$p(\beta|D,I_a) = p(\beta|D,I_b,\phi = 0) \tag{40}$$

$$p(\beta|D,I_b) = \int p(\beta|D,I_b,\phi)\, p(\phi|D,I_b)d\phi \tag{41}$$

respectively. The model (a) result appears, in the context of model (b), as a conditional distribution conditional on $\phi = 0$; while the model (b) result is instead a weighted average of distributions conditional on all possible values of ϕ.

Evidently, then, considering ϕ unknown will in general make our estimate of β worse, in agreement with that intuitive folk-theorem. But the folk-theorem can also be false in some cases, since it is possible for a weighted average of distributions to be more sharply peaked than some particular one of those distributions. In this model, in order to cause appreciable deterioration in our estimate of β, two conditions must be present: our estimate of β from the conditional posterior distribution $p(\beta|DI_b\phi)$ must depend appreciably on ϕ; and ϕ must itself be not well determined by the prior information I_b and the data D.

One stated, this seems so obvious that intuition should have seen it long ago; but we can point to no record indicating that it actually did. It would be hard for sampling theory to state such a result because the deterioration is caused by the old parameter becoming correlated, in the posterior distribution, with an unknown quantity; and sampling theory does not recognize such notions.

Of course, the situation may be different if introducing the new parameter causes a drastic change in the model, not just adding a new dimension to the parameter space. The two problems might be so different that there is no meaningful comparison at all.

In our seasonal adjustment problem, there is no such drastic change, and adding A_0 to the Example 2 problem affects our conclusions thus: our estimate of the oscillating seasonal component is not changed at all, because the added constant term is orthogonal to the seasonal sinusoids $C(kt)$, $S(kt)$. But our estimates of the irregulars (e_1 ... e_N) are shifted by an amount proportional to our estimate of A_0, and their posterior correlations are increased by an amount corresponding to the probable error of our estimate of A_0. But actually A_0 can be estimated quite accurately from the data of a few years, and so the last factor in (41) is sharply peaked and there is very little effect on the accuracy of our results.

The rules (40), (41) also enable us to judge how various extensions of our seasonal adjustment model will affect our conclusions. In particular, detrending would have to be included in many real problems by adding to the model equations (1), (2) a term Ct where C is a new trend rate parameter to be estimated from the data. But the Bayesian detrended seasonal adjustment will differ from what is commonly done on intuitive grounds. Our new joint posterior distribution of $(A_0 \dots B_5)$ will not be an estimate from the detrended data; but rather as indicated by (41) we should take a weighted average of the joint distributions conditional on all possible values of C. The Bayes estimates would approach an estimate from detrended data if C were itself accurately determined by the data.

We do not go into the Bayesian detrending analysis here, but it has been carried out and we find astonishingly little change in our seasonal adjustment conclusions. The reason is that the new linear term Ct is nearly orthogonal to the seasonal sinusoids. The estimates of $(A_1 \dots B_5)$ are slightly changed by amounts proportional to our estimate of the trend rate C, their joint posterior distribution is no longer quite independent, and their probable errors are slightly increased, by a factor of $(1-6N^{-2})^{-1}$, which is only 4 % for $N = 12$, and quite negligible if we have data for two or more years.

Thus Bayesian seasonal adjustment —contrary to what a naive application of that folk-theorem might suggest— accommodates detrending easily. Of course, estimates of the irregulars will be corrected if there is evidence for a strong trend; but there is very little change in their posterior correlations or accuracy.

8. RANK OF THE COVARIANCE MATRIX

The $(N \times N)$ matrix R defined in (25) is real and symmetric; therefore it has a full set of N orthonormal eigenvectors $\mathbf{h}_i = (h_{1i} \dots h_{Ni})$ and eigenvalues β_i:

$$\sum_{r=1}^{N} R_{tr} h_{ri} = \beta_i h_{ti}, \qquad 1 \le t,i \le N \tag{42}$$

Let $\mathbf{y}$ be any real $(N \times 1)$ vector with components $(y_1 \dots y_N)$, not all zero and denote scalar products of real vectors by $(x,y) = \Sigma\, x_t y_t$. Since $\mathbf{y}$ has an expansion $\mathbf{y} = \sum_i (h_i,y)\mathbf{h}_i$, the quadratic form

$$F(\mathbf{y}) = (y'Ry) = \sum_{t,r=1}^{N} R_{tr}\, y_t\, y_r \tag{43}$$

can be decomposed as

$$F(\mathbf{y}) = \sum_{i=1}^{N} |(h_i,y)|^2 \beta_i\,. \tag{44}$$

But in our case, from (25) this is also equal to

$$F(\mathbf{y}) = \sum_{k=0}^{6} M_k^{-1} |\sum_t y_t \exp(i\pi kt/6)|^2 \ge 0 \tag{45}$$

therefore R is positive semidefinite. But then if $F(\mathbf{y}) = 0$, from (44) $\mathbf{y}$ must be orthogonal to all $\mathbf{h}_i$ with $\beta_i > 0$; i.e. $\mathbf{y}$ is itself an eigenvector with zero eigenvalue; call it a *zector* for short. The number of linearly independent zectors is $(N-r)$, where r is the rank of R.

Consider, then, the prior information I_1 of Sec. 5, that there is no oscillating seasonal component but there may be a DC offset A_0, which led in (28) to $R_{tr} = M_0^{-1}$, $1 \le t,r \le N$. This gives

$$F(\mathbf{y}) = M_0^{-1} |\sum_t y_t|^2 = M_0^{-1}(u_0,y)^2 \tag{46}$$

where the vector $\mathbf{u}_0$ with components $(1 \dots 1)$ is an eigenvector of R with eigenvalue $\beta = N/M_0 > 0$. But from (46) every vector $\mathbf{z}$ orthogonal to $\mathbf{u}_0$ yields $F(\mathbf{z}) = 0$, and is therefore a zector. In an N-dimensional space there are $N-1$ linearly independent vectors orthogonal to $\mathbf{u}_0$, therefore the positive eigenvalue is nondegenerate, and the rank of R is one, as stated in Sec. 5.

The prior information I_2 of Sec. 6, that only a fundamental seasonal component, of period 12, is present, led to the covariance matrix (36) in which only the term in M_1^{-1} is present. For this we have

$$F(\mathbf{y}) = M_1^{-1}|\sum_t y_t \exp(i\pi t/6)|^2 = M_1^{-1}|(u_1 + iv_1, y)|^2$$
$$= M_1^{-1}[(u_1,y)^2 + (v_1,y)^2] \tag{47}$$

where $\mathbf{u}_k$, $\mathbf{v}_k$ denote the linearly independent vectors

$$\mathbf{u}_k \text{ with components } (u_{tk} = \cos\pi kt/6, \quad 1 \le t \le N)$$
$$\mathbf{v}_k \text{ with components } (v_{tk} = \sin\pi kt/6, \quad 1 \le t \le N) \tag{48}$$

Evidently, $F(\mathbf{u}_1) > 0$ and $F(\mathbf{v}_1) > 0$; and any vector $\mathbf{z}$ that is orthogonal to both $\mathbf{u}_1$ and $\mathbf{v}_1$ is a zector. There are N-2 linearly independent zectors, so $R^{(2)}$ is of rank 2.

For the prior information I_3 which allowed the possibility of both the DC offset and the first harmonic component, R yields the quadratic form

$$F(\mathbf{y}) = M_0^{-1}(u_0,y)^2 + M_1^{-1}[(u_1,y)^2 + (v_1,y)^2] \tag{49}$$

from which it is now evident that R is of rank 3, since every vector orthogonal to $\mathbf{u}_0$, $\mathbf{u}_1$, and $\mathbf{v}_1$ is a zector.

In general, then, we have

$$F(\mathbf{y}) = \sum_{k=0}^{6} M_k^{-1}[(u_k,y)^2 + (v_k,y)^2] \tag{50}$$

and the rank of R is the number of scalar products appearing. Note that the cases $k = 0$, $k = 6$ are special, since $v_0 = v_6 = 0$. Every nonzero M_k^{-1} contributes 1 or 2 to the rank. and when all are non-zero the maximum possible rank of R is 12.

Stated differently, every seasonal parameter in the set $(A_0 \ldots B_5)$ that is initially unknown contributes one to the rank of R.

9. CONCLUSION

Our analysis has shown how prior information about the seasonal parameters can have a major effect on our estimates of the irregulars or functions of them. In the extreme case where we know that a particular harmonic component is zero, the dimensionality of the posterior distribution is reduced. A real application will probably never be so utopian, but if we know that a particular harmonic component must be very small, R will approach a matrix of reduced rank, and as a result it will be possible to estimate some functions of the irregulars much more accurately than one would have thought from the sampling theory result (27).

Consider, briefly, what kind of prior information one might have in the case of a real economic time series. From prior knowledge familiar to all of us, we expect that department store sales will peak rather sharply in December, while sales of beer and ice cream will peak more broadly in July, bank loans to individuals in the U.S.A. may peak just before April 15, agricultural employment will peak at harvest time, etc.

In all these cases the first harmonic seasonal component is clearly the major one. Indeed, it is hard to think of any case where there is a repetitive mechanism tending to generate a second harmonic (i.e. a reason why anything should go from maximum to minimum in alternate quarters), much less any higher harmonic. That is, virtually every economic time series surely has a "driving mechanism" with a basic period of one year, the appearance of harmonics being due only to the nonsinusoidal nature of the variation, rather than to any influence that encourages repetition after a shorter period.

There are two basic kinds of periodic but nonsinusoidal behavior: (a) high-low asymmetry making, for example, sharp peaks but broad troughs, as we conjecture to be the case for department store sales; and (b) up-down asymmetry tending, for example, to make the falling portion of a curve steeper than the rising portion. It seems plausible that sales of bathing suits might rise slowly throughout the Spring and Summer, but drop precipitously in the Fall.

Type (a) behavior is represented by a Fourier series with only even order harmonics, while type (b) has both odd and even. Doubtless, both effects are present to some extent in most time series; but it seems highly unlikely that any economic quantity would exhibit only odd harmonics, making the peaks and troughs mirror images of each other (although that is the usual case for the electrical engineer's seasonal adjustment problem, elimination of complex and changing hum interference waveforms from sensitive circuits). It appears that in most cases the economist may expect the sinusoidal components of order $k = 1,2,4$ to predominate, while $k = 3,5,6$ should be much weaker.

We suggest that putting this information into our priors may make a noticeable improvement in seasonal adjustment, just as Litterman's use of priors that express our common-sense judgment that high order autoregressive coefficients are small, improves the forecasting of time series.

Clearly, what is needed now is to put these ideas to the test by analysis of real data for which conventional seasonal adjustments have been made by X-11, SABL, or some other current program. The writer will undertake to do this computation but, not being a professional economist, feels the need of help in choosing samples of data that appear to be promising for this purpose. Case where hindsight was able to make a significant correction of the first seasonal adjustment would be particularly valuable.

APPENDIX A: WHY A GAUSSIAN ERROR DISTRIBUTION?

Stigler's Law of Eponymy (1980), illustrated by its name, states that "No scientific discovery is named after its original discoverer". Thus we find that the distribution $f(x) = \exp(-x^2/2)$ was used by Laplace in 1774 three years before Gauss was born, and by de Moivre in 1733 sixteen years before Laplace was born. So we are well within the Letter of the Law if we continue to call $f(x)$ a "Gaussian distribution".

The usual rationale for assigning iid Gaussian prior probabilities to "random errors" is that if the real errors are the resultant of many small independent contributions, then by the Central Limit Theorem the total error will have nearly a Gaussian frequency distribution whatever the distributions of the individual small components. Such an argument, although of course correct as far as it goes, does not take note of other reasons that may be equally cogent or more so.

Our use of the term "prior probabilities" in this context may seem unusual; conventionally, $p(e|I)$ would be called a sampling distribution. Note, however, that "sampling distributions" are from a Bayesian standpoint simply the prior probabilities we assign to the errors, or "noise", not different in logical status from prior probabilities assigned to parameters or hypotheses. Indeed, for seasonal adjustment the term seems particularly called for, since our aim is just to convert the prior probability $p(e|I)$ assigned to the noise, into its posterior distribution $p(e|yI)$.

In Bayesian inference, a sampling distribution is no more required to be a frequency distribution than is any other prior. Our aim in writing a prior distribution for a parameter is to represent our state of prior knowledge about the range of possible values that parameter may have in the specific case at hand; and this is not necessarily a frequency in any real or conceivable set of repetitions of the problem. Indeed, it will not be a frequency except in the special case (of which no example is known to us) where our prior information about the parameters consists solely of frequencies with which various values have occurred in other cases.

But in almost every real problem there are special circumstances, which make the present instance unique and not comparable to others. Then whether our parameter would would or would not have the same value in some other problem that we are not reasoning about, is irrelevant for our problem.

Likewise, in writing a sampling distribution we are representing our state of prior knowledge about the possible errors that may occur in the specific case at hand --which, depending on special circumstances, may or may not be related to the frequencies of those errors in some class of other situations that we are not reasoning about.

In fact, there is almost no real problem in which we actually have prior knowledge of the frequency distribution of errors; and if we did have such knowledge, it would not in general suffice to determine a reasonable sampling distribution. For in the case of errors as well as that of parameters, it is typical of real problems that we have prior information which does not happen to consist of frequencies; but is none the less cogent.

A rational prior error distribution, or sampling distribution $p(e|I)$, should incorporate all our prior information about the errors; not just the part of it that happens to refer to frequencies. Indeed, the frequency part is usually missing altogether or incomplete, consisting of only one or two moments of the error distribution. An experimental physicist or electrical engineer usually knows the average power level of his noise, less often something about its spectral distribution; and seldom anything else. An economist willing to use past experience about the general magnitude of the irregular fluctuations; but either lacking, or distrusting the relevance of, past frequency data, would be in much the same position.

As we argued extensively elsewhere, the prior distribution that most honestly represents our state of knowledge (i.e. that agrees with what we know but does not assume anything beyond that) is the one with maximum entropy subject to the constraints imposed by what we know. If the prior information fixes (or can be reasonably thought of as fixing) the first two moments, this principle will lead to the Gaussian distribution, as has been observed countless times.

However, this is not the whole story. Another of the common folk-theorems of statistics is that if we use an iid Gaussian prior for the errors, but the errors do not in fact have an iid Gaussian frequency distribution, then something terrible will happen to us and we shall be led to draw all manner of wrong and misleading conclusions. Like the other folk-theorems mentioned above, this one is in need of more careful statement.

We learn form the Asymptotic Equipartition Theorem of Information Theory (Feinstein, 1958) that the entropy of a distribution is essentially a measure of its "size"; i.e. over how large a volume W of the sample space is the probability density of the errors, or "noise", reasonably large? As $N \to \infty$ the iid Gaussian distribution is the one that occupies, asymptotically, the greatest possible volume of N-dimensional sample space for given first and second moments. As it is usually put in the literature of Information

Theory, entropy is an asymptotic measure of the size of the "basic support set" of the distribution.

From this we see that the Central Limit Theorem is, in a very fundamental sense, a special case of the principle of maximum entropy. For first and second moments, being additive under convolution, constrain the possible distributions that can be reached by convolution. The CLT thus tells us that, barring very unusual circumstances, no other constraints exist; a distribution with finite first and second moments will expand under repeated convolution until it fills up the entire volume allowed by those two constraints. In a class of generalized CLT's, distributions could be merged in other ways than convolution, and would expand asymptotically into the one with maximum entropy subject to whatever constraints are imposed by the new method of merging.

If we assign an iid Gaussian prior $N(0,\sigma)$ to our noise sequence $(e_1 \ldots e_N)$, then the high-probability volume, or basic support set, is an N-dimensional sphere R of radius about $\sigma N^{1/2}$. Any systematic effect that one is trying to detect by a significance test will be effectively obscured by this noise (i.e. we will not be able to distinguish it from the noise) if the effect is so small that the sample values remain inside the noise sphere R. It will appear statistically significant if its effect is enough to carry the sample values outside R.

Thus the iid Gaussian prior assignment is not a "physical assumption" that might lead us to errors if wrong; quite the contrary, it is the safest, most conservative assignment we can make if we know only the first two moments, because its high-probability region R takes into account every possible noise vector that is allowed by that information. To use any other prior assignment in this state of knowledge would amount either to contradicting our prior knowledge (if our prior has different moments); or to make additional assumptions, not warranted by our information, that contract R to some arbitrary subset $R' \subset R$; and thus invite erroneous conclusions. For if the noise vector happens to lie in the complementary set $R - R'$, there will appear to be a real, statistically significant effect that is actually only an artifact of our particular choice of R'.

Put differently, if we assign an iid Gaussian error distribution but the frequency distribution of errors in N measurements is not, in fact, iid Gaussian, then for a given mean square error the result will not be that we shall see spurious effects that are not there; but only that our discriminating power to see small effects is not as sharp as it could be, and the accuracy of our parameter estimates is not as high as it could be.

If we had additional prior information, beyond the mean square error, about the specific way in which the noise values depart from iid Gaussian, then we could use that information to define a subset $R'' \subset R$ within which we know that the noise almost certainly lies; and then any sample that falls in $R - R''$ will be statistically significant. Thus additional prior information about the noise can be crucial for the conclusions we are able to draw from a given data set.

But this information need not consist of frequencies. Any information that constrains the possible N-dimensional noise vectors to a subset $R'' \subset R$ will have this effect of increasing the discriminating power of our signficance tests and the accuracy of our estimates. In particular, information about non-independence (the noise is not "white") makes the noise sequence in part predictable, and is thus highly valuable for extracting systematic "signals" from the noise.

Maximum-entropy spectrum analysis (Burg, 1975; Childers, 1978; Jaynes, 1982) is just the means by which we exploit the increased predictability of a time series that results from information about a few values of its autocovariance function. The maximum entropy formalism is the analytical means for locating the subset R'' defined by this information.

Likewise, maximum-entropy reconstruction of images, crystallographic, or molecular structure is the analytical means for locating the high-probability region R'', in the space of possible true states of nature, defined by the incomplete data of a blurred image. We think that future advances in pattern recognition will come from a similar taking into account of information that is ignored by sampling theory because it does not consist of frequencies.

There is no reason other than historical precedent why these methods should be confined to physics and engineering; they should find use in any application where there is cogent informtion that sampling theory finds indigestible.

Thus we applaud the custom of assigning iid Gaussian priors. Nothing better could have been done by the Bayesian or anyone else, unless he had additional, quite specific prior information beyond the first two moments of the noise. Whatever the "true" frequency distribution of the noise, if it is unknown then we cannot make use of it for inference; and if it is known, additional information may be equally cogent.

But what if our prior information is so meager that we do not even know the second moment σ^2? This is perhaps the most common situation outside of physics. We do not claim it as the optimal thing to do in every case, but it is useful and computationally feasible —a kind of Bayesian jackknife— to reason as follows: if we did know σ, the Gaussian assignment would be indicated, so σ is a relevant hyperparameter and the problem reduces to assigning a prior to σ.

APPENDIX B: THE PRIOR FOR SIGMA

In discussing our choice of a prior $p(\sigma|I)$, we can answer some common misgivings by noting that the Bayesian formalism automatically provides a diagnostic check on our priors and model. Of course, their adequacy cannot be tested by Bayesian or any other methods if we have only a small amount of data. But with enough data this becomes possible; for then the posterior density of σ from the Jeffreys prior becomes sharply peaked, the data alone pointing to a well-defined value of σ. A highly informative prior sharply peaked at a very different value thus stands in conflict with the evidence of the data; intuitively, we would be led to doubt the validity of our prior information or model.

This observation leads to the formal significance tests of Jeffreys, used recently by Zellner and co-workers, if we convert that intuitive feeling into a well-posed question. In our view, then the "diagnostic phase" is indeed an essential part of inference; but it requires no departure from Bayesian methods. Quite the contrary, Bayesian methods are required for a full treatment of it. This is discussed further in Appendix C.

If we have very little data, then of course our prior distributions can matter a great deal for the conclusions we are able to draw, since our prior and posterior states of knowledge are not very different. But a t-distribution with many degrees of freedom goes asymptotically into a Gaussian, and so if we have a reasonably large amount of data, even the aforementioned extremes in the prior $p(\sigma|I)$ cannot lead us to very different conclusions about seasonal adjustment unless the data sharply contradict our informative prior, in which case we would start over again anyway. So we made the simplest choice in the text.

APPENDIX C: THE "DIAGNOSTIC PHASE" OF INFERENCE

G.E.P. Box (1982) noted the Bayesian significance test for comparing different models, but criticized it on the grounds that it could lead us to misleading conclusions if the class of alternatives did not happen to include the true one. At first glance it may appear that a test that does not refer to any specific class of alternatives is free from this objection. Such tests are indeed useful, and we do not mean to argue against them as easy approximations, often good enough for the purpose. However, we note three reasons why a full, well-posed diagnostic test must be Bayesian.

In the first place, not specifying a class of alternatives does not mean that the alternatives are not there; it means only that the test has not been fully defined. Any choice of test statistic, whatever rationale is given for it, is necessarily an implicit assumption about some class of alternatives. That is, given any null hypothesis H_0, data D, test statistic $t(D)$, and rule such as "reject H_0 if $t > t_0$", we are judging H_0 against a class of alternatives for which one expects large values of t.

If one fails to specify what that class is, he is not thereby prevented from applying the test; but having done so, he does not know what the test has accomplished, or what to do if H_0 is rejected. As Jeffreys (1939) put it, there is not the slightest use in rejecting any hypothesis unless we can do it in favor of some alternative known to be better.

Of course, not all "frequentist" significance tests have a Bayesian interpretation, for there are still frequentists who do not believe in the likelihood principle. For example, suppose that the point at issue is the value of some parameter β. If D and D' denote two different data sets that give the same likelihood function $L(\beta)$, but D is in the "accept" region and D' in the "reject" region, then in two situations where we have the same state of knowledge about β we are drawing different conclusions, a violation of our basic "Desideratum of Consistency". Such an irrational test cannot have —nor would we wish it to have— any Bayesian interpretation.

But a test that respects the likelihood principle (data sets that give the same likelihood function also lead to the same decision) necessarily partitions the class of possible posterior distributions $p(\beta|DI)$ into "accept" and "reject" subclasses. Such a test can be interpreted —and defined— in Bayesian terms.

For the most common significance tests, Chi-squared, or the one-sided t-test and F-test, it is straightforward mathematics to construct a Bayesian significance test that leads to the same test statistic and the same procedure; but does refer to a specific class of alternatives, and therefore does tell us what the test has accomplished.

Given a null hypothesis H_0 and some class C of alternative hypotheses $(H_1, H_2, \ldots)$, a usable test, that goes at least part way toward a full Bayesian posterior odds ratio test, is to search out class C for the best alternative to H_0 by the likelihood criterion; i.e. given the data D, calculate the test statistic

$$t(D;C) = \max_{H \epsilon C} \log[p(D|H)/p(D|H_0)] \tag{C1}$$

which tells us how much the data could support an alternative in C, relative to H_0.

For example, the Chi-squared test with n categories is commonly cited as a test without alternatives; yet Chi-squared is readily interpreted in Bayesian terms, as the statistic that searches out the class B_n of Bernoulli alternatives (i.e. n possible results at each trial, independence of different trials). Chi-squared is a two-term Taylor series approximation to $2t(D;B_n)$. Thus the numerical value of Chi-squared has a definite meaning: it tells us how much improvement in fit could be obtained within the class B_n of alternatives.

This brings up the second reason for using a Bayesian test, the logic of the decision criterion, or choice of the critical value t_0. In the traditional Chi-squared test the decision is based, not on the numerical value of Chi-squared, but on tail areas of the Chi-squared distribution conditional on H_0. But the illogical nature of any test that tries to decide solely on grounds of probabilities conditional on the null hypothesis, while simply ignoring the probabilities conditional on the alternatives, is too obvious to dwell on.

The situation is particularly embarrasing if H_0 is rejected; for surely, if we reject H_0, then we must also reject probabilities conditional on H_0; but then if no other probabilities have been used, what was the justification for the decision? Orthodox logic seems to saw off its own limb. A Bayesian test is free of this dilemma, since it decides on grounds of all the probabilities involved.

Another difficulty with interpreting the Chi-squared test as concerned with tail areas but not with alternatives, is that there is then no reason why one tail should be better than another; one ought to reject H_0 just as readily if Chi-squared turns out to be much smaller than expected. But then it would be clearly illogical to reject H_0 in favor of any alternative in B_n, for H_0 is supported by the data more than any such alternative. An unexpectedly small Chi-squared could only be grounds for rejecting H_0 in favor of an alternative for which Chi-squared is expected to be small (such as one with negative correlations or some mechanism that constrains the data).

But this brings home to us the third reason why a fully satisfactory diagnostic test must be Bayesian; whether we should actually reject our model cannot be judged reasonably until we take into consideration not only the class of alternatives in some quantity like $t(D;C)$, but also their prior probabilities. Usually, one is willing to reject H_0 when Chi-squared is large, because the indicated class of alternatives B_n is judged to have a reasonably high prior probability. But one would seldom reject H_0 when Chi-squared is unexpectedly small, because the alternatives for which one expects a small Chi-squared have very low prior probability. Tests that ignore alternatives and their prior probabilities could be —and we think have been— very misleading.

A Bayesian test with a set of alternatives so poorly chosen that the true one is not even in it could also be misleading, in the sense that it could not lead us to the true one; but would a test that ignores alternatives leave us in any better position? At least, the Bayesian would know what class of alternatives had been searched out, and the most likely one in that class. Surely, an antiBayesian who does not know even that much is not in less danger of falling into error; but more.

It would be very nice to have a formal apparatus that gives us some "optimal" way of recognizing unusual phenomena and inventing new classes of hypotheses that are most likely to contain the true one; but this remains an art for the creative human mind. In trying to practice this art, the Bayesian has the advantage because his formal apparatus already developed gives him a clearer picture of what to expect; and therefore a sharper perception for recognizing the unexpected. To one who expects nothing in particular, nothing can be unexpected. This applies especially to methods of data analysis that decline to use any formal apparatus at all.

REFERENCES

BOX, G.E.P. (1982). An Apology for Ecumenism in Statistics, *Tech. Rep.* **2408**, Math. Research Center, University of Wisconsin.

BRICOGNE, G. (1982). In *Computational Crystallography*, D. Sayre, Ed., Oxford University Press, pp. 258-264. Also Maximum Entropy and the Foundations of Direct Methods (1983), Dep. of Biochemistry Report, Columbia University, New York.

BRYAN, R.K., BANSANL, M., FOLKHARD, W., NAVE, C. and MARVIN, D.A. (1983). Maximum-Entropy calculation of the electron density at 4A resolution of Pf1 filamentous bacteriophage, *Proc. Nat. Acad. Sci.* USA, **80**, pp. 4728-4731.

BURG, J.P. (1975). Maximum Entropy Spectral Analysis. Stanford University. Thesis.

CHILDERS, D. (1978). *Modern Spectrum Analysis*. New York: IEEE Press.

CURRIE, R.G. (1981). Solar Cycle Signal in Earth Rotation, *Science,* **211**, 386-389; See also Evidence for 18.6 years M_N Signal in Temperature and Drought Conditions in North America since AD 1800, *J. Geo. Res.* **86**, 11055-11064.

FEINSTEIN, A. (1958). *Information Theory*, New York: Wiley.

FRIEDEN, B.R. (1980). Statistical Models for the Image Restoration Problem, *Computer Graphics and Image Processing,* **12**, 40-59.

GIBBS, J.W. (1902). *Elementary Principles in Statistical Mechanics*, reprinted in *The Collected Works of J. Willard Gibbs*, Vol. 2, New Haven, Conn. University Press, 1948 and New York: Dover Publications, Inc., 1960.

GULL, S.F. and DANIELL, G.J. (1978). Image Reconstruction from Incomplete and Noisy Data, *Nature,* **272**, 686-690; See also The Maximum Entropy Algorithm Applied to Image Enhancement, *Proc. IEEE,* **5**, 170-173 (1980).

JAYNES, E.T. (1968). Prior Probabilities *IEEE Trans. on Systems Science and Cybernetics,* **SSC-4**, 227-241.

— (1981). Marginalization and Prior Probabilities, *Bayesian Analysis in Econometrics and Statistics*, North-Holland, Amsterdam, A. Zellner, Editor.

— (1982). On the Rationale of Maximum-Entropy Methods, *Proc. IEEE,* **70**, 939-952.

— (1983). *Papers on Probability, Statistics and Statistical Physics,* Dordrecht-Holland: D. Reidel.

— (1984). The Intuitive Inadequacy of Classical Statistics, in *Proceedings of the International Convention on Fundamentals of Probability and Statistics,* Luino, Italy, D. Costantini, Editor (in press).

JEFFREYS, H. (1939). *Theory of Probability*, London: Oxford University Press.

KADANE, J.B. (1980). Predictive and Structural Methods for Eliciting Prior Distributions, *Bayesian Analysis in Econometrics and Statistics,* North-Holland, Amsterdam, A. Zellner, Editor.

KASS, R. (1982). A Comment on 'Is Jeffreys a Necessarist?', *Am. Stat.* **36**, 390-392.

LAPLACE, P.S. (1814). *Essai Philosphique sur les Probabilities*, Courcier Imprimeur, Paris; reprints of this work and of Laplace's much larger *Theorie Analytique des Probabilities* are available from Editions Culture et Civilisation, 115 Ave. Gabriel Lebron, 1160 Brussels, Belgium.

LITTERMAN, R. (1980). A Bayesian Procedure for Forecasting with Vector Autoregression, Manuscript dated September 1980 and Ph.D. Thesis, University of Minnesota, 1979.

MEAD, L.R. and PAPANICOLAOU, N. (1984). Maximum Entropy in the Problem of Moments *J. Math. Phys.* **25**, 2404-2417.

POPPER, K.R. (1959). *The Logic of Scientific Discovery*, Basic Books, Inc., New York.

RISSANEN, J. (1983). A Universal Prior for the Integers and estimation by minimum description length. *Ann. Stat.* **11**, 416-431.

SAVAGE, L.J. (1954). *The Foundations of Statistics,* New York: J. Wiley & Sons.

— (1981). *The Writings of Leonard Jimmie Savage - A Memorial Selection,* Published by The American Statistical Association and the Institute of Mathematical Statistics.

STIGLER, S.M. (1980). Stigler's Law of Eponymy, *Trans. New York Acad. Sci.* **39**, 147-159.

TUKEY, J.W. (1978). Granger on Seasonality, in *Seasonal Analysis of Time Series,* U. S. Government Printing Office: A. Zellner, Editor, Washington.

VAN DANTZIG, D. (1957). Statistical Priestcraft: Savage on Personal Probabilities, *Statistica Neerlandica,* **11**, 1-16.

WILKINS, S.W., VARGHESE, J.N. and LEHMANN, M.S. (1983). Statistical Geometry I. A Self-Consistent Approach to the Crystallographic Inversion Problem Based on Information Theory, *Acta Cryst.* **A39**, pp. 47-60.

WINKLER, R.L. (1980). Prior Information, Predictive Distributions, and Bayesian Model-building, in *Bayesian Analysis in Econometrics and Statistics*, North-Holland, Amsterdam: A. Zellner, Editor.

ZELLNER, A. (1982). Is Jeffreys a Necessarist?, *Am. Stat.* **36**, 28-30.

DISCUSSION

J.J. DEELY (*University of Canterbury, NZ*)

Firstly, I must apologise to Professor Jaynes for not having enough time to verify the details of his paper. I had to leave New Zealand early in August and have been on the go ever since. I should also point out to him that his paper is the only one presented during this conference in which *two* discussants were devoted solely to one paper. Perhaps the fact that his first two initials are "E.T." had some bearing.

Professor Jaynes is to be commended for his labour of love in reminding us again that the Maximum Entropy Principle (MEP) provides us with a method which converts prior information into highly informative priors. This paper provides a new orchestration for that old song. He uses a simple forecasting model:

$$y_t = s_t + \epsilon_t, \qquad 1 \leq t \leq N$$

where

$$s_t = A_0 + \sum_{k=1}^{6} \left(A_k \cos \frac{2\pi kt}{12} + B_k \sin \frac{2\pi kt}{12} \right)$$

with $B_6 = 0$. This implies a seasonal model with $s_t = s_{t+12}$ for every t. The coefficients are assumed to have prior distributions, $A_k \sim N(a_k,\sigma_k)$, $B_k \sim N(b_k,\sigma_k)$ and the errors ϵ_t are independent and $N(0,\sigma)$. It is the choice of these prior distributions which causes me concern. Professor Jaynes has stated "... a high priority research problem must be the development of the formal apparatus that can realise the aforementioned desideratum, i.e. to have a theory which deals seriously with prior knowledge by converting specific prior information into specific prior probability assignments, in a wider variety of problems," and "... problems of inference are basically ill-posed if prior information is not considered". He then adds "... it (MEP) is a rule for constructing informative priors when we have partial prior information that restricts the possibilities significantly but not completely". I am in complete agreement with these statements, but what has not been done is to convince me that MEP is a *good* way to deal with prior information. This probably sounds like heresy but I feel the subjective atmosphere here is so healthy that diverse opinions can be openly expressed. (I hope that future Bayesian Symposiums will zealously guard this attitude). The reasons for my saying that I am not convinced about MEP begin with a lecturer in my undergraduate days at Georgia Tech. who introduced the subject of Entropy in a thermodynamics course, by saying, "you will never use this again" and I believed him! More importantly my statistical training brought me through classical procedures and least favourable prior distributions, and to use Bayes only for admissibility. Hence it is not surprising that a paper, Deely, Zimmer and Tierney (1970) appeared, which showed that MEP does not correspond to least favourable priors, not even in the restricted minimax sense.

I give this bit of history to show how far off the MEP track I am. To reach me and others like me (if there are any?) a decision theoretic structure will have to be used or something similar using beautiful mathematics —even hand waving— but not repeating over and over again that MEP is *good*. Jim Berger has already suggested in his paper that using the decision theory structure as a check for various procedures may not be a bad thing. Furthermore, Professor Jaynes has given me a little more insight into MEP during a conversation this week and that word is "neutral". In the paper he describes it in the following way, "As we have argued extensively elsewhere, the prior distribution that most honestly represents our state of knowledge (i.e. that agrees with what we know but does not assume anything beyond that) is the one with maximum entropy subject to the constraints imposed by what we know". Thus all that remains to do is to give a nice intuitive decision theoretic definition of "neutral" or "honestly represents" and then derive necessary and sufficient conditions for MEP to be neutral.

Now concerning the specific applications in this paper, I'm not surprised at the nice results. However as Professor Jaynes admits, his examples and results for them are not realistic. "Clearly, what is needed now is to put these ideas to the test by analysis of real data..." I certainly concur with this and would like to point out one definite area where I think his model is unrealistic. To allow for changes in the seasonal factors over time, it seems to me that the coefficients A_k and B_k $(k=1,\ldots,6)$ should be allowed to change. One way to do this is to imagine the coefficients being drawn repeatedly each 12 months from the same distribution. Thus given data for $N/12 = n$ years, forecasting the future requires updating the prior information using the past data for $n-1$ years. In this context I believe MEP would be incoherent, and possibly I can prove this by the next Symposium.

A.F.M. SMITH (*University of Nottingham*):

I believe that Jaynes' paper contains a very important message but I'm not sure that it has anything to do with Time Series or Entropy. However, it does concern a topic which Jaynes emphasized at our previous conference: namely, the notion of "question posing" as a dual activity to "inference" (formalized, to some extent, in the work of R.T. Cox). In particular, the business of deciding which questions can be well-answered by a particular data set (given a parametric model and a prior specification) can be thought of as corresponding to a principal components analysis in parameter space, performed on the posterior covariance matrix (perhaps following various exploratory non-linear transformations of individual parameters). Part of Jaynes' analysis seems close in spirit to this idea, with the added refinement that —with all other ingredients fixed— one can give an operational meaning to "highly informative priors" as those which lead to a simplifying principal components analysis.

J.R.M. AMEEN (*University of Warwick*):

It is appealing that prior probabilities should be considered as highly informative relative to the hypotheses under which the data is collected, when the data have almost nothing to tell on the basis of some discrimination measure between prior and posterior probabilities. Is this the procedure under which Professor Jaynes' prior specifications turn out to be highly informative?

Regarding model specifications, although the simplicity and restrictedness seem to be for the sake of argument, the Dynamic Linear Models of Harrison and Stevens (1976) may have been more justified.

Prior probabilities need not be completely ruined by the occurence of a sharp change in the data as it is stated at the end of appendix B of the paper. This point is emphasized with practical examples in Ameen and Harrison (1983).

J. BERGER (*Purdue University*):

This was a very enjoyable paper to read, not only because of the interesting Bayesian analysis of seasonal adjustment, but also because of the many philosophical meanderings that are sprinkled throughout. With only two of these meanderings did I not feel immediate agreement.

One was in Appendix C, in the discussion of significance testing "sawing off its own limb". The "justification" for the decision to reject H_0, if the tail area probability of the significance test is small, is surely just that data inconsistent with the hypothesis casts doubt on the hypothesis. I fully agree, however, that any attempt to treat a tail area significance level as a measure of doubt in any quantifiable sense is generally meaningless. It is perhaps this which Professor Jaynes is calling "limb-sawing" illogic: how can a significance level provide any *absolute* measure of doubt for the hypothesis if it is calculated under the hypothesis and the hypothesis seems wrong. Of course, the comments about the need to consider alternatives raise the main concern with significance testing.

The only real disagreement I had was with the justification in Appendix A of the Gaussian error distribution. I view maximum entropy as a wonderful tool but, as with all wonderful tools, fear the tendency to make the problem fit the tool. In particular, maximum entropy, as here, works best when one assumes that prior knowledge provides moments of the distribution. Now I can conceive of many situations where such may be the case but, in the majority of situations that I see, prior knowledge is more likely to be in the form of, say, medians and quartiles of a distribution. Medians and quartiles can be assessed by consideration of sets with large probability, while moments depend on delicate features of a distribution (such as its tail behavior) which are hard to assess. It may be a mistake to assume that moments even exist (i.e., the error distribution could have a fat enough tail that, as an approximation, it would better to proceed as if the moments were infinite). The rather unsatisfactory behavior of maximum entropy distributions for specified medians and quartiles (and continuous distributions) is then a source of concern. The justification of normal error distributions (or, for that matter, normal prior distributions) by maximum entropy thus leaves me somewhat uneasy.

W.H. DuMOUCHEL (*M.I.T.*)

I agree strongly with Professor Jaynes that the real opportunities for Bayesians lie in the use of informative priors. How else could we hope to do better than frequentists? To use noninformative priors is, basically, to play on their turf.

The author's discussion of the special problems involved in seasonal adjustment is interesting, but, in spite of his apology, I regret the absence of real data in his presentation. One wants to see the methods in action --which harmonics are used? Why? What values of σ and b are used? Quite possibly Professor Jaynes might develop more sympathy for those who allow arbitrary harmonics to enter their models if he had experience with a wide variety of real series, only a moderate proportion of which follow idealized seasonal models. For example, there may be a spike in which one month stands out from its adjacent months, or there may be more than one harvest time, several months apart, etc. If simple fits to trigonometric series work consistently well, frequentists should be able to profit from them as well as Bayesians. Show us the data!

The assumption that the prior distributions of the size of the first and second harmonics are independent doesn't seem reasonable to me. If one switches to polar coordinates, so that R_1^2 and R_2^2 are $(A_1^2+B_1^2)$ and $(A_2^2+B_2^2)$, respectively, then I would be more willing to believe that R_1 is independent of R_2/R_1.

Finally, I am afraid I don't see the point of the author's elaborate discussion of the singular posterior distribution of the $\{e_i\}$, at least in the context of the Bayes/non-Bayes controversy. Classical statisticians at least since Fisher have known that the residuals and the fitted values from a regression model each have singular distributions. Isn't that all you are saying?

I.J. GOOD (*Virginia Polytechnic Inst. and State Univ.*)

Regarding minimax entropy I'd like to draw attention to my comment on the paper by Bernardo and Bermudez.

A previous discussant said that the principle of maximum entropy is not a minimax principle with the usual loss functions. But it seems to me that the minimization of expected weight of evidence (minimum discrimination information), which is a generalization of the maximization of entropy, must be a minimax procedure if weight of evidence is regarded as a utility or quasi-utility (Good, 1969). This is so by virtue of Wald's theorem that the least favourable prior distribution is minimax. The minimax property by itself would not be of much interest if weight of evidence were not a reasonable quasi-utility having the property of invariance under transformations of the independent variable for continuous distributions, and having the analogous "splitative" property for discrete distributions. That is, weight of evidence, in the discrete case, is unchanged if categories are split, such as by tossing a coin (Good, 1973).

REPLY TO THE DISCUSSION

It is I who must apologize to John Deely for failure to get my completed manuscript to him before this meeting. Desperate attempts to meet three publication deadlines simultaneously led also to omission of some of the results (particularly numerical analysis of real data) that I had hoped to present.

But all heard my vow not to speak at any more meetings until I could present such an analysis. This should go a long way toward meeting the very valid criticism of Deely and others, that one does need to judge performance in the arena of real applications, and not just philosophy and theorems. Nobody feels that need more strongly than I, and only the pressure of other prior committments has deterred me.

However, we are not operating in a complete vacuum: my paper gives several references to recent work where useful numerical results, not duplicated by other statistical methods, have been obtained with MEP. Applications in several different fields are now growing so rapidly, the number of workers appearing to double every year, that it is no longer possible for one person to keep up with what is being done.

And this growth owes nothing to philosophy or theorems. In every field of application, nobody will believe that the method will work (the philosophy is attacked and the theorems are ignored) until somebody actually applies it and shows that it does work. It is the computer printouts —the sharp detail, uncluttered by the spurious artifacts that were generated by previous methods of data analysis like lag windows and inverse filters—that have produced the converts. A method that gives such results would be used just as much if it had no theoretical justification at all.

Today, we are rather far beyond the stage of "repeating over and over again that MEP is good" which would have been a valid criticism in 1965. Even at the time of Deely's 1970 criticism of MEP, Burg's Maximum-Entropy Spectrum Analysis (MESA) had been available in the literature for three years. By 1978 the literature had grown to the point where the IEEE issued a special volume (Childers, 1978) of reprints on MESA. The September 1982 IEEE Proceedings is a special issue devoted largely to it. But there is still a serious shortage of such numerical results outside of physics and engineering. This can and will be corrected.

In a recent talk George Tiao noted that: "In the marketplace of statistical ideas, there are many sellers but few buyers" —a profoundly true observation in spite of the seeming paradox that before one can become a seller he must first have been a buyer. The explanation is that one must be a seller for a very long time, because of the difficulty —more acute in statistics than in any other field— of clarifying to potential buyers exactly what it is that one is trying to sell.

Nothing could be further from our intention than to construct a "least favorable" prior. We seek rather a prior that deals most honestly with our prior information by representing the whole truth and nothing but the truth, thus enabling one to separate the truth from the artifacts. The philosophy, the theorems, and the computer printouts all support the view that MEP is accomplishing this. On the other hand, "least favorable" priors ignore —and therefore in general contradict— our prior information, and could be disastrous in these problems.

I cannot imagine what one could mean by a decision theoretic justification for MEP. Wald's decision theory makes no contact with any principles for assigning priors. Various theoretical justifications for MEP are in the literature, but they appeal rather to requirements of logical consistency, which are neutral toward the value judgments underlying decision theory. This neutrality appears to me the very essence of "scientific objectivity".

Similarly, it was not my intention to discuss all the real problems of seasonal adjustment. My paper formulated and solved one specific problem in order to demonstrate one point, the effect that prior information can have in seasonal adjustment. Believing that point to be new, we did not want to obscure it with unnecessary details. In various situations the model chosen may indeed be "not realistic" in many different ways; and so we noted the possibility, and the ultimate necessity, of extending it. But, having seen this one solution, the extensions are straightforward exercises that John Deely could assign to his students as homework.

In reply to Adrian Smith:

My presentation had, indeed, very little to do with entropy which was, after all, invoked only to justify what everybody has been doing all along. It is surprising that some other comments are so preoccupied with entropy, since if the same Gaussian priors had been called simply "assumptions" nobody would have questioned them. It seems that new rationales are disturbing even when they support previous practice, as Harold Jeffreys and Jimmie Savage found also.

There was, however, something to do with time series. Just how much will be clearer when those promised but still undelivered analyses of real data are at hand.

I am in full —indeed, enthusiastic— agreement with Adrian Smith on the importance, for practice and future theoretical development, of recognizing that the "design of models'' or ''design of hypotheses'' cannot be separated from the ''design of experiments''. Now that we are happily beyond the stage of mysticism (was the specific Latin square generated by a *truly* random selection process?), a new attack on these problems is possible, in which we give just as much attention to: ''Which questions can be well-answered by a given kind of data?'' as to the converse: ''Which kind of data can best answer a given question?''

The instincts of good scientists have always told them that it is idle to invent hypotheses which cannot be tested by the data that it is feasible to get. Progress is always made by asking the questions that are answerable at the time. For Isaac Newton it would have been foolish to ask questions which were not foolish for Erwin Schrödinger 250 years later. For Gregor Mendel it would have been foolish to ask questions which are not foolish for Francis Crick today. The same point was made in Arnold Zellner's call for ''sophisticatedly simple'' models in econometrics.

With the good beginnings made by R.T. Cox, it should be possible to formalize this intuitive wisdom in a new theory of scientific inference, which includes the ''diagnostic phase'' and hypothesis formulation automatically, and will prove to be a far more powerful tool than our intuition; just as our present Bayesian principles formalize and strengthen our intuition about plausible reasoning from given hypotheses. In both, prior information will be of crucial importance. The principal components analysis in parameter space suggested by Adrian Smith may be a very good starting point for this quest.

In reply to Jack Good:

Jack Good's remarks illustrate an important point that deserves to be emphasized; in statistics it is the rule that a given procedure may be interpreted and justified in various ways.

The maximum entropy *principle*, as a rationale for assigning priors, was questioned by John Deely who wanted to see instead a minimax decision theory type argument; and by Jim Berger, who left us wondering what he would like instead. Deely's goal seemed to me dubious because priors and decisions lie at opposite ends of the inference-action chain.

In contrast, the maximum entropy *procedure* exists independently of any rationale; and my comments on Professor Csiszar's paper noted one conceivable way of interpreting it in decision-theory terms. Jack Good's comments seem to offer a rationale for thinking of entropy as related to utility rather than priors, and so tend to support this view.

But we are not compelled to choose one view and abandon the other; perhaps they are both appropriate in different circumstances, and this shows only that the procedure solves more than one problem. Such a situation is not new in statistics.

Indeed, as John Tukey has noted, a procedure does not have hypotheses, and really needs no rationale at all; its justification lies in the results it gives. In this connection, nothing in the procedure requires the quantity whose entropy is maximized to be even a probability. As far as the mathematics is concerned, "any" non-negative integrable function $f(x)$ has an entropy $H = -\int f(x) \log f(x)\, dx$, whose maximization is a well-posed problem under very general conditions. The solution will be a purely mathematical result, independent of whatever conceptual meaning you or I might attach to $f(x)$.

Some of the current applications take advantage of this flexibility; the cited work of Papanicolaou and Mead finds the maximum entropy procedure a surprisingly powerful method for approximating functions $f(x)$ about which only a few moments are known, with advantages in both accuracy and stability over other common schemes such as Padé approximants.

Other interesting examples of the procedure escaping from the confines of its original rationale (such as filling gaps in incomplete contingency tables) are in the articles cited and in others shortly to appear.

Jim Berger's comments came as welcome relief because, unlike most of my recent discussants and commentators, he does not ask me to defend statements that I did not make. Instead, I am able to note that I did say, in different words, the things that he mentions. But his remarks end abruptly just at the most interesting point; I hope he will end the suspense by continuing his line of thought elsewhere.

On the trivial "limb-sawing" matter, surely we agree that probabilities conditional on an hypothesis that we have rejected stand in a rather peculiar logical position. Would you use them for future prediction? In principle, it must be the probability of the hypothesis, conditional on the data, that justifies our decision to reject it. Pragmatically, it doesn't matter because the two probabilities are related mathematically through Bayes' theorem. But orthodox statistics cannot say it that way, and so puts up a logically puzzling rationale.

The important matter here is the status of iid Gaussian sampling distributions and of maximum entropy as a principle for assigning probability distributions. As Jim correctly notes, the former is only a special case of the latter. I too stated that this rationale for the Gaussian assignment holds when the prior information consists of the first two moments (or just the second moment) of the noise; and that when our prior information is different, a different prior distribution will be appropriate. So we are in agreement here.

But Jim and I seem to work on different kinds of problems; my experience is that in the great majority of those arising in physics and engineering the first two moments are exactly the prior information we have, because they are connected directly to the noise energy. I have never seen a real problem in which we had prior information about percentiles, and would need to know more about them (are percentiles the only information?) before deciding what to recommend.

Now we come to the serious point that calls for more explanation. He suggests that if the prior information does not consist of moments, then not only will the iid Gaussian distribution not be appropriate, but also that the maximum entropy principle may not be appropriate to find the new distribution. This is tantalizing because it seems to imply either that he proposes to leave us with no principle, or that some other principle will then be more appropriate.

All I can say to this is that, if Jim Berger or anyone else is in possession of a principle for assigning priors that can, while using the same prior information, give more satisfactory results than does maximum entropy in even one case, then I and a few thousand others are breathlessly eager to hear what it is. If I knew a better principle, I would be expounding it and using it in my current research.

In reply to Dr. Ameen:

The point of my work was to show the surprisingly large effect that prior information can have in seasonal adjustment. To do that, one naturally chooses one specific model to analyze, and keeps it as simple as possible. Anyone who wishes to analyze a different model is perfectly free to do so. Presumably, the same sensitivity to prior information will be found in any model of seasonal adjustment —or indeed any problem where we want to estimate the "noise"— if it is given a full Bayesian analysis.

Finally, in response to Prof. DuMouchel:

Professor DuMouchel calls on me to defend statements that I did not make. As in other replies, I must emphasize that the purpose of my presentation was to demonstrate, by analyzing a simple case, the large effect that prior information can have in seasonal adjustment. It is indeed true that in the real world one can find other cases than the one I chose. Therefore extensions of my model, in a dozen different ways, are also of interest; by all means, let us study them.

Let me assure everybody that, being one of them, I have the greatest sympathy for "those who allow arbitrary harmonics to enter their models". Nothing in my argument forbids one to model whatever harmonics he pleases. But in the model I chose to analyze (known period of one year, monthly data) there are no higher harmonics than the sixth.

Likewise, nothing forbids one to consider different prior probabilities than the ones I chose. If one has evidence supporting a non-independent prior distribution for the first and second harmonics, by all means use it. Presumably, still more interesting things will then be found. I would be very interested to learn how these different kinds of prior information interact; perhaps Professor duMouchel might tell us about this at a future meeting.

Have statisticians since Fisher known about the singular posterior distributions for the $\{e_i\}$ that I pointed out? How could they even recognize the existence of correlations in any posterior distribution, without becoming Bayesians?

REFERENCES IN THE DISCUSSION

AMEEN, J.R.M. and HARRISON, P.J. (1984). Normal discount Bayesian models. This volume.

DEELY, J.J., ZIMMER, W.J. and TIERNEY, M.S. (1970). On the usefulness of the maximum entropy principle in the Bayesian estimation of reliability. *IEEE Trans. Rel. R*-**19, 3,** 110-115.

GOOD, I.J. (1969). What is the use of a distribution. *Multivariate Analysis II,* P.R. Krisnaiah (ed). New York: Academic Press, 183-203.

— (1973). Information, rewards, and quasi-utilities. *Science, Decision, and Value.* J.J. Leach, R. Butts & G. Pearce (eds.). Dordrecht: D. Reidel, 115-127.

HARRISON, P.J. and STEVENS, C.S. (1976). Bayesian forecasting. *J. Roy. Stat. Soc. B,* **38,** 205-247 (with discussion).

BAYESIAN STATISTICS 2, pp. 361-374
J.M. Bernardo, M.H. DeGroot, D.V. Lindley, A.F.M. Smith (Eds.)

Average Case ϵ-Complexity in Computer Science - A Bayesian View

J.B. KADANE
Carnegie-Mellon University

G.W. WASILKOWSKI
Columbia University
and University of Warsaw

SUMMARY

Relations between average case ϵ-complexity and Bayesian statistics are discussed. An algorithm corresponds to a decision function, and the choice of information to the choice of an experiment. Adaptive information in ϵ-complexity theory corresponds to the concept of sequential experiment. Some results are reported, giving ϵ-complexity and minimax-Bayesian interpretations for factor analysis. Results from ϵ-complexity are used to establish that the optimal sequential design is no better than optimal nonsequential design for that problem.

Keywords: ADAPTIVE INFORMATION; ALGORITHM; COMPLEXITY; EXPERIMENT; FACTOR ANALYSIS; INFORMATION; SEQUENTIAL ANALYSIS.

1. INTRODUCTION

This paper shows that average case analysis of algorithms and information in ϵ-complexity theory is related to optimal decisions and experiments, respectively, in Bayesian Theory. Finding such relations between problems in previously disjoint literatures is exciting both because one discovers a new set of colleagues and because results obtained in each literature can illuminate the other.

Sections 1 and 2 explain, respectively, the worst-case and average-case analysis of algorithms. Section 3 establishes the correspondence mentioned above. Finally, Section 4 discusses some results from the average case ϵ-complexity literature, and its interpretation for Bayesians. We hope that the relation reported here can lead to further fruitful results for both fields.

1.1. *Worst Case Analysis*

In this section we briefly present some major questions addressed in ϵ-complexity theory. We first discuss the worst case model which is conceptually simpler than the average case model, discussed in Section 2.

An expository account of ϵ-complexity theory (which is also known as information-centered theory) may be found in Traub and Wozniakowski (1984). The general worst case model is presented in two research monographs: Traub and Wozniakowski (1980) and Traub, Wasilkowski and Wozniakowski (1983). The first of these has an extensive annotated bibliography. Reports on the average case model are cited in Section 2.

A simple integration problem provides a suggestive illustration. We wish to approximate $\int_0^1 f(t)dt$ knowing n values of f at points t_i, $N(f) = [f(t_1),\dots,f(t_n)]$, and knowing that f belongs to a given class F of functions where F is a subclass of a linear space F_1. This means that, for given information value $y = N(f)$ we approximate the integral of f by $\phi(y)$ where $\phi:\mathbb{R}^n \to \mathbb{R}$ is a mapping called an algorithm. In the worst case model discussed in this section, the error of ϕ is determined by its performance for the "hardest" element f, i.e.

$$e^w(\phi,N) = \sup_{f \in F} \left| \int_0^1 f(t)dt - \phi(N(f)) \right| .$$

The radius of information $r^w(N)$, is defined as the minimal error among all algorithms that use N, and the optimal algorithm ϕ^* is defined so that its error is minimal, i.e.,

$$e^w(\phi^*,N) = r^w(N) = \inf_{\phi} e^w(\phi,N).$$

Suppose now that the points t_i may be varied. Then $N^*(f) = [f(t_1^*),\ldots,f(t_n^*)]$ is nth optimal iff the points t_i^* are chosen to minimize the radius r^w. Hence, roughly speaking, an optimal algorithm ϕ^* that uses nth optimal information N^* approximates $\int_0^1 f(t)dt$ for every $f \in F$, with minimal error among all algorithms that use n function evaluations. Observe that these concepts are independent of a model of computation.

In ϵ-complexity theory we are interested in minimizing errors as well as cost. More precisely, suppose we are given $\epsilon > 0$. Then the problem is to find information N, and an algorithm ϕ that uses N, so that ϕ approximates $\int_0^1 f(t)dt$, $\forall f \in F$, with error not greater that ϵ and the cost of computing $\phi(N(f))$ is minimized. Observe that this cost (denoted by comp(ϕ,f)) is the sum of two terms: the cost of computing $y = N(f)$ (denoted by comp(N,f)) and the cost of computing $\phi(y)$ given y (denoted by comb(ϕ,y)). Of course, comp$(\phi,f) \geq$ comp(N,f). A major problem of ϵ-complexity can be stated as follows: find N^{**} and ϕ^{**} that uses N^{**} such that $e^w(\phi^{**},N^{**}) \leq \epsilon$ and

$$\sup_{f\epsilon F} \text{comp}(\phi^{**},f) = \min\{ \sup_{f\epsilon F} \text{comp}(\phi,f) : e^w(\phi,N) \leq \epsilon\}.$$

Of course, the choice of N^{**} and ϕ^{**} depend strongly on the model of computation, i.e., how much various operations cost. In this section we discuss a very simple model, assuming that comp(N,f) is proportional to n, say comp$(N,f) = cn$, where c is so large that comb$(\phi^*,N(f))$ is negligible for some optimal algorithm ϕ^* that uses N. Then to choose N^{**} and ϕ^{**} we must find information with the minimal number of function evaluations such that $r^w(N^{**}) \leq \epsilon$. Then for ϕ^{**} we can take the optimal error algorithm ϕ^* that uses N^{**}.

We now comment on the assumption that comb$(\phi,N(f)) \ll cn$. For many problems there exists an optimal algorithm ϕ^* which is linear, i.e., $\phi^*(y) = \sum_{i=1}^{n} y_i g_i$, $g_i \in \mathbb{R}$. Since an arithmetic operation (we take its cost to be unity) is less expensive than a function evaluation,

$$\text{comb}(\phi,y) = 2n-1 \ll cn.$$

Hence the above assumption is satisfied whenever ϕ^* is linear. This also explains one reason why we are particularly interested in linear optimal algorithms.

We now indicate another important question studied in ϵ-complexity theory. Recall that in our example the information N is of the form, $N(f) = [f(t_1),\ldots,f(t_n)]$. If the points t_i are given independently of f, then N is called nonadaptive. If the t_i depend on previously computed information values, N is called adaptive. Nonadaptive information is desirable on a parallel computer and for distributed computation since the information can then be computed on various processors at the same time. This lowers significantly the cost comp(N,f). Adaptive information has to be computed sequentially which means that comp(N,f) remains nc. Hence, if N^{non} is nonadaptive and N^a is adaptive, we prefer N^{non} unless $r^w(N^a) \ll r^w(N^{non})$. This explains why we are interested in the following question: when is adaptive information more powerful than nonadaptive information?

We described some of the major questions addressed in ϵ-complexity theory by using integration as an example. The same questions can be asked in great generality where, for example, different operators are considered instead of $\int_0^1 f(t)dt$, and different information operators N are studied instead of function evaluations. For many problems optimal information and optimal algorithms are known (See Traub and Wozniakowski (1980)). Sometimes this information and these algorithms are new. Furthermore, for many problems (including the integration problem) adaptive information is not more powerful than nonadaptive information. The significance of this result is that adaptation is widely used by practitioners.

2. AVERAGE CASE ANALYSIS

In the previous section we briefly discussed the worst case model where the error of an algorithm was defined by its performance for the "hardest" f. For some problems this model might be too pessimistic. Researchers in ϵ-complexity theory also analyze average case models, three of which we presente in this section. For simplicity we discuss only problems defined on Hilbert spaces. This presentation is based on Traub, Wasilkowski and Wozniakowski (1982, 1984), Wasilkowski and Wozniakowski (1984), and Wasilkowski (1983).

Let F_1 and F_2 be two real separable Hilbert spaces and let S,

$$S: F_1 \to F_2, \tag{2.1}$$

be a continuous operator. We call S a *solution operator*. For instance, we might take $S(f) = \int_0^1 f(t)dt$ which corresponds to the integration problem discussed above, $S(f) = f$ which corresponds to the approximation problem or $S = \Delta^{-1}$ where $\Delta u = -\sum_{i=1}^{p} \partial^2 u/\partial x_i^2$ which corresponds to a differential equation problem.

As in Section 1 we want to approximate $S(f)$ for every $f \epsilon F_1$ but now with the average error as small as possible. In order to define average error we assume that the space F_1 is equipped with a probability measure, μ, $\mu(F_1) = 1$, defined on the Borel sets of F_1.

To find an approximation to $S(f)$ we must know something about f. We assume $y = N(f)$ is known, where now N is defined as follows:

$$N(f) = [L_1(f), L_2(f,y_1), \ldots, L_n(f,y_1,\ldots,y_{n-1})] \, \epsilon \, \mathbb{R}^n\,, \tag{2.2}$$

where $y_1 = L_1(f)$, $y_i = L_i(f,y_1,\ldots,y_{i-1})$ $(i=2,\ldots,n)$, and for every $y \,\epsilon\, \mathbb{R}^{i-1}$ the functionals $L_i(\bullet,y)$: $F_1 \to \mathbb{R}$ belong to some given class L of measureable functionals. Such an operator N is called an *adaptive information operator* and the number n of functional evaluations is called the *cardinality of* N, card $(N) = n$. In general, the choice of the ith evaluation depends on the previously computed information $y_1,\ldots,y_{i-1}$. If $L_i(\bullet,y) \equiv L_i$ for every y then N is called *nonadaptive*. Of course, for nonadaptive information N, $N(f)$ can be very efficiently computed in parallel.

To illustrate this concept assume, as in Section 1, that F_1 is a space of functions. Then $N(f)$ might consist of function evaluations, $N(f) = [f(t_1),\ldots,f(t_n)]$, i.e., $L_i(f) = f(t_i)$. If the

t_i's are fixed then N is nonadaptive. Otherwise, if the selection of t_2 depends on the value $f(t_1)$ and so on, then N is adaptive.

Knowing $N(f)$ we approximate $S(f)$ by $\phi(N(f)$ where ϕ is an *algorithm that uses N.* By an algorithm we mean any mapping from $N(F_1) = \mathbb{R}^n$ into F_2. Then the *average error of* ϕ is defined as

$$e^{avg}(\phi,N) = \{ \int_{F1} \| S(f) - \phi(N(f)) \|^2 \mu(df) \}^{1/2} , \tag{2.3}$$

where the integral in (2.3) is understood as the Lebesgue integral.

We pause to comment on definition (2.3)

- The average error of an algorithm ϕ is conceptually the same as the error $e^w(\phi,N)$ in the worse case model, except the supremum is replaced by an integral.
- The average error of ϕ is well defined only if ϕ is "measurable" (or more precisely only when $\| S(\cdot) - \phi(N(\cdot)) \|^2$ is measurable in f). Since "nonmeasurable" algorithms might be as good as "measurable" ones, we would like to have the concept of average error for *every* algorithm. It is possible to extend the definition (2.3) so that average error is well defined for every ϕ (see Wasilkowski (1983)) but for the purpose of this paper we shall assume that ϕ is chosen such that (2.3) is well defined.
- The average error is defined as an average value of $\| S(f) - \phi(N(f)) \|^2$. Of course, in general $\| S(f) - \phi(N(f)) \|^2$ can be replaced by a different error function $E(S(f),\phi(N(f)))$ (see, e.g. Traub, Wasilkowski and Wozniakowski (1981) and Wasilkowski (1983) and the Appendix).

Let

$$r^{avg}(N) = \inf_{\phi} e^{avg}(\phi,N) \tag{2.4}$$

be the *average radius of information N.* Then by an *optimal average error algorithm* we mean an algorithm ϕ^* that uses N and enjoys the smallest error, i.e.,

$$e^{avg}(\phi^*,N) = r^{avg}(N) . \tag{2.5}$$

For a given integer n, let Ψ_n be the class of all information operators N of the form (2.2) with cardinality not greater than n. We shall say that $r^{avg}(n)$ is an *nth minimal average radius* if

$$r^{avg}(n) = \inf_{N \in \Psi_n} r^{avg}(N) . \tag{2.6}$$

We shall say that N_n^* from Ψ_n is *nth optimal* if the radius of N_n^* is minimal, i.e.,

$$r^{avg}(N_n^*) = r^{avg}(n) . \tag{2.7}$$

Using this notation, we now describe some of the major questions addressed in the average case model. Given the problem, i.e., solution operator S, probability measure μ, class of information operators N and error tolerance ϵ, find N^{**} and ϕ^{**} with minimal (or almost minimal) complexity such that

$$e^{avg}(\phi^{**},N^{**}) \leq \epsilon.$$

The complexity of the average case model can be measured in different ways. Depending on the problem, sometimes it is defined as complexity of ϕ in the worst case, i.e.,

$$\sup_{f \in F_1} \text{comp}(\phi,f) .$$

where comp(ϕ,f) is discussed in Section 1.

We call this Case A. Sometimes complexity is defined by the average case complexity, which is

$$\int_{F_1} \text{comp}(\phi, f)\mu(df),$$

which we call Case B. However, if we agree that the assumptions in Section 1 are satisfied the search for optimal ϕ^{**} and N^{**} can be simplified, namely, to find ϕ^{**} and N^{**} we need only to find an n^*th optimal information operator with the minimal n^* such that

$$r^{avg}(n^*) \leq \epsilon,$$

here called Case C. Then the optimal N^{**} is $N^{**} = N^*_n$ and the optimal algorithm ϕ^{**} is an optimal average error algorithm ϕ^* that uses N^{**}. The conditions which guarantee that ϕ^* is linear, i.e., $\phi^*_n((y_1,\ldots,y_n)) = \sum_{i=1}^{n} y_i\, g_i, g_i \in F_2$, are also studied (see the Appendix). Another important question posed in ϵ-complexity theory is when adaptation is more powerful than nonadaptation.

We end this section by presenting the concepts of local errors and local radii.

Let N be an information operator. For simplicity assume that $N(F_1) = \mathbb{R}^n$. Define a measure $\mu_{1,N}$ on Borel sets from $\mathbb{R}^n$ as

$$\mu_{1,N}(A) = \mu(N^{-1}(A)) = \mu(\{f \epsilon F_1, N(f) \epsilon A\}). \tag{2.8}$$

Of course, $\mu_{1,N}$ is a probability measure. Then there exists a family of probability measures $\mu_{2,N}(\bullet\,|y)$ on F_1 such that $\mu_{2,N}$ are concentrated on $N^{-1}(y)$, i.e., $\mu_{2,N}(N^{-1}(y)\,|y) = 1$, for almost every y and

$$\mu(B) = \int_{\mathbb{R}^n} \mu_{2,N}(B|y)\mu_{1,N}(dy). \tag{2.9}$$

(For more detailed discussion see Parthasarathy (1967, p. 197). Such measures $\mu_{2,N}(\bullet\,|y)$ are called conditional measures. Then the average error of an algorithm ϕ can be rewritten as

$$e^{avg}(\phi,N)^2 = \int_{\mathbb{R}^n} \{\int_{F_1} \|S(f) - \phi(y)\|^2 \mu_{2,N}(df|y)\}\, \mu_{1,N}(dy)$$

$$= \int_{\mathbb{R}^n} e^{avg}(\phi,N,y)^2\, \mu_{1,N}(dy). \tag{2.10}$$

where

$$e^{avg}(\phi,N,y) = \{\int_{F_1} \|S(f) - \phi(y)\|^2 \mu_{2,N}(df|y)\}^{1/2} \tag{2.11}$$

is called a *local average error of* ϕ. Let

$$r^{avg}(N,y) = \{\inf_{f \epsilon F_2} \int_{F_1} \|S(f) - g\|^2 \mu_{2,N}(df|y)\}^{1/2} \tag{2.12}$$

be the *local average radius of* N. It is proven in Wasilkowski (1983) that $r^{avg}(N,y)^2$, as a function of y, is μ_1-measurable (see the Appendix). Hence, we have

$$r^{avg}(N)^2 = \int_{\mathbb{R}^n} r^{avg}(N,y)^2\, \mu_{1,N}(dy) \tag{2.13}$$

and ϕ^* is optimal iff

$$e^{avg}(\phi^*,N,y)^2 = r^{avg}(N,y)^2 = \inf_{g \epsilon F_2} \int_{F_1} \|S(f) - g\|^2 \mu_{2,N}(df|y),\ \forall\, y,\ a.e. \tag{2.14}$$

Finally, N_n^* is nth optimal iff

$$r^{avg}(N_n^*)^2 = \inf_{N \epsilon \Psi n} \int_{\mathbb{R}^n} \{ \inf_{g \epsilon F_2} \int_{F_1} \| S(f) - g \|^2 \mu_{2,N}(df|y) \} \mu_{1,N}(dy). \tag{2.15}$$

3. A BAYESIAN INTERPRETATION OF THE AVERAGE CASE MODEL

Recall that the Bayesian scheme for the design of experiments comes in two equivalent forms, normal and extensive (Lindley, 1971). In the normal form, one chooses a decision function $\delta(x)$, depending on the data x, to minimize expected loss over both the sample space X and parameter spaces Ω:

$$\min_{\delta} \int_{\Omega} \int_{X} L(\delta(x),\theta)\, p(x|\theta)dxp(\theta)d\theta. \tag{3.1}$$

where L is a loss function and the integrals are with respect to the Lebesgue measure.

In the extensive form, a Bayesian chooses an experiment e and a decision d, after observing x but not observing θ, to

$$\min_{e} \int_{X} \min_{d} \int_{\Omega} L(d,\theta)p(\theta|x,e)d\theta p(x|e)dx. \tag{3.2}$$

Comparing (3.1) with (2.3) and (2.4) for the normal forms, and (3.2) with (2.15) for the extensive forms, we see identical forms, leading to the correspondence exhibited in Table 1.

TABLE 1

Correspondence of Notation Between Bayesian Decision Theory and ϵ-Complexity Theory (Case C)

Language of Bayesian Decision Theory		*Language of* ϵ*-Complexity Theory*	
Ω	Finite dimensional parameter space	F_1	Space of problem elements
θ	parameter	f	problem element
e	experiment	N	information
X	space of data	$N(F_1)$	space of information values
x	data	y	information value
$\delta(\cdot)$	decision function	$\phi(\cdot)$	algorithm
d	decision	g	value of algorithm
L	loss	$\|S(f)-g\|^2$	algorithm error
$p(\theta)d\theta$	prior distribution	μ	probability measure
$p(\theta|x,e)$	posterior distribution	$\mu_{2,N}(\cdot|y)$	conditional probability on F_1, given that $N(f)=y$
$p(x|e)$	marginal distribution of data	$\mu_{1,N}(\cdot)$	probability measure induced by N
expression (3.1)	Bayes risk of experiment	$[r^{avg}(N)]^2$	squared radius of information

In both cases A and B, researchers in ϵ-complexity theory keep the error and the cost of computation separate. They want to guarantee that the error is not greater than ϵ, and then they ask about minimal cost of computation. A Bayesian might prefer a formulation in which cost and error were represented in the loss function L so that the optimal information N and algorithm ϕ would minimize

$$\int_{F_1} L(\phi,N,f)\,\mu(df)$$

unconditionally. Observe that letting the loss function

$$L(\phi,N,f) = \text{comp}(\phi,f) + \frac{1}{\psi_\epsilon(\phi,N)} - 1 , \tag{3.3}$$

$$\text{where } \psi_\epsilon(\phi,N) = \begin{cases} 1 & \text{if } e^{avg}(\phi,N) \leq \epsilon \\ 0 & \text{otherwise} . \end{cases}$$

will yield a Bayesian formulation of Case B. However, Case A appears not to have a correspondence with Bayesian statistics.

With the above as background, the role of adaptive information becomes clearer to statisticians. Adaptive information is defined to be an N dependent on past values of y, that is, an experiment dependent on past data. Thus what researchers in ϵ-complexity mean by adaptive information is related to what statisticians mean by the sequential design of experiments. To ask whether adaptation helps in the average case is to ask whether sequential experimentation yields greater expected utility.

The average case models presented in this paper are not the only ones studied in ϵ-complexity theory. We now present very briefly another model which has no correspondence in the Bayesian Decision Theory and whose conclusion (that probabilistic algorithms are better than nonprobabilistic algorithms) is contrary to Bayesian Decision Theory.

In the model to be presented here we have a class Ψ of pairs (N,ϕ) where, as always N is an information operator and ϕ is an algorithm that uses N. In general, Ψ is uncountable. Consider a random variable R with values in Ψ. This random variable defines the following probabilistic method: according to R, randomly choose $(N_R,\phi_R) \in \Psi$ and then approximate $S(f)$ by $x = \phi_R(N_R(f))$. Observe that if R is constant, then this probabilistic method is an ordinary algorithm discussed in this paper. For given $\epsilon > 0$ and $p \in [0,1]$ let $\mathbb{R}(\epsilon,p)$ be the set of all random variables R such that

$$\text{prob}(\| S(f) - \phi_R(N_R(f)) \| \leq \epsilon) \geq p.$$

i.e., $\mathbb{R}(\epsilon,p)$ is the set of all probabilistic methods which, with probability at least p, yield an approximation with error at most ϵ. Now the problem is to find an optimal method, or equivalently and optimal R^*, with minimal complexity among all methods from $\mathbb{R}(\epsilon,p)$. Here the complexity of R is defined as the average complexity.

It turns out that for many problems the optimal R^*, although discrete, is not constant. This means that in this model, probabilistic methods are better than nonprobabilistic methods. This will be reported in a future paper by Wasilkowski.

4. AN APPLICATION TO FACTOR ANALYSIS

In this section we report some results from the average case ϵ-complexity literature giving an interpretation of factor analysis. This and the correspondence between Bayesian statistics and ϵ-complexity yields a Bayesian interpretation of factor analysis with some minimax elements, much along the lines suggested by Manski (1981), Lambert and Duncan (1981) and Berger (1984).

Consider the problem of representing a linear space by a few vectors capturing most of the variability in a particular sense. In a factor analysis one chooses any system of vectors spanning the space spanned by the eigenvectors corresponding to the largest eigenvalues of the covariance matrix V. This problem can be displayed in the language of ϵ-complexity as follows.

Let F_1 be a separable Hilbert space (although statisticians will be more familiar with the large, finite dimensional case $F_1 = \mathbb{R}^m$, we prefer to talk about not necessarily finite spaces, since for many interesting problems in ϵ-complexity the spaces are infinite dimensional). To represent this space means to approximate the solution operator S, which for this case is the identity operator,

$$S = I,$$

possessing partial information N,

$$N(f) = [(f,z_1),\ldots,(f,z_n)] \,. \tag{4.1}$$

Here z_j are some vectors from F_1 and $(\bullet,\bullet)$ denotes the inner product in F_1. Assume now that the probability measure μ on F_1 is unknown and what we know is its covariance operator V. Recall that V is defined by

$$(Vg,h) = \int_{F_1} (f,g)\,(f,h)\,\mu(df),\ \forall\, g,h \in F_1 \tag{4.2}$$

and for $F_1 = \mathbb{R}^m$, V is the covariance matrix of μ. Of course, V is symmetric and nonnegative definite. Without loss of generality we can assume that V has finite trace (this is equivalent to the assumption that $\int_{F_1} \|f\|^2\,\mu(df) < +\infty$) since otherwise we cannot approximate $S = I$ with finite error. Since μ is unknown we replace the problem (2.15) by the following one. Find N^* and ϕ^* that uses N^* such that

$$\sup_{\mu} e^{avg}(\phi^*,N^*;\mu) = \inf_{N}\inf_{\phi}\sup_{\mu} e^{avg}(\phi,N;\mu)\,. \tag{4.3}$$

where $e^{avg}(\phi,N;\mu)$ denotes the average error of ϕ for a measure of μ, and the supremum is taken over all measures with fixed covariance operator V.

This is, we believe, the ϵ-complexity formulation of the problem studied in factor analysis. To solve this problem we first fix N. From Wasilkowski and Wozniakowski (1982) it follows immediately that

$$\inf_{\phi}\sup_{\mu} e^{avg}(\phi,N;\mu) = e^{avg}(\phi^*;N,\mu^*) = r^{avg}(N;\mu^*) \tag{4.4}$$

where ϕ^* is the *spline algorithm* that uses N and μ^* is any *orthogonally invariant measure* (see the Appendix for explanation). The formal definitions of spline algorithms and of orthogonal invariance can be found in the paper cited above as well as in the Appendix. We only stress that the spline algorithm is linear (i.e., is simple) and that Gaussian measures are examples of orthogonally invariant measures. Hence from (4.4) we have that the spline algorithm ϕ^* that uses N is optimal in the sense of (4.3). Furthermore, the Gaussian measure with covariance operator V is the "least favorable" measure for every information N.

We now exhibit the optimal information N^*. Let $\eta_1^*, \eta_2^*, \ldots,$ $(\|\eta_j^*\| = 1)$ be the eigenvectors of the operator V, i.e.,

$$V\eta_j^* = \lambda_j^*\eta_j^*, \quad \lambda_1^* \geq \lambda_2^* \geq \ldots \geq 0. \tag{4.5}$$

Let

$$N^*(f) = [(f,\eta_1^*),\ldots,(f,\eta_n^*)]\,. \tag{4.6}$$

From (4.4) and Wasilkowski and Wozniakowski (1982) (see also Appendix) it follows that

$$\inf_N \inf_\phi \sup_\mu e^{avg}(\phi,N,\mu) = e^{avg}(\phi^*,N^*,\mu^*) = r^{avg}(N^*,\mu^*) = \{\sum_{i=n+1}^{\infty} \lambda_i^*\}^{1/2}\,. \tag{4.7}$$

This means that N^* defined by (4.6) is optimal and that the spline algorithm ϕ^*, which for this information has a very simple form

$$\phi^*(y) = \sum_{i=1}^{N} y_i\,\eta_i^*, y = [y_1,\ldots,y_n] \in \mathbb{R}^n,$$

is the unique optimal algorithm for the problem (4.3).

We now comment on the choice of information N^*. Suppose that instead of N^* one chooses N,

$$N(f) = [(f,z_1),\ldots,(f,z_n)]\,,$$

where $z_1,\ldots,z_n$ spans the same space as $\eta_1^*,\ldots,\eta_n^*$, i.e.,

$$\mathrm{lin}\{z_1,\ldots,z_n\} = \mathrm{lin}\{\eta_1^*,\ldots,\eta_n^*\}$$

Then information N^* and N are equivalent. More precisely, if $\phi_{N^*}^*$ and ϕ_N^* are optimal algorithms that use N^* and N, respectively, then

$$\phi_{N^*}^*(N^*(f)) = \phi_N^*(N(f)), \forall f \in F_1\,,$$

and

$$r^{avg}(N^*;\mu) = r^{avg}(N;\mu), \quad \text{for every } \mu.$$

Observe that N^* defined above is nonadaptive. It is natural to ask whether N^* remains optimal among all adaptive information operators. From Walsilkowski and Wozniakowski (1984), we know that adaptation is not more powerful in the average if the measure μ is orthogonally invariant (see Appendix). Since for every N the supremum in (4.3) is attained for such measures, this implies that adaptation does not help in our problem.

In the language of statistics (we refer the reader to Table 1 as necessary), $Sf = f$ is a random variable, N is an experiment which gives ω, for n chosen vectors z_j $(j = 1,\ldots,n)$, the value of the random variables $(S(\bullet),z_j)$, which can be written in the finite dimensional case, $z_j^T S(\bullet)$. Note that the covariance matrix V of S in the finite dimensional case is the covariance operator of μ. Knowing the matrix V, we wish to find, for fixed n, optimal vectors $\eta_1^*,\ldots,\eta_n^*$ to satisfy (4.3), that is to minimize, over experiments N and estimates ϕ^*, the loss against the least favorable distribution μ for S. (This latter aspect gives rise to the minimax character of the criterion). The nature of this optimal choice of vectors η_j^* is that they span the space spanned by the eigenvectors of V corresponding to the n largest eigenvalues. This is exactly the space of possible factor analysis of V. We therefore have a Bayes-minimax interpretation of factor analysis, and one that does not appear to be available in the statistical literature.

Another conclusion is that the experiment N^*, which is nonsequential, is optimal among all sequential experiments.

5. CONCLUSION

We believe that the relations between Bayesian decision theory and average case ϵ-complexity theory may have important consequences for both groups of researchers. We have found in our discussions that despite the similarities reported here, the perspectives of the two fields are rather distinct. Only as we further explore the connections between these two areas can we determine how much progress can be made in each by exploiting the relations reported here.

I. APPENDIX

We present very briefly results from the ϵ-complexity literature which have been mentioned in Sections 2 and 4.

Result 1 (Wasilkowski (1983))

Let F_1, F_2 be separable metric spaces and let μ be a probability measure defined on the σ-field of Borel sets from F_1. Let

$$E : F_1 x F_2 \rightarrow \mathbb{R}_+$$

be a function such that

for every $g \in F_2$, $E(\bullet,g)$ is measurable, and

for almost every $f \in F_1$, $E(f,\bullet)$ is continuous on F_2.

Let N,

$$N : F_1 \rightarrow H,$$

be measurable where $H = N(F_1)$ is a separable metric space.

Then the function $r : H \rightarrow \mathbb{R}_+$,

$$r(y) = \inf_{g \in F_2} \int_{F_1} E(f,g)\mu_{2,N}(df|y)$$

is $\mu_{1,N}$-measurable, where $\mu_{1,N}(\bullet) = \mu(N^{-1}(\bullet))$ and $\mu_{2,N}(\bullet\,|y)$ is the conditional measure.

Letting, as in Section 2, F_1, F_2 be Hilbert spaces, $E(f,g) = \|S(f)-g\|^2$ and $N : F_1 \rightarrow \mathbb{R}^n$, we have that $r(y) = r^{avg}(N,y)$, $\forall\, y \in \mathbb{R}^n$. Hence the measurability of the squared local radius $r^{avg}(N,y)$ follows from the result cited above.

Result 2 (Wasilkowski and Wozniakowski (1982))

Let $S : F_1 \rightarrow F_2$ be a continuous operator and let F_1, F_2 be separable Hilbert spaces. Let μ be an *arbitrary* Borel measure on F_1 with covariance operator V. Let

$$N(f) = [(f,z_1),\ldots,(f,z_n)] .$$

Without loss of generality we can assume that $(Vz_i,z_j) = \delta_{ij}$. Then the algorithm ϕ^*,

$$\phi^*(y) = \sum_{i=1}^{n} y_i\, SVz_i,$$

called the *spline algorithm,* is *optimal among all linear algorithms* and its average error is given by

$$e^{avg}(\phi^*,N;\mu) = \{\text{trace}(V^{1/2} S^* SV^{1/2}) - \sum_{i=1}^{n} \|SV^{1/2} z_i\|^2\}^{1/2} .$$

The measure μ is *orthogonally invariant* iff

$$\mu(B) = \mu(D(B)), \forall B \quad F_1, \quad B-\text{Borel set},$$

where D is an operator of the following form

$$D(f) = 2(f,z)Vz - f$$

and z is an arbitrary element from F_1 so that either $z = 0$ or $(Vz,z) = 1$ (for a more detailed discussion see Wasilkowski and Wozniakowski (1984)).

If μ^* is orthogonally invariant then the spline algorithm is *optimal among all algorithms*, i.e.,

$$r^{avg}(N;\mu^*) = e^{avg}(\phi^*,N;\mu^*) = \{\text{trace}(V^{1/2} S^* SV^{1/2}) - \sum_{i=1}^{n} \|SV^{1/2} z_i\|^2\}^{1/2}.$$

Observe that the error of the spline algorithm does not depend on a particular measure μ. It depends only on the covariance operator V. Hence for an arbitrary μ with covariance operator V,

$$r^{avg}(N;\mu) \le e^{avg}(\phi^*,N;\mu) = e^{avg}(\phi^*,N;\mu) = r^{avg}(N;\mu^*).$$

This explains why orthogonally invariant μ^* is the least favorable measure for factor analysis. ■

We now present the result concerning adaptive information.

Result 3 (Wasilkowski and Wozniakowski (1984))

Let S, F_1, F_2 be as in Result 2. Let μ be an orthogonally invariant measure. Then for arbitrary adaptive information N^a,

$$N^a(f) = [(f,z_1), (f,z_2(y_1)),\ldots,(f,z_n(y_1,\ldots,y_{n-1})]$$

where $y_i = y_i(f) = (f,z_i(y_1,\ldots,y_{i-1}))$ are measurable, there exists an element $y^* \in \mathbb{R}^n$ such that

$$r^{avg}(N^a) \ge r^{avg}(N^{non}_{y*}).$$

Here N^{non}_{y*} denotes the nonadaptive information which consists of the same evaluations as N^a for fixed $y_n = y^*_n,\ldots,y_n = y^*_n$, i.e.,

$$N^{non}_{y*}(f) = [(f,z^*_1),\ldots,(f,z^*_n)]$$

with $z^*_1 = z_1$, $z^*_2 = z_2(y^*_1),\ldots,z^*_n = z_n(y^*_1,\ldots,y^*_{n-1})$.

ACKNOWLEDGEMENTS

The authors are grateful to H.T. Kung, M.I. Shamos and J.F. Traub for their roles in bringing them together. Joseph B. Kadane was supported in part by ONR Contract 014-82-K-0622 and G.W. Wasilkowski was supported in part by the National Science Foundation under Grant MCS-7823676.

REFERENCES

BERGER, J.O. (1984). The Robust Bayesian Viewpoint. *Robustness in Bayesian Statistics.* (J.B. Kadane, ed.) Amsterdam: North-Holland, 63-144, (with discussion).

LAMBERT, D. and DUNCAN, G. (1981). Bayesian learning based on partial prior information. *Tech. Rep.* **209**, Department of Statistics, Carnegie-Mellon University.

LINDLEY, D.V. (1971). *Bayesian Statistics: A Review.* Philadelphia: SIAM.

MANSKI, C.F. (1981). Learning and decision making when subjective probabilities have subjective domains. *Annals of Statistics,* **9**, 59-65.

PARTHASARATHY, K.R. (1967). *Probability Measures on Metric Spaces.* New York: Academic Press.

TRAUB, J.F., WASILKOWSKI, G.W. and WOZNIAKOWSKI, H. (1984). Average Case Optimality for Linear Problems. *Theory Comp. Sci.* **29**, 1-25.

— (1983). *Information, Uncertainty, Complexity. Reading, MA: Addison-Wesley.*

TRAUB, J.F. and WOZNIAKOWSKI, H. (1980). *A General Theory of Optimal Algorithms.* New York: Academic Press.

— (1984). Information and Computation. *Advances in Computers,* **23**, (M. Yovitz, ed.) New York: Academic Press.

WASILKOWSKI, G.W. (1983). Local Average Errors. *Tech. Rep.* Columbia University, New York.

WASILKOWSKI, G.W. and WOZNIAKOWSKI, H. (1982). Average Case Optimal Algorithms in Hilbert Spaces, *Tech. Rep.* Columbia University, New York.

WOZNIAKOWSKI, H. (1984). Can Adaption Help on the Average? *Numer. Math.* (to appear).

DISCUSSION

SMITH, J.Q. (*University College, London*)

The authors are to be congratulated for making a very interesting connection between theoretical statistics and theoretical computer science. Papers that bridge two disciplines take a more than average amount of effort to unite and the authors' attempt is all the more laudable for that. I look forward to substantial cross-fertilisation of disciplines when the ϵ-complexity literature becomes more accesible.

I am interested by the factor analysis example. Just before equation (2.4) the authors suggest that error/loss functions other than squared error might be used. It would be a valuable addition to statistical theory to know how such different error criteria change the optimal solution for this problem.

Since Hilbert spaces seem to be the natural domain of ϵ-complexity theory it might be fruitful to consider connections here with time series. Time series in their conventional statistical settings are assumed to have been generated from an infinite past data set. In particular by analogy with the factor analysis example given in the paper it might be possible to use epsilon-complexity theory to find the subset of past n data points which allows the construction the "best" approximation to, for example, the one step ahead forecast distribution.

There is a danger of course that cross-fertilisation of these two disciplines might give misleading results because what is a reasonable assumption in one field is not reasonable in the other. For example, statisticians, might want to use results from epsilon-complexity to show that certain sequential designs were no better than fixed design. Whereas it is plausible that in applications pertaining to computer science the prior family F measured by μ is fixed, in statistical applications I believe this will rarely be the case. For example if F were a parametric family of distributions it would be unusual to think that it was the true family in any sense, F would just be a convenient modelling assumption. We may well expect to adjust or extend F after collecting a large amount of data. If we envisaged such adjustments then a sequential design would allow such adjustments whereas the fixed design would not.

Finally, I would have found it helpful for theorems mentioned in the paper to be stated at least in an appendix since the research reports cited are not easily referenced. (Ed. note: The authors have adopted this suggestion in their revision).

R. VIERTL (*Techn. Univ. Wien*)

The presented paper is an interesting link between computer science and statistics. The formal similarity of elements of ϵ-complexity theory and Bayesian statistics is pointed out very nicely. But I think for the open-minded reader with future perspectives, the paper could tell much more. In my opinion the connection between computer science and statistics is essential for modern applied statistics, because today really applied statistics cannot be done without computers.

The paper shows that statisticians could cooperate with computer scientists in a way I didn't realize before, and so progress could be made in both fields.

Another point which came to my mind when reading the paper was that probabilists and statisticians should be reminded to look at problems more from the computational-complexity approach to probability, which could be done from a Bayesian viewpoint. For references, see the book on Theories of Probability by T. Fine. Also good is the idea of incorporating costs, since it makes no sense just to deny research costs. The result that adaptive information -in the sense of the paper and cited references- is often not more powerful than nonadaptive information is an interesting conclusion. But the proofs could be given and further Bayesian analysis should be done in the future. The final remarks on factor analysis and the Bayesian interpretation should be read by all statisticians.

M. GOLDSTEIN (*Hull University*)

I was interested to read that a Bayesian interpretation of factor analysis may follow from ideas suggested by Manski and others. I have found it quite challenging to modify Manski's type of approach into a working method, even for apparently simpler problems, as guidance as to the sequential choice of action is not usually explicit in the criterion. I would welcome any comments from the authors concerning the relationship between their criterion and those that they cite.

REPLY TO THE DISCUSSION

The authors thank the commentators for their kind thoughts about our attempt to demonstrate a connection between theoretical computer science and theoretical statistics.

Smith asks how our remark before equation (2.4) about different loss functions might apply to the factor analysis example. That remark was intended only to remind the reader of the possibility of different loss functions. For example, if F_2 is a linear topological space, $S: F_1 \to F_2$ is linear and continuous, and the probability measure μ is orthogonally invariant, then the spline algorithm remains optimal when squared error is replaced by an arbitrary convex, symmetric function. We do not know, however, how such different loss functions reflect on optimal algorithms in the factor analysis setting.

Smith also raises the question of the possible application of our approach to time series problems. One time series problem can be formulated as follows: Given a space F_1 of functions f defined on $[0, \infty]$ (or on a discrete subset of $[0, \infty]$), we wish to predict the value $f(t^*)$ of $f \in F$, at time t^* where only the n values of f at times $t_1,\ldots,t_n$ have been observed. Then set $S(f) = f(t^*)$ and $N(f) = [f(t_1),\dots,f(t_n)]$. Thus we have a linear operator S and linear information operator N, so formally the results of epsilon-complexity theory apply. Whether this may lead to new results in time-series prediction we do not know.

Smith's remark about inflexible models applies equally to both epsilon-complexity and statistics. If the measure μ does not represent your opinion adequately, then it should be modified until it does. If, having done that, some theorem applies (such as that sequential sampling does not benefit one compared to a sample of fixed size), then it gives valid information. Perhaps what Smith intends is that he suspects that the conditions under which such a theorem applies are very special. But this does not have to do particularly with the relation of the two disciplines.

We have adopted Smith's suggestion of an appendix, and thank him for it.

We agree with much of the spirit of Viertl's remarks. Our current joint efforts are directed toward exploring further the relationship between results on when adaptive information is no more powerful than non-adaptive information in epsilon-complexity theory, and when sequential designs are no more powerful than non-sequential designs in statistics.

Finally, Goldstein asks us to say more about the relationship of Manski's ideas to our treatment of factor analysis. Manski's work, and the other references cited, are relevant to us because, like us, he uses a combination of Bayesian and minimax ideas together.

We thank all these commentators for their thoughful remarks.

BAYESIAN STATISTICS 2, pp. 375-390
J.M. Bernardo, M.H. DeGroot, D.V. Lindley, A.F.M. Smith (Eds.)

Reconciliation of Discrete Probability Distributions

DENNIS V. LINDLEY
Somerset, UK

SUMMARY

Several subjects each provide a discrete, probability distribution over the same set. This paper addresses the problem of how these distributions should be reconciled to form a single distribution incorporating all the subjects' knowledge. Section 1 provides the formal structure, develops a special case in which the stated log-probabilities are normally distributed and discusses a technical problem that arises when using log-probabilities. Section 2 discusses the normal model when there is symmetry amongst the set of events. The methods of the paper are shown in sections 3 and 4 not to have the properties of marginalization or external Bayesianity that have been proposed: it is argued that neither property is desirable. Correlation between subjects is the topic of section 5. In section 6 there is a discussion of possible applications of the methods of pooling and reconciliation.

Keywords: POOLING; RECONCILIATION; LOG-ODDS; LOG-PROBABILITIES; BAYES THEOREM; MULTIVARIATE NORMAL; DISCRIMINATION; MARGINALIZATION; EXTERNAL BAYESIANITY; CORRELATION.

1. THE NORMAL MODEL

A decision-maker, or investigator N, is interested in a quantity or set of quantities about which he is uncertain and consults a number m of experts, or subjects $S_1, S_2, \ldots S_m$ each of whom provides N with his personal probability distribution for the quantities. How should N combine these distributions to form his considered distribution in the light of the information provided by the subjects? In this paper the case where the distributions are discrete will be considered. An example is provided by a product that a firm might manufacture. The product is assessed on several attributes such as, acceptability of the product, cost of manufacture etc... Each attribute is classified as low, medium or high. With r attributes there is a probability distribution on 3^r values. The firm, N, asked several experts S_i to assess the product: that is, to provide a probability distribution over the 3^r values, and then needed to combine these into a single distribution.

Generally then there are n exclusive and exhaustive events $A_1, A_2, \ldots A_n$ (in the example $n=3^r$) and each S_i provides his probability for A_j $(1 \leq i \leq m,$ $1 \leq j \leq n)$. We shall find it convenient to work in terms of the logarithms of these probabilities and write $q_{ij} = \log pr_i(A_j)$ where $pr_i(A_j)$ is the probability for A_j given by S_i. These form an $m \times n$ matrix Q with typical element q_{ij}. In the rest of the paper we shall always use the notation $p(\cdot \mid \cdot)$ to denote N's probability for the quantities that appear before the vertical line conditional on the values of those given after the line. Then the rule of combination adopted by N should be

$$p(A_j \mid Q, H) \propto p(Q \mid A_j, H) p(A_j \mid H). \tag{1.1}$$

Here H denotes N's knowledge of the situation before he consults the subjects, so that $p(A_j|H)$ ($j = 1, 2, \ldots n$) is his distribution assessed without their expertise and $p(A_j|Q, H)$ is that with it. The likelihoods $p(Q|A_j, H)$ express N's opinion of the subjects, giving his probability for what they would say were the true situation to be A_j: for example if N had a very high opinion of S_i he would assess $p(q_{ij}|A_j, H)$ to concentrate around near-zero values of q_{ij}, where $pr_i(A_j)$ is nearly one, whereas $p(q_{ik}|A_j, H)$ with $j \neq k$ would be high for numerically very large values of q_{ik}, where $pr_i(A_k)$ is near zero. Alternatively expressed N would, if A_j were the true value, expect S_i to give high probability for A_j and low probability for other possibilities A_k, $k \neq j$. (1.1) is Bayes formula and expresses how N coherently changes his opinion of $A = (A_1, A_2, \ldots A_n)$ in the light of the information Q provided by the subjects.

To make significant progress in using (1.1) we take a special form for the likelihoods and make

Assumption 1. The mn quantities q_{ij} have, given A_k, a multivariate, normal distribution with means

$$E(q_{ij}|A_k, H) = \mu_{ijk} \qquad (1 \leq i \leq m;\ 1 \leq j, k \leq n) \tag{1.2}$$

and covariances (variances when both $i = k$ and $j = l$)

$$\operatorname{cov}(q_{ij}, q_{kl}\ |A_s, H) = \sigma_{ijkl} \qquad (1 \leq i, k \leq m;\ 1 \leq j, l \leq n). \tag{1.3}$$

(Remember these are judgments made by N about the subjects' expertise and are N's probabilities and therefore N's means and covariances.) Notice that the means are quite general but that the covariances are not supposed to depend on the true event A_s. The latter, expressed in (1.3), says that N thinks that, for each subject, his precisions in evaluating his probabilities do not depend on the true event: he is equally precise whatever be A_s. Notice that with $k = i$, σ_{ijil} expresses the covariance between S_i's opinion of A_j and A_l; and with $j = l$, σ_{ijkj} expresses the covariance between two subjects', S_i and S_k, opinions of the same event A_j. If N, as suggested above, has a high opinion of S_i then μ_{ijk} will be larger when $j = k$ than otherwise. There are some further, important points to be noted about Assumption 1 but it is convenient to delay discussion of them until after we have studied the following consequence of the assumption.

Theorem 1. If Assumption 1 obtains the logarithm of the likelihood in (1.1) is given by

$$\log p(Q|A_s, H) = c + \sum_{i,j} \beta_{ijs} q_{ij} + \alpha_s \tag{1.4}$$

where

$$\beta_{ijs} = \sum_{k,l} \sigma^{ijkl} \mu_{kls} \quad , \tag{1.5}$$

$$\alpha_s = -\frac{1}{2} \sum_{i,j,k,l} \mu_{ijs} \sigma^{ijkl} \mu_{kls} \quad , \tag{1.6}$$

σ^{ijkl} are the elements of a matrix which is inverse to that with elements σ_{ijkl}, and c is a constant* (that is, which does not depend on s) whose value is irrelevant.

Proof. The normality of the q's in Assumption 1 means that the logarithm of the probability is a quadratic form in the q's plus a normalizing constant which depends only

* c is a generic symbol for a constant; not necessarily the same constant each time it appears.

on the covariances, and therefore not on A_s. Hence

$$\log p(Q|A_s,H) = c - \frac{1}{2} \sum_{i,j,k,l} (q_{ij} - \mu_{ijs})\sigma^{ijkl}(q_{kl} - \mu_{kls})$$

which immediately reduces to (1.4) on using the notation of (1.5) and (1.6).

The important, simplifying feature of (1.4) is that it is linear in the elements q_{ij} of Q, the values provided by the subjects. The coefficients, (1.5) and (1.6), can be calculated by N in advance of the subjects' opinions being expressed and depend only on N's views of the subjects. Readers familiar with the calculations for discrimination between two multivariate normal distributions having the *same* dispersion matrix but different means will recognize those in the proof of Theorem 1 to be essentially the same: A_s and A_t have to be discriminated and the resulting discriminant (1.4) is linear. The values σ^{ijkl} appearing in (1.5) and (1.6) are the elements of the precision matrix of the mn values q_{ij} corresponding to the dispersion matrix given by (1.3).

Theorem 2. If Assumption 1 obtains and

$$\gamma_s = \log p(A_s|H), \tag{1.7}$$

being N's log-probability for A_s before consulting the subjects, then his same log-probability after receiving the subjects' opinions is

$$\log p(A_s|Q,H) = c + \sum_{i,j} \beta_{ijs} q_{ij} + \alpha_s + \gamma_s. \tag{1.8}$$

This is immediate on inserting (1.4) and (1.7) into Bayes theorem (1.1) in logarithmic form. We now have a simple answer to our question as to how N should combine the subjects' views, Q, to reach an opinion of his own: it is simple because, on the logarithmic scale, the combination is linear.

Before considering this further and giving some examples, let us return to discuss Assumption 1. Care needs to be exercised in its interpretation. The probabilities $pr_i(A_j)$ for any S_i are constrained to be positive* and add to one. Thus the q_{ij} are negative and themselves are awkwardly constrained by $\sum_j e^{q_{ij}} = 1$. In particular, they cannot be normally distributed. The difficulty can be resolved by ignoring the constraint and only using the results for *contrasts*: that is, linear forms $\sum_j q_{ij}c_j$ with $\sum_j c_j = 0$. These are unaffected by the constraint since if all the probabilities (for a single subject) are multiplied by a, all the log-probabilities have log a added to them, $\sum_j (q_{ij} + \log a)c_j = \sum q_{ij}c_j$ and a contrast is unaffected. The familiar case is $n = 2$ when $c_1 = 1$, $c_2 = -1$ gives $q_{i1} - q_{i2}$, the log-odds for A_1 against A_2. Then Assumption 1 essentially says that the log-odds are normally distributed. For $n > 2$ we may take $q_{ij} - q_{in}$ $(j < n)$ with some reference event, here A_n.

However, even with this convention of only using contrasts, our difficulties are not over because the log-likelihood (1.4) and the resulting expressions for N's log-probabilities (1.8) are not, in general, contrasts: specifically $\sum_j \beta_{ijs} \neq 0$. How then are we to interpret these results? We discuss (1.4): (1.8) differs only in the addition of N's prior opinion. We shall work in terms of the contrasts $q_{ij} - q_{in}$ $(j < n)$, but the argument will apply to any full set of $(n - 1)$ contrasts. Were these contrasts to be used no difficulties would arise. They can assume any values, are not constrained in any way and the normal distribution for them implied by Assumption 1 is sensible. The only objection to their use is their lack of symmetry in singling out A_n: and asymmetrical constraints are difficult to interpret.

Denote by Q^* the set $\{q_{ij} - q_{in} : 1 \le i \le m, 1 \le j < n.\}$ and by q the set $\{q_{in}: 1 \le i \le m\}$: so that $Q = (Q^*, q)$. Now

$$p(Q | A_l, H) = p(Q^* | A_l, H) p(q | Q^*, A_l, H).$$

Our analysis has been in terms of the left-hand side: an analysis in terms of the first term on the right-hand side would, as has just been remarked, be sensible. But these two analyses would be the same were the remaining term $p(q | Q^*, A_l, H)$ not to depend on A_l, for then the two sets of log-probabilities would differ only by a constant which may be absorbed into the constant already present in (1.4). Now, by Assumption 1, $p(q | Q^*, A_l, H)$ is normal so only the mean and variance need be considered. But by (1.3) the variance does not depend on A_l. Consequently all that is required is that $E(q | Q^*, A_l, H)$ does not depend on l. We therefore add

Assumption 2. $E(q | Q^*, A_l, H)$ does not depend on l.

With this holding, Theorem 1 has its obvious, satisfactory interpretation.

We now show how this can be arranged to be true. Because of the normality the expectations are linear functions of the elements of Q^* which we may write

$$E(q_{in} | Q^*, A_l, H) = \sum_{j \le m, k < n} \gamma_{ijk}(q_{jk} - q_{jn} - \mu_{jkl} + \mu_{jnl}) + \mu_{inl} \tag{1.9}$$

for suitable coefficients γ_{ijk} which, because they depend on second moments (1.3), do not depend on l. Multiply both sides of this equation by $q_{rs} - q_{rn} - \mu_{rsl} + \mu_{rnl}$ for some r, s, $1 \le r \le m$, $1 \le s < n$ and take expectations. The result is

$$\sigma_{inrs} - \sigma_{inrn} = \sum_{j \le m, k < n} \gamma_{ijk}(\sigma_{jkrs} - \sigma_{jkrn} - \sigma_{jnrs} + \sigma_{jnrn}). \tag{1.10}$$

Whereas the condition that the expectation (1.9) does not depend on l is that

$$c + \mu_{inl} = \sum_{j \le m, k < n} \gamma_{ijk}(\mu_{jkl} - \mu_{jnl}). \tag{1.11}$$

Thence the γ's may be determined from (1.10) and then μ_{inl} be chosen to satisfy (1.11).

2. SYMMETRY AMONGST EVENTS

Even within the constraints of the normal distributions with covariance structure that is the same for all events (Assumption 1) the number of quantities N has to assess is very large: in fact mn^2 means and $mn(mn-1)/2$ covariances. Some simplification seems called for. To achieve this suppose

Assumption 3.

$$\begin{aligned} \mu_{ijk} &= \mu_{i1} && \text{when } j = k, \text{ for all } k \\ &= \mu_{i0} && \text{otherwise,} \end{aligned} \tag{2.1}$$

and

$$\begin{aligned} \sigma_{ijkl} &= \sigma_{ik1} && \text{when } j = l, \text{ for all } l \\ &= \sigma_{ik0} && \text{otherwise.} \end{aligned} \tag{2.2}$$

(2.1) says that of any subject S_i, the value N expects him to give for an event which is true is the same for all events, μ_{i1}; and the value N expects him to give for any event, when some

* It is supposed that no subject assigns zero for the probability of any event. This is Cromwell's rule (Lindley, 1982a).

different event is true, is the same for all pairs of differents events, μ_{i0}. These values can depend on the subject, allowing N to rate some subjects better than others. The principle is that the events are all placed on an equal footing; only rightness or wrongness matters. (2.2) says essentially the same thing about the covariances: σ_{ik1} is the covariance between S_i and S_k when assessing the same event, σ_{ik0} that for different events. (By (1.3) these do not depend on the true event.)

Theorem 3. The values given in (2.1) and (2.2) satisfy the coherence requirements (1.10) and (1.11) if and only if

$$\mu_{i1} = -(n-1)\mu_{i0} + c. \tag{2.3}$$

Inserting the values given by (2.2) into (1.10) with $s \neq n$

$$\sigma_{ir0} - \sigma_{ir1} = \sum_{j \le m} \gamma_{ijs}(\sigma_{jr1} - \sigma_{jr0} - \sigma_{jr0} + \sigma_{jr1})$$

$$+ \sum_{j \le m,\ k \neq s,n} \gamma_{ijk}(\sigma_{jr0} - \sigma_{jr0} - \sigma_{jr0} + \sigma_{jr1})$$

$$= \sum_{j \le m,\ k < n} \gamma_{ijk}(\sigma_{jr1} - \sigma_{jr0}) + \sum_{j \le m} \gamma_{ijs}(\sigma_{jr1} - \sigma_{jr0}).$$

It follows that γ_{ijs} does not depend on s; so write $\gamma_{ijk} = \gamma_{ij}$ when we have

$$(\sigma_{ir0} - \sigma_{ir1}) = n \sum_{j \le m} \gamma_{ij}(\sigma_{jr1} - \sigma_{jr0})$$

and $\gamma_{ij} = -n^{-1}$ when $i = j$ and zero otherwise.
Substituting these values for γ_{ij} into (1.11) gives for $l \neq n$

$$c + \mu_{i0} = \sum_{j \le m} \gamma_{ij}(\mu_{j1} - \mu_{j0}) = -n^{-1}(\mu_{i1} - \mu_{i0})$$

and for $l = n$

$$c + \mu_{i1} = (n-1) \sum_{j \le m} \gamma_{ij}(\mu_{j0} - \mu_{j1}) = -\left(\frac{n-1}{n}\right)(\mu_{i0} - \mu_{i1}).$$

These equations are identical and yield (2.3).

Since the q's (or log p's) and hence the μ's are essentially unaffected by a change of origin we can easily arrange for (2.3) to be satisfied. Starting with arbitrary μ_{i1} and μ_{i0} add to each $a_i = -[\mu_{i1} + (n-1)\mu_{i0}]/n$; the new values will satisfy (2.3) with $c = 0$. This is essentially centering all the subjects at a common origin.

The consequences of the special model satisfying (2.1) and (2.2) (and (2.3)) are now explored. It is first necessary to evaluate the α's and β's, equations (1.5) and (1.6). These involve the values σ^{ijkl} of the matrix inverse to that with elements σ_{ijkl}. The latter matrix may be regarded as m^2 submatrices, the submatrix in "row" i and "column" k referring to the covariances between S_i and S_k and the rows and columns within submatrices referring to the events. Each of these submatrices, by (2.2), has one value for all the diagonal elements, namely σ_{ik1}, and another, σ_{ik0}, for all off-diagonal elements. Such matrices form a group. It therefore follows that the inverse matrix has the same structure; namely

$$\begin{aligned} \sigma^{ijkl} &= \sigma^{ik1} \quad \text{when } j = l, \text{ for all } l \\ &= \sigma^{ik0} \quad \text{otherwise.} \end{aligned} \tag{2.4}$$

(The evaluation of these elements in terms of σ_{ik1} and σ_{ik0} is tedious but straightforward.)

Insertion of the values in (2.4) into (1.5) gives

$$\beta_{ijs} = \sum_k \sigma^{ik1}\mu_{kjs} + \sum_{k,l\neq j} \sigma^{ik0}\mu_{kls}.$$

If $j \neq s$ this is

$$\beta_i = \sum_k \sigma^{ik1}\mu_{k0} + \sum_k \sigma^{ik0}\mu_{k1} + (n-2)\sum_k \sigma^{ik0}\mu_{k0}$$

where we have written β_{ijs} ($j \neq s$) as β_i, so defining β_i, since the right-hand side depends neither on j nor s. When $j = s$

$$\beta_{iss} = \sum_k \sigma^{ik1}\mu_{k1} + (n-1)\sum_k \sigma^{ik0}\mu_{k0}.$$

But using (2.3) with $c = 0$ we easily have that

$$\begin{aligned} \beta_{ijs} = \beta_i &= \sum_k \mu_{k0}(\sigma^{ik1} - \sigma^{ik0}) && \text{for } j \neq s \\ &= -(n-1)\beta_i && \text{for } j = s. \end{aligned} \tag{2.5}$$

A similar calculation for α_s, equation (1.6), establishes that

$$\alpha_s = \frac{1}{2}(n-1)\sum_i(\mu_{i0} - \mu_{i1})\beta_i = \alpha, \text{ say} \tag{2.6}$$

which does not depend on s. Inserting these values for the α's and β's into the results of Theorem 1 we have

Theorem 4. If Assumptions 1 and 3 obtain, then the logarithm of the likelihood in (1.1) is given by

$$\log p(Q|A_s,H) = c - n\sum_i \beta_i q_{is}. \tag{2.7}$$

There is a corresponding simplification in Theorem 2. The attractive feature of this analysis is that when N comes to consider his probability for A_s, in the light of all the information provided by the subjects, he need only consider the subjects' assessments for A_s, namely $\{q_{is}, 1 \leq i \leq m\}$, and not the assessments for other events. The coefficients of the relevant q's, $\{\beta_i, 1 \leq i \leq m\}$, are available from (2.5). Only the elements σ^{ik1} and σ^{ik0} are tedious to compute.

Some special cases are now considered: firstly, that with only one subject. The suffix i in (2.1), referring to subjects, may be dropped and we have simply μ_1 and μ_0 reflecting what N expects S to announce for true and false events respectively. From (2.3) $\mu_1 = -(n-1)\mu_0$, using $c = 0$. Similarly in (2.2), i and k may be dropped, leaving σ_1 and σ_0, the variance and common covariance for S's statements as assessed by N. It is immediate from (2.5) that β (the suffix i being omitted) is $\mu_0(\sigma^1 - \sigma^0)$. Simple calculations show that $\sigma^1 - \sigma^0 = (\sigma_1 - \sigma_0)^{-1}$ and hence (2.7) says simply that ($c = 0$)

$$\log p(Q|A_s,H) = \frac{-n\mu_0 q_s}{(\sigma_1 - \sigma_0)} = \frac{n}{n-1}\,\frac{\mu_1 q_s}{\sigma_1 - \sigma_0}.$$

In the case of only two events, $n = 2$, this agrees with the results of Lindley (1982b: Example 3) where the calculations were all in terms of the log-odds, here $q_1 - q_2$. On A_1 this has mean $\mu_1 - \mu_0 = 2\mu_1 = \mu$ in the notation of the earlier reference, and variance $2(\sigma_1 - \sigma_0) = \sigma^2$ there.

A second simple case is that in which N judges the subjects to be independent: that is, $\sigma_{ijkl} = 0$ whenever $i \neq k$. The second-moment structure is therefore described in terms of σ_{i1} and σ_{i0} for S_i. The inverse elements will have the same structure and hence, from (2.5), $\beta_i = \mu_{i0}(\sigma^{i1} - \sigma^{i0})$ which, exactly as with a single subject, reduces to $-\mu_{i1}/(n-1)(\sigma_{i1} - \sigma_{i0})$.

Finally, from (2.7),

$$\log p(Q|A_s,H) = \frac{n}{n-1}\sum_i \frac{\mu_{i1}q_{is}}{\sigma_{i1}-\sigma_{i0}}. \tag{2.8}$$

Notice that $\mu_{i1}/(\sigma_{i1}-\sigma_{i0})$ is the weight attached by N to q_{is}, S_i's opinion about A_s and is a measure of S_i's worth as seen by N.

The above model with its symmetry between events is too simplistic in many situations, particularly those in which the events $A_1, A_2, \ldots A_n$ have a natural order. For example, they may correspond to grouping of a continuous underlying quantity X, A_i being $a_i < X < a_{i+1}$: or they may be ordered as in the example of section 1; low, medium and high. In such cases the correlation between adjacent events A_i and A_{i+1} is likely to be stronger than between distant events A_i and A_{i+4}, say. One possibility is to suppose σ_{ijkl} to be of the form $\sigma_{ik}\varrho_{ik}{}^{|j-l|}$ with $0<\varrho_{ij}<1$, corresponding to a Markovian relationship between the events. This is technically difficult to handle due to the implied, complicated form for σ^{ijkl} which, as we have seen, plays an important role in the calculations. A simpler possibility is to replace the Markovian character, which is akin to linear autoregression, by supposing the structure to be like a moving average. This is most easily described through the precision matrix directly, supposing $\sigma^{ijkl} = \sigma^{ik}$ when $j = l$, $\sigma^{ik}\varrho^{ik}$ when $j = l-1$ or $l+1$ and otherwise zero. The values of the ϱ's reflect the adjacent correlations. This matrix has submatrices with zeroes everywhere except in the leading diagonal and immediately adjacent diagonals so that β_{ijs}, for example, (1.5) is easy to evaluate. Unfortunately such matrices do not form a group and the computation of σ_{ijkl} is complicated. The details are not pursued here.

3. MARGINALIZATION

McConway (1981) and others earlier have considered the concept of marginalization in connection with the pooling of subjects' opinions. It arises in the following way. Suppose that the original events $A_1, A_2, \ldots A_n$ are reduced by combining some events together to form a reduced set $B_1, B_2, \ldots B_r$ $(r<n)$. A common case is where the events are defined in terms of two, discrete quantities X and Y or grouped, continuous ones. Employing two suffixes, where previously one has been used, A_{rs} is the event that both $X = r$ and $Y = s$. If only X is considered then B_r is the event that $X = r$ and is the union of the events $A_{r1}, A_{r2}, \ldots A_{rd}$ where Y takes d values. Clearly the B's form the margins of the array of events A_{rs}. Generally, the reduction from any set of events by unions will be referred to as marginalization. McConway in his paper uses the concept of a σ-algebra but this abstraction is unnecessary in our finite case.

Suppose now that although the subjects provide all the probabilities $pr_i(A_j)$ $(i = 1, 2, \ldots m; j = 1, 2, \ldots n)$ N is only interested in the margins $p(B_k|Q,H)$. There are two ways that N can proceed.

First method. Calculate the marginals for each subject by $pr_i(B_k) = \Sigma pr_i(A_j)$, where the summation is over all events A_j whose union is B_k, and then pool the results.

Second method. Pool the original statements about the A's to obtain $p(A_j|Q,H)$ and then construct the marginals by summation over all A_j in B_k.

(The methods only differ in the order in which the two operations of pooling and marginalization proceed.)

It has been suggested that the two methods ought to give the same results if the pooling mechanism is sensible (the marginalization mechanism being dictated by the rules of probability). McConway shows that they do if, and only if, the pooling is a linear operation on the original subject probabilities: that is, if

$$p(A_j|Q,H) = \sum_i \omega_i pr_i(A_j)$$

for weights ω_i for subject S_i.

The method proposed here is linear in the logarithms of the probabilities (equation (1.8)) and also may use subject judgments of A_k, $k \neq j$, when assessing $p(A_j|Q,H)$. It therefore does not satisfy the requirement of marginalization that the two methods lead to the same result. The questions therefore arise of whether our pooling method is satisfactory and whether marginalization is a sensible desideratum. The questions cannot both be answered in the affirmative. We argue that it is the marginalization argument that is at fault. The reason is that when N does the pooling in the first method he has less information than in the second: in the former he has subjects' opinions only about the margin, in the latter he has opinions about the full distribution. It is therefore to be expected that, with different information, the results will differ. Let us analyze it in detail.

Write Q_B for the set of log-probabilities[1] for the margin and Q_{AB} for the set of log-probabilities for the conditional distribution of the A's given the marginal B's, so that $Q = (Q_B, Q_{AB})$. In what follows summation is over all events A_j whose union is B_k. The calculations in the second method are as follows:

$$\begin{aligned} p(B_k|Q,H) &= \Sigma p(A_j|Q,H) \\ &\propto \Sigma p(Q|A_j,H)\, p(A_j|H) \\ &= \Sigma p(Q_B|A_j,H)\, p(Q_{AB}|Q_B,A_j,H)\, p(A_j|H) \end{aligned} \tag{3.1}$$

whereas with the first method

$$p(B_k|Q_B,H) \propto p(Q_B|B_k,H)\, p(B_k|H). \tag{3.2}$$

The two right-hand sides of (3.1) and (3.2) are different, particularly because of the occurrence of $p(Q_{AB}|Q_B,A_j, H)$ in (3.1). If this genuinely depends on the true event A_j, N will have more information with the second method. Even if it does not the two expressions will, in general, be different unless $p(Q_B|A_j,H)$ is the same for all A_j included in B_k. We return to discuss marginalization in the commentary of section 6 below.

4. EXTERNAL BAYESIANITY

One way of thinking about the topic of this paper, as in §3, is to regard it as a method of pooling the opinions of the subjects to reach a single opinion; the pooling being done by N. In effect, there is a transformation $T(\{q_{ij}\})$ from the set of q's to a single set of log-probabilities over the events. The exact form of T is given by (1.8). Mandansky (1964, 1978) has suggested that the transformation should have a property that he calls "externally Bayesian", which essentially says that, to an external observer, the group of subjects should look like a single Bayesian when assessed on their pooled values. To make this more precise, suppose each subject learns of a new piece of information and that they all agree that on A_j it has log-likelihood ℓ_j and that, given A_j, it is independent of their earlier information. Then subject i will update q_{ij} to $q_{ij} + \ell_j$ assuming, as we do, that he is coherent. The transformation will then give $T(\{q_{ij} + \ell_j\})$. But if the pooled opinion is to look externally Bayesian it too should update by adding ℓ_j to the distribution in logarithm

[1] In the present context the rules of combination of the q's are not used and probabilities could replace the logarithms.

form. That is, $T_s(\{q_{ij} + \ell_j\}) = T_s(\{q_{ij}\}) + \ell_s$ where T_s is the log-probability for A_s. The methods proposed here are not externally Bayesian for, on replacing q_{ij} in (1.8) by $q_{ij} + \ell_j$, $T_s(\{q_{ij}\})$ or $\log p(A_s|Q,H)$ is increased by $\sum_{ij} \beta_{ijs}\ell_j$, not ℓ_s. (In the special case of section 2, the increase, from (2.7) is $-n\Sigma\,\beta_i\ell_s$, only equal to ℓ_s if $\Sigma\beta_i = n^{-1}$ where β_i is given by (2.5).

The methods proposed fail the criterion of external Bayesianity as well as that of marginalization. Again we argue that it is the criterion that is at fault. It is enough if we take the case of a single subject who learns of the truth of an event E and assigns it log-likelihood ℓ_j on A_j. He will then change from q_j to $q_j + \ell_j = q_j^*$, say. Let E^* denote the event that N knows E and also knows that *S knows E and assigns log-likelihood* ℓ_j: so that $E^* = (E, E_s)$, where E_s is the event in italics. Then N has q^* (the set of q_j^*s) and E^* so considers (omitting H for clarity)

$$p(A_j|q^*,E^*) \propto p(q^*,E^*|A_j)p(A_j)$$

$$= p(E|A_j)\,p\,(E_s|E,A_j)\,p\,(q^*|E^*,A_j)p(A_j). \tag{4.1}$$

Suppose $p(E_s|E,A_j)$ does not depend on j: that is knowing that S knows E, over and above N's knowing E, does not change N's opinion about any A_j. (There are occasions where the mere knowledge that someone knows something, without knowing what that something is, can change one's opinion. "X was known to the police" can influence one's view of X. An insightful discussion is given by Aumann (1976). We return to the point in section 5). If so $p(E_s|E,A_j)$ may be omitted from (4.1). Since $p(q^*|E^*,A_j) = p(q|A_j)$ where q is the set of q_j's, (4.1) gives on taking logarithms

$$\log p(A_j|q^*,E^*) = c + \ell_j + \log p(q|A_j) + \log p(A_j).$$

This is exactly $\log p(A_j|q) + \ell_j$, so that N will update coherently, as did S. However, N will not process q^* as he did q because his attitude to S has changed because of E^*. His opinion of S has changed because he knows that S knows E. Hence the transformation T changes. With several subjects, the pooling mechanism will change because of the new knowledge shared by all the subjects. As we see it, the fallacy in external Bayesianity is not to recognize that T changes with new information. Genest (1982) has shown that external Bayesianity leads to dictatorships.

5. CORRELATIONS BETWEEN SUBJECTS

A difficulty in the implementation of the methods of this paper is the substantial assessment problem that N has. Yet there seems no escape from it since N has to describe his opinion of the subjects in some way, for their perceived expertise is an essential ingredient of the problem. Perhaps the most difficult part of the assessment is the evaluation of the correlations between the subjects, which are now discussed. Related work on this topic is contained in French (1981) and Winkler (1981). The most important source of correlation is the knowledge held in common by two subjects. Let us begin by considering the situation where there is no such common knowledge and N judges the S's to be independent: $\sigma_{ijkl} = 0$ whenever $i \neq k$. This might arise when they are experts in different fields, as when a politician consults an economist and a soldier about the prospects of a conflict between two states. To simplify the discussion, suppose N judges each of the independent subjects to be equally good. That is,

$$\begin{aligned} \mu_{ijk} &= \mu_{jk}, \text{ say, for all } i \\ \sigma_{ijkl} &= \sigma_{jl}, \text{ when } i = k, \text{ and zero otherwise.} \end{aligned} \tag{5.1}$$

Then
$$\beta_{ijs} = \sum_l \sigma^{jl}\mu_{ls} = \beta_{js}, \text{ say,}$$
$$\alpha_s = \frac{1}{2} m \sum_{j,l} \mu_{js}\sigma^{jl}\mu_{ls} \tag{5.2}$$
and
$$\sum_{i,j}\beta_{ijs}\, q_{ij} = \sum_j \beta_{js}\, q_{\cdot j},$$
where $q_{\cdot j} = \sum_i q_{ij}$. In particular, if each subject gives the same assessment, $q_{ij} = q_j$, then $q_{\cdot j} = mq_j$ and the log-likelihood, (1.4), is, for A_s,
$$c + m\sum_j \beta_{js}q_j + \frac{1}{2} m \sum_{j,l} \mu_{js}\,\sigma^{jl}\mu_{ls},$$
of the form $c + mf_s$ where f_s is a function of s. As m increases the largest f_s will dominate —remember it is only differences of log-likelihoods that are relevant— and hence for the event corresponding to that f_s its probability will tend to one. For example, if each of m equally competent, independent subjects each gives probability 0.7 to an event, N will conclude that the event has probability near 1.0. At first surprising, on reflection the conclusion seems correct, for if the event were false it is unlikely[1] that each subject would assign the value 0.7. As Richard Savage commented to me, the heart of the paradox lies in the concept of independent subjects. Only in extreme cases is it reasonable to suppose independence, generally some knowledge will be shared. Let us therefore consider how shared knowledge can affect the covariances between subjects.

Let m subjects be assessed as in section 1 and then suppose that each subject observes a random quantity X: that is, they all share the experience of X. Suppose that N does not observe X but does know that each subject has seen it. (He knows that they know but not what they know, in contrast to the case considered in section 5.) Suppose that, given A_s ($s = 1, 2, \ldots n$), each S_i judges X to be independent of his other information (and N knows this) so that S_i will add his perceived log-likelihood of A_s for X to his earlier value. It then only remains to describe this log-likelihood. Suppose that

$$\text{on } A_s\text{:} \quad S_i \text{ judges } X \text{ to be } N(\nu_{is}, \tau_i), \tag{5.3}$$

(normal with mean ν_{is} and variance τ_i: again as with (1.3) the variance is supposed not to depend on the true event). Suppose also that

$$\text{on } A_s\text{:} \quad N \text{ judges } X \text{ to be } N(\nu_s, \tau). \tag{5.4}$$

From (5.3) S_i, having seen $X = x$, will have log-likelihood $-(x-\nu_{is})^2/2\tau_i$ or, omitting constants, $\nu_{is}(x - \frac{1}{2}\nu_{is})/\tau_i$ and will therefore update q_{ij} to q^*_{ij} given by

$$q^*_{ij} = q_{ij} + \nu_{ij}(x - \tfrac{1}{2}\nu_{ij})/\tau_i. \tag{5.5}$$

N, not knowing x, will perceive it to be governed by (5.4). Thus on taking expectations

$$\mu^*_{ijs} = E(q^*_{ij}|A_s, H)$$
$$= \mu_{ijs} + \nu_{ij}(\nu_s - \tfrac{1}{2}\nu_{ij})/\tau_i. \tag{5.6}$$

1 It is here being assumed that N judges that a stated probability above $\frac{1}{2}$ makes the corresponding event more likely: that is $\mu_{ss} > \mu_{js}$. The case of the meteorologist who gives 70 % chance of rain (dry) whenever it is dry (wet) is ignored.

Similarly for the covariances

$$\sigma^*_{ijkl} = \text{cov}\,(q^*_{ij}, q^*_{kl} \mid A_s, H)$$

$$= \sigma_{ijkl} + \nu_{ij}\nu_{kl}\tau/\tau_i\tau_k. \quad (5.7)$$

The special case where all agree on the distribution of X, so that $\nu_{is} = \nu_s$ and $\tau_i = \tau$, gives

$$\mu^*_{ijs} = \mu_{ijs} + \frac{1}{2}\nu_j\nu_s/\tau$$

and (5.8)

$$\sigma^*_{ijkl} = \sigma_{ijkl} + \nu_j\nu_l/\tau,$$

with $\nu_s/\tau^{1/2}$ as the relevant quantities affecting the change in N's assessment.

This analysis suggests a way in which N might go about the assessment procedure: see Freeling (1981). He might consider each subject and think of the quantities that subject will consider when he arrives at his judgment. For example, with three subjects he might suppose S_1 takes into account X_1, X_2 and X_3: S_2 uses X_1 and X_4 and S_3, X_2, X_4 and X_5. Next he will consider how he thinks each subject will view the quantities; and also how he, N, will view them. Using formulae like (5.6) and (5.7), the required pattern of μ's and σ's can be built up. In the example, there will be correlation between S_1 and S_2 because of the shared X_1; S_1 and S_3 because of X_2; S_2 and S_3 because of X_4; yet S_1 and S_3 (but not S_2) have something additional to contemplate: S_1, X_3 and S_3, X_5. It seems idle to theorize further until some practical experience of the method is available.

6. DISCUSSION

The situation described in this paper where one person, N, coordinates the views of others is one that surely occurs in practice, as when a company seeks advice from many different sources. Then the Bayesian argument expressed in (1.1) is normatively the only reasonable technique leading, when normality obtains, to the calculations displayed here. We suggest that there are other situations where the results of this paper might apply.

One of them has already been mentioned, namely where the problem is that of m subjects pooling their separate opinions to reach a concensus. This might arise when the subjects form a board of directors and the board has to reach a judgment. Several methods have been proposed: DeGroot (1974) provides an example. This paper provides another, for Theorem 2 still makes sense as a pooling mechanism without any overt mention of N. It is true that the judgments have to be made about the μ's and σ's but extra judgments beyond the q's are required in any method of pooling. Furthermore, we shall now argue that this is the only normative method of pooling. The reason is that the only normative theory of decision-making that is available is that for a single decision-maker: there is no parallel theory for more than one decision-maker. Valuable as game theory is, it has not yet produced a satisfactory solution to other than the two-person, zero-sum game (and even that solution has its defects as Kadane and Larkey (1982) have pointed out). Consequently the only available normative approach requires reduction to a single decision-maker to produce the probabilities that, with the utilities, are the basic ingredients of the normative view. We therefore contend that the introduction of N is not merely an artifice but essential, at least at the present stage of development. Any other approach has an element of adhockery about it because it does not derive from the inherent logic of the Bayesian paradigm. This is most clearly brought out in the two requirements of marginalization and external Bayesianity. Both of these requirements are essentially adhockeries, suggested as plausible requirements of a pooling scenario. At a basic level, neither have the elemental simplicity of the axioms of decision analysis and, as we have seen, fail under the close scrutiny of that analysis. Marginalization fails because it forgets vital information: external Bayesianity fails because it omits to notice that the pooling function changes when the subjects acquire a new fact. Only an analysis within the Bayesian view can ensure sound logic: hence the introduction of N.

There is another possible application of the results of this paper and that is to the case where N and all the subjects are the same person. A single individual is contemplating a distribution over the events A_i. He looks at it one way and reaches the judgment $\{q_{1j}, j = 1, 2, \dots n\}$: another way produces $\{q_{2j}\}$, and so on. Brown and Lindley (1982) provide examples. There S_i is replaced by the ith way of looking at the problem. The task is then to reconcile the different views: pooling becomes reconciliation. We argue that the methods of this paper are appropriate to the reconciliation problem and indeed provide the only normative solution. In this problem N enters naturally as the single person reconciling the different approaches he has considered.

In their paper on reconciliation Lindley, Tversky and Brown (1979) suggest another way of looking at the likelihood $p(Q|A_j,H)$. They suggest it might be useful to introduce true probabilities (or log-probabilities) π_{ij} for subject S_i contemplating event A_j corresponding to stated values q_{ij} and to suppose that (omitting H for clarity)

$$p(Q|A_j) = \int p(Q|\Pi)p(\Pi|A_j)d\Pi,$$

where $\Pi = \{\pi_{ij} : 1 \le i \le m,\ 1 \le j \le n\}$. Here $p(Q|\Pi)$ is the distribution for Q given Π, supposed not to depend on the event A_j, and to describe errors of probability assessment, whereas $p(\Pi|A_j)$ measures the subjects, real knowledge of the events. If both these distributions are normal, with Q having mean a linear function of Π with variance that does not depend on Π, then $p(Q|A_j)$ will also be normal in accord with Assumption 1. It would be possible for N to make statements about Π in the light of Q. Lindley et al. called this the internal problem, whilst the one discussed here is external (to the subjects).

ACKNOWLEDGMENT

The work on this paper was partly supported by the Office of Naval Research under contract number N00014-81-C-0330 through Decision Science Consortium, Inc., Falls Church, Virginia and partly by Los Alamos National Laboratory under P.O. number 9-L42-1028Z-1 through the George Washington University, Washington, D.C. I am grateful to Rex Brown and Nozer Singpurwalla for arranging these and for many, useful discussions; and to Christian Genest (1982) for letting me see an advance copy of his paper which stimulated the results of section 4. An unknown discussant at a Yale seminar drew my attention to the unexpected behaviour when all subjects agree.

REFERENCES

AUMANN, R.J. (1976). Agreeing to disagree. *Ann. Statist.* **4**, 1236-1239.

BROWN, R.V. and LINDLEY, D.V. (1982). Improving judgment by reconciling incoherence. *Theory and Decision* **14**, 113-132.

DEGROOT, M.H. (1974). Reaching a consensus. *J. Amer. Statist. Ass.* **69**, 118-121.

FREELING, A.N.S. (1981). Reconciliation of multiple probability assessments. *Organizational Behavior and Human Performance.* **28**, 395-414.

FRENCH, S. (1981). Consensus of opinion. *Eur. J. Opl. Res.* **7**, 332-340.

GENEST, C. (1982). External Bayesianity: an impossibility theorem. Tech. report 82-8: Institute of Applied Mathematics and Statistics; University of British Columbia.

KADANE J. and LARKEY, P. (1982). Subjective probability and the theory of games. *Management Science* **28**, 113-120.

LINDLEY, D.V. (1982a). The Bayesian approach to statistics. In *Some Recent Advances in Statistics*: ed, J.T. de Oliviera and B. Epstein. London: Academic Press, 65-87.

— (1982b). The improvement of probability judgments. *J. Roy. Statist. Soc. A* **145**, 117-126.

LINDLEY , D.V., TVERSKY, A. and BROWN, R.V. (1979). On the reconciliation of probability assessments. *J. Roy. Statist. Soc. A* **142**, 146-180 (with discussion).

MADANSKY, A. (1964). Externally Bayesian groups. Rand Memo. RM-4141-PR.

— (1978). Externally Bayesian groups. Unpublished ms. University of Chicago.

McCONWAY, K.J. (1981). Marginalization and linear opinion pools. *J. Amer. Statist. Ass.* **76**, 410-414.

WINKLER, R.L. (1981). Combining probability distributions from dependent information sources. *Management Science* **27**, 479-488.

DISCUSSION

K.J. McCONWAY (*Open University, UK*)

Professor Lindley's paper is valuable, particularly in that it provides us with a greater understanding of the workings of his approach to what French (elsewhere in this volume) calls the expert problem. In particular, it provides a clear discussion of what can be meant by the intuitive idea of correlated experts. I shall not comment on the technical details of the paper, interesting though they are.

I shall start with the problem of marginalization. In Section 3, Lindley shows that even if the decision maker is interested only in the probabilities of a reduced set of events (the B's), then he has less information on which to base his analysis if he learns only the experts' probabilities for the B's than if he learns their probabilities for the full set of events (the A's). This seems to imply that it would be worth in *any* case asking the experts to state probabilities over a much larger set of events than those in which the decision-maker is really interested. Perhaps this should be implemented in practical application of this work.

One subject Professor Lindley does not really touch on could be described as the logical structure of the set of events. Suppose all the experts give distributions in which two particular events are independent. Some authors (see Laddaga, 1977) recommend that in this case the decision maker should also make these events independent. This is probably unsatisfactory as a general rule. However, if the experts have given these assessments of stochastic independence because they believe the two events are logically independent, it would seem reasonable to take this into account. Lindley's method as described in the paper does not preserve independence in this way. I should like to ask whether he considers this important, and if so, how it could be dealt with.

At a more general level, the methods described proceed by turning the usual Bayesian crank from prior via likelihood to posterior. In most Bayesian analysis, the conceptually difficult part is finding the prior; the likelihood involves probability modelling that has much in common with all approaches to inference. But here things are reversed. The prior is given, and it is the likelihood that creates the difficulty. Do we really want to turn the

Bayesian crank in the usual direction, or does it just complicate things? Why not invent the posterior directly, rather than the likelihood? One could then go back to deduce the likelihood, and see if it was sensible. (This approach has parallels with the method of imaginary results for assessing prior distributions, where intuitive prior - to posterior updating is used to deduce a prior via an agreed likelihood.)

Finally, I would like to ask whether the expert problem is of any great practical significance. The analysis here, and most others, take the stated probabilities as the only formal data. But in general there will usually be discussions and interaction between the experts and the decision-maker, and probably amongst the experts too. How should we take these into account?

A. O'HAGAN (*University of Warwick, UK*)

It is always a pleasure to read a paper by Dennis Lindley. As usual, I find myself in complete agreement with him. His approach to the problem is obviously correct, and his treatment of the marginalisation and external Bayesianity principles is definitive.

My one criticism is from a surprising quarter. Lindley is usually so lucid that the obscurity of his justification for using the multivariate normal model is quite uncharacteristic. Having wrestled with his argument leading to Assumption 2, it seemed to me that it might help to phrase it in terms of sufficiency, as follows.

1. The set Q^*, of contrasts between the experts' log probabilities, is sufficient. To prove sufficiency formally, we must show that the distribution of the rest of the data, given the sufficient statistics, does not depend on the parameter. In this case, the parameter is the suffix s indexing the set $\{A_s\}$. Given Q^*, all the expert data $Q = (Q^*,q)$ are determined precisely. Therefore $p(q \mid Q^*,H)$ is degenerate, and so $p(q \mid Q^*,A_s,H)$ is the same degenerate distribution.
2. Because of the sufficiency of Q^*, the actual form of $p(q \mid Q^*,A_s,H)$ is irrelevant for posterior inference, since it is absorbed into the proportionality constant. Therefore we can "cheat" by giving it any other form we like, so long as we preserve sufficiency by making it independent of s. The posterior, and any inferences derived from it, will be unaffected.
3. Lindley does just this: $p(q \mid Q^*,A_s,H)$ is assigned a non-degenerate normal distribution for convenience, but he ensures through Assumption 2 that sufficiency of Q^* is retained.
4. For a given expert, *any* set of n-1 linearly independent contrasts in his log probabilities is sufficient. Therefore Lindley's claim, that his analysis does not depend on his particular choice of Q^* and q, is also justified.

Having put the argument like this, I am convinced that Lindley's sleight-of-hand is not only valid but also a very clever and useful device.

J. M. BERNARDO (*University of Valencia, Spain*)

I think it is now widely accepted that treating the information provided by the experts as data for the investigator and using Bayes theorem is indeed the only normative approach to the problem discussed. Moreover, this paper clearly shows that superimposed requirements such as marginalization or external Bayesianity do not stand up to close examination. Consequently, the problem is reduced to the provision of a technically feasible procedure of turning Bayes' handle.

The author's choice has been to work in terms of the log-probabilities of the experts and to assume specific forms of multinormality for them. I wonder whether reasons, other than the mathematical convenience demonstrated in Theorem 1 could be invoked to defend this choice. The fact that log-probabilities are the natural expression for additive information measures suggests that log-probabilities may indeed be the natural parametrization of a problem where the key issue is combining information; however, since they are negative quantities, it is rather akward to have to make the normality assumption. It may be more natural and more general to assume that the $-q_{ij}$'s have a multivariate Gamma distribution (c.f. Johnson and Kotz, 1972, pp. 216-231). Alternatively, one could assume a multivariate Student, or even a multivariate Normal, for the log-odds of the experts. Naturally, the final answer rests on whether the investigator's opinions are well described by one model or another, but I suspect that the last possibility would provide better fits.

Finally, I wonder whether it would be feasible to model the investigator's opinions of the experts through a hierarchical prior structure on the parameters involved in the family chosen, rather than assuming them directly assessed.

REPLY TO THE DISCUSSION

One feature of the paper that bothers O'Hagan and Bernardo, as well as myself, concerns the constraints on the log-probabilities and the "sleight-of-hand" that avoids them by using only contrasts. When I appreciated that this might be possible I felt an unease that was not entirely eliminated by my proof, which O'Hagan correctly says is not lucid. It is therefore good to have his better proof which links the ideas to sufficiency. If only the same trick was available with multinomial inferences.

Bernardo goes on to suggest alternative assumptions about the log-probabilities. My own attitude is that anything that obeys the laws of probability is possible. (The same remark applies to McConway's point about going from posterior to likelihood.) As de Finetti says, each problem should be considered on its merits. The choice of the multivariate normal was entirely dictated by mathematical convenience. The other possibilities he mentions would repay investigation though they would, I think, still have difficulties over the constraint that the probabilities must add to one.

McConway makes the point that "it would be worth in *any* case asking the experts to state probabilities over a much larger set of events than those in which the decision-maker is really interested". I agree completely: indeed I think that the point is tremendously important. Probability is about how judgments fit together and it is the subject's ability to fit that matters. Experimental psychologists who have studied probability assessment have all missed this point. Do not ask a subject only about an event A of interest: ask him about related events B, C, ...etc. Being forced to think about these as well and therefore being forced to use the laws of probability (and not just that $p(A)$ lies between zero and one) will sharpen the assessment of $p(A)$.

Until McConway pointed it out, I had not realized that if all the S_i think A and B are independent, N might not. But could logic make two events probabilistically independent?

I do think the expert problem is important but McConway is right when he says the experts might provide more than probabilities - for example, confidence in their stated values as well. The treatment is the same: if X is the totality of what S says, then $p(X|A)$ and $p(X|\bar{A})$ are the essential ingredients. The paper under discussion deals with a special case of expert opinion.

REFERENCES IN THE DISCUSSION

JOHNSON, N.L. and KOTZ, S. (1972). *Distributions in Statistics: Continuous Multivariate Distributions.* New York: Wiley.

LADDAGA, R. (1977). Lehrer and the consensus proposal. *Synthese,* **36**, 473-477.

BAYESIAN STATISTICS 2, pp. 391-392
J.M. Bernardo, M.H. DeGroot, D.V. Lindley, A.F.M. Smith (Eds.)

The Bayesian Approach to Global Optimization

J. B. MOCKUS
Academy Sciences, Vilnius, URSS

SUMMARY

One of the important and difficult problems addressed in computer-aided design and planning is how to find the best decision. In many cases the problem can be reduced to one of global optimization, where the minimum of multimodal functions is to be found. Usually it cannot be solved using conventional numerical methods of optimization, which guarantee some fixed accuracy level. Therefore, a heuristic approach is often used which works reasonably well in many practical situations, but which has no mathematical justification.

In this paper a new approach is discussed, which considers average deviation. It was developed as a result of generalization and mathematical justification of the heuristic approach. In this approach the function to be minimizes is regarded as a sample of a random function. This allows the possibility of defining the search procedure which minimized the average deviation from the global minimum and, which therefore is optimal in the Bayesian sense.

The main objection to the Bayesian approach is doubt about the existence and the arbitrary character of the *a priori* probability distribution. Therefore, in this paper conditions are given in which Bayesian methods converge to the global minimum of any continuous function. It is also proven that under some simple and natural assumptions concerning the relations of subjective likelihood, there exist probability distributions which agree with these likelihood relations.

A package of FORTRAN programs implementing the Bayesian approach has been developed and used to optimize a set of different test functions. The results are compared with alternative approaches.

The results of engineering applications in the design of optimal vibromotors and the most thermostable polymeric compositions are mentioned.

Keywords: AVERAGE DEVIATION; NUMERICAL METHODS.

This paper has been published in the volume of Proceedings of the Indian Statistical Institute Golden Jubelee, *Statistics Applications and New Directions,* (1984).

BAYESIAN STATISTICS 2, pp. 393-424
J.M. Bernardo, M.H. DeGroot, D.V. Lindley, A.F.M. Smith (Eds.)

Repetitive Assessment of Judgmental Probability Distributions: A Case Study

JOHN W. PRATT and ROBERT SCHLAIFER
Harvard University

SUMMARY

An executive assessed three quartiles and two "astonishing" points of her judgmental probability distribution of gross receipts for each of 190 programs at two movie theaters over 16 months, learning the actual values soon after each assessment. Analysis of the data shows that: after a very short learning period her lower quartile and median were extremely well calibrated over all but 37% of the actual grosses exceeded her upper quartile; her relative errors (actual/median) were very nearly lognormal with larger variance in the larger, less familiar theater; the ratios among her fractiles varied little; the relation of their variation to relative error was statistically significant in each tail and not badly calibrated but of little practical importance; there was no clear relation between forecast errors and overall measures of subjective spread and skewness.

Keywords: ASSESSED QUARTILES; CALIBRATION; DINEMA DATA; JUDGMENTAL PROBABILITY; PROBABILISTIC FORECASTING; SUBJECTIVE PROBABILITY.

1. INTRODUCTION

In 1962-3 a small, somewhat informal experiment was done on the behavior of a real business person forecasting a real business quantity, probabilistically, with feedback. The results were never published because the experiment was informal, because few in the authors' fields were interested then, and for all the other, less good reasons some people don't get around to publishing things. They are of interest still because little of the kind is available even now for real business situations and because they are not what one might casually expect from the extensive evidence which has been published, most prominently on nonbusiness people in unfamiliar situations quite unlike business. Many of these situations have also embodied just those things we all know are counterintuitive, and cried out for the use of probabilistic and statistical calculations in the assessment process. Indeed, their purpose has often been to elicit and elucidate natural judgmental biases. The only other "real tasks with experts" noted in a report (Lichtenstein *et al.*, 1982) on the State of the Art to 1980 (which was so thorough as to include the experiment discussed here) were in meteorology and dichotomous medical diagnosis.

This paper is mainly an analysis of the data, preceded by a brief description of the experiment and followed by some discussion and conclusions.

(1) Research support by the Associates of the Harvard Business School and the Second Valencia International Meeting on Bayesian Statistics (held Sept. 1983) is gratefully acknowledged.

2. THE EXPERIMENT

For several years before data-collection began, Joy (Mrs. John W.) Pratt had been doing publicity, advertising, programming, etc. for first one and then both of the moving-picture theaters near Harvard University. Her knowledge was similar to that of those choosing the films, though she was little involved in the final choice. For each scheduled run of a program (movie or double feature), she forecast the gross receipts, this being the naturally available and customarily used measure of attendance.

Her forecasts were in the form of quartiles and ''astonishingly'' high and low values. We shall call all five assessments ''fractiles'', although ''astonishing'' was not interpreted probabilistically, it being already local folk-knowledge that assessments of extreme probabilities or fractiles are difficult to make or interpret exactly. She made each forecast 0-8 days beforehand. To do it much earlier would often have been impossible because the films were not yet selected. She usually learned the actual gross within days. (In the absence of the experiment she might well have learned it less precisely and later.) Thus at the time of each forecast she had typically received feedback from all but 0-4 previous occasions. (Irregularities shouldn't affect calibration because subjective distributions apply given information at the time assessed.) Quartiles were explained to her both as successive equally-likely-division points and as producing four equally likely intervals. She had some but not great difficulty understanding her task, although not particularly trained or talented in any mathematical field. She had little difficulty carrying it out after the first few times. She took into account a great many factors, especially those affecting the substantial student portion of the audience, including day of the week, exams, holidays, vacations, major arts and sports events, the popularity of the directors and actors, and the time and success of previous showings of the films at the two theaters and to some extent at others. Calibration was never mentioned to her, however, and she saw only the list on which she entered the data, reproduced in Appendix 1. Data were collected from March 1962 until a price increase in July 1963, except during summer 1962. The reasons for the gap, good or bad, included her lesser availability then, a quite different audience, more frequent changes of program, and fewer showings per day.

3. OVERVIEW OF THE STATISTICAL ANALYSIS

A Preliminary Look. A graphical display of the first portion of data (made in 1962 in ignorance of Spear's 1952 ''range-bar'' charts; cf. Tufte 1983) is reproduced as Figure 1. It gives a clear impression of initial overconfidence rapidly dispelled. A similar display of the rest of the data shows no obvious further learning, and other simple plots and counts made in 1962-3 indicate somewhat imperfect calibration in a pessimistic direction, about what one would expect given the form of the feedback except that the pessimism is mainly in the upper quartile. Interactive regression done in 1969 confirmed and clarified these features of the data and uncovered the proportionality mentioned in the summary. The analysis presented here is in agreement, but was begun *de novo*, done more carefully, and carried further.

Outline of Analysis. Before examining the behavior of the forecast fractiles and errors, we consider in Section 4 transforming the data, eliminating an initial time-period, and pooling the two theaters. We next discuss the use and usefulness of the assessments in decision making, the median alone in Section 5, all five fractiles in Section 6, allowing but not finding autocorrelated errors. In Section 7 we investigate the statistical relations among the assessments, and in Section 8 their calibration, both marginal and conditional on the information available. Section 9 contains some further discussion and conclusions.

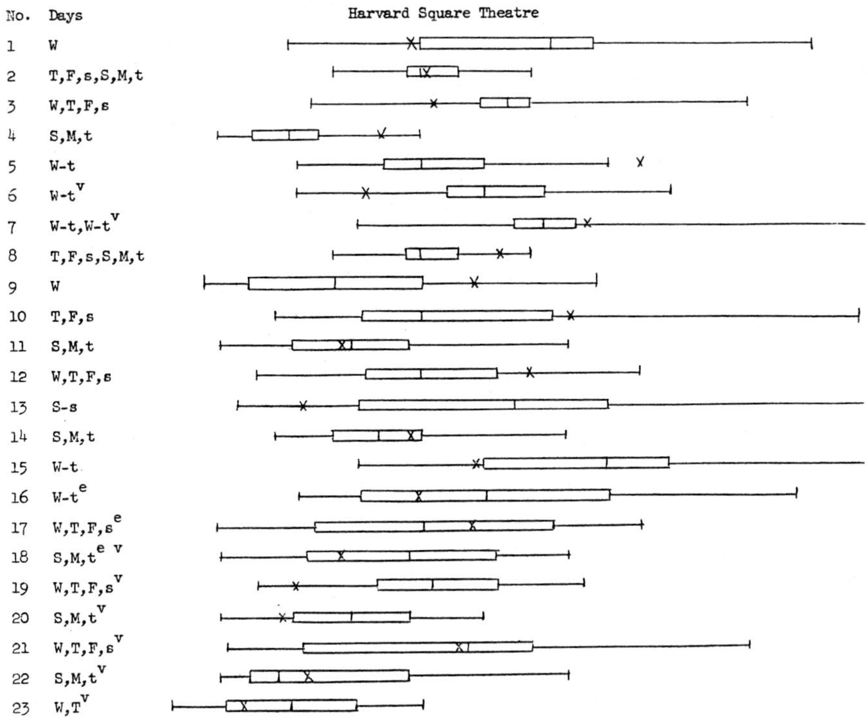

v Vaction periods: s,S,M,t of 6, first W,T,F,s,S of 7, t of 18, all of 19-23

e Examination period: all of 16 and 17, S,M of 18

Note (1984). Data for March-July 1962. Plots for later dates give only the number of days in the column headed "Days". Similar plots were made for the Brattle Theater.

Figure 1

Prior distribution of gross and actual gross, per day basis

Prior:

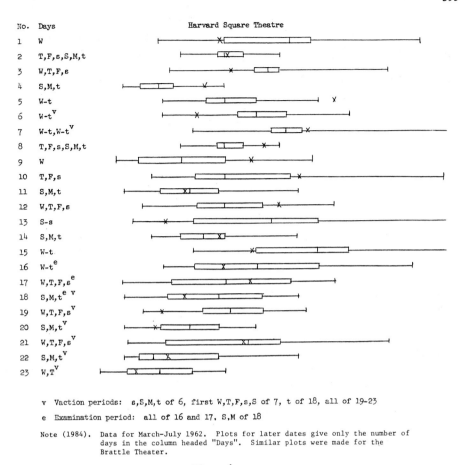

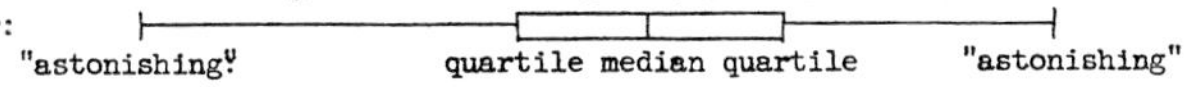

Actual indicated by X.

4. TRANSFORMATION AND SELECTION OF DATA

To keep analysis simple, we want to describe the relations of interest by homoscedastic models which fit as much of the data as possible homogeneously (with constant parameter values). In this section we consider the relation of the actual gross to the assessed median and interquartile range in a preliminary way with the limited objectives of transforming to approximate homoscedasticity and selecting data which can be fit homogeneously by a simple model. The choices we make will be seen later to make most other relations as well close to simply homogeneous and homoscedastic.

Tranformation of Variables. Instead of being (approximately) additive, the assessor's forecast errors (actual gross vs. median) and subjective uncertainty about the popularity of each program might be more nearly multiplicative (proportional). In addition, independent daily variability might contribute to variance in proportion to the length (number of days) of the run, given the popularity of the program. All this suggests that a transformation having curvature somewhere between logarithmic and linear may produce approximate homoscedasticity.

Absolute forecast errors were plotted against the (median) forecast on the natural and logarithmic scales for the two theaters combined, and inspection of the plots shows that whereas absolute errors increase with the forecast, there is little if any heteroscedasticity on the logarithmic scale. Ordinary least-squares (OLS) regressions of absolute errors on forecasts confirm that they are materially correlated on the natural scale (slope = .16 with standard error .02, $P<.0005$; $R^2=.29$) but almost uncorrelated on the logarithmic scale (slope = $-.022$ with standard error .015, $P=.075$; $R^2=.01$).(2) We shall see later that taking logarithms also eliminates almost all heteroscedasticity related to other variables and produces a very close to normal error distribution.

Since a logarithmic transformation of the data will so greatly simplify analysis and also allow simple interpretation in terms of relative error, other transformations seem not worth considering and *we work henceforth entirely in terms of the logarithms of actual gross receipts and all assessments.* Even if the word is omitted, logarithms will be intended unless the contrary is explicitly stated. In particular, "absolute error" is the absolute logarithm of the actual/forecast ratio in natural units, and the many regression slopes of logs on logs are elasticities, ratios of proportional changes in natural units when the changes are small.

Fractile Differences instead of Fractiles. The five assessments are highly correlated with one another, and this leads to great uncertainty about regression coefficients. To reduce the multicollinearity and at the same time facilitate interpretation of the signs of regression coefficients and perhaps follow the assessment process better, we shall replace the four assessments other than the median by "fractile differences" using the notation

$D_{01} = F_{25}-F_{01}$ = first quartile minus lower astonishing point,
$D_{25} = F_{50}-F_{25}$ = median minus first quartile,
$D_{75} = F_{75}-F_{50}$ = third quartile minus median,
$D_{99} = F_{99}-F_{75}$ = upper astonishing point minus third quartile.

Transformation to Daily Basis. One might convert all actual grosses and assessments to a daily basis, but nothing essential in our analyses would change. Because division of natural values by N (the number of days) simply subtracts log N from their logarithms, it does not change differences such as the forecast error and the interquartile range, nor, in regressions like all our main ones that include log N as an independent variable, does it change the coefficients, standard errors, or P-values, except perhaps those of log N itself. The residual standard deviation is also unchanged, but R^2 is typically much larger for the dependent variable gross than it would be for gross per day.

Stability. Figure 1 strongly suggests that learning about the magnitude of forecast errors occurred, but not when it ended; and it is quite possible that the forecasts themselves become more accurate during a learning period, although it would be hard to tell whether this was true by simply inspecting Figure 1. It is obviously desirable to eliminate data for an

(2) Throughout this paper s^2 is the usual (unbiased) estimate of residual variance, but R^2 is the sample fraction of variance explained (or, occasionally, an analog), not "corrected" or "adjusted" for degrees of freedom. $F(m,n)$ has m and n degrees of freedom. All P-values are one-tailed; those for individual regression coefficients are based on t and conversion to $F(1,n)$ would double them.

initial time period which includes essentially all learning so that statistical relations in the remaining data will be stable (temporally homogeneous), but at the same time it is obviously impossible to date the end of the learning period exactly. We shall consider a few possible dividing points, comparing first the before and after regressions of actual on median and then, if this comparison would permit pooling, the regressions of absolute error on median and interquartile range. The two comparisons probe the stability of very simple relations of location and scale respectively. More complicated models could be tried if stability could not be obtained simply, but here it can be.

Table 4-1: Regressions of Gross Receipts on Median Forecast

A: Pre-Summer vs Post-Summer

	con	F_{50}	R^2
Pre-summer	1.607(.451)<.000>	.803(.057)<.000>	.812
Post-summer	.043(.227)<.426>	.997(.029)<.000>	.892
Difference	-1.564(.513)<.001>	.194(.065)<.002>	

B: Last Half Pre-summer vs Post-Summer

	con	F_{50}	R^2
Pre-summer	1.795(.508)<.001>	.773(.065)<.000>	.835
Post-summer	.043(.227)<.426>	.997(.029)<.000>	.892
Difference	-1.752(.669)<.005>	.224(.085)<.005>	

C: First 1/3 Post-Summer vs Last 2/3

	con	F_{50}	R^2
First 1/3	-.467(.658)<.241>	1.051(.082)<.000>	.810
Last 2/3	.040(.234)<.432>	1.002(.031)<.000>	.915
Difference	.507(.630)<.211>	-.049(.080)<.269>	

D: First 1/3 vs Last 2/3, Harvard Program 47 Eliminated

	con	F_{50}	R^2
First 1/3	.313(.626)<.310>	.956(.078)<.000>	.802
Last 2/3	.040(.234)<.432>	1.002(.031)<.000>	.915
Difference	-.273(.644)<.336>	.046(.081)<.287>	

The summer gap is a natural dividing point, and we consider it first. That pre- and post-summer data are *not* homogeneous is shown by Table 4-1A, which gives separate pre- and post-summer regressions and the interaction terms (pre-post differences) in a combined regression, with standard errors in parentheses and one-tail P-values in angle brackets. The regression of actual gross receipts on the median forecast F_{50} has a much smaller constant term and a slope much closer to 1 after the summer than before. The interaction terms contribute only .006 to an R^2 of .881 (that is, dropping them from the combined regression reduces R^2 but they are individually statistically highly significant and together have $F(2,186) = 4.81$, $P = .009$, and these results change very little even when the four observations with influence greater than 3.0 are eliminated from the data. (Roughly, an observation of influence k has the same root mean square effect on the fitted values of y as k "average" observations; thus the influence here is on the full combined regression, not the differences alone.)

Can we use some if not all pre-summer data? As shown by Table 4-1B, the regression of gross on median in the last half (approximately) of the pre-summer data still differs materiallv from the post-summer regression (3.) Indeed, it is a little further than the whole pre-summer regression both from the post-summer regression and from constant 0 and slope 1. It becomes a little closer to both than the whole pre-summer regression if the two observations with influence greater than 3.0 are dropped, but the change is too little to justify pooling.

Is all post-summer data homogeneous? Table 4-1C compares the regression of gross on median in the first third (approximately) of the post-summer data with that in the remaining two thirds (4.) The differences in the coefficients are individually statistically insignificant and together contribute only .004 to an R^2 of .896, but they have $F(2,138) = 2.54$, $P = .082$. Harvard program 47 has influence 9.5, however, and when it is eliminated as in Table 4-1D, the differences become individually still less significant and together contribute only .002 to an R^2 of .900 and give $F(2,137) = 1.53$, $P = .221$.

Regression of absolute error on median and interquartile range (IQR) tells a rather similar story about homogeneity of the post-summer data. The difference between the coefficients of interquartile range is large and statistically significant (Table 4-2A), and the three differences together contribute .053 to an R^2 of .094 and have $F(3,136) = 2.67$, $P = .050$; but Harvard program 47 has influence 8.6 and when it is eliminated as in Table 4-2B the differences become individually insignificant and together contribute only .010 to an R^2 of .030 and have $F(3,135) = .48$, $P = .695$.

Not only is the same observation most influential in both post-summer combined regressions, with the largest residual, high leverage, and unusually high influence; it also has the smallest gross and largest absolute error in all the data. In fact, it was a Christmas-day-only showing of two opera films at the larger and more commercial of the two theaters, and its gross was astonishingly low even though the forecast distribution was low and broad (having the second-highest interquartile range post-summer, sixth highest in all data). Because this maverick observation would heavily influence and confuse many analyses to follow, we eliminate it henceforth (but not the December 25-26 showing of "Jules and Jim" at the Brattle).

(3) In each theater, the first half of the pre-summer data included 50 days of movies, namely the first 9 programs.

(4) The split followed the 200th day of movies in each theater, namely Brattle program 39 and Harvard program 49.

Table 4-2: Absolute Error on Median and Interquartile Range

A: First 1/3 vs Last 2/3 Post-Summer; All Observations

	con	F_{50}	IQR	R^2
First 1/3	-.053(.478)<.456>	.007(.053)<.451>	.835(.318)<.006>	.188
Last 2/3	.115(.169)<.249>	.004(.019)<.420>	.215(.147)<.073>	.022
Difference	.168(.460)<.357>	-.003(.051)<.478>	-.620(.322)<.028>	

B: First 1/3 vs Last 2/3 Post-Summer; Harvard Program 47 Eliminated

	con	F_{50}	IQR	R^2
First 1/3	-.148(.407)<.359>	.034(.046)<.227>	.375(.295)<.106>	.046
Last 2/3	.115(.169)<.249>	.004(.019)<.420>	.215(.147)<.073>	.022
Difference	.264(.439)<.274>	-.031(.049)<.266>	-.160(.328)<.313>	

With this one observation eliminated, simple homogeneous relations appear to hold in the post-summer data for both location and scale, while even the latter half of the pre-summer data appear different. The data are not adequate to date the end of instability more precisely, and the summer is a natural break-point. Accordingly, *all analyses in the remainder of this paper will be based on the 141 post-summer observations, excluding the Christmas-day opera program at the Harvard Square Theater.*

Pooling Theaters. Because the two theaters differ in size and type, they might also differ in the relations between actual and assessed gross, but no such difference appears in the regressions of log actual on log median in Table 4-3A. The differences in intercepts and slopes for the two theaters are less than 1/3 of their standard errors and together they contribute only .0005 to an R^2 of .898 and have $F(2,137) = .32$, $P = .73$. In the regressions of absolute error on log median and interquartile range in Table 4-3B, the two coefficients differ negligibly between theaters. The coefficient of interquartile range is larger and more appropriate at the Harvard than the Brattle, but the difference is statistically not definitive ($P = .125$), and all three differences together contribute .041 to an R^2 of .060 and have $F(3,135) = 1.94$, $P = .125$. Eliminating the three observations with influence greater than 3.0 would leave all differences in coefficients with $P \geq .40$, a combined contribution of .025 to an R^2 of .030, and $F(3,132) = 1.11$, $P = .346$, but for better or worse we shall not eliminate these observations.

It thus appears that both theaters can be described by a simple homogeneous model when the median and interquartile range are considered. When the first and third quartiles are considered separately, however, there is evidence that the relations are no longer homogeneous. This is curious because it means that the heterogeneity of the theaters with respect to the individual quartiles essentially cancels out for the interquartile range. We show in Table 4-3C the regression of absolute error on the quartile-median differences D_{25} and D_{75} and the median F_{50}, omitting the constant which differs negligibly between theaters ($P = .44$). The coefficients of D_{25} and D_{75} each differ significantly between theaters, with $F(2,133) = 4.09$, $P = .019$. (Dropping influential observations can increase or decrease the differences, depending on the cutoff level used, but dropping all 6

Table 4-3: Brattle vs Harvard Theaters, Post-Summer

A: Gross Receipts on Median Forecast

	con	F_{50}	R^2
Brattle	.294(.279)<.148>	.962(.038)<.000>	.894
Harvard	.174(.505)<.366>	.983(.061)<.000>	.806
Difference	-.120(.548)<.413>	.021(.070)<.383>	

B: Absolute Error on Median and Interquartile Range

	con	F_{50}	IQR	R^2
Brattle	.307(.202)<.067>	-.016(.023)<.243>	.056(.140)<.345>	.017
Harvard	.291(.319)<.183>	-.019(.036)<.295>	.356(.232)<.065>	.048
Difference	-.016(.366)<.483>	-.003(.041)<.471>	.300(.260)<.125>	

C: Absolute Error on Median and Quartile Differences

	F_{50}	D_{25}	D_{75}	R^2
Brattle	-.012(.024)<.300>	.254(.241)<.148>	-.211(.299)<.242>	.031
Harvard	-.017(.033)<.303>	-.518(.352)<.073>	.865(.270)<.001>	.183
Difference	-.005(.040)<.452>	-.772(.416)<.033>	1.076(.420)<.006>	

observations with influence above the very low cutoff 2.5 leaves $F(2,127)=3.34$, $P=.039$, and coefficients significant at $P=.014$ and .028.) It is true that the negative coefficients of D_{75} at the Brattle and D_{25} at the Harvard are contrary to intuition with uncompelling P-values, and it is tempting to assume that in reality both have small coefficients at both theaters and thus homogeneity holds. According to the evidence, however, the two differences between theaters are very probably real. Although they are very uncertain and might be small, they are probably material and could well be large.

Since this difference between theaters is strange and unilluminating and the other differences we have found are at most small relative to the effects and disturbances in the relations we examine, and since pooling the theaters effectively doubles the available data and halves the statistical displays, we shall generally continue to pool the two theaters henceforth.

5. USE OF ASSESSED MEDIAN BY A DECISION MAKER

We are now ready to investigate the behavior of the assessments as a stochastic process. This is of interest from two points of view: the psychology of judgmental probability, and the use of assessments of this kind by a decision maker other than the assessor. Effects may be scientifically interesting even if they are too small to matter practically, and in due course we shall seek all we can find. First, however, we shall forget psychology and treat the assessments like any other data a decision maker might use. In this section, we ask how much can be learned from the assessed median alone; it presumably represents the kind of point forecast one might ordinarily obtain or seek. The next section examines how much the remaining assessments add and whether the more elaborate game would be worth the candle in practice. In both sections we discuss the choice and estimation of regression models on which the decision maker's forecast distribution might be based, but we omit the standard Bayesian derivation of a forecast distribution given a model and we do not consider weighting alternative models Bayesianly rather than choosing just one. As we have already said, we work entirely with logarithms and with the 141 observations on the two theaters that remain after elimination of all pre-summer data and Harvard program 47.

Simple Adjustment of the Median. The simplest approach would be to treat the forecast errors as independently, identically distributed, find their sample mean (.027) and standard deviation (.278), and examine the shape of their sample distribution. For comparability with later regressions, the estimates are described in the first part of Table 5-1 as a regression of gross on median with the coefficient of median constrained to be 1; its mean squared error is 90% smaller than the variance of log gross. The empirical cumulative distribution of the residuals (forecast errors −.027) is close to normal with mean 0 and standard deviation .278, differing from it by at most .080 (the Kolmogorov-Smirnov statistic; $P=.026$ roughly adjusted for fitting as in Lilliefors, 1967). When graphed as in Figure 2, the empirical distribution appears well fitted by the normal except for a suggestion of bimodality, as if two distributions with different means were mixed in the data. Theater is the dichotomy it is natural to suspect, but the difference between the means in the two theaters is far too small to provide an explanation, being less than 8% of

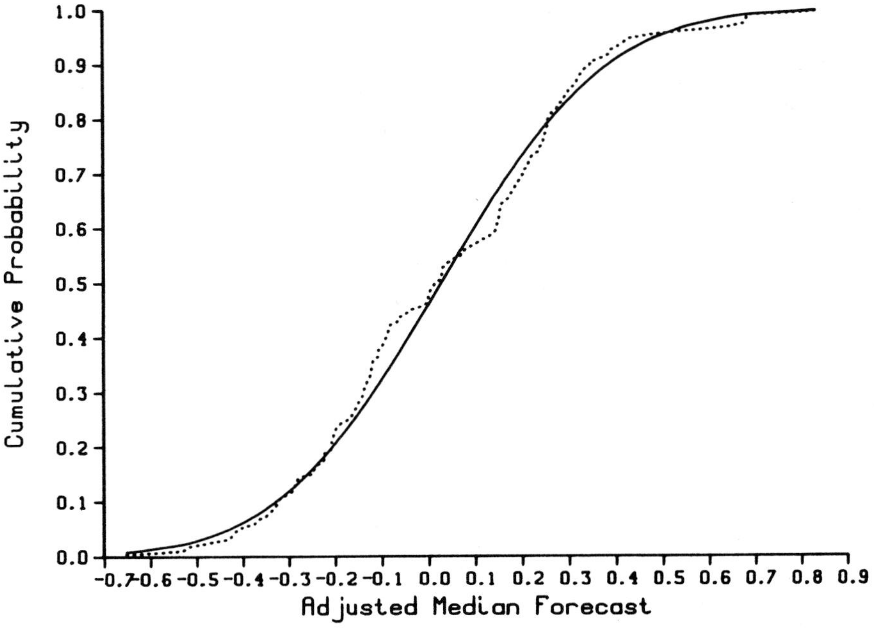

Figure 2

the standard deviation within either theater; and since differences in standard deviation could not produce bimodality, it seems very likely that the observed bimodality reflects nothing but sampling error.

Regression on Median and Length of Run. The decision maker might expect to improve upon simple adjustment by adopting a regression model in which the mean and standard deviation of the gross are functions of the assessor's median and the length of run, and he might also ask whether the errors are autocorrelated or otherwise related. We shall first derive an estimate of the mean and examine autocorrelation by OLS regression, then derive an estimate of the standard deviation by a similar regression, then derive a revised estimate of the mean by WLS regression, and finally examine the normality of the residuals.

OLS Estimates of Mean and Residual SD. OLS regression of G = log(gross) on F_{50} = log(median) and L = log N = log(length of run) is shown second in Table 5-1. The difference between the full regression and simple adjustment of the median is completely negligible, increasing R^2 by only .0007 and having $F(2,138) = .56$, $P = .58$, but we shall nevertheless use the full regression, partly because its coefficients are not unreasonable and an uncertain estimate seems better than none, but primarily to discover how much difference more refined analysis can make.

No available variables other than L are convenient to use in conjunction with F_{50}. One could construct other variables from the available information such as day of the week, but like L they are already reflected judgmentally in the median. Since they would be very hard to construct and fit appropriately into a model and seem less likely than L to be informative or better handled by statistics than judgment, and since even L added negligibly to the median, we shall not pursue this line here.

When first-order autocorrelation of the residuals is allowed, the standard Cochrane-Orcutt estimate is only .035 and R^2 increases by only .0001. Again it appears that the complications far outweigh the likely gains of considering other possible relations among residuals, and we shall not do so.

Heteroscedasticity. If there were appreciable heteroscedasticity related to the median or the length of run, we would expect it to be monotonic and hence to see at least some reflection of it in a linear regression of the absolute residual A on these variables. The OLS

Table 5-1: Regressions on Median and Length of Run

	con	F_{50}	L	s	R^2
G (ADJ)	.027 (.023)	1	0	.278	.898
G (OLS)	.416 (.371) <.132>	.939 (.058) <.000>	.069 (.076) <.185>	.279	.898
A (OLS)	.005 (.209) <.491>	.037 (.033) <.131>	−.051 (.043) <.119>	.158	.010
G (WLS)	.335 (.368) <.182>	.953 (.058) <.000>	.045 (.076) <.278>	1.22$\hat{A}$	.897

regression of A on F_{50} and L, shown third in Table 5-1, has only $R^2 = .010$, $F(2,138) = .72$, $P = .49$, and the sample standard deviation of the predicted absolute residual is only .022. (Indeed similar results in the previous section were the major reason for taking logarithms.) Nevertheless the predicted absolute residual runs from .20 to .26 over the range of the data, so it is sometimes appreciably different from its mean (.23); and again we use the full regression partly because the coefficients are not unreasonable (indeed that of L looks "right"), much more because we want something close to an upper bound on the effect of refinement.

For those who would routinely weight by N (scale by $\sqrt{N}$) after taking logarithms in situations like ours, we note that if the model fit perfectly with only independent and homoscedastic daily disturbances, the residual variance would be approximately proportional to $1/N$, justifying their weighting; but the regression of the OLS squared residual on $1/N$ is against them. It is $.072 + .010/N$, so that even for $N = 1$ the estimated daily variance is less than 1/8 of the total, and R^2 is tiny (.0007). Although the estimates have relatively large standard errors (.015 and .033 respectively), they are incompatible with proportionality to $1/N$.

WLS Estimates of Mean and Residual SD. The heteroscedasticity means that more efficient fitting of the mean is theoretically possible. Although the difference here is minor, Table 5-1 shows finally an appropriate WLS regression where the scale factor is the predicted value $\hat{A}$ of the absolute residual obtained in the way described just above. A WLS residual divided by the associated scale factor $\hat{A}$ will be called a scaled residual.

The account taken of heteroscedasticity in the WLS regression has brought us even closer than before to simple adjustment of F_{50}, and the difference is even less significant: $F(2,138) = .36$, $P = .70$. The revision of the formula for the mean would in principle also call for a revision of the formula for the variance, etc., but the effort in refinement has long since passed the point of diminutive returns.

For the record, we report that the standard deviation of the scaled Brattle residuals is about 10% smaller and that of the Harvard residuals about 10% larger than the over-all standard deviation, and the means are about one-tenth of a standard deviation below 0 for the Brattle, above for the Harvard. We have already said that we would ignore theater differences in our analysis.

Normality of Residuals. The empirical cumulative distribution of the scaled residuals of the WLS regression is very close to normal with mean 0 and standard deviation 1.22, differing from it by at most .061 ($P = .22$ adjusted as before for fitting a constant mean and standard deviation, but not adjusted for further fitting). The graphical fit, in Figure 3, again seems very good except for observed bimodality not attributable to theaters, the means within theaters differing this time by about 21% of the smaller standard deviation.

Conclusion. Using only the assessed median and the length of run, a decision maker might determine his forecast distribution of log gross from a heteroscedastic normal linear regression model with mean and *standard deviation* estimated by WLS in accordance with Table 5-1, but the regression refinements will affect him little and buy him almost nothing. Simple adjustment (ADJ) and OLS differ negligibly in estimated root mean squared error (= standard deviation), and they have almost the same performance not only on average but even on almost all occasions. The first line of Table 5-2 shows the minimum, maximum, mean, and standard deviation of the difference in estimated mean for the 141 observations in our data; the standard deviations estimated by simple adjustment and OLS are constant and differ by less than .5%.

By definition, WLS cannot reduce the mean squared error of OLS on the same data, or OLS the weighted mean squared error of WLS. But again we can see from Table 5-2

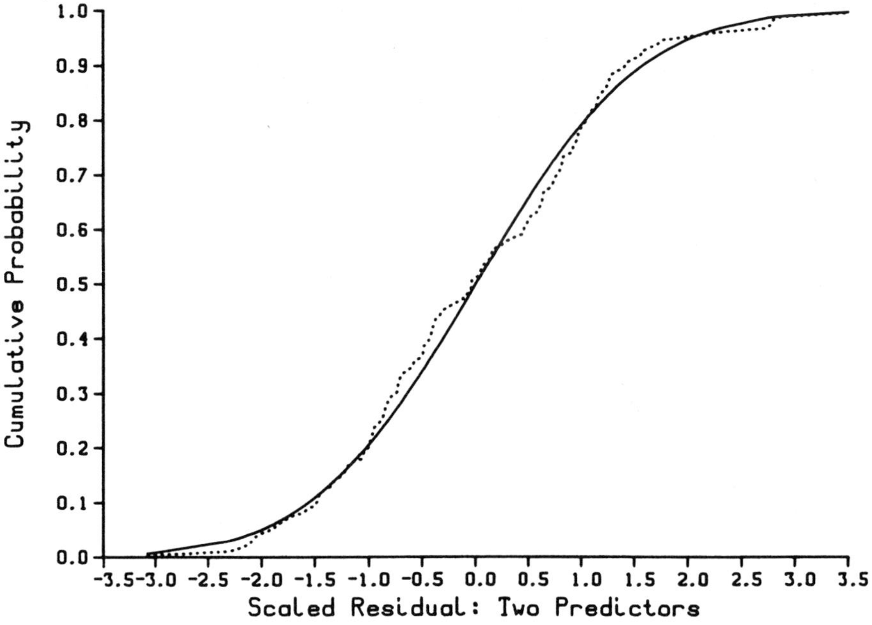

Figure 3

Table 5-2: Differences between Estimated Regressions

	min	max	mean	sd
Mean				
$OLS(F_{50},L)-ADJ(F_{50})$	-.053	+.047	.000	+.025
$WLS(F_{50},L)-OLS(F_{50},L)$	-.014	+.018	.000	+.008
$WLS(F_{50},L)-ADJ(F_{50})$	-.045	+.037	.000	+.020
Residual SD				
$WLS(F_{50},L)-OLS(F_{50},L)$	-.036	+.039	-.001	+.019
$WLS(F_{50},L)-ADJ(F_{50})$	-.035	+.039	-.001	+.019

that WLS rarely differs materially from either OLS or simple adjustment as regards either the estimated mean or the estimated standard deviation. The importance of accuracy about spread of course varies greatly from one decision problem to another, but is usually minor compared to accuracy about the mean.

6. USE OF OTHER ASSESSMENTS BY A DECISION MAKER

We now investigate what the decision maker can gain by bringing in the four assessments other than the median, or equivalently, the four fractile differences defined in Section 4. Since our approach will be the same as in Section 5 and the results are not startling, we shall abbreviate the exposition.

Analysis. The OLS regression of G on all assessments and L appears in Table 6-1. The extra assessments add only .011 to the $R^2 = .898$ of the regression of G on F_{50} and L alone (Table 5-1), a fact which is not surprising since the assessed median is attempting to capture all information relevant to the location of a practically symmetric distribution. All signs are "right", however, and to obtain an upper bound we again assume that the decision maker uses all available variables.

The estimated first-order autocorrelation of the residuals is only −.046, and taking account of it increases R^2 by only .0002; and therefore we again ignore intertemporal dependence.

Information about heteroscedasticity should be provided by the four assessments other than the median, but regression of the absolute residual A of the regression of G on all assessments and L indicates that little is. The full regression shown in Table 6-1 has one statistically highly significant coefficient, that of D_{75}, but overall only $R^2 = .060$, $F(6,134) = 1.44$, $P = .205$. The coefficients of the other D's have "wrong" signs and dropping them reduces R^2 by only .014, with $F(4,134) = .65$, $P = .59$. We nevertheless use the full regression for consistency with our treatment of the mean and because we want something close to an upper bound on the effect the extra assessments could have on the decision maker's distribution. For the record we mention that $1/N$ explains an even smaller and statistically less significant fraction of the squared residuals here than in Section 5.

Finally we scale by the predicted absolute residual given by the regression in Table 6-1 and obtain the WLS regression also shown there.

Table 6-1: Regressions on all Predictors

	con	F_{50}	D_{01}	D_{25}	D_{75}	D_{99}	L	s	R^2
G	.362	.950	-.148	-.620	1.157	.033	.017	.268	.909
(OLS)	(.429)	(.060)	(.095)	(.362)	(.346)	(.155)	(.080)		
	<.200>	<.000>	<.061>	<.045>	<.001>	<.416>	<.415>		
A	.121	.015	-.011	-.246	.488	-.027	-.032	.139	.060
(OLS)	(.222)	(.031)	(.049)	(.188)	(.179)	(.081)	(.042)		
	<.294>	<.314>	<.415>	<.096>	<.004>	<.371>	<.224>		
G	.433	.945	-.118	-.657	1.095	-.034	.020	1.21$\hat{A}$	.914
(WLS)	(.408)	(.058)	(.093)	(.344)	(.387)	(.143)	(.078)		
	<.146>	<.000>	<.102>	<.029>	<.003>	<.406>	<.398>		

The difference between theaters, which we are ignoring, is very small over all, in the same direction as in Section 5 but less than half as great. Separating the theaters might nevertheless increase the measurable heteroscedasticity appreciably because of the peculiar interactions mentioned at the end of Section 4, but we have not investigated this.

The empirical distribution of the scaled residuals and a normal distribution with the same mean (0) and standard deviation (1.21) are plotted in Figure 4. The difference is at most .077 (partly adjusted $P = .038$).

Conclusion. With all assessment available, a decision maker might determine his forecast distribution of log gross from a heteroscedastic normal linear regression model

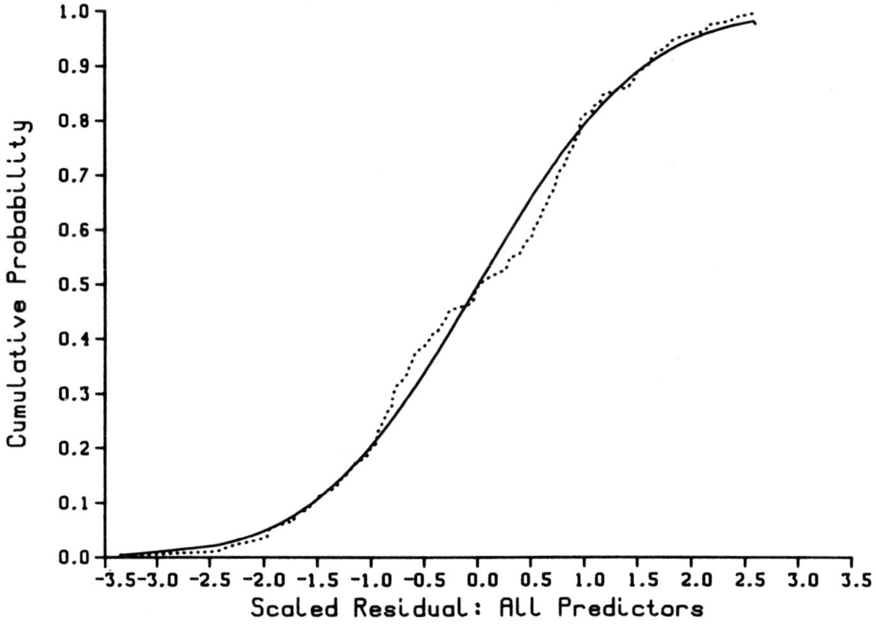

Figure 4

with mean and standard deviation determined by the WLS regression. The reduction in root mean square error bought by the extra assessments is virtually nil, however, and as shown by Table 6-2 it is only occasionally that a mean and standard deviation so calculated would differ materially from their counterparts based only on F_{50} and L, or even from simple adjustment of F_{50} alone. Nevertheless, these differences, though small, are much larger than the differences between F_{50} alone and F_{50} with L or between OLS and WLS applied to the same assessments.

Table 6-2: Differences between Estimated Regressions

	min	max	mean	sd
Mean				
WLS(all)-OLS(all)	-.037	+.043	-.000	+.013
WLS(all)-WLS(F_{50},L)	-.164	+.135	+.000	+.089
WLS(all)-ADJ(F_{50})	-.195	+.298	-.000	+.086
Residual SD				
WLS(all)-OLS(all)	-.076	+.161	-.000	+.042
WLS(all)-WLS(F_{50},L)	-.111	+.148	-.010	+.042
WLS(all)-ADJ(F_{50})	-.086	+.151	-.010	+.042

7. RELATIONS AMONG ASSESSED FRACTILES

Having examined the value to a decision maker of the expert's assessments, we now turn to the stochastic behavior and calibration of the assessments as a mental process, without concern for practical importance. In this section we investigate how assessed fractiles vary in relation to one another and length of run, regardless of the actual gross. Their variation's informativeness about the actual gross was the subject of Sections 5 and 6; calibration will be examined in Section 8.

If one thinks of errors proportionally, one might assess fractiles whose ratios are always the same on the natural scale and whose differences are therefore constant on the logarithmic scale. Once these ratios are known, all information is available from the median. If the other fractiles can be quite accurately predicted from the median (linearly or otherwise), then although the deviations from prediction could in principle be extremely informative and/or well calibrated, in practice they may for example reflect little more than rounding of assessments to multiples of 5 or 10.

In our data, the differences among the fractiles are much less variable than the fractiles themselves, as shown by Table 7-1, where F_{25} and F_{75} denote the first and third quartiles and F_{01} and F_{99} the astonishing points. The variance of D_{01}, for example, is only .0751 times the variance of F_{01}. As we have already said, a large common component of the variation of the fractiles is eliminated by replacing all but F_{50} by fractile differences.

Quartiles. Regressions of the differences D_{25} and D_{75} on F_{50} and L (Table 7-2) explain about 23 % of the variance of D_{25} but only about 5 % of that of D_{75}, and the coefficient of F_{50} is extremely small. Given the median, the quartiles move closer to the median as L increases, consistent with some less-than-proportional variability arising from evening out daily fluctuations. Given L, the quartiles do not materially change distance from the median as the median increases.

Table 7-1: Ratios of Variances and SD's

	$D_{01}:F_{01}$	$D_{25}:F_{25}$	$D_{75}:F_{75}$	$D_{99}:F_{99}$
variances	.0759	.0071	.0070	.0508
sd's	.2755	.0841	.0835	.2254

Table 7-2: Regressions of Quartile Differences on F_{50} and L

	con	F_{50}	L	s	R^2
D_{25}	.232	.004	-.059	.065	.231
	(.086)	(.013)	(.018)		
	<.004>	<.391>	<.001>		
D_{75}	.220	-.003	-.020	.068	.045
	(.090)	(.014)	(.019)		
	<.008>	<.426>	<.146>		

The fact that the residual standard deviations of the regressions of D_{25} and D_{75} on F_{50} and L are only .065 and .068 respectively whereas the marginal means are +.189 and +.175 shows that any factors other than median and length of run that were taken into account in assessing their values seemed of very little importance to the assessor. Quite possibly, the residual variation represents nothing but noise, part if not all of which is round-off.

One might also ask whether D_{25} and D_{75} are related to one another except through F_{50} and L, and a natural measure of this relation is their partial correlation, with F_{50} and L partialed out. This correlation is only +.19, indicating that the factors other than L which most influenced D_{25} and D_{75} were not common, and thus tending to confirm the view that the residual variation in the regressions of D_{25} and D_{75} on F_{50} and L may be mostly noise. (However this may be, the correlation shows that adding either of the differences to the regression of the other on F_{50} and L would reduce the residual variance by only 4%.)

Are D_{25} and D_{75} equivalent measures of nondirectional uncertainty? The standard deviation of $D_{25}/(D_{25}+D_{75})$ is only .107, indicating not much variability in the apportionment of uncertainty upward and downward. On the other hand, what variability D_{25} and D_{75} have is largely unrelated: their marginal correlation is only .27, and the part of their variability not explained by their relations to F_{50} and L has only the partial correlation .19 just mentioned.

Lower Astonishing Point. As shown by the first of the two regressions in Table 7-3, the relation of the lower fractile difference D_{01} to F_{50}, D_{25}, D_{75}, and L is much like that of D_{25} and D_{75} to F_{50} and L except that the residual standard deviation is nearly four times as great. About one quarter of the variance of D_{01} is explained.

Upper Astonishing Point. As also shown in Table 7-3, regression on F_{50}, D_{25}, D_{75}, and L explains about the same fraction of the variance of D_{99} as of the variance of D_{01}, but the regressions have strikingly different coefficients. Whereas D_{01}, D_{25}, and D_{75} decrease as L increases, D_{99} increases as L increases and decreases as F_{50} increases. The residual standard deviation of D_{99} is somewhat over half that of D_{01} but still more than twice that of D_{25} and D_{75} in Table 7-2. The coefficient of partial correlation between D_{01} and D_{99} with all other predictors partialed out is +.18, just about equal to the coefficient of partial correlation +.19 between D_{25} and D_{75} with F_{50} and L partialed out. The standard deviation of $D_{01}/(D_{01}+D_{99})$ is .118 and the marginal correlation of D_{01} and D_{99} is .31: they are nondirectional measures of uncertainty to about the same degree as D_{25} and D_{75} are.

Table 7-3: Regressions of Extreme Differences on all other Predictors

	con	F_{50}	D_{25}	D_{75}	L	s	R^2
D_{01}	.615	.011	.344	.035	-.218	.246	.266
	(.340)	(.051)	(.331)	(.314)	(.070)		
	<.036>	<.417>	<.150>	<.456>	<.001>		
D_{99}	1.408	-.140	-.070	.294	.066	.150	.254
	(.208)	(.031)	(.202)	(.192)	(.043)		
	<.000>	<.000>	<.366>	<.064>	<.062>		

Effect of Theater Capacity. The explanation for the special dependence of the upper astonishing point on the median may lie in the influence of theater capacity. When demand for a particular showing of a program exceeds capacity, some of the excess may be displaced to other showings, but much of it will be lost. This would naturally affect F_{99} far more than the other assessments, and would reduce it when the median is high relative to the length of run. When assessments are converted to a per-day basis, the coefficients in Table 7-3 remain the same except for the coefficient of L, which for D_{99} becomes negative ($-.074$ with standard error .023): the upper astonishing point (like the lower) comes closer to the median as the length of run increases and daily fluctuations average out.

Unequal and Extreme Odds. Another rather peculiar property of the assessments is the fact that the variation in D_{01} and D_{99} that is not explained by all the other predictors is much greater than the variation in D_{25} and D_{75} that is not explained by F_{50} and L. This could be because an astonishing point is not quantitatively defined or because of its extremeness *per se*. We tentatively suggest another explanation. A good deal of folk knowledge supports the notion that odds of 1 to 1 or probabilities equal to .5 are much easier to internalize than other odds or probabilities. Nonacademic evidence in support of this proposition can be found in the fact that the bets on winners at unequal odds which were common years ago have where possible been largely superseded by even-money bets on point spreads. Assessment of a median calls for definition of two events with equal odds, and quartiles at least *can* be assessed as conditional medians, but such an approach becomes more difficult for octiles, sextodeciles, etc., and impossible for "astonishing points" or .1 or .01 fractiles.

8. CALIBRATION OF THE ASSESSED QUARTILES

Finally we examine the calibration of the three assessed quartiles, comparing an estimate of the actual probability that gross receipts will be less than each quartile with the probability that defines the quartile. The actual probability may be either marginal or conditional on specified information, i.e., on the values of a specified set of variables, and therefore we shall distinguish between marginal and conditional calibration. One actual gross was exactly equal to an assessed quartile (the median). We count it as smaller in our computations, but for simplicity our discussion ignores the possibility of such equalities.

Marginal Calibration. The marginal fractions of the 141 observed grosses falling below the five assessed "fractiles", exhibited in Table 8-1, show excellent calibration of the first two quartiles but distinctly too few observations below the upper quartile (fraction .631 with standard error .041). The assessor may have hesitated to be optimistic in the optimistic half of the distribution, but the upper astonishing point shows no such hesitation, and we would have expected it more below the median. Increasing the upper quartile by 0.06 on every observation (6% on the natural scale) would correct its marginal calibration. Its calibration was worse in the more familiar theater, but the theaters differed in fraction below by only .046 with standard error .081. Including Harvard program 47, whose actual gross was less than F_{01}, would multiply the fraction *above* each fractile by 141/142.

Table 8-1: Marginal Calibration of Assessments

	F_{01}	F_{25}	F_{50}	F_{75}	F_{99}
Fraction below	.000	.262	.482	.631	.972

Conditional Calibration of the Median. We shall say that the median is conditionally well calibrated with respect to variables x if given any value of x whatever, the long-run relative frequency of actual values below the median would be .5. The most obvious way to investigate this is by binary regression with a dichotomous dependent variable for above/below the median, but one might hope to obtain a more powerful procedure by using the actual values of G rather than reducing them to two categories. Since the median uniquely minimizes the expected absolute deviation, least-absolutes (LA) regression can be used to estimate the conditional median. OLS gives the same slope coefficients as LA except for sampling and model error if the difference between the conditional mean and median of G given the predictors is constant; it is more powerful unless the error distribution is very high-tailed; and it provides standard errors unavailable by LA. Thus LA and OLS complement one another and binary regression. We ignore constant terms because they relate primarily to marginal calibration which we have already examined directly.

We use a probit model for binary regression because it agrees with a normal regression model. (Logit results are very similar throughout except for R^2, as are results with Harvard program 47 included.) For comparability and interpretability of the slope coefficients, *in this section we standardize all independent and dependent variables* in LA and OLS regressions and scale the probit coefficients correspondingly (5.)

Ideally we would examine the conditional calibration of the median with respect to all information available to the assessor. Most of what is most relevant in the available information is reflected in the median itself, and some more may be reflected in the other assessed fractiles. There may, however, be relevant information not captured by the assessments, at least in the way we shall bring them into our models. Since log length of run L is the most promising and only convenient other information available, we shall use only L in addition to the five assessments and inquire whether the median is conditionally well calibrated with respect to the five assessments and L, or can be improved by using them somehow.

Table 8-2 shows the probit, LA, and OLS regression slopes of $G-F_{50}$ on these

Table 8-2: Standardized Regressions of $G-F_{50}$ on all Predictors

	F_{50}	D_{01}	D_{25}	D_{75}	D_{99}	L	R^2
PRB	-.406	-.184	-.014	.172	.039	.325	.100
	(.228)	(.128)	(.121)	(.111)	(.119)	(.231)	
	<.037>	<.075>	<.452>	<.060>	<.371>	<.080>	
LA	-.311	-.204	-.090	.372	.121	.098	.118
OLS	-.149	-.150	-.163	.287	.020	.040	.111
	(.181)	(.097)	(.095)	(.086)	(.096)	(.184)	
	<.206>	<.061>	<.045>	<.001>	<.416>	<.415>	

(5) Thus our OLS slopes are so-called "beta" coefficients. In our probit analyses, standard errors and P-values are based on asymptotic theory of unknown accuracy here; $R^2 = V/(V+1)$ where V is the sample variance of the linearly fitted values and hence is the explained variance of the latent continuous dependent variable y with unexplained variance 1 underlying the probit model; and the coefficients of the probit regression on standardized predictors are divided by $\sqrt{(V+1)}$, the estimated marginal standard deviation of y.

variables. Perfect calibration would imply that except for sampling error, the slopes would all be 0 (under the additional assumption mentioned for OLS). The regressions give evidence of a statistically significant but weak relation: OLS has $F(6,134)=2.78$, $P=.014$, but by all three methods R^2 is only about .10. (The average absolute deviation of the LA regression is 94% of that of $G-F_{50}$, and one minus the square of this is shown as R^2 in Table 8-2.) The slopes obtained by the three methods are in reasonable agreement in relation to their standard errors. The slopes of F_{50} and D_{75} are generally largest in size and those of F_{50} are negative, suggesting that variations in F_{50} given the shape of the assessed distribution should be somewhat discounted but that D_{75} contains some information not fully incorporated in F_{50} by the assessor.

Conditional Calibration of the Lower Quartile. The lower quartile is of course the median of the lower half of the distribution, that is, the conditional median given that the actual gross is below its median. To investigate its conditional calibration we may therefore apply the methods used above for the median to the 68 observations where the actual gross was below the assessed median, although the coefficients (and power) of OLS are affected by the skewness of the lower half of the distribution. If the ratio of conditional mean to median of $G-F_{50}$ given the predictors and given $G<F_{50}$ is a constant k (as when the conditional distribution has always the same shape), then since D_{25} is among the predictors, the mean of $G-F_{25}$ is k times the median plus $(1-k)D_{25}$, and except for sampling fluctuation the OLS (mean) regression will equal k times the LA (median) regression plus $(1-k)D_{25}$, or simply $(1-k)D_{25}$ with all other slopes 0 in the case of perfect calibration. Any $k>.5$ is possible; $k\geq 1$ for unimodal distributions and $k=1$ for the rectangular. For the half-normal, $k=1.183$ and hence the expected OLS regression under perfect calibration is $-.183D_{25}$.

Table 8-3 shows the probit, LA, and OLS regressions of $G-F_{25}$ on all assessments and L in the data where $G\leq F_{50}$. In the probit regression, the differences of the coefficients from the value 0 implied by perfect calibration show little statistical significance, but there is some indication of imperfect conditional calibration ($R^2=.19$) and the coefficient of D_{25} has $P=.02$. The effects of D_{25} and D_{01} appear about 60% larger in the LA regression and suggest that the assessed variation in D_{25} is too great, that is, when F_{25} is low relative to F_{50} (and F_{01}), the down-side uncertainty is overstated. This may reflect the perversity of D_{25} at the Harvard Square Theater shown in Table 4-3C and discussed in Section 4. The estimated miscalibration is less by OLS than by the other methods. R^2 is smaller, and the coefficient of D_{25}, .332, is smaller in relation to the LA coefficient than the value $1.183\times.509\ \ -.183=.419$ expected under normality although substantially and

Table 8-3: Standardized Regressions of $G-F_{25}$ given $G\leq F_{50}$

	F_{50}	D_{01}	D_{25}	D_{75}	D_{99}	L	R^2
PRB	.022	-.203	.318	-.029	-.099	-.263	.186
	(.306)	(.174)	(.157)	(.148)	(.170)	(.304)	
	<.471>	<.122>	<.021>	<.422>	<.280>	<.194>	
LA	-.300	-.300	.509	.080	.004	.017	.266
OLS	.015	-.117	.332	.059	.000	-.096	.135
	(.252)	(.143)	(.135)	(.125)	(.144)	(.252)	
	<.476>	<.209>	<.009>	<.320>	<.499>	<.352>	

statistically significantly greater than $-.183$ ($P<.001$) and even than 0 ($P=.009$). The other coefficients are also generally smaller in magnitude by OLS than by the other methods. This is somewhat surprising and could be just sampling fluctuation, but all OLS coefficients together have only $F(6,61)=1.58$, $P=.168$, and the difference of the OLS regression from $-.183D_{25}$ has $F(6,61)=2.40$, $P=.038$.

Conditional Calibration of the Upper Quartile. Proceeding exactly as we did in discussing conditional calibration of the lower quartile except that now we expect the LA regression to be k times the OLS regression plus $(k-1)D_{75}$, we show in Table 8-4 the probit, LA, and OLS regressions of $G-F_{75}$ on all predictors in the 73 observations where $G>F_{50}$. The evidence of imperfect calibration is about as strong as for the lower quartile, with R^2 in the same range, and the three regressions are generally in close agreement, with one surprising result. The negative coefficient of D_{25} is far larger in magnitude than that of any other fractile difference and is the only one that is statistically significant or even any where near significance. Thus the over-variability of the assessment of down-side uncertainty seems to apply to up-side as well as down-side outcomes. The coefficients of F_{50} and L are questionable by OLS but less so by probit and LA. The difference of the OLS regression from 0 has $F(6,66)=2.13$, $P=.061$; the difference from $.183D_{75}$ has $F(6,66)=2.34$, $P=.042$.

Table 8-4: Standardized Regressions of $G-F_{75}$ given $G>F_{50}$

	F_{50}	D_{01}	D_{25}	D_{75}	D_{99}	L	R^2
PRB	.156	.114	-.388	-.043	-.112	-.298	.137
	(.377)	(.187)	(.182)	(.164)	(.178)	(.396)	
	<.339>	<.271>	<.016>	<.397>	<.264>	<.225>	
LA	.230	.125	-.367	.004	-.027	-.303	.149
OLS	.391	.081	-.425	.024	-.069	-.525	.162
	(.283)	(.137)	(.138)	(.121)	(.131)	(.288)	
	<.086>	<.278>	<.002>	<.421>	<.301>	<.037>	

Other Methods of Analysis. Two methods of using all the data to investigate the conditional calibration of the quartiles may be worth mentioning. First, one can do binary regression for the probability that $G>F_{25}$ or $G>F_{75}$ in all data, perfect calibration implying probability .75 or .25 given any set of values of the independent variables and hence 0 slopes again. The results for probit regression do not differ greatly from the probit results in Tables 8-3 and 8-4. Second, one could generalize LA regression to the criterion whose expectation is minimized by the k'th fractile, namely k times the absolute error for underestimates, $1-k$ times the absolute error for overestimates. We have no program to do this.

9. DISCUSSION AND CONCLUSIONS

Our main analysis is confined to a stable period of sufficient duration so that, especially by combining the two theaters, we have a substantial amount of data bearing on the questions we ask. Even this amount of data could not provide firm evidence on many details, because the independent variables varied too little in the relevant directions, but it follows that these details have little practical importance in this particular context. To summarize our main conclusions:

1. After the initial learning period, the lower quartile and median had excellent marginal calibration but the upper quartile was 6% too low. Their ratios varied little but the median and the ratio of the lower quartile to the median were nevertheless somewhat over-variable and thus imperfectly calibrated conditionally, at least with respect to the most salient available information.

2. Once enough data were available for even rough calibration, a forecast distribution derived in the usual way from the OLS regression of log actual on log forecast and log length of run could not be materially improved by use of additional probabilistic assessments or analytical refinements, and even this regression was negligibly better than simple adjustment of the median.

Redoing the analysis in terms of total grosses in natural units would of course add well calibrated variability to the fractiles and from this angle the conclusions would appear different, but conclusions for total gross in natural units seem almost irrelevant since one would almost inevitably convert to ratios, logarithms, or a daily basis.

The over-variability mentioned in conclusion 1 may possibly have derived from a feeling on the part of the assessor that she should provide variation to give meaning to the exercise, or some desire to experiment on her own, but this seems unlikely since she was working in terms of total grosses in natural units and was not calculating ratios frequently let alone systematically and thus could scarcely have been aware of the absence of relative variability.

We made no attempt to analyze the initial, transient period quantitatively. The question, from the point of view of either the psychology of assessment or decision making, is when statistical analysis starts to contribute materially to which of the assessor's judgments, and how. Qualitatively speaking, the very early assessments were greatly overconfident, at least as much so as the literature would lead one to expect. Merely gathering the actual values for the record sufficed to induce very rapid learning, however, even in the absence of any processing for feedback saliency. Learning was not forgotten over a summer gap of 13 weeks. We have unfortunately no direct evidence on its transferability to other tasks – how similar another task would have to be for the assessor to be how much better calibrated than she would have been without the experience. Although the difference between theaters could be further analyzed, it would not bear directly on this question since she began their assessments simultaneously.

And we have no evidence whatever, nor is any known to us, on the variability between forecasters as regards speed of learning, calibration after learning, or forecast accuracy in tasks of this kind. We would very much like to know how universally one could without serious loss take logarithmic errors to be normal with standard deviation .25.

Finally we relate our findings directly to the traditional framework for discussing experts' assessments, beginning with information content vs. calibration. The information we found here was mostly contained in the assessed median and length of run and obtainable from them by straightforward statistical analysis; additional assessments would not be worth the bother in practice.

Calibration of a continuous forecast distribution is traditionally investigated via median and spread, marginally. But conditional calibration is also relevant if forecast distributions do or could vary, as is skewness especially if shape varies. We first used interquartile range to measure spread and $D_{75}-D_{25}$ to measure skewness when investigating conditional calibration, but this made the results difficult to interpret. It appears that the two halves of the distribution should be considered separately when assessments have been obtained as these were, at least for this assessor.

Two real peculiarities of the assessments emerged in our investigation. They are reflected in conclusion 1 above, but to amplify: First, the marginal calibration of the lower quartile and median deviated from expectation by less than 3 observations in over 140, but the upper quartile was low by more than 16 observations. The assessor took a pessimistic stance on average about the better outcomes while not exhibiting pessimistic about either the number or size of the worse outcomes.

Second and last, and curiousest, was the conditional miscalibration of the ratio between the lower quartile and the median. Actual grosses both below and above the median were less likely to fall beyond the quartiles when the lower quartile and median were relatively far apart then when they were close together.

REFERENCES

LICHTENSTEIN, S., FISCHHOFF, B. and PHILLIPS, L.D. (1982). Calibration of Probabilities: The State of the Art to 1980, *Judgment Under Uncertainty: Heuristics and Biases,* Daniel Kahneman, Paul Slovic and Amos Tversky, Eds., Cambridge: University Press.

LILLIEFORS, H.W. (1967). On the Kolmogorov-Smirnov Test for Normality with Mean and Variance Unknown. *J. Amer. Statist. Assoc.* **62**, 399-402.

SPEARE, M.A. (1952). *Charting Statistics.* New York.

TUFTE, E.R. (1983). *The Visual Display of Quantitative Information,* Chesire, Connecticut: Graphics Press.

APPENDIX 1--DATA

Given here are all original data except film names, show dates (replaced here by length of run), and assessment dates. The Brattle show dates cover March 4 to June 23, 1962 and September 23, 1962 to July 23, 1963 except April 16 and December 24, 1962 and April 22, 1963. The Harvard Square show dates cover March 7 to June 21, 1962 and September 26, 1962 to July 23, 1963 except April 18, October 24, November 15, November 29, December 5, and December 24, 1962, and January 23, February 24, April 24, and June 9-11, 1963. December 29, 1962 was originally recorded and is here counted as both the last day of Harvard program 48 and the first day of Harvard program 49. Harvard program 74 was shortened from 7 days to 5 after the assessments had been made; its actual gross of 3794 is multiplied here by 7/5 and the two cancelled days are counted in both it and the next program. All the original data are available from the authors uncoded. The data tabled here and much derived data are available in machine-readable form.

Brattle Theater

Program Number	Length of Run	Assessed Fractiles Low	.25	.50	.75	High	Actual Gross
1.	3.	700.	870.	900.	943.	1000.	2400.
2.	4.	2000.	2200.	2300.	2400.	2500.	2755.
3.	7.	1400.	3000.	3200.	3500.	5000.	2379.
4.	7.	2800.	3600.	4000.	4300.	6500.	3422.
5.	7.	1800.	2000.	2200.	2500.	4000.	3096.
6.	3.	499.	700.	930.	1400.	2500.	1425.
7.	4.	900.	1100.	1600.	2200.	4000.	2673.
8.	7.	1400.	2800.	3400.	4000.	7000.	3325.
9.	6.	1400.	2200.	3100.	3900.	5000.	1539.
10.	7.	1400.	2200.	3300.	4000.	6000.	2231.
11.	7.	1400.	2500.	3500.	4000.	6000.	2969.
12.	7.	1400.	2300.	2900.	3400.	6000.	3940.
13.	3.	500.	700.	900.	1200.	1500.	1039.
14.	4.	1000.	1600.	2400.	3000.	4000.	1923.
15.	3.	600.	1100.	1600.	2000.	3000.	2105.
16.	4.	1500.	1800.	2700.	3200.	5000.	2643.
17.	2.	500.	800.	1250.	1500.	1800.	1523.
18.	2.	800.	900.	1400.	1700.	2000.	1952.
19.	3.	1000.	1400.	2100.	2500.	3100.	2483.
20.	7.	1600.	2100.	3100.	4100.	5000.	4680.
21.	3.	500.	800.	1000.	1400.	2500.	1286.
22.	4.	1000.	1200.	2000.	2700.	3500.	1500.
23.	3.	1000.	1200.	1500.	1800.	2000.	1248.
24.	2.	400.	600.	800.	1000.	1500.	1264.
25.	2.	400.	600.	800.	1000.	1500.	1278.
26.	7.	2800.	3500.	4000.	4700.	6000.	5071.
27.	7.	2000.	2900.	3500.	4300.	6000.	2627.
28.	7.	2400.	3100.	3500.	4000.	6000.	3611.
29.	7.	2400.	3000.	3600.	4000.	6000.	2597.
30.	7.	2000.	2400.	2800.	3200.	4000.	2620.
31.	7.	2600.	3500.	4000.	4300.	6000.	2676.
32.	7.	2000.	2600.	2800.	3100.	4000.	3421.
33.	7.	2000.	3300.	3600.	4000.	6000.	3704.
34.	7.	2600.	3700.	4000.	4500.	6000.	4797.
35.	7.	2000.	3100.	3500.	4000.	6000.	2988.
36.	3.	700.	1000.	1200.	1500.	3000.	1549.
37.	4.	1200.	2000.	2500.	2800.	3000.	1858.
38.	3.	700.	900.	1000.	1200.	1800.	1074.
39.	4.	800.	1400.	1700.	2000.	4000.	2079.
40.	8.	2800.	3500.	4000.	5000.	7000.	3595.
41.	2.	350.	500.	700.	850.	1200.	901.
42.	3.	600.	900.	1100.	1300.	2000.	1231.
43.	7.	2500.	3200.	3500.	4000.	6000.	3008.
44.	3.	900.	1200.	1500.	2000.	3000.	1124.
45.	4.	1000.	1500.	2000.	2500.	4000.	2534.
46.	3.	900.	1200.	1500.	2000.	3000.	1519.
47.	4.	1000.	1500.	2200.	2700.	4000.	2262.
48.	7.	2800.	3500.	4000.	4700.	6000.	4823.
49.	2.	800.	1100.	1400.	1700.	2500.	1749.
50.	2.	800.	1000.	1200.	1400.	2100.	1490.

Brattle Theater (continued)

Program Number	Length of Run	Assessed Fractiles Low	.25	.50	.75	High	Actual Gross
51.	3.	1000.	1600.	2000.	2500.	4000.	2668.
52.	7.	2000.	3000.	3800.	4300.	6000.	4898.
53.	7.	2500.	3500.	4000.	4800.	8000.	3372.
54.	7.	1700.	2300.	2800.	3200.	4000.	3578.
55.	2.	500.	900.	1200.	1800.	3000.	1284.
56.	2.	400.	700.	1000.	1200.	3000.	522.
57.	1.	200.	350.	500.	600.	1000.	581.
58.	1.	300.	600.	800.	950.	1500.	841.
59.	1.	300.	600.	900.	1100.	1600.	1152.
60.	1.	300.	500.	600.	700.	1200.	591.
61.	1.	100.	300.	400.	500.	1000.	554.
62.	1.	100.	300.	400.	500.	1000.	484.
63.	1.	100.	250.	350.	450.	800.	524.
64.	1.	100.	300.	400.	500.	1000.	277.
65.	1.	100.	500.	600.	700.	1000.	521.
66.	1.	100.	600.	700.	800.	1500.	525.
67.	1.	300.	500.	600.	800.	1700.	552.
68.	1.	100.	300.	400.	500.	1000.	579.
69.	1.	100.	300.	400.	500.	1000.	262.
70.	1.	100.	400.	500.	600.	1200.	415.
71.	3.	300.	1500.	2000.	2500.	4000.	1463.
72.	7.	2500.	3000.	3500.	4000.	6000.	4159.
73.	7.	2500.	3000.	3500.	4000.	6000.	3364.
74.	3.	500.	750.	900.	1100.	3000.	619.
75.	4.	600.	1200.	1700.	2200.	3000.	1455.
76.	3.	400.	700.	900.	1200.	3000.	899.
77.	4.	600.	1500.	1800.	2200.	4000.	2226.
78.	3.	600.	1000.	1300.	1600.	3000.	1516.
79.	4.	600.	1700.	2000.	2400.	4000.	1595.
80.	3.	600.	900.	1000.	1100.	1900.	1138.
81.	3.	600.	1000.	1200.	1500.	2100.	2271.
82.	7.	2000.	2900.	3100.	3500.	5000.	2476.
83.	7.	2000.	2900.	3200.	3600.	5000.	2276.
84.	7.	2000.	2900.	3200.	3600.	5000.	2824.
85.	7.	2000.	3000.	3500.	3800.	6000.	2839.
86.	7.	2500.	3500.	4200.	5000.	9000.	5372.
87.	7.	2400.	3200.	3600.	4200.	9000.	5077.
88.	2.	400.	750.	900.	1000.	1600.	1042.
89.	3.	600.	900.	1200.	1350.	1800.	1305.
90.	2.	400.	850.	1000.	1200.	1600.	1347.
91.	3.	600.	1100.	1300.	1500.	2100.	1550.
92.	2.	400.	600.	800.	900.	1600.	679.
93.	2.	300.	600.	800.	1000.	2000.	719.
94.	3.	600.	900.	1100.	1300.	2100.	975.
95.	2.	400.	600.	800.	1000.	1800.	1054.
96.	2.	400.	700.	850.	1000.	1800.	938.
97.	3.	900.	1300.	1500.	1700.	2400.	1332.
98.	2.	400.	650.	800.	900.	2400.	926.
99.	2.	400.	700.	800.	900.	2000.	1035.
100.	3.	900.	1300.	1500.	1700.	2400.	1328.
101.	2.	400.	700.	800.	900.	1800.	654.
102.	2.	500.	800.	900.	1000.	2000.	1328.
103.	3.	900.	1400.	1600.	1700.	2400.	NA
104.	2.	500.	700.	800.	950.	2000.	NA
105.	2.	500.	900.	1000.	1200.	2000.	NA
106.	3.	900.	1800.	2000.	2200.	3000.	NA

Harvard Square Theater

Program Number	Length of Run	Assessed Fractiles Low	.25	.50	.75	High	Actual Gross
1.	1.	700.	1000.	1300.	1400.	1900.	988.
2.	6.	4800.	5800.	6000.	6500.	7500.	6083.
3.	4.	3000.	4550.	4800.	5000.	7000.	4127.
4.	3.	1600.	1850.	2100.	2300.	3000.	2766.
5.	7.	5000.	6400.	7000.	8000.	10000.	10559.
6.	7.	5000.	7400.	8000.	9000.	11000.	6159.
7.	14.	12000.	17000.	18000.	19000.	30000.	19432.
8.	6.	4800.	5800.	6000.	6500.	7500.	7083.
9.	1.	500.	600.	800.	1000.	1400.	1128.
10.	3.	2000.	2600.	3000.	3900.	6000.	4029.
11.	3.	1600.	2100.	2500.	2900.	4000.	2453.
12.	4.	2500.	3500.	4000.	4700.	6000.	5009.
13.	7.	4000.	6000.	8500.	10000.	15000.	5072.
14.	3.	2000.	2400.	2700.	3000.	4000.	2931.
15.	7.	6000.	8000.	10000.	11000.	15000.	7965.
16.	7.	5000.	6000.	8000.	10000.	13000.	6983.
17.	4.	2100.	3000.	4000.	5200.	6000.	4470.
18.	3.	1600.	2200.	2900.	3500.	4000.	2421.
19.	4.	2500.	3600.	4100.	4700.	5500.	2887.
20.	3.	1600.	2100.	2500.	2900.	3400.	2031.
21.	4.	2200.	2900.	4400.	5000.	7000.	4324.
22.	3.	1600.	1800.	2000.	2900.	4000.	2208.
23.	2.	850.	1100.	1400.	1700.	2000.	1192.
24.	4.	2900.	3600.	4100.	4600.	6000.	3755.
25.	3.	1400.	2100.	2500.	3000.	4000.	1470.
26.	4.	2600.	3200.	3600.	4000.	6000.	6244.
27.	3.	1200.	1600.	1800.	2200.	3200.	2098.
28.	1.	600.	800.	1000.	1150.	1600.	907.
29.	6.	5000.	6000.	7500.	8500.	10000.	6484.
30.	4.	2500.	3600.	4000.	4500.	6000.	5652.
31.	3.	1000.	1600.	2000.	2200.	4000.	2006.
32.	6.	3000.	5000.	6000.	6500.	8000.	6994.
33.	4.	3000.	5000.	6000.	6500.	8000.	5991.
34.	3.	1800.	2800.	3000.	3200.	5000.	2116.
35.	1.	500.	700.	900.	1000.	1200.	536.
36.	6.	3000.	5500.	6000.	6400.	8000.	7809.
37.	3.	1000.	2400.	3000.	3900.	5000.	1861.
38.	3.	1000.	2400.	3000.	3300.	5000.	2447.
39.	1.	600.	800.	900.	1100.	1500.	1253.
40.	6.	3000.	5400.	6000.	6400.	8000.	8254.
41.	3.	2000.	2500.	3000.	3500.	5000.	3529.
42.	3.	1600.	2000.	2200.	2900.	4000.	1918.
43.	7.	5000.	6500.	8000.	9500.	12000.	6526.
44.	6.	4000.	5200.	6000.	6800.	10000.	4337.
45.	4.	2500.	3800.	4400.	5500.	7000.	2853.
46.	1.	400.	700.	950.	1100.	1500.	950.
47.	1.	300.	500.	700.	950.	1300.	246.
48.	4.	3000.	4000.	5500.	6000.	7500.	3628.
49.	4.	1000.	1700.	2000.	3000.	5000.	3212.
50.	4.	2000.	2600.	3000.	4000.	6000.	4592.

Harvard Square Theater (continued)

Program Number	Length of Run	Assessed Fractiles Low	.25	.50	.75	High	Actual Gross
51.	3.	1000.	1800.	2000.	3000.	5000.	4579.
52.	4.	1000.	2800.	3400.	4000.	7000.	5154.
53.	3.	1000.	1800.	2100.	2700.	4000.	1919.
54.	7.	5000.	7000.	8000.	9000.	13000.	10783.
55.	3.	1500.	3000.	4000.	5000.	8000.	3591.
56.	3.	1200.	1650.	1900.	2200.	4000.	2186.
57.	7.	4000.	6000.	7000.	9000.	12000.	9205.
58.	4.	2500.	3600.	4000.	4700.	7000.	5471.
59.	3.	1500.	2500.	3000.	4000.	6000.	2731.
60.	4.	2000.	3500.	4000.	5000.	8000.	5349.
61.	3.	1400.	1900.	2400.	3000.	5000.	2206.
62.	2.	1000.	1500.	1800.	2700.	5000.	3558.
63.	8.	4000.	6500.	8000.	10000.	15000.	8081.
64.	3.	1000.	1700.	1900.	2100.	4000.	1811.
65.	4.	2000.	3300.	4100.	5300.	8000.	5023.
66.	3.	1000.	1800.	2000.	2300.	4500.	2478.
67.	4.	2000.	3300.	4400.	5000.	8000.	3884.
68.	3.	1000.	2000.	2500.	3000.	5000.	3428.
69.	4.	2000.	3000.	3800.	4100.	8000.	3305.
70.	3.	2000.	2500.	3000.	3500.	5000.	2344.
71.	7.	4000.	7000.	8000.	9000.	12000.	7495.
72.	7.	4000.	6600.	7000.	8000.	12000.	5499.
73.	2.	900.	1900.	2500.	3000.	4000.	2029.
74.	7.	4000.	6000.	7000.	8500.	12000.	5312.
75.	2.	900.	1700.	2000.	2200.	4000.	1620.
76.	5.	4000.	5000.	6000.	7500.	10000.	11795.
77.	6.	6000.	9000.	10000.	13000.	20000.	8755.
78.	2.	600.	1200.	1400.	1800.	2500.	1118.
79.	9.	5000.	7000.	9000.	12000.	20000.	11604.
80.	3.	1000.	1800.	2200.	2600.	4000.	2288.
81.	7.	4000.	6000.	7500.	9000.	12000.	11040.
82.	4.	2000.	4000.	5500.	6500.	9000.	4820.
83.	3.	1500.	2500.	3000.	3600.	6000.	2468.
84.	7.	5000.	7500.	9000.	11000.	15000.	6465.
85.	4.	2000.	2600.	3000.	3800.	6000.	5920.
86.	7.	5000.	7000.	8000.	9000.	15000.	9299.
87.	7.	5000.	8000.	9200.	10500.	20000.	7184.
88.	14.	9000.	15000.	17000.	20000.	35000.	17499.
89.	4.	3000.	5000.	6000.	7000.	10000.	NA
90.	3.	1000.	2200.	2500.	3000.	6000.	NA
91.	7.	5000.	7000.	8000.	9500.	12000.	NA

DISCUSSION

W.H. DuMOUCHEL (*M.I.T.*)

This paper by Professors Pratt and Schlaifer provides the Bayesian community with a most interesting set of data – a lengthy history of subjective forecasting in a real business environment. This data will surely be analyzed and reanalyzed by others with various objectives in mind. The authors' objectives seem to be first, to provide a basic description of the stochastic model which the forecasts seem to follow; second to assess whether the subjective forecasts could be profitably used as is, or whether a statistically sophisticated decision maker would gain much by using the forecasts and observations as a training set to produce a method of adjusting future forecasts; and, third, to understand in some detail just how well calibrated this particular subjective forecaster became, and in what ways can the calibration be shown to be less than perfect. The authors have succeeded, I think, in all three of these objectives. For me, the most interesting result was how little the "Decision Maker" can hope to gain by use of a statistical model, compared to just using the raw forecast, at least in the mean of his posterior distribution, although for some applications the expected improvement in the spread of the forecasts might be valuable.

There was one topic not much mentioned in the paper, which I would be anxious to see developed. I would like to take a closer look at just how the forecaster became calibrated during the first few months of the series. One conceivable model for her behavior is that she was intuitively using Bayes Theorem to update her opinion about the parameters of the stochastic process generating the observations. Is it possible to test this possibility from the data? That is, how well do these series of predictive distributions follow the Bayesian Paradigm?

I have just two technical suggestions for the data analysis. First, there were several regressions in which the dependent variable was the absolute value of a residual, and therefore almost certain to have a distribution skewed to the right. A square root transformation or a regression which assumed an exponential error term would seem to be more appropriate than simple least squares. Second, since the forecaster received feedback several days late, it might be more powerful to test for auto-correlation in the forecast errors with a time lag of one week or so, rather than just the one-day lag presented here.

P.J. BROWN (*Imperial College, London*)

This is an exercise in analysing the subjectively assessed fractiles of receipts at two cinemas compared to the actual receipts. The presentation of data in Exhibit 1 has meant that I have been able to do some analysis as well. Unfortunately the important background data, day of week, vacation indicator, type of film, etc., from which a statistical analysis of the accuracy of assessment might be performed is unavailable. Hence one is essentially limited to analysing the fractiles as surrogates for the detailed missing background data; forgoing an analysis in which daily receipts might be expected to be log-linear in day effect, vacation effect, length of run and some measure of film popularity. In such modelling the capacity of the cinema would also be important.

It is natural to think in terms of proportional effects and to attempt to model in terms of logarithms, an empirical conclusion of the paper. What I find surprising is that despite the declared intention to treat the data as a stochastic process, Section 5, time is not considered, except to remove some early data.

I contend that a few time series plots are far more illuminating as a preliminary analysis than the numerous regression analyses. Subsequent analysis might look for a learning model of the time dependent data, perhaps in terms of a multinomial reduction of the observations. The difficulty in making coherent sense in a plethora of p-values is evident from the authors' regression analyses.

Median Assessment

A time series plot of log(actual/median) reveals no special patterns, a histogram confirms $N(0,0.297)$, with Harvard having a slightly larger standard deviation than Brattle, 0.317 to 0.279. About 90% variation of log(actual) is explained by log(median), so that as Professors Pratt and Schlaifer have noted, median assessments at least explain a large proportion of the variation. Whether there was still room for improvement would depend on the nature of the missing "independent variables".

Quartile Assessment

We may use the above normal distribution of log(actual/median) to predict an upper quartile from the upper 75% point of the standard normal distribution, 0.675, as

$$\log(\text{upper quartile}) = \log(\text{median}) + 0.675 \times 0.297.$$

This reflects the quartile assessment implied by the homoscedastic and patternless time series of log(actual/median). A time series plot of log(quartile/pred quartile) reveals some interesting patterns. Aside from there being only one fifth of the values greater than zero, as opposed to a well calibrated 50 %, there seems to be considerable cyclical behaviour. This behaviour is perhaps even more marked for a time series plot of log(upper/lower quartile), see Figure 1. Calculation of the autocorrelation function corroborates the extent of the tracking, the first five lagged autocorrelation being 0.48. 0.40, 0.32, 0.32, 0.19.

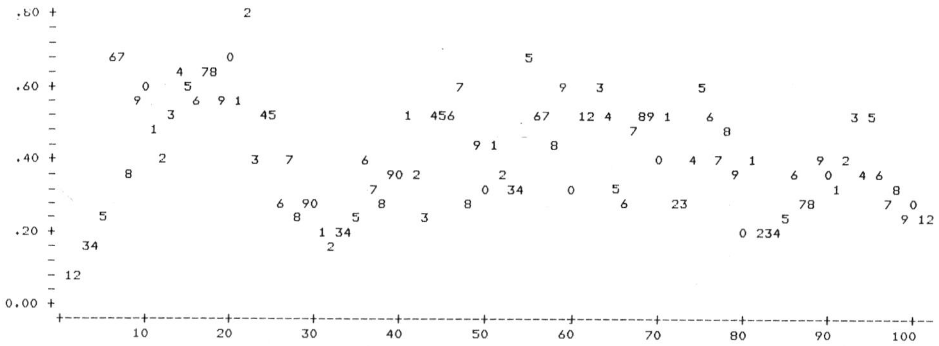

I know the exercise is now 20 years old but it might be illuminating to hear how Mrs. Pratt went about her specification of fractiles; though still the unanswered question remaining is whether a proper statistical analysis of the receipts in terms of the concomitant data would have proved more accurate.

D.V. LINDLEY (*Somerset, UK*)

This is an admirable paper because it discusses, with good data, an important Bayesian problem; namely how an expert in a field can communicate her uncertainty in the probabilistic terms necessary for a coherent argument. We need more data sets of the high quality of this one. My main point is to connect this paper with mine given at the conference. In the latter it was suggested that the key to the analysis lay in the likelihood for the true value (in this case, cinema attendance θ) given the data (five fractiles x_i of the subject's distribution for attendance). This is equivalent to the density $p(x_1,\ldots,x_5|\theta)$. This describes the expert's knowledge and assessment skills. Is is a pity that an analysis with this function was not used instead of the plethora of tail-areas, significance tests and other sampling-theory devices that do occur.

REPLY TO THE DISCUSSION

First we should note that the discussants had only a very preliminary draft of the paper. We apologize for any missed connections, misunderstandings, or redundancies resulting, although our revisions were independent of the discussion and would have affected it little in essentials, we believe.

Dr. DuMouchel has stated our objectives correctly and well. Technical improvements in our analysis could surely be made, but we doubt that they would change our preliminary transformation and selection of data in Section 4, and the logarithmic data used thereafter were very close to symmetric and normal, skewed slightly to the left if at all, except when actual grosses on one side of the median only were used in Section 8. Here we did regret not having more powerful technology handy, especially since we had less data and some curious results.

Analyzing the learning period, discussed by both Dr. DuMouchel and Dr. Brown, would indeed be valuable and difficult, we think.

We are less sanguine than Dr. Brown (and others) about improving the assessed median by incorporating background data in the analysis. Regarding availability of data, day of the week can be determined using a calendar from the information in the Appendix or the dates on the original sheets. Vacations, holidays, and exam periods could also be looked up but affect only a fraction of the data. Type and popularity of film would have to be defined quantitatively or categorically somehow and determined starting from the title alone. Daily receipts are not available. Theater capacity constrained only occasional showings, not whole runs or even days, so its numerical value would be little help in analyzing its effect. The assessor took into account all this information and more, mentally adding new variables and responding to changes as they arose. Thus the question is whether she misused certain information in a way that statistical analysis could disentangle and amend even without using all her information. We would welcome any attempt in this direction.

We would also welcome a search for time dependencies of all kinds. We looked only for the simplest type of statistical dependence of forecast errors that would suggest (conditional) miscalibration. Well calibrated assessments may of course be time dependent in themselves. Thus the interesting autocorrelation of log interquartile range discovered by Dr. Brown could reflect autocorrelation of background variables, not miscalibration. The marginal miscalibration of the upper quartile Dr. Brown points to is described in another way in the paper.

Our statistical devices were intended to be convenient and readily understood and hence conventional, not pioneering. Most of them we believe have useful interpretations, conventional or unconventional, descriptive if not inferential. We hope the twice-deplored plethora of P-values is unobstrusive enough to be ignored where it is not helpful. Personally, we find t tail-probabilities helpful and F tail-probabilities usually not helpful, interpreting both simply as approximate posterior probabilities when models are reasonable and data not too weak. Fully Bayesianizing our analysis would be harder than it may appear, for reasons indicated below.

Professor Lindley's brief remarks ignore and even his paper scants essential complexities of a Bayesian analysis of both our problem and the in fact different problem his remarks and paper apply to directly. If the successive values of the assessed fractiles $\mathbf{x} = (x_1,\ldots,x_5)$ and attendance θ were so helpful as to be jointly iid in classical language (to be "exchangeable" Professor Lindley would say), then one could study their joint distribution equally validly by way of $p(\theta\,|\,\mathbf{x})$ and the marginal distribution of $\mathbf{x}$, as we did, or by way of $p(\mathbf{x}\,|\,\theta)$ and the marginal distribution of $\mathbf{x}$, as Lindley prefers, or in other ways. To handle background variables our approach was simply to condition on them, but a fully Bayesian or Lindleyesque approach would be more complicated. The distributions we chose are the most directly relevant to calibration, and also the most directly usable by a decision maker lacking other information. We view much of our analysis as looking at these distributions through a "flat" prior.

A full and proper Bayesian analysis would require a proper prior distribution on the "true" joint distribution of all observed variables (ours would *not* be Dirichlet) or on the parameters of some model (perhaps hierarchical). This looks beyond sensible bounds to us, even in the iid case without background variables. A partial, proper Bayesian analysis of $p(\theta\,|\,\mathbf{x})$ using a regression model with a proper prior distribution on the parameters would have some appeal to us —more so if there were other studies of experts available on which to base the prior. Professor Lindley prefers to look at $p(\mathbf{x}\,|\,\theta)$, but since it doesn't speak directly to either the calibration question or the decision maker's question, a full and proper Bayesian analysis appears necessary to any progress along this route.

Here and usually it could not be assumed that $(\mathbf{x},\theta)$ is iid and the full analysis would be even more impossible. It now becomes an essential question whether either $p(\mathbf{x}\,|\,\theta)$ or $p(\theta\,|\,\mathbf{x})$ is approximately the same on all occasions. We believe that $p(\mathbf{x}\,|\,\theta)$ clearly is not whatever is done with the background variables, while $p(\theta\,|\,\mathbf{x})$ may well be when the background variables are adjoined to $\mathbf{x}$. If the process generating θ changes (easier said than defined since remodeling may subsume change) a knowledgeable expert should change $p(\mathbf{x}\,|\,\theta)$. An expert is not a measuring instrument whose conditional error distribution $p(\mathbf{x}\,|\,\theta)$ can be expected to be the same however θ is generated or whatever population it comes from. Indeed, if the marginal distribution of θ changes, then either $p(\theta\,|\,\mathbf{x})$ or $p(\mathbf{x}\,|\,\theta)$ must change, and an expert trying to be well calibrated will change $p(\mathbf{x}|\theta)$ so as to maintain the calibration of $p(\theta\,|\,\mathbf{x})$ to the best of his or her ability. So unless you know a lot more than the expert, which here and typically you do not, you should study $p(\theta\,|\,\mathbf{x})$ rather than $p(\mathbf{x}\,|\,\theta)$ not only because it is more directly relevant (to both calibration and decision making) but more importantly because it is more plausibly stable. Furthermore our approach should be quite robust, that is, even if the process generating the data is quite unstable in some respects, our approach should detect and estimate the kinds of miscalibration anticipated as well as can be hoped for. We confess we do not see how to study our questions via $p(\mathbf{x}\,|\,\theta)$ except by an indirect route to at best similar answers.

Our decision maker was more a device to separate informativeness from calibration than an exemplary Bayesian. If, however, a real decision maker had no information not available to the expert, we doubt that he could improve much on the forecast distributions we proposed in Sections 5 and 6, although the partially proper Bayesian analysis mentioned above would be more satisfying and perhaps more "coherent" or "rational". If the decision maker wants to take into account his own information H about θ on a particular occasion, he cannot use the standard prior-posterior analysis blindly unless his information is entirely independent of the expert's, an extremely unlikely possibility. He could assess his distribution for θ given H and assess the likelihood of $\mathbf{x}$ given θ and H and multiply them, but neither our data nor any other likely type of data would bear directly on that likelihood, so considerable and difficult assessment would probably be required. Alternatively, he could assess a distribution for θ prior to H and a joint likelihood of $\mathbf{x}$ and H given θ, allowing for the dependence between his information and the expert's. This amounts to treating himself as another expert in the framework of Professor Lindley's paper. Either way he could not reasonably work only in terms of the assessed distributions of θ above (decision maker's and expert's or several experts'), let alone combine them in any automatic way. It is important to probe and worry about what information, background, prejudices, etc. are shared by the assessors. Combining assessed distributions in practice looks very hard to do formally, sensibly, and validly by any method. We do not question the validity of either Professor Lindley's preferred approach provided it is properly understood, or his rejection of other proposed pooling methods. That these proposals are entertaining but not to be seriously entertained already seemed obvious, however, from a simple example that occurred to one of us (JWP) long ago. It is nicely described by Raiffa (1968, pp. 231-2), but in brief, when all assessors agree that two events are independent but disagree about their probabilities, non-Bayesian proposals typically make them dependent in the pooled distribution even when one wants them to be independent, as one easily might.

REFERENCE IN THE DISCUSSION

RAIFFA, H. (1968). *Decision Analysis*. New York: Addison Wesley.

BAYESIAN STATISTICS 2, pp. 425-462
J.M. Bernardo, M.H. DeGroot, D.V. Lindley, A.F.M. Smith (Eds.)

Multivariate Group Assessment of Probabilities of Nuclear War

S. JAMES PRESS
University of California, Riverside

SUMMARY

This paper describes a new method for assessing proper multivariate prior distributions. The method involves first assembling an informed, homogeneous, group of people whose opinions you would like to be reflected in the prior distribution and then obtaining numerical assessment of their views about a set of issues. In many situations it is reasonable to substitute the distribution of the single point opinions of the group for the distribution of opinions of a single individual. The approach described here is appropriate in such contexts. The group point assessments are then merged into a single representative prior distribution by density estimation. The method is illustrated with an empirical example involving the views of experts regarding the chances of nuclear war between the United States and the Soviet Union.

Keywords: BAYES; GROUP JUDGMENT; MULTIVARIATE PRIOR DISTRIBUTION; NUCLEAR WAR; QUALITATIVE CONTROLLED FEEDBACK; SUBJECTIVE PROBABILITY ASSESSMENT.

1. INTRODUCTION

Suppose you are a decision maker who would like to be helped to make a meaningful judgment. There are, in general, several correlated observable variables and several correlated (unknown) unobservables. You can make inferences about the unobservables by using Bayes theorem for the posterior distribution of the unobservables, conditional on the observables. To implement the procedure you will need a sampling distribution for the observables conditional on the unobservables, and you will need a joint (multivariate) prior distribution for the unobservables. In this paper we confine our attention to the assessment of the multivariate prior distribution.

We propose to assess the multivariate prior distribution by using the combined judgments of an informed, homogeneous group. In Sections 2, 3, 4, we will describe elicitation procedures for generating a vector of assessments from each group member; we will show how to merge the vectors into a multivariate prior density; and we will discuss a model for grouping opinions. For illustration, in Section 5 we report on some results of an empirical application of this procedure to the problem of assessing a prior distribution for the probability that the U.S. or the Soviet Union will detonate a nuclear weapon in the other's homeland during each year of the next decade. Finally, in Section 6, we discuss some psychological aspects of the subjective assessment problem that suggest additional research.

2. MODEL FOR CONVERGENCE OF THE ASSESSED MEAN

Suppose $\theta:(p\times 1)$ denotes a continuous, unobservable, random vector which is the mean of some p-dimensional sampling distribution with density $f(\mathbf{z}\,|\,\theta)$. You would like to assess your prior probability density for θ, and you would like your prior density to reflect the combined judgments of some informed population. We assume the informed population consists of "experts", in that they possess contextual knowledge. You could try to assess your own p-dimensional prior distribution for θ. Alternatively, you might be willing, because of the greater degree of expertise possessed by a readily available group of experts, to accept "best guess" point assessments of θ given by the panel of experts and then use as your prior the distribution of these point assessments. We take a probability sample from the population of experts and ask each subject, for a "best guess" opinion about θ: $(p\times 1)$. Let $\theta_j:(p\times 1)$ denote the vector of opinions about θ given by the jth subject; $j=1,\ldots,N$.

Assume[1]:

$$\underset{(p\times 1)}{\theta_j} = \underset{(p\times 1)}{\theta^*} + \underset{(p\times 1)}{b_j} + \underset{(p\times 1)}{\varepsilon_j}, \qquad j = 1,\ldots,N, \tag{1}$$

where θ^* denotes some hypothetical, true value of θ, as perceived by the jth group member (that is, θ^* denotes what his assessment would be if all of his errors and biases were removed, and if he were exposed to all available information on the subject); the ε_j denote independent, identically distributed error vectors; and $\mathbf{b}_j$ denotes the bias in the perception of θ by the jth panelist.

2.1. *General Case: Inhomogeneous Populations*

In general, when panelists are selected from inhomogeneous populations, the b_j's will all be different from one another. Define $\bar{\theta}$ the average opinion of θ in the sample group. (Mean judgments have been found to be surprisingly accurate compared to other aggregation schemes; see Hogarth, 1978, for a summary of this literature.)

$$\bar{\theta} = \frac{1}{N}\sum_{j=1}^{N} \theta_j .$$

In general, with such inhomogeneous populations, as the group size increases, $\bar{\theta}$ need not approach θ^* in probability. In general, $\bar{\theta}$ will be an inconsistent assessment of θ^*. Since $E(\theta_j) = \theta^* + \mathbf{b}_j$, a sufficient condition for consistent assessment of θ^* is that[2], if $\bar{\mathbf{b}} = \frac{1}{N}\sum_{j=1}^{N} \mathbf{b}_j$ $\bar{\theta} \xrightarrow{P} \theta^*$, if $\bar{\mathbf{b}} \to \mathbf{0}$, and if for some δ, $0 < \delta \leq 1$,

$$\sum_{j=1}^{N} \left(\frac{\theta_j^{(i)}}{N}\right)^{1+\delta} \xrightarrow[N\to\infty]{P} 0, \; i = 1,\ldots p,$$

where $\theta_j^{(i)}$ denotes the *ith* component of $\underset{(p\times 1)}{\theta_j}$.

[1] This model is discussed in greater detail in Press (1983).

[2] See Rao, (1965).

2.2. *Case of Homogeneous Populations*

From here on we assume our sample panel is homogeneous, in that $\underset{(p\times 1)}{\mathbf{b}_j} = \underset{(p\times 1)}{\mathbf{b}} = \underset{(p\times 1)}{\text{constant}}$, for all panelists. Then

$$\underset{(p\times 1)}{\theta_j} = \underset{(p\times 1)}{\theta^*} + \underset{(p\times 1)}{\mathbf{b}} + \underset{(p\times 1)}{\varepsilon_j} \quad ,j = 1,\ldots,N \ .$$

Now it is still true that $\bar{\theta}$ need not be consistent, but now the θ_j's are independent and identically distributed. So $\bar{\theta}$ is a consistent assessment of θ^* if and only if $\mathbf{b} = \mathbf{0}$. Increasing the group size in a sample drawn from a homogeneous population will yield a consistent assessment only if $\mathbf{b} = \mathbf{0}$. We note that the assumption that $\mathbf{b}_j$ = constant for all panelists is a major simplification of a very complex problem. It is an assumption that may not be satisfied in many situations, but we are using the assumption to develop what we hope is a first approximation to many situations. The more homogeneous is the group, in as many variables as possible, the more likely is the assumption to be valid.

2.3. *Qualitative Controlled Feedback*

One method of driving the bias in the model toward zero (see below in this paragraph) is to use Qualitative Controlled Feedback[3] (QCF). The QCF procedure is similar to Delphi, but differs in the type of information fed back on each round. A Delphi procedure is designed to drive the panel toward consensus ($\theta_1 = \theta_2 = \ldots = \theta_N$), but the common assessed value need not be θ^*. In QCF we have the panel sharing "reasons" and information relating to the substantive issues, so they jointly tend to drive the bias to zero on successive rounds. There is no pressure to move towards consensus. It should be noted that even though the panel members are assumed to have the same average biases, they might not all think of the same reasons on the first round. They might each develop a slate of reasons that might or might not have an intersection with the slates of other panel members, but they still have the same average basic biases. On subsequent rounds the reasons are combined into a composite and then shown to panel members in composite form so all panelists can benefit from the jointly developed composite on subsequent rounds of assessments. In a homogeneous group, as panel members keep providing additional reasons on subsequent rounds, more and more information is revealed about the issue, and as a result, the common bias vector tends to be driven toward zero. We note that the model introduced in equation (1) will be used only to assess whether or not a sample mean will converge to a population mean. We will not need this structure to use QCF for data collection, nor to establish the prior density.

2.4. *Multivariate Density Assessment*

Now assume we have a homogeneous panel of N experts (randomly sampled) who provide assessments

$$\theta_1{:}(p\times 1),\ldots,\theta_N{:}(p\times 1).$$

[3] In QCF a panel is asked a battery of questions, and each panelist must respond independently but must give reasons for his responses. The reasons from all panelists are merged and fed back as a composite to the panel, which is asked the same battery again. The process is repeated. See Press (1978).

Assume the N assessments of the expert panel, $\theta_1:(p \times 1),\ldots,\theta_N:(p \times 1)$ are independent and identically distributed, all with multivariate density $g(\theta)$. (If the assessments are correlated with common intraclass correlation induced by QCF feedback, their covariance matrix must first be diagonalized. The final $(p-1)$ elements of the resulting vectors will be mutually uncorrelated and identically distributed).

We adopt the Parzen-Cacoullos approach to density assessment and permit the density to assume the form

$$\underset{(p\times 1)}{g_N(\theta)} = \frac{1}{N}\sum_{j=1}^{N} \underset{(p\times 1)}{\phi_j(\theta)} ,$$

where

$$\phi_j(\theta) = \frac{1}{\delta^p(N)} K\left(\frac{\theta - \theta_j}{\delta(N)}\right) ,$$

$K(\theta)$ is a kernel chosen to satisfy suitable regularity conditions, and $\delta(N)$ is a sequence of positive constants satisfying

(1) $$\lim_{N\to\infty} \delta(N) = 0 ,$$

and

(2) $$\lim_{N\to\infty} N\delta^p(N) = \infty .$$

Normal Kernel

For simplicity, we adopt the normal density kernel

$$K(\theta) = \frac{1}{(2\pi)^{p/2}} e^{(-1/2)\theta'\theta} ,$$

and we take

$$\delta(N) = \frac{C}{N^{1/(p+1)}} ,$$

where C is any preassigned constant (fixed in any given sample to smooth the density).

Thus, if $\mathbf{X}_j$ has density $\phi_j(\mathbf{x})$,

$$\underset{(p\times 1)}{\mathbf{X}_j} \sim N(\theta_j, \delta^2(N)\mathbf{I}_p).$$

Therefore, the group, multivariate prior density becomes

$$g_N(\theta) = [\mathbf{C}^p N^{1/(p+1)}(2\pi)^{p/2}]^{-1} \sum_{j=1}^{N} \exp\{-\frac{N^{2/(p+1)}}{2\mathbf{C}^p}(\theta-\theta_j)'(\theta-\theta_j)\} .$$

The multivariate prior density generated by the group has the properties that $g_N(\theta)$ is assymptotically unbiased, and

$$\lim_{N\to\infty} E\,[g_N(\theta) - g(\theta)]^2 = 0$$

at every point of continuity θ of $g(\theta)$.

Note: Kernal density estimation is only one smoothing technique that could be used to merge the assessed points (vectors). An interesting alternative is Projection Pursuit density estimation (see Friedman et al., 1981) which has the virtue of requiring very few assumptions about the structure.

3. EXAMPLE - ELICITATION OF THE PRIOR DISTRIBUTION FOR A MEAN VECTOR

Suppose θ: $p\times 1$ denotes a vector of means of a distribution in p dimensions and we wish to develop a prior distribution for θ under the assumptions of Sections 1 and 2. We assume there is a panel of N experts who have been sampled, and who will each give a point assessment of $\theta^{(i)}$, the ith component of θ, for every i. Each subject could be asked a battery of p questions of the form: give a value $\widetilde{\theta^{(i)}}$ so that it is equally likely that $\widetilde{\theta^{(i)}} > \theta^{(i)}$, and $\widetilde{\theta^{(i)}} < \theta^{(i)}$, $i = 1,2,\ldots,p$. This yields a vector $\widetilde{\theta_j} \equiv (\widetilde{\theta_j^{(i)}})$, $i = 1,\ldots,p$, for the jth panelist; for all $j = 1,\ldots,N$. The $\widetilde{\theta_j}$ vectors are then merged into a prior density, $j = 1,\ldots,N$.

This procedure could also be applied to assessment of a prior distribution for a covariance matrix, but it is somewhat more complicated in that the procedure requires that correlation coefficients be assessed as well as marginal parameters (variances). Moreover, positive definiteness conditions would need to be satisfied (see Gokhale and Press, 1983).

4. CHARACTERISTICS OF THE MULTIVARIATE PRIOR DENSITY. ASSESSMENT PROCEDURES PROPOSED

We note the following aspects of the procedures proposed:

(i) *Consensus of a group of experts is not required*, as is the case with some group procedures. It is the diversity of opinions of experts that is interesting. When a homogeneous group of experts is given the same battery of questions independently, its degree of disagreement reflects the amount of inherent uncertainty in the underlying issue.

(ii) *No need for an assumed functional form for the prior.* We don't use natural conjugates, or any other artificially structured family of priors, and then assess hyperparameters. We merely let the actual empirical distribution of expert point assessment determine the prior. (There are some parametric assumptions made about the form of the density kernel, but these do not predetermine the form of the prior).

(iii) *Assessment ease.* It is not necessary for subjects to think about the likelihood of certain events occuring simultaneously, in order to assess a prior distribution for a p-vector; alternatively, we replace the problem of assessing a p-dimensional prior distribution for a single individual with a problem of assessing the p-dimensional means of N individuals and letting the distribution of these mean vectors be a surrogate for the single individual's prior. This compromise makes it possible to circumvent the "curse of dimensionality". We need only administer a battery of p-questions to N people, and each person's p-responses are correlated. Subjects need only to think about one-dimensional (marginal) events as long as priors for means are being assessed. In such cases, it is as easy to assess a p-dimensional prior, as a one-dimensional prior. In cases where priors for covariance matrices are being assessed these ideas are still valid, with the small modification that correlation coefficients must be assessed as well.

(iv) *Additional fractiles*. It may be of interest sometimes to assess several fractiles of $\theta:(p\text{x}1)$ from each subject (such as medians and other quartile points).

(v) *Convergence to "truth"*. If there is indeed "truth", in some absolute sense, increasing the size of the group will not necessarily cause convergence to truth. In inhomogeneous groups, such convergence is highly unlikely. In homogeneous groups, convergence may occur if group members are permitted to interact (in face-to-face contact, using QCF, or whatever), to exchange information, and to argue rationally for their opinions. Then, the bias in the group response may be driven to zero (because people with the same average biases are still likely to generate different reasons, but when they share a composite of reasons the common bias might be reduced to zero).

5. EMPIRICAL APPLICATION

How do experts in the United States feel about the chances of nuclear war between the United States and the Soviet Union during the 1980's? We were motivated to raise this question in an effort to establish a barometer of expert opinion on this subject that could be reexamined every few years to see whether expert perception was changing. To answer the basic question a mail survey of subjective opinions was carried out among American experts on such issues using Qualitative Controlled Feedback. We next describe the nature of the survey, the tasks the subjects carried out, and some results of our analysis of the survey responses.

It might be noted that while the empirical application of our assessment procedure relates to a problem in which one can never know a precise answer, we could as easily create a scenario in which the problem is that there are many weather forecasters who are predicting probability-of-rain in a flooded area for the next several weeks. We could generate a representative prior for the group, but we could also wait for the correct answer, and then calibrate the group.

5.1 *Nature of the Survey*

A list was prepared for our problem containing the names of all experts in strategic policy who were employed at The Rand Corporation, Santa Monica, California, or who were members of the California Arms Control Seminar. A stratified random sample of 305 experts was selected from this list for solicitation to participate in the survey. There were also some non-experts who were randomly sampled for the survey, but they were used only to help generate reasons (see below); their numerical subjective assessments were not used in the results described below. A letter of request for participation in the (three stage) survey was sent out to all prospective survey panelists under the signature of a Rand Vice President who emphasized the importance of participating (we did it this way to reduce non-response). These subjects were then asked to provide written answers to a mailed questionaire.

Earlier research has shown (see Harman and Press, 1978) that an important component is improving the validity of group judgments involves having a wide diversity of disciplines represented on the panel. This is because repondents from different disciplines see the world differently, and ofte~ think about "reasons" for their responses that are suggested by their own discipline (and not viewed as important or relevant in other disciplines). So using only experts from a single discipline could generate a composite of reasons that did not include what could be the most important reasons. Accordingly, in our panel we included both experts (people who had been working in the strategic area) and non-experts (well educated people at The Rand Corporation who did not work in the strategic area, but were mostly people with advanced degrees in psychology, education, sociology, law, etc.). We ended up with a total of 186 subjects who responded to the first round questionnaire (51% response rate). Within this group there were 149 "experts" and 37 "non-experts". Moreover, of the experts, 79 were employed at The Rand Corporation and 70 were employed at other organizations. On the second round, 151 people responded (81%), of whon 124 were experts; and on the third round 130 people responded (86%), of whom 110 were experts. Thus, there were 130 people who completed all three questionnaires. The three survey instruments for the three rounds of questions were first pilot-tested on a panel of 22 subjects in the Washington, D.C. office of Rand (and the survey instrument was adjusted accordingly).

5.2. *Tasks Carried Out by the Subjects*

Each subject in our final sample survey was independently asked in early 1981 (in October, 1980 for Round 1, in December, 1980 for Round 2, and in January, 1981 for Round 3):

> "Please express your subjective probability assessment that the United States or the Soviet Union will detonate a nuclear weapon in the other's homeland during the year 1981"

The question was then repeated for each of the ten years in the decade (vectors of the ten-dimensional assessments were merged to form a ten-dimensional prior distribution). Subjects were advised to assume for every year separately that no such detonations had occurred in previous years.

Scale Design

The subjects were first asked to respond to this basic question for 1981 on three different scales (to see if they quantified their judgments consistently on these scales). They were also asked many subsidiary background questions of a demographic, socioeconomic, and attitudinal nature, so that later we would be able to relate a panelist's responses to his (her) background characteristics. Subjects were also asked to respond to the basic question for each year separately during the decade of the 1980's. To assist respondents in understanding the meanings of extremely low probabilities (see Section 6 for a discussion of possible biases that are introduced when subjects have an incomplete understanding of extremely low probabilities), various points on the probability scales (for the basic question) were given word interpretations, (see below) which could also serve as anchors (see Section 6 for a discussion of Anchoring).

In order to accustom the subjects to scaling their opinions on this issue they were asked first to respond to the basic question on a simple ordinal scale in which they were asked to evaluate the event in question as "very unlikely", "likely", etc.; then they were asked to respond on a more complex ordinal scale (by circling one + point on the scale below).

Ordinal Scale

+ The probability is greater than the probability of getting a head by flipping a fair coin once. (The chance of getting a head is 0.5).
+ The probability is less than the probability of getting a head by flipping a fair coin once, but greater than the probability that a randomly selected state in the United States would be the state in which you live. (The actual chance that a randomly selected state would be yours is 2 %).
+ The probability is less than the probability that a randomly selected state would be the one in which you live, but greater than the probability of throwing boxcars (double sixes) in two successive throws of the dice. (The actual chance of throwing double boxcars is 1 in 1,296).
+ The probability is less than the probability of throwing double boxcars in two successive throws of the dice, but greater than the probability that a randomly selected person in Los Angeles would be you if you lived in Los Angeles. (There are about 2 million people living in Los Angeles).
+ The probability is less than the chance that a randomly selected person in Los Angeles would be you if you lived in Los Angeles, but greater than the probability that a randomly chosen American will be you. (There are about 220 million Americans).
+ The probability is less than the probability that a randomly chosen American will be you.

Subjects were then asked the same basic question, but were asked to respond on a *linear scale* (by circling one + point on the scale below).

Linear Scale

A sure thing ⇒	100 % + 1.00	⇒ Certain
	+	
	95 % + .95	
	+	
	90 % + .90	⇒ Very Likely
	+	
	85 % + .85	______________
	+	
	80 % + .80	
	+	
	75 % + .75	
	+	⇒ Somewhat Likely
	70 % + .70	
	+	
	65 % + .65	
	+	
	60 % + .60	______________
	+	
	55 % + .55	
	+	
	50 % + .50	⇒ Chances about Even
	+	
	45 % + .45	
	+	
	40 % + .40	______________
	+	
	35 % + .35	
	+	⇒ Somewhat Unlikely
	30 % + .30	
	+	
	25 % + .25	
	+	
	20 % + 20	______________
	+	
	15 % + .15	
	+	
	10 % + .10	⇒ Very Unlikely
	+	
	5 % + .05	
	+	
	0 % + 0.00	⇒ Impossible

Subjects were then asked the same basic question again, but were asked to respond on a *log-scale* (by circling one + point on the scale below).

Log Scale

A sure thing ⇒	1 +		*The Probability of:*
	1 / 3 +		
	+		
A 10 % chance ⇒	1 / 10 +	⇐	a randomly chosen
	+		American being Black
	1 / 30 +		
	+		
A 1 % chance ⇒	1 / 100 +		
	+		
	1 / 300 +	⇐	a randomly chosen American
One chance	+		male age 35-44 dying during a
in a thousand ⇒	1 / 1,000 +		one year period.
	+		
	1 / 3,000 +		
	+	⇐	a randomly chosen American
	1 / 10,000 +		being killed in a motor vehicle
	+		accident during a one year period
	1 / 30,000 +		
	+		
	1 / 100,000 +	⇐	getting all heads while flipping a
	+		fair coin 17 times.
	1 / 300,000 +		
One chance	+		
in a million ⇒	1 / 1,000,000 +		
	+	⇐	being killed in an accident during
	1 / 3,000,000 +		a scheduled domestic airline
	+		flight.
	1 / 10,000,000 +		
	+	⇐	selecting a given Californian
	1 / 30,000,000 +		from among all Californians.
	+		
	1 / 100,000,000 +		
	+	⇐	a randomly chosen American will
One chance	1 / 300,000,000 +		be you.
in a	+		
billion ⇒	1 / 1,000,000,000 +		
less than	1 / 1,000,000,000 +		

Finally, subjects were asked to respond to the same basic question, and to answer on a log-scale, but for *all of the years separately* in the 1980's (see scale below). This yielded ten correlated assessments for each subject.

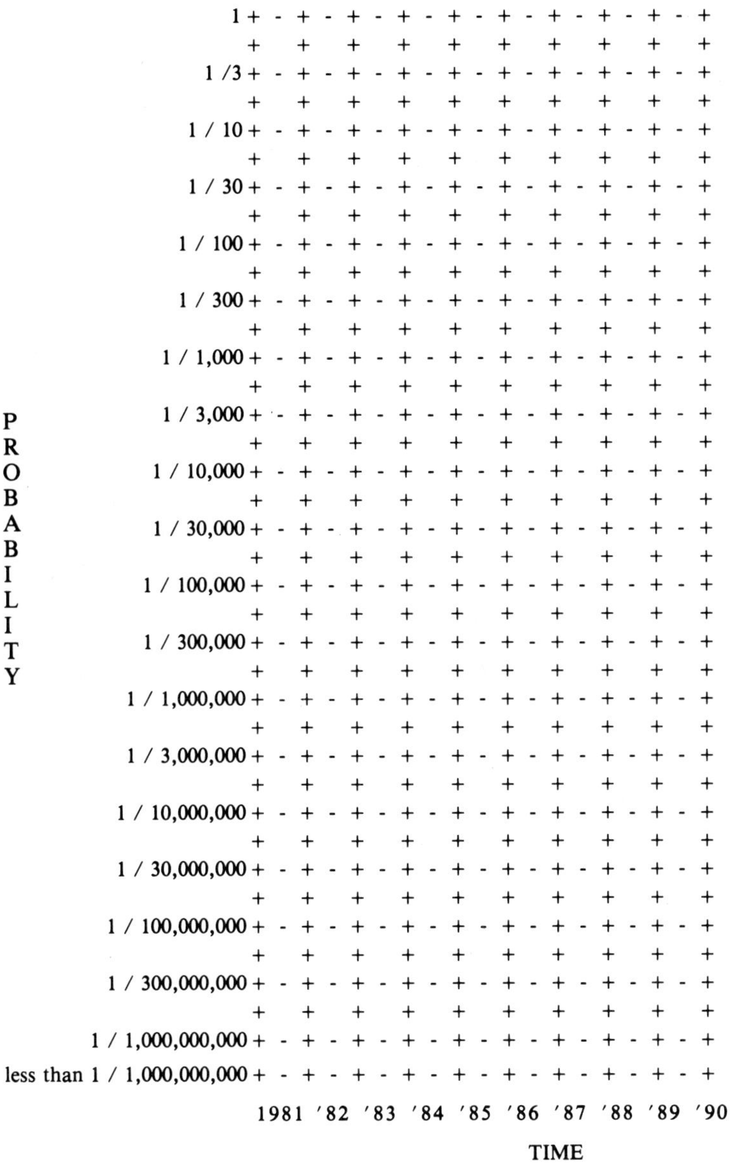

Note that subjects were instructed to assume for their assessments in any given year that no detonations had occurred up to that time. We also note that we might have randomized the order of presentation of the scales to try to minimize any effect that order might have, but we didn't. Note also that on reflection we find that it would have been useful to advise subjects about the implications of their probability assessments (in terms of the Boolian algebra of those assessments) to see whether or not they would be coherent (this suggestion was made by Dennis Lindley). Unfortunately, we cannot go back to do this.

Open Ended Questions

In addittion to being asked this battery of questions, independently on the first round, subjects were given an open ended question; namely, to list all of the reasons they had for answering the basic question the way they did. The reasons given by all subjects were merged into a composite (by eliminating reasons which were paraphrases of one another, non-factual reasons, or reasons which were specific examples of the same issue raised by someone else, in more general terms). In accordance with the procedures suggested by QCF, the composite of reasons was then presented to the same panel along with a second round survey instrument containing the same basic question (for the entire decade), plus other questions suggested by the pool of reasons. The procedure was repeated for a total of three rounds. There were 109 distinct reasons generated by the panel on the first round. Subjects were also asked on the second round to give any new reasons that occurred to them (some 14 new reasons emerged). In addition, they were asked the extent to which they agreed with each of the reasons listed in the composite, and they were asked how important each reason in the composite was in shaping their thinking about the basic question. On the third round subject were presented with the composite of reasons from the first round, plus a composite of the new reasons from the second round (all reasons from the first two rounds are listed in the Appendix, for convenience). Subjects were also asked attitudinal questions about strategically related issues. They were asked the same basic question a third time. In accordance with the procedures of the QCF data collection protocol, in no case was any information about the quantitative probability assessments in the basic question fed back to subjects on a subsequent round. Subjects received only a composite of reasons.

We decided to use a log-scale for responses to the basic question because such a scale affords subjects an opportunity to distinguish among very small probabilities more easily than would be possible with a conventional linear scale. We decided to use multiple scales for responses in order to rule out subjects who could not give the same answer on different scales (they were classified as "inconsistent"). By eliminating inconsistent subjects from our analysis we minimized the effect on results that might otherwise be incorrectly attributed to the nature of the scale. After eliminating inconsistent subjects, 107 subjects remained who were both experts, and who responded with the same answer on all three scales.

Meaning of Subjective Probability

It was explained to the subjects that the subjective probability they were assessing represented their own personal degree of belief about the event. This helped many subjects to understand the inherent nature of the number they were being asked to submit (some subjects at first felt the number should be a frequency for the event, and how could they give a frequency for an event that had never occurred).

Credibility Intervals

Another question in all three survey instruments attempted to measure the subject's confidence in his subjective probability assessment. The subject was instructed:

> Draw the upper line so you feel that for each year, there is only a 5 % chance that the true probability for that year is greater. Draw the lower line so you feel there is only a 5 % chance that the true probability for each year is less. The interval between these lines is called a credibility interval. (Remember that the scale on which you are recording your probability assessment is logarithmic, and thus non-linear. Your lower limit may or may not be the same distance from your original assessment as your upper limit).

Average credibility bounds for the group for Rounds 1 and 3 may be found in Figure 7. Thus, not only do we have measures of location of each of the expert's prior distributions, we also have the .05, and .95 fractiles of their priors.

5.3. *Analysis of Survey Responses*

In this section we develop the prior distributions (and their characteristics) generated by the survey responses for the probability of the nuclear detonation event under consideration, over the decade. We also provide the most frequently given reasons (the complete composite is given in the Appendix).

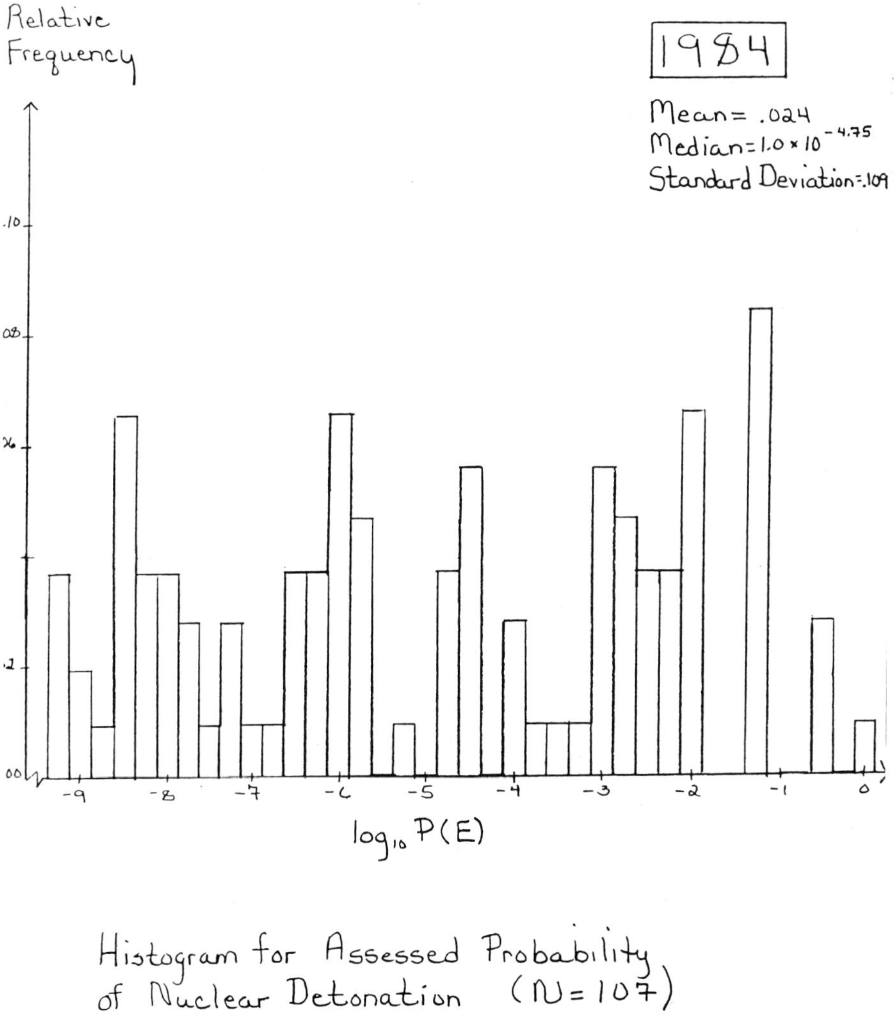

FIGURE 1

Distributions of Responses

Figures 1, 2, 3 are histograms for the Round 1 distributions of responses of consistent experts for the years 1984, 1987, 1990 (for simplicity we present histograms for only three of the ten years of assessments). The abscissas are logarithms of the probabilities. The figures also contains means, medians, and standard deviations. Note that medians are in the vicinity of 10^{-5}, by comparison with means of about .03, while standard deviations are about .1. We note a tendency of subjects to select (on a log scale) responses that are of the form 10^{-n}, $n = 2,3,\dots$.

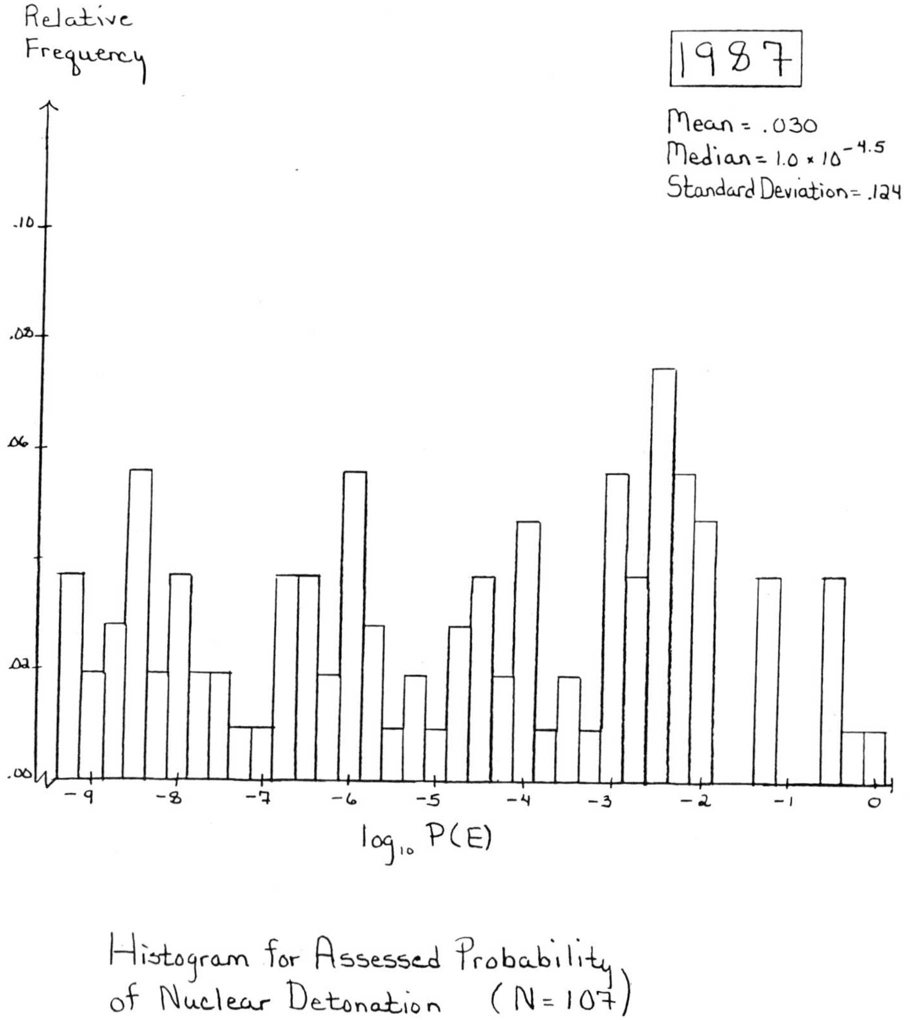

FIGURE 2

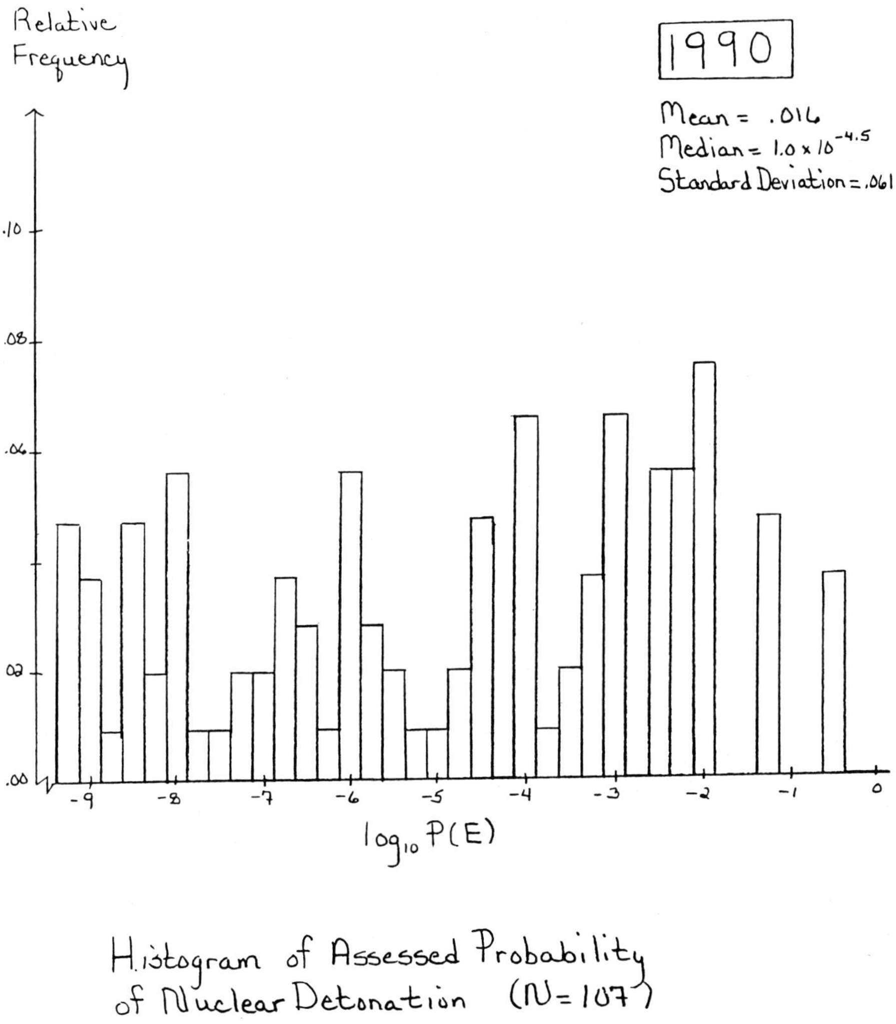
Relative Frequency
1990
Mean = .016
Median = 1.0 x 10^-4.5
Standard Deviation = .061
.10
.08
.06
.02
.00
-9
-8
-7
-6
-5
-4
-3
-2
-1
0
log10 P(E)
Histogram of Assessed Probability of Nuclear Detonation (N=107)

FIGURE 3

We formed ten-dimensional vectors of responses for the 107 consistent experts. These response vectors contained the subjective probability assessments of each subject, for each of the ten years in the decade of the 1980s. The 10-vectors were then merged into a single multivariate prior density by the density estimation technique defined in Sectio 2.4.

Define θ_j:(10×1) as the response vector for panelist j; each component of θ_j is the assessed probability for each of the ten years in the decade, from 1981-1990.

The empirical (multivariate) density of responses (the prior degsity) is given by

$$g_{107}(\theta) = [C^{10}(107)^{1/11}(2\pi)^5]^{-1}$$

$$\cdot \sum_{j=1}^{107} \exp\{- \frac{(107)^{2/11}}{2C^2} (\theta-\theta_j)'(\theta-\theta_j)\} \quad . \tag{2}$$

The 1-dimensional marginal density for the ith component is given by

$$g_{107}(\theta^{(i)}) = [C(107)^{10/11}\sqrt{2\pi}]^{-1}$$

$$\cdot \sum_{j=1}^{107} \exp\{- \frac{(107)^{2/11}}{2C^2} (\theta^{(i)} - \theta_j^{(i)})^2 \quad . \tag{3}$$

The smoothing constant C was taken to be C = .1 because that choice yielded a smooth, well behaved density on a scale that was readily interpretable[4]. We have plotted three of the Round 1 marginal densities for 1984, 1987, 1990, in Figure 4. The abscissa is logarithm of the probability. Note that on this scale the distribution is approximately uniform over a wide range (a range of $10^{-9.25}$ to 10^{-2}). Figure 5 is a plot of the empirical density for the three illustrative years, for both Round 1 and Round 3, on a conventional linear probability scale, for the panel of consistent experts. It should be noted that the Round 1 prior density is more peaked than the Round 3 prior density and is uniformly greater for all probabilities in (0,.1). In Figure 6 we see the same curves plotted (but for smaller samples) for the panel of consistent non-experts, just for comparison. The greater peakedness of the Round 1 density is still manifest as is the dominance of the Round 1 density in (0,.1). Note that in Figures 5 and 6 the sample sizes differ from Rounds 1 to 3. Because we were concerned that perhaps the people who participated in Round 1 but not in Round 3 might be more pessimistic about war (and would therefore be inclined to assess higher probabilities than other subjects), we reran the analysis of Round 1 using just the Round 3 subjects. We found that there was very little difference in the results shown in Figures 5 and 6, reflecting the fact that the subjects who did not complete Round 3 questionnaires were a representative sample (the non-respondents were relatively unbiased). The multivariate prior density in equation (1), and the marginal densities in equation (2) may now be used in Bayes theorem with likelihoods of data obtained from strategic policy planners to obtain posterior densities of the nuclear detonation probabilities.

Averaging the Expert Opinions

It may be of interest to examine the mean assessed probability of nuclear detonation for all of the consistent experts as a function of year in the decade. We caution, however,

[4] By plotting the fitted density for values of C from 10^{-5} γ 10^2 it was found that the density become extremely noisy for $10^{-5} < C < 10^{-4}$, and seemed too smooth for $1 < C \cdot C = 10^{-1}$ yielded the required structural form without overfitting.

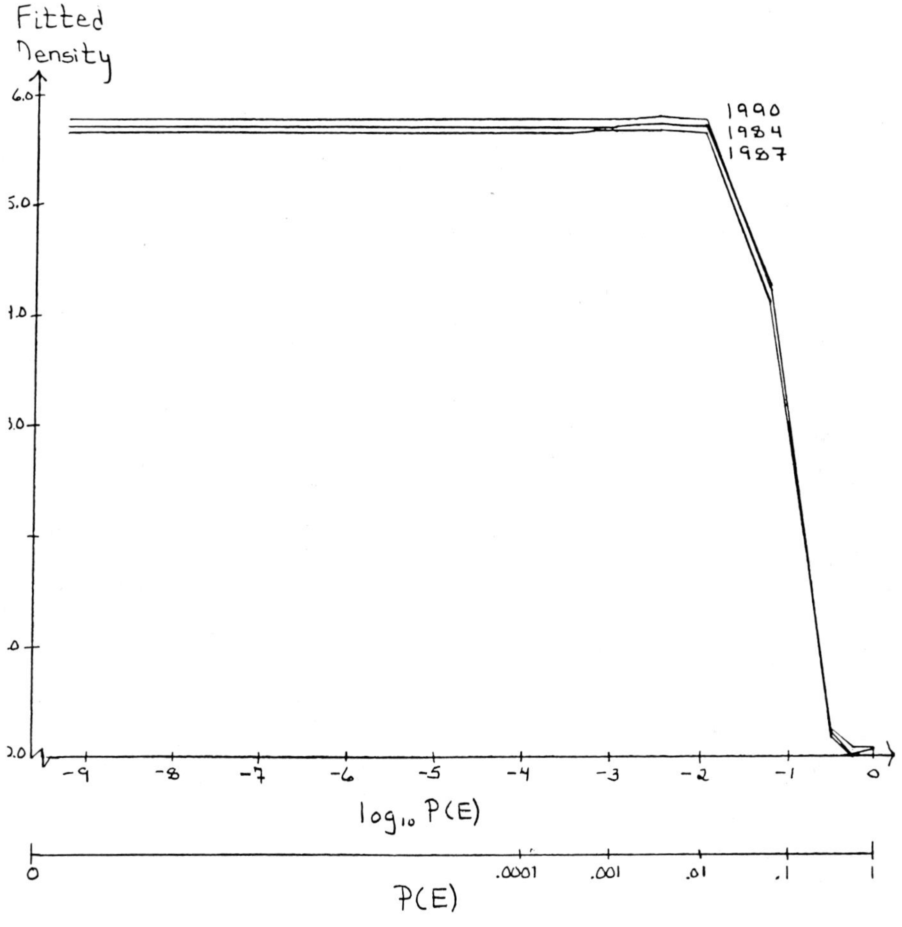

Fitted Density for Assessed Probability of Nuclear Detonation (N = 107) (Log Scale)

FIGURE 4

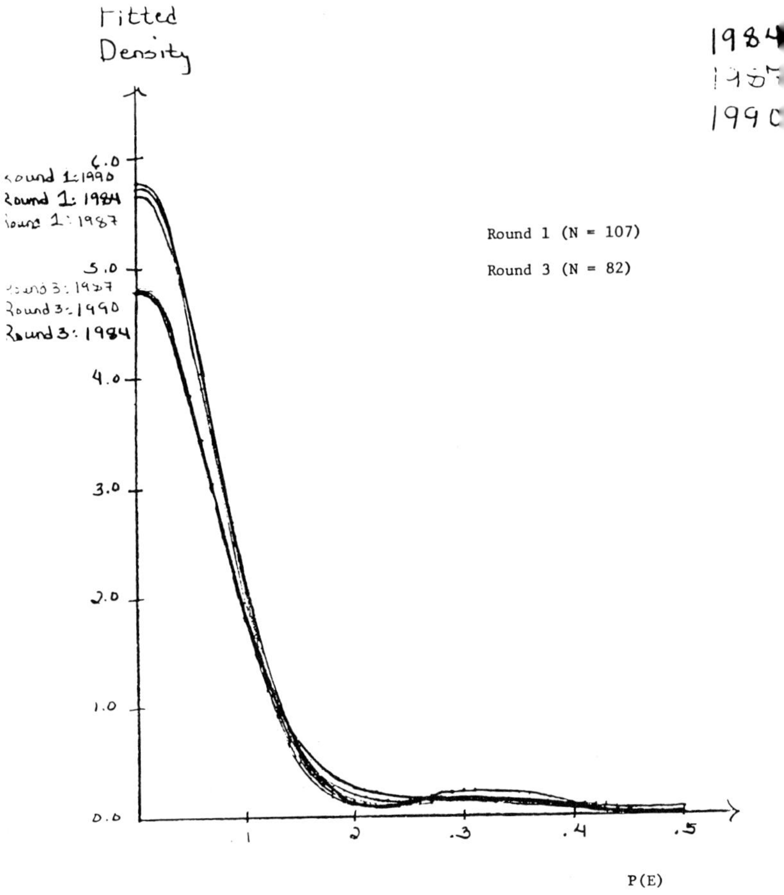

Fitted Prior Density of Consistent Experts

FIGURE 5

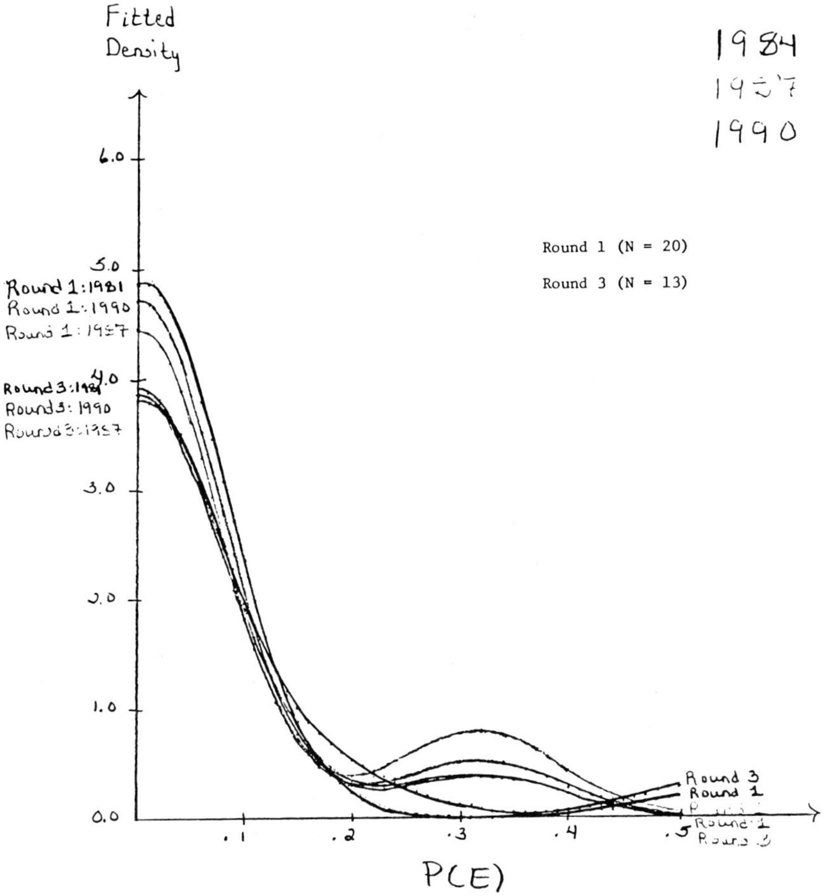

Fitted Prior Density of Consistent Non-Experts

FIGURE 6

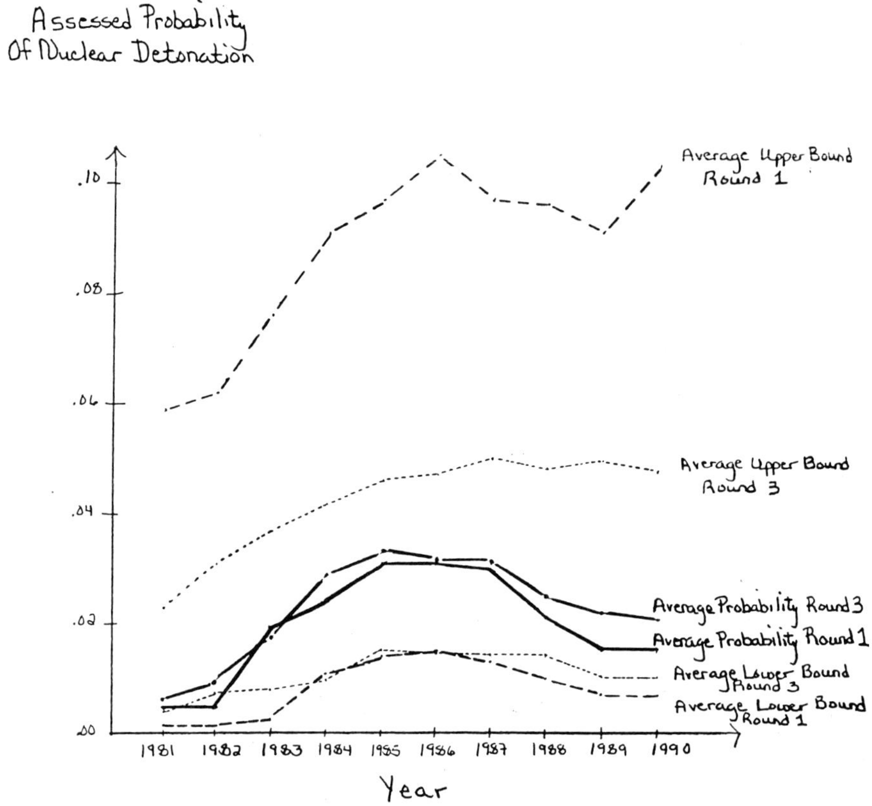

Assessments of Consistent Experts
Round 1 : n = 107
Round 3 : n = 82

FIGURE 7

that such a mean assessment should be constructed only in terms of an underlying *homogeneous* population (since then the sample mean can be expected to be consistent). In the face of inhomogeneity, such an averaging of beliefs can be misleading. We have done our best to achieve homogeneity in terms of expertise, as measured by education and contextual knowledge, but there are still uncertainties about other factors (attitudinal, discipline-based variations, etc.). In Figure 7 we present the average assessed probability of nuclear detonations as a function of year. We have also taken the confidence (credibility) statements of each of the respondents and averaged them to obtain average upper (95 %) and lower (5 %) bounds for the assessed probability of nuclear detonation. Note in Figure 7 that from Rounds 1 to 3 the credibility intervals tighten substantially, reflecting the fact that respondents felt increasingly confident about their responses after three rounds. It should be noted that the mean assessed probability (for the consistent experts) never exceeds about 3 % (it peaks in 1985) throught the decade on Round 1. It is difficult to know whether the absolute assessed probabilities are really meaningful, but it would seem reasonable to conclude that the 1985 peaking of the mean assessed belief of experts reflects a real perception of this group. The medians, by sharp contrast, never exceed about .003 % on Round 1 (three of them are given in Figures 1-3). The same is true for Round 3, and for the Round 3 totals of consistent experts and non-experts. Round 3 medians of consistent non-experts were uniformly smaller than those of consistent experts.

Suppose we were to make the simplifying, but obviously incorrect, assumption that for each subject, every year's assessment is independent of every other year's and that all years have the same assessed probability value of .03 (this value is the maximum of the averages for the group, over the decade, on Round 3). Then the probability of exactly one such nuclear detonation taking place during the decade of the 1980's would be given by the binomial probability

$$\binom{10}{1}(.03)(.97)^9 \cong 23\ \%\ ,$$

a startlingly large number to many people for such a horrendous event. Perhaps if we had informed subjects of such implications of their assessments they would have reduced their magnitudes; there is no way to tell a posteriori.

The Most Frequently Given Reasons

Our pool of reasons was generated by the sample of 186 respondents (combined experts and non-experts) on Round 1. Three judges were used (using majority rule) to classify all reasons that were questionable regarding whether two reasons were equivalent, and to resolve other ambiguities. On Round 1, the reason given most often (14 % of the respondents) was #97. This reason centers around the fear of respondents that third party nations (other than the U.S. or the U.S.S.R.) will become involved in a nuclear conflict, because of nuclear proliferation. Respondents were also concerned about nuclear involvements occurring over Middle East oil (Reason # 76) (13 % of respondents gave this reason), and occurring because of accidents (12.9 % of respondents gave Reason #83). The three most frequently given reasons are listed in Figure 8. A complete listing of all of the reasons in the composite, along with their response frequencies, is given in the Appendix. The reasons are listed there in the order and format in which they were presented to the panel of respondents (but response frequencies were not given).

It would be of interest, from a substantive point of view, to relate the reasons given by a particular respondent to the background of that particular respondent. It would also be of interest to study whether specific schools of strategic thought emerge; to develop some of the policy implications of the reasons generated by the panel; to correlate the attitudes of respondents with their probability assessments; and to study how these same experts viewed a variety of other related issues. While these questions are of paramount importance substantively, they have only peripheral interest in this methodological research; such issues will therefore be treated elsewhere (see Gallavan, 1985).

Reason #	Reason	*Response* Frequency	*Response Frequency* Rank
97	Increasing world nuclear proliferation increases the chances of nuclear war among other third party nations facing critical economic, political, and demographic problems, and makes third parties much more likely to get into direct nuclear conflict than the U.S. and the U.S.S.R.	14 %	1
76	The U.S. and the U.S.S.R. both view Middle East oil as vital to their economies and will find their oil contention peaking over the next decade causing them to become more directly involved in unstable Third World countries, perhaps using military force to secure applies.	13 %	2
83	Probability of accident (computer, radar, or human failure) causing a nuclear strike is greater than the probability of U.S. or U.S.S.R. intentionally striking the other side.	12.9 %	3

Principal Reasons for Assessments Given By 186 Respondents

FIGURE 8

6. PSYCHOLOGICAL FACTORS RELATING TO SUBJECTIVE PROBABILITY ASSESSMENT

This paper has been concerned with assessing multivariate prior distributions by combining the judgments of a group of human subjects. The methodology proposed is statistical, but there are also many important aspects of the assessment problem that originate in cognitive and social psychology. Such factors must be understood if any reasonable degree of closeness to "truth" is to be claimed. We briefly comment on some of these factors below, as they relate to the empirical problem of this paper.

Major heuristics and biases that exist in subjective probability assessments were addressed in an important work by Tversky and Kahnemann (1974). Related work has been carried out by many others; e.g., Fujii et al. (1977); Nisbett and Ross (1980); and Lichtenstein et al. (1978). The literature on the subject is currently large, and is growing.

The basic Tversky and Kahneman work described three heuristics that are employed in making judgments under uncertainty. The first is "representativeness", which is usually employed when people are asked to judge the probability that an event A belong to class B. In making judgments in such instances, subjects generally evaluate probabilities by the degree to which A is perceived to be similar to, or representative of B. Such perceptions ignore prior information that includes known relative frequencies of such events, and therefore, such judgments are generally biased. The second heuristic they described is "availability", in which judgments of frequency are made on the basis of instances, or scenarios that the subject can readily construct, in which similar events occurred. Such judgments can be grossly inaccurate in many ways. For example, recent events are more likely to be remembered than distant events, lack of interest in, or exposure to, certain phenomena can mask their real prevalence; etc. Their third heuristic is "anchoring", which is the tendency to start a numerical prediction from a fixed (but often irrelevant) point, and then adjusting the final value in minor ways from the fixed anchoring point. Clearly, if the anchor is greatly in error, the numerical prediction is likely to be poor.

We next enumerate some heuristics that may have been used by the subjects in our survey on the probability of nuclear detonation in the decade of the 1980's. It is not known whether such heuristics were actually employed; moreover, if they were employed, it is not known how much bias would have been introduced into the subjective assessments of our subjects as a result. The following conclusions derive from research in cognitive and social psychology; sources may be found in the References.

Potential Sources of Assessment Biases

(1) *Desirable/undesirable events* - it is known that subjects tend to underassess the probability of an undesirable event (such as nuclear war), and to overassess the probability of a desirable one. In our study this effect should have given rise to probability assessments that are too low.

(2) *Availability heuristic* - subjects tend to underassess the probability of events that occur infrequently, because they cannot readily construct scenarios for the event in question. For this reason, subjects might tend to underassess the probability of an event that has never occurred (such as nuclear war). On the other hand, because subjects were asked to think hard about scenarios (reasons) for why nuclear detonations might or might not take place, their assessments are perhaps thereby inclined upward overall. However, we feel that use of this heuristic would again give rise to probability assessments that are too low.

(3) *Low/high probability events* - it is known that subjects generally tend to overassess very low probability events, and to underassess very high probability events. This may be a numerical phenomenon that derives from a person's inability to conceptualize and distinguish among extremely low or extremely high frequencies. Since the event of nuclear detonation in our problem is generally perceived to be a very low probability event, the effect attributable to this source of bias is surely a set of assessments that are too high.

(4) *Anchoring heuristic* - one effect of the anchoring heuristic is for subjects to underassess the probability of disjunctive events. We consider the carefully nurtured delicate balance among nations with respect to their differing economic, political and social systems as a complex construct involving many built-in safeguards against the breakdown of the construct, and the exercise of the ultimate option of disagreement-war. Since complex systems tend to break down when any of their essential components fails (disjunctive events), anchoring in our problem would most likely lead to assessments that are generally too low. (There are of course both conjunctive and disjunctive events involved in the compounding of events leading to nuclear war, but our belief is that the disjunctive effects would dominate).

Another effect attributable to anchoring is that once subjects state their first round subjective assessments, they are inclined to retain those positions and beliefs on subsequent rounds regardless of rational arguments presented that should logically dissuade them from those positions. This effect was actually - bserved in the round-to-round data obtained. It was found that most subjects held to their original positions on subsequent rounds, and those who did change, altered their first round positions only minimally (there were a few exceptions to this but they were rare and usually attributable to limited understanding of the numerical assessment problem).

We have listed above some possible sources of bias in our subjective probability assessments of nuclear war. There are undoubtedly other sources of bias we have not mentioned. It would be extremely useful to know more about these sources of bias in quantitative terms. We would like to know how to measure the amount of bias likely to be introduced by use of a given heuristic; how to determine whether or not a given heuristic has been used (perhaps we should ask the respondent?); and how to evaluate situations in which several competing heuristics are interacting in a given subject, and across subjects. These are all topics for further research. For our problem, we must be content at this time to know that these biases are very likely to exist in the data, but we don't know how to evaluate their combined effects. Most of the effects discussed here appear to lead to underassessment. This suggests that our final subjective probability assessments are somewhat low, and therefore perhaps they should be viewed as lower bounds to the underlying beliefs of our group of experts.

REFERENCES

FRIEDMAN, J.H., STUETZLE, W. and SCHROEDER, A. (1981). Projection Pursuit Density Estimation, *Tech. Rep. ORION* **002**, Dept. of Statist. Standford University.

FUJII, T., SEAVER, D.A. and EDWARDS, W. (1977). New and Old Biases in Subjective Probability Distributions: Do They Exist and are They Affected by Elicitation Procedures, *SSRI Res. Rep.* **77-4**, University of Southern California, Social Science Research Institute.

GALLAVAN, R. (1985). An Experiment in Collective Strategic Assessment, Dissertation in the Dept. of Statistics, University of California, Riverside.

GOKHALE, D.V. and PRESS, S.J. (1982). Assessment of a Prior Distribution for the Correlation Coefficient in a Bivariate Normal Distribution, *J. Roy. Statist. Soc. A,* **145**(2), 237-249.

HARMAN. A.J. and PRESS, S.J. (1978). Assessing Methods in Policy Formulation, Basel: Birkhauser Verlag, edited by D.W. Bunn and H. Thomas, pp. 123-147.

HOGARTH, R.M. (1978). A Note on Aggregating Opinions, *Organizational Behavior and Human Performance,* **21**, 40-46.

LICHTENSTEIN, S., SLOVIC, P., FISCHHOFF, B., LAYMAN, M. and COMB, B. (1978). Judged Frequency of Lethal Events, *Jour. of Experimental Psych: Human Learning and Memory,* **4**, n° 6, 551-578.

NISBETT, R. and ROSS, L. (1980). *Human Inference: strategies and short-comings of social judgment.* Englewood Cliffs, New Jersey: Prentice-Hall.

PRESS, S.J. (1978). Qualitative controlled feedback for forming group judgments and making decisions, *J. Amer. Statist. Assoc.,* **73, 363**, 526-535.

— (1983). Group assessment of multivariate prior distributions, *Technical Forecasting and Social Change,* **23**, 247-259.

RAO, C.R. (1965). *Linear Statistical Inference and its Applications,* New York: John Wiley and Sons, Inc., 118.

TOURANGEAU, R. (1983). Cognitive Science and Survey Methods, paper prepared for the Advanced Research Seminar on Cognitive Aspects of Survey Methodology, Committee on National Statistics, U.S. National Academy of Sciences, Wash. D.C.

TVERSKY, A. and KAHNEMAN, D. (1974). Judgments under Uncertainty: Heuristics and Biases, *Science,* **185**, 1124-1131.

ACKNOWLEDGEMENTS

I am grateful to: Rosemary Gallavan for her assistance in performing numerical calculations and for helping to prepare graphs; to The Rand Corporation for its financial support to collect the data for the empirical work reported; to Dr. John Rolph for making the empirical data available for analysis; to Dr. Carl Hensler, and Michael Kanzelberger, and to Col. Michael Seaton of the U.S. Air Force, for helping to design the survey intruments; to Dr. Alvin Harman for suggesting that the empirical study be undertaken; and to Professor Judith Tanur for her suggestions for improving the exposition. R. Tripathi, L.C. Chang, and C. Yang assisted in classifying the reasons given by the respondents. Finally, I am grateful to my two Discussants, Dr. Philip Brown and Dr. William DuMouchel and to other participants in the Second Bayesian Valencia Conference, September 6-10, 1983, for raising meaningful questions which helped me to formulate my thinking about my final draft of this paper.

APPENDIX

Below, we give the composite pool of 109 reasons given by the panel of QCF respondents for their beliefs about the basic nuclear detonation event. The composite was generated by the 186 respondents on Round 1. For convenience, the reasons have been grouped by category. The composite is followed by a composite of 14 new reasons generated by the panel on Round 2. The column on the right gives the percent of respondents who gave each of the reasons on Round 1.

REASON ELICITED BY ROUND 1
UNITED STATES

LEADERSHIP		*Response Frequency*
1.	US election of Reagan will reduce chances of nuclear exchange.	1.1 %
2.	US President Reagan will initiate a hardline militaristic policy that will gradually cause deterioration of US/SU relationship.	3.8
3.	US is not likely to get strong leadership in the future.	2.7
4.	Additions to US military capability such as new rapid deployment forces will cause US leadership's confidence to increase and US to increasingly challenge the SU.	0.5
5.	US leadership that regains confidence in its own military power and stands up to the Soviets in the international arena will contribute to reduced tensions in US/SU relationship.	1.1
6.	Never before has such a small group of people been able both to make the decision to go to general war and to execute that decision.	0.5
7.	US defense policymakers who have a tendency to look for acceptable and controllable applications for nuclear weapons have an image of nuclear war that is too neat and sterile, lacking appreciation of the likely horrible societal and individual human consequences.	1.1
8.	US attention to the vulnerability issue of US land-based missiles has increased slightly the uncertainty of US decisionmakers about the viability of the US nuclear deterrent and increases the likelihood that they will behave rashly.	3.8

PUBLIC OPINION		*Response Frequency*
9.	The US public is becoming more hawkish - more supportive of military expenditure and military intervention abroad.	3.2

10.	US national character and moral principles precludes a nuclear first strike and supports a reluctance to use nuclear weapons under any circumstances.	5.4
11.	US is developing a non-rational disillusionment with the Soviet Union at the same time as it develops a non-rational romanticized image of China.	1.1
12.	US public does not have the will to compete with SU militarily.	0.5

POLICY

13.	US mutual assured destruction (MAD) doctrine remains the real cornerstone of US strategic defense despite doctrinal flirtations with limited nuclear retaliation concepts as in Presidential Directive (PD) 59.	1.1
14.	US policy of detente as practiced had the critical flaw of not allowing for US efforts to redress the military balance as required.	1.1
15.	To protect US national interests in sensitive areas, US should rely principally on the threatened use of tactical nuclear weapons and the concomitant risk of escalation for reasons of economy.	0.5
16.	US suffers from a breakdown of a broad foreign policy consensus that endured from WWII to the Vietnam War.	1.6
17.	US inability to develop a foreign policy consensus has produced defense policies permeated with wishful thinking and an unwillingness to use military force in regional disputes.	0.5
18.	US involvement in the internal politics of underdeveloped countries will increase.	1.1
19.	US is in a period of economic malaise that will continue.	2.7

FORCES		*Responses Frequency*
20.	US lacks effective civil defense and post-attack recovery and reconstitution capabilities which temps the Soviets to think that the US will fare much worse in a nuclear exchange.	0.5 %
21.	US land-based missile vulnerability is not sufficient to warrant massive remedy by MX deployment.	0.5
22.	The invulnerability of US strategic submarines can be expected to be severely threatened in this decade.	0.5
23.	US conventional military capabilities are inadequate to counter the SU in areas in which confrontations are likely to occur.	1.1
24.	US scrapping of B-1 bomber and neutron bomb programs, and delays in other military programs, will give the Soviets an exploitable political advantage in strategic armaments without their having to take any action of their own.	0.5
25.	US arms increases and improvements will lessen SU willingness to resort to use of military force.	0.5

SOVIET UNION
LEADERSHIP

26.	Brezhnev's death will cause instability in Soviet leadership.	2.7
27.	SU next leadership is likely to be dominated by younger, more aggresive and less cautious men.	3.2
28.	SU new leaders will be less doctrinaire and more pragmatic.	2.7
29.	Worldwide negative public opinion acts as a significant deterrent to the SU using nuclear weapons.	1.6

POLICY

30.	SU will feel less inhibition than US to detonating a nuclear weapon to demonstrate resolve in a crisis and would not let moral considerations impede use of nuclear weapons.	0.5
31.	SU places genuine credence in the eventual triumph of socialism and can be expected to use nuclear weapons to eliminate the US as a bastion of capitalism if it perceives a favorable balance of forces for this purpose.	1.6
32.	SU will become more conservative and committed to the status-quo over time.	2.7
33.	SU is arming out of fear and determination not to become the victims in war again as memory of WWII calamity must still be vivid in minds of SU leadership.	1.1
34.	SU understands that the industrial democracies are extremely vulnerable to an oil and raw material shut-off and has a long term strategy to deny the West control or access to Middle Eastern oil and central and southern African mineral wealth.	4.3
35.	SU successful invasion of Afghanistan demonstrates SU willingness to use military capability to pursue international objectives.	3.2
36.	SU believes its military doctrine that nuclear war can be fought and won.	0.5
37.	SU has an increasing tendency to sign «Friendship Agreements» which have the appearance of military alliance as well as economic alliance and if invoked, would draw member nations into a conflict with a greater momentum than might occur otherwise, as in WWI.	0.5
38.	SU leaders will be tempted to divert attention of domestic population from chronic and persistent food shortages and other economic problems with military encroachment on more fertile lands.	4.8
39.	SU first strike on US, if it occurs, will probably be caused by Soviet domestic problems such as a disintegration of Soviet society, a leadership crisis, or a last ditch effort by a Soviet leader to keep power.	1.1
40.	SU is jockeying for position in the Middle East in anticipation of it own oil needs and can be expected to make a move in the region within this decade that is likely to lead to a US/SU conflict.	3.2
41.	SU is not content with its world position and is driven by an imperialistic dynamic consistent with the czarist tradition, thinly veiled by communist ideology that views the US as the major obstacle.	0.5
42.	SU will compete in the world market to offset shortfalls in petroleum rather than use military force in the Middle East.	0.5
43.	SU does not need to engage in nuclear exchange because they are making sufficient progress in subverting the world to their will, maintaining a comfortable rate of expansion, and achieving strategic objectives.	4.8
44.	SU prefers slow but progressive adventurism to direct confrontation with the West.	1.1
45.	SU will become a net importer of oil, less self-sufficient in energy overall, and less able to afford the cost of energy required by its Eastern European allies within this decade.	3.2

FORCES

46.	SU can be expected to question the invulnerability of its own strategic forces by late in the decade under current trends of US weapons deployment including the MX missile.	0.5
47.	The SU has civil defense programs which would be of political advantage in a serious crisis and might make a significant difference to the outcome of a nuclear exchange with the US.	0.5

48.	Because of the superior strength of Soviet conventional forces, they are less likely than the US to escalate to tactical nuclear weapons.	2.1
49.	A SU technological breakthrough in defense against US nuclear retaliation would bring about a Soviet preemptive strike against the US.	1.1
50.	SU relative nuclear advantage and conventional superiority in many likely areas of confrontation will embolden Soviets to undertake other geopolitical risks like those taken in Afghanistan.	4.3
51.	SU will endeavor to keep any armed conflict within the conventional realm because of their superiority in this category of armaments.	0.5
52.	SU has demonstrated resolve to overpower US and NATO by military force by their vigorous military build-up in all categories of armaments.	2.7
53.	SU main and most durable resource is military power which it can be expected to use to stem reverses in international political influence and energy situation, and to stifle political and economic dissent at home.	3.8
54.	SU economic and agricultural development needs will divert attention and resources from military build-up and increase SU industrial, agricultural, and economic interaction with the West.	2.1
55.	US/SU relations are worsening and will continue to erode, increasing the number of risky confrontations.	1.1
56.	US and SU find the political utility of nuclear weapons to be in their ability to project convincing threats without using them.	3.2
57.	US and SU fears about the use of nuclear weapons are being dulled by so much discussion about their potential use.	0.5
58.	Neither US nor SU have a reasonable chance of achieving effective strategic nuclear superiority in this decade that could be used to disarm the adversary or hold significant political advantage over him.	4.8
59.	Even if US/SU nuclear forces are mismatched or not at parity by all measures, the US and NATO allies could cause large scale strategic damage to the SU, guaranteeing that the probability of nuclear exchange will remain low throught the decade.	2.1
60.	US/SU previous crisis behavior indicates that each is smart enough not to cause the other to suffer a sizable and historically unprecedented diplomatic or political defeat.	2.1
61.	US efforts to redress self-perceived deficits in military balance will tempt Soviets to use depreciating military assets before they lose too much of their relative advantage.	3.2
62.	US and SU will find alternative sources of energy to take the pressure off Middle Eastern oil within this decade.	1.6
63.	US and SU leadership change is bound to create tension at first, but will be ameliorated as new leaders come to understand positions of others.	2.1
64.	Despite the pronounced variations in US and SU declaratory strategic deterrent doctrine, both nations have accepted the concept of mutually assured destruction because neither has an incentive to strike first.	10.2
65.	Both US and SU leadership feel strong antipathy toward use of nuclear weapons.	3.2
66.	Probability of war between US and SU is inversely related to the degree of economic integration between the two countries.	1.6
67.	US and SU common interests are increasing making it reasonable to expect that they will return to a policy of detente within this decade.	3.8
68.	US and SU safeguards against accidental or unauthorized launch, including emergency communications to explain unusual situations, are sufficient to make	

the probability of accidental nuclear war so low that it can be ignored. 7

69. US and SU development of more diversity and precision in strategic weaponry and defensive technologies will further increase the stability of mutual deterrence, make it less likely that there will be any worthwhile payoff from nuclear war, and thus lessen the chances of a strategic nuclear exchange. 10.8

70. US/SU use of limited nuclear strikes with small-yield nuclear weapons carries too great a risk of uncontrolled escalation to all-out war and societal destruction to be seriously considered as a viable policy option by either nation. 7

71. **US and SU are both highly uncertain about the other's intentions and behavior, the performance of both side's weapons, and the outcomes of the use of nuclear weapons. This deters use of nuclear weapons. 4.8**

72. US and SU national interests will not clash sufficiently to warrant nuclear war and in any case, both have sufficient alternatives to exert force on one another to obviate the need to use nuclear weapons for gaining social, political, and economic goals. 5.4

73. US military strength is declining while SU military strength is increasing, resulting in a general military balance in the 80's that favors the Soviets. 8.1

74. SU use of nuclear weapons is most likely in the middle of this decade when SU has the greatest net advantage over the US in strategic arms and hence can expect the smallest relative adverse consequences of such action. 4.8

75. The way conventional hardware costs are escalating and technological improvements are eroding the distinctions between nuclear and conventional weapons, it is becoming easier to justify the application of limited nuclear force. 1.1

76. US and SU both view Middle East oil as vital to their economies and will find their oil contention peaking over the next decade causing them to become more directly involved in unstable Third World countries, perhaps using military force to secure supplies. 13

77. **US/SU mutual deterrence relationship is being destabilized by Soviet's continual attempts to achieve a disarming first strike capability through continually expanding strategic counterforce programs. 2.7**

78. US and SU are both capable of achieving a disarming first strike nuclear capability in this decade. 1.1

79. US and SU improvements in accuracy and yield-to-weight ratios of strategic nuclear weapons make it thinkable to use nuclear weapons in a preemptive counterforce strike that attempts to avoid the mass destruction of civilians and collateral damage and therefore makes some kind of nuclear detonation more likely. 1.6

80. US/SU nuclear exchange is most likely to come from escalation from a lower level of conflict, such as from the Soviet response to US tactical nuclear defense against Warsaw Pact invasion of Western Europe. 8.6

81. **US and SU use of nuclear weapons is more likely to be initially in some third country where each can watch the reaction of its counterpart and the world. 2.7**

82. US and SU development and deployment of smaller, more accurate nuclear weapons makes deterrence threats more credible and potent by making limited nuclear responses to provocation possible. 3.2

83. Probability of accident (computer, radar, or human failure) causing a nuclear strike is greater than the probability of US or SU intentionally striking the other side. 12.9

84. Random events (like the Iran-Iraq war) occurring at the same time as a general deterioration of US/SU relations represents the most likely causal path to an intentional nuclear exchange. 2.1

85. The longer there has been peace, the more the US and SU strategic capabilities will have a chance to equalize, and so the chances of war will decline. 1.6

86. When US capabilities are at parity with, or superior to, SU strategic capabilities, the chances of war will decline. 5.9

87. US efforts to redress self-perceived deficits in military balance by end of decade will cause US and SU to engage in much dangerous military posturing to demonstrate relative positions. 2.7

88. US nuclear alliance with China will aggrevate SU temptation to preempt in a crises, especially as Chinese nuclear threat increases. 3.8

89. US strategic arms superiority gained through superior production capability will provide the Soviets with an incentive to negotiate future strategic arms limitations. 0.5

90. US/SU abandonment of SALT process will lead to an intensified arms including an increase in the number of SU nuclear weapons threatening the US in the next five years. 4.3

91. If US or SU develops a fairly effective anti-ballistic missile defense system within this decade, the SALT ABM protocol will be set aside, officially or not. 2.1

92. US and SU will negotiate further strategic arms controls. 4.3

93. Unfavorable public opinion will be potent enough in both US and SU to prevent either nation's leader from embarking on a large-scaled armed conflict or from using nuclear weapons. 5.4

94. The use of nuclear weapons is a far scarier act to contemplate for real political leaders than it appears to strategic analysts who view it as an abstract intellectual exercise, and even the analysts themselves often find it difficult to use nuclear weapons in politico-military gaming exercises. 4.8

THIRD WORLD

95. World economic situation will improve and stabilize over the next decade and increse political stability. 1.6 %

96. In spite of limited conflicts throughout the world, humanity has become more rational and large scale wars appears less likely. 4.3

97. Increasing world nuclear proliferation increases the chances of nuclear war among other third party nations facing critical economic, political and demographic problems, and makes third parties much more likely to get into direct nuclear conflict than US and SU. 14

98. Increasing population, urbanization, and external debt generated by oil imports is putting a strain on the social and economic systems of Third World countries that will lead to breakdowns, social unrest, and the temptation to seek military solutions. 5.4

99. Worldwide terrorism and violence is on the increase and terrorists can be expected to intensify efforts to obtain a nuclear weapon this decade. 3.8

100. World education and communications have improved dramatically, improving the prospects for peace. 1.1

101. Regional conflicts can be expected to be resolved using surrogates or proxies (smaller, less powerful client states) at which the Soviets have demonstrated a superior aptitude. 1.1

102. Western European energy scarcity will cause these nations to display more aggressive stance with respect to the Middle East. 2.1

103. NATO does not have a credible nuclear deterrent. 0.5

104. That an antagonistic mentality and rationality dominates relations among nations has been demonstrated historically. 0.5

105. The Middle East situation will continue to deteriorate with the continued rise of Islamic nationalism, perhaps even including the fall of Saudi Arabia to non-traditional leadership within this decade. 2.7

106. Yugoslavian internal politics and national problems can reasonably be expected to erupt at any time and produce conflict between US and SU. 0.5

107. Improved communications among world's nations exacerbates social and political tensions by offering vivid and immediate comparison of different material conditions, laying bare the increasing gap between rich and poor nations. 0.5

108. SU Eastern European satellites will become more restive over time with current SU economic policies and can be expected to use leadership succesion period to test limits of Soviet control and increase relative autonomy, perhaps seriously affecting the cohesion of the Warsaw Pact. 2.1

109. US and SU will be put on the defensive by a power shift to the Third World holders of critical materials and driven closer together to the point where they will develop ways of resolving their conflicts peacefully at the expense of Third World countries. 2.1

NEW REASONS ELICITED BY ROUND 2

1. Ronald Reagan has been elected President of the United States.
2. Senate elections have put more Republicans in power.
3. Unrest continues in Poland.
4. The Soviet Union announced discovery of huge oil reserves in Siberia.
5. There has been a rapid rise of anti-semitism which indicates growing international unrest and pressures for war.
6. The United States is not content with its world position, will seek greater wealth, and come into conflict with more nations in the future.
7. The uncertainty surrounding the outcome of a strategic nuclear exchange is more important than the actual strategic force balance for determining the likelihood of such an event.
8. American military leaders exhibit conservatism when confronted with the actual prospect of going to war.
9. The US economy will not permit the military build-up necessary to off-set Soviet advantages in the strategic balance of forces.
10. The nature of Reagan's advisors will make American entry into some kind of war more likely.
11. New technology will permit some third party to deliver nuclear weapons by radio-controlled cruise missiles that would be difficult to stop.
12. The United States and the Soviet Union will only initiate conflicts with less capable opponents.
13. Assessment of the subjective probability of a strategic nuclear exchange should include consideration of political context, and the interplay of many factors over time.
14. One respondent wished to be paraphrased exactly this way: «The Afghanistan invasion and the SU failure to invade Iran since the Shah's departure strongly confirm the thesis that the Soviet Union will only take overt military action to support a native communist regime or keep a communist-run country from backsliding».

DISCUSSION

J.R.M. AMEEN (*University of Warwick, U.K.*)

Professor Press's paper deals with an interesting and very serious matter namely the probability of detonating a nuclear weapon by either the US or SU in the other's homeland. It would have been of more interest for the study to be world-wide and some degree of confidence for the reported results in Fig. 7 should have been given.

As admitted by the author at the very end of the paper, the whole of the battery is deceived into under estimating the probability assessments by the ordinal scale and the very sharp log scale. The inconsistent assessors seem to be among those who have assessed high probabilities on the linear scale and had little chance to perform a close assessment on the log scale since the density of the consistency is not uniform on the full scale. It is likely that the types of questions and their ordering implicitly educates but has a critical effect on the assessed probabilities. For example the experts should be reminded whether their very small probabilities of this year will lead to an increase in arms production and deployment which will lead to more instability in the war zones which in turn increases the experts probability assessments for the next year. This and many of the reasons listed at the end of the paper emphasize the need for a dynamic structure. Further conditionality should be introduced as pointed out by Dr. M. Goldstein. At least the reported results should have been calibrated in the sense of De Groot.

P.J. BROWN (*Imperial College, London*)

I enjoyed this interesting and thought provoking paper. Especially as it is concerned about thoughts about "the unthinkable", and illustrating as it does the fundamental personalistic standpoint that all subjective probabilities can be specified, at least that is for any individual. It is sociologically interesting in the diversity of assessments given by a "homogeneous" group of experts. Statistically, I suppose there is little new, Professor Press has described models for qualitative controlled feedback before and in particular back at the First Valencia Conference of 1979. A development is in the use of a kernel estimate of the multivariate prior, thus avoiding dependence on somewhat arbitrary natural conjugate priors. I did wonder here why the kernel method had been applied to the probabilities rather than the log probabilities. The latter would seem to be more suited. However, I think technical questions concerning the kernel are to my mind secondary to the meaning to attach to the average assessed probabilities and even the individual probabilities in the context of the wider issues generated by the application and the divorce of evidence, of which reasons are one manifestation, and probabilities.

I have experienced related misgivings in the area of long range energy demand forecasting in the U.K. and U.S. There earlier simple extrapolations of growth have been replaced by more sophisticated econometric models relating energy consumption to economic growth. Various scenarios of economic growth are posited and fed into the relationship as fitted to past data to produce a forecast of the future for each scenario. What is explicitly ignored is the theory laden aspect of the data, that new policy can change the future development and that the fitted relationship is dependent on past policies. See Brown (1983) for an extended critique.

Jim Press's experts seem to be predicting both what is uncontrollable and at the same time what some would argue is well within the ambit of government to influence by way of arms limitation and other measures. Is passive prediction of policy good enough or should the experts be attempting more to assess probabilities conditional on various policies together with their assessed probabilities of success. This they may already do but my overiding query is what do these unconditional and average probabilities mean.

In a similar vein I wonder how the listed reasons relate to probabilities? Why should qualitative controlled feedback drive the "bias" to zero(p3). Later (p31) it is even suggested that the change by round is minimal. If it is not minimal does the direction of movement reflect a common assessment of the importance of the added reasons amongst the "homogeneous" experts.

Finally, I suppose if I was trying to specify my probability of detonation conditional on a set of achieved policy objectives I would consider a time span of ten years as well as ten individual years, with perhaps a few further checks on coherence. The ten year span would have the effect of scaling up the small probabilities to sizeable probabilities which I would feel more familiar in specifying. However such probabilities are not derivable from the between individual multivariate distribution of the one year marginal probabilities assessed here.

M. GOLDSTEIN (*Hull University*)

This paper concerns such a chilling issue that technical queries are perhaps irrelevant. However, there is one point that puzzles me. Precisely what were theevents for which the experts were specifying probabilities? At first sight, it appears that the unconditional events are being considered, namely that the experts considers each year separately and assesses for each the probability of nuclear detonation. However, it seems from the discussion that many of the respondants considered for each year a situation where no nuclear detonation had occurred before the beginning of the year, so that they were in fact considering the probability of nuclear detonation in each year conditional on the event that there had been no such detonation in the preceding years. The numerical difference between these two assessments becomes considerable by the end of the decade, and I wonder what guidance was given to ensure that all experts specified probabilities for the same conditional or unconditional events.

I.J. GOOD (*Virginia Polytech. Inst. and State University*)

Probabilities close to 0 or 1 are not easy to judge so it might have helped to ask for single judgements over the next ten years instead of, or as well as, judgements for each year.

A paper that might be worth citing is Good (1966) because I think it was an early paper on the probability of war, although it did not depend on a sample.

K. McCONWAY (*The Open University, U.K.*)

I welcome this paper, and it is particularly good to see some real data being produced and analysed in this field.

I would like to raise a few points about the details of Professor Press's survey. Firstly, were all the respondents given the three scales in the same order? If they were, then presumably their responses on the second two scales might be consciously adjusted to fit with the first response. This procedure might produce a more *constant* bias than would be the case if different people had the scales in different orders, but it would seem unlikely that the value of the constant would be zero.

Secondly, while I agree that the use of words rather than just numbers on the scales is useful, there may be problems with the words used. It is well known that different people mean different things by words like "somewhat likely". Lifelong inhabitants of Los Angeles would be affected by availability biases on the ordinal scale statements than people who were less familiar with that city. On the log scale, probabilities of accidents are used. Work by Lichtenstein *et al* (1978) showed that people tend to overrate the chances of accidental death, and this was attributed to availability bias. Perhaps further research is needed on the best forms of words to use in scales of this type.

R.L. WINKLER (*Indiana University*)

The assessment of probabilities for future events which are generally regarded as reasonably unlikely is a very difficult task. In this study, the assessment procedure was carefully designed to help the subjects express their judgments in probabilistic form. The use of different scales and reference events is particularly valuable, as is the awareness of potential cognitive heuristics and biases. Repeating the process with feedback concerning reasons allows some controlled sharing of information among subjects.

The consideration of more than one event and more than one assessor complicates matters considerably. When assessments are made for multiple events or variables by a number of subjects and the assessments from different subjects are combined to form a single distribution, two types of dependence are likely to be encountered. Individual subjects may judge the different events or variables to be dependent, and the errors of estimation of different subjects may be correlated. In terms of the model, the elements of ϵ_j may be correlated for a fixed j, and error vector $\epsilon_1,\ldots,\epsilon_N$ may be correlated.

Dependence among the various events of variables is handled by having each expert assess a complete multivariate distribution instead of only providing information about marginal probabilities. This possibility, which is mentioned but not pursued in the paper, increases the assessment burden considerably (by requiring a covariance matrix in the normal case, for example).

Dependence among subjects does not increase the assessment burden for the subjects, but it requires the analyst to make judgments about such dependence and incorporate it into the model (perhaps using some simplifying assumption such as an intraclass correlation structure). I suspect that the error vectors $\epsilon_1,\ldots,\epsilon_N$ are likely to be dependent with reasonably high positive correlations. The emphasis on using an informed, homogeneous group should lead to particularly high correlations. Moreover, the fact that the homogeneity may provide similar bias terms would seem to make it less likely that the average bias will be driven to zero and that the amount of disagreement will reflect the amount of inherent uncertainty in the underlying issue, as claimed in the paper. As the group becomes more homogeneous, the amount of disagreement is likely to be reduced and some important sources of uncertainty concerning the underlying issue may be ignored. If potential subjects are divided into homogeneous groups, sampling entirely within one group is analogous to stratifying a population as if for stratified sampling but then using cluster sampling instead. It is recognized in the paper that "an important component in improving the validity of group judgments involves having a wide diversity of disciplines represented on the panel", which suggests that the group should be heterogeneous.

On a different note, how might the choice of a time span affect the probabilities? Weather forecasters giving separate forecasts for the probability of precipitation in each of two consecutive six-hour periods instead of a single probability of precipitation for the entire twelve-hour period appear to have some difficulty in taking into account the possible relationship between the six-hour periods (Winkler and Murphy, 1968). What would happen if the subjects were asked for their subjective probabilities of nuclear war in six-month periods, two-year periods, and so on? Would the results be consistent?

The method described in this paper has some very appealing features. I would be interested in seeing the impact upon the results of the formal incorporation of dependence, the use of a heterogeneous set of subjects, the consideration of different time periods, and the provision of slightly different types of feedback (e.g., the assessed probabilities as well as the reasons) or even some interaction among subjects. There are many situations in which the information available is primarily subjective in nature, and the development of methods to obtain, summarize, and analyze information from knowledgeable individuals is an important area of study.

REPLY TO THE DISCUSSION

While Professor J.R.M. Ameen of the University of Warwick suggests that the study would have been of more interest if it had been worldwide, support for the data collection in this project was predicated upon our collection opinions of experts in the United States. Contrary to Professor Ameen's belief, some "inconsistent" assessors assessed high probabilities and other assessed low probabilities. That is, over, the reported results could not have been "calibrated in the sense of DeGroot", since there was no way for true values to be known, and, therefore, assessors could not be scored. So, it is not clear what Professor Ameen had in mind with respect to this suggestion.

Professor Philip J. Brown of Imperial College, London, raises some thought-provoking questions. He asks in particular, what do our probabilities mean? The conditional probabilities are very realistic, actually, in the sense that all respondents feel, as most of us do, that once a nuclear detonation has taken place, it's all over anyhow; moreover, most of us really don't care much about whether there would be new nuclear detonations during the following year. So, it made most sense in designing the basic question, to have respondents assume that no nuclear detonation had taken place in either of the antagonists' homelands up until that point in time, no matter which year it was, and for the respondent then to assess the probability that that year would be the one when the first nuclear detonation would take place. The average probabilities have meaning for me only to the extent that they represent the average belief of a group of experts. The issue of whether or not the average converges to something meaningful has to do with our conceiving of the panel size steadily increasing, and the sample mean converging to some underlying population mean. In fact, if we restrict our thinking to the sample of experts currently available, and ignore notions of convergence, we don't run into these issues. Professor Brown also asks about why qualitative control feedback should drive the bias to zero. The answer is a qualitative rather than a mathematical one. Even though panel members in a homogeneous group are assumed to have the same biases, they might not all think of the same reason in the first round. They might each develop a slate of reasons that might or might not have an intersection with the slates of other panel members, but they would still have the same basic biases. On subsequent rounds, the reasons are combined to form a composite, and then the reasons are shown to the panel members in composite form, so that all panelists can benefit from the jointly-developed composite on subsequent rounds of assessment. As panel members keep providing additional reasons on subsequent rounds, more and more information is revealed about the issue. As panel members become increasingly well informed and have the opportunity to argue their positions on paper, beliefs they may have about events taking place which will affect the probabilities one way or the other, are also affected. As a consequence, the more their beliefs are aired, the more likely it is that all panelists will develop a better understanding of the underlying processes governing the likelihood of our basic event, which is to say apparent biases and differences in beliefs will tend to be minimized. Fundamental differences in philosophy that may exist in very inhomogeneous groups will probably tend to remain, even after feedback.

Professor Michael Goldstein, of the University of Hull, points out that our respondents were providing assessments for the probability of nuclear detonation in each year, conditional on the event that there had been no such detonation in the preceding years, and in that respect, he is correct. In all of our interactions with the respondents in this exercise, both by mail and by telephone, in addition to any personal contact we had with them, the respondents were cautioned to assess their probabilities on this basis, to insure that all experts specified probabilities for the same conditional events.

Professor Jack Good of Virginia Polytechnic Institute and State University, points out that probabilities close to 0 or 1 are not easy to judge, so it might have helped to ask for single judgements over the next 10 years instead of, or as well as, judgements for each year. I totally concur with Professor Good in his belief that probabilities close to 0 or 1 are not easy to judge. In fact, because I think they are extremely difficult to judge, we introduced many schemes for assisting the probability assessors in quantifying their judgements: using multiple scales; using probabilities of almanac types of events with which they might to be able to identify, to provide anchoring points; using words interpretations of probabilities; etc. Because it was our belief, however, that people do better with probability assessments by breaking them down into pieces, we did not ask for single judgements over the next ten years. Professor Good has also reminded us of his 1966 paper, "The probability of war". While this paper did not depend upon a sample of subjective assessments of individuals about the probability of war, it, nevertheless, was concerned with the probability of all-out atomic war, and raised some very interesting philosophical questions regarding the nature of such a probability. The paper should be of as much interest to others who are interested in the meaning of the probability of war, as the paper was to me.

Professor Kevin McConway, from the Open University in England, points out that we might have randomized the presentation of our various scales to the respondents. From some points of view, such a procedure would seem to be a reasonable one, but from the point of view of training the probability assessor to slowly adjust to thinking in terms of a log scale (when, in fact, many of the assessors were really quite unaccustomed to using log scales at all), I think the actual procedure we used was the appropriate one. I applaud Professor McConway's suggestion that further research is probably needed on the best forms of words to use in scales of this type to assist the probability assessor in understanding what the various degrees of belief really mean. We had great difficulty selecting the appropriate words to be used, and selecting the appropriate events to use. In fact, I pilot tested the words and the events on various people and found substantial variation among subjects regarding their ability to identify with the various events which were used as benchmark events to assess probabilities. All people on whom I pilot tested words, such as "very unlikely", agreed that the associated probability should be something quite "remote", but, of course, they disagreed on what numerical values should be associated with a very remote probability. This seems to be the best that we can do at this time. My belief is that by superimposing such verbal expressions of degree of belief on the scale, we provided a collection of anchoring points to the subjects, which served as uniform benchmarks for all subjects. I'm not sure whether this was a good thing or a bad thing to do.

Professor Robert L. Winkler, of Indiana University, has raised some questions about the advisability of having a homogeneous group for our panel of experts and points out that while in the paper we talk about the usefulness of the panel being homogeneous, at the same time we say, "an important component in proving the validity of judgements involves having a wide diversity of disciplines represented on the panel", which suggests that the group should be heterogeneous. In fact, these notions are not mutually contradictory. In constituting our own panel, we included people from a wide diversity of disciplines purposely, in order to have people on the panel who would address the issues with totally different perspectives. Our hope in doing so was to increase the totality of reasons, arguments, and justifications regarding the state of the world that would affect the probability of nuclear war, and such a composite of information would be likely to be more broadly based if people from many disciplines were included than it would be if our panel were a monolith. Reasons given on the various rounds of qualitative control feedback by all panel members were used in developing the total composite fed back to all panel members. The actual group of people classified as experts did not actually come from many disciplines. In fact, they came from only a few disciplines. The people who came from the other disciplines were mostly non-experts. While the non-experts were permitted to engage in the exercise, and to provide reasons for their judgements, their final numerical assessments were not included in the averages or probability distributions assessed. When those computations were made, only numerical assessments of the experts were used. Furthermore, the only issue that arises regarding whether the panel was homogeneous or inhomogeneous involves the interpretation of our sample mean of probability assessments. If we are not concerned about convergence of such a sample mean, in a probabilistic sense, then homogeneity of the panel is an irrelevant issue. The questions Professor Winkler raises regarding "what would happen if the subjects were asked for their subjective probabilities of nuclear war in 6 month periods, 2 year periods, and so on", is an interesting one, but one we didn't address. Finally, Professor Winkler has asked about what would happen with slightly different types of feedback, for example, feeding back the assessed probabilities as well as the reasons. In fact, we carried out such a sub-experiment in addition to the one described in the paper, and so we have data bearing on that question, but so far, we have not analyzed it. One of these days, we hope to do so.

I am grateful to all of the people who attended the Second Bayesian Valencia Conference and took the time, effort, and interest in this paper to detail their comments in writing. I'm also grateful to other participants who raised very stimulating questions about the paper verbally at the conference. The paper which appears in these Proceedings is a somewhat revised version of the paper presented at the meeting, and in the revision, I have attempted to improve the paper by reacting to the verbal discussion at the conference in a meaningful way. I hope I have been succesful, and that the paper is a better product for it.

REFERENCES IN THE DISCUSSION

BROWN, P.J. (1983). Forecasting and Public Policy. *The Statistician*. (To appear).

GOOD, I.J. (1966). The probability of war. *J. Roy. Statist. Soc. A,* **129**, 268-269.

LICHTENSTEIN, S. SLOVIC, P. FISCHHOFF, B., LAYMAN, M. and COMBS, B. (1978). Judged frequency of lethal events. *Journal of Experimental Psychology: Human Learning and Memory*, **4**, 551-578.

WINKLER, R.L. and MURPHY, A.H. (1968). Evaluation of subjective precipitation probability forecasts. *Proceedings of the First National Conference on Statistical Meteorology*. Boston: American Meteorological Society, 148-157.

BAYESIAN STATISTICS 2, pp. 463-472
J.M. Bernardo, M.H. DeGroot, D.V. Lindley, A.F.M. Smith (Eds.)

The Use of Propensity Scores in Applied Bayesian Inference

D.B. RUBIN
University of Chicago, USA

SUMMARY

The propensity score is the conditional probability, given a vector of covariates, of inclusion in a sample survey, or assignment to a treatment in an experiment; in a randomized survey or experiment, they are simply the a priori randomization probabilities for the units. Although propensity scores are used by frequentists to obtain unbiased estimates over randomization distributions, they are widely regarded as irrelevant to Bayesian inference. Despite this view, they do in fact play an important role in applied Bayesian inference. Generally, they are the coarsest possible summary of the information in the covariates such that given the summary, the mechanism that assigns units to treatments or selects units for inclusion in the sample can be ignored. Since modelling simple structures is usually an easier task than modelling more complex structures, models that use the propensity score as a scalar summary of covariate information for each unit may often be more accurate reflections of reality than models that incorporate all covariate information. Thus, the applied Bayesian who concentrates on models that summarize covariates by propensity scores may be more likely to be calibrated to reality than the Bayesian who builds models using all covariates. This conclusion does not mean that the applied Bayesian should necessarily ignore covariate information other than the propensity score —efficiency considerations and the existence of subgroups of particular interest often lead to models using other covariate information— but it does mean that propensity scores, the randomization probabilities, can play an important role in a valid and scientific Bayesian analysis of data.

Keywords: ADEQUATE SUMMARIES; CALIBRATION; EXPERIMENTS; IGNORABLE MECANISMS; RANDOMIZATION PROBABILITIES; SURVEYS.

1. INTRODUCTION

It has often been argued that randomization probabilities in surveys or experiments are irrelevant to a Bayesian statistician. In contrast to this common claim, these probabilities can be demonstrated to be of substantial importance to the applied Bayesian statistician, even though the Bayesian does not use these probabilities in data analysis in the way that frequentists do. Modelling the conditional distribution of outcome variables given the randomization probabilities in many cases can be a primary task in a valid Bayesian analysis of data. In senses to be made precise, the randomization probabilities are the coarsest possible summary of information that allows the Bayesian statistician to ignore sampling and treatment assignment mechanisms, albeit in some cases too coarse. Since the specification of a distribution for outcomes given a simple scalar covariate is usually an easier task than the specification of a distribution for outcomes given a complex multivariate covariate, Bayesian models using the randomization probabilities to summarize covariate information may often be a better reflection of reality than Bayesian models that try to model multivariate covariate information. Thus Bayesian models that focus attention on the modelling of outcomes given randomization probabilities may often be better calibrated to the real world.

This argument will be developed and illustrated for the case of survey sampling. Parallel development for the case of experiments for causal effects will then be briefly indicated. Practical issues concerning modelling using randomization probabilites and additional covariate information will also be briefly noted, as will problems of nonresponse and nonrandomized studies. Throughout, measure theoretic details are ignored.

2. NOTATION FOR SAMPLE SURVEYS

Let N be the number of units in a population, and let $\mathbf{Y} = (Y_1,\ldots,Y_N)$ be an outcome variable whose values are the object of a sample survey; a common objective is to estimate the total Y_+ or mean $\bar{Y}$ in the population, $Y_+ = \Sigma\, Y_i$ and $\bar{Y} = Y_+/N$ respectively. Each Y_i could be a vector, but to convey essential ideas it is simpler to consider each Y_i to be a scalar. Let $\mathbf{X} = (X_1,\ldots,X_N)$ be the covariate whose values are available for making sampling decisions; each X_i generally is a vector. Finally, let $\mathbf{I} = (I_1,\ldots,I_N)$ be the indicator for inclusion in the sample: if $I_i = 1$, Y_i is observed, and if $I_i = 0$, Y_i is not observed; $\mathbf{I}$ itself is entirely observed.

Unconfounded probability sampling mechanisms

The conditional distribution of a sampling indicator $\mathbf{I}$ given $\mathbf{Y},\mathbf{X}$ will be called the sampling mechanism and written $Pr(\mathbf{I}|\mathbf{Y},\mathbf{X})$. In order to avoid confusion about the inferential content of indices, they will be assumed to a random permutation of $1,\ldots,N$; all information to be modelled is to be coded in $\mathbf{Y}$ or $\mathbf{X}$, not in the indices. As a consequence, all probability specifications will be exchangeable in the unit indices.

In most scientific survey applications, the survey is carefully designed and documented to support the assumption that the sampling mechanism has $\mathbf{I}$ unconfounded with $\mathbf{Y}$:

$$Pr(\mathbf{I}|\mathbf{Y},\mathbf{X}) = Pr(\mathbf{I}|\mathbf{X})\,. \tag{1}$$

Typically, some randomization is employed in the sense that in addition to (1) the accepted specification for the sampling mechanism asserts that it is possible to observe each value of Y_i:

$$Pr(I_i = 1|\mathbf{Y},\mathbf{X}) > 0 \qquad \text{for all } i\,. \tag{2}$$

For example, with simple random sampling, all samples of Y_i of size n are considered to be equally likely given $\mathbf{Y},\mathbf{X}$;

$$P(\mathbf{I}|\mathbf{Y},\mathbf{X}) = P(\mathbf{I}|\mathbf{X}) = \begin{cases} \binom{N}{n}^{-1} & \text{if} \quad \sum_i^N I_i = n \\ 0 & \text{otherwise.} \end{cases}$$

Of course, if the random numbers associated with each unit and used to make sampling decisions (choose n units with the smallest random numbers) were included in $\mathbf{X}$, the sampling mechanism would be deterministic in the sense that $Pr(I_i|\mathbf{Y},\mathbf{X}) = 1$ or 0 for all i. Consequently, in order for $\mathbf{X}$ to be useful in (2), it must exclude variables known a priori to be unrelated to $\mathbf{Y}$.

If both (1) and (2) hold, the sampling mechanism will be said to be, given **X**, an unconfounded probability sampling mechanism, or for brevity, a probability sampling mechanism. Throughout our discussion of survey sampling, probability sampling, defined by (1) and (2) will be assumed unless explicitly stated otherwise.

Propensity scores

Let $e_i = Pr(I_i = 1 | \mathbf{Y},\mathbf{X}) = Pr(I_i = 1 | \mathbf{X})$, a function of **X**. The vector $\mathbf{e} = (e_1,...,e_N)$ will be called the vector of propensity scores, that is, of propensities for inclusion in the sample. This terminology is taken from Rosenbaum and Rubin (1983) and Rubin (1983).

The propensity scores are the randomization probabilities. Frequentists commonly use them as weights to create unbiased estimates. For example, the Horvitz-Thompson estimator of Y_+ is $\sum_1^N Y_i I_i / e_i$, which can be shown to be unbiased for Y_+:

$$E(\sum_1^N Y_i I_i / e_i | \mathbf{Y},\mathbf{X}) = \sum_1^N (Y_i/e_i)\, E(I_i | \mathbf{Y},\mathbf{X}) = Y_+ .$$

Propensity scores play a role in Bayesian inference too, but not in so simple a manner. In many cases, the propensity score vector can be an "adequate" summary of **X**.

Adequate summaries of **X**

A function of **X**, $\mathbf{a} = (a_1,...,a_N)^T = \mathbf{a}(\mathbf{X})$, will be said to be an *adequate* summary of **X** for an unconfounded sampling mechanism if

$$Pr(\mathbf{I} | \mathbf{X}) = Pr(\mathbf{I} | \mathbf{a}) . \tag{3}$$

When the reference to the sampling mechanism is clear and (3) holds, we will simply say that **a** is an adequate summary of **X**, or when both the reference to the sampling mechanism and **X** are clear and (3) holds, we will simply say that **a** is adequate. When **a** is adequate, it provides a complete summary of the data used to make sampling decisions. Trivially, with all unconfounded sampling designs, **X** is adequate. Of far more interest is the fact that with common probability sampling designs —such as simple random sampling, stratified sampling with different sampling fractions across strata defined by X_i, probability proportional to X_i sampling, and independent Bernoulli sampling with probabilities a function of X_i— the propensity scores are adequate. Moreover, in such cases, the propensity scores are the coarsest possible summary of **X** that is adequate.

The propensity score vector as the coarsest possible adequate summary of **X**

One function of **X**, $\mathbf{a}_1(\mathbf{X})$, will be said to be *coarser* than another function of **X**, $\mathbf{a}_2(\mathbf{X})$, if $\mathbf{a}_1$ is a function of $\mathbf{a}_2$ but $\mathbf{a}_2$ is not a function of $\mathbf{a}_1$, where both $\mathbf{a}_1$ and $\mathbf{a}_2$ are vectors of length N.

The propensity score is the coarsest possible adequate summary of **X** in the sense that:

(a) If a function of **X** is coarser than the propensity score, it cannot be an adequate summary of **X**;

and

(b) There exist examples in which the propensity score is an adequate summary of **X**.

Only (a) requires proof since we have already described common examples in which the propensity score is an adequate summary of **X**.

Theorem 1 Any function of $\mathbf{X}$ that is coarser than the propensity score cannot be adequate.

Proof Suppose $\mathbf{a}$ is an adequate summary of $\mathbf{X}$. Then $Pr(\mathbf{I}|\mathbf{X}) = Pr(\mathbf{I}|\mathbf{a})$, and thus since $\mathbf{e} = E(\mathbf{I}|\mathbf{X})$, the propensity score vector also satisfies

$$\mathbf{e} = E(\mathbf{I}|\mathbf{a})\,. \tag{4}$$

Hence, $\mathbf{e}$ must be a function of $\mathbf{a}$, which means that $\mathbf{a}$ cannot be coarser than $\mathbf{e}$.

Of course, the propensity score is not the unique coarsest possible adequate summary of $\mathbf{X}$. Any one-one function of $\mathbf{e}$, for example, will be adequate if $\mathbf{e}$ is adequate; but $\mathbf{e}$, or a monotone function of $\mathbf{e}$, may be particularly appropriate for modelling since sampling mechanisms often are designed with the hope that Y_i is approximately proportional to e_i so that Y_+ will be easy to estimate precisely.

3. BAYESIAN INFERENCE FROM AN ADEQUATE SUMMARY OF X

Suppose $\mathbf{a} = (a_1,\ldots,a_N)^T$ is a function of $\mathbf{X}$ that is observed for data analysis but that $\mathbf{X}$ itself is unobserved for data analysis. We now show that if $\mathbf{a}$ is an adequate summary of $\mathbf{X}$, the conditional distribution of $\mathbf{Y}$ given $\mathbf{a}$ is all that the Bayesian needs to specify in order to draw valid inferences for population values of $\mathbf{Y}$.

For notational convenience, partition $\mathbf{Y}$ so that $\mathbf{Y}_{obs}$ is observed ($\mathbf{Y}_{obs}$ consists of the Y_i such that $I_i = 1$) and $\mathbf{Y}_{nob}$ is missing ($\mathbf{Y}_{nob}$consists of the Y_i such that $I_i = 0$), where $\mathbf{Y} = (\mathbf{Y}_{obs},\mathbf{Y}_{nob})$ for an appropriate arrangement of indices. Bayesian inference for a population summary of $\mathbf{Y}$, such as Y_+, follows from the conditional distribution of $\mathbf{Y}_{nob}$ given all observed values, that is, from the posterior distribution of $\mathbf{Y}_{nob}$: $Pr(\mathbf{Y}_{nob}|\mathbf{Y}_{obs},\mathbf{a},\mathbf{I})$ evaluated at the observed values of $\mathbf{Y}_{obs}$, $\mathbf{a}$, $\mathbf{I}$.

Ignorable sampling mechanisms

The sampling mechanism is said to be ignorable (Rubin, 1976, 1978) at $(\mathbf{Y}_{obs},\mathbf{a},\mathbf{I})$ if the posterior distribution of $\mathbf{Y}_{nob}$ does not explicitly depend on $\mathbf{I}$:

$$Pr(\mathbf{Y}_{nob}|\mathbf{Y}_{obs},\mathbf{a},\mathbf{I}) = Pr(\mathbf{Y}_{nob}|\mathbf{Y}_{obs},\mathbf{a})\,.$$

When the sampling mechanism is ignorable, inference for $\mathbf{Y}_{nob}$ follows simply by Bayes Theorem from the specification for the conditional distribution of $\mathbf{Y}$ given $\mathbf{a}$:

$$Pr(\mathbf{Y}_{nob}|\mathbf{Y}_{obs},\mathbf{a}) = Pr(\mathbf{Y}|\mathbf{a})/\textstyle\int Pr(\mathbf{Y}|\mathbf{a})\,d\mathbf{Y}_{nob}\,.$$

Nonignorable sampling mechanisms, such as probability proportional to X_i with no functions of $\mathbf{X}$ observed, are generally difficult to handle in practice. The explicit dependence of the posterior distribution of $\mathbf{Y}_{nob}$ on $\mathbf{I}$ means that the conditional distribution of Y_i given $I_i = 0$ must be estimated, and it generally differs from the conditional distribution of Y_i given $I_i = 1$. But there are no observed Y_i values when $I_i = 0$ from which to estimate this distribution. Thus any inference for quantities involving unobserved Y_i values must rely on assumptions that are not directly confronted by the observed data. Obviously, nonignorable sampling mechanisms are to be avoided in general.

Adequate summaries imply ignorable sampling mechanisms

Theorem 2 Suppose $\mathbf{a} = \mathbf{a}(\mathbf{X})$ is an adequate summary of $\mathbf{X}$ for the unconfounded sampling mechanism $Pr(\mathbf{I}|\mathbf{X})$. Then for all possible observed values and specifications for $\mathbf{Y}$ given $\mathbf{X}$, the sampling mechanism is ignorable given $\mathbf{a}$:

$$Pr(\mathbf{Y}_{nob}|\mathbf{Y}_{obs},\mathbf{a},\mathbf{I}) = Pr(\mathbf{Y}_{nob}|\mathbf{Y}_{obs},\mathbf{a}) .$$

Proof By Bayes Theorem

$$Pr(\mathbf{Y}_{nob}|\mathbf{Y}_{obs},\mathbf{a},\mathbf{I}) = Pr(\mathbf{Y}_{nob}|\mathbf{Y}_{obs},\mathbf{a}) \frac{Pr(\mathbf{I}|\mathbf{Y},\mathbf{a})}{Pr(\mathbf{I}|\mathbf{Y}_{obs},\mathbf{a})} . \tag{5}$$

The ratio on the right hand side of (5) is one if $\mathbf{a}$ is adequate.

If $\mathbf{a}$ is not an adequate summary, then examination of the ratio on the right hand side of (5) shows that whether the sampling mechanism is ignorable generally depends on the implied specification for $Pr(\mathbf{I},\mathbf{Y}|\mathbf{a})$ as well as on the observed values of $\mathbf{I}$, $\mathbf{a}$ and $\mathbf{Y}_{obs}$.

4. FITTING MODELS AND BEING WELL-CALIBRATED

Thus far we have shown that

(1) Using ignorable sampling mechanisms is desirable;

(2) The sampling mechanism is ignorable given any adequate summary of covariates and is generally not ignorable given an inadequate summary; and

(3) The coarsest possible adequate summary of covariates is the vector of propensity scores, though it may be too coarse.

Points (2) and (3) imply that the vector of propensity scores is the coarsest possible summary of $\mathbf{X}$ such that the sampling mechanism is generally ignorable given the summary. We have yet to show why it may be of interest to use the coarsest possible adequate summary.

The basic ideas are that (a) the task of specifying realistic models for $Pr(\mathbf{Y}|\mathbf{a})$ is in principle simpler the coarser the summary $\mathbf{a}$, (b) there may be effectively more data to estimate essential parameters the coarser the summary, and (c) when the specification for $Pr(\mathbf{Y}|\mathbf{a})$ is accurate, the Bayesian will be well-calibrated in general and in all statements conditional on $\mathbf{a}$.

Calibration

A Bayesian is well-calibrated if his probability statements have their asserted coverage in repeated experience. For example, if $C_1,C_2,\ldots$ represents a series of 95 % Bayes interval estimates for unknown $U_1,U_2,\ldots$ from known data sets $D_1,D_2,\ldots$, then these statements are well-calibrated if 95 % of them cover their unknowns and 5 % do not. A subsequence of $C_1,C_2,\ldots$ is well-calibrated if 95 % of those C_i in the subsequence cover their unknowns. For interesting discussion of this idea, see Dawid (1982). Clearly, it is desirable for a Bayesian to be well-calibrated overall and for all subsequences defined by characteristics of the data sets, and he will obviously be well-calibrated overall and in such subsequences if his models are correct.

That is, if his model are correct, $Pr(U_i \in C_i|D_i) = .95$ and thus averaging over all data sets

$$Pr(U_i \in C_i) = .95 ,$$

or averaging over all data sets with observed characteristic Q,

$$Pr(U_i \in C_i | D_i \text{ satisfies } Q) = .95 .$$

Thus, the Bayesian who uses realistic models can be expected to be well-calibrated generally and in any collection of cases with common observed characteristics.

Because it is desirable to be calibrated in as many subsequences as possible, there is an attraction for the Bayesian to record and condition upon as many features of the data as possible (i.e., make each D_i as extensive as possible). If the resultant models are correct, the Bayesian will be well-calibrated in more subsequences than if he had not made each data set D_i as extensive. However, complicated data sets are more difficult to model well, and if the Bayesian's model is poor, he might not only be uncalibrated on many subsequences where he thought he was calibrated, he may not even be well-calibrated in the overall sequence. The coarser the data set D_i, the more likely it may be that the Bayesian is well-calibrated overall and in all subsequences definable by the observed coarser characteristics. There are two related reasons for this: first, it will often be easier to estimate essential parameters with coarser summaries, and second, the coarser the summary, the easier it may be to provide specifications.

Easier to estimate essential parameters with coarser summary

Suppose that there are K distinct values of the adequate covariate $\mathbf{a}$ represented among the unobserved Y_i in $\mathbf{Y}_{nob}$. Choose one of these distinct values, for concreteness, say the one having the fewest components in $\mathbf{Y}_{obs}$. That is, suppose that for $a_i = a^*$, there exist $m_{nob} > 0$ unobserved values of Y_i and $m_{obs} \geq 0$ observed values of Y_i. Since (i) $\mathbf{a}$ but not $\mathbf{X}$ is observed, (ii) the indexing of units is random, and (iii) the sampling mechanism is ignorable given $\mathbf{a}$, the distribution of $\mathbf{Y}$ for the $m_{obs} + m_{nob}$ units with $a_i = a^*$ is exchangeable. If $m_{obs} = 0$, then in order to draw inferences about the m_{nob} unobserved values of Y_i we must rely on prior assumptions relating these m_{nob} units with $a_i = a^*$ to the other units with observed values of Y_i but $a_i \neq a^*$. If $m_{obs} > 0$, then there are data available to directly estimate the m_{nob} unobserved values of Y_i with $a_i = a^*$, namely, the m_{obs} observed values of Y_i with associated $a_i = a^*$. Obviously, for fixed $Pr(\mathbf{Y}|\mathbf{a})$, the larger m_{obs}, the less sensitive the estimation of the m_{nob} unobserved values of Y_i will be to prior assumptions.

For fixed $(\mathbf{Y},\mathbf{X},\mathbf{I})$, in order to maximize the number of Y_i observations at each distinct value of an adequate summary, use the coarsest possible adequate covariate. And the coarsest possible (albeit sometimes too coarse) adequate summary is the vector of propensity scores.

Of course, this argument has to be carefully considered in cases with strong prior information because then reduction to coarser $\mathbf{X}$ may not simplify good modelling efforts but actually complicate them.

Note that if the sampling mechanism is systematic, no Y_i in Y_{nob} has an a_i value found among the Y_i in Y_{obs}, and estimation of Y_{obs} is entirely reliant upon prior assumptions (which is not necessarily bad when prior information about relationships is strong and realistic).

The coarser the summary, the easier it often is to provide a specification for $Pr(\mathbf{Y}|\mathbf{a})$

Intuition suggests that it is often easier to provide a realistic specification for $Pr(\mathbf{Y}|\mathbf{a})$ when $\mathbf{a}$ is coarser. This intuition is partially related to the previous discussion concernng the increase in the number of observations at each distinct value of a coarser summary. A related but different point avoids considerations of the pattern of observed values in the data. If $Pr(\mathbf{Y}|\mathbf{a})$ is a correct specification, then we can obtain by direct calculation the correct specification for $Pr(\mathbf{Y}|\mathbf{F}(\mathbf{a}))$ where $\mathbf{F}(\ )$ is a function of $\mathbf{a}$; in contrast, given the correct specification for $Pr(\mathbf{Y}|\mathbf{F}(\mathbf{a}))$, we cannot calculate the correct specification for $Pr(\mathbf{Y}|\mathbf{a})$ unless $F(\ )$ is one−one. In this sense, it takes less work to provide a correct specification for $Pr(\mathbf{Y}|\mathbf{a})$ given a coarser summary for $\mathbf{a}$.

5. ILLUSTRATION

Suppose that X_i is scalar and measures the size of the ith unit (e.g., units = cities, X_i = number of inhabitants). If sampling is probability proportional to size, X_i, then $e_i \propto X_i$, and $\mathbf{e}$ is an adequate covariate. Since $\mathbf{e}$ is a one-one function of $\mathbf{X}$, no reduction in complexity is achieved by modelling $\mathbf{Y}$ given $\mathbf{e}$ rather than $\mathbf{Y}$ given $\mathbf{X}$.

Suppose instead that $\mathbf{X}$ is used to create ten size-strata such that the total of the X_i in each stratum is approximately the same, with the result that the numbers of units in the strata vary; a simple random sample of size m is taken within each stratum. Now the propensity scores are constant within each stratum and vary across strata, and are thus coarser than the size vector, yet adequate. If $m = 1$, the difference between the complexity of modelling $\mathbf{Y}$ given $\mathbf{X}$ and $\mathbf{Y}$ given $\mathbf{e}$ is trivial. If m is larger, however, say five or ten, then modelling $\mathbf{Y}$ given $\mathbf{e}$ may be substantially easier than modelling $\mathbf{Y}$ given $\mathbf{X}$ because of the exchangeability of units within each value of e_i. Consequently, we might expect answers based on obvious models for $\mathbf{Y}$ given $\mathbf{e}$ to be better calibrated than answers based on obvious models for $\mathbf{Y}$ given $\mathbf{X}$.

Hansen-Madow-Tepping example

Hansen, Madow and Tepping (1983) (HMT) present such an example. Their purpose is to show how the model-based perspective on survey sampling is not as safe as the randomization-based perspective. They compare estimators under a superpopulation model; one of their tables presents calibration results for interval estimators. They find that estimators whose model-based justifications focus attention on the accurate modelling of $\mathbf{Y}$ given $\mathbf{e}$ fare better than estimators that dissipate modelling energy by attending to the full distribution of $\mathbf{Y}$ given $\mathbf{X}$ (e.g., a linear model for Y given $\mathbf{X}$). Thus, their example helps to illustrate that it can be advantageous for the applied Bayesian (i.e., one with finite resources who wants to be calibrated) to record for analysis the probabilities of selection and possibly forego the extra information coming from the available covariate.

Residual covariation between $\mathbf{Y}$ *and* $\mathbf{X}$ *given* $\mathbf{e}$

Of course, the above discussion should not be read as suggesting that the applied Bayesian should systematically discard all covariate information not contained in the propensity scores even when they are adequate. Although accurate modelling of $\mathbf{Y}$ given $\mathbf{e}$ is needed in order to be well-calibrated, modelling the residual covariation between $\mathbf{Y}$ and $\mathbf{X}$ at each fixed value of $\mathbf{e}$ is often extremely desirable in order to increase precision or to obtain estimates for groups of units with specific values of $\mathbf{X}$. Thus, it is generally desirable to focus extra attention on the accurate modelling of Y_i given e_i in order to insure calibration overall (and given $\mathbf{e}$), and then to model the regression of Y_i on X_i within the fixed levels of e_i in order to increase precision and increase the likelihood of being calibrated in subsequences defined by $\mathbf{X}$. In the HMT example with several observations in each of ten strata, this perspective suggests, for example, fitting a separate normal linear regression of Y_i on X_i within each level of e_i. With such a model, the specification for Y_i

given e_i is likely to be a reasonable reflection of reality, thereby leading to overall calibration, whereas the specification for the distribution of the Y_i given e_i residuals on the X_i given e_i residuals is more questionable, but is being relied upon primarily to increase precision rather than guarantee calibration.

More development of this issue is needed to address questions such as the extent to which the Bayesian remains well-calibrated whenever the specification for Y_i given e_i is correct, even if the specification for Y_i given X_i is grossly incorrect.

6. EXTENSION TO OTHER SITUATIONS

With only minor notational changes, the essential notions in this paper can be applied to other contexts. We simply note some extensions here.

Two treatment studies for causal effects

In a two treatment study for causal effects, there exist two versions of the outcome variable; Y_{i1}, the value for the ith unit that would be observed under treatment 1, and Y_{i0}, the value for the ith unit that would be observed under treatment 0 (Rubin, 1978). If $I_i = 1$, Y_{i1} is observed and Y_{i0} is not observed; if $I_i = 0$, Y_{i0} is observed and Y_{i1} is not observed. Letting $Y_i = (Y_{i1}, Y_{i0})$, the notation and discussion of the previous sections needs only to be trivially modified (e.g. instead of $Pr(\mathbf{I} | \mathbf{Y},\mathbf{X})$ being called the sampling mechanism, it should now be called the treatment assignment mechanism).

One caveat, however, is that in the *experimental* treatment assignment context, the propensity scores are often not adequate; simply consider the standard paired comparison experiment - each unit is equally likely to receive either treatment so $e_i = .5$. Nevertheless, the spirit of much of the previous discussion still has force: often it may be wise to focus modelling energy on the coarsest possible adequate summary of the covariate. The propensity score was first defined in the context of two treatment observational studies for causal effects (Rosenbaum and Rubin, 1983), and the notion seem to continue to be a fruitful one (Rosenbaum and Rubin, 1984, Rosenbaum, 1983).

Nonresponse in surveys and other examples

Another situation in which propensity scores may be useful is in modelling nonresponse in sample surveys. Considering the intended sample to be the population, Y_i are observed for respondents ($I_i = 1$) and Y_i are missing for nonrespondents ($I_i = 0$). In this case, the propensity scores are the probabilities of being a respondent.

Of course, all of these examples can be combined in one encompassing example with propensity scores defined for each type of mechanism that is operating. In such cases, there will generally exist multiple propensity score vectors, one for each mechanism, i.e., e_i is a vector. Nonetheless, our basic message still holds: accurate modelling of outcomes given propensities is an important aspect of a valid and scientific applied Bayesian analysis.

REFERENCES

DAWID, A.P. (1982). The Well-Calibrated Bayesian. *J. Amer. Statist. Assoc.*, **77**, 605-613. (with discussion).

HANSEN, M.H., MADOW, W.G. and TEPPING, B. (1983). An Evaluation of Model-Dependent and Probability-Sampling Inferences in Sample Surveys. *J. Amer. Statist. Assoc.*, **77**, 776-807 (with discussion).

ROSENBAUM, P.R. (1983). Conditional Permutation Tests and the Propensity Score in Observational Studies. *Tech. Rep.* **21**, Wisconsin Clinical Cancer Center.

ROSENBAUM, P.R. and RUBIN, D.B. (1983). The Central Role of the Propensity Score in Observational Studies for Causal Effects. *Biometrika,* **70**, 41-55.

— (1984). Balanced Subclassification in Observational Studies Using the Propensity Score: A Case Study. *J. Amer. Statist. Assoc.* To appear.

RUBIN, D.B. (1976). Inference and Missing Data. *Biometrika,* **63**, 581-592.

— (1978). Bayesian Inference for Causal Effects: The Role of Randomization. *Ann. Statist.*, **6**, 34-58.

— (1983). Discussion of Hansen, Madow, Tepping. *J. Amer. Statist. Assoc.*, **77**, 803-805.

DISCUSSION

R.A. SUGDEN (*Goldsmiths' College, London*)

Rubin shows that a survey analyst who knows the sampling covariates **X**, or *design* variates (Scott, 1977), may be better calibrated by using a coarser adequate summary. He also shows that an analyst who does not know **X** but has an adequate summary **a** can draw valid inferences ignoring the selection mechanism and using only the conditional distribution of unsampled values given the sampled ones and **a**.

I would like to comment only on this second aspect of the paper. Theorem 1 gives the impression that the propensity scores or inclusion probabilities are the most fundamental design information that might be available. In fact the function of **X** which actually specifies the selection mechanism itself namely $pr[\mathbf{I}\,|\,\mathbf{X}]$ may be needed. Are the observed values of this always an adequate summary? I think so, for we can write

$$pr[\mathbf{I}\,|\,pr[\mathbf{I}\,|\,\mathbf{X}]] = pr[\mathbf{I}\,|\,\mathbf{X}]\,.$$

A difficulty arises as Rubin restricts himself (artificially in my opinion) to summaries which are vectors with N components whereas there are up to N_{C_n} distinct values here. However, for stratified *random* sampling, whether for unequal or equal sampling fractions, $pr[\mathbf{I}\,|\,\mathbf{X}]$ is in one-one correspondence with the stratum identifier. Inference therefore proceeds conditional on the strata.

A general but related point is that, although in some experiments treatment assignments may be independent given **X**, in sampling there is usually information in the joint inclusion probabilities, i.e. covariances as well as expectations of components of **I** must be considered. Propensity scores are simply not enough, except in a few simple cases that Rubin lists.

I am rather worried about Rubin's assumption of exchangeability in section 2. Many practical situations are of course non-exchangeable and I wonder if his conclusions are still valid. It should be possible to extend Rubin's approach to include summaries which can be functions of **I**, the sampling indicator. Commonly covariate information may only be available to the analyst for sampled units, for example in poststratification the strata are only known for these units. An important case, excluded in Rubin's paper, where the selection mechanism is ignorable, arises from poststratification where the only other covariate information is knowledge of the stratum sizes.

T.M.F. Smith and I are working on these and related problems to the above discussion and some preliminary results appear in Sugden, (this volume).

I think this paper is important because it opens up discussion of the rôle of randomisation in survey sampling inference, which for Bayesians had become rather sterile.

REPLY TO THE DISCUSSION

The author declined to reply.

REFERENCES IN THE DISCUSSION

SCOTT, A.J. (1977). Some comments on the problem of randomisation in surveys. *Sankhyā C* **39**, 1-9.

BAYESIAN STATISTICS 2, pp. 473-494
J.M. Bernardo, M.H. DeGroot, D.V. Lindley, A.F.M. Smith (Eds.)

Outliers and Influential Observations in Linear Models

L.I. PETTIT and A.F.M. SMITH
Goldsmiths' College, London, U.K. *University of Nottingham, U.K.*

SUMMARY

Previous Bayesian approaches to modelling outliers in linear models are reviewed and reconsidered. General Bayesian measures of outlyingness and influence are presented and illustrated, and comparisons are made with non-Bayesian measures proposed by other authors. Two main conclusions are drawn: first, that the computational explosion implicit in many Bayesian approaches to outliers can be avoided; secondly, that the Bayesian influence measure proposed can differ in important ways from classical measures.

Keywords: INFLUENCE; LINEAR MODELS; OUTLIERS; SENSITIVITY ANALYSIS

1. INTRODUCTION

Operational forms of sensitivity analysis are an essential feature of Bayesian statistics motivated either by an individual need to explore personal preferences for coherent prior/data/posterior combinations (no chronology intended), or by the group need of a community of scientific report readers for a range of coherent prior/data/posterior "mappings" to aid the process of individual assessment.

In this paper, we shall re-examine and further develop methodology for two types of Bayesian sensitivity analysis. The first of these relates to the situation where a data set is regarded as "fixed" and we wish to examine the plausibility of a range of *a priori* specifications in the light of the data and to study the variation in the *a posteriori* statements corresponding to these individual *a priori* specifications. Specifically, we shall deal with univariate location-scale and multivariate normal models incorporating a range of assumptions about the number of possible "outliers" in the data set, such assumptions constituting, in a sense, departures from the "optimistic" assumption of "no outliers". The second form of sensitivity analysis relates to the situation where an *a priori* specification is "fixed" and we wish to examine the changes in *a posteriori* statements that would result from a range of "perturbations" to the data set. In particular, we shall deal —within the general linear model framework— with "perturbations" corresponding to the omission from the data set of individual data points, pairs of points, etc. Measures of changes in *a posteriori* inferences can then be regarded as measures of the "influence" of the omitted observations.

2. OUTLIERS

2.1. *Bayesian sensitivity analysis via model elaboration*

A general philosophy and methodology for carrying out, for a "fixed" data set, an analysis of the sensitivity of model specification in the neighbourhood of a "standard", "simplified" or "optimistic" model has been described in detail in Box (1980), and consists in applying the Bayesian paradigm to a "judicious and grudging elaboration of the currently proposed model to ensure against particular hazards". (See, also, Simar, 1983, and Smith, 1983, for further discussion and references.)

With a minimum of notational complication, we suppose that the currently proposed parametric model is represented, for data vector x, by the density $p(x|\theta)$, where θ is an unknown parameter vector, and that the elaborated neighbourhood of models is represented by the family of densities $p(x|\theta,\lambda)$, $\lambda \epsilon \Lambda$, where Λ is some form of labelling set describing the neighbourhood, with $p(x|\theta) = p(x|\theta,\lambda_0)$ for some $\lambda_0 \epsilon \Lambda$. Thus, for example, in the case of model elaboration to protect against outliers, Λ describes the possible numbers and types of outlier, with λ_0 denoting no outliers.

Applying the Bayesian paradigm to the elaborated framework, inferences about θ are summarized by

$$p(\theta|x) = \int p(\theta|x,\lambda)p(\lambda|x)d\lambda \tag{2.1}$$

where

$$p(\theta|x,\lambda) \propto p(x|\theta,\lambda)p(\theta|\lambda) \ , \tag{2.2}$$

$$p(\lambda|x) \propto p(x|\lambda)p(\lambda) \tag{2.3}$$

and

$$p(x|\lambda) = \int p(x|\theta,\lambda)p(\theta|\lambda)d\theta. \tag{2.4}$$

As was pointed out by Box and Tiao (1964), the individual elements appearing in (2.1)-(2.4) provide a comprehensive framework for conditional and overall inferences for θ, as well as for sensitivity analysis within the neigbourhood described by Λ:

- for given $p(\theta,\lambda) = p(\theta|\lambda)p(\lambda)$ and x, the sensitivity of conclusions to assumptions is revealed by examining the variation in $p(\theta|x,\lambda)$, considered as a function of $\lambda \epsilon \Lambda$;
- since Λ specifies the forms of potential departures of interest from the presumed model ($\lambda = \lambda_0$), the form of $p(\lambda)$ can be chosen (or varied) to reflect actual beliefs about such departures (or as a means of understanding the effects of such beliefs on overall inferences);
- for given $p(\lambda)$, $p(\theta|\lambda)$, the form of $p(\lambda|x)$ provides information about the relative plausibilities of the various model elaborations described by Λ.

If simple summary estimates and measures of uncertainity are required, it may suffice to quote the posterior mean and dispersion matrix for θ, which are given by

$$E(\theta|x) = E[E(\theta|x,\lambda)|x] \tag{2.5}$$

$$V(\theta|x) = E[V(\theta|x,\lambda)|x] + V[E(\theta|x,\lambda)|x] \tag{2.6}$$

The former is seen to be an adaptive weighted average of the posterior means corresponding to particular choices of λ; the latter expresses overall uncertainty as a combination of average within- and between- model uncertainty. In the case where Λ consists of a finite list, ranging, say, over all posible identifications of particular subsets of the observations as outliers (up to some maximum number) expectation with respect to

$p(\lambda \mid x)$ reduces to summation, so that, for example, (2.5) becomes.

$$E(\theta \mid x) = \sum_{\lambda \epsilon \Lambda} E(\theta \mid x,\lambda) p(\lambda \mid x). \tag{2.7}$$

In the remainder of Section 2, we shall re-examine the use of this elaborated neighbourhood approach to the Bayesian formulation and analysis of linear models incorporating outliers, where —following Freeman (1981) —

> "the word 'outlier' here will mean any observation that has not been generated by the mechanism that generated the majority of observations in the data set"

2.2 *Problems with previous Bayesian approaches to outlier models*

Let us consider first, for simplicity, the one-sample location-scale situation in which the "standard" model assumes observations to be described by the normal density $\phi(x \mid \mu,\sigma)$, conditionally independent given μ and σ. Typical forms of model elaboration to include the possibility of outliers assume that aberrant observations follow an alternative normal density $\phi_\delta(x \mid \mu,\sigma)$, the latter taken to be either of the form

$$\phi_\delta(x \mid \mu,\sigma) = \phi(x \mid \mu + \delta,\sigma), \text{ the } \textit{location-shift} \text{ model, or, with } \delta > 1;$$

$$\phi_\delta(x \mid \mu,\sigma) = \phi(x \mid \mu,\delta\sigma), \text{ the } \textit{inflated-variance} \text{ model.}$$

Further scope in the precise form of elaboration used is provided by the choice of the range of the number of possible outliers to be entertained. In addition, models differ according as δ is assumed fixed, or varying from one aberrant observation to another, and —for the latter, in the location-shift case— having possibly different signs. Various such elaborations are considered in detail by Box and Tiao (1968), Abraham and Box (1978), Guttman, Dutter and Freeman (1978) and Freeman (1981).

There are, however, a number of problems with previous Bayesian approaches to outliers via model elaboration:

- the first of these relates to the basic idea of an aberrant observation as one which is generated by a mechanism which differs from that generating the majority of the observations; without further elaboration, this concept does not embrace the intuitive notions of "surprisingness" or "outlyingness" which seem to underlie most people's understanding of the term "outlier";

- this point is closely linked to the issues arising from the confusion of the statements "some observation is an outlier" and "the actually outlying observation is an outlier" (see, for example, Barnett and Lewis, 1978, p.33);

- it may be of interest to model and to distinguish the direction of outlyingness; however, on the basis of numerical studies of various previous Bayesian formulations of the problem, Freeman (1981) concluded "There is at present no known prior structure that permits large positive and negative contaminations to show themselves simultaneously";

- finally, the number of elements in Λ may become prohibitively large (consider, for example, the combinatorial explosion resulting from a consideration of all possible subsets of up to 10 % outliers in a sample of size 100), so that, computationally, the routine calculation of quantities like (2.7) may be impossible.

Similar problems arise with more general forms of outlier model, including the multivariate normal case, where the inflated-variance form assumes that most observations follow an $N_p(\mu,\Sigma)$ distribution, but that aberrant observations follow an $N_p(\mu,\delta\Sigma)$ distribution, with $\delta > 1$.

In Section 2.5, we shall consider a unified framework for the analysis of such models. First, however, we introduce a further element into the usual Bayesian form of model elaboration for outliers. This modified approach will be seen to resolve the difficulties mentioned above and open the way to a Bayesian approach to outliers which is both computationally feasible and more in accord with common sense ideas.

2.3 *A modified approach to outlier models: location-scale case*

In order to provide a clear statement of the basic ideas, we shall consider, for simplicity, the situation where there is assumed to be at most one aberrant observation, described by the location-shift model $\phi(x\,|\,\mu+\delta,\sigma)$, with $\delta > 0$ (which we shall call an *upper* outlier; $\delta < 0$ would correspond to a *lower* outlier).

Consider now a sample $x = (x_1,\ldots,x_n)$ and define the following events:

A_1 = "x contains one aberrant observation"
$A_1(i)$ = "observation x_i is aberrant, the others are not"
L_i = "observation x_i is the largest in x".

The key idea in our modification of previous Bayesian analyses is as follows. For "outlyingness" to be an appropriate interpretation of "aberrant", we require that $p(\delta)$, the prior specification encapsulating assumptions about the magnitude of the shift producing the aberrant observation, be such that, for $i=1,\ldots,n$,

$$P(A_1(i) \cap \bar{L}_i) \approx 0. \tag{2.8}$$

In other words, for a genuine *outlier* model, the component $p(\delta)$ should specify that, with high probability, an aberrant observation will, in fact, "outlie".

It now follows that:

$$\begin{aligned} P(A_1|x) &= \sum_i P(A_1(i)\,|x) = \sum_i P(A_1(i)\cap L_i|x) + \sum_i P(A_1(i)\cap \bar{L}_i|x) \\ &= \sum_i P(A_1(i)\,|\,L_i,x)P(L_i|x) + \sum_i P(A_1(i)\cap \bar{L}_i|x), \end{aligned} \tag{2.9}$$

and so, noting that (2.8) implies that the final summation is (approximately) zero and that if $x_{i*} = \max_i\{x_i\}$ then $P(L_i|x) = 1$ for $i = i*$, $= 0$ otherwise, (2.9) reduces to

$$P(A_1|x) \approx P(A_1(i*)\,|x). \tag{2.10}$$

This result —an immediate consequence of the specification of a $p(\delta)$ such that (2.8) is satisfied— is the key to avoiding the computational problem arising from the implicit combinatorial explosion in weighted average forms such as (2.7), which characterize inferences from elaborated neighbourhoods of models. The approach carries over in an obvious way to situations involving both upper and lower outliers and to the modelling of more than one of each. Thus, if $M(r_1,r_2)$ denotes a model which assumes there to be r_1 lower and r_2 upper outliers then —assuming an obvious generalization of (2.8)— $P(M(r_1,r_2)\,|x)$ is approximately equal to the posterior probability of a model which *a priori* pin-points the *actual* r_1 lower and r_2 upper outliers occurring in the sample.

In essence, the modified approach says that, when we consider possible sets of "outliers", we are justified in confining our attention in *a posteriori* calculations to those subsets which actually do "outlie". This is both pragmatically helpful in implementing a feasible sensitivity analysis and is also somewhat in accord with common sense. Using this approach, the difficulties described in Section 2.2 largely disappear.

However, there remains the rather important task of examining the implicit forms of $p(\delta)$ required in order that (2.8) —and it obvious generalizations— be satisfied. Again, for simplicity, we shall confine attention in our discussion to the case of one upper outlier. The question we need to answer is the following: if $\tilde{y}_i \sim N(\mu,\sigma^2)$, $i = 1,\ldots,n-1$, and $\tilde{y} \sim N(\mu+\delta,\sigma^2)$, with $\delta > 0$, what is the value of $P(\tilde{y} > \max_i\{\tilde{y}_i\})$? Writing $\tilde{z}_i = \tilde{y} - \tilde{y}_i$, $i = 1,\ldots,n-1$, the vector $\tilde{z} = (\tilde{z}_1,\ldots,\tilde{z}_{n-1})$ has a multivariate normal distribution with mean $\delta 1_{n-1}$ and dispersion matrix $\sigma^2(I_{n-1}+J_{n-1})$, where J_{n-1} is a square matrix of unities. The quantity we require is then equal to $P(\tilde{z}_1 \geq 0,\ldots,\tilde{z}_{n-1} \geq 0)$, which cannot be found analytically, but is bounded below by the corresponding probability obtained by replacing the actual dispersion matrix by $2\sigma^2 I_{n-1}$ (see Das Gupta *et al*, 1972, p. 261, Remark 5.1). We thus obtain

$$P(\tilde{y} > \max_i\{\tilde{y}_i\}) \geq \left[\Phi\left(\frac{\delta}{\sigma\sqrt{2}}\right)\right]^{n-1}, \tag{2.11}$$

where Φ is the standard normal distribution function. Table 2.1 gives values of the right-hand side of (2.11) for various values of n and δ/σ.

δ/σ \ n	10	15	20	30	50
3	.857	.787	.722	.608	.432
3.5	.942	.911	.811	.824	.721
4	.979	.968	.956	.934	.891
5	.999	.998	.998	.997	.995

TABLE 2.1 *Values of the lower bound in (2.11)*

An alternative way of investigating whether (2.8) is a satisfactory approximation is to find the probability that the aberrant observation exceeds the expected value of the extreme order statistic, values of which are given, for example, in Neave (1978, Table 2.4). For the case of a single upper outlier, the values of $P(\tilde{y} > E(\max_i \tilde{y}_i))$ are given in Table 2.2.

δ/σ \ n	10	15	20	30	50
3	.935	.903	.876	.834	.776
3.5	.978	.964	.951	.929	.896
4	.994	.989	.984	.976	.960
5	.999	.999	.999	.998	.997

TABLE 2.2 *Values of* $P(\tilde{y} > E(\max \tilde{y}_i))$

In the case of the inflated-variance model for aberrant observations, we may not be interested in distinguishing upper and lower outliers, so that the event L_i might be redefined as "observation x_i is the largest or smallest in x". The reasonableness of assumption (2.8) could then be investigated by calculating $P(\tilde{y} < E(\min_i \tilde{y}_i)) + P(\tilde{y} > E(\max_i \tilde{y}_i))$, where $\tilde{y}_i \sim N(\mu,\sigma^2)$, $i,\ldots,n$-1, and $\tilde{y} \sim N(\mu,\delta^2\sigma^2)$, with $\delta > 1$. Values of this quantity, for various choices of n and δ, are given in Table 2.3.

δ \ n	10	15	20	30	50
3	.620	.568	.526	.498	.454
5	.764	.734	.710	.684	.654
10	.880	.864	.852	.840	.820
15	.921	.910	.902	.893	.881

TABLE 2.3. *Values of* $P(\tilde{y} < E(\min \tilde{y}_i)) + P(\tilde{y} > E(\max \tilde{y}_i))$

Using such tables, we can check, in any specific application, whether the number of observations and our prior opinions about δ are such as to enable us to feel comfortable in using the approximation (2.10), or its generalization to more complex models. Broadly speaking, and bearing in mind the large potential computational saving and thus the possibility of implementing analyses for much larger data sets, these results suggest to us that the approximation will often provide a satisfactory operational framework for flexible Bayesian modelling and analysis of outliers.

2.4. *A modified approach to outlier models: multivariate case*

In a univariate sample there is a natural ordering which we exploited in the modified approach to outlier modelling described in the previous section. For multivariate data, or observations from a general linear model, there is no such obvious ordering. However, to extend our suggested approach to these more complicated situations we need an "ordering" principle which indicates the extent to which subsets of the data appear "aberrant" in comparison with other observations in the data set.

The ordering principle which we shall adopt is based on what we shall call the "predictive ordinate measure". Given a sample $x = (x_1,...,x_n)$, let us suppose that the "standard" model generating the non-aberrant observations has the form $p(x_i|\psi)$, where ψ represents all unknown parameters. Let us also denote by $x(S)$ the elements of x whose labels are in $S \subseteq \{1,...,n\}$, and define

$$p(x(S)|x(\bar{S})) = \int p(x(S)|\psi)p(\psi|x(\bar{S}))d\psi \qquad (2.12)$$

to be the predictive density for $x(S)$ given $x(\bar{S})$, where $\bar{S}$ denotes the complement of S in $\{1,...,n\}$. Small values of (2.12) indicate that observations $x(S)$ are "surprising" in relation to $x(\bar{S})$ (and the prior specification for ψ), and this predictive ordinate measure can be used to order individual observations, pairs, triples, etc., on the basis of their aberrant nature compared with the other observations. (See Also, Geisser, 1980.)

It is easy to see that this measure reduces, in the univariate case, to the "natural" ordering, in that "surprisingness" is synonymous with "outlyingness". More general properties of this measure are easily established in the multivariate case where non-aberrant observations are assumed to be $N_p(\mu,\Sigma)$, and where, conditional on Σ, μ is assigned an $N_p(\mu_0,a^{-1}\Sigma)$ distribution and Σ, marginally, is assigned an inverse-Wishart distribution with parameters q and Q (see, for example, Box and Tiao, 1973, p. 460). Then, if S contains r elements, (2.12) is proportional to the determinant of

$$Q + \sum_{j\notin S} x_j x_j^T + a\mu_0\mu_0^T - \left[\sum_{j\notin S} x_j + a\mu_0\right](n+a-r)^{-1}\left[\sum_{j\notin S} x_j + a\mu_0\right]^T \qquad (2.13)$$

Using (2.13), it can be shown that the individual observation ($r = 1$) with the minimum predictive ordinate must lie at one of the vertices of the convex hull of the data set. Similarly, the pair of observations ($r = 2$) with the minimum predictive ordinate among all pairs will be such that if one of them is deleted from the data set the other will be at a vertex of the convex hull of the remaining observations. This latter result then extends inductively in an obvious manner to any value of r (see Pettit, 1983, Section 3.2, for proofs of these results).

The results just quoted imply that the computations required to identify, for various r, the subsets of size r with minimum predictive ordinate are not excessive, since minimization reduces to certain systematic searches of vertices of convex hulls of sets of points (for which efficient algorithms exist: see, for example, Green and Silverman, 1979). The amount of computation required in any particular case cannot, of course, be specified exactly, since the number of observations lying at vertices of the convex hull is a random quantity. However, Efron (1965) has obtained the expected number of points lying on the convex hull of a set of observations with a common bivariate normal distribution and we can use his results in order to illustrate the computational savings involved. For example, if $n = 50$, the expected number is 8.0 and we would anticipate having to work out the predictive ordinate of about 8 points in order to identify the most aberrant single points, $36 = \binom{9}{2}$ points in order to identify the most aberrant pair of observations and $120 = \binom{10}{3}$ points in order to identify the most aberrant triple (assuming the expected numbers of vertices in the convex hull for $n = 49, 48$ also to be about 8). This compares with 50, 1225 and 19600 evaluations if all $\binom{n}{r}$ predictive ordinates had to be calculated.

Moreover, these figures are for a common bivariate normal. If there are, in fact, aberrant observations, from an $N_p(\mu,\delta\Sigma)$ distribution, with $\delta > 1$, the number of points in the convex hull may be greatly reduced.

By analogy with the key step in the approach of Section 2.3, we shall now argue that for the "aberrant" observation model to be a genuine "outlier" model, we require $p(\delta)$ such that if r aberrant observations are assumed then, with high probability, these correspond to the actual subset of r observations having minimal predictive ordinate among all subsets of r. By a similar argument to that lending to (2.10), we can then confine attention in *a posteriori* calculations to the subsets of potential outliers identified by the predictive ordinate ranking. As we have just seen, this results in a computationally feasible methodology which restricts attention in assessing potential outliers to those subsets of observations which actually do "outlie", in the sense of being the vertices of the convex hull of the complete data set, or the convex hull of successively "trimmed" versions of the data set.

As an indication of the form of analysis required in order to understand the kinds of assumptions about δ required for this to be a reasonable approximation, we shall examine the simple case of one outlier ($r = 1$), and assume that $\tilde{y}_1,\ldots,\tilde{y}_{n-1}$ have independent $N_p(0,I)$ distributions and independent $\tilde{y}$ has an $N_p(0,\delta I)$ distribution. If $\|\cdot\|$ denotes the sum of squares of the components of the vector argument, the probability that $\tilde{y}$ turns out to have the minimum predictive ordinate is given by

$$P(\|\tilde{y}\| > \|\tilde{y}_i\|, i = 1,\ldots,n-1) = P\left(\frac{\|\tilde{y}_i\|}{\|\tilde{z}\|} < \delta, i = 1,\ldots,n-1\right) \tag{2.14}$$

where $\tilde{z} = \delta^{-1/2}\,\tilde{y}$. But the joint distribution of $\tilde{v}_i = \|\tilde{y}_i\|/\|\tilde{z}\|$, $i = 1,\ldots,n-1$, is multivariate-F (Johnson and Kotz, 1972), so that (by a result of Hewett and Bulgren, 1971)

$$P(\tilde{v}_i < \delta, i = 1,\ldots n-1) \geq [P(\tilde{u} < \delta)]^{n-1}, \tag{2.15}$$

where $\tilde{u}$ has an $F_{p,p}$ distribution. For given n and δ, we thus have a lower bound for the required probability. In Table 2.4, we present a selection of required values of δ, for given values of n, to obtain the bound, α, say, defined by the right-hand side of (2.15).

		α \ n	15	20	30
p	2	.85	85	117	180
		.90	130	180	270
	3	.85	27	33	45
		.90	36	45	59
	4	.85	15	18	22
		.90	19	22	28

TABLE 2.4. *Values of δ such that the probability α is exceeded, for given n and p.*

We have focussed, in Sections 2.3 and 2.4, on the univariate and multivariate location-scale models. The predictive ordinate measure can be applied equally to the general linear model case, but we shall postpone the details of this until Section 3, where a comparative discussion of "outlyingness" and "influence" will be given.

2.5. *A general linear model framework for analysing Bayesian outlier models*

We shall show in this section that many of the outlier models we wish to consider in the linear framework can be formulated within the structure

$$x \sim N(A_1\theta_1, C_1),\ \theta_1 \sim N(A_2\theta_2, C_2). \tag{2.16}$$

for suitable choices of the matrices involved.

It then follows (see, for example, Lindley and Smith, 1972), that if C_1, C_2 are known, the posterior probability for a model thus specified is proportional to the prior probability multiplied by

$$|C_1 + A_1C_2A_1^T|^{-1/2} \exp\{-\frac{1}{2}(x - A_1A_2\theta_2)^T(C_1 + A_1C_2A_1^T)^{-1}(x - A_1A_2\theta_2)\}. \tag{2.17}$$

Alternatively, if, conditional on σ^2, $C_1 = \sigma^2 I_n$, $C_2 = \sigma^2 V$, with V known, and the prior for the unknown σ^2 is specified in the form $\nu\lambda/\sigma^2 \sim \chi^2_\nu$, (2.17) is replaced by

$$|I_n + A_1VA_1^T|^{-1/2}[\nu\lambda + (x - A_1A_2\theta_2)^T(I_n + A_1VA_1^T)^{-1}(x - A_1A_2\theta_2)]^{-1/2(n+\nu)}. \tag{2.18}$$

If interest centres on inferences for θ_1, we note (see, for example, Lindley and Smith, 1972) that, under the particular model specification (2.16), the distribution of θ_1, given x, C_1, C_2 is $N(Bb, B)$, where

$$B^{-1} = A_1^TC_1^{-1}A_1 + C_2^{-1}, \qquad b = A_1^TC_1^{-1}x + C_2^{-1}A_2\theta_2. \tag{2.19}$$

Under the alternative specification for unknown σ^2 given above, the posterior distribution for θ_1 is Student-t with degrees of freedom $n + \nu$ and mean and dispersion matrix given by

$$(A_1^TA_1 + V^{-1})^{-1}(A_1x + V^{-1}A_2\theta_2) \text{ and } (A_1^TA_1 + V^{-1})^{-1}, \tag{2.20}$$

respectively.

Overall inference for θ_1 has the form of a weighted-average of such normal or Student-t densities, with weights given by (2.17) or (2.18). Moreover, in the light of the discussion of Sections 2.3 and 2.4, the only non-negligible posterior weights may be assumed to be those corresponding to models of the form (2.16) which label as outliers those subsets of the observations which actually turn out to "outlie" (in accordance with the predictive ordinate measure).

In order to exemplify the ways in which various proposed outlier models can be formulated within the framework of (2.16), we shall consider, for notational convenience, some particular univariate location-scale models which specify that $x_1,\dots,x_{r_1}$ are lower outliers and $x_{n-r_2+1},\dots,x_n$ are upper outliers, with $r_1+r_2 = r$. The first two examples are location-shift models, with δ_i denoting the shift if x_i is an aberrant observation; the third example is the inflated-variance model.

Location-shift model; δ_i's different

We consider, as an example, the model of Guttman, Dutter and Freeman (1978), with the additional assumption that, conditional on σ^2, the δ_i's are $N(\delta_0, \eta^2\sigma^2)$ in the case of upper outliers and $N(-\delta_0,\eta^2\sigma^2)$ in the case of lower outliers. If the location μ is assigned, conditional on σ^2, an $N(\mu_0,\varkappa^2\sigma^2)$ distribution, then, in the notation of (2.16), we have

$$A_1 = \begin{bmatrix} 1_n & \begin{matrix} I_{r1} & 0 \\ 0 & 0 \\ 0 & I_{r2} \end{matrix} \end{bmatrix}, \quad C_1 = \sigma^2 I_n, \quad \theta_1^T = (\mu,\delta_1,\dots,\delta_{r1},\delta_{n-r2+1},\dots,\delta_n),$$

$$A_2 = \begin{bmatrix} 1 & 0 \\ 0 & -1_{r1} \\ 0 & 1_{r2} \end{bmatrix}, \quad C_2 = \sigma^2 \begin{bmatrix} \varkappa^2 & 0 \\ 0 & \eta^2 I_r \end{bmatrix}, \quad \theta_2 = \begin{bmatrix} \mu_0 \\ \delta_0 \end{bmatrix}.$$

Location-shift model: $|\delta_i|$'s equal

We consider, as an example, the model of Abraham and Box (1978) extended to include upper and lower outliers, with the same prior specification for μ as in the previous model and with the assumption that the common shift is normally distributed with positive mean value δ_0 and variance $\eta^2\sigma^2$. In the notation of (2.16), we then have

$$A_1 = \begin{bmatrix} 1_n & \begin{matrix} -1_{r1} \\ 0 \\ 1_{r2} \end{matrix} \end{bmatrix}, \quad C_1 = \sigma^2 I_n, \quad \theta_1^T = (\mu,\delta)$$

$$A_2 = I_2, \quad C_2 = \sigma^2 \begin{bmatrix} \varkappa^2 & 0 \\ 0 & \eta^2 \end{bmatrix}, \quad \theta_2 = \begin{bmatrix} \mu_0 \\ \delta_0 \end{bmatrix}$$

Inflated-variance model

With the same prior specification for μ as in the previous models, the inflated-variance model of Box and Tiao (1968) corresponds, in the notation of (2.16), to

$$A_1 = 1_n, \; C_1 = \sigma^2 \begin{bmatrix} \delta^2 I_{r1} & 0 & 0 \\ 0 & I_{n-r} & 0 \\ 0 & 0 & \delta^2 I_{n-r2} \end{bmatrix}, \quad \theta_1 = \mu$$

$$A_2 = 1, \; C_2 = \varkappa^2\sigma^2, \; \theta_2 = \mu_0.$$

2.6. *Links with non-Bayesian analyses of outlier models*

Using (2.19), the form of posterior mean for any model specified in terms of (2.16) can be found explicitly. In general, we obtain complex weighted-averages of the mean of the non-aberrant observations (as indicated by the particular model) and means of the other observations suitably "adjusted" (in line with the form of location-shift or inflated-variance assumed; see Pettit, 1983, Section 2.3, for details).

However, if $\varkappa^2$, $\eta^2 \to \infty$ in the location-shift models and $\varkappa^2, \delta^2 \to \infty$ in the inflated-variance model —representing, in some sense weak prior information about the location parameter and the magnitude of the outliers then the posterior mean can be shown to reduce to the ordinary mean of the "good" observations (as defined by the model in question). The overall posterior mean —corresponding to (2.7)— is thus seen to be an adaptive weighted average of "trimmed" means, each of the latter corresponding to the deletion of the particular observations specified as outliers by the particular outlier model elaboration under consideration. Moreover, as we have argued in Sections 2.3 and 2.4, in *a posteriori* calculations we need only consider as non-negligible the posterior weights on models which pin-point as potential outliers those actual subsets of the observations which do "outlie".

In order to obtain greater insight into the nature of these posterior weights, it is convenient to think in terms of the ratio of posterior to prior odds (or Bayes factor) for the "no outlier" model versus the model, $M(r_1,r_2)$, which specifies r_1 lower and r_2 upper outliers. In the case of location-shift models, we then have a special case of Bayes factors for nested linear models (see, for example, Smith and Spiegelhalter, 1980, and Spiegelhalter and Smith, 1982), which can be shown to be monotone functions of conventional F-test statistics. Many classical test statistics for outliers versus the null hypothesis of no outliers (see Barnett and Lewis, 1978) can be related in this way to the posterior weights appearing in our modified Bayesian elaborated model approach. The latter thus enables us to provide a unified overview of the many *ad hoc* test proposals in the non-Bayesian literature by making explicit the form of outlier model for which the statistic appears as part of the posterior weight function. A selection of these correspondences are listed below (where $B_{0,(r_1,r_2)}$ denotes the Bayes factor on "no outliers" versus model $M(r_1,r_2)$; z denotes the ordered version of the sample x; $\bar{z}$ is the sample mean; $\bar{z}_*$ is the trimmed mean excluding z_n; $\bar{z}_{**}$ is the trimmed mean excluding z_1 and z_n).

	Classical test statistics	*Bayes factors in which they appear*
1.	$\dfrac{\sum_1^n (z_i-\bar{z})^2}{\sum_1^{n-1} (z_i-\bar{z}_*)^2}$	$B_{0,(0,1)}$ δ_i's different
	Grubbs (1950) test for 1 upper outlier	
2.	$\dfrac{\sum_1^n (z_i-\bar{z})^2}{\sum_2^{n-1} (z_i-\bar{z}_{**})^2}$	$B_{0,(1,1)}$ δ_i's different
	Grubbs (1950) test for 1 upper and 1 lower outlier	
3.	$\dfrac{\sum_1^n (z_i-\bar{z})^2}{(z_n-z_1)^2}$	$B_{0,(1,1)}$ $\lvert\delta_i\rvert$ constant
	David *et al* (1954) test for 1 upper and 1 lower outlier	

4.	$\dfrac{\sum_{1}^{n} (z_i - \bar{z})^2}{(z_{n-1} + z_n - 2\bar{z})^2}$	$B_{0,(0,2)}$ $\lvert\delta_i\rvert$ constant
	Murphy (1951) test for 2 upper outliers	

Extensive numerical studies (see Pettit, 1983, Section 2.4) demonstrate that the posterior weights obtained from the models specified in Section 2.5 have intuitively desirable behaviour and avoid the difficulties with previous Bayesian outlier models highlighted in Freeman (1981).

2.7. *Two-sample and multivariate problems*

The location-scale models presented in Section 2.5 can easily be extended to deal with two-sample problems: for details, see Pettit and Smith (1984). In particular, inferences about differences in location in the two samples are based on weighted averages of Student-t distributions, with weights ranging over model elaborations which take into account various possibilities for the numbers and types of outlier in each sample. Similarly, inferences for variance ratios are based on weighted averages of F-distributions.

A detailed analysis of the multivariate outlier elaboration described at the end of Section 2.2 is also given in Pettit and Smith (1984). In particular, it is shown that in the bivariate case ($p = 2$), a robustified estimate of correlation is obtained in the (approximate) form of a weighted-average of standard correlation estimates based on successive "convex-hull trimming" of the sample.

3. INFLUENTIAL OBSERVATIONS

3.1. *A measure of the influence of a subset of observations*

As a measure of the influence of a subset of the observations in a data set, we shall use the symmetric form of the Kullback-Leibler distance, first proposed by Jeffreys (1946) in a somewhat different context.

Using the general notation of Section 2.4, we define the influence of the subset of observations with labels in $S \subseteq \{1,\ldots,n\}$ to be

$$I(S) = \int \log \frac{p(\psi|x)}{p(\psi|x(\bar{S}))} [p(\psi|x) - p(\psi|x(\bar{S}))] d\psi. \tag{3.1}$$

If $I(S)$ is small, observations with labels in S have little influence (in the intuitive sense) in that if they are omitted from the data set the posterior distribution for ψ is little changed.

If we are dealing with linear models which can be specified within the framework of (2.16), and if C_1, C_2 are known, so that $\psi = \theta_1$, then $p(\theta_1|x)$ is the density of an $N(Bb,B)$ distribution (with B, b given by (2.19)) and $p(\theta_1|x(\bar{S}))$ corresponds to an $N(Dd,D)$ distribution for some easily identified D,d. Straightforward manipulations then establish that

$$I(S) = -p + \tfrac{1}{2}[tr(D^{-1}B + B^{-1}D) + (Bb\text{-}Dd)^T(D^{-1} + B^{-1})(Bb - Dd)], \tag{3.2}$$

where p is the dimensionality of θ_1.

Johnson and Geisser (1982) have also considered the problem of influential observations from a Bayesian viewpoint, but the measure they adopt is the Kullback-Leibler distance between the *predictive* distributions of a new set of data at the given design points conditional on all the data and conditional on all the data with observations omitted one at a time, two at a time, etc. This approach may be appropriate if we are interested in prediction, but if we are interested in the influence of a point on inferences from a given sample it would seem more natural to use a measure of the effect of that point on the *posterior* distribution rather than the predictive (although technically, of course, many of the results are similar).

3.2. *Influence and outlyingness measures for a single observation*

In the case where $S=\{i\}$, we note that, if, for $j=1,\ldots,n$, we have independent observations such that

$$x_j \sim N(a_j^T\theta_1,\sigma^2)$$
$$\theta_1 \sim N(A_2\theta_2,C_2) \tag{3.3}$$

then writing $A_1^T = [a_1,\ldots a_n]$ and using results from Lindley and Smith (1972) the quantities occurring in (3.2) are given by

$$B^{-1}=\sigma^{-2}A_1^TA_1+C_2^{-1}$$
$$b=\sigma^{-2}A_1^Tx+C_2^{-1}A_2\theta_2 \tag{3.4}$$

and

$$D^{-1}=B^{-1}-\sigma^{-2}a_ia_i^T$$
$$d=b-\sigma^{-2}a_ix_i\ . \tag{3.5}$$

Using the relations in (3.5) we may simplify (3.2) and rewrite it in two forms. First, eliminating D and d, we have

$$I(\{i\})=\frac{1}{2}\,\frac{u_i}{1-u_i}\left[u_i+\frac{(2-u_i)(x-a_i^TBb)^2}{\sigma^2(1-u_i)}\right] \tag{3.6}$$

where

$$u_i=\frac{a_i^TBa_i}{\sigma^2}\ .$$

Alternatively, eliminating B and b, we have

$$I(\{i\})=\frac{1}{2}\,\frac{w_i}{1+w_i}\left[w_i+\frac{(2+w_i)(x_i-a_i^TDd)^2}{\sigma^2(1+w_i)}\right] \tag{3.7}$$

where

$$w_i=\frac{a_i^TDa_i}{\sigma^2}\ .$$

We have assumed so far that σ^2 is known. If we consider σ^2 unknown, the posterior distributions for θ_1 are t-distributions and we cannot find an analytic expression for the Kullback-Leibler distance. Johnson and Geisser (1982) consider approximations to this quantity, but, for simplicity, we prefer to consider the sensitivity of the measure for a range of plausible values of σ^2. In general we have found that the ordering of the most influencial observations is not much affected by different choices of σ^2.

The measure proposed for potential "outlyingness" is the predictive ordinate measure defined by (2.12). In the particular case considered above, where $S = \{i\}$, the predictive ordinate measure for x_i is easily seen to be equal to

$$[(2\pi\sigma^2)^{-1/2}(1-u_i)]^{1/2}\exp\left[-\frac{1}{2}\,\frac{(x_i-a_i^TBb)^2}{\sigma^2(1-u_i)}\right]. \tag{3.8}$$

In this case, it is possible to integrate out σ^2, but to facilitate the comparison of influence and outlyingness measures we can in practice consider the expression (3.8) for a range of plausible values of σ^2. Again, we have found that in general the ordering of the most outlying observations is not affected by different choices of σ^2.

3.3. *Links with non-Bayesian measures of influence and outlyingness*

Recent books by Belsley, Kuh and Welsch (1980), and Cook and Weisberg (1982), have reviewed a number of non-Bayesian proposals for regression diagnostics, based on *ad-hoc* combinations of statistics which measure —in some intuitively appealing sense— the influence and outlyingness of subsets of the observations.

We shall not attempt a comprehensive overview of such proposals and their links —or otherwise— with Bayesian ideas. Instead, we shall concentrate on some of the more widely discussed ideas which are based on the information contained in the least-squares residuals and their variances:

$$r_i = x_i - a_i^T \hat{\theta}_1, \qquad V(r_i) = \sigma^2(1-v_i) \tag{3.9}$$

where $v_i = a_i^T (A_1^T A_1)^{-1} a_i$, a circumflex denotes the least-squares estimate and, for simplicity, σ^2 is assumed known.

In particular, when considering individual points, large values of the "Studentised residuals"

$$t_i = (x_i - a_i^T \hat{\theta}_1)/[\sigma^2(1-v_i)]^{1/2} \tag{3.10}$$

can be considered as an intuitive indication of outlyingness, and large values of v_i typically indicate an observation with an aberrant a_i (i.e. an "outlying" value of the independent variable vector, or design point). Much discussion has then centred on how to combine these values into some overall measure (but with σ^2 replaced by some standard point estimate). For example, the following alternatives have been proposed and their merits and demerits analysed:

$\left(\frac{t_i^2}{p}\right)\left(\frac{v_i}{1-v_i}\right)$, Cook (1977); $\left(\frac{n-p}{p}F_i\right)\left(\frac{v_i}{1-v_i}\right)$, Welsch and Kuh (1977);

$\left(1-\frac{t_i^2}{n-p}\right)(1-v_i)$, Andrews and Pregibon (1978);

$\left(\frac{t_i^2}{p}\right)(v_i)$, Cook and Weisberg (1980); $\left(\frac{F_i}{p}\right)(v_i)$, Cook and Weisberg (1980);

where $F_i = t_i^2(n-p-1)/(n-p-t_i^2)$.

To compare the Bayesian influence measure with these *ad hoc* suggestions, we let $C_2^{-1} \to 0$ in (3.6), representing vague prior information about θ_1. Noting that $u_i \to v_i$, we obtain

$$I(\{i\}) = \frac{1}{2}\frac{v_i}{1-v_i}[v_i + (2-v_i)t_i^2], \tag{3.11}$$

and, after some manipulation, $I(\{i\})$ can be examined as a function of the above forms. Similarly, the predictive ordinate measure (3.8) reduces to

$$[(2\pi\sigma^2)^{-1/2}(1-v_i)]^{1/2}\exp(-\frac{1}{2}t_i^2). \tag{3.12}$$

To examine the "trade-off" in (3.11) between outlyingness in observation-space, which is measured directly by t_i^2, and outlyingness in design-space, which is closely related to v_i, we define (dropping the label i, for convenience)

$$I_1 = t^2, \qquad I_2 = \frac{v}{1-v} = \frac{V(\hat{y})}{V(r)}, \tag{3.13}$$

so that

$$I = \frac{1}{2}\frac{I_2}{1+I_2}[I_2 + I_1(2+I_2)]. \tag{3.14}$$

Contour plots of I against I_1, I_2, with observations in a data set superimposed then provide a possible Bayesian diagnostic summary, indicating, for each data point, both the individual measures, I_1, I_2 and the overall influence measure I (which is *not* necessarily a maximum for points having maximum values of either I_1 or I_2).

We shall illustrate these ideas with an example, which further highlights differences between Bayesian and non-Bayesian overall measures of influence.

3.4. *An illustrative example*

We reconsider the data presented by Mickey, Dunn and Clark (1967), which has been much used to compare suggested regression diagnostics. The data are from a study on cyanotic heart disease and Figure 3.1 shows a scatter plot of Gessel adaptive score (y) versus age at first word (x), for each of 21 children. A straight line model is thought to be plausible for these data, but it is possible that the observations labelled 2, 18 and 19 are dominating any visual reaction to the plot.

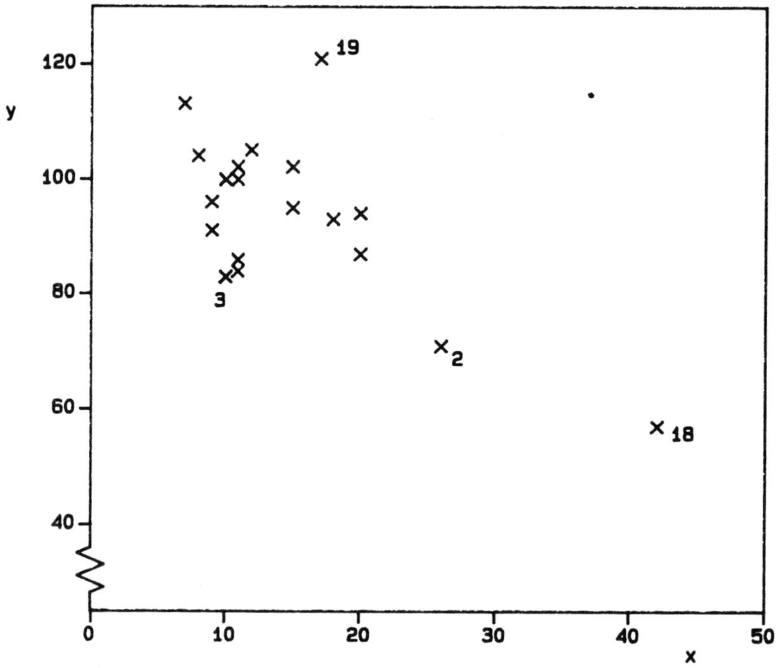

FIGURE 3.1. *Scatter plot of the Gessell adaptive score data*

Various of the measures we have discussed above —(3.12), (3.13), (3.14)— were calculated for this data set (using a value of $\sigma^2 = 50$: other reasonable choices do not lead to essentially different conclusions, although we shall return to this point briefly in Section 4). Table 3.1 gives the results obtained for the data shown in Figure 3.1.

Observation	I_1	I_2	I	$(2\pi)^{1/2}\times(3.12)$
1	0.09	.05	.005	.132
2	4.53	.10	.424	.014
3	4:67	.06	.280	.013
4	1.64	.08	.122	.060
5	1.71	.05	.085	.059
6	0.00	.08	.003	.136
7	0.25	.06	.017	.121
8	0.14	.06	.010	.128
9	0.22	.09	.021	.122
10	0.96	.08	.075	.084
11	3.55	.08	.271	.023
12	0.30	.08	.024	.117
13	5.19	.07	.338	.010
14	3.85	.06	.226	.020
15	0.44	.06	.027	.110
16	0.04	.07	.005	.134
17	1.58	.05	.086	.062
18*	0.00	5.32	2.239	.056
19	17.22	.05	.859	.000
20	2.79	.06	.164	.034
21	0.04	.07	.005	.134

TABLE 3.2. *Values of I_1, I_2, I and $(2\pi)^{1/2}\times(3.12)$ for the amended data (with point 18 replaced by 18* as described in the text).*

The results show that the overall influence of observation 18 is due to its outlying position in the design-space (high value of I_2), whereas observation 19 is outlying in the observation-space. Figure 3.2 provides an overall summary plot (with observation 19 omitted for convenience of presentation).

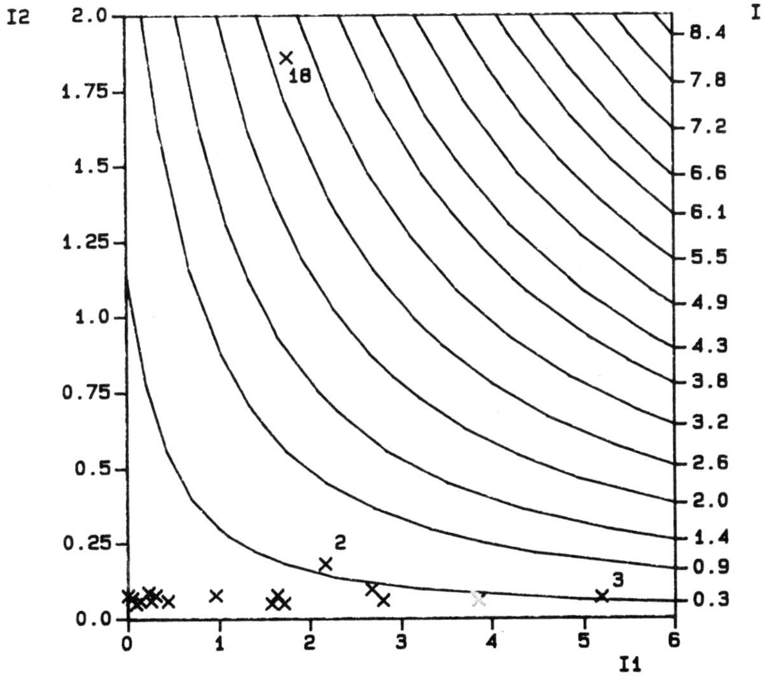

FIGURE 3.2. *Contours of I against I_1, I_2 for the Gessell adaptive score data*

To explore further the way in which I compares with other influence measures, we follow Draper and John (1981) in considering the data with the point 18 replaced by 18* in which the x-value is changed to 62.43 so that 18* lies on the least squares line fitted to the observations omitting 18. Since we have used a vague prior this will also be very close to the line given by the posterior means of the parameters. We see from Table 3.2 that, in contrast to the ordering given by Cook's (1977) statistic, 18* is still the most influential point as measured by I. *This is because although removal of 18* will not affect posterior means, it will affect posterior variances.* The posterior distribution provides inferences which go wider than just point estimates; they also indicate measures of uncertainty. The classical influence measure is concerned with the effect of an observation on the point estimate; the Bayesian measure of influence has greater content in that it reflects the effect on the overall inference.

Observation	I_1	I_2	I	$(2\pi)^{1/2}\times(3.12)$
2	4.53	.10	.424	.014
3	4:67	.06	.280	.013
11	3.55	.08	.271	.023
18*	0.00	5.32	2.239	.056
19	17.22	.05	.859	.000

TABLE 3.2. *Values of I_1, I_2, I and $(2\pi)^{1/2}\times(3.12)$ for the amended data (with point 18 replaced by 18* as described in the text).*

If we replace 18 by 18**, in which the y value is changed to 72.91, which still lies on the least squares line fitted to the observations omitting 18, then we see from Table 3.3 that 18** is not as influential as observation 19 or 2 but that it does have some effect.

Observation	I_1	I_2	I	$(2\pi)^{1/2}\times(3.12)$
2	4.89	.18	.838	.011
3	4.69	.07	.306	.013
11	3.62	.10	.349	.022
18**	0.00	1.86	.607	.084
19	17.30	.06	.945	.000

TABLE 3.3. *Values of I_1, I_2, I and $(2\pi)^{1/2}\times(3.12)$ for the amended data (with point 18 replaced by 18** as described in the text).*

Finally, we note that although we have not had space here to develop the algebraic forms for influence and outlyingness measures for pairs of data points, triples, etc., these are easily derived. Table 3.4 shows the numerical results for the overall influence and predictive ordinate measures for selected pairs of points from the data set shown in Figure 3.1.

Observation	I	$(2\pi)\times(3.2)\times 10^4$
18,2	20.161	1.082
18,3	2.313	4.417
18,11	5.629	6.874
18,19	1.392	.006
19,2	.498	.006
19,3	.591	.001
19,11	1.902	.002

TABLE 3.4. *Values of I and $(2\pi)\times(3.2)\times 10^4$ for pairs of data points*

4. DISCUSSION

Models and data are both outcomes of processes and activities about which we need to be cautious and circumspect —all the more so if Bayes'theorem is being interpreted in an "input-output" sense. There is thus a pressing need for easily implementable forms of sensitivity analysis both in relation to model choice and data acceptance— motivated and developed from a Bayesian standpoint.

There are many problems: *conceptual* (how to define "outliers", "influence", etc.); *computational* (how to deal with the combinatorial explosion implicit in most elaborated neighbourhood approaches), and *presentational* (how to summarize investigations of many alternative models, or assessments of many subsets of a large data set).

In this paper we have suggested some possible solutions -but our suggestions are very much to be regarded as tentative, and requiring considerable further exploration to map out the territorial gap between theoretical musing and practical application. In particular, we need to understand whether real data sets *can* be routinely and adequately modelled using these kinds of ideas, and if so, whether the information about δ and σ^2 in the data can be extracted without introducing yet further computational problems.

ACKNOWLEDGEMENTS

Much of this material is taken from the first author's Ph.D. thesis, written at Nottingham University under the supervision of the second author and supported by a Science and Engineering Research Council Studentship. Our interest in rethinking the modelling of outliers was stimulated by the results presented at the 1st. International Meeting on Bayesian Statistics by P.R. Freeman.

REFERENCES

ABRAHAM, B. and BOX, G.E.P. (1978). Linear models and spurious observations. *Appl. Statist.*, **27**, 120-130.

ANDREWS, D.F. and PREGIBON, D. (1978). Finding the outliers that matter. *J. Roy. Statist. Soc. B.* **40**, 85-93.

BARNETT, V. and LEWIS, T. (1978). *Outliers in Statistical Data.* Chichester: Wiley.

BELSLEY, D.A., KUH, E. and WELSCH, R.E. (1980). *Regression Diagnostics.* New York: Wiley.

BOX, G.E.P. (1980). Sampling and Bayes inference in scientific modelling and robustness *J. Roy. Statist. Soc. A.*, **143**, 383-430 (with discussion).

BOX, G.E.P. and TIAO, G.C. (1964). A Bayesian approach to the importance of assumptions applied to the comparison of variances. *Biometrika,* **51**, 153-167.

— (1968). A Bayesian approach to some outlier problems. *Biometrika,* **55**, 119-129.

— (1973). *Bayesian Inference in Statistical Analysis.* Massachusetts: Reading: Addison-Wesley.

COOK, R.D. (1977). Detection of influential observations in linear regression. *Technometrics,* **19**, 15-18.

COOK, R.D. and WEISBERG, S. (1980). Characterizations of an empirical influence function for detecting influential cases in regression. *Technometrics,* **22**, 495-508.

— (1982). *Residuals and Influence in Regression.* New York: Chapman and Hall.

DAS GUPTA, S., EATON, M.L., OLKIN, I., PERLMAN, M., SAVAGE, L.J. and SOBEL, M. (1972). Inequalities on the probability content of convex regions for elliptically contoured distributions. *Proc. 6th. Berk. Symp.* **2**. 241-265. California: University Press.

DAVID, H.A., HARTLEY, H.O. and PEARSON, E.S. (1954). The distribution of the ratio, in a single normal sample, of range to standard deviation. *Biometrika,* **41**, 482-493.

DRAPER, N.R. and JOHN, J.A. (1981). Influential observations and outliers in regression. *Technometrics,* **23**, 21-26.

EFRON, B. (1965). The convex hull of a random set of points. *Biometrika*, **52**, 331-343.

FREEMAN, P.R. (1981). On the number of outliers in data from a linear model. *Bayesian Statistics* (Bernardo, J.M., *et. al* eds.), 349-365. Valencia: University Press.

GEISSER, S. (1980). In discussion of Box (1980).

GREEN, P.J. and SILVERMAN, B.W. (1979). Constructing the convex hull of a set of points in the plane. *The Computer Journal*, **22**, 262-266.

GRUBBS, F.E. (1950). Sample criteria for testing outlying observations. *Ann. Math. Statist.,* **21**, 27-58.

GUTTMAN, I., DUTTER, R. and FREEMAN, P.R. (1978). Care and handling of univariate outliers in the general linear model to detect spurosity —a Bayesian approach. *Technometrics,* **20**, 187-193.

HEWETT, J, and BULGREN, W.G. (1971). Inequalities for some multivariate F-distributions with applications. *Technometrics*, **13**, 397-402.

JEFFREYS, H. (1946). An invariant form for the prior probability in estimation problems. *Proc. Roy. Soc. A.,* **186**, 453-461.

JOHNSON, N.L. and KOTZ, S. (1972). *Distributions in Statistics: Continuous Multivariate Distributions.* New York: Wiley.

JOHNSON, W. and GEISSER, S. (1982). Assessing the predictive influence of observations. In *Statistics and Probability: Essays in Honor of C.R. Rao* (Kallianpur, G., *et al* eds.), 343-358. Amsterdam. North-Holland.

LINDLEY, D.V. and SMITH, A.F.M. (1972). Bayes estimates for the linear model. *J. Roy. Statist. Soc. B.,* **34**, 1-41 (with discussion).

MICKEY, M.R., DUNN, O.J. and CLARK, V. (1967). Note on use of stepwise regression in detecting outliers. *Computers and Biomed. Res.,* **1**, 105-111.

MURPHY, R.B. (1951). *On tests for outlying observations.* Unpublished Ph.D. thesis. Princeton University.

NEAVE, H.R. (1978). *Statistics Tables.* Allen and Unwin: London.

PETTIT, L.I. (1983). *Bayesian approaches to outliers.* Unpublished Ph.D. thesis. Nottingham University.

PETTIT, L.I. and SMITH, A.F.M. (1984). Bayesian model comparison in the presence of outliers. *Bull. 44th Int. Statist. Inst. Meeting, Madrid.* (To appear).

SIMAR, L. (1983). Protecting against gross errors: the aid of Bayesian methods. *Specifying Statistical Models* (Florens, J.P., *et al* eds.), 1-12. New York: Springer Verlag.

SMITH, A.F.M. (1983). Bayesian approaches to outliers and robustness. *Specifying Statistical Models* (Florens, J.P., *et al* eds.), 13-35. New York: Springer-Verlag.

SMITH, A.F.M. and SPIEGELHALTER, D.J. (1980). Bayes factors and choice criteria for linear models. *J. Roy. Statist. Soc. B.,* **42**, 213-220.

SPIEGELHALTER, D.J. and SMITH, A.F.M. (1982). Bayes factors for linear and log-linear models with vague prior information. *J. Roy. Statist. Soc. B.,* **44**, 377-387.

WELSCH, R.E. and KUH, E. (1977). Linear regression diagnostics. *Working Paper N° 173, National Bureau of Economic Research.* Cambridge, Massachusetts.

DISCUSSION

P.R. FREEMAN (*University of Leicester*)

It is a pleasure to discuss two very good papers in this session, both containing analyses of real data. They both provide clear advances in the practical application of

Bayesian methods, of exactly the kind we hoped for at the end of the previous conference when we peered into the future. In particular they provide nicely contrasting views of how to treat outliers and this can only be good since no single answer can ever be adequate for such a multi-facetted problem.

Pettit and Smith go straight to the heart of the problem by declaring "outliers must outlie". The absence of this from my own formulation (Freeman (1981)) led to peculiar results, principally because Bayes theorem doesn't push down to zero the posterior probability that a nice centrally-positioned observation was in fact generated by a different mechanism. By forcing this to be negligible, Pettit and Smith are in effect saying, quite rightly, that such a possibility is of no interest and they refuse to let it spoil their analysis. All is not plain sailing, however, since a prior for δ has to be chosen and this cannot simply be made arbitrarily large in order to satisfy condition (2.8). Inferences about location are probably pretty robust but posterior probabilities of particular *numbers* of outliers (r_1 and r_2 in section 2.5) may well do funny things.

It is good to see all the web of ad-hockery that has been woven (almost entirely in the USA) around outliers and influential observations being swept aside and replaced by a single piece of Bayesian ad-hockery: the predictive ordinate measure. At least, I can only think of very artificial justifications for it on decision-theoretic grounds. Its elegant consequences remind me of Bill Eddy's work on convex hull peeling (Eddy and Gale, (1981)). Have the authors explored any of the possible links with this?

Influence is just as much prone to the combinatorial explosion as are outliers, since a pair of observations may be jointly influential without either individually being so, and vice-versa. The existence of several different sources of influences seems to me to preclude such a single simplification as condition (2.8) for outliers, but I'm sure the authors have thought of one.

A.P. DEMPSTER (*Harvard University*)

Pettit and Smith open with an accurate and wise statement on the need for Bayesian diagnostics which serve as data-generated stimuli to careful thought about prior inputs to formal models. They review and extend diagnostics which examine single sample points or units. These they subdivide into assessing "outlyingness" or "surprisingness", and assessing "variation in the *a posteriori* statements". I have a terminological quibble over their application of the label "Bayesian sensitivity analysis" to both types of diagnostic, since only the latter is directly concerned with the sensitivity of Bayesian conclusions, while the former has a more pragmatic flavor of wanting to bring Bayesian practice into conformity with the non-Bayesian practice of downweighting extreme observations.

Pettit and Smith make some claims about their approach to outliers which I regard as logically shaky. In fact, I wonder if they have a new approach at all, since the specific model elaborations discussed in §2 are of the familiar types which postulate shifts in location or scale associated with single observations. The probabilities displayed in Tables 2.1 to 2.4 are interesting, but they hardly imply that the new approach "will be seen to resolve the difficulties" cited in §2.2. Only if reasonable priors fall in ranges which the tables indicate will the difficulties go away. By omitting this qualification, Pettit and Smith manage to sound like Neyman, whose program of inductive behavior was to contemplate operating characteristics such as those in the tables, and to choose procedures which are nice in some desirable sense. Unfortunately, desirable in this sense often means undesirable in a Bayesian sense, such as making the prior depend on sample size. I like the definition of outlier quoted from Freeman at the end of §2.1, but to follow it I need models which reflect genuine prior probabilities concerning "aberrant mechanisms", which will conform to desirable operating characteristics only sometimes.

Influence diagnostics are important, for the reason cited in the opening sentence of §1, but why should Bayesians emulate frequentists by debating one *ad hoc* criterion as opposed to another. Bayesian methods are problem-solving tools, and Bayesian statisticians should set up criteria for problem definition which include a target of analysis (e.g., estimate or decision) and a measure of the cost of an error in a reported analysis such as might be due to an inadequate specification of prior probabilities. My question about §3 is therefore: when or how do information-theoretic portmanteau measures of shifts in posterior relate to problem-specific measures of such shifts?

S. GEISSER (*University of Minnesota*)

I will comment on two aspects of this paper. First the authors have nicely implemented a suggestion I made, Geisser (1980) and demonstrated its usefulness. I called it a diagnostic based on the conditional predictive probability function. The authors referred to it as a predictive ordinate - perhaps as a compromise and to distinguish it from the Box criticism technique, I would suggest "conditional predictive ordinate" (CPO) diagnostic. As I had mentioned previously, Geisser (1980), this CPO diagnostic is defined even for certain quasi-priors and more recently Geisser (1983), that it obeys the likelihood principle - which is generally not the case with Box's checking procedure.

The work of the authors on influential observations has taken the same turn as a technical report by Johnson and Geisser (1981), to be published shortly in revised form and title, in the Journal of Statistical Planning and Inference. There we developed a Bayesian approach to estimative influence measures in the Multivariate Normal Regression situation. It is no doubt a tribute to the Bayesian approach to see some of the same ideas surfacing somewhat independently. In our work, we did find, however, that exact expressions can be obtained for the Estimative Influence measures that jointly take into account the matrix of regression coefficients and the covariance matrix, and also obtained a good approximation for the influence measures that consider only the regression matrix. In the Gessell data which was also featured in our paper, there is a reversal in the ordering of observations 18 and 19 depending on whether influence is measured for regression coefficients alone or jointly with the variance.

A review featuring only the univariate case of some of this material appears in my paper in this symposium. Finally, although in Johnson and Geisser (1981, 1982, 1983) we listed the three possible K-L divergences that could be used as influence measures and used the symmetric one in the estimative case, we have come to suspect that the most natural measure of the three is the one in which the expectation is taken over the parametric distribution containing only the included set of observations. This measure is featured in my symposium paper here, as well as other aspects of influence measures.

J.M. BERNARDO (*Universidad de Valencia*)

The authors are to be congratulated for their important contributions to the two main approaches to the problem of identifying atypical or aberrant observations. I personally believe that the influential observations approach is better for at least two reasons: (i) it is more general in that it does not require one to specify the particular model by which aberrant observations are generated and (ii) it may be given a full decision-theoretic justification which I now briefly describe.

Let y be the quantity we are interested in; this may either be a function $y=y(\psi)$ of the parameters ψ of the model $p(x|\psi)$ which generate non-aberrant observations or else a function $y=y(x_{n+1})$ of the next observation x_{n+1}. We are thus in a decision-making situation where the action space is the class of probability distribution of y. Moreover, let $u\{p_y(\cdot), y_0\}$ be the utility of describing one's opinion about y by $p_y(\cdot)$ where y_0 is the true value of y. We shall naturally require u to be a *proper* utility function. Finally, let $\hat{p}_y(\cdot)$ be an *approximation* to $p_y(\cdot)$. Then, the *expected loss* of using $\hat{p}_y(\cdot)$ rather than $p_y(\cdot)$ is given by

$$\delta\{\hat{p}_y(\cdot)|p_y(\cdot)\} = \int p_y(y)[u\{p_y(\cdot),y\} - u\{\hat{p}_y(\cdot),y\}]dy\,.$$

Now, with the notation of Section 2.4, if $x(S)$ were the set of aberrant observations in a sample x, we could think of $p(y|x) = p(y|x(S),x(\bar{S}))$ as the approximation to the posterior $p(y|x(\bar{S}))$ based on the non-aberrant observations obtained by *contaminating* $x(\bar{S})$ with the addition of $x(S)$. Thus, a general measure of the influence of $x(S)$ in our inferences about y would be

$$\int p(y|x(\bar{S}))[u\{p_y(\cdot|x(\bar{S})), y\} - u\{p_y(\cdot|x), y\}]dy$$

where u is any proper utility function, and this is to be interpreted as the *expected loss* due to the use of the aberrant observations $x(S)$.

If u is local, i.e. if $u\{p_y(\cdot),y\}=u\{p_y(y),\}$ and proper then, necessarily, (Savage, 1971, Bernardo, 1979) $u\{p_y(\cdot),y\} = A\log p_y(y) + B(y)$ and hence our measure reduces to

$$\int p(y|x(\bar{S}))\log\left(\frac{p(y|x(\bar{S}))}{p(y|x)}\right)dy\,.$$

If $y=x_{n+1}$ this is the Johnson and Geisser (1982) suggestion, if $y=\psi$ this is a non-symmetric version of $I(S)$ in 3.1. I wonder about the arguments of the authors in favour of a symmetric definition, and I would like to know the differences which this makes.

W. POLASEK (*Universität Wien*)

To my knowledge the concept of sensitivity analysis in Bayesian Statistics dates back to Leamer (1972) and Leamer (1978). The first type of sensitivity analysis, where the data set is regarded as fixed, was used in Leamer (1972) for a Bayesian distributed-lag analysis and was extended by Polasek (1981) using matrix by matrix derivatives to a general local "prior-posterior" sensitivity analysis. The second type of sensitivity analysis, where the *a priori* specification is fixed, is also called "local resistance" analysis. This can be used for general Bayesian regression diagnostics (see Polasek, 1983) or for special models as in Polasek (1982).

A Bayesian type of outlier detection program was also developed by Kitagawa and Akaike (1982) and Kitagawa (1983). The program package is called OUTLAP and is available from the authors. They provide the calculation of the probability of k outlying observations, where k is the prespecified number of observations, and an outlier correction method using the mean of the posterior distribution.

REPLY TO THE DISCUSSION

Professor Freeman is quite right to emphasize that the usefulness of our approximate approach depends on the "realism" of the implied prior for δ. We certainly did not intend to suggest that our proposal be adopted routinely, without careful thought. Perhaps our presentation did not emphasize this enough though, since Professor Dempster has also clearly picked up some bad vibrations. Our intended perspective is the following: we see the approach as potentially useful if (a) we are prepared to use these kinds of "aberrant observation" model elaborations, and (b) find the implicit forms of $p(\delta)$ "reasonable". The latter is to be interpreted in the following sense: (approximate) specification of $p(\delta)$ comes first; then, for a given or anticipated n, its "reasonableness" can be checked using the kinds of calculation summarized in Tables 2.1 - 2.4. So far as the suggested links with convex hull peeling are concerned (Freeman), further discussion of this can be found in Pettit and Smith (1984). The Kitagawa and Akaike work (Polasek) is based on maximum likelihood approximations to proper Bayesian methodology and, in fact, does not perform well on the second test data set given in Freeman (1981).

The question of measuring the "influence" of observations —and of developing approaches to Bayesian "diagnostics" in general— seems to us an extremely important one, but, as yet, perhaps not well-formulated. We therefore sympathize with the general tenor of Professor Dempster's remarks and are intrigued by their more concrete elaboration in Professor Bernardo's contribution. However, the issues and methodological needs are pressing and Bayesians must surely start exploring these relatively unchartered waters with more energy than hitherto.

Professor Geisser has gently reminded us that other voyages of discovery are already under way, but all of us are currently setting sail in rather *ad hoc* directions. We hope that clearer charts will be available before our 1987 meeting.

REFERENCES IN THE DISCUSSION

BERNARDO, J.M. (1979). Expected utility as expected information. *Ann. Statist.* **7**, 686-690.

EDDY, W.F. and GALE, J.D. (1981). The convex hull of a spherically symmetric sample. *Adv. Appl. Prob.* **13**, 751-763.

FREEMAN, P.R. (1981). On the number of outliers in data from a linear model. *Bayesian Statistics,* (Bernardo, J.M. *et al.* eds.), 349-365. Valencia: University Press.

GEISSER, S. (1983). A remark on a model criticism technique. *Bull. Int. Statist. Inst. 44th Session,* **2**, 925-927.

JOHNSON, W. and GEISSER, S. (1981). On the architecture of estimative influence functions. *Tech. Rep.,* **23**, University of California, Davis.

— (1983). A predictive view of the detection and characterization of influential observations in regression analysis, *J. Amer. Statist. Assoc.* **78**, 381, 137-144.

KITAGAWA, G. and AKAIKE, H. (1982). A quasi Bayesian approach to outlier detection, *Ann. Inst. Stat. Math. B.* **34**, 389-398.

KITAGAWA, G. (1980). OUTLAP, An outlier analysis program. Computer Science Monographs, *The Institute of Statistical Mathematics,* Tokyo.

— (1983). Bayesian analysis of outliers via Akaike's predictive likelihood of a model Mimeo; *The Institute of Statistical Mathematics,* Tokyo.

LEAMER, E.E. (1972). A class of informative priors and distributed lag analysis. *Econometrica*, **40**, 1059-1081.

— (1978). *Specification Searches.* New York: Wiley.

POLASEK, W. (1981). Matrix Derivatives and Sensitivity Analysis, in *G. Feichtinger et al. (eds). Operation Research in Progress,* Reidel Pub. Co.

— (1982). Local Resistance in Distributed Lag Models, *Statistische Hefte,* **23**(1), 44-51.

— (1983). Bayesian regression diagnostics. (To appear).

SAVAGE, L.J. (1971). Elicitation of personal probabilities and expectation. *J. Amer. Statist. Assoc.* **70**, 271-294.

BAYESIAN STATISTICS 2, pp. 495-510
J.M. Bernardo, M.H. DeGroot, D.V. Lindley, A.F.M. Smith (Eds.)

Multiparameter Bayesian Inference Using Monte Carlo Integration — Some Techniques for Bivariate Analysis

LELAND STEWART
Lockheed Research Lab., Palo Alto

SUMMARY

Bayesian analysis using Monte Carlo integration is an effective approach for handling rich multiparameter families of distributions; nonconjugate priors; extrapolation uncertainty; and the computation of posterior distributions for prameters or predictions of interest. This paper presents some techniques for bivariate analysis. In the example, a 20-parameter family is used that includes distributions that deviate moderately from bivariate normals. Posterior distributions can be computed for parameters of interest such as the correlation and the probability contained in a specified region of the *x-y* plane.

Keywords: BAYESIAN STATISTICS; MONTE CARLO INTEGRATION; NUMERICAL INTEGRATION; BIVARIATE DISTRIBUTIONS; POSTERIOR DISTRIBUTIONS.

1. INTRODUCTION

Bayesian statistics and numerical methods form a particularly effective partnership. There are at least two reasons for this:

(i) It is often easy to compute the values of the likelihood function and the prior density (and therefore the posterior density up to a multiplicative constant) on any specified set of points in the parameter space. This is true even for many models and problems that would be very difficult to handle with conventional methods.

(ii) The principal computational problem in making a Bayesian inference or decision is the evaluation of integrals involving the posterior density.

Thus the solution to many difficult inference and decision problems can be obtained by numerical integration using only the values of the likelihood function and the prior density on a finite set of parameter points. The suitable choice of this set is the most important, and sometimes the most difficult, aspect of the computation.

2. BAYESIAN ANALYSIS USING MONTE CARLO INTEGRATION

Bayesian inference and decision making usually require the computation of posterior cumulative distribution functions and/or posterior moments of *functions* of the parameter vector

$$\theta = (\theta_1,\theta_2,\ldots,\theta_k)'\epsilon\Theta$$

This requires evaluation of integrals of the form

$$I = \int_{\Theta} h(\theta)\, \xi(\theta \mid D)\, d\theta \qquad (1)$$

where h is a real valued function. The posterior density $\xi(\theta \mid D)$ is given by Bayes' theorem,

$$\xi(\theta \mid D) \propto \xi(\theta)\, L(\theta \mid D)$$

where $\xi(\theta)$ is the prior density and $L(\theta \mid D)$ is the likelihood function corresponding to the observed data D.

The Monte Carlo approximation to I is given by

$$\hat{I} = \sum_{m=1}^{M} h(\theta_m)\, W(\theta_m) \Big/ \sum_{m=1}^{M} W(\theta_m) \qquad (2)$$

The "weights" $W(\theta_m)$, $m = 1,2,\ldots,M$, are computed from

$$W(\theta) = \xi(\theta)\, L(\theta \mid D)/g^*(\theta). \qquad (3)$$

The θ_m, $m = 1,2,\ldots,M$, are generated independently from the K-variate density $g^*(\theta)$. This is known as "importance sampling". The density $g^*(\theta)$ is called the "importance function" or "generating density" and is usually chosen to approximate the posterior density, with the restriction that its form must be such that the θ_m's can be easily generated.

Details about the techniques used in applying this methodology and discussions of Monte Carlo integration error and its reduction can be found in Stewart, 1968, 1970, 1977, 1983; Stewart and Johnson, 1971, 1972; Johnson and Stewart, 1971; Kloek and van Dijk, 1975, 1978; van Dijk and Kloek, 1978, 1980, 1983, 1984; McGhee and Walford, 1965, 1967, 1968; Heiberger, 1976; Zellner and Rossi, 1982.

3. BIVARIATE ANALYSIS

This paper presents some techniques that can be used to carry out bivariate analyses using multiparameter families that include densities that deviate moderately from normal densities. In some respects, the approach presented here represents an extension of the methodology for multiparameter univariate Bayesian analysis using Monte Carlo integration described in Stewart (1979). An outline of the approach follows:

1. Densities are defined by

$$f(x,y) = \exp[\phi(x,y)] \Big/ \iint \exp[\phi(x',y')]\,dx'\,dy' \qquad (4)$$

2. $\phi(x,y)$ is a continuous Gaussian stochastic process on the x, y plane whose probability structure represents the prior.

   Up to this point we have used the same method as Leonard (1978).

3. The mean, $E[\phi(x,y)]$, and covariance, $\mathrm{cov}[\phi(x_1,y_1), \phi(x_2,y_2)]$, functions of $\phi(x,y)$ are chosen to define an operational prior that approximates the actual prior in the region of high likelihood. $\phi(x,y)$ is the sum of a "random" component and a "structured" component. In the example, the random component will be chosen to be stationary in x and y. The mean and covariance functions of the structured component will be computed to approximate the mean and covariance functions of $\ln \lambda(x,y)$ where $\lambda(x,y)$ is an "asymmetric bivariate logistic" density with a relatively diffuse prior.

4. Numerical integration is used to compute the denominator of (4) as well as the

properties of $f(x,y)$ that are of interest, e.g., means, variances, correlation, probability in a specified region, etc. We will need, therefore, only the values of $f(x,y)$ on the integration points $\{x_i, y_i: i = 1, 2, \ldots, IM\}$ and the data points $\{x_i^*, y_i^*, i = 1, 2, \ldots, ID\}$.

Because of this, the continuous stochastic process, $\phi(x,y)$, can be replaced by the vector

$$\phi = [\phi(x_1,y_1), \phi(x_2,y_2), \ldots, \phi(x_{IM}, y_{IM}), \phi(x_1^*, y_1^*), \ldots, \phi(x_{ID}^*, y_{ID}^*)] \tag{5}$$

whose distribution under the prior is multivariate normal of dimension $IM + ID$.

5. The number of parameters at this stage can be very large, for example $IM + ID = 420$ in the example. To reduce this number we first rewrite ϕ as

$$\phi = \sum_{i=1}^{IM+ID} \theta_i \mathbf{e}_i + E\phi \tag{6}$$

where the $\mathbf{e}_i$ are the eigenvectors of the covariance matrix of ϕ and the θ_i are independent normal with means equal zero and variances equal to the eigenvalues. Finally, since the eigenvalues decrease in magnitude so that the contribution of most of the terms is small, we use

$$\phi = \sum_{k=1}^{K} \theta_k \mathbf{e}_k + E\phi \tag{7}$$

as our formula for ϕ. K will equal 20 in the example. (In order to appropriately weight the importance of the different components of ϕ, we may want to transform ϕ before computing eigenvectors and eigenvalues and then retransform the results).

6. All numerical evaluations of integrals involving the bivariate densities are carried out in a manner analogous to the Monte Carlo evaluation of integrals involving the posterior density. Here the weights $w(x,y)$ are defined by

$$w(x,y) = \exp[\phi(x,y)]/g^{**}(x,y) \tag{8}$$

The likelihood is given by

$$L(\theta \mid D) = \prod_{i=1}^{ID} f(x_i^*, y_i^* \mid \theta)$$

$$\cong \exp\left[\sum_{i=1}^{ID} \phi(x_i^*, y_i^*)\right] \Big/ \left[(1/IM) \sum_{j=1}^{IM} w(x_j, y_j)\right]^{ID} \tag{9}$$

4. EXAMPLE

Data. The data for the example are shown in Fig. 1. The $ID = 20$ observations were generated from the density

$$f(x,y) = f(x) f(y \mid x) \tag{10}$$

where $f(x)$ is the density corresponding to the asymmetric distribution function, F, defined by

$$F^{-1}(u) = \ln u - 0.2 \ln(1-u) \tag{11}$$

and $f(y|x)$ is normal with mean equal to x and standard deviation equal to $0.8 \exp(-x/3)$. The sample correlation is 0.75.

Integration Points. Also shown in Fig. 1 are the $IM = 400$ integration points x_i, y_i used for integration involving the bivariate densities. Let (r_i, α_i) denote polar coordinates. The integration points were generated from a bivariate Student's t distribution with four degrees of freedom by choosing the r_i to be evenly spaced in probability and

$$\alpha_i = i \cdot 2\pi \cdot 0.36964736 \tag{12}$$

The constant was chosen by trying 20 different constants and choosing the case that gave the best spacing of the integration points. One would expect improvements in the choice of the integration points with further study.

The final step was to transform the points just generated in accordance with the means, variances, and correlation of the data and expand by a factor of two. The g^{**} in Eq. (8) is a bivariate t with four degrees of freedom transformed as just described.

DATA AND INTEGRATION POINTS FOR THE EXAMPLE

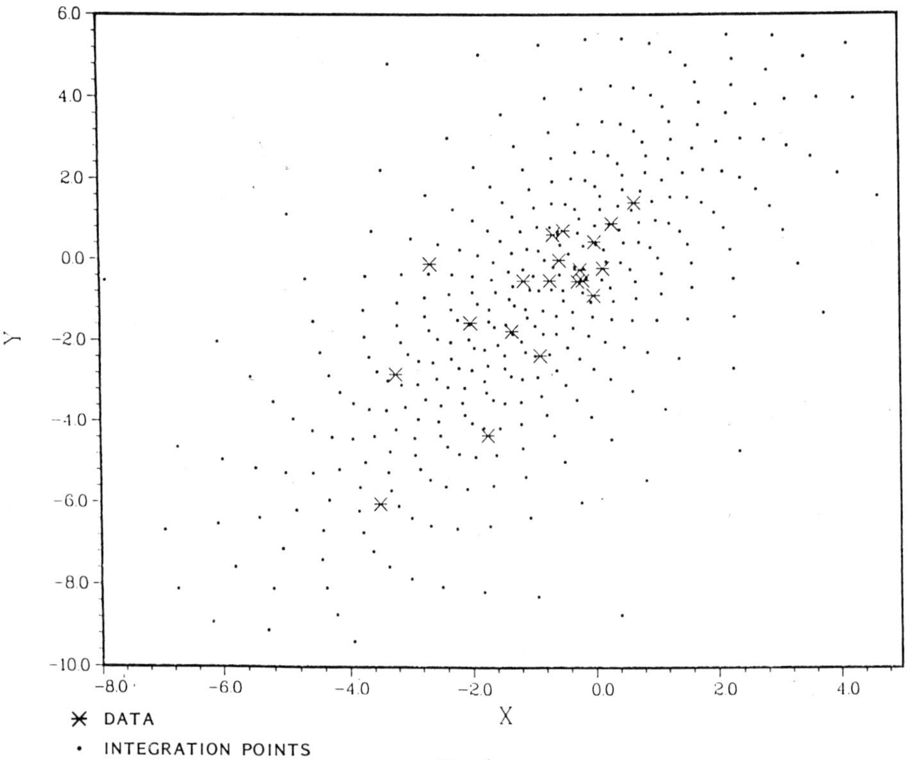

Fig. 1

Prior. To describe the construction of the prior we first consider an "asymmetric bivariate logistic" density defined by

$$\lambda(x,y) \propto [\exp(r/2\Psi(\gamma)) + \exp(-r/2\Psi(\gamma + \pi))]^{-2} \tag{13}$$

where (r,γ) are the polar coordinates of $(x-C_x, y-C_y)$, (C_x,C_y) is the location of the "center" of the distribution and Ψ is a positive function defined on the circle.

Under the prior used in the example, C_x and C_y are independent with mean zero and variance one, and $\psi(\gamma)$ is a positive stochastic process with mean equal to $3^{1/2}/\pi$ and

$$\text{cov}[\psi(\gamma), \psi(\gamma+\tau)] = 4(4-3\cos\tau)\exp[3(\cos\tau-1)] + \tfrac{1}{8} + \frac{\cos 2\tau}{16} \tag{14}$$

Under this prior, $\ln \lambda(x,y)$ is a random function whose covariance function, $\text{cov}[\ln \lambda(x_i,y_i), \ln \lambda(x_j,y_j)]$, was approximated using a first order Taylor's expansion for $\ln \lambda(x,y)$ that was linear in C_x, C_y, $\psi(\gamma)$ and $\psi(\gamma+\pi)$. This is the structured component of the prior described in "3" in Section 3. Note that there are two approximations here. The non-Gaussian random function $\ln \lambda(x,y)$ is replaced by a Gaussian process whose covariance function is computed from a Taylor's approximation. Determining the range of applicability of these approximations requires further study.

The random component of the prior was chosen to have covariance function

$$(d^2/6 + d + 2)\exp(-d/2) \tag{15}$$

where

$$d = [(x_i-x_j)^2 + (y_i-y_j)^2]^{1/2}$$

The covariance functions, Eqs. (14) and (15), were chosen from theoretical considerations and by looking at sample functions generated from processes with these and other covariance functions.

The operational prior just constructed would be appropriate if we wanted to approximate our actual prior in a neighborhood (in the set of bivariate distributions) centered on a bivariate distribution with means and correlation approximately the same as the data. The final step, therefore, is to transform the prior defined above accordingly, to obtain our operational prior, but the data are used only to direct us to the region in the set of bivariate distributions over which the operational prior should approximate the actual prior.

Results. The bivariate density $f(x,y|\theta_{max})$, where θ_{max} maximizes the posterior density, is shown in Fig. 2. The contours of this density are shown in the upper left of Fig. 3.

Figure 3 also shows contours of five densities generated from the posterior. Since a sample cannot be generated directly from the posterior distribution, we use an acceptance-rejection method which accepts θ_m's for which

$$U_m \leq W/\theta_m)/W^* \tag{16}$$

where the U_m are independent uniform [0,1] random numbers and

$$W^* \geq \max \quad W(\theta_m): m = 1, 2, \ldots, M \tag{17}$$

$W(\theta)$ is defined in Eq. (3).

The posterior distribution for the correlation, ϱ, is shown in Fig. 4. This was computed using Eq. (2) with

$$\begin{aligned} h(\theta) &= 1 \text{ when } \varrho(\theta) \leq \varrho_0 \\ h(\theta) &= 0 \text{ otherwise} \end{aligned} \tag{18}$$

for $\varrho_0 = 0, .01, .02, \ldots, .99, 1.00$

The Monte Carlo sample size, M, equalled 2000 for this example.

The posterior distribution of $P[x < -3.8]$ is shown in Fig. 5. The posterior

THE DENSITY $f(x,y \mid \theta_{MAX})$

θ_{MAX} MAXIMIZES THE POSTERIOR DENSITY

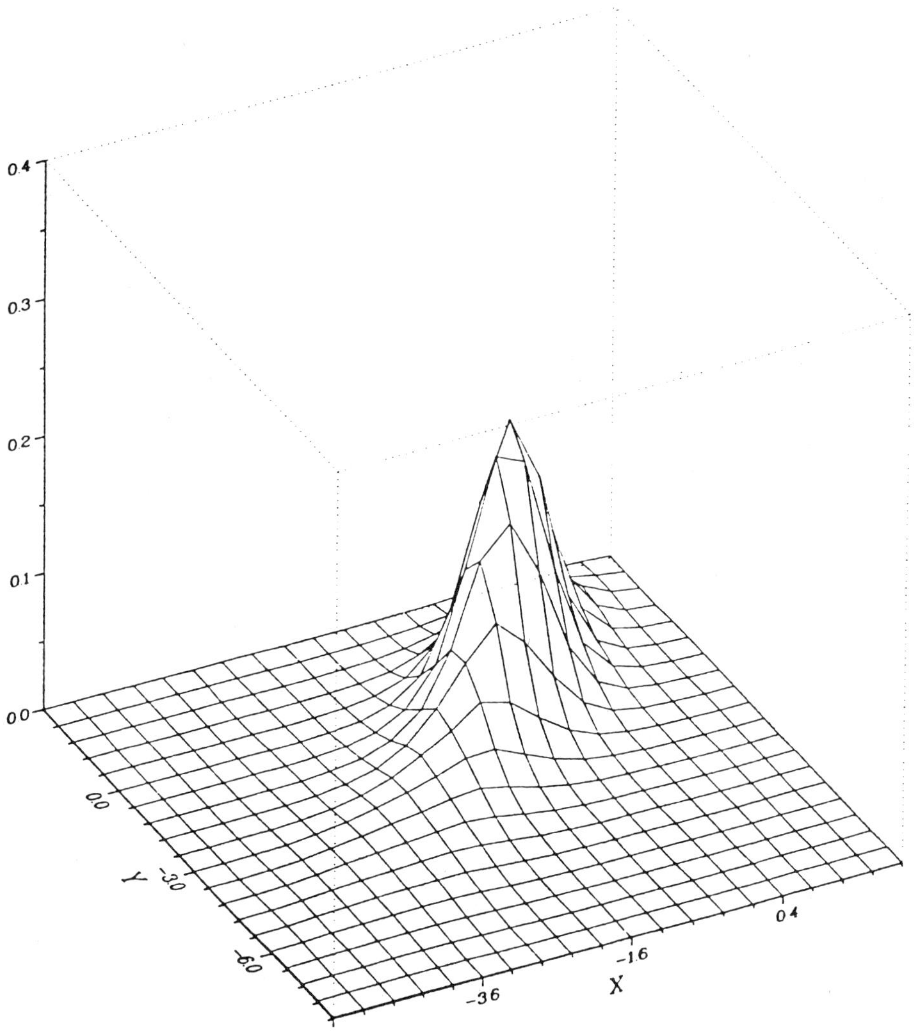

Fig. 2

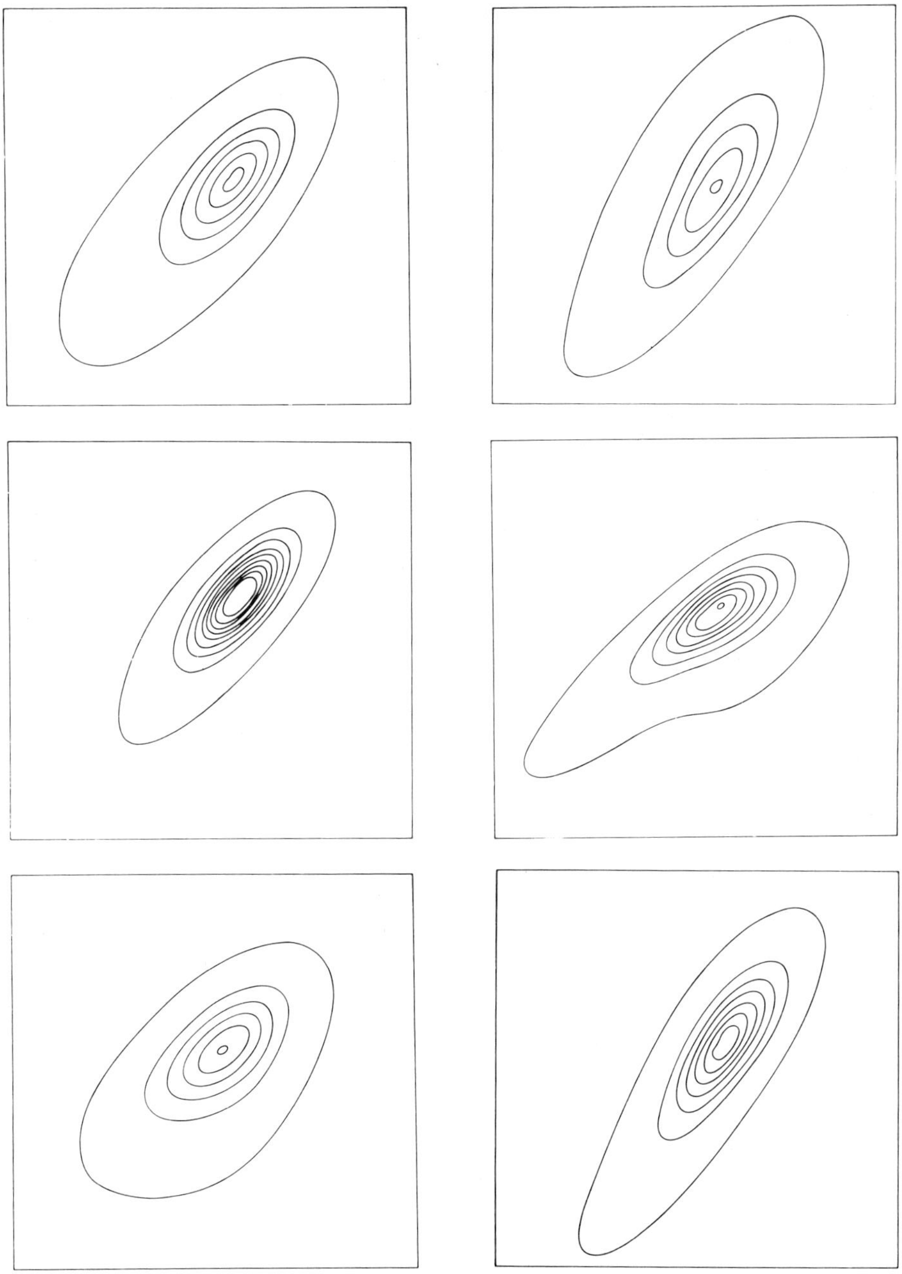

- UPPER LEFT – MAXIMUM POSTERIOR
- ALL OTHERS GENERATED FROM THE POSTERIOR DISTRIBUTION
- CONTOURS CORRESPOND TO VALUES $0.005 + 0.04\,n$ FOR $n = 0, 1, 2, \ldots$

CONTOURS OF DENSITY FUNCTIONS

Fig. 3

distribution for the probability of a more complex region can also be easily computed. Furthermore, in a Bayesian analysis, the posterior distribution is the same whether the region is specified in advance or is determined from the data (providing the choice of region is noninformative). The latter case is of particular interest in classification and discrimination problems. Note that choice of an effective set of integration points may depend on the region and therefore those points should be chosen after the region is determined.

Figure 6 shows the logarithm of conditional densities (unnormalized) of y given that $x = -3.2$. These were generated from the posterior as described earlier. The array of integration points shown in Fig. 1 was replaced by a rectangular array to produce Figs. 2 and 6.

5. CONCLUSIONS

The methodology described here represents an effective approach for handling bivariate problems when the distributions can deviate moderately from normal. There are, however, several areas that deserve further attention. These include:

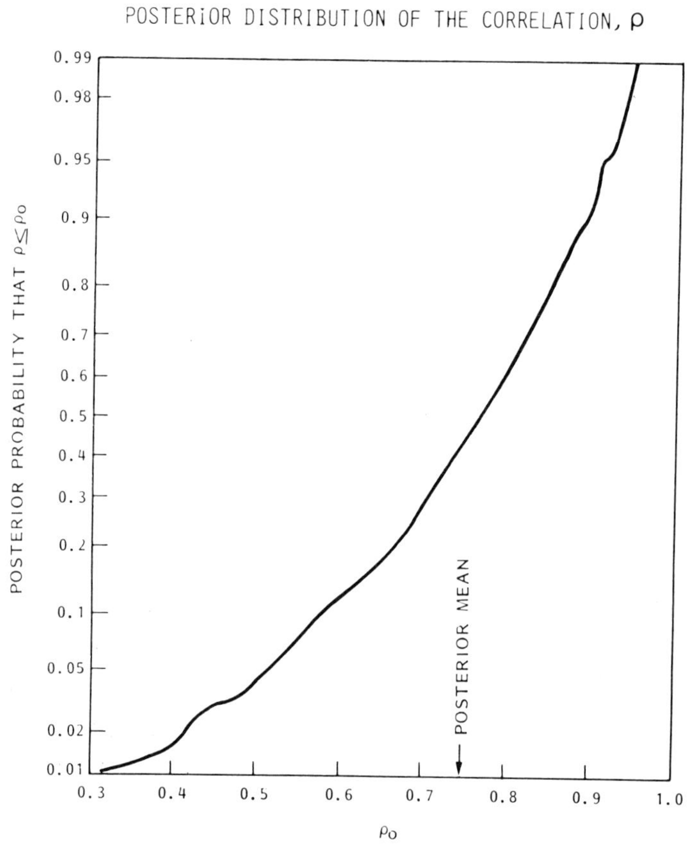

Fig. 4

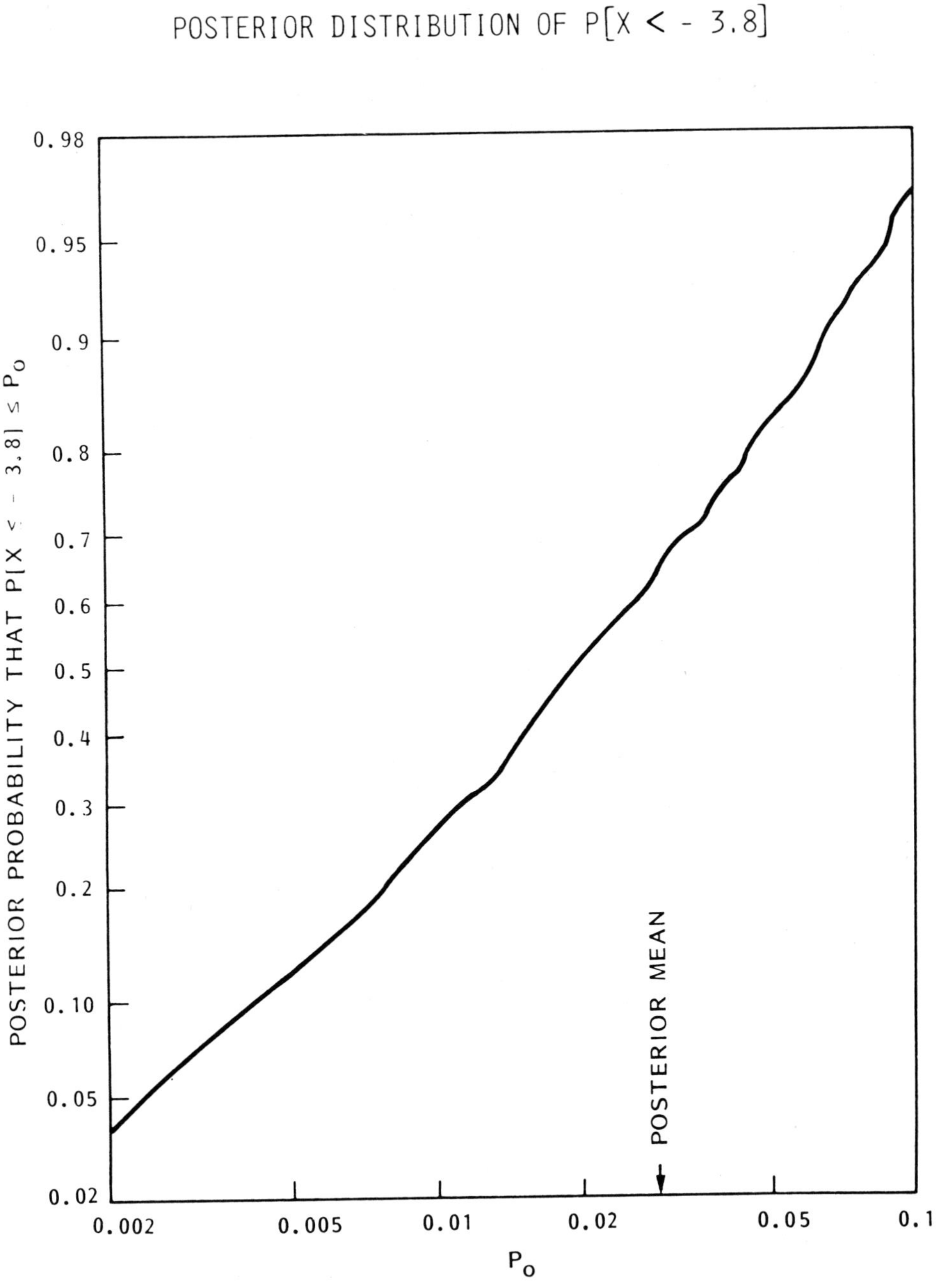
POSTERIOR DISTRIBUTION OF P[X < - 3.8]
0.98
0.95
0.9
0.8
0.7
0.6
0.5
0.4
0.3
0.2
0.10
0.05
0.02
POSTERIOR PROBABILITY THAT P[X < - 3.8] ≤ Po
POSTERIOR MEAN
0.002
0.005
0.01
0.02
0.05
0.1
Po

Fig. 5

LOG CONDITIONAL DENSITIES (UNORMALIZED),
ln cf(y|x = -3.2), GENERATED FROM
THE POSTERIOR DISTRIBUTION

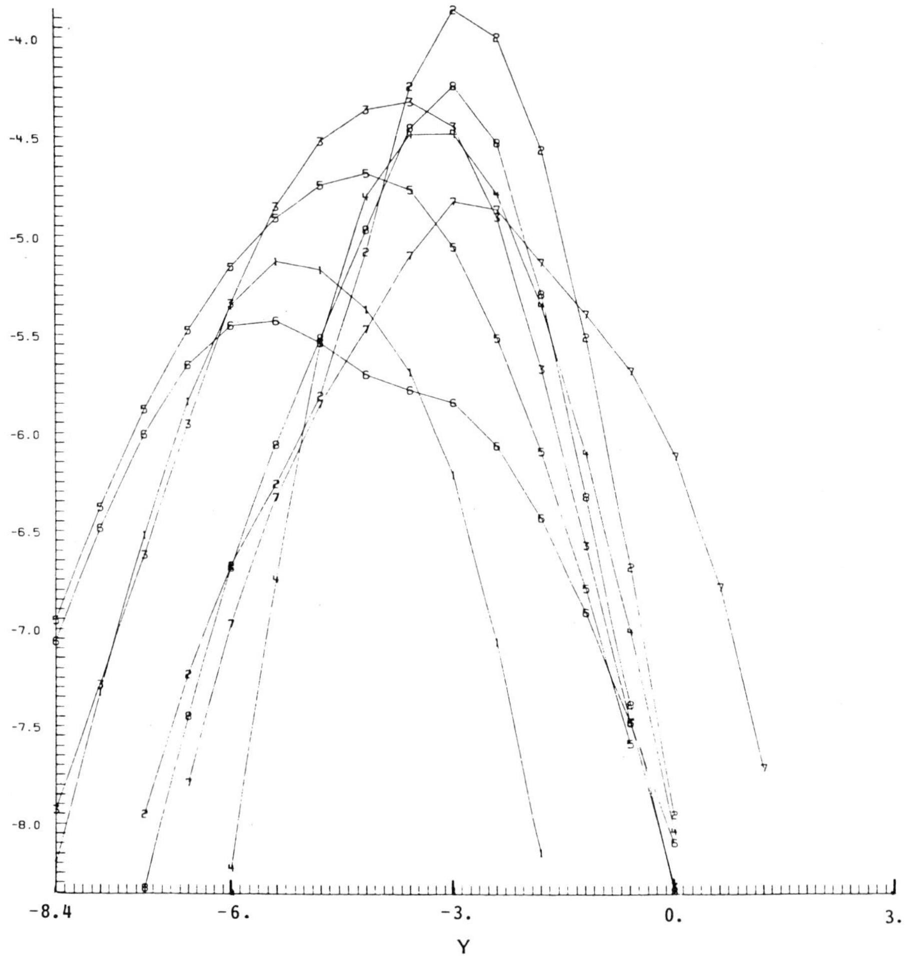

Fig. 6

1. The study of numerical integration accuracy.
2. The study of the properties of the prior used here and alternative ways of stating priors.
3. The determination of the range of validity of the approximations used in defining the prior.

6. ACKNOWLEDGEMENT

This work was supported under the Lockheed Independent Research Program.

REFERENCES

HEIBERGER, R.M. (1976). Posterior moments of scalar functions of a covariance matrix, *Communications in Statistics - Simulation and Computation,* Vol. B5 (2 and 3), 107-140.

JOHNSON, J.D. and STEWART, L. (1971). Failure prediction from interval data, *Annals of Assurance Sciences - Proceedings 1971 Annual Symposium on Reliability*. The Institute of Electrical an Electronics Engineers, Inc., 356-61.

KLOEK, T., and VAN DIJK, H.K. (1975). *Bayesian estimates of equation system parameters - an unorthodox application of Monte Carlo,* Report. 7511/E, Econometric Institute, Erasmus University, Rotterdam.

— (1978). Bayesian estimates of equation system parameters; an application of integration by Monte Carlo, *Econometrica* **46**, p. 1-19. Reprinted in: A. Zellner, ed., (1980). *Bayesian analysis in econometrics and statistics, essays in honor of Harold Jeffreys,* Amsterdam: North Holland.

LEONARD, T. (1978). Density estimation, stochastic processes and prior information, *J. Roy. Statist. Soc. B,* **40**, 113-146.

MCGHEE, R.B. and WALFORD, R.B. (1965). *A Bayesian solution to the reentry tracking problem, USCEE Report* **151**, Dep. Elec. Eng. University of Southern California, Los Angeles.

— (1967). *A Bayesian approach to nonlinear system identification, USCEE Report* **193**, Dep. of Elec. Eng. University of Southern California, Los Angeles.

— (1968). A Monte Carlo approach to the evaluation of conditional expectation parameter estimates for nonlinear dynamic systems, *IEEE Trans. on Automatic Control,* AC-**13**, 29-37.

STEWART, L.T. (1968). Inferences about the tail of a function using Bayesian extrapolation, handout for presentation at the Annual Meeting of the Amer. Statist. Assoc.,Lockheed Palo Alto Research Laboratories, Palo Alto, CA.

— (1970). Bayesian analysis of multiparameter univariate problems using Monte Carlo integration, handout for presentation at the Conference on Bayesian Inference, Oregon State University, Lockheed Palo Alto Research Laboratories, Palo Alto, CA.

— (1977). *Bayesian analysis using Monte Carlo integration,* Lockheed Palo Alto Research Laboratories, Palo Alto, CA.

— (1979). Multiparameter univariate Bayesian analysis, *J. Amer. Statist. Assoc.,* **74**, 684-93.

— (1983). Bayesian analysis using Monte Carlo integration - a powerful methodology for handling some difficult problems, *The Statistician,* **32**, 195-200.

STEWART, T. L. and JOHNSON, J.D. (1971). An example of the use of Monte Carlo integration in Bayesian decision problems, *Annals of Reliability and Maintainability,* **10**, 1971, New York: The Amer. Soc. of Mech. Eng., 19-23.

— (1972). Determining optimum burn-in and replacement times using Bayesian decision theory. *IEEE Transactions on Reliability, R*-**21**, 170-175.

VAN DIJK, H.K. and KLOEK, T. (1978). *Posterior analysis of Klein's Model I, Report* 7824/E, Econometric Institute, Erasmus University Rotterdam. Rotterdam.

— (1980). Further experience in Bayesian analysis using Monte Carlo integration, *Journal of Econometrics,* **14**, 307-328.

— (1983). Monte Carlo analysis of skew posterior distributions; an illustrative econometric example, *The Statistician,* **32**, 216-223.

— (1984). Experiments with some alternatives for simple importance sampling in Monte Carlo integration, in this volume.

ZELLNER, A. and ROSSI, P.E. (1982). Bayesian analysis of dichotomous qualitative response models. H.G.B. Alexander Research Foundation paper, Graduate School of Business, University of Chicago. Presented at the 24th meeting of NSF-NBER Seminar in Bayesian Inference in Econometrics, SUNY-Albany, May 1982.

DISCUSSION

J.C. NAYLOR (*Derby Lonsdale College of Higher Education*)

The paper by Stewart presents a technique for making inferences about a bivariate density using a rich class of such densities. The success of this technique lies in the use of Monte Carlo methods to evaluate integrals over a high dimensional parameter space (20 parameters in the example presented). The method also involves numerical integration in the computation of the likelihood function, and it is this aspect that I should like to discuss in a little more detail.

The computation of the integral implicit in the likelihood function seems to me to be critical to the operation of the method in two ways. Firstly, any errors in the integration will give rise to a noise component in the likelihood function, especially if Monte-Carlo methods are used. In any numerical computation of a likelihood function, we may expect rounding errors and the use of approximations to produce such a noise component, and so this example is simply a potentially extreme case of a general situation. However this example does offer the opportunity to vary the noise contribution (by using integration rules of varying precision) and I would like to ask Dr. Stewart if he has carried out any investigation of this type.

Secondly, the choice of method for this integration is important in relation to computational cost, since this bivariate integral must be evaluated for each point in the main Monte Carlo sample used for integrations over the parameter space. I would like to ask if the rule actually chosen is in any sense best. Is there some good reason for choosing g^{**} to be a bivariate t density on 4 degrees of freedom? Why should a pseudo-Monte Carlo rule be preferred to a classical rule for an integration over only two dimensions? A classical rule related to the student t form is suggested by Harper (1962) and a cartesian product based on this rule would be one possible alternative. The use of a rectangular grid in constructing figures 2 and 6 implies the use of a product type rule; assuming some of the other calculations were repeated a comparison of results could give further insight into effective choice of rule for this part of the computation.

One further remark, the use of bi-splines may have seemed the more natural extension of Dr. Stewart's earlier work on univariate densities, Stewart (1979). However this approach using a Gaussian process would seem to be more readily extended to multivariate analysis of this type.

T.J. SWEETING (*University of Surrey*)

Dr. Stewart's paper emphasized the application of Monte Carlo integration in the area of Bayesian nonparametric statistics, and my remarks are confined to this application. I have two main comments, one of a general nature and one technical.

First of all, I think that the general approach to the Bayesian estimation of continuous functions as exemplified this evening has much potential for practising Bayesian statisticians. My own assessment of where this potential lies may differ however from Dr. Stewart's. I believe that Bayesian nonparametric methods will be most used for (a) preliminary data analysis and (b) model criticism and model revision. The example discussed in the paper provides a convenient illustration. Although the method of generating the prior distribution is ingenious, I feel that the analysis of the example data is incomplete as no attempt has been made to identify and investigate a parsimonious model for the data. Supposing for example that $y|x$ has some meaningful interpretation, on the basis of Fig. 1, or from the first plot in Fig. 3, one could presumably come up with a tentative model of the form $y_i = \mu(x_i) + \sigma(x_i)\epsilon_i$, given x_i with μ linear, σ decreasing, the ϵ_i independent standard normal, and x having a distribution in some given parametric class. Furthermore, it would not be unreasonable to take $\log \sigma(x)$ to be linear. *Now* one could construct a two-dimensional Gaussian process to express prior uncertainty about this model, and use it to check the robustness of parametric inferences and/or predictions to departures from the model by comparing appropriate posterior distributions. It would be of further interest to separate out in some way the effects of departures from each of the separate distributional and linearity assumptions contained in the above model.

My second point concerns the parametric approximation in Section 3.5. The method here struck me as rather cumbersome, and presumably involves a fair amount of computational effort. As an alternative one could remain with a continuous Gaussian process, but one having a finite spectrum approximating the desired continuous spectrum. For simplicity, suppose that a random function ϕ on (0,1) is generated by a stationary Gaussian process with absolutely continuous spectral distribution $F\ (\omega|\beta)$ and autocorrelation function $\varrho(\beta h)$ for some given $\beta > 0$. The larger the value of β, the more terms are needed in the spectral representation to approximate the process. Omitting the details, consider a representation of the form

$$\hat{\phi}(t) = \sigma \sum_{|j| \leq k} e^{ijct} f_j^{1/2}(c/\beta)\theta_j \tag{1}$$

where θ_j, $j=0, \pm 1, \ldots, \pm k$ are independent $N(0,1)$, $k = [A\beta/c]$ and $f_j(\delta) = {}_\gamma (\{j+1\}\delta) - F(j\delta)$. For suitably chosen values of c and A, (1) generates a stationary Gaussian process with autocorrelation function approximating $\varrho(\beta h)$, the approximation being uniform in $|h| < 1$ and all $\beta > 0$. (A multivariate generalization is straightforward). The representation (1) has a number of advantages.

1. The number of parameters required, k, is readily calculated.
2. The coefficients are generally simple to calculate, and $\hat{\phi}(t)$ easily computed for any t.
3. No further smoothing is required to plot $\hat{\phi}(t)$. (All realizations of ϕ will be infinitely differentiable, even when this is not true of the original process approximated).

It would be of interest to generate results based on (1).

J. BACON-SHONE (*University of Hong Kong*)

One of the main reasons cited for using Monte Carlo integration techniques is that deterministic methods are infeasible because they require large number of points. I am not convinced yet that this is true. Naylor and Smith (1982) have shown that product-rule Hermitian integration is very effective for problems of 5 parameters and more where all the marginals are required. I have been investigating the behaviour of spherical rules which are essentially the combination of a Laguerre rule for r and a trapezoidal rule over the solid angle. The crucial aspects are that very few points are required to start with and that unless the Laguerre rule is changed, all points are fully utilized on increasing the order of the method. It will be of interest to see how well this method can compare with existing methods.

J.M. BERNARDO (*Universidad de Valencia*)

Monte Carlo integration is usually employed to derive posterior expectations of functions of the parameters which label the sampling model, as briefly described by the author in Section 2. I would like to draw attention to the fact that in a large class of Bayesian problems where a pre-posterior analysis is necessary, such as model choice, experiment design, sensitivity analysis or evaluation procedures, integrals of the type

$$I = \int_X p(x|D)\, I(x|D)\, dx\,, \qquad (1)$$

are required, where usually $I(x|D)$ is the result of another integral of the type described in equation (1). Monte Carlo integration procedures then become extremely simple; indeed, the usually available random sample from the predictive distribution $p(x|D)$, $z = \{x_1,\ldots,x_n\}$, directly provides the integration points, the weights become constants, and the evaluation of I simply reduces to $\Sigma I(x_i|D)/n$. A succesful application of this technique occurs in Bernardo and Bermudez (1983, Section 2.4).

REPLY TO THE DISCUSSION

I thank the discussants for their thoughtful comments and suggestions.

Dr. Bacon-Shone mentioned the use of deterministic methods to evaluate the posterior integrals. One way of carrying out Monte Carlo integration is to generate the θ_m from quasi-random sequences rather than random or pseudo-random sequences. Quasi-random sequences are usually generated deterministically with the sole intent of minimizing integration error rather than to simulate true random sequences as is the intent with pseudo-random sequences. Theoretical studies show that under certain conditions integration error can be significantly reduced by use of quasi-random sequences. It appears that this may also be true in the practical application of the methods described in this paper although problems can arise when integrating over subsets of Θ (see Zaremba, 1968).

I agree with Dr. Naylor that we must be concerned with integration error. As mentioned at the end of the paper, a study of computational accuracy is planned. For numerical methods to be satisfactory, we must be able to achieve the desired accuracy for acceptable computational costs. Based on the few runs that have already been made, there has been no indication that this cannot be accomplished.

The bivariate integration rule that was used was chosen because it is simple and reasonably efficient and it illustrates quasi-Monte Carlo integration. Student t densities, with their heavy tails, often make good importance functions, but the choice of integration points could certainly be improved. The possibility of extending this bivariate methodology to higher dimensional multivariate problems was considered in choosing the techniques that were presented in this paper.

Dr. Sweeting believes that "Bayesian nonparametric methods will be most used for (a) preliminary data analysis and (b) model criticism and model revision". I hope that Bayesian approaches using Monte Carlo integration will find wide use for such problems. Much of the motivation for my work, however, came from seeing so many cases where the use of parsimonious models has been abused. For example, in problems involving extrapolation, parsimonious models have usually yielded overly optimistic (i.e., too small) posterior uncertainties. I think that there are many problems that do not call for the formulation and use of parsimonious models. Bayesian multiparameter (i.e., approximately nonparametric) analysis will provide a powerful approach to these problems and I believe it will be frequently used when programs become generally available.

I think that we still have much to learn about how to specify good priors in multivariate problems. With large samples we may be able to get by, except in the tails, with priors that essentially just smooth the data. But for small samples, or for inferences beyond the range of the data, the prior may have to be chosen with much greater care. Just transforming a standard stochastic process, for example, may define a nonparametric prior that is not obviously unreasonable. But this prior may, in fact, have most of its prior probability on multivariate distributions that rarely, if ever, represent real population or would have been given a very low subjective prior probability had the prior been carefully chosen. It was this general concern that led to the use of the "structured" component of the prior in the paper.

Dr. Sweeting suggests an alternative parametric approximation to the $\phi(x,y)$ process. In light of the comments of the preceding paragraph, I am not convinced that the specific form he presents could be used. His suggestion, however, could be applied to the stationary "random" component of $\phi(x,y)$ and to the stochastic process $\psi(\gamma)$, Eq. (14), that defines the "structured" component. With this approach we could define a (non-Gaussian) random function $\phi(x,y)$ in terms of a parameterized set of continuous, easily computable functions instead of the eigenvectors in Eq. (7). The advantages of this would have to be weighted against possible increased difficulties in constructing good importance functions for Monte Carlo integration and in extending this methodology from the bivariate case to higher dimensions.

One final note of caution. In the example in the paper it was easy to carry out the Monte Carlo integration efficiently over the 20-dimensional-parameter space. In that example, as in many real problems, the posterior was well behaved in the sense that it was easy to construct an importance function that was an adequate approximation to the posterior density. The reader should not conclude, however, that all problems of 20 parameters or less can be *easily* handled by Monte Carlo integration. Van Dijk and Kloek, in their recent work, illustrate some of the difficulties that have to be overcome to do efficient Monte Carlo integration when posterior distributions are less well behaved.

REFERENCES IN THE DISCUSSION

BERNARDO, J.M. and BERMUDEZ, J.D. (1983). The choice of variables in probabilistic classification. *This volume*.

HARPER, W.N. (1962). Quadrature formulas for infinite integrals. *Math. Comp,* **16**, 170-175.

NAYLOR, J.C. and SMITH, A.F.M. (1982). Applications of a method for the efficient computation of posterior distributions. *Appl. Statist.* **31**, 214-225.

STEWART, L. (1979). Multiparameter univariate Bayesian analysis. *J. Amer. Statist. Assoc.,* **74**, 684-693.

ZAREMBA, S.K. (1968). The Mathematical Basis of Monte Carlo and Quasi-Monte Carlo Methods, *SIAM Review,* **10**, 303-14.

BAYESIAN STATISTICS 2, pp. 511-530
J.M. Bernardo, M.H. DeGroot, D.V. Lindley, A.F.M. Smith (Eds.)

Experiments with Some Alternatives for Simple Importance Sampling in Monte Carlo Integration

H.K. VAN DIJK and T. KLOEK
Econometric Institute, Erasmus University, Rotterdam

SUMMARY

Some alternatives for simple importance sampling [compare Kloek and van Dijk (1978) and van Dijk and Kloek (1980)] are investigated for the computation of posterior moments and densities. An importance sampling method that is based on a mixture of a finite number of multivariate normal densities is compared with simple importance sampling and with a method that is based on a combination of Monte Carlo and classical numerical integration. These methods are intended to handle econometric applications where a simple importance function that is a reasonable approximation to the posterior density is difficult to find. For illustrative purposes use is made of a small econometric model. The results include bivariate marginal densities of the importance functions and the posterior plotted in three dimensional figures.

Keywords: MONTE CARLO; NUMERICAL INTEGRATION; SIMULTANEOUS EQUATION MODEL; POSTERIOR MOMENTS; BIVARIATE POSTERIOR DENSITIES; IMPORTANCE SAMPLING; COMPOSITION METHOD; FINITE MIXTURES; MIXED INTEGRATION.

1. INTRODUCTION

In two earlier papers [Kloek and van Dijk (1978) and van Dijk and Kloek (1980)], we applied Monte Carlo integration (in particular, importance sampling) for the purpose of finding posterior moments and densities for parameters of econometric simultaneous equation models. (A simultaneous equation model is nonlinear in the sense that the expected values of the endogenous variables are nonlinear functions of the parameters of interest.) We made use of the multivariate Student density as an importance function. In these papers we emphasized the use of importance functions that fit reasonably well to the posterior distribution. So there exists a problem when the posterior is skew. In van Dijk and Kloek (1983a) we addressed this problem, but in a rather *ad hoc* way. For that reason we have started to investigate alternative approaches. We shall use the term simple importance sampling (SIS) for the approach used in our 1978 and 1980 papers.

The present paper contains some experiments with two alternative approaches to SIS. First, we propose a flexible importance function which consists of a mixture of a finite number of multivariate normal densities. This importance function is intended for cases where the dominating skewness is in one or two directions. More details are given in Section 2. Second, we make use of an alternative to simple importance sampling, where the basic idea is to transform the s-dimensional space of parameters of interest θ into another s-dimensional space of vectors $(\eta'\varrho)$ where ϱ is a scalar and η an $(s-1)$-vector, and the prime denotes transposition. The transformed posterior density of $(\eta'\varrho)$ is decomposed as a conditional density of ϱ given η and a marginal density of η. For ϱ we take $\pm\, d$, where d is a measure of the distance between a point $\theta^{(i)}$, generated at random, and a central point (θ^0) such as the posterior mode or a preliminary estimate of the posterior mean. For η we take the direction $(\theta^{(i)} - \theta^0)/\varrho^{(i)}$ with one coordinate deleted in order to avoid degeneracy. After having performed the transformation described we draw a vector $\eta^{(i)}$ and apply classical numerical integration with respect to ϱ given $\eta^{(i)}$. So, this method amounts to a combination of classical numerical integration and Monte Carlo, which we call *mixed integration*. For a motivation and details of this approach we refer to van Dijk and Kloek (1983b). This method is intended to handle cases where the posterior is skew in several directions.

In Section 3 we discuss the results that have been obtained with some experiments using the alternative approaches. As an example we make use of the same econometric model as in Kloek and van Dijk (1978). For details with respect to the choice of the prior density we refer to van Dijk and Kloek (1983a). Our conclusions are given in Section 4.

2. A NORMAL-PIECEWISE-UNIFORM IMPORTANCE FUNCTION

In this section we describe a flexible class of importance functions, which may be useful in cases where the surface of the posterior density of a nonlinear model is skew predominantly in one or two directions. The present approach was inspired by a technique for generating univariate random variables, called *composition* [compare Rubinstein (1981, Chapter 3), and the references cited there]. In this technique a composite density function is defined as a convex linear combination of a number of more elementary density functions in the following way. Let $p_1,\ldots,p_h$ be a set of nonnegative constants that sum to unity and let $\theta_1,\ldots,\theta_h$ be a set of random variables with distribution functions $F_1(\theta_1),\ldots,F_h(\theta_h)$ and density functions $I_1(\theta_1),\ldots,I_h(\theta_h)$. Define the random variable θ as equal to θ_j with probability p_j for $j = 1,\ldots,h$. The composite distribution function of θ is given in the point θ^* as

$$F(\theta^*) = P\,[\theta \leq \theta^*] = \sum_{j=1}^{h} p_j P[\theta_j \leq \theta^*] = \sum_{j=1}^{h} p_j F_j(\theta^*) \tag{1}$$

and the composite density function is given as

$$I(\theta) = \sum_{j=1}^{h} p_j I_j(\theta) \tag{2}$$

where we deleted the asterisk for notational convenience. In certain univariate cases this approach is very efficient [see Atkinson and Pearce (1976)]. In the literature the density (2) is also known as a mixture of a finite number of density functions with mixing parameters $p_1,\ldots,p_h$ [see Everitt and Hand (1981)].

One may extend the use of the composite density (2) to cases where θ is a vector of parameters. We have experimented with a composite density as an importance function where the elementary densities are multivariate normals. However, there are at least two problems with this approach. First, the number of parameters of a composite importance function $I(\theta)$ is very large. The estimation of the parameters of the multivariate normal densities $I_1(\theta),\ldots,I_h(\theta)$ and the estimation of the mixing parameters $p_1,\ldots,p_h$ is a far from trivial matter. Second, equation (2) implies that in order to evaluate $I(\theta)$ in a point $\theta^{(i)}$, every elementary function $I_j(\theta)$ has to be evaluated in $\theta^{(i)}$. This may be computationally expensive. *So, there exists a need for simplification of the estimation of the large number of parameters without affecting the flexibility of the composite density more than marginally.* Below we present a first attempt in this direction.

First, we note that the index variable j in equation (2) can be interpreted as a discrete random variable with probability mass function p_j and the elementary density functions can be interpreted as conditional density functions $I(\theta | j)$ with the index j as a conditioning variable. One may generalize this by assuming more general distributions of the conditioning variable. In the continuous case, for instance, one has

$$I(\theta) = \int I^C(\theta | z) I^M(z) dz \tag{3}$$

The composite density (3) is an infinite mixture of conditional densities $I^c(\theta | z)$ given the values of the continuous random variable z. The marginal density $I^M(z)$ of z, which is an auxiliary random variable, is called the mixing density. The Student density is an example of an infinite mixture of normal densities where the gamma-2 is the mixing density.

In our example we noted from a preliminary diagnostic analysis that the posterior density $p(\theta)$ is, roughly speaking, very skew in the direction of one component of θ. We make use of this information in the following way. Let θ be partitioned as

$$\theta = \begin{bmatrix} \theta_1 \\ \theta_2 \end{bmatrix} \tag{4}$$

where θ_1 contains s-1 components and θ_2 is a scalar and the skewness is supposed to be predominant in the direction of θ_2. The importance function $I(\theta)$ is factorized as the product of a conditional multivariate normal (CMN) density of θ_1 conditional upon θ_2 and a marginal piecewise-uniform (MPU) density of θ_2. The MPU distribution of θ_2 is defined as

$$F(\theta_2) = \begin{cases} 0 & \text{if } \theta_2 \leq a_0 \\ P_{j-1} + \dfrac{p_j}{a_j - a_{j-1}} (\theta_2 - a_{j-1}) & \text{if } a_{j-1} \leq \theta_2 \leq a_j \quad (j = 1,\ldots,h) \\ 1 & \text{if } a_h \leq \theta_2 \end{cases} \tag{5}$$

where

$$P_0 = 0, \quad P_j = P_{j-1} + p_j (j = 1,\ldots,h), \quad \sum_{j=1}^{h} p_j = P_h = 1$$

and the MPU density function of θ_2 is defined as

$$I^M(\theta_2) = \begin{cases} 0 & \text{if } \theta_2 < a_0 \\ \dfrac{p_j}{a_j - a_{j-1}} & \text{if } a_{j-1} < \theta_2 < a_j \quad (j = 1,\ldots,h) \\ 0 & \text{if } a_h < \theta_2 \end{cases} \tag{6}$$

Random drawings θ_2 can easily be generated by an inversion method because the distribution of the random variable $F(\theta_2)$ is uniform on [0,1]. The CMN density of θ_1 conditional upon θ_2 is (for $j = 1,\ldots,h$) given as

$$I_j^C(\theta_1 | \theta_2) = (2\pi)^{-(s-1)/2} |V_j|^{-1/2} \times \exp[-\tfrac{1}{2}(\theta_1 - \mu_j)' V_j^{-1} (\theta_1 - \mu_j)] \quad \text{if } a_{j-1} < \theta_2 < a_j \tag{7}$$

where μ_j and V_j are the well known parameters of a multivariate normal density. Therefore the CMNMPU importance function can be written as

$$I(\theta) = I_j^C(\theta_1|\theta_2)I^M(\theta_2)$$
$$= I_j^C(\theta_1|\theta_2)\ \frac{p_j}{a_j - a_{j-1}} \qquad \text{if } a_{j-1} < \theta_2 < a_j \tag{8}$$

The parameters of (8) are μ_j, V_j, p_j, $(j = 1,\ldots,h)$ and the interval bounds $a_0, a_1,\ldots,a_h$.

One may compare the case of the CMNMPU importance function (8) of the s-vector θ with the case where the composite importance function of θ, given in equation (2), contains s-dimensional conditional multivariate normal densities as elementary functions and the conditioning index variable is an auxiliary random variable. In contrast, in the CMNMPU approach we have partitioned θ [see (4)] and our conditional densities are $(s-1)$-dimensional rather than s-dimensional. As a result we have a composition (or mixture) of a finite number of conditional multivariate normal densities of the $(s-1)$-vector θ_1 conditional upon the scalar θ_2. The marginal piecewise-uniform density of the continuous random variable θ_2 plays the role of the mixing density.

The CMNMPU approach has three advantages.

First, it simplifies the evaluation of the importance function $I(\theta)$. In a particular point $\theta^{(i)}$ one has to evaluate only one elementary normal density $I_j^C(\theta_1^{(i)}|\theta_2^{(i)})$ for a given value $\theta_2^{(i)}$ [compare equation (8)].

Second, the number of parameters of the CMNMPU density is smaller than the number of parameters of a composition of s-dimensional multivariate normal densities. This may simplify the estimation of the parameters, see below. However, the number of parameters of the CMNMPU density is larger than the number of parameters of a multivariate Student density. This gives the CMNMPU density flexibility. In our example of a three dimensional posterior density we make use of a CMNMPU density with $h = 20$ conditional bivariate normal densities and a total of 138 unrestricted parameters. [20×5 parameters (μ_j, V_j) of the bivariate normals; 19 unrestricted bounds a_j $(j = 1,\ldots,h-1)$ since the end points a_0 and a_h are determined a priori; and 19 unrestricted parameters p_j, $j = 1,\ldots,h-1$]. More generally we make use of $h(s-1 + (1/2)s(s-1)) + 2h-2$ parameters if the model contains s parameters of interest and we use h elementary densities. By contrast, a composition of s-dimensional normal densities contains an additional $h(s+1)$ parameters.

Third, there are relatively simple algorithms to estimate the parameters of the CMNMPU importance function. This advantage is the most important one. One of these algorithms is described in the following four steps.

Step 1. Generate a random sample for θ of size N ($N = 2000$, say), by making use of the simple importance sampling method (SIS) described in van Dijk and Kloek (1980). Estimates of the parameters of the importance function of SIS have been discussed in Section 4 of that paper;

Step 2. Arrange the drawings of θ_2 in ascending order and divide them into h ($h = 20$, say) groups. Then the interval bounds a_j, $j = 1,\ldots,h-1$ are given by the random drawings $\theta_2^{(100)}, \theta_2^{(200)},\ldots,\theta_2^{(1900)}$. The bounds a_0 and a_h are given *a priori* (or by a preliminary diagnostic analysis);

Step 3. Compute posterior first and second order moments of θ_1, by a standard Monte Carlo method, for the twenty regions constructed in Step 2. Take these moments as estimates for μ_j, V_j, $j = 1,\ldots,20$;

Step 4. Estimate p_j, $j = 1,\ldots,20$, by a linear regression of the posterior density $p(\theta)$ on the importance function given in the right hand side of (8). The regression coefficients are

$$\hat{p}_j = \frac{\sum_{i=100(j-1)+1}^{100j} \frac{1}{a_j - a_{j-1}} I_j^c(\theta_1^{(i)} | \theta_2^{(i)}) p(\theta^{(i)})}{\sum_{i=100(j-1)+1}^{100j} [\frac{1}{a_j - a_{j-1}} I_j^c(\theta_1^{(i)} | \theta_2^{(i)})]^2} \tag{9}$$

Note that the summation is on points satisfying $a_{j-1} < \theta_2^{(i)} < a_j$.

One may perform a second round of parameter estimation by generating a sample of 2000 random drawings from the CMNMPU density (8) and reiterating Steps 2 to 4. Since our starting point was a composition approach we shall call the approach consisting of Steps 1-4 COM1 and the approach where a second round of parameter estimation has been performed COM2. In the next section we report some results using COM1 and COM2.

Note that as an alternative to Step 4 one may take the integral of the posterior density on the region $-\infty < \theta_1 < \infty$ and $a_{j-1} < \theta_2 < a_j$, and use this integral as estimator for the parameter p_j. Estimates of these integrals for $j = 1,\ldots,h$ can be computed by the same standard Monte Carlo method as mentioned in Step 3.

Finally, we emphasize that our approach of estimating parameters of a multivariate composite density is only a first step. More research is needed in this area.

3. SOME EXPERIMENTAL RESULTS

In this section we discuss the results of some experiments with alternative Monte Carlo methods using a simple example. We take as an example the three dimensional marginal posterior density of the structural parameters β_1, β_2 and γ_2 of the Johnston model [see Kloek and van Dijk (1978) and van Dijk and Kloek (1983a)]. The prior on all structural parameters is the same as described in the latter reference. In this particular case we have taken a uniform prior on the interval (—2, +2) for the three parameters $(\beta_1, \beta_2, \gamma_2)$. Further, we note that the marginal posterior density of $(\beta_1, \beta_2, \gamma_2)$ is in our case of two stochastic equations equal to the concentrated likelihood function apart from a constant factor. We shall consider three methods: simple importance sampling (SIS); simplified composition in one and two rounds (COM1 and COM2), compare Section 2; and mixed integration (MIN).

Next, we describe briefly some problems that had to be solved in the design of the experiments. First, there exists the problem of the comparability of the Monte Carlo rounds for the different methods. We make use of the same starting value of the random number generator for all three methods. But the methods differ with respect to the importance function and as a consequence with respect to the way the random numbers are generated. Therefore we opted for the (crude) criterion of approximately equal CPU-time. All results of Tables 2 and 3 were taken after approximately 40 seconds CPU-time on a DEC 2060 computer. The computer programs were executed at different times of the day and night in order to verify the sensitivity with respect to the workload of the computing system. The results were only marginally affected. The 40 CPU-seconds gave 10,000 accepted random drawings for SIS and COM1 and COM2. The parameters of the CMNMPU-importance function were estimated with a sample of 2000 random drawings. The program for MIN was stopped after the 40 CPU-seconds, mentioned above, had been reached.

Second, we performed a preliminary diagnostic analysis. This analysis indicated that in our case several prior bounds could be reduced in absolute value without affecting the posterior results substantially. So we changed the lower bound of -2 for the three parameters $(\beta_1, \beta_2, \gamma_2)$ into the vector $(-2.0, -1.7, -.4)$ and the upper bound of $+2$ for the three parameters into (.8, .25, 1.0). In this way we reduce the numerical integration on the region where the posterior density is almost zero. This is an advantage in particular for MIN where one-dimensional integrals are computed on the intervals, bounded by the upper and lower bounds, mentioned above. Further, the diagnostic results indicated that the posterior is very skew in the direction of the parameter β_1. So β_1 was chosen as the parameter θ_2 of equation (4). We note that there is a minor notational problem with respect to β_1. In reporting the results we have taken the mixing parameter β_1 as the first element of the vector $\theta' := (\beta_1, \beta_2, \gamma_2)$, while in the theoretical discussion of Section 2 the last element of θ, denoted by θ_2, was taken as the mixing parameter.

Third, we had to specify the values of the parameters of the importance functions in the different Monte Carlo methods. As mentioned before we make use of a multivariate Student density as importance function in SIS and in the first step of COM1 and COM2. Further, we make use of a multivariate normal importance function in MIN. The location parameters (μ_1, μ_2, μ_3) correspond with the parameters of interest $(\beta_1, \beta_2, \gamma_2)$. We shall consider two cases of parameter estimates. The degrees of freedom parameter of the Student function is fixed at unity in both cases. The location parameters μ and the covariance matrix V are different in the two cases. Case I consists of taking the posterior mode for μ and minus the inverse of the Hessian of the log posterior for V. We call this case the *local* approximation case. Case II consists of taking a preliminary estimate of the posterior mean for μ and a preliminary estimate of the posterior covariance matrix as estimate for V. This we call the *global* approximation case.

Fourth, there exists a problem with respect to the choice of the parameter estimates of μ and V in the global case. One may apply the different Monte Carlo methods in a two-stage approach in the sense that the posterior moments obtained in the first (local) round of each different method are used as parameter estimates of the importance functions in the second (global) round. As a consequence a poor approximation of the posterior moments in the first round of a particular approach influences the posterior results in the second round. In order to avoid different effects of large sampling errors in different Monte Carlo methods, we decided to take the same set of posterior estimates in the second round for all alternative Monte Carlo methods. Further, we decided to take a large sample of 100,000 Monte Carlo drawings, using the COM2 approach, in the first round. This is a very large sample and we performed a sensitivity analysis with respect to the sample size in the following sense. The posterior first and second order moments from the COM2 (local) approach, using 10,000 random drawings (instead of 100,000 drawings) were taken as parameter estimates of μ and V for all alternative methods in the global case. The results of Tables 2 and 3 were not very much affected. We emphasize that it is attractive to have a rather small sample of random drawings (say $N = 1000$) in the first round of Monte Carlo. Further research is needed to decide upon the trade-off between sample size and desired accuracy of the preliminary estimates of μ and V in the first round.

Fifth, Table 1 gives the two cases of estimates of importance function parameters. These estimates indicate three major differences between the local and global case. First, the modes and mean of β_1 and β_2 differ considerably, which indicates skewness. Second, the posterior standard deviations of β_1 and β_2 are for the global case roughly eight times as large as for the local case. This indicates leptokurtosis. Third, the correlation between β_2 and γ_2 is positive in the global case but negative in the local case. This is an indication that the contours of the posterior density are not concentric ellipsoids as is the case in linear models. This concludes our discussion of the main points of the experimental design.

The results on the numerical errors of the posterior mean estimates of β_1, β_2 and γ_2 are presented in Tables 2 and 3 using the alternative Monte Carlo methods. We take two measures of numerical error. First, we take the ratio (× 100) of the standard deviation of the Monte Carlo estimate of the posterior mean and the posterior standard deviation. This relative numerical error is chosen because we are more interested to estimate a posterior mean accurately if the posterior variance is small, than if it is large. Second, a quadratic loss function (× 100) is evaluated around the large sample estimate of the posterior mean reported in Table 1, with the inverse of the posterior covariance matrix as the matrix of weights. This function has been tentatively chosen as a summary statistic for high dimensional cases.

TABLE 1

Estimates of Importance Function Parameters

Means	μ_1	μ_2	μ_3
Posterior mode (I)	.46	.09	.36
Posterior* mean (II)	-.60	-.31	.31
Standard deviations	ν_1	ν_2	ν_3
Local approximation in mode (I)	.10	.04	.11
Posterior standard deviations (II)	.79	.33	.15
Correlations	r_{12}	r_{13}	r_{23}
Local approximation in mode (I)	.88	.17	-.16
Posterior correlations (II)	.92	.19	.33

* These posterior moments have an absolute numerical error which is less than .005 at the five per cent significance level.

TABLE 2

Relative Numerical Errors of Posterior Mean Estimates

	SIS	MIN	COM1	COM2	Best method
			Case I (Local approximation)		
β_1	9.92	5.51	1.95	1.52	COM2
β_2	10.02	6.22	2.19	1.88	COM2
γ_2	5.56	4.47	2.96	2.30	COM2
			Case II (*Global approximation*)		
β_1	2.88	2.05	1.63	1.62	COM2
β_2	2.64	1.90	1.75	1.69	COM2
γ_2	2.12	1.71	1.55	1.57	COM1

TABLE 3

*Squared error losses of posterior mean estimates**

	SIS	MIN	COM1	COM2	Best Method
Case I	22.23	8.62	.48	1.27	COM1
Case II	.12	.62	.03	.01	COM2

* The squared error loss function $L(\bar{\theta}, E(\theta)) = (\bar{\theta}-E(\theta))' W(\bar{\theta}-E(\theta))$, where $\bar{\theta}$ is the Monte Carlo estimate of $E(\theta)$ and W is the inverse of the posterior covariance matrix.

We make the following remarks on the numerical results in Tables 2 and 3.

1) According to the results of the relative numerical error COM2 performs best with one exception, but the differences with COM1 are very small. The results for MIN are between those for COM and SIS for all cases. Note that the results of the squared error loss functions are in some cases different from the results of the relative error criterion. For instance, in Table 3 MIN does poorly in case II. Since the summary statistics give sometimes conflicting evidence we shall investigate the approximation of the marginal posterior densities in some cases.

2) The local approximations of μ and V are poor starting values for SIS and MIN. There is a substantial gain in computational efficiency for these methods when the parameter estimates of the global case are taken as values of the importance function parameters.

3) The results of the different methods are sensitive for different reasons. With respect to SIS we note that the results are sensitive for the prior bound of $\beta_1 = -2.0$. When the value of this bound is relaxed SIS performs worse. That is, SIS is sensitive to the degree of skewness of the problem. We shall comment on this below. The computational efficiency of MIN is sensitive for the accuracy level of the one-dimensional classical numerical integration. We took an iterative 16-point Gauss-Legendre formula for the numerical integration with an accuracy level of three digits, because the Monte Carlo estimates of the posterior means have usually no more than a two-digit accuracy. Further, it has been mentioned before (see the second problem of the design of the experiments) that it is an advantage for MIN to change the upper and lower bounds of the region of integration in such a way that one-dimensional integration on intervals where the posterior density is almost zero can be avoided. The results of COM1 and COM2 are, of course, dependent upon the preliminary diagnostic analysis that indicated that in our case the skewness was in the direction of β_1. Here MIN appears rather robust since it is not dependent on *a priori* or diagnostic knowledge with respect to the direction of skewness (compare COM1 and COM2) and it is not dependent on the degree of skewness (compare SIS). Only a few hundred random drawings were sufficient to indicate the direction(s) of skewness.

Next, univariate and bivariate marginal posterior densities and univariate and bivariate marginal importance functions are shown in Figures 1 to 6 for simple importance sampling based on the local and global cases and for the two-step composition method for the global case. The marginal posterior densities from the mixed integration method require additional numerical calculations. For diagnostic purposes one may use a rough approximation procedure. We have deleted these results in order to save space.

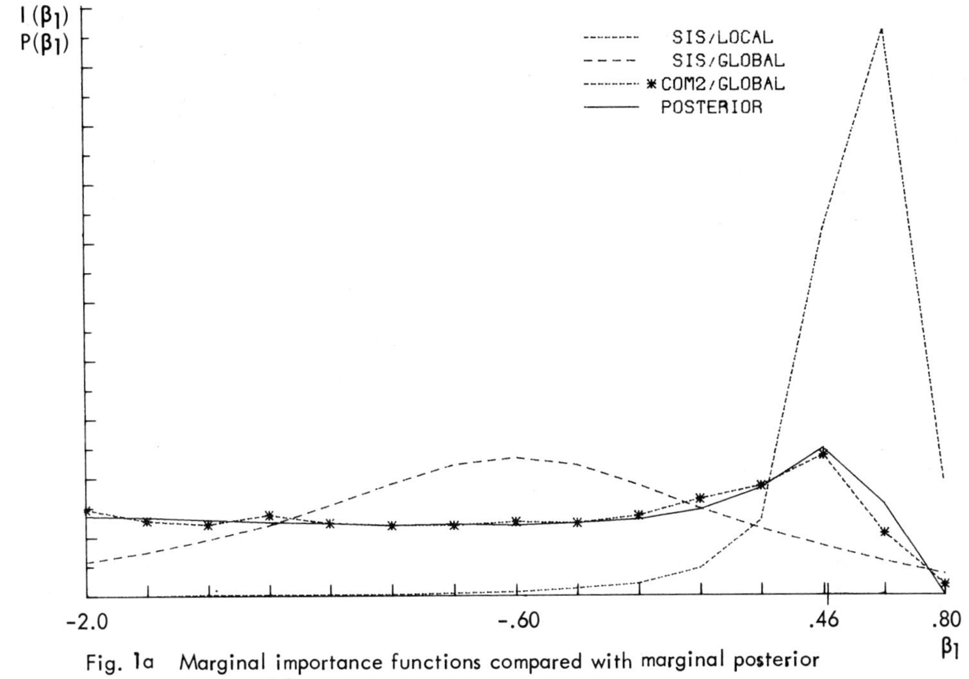

Fig. 1a Marginal importance functions compared with marginal posterior density of β_1

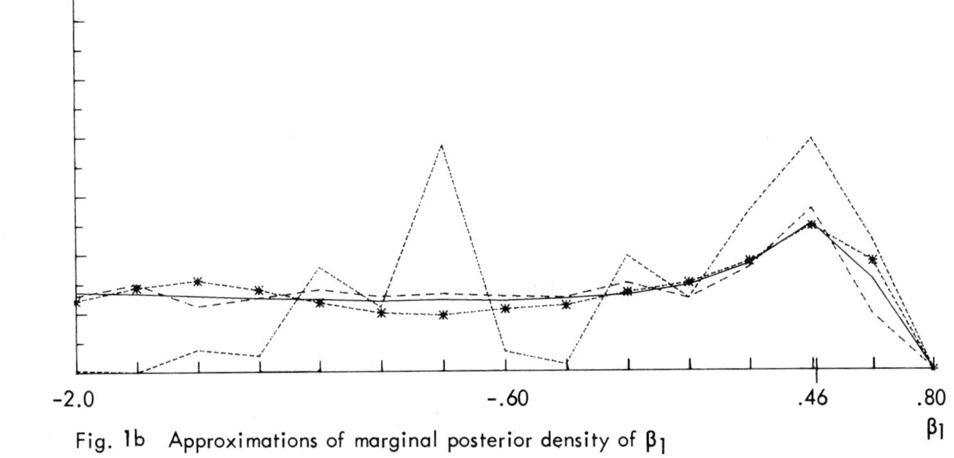

Fig. 1b Approximations of marginal posterior density of β_1

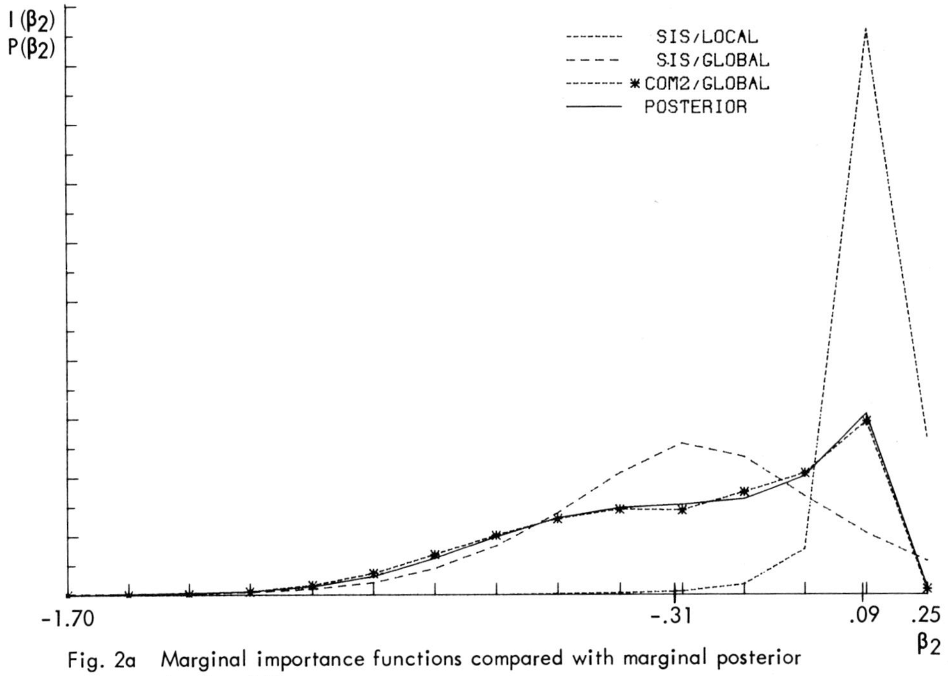

Fig. 2a Marginal importance functions compared with marginal posterior density of β_2

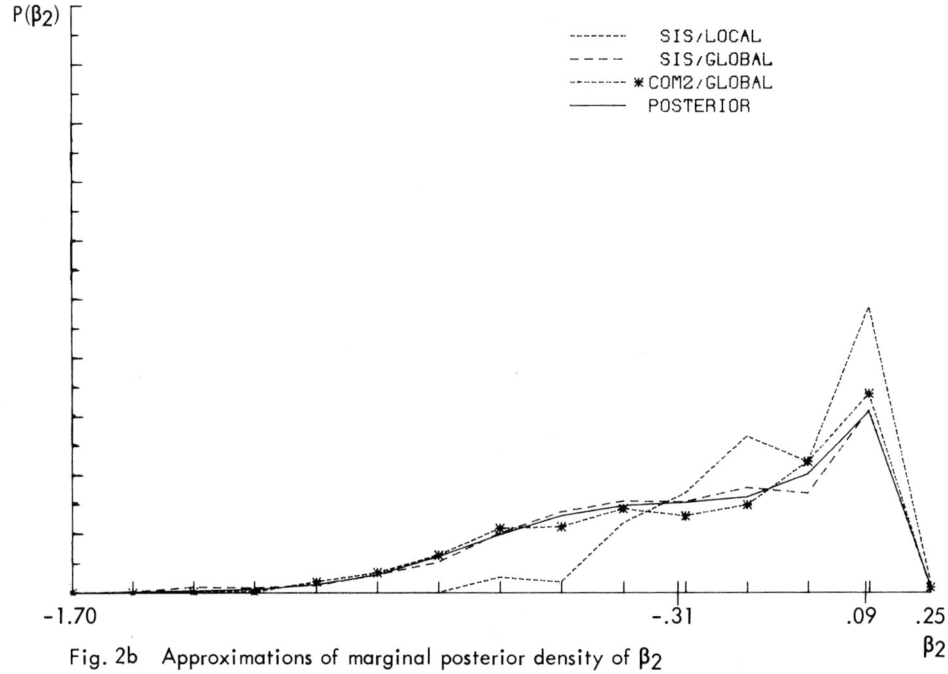

Fig. 2b Approximations of marginal posterior density of β_2

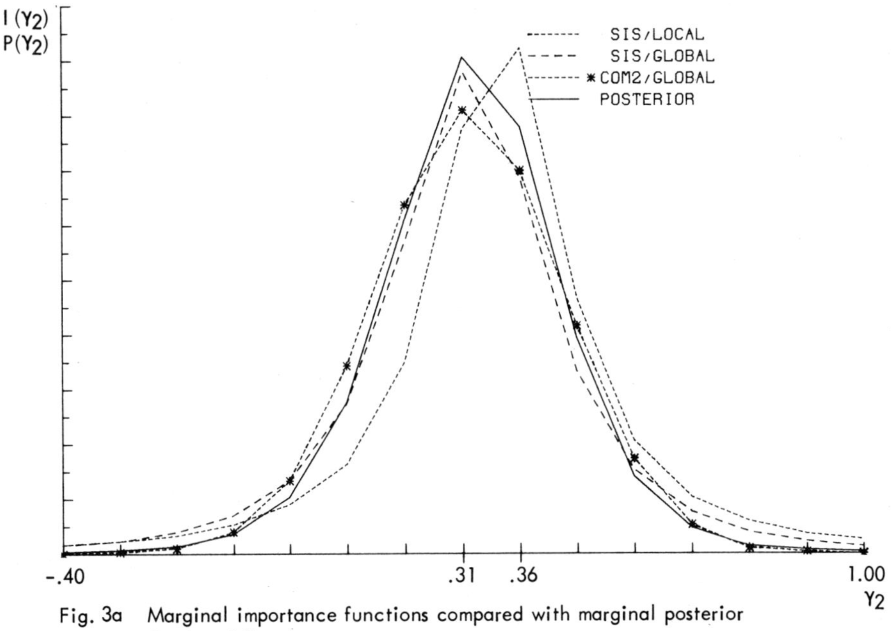

Fig. 3a Marginal importance functions compared with marginal posterior density of Y_2

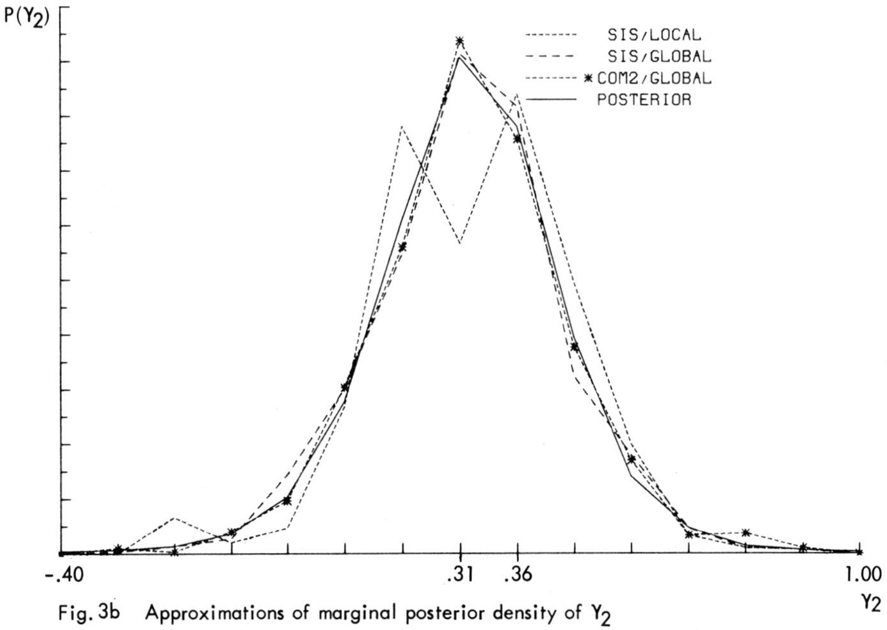

Fig. 3b Approximations of marginal posterior density of Y_2

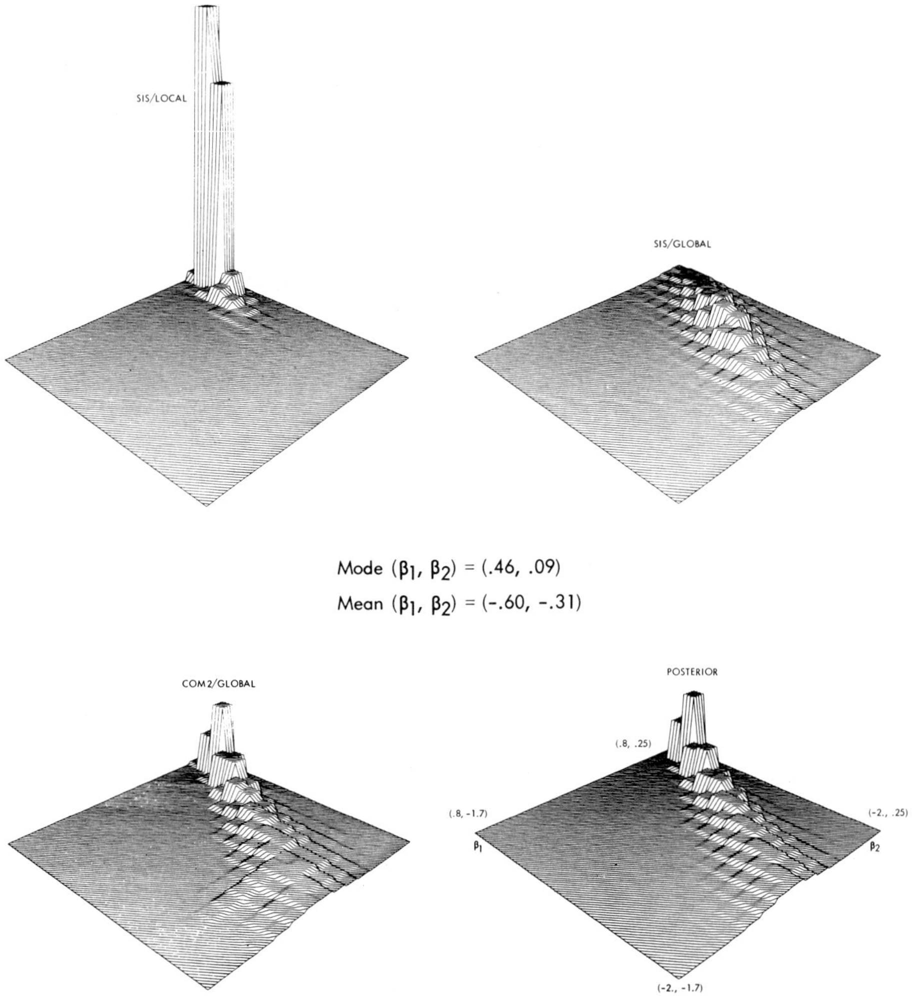

Fig. 4 Bivariate marginal importance functions and bivariate marginal posterior density of (β_1, β_2)

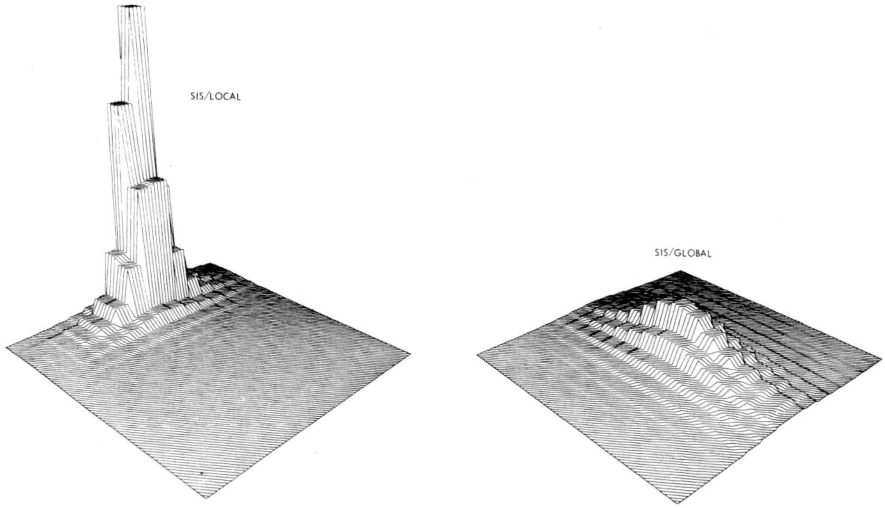

Mode $(\beta_1, \gamma_2) = (.46, .36)$

Mean $(\beta_1, \gamma_2) = (-.60, .31)$

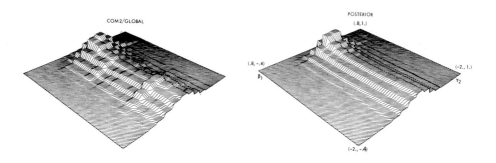

Fig. 5 Bivariate marginal importance functions and bivariate marginal posterior density of (β_1, γ_2)

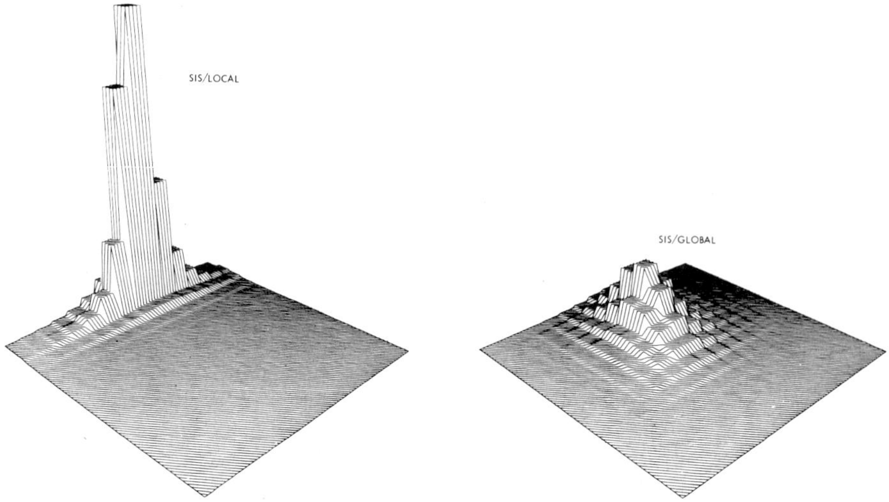

Mode (β_2, γ_2) = .09, .36

Mean (β_2, γ_2) = (-.31, .31)

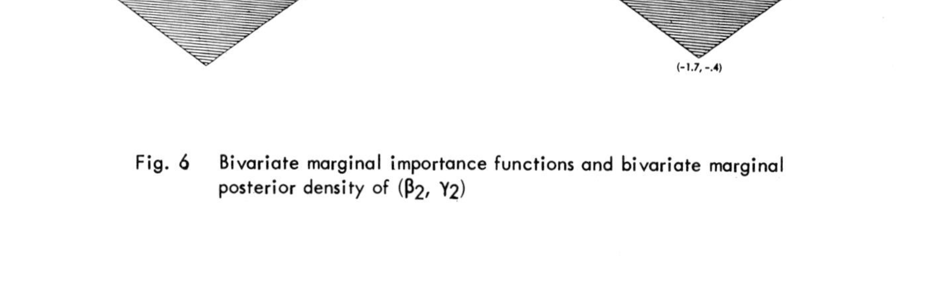

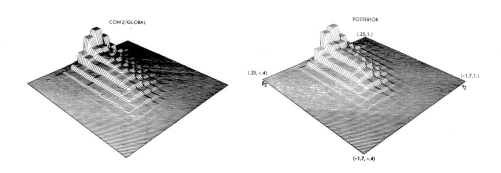

Fig. 6 Bivariate marginal importance functions and bivariate marginal posterior density of (β_2, γ_2)

The approximations to the marginal poseterior densities of β_1, β_2 and γ_2 are presented in Figures 1b to 3b for a sample of 2000 random drawings. Especially Figure 1b illustrates the poor approximation of the SIS/LOCAL importance function for the posterior of β_1. The spikes in the middle of this figure are caused by very large values of the ration $p(\theta)/I(\theta)$. The Student importance function in the SIS/LOCAL approach is considerably below the posterior density in the long tail for β_1. This phenomenon has shown up in several other experiments. The bivariate densities illustrate clearly the poor approximation of the Student density in the local case. The CMNMPU density appears to approximate the posterior very well in the global case. Finally, we note that the shapes of the different bivariate posterior densities confirm the remarks about skewness, leptokurtosis and nonlinearity which we made from an inspection of the point estimates of Table 1 above.

4. CONCLUSIONS

In this paper we have reported on the results of some experiments with alternative Monte Carlo methods for simple importance sampling. The results must be interpreted with care since the experiments were limited and were performed on one model only. However, preliminary experiments with Klein's Model I, which involves nine-dimensional numerical integration, using the same methods appear to confirm the following. First, considerable skewness causes gross approximation errors for simple importance sampling, but a minor case of skewness can be dealt with. Second, poor values of the importance function parameters yield also gross approximation errors for simple importance sampling. Mixed integration appears more robust and can be used in a mechanical way. Our experience with the computation of the posterior moments of the Klein-Goldberger model which involves thirty-dimensional numerical integration indicates that mixed integration is feasible in high dimensional cases where skewness occurs in several directions. In contrast, simple importance sampling does not appear to give reliable results in this case [compare van Dijk and Kloek (1982)]. Third, the simplified composition approach, which makes use of the CMNMPU importance function, is computationally efficient once the direction of skewness is known.

Finally, we emphasize that more research is needed in diagnostic analysis in order to investigate which Monte Carlo alternative is suitable and that more research is needed in the area of estimating the parameters of mixtures of multivariate density functions.

ACKNOWLEDGEMENT

We wish to thank A.S. Louter for his assistance with the preparation of the necessary computer programs.

REFERENCES

ATKINSON, A.C. and PEARCE, M.C. (1976). The computer generation of Beta, Gamma and Normal random variables. *J. Roy. Statist. Soc. A,* **139**, 431-448.

EVERITT, B.S. and HAND, D.J. (1981). *Finite mixture distributions.* London: Chapman and Hall.

KLOEK, T. and VAN DIJK, H.K. (1978). Bayesian estimates of equation system parameters; an application of integration by Monte Carlo. *Econometrica,* **46**, 1-19. Reprinted in: A. Zellner, ed. (1980). *Bayesian analysis in econometrics and statistics.* Amsterdam: North-Holland.

RUBINSTEIN, R.Y. (1981). *Simulation and the Monte Carlo method.* New York: Wiley.

VAN DIJK, H.K. and KLOEK, T. (1980). Further experience in Bayesian analysis using Monte Carlo integration. *J. of Econometrics,* **14**, 307-328.

— (1982). Posterior moments of the Klein-Goldberger model. *Rep.* **8223**/E, Econometric Institute, Erasmus University, Rotterdam. To appear in: P.K. Goel and A. Zellner eds. *Bayesian inference and decision techniques with applications.* Amsterdam: North-Holland.

— (1983a). Monte Carlo analysis of skew posterior distributions: An illustrative econometric example. *The Statistician,* **32**, 216-223.

— (1983b). Posterior moments computed by Mixed Integration. *Rep.* **8318**/E, Econometric Institute, Erasmus University, Rotterdam.

DISCUSSION

J.C. NAYLOR (*Derby Lonsdale College of High Education*)

The paper by van Dijk and Kloek considers Monte Carlo integration methods. Here a number of methods are compared and the results clearly show the need for careful selection or construction of an appropriate importance function. Again non Monte Carlo or classical methods have been largely ignored. Since, in the example used, the number of dimensions of the parameter space is not large (3), there is opportunity to compare a wider range of methods.

The only rule considered with some classical component in it was MIN, which has not performed well in this experiment. I wonder if the apparently arbitrary limit on *cpu* time might have worked against this rule? The choice of a single time limit involves an assumption that relative performance is a monotonic function of computation time. Another possible approach would be to determine the *cpu* time needed to achieve some desired (average) accuracy.

As an alternative approach, I applied the methods of Naylor and Smith (1982) to this example. The use of finite ranges for the parameters (which is surely an advantage to the Monte Carlo method) produced some difficulty, especially since the marginal posterior for β_1 is almost uniform for most of the chosen range. First results actually produced a marginal density for β_1 *which increased as* β_1 became more negative. In this paper the authors state that the concentrated likelihood is proportional to the marginal posterior density for the parameters of interest. From Kloek and van Dijk (1978) these two functions are defined as being of the form

$$\|\Gamma\|^n |n\Gamma'\Omega'\Gamma + (\beta_1-\hat{\beta}_1)'Z_1'NZ_1(\beta_1-\hat{\beta}_1)|^\alpha$$

where $\alpha = -(n-1)/2$ for the marginal posterior and $\alpha = -n/2$ for the concentrated likelihood. Using the latter value for α I obtained a plot of the marginal for β_1 in good agreement with the results presented here. I would like to ask Dr. van Dijk if he could explain this apparent discrepancy and perhaps confirm that the results I obtained were of the form he would have expected. I think that for the purpose of such a comparison the actual correctness of the likelihood is less important than agreement in the function used by the different methods.

The example problem presented actually involves 8 parameters, 5 of which have been integrated out analytically. A further interesting test of the Monte Carlo method is then made possible by ignoring these analytic integrations and working with the full 8 parameter problem. This would give some measure of the performance of the methods in more dimensions while having the results of the 3 parameter analysis for comparison.

T.J. SWEETING (*University of Surrey*)

Drs. van Dijk and Kloek have a great deal of experience in using Monte Carlo integration techniques in Bayesian analysis, and their results allow us to gain a quick appreciation of the problems. It is clear from their work that in general one cannot rely on simple importance sampling (SIS) alone, and one needs to combine this with analytical integration, approximations, and classical numerical integration.

For the practical application of the Monte Carlo method one needs to ensure that the location, spread and tail behaviour of the importance function approximately concur with those of the posterior density. In the case where there is no significant skewness, this can be effectively achieved by the mixed integration (MIN) method introduced by the authors (Van Dijk and Kloek (1983b)). The more difficult cases arise where there is a marked degree of skewness in which case, as was demonstrated this evening, SIS based on a local approximation to the posterior density will perform very poorly. The method presented is designed for cases where there are only one or two principal directions of skewness, θ_2 say. The authors propose using a normal importance function for $\theta_1|\theta_2$, and a piecewise-uniform importance function for θ_2 (their CMNMPU method), and clearly very good results are obtained in the example given. One obvious refinement of the method more appropriate for general use would be to use mixed integration for $\theta_1|\theta_2$ i.e. CMINMPU. This would permit arbitrary tail behaviour in the distribution of $\theta_1|\theta_2$, and may also alleviate problems caused by any residual skewness. Secondly, ordinary numerical integration could be used instead for θ_2, when the dimension is only one or two. Thus a mixed integration method could be used, with two numerical integrations taking care of the tail behaviour and skewness problems.

A.F.M. SMITH (*University of Nottingham*)

Have the authors experimented with the use of mixtures of conjugate families as importance sampling function for any families other than multivariate normal? For example, mixtures of gammas in the "skew direction"?

J. BACON-SHONE (*University of Hong Kong*)

If I may quote Professor D.V. Lindley out of context, 'Bayesians don't randomize'. This implies to me that Monte Carlo integration is in general a bad idea as it involves introducing extra random variation. As I understand importance sampling, we are trying to find

$$\int h(x)dF(x)$$

so we look for G such that

$$G^{-1}[F(x)] \cong x$$

and such that it is quick and easy to sample from G. That this is a good idea would seem to be related to the surprisingly good results obtained when using trapezoidal integration (see Walley and Fearn (1979)). This in turn implies that transformation of variables is crucial in allowing the use of deterministic methods.

REPLY TO THE DISCUSSION

Dr. Naylor questions us about the use of a certain CPU-time limit as a stopping rule for all methods considered. The assumption that relative performance is a monotonic function of computation time is in accordance with the classical formula σ^2/N for the variance of a posterior distribution if the prior is locally uniform and σ^2 is (sufficiently well) known. The question whether our 40 CPU-seconds are sufficient in the sense that they give a reasonable accuracy of the posterior moments and densities require further research. When the sample is very small one has a large and unreliable estimate of σ^2; when the sample is extremely large one has excessive computational costs. There is apparently a need for a sensitivity analysis and for several other diagnostic checks. Given our experience, we recommend the strategy that after a relatively small number of drawings (500, 1000, 1500,...) one evaluates (relative) error estimates, plots of the univariate and bivariate posterior densities and the empirical distribution of the ratio of the posterior density and the importance function.

With respect to the proper value of α, which is specified in the equation given by Dr. Naylor, we remark that $\alpha = n-1$ in our 1978 paper and that $\alpha = n$ in our present paper. The value of α depends on the particular noninformative prior that one specifies with respect to the covariance matrix Σ of the structural disturbances. [For details, see, e.g., van Dijk and Kloek (1977a)]. From preliminary experiments we learnt that our posterior density is sensitive to the choice of the noninformative prior on Σ. Similar results have been reported for the case of Klein's model I by van Dijk and Kloek (1977b). The problem of the sensitivity of posterior results to the choice of the noninformative prior on Σ, which we consider as nuisance parameters, is a topic of further research.

Finally, Dr. Naylor suggested that in our example the nuisance parameters may be handled by means of Monte Carlo rather than by means of analytical integration. One deals then with an eight-dimensional numerical integration problem. An inverted Wishart density may be used in this case as a factor in the importance function. The specification of all the parameters of this density is, however, not a trivial matter.

Dr. Sweeting's remark that mixed integration (MIN) can be used effectively in cases without significant skewness seems to suggest that MIN is not suitable for cases with considerable skewness. We emphasize that MIN can deal with an *arbitrary* degree of skewness in *any* direction of a unimodal multivariate posterior density.

Dr. Sweeting has made two suggestions with respect to our CMNMPU-importance sampling approach. These may be summarized as follows. The posterior mean of θ can be written as

$$E\theta = E[E(\theta_1|\theta_2)]$$

where the first expectation on the right hand side of the equality sign is taken with respect to θ_2 and the second expectation is with respect to θ_1 given a value of θ_2. The first suggestion is to perform mixed integration with respect to θ_1 given a value of θ_2 (the so-called CMINMPU-approach). The second suggestion is to use classical numerical integration with respect to θ_2. As a consequence, one has to perform mixed integration with respect to θ_1 for several different values of θ_2.

Next, consider two extreme cases of conditional skewness of θ_1. In the first case the conditional posterior density of θ_1 is (almost) symmetric given a value of θ_2. Then the CMNMPU-approach appears more efficient than the approach suggested above since one-dimensional classical numerical integration is avoided in the CMNMPU-approach. In the second case the conditional posterior density of θ_1 is still rather skew. Then the CMNMPU-approach is not very efficient (except for the case where this skewness can be identified with one element of θ_1 or with a linear combination of a subset of θ_1). We recommend the used of mixed integration with respect to the complete vector θ in this case rather than the repeated use of mixed integration with respect to the subvector θ_1.

Professor A.F.M. Smith asked us about the use of the natural conjugate family as a class of importance functions. Rothenberg (1963, 1973) has shown the mathematical restrictiveness of the conjugate family for the linear simultaneous equation model [see also Harkema (1971)]. An extension of the conjugate family has been suggested by Drèze. This extended natural conjugate family may be useful as an importance function [see Drèze and Richard (1983), and Bauens (1983)]. More experience is needed to determine the usefulness of such an approach. Finite mixtures of conjugates appear very flexible but, as we indicated in Section 2, the estimation of all the parameters is another problem for research.

Dr. Bacon-Shone quotes Lindley who said "Bayesians don't randomize" and suggests that this is a categorical imperative. (At least, that is our interpretation of his comment). This reminds us of another categorical imperative which states "Muslims don't charge interest". In our opinion, the latter rule reveals lack of knowledge of economics. The desire to prevent usury does not justify such a strict rule. The analogy suggests that the former categorical imperative might reveal lack of knowledge of multidimensional numerical integration: there exists an important class of multidimensional numerical integration problems that can be much more efficiently solved by using Monte Carlo than by means of traditional integration methods. We refer to Kloek and van Dijk (1978) for an extensive discussion of the advantages of Monte Carlo integration for the computation of posterior moments and densities of the parameters of the simultaneous equation model in high dimensional cases (say, more than five or six). Recently, we reported on the computation of the first and second order moments of a thirty-dimensional posterior density using the mixed integration method [van Dijk and Kloek (1982)].

Dr. Bacon-Shone's remark about the importance of transformations of random variables for classical numerical integration applies also to Monte Carlo integration. The problem is that in the multivariate case it is far from trivial to find transformations from the uniform density into such a flexible functional form that a good approximation to the posterior density is obtained.

Finally, we thank all the discussants for their stimulating comments.

REFERENCES IN THE DISCUSSION

BAUENS, L. (1983). *Bayesian full information analysis of simultaneous equation models using integration by Monte Carlo.* Doctoral dissertation, Université Catholique de Louvain.

DRÈZE, J.H. and RICHARD, J.F. (1983). Bayesian analysis of simultaneous equation systems. *Handbook of Econometrics,* **1.** (Z. Griliches and M. Intriligator, eds.) Amsterdam: North Holland.

HARKEMA, R. (1971). *Simultaneous equations, a Bayesian approach.* Rotterdam: Rotterdam University Press.

KLOEK, T. and VAN DIJK, H.K. (1978). Bayesian estimates of equation system parameters: an application of integration by Monte Carlo. *Econometrics,* **46,** p. 1-19.

NAYLOR, J.C. and SMITH, A.F.M. (1982). Applications of a method for the efficient computation of posterior distributions. *Appl. Statist.* **31,** 214-225.

ROTHENBERG, T.J. (1963). A Bayesian analysis of simultaneous equation systems. *Rep.* **6315,** Econometric Institute, Erasmus University, Rotterdam.

— (1973). *Efficient estimation with a priori information.* New Haven: Yale University Press.

VAN DIJK, H.K. and KLOEK, T. (1977a). Predictive moments of simultaneous econometric models; A Bayesian approach. *New developments in the applications of Bayesian methods.* (A. Aykaç and C. Brumat eds.) Amsterdam: North Holland.

— (1977b). Likelihood diagnostics and posterior analysis of Klein's model I. *Tech. Rep.*, Econometric Institute, Erasmus University, Rotterdam, presented at the fifteenth NBER-NSF seminar on Bayesian inference in econometric, 1977, Madison, Wisconsin.

WALLEY, D. and FEARN, T. (1979). Trapezium rules, O.K. *Appl. Statist.* **28**, 296-297.

BAYESIAN STATISTICS 2, pp. 531-558
J.M. Bernardo, M.H. DeGroot, D.V. Lindley, A.F.M. Smith (Eds.)

Generalized Linear Models: Scale Parameters, Outlier Accommodation and Prior Distributions

MIKE WEST
(*University of Warwick*)

SUMMARY

Bayesian data analysis and inference for regression models in the exponential family are considered with particular emphasis on the modelling of outliers and the specification and use of flexible prior distributions. It is shown how scale parameters can be used to detect, highlight and discount particular model components that are in some way inadequate or in disagreement with many other components. This general approach covers simple goodness of fit of the whole model, outlying observations or groups of observations, and discordant components of prior distributions, and is illustrated in several examples.

Keywords: EXPONENTIAL FAMILY; GOODNESS OF FIT; INFLUENCE FUNCTIONS; MEAN VARIANCE FUNCTIONS; NON-LINEAR REGRESSION; OUTLIERS; SCALE MIXTURE MODELS; SCALE PARAMETERS; SHRINKAGE; STRUCTURED PRIORS.

1. INTRODUCTION

The general form of the density of the exponential family distribution for an observation x is taken as

$$p(x|\theta,\phi) = \exp[\phi\{x\theta - a(\theta)\}]\, b(x,\phi) \tag{1.1}$$

where the natural parameter θ and the scale parameter ϕ determine the mean and variance of x as

$$E[x|\theta,\phi] = a'(\theta) \text{ and } V[x|\theta,\phi] = a''(\theta)/\phi\,. \tag{1.2}$$

As a likelihood for θ given x and ϕ (1.1) is located at the maximum likelihood value t, satisfying

$$x = a'(t)\,, \tag{1.3}$$

and the corresponding observed information is $\phi.a''(t)$.

The notation for some special cases of interest is as follows:

(i) $(x|\theta,\phi) \sim N(\theta,\phi)$ when x is normally distributed with mean θ and precision ϕ.

(ii) $(x|\theta,\phi) \sim G(\theta,\phi)$ when x has a gamma distribution with mean θ^{-1} and index ϕ; $p(x|\theta,\phi) = (\theta\phi)^{\phi} x^{\phi-1} e^{-\theta\phi x}/\Gamma(\phi)$, $\quad x>0$.

(iii) $(y|\theta,\phi) \sim B(n,\pi)$ when $y = nx$ is binomially distributed with $\phi=n$ and $\theta = \log[\pi/(1-\pi)]$, $0<\pi<1$, $y=0,1,\ldots,n$.

(iv) $(x|\theta,\phi) \sim P(\mu)$ when x has a Poisson distribution with mean $\mu = e^{\theta}$ and $\phi=1$.

The class of generalized linear models for a set of n observations $\mathbf{x} = \{x_1,...x_n\}$ supposes a joint density of the form

$$p(\mathbf{x}|\beta,\phi,H) = \prod_{r=1}^{n} p(x_r|\theta_r,\phi_r), \qquad (1.4)$$

where the scale parameters $\phi = \{\phi_1,...,\phi_n\}$ are known, H is an $n \times p$ matrix of known regressors with rows $\mathbf{h}_r^T$, $r=1,...,n$, β is a p-vector of regression parameters and $\theta_r = f(\mathbf{h}_r^T\beta)$, $r=1,...,n$, for some known function $f(\bullet)$. (It is often reasonable to specify the regression function f implicitly by linking the mean function $a'(\theta_r)$ directly to a function of $\mathbf{h}_r^T\beta$ as in Nelder and Wedderburn, 1972). In practice it is often satisfactory to take f to be the identity function for the normal, binomial and Poisson models, and $f(x) = e^x$ for the gamma, although there may be several candidate functions to be considered.

In practice, the overall goodness of fit of a particular model M_0, say, can often be assessed by embedding it in a more general class $M(\alpha)$, say, indexed by a single parameter α such that $M(\alpha_0) = M_0$ for some α_0. Concerning the error model (1.1) a particular example is that of the exponential distribution embedded in the gamma family with scale parameter α. Section 2 of this paper is concerned with the extension of this use of scale parameters to the general exponential family model; in particular, the discrete distributions. Use of an approximate scaled likelihood provides a simple method of assessment of the goodness of fit of binomial and Poisson models, for example.

Section 3 is concerned with outliers and outlier modelling in nominal exponential family models and is based on the use of alternative heavy-tailed error distributions as in standard linear models (West, 1983). This approach involves the discounting of components of the likelihood essentially by allocating individual scale parameters to the observations and draws on the developments of section 2 concerning scaled likelihoods.

Finally, the use of multivariate normal prior distributions for regression parameters in generalized linear models is discussed in section 4,, along with some problems that arise as a consequence of the normal form. In particular, the use of standard shrinkage priors may lead to distorted conclusions due to under —or over— shrinkage in many cases. The use of heavy-tailed priors, such as the Student t distributions, as alternatives to normality alleviates this problem and several illustrative examples are discussed.

2 .SCALE PARAMETERS AND THE EXPONENTIAL FAMILY

2.1. *Variance functions*

The specification of a model for the systematic variation in the data x in terms of the regression discussed above determines the variance function $a''(\theta_r)$ of (1.2). When the data are continuous (e.g. normal or gamma) then it is often the case that $\phi_r = \phi$ for each r and ϕ is unknown. In this case, learning about ϕ will provide information about variation in the data that it is not accounted for by the regression of the θ_r on the rows of H, whether it is residual variation due to the inadequacy of the systematic component or (apparently) purely random. If the systematic component of the model is satisfactory (perhaps on non-statistical grounds), the adequacy of particular cases of (1.1) for the random component can be examined by making inferences about ϕ. For example, a nominal exponential distribution can be assessed by using a gamma model and examining the posterior support for $\phi = 1$.

By contrast the discrete distributions do not have this flexibility for modelling purely random variation; the binomial and Poisson models have ϕ fixed. This restriction is sometimes felt in applications when observed random variation in discrete data appears to be greater than that provided by a nominal binomial or Poisson sampling model although such models may be attractive for other reasons. One approach to modelling extra variation of this kind is to replace the nominal distribution with a more diffuse form derived as a mixture with respect to some prior distribution for the parameters θ_r. For example, in biological bio-assay problems extra-binomial variation is often adequately modelled using beta-binomial models (Crowder, 1978). A similar approach (Williams, 1982) involves an approximate analysis that requires only the mean and variance of the prior distribution. Thus if $(x|n,\pi) \sim B(n,\pi)$ with $E[\pi|\mu] = \mu$ and $V[\pi|\mu] = \alpha\mu(1-\mu)$ for some μ and α, $(0<\mu<1, 0<\alpha)$, then $E[x|n,\mu] = n\mu$ and $V[x|n,\mu] = n\mu(1-\mu)\phi^{-1}$ where $\phi^{-1} = 1 + (n-1)\alpha$. Notice that the relationship between the mean and variance of $p(x|n,\mu)$ is similar to that of the binomial model but with an additional scale parameter ϕ. When fitting $p(x|n,\mu)$ to data, ϕ can be viewed as measuring the goodness of fit of the nominal binomial distribution with ϕ near unity indicating a very tight prior for $(\pi|\mu)$, ϕ far from unity giving a more diffuse form.

An alternative approach is required, of course, in modelling variation less than that predicted by a nominal binomial or Poisson model. Altham (1978) discusses generalizations of the binomial distribution in which the variance is proportional to that of the binomial with a constant of proportionality now able to take values less than unity, as well as greater than unity. This provides models for distributions that are more peaked or more diffuse than binomial (and can be extended to generalizations of the Poisson with similar properties), although the complex forms of these models lead to intractable likelihoods.

A simple, tractable, alternative model having the same mean/variance relationship is the density (1.1) with the scale parameter ϕ now unrestricted. This follows since, whatever the function $b(x,\phi)$ may be, the form of (1.1) as a function of θ determines the properties in (1.2). This idea extends to the problems discussed by Wedderburn (1974) in which the relationship between the mean and variance of x is used to construct an approximate likelihood based on the connection between (1.1) and (1.2). A particular example concerns data in the form of essentially continuous proportions or rates $(0<x_r<1, r=1,...,n)$. A density for x_r of the form

$$g(x_r,\phi_r).\pi_r^{\phi_r x_r}(1-\pi_r)^{\phi_r(1-x_r)}, 0<x_r<1,$$

leads to $E[x_r|\pi_r,\phi_r] = \pi_r$ and $V[x_r|\pi_r,\phi_r] = \pi_r(1-\pi_r)\phi_r^{-1}$ and may provide a suitable model for studying the relationships between the π_r even though the x_r are not derived as binomial proportions. Here the likelihood has the form of a nominal $B(1,\pi_r)$ model with an arbitrary scale parameter ϕ_r, whereas in the case of extra (or under) binomial variation the required variance $n_r\pi_r(1-\pi_r)\phi_r^{-1}$ is obtained by replacing the nominal $B(n_r,\pi_r)$ density for $n_r x_r$ by a likelihood of the form

$$g(x_r,\phi_r).\pi_r^{\phi_r n_r x_r}(1-\pi_r)^{\phi_r n_r(1-x_r)}, 0<x_r<1,$$

for some function g. In both cases the function g is not determined but the dependence on the scale parameter ϕ_r is required to provide a full learning model. More generally the function $p(x|\theta,\phi)$ of (1.1) defines an exponential family density only for certain values of ϕ yet for any $\phi>0$ has the same form, in terms of location and shape, as a likelihood in θ; the addition of a general scale parameter simply discounts the likelihood when $\phi<1$ and enhances it when $\phi>1$. In general the function $b(x,\phi)$ is unspecified but the dependence on ϕ must be considered in order to provide a joint likelihood for θ and ϕ. This is now examined.

2.2. *Scaled exponential family likelihoods*

For a standard exponential family distribution having a fixed scale parameter ϕ_0, say, the deviance function $D(x|\theta)$ is defined as

$$D(x|\theta) = -2\log[p(x|\theta,\phi_0)/p(x|t,\phi_0)] \tag{2.1}$$

where t is the maximum likelihood value in (1.3). Clearly, from (1.1),

$$D(x|\theta) = -2\phi_0[x(\theta-t)-a(\theta)+a(t)], \tag{2.2}$$

and $D(x|\theta) \geq 0$, with equality only at $\theta = t$. The deviance plays a central role in current (non-Bayesian) analyses of exponential family models (Baker and Nelder, 1978) that is clearly based on the parallel between $D(x|\theta)$ in general models and the quadratic form $\phi_0(x-\theta)^2$ in the normal case. The identity

$$p(x|\theta,\phi_0) = p(x|t,\phi_0)\exp[-D(x|\theta)/2] \tag{2.3}$$

illustrates this point. For the normal model there is, of course, a natural scaled likelihood obtained simply by changing the normal precision from ϕ_0 to $\phi\phi_0$, for any $\phi > 0$, and this has the form

$$p(x|t,\phi_0)\,\phi^{1/2}.\exp[-\phi.D(x|\theta)/2] \tag{2.4}$$

in this special case.

Define $m(x|\theta,\phi) = \phi^{1/2}.\exp[-\phi D(x|\theta)/2]$. As a function of θ in the general case, $m(x|\theta,\phi)$ is a natural power transformation of the standard density,

$$m(x|\theta,\phi) \propto p(x|\theta,\phi_0)^{\phi}, \tag{2.5}$$

with the power ϕ scaling the information content of the original model. The function in (2.4) is proportional to (1.1) as a function of θ and equality is established for any $\phi > 0$ if $b(x,\phi) = p(x|t,\phi_0)^{1-\phi}\phi^{1/2}$.

As noted above, (2.4) is the exact density of the scaled normal distribution when $D(x|\theta) = \phi_0(x-\theta)^2$. In the case of a gamma model, for which

$$p(x|\theta,\phi_0) = (\theta\phi_0)^{\phi_0}.x^{\phi_0-1}e^{-\theta\phi_0 x}/\Gamma(\phi_0), \tag{2.6}$$

(2.4) gives

$$p(x|t,\phi_0)\,m(x|\theta,\phi) = c(\phi,\phi_0)\,p(x|\theta,\phi\phi_0) \tag{2.7}$$

where $p(x|\theta,\phi\phi_0)$ is the natural scaled density (2.6) with ϕ_0 replaced by $\phi\phi_0$ and

$$c(\phi,\phi_0) = \phi^{1/2}\Gamma(\phi\phi_0)\,e^{\phi\phi_0}\,\phi_0^{\phi_0}\,e^{-\phi_0}/[\Gamma(\phi_0)(\phi\phi_0)^{\phi\phi_0}].$$

Clearly the form of (2.7) is correct as a function of both x and θ; the exact density being given only if $c(\phi,\phi_0) = 1$. In fact approximating $\Gamma(\phi_0+1)$ and $\Gamma(\phi\phi_0+1)$ using Stirling's formula gives $c(\phi,\phi_0) = 1$. Otherwise for a wide range of values of ϕ and ϕ_0, c is close to unity.

In general (2.4) may be viewed as an approximate sampling model having the mean/variance relationship (1.2) providing a scaled version of the original exponential family likelihood for θ and an approximate likelihood for ϕ that reduces to the original model when $\phi = 1$. It will be shown in the next section that the usual exponential family analyses can be performed within this model and that inferences about ϕ provide a check of the goodness of fit of the standard (binomial, Poisson etc.) models.

2.3. *Analyses of random samples using* $m(x|\theta,\phi)$

Suppose the likelihood of θ and ϕ based on independent observations $x_1,\ldots,x_n$ is proportional to

$$m(\mathbf{x}|\theta,\phi) = \prod_{r=1}^{n} m(x_r|\theta,\phi)$$

$$= \phi^{n/2}\exp[-\phi D(\mathbf{x}|\theta)/2]\,, \tag{2.8}$$

where

$$D(\mathbf{x}|\theta) = \sum_{r=1}^{n} D_r(x_r|\theta)\,, \tag{2.9}$$

with

$$D_r(x_r|\theta) = -2n_r[x_r(\theta-t_r) - a(\theta) + a(t_r)] \tag{2.10}$$

and $x_r = a'(t_r)$ with the scale parameters n_r known for each r.

Sweeting (1982) discusses Bayesian analyses of continuous models using improper prior distributions for θ and ϕ that will be considered briefly at the end of this section. Proper conjugate priors for θ and ϕ are available, given by

(i) $$p(\theta|\phi) \propto \exp[\phi\{\alpha_0\theta - \beta_0 a(\theta)\}]\,, \tag{2.11}$$

which is a power transformation of the standard prior $p(\theta|\phi=1)$ with power ϕ; and

(ii) $$p(\phi) \propto \phi^{\nu_0/2-1}e^{-\phi\delta_0/2}\,, \quad \phi>0, \tag{2.12}$$

so that $\delta_0\phi \sim \chi^2_{\nu_0}$.

Let t be the mode of (2.11), so $\alpha_0 = \beta_0 a'(t_0)$. Normalization of (2.11) introduces a term in ϕ that is generally complex. For example with a Poisson based model with $a(\theta) = e^\theta$,

$$p(\theta|\phi) = \exp[\phi\{\alpha_0\theta - \beta_0 e^\theta\}]\cdot(\beta_0\phi)^{\alpha_0\phi}/\Gamma(\alpha_0\phi),$$

and the resulting analysis involves gamma functions in ϕ. A useful approximation obtains when Stirling's formula for $\Gamma(\alpha_0\phi+1)$ is applied, giving

$$p(\theta|\phi) \propto \phi^{1/2}\cdot\exp[\phi\{\alpha_0(\theta-t_0) - \beta_0(e^\theta - e^{t_0})\}]$$

where $t_0 = \log(\alpha_0/\beta_0)$ in this case. This expression can be written as

$$p(\theta|\phi) \propto \phi^{1/2}\cdot\exp[-\phi D(\theta)/2] \tag{2.13}$$

where

$$D(\theta) = -2[\alpha_0(\theta-t_0) - \beta_0(e^\theta - e^{t_0})]$$

$$= -2\log[p(\theta|\phi=1)/p(t_0|\phi=1)] \tag{2.14}$$

may be called the deviance of the standard prior $p(\theta|\phi=1)$ by analogy with (2.1).

The form (2.13) is also obtained when $\log p(\theta|\phi=1)$ is well approximated by a quadratic expansion about t_0, since then

$$p(\theta|\phi=1) = p(t_0|\phi=1)\exp[-a''(t_0)\cdot(\theta-t_0)^2/2]$$

and so $p(\theta|\phi) \propto p(\theta|\phi=1)^\phi$ has the approximately normalized form

$$[\phi a''(t_0)/2\pi]^{1/2}\cdot[p(\theta|\phi=1)/p(t_0|\phi=1)]^\phi$$

which is proportional to (2.13) as a function of θ and ϕ. Clearly as an alternative prior, a normal distribution to begin with is exactly of the form (2.13) and may provide a suitable model, particularly with a view to extending the analysis to more complex regression models. This would involve a lognormal, rather than gamma, prior in Poisson models and logistic-normal, rather than beta, in binomial models. Of course, in many cases these models will be similar (Lindley, 1965, pp. 148 & 156, Aitchison and Shen, 1980).

The following analysis is based on the approximate prior (2.12) and (2.13), some further features of which are now easily obtained as

(iii) $$p(\phi\,|\,\theta) \propto \phi^{(\nu_0+1)/2-1}\exp[-\phi\{\delta_0+D(\theta)\}/2]\,, \tag{2.15}$$

$$\text{or } \{\delta_0+D(\theta)\}\cdot(\phi\,|\,\theta) \sim \chi^2_{\nu_0+1};$$

(iv) $$p\,(\theta) \propto [\nu_0+\phi_0\,D(\theta)]^{-(\nu_0+1)/2}\,, \tag{2.16}$$

when $\phi_0 = \nu_0/\delta_0 = E[\phi]$.

Clearly these results parallel those of the joint normal/chi-square prior for normal mean and precision parameters with $(\theta\text{-}t_0)^2$ replaced by $D(\theta)$.

Now the standard analysis leads to a posterior for θ of the form

$$p\,(\theta\,|\,\phi=1,\mathbf{x}) \propto \exp[\alpha_1\theta-\beta_1 a(\theta)]$$

where $\alpha_1 = \alpha_0 + \sum_{r=1}^{n} n_r x_r$ and $\beta_1 = \beta_0 + \sum_{r=1}^{n} n_r$. The mode for θ is t_1 satisfying $\alpha_1 =$ $= \beta_1 a'(t_1)$.

Define the posterior deviance $D(\theta\,|\,\mathbf{x})$ by

$$D(\theta\,|\,\mathbf{x}) = -2\log[p\,(\theta\,|\,\phi=1,\mathbf{x})/p\,(t_1\,|\,\phi=1,\mathbf{x})] \tag{2.17}$$

It can be shown that, whatever form $D(\theta)$ has,

$$D(\theta) + D(\mathbf{x}\,|\,\theta) = D(\theta\,|\,\mathbf{x}) + D(\mathbf{x}) \tag{2.18}$$

where

$$D(\mathbf{x}) = [D(\theta)+D(\mathbf{x}\,|\,\theta)]_{\theta=t_1}\,, \tag{2.19}$$

may be called the residual deviance.

Hence the joint posterior for θ and ϕ has the form

$$p\,(\theta,\phi\,|\,\mathbf{x}) \propto \phi^{(\nu_1+1)/2-1}\exp[-\phi\{\delta_1+D(\theta\,|\,\mathbf{x})\}/2] \tag{2.20}$$

where $\nu_1 = \nu_0 + n$ and $\delta_1 = \delta_0 + D(\mathbf{x})$, and so, by analogy with the prior,

(i) $p\,(\theta\,|\,\phi,\mathbf{x}) \propto p\,(\theta\,|\,\phi=1,\mathbf{x})^{\phi}$, a natural power transformation or scaling of the standard posterior;

(ii) $\delta_1(\phi\,|\,\mathbf{x}) \sim \chi^2_{\nu_1}$ with mean $\phi_1 = (\nu_0+n)/[\delta_0+D(\mathbf{x})]$;

(iii) $[\delta_1+D(\theta\,|\,\mathbf{x})]\cdot(\phi\,|\,\theta,\mathbf{x}) \sim \chi^2_{\nu_1+1}$;

(iv) $p(\theta\,|\,\mathbf{x}) \propto [\nu_1+\phi_1\cdot D(\theta\,|\,\mathbf{x})]^{-(\nu_1+1)/2}$, a heavy-tailed analogue of $p\,(\theta\,|\,\phi=1,\mathbf{x})$ comparable to the Student t posterior in the normal location/scale model.

By contrast the result of Sweeting (1982) depend on the use of the improper prior $p\,(\theta,\phi) \propto \phi^{-1}$. Such a prior would lead to (2.20) with ν_1 replaced by n-1, $\alpha_0 = b_0 = 0$ and $D(\theta) = 0$ for all θ. In this case t_1 is the maximum likelihood estimate of θ and $D(\mathbf{x})$ is just $D(\mathbf{x}\,|\,t_1)$, the residual deviance of Nelder and Wedderburn (1972) and Baker and Nelder (1978).

Now, for any ϕ, $p\,(\theta\,|\,\phi,\mathbf{x})$ is unimodal at t_1, the mode of the standard posterior and of $p\,(\theta\,|\,\mathbf{x})$. The observed information of the latter is

$$\left[-\frac{d^2}{d\theta^2}\log p\,(\theta\,|\,\mathbf{x})\right]_{\theta=t_1} = \bar{\phi}(t_1)\cdot\beta_1\cdot a''(t_1) \tag{2.21}$$

where $\bar{\phi}(\theta) = (\nu_1+1)/[\delta_1+D(\theta|\mathbf{x})] = E[\phi|\theta,\mathbf{x}]$, and $\beta_1\, a''(t_1)$ is the information of $p(\theta|\phi=1,\mathbf{x})$. Further features of (2.20) are illustrated now by example.

Example 2.1. Suppose the standard model to be Poisson with rate $\mu=e^\theta$ so that $a(\theta) = e^\theta$ and $n_r = 1$ for each r, with $x_r = 0,1,2,\ldots$. Then transforming from θ to μ gives

$$(\mu|\phi,\mathbf{x}) \sim G(\beta_1/\alpha_1,\phi\alpha_1)$$

where $\alpha_1 = \alpha_0 + n\bar{x}$ and $\beta_1 = \beta_0 + n$. So, if $m_1 = \alpha_1/\beta_1$, $E[\mu|\phi,\mathbf{x}] = E[\mu|\mathbf{x}] = m_1$ as in the standard model, and $V[\mu|\phi,\mathbf{x}] = m_1^2/(\phi\,\alpha_1)$ with $V[\mu|\mathbf{x}] = m_1^2 E[\phi^{-1}|\mathbf{x}]/\alpha_1$ where $E[\phi^{-1}|\mathbf{x}] = \delta_1/(\nu_1-2)$ is the reciprocal of the posterior mode of ϕ.

Lindley (1965, p. 189) gives an example concerning eight nominally Poisson observations given by 50, 65, 52, 63, 56, 49, 60, 45. Taking α_0, β_0, ν_0 and δ_0 to near zero, the above analysis leads to $\alpha_1 = 440$, $\beta_1 = 8$, $\nu_1 = 8$ and $\delta_1 = 6.53$. So $p(\phi|\mathbf{x})$ has mean 1.23 and mode 0.92 indicating consistency with the nominal Poisson model. Also $E[\phi^{-1}|\mathbf{x}] = 1.09$ so $V[\mathbf{x}|\mu]$ is only about 9 % greater than the variance of the usual model.

Example 2.2. Suppose the standard model to be binomial with $x_r = y_r/n_r$ where $(y_r|\theta,n_r) \sim B(n_r,\pi)$ and $\theta = \log[\pi/(1-\pi)]$. Then $(\pi|\phi,\mathbf{x})$ is beta distributed

$$p(\pi|\phi,\mathbf{x}) \propto \pi^{\phi\alpha_1-1}(1-\pi)^{\phi(\beta_1-\alpha_1)-1}, \qquad 0<\pi<1,$$

where $\alpha_1 = \alpha_0 + \sum_{r=1}^{n} n_r x_r$ and $\beta_1 = \beta_0 + \sum_{r=1}^{n} n_r$. So, again, $E[\pi|\phi,\mathbf{x}] = E[\pi|\mathbf{x}] = \alpha_1/\beta_1$ for all ϕ and the variance $V[\pi|\mathbf{x}]$ is the standard value inflated by $E[\phi^{-1}|\mathbf{x}]$.

Smith (1979) analyzed in depth a sequence of thirteen nominally binomial observations relating to the Lindisfarne scribes problem. The data are

n_r:	21	36	44	30	52	45	48	57	48	22	20	21	20
x_r:	0.57	0.72	0.71	0.80	0.54	0.76	0.81	0.81	0.85	0.86	0.85	0.81	0.80

A first step in an analysis of these data might be to consider a simple binomial model with common parameter π, and to assess the goodness of fit by looking at $p(\phi|\mathbf{x})$. The calculations, with α_0, β_0, ν_0, δ_0 near zero, give the approximate values $\alpha_1 = 350$, $\beta_1 = 464$, $n_1 = 13$ and $d_1 = 23.89$. So $p(\phi|\mathbf{x})$ has mean 0.54 and mode 0.46 indicating that the dispersion in the data is greater than that provided by the simple binomial model. Further $E[\phi^{-1}|\mathbf{x}]=2.17$, so the variance of $p(\pi|\mathbf{x})$ is more than twice that of the standard beta posterior. Correspondingly, the information (2.22) is 59.17 compared with 100.96 in the standard model.

Clearly the binomial model is untenable in this example. In such cases the individual components of the residual deviance $D(\mathbf{x})$ indicate where the lack of fit arises; $D(\theta)|_{\theta=t_1}$ measures disagreement between the prior and posterior and $D_r(x_r|t_1)$ is the lack of fit contribution of the rth observation, playing a role similar to that of the squared residual in normal models. In this example, $D(\theta) \cong 0$ for all θ and the $D_r(x_r|t_1)$ are given by

8.39, 0.19, 0.56, 0.35, 11.39, 0, 0.93, 0.90, 2.90, 1.61, 1.10, 0.37, 0.24

indicating that the main problem lies with x_5 and, to a lesser extent, x_1. Further elaboration of models to account for such discrepancies is the subject of the remainder of the paper.

3. OUTLIERS AND OUTLIER MODELS

3.1. *Outliers in the exponential family*

Consider a single observation x drawn from the distribution with density (1.1). The behaviour of the posterior distribution for θ given x as x varies across the sample space is qualitatively similar to the behaviour of the posterior when (1.1) is normal (and θ is the normal mean), in the sense that positive account is taken of x no matter how extreme it may be. For instance with the standard conjugate prior $p(\theta) \propto \exp[\alpha_0\theta - \beta_0 a(\theta)]$ all posterior inferences are functions of the updated parameters $\alpha_1 = \alpha_0 + \phi x$ and $\beta_1 = \beta_0 + \phi$ and extreme values of α_1 are obtained when x is extreme and should perhaps be considered an outlier. The posterior mode, for example, is t_1 where $\alpha_1 = \beta_1 a'(t_1)$ and can be written in terms of the prior mode t_0 using

$$a'(t_1) = a'(t_0) + \phi[x - a'(t_0)]/\beta_1.$$

Hence, since $a(\theta)$ is convex in general, t_1 is an increasing function of x indicating that $p(\theta|x)$ favours larger values of θ as x increases.

This can be formalized using the ideas developed by O'Hagan (1979) for outlier analysis when the observations are drawn from a symmetric location distribution. By analogy with O'Hagan, the exponential family distribution with density (1.1) can be shown to be outlier-resistant, meaning that, for any proper prior, the posterior probability

$$F_c(x) = \text{Prob}(\theta \leq c | x)$$

is a decreasing function of x for all values c that θ may take. Clearly the definition of outlier resistance is symmetric since it also implies that $1 - F_c(x)$ is a decreasing function of any decreasing function of x so that smaller values of θ are favoured as x decreases. The proof of the outlier resistance of the exponential family distribution in general is a simple extension of Theorem 2 of O'Hagan's paper and requires only a minor modification to cover discrete models, as follows:

$$F_c(x) = \int m_c(\theta)\exp[\phi\{x\theta - a(\theta)\}]p(\theta)d\theta / \int \exp[\phi\{x\theta - a(\theta)\}]p(\theta)d\theta$$

where

$$m_c(\theta) = \begin{cases} 1 & , \text{if } \theta \leq c, \\ 0 & , \text{if } \theta > c, \end{cases}$$

and $p(\theta)$ is the prior.

In the discrete models $F_c(x)$ takes discrete values as a function of x but is defined also when x is viewed as continuous. For example in the binomial and Poisson cases, $F_c(x)$ is defined and differentiable for $0 < x < 1$ and $0 < x$ respectively. O'Hagan's Theorem 2 can now be followed to conclude that

$$\frac{d}{dx} F_c(x) = \text{cov}[\theta, m_c(\theta) | x]$$

and this is less than or equal to zero for each c. The result follows.

Clearly if extreme observations are to be protected against, then the exponential family likelihood must be either abandoned or modified to accomodate them. In generalized linear regression models the likelihoods suffer a similar sensitivity to outliers and anomalous observations can be obscured by the complex nature of the model making them difficult to detect by simple examination of the data. Formal methods of outlier detection and accommodation, so well developed and regularly utilised in standard linear models, are required for general exponential family models and a simple model is now discussed.

3.2. *Outlier modelling*

The analysis of linear, nominally normal regression models using heavy-tailed symmetric location/scale error distributions is discussed in West (1983). In particular the use of Student t models leads to "good" observations being treated as they would be in a normal model with outliers being downweighted and, ultimately, contributing little to the posterior distributions of interest. To motivate the generalization to exponential family models, consider the influence function (or score function) of a general likelihood $p(x|\theta)$, given

$$g(x|\theta) = \frac{d}{d\theta} \log p(x|\theta). \tag{3.1}$$

The influence function occupies a central role in both Bayesian and non-Bayesian studies of outliers (Ramsay and Novick, 1980) and enters into the additive re-expression of Bayes theorem which, assuming differentiability, defines the posterior score function by

$$\frac{d}{d\theta} \log p(\theta|x) = \frac{d}{d\theta} \log p(\theta) + g(x|\theta) \tag{3.2}$$

where $p(\theta)$ is the prior. So $g(x|\theta)$ determines the first order influence of the observation x on $p(\theta|x)$ at any point θ and protection against extreme observations requires that $g(x|\theta)$ be bounded in x for all θ and, for fixed θ, that $g(x|\theta)$ tends to zero at the extremes if such outlying values are to be totally ignored.

The exponential family density (1.1) with $\phi = 1$ has influence function $x - a'(\theta)$. In the case of nominally normal models for which $a'(\theta) = \theta$, outliers are accommodated using an alternative heavy-tailed distribution with a bounded and redescending score function such as the Student t model with k degrees of freedom, for which

$$g(x-\theta) = (k+1)(x-\theta) / [k + (x-\theta)^2]. \tag{3.3}$$

In West (1983), the construction of such heavy-tailed models as scale mixtures of normal distribution is discussed. In the Student t example, $p(x|\theta)$ is a mixture of the conditional normal distributions, $(x|\theta,\phi) \sim N(\theta,\phi)$, with respect to the precision ϕ viewed as a random quantity such that $k\phi \sim \chi_k^2$. The influence function in (3.3) is then given by

$$g(x-\theta) = E[\phi(x-\theta)|x,\theta] = \bar{\phi}(x-\theta).(x-\theta)$$

where $\bar{\phi}(x-\theta)$ is the mean of the posterior for ϕ given x and θ, which is given by

$$\{k + (x-\theta)^2\}.(\phi|x,\theta) \sim \chi_{k+1}^2.$$

Suppose, by analogy, that the parameter ϕ of the general exponential family distribution (1.1) is viewed as a random quantity with some prior distribution and that $p(x|\theta)$ is the mixture with respect to this prior. Then $p(x|\theta)$ has influence function

$$\begin{aligned} g(x|\theta) &= E[\phi\{x - a'(\theta)\}|x,\theta] \\ &= \bar{\phi}(x,\theta).\{x - a'(\theta)\} \end{aligned} \tag{3.4}$$

where, again, $\bar{\phi}(x,\theta)$ is the mean of the posterior for ϕ given x and θ. Following the discussion of section 2, suppose that the approximate likelihood of (2.4),

$$m(x|\theta,\phi) \propto \phi^{1/2} \exp[-\phi D(x|\theta)/2] \tag{3.5}$$

is adopted. A convenient prior for ϕ is a chi-square distribution and, by analogy with the Student t model, if $k\phi \sim \chi_k^2$, $k > 0$, then

$$m(x|\theta) = \int_0^\infty m(x|\theta,\phi)\, p(\phi) d\phi$$
$$\propto [k + D(x|\theta)]^{-(k+1)/2} \tag{3.6}$$

with influence function as in (3.4) where, now,

$$\bar{\phi}(x,\theta) = (k+1)/[k+D(x|\theta)] \,. \tag{3.7}$$

The likelihood $m(x|\theta)$ is a heavy-tailed alternative to the standard exponential family model and, in the normal, gamma, binomial and Poisson cases, the corresponding influence functions have the desirable bounded and redescending forms. As an example, if the deviance is that of the binomial model, $x \sim B(20, 0.25)$, the influence function $g(x|\theta)$ and the weight function $\bar{\phi}(x,\theta)$ are as shown in Figure 1 in the case $k=5$. Notice that the forms of these functions are similar to the corresponding functions of the Student t distribution when $D(x|\theta) = (x-\theta)^2$ although, of course, in the binomial model they are both skewed and discrete. Values of x near the sampling mean of 5 are treated approximately as in the standard binomial model, with $g(x|\theta)$ roughly linear in x and $\bar{\phi}(x,\theta)$ near unity. As $|x-5|$ increases the weight $\bar{\phi}(x,\theta)$ decays and $|g(x|\theta)|$ reaches a maximum before redescending. Similar influence and weight functions are obtained when $D(x|\theta)$ is the Poisson or gamma deviance, corresponding to heavy-tailed alternatives to the most commonly used exponential family models. Further, the use of alternative prior distributions for ϕ will provide a range of models analogous to the range of unimodal and symmetric distributions constructed as scale mixtures of normal distributions (West 1982, 1983). The chi-square prior is particularly convenient and is used as an example in the next section.

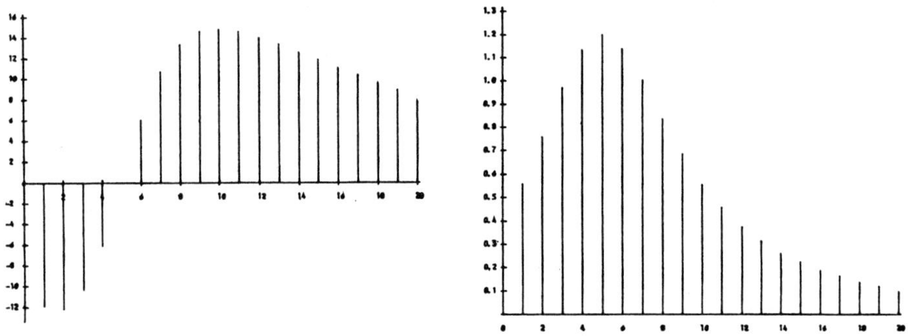

1.a 1.b

FIG. 1. *Influence function (a) and corresponding weight function (b) of the B(20, 0.25) based heavy-tailed model with $k=5$.*

3.3. Accommodation of outliers

Suppose that the likelihood for θ based on observations $\mathbf{x} = \{x_1,\ldots,x_n\}$ is

$$m(\mathbf{x}|\theta) \propto \prod_{r=1}^{n} [k + D_r(x_r|\theta)]^{-(k+1)/2} \tag{3.8}$$

where $D_r(x|\theta)$ is the deviance of (2.10) with known scale parameter n_r, and let the prior for

θ be the usual conjugate form $p(\theta) \propto \exp[\alpha_0\theta - \beta_0\, a(\theta)]$. The posterior $p(\theta\,|\,\mathbf{x})$ is, of course, somewhat intractable in general. The modal equation is given by

$$\alpha_1(\theta) = \beta_1(\theta).a'(\theta)\,, \tag{3.9}$$

where

$$\alpha_1(\theta) = \alpha_0 + \sum_{r=1}^{n} \bar{\phi}_r(x_r,\theta).n_r x_r \tag{3.10}$$

and

$$\beta_1(\theta) = \beta_0 + \sum_{r=1}^{n} \bar{\phi}_r(x_r,\theta).n_r\,, \tag{3.11}$$

with $\bar{\phi}_r(x,\theta)$ defined by (3.7) with $D(x\,|\,\theta)$ replaced by $D_r(x\,|\,\theta)$. If $p(\theta\,|\,\mathbf{x})$ is unimodal at t_1, say, then t_1 is also the mode of a posterior based on observations from the standard model (1.1) with each observation having its own scale parameter $\bar{\phi}_r(x_r,t_1).n_r$. The discounting of extreme observations is naturally achieved with the contribution of x_r to $\alpha_1(t_1)$ and $\beta_1(t_1)$ decaying accordingly. In many applications with a small proportion of extreme observations, most of the weights $\bar{\phi}_r(x_r,t_1)$ will be near unity with the outliers receiving much smaller weights. As a consequence $p(\theta\,|\,\mathbf{x})$ will be unimodal at t_1 and the distribution proportional to $\exp[\alpha_1(t_1)\theta - \beta_1(t_1)a(\theta)]$ may be a useful, tractable, approximation.

Actually the posterior information at t_1 is given by

$$\beta_1(t_1).a''(t_1) - 2(k+1)^{-1} \sum_{r=1}^{n} \{\bar{\phi}_r(x_r,t_1)n_r[x_r - a'(t_1)]\}^2 \tag{3.12}$$

which is always less than or equal to $\beta_1(t_1).a''(t_1)$ reflecting the fact that the above approximation underestimates the dispersion of $p(\theta\,|\,\mathbf{x})$; a better approximation is obtained by choosing α_1 and β_1 so that both the mode and the information coincide.

Following West (1983), the structure of $p(\theta\,|\,x)$ can be examined as follows.

The model (3.8) can be written as

$$m(\mathbf{x}\,|\,\theta) \propto \sum_{r=1}^{n} m(x_r\,|\,\theta)$$

where

$$m(x_r\,|\,\theta)\; \alpha \int_0^\infty \phi_r^{1/2}.\exp[-\phi_r D_r(x_r\,|\,\theta)/2].p(\phi_r)d\phi_r \tag{3.13}$$

and $p(\phi_r)$ is the density of ϕ_r when $k\phi_r \sim \chi^2_k$. Conditional on $\phi = \{\phi_1,\ldots,\phi_k\}$ then, the likelihood is similar to the standard exponential family form with each x_r having a scale parameter ϕ_r and the ϕ_r are implicitly independent and independent of θ. The joint posterior for θ and ϕ is then

$$p(\theta,\phi\,|\,x) \propto p(\theta).\left[\prod_{r=1}^{n} \phi_r^{(k+1)/2-1}\right].\exp[-\sum_{r=1}^{n} \phi_r\{k+D_r(x_r\,|\,\theta)\}/2] \tag{3.14}$$

with the properties

(i) $$p(\theta\,|\,\phi,\mathbf{x}) \propto \exp\left[(\alpha_0 + \sum_{r=1}^{n} \phi_r\, n_r\, x_r)\,\theta - (\beta_0 + \sum_{r=1}^{n} \phi_r\, n_r)a(\theta)\right] \tag{3.15}$$

(ii) $$p(\phi\,|\,\theta,\mathbf{x}) = \prod_{r=1}^{n} p(\phi_r\,|\,\theta,\mathbf{x}) \text{ and } [k+D_r(x_r\,|\,\theta)](\phi_r\,|\,\theta,\mathbf{x}) \sim \chi^2_{k+1}\,; \tag{3.16}$$

(iii) $p(\theta\,|\,\mathbf{x})$ is then a mixture of the conditionally conjugate distributions (3.15) with respect to $p(\phi\,|\,\mathbf{x})$ and approximating by a conjugate form with the same mode/information for convenience is one of several possible approximate schemes for mixtures.

Calculation of t_1 can easily be done iteratively. Starting with $\phi_r = 1$ the prior mean of ϕ_r, for each r, the values of α_1 and β_1 in (3.10) and (3.11) substituted in (3.9) gives the first approximation to t_1 as the posterior mode for the standard analysis. Using this to calculate new weights using (3.7) and iterating generally achieves convergence in three or four steps. West (1983) discusses the significance of the weights in the symmetric location problem in which the influence function reaches the maximum and begins to redescend when the weight $\bar{\phi}_r(x_r,t_1)$ equals $(k+1)/2k$. At this point the contribution of x_r to the information function (3.12) becomes negative and the observation begins to be discounted. It is convenient to use this value as a rule of thumb in judging the results of an analysis; in the symmetric location problem the cut off corresponds to a residual x_r-t_1 of $\pm\sqrt{k}$. So with $k=9$ a residual of ± 3 is judged extreme and the corresponding weight is 0.56. With $k=5$ the residual value is 2.24 and the weight is 0.6.

Example 3.1. Consider the binomial case with $x_r = y_r/n_r$ where $(y_r|\pi,n_r) \sim B(n_r,\pi)$ and $\theta = \log[\pi/(1-\pi)]$. The data of example 2.2 was analysed using the above model with $k=9$; the likelihood is then very similar to binomial just as the Student t distribution with 9 degrees of freedom is similar to the normal. Taking α_0 and β_0 to be near zero, the value of $e^{t_1}/[1+e^{t_1}]$ is 0.771 compared with 0.754 from the standard model and the posterior information (3.12) is 62.43 whereas $\beta_1.a''(t_1) = 78.53$. The weights $\bar{\phi}_r(x_r,t_1)$ are all greater than 0.9 but for the cases $r=1$ and 5, when the values are 0.77 and 0.45 respectively. The fifth observation is clearly an outlier, $x_5 = 0.54$ with $n_5 = 52$; whereas $x_1 = 0.57$ and $n_1 = 21$, though somewhat discounted, is not judged too extreme. Although x_1 and x_5 are similar, the precision of x_5 is proportional to $n_5 = 52$ which is roughly 2.5 times that of x_1, so a smaller weight is required for the more precise, hence more influential, contribution to the likelihood.

The simple outlier accommodating analysis has identified why the binomial model of example 2.2 is untenable and provides a guide to choosing a more elaborate model, if desirable. The change point analysis of Smith (1979) confirms the conclusion that the data is essentially a homogeneous binomial sample if x_5 is omitted. In more complex regression models this type of investigation is an invaluable aid to identifying strange and interesting observations or groups of observations, discounting their influence in the current model and indicating appropriate modifications to that model. Pregibon (1982) has recently investigated a maximum likelihood approach to outlier accommodation in logistic regression and this has features similar to the above analysis when extended to more general regression problems; this extension is discussed in the next section.

4. PRIOR DISTRIBUTIONS AND MODELLING IN REGRESSION

4.1. *Analyses with normal priors*

In the special case of linear regression on the natural parameter, $\theta_r = \mathbf{h}_r^T\beta$ in the model of (1.4), the likelihood is given by

$$p(\mathbf{x}|\beta,\phi_0) \propto \exp\left[\phi \sum_{r=1}^{n} n_r\{x_r(\mathbf{h}_r^T\beta) - a(\mathbf{h}_r^T\beta)\}\right] \tag{4.1}$$

where $\phi>0$ and $n_r>0$, $r=1,\ldots,n$, are known scale parameters. The analysis of this special case illustrates the essential features of more general models for which $\theta_r = f(\mathbf{h}_r^T\beta)$ with $f(.)$ a given non-linear function. Of course the special models (log linear and logistic, for example) are often adequate; they are also attractive for their simplicity and by virtue of the unimodality properties of (4.1) in β when H has full rank p (Wedderburn, 1976).

Simple conjugate prior distributions for β are available only in very special cases; for example with one-way classifications when the data forms a set of independent random samples and the corresponding independent likelihood components each depend on a single β_j. Even in this case the conjugacy is only useful if the hyperparameters of the prior for β are known. With unknown hyperparameters the analysis is, in general, intractable and particularly so when the dimension p of β is relatively large. As an alternative, use will be made of normal prior distributions that lead to posteriors that, though still complex, have some useful, simplifying features. Leonard (1972) has used normal prior distribution for the natural parameters in simple binomial models with some success. Clearly the specification of a prior mode and covariance structure or information matrix for β determines the normal prior; these two quantities alone may embody much of the information required for the prior and often the actual form will be unimportant. In such cases the normal model is a convenient and tractable approximation.

Suppose then, that the prior is taken as

$$p(\beta) \propto \exp\left[-(\beta-\mathbf{b}_0)^T B_0(\beta-\mathbf{b}_0)/2\right] \tag{4.2}$$

denoted by $\beta \sim N(\mathbf{b}_0, B_0)$, where $\mathbf{b}_0$ and B_0 are known. The posterior $p(\beta|\mathbf{x},\phi)$ will be unimodal when the likelihood (4.1) is unimodal as it is in most special cases of interest (Wedderburn, 1976), In such cases, let $\mathbf{b}_1$ be the posterior mode. Define the posterior score and information functions by

$$g(\beta|\mathbf{x},\phi_0) = \frac{d}{d\beta} \log p(\beta|\mathbf{x},\phi_0) \tag{4.3}$$

and

$$G(\beta|x,\phi_0) = -\frac{d}{d\beta^T} g(\beta|\mathbf{x},\phi_0) \tag{4.4}$$

respectively. Then

$$\mathbf{g}(\beta|x,\phi_0) = \phi_0 \sum_{r=1}^{n} n_r.\,[x_r - a'(\mathbf{h}_r^T\beta)].\mathbf{h}_r - B_0(\beta-\mathbf{b}_0) \tag{4.5}$$

and

$$G(\beta|\mathbf{x},\phi_0) = \phi_0 \sum_{r=1}^{n} n_r.a''(\mathbf{h}_r^T\beta).\mathbf{h}_r\,\mathbf{h}_r^T + B_0 \tag{4.6}$$

$$= \phi_0 H^T A(\beta) H + B_0, \tag{4.7}$$

where $A(\beta)$ is the $n \times n$ diagonal matrix with diagonal elements $n_r\, a''(\mathbf{h}_r^T\beta), r=1,\ldots,n$. The mode $\mathbf{b}_1$ may be calculated iteratively as the limit of the second order gradient algorithm given by

$$\beta_{i+1} = \beta_i + G^{-1}(\beta_i|\mathbf{x},\phi_0).g(\beta_i|\mathbf{x},\phi_0) \tag{4.8}$$

A similar method was used by Leonard (1972) in the case of grouped binomial data and experience with (4.8) has shown that convergence is achieved rapidly in many common models.

Using (4.5 - 4.8) it can be shown that

$$\beta_{i+1} = [\phi_0 H^T A(\beta_i) H + B_0]^{-1}\,[\phi_0 H^T A(\beta_i)\mathbf{y}(\beta_i) + B_0\,\mathbf{b}_0], \tag{4.9}$$

where $\mathbf{y}(\beta)$ is an n-vector with elements

$$y_r(\beta) = \mathbf{h}_r^T\beta + [x_r - a'(\mathbf{h}_r^T\beta)]/a''(\mathbf{h}_r^T\beta) \tag{4.10}$$

for $r=1,\ldots,n$. Then the calculation of $\mathbf{b}_1$ involves the iterative use of standard normal theory recursions; β_{i+1} is the posterior mode for β from a regression of the "normal observations" $\mathbf{y}(\beta_i)$ on $H\beta$ with a "covariance" matrix $A(\beta_i)^{-1}$. This is a direct generalization of the standard maximum likelihood method of Nelder and Wedderburn (1972) and, if B_0 is small, $\mathbf{b}_1$ is approximately equal to the usual maximum likelihood estimate of β. In many cases inferences about β may be based on the asymptotic normal distribution, $(\beta\,|\,\mathbf{x},\phi) \sim N(\mathbf{b}_1,B_1)$, where $B_1 = G(\mathbf{b}_1|\mathbf{x},\phi)$.

The extension of this analysis to the usual hierarchical prior structure of normal models (Lindley and Smith, 1972) is immediate. The general form is given by $\mathbf{b}_0 = F\mu$, where μ is a $q-$vector of unknown hyperparameters and F is a known $p \times q$ transfer matrix $(q<p)$. Supposing that the prior for μ is an improper reference form $p(\mu) \propto$ constant, then the marginal prior for β is approximately given by

$$p(\beta) \propto \exp[\,-\,(\beta - F\mathbf{m})^T B_0 (\beta - F\mathbf{m})/2] \tag{4.11}$$

where $\mathbf{m} = (F^T B_0 F)^{-1} F^T B_0\, \beta$ is the posterior mode of μ given β. Rewriting (4.11) gives

$$p(\beta) \propto \exp\,[\,-\beta^T E^T B_0 E \beta/2] \tag{4.12}$$

where $E = I - F(F^T B_0 F)^{-1} F^T B_0$. Hence the equations (4.5-9) hold with B_0 replaced by $E^T B_0 E$ and $\mathbf{b}_0$ replaced by $\mathbf{0}$ throughout. Some special cases of this model are considered later. Applications, of course, require either the specification of B_0 or an analysis involving learning about unknown scale and/or covariance parameters and some particular cases are now examined.

4.2. *Unknown scale parameters in prior and likelihood*

Suppose initially that $\mathbf{b}_0$ and B_0 of (4.2) are specified but that the scale parameter of the likelihood is unknown. Following the discussion of section 2 the approximate scaled likelihood is taken as

$$m(\mathbf{x}\,|\,\beta,\phi) \propto \phi^{n/2} \exp\,[-\phi D(\mathbf{x}\,|\,\beta)/2] \tag{4.13}$$

where

$$D(\mathbf{x}\,|\,\beta) = \sum_{r=1}^{n} n_r . D_r(x_r\,|\,\theta_r) \tag{4.14}$$

and $D_r(x_r\,|\,\theta_r)$ is given by (2.10) with $\theta_r = \mathbf{h}_r^T\beta$. A conjugate analysis is obtained as in section 2.3 by scaling the normal prior with ϕ, giving a prior

$$p(\beta\,|\,\phi) \propto \phi^{p/2} \exp[-\phi D(\beta)/2] \tag{4.15}$$

where $D(\beta) = (\beta - \mathbf{b}_0)^T B_0 (\beta - \mathbf{b}_0)$ is the prior deviance. In this case

$$p(\beta\,|\,\mathbf{x},\phi) \propto \phi^{p/2} \exp[-\phi D(\beta\,|\,\mathbf{x})/2] \tag{4.16}$$

where $D(\beta\,|\,\mathbf{x}) = -2 \log[p(\beta\,|\,\mathbf{x},\phi=1)/p(\mathbf{b}_1|\,\mathbf{x},\phi=1)]$ is the posterior deviance as in the special case of (2.17). Again the deviance identity

$$D(\beta) + D(\mathbf{x}\,|\,\beta) = D(\beta\,|\,\mathbf{x}) + D(\mathbf{x}) \tag{4.17}$$

holds, with $D(\mathbf{x}) = [D(\beta) + D(\mathbf{x}\,|\,\beta)]_{\beta=b_1}$ being the residual deviance of the standard model for which $\phi = 1$. Often a quadratic approximation to $D(\beta\,|\,\mathbf{x})$ about $\mathbf{b}_1$ is adequate in which case (4.16) is normal.

The conjugate analysis is completed if ϕ has a chi-square prior, $\delta_0\phi \sim \chi^2_{\nu_0}$, say, as in section 2.3. For then

(i)
$$p(\beta\,|\,\mathbf{x}) \propto [\delta_1 + D(\beta\,|\,\mathbf{x})]^{-(\nu_1+p)/2} \tag{4.18}$$

where $\nu_1 = \nu_0 + n$ and $\delta_1 = \delta_0 + D(\mathbf{x})$. If $D(\beta\,|\,\mathbf{x})$ is approximately quadratic, (4.18) is a multivariate t distribution with ν_1 degrees of freedom and precision matrix scales by $\bar{\phi} = (\nu_0+n)/[\delta_0+D(\mathbf{x})]$.

(ii) $p(\phi|\mathbf{x}) = p(\beta,\phi|\mathbf{x})/p(\beta|\phi,\mathbf{x})$ gives $\delta_1(\phi|\mathbf{x}) \sim \chi^2_{\nu_1}$ so that $\bar{\phi}$ is the posterior mean of ϕ.

Clearly this analysis parallels the standard normal linear modelling results with deviances replacing quadratic forms throughout. Similar results hold if the standard improper prior $p(\beta,\phi) \propto \phi^{-1}$ is adopted as in Sweeting (1982); simply replace ν_1 by $n-p$ and set $\delta_0 = D(\beta) = 0$, for all β, in equations (4.16) onwards.

Of course the assumption that B_0 is completely known is somwhat unrealistic. The simplest useful extension is to suppose that the prior has a single unknown scale parameter γ, say, independent of the scale of the likelihood, thus

$$p(\beta|\gamma) = p(\beta|\gamma,\phi) \propto \gamma^{p/2}\exp[-\gamma D(\beta)/2]\,, \tag{4.19}$$

with $D(\beta)$ as in (4.15). Now, of course, there is no conjugate analysis. If ϕ is also unknown, a useful prior for ϕ and γ is given by taking them independent, with $\delta_0\,\phi \sim \chi^2_{\nu_0}$ and $\tau_0\,\gamma \sim \chi^2_{\sigma_0}$, for some δ_0, ν_0, τ_0 and σ_0. Then, integrating $m(\mathbf{x}|\beta,\phi)$ of (4.13) with respect to $p(\phi)$ gives

$$m(\mathbf{x}|\beta) \propto [\delta_0+D(\mathbf{x}|\beta)]^{-(\nu_0+n)/2} \tag{4.20}$$

and, similarly, integrating (4.19) with respect to $p(\gamma)$ gives

$$p(\beta) \propto [\tau_0+D(\beta)]^{-(\sigma_0+p)/2}\,. \tag{4.21}$$

Hence $p(\beta|\mathbf{x})$ has score function

$$\bar{\phi}(\beta)\,.\sum_{r=1}^{n} n_r[x_r-a'(\mathbf{h}_r^T\beta)]\mathbf{h}_r - \bar{\gamma}(\beta).B_0(\beta-\mathbf{b}_0) \tag{4.22}$$

where

$$\bar{\phi}(\beta) = (\nu_0+n)/[\delta_0+D(\mathbf{x}|\beta)] = E[\phi|\mathbf{x},\beta]\,, \tag{4.23}$$

and

$$\bar{\gamma}(\beta) = (\sigma_0+p)/[\tau_0+D(\beta)] = E[\gamma|\mathbf{x},\beta]\,. \tag{4.24}$$

This score is similar to that in (4.5), with ϕ replaced by the conditional mean given $\mathbf{x}$ and β and B_0 scaled by the conditional mean of γ given β (since $p(\gamma|\mathbf{x},\beta) = p(\gamma|\beta)$). The corresponding information function is given by

$$[\bar{\phi}(\beta).H^TA(\beta)H + \bar{\gamma}(\beta).B_0] - S(\beta) \tag{4.25}$$

where

$$S(\beta) = 2.\bar{\phi}(\beta)^2(\nu_0+n)^{-1}\mathbf{u}(\beta).\mathbf{u}(\beta)^T + 2.\bar{\gamma}(\beta)^2(\sigma_0+p)^{-1}\mathbf{v}(\beta)\mathbf{v}(\beta)^T \tag{4.26}$$

with $\mathbf{u}(\beta) = \sum_{r=1}^{n} n_r[x_r-a'(\mathbf{h}_r^T\beta)]\mathbf{h}_r$ and $\mathbf{v}(\beta) = B_0(\beta-\mathbf{b}_0)$.

So although the posterior mode $\mathbf{b}_1$ coincides with that from an analysis with known scale parameters $\bar{\phi}(\mathbf{b}_1)$ and $\bar{\gamma}(\mathbf{b}_1)$, the corresponding information matrix (in square brackets in (4.25) when $\beta=\mathbf{b}_1$) must be corrected by the negative definite $S(\mathbf{b}_1)$ to account for the uncertainty about ϕ and γ. Finally note that, if ϕ is known, as in the standard binomial and Poisson models, then the above analysis can be used with the known value of ϕ specified and the first term of $S(\beta)$ in (4.26) set to zero; formally, let ν_0 and δ_0 tend to infinity subject to $\nu_0/\delta_0 = \phi$.

The simplest example is that of grouped data, p groups with r_j observations in the jth group, $(j=1,\ldots,p)$, $\sum_{j=1}^{p} r_j=n$. Relabelling the observations x_{jk} and the known scale parameters n_{jk}, gives

$$p(\mathbf{x}|\beta,\phi) \propto \prod_{j=1}^{p} \exp[\phi\{y_j\beta_j - z_j a(\beta_j)\}] \tag{4.27}$$

where $y_j = \sum_{k=1}^{r_j} n_{jk} x_{jk}$ and $z_j = \sum_{k=1}^{r_j} n_{jk}$.

Now, taking the β_j to be independently normally distributed with mode μ and precision γ and adopting an improper uniform reference prior for μ gives a marginal prior $p(\beta|\gamma)$ of the form of (4.12) where now $E = I - \mathbf{1}\mathbf{1}^T p^{-1}$. In this case, the score function (4.22), with $B_0 = E$ and $\mathbf{b}_0 = 0$, can be used to show that $\mathbf{b}_1$ is the solution in $\beta = (\beta_1,\ldots,\beta_p)^T$ of

$$\bar{\phi}(\beta).z_j.[a'(\beta_j) - a'(\hat{\beta}_j)] = \bar{\gamma}(\beta).(\bar{\beta}_0 - \beta_j) \tag{4.28}$$

where $y_j = z_j.a'(\hat{\beta}_j)$ so that $\hat{\beta}_j$is the maximum likelihood estimate of β_j, $(j=1,\ldots,n)$. This equation illustrates the shrinkage effect induced by the prior; since $a'(\beta_j) - a'(\hat{\beta}_j) = a''(\eta_j).(\beta_j - \hat{\beta}_j)$, where η_j lies between β_j and $\hat{\beta}_j$, (4.27) becomes

$$\beta_j = w_j.\hat{\beta}_j + (1-w_j)\bar{\beta} \tag{4.29}$$

where $w_j = \bar{\phi}(\beta).z_j.a''(\eta_j)/[\bar{\phi}(\beta)z_j.a''(\eta_j) + \bar{\gamma}(\beta)]$, so that $0 < w_j < 1$ for each j.

Example 4.1. The model above reduces to that of Leonard (1972) when the data is binomial $(y_j|\beta_j) \sim B(z_j,\pi_j)$ where $\beta_j = \log[\pi_j/(1-\pi_j)]$, and $\phi = 1$. This model was fitted to the Lindisfarne scribes' data discussed earlier, in an elaboration of the simple model of example 2.2. Taking σ_0 and τ_0 to be near zero leads to $\bar{\gamma}(\mathbf{b}_1) = 5.95$ and, in this case, there is actually very little shrinkage of the elements of $\mathbf{b}_1$ from the maximum likelihood values. A larger value of $\bar{\gamma}$ would lead to more shrinkage, of course, and this is discussed further in a later example. Also, in this case, $\bar{\beta} = 1.203$ with $e^{\bar{\beta}}/(1+e^{\bar{\beta}}) = 0.769$.

Example 4.2. Cox and Snell (1981, p. 143) discuss data concerning the intervals between failures of air-conditioning equipment in ten aircraft and fit ten separate gamma distributions, one for each aircraft. This can be viewed as a special one-way classification and the above analysis can be used with a minor modification to include separate scale parameters for each of the $p=10$ groups. This simply replaces ϕ in (4.27) by ϕ_j, $n_{jk} = 1$ $z_j = r_j$ and $y_j = \sum_{k=1}^{r_j} x_{jk}$, $(j=1,\ldots,p)$. Then, assuming the ϕ_j to be independent with common prior given by $\delta_0\phi_j \sim \chi^2_{\nu_0}$, $j=1,\ldots,p$, the modal equations for the β_j are (4.28) with $\bar{\phi}(\beta)$ replaced by

$$\bar{\phi}_j(\beta) = (\nu_0 + r_j)/[\delta_0 + \sum_{k=1}^{r_j} D(x_{jk}|\beta_j)]\,,$$

for each j.

For this data, the nominal model was taken as exponential with μ_j the exponential mean of the jth group, and the scaled likelihoods can then be viewed as approximations to the gamma models used by Cox and Snell. The results reported below are based on a normal prior for the real valued parameters $\beta_j = -\log(\mu_j)$ rather than the natural parameters $-\mu_j^{-1}$, with the appropriate minor modifications to the analysis outlined above. The columns of Table 1 headed μ_j give the values of $\exp(-b_j)$ where b_j is the posterior mode of β_j from the relevant analysis; the columns headed ϕ_j give the corresponding values of $\bar{\phi}_j$. In both examples (i) and (ii), ν_0 and δ_0 were close to zero. In (i) there is no shrinkage ($\gamma = 0$, $\sigma_0 = \tau_0 = \infty$) so that the μ_j are the maximum likelihood estimate given by Cox and Snell.

The figures in brackets following the ϕ_j column are the exact maximum likelihood estimates of the group scale parameters from the gamma model for comparison with the scaled likelihood analysis. In (ii) there is shrinkage with $\sigma_0=0.01$ and $\tau_0=1$ resulting in $\bar{\gamma}=7.34$. There is marked shrinkage in the μ_j column, particularly for the final group where μ_{10} is shrunk from the maximum likelihood value of 200 down to 118-8.

TABLE 1

Analyses of aircraft air-conditioning data

(i) No shrinkage					(ii) Shrinkage	
j	r_j	μ_j	ϕ_j		μ_j	ϕ_j
1	30	59.6	0.69	(0.81)	66.9	0.68
2	24	64.1	0.92	(1.06)	70.1	0.92
3	27	76.8	1.00	(1.13)	79.6	1.00
4	16	82.0	1.60	(1.75)	83.9	1.58
5	19	83.5	1.52	(1.67)	84.5	1.54
6	23	95.7	0.83	(0.97)	94.0	0.64
7	12	108.1	0.59	(0.71)	99.0	0.58
8	15	121.3	0.78	(0.83)	108.8	0.78
9	14	130.9	1.46	(1.61)	119.1	1.43
10	9	200.0	0.36	(0.46)	118.8	0.32

It is of interest to compare the shrinkage effects in examples 4.1 and 4.2. In the former very little shrinkage occurs corresponding to a large value of the sum of squares $D(\beta)$ = $= \sum_{j=1}^{p} (\beta_j-\bar{\beta})^2$ when $\beta=\mathbf{b}_1$. For the binomial data of this example, observation number five is quite separate from the majority of the data but is also very precise, $n=52$, (see example 2.2) and so has a large influence on the analysis. This, and the relatively large spread of the other groups, leads to an inflated value of $D(\beta)$. Hence the lack of shrinkage, or under-shrinkage. By comparison in example 4.2, although group 10 is extreme as indicated by the maximum likelihood estimate of μ_{10}, it plays only a restricted role in the analysis due to both the small number of observations in the group ($r_{10}=9$) and the comparatively large spread ($\bar{\phi}_{10}=0.36$). This results in a relatively small value of $D(\mathbf{b}_1)$ and marked shrinkage, particularly for the imprecise group 10. In this case, the shrinkage of the extreme group is desirable since there is a relatively large degree of uncertainty associated with that group; the posterior for β_{10} given data from group ten alone is much more diffuse than the corresponding posteriors for the other β_j. In cases where an extreme, or outlying, group is more precise it is important to consider that group separately and apply shrinkage to the other, homogeneous, groups. In example 4.1 there is already a relatively large spread amongst the groups without the outlying groups 1 and 5 and thus $\bar{\gamma}$ is rather small resulting in under-shrinkage. If, however, the homogeneous groups are very close together the above analysis can result in a rather small value of $\Sigma(\beta_j-\bar{\beta})^2$ and hence a large value of $\bar{\gamma}$ leading to overshrinkage as discussed in West (1983). In such a case the weights w_j of (4.29) are relatively small so there is marked shrinkage towards the group mean $\bar{\beta}_0$ which is itself over-influenced by the outlying group. Thus differences between groups, particularly concerning extreme groups, tend to be obscured as a result.

Those problems can be overcome using a slightly more flexible prior model and this is discussed in the next section.

4.3. *Scale parameters for individual model components*

The normal prior is clearly inadequate in the one-way model just discussed when one, or a small proportion, of the groups are distinct from the rest. The problem concerning under and over-shrinkage are induced by the normal prior and can be avoided by using a prior with more weight in the tails as in modelling outlying observations. In the simple one-way model this can be done simply by replacing the nominal normal prior, $(\beta_j|\mu,\gamma) \sim N(\mu,\gamma)$ independently, with

$$p(\beta|\mu,\gamma) = \prod_{j=1}^{p} \gamma^{1/2} p[\gamma^{1/2}(\beta_j-\mu)] \, . \tag{4.30}$$

where $p(.)$ is the density of a suitable heavy-tailed, unimodal and symmetric distribution. Again the Student t distributions are particularly convenient and will be used throughout this section; the analysis extends easily to general scale mixtures of normal distributions, of course (West, 1982, 1983).

Suppose then, that $p(\beta) \propto [k+\beta^2]^{-(k+1)/2}$, $k>0$. Then (4.30) can be written as

$$p(\beta|\mu,\gamma) \propto \prod_{j=1}^{p} \int_0^\infty \gamma^{1/2}\gamma_j^{1/2}. \exp[-\gamma\gamma_j(\beta_j-\mu)^2/2].p(\gamma_j)d\gamma_j$$

where $p(\gamma_j)$ is the density of the scale parameter γ_j when $k\gamma_j \sim \chi_k^2$. Conditional on $\gamma = \{\gamma_1,\ldots,\gamma_p\}$, the analysis follows that of section 4.2 with $\gamma_j\bar{\gamma}(\beta)$ replacing $\bar{\gamma}(\beta)$ in equations (4.28-4.29) so that the mode of $p(\beta|\mathbf{x},\gamma_0)$ satisfies

$$\beta_j = w_j.\hat{\beta}_j + (1-w_j)\beta^*, \tag{4.31}$$

where

$$w_j = \bar{\phi}(\beta).z_j.a''(\eta_j)/[\bar{\phi}(\beta)z_ja''(\eta_j)+\gamma_j.\bar{\gamma}(\beta)], \tag{4.31}$$

for some η_j lying between β_j and $\hat{\beta}_j$; also

$$\beta^* = \sum_{j=1}^{p} \gamma_j.\beta_j / \sum_{j=1}^{p} \gamma_j \tag{4.32}$$

and

$$\bar{\gamma}(\beta) = (\sigma_0+p)/[\tau_0+D(\beta)], \tag{4.33}$$

with

$$D(\beta) = \sum_{j=1}^{p} \gamma_j(\beta_j-\beta^*)^2, \tag{4.34}$$

for each j. Thus the scale parameters γ_j can be used to discount the effect of outlying or extreme groups on the overall mean β^* and the deviance $D(\beta)$ in the same way that observational outliers are discounted in section 3.

Notice also that, if γ_j is relatively small, then the shrinkage of the corresponding group mean is reduced since w_j is increased. Thus in cases where one group is separated from the rest, little shrinkage is applied to that group by using a suitably small scale parameter, and the other groups will be shrunk towards β^* which is relatively unaffected by the outlying β_j. The degrees of shrinkage and the values of the γ_j will, of course, be data dependent.

Now, following the development of section 3.3, define

$$\bar{\gamma}_j(\beta) = E[\gamma_j|\beta,\bar{\gamma}(\beta)] = (k+1)/[k+\bar{\gamma}(\beta)(\beta_j-\beta^*)^2] \tag{4.35}$$

and $\bar{\gamma}(\beta) = \{\bar{\gamma}_1(\beta),\ldots,\bar{\gamma}_p(\beta)\}$. Since the calculation of marginal posteriors for β is impossible to perform analytically, suitable conditional distributions must be considered. Inferences based on $p(\beta|\mathbf{x},\mathbf{y})$ for suitable values of $\mathbf{y}$ will provide useful information about β and will generally highlight any strange or interesting features that a simple normal prior would obscure. By analogy with the results of section 3.3 a suitable choice of γ appears to be $\bar{\gamma}(\mathbf{b}_1)$ where $\mathbf{b}_1$ satisfies (4.31) when $\gamma = \bar{\gamma}(\mathbf{b}_1)$. The calculation of $\mathbf{b}_1$ is performed iteratively; for given γ_j, starting with the prior mean of unity, a first approximation to $\mathbf{b}_1$ is calculated via iterative solution of (4.28), corresponding to (4.31). This also gives values of β^* and $D(\beta)$ and so new values of $\bar{\gamma}_j$ may be found using (4.35). In practice, this routine has converged rapidly in many examples, some of which are discussed below.

Example 4.3. Returning to the binomial data of Example 4.1, the above analysis was performed with $k=5$ and, again, $\sigma_0=\tau_0\cong 0$. This gave $\beta^* = 1.277$ with $e^{\beta^*}/[1+e^{\beta^*}] = 0.782$ compared with 0.769 from the unweighted analysis, and $\bar{\gamma}(\mathbf{b}_1) = 10.10$ compared with 5.95. Clearly there is much more shrinkage in this analysis; the weights $\bar{\gamma}_j(\mathbf{b}_1)$ are given by

0.55. 1.08, 0.99 1.19, 0.38, 1.17, 1.15, 1.17, 0.93, 0.98, 1.06, 1.18, 1.19

and so, recalling the interpretation of scale parameter estimates in section 3.3, β_5 and, to a lesser extent, β_1 are indicated as outlying.

Example 4.4. The air-conditioning data of Example 4.2 was analysed using $k=5$ and the same prior for γ, $\sigma_0=0.01$ and $\tau_0=1$. In this case there was very little difference between the normal and Student t based analyses with all the weight $\bar{\gamma}_j(\mathbf{b}_1)$ being greater than unity in the latter. Thus the large amount of shrinkage is justified and the normal prior suitable.

In this simple one-way model, the individual components of the nominally normal prior for $(\beta|\mu,\gamma)$ are replaced by heavy-tailed components which, as scale mixtures of normals, can be viewed as power transformations of the original independent normal components. In more general models there may not be a simple prior with the β_j conditionally independent; the matrix B_0 of (4.2) may not be diagonal. In such cases the parameter vector $\beta' = A\beta$, where $B_0 = A^TA$, has a nominally normal prior with identity precision matrix and so the componentwise weighting of the prior can be applied to β' rather than β (West, 1983).

Finally complete weighting of all the model components simply involves the modelling of both likelihood and prior components with heavy-tailed distributions in order to accommodate observational outliers and protect against misspecification of the nominally normal prior. In the most general case the likelihood has the form

$$m(\mathbf{x}|\beta,\phi,\phi) \propto \prod_{r=1}^{n} \phi^{1/2}\phi_r^{1/2}\exp[-\phi\phi_r D_r(x_r|\theta_r/2]$$

where $\theta_r = \mathbf{h}_r^T\beta$, $r=1,\ldots,n$, and the prior, nominally $N(\mathbf{b}_0,\gamma B_0)$, is replaced by

$$p(\beta|\mathbf{b}_0,B_0,\gamma,\gamma) \propto \prod_{j=1}^{p} \gamma^{1/2}\gamma_j^{1/2}\exp[-\gamma\gamma_j\{\mathbf{a}_j^T(\beta-\mathbf{b}_0)\}^2/2],$$

where $\mathbf{a}_j^T$ is the jth row of A, $B_0 = A^TA$. Inferences can then be based on conditional posteriors for β given $\mathbf{x}$, ϕ and γ, for suitable values of ϕ and γ. Taking $\bar{\gamma}$ as defined in (4.35), the iterative procedure outlined there is directly extended to ϕ, using $\bar{\phi}(\beta) = \{\bar{\phi}_1(\beta),\ldots,\bar{\phi}_n(\beta)\}$, where

$$\bar{\phi}_r(\beta) = (k+1)/[k+\bar{\phi}(\beta).D_r(x_r|\theta_r)], \qquad r=1,\ldots,n,$$

and

$$\bar{\phi}(\beta) = (\nu_0+n)/\left[\delta_0 + \sum_{r=1}^{n} \bar{\phi}_r(\beta).D_r(x_r|\theta_r)\right]$$

Further details can be derived as necessary from the earlier analyses and are not discussed here. The final example below concerns a more complex model using full weighting of individual components as discussed above.

Example 4.5. The data plotted in Figure 2 are the half-hourly counts of incoming telephone calls at the University of Warwick during the week 6th - 12th September 1982, given in the Appendix.

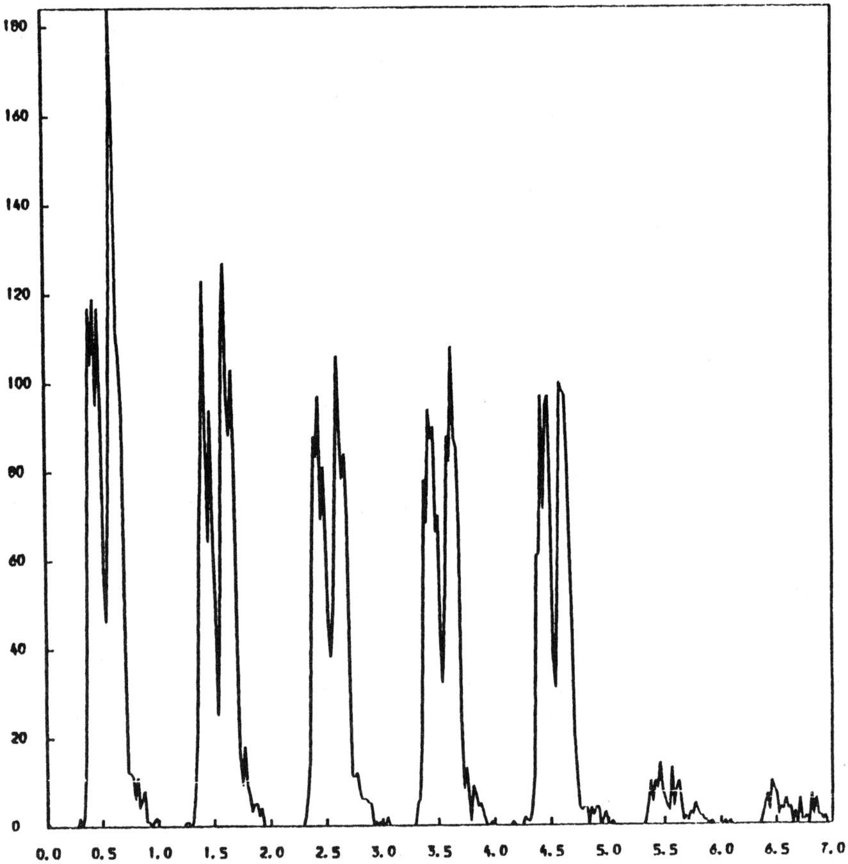

FIG. 2. *Incoming telephone calls (per ½ hour), University of Warwick, September 6th-12th, 1982*

The observations consist of 48 non-negative integers for each of the 7 days, Monday to Sunday inclusive, with more than 20 % being zeroes. Let x_{jk} be the kth observation on day j, ($k=1,\dots,48$, $j=1,\dots,7$). The model fitted for this example supposes that the data are Poisson distributed $x_{jk} \sim P(\mu_{jk})$, where $\theta_{jk} = \log(\mu_{jk})$ is the sum of two components. Firstly, to model the basic form of evolution over the day ignoring the lunchtime dip, a quadratic component,

$$\beta_{0,j} + \beta_{1,j}k + \beta_{2,j}k^2 .$$

Secondly, to model the lunchtime dips and the increased number of calls in late afternoon, a Fourier component,

$$\sum_{t=1}^{T} \{\beta_{2t+1,j}.\cos(2\pi kt/48) + \beta_{2t+2,j}.\sin(2\pi kt/48)\}$$

where T is taken as 5 in this example. The prior for the coefficients of this loglinear model is

$$(\beta_{i,j}|\mu_i,\gamma_i,\gamma_{ij}) \sim N(\mu_i,\gamma_i\gamma_{ij}), \qquad (j=1,\ldots,7)$$

with $5\gamma_{ij} \sim \chi_5^2$, $p(\mu_i,\gamma_i) \propto \gamma_i^{-1}$, with the γ_{ij}, γ_i, and μ_i mutually independent, $(i=0,1,\ldots,2T+2)$. So, unconditional on the γ_{ij}, the $\beta_{i,j}$ are exchangeable across days indexed by j, for each i. The marginal Student t prior is used to detect and highlight major differences if they exist and protect the analysis from under or over-shrinkage. Similarly, outliers are modelled using the outlier accommodating model in the Poisson case, with x_{jk} having scale parameter ϕ_{jk} and $5\phi_{jk} \sim \chi_5^2$ with the ϕ_{jk} independent and independent of all other hyperparameters.

The model fits rather well. The peak in the afternoon on Monday ($x_{1,30} = 184$ and $x_{1,31} = 155$) is considered atypical; $x_{1,30}$ is an outlier with $\bar{\phi}_{1,30} < 0.6$. Concerning the Fourier component, the values of $\bar{\gamma}_{ij}$ are all reasonably close to unity with most of them greater than unity. Thus the assumption that, across days, the coefficients of the Fourier term are exchangeable and normally distributed is satisfactory. Concerning the quadratic term, again most of the $\bar{\gamma}_{ij}$ are relatively large with the exception of those on Sunday, when $j=7$, $i=1,2$. Here $\bar{\gamma}_{1,7} = 0.58$ and $\bar{\gamma}_{2,7} = 0.6$ indicating that the coefficients of the linear and quadratic terms on Sunday are quite different from the rest of the week. The difference between Sunday and the weekdays is not due to the overall level of the number of calls (the form of Saturday is consistent with the weekdays); rather it is due to a displacement of the daily trend and an increase in the spread of the quadratic form.

To see this note that the quadratic term contributes a multiplicative "normal" form,

$$\exp[\beta_{0,j} - \beta_{1,j}^2/(4\beta_{2,j})].\exp[+\beta_{2,j}\{k+\beta_{1,j}/(2\beta_{2,j})\}^2]$$

to the mean of $x_{j,k}$. The second component of this function determines the peak of the trend at $p_j = -\beta_{1,j}/(2\beta_{2,j})$ and the spread is measured by $\beta_{2,j}$. From the fitted model, the values of $\beta_{2,j}$ are, of course, negative, and the fitted values of p_j are given by

30.45, 28.68, 30.0, 29.70, 29.41, 31.33, 32.5

corresponding to peaks at approximately

3.15 p.m., 2.21 p.m., 3.00 p.m., 2.51 p.m., 2.42 p.m., 3.39 p.m., 4.15 p.m.

Secondly the value of the fitted quadratic coefficients $\beta_{2,j}$ are

−0.022, −0.019, −0.018, −0.017, −0.015, −0.008.

The extremeness of the form on Sunday is evident. Due to the weighting with individual scale parameters, Sunday has a restricted influence on the estimation of the hyperparametes μ_i and γ_i; the fitted quadratic term using the μ_i peaks at 2.53 p.m. and has spread measured by the fitted value of $\mu_2 = -0.018$.

Finally, a further analysis with seven additional scale parameters $\phi_1,\ldots,\phi_7$, one for each day, to assess the goodness of fit of the model produces very similar results. Using independent reference priors $p(\phi_j) \propto \phi_j^{-1}$ for each j leads to values of $\bar{\phi}_j$ given by

0.76, 0.89, 2.04, 0.82, 0.76, 1.82, 0.80.

Values $\phi_j = 1$ correspond to the standard Poisson model. The conditional posteriors for the ϕ_j are χ^2_{48} with means $\bar{\phi}_j$ so that for Wednesday and Saturday values of ϕ_j greater than unity are indicated suggesting that there is somewhat less variation than provided for by the model on those days. On other days the $\bar{\phi}_j$ are less than unity indicating a poorer fit although in each case the values $\phi_j = 1$ are still fairly well supported. There is room for improvement in the model; an alternative analysis of this data using a time dynamic extension of this model appears in West, Harrison and Migon (1983).

APPENDIX

Half-hourly counts of incoming telephone calls at the University of Warwick, Mon. Sept. 6th - Sun. Sept. 12th., 1982. (I am grateful to Mr. Keith Halstead of the University Computer Unit for providing this data).

	1	0	0	0	0	0	0	0	0	0	0	0
	0	0	2	0	2	17	67	117	104	119	95	117
Monday	94	57	46	73	121	184	155	111	106	93	44	12
	12	11	6	11	4	6	8	1	1	0	1	2
	0	0	0	0	0	0	0	0	0	0	0	0
	1	1	0	1	10	21	64	86	123	87	64	94
Tuesday	63	49	25	56	112	127	96	88	103	80	42	16
	9	18	9	7	3	5	5	2	4	0	0	0
	0	0	0	0	0	0	0	0	0	0	0	0
	0	0	0	2	7	15	59	88	83	97	69	81
Wednesday	64	45	38	50	78	106	89	78	84	70	37	11
	11	12	8	6	6	6	5	5	0	1	0	1
	0	0	2	0	0	0	0	0	0	0	0	0
	0	0	0	5	6	20	78	68	94	87	90	66
Thursday	70	42	32	54	88	82	108	87	85	68	25	8
	13	7	1	9	7	4	5	3	1	0	0	1
	0	0	0	0	0	0	0	0	1	0	0	0
	0	2	1	1	3	11	61	61	97	71	95	97
Friday	72	38	31	65	100	98	97	82	62	44	20	11
	4	3	4	4	0	4	2	4	4	0	1	3
	0	0	0	0	0	0	0	0	0	0	0	0
	0	0	0	0	1	5	10	5	10	8	14	8
Saturday	6	4	3	13	4	8	10	5	1	2	1	3
	2	5	3	2	2	1	1	0	1	0	0	0
	0	0	1	0	1	0	0	0	0	0	0	0
	0	0	0	0	0	0	2	4	7	3	10	8
Sunday	7	2	4	3	6	3	4	0	3	0	6	1
	1	2	0	7	2	6	2	-2	1	2	0	0

REFERENCES

AITCHISON, J. and SHEN, S.M. (1980). Logistic-normal distributions: some properties and uses. *Biometrika,* **67**, 261-272.

ALTHAM, P.M.E. (1978). Two generalizations of the binomial distribution. *Appl. Statist.,* **27**, 162-167.

BAKER, R.J. and NELDER, J.A. (1978). *GLIM Release* **3**. N.A.G. Oxford.

COX, D.R. and SNELL, E.J. (1981). *Applied Statistics.* Chapman and Hall.

CROWDER, M.J. (1978). Beta-binomial anova for proportions. *Appl. Statist.,* **27**, 34-37.

LEONARD, T. (1972). Bayesian methods for binomial data. *Biometrika,* **59**, 581-590.

LINDLEY, D.V. (1965). *Introduction to probability and statistics from a Bayesian viewpoint: part 2, Inference.* Cambridge: University Press.

LINDLEY, D.V. and SMITH, A.F.M. (1972). Bayes estimates for the linear model. *J. Roy. Statist. Soc. B.,* **135**, 370-384.

O'HAGAN, A. (1979). On outlier rejection phenomena in Bayes inference. *J. Roy. Statist. Soc. B,* **41**, 358-367.

PREGIBON, D. (1982). Resistant fits for some commonly used logistic models with medical applications. *Biometrics* **38**, 485-499.

RAMSEY, J.O. and NOVICK, M.R. (1980). PLU Robust Bayesian decision theory: point estimation. *J. Amer. Statist. Ass.,* **75**, 901-907.

SMITH, A.F.M. (1979). Change-point problems: approaches and applications. In *Bayesian Statistics.* (J.M. Bernardo et al., eds). Valencia: University Press.

SWEETING, T. (1981). Scale parameters: a Bayesian treatment. *J. Roy. Statist. Soc. B,* **43**, 333-338.

WEDDERBURN, R.W.M. (1974). Quasi-likelihood functions, generalized linear models, and the Gauss-Newton method. *Biometrika,* **61**, 439-448.

— (1976). On the existence and uniqueness of the maximum likelihood estimates for certain generalized linear models. *Biometrika,* **63**, 27-32.

WILLIAMS, D.A. (1982). Extra-binomial variation in logistic linear models. *Appl. Statist.,* **31**, 144-148.

WEST, M. (1982). Aspects of recursive Bayesian estimation. Unpublished PH.D. thesis, University of Nottingham.

— (1983). Outlier models and prior distributions in Bayesian linear regression. University of Warwick *Res. Rep.* N° **37**.

WEST, M., HARRISON, P.J. and MIGON, H.S. (1983). Dynamic generalized linear models and Bayesian forecasting. University of Warwick *Res. Rep.* N° **41**.

DISCUSSION

A.P. DEMPSTER (*Harvard University*)

This is a very interesting paper, packed with details of new models and associated techniques. Despite the clarity of style and generosity with numerical examples, however, and despite several attempts at reading both front to back and back to front, I still find the details very hard to follow. I feel sure that if I could persuade Dr. West to explain slowly to me his premises, goals, and notation, in terms I understand fully, I would be a better equipped Bayesian. As it is, I see that new models for discrete data are being provided which can allow for increased or decreased variance over binomial or Poisson models. And the idea of modulating shrinkage to allow for random effects models with other-than-normal tails is important.

To initiate some dialogue, here are two questions. My experience with implementation of Bayesian modelling techniques is that the computations are very time-consuming to specify and program. Yet no hint of this appears in the examples. Are there easy and justifiable approximations for everything? Secondly, I believe that Bayesian applications are necessarily tied to externally defined problems, and inputs of nontrivial prior knowledge are needed. Yet Dr. West presents Bayesian modelling as an exercise in pure data analysis. How is this possible?

C.A. DE B. PEREIRA (*Universidade de Sào Paulo*)

Suppose that you postulate a continuous model and get a set of observations that are close to each other. Three of them, unfortunately, are so close that you can assume that they have the same value. Would you consider any one of these three as an outlier?

P.R. FREEMAN (*University of Leicester*)

Dr. West's paper is certainly one of the meatiest (or perhaps in deference to the local cuisine, fishiest) of the conference. It will be seen in retrospect, I think, as one of an important series that shed Bayesian light on the very powerful area of generalised linear models that has hitherto been the domain only of frequentist darkness. The paper is full of elegant results, with squared differences replaced by deviances and normals by Student t's etc. It is all achieved at the expense of approximations and ad-hockery and I would need much more time to convince myself of the justification of every step. Let me encourage Dr. West to tackle many more datasets so that we can see how it all works out in practice. The one area where I've had some experience makes me preach more caution and care in looking at posterior distributions than Dr. West had time or space to present in his paper. Teather (1984) used Leonard's (1972) model, itself a special case of equation (4.27), to get smoothed estimates of probabilities for use in Bayesian diagnosis. This turned up very strange properties when data were fairly sparse, in that the amount of shrinkage was very sensitive to small changes in the data. Careful numerical work revealed that this was due to the joint posterior being peculiar, with a ridge and several local modes. Modal estimates such as (4.29) may seduce by their elegance but conceal the truth.

Beethoven once described his late string quartets as "showing slightly less lack of imagination than hitherto". This could apply with almost as much understatement to the two papers in this session.

J.R.M. AMEEN (*University of Warwick*)

I would like to make some comments on Dr. West's interesting paper concerning dynamic rather than static models. In this context the use of power transformations in describing uncertainty has gained attention in the work of Smith (1979) and Souza (1978). These were concerned with Bayesian steady forecasting systems which accommodate non-normal dynamic models parallel to the normal dynamic linear models of Harrison and Stevens (1976).

Although the ideas of this paper deal with "Stationary Statistics", i.e. the study of time independent random variables, they are closely related to those of Ameen and Harrison (1983 a,b) which *could* have been referred to. However, it is interesting to note that the goal of the present paper can be achieved in a wider sense with less penalty:

restriction to the exponential family and natural conjugacy of the prior distributions are not required.

The following is a simple generalised dynamic reconstruction of the structural ideas in the present paper:

Let the observation *pdf* be $f(y_t|\vartheta_t)$ and $f(\vartheta_{t-1}|D_{t-1})=f_{t-1}(\vartheta_{t-1})$ be the posterior *pdf* for the state variable ϑ at time t-1 given D_{t-1} which represents all the available information at that time. Moreover, let φ be a scaling random variable and its posterior *pdf* at time $t-1$ be

$$f(\varphi_{t-1}|D_{t-1})=f_{t-1}(\varphi_{t-1}) \propto \varphi_{t-1}^{\nu_t-1} e^{-\delta_{t-1}\varphi_{t-1}}$$

Let the prior scaling *pdf* for time t be given by

$$f_{t-1}(\varphi_t) \propto \varphi_t^{\lambda(\nu_{t-1})} e^{-\mu(\delta_{t-1})\varphi_t} \tag{1}$$

where $\lambda(\bullet)$ and $\mu(\bullet)$ are feasible functions of the posterior parameters ν_{t-1} and δ_{t-1}. Now given x_t as the mode of ϑ_t calculated from the observation information $f(y_t|\vartheta_t)$, and m'_{t-1} the prior mode of $\vartheta_t|D_{t-1}$, with the same approximation that Dr. West has used paralleling Normal theory, the link between Y_t, ϑ_t and φ_t will be given as follows:

$$f(y_t|\vartheta_t,\varphi_t) \propto \varphi_t^{1/2} f^{1-\varphi_t}(y_t|x) f^{\varphi_t}(y_t|\vartheta_t)\,.$$

$$f_{t-1}(\vartheta_t|\varphi_t) \propto \varphi_t^{1/2} f_{t-1}^{1-\varphi_t}(m'_{t-1}) f_{t-1}^{\varphi_t}(\vartheta_t) \tag{2}$$

where $f_{t-1}(\varphi_t)$ is the prior *pdf* for $\vartheta_t|D_{t-1}$.

(1) and (2) gives

$$f_{t-1}(\vartheta_t,\varphi_t) \propto \varphi_t^{\lambda(\nu_{t-1})+1/2} e^{-\mu(\delta_{t-1})\varphi_t} f_{t-1}^{1-\varphi_t}(m'_{t-1}) f_{t-1}^{\varphi_t}(\vartheta_t)\,.$$

Accordingly, the approximate posterior density function for $\vartheta_t,\varphi_t|y_t,D_{t-1}$ is derived as follows:

$$f(\vartheta_t,\varphi_t|y_t,D_{t-1})=f(y_t|\vartheta_t,\varphi_t)f_{t-1}(\vartheta_t,\varphi_t)/f(y_t)$$

$$\propto \varphi_t^{\lambda(\nu_{t-1})+1} e^{-[\mu(\delta_{t-1})+\log \frac{f_{t-1}(m'_{t-1})f(y_t|x_t)}{f_{t-1}(m_t)f(y_t|m_t)}]\varphi_t} f_t^{1-\varphi_t}(m_t) f_t^{\varphi_t}(\vartheta_t)$$

where m_t is the posterior mode of $\vartheta_t|D_t$. In comparison with the posterior *pdf*

$$f_t(\vartheta_t,\varphi_t) \propto \varphi_t^{\nu_t+1/2} e^{-\delta_t\varphi_t} f_t^{1-\varphi_t}(m_t) f_t^{\varphi_t}(\vartheta_t)$$

we have

$$\nu_t = \lambda(\nu_{t-1})+1/2$$

$$\delta_t=\mu(\delta_{t-1})+\ln\frac{f_{t-1}(m'_{t-1})f(y_t|x_t)}{f_{t-1}(m_t)f(y_t|m_t)}\,.$$

For independent observations $y_t=(y_1,y_2,\ldots,y_n)_t$, $f_t(\vartheta_t|y_t)$ is replaced by

$$f_t(\vartheta_t|y_t) \propto \prod_{i=1}^{n} f_t^{\varphi_t}(\vartheta_t|y_{i,t}) \text{ and}$$

$$f_t(m_t|y_t) \text{ by } f_t(m_t|y_t).$$

It is thus seen that the static models are particular cases of the dynamic models. Further, multivariate Y_t and/or ϑ_t cases are straightforward with no reference to the exponential family and conjugacy principle.

Since the preparation of this note I have seen a paper by West, Harrison and Migon (1983) in which the ideas are related to the paper I discuss.

REPLY TO THE DISCUSSION

The invited discussants raise several important points about hierarchical models in general and those of my paper in particular.

(i) The estimation of hierarchical variances/scale parameters in structured prior models is an old and important practical problem. In applications, the fact that the model + data combination is often incapable of determining unique and appropriate degrees of shrinkage is evidenced by the occurrence of local modes and ridges in joint posteriors as mentioned by Professor Freeman and discussed by Tony O'Hagan (1983) at this conference. Often this is due to the use of unsuitable priors, in particular, the standard 'shrinkage' priors; the model does not fit and the use of heavy-tailed priors may help us to understand why as discussed in section 4.3 of the paper.

(ii) The computations associated with these models can, as pointed out by Professor Dempster, be time consuming. The examples in the paper are relatively straightforward although the iterative solution of modal equations requires some care in application. A fuller understanding of such complex joint posteriors is certainly needed; current joint work at Warwick with Tony O'Hagan is directly concerned with the development of relatively simple numerical techniques and graphical displays for hierarchical and heavy-tailed models. An alternative approach, which avoids some of the complexity inherent in static models, is to reformulate them as dynamic (non-linear, non-normal) models and operate sequentially (West, Harrison and Migon, 1983).

(iii) Concerning Professor Dempster's final point, the examples in the paper (including the priors) are provided simply to illustrate general methods that are applicable in a wide variety of problems; certainly each real application requires the input of non-trivial priors and prior structure. The use of scale parameters for individual model components should be viewed as an exploratory device that indicates possible deficiencies in the model; further elaboration of the model (including the prior structure) may then be performed to remedy these defects if required.

To Dr. Pereira the answer is that equal observations are treated equally in the simple independent sample case.

Mr. Ameen discusses sequential application of the basic deviance decomposition (2.18). This decomposition is just Bayes' Theorem in disguise and so is, of course, quite general; I look forward to further details of the multivariate, non-exponential family and non-conjugate extensions mentioned by Mr. Ameen. Sequential analyses for dynamic generalized linear modesl are discussed in West, Harrison and Migon (1983).

REFERENCES IN THE DISCUSSION

AMEEN, J.R.M. and HARRISON, P.J. (1983). Information statistics. *Warwick Statist. Res. Rep.* **31**.

— (1984). Normal discount Bayesian models. This Proceedings.

HARRISON, P.J. and STEVENS, C.F. (1976). Bayesian forecasting. *J.R. Statist. Soc. B.* **338**, 205-247 (with discussion).

LEONARD, T. (1972). Bayesian methods for binomial data. *Biometrika,* **59**, 581-590.

O'HAGAN, A. (1983). Shoulders in hierarchical models. These Proceedings.

SMITH, J.Q. (1979). A generalisation of the Bayesian steady forecasting model. *J.R. Statist. Soc. B*, **41**, 375-387.

SOUZA, R.C. (1978). A Bayesian Entropy approach to forecasting. Ph.D. thesis, University of Warwick, Coventry, U.K.

TEATHER, D. (1984). The estimation of exchangeable binomial parameters. *Commun. Statist. A*. **13**.

WEST, M., HARRISON, P.J. and MIGON, H.S. (1983). Dynamic generalised linear models and Bayesian forecasting. *Warwick Res. Rep.* **41**.

BAYESIAN STATISTICS 2, pp. 559-570
J.M. Bernardo, M.H. DeGroot, D.V. Lindley, A.F.M. Smith (Eds.)

Information Loss in Noisy and Dependent Processes

ROBERT L. WINKLER
Indiana University and Duke University

SUMMARY

When information is gathered in the real world, the data-generating process is often not independent and free of noise. This usually results in less informative data, and such information loss is investigated in this paper. Two specific models, a dichotomous process with noise and a normal process with dependence, are studied. The results suggest that the loss of information can be quite extensive, particularly with dependence. In a Bayesian analysis, the information loss leads to more emphasis on the prior distribution and less on the sample evidence and can lead to reductions in the expected value of sample information in decision-making problems.

Keywords: DEPENDENT PROCESSES; INFORMATION; NOISY PROCESSES; VALUE OF INFORMATION.

1. INTRODUCTION

Although methods are readily available for modeling dependence and noise in data, data-generating processes are usually modeled in practice as if they were independent and free of noise. In the context of carefully controlled experiments, a context within which many statistical methods have been developed and taught, this may not be unreasonable. When information is gathered in the real world, however, the process is often noisy and/or dependent. For example, in some situations the only information available is subjective information. This opens up the strong possibility of noise due to factors such as lying or differences among individuals and of dependence due to overlapping information. Of course, processes generating so-called "objective" data can also be noisy and dependent.

If the noise or dependence in a data-generating process is not modeled appropriately, the information may be perceived as being more accurate than it actually is. That is, the likelihood function and posterior distribution may be tighter than they should be, since the information loss due to the noise or dependence is ignored. As a result, inferences based on the likelihood function or posterior distribution may have systematic errors. Moreover, a decision maker contemplating information acquisition may overstate the expected value of the information, and this may lead to suboptimal information-purchasing decisions.

The purpose of this paper is to study information loss in noisy and dependent processes. The extent of such information loss and the resulting implications for inference and decision making are investigated for two selected models. In Section 2, a dichotomous process with noise is studied. In Section 3, a normal process with dependence is investigated. Some implications for the value of information are considered in Section 4, and Section 5 contains a brief summary and discussion.

2. A DICHOTOMOUS PROCESS WITH NOISE

Data are frequently generated from dichotomous processes, with the parameter of interest being a proportion. In practice, inferences about a proportion π are almost always based on the assumption that the data-generating process is Bernoulli with parameter π. The maximum likelihood estimator is then the sample proportion, and a conjugate Bayesian analysis involves beta prior and posterior distributions.

Noise in the data-generating and data-handling processes can render the Bernoulli assumption inappropriate. For example, respondents may lie when answering yes-no questions, the classification of items into two classes (e.g., defective and good) may not be perfect, and errors may occur when data are recorded. The possibility of such noise has been considered in the literature, as illustrated by Press (1968), who considers estimation from misclassified data. Also, randomized response sampling (Warner, 1965) represents an attempt to systematize the noise.

In most applications involving dichotomous processes, potential noise is ignored and inferences about π are based on the Bernoulli model. When noise is present, ignoring it will generally lead to systematic errors of estimation and overstatements of the accuracy of estimation. In a Bayesian analysis, this means that the location of the posterior distribution is shifted and the dispersion is understated. In this section, the systematic estimation errors and loss of information associated with noise in the process are studied.

Suppose that each member of a large population is in either Group A or Group B, but not both, and let π denote the proportion of the population in Group A. Furthermore, assume that the prior distribution of π is a member of the beta family, with density

$$f(\pi) = f_\beta(\pi \mid \alpha_o, \beta_o) = [B(\alpha_o,\beta_o)]^{-1}\pi^{\alpha_o-1}(1-\pi)^{\beta_o-1}, \tag{1}$$

where $\alpha_o > 0$, $\beta_o > 0$, and $B(\alpha_o,\beta_o) = \Gamma(\alpha_o)\Gamma(\beta_o)/\Gamma(\alpha_o+\beta_o)$ is the beta function. The beta family is conjugate with respect to a Bernoulli process, and beta distributions seem to be viewed as reasonable approximations for a wide variety of types of prior information concerning a proportion.

A random sample is taken from the population, and each individual in the sample answers "yes" or "no" to the question "Do you belong to Group A?" Let p and q denote the probabilities of lying for members of Groups A and B, respectively, and to simplify the analysis, assume that p and q are known. The probabilities of receiving "yes" and "no" answers are then

$$\lambda = P(\text{yes}) = (1-p-q)\pi + q \tag{2}$$

and

$$1-l = P(\text{no}) = (p+q-1)\pi + 1-q. \tag{3}$$

Instead of being Bernoulli in π, the data-generating process is Bernoulli in λ. The case with $p=q=0$ corresponds to the noise-free Bernoulli process. To avoid extreme cases in which the noise is so great that inferences run counter to the impression given by the sample evidence because λ is a decreasing function of π, it is assumed that $p+q < 1$.

If a sample of size n results in r "yes" answers, the likelihood function is

$$\ell(r,n \mid \pi) = [(1-p-q)\pi + q]^r[(p+q-1)\pi + 1 - q]^{n-r}. \tag{4}$$

The maximum likelihood estimator of π is

$$\hat{\pi} = \begin{cases} 0 & \text{if } r/n < q, \\ (\hat{\pi}_B - q)/(1-p-q) & \text{if } q \le r/n \le 1-p, \\ 1 & \text{if } r/n > 1-p, \end{cases} \tag{5}$$

where

$$\hat{\pi}_B = r/n \qquad (6)$$

is the sample proportion. Ignoring the noise amounts to setting $p = q = 0$ and taking $\hat{\pi}_B$ as the maximum likelihood estimator of π. The difference between $\hat{\pi}_B$ and π is

$$\hat{\pi}_B - \hat{\pi} = \begin{cases} r/n & \text{if } r/n < q, \\ [q-(p+q)(r/n)]/(1-p-q) & \text{if } q \le r/n \le 1-p, \\ (r-n)/n & \text{if } r/n > 1-p. \end{cases} \qquad (6)$$

This difference is graphed as a function of r/n in Figure 1 for three cases: $p = q = 0.10$; $p = q = 0.30$; and $p = 0.30$, $q = 0.10$. Note that $\hat{\pi}_B - \hat{\pi}$ is a piecewise linear function of r/n with the greatest positive systematic error at $r/n = q$, the greatest negative systematic error at $r/n = 1-p$, and systematic errors of zero at 0, $q/(p+q)$, and 1. If π is small, the overall estimation error associated with the use of $\hat{\pi}_B$ tends to be positive and is driven primarily by q; if π is large, the overall estimation error tends to be negative and is driven primarily by p.

FIGURE 1

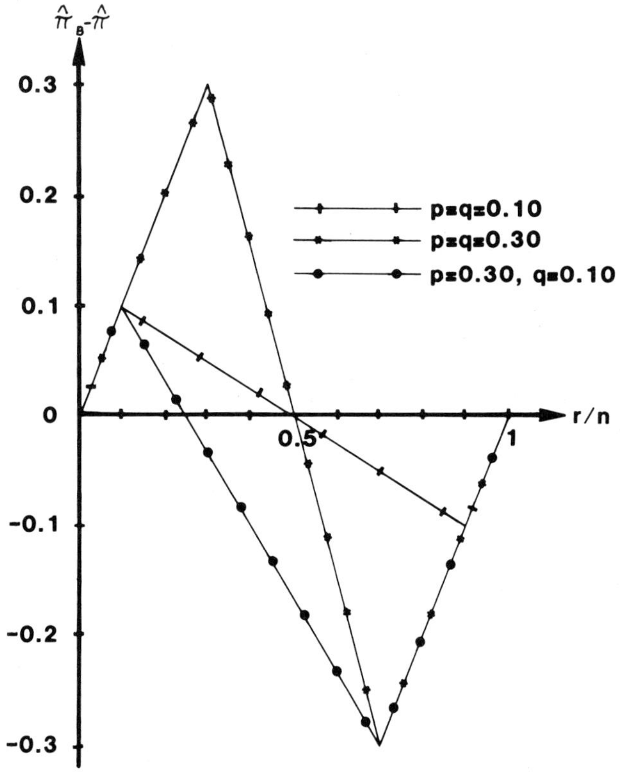

The likelihood function in (4) can be expressed as a mixture of likelihood functions Bernoulli in π, with a natural Bayesian interpretation involving the uncertainty about how many respondents in the sample are actually in Group A. Approximating this mixture by a single likelihood function Bernoulli in π,

$$\ell^*(r^*,n^* \mid \pi) = \pi^{r^*}(1-\pi)^{n^*-r^*}, \tag{7}$$

simplifies the analysis of the information loss due to the noise. If

$$q < r/n < 1-p, \tag{8}$$

equating modes and the curvature of the modes of $\log \ell(r,n \mid \pi)$ and $\log \ell^*(r^*,n^* \mid \pi)$ yields

$$n^* = n\hat{\pi}(1-\hat{\pi})/(\hat{\pi}+c_1)(1-\hat{\pi}+c_2) \tag{9}$$

and

$$r^* = n^*\hat{\pi}, \tag{10}$$

where

$$c_1 = q/(1-p-q) \tag{11}$$

and

$$c_2 = p/(1-p-q). \tag{12}$$

The approximation should be adequate for most purposes unless n is extremely small. It has been shown to be quite accurate for the case with $p=q$ (Winkler and Franklin, 1979).

If the sample results can be thought of as approximately equivalent to a noise-free sample of size n^* with r^* "yes" answers, then n^* can be interpreted as the equivalent number of noise-free observations. From (9),

$$n^*/n = \hat{\pi}(1-\hat{\pi})/(\hat{\pi}+c_1)(1-\hat{\pi}+c_2). \tag{13}$$

As long as (8), the initial condition for the approximation, is satisfied,

$$n^*/n < 1. \tag{14}$$

As anticipated, then, the noise leads to an effective reduction in sample size. In Figure 2, n^*/n is graphed as a function of r/n for $p=q=0.10$; $p=q=0.30$; and $p=0.30$, $q=0.10$.

FIGURE 2

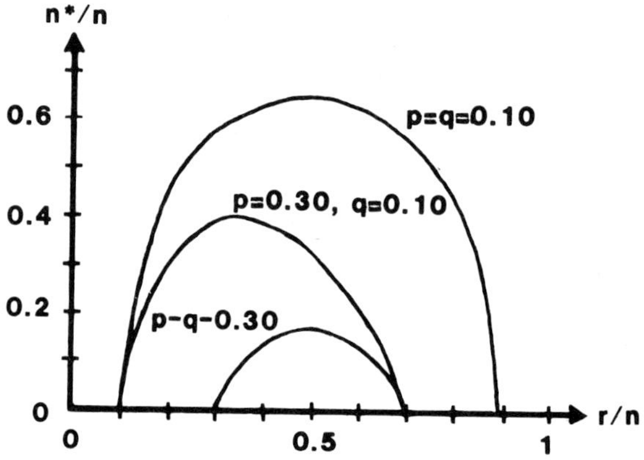

As p or q increases, the curve is lower, implying that less information is retained after the noise is taken into account. A rough measure of the proportion of information lost as a result of the noise is $1-(n^*/n)$. The information loss tends to be greatest when π is near zero or one, with n^*/n approaching zero as $r/n \to q^+$ or $r/n \to (1-p)^-$. Furthermore, the sensitivity of the information loss to changes in r/n is greatest when r/n is near q or $1-p$.

Through the likelihood function, noise in the data-generating process affects Bayesian inference as well as likelihood-based inference. Given the prior distribution in (1), the approximate posterior distribution for π is a beta distribution with parameters

$$\alpha_n = \alpha_o + r^* \tag{15}$$

and

$$\beta_n = \beta_o + n^* - r^*. \tag{16}$$

The posterior mean and variance are

$$E(\pi \mid r^*, n^*) = (\alpha_o + r^*)/(\alpha_o + \beta_o + n^*) \tag{17}$$

and

$$V(\pi \mid r^*, n^*) = (\alpha_o + r^*)(\beta_o + n^* - r^*)/(\alpha_o + \beta_o + n^*)^2(\alpha_o + \beta_o + n^* + 1). \tag{18}$$

The impact of noise is generally to shift the posterior distribution toward the prior distribution and to increase its dispersion. The reduction in the effective sample size results in more weight being given to the prior mean in the computation of the posterior mean. Of course, if the noise is ignored, the prior information receives less weight than it should and more influence is exerted by an estimate, r/n, which has the systematic error given by (6).

3. A NORMAL MODEL WITH DEPENDENCE

For inferences about a continuous variable, observations often take the form of numerical estimates or forecasts of the variable of interest. Typically the errors in these estimates or forecasts are modeled with a normal distribution. Inferences are then based on standard procedures developed for an independent normal data-generating process. The sample mean is a maximum likelihood estimator, and a conjugate Bayesian analysis involves normal or normal-gamma prior and posterior distributions.

Independence is a key assumption in standard normal theory. In some instances the observations might be expected to be somewhat redundant. For example, the estimates or forecasts may be based at least in part on the same data, on similar underlying assumptions or theories, and on common methods of data analysis. The redundancy can be represented in a normal model in terms of positive correlations among errors of estimation. A normal model with correlated errors has been studied from a Bayesian viewpoint by Geisser (1965), and Clemen and Winkler (1985) investigate some limits for the precision and value of information when observations are dependent. Information loss associated with dependence is discussed in this section.

Denote the variable of interest, which could be a parameter of a statistical model or a future observation, by θ, and suppose that the prior distribution of θ is a normal distribution with mean m_o and variance σ_o^2. The sample information consists of estimates $x_1,\ldots,x_n$ of θ, with

$$x_i = \theta + u_i \tag{19}$$

for $i=1,\ldots,n$. The errors of estimation $u_1,\ldots,u_n$ are normally distributed with mean zero, variance σ^2, and pairwise correlation ϱ, with $\sigma^2 > 0$ and $\varrho \geq 0$. To simplify the analysis, assume that σ^2 and ϱ are known. Note that the case with $\varrho = 0$ corresponds to the typical independent normal data-generating process.

The likelihood function for this process can be expressed in the form

$$\ell(m,n|\theta) = \exp\{-n(\theta-m)^2/[1+(n-1)\varrho]\sigma^2\}, \tag{20}$$

where

$$m = \sum_{i=1}^{n} x_i/n \tag{21}$$

is the sample mean. The maximum likelihood estimator is simply $\hat{\theta}=m$, as in the independent case. If the variances or correlations differed, then the maximum likelihood estimator would be a weighted average of the observations instead of a simple average.

The variance of the sampling distribution of m is

$$V(m|\theta) = [1+(n-1)\varrho]\sigma^2/n, \tag{22}$$

which is larger than the variance in the independent case by a factor of $1+(n-1)\varrho$. To achieve the variance in (22) in the independent case would require

$$n^* = [1+(n-1)\varrho]^{-1}n \tag{23}$$

observations. The likelihood function in (20) is equivalent to that from an independent sample of n^* observations with sample mean m.

As long as $\varrho > 0$,

$$n^*/n = [1+(n-1)\varrho]^{-1} < 1. \tag{24}$$

In Figure 3, n^*/n is graphed as a function of n for selected values of ϱ. It is a decreasing convex function of n, indicating a rapid loss of information beyond the first observation. As $n \to \infty$, $n^*/n \to 0$ because n^* approaches a finite upper bound, ϱ^{-1}. This means, for example, that even with a correlation as low as $\varrho = 0.10$, any sample, no matter how large, is equivalent to less than ten independent observations.

FIGURE 3

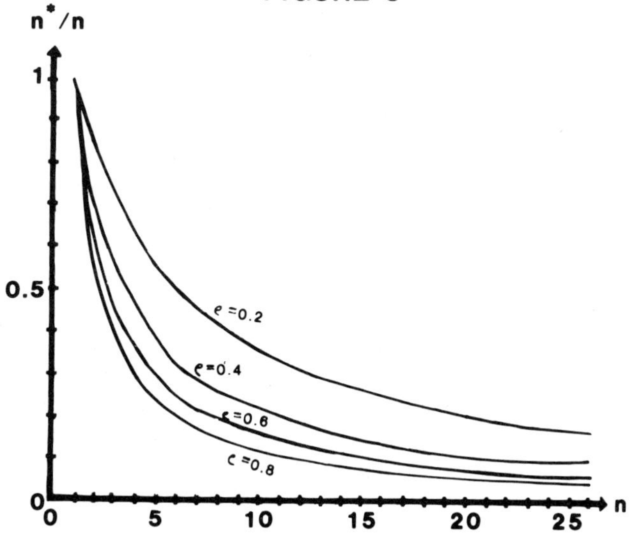

The loss of information due to dependence has a major impact on the posterior distribution in a Bayesian analysis. Given a normal prior distribution for θ with mean m_o and variance σ_0^2, the posterior distribution after observing $x_1,\ldots,x_n$ is also normal with mean

$$E(\theta\,|\,x_1,\ldots,x_n) = (n_o m_o + n^* m)/(n_o + n^*) \tag{25}$$

and variance

$$V(\theta\,|\,x_1,\ldots,x_n) = \sigma^2/(n_o + n^*), \tag{26}$$

where

$$n_o = \sigma^2/\sigma_o^2 \tag{27}$$

is the equivalent sample size represented by the prior information. Dependence results in less weight being given to m and more to m_o. Also, it causes the posterior variance to be higher than it would be after n independent observations. In particular, since n^* is bounded from above, the posterior variance is bounded from below:

$$V(\theta\,|\,x_1,\ldots,x_n) > \sigma^2/(n_o + \varrho^{-1}). \tag{28}$$

This analysis implicitly assumes that $u_1,\ldots,u_n$ are independent of the prior error of estimation,

$$u_o = m_o - \theta\,. \tag{29}$$

As a result, the effect of the dependence among $u_1,\ldots,u_n$ is only felt with the second and later observations. However, for the same reason that $u_1,\ldots,u_n$ may be dependent, there may be some degree of dependence between these errors of estimation and u_o. If this is the case, then there is some information loss even with the first observation.

4. IMPLICATIONS FOR THE VALUE OF INFORMATION

The results in Sections 2 and 3 concerning information loss for a noisy dichotomous process and a dependent normal process involve only the inferential aspects of the information loss. Information is often gathered in order to help a decision maker make a better, more informed decision. A measure of interest to the decision maker who is contemplating sampling is the expected value of sample information.

The expected value of sample information (EVSI) to a decision maker is defined as the maximum amount that the decision maker could pay for the information without being worse off (in an expected value sense, before the sample is observed) than without the information. As a function of the sample size n, EVSI is a non-negative, non-decreasing function; additional information cannot have negative value. But, as shown in Sections 2 and 3, noise and dependence generally result in a reduction in the effective sample size (i.e., $n^* < n$). This, in turn, will lead to reductions in EVSI in most cases.

The magnitudes of reductions in EVSI associated with noise and dependence are functions of the decision maker's prior distribution and the details of the decision-making problem. Examples of reductions are illustrated in Figures 4 and 5. In Figure 4, EVSI is a concave function of n, as is the case in many decision-making problems. This example happens to be based on a point estimation problem with linear loss. In Figure 5, EVSI is convex for small n and concave for larger sample sizes. This form of EVSI is encountered, for instance, in two-action problems where small samples are unlikely to cause the decision maker to switch actions.

Both examples (Figures 4 and 5) involve a normal data-generating process. The value of information is graphed as a function of n for three cases:

FIGURE 4

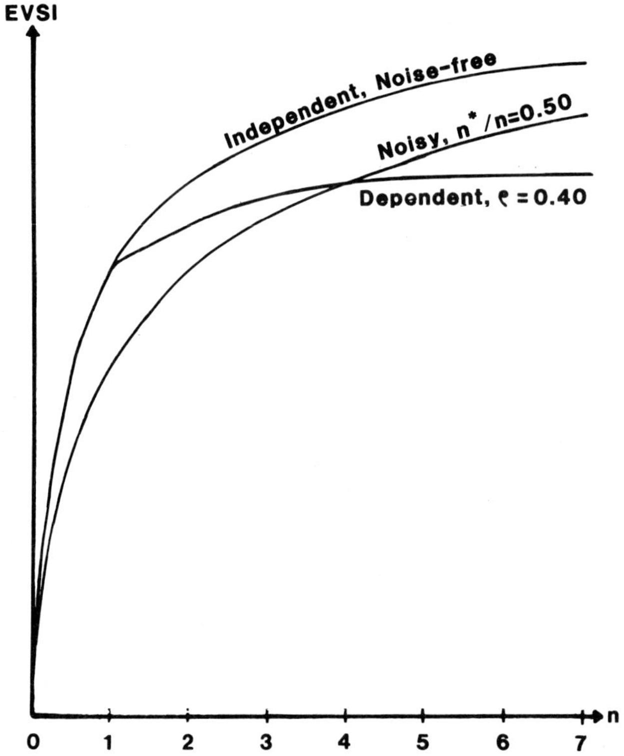

FIGURE 5

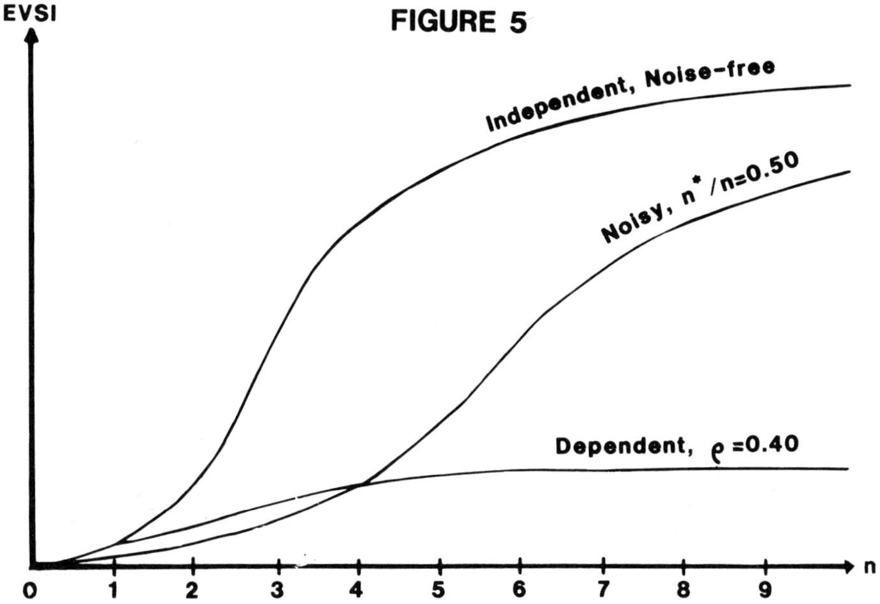

1. an independent, noise-free process,
2. a noisy process with $n^*/n = 0.50$,
3. a dependent process with $\varrho = 0.40$.

For a noisy dichotomous process, Figure 2 suggests that the information loss is often more serious in magnitude than $n^*/n = 0.50$. However, for a normal process with noise, $n^*/n = 0.50$ implies that the variance of the noise equals the process variance. The correlation of 0.40 used for the dependent process seems moderate in view of much higher correlations observed among football predictions (Winkler, 1981) and among economic forecasts currently being studied. Different choices of n^*/n in the noisy process and ϱ in the dependent process would shift the curves in Figures 4 and 5, but the essential results would not be changed.

Because the first observation suffers no loss in information value, the EVSI for the dependent process is greater than the EVSI for the noisy process at first. The incremental increases in value as n increases drop off rapidly for the dependent process, however, with the limiting EVSI equalling the EVSI for $n = \varrho^{-1}$ with an independent, noise-free process. In contrast, the noisy process can overcome information loss through increases in sample size, and its limiting EVSI is identical to that for the independent, noise-free process.

Shifts in EVSI curves have implications for information gathering, of course. Reductions in EVSI often can be expected to lead to smaller optimal sample sizes. If the cost of sampling is relatively large and the optimal sample size is small when no noise or dependence are present, then the introduction of noise or dependence may make it optimal not to sample at all (or perhaps to use $n = 1$ with a dependent process). On the other hand, if sampling is cheap, then the optimal sample size may actually increase for a noisy process, with the low cost making it worthwhile to overcome some of the information loss by taking a larger sample. With dependence, larger samples are inadvisable because the loss of information cannot be overcome.

5. SUMMARY AND CONCLUSIONS

When information is gathered in the real world, the data-generating process is often not independent and free of noise. The two simple models studied in this paper suggest that the loss of information in noisy and dependent processes can be quite extensive. In the noisy dichotomous process, the effective sample size is sometimes a small percentage of the actual sample size and estimates of the proportion can have substantial systematic errors if the noise is ignored. These results are most pronounced when the proportion is near zero or one. In the dependent normal process, the information loss escalates rapidly as the sample size increases and the loss of information cannot be overcome by increasing the sample size. There is a very restrictive bound on the total amount of information available, regardless of sample size, from the process. Because of this bound, dependence has potentially more serious implications for inference and decision than does noise.

In a Bayesian analysis, the information loss results in more emphasis being placed on the prior distribution and less on the sample evidence. This can make the specification of the prior distribution more crucial and can lead to reductions in the expected value of sample information in decision-making problems. In some situations with a noisy process, the decision maker should take a larger sample than would be taken with a noise-free process, thereby spending more on sampling to make up for the information loss. In other situations, the sample size should be reduced and the decision maker should simply base the decision on less information. For the dependent normal process investigated in Section 3, the sample size should never be increased in response to the dependence.

Extreme cases which could lead to information gains instead of losses have not been considered here. For example, in a normal model with unequal variances, high correlations can lead to a more informative sample than independent sampling, but the inferences are non-robust in the sense that they are extremely sensitive to estimates of the variances and correlations. Moreover, the emphasis in this paper has not been on the models, which are relatively simple with parameters such as p, q, σ^2, and ϱ known, but on the extent of the information loss. A full Bayesian analysis would generally involve a joint prior distribution for π, p, and q in the noisy dichotomous process and for θ, σ^2, and ϱ in the dependent normal process. Nonetheless, simple models such as those used in this paper can serve a useful purpose by providing easy-to-understand guidelines concerning the informational impact of noise and dependence. Such guidelines can help practitioners arrive at more realistic assessments of the degree of sampling variability, the dispersion of their posterior distributions, the value of sample information, and the sensitivity of measures of these characteristics to the degree of noise or dependence. In terms of information gathering, the guidelines may be helpful in considering tradeoffs between the cost of sampling and the degree of noise or dependence in the process.

ACKNOWLEDGEMENTS

This research was supported in part by the National Science Foundation under Grant IST8018578.

REFERENCES

CLEMEN, R.T. and WINKLER, R.L. (1983). Limits for the precision and value of information from dependent sources. *Operations Research*, **33**, (in press).

GEISSER, S. (1965). A Bayes approach for combining correlated estimates. *J. Amer. Statist. Assoc.* **60**, 602-607.

PRESS, S.J. (1965). Estimating from misclassified data. *J. Amer. Statist. Assoc.* **63**, 123-133.

WARNER, S.L. (1965). Randomized response: A survey technique for eliminating evasive answer bias. *J. Amer. Statist. Assoc.* **60**, 63-69.

WINKLER, R.L. (1981). Combining probability distributions from dependent information sources. *Management Science*, **27**, 479-488.

WINKLER, R.L. and FRANKLIN, L.A. (1979). Warner's randomized response model: A Bayesian approach. *J. Amer. Statist. Assoc.* **74**, 207-214.

DISCUSSION

J.Q. SMITH (*University College, London*)

Professor Winkler has presented some simple but important illustrations of how common model assumptions (usually made for convenience or by convention) can be very unrobust. It is not until the consequences of deviating slightly from model assumptions are calculated explicitly that the extent of this lack of robustness is realised. It is doubly worrying since I can think of no realistic undesigned statistical application where it is reasonable to blindly assume that observations are not noisy or dependent in some way. Of course, problems with lack of independence can be conditioned away once we *know* the correct model, but how do we find it?

The pitfalls in assuming independence in the most innocuous of experiments are illustrated through the results of this classroom coin tossing experiment. Each student in the class was asked to toss a coin a large number of times and record the sequence of heads and tails. I blithely assumed that this was a Binomial experiment. However observing the students I became aware that after the half-way point of the experiment:

(a) the flipping finger of many of the students was becoming sore so that the coin only spun in the air a small number of times;

(b) to be effecient most students maintained a mechanical rhythm of tossing.

The combination of these two factors meant that the coins spun approximately the same number of half revolutions each toss so that adjacent observations were highly dependent.

I do not believe I could have thought of this dependence before performing the experiment and I certainly believed my independent assumptions were a priori reasonable. Yet the dependence I missed would be crucial to how I should bet on the next toss on the evidence of past tosses, for example. If independence assumptions do not work in applications of coin tossing experiments when will they work!

R. VIERTL (*Technische Universität Wien*)

Compared to some semi-Bayesian papers presented at this meeting this is real Bayesian work - in my subjective opinion at least. But for all statistical inference, it is important for the following reasons: Looking at inferential problems in interesting applications, observations are usually not identically distributed. But they are in a certain way similar. Therefore, a theory for similar observations, suitable for applications, should be developed. Now Prof. Winkler's paper in the first part is the development of this idea for a special case and so it is an important step in that direction. In this connection, a paper by Basu and Pereira (1982) where Bayesian methods are used on the problem of nonresponse should be noted. The second part of the paper is a very informative approach pointing out the influences of dependencies in observations. This is again a very important step in order to develop a theory for non-independent observations. Especially it shows how substantial the use of prior information for dependent processes is. One thing which needs more future research is the way to measure uncertainty and information. In my opinion entropy and Fisher-information are not very good measures of uncertainty in probability distributions, especially for applied statistics. It would be necessary to do more research in stochastic analysis of information processes but not in the sense of present information theory which should better be named communication theory. The paper can be recommended for reading also by non-Bayesians.

M. GOLDSTEIN (*Hull University*)

This paper argues that statisticians often perform badly because they use their simple independence models to analyse dependent data. There is some discussion, in the "normative" psychological literature, of precisely the opposite argument for how people actually analyse information (see, for example, Goldstein (1979)). That is, they are so used to considering situations in which they receive complicated, interdependent data that when they see data which actually is independent (for example in psychological testing), they make unappropriate judgements, because they have rarely encountered such situations before. For example, if a simple Bernoulli process is analyzed instead as a one stage Markov process, then all the usual "conservatism" behaviour will be observed.

REPLY TO THE DISCUSSION

I am grateful to the discussants for their kind comments. The coin-tossing experiment of Dr. Smith illustrates that dependence can occur in the most "seemingly independent" situations. I believe that a study of various ways in which dependence can arise and of the corresponding inferential models would be very useful in expanding the storehouse of models of dependent processes as well as in helping us better understand dependence.

I agree with Professor Viertl's concern about the measurement of uncertainty and information. In decision theory, EVSI is a suitable measure, but it is problem-dependent. In the context of this paper, n^* seems reasonable for inferential purposes but is still not entirely satisfactory. I have not seen a convincing argument for a measure that is ideal in all situations.

Dr Goldstein's point is very well-taken. Individuals seem to learn on a purely intuitive level to handle complex situations with noisy or dependent data and may, as a result, have trouble dealing with independent, noise-free data (Winkler and Murphy, 1973). It is indeed ironic, then, that since statisticians are trained to model data-generating processes with strong emphasis on independent, noise-free models, they may not fully appreciate the serious ramifications of noise or dependence for inferences and predictions generated by the model.

REFERENCES IN THE DISCUSSION

BASU, D. and PEREIRA, C. (1982). On the Bayesian analysis of categorical data: the problem of non-response. *J. Statist. Planning and Inference*, **6**, 345-362.

GOLDSTEIN, M. (1979). Interdependent data and human conservatism. *Brit. J. Math. Statist. Psychol.* **32**, 72-79.

WINKLER, R.L. and MURPHY, A.H. (1973). Experiments in the laboratory and the real world. *Organizational Behavior and Human Performance,* **10**, 252-270.

BAYESIAN STATISTICS 2, pp. 571-586
J.M. Bernardo, M.H. DeGroot, D.V. Lindley, A.F.M. Smith (Eds.)

Bayesian Statistics in Econometrics

ARNOLD ZELLNER
University of Chicago

SUMMARY

After recognizing the considerable progress of Bayesian econometrics since the early 1960s, the following research topics are discussed: (1) development of Bayesian computer programs, (2) diagnostic procedures for checking models, prior distributions and data, (3) procedures for assessing prior distributions, (4) numerical integration and analysis of non-standard models, and (5) simplification of some econometric and time series models. It is concluded that further work on these topics will contribute substantially to promote more applications of Bayesian methods in econometrics and other areas of science.

Keywords: BAYESIAN COMPUTER PROGRAMS; BAYESIAN ECONOMETRICS; ECONOMETRIC MODELING; PRIOR DISTRIBUTIONS; REGRESSION DIAGNOSTICS.

1. INTRODUCTION

In my 1980 paper, "The Current State of Bayesian Econometrics", I described the remarkable progress that had occurred in Bayesian Econometrics since the early 1960s. As many of you know, there has been substantial growth in the numbers of Bayesian researchers, publications and university courses. Many inference and decision problems in econometrics have been solved using Bayesian techniques and these solutions have compared very favorably with non-Bayesian solutions. Bayesian learning and optimization models have been widely employed in economic theory. A number of applied Bayesian econometric studies have been published. Bayesian econometric texts and monographs have been published and many current general econometrics textbooks include Bayesian material. Thus, in general, progress in Bayesian Econometrics since the 1960s has been substantial. However, in my 1980 paper, I noted several topics requiring additional work in order to facilitate more wide-spread applications of Bayesian techniques in econometrics, namely:

(1) Bayesian computer programs,

(2) diagnostic checking of models, prior distributions and data,

(3) procedures for assessing priors,

(4) analysis of "non-standard" models, e.g. non-normal and non-linear models, models for mixtures of discrete and continuous variables, time series models, etc., and

(5) simplification of some current econometric and time series models.

In the present paper, I shall discuss these five topics and refer to some recent work relating to each of them.

2. BAYESIAN COMPUTER PROGRAMS

In connection with activities of our NSF-NBER Seminar on Bayesian Inference in Econometrics, S.J. Press (1980) published a paper listing and describing a number of publicly available Bayesian computer programs. Many of these programs provide users with the capability of analyzing multiple regression and other standard econometric and statistical models using diffuse and/or certain types of informative prior distributions. While it may be dangerous to generalize about these computer programs, it is my impression that many, if not all of them, require expansion to permit (a) improved diagnostic checking of models, data and prior distributions, (b) easier assessment and analysis of prior distributions, and (c) enhanced numerical integration capabilities. Under (a), I have in mind Bayesian analyses of and tests for possible autocorrelation (see, e.g. Fomby and Guilkey (1978) and Griffiths and Dao (1980)), heteroscedasticity (see, e.g. Surekha and Griffiths (1983)), appropriate functional forms (Box-Cox and other transformations), plots and analyses of residuals that can be given a Bayesian interpretation (see Zellner (1975)), outlier analyses, etc. Some further diagnostic checks will be described below.

As regards (b), it is highly probable that computer programs providing easier assessment and analysis of prior distributions will facilitate applications of Bayesian techniques. It may be worthwhile to have several different assessment techniques on call to allow users flexibility in approaching different problems. Last, by having enhanced numerical integration techniques, including multivariate integration techniques available, greater flexibility is provided in choosing prior distributions' and likelihood functions' forms.

While much more could be said about the design of Bayesian computer programs, at present I shall just emphasize the extreme importance of having them include a satisfactory range of diagnostic checks of models' assumptions, data quality and prior distributions' forms and adequate numerical integration capabilities.

3. DIAGNOSTIC CHECKING OF MODELS, PRIORS AND DATA

As mentioned above, Bayesian analyses of departures from assumptions of the standard normal linear multiple regression model are available. These usually involve expanding the standard model to allow for a possible departure or departures. Posterior distributions for the added parameters can be obtained. Conditional posterior distributions can be utilized to determine how sensitive inferences about regression coefficients or future values of the dependent variable are to departures from standard assumptions. Also, posterior odds ratios can be computed for, say hypotheses about an autocorrelation parameter, namely H_0: $\varrho = 0$ and H_1: $\varrho \neq 0$ (see, e.g. Griffiths and Dao (1980)). To perform these analyses of possible departures from standard assumptions, usually good numerical integration programs are needed since many integrations in this area cannot be performed analytically. Also the capability of averaging results over alternative hypotheses using posterior probabilities (see Zellner and Vandaele (1975)) is an option that is worth considering.

One type of diagnostic check that is frequently employed in applied regression analyses is the assessment of the possible effects of left-out variables on analyses, the so-called left-out variable (LOV) problem. As regards a standard form of the LOV problem, assume that $\mathbf{y} = X_1\beta_1 + X_2\beta_2 + \mathbf{u}$ is thought to be an appropriate formulation of a regression relation linking a dependent variable to two sets of independent variables, X_1 and X_2 and a standard normal, zero-mean, scalar covariance matrix error vector $\mathbf{u}$. Suppose that for some reason, e.g. ignorance, lack of data on X_2, etc., an investigator does not or cannot include X_2 in his regression analysis. How will this omission affect his inferences about β_1? This question can be answered by computing the conditional posterior pdf for β_1 given X_2 and β_2 and examining it properties. Assume that we have a normal natural conjugate prior for $\beta'=(\beta_1',\beta_2')$ with mean vector $\bar{\beta}'=(\bar{\beta}_1',\bar{\beta}_2)$ and conditional covariance matrix $V(\beta|A,\sigma) = A^{-1}\sigma^2$. With A positive definite symmetric, we can write $A=K'K= (K_1:K_2)''(K_1:K_2)$, where the partitioning of K is conformable with the partitioning of $\beta'=(\beta_1',\beta_2')$. Then the conditional posterior mean of β_1 given β_2, X_2 and the sample and prior information is

$$\tilde{\beta}_1^c=\tilde{\beta}_{y\cdot 1}+(X_1'X_1+K_1'K_1)^{-1}[K_1'K_2(\bar{\beta}_2-\beta_2)-X_1'X_2\beta_2] \tag{1}$$

where $\tilde{\beta}_{y\cdot 1} = (X_1'X_1 + K_1'K_1)^{-1}(X_1'\mathbf{y}+K_1'K_1\bar{\beta}_1)$, the posterior mean for β_1 if just X_1 enters the model and if the prior on β_1 is $N[\bar{\beta}_1, (K_1'K_1)^{-1}\sigma^2]$. The second expression on the right side of (1) can be used to appraise differences between $\tilde{\beta}_1^c$ and $\tilde{\beta}_{y\cdot 1}$ for given β_2 and $(X_1'X_1 + K_1'K_1)^{-1}X_1'X_2$. Clearly if $\beta_2 = \bar{\beta}_2=\mathbf{0}$, then $\tilde{\beta}_1^c=\tilde{\beta}_{y\cdot 1}$. Also this equality holds if $K_1'K_2=0$ (elements of β_1 and β_2 are a priori uncorrelated and $\beta_2=\mathbf{0}$ even if $\bar{\beta}_2\neq\mathbf{0}$). Last, if our prior information is vague, i.e. $A \to \mathbf{0}$, then (1) becomes

$$\tilde{\beta}_1^c = (X_1'X_1)^{-1}X_1'\mathbf{y}-(X_1'X_1)^{-1}X_1'X_2\beta_2 \tag{2a}$$

or

$$\tilde{\beta}_{y\cdot 1} = (X_1'X_1)^{-1}X_1'\mathbf{y} = \tilde{\beta}_1^c + (X_1'X_1)^{-1}X_1'X_2\beta_2\,. \tag{2b}$$

In (2a) the conditional posterior mean for β_1 given X_2 and β_2 is equal to the least squares quantity, $\tilde{\beta}_{y\cdot 1} = (X_1'X_1)^{-1}X_1'\mathbf{y}$, the posterior mean when just X_1 enters the regression and a diffuse prior pdf is employed, minus $(X_1'X_1)^{-1}X_1'X_2\beta_2$. By assigning values for $(X_1'X_1)^{-1}X_1'X_2$ and β_2 it is possible to assess the extent to which $\tilde{\beta}_1^c$ departs from $\tilde{\beta}_{y\cdot 1}$. Here if $X_1'X_2=0$ and/or $\beta_2=\mathbf{0}$, $\tilde{\beta}_1^c=\tilde{\beta}_{y\cdot 1}$. Also, it is interesting to note that sampling theorists in econometrics consider the sampling mean of $\tilde{\beta}_{y\cdot 1}= (X_1'X_1)^{-1}X_1'\mathbf{y}$ given the model $\mathbf{y}=X_1\beta_1 + X_2\beta_2+\mathbf{u}$ to obtain

$$E\tilde{\beta}_{y\cdot 1} = \beta_1 + (X_1'X_1)^{-1}X_1'X_2\beta_2\,. \tag{3}$$

Thus the bias of $\tilde{\beta}_{y\cdot 1}$ as an estimator for β_1 is $(X_1'X_1)^{-1}X_1'X_2\,\beta_2$. They then guess the values (or algebraic signs) of the elements of $(X_1'X_1)^{-1}X_1'X_2$ and of β_2 in an effort to determine the sizes (or algebraic signs) of the biases. The similarity in form of (2b) and (3) should be noted. However (1) and (3) are of somewhat different form since (1) reflects the impact of an informative prior distribution for β_1 and β_2. Also, while just conditional posterior means have been considered above, the analysis can be extended to consider the entire conditional posterior pdf for β_1 given β_2 and X_2. Finally, the above analysis has been considered in the framework of an extended model, (i) $\mathbf{y} = X_1\beta_1+X_2\beta_2+\mathbf{u}$ and (ii) $X_2=X_1\Pi+V$, the latter equation representing a standard multivariate normal regression system. On substituting from (ii) in (i), $\mathbf{y}=X_1(\beta_1+\Pi\beta_2)+\mathbf{u}+V\beta_2$ or $\mathbf{y}=X_1\eta + \epsilon$, where $\eta=\beta_1+\Pi\beta_2$ and $\epsilon=\mathbf{u}+V\beta_2$. Then a posterior distribution for η can be transformed to a conditional posterior distribution for β_1 given $\Pi\beta_2$. Also prior information about Π and β_2 can be introduced into this type of LOV analysis which is very important in applied work.

Another type of analysis that seems valuable in applied econometric work is a Bayesian version of regression influence and residual analysis (see, e.g. Johnson and Geisser (1980), Barnett and Lewis (1978), Cook and Weisberg (1982) and Belsley, Kuh and Welsch (1980)). Here measures of influence that reflect not only the sample data's impact on analyses but also the prior distribution's impact will be presented. We consider the usual linear normal regression model $\mathbf{y}=X\beta+\mathbf{u}$ with a natural conjugate prior in the normal, inverted gamma form, $p_N(\beta|\sigma,\bar{\beta},A)p_{IG}(\sigma|\nu_0,s_0)$. The posterior pdf for β and σ is given by:

$$p(\beta,\sigma|D) \propto \sigma^{-(n+k+\nu_0+1)} \exp\{-[\nu_0 s_0^2 + (\mathbf{y}-X\beta)'(\mathbf{y}-X\beta) + (\beta-\bar{\beta})'A(\beta-\bar{\beta})]/2\sigma^2\} \tag{4}$$

where D denotes the prior and sample information. Writing the positive definite symmetric matrix $A=K'K$, the two quadratic terms in the exponential of the posterior distribution in (4) can be written, with K selected as discussed in Zellner (1983), as

$$\left[\begin{pmatrix}\mathbf{y}\\ K\bar{\beta}\end{pmatrix}-\begin{pmatrix}X\\ K\end{pmatrix}\beta\right]'\left[\begin{pmatrix}\mathbf{y}\\ K\bar{\beta}\end{pmatrix}-\begin{pmatrix}X\\ K\end{pmatrix}\beta\right] \tag{5a}$$

or

$$(\mathbf{z}-W\beta)'(\mathbf{z}-W\beta) \tag{5b}$$

where $\mathbf{z}'=[\mathbf{y}'\ (K\bar{\beta})']$ and $W'=(X'\ K')$. On completing the square on β in (5b), we have

$$(\mathbf{z}-W\tilde{\beta})'(\mathbf{z}-W\tilde{\beta}) + (\beta-\tilde{\beta})'W'W(\beta-\tilde{\beta}) \tag{6}$$

where

$$\tilde{\beta}=(W'W)^{-1}W'z=(X'X+K'K)^{-1}(X'y+K'K\bar{\beta})$$

is the posterior mean. Also, the "sum of squared residuals",

$$(\mathbf{z}-W\tilde{\beta})'(\mathbf{z}-W\tilde{\beta}) = (\mathbf{y}-X\tilde{\beta})'(\mathbf{y}-X\tilde{\beta}) + (K\bar{\beta}-K\tilde{\beta})'(K\bar{\beta}-K\tilde{\beta})$$

breaks up into two parts, one showing goodness of fit of $X\tilde{\beta}$ to $\mathbf{y}$ and the other goodness of fit of $K\tilde{\beta}$ to $K\bar{\beta}$. It is recommended that the "augmented" residual vector $\tilde{\epsilon}_A=\mathbf{z}-W\tilde{\beta}$ be computed and studied carefully. Also, a plot of the elements of $\tilde{\epsilon}_A$ against the "fitted values" $W\tilde{\beta}$ will be useful.

Next, if a row or several rows of $(\mathbf{z},W)$ are deleted and a posterior mean $\tilde{\beta}_a$ computed, a slightly generalized Cook's distance measure can be computed, namely

$$(\tilde{\beta}_a-\tilde{\beta})'W'W(\tilde{\beta}_a-\tilde{\beta})/k\hat{\sigma}^2 \tag{7}$$

where k is the number of regression coefficients and $1/\hat{\sigma}^2$ is the posterior mean of $1/\sigma^2$. The numerator of the distance measure in (7) can be expressed as

$$(W\tilde{\beta}_a-W\tilde{\beta})'(W\tilde{\beta}_a-W\tilde{\beta}) = (X\tilde{\beta}_a-X\tilde{\beta})'(X\tilde{\beta}_a-X\tilde{\beta}) + (K\tilde{\beta}_a-K\tilde{\beta})'(K\tilde{\beta}_a-K\tilde{\beta})$$

and it is recommended that the elements of the vectors $X\tilde{\beta}_a-X\tilde{\beta}$ and $K\tilde{\beta}_a-K\tilde{\beta}$ be studied very carefully. Also the elements of the residual vector, $\mathbf{z}_a-W_a\tilde{\beta}_a$, where $(\mathbf{z}_a\ W_a)$ is the matrix $(\mathbf{z}\ W)$ with one or more rows deleted deserves close attention.

As regards influential points, let $\mathbf{w}_i'$ denote the i'th row of W. Of course, $\mathbf{w}_i'$ can be a row of X or a row of K since $W'=(X'\ K')$. Then the posterior moments of $\mathbf{w}_i'\beta$ are $E\mathbf{w}_i'\beta=\mathbf{w}_i'\tilde{\beta}$ and $\mathrm{Var}(\mathbf{w}_i'\beta)=\mathbf{w}_i'(W'W)^{-1}\mathbf{w}_i\,E\sigma^2$. Thus it is clear that $\mathbf{w}_i'(W'W)^{-1}\mathbf{w}_i$ is an important factor in determining the posterior variance of $\mathbf{w}_i'\beta$. Letting $q_{ii}\equiv\mathbf{w}_i'(W'W)^{-1}\mathbf{w}_i$, we have

$$\mathrm{Var}(\mathbf{w}_i'\beta)/\mathrm{Var}(\mathbf{w}_j'\beta)=q_{ii}/q_{jj}\,. \tag{8}$$

It is seen that the q_{ii} are slightly generalized versions of the diagonal elements of the "hat" matrix, $X(X'X)^{-1}X'$ that yields $\hat{\mathbf{y}}=X(X'X)^{-1}X'\mathbf{y}$, the vector of fitted values in a diffuse prior Bayesian analysis. They will be helpful in understanding which rows of W probably represent influential points. In the case of a single independent variable measured as a deviation from its mean, $q_{ii} = x_i^2 / (\Sigma_1^n x_i^2 + a)$ and for the last row of W, $q_{n+1,n+1} = a/(\Sigma_1^n x_i^2 + a)$, a measure of the influence of the prior information.

The above measures and residuals will probably be helpful in understanding in part how analyses are affected by particular data and prior constellations. On prior distributions, there is often some uncertainty in assigning values to prior parameters. One might put a prior distribution on such a parameter, say A in the regression natural conjugate prior distribution and integrate it out. In the present example, one might attempt to assess an inverted Wishart distribution for A, a rather formidable task. Perhaps it may be better to resort to a "mixed prior", that is several variants of a prior, say $f_i(\theta \mid \mathbf{a}_i)$, $i = 1,2,\ldots,m$ where $\mathbf{a}_i$ is a vector of prior parameters, each with an associated prior probability Π_i. Then the mixed prior is $f(\theta \mid \bullet) = \Sigma_{i=1}^m \Pi_i f_i(\theta \mid \mathbf{a}_i)$. Given a likelihood function $\ell(\theta \mid \mathbf{y})$, posterior odds ratios can be computed,

$$K_{ij} = \frac{\Pi_i}{\Pi_j} \frac{\int \ell(\theta \mid \mathbf{y}) f_i(\theta \mid \mathbf{a}_i) d\theta}{\int \ell(\theta \mid \mathbf{y}) f_j(\theta \mid \mathbf{a}_j) d\theta} \tag{9}$$

and from the K_{ij}, posterior probabilities, $P_1, P_2, \ldots, P_m$ can be obtained. For example if $m=2$, $K_{12} = P_1/P_2$ and if the possible priors are exhaustive $P_2 = 1 - P_1$, then $P_1 = K_{12}/(1 + K_{12})$. Further, the posterior distribution is given by

$$p(\theta \mid D) = \sum_{i=1}^{m} P_i p_i(\theta \mid D_i)$$

where $p_i(\theta \mid D_i)$ is the normalized posterior pdf for θ based on data $\mathbf{y}$ and prior distribution $p_i(\theta \mid \mathbf{a}_i)$. When $\ell(\theta \mid \mathbf{y})$ is the likelihood function for a normal linear multiple regression model and the prior distributions are all in the natural conjugate form with different prior parameters, the above integrations can be done analytically and differences in the individual posterior distributions, the $p_i(\theta \mid D_i)$ appraised using some of the measures introduced above since varying the natural conjugate priors for β can be viewed as changing $K\beta$ and K. For non-standard models, integrations will probably have to be done numerically. Finally, if both the prior distribution's form and the likelihood function's forms are varied, then a more complete sensitivity analysis will have to be performed.

4. PROCEDURES FOR ASSESSING PRIOR DISTRIBUTIONS

As mentioned in the introduction, it is important to have procedures that permit convenient and accurate assessment of prior distributions. As regards diffuse or vague prior distributions, it is my impression that convenient reference non-informative prior distributions are available for a number of standard models. Of course when there are many parameters in a model, such "vague" or "diffuse" priors have to be considered very carefully --see e.g. Hill's (1975) cogent discussion of this range of issues.

As regards assessing informative prior distributions, several issues arise. First, should assessment procedures involve questions about possible values of parameters or just be restricted to questions about observables, say possible values of the dependent variable? (See Kadane (1980) for a discussion of these issues). Given the fact that in many sciences including physics, biology, chemistry, astronomy, economics, etc., scientists think in terms of models' parameters as well as in terms of predictions, I put forward a combined predictive-structural procedure for assessing prior distributions for regression parameters (see Appendix for a brief description of this approach). The procedure not only provides a prior distribution for the parameters but also involves very important diagnostic checks on regression models' properties and the consistency of prior beliefs. It also forces users to become very familiar with their models' properties. While I believe that this approach will be useful for certain types of problems, I recognize that there are others for which it may not be suitable (see, e.g. Zellner and Williams (1973), where a completely different approach was employed to assess an informative prior distribution). Thus there is a need for a multiplicity of convenient and accurate assessment methods.

Second, there is the issue of whether reference informative priors (RIPs) are useful. I regard the class of natural conjugate prior distributions as a very useful important class of RIPs. In regression models, as well as in other models, the form of the natural conjugate prior is the form of the posterior distribution as the sample information increases, provided of course that there is no change in the form of the model, a feature of natural conjugate priors that several have noted in the past. Also, as the sample size grows, the posterior precision matrix becomes proportional to $X'X$ in multiple regression models, a property that has led me to investigate assessment procedures that lead to a natural conjugate prior distribution with prior precision matrix proportional to $X'X$ (see Zellner (1983), for a description of this assessment procedure), a class of prior distribution that I have called "g-priors". These priors are relatively easy to assess, requiring just a prior mean vector $\bar{\beta}$, a prior mean for the error term variance, $\bar{\sigma}^2$ and a choice of the value of g. With this type of prior, the posterior mean for a regression coefficient vector is $\tilde{\beta} = (\hat{\beta} + g\bar{\beta})/(1+g)$, where $\hat{\beta} = (X'X)^{-1}X'\mathbf{y}$. It is seen that $\tilde{\beta}$ is a simple average of $\hat{\beta}$ and the prior mean $\bar{\beta}$ (further details are provided in Zellner (1982), (1983), including discussion of a possible dependence of g on n, the sample size and extensions of the approach).

As more procedures for assessing prior distributions become available (see, e.g. Kadane et al. (1980) for a recent contribution), their relative properties can be appraised and I believe it will be the case that a multiplicity of assessment procedures will emerge. This should provide Bayesian analysts with needed tools for structuring and solving applied problems.

5. ANALYSES OF NON-STANDARD PROBLEMS

By non-standard problems, I mean problems in which it is difficult, if not impossible to analyze posterior and predictive distributions and posterior odds ratios completely analytically. For examples of such analyses see Kloek and van Dijk (1980), Monahan (1983) and Zellner and Rossi (1984). In these studies, the likelihood functions' forms are very complicated. Also, in Monahan's study of ARMA processes, the parameter space was restricted to reflect stationarity and invertibility assumptions. Thus, in this case, and in the Kloek-van Dijk and Zellner-Rossi studies, posterior distributions' normalizing constants, moments and marginal distributions could not be evaluated analytically. Numerical integration methods were employed in these studies that proved to be very useful. In Monahan's analysis, exact posterior and predictive distributions for low order ARMA processes were computed as well as posterior odds ratios helpful in approaching model selection problems. In the Zellner-Rossi paper, exact finite sample posterior distributions for logit-models were compared with approximate asymptotic posterior distributions. There can be no question but that for many non-standard problems numerical integration techniques have been shown to be invaluable. Further work to improve numerical integration techniques and to incorporate them in Bayesian computer programs will be very fruitful.

6. SIMPLIFICATION OF CURRENT MODELS

Many current econometric models and time series models are very complicated and involve extremely large numbers of parameters. For example, some macroeconometric models involve *hundreds* of non-linear stochastic difference equations with many parameters. These models are difficult to understand and cannot be analyzed satisfactorily from the Bayesian or any other point of view. There is a need for model simplification in such cases. Similarly in the time series area, some have entertained multivariate or vector autoregressive models in the form $\mathbf{z}_t = \alpha + A_1\mathbf{z}_{t-1} + A_2\mathbf{z}_{t-2} + \ldots + A_m\mathbf{z}_{t-m} + \epsilon_t$ where $\mathbf{z}_t$ is a $p \times 1$ vector of variables, α is a $p \times 1$ parameter vector, the A's are $p \times p$ coefficient matrices and ϵ_t is a zero-mean white noise error vector with $p \times p$ covariance matrix Σ. There are $p + mp^2 + p(p+1)/2$ parameters in this system. If $p = 6$ and $m = 10$, a "small" system of 6 variables, the number of parameters is 387. With 20 years of quarterly data, the number of observations is $6 \times 4 \times 20 = 480$. Thus the observation-parameter ratio, 480/387 is abysmally low which usually means that estimation and prediction will not be very precise and that tests of hypotheses about parameters' values will not be very powerful (for empirical evidence supporting these statements, see Litterman (1980)). There is thus a need for model simplification in such cases.

As regards multivariate ARMA processes, they may be more parsimonious than vector autoregressions but they too involve many parameters. For example, the ARMA system for the $p \times 1$ vector $\mathbf{z}_t$, $H_0\mathbf{z}_t + H_1\mathbf{z}_{t-1} + \ldots + H_r\mathbf{z}_{t-r} = \alpha + \epsilon_t + F_1\epsilon_{t-1} + \ldots + F_q\epsilon_{t-q}$, with the H's and F's $p \times p$ matrices, α a $p \times 1$ parameter vector and Σ the $p \times p$ covariance matrix of the white noise error vector ϵ_t, contains $(r+1+q)p^2 + p + p(p+1)/2$ parameters. With $p = 6$, $r = 3$ and $q = 4$, the number of parameters is 315, still very large for this "small" system involving just 6 variables. There is clearly a need to utilize prior information, that is subject matter information and theory, in an effort to simplify these models.

Litterman (1980) used a special prior distribution for parameters of a seven variable vector autoregressive model. His prior reflects the information that processes for individual variables are somewhat close to being random walks. Using his flexible prior distribution that roughly incorporates this information, he has been successful in deriving predictive means for seven quarterly, seasonal adjusted macrovariables that have been relatively reliable in forecasting (see Litterman's paper for explicit comparisons of his forecasting results with those provided by other methods).

Subject-matter information sometimes suggests that some variables in $\mathbf{z}_t$ are exogenous, that is $\mathbf{z}_t' = (\mathbf{y}_t' \mathbf{x}_t')$ with $\mathbf{x}_t$ a vector of exogenous variables. The assumption that $\mathbf{x}_t$ is exogenous places a number of restrictions on a multivariate ARMA process (see, e.g. Zellner and Palm (1974)) and this reduces the number of parameters considerably. Also, further subject-matter considerations usually lead to other reasonable restrictions on models' parameters. These considerations are extremely important in producing a tentative model that is relatively simple, understandable and capable of being analyzed well from a Bayesian statistical point of view. This point of view seems to be reasonable to the present writer. However, there are some in econometrics who dogmatically assert that, "The world is complicated and therefore we need complicated models" or that "A model with an infinite number of parameters is realistic". The differences in actual analyses produced by these divergent prior views are enormous. While some may wish to try to settle these issues philosophically, I take a more pragmatic approach. I know of many sophisticatedly simple models in economics, physics, etc. that work reasonably well in explaining past data and making predictions about new data (Newton's Laws, Einstein's Laws, the Gas Laws,

Schrödinger's wave equation, Laws of Demand and Supply, Friedman's consumption model, etc.). However, I do not know of any large, complicated models in any area of science that have performed well in explaining past data and predicting as yet unobserved new data. Thus, I am of the opinion (hardly novel) that entertaining sophisticatedly simple models is an excellent point of departure for work in econometrics and other areas of science.

7. CONCLUDING REMARKS

In summary, having more and better Bayesian computer programs, diagnostic checking procedures, procedures for assessing prior distributions, numerical integration techniques for analyzing non-standard models and sophisticatedly simple subject-matter models will contribute substantially to promote more widespread applications of Bayesian analysis in econometrics and other areas. As noted in my paper, substantial progress is being made on these and other Bayesian research topics and thus I agree with Jack Good that he is winning his bet with Bartlett that he described as follows in his address at the 21st meeting of the NBER-NSF Seminar on Bayesian Inference in Econometrics in 1980:

> In about 1946 I had a long argument about Bayesian statistics with Maurice Bartlett in the Faculty Lounge at Manchester University. At the end of our argument Bartlett proposed the bet or test of waiting for a hundred years to see which was dominant, a classical or a neo-Bayesian one. After a third of a century I think Bartlett is losing the bet.

REFERENCES

BARNETT, V. and LEWIS, T. (1978). *Outliers in Statistical Data.* New York: John Wiley & Sons.

BELSLEY, D.A., KUH, E. and WELSCH, R.E. (1980). *Regression Diagnostics.* New York: Wiley.

COOK, R.D. and WEISBERG, S. (1982). *Residuals and Influence in Regression.* New York: Chapman and Hall.

FOMBY, T.B. and GUILKEY, D.K. (1978). On Choosing the Optimal Level of Significance for the Durbin-Watson Test and the Bayesian Alternative. *J. of Econometrics,* **8**, 203-213.

GRIFFITHS, W.E. and DAO, D. (1980). A Note on a Bayesian Estimator in an Autocorrelated Error Model. *J. of Econometrics,* **12**, 389-392.

HILL, B.M. (1975). On Coherence, Inadmissibility and Inference about Many Parameters in the Theory of Least Squares. *Studies in Bayesian Econometrics and Statistics in Honor of Leonard J. Savage,* Fienberg, S.E. and Zellner, A. (eds.) Amsterdam: North-Holland, 555-584.

JOHNSON, W. and GEISSER, S. (1980). A Predictive View of the Detection and Characterization of Influential Observations in Regression Analysis. School of Theoretical Statistics, *Tech. Rep.* **365**. University of Minnesota.

KADANE, J.B. (1980). Predictive and Structural Methods for Eliciting Prior Distributions. *Bayesian Analysis in Econometrics and Statistics: Essays in Honor of Harold Jeffreys.* Amsterdam: North-Holland, 89-93.

— et al. (1980). Interactive Elicitation of Opinion for a Normal Linear Model, *J. Amer. Statist. Assoc.,* **75**, 845-854.

KLOEK, T. and VAN DIJK, H.K. (1980). Bayesian Estimates of Equation System Parameters: An Application of Integration by Monte Carlo. *Bayesian Analysis in Econometrics and Statistics: Essays in Honor of Harold Jeffreys.* Amsterdam: North-Holland, 311-329.

LITTERMAN, R. (1980). A Bayesian Procedure for Forecasting with Vector Autoregressions. ms., Dept. of Economics, MIT, Cambridge, MA.

MONAHAN, J.F. (1983). Fully Bayesian Analysis of ARMA Time Series Models. *J. of Econometrics,* **21**, 307-331.

PRESS, S.J. (1980). Bayesian Computer Programs. *Bayesian Analysis in Econometrics and Statistics: Essays in Honor of Harold Jeffreys.* Amsterdam: North-Holland, 429-442.

SUREKHA, K. and GRIFFITHS, W.E. (1983). A Monte Carlo Comparison of Some Bayesian and Sampling Theory Estimators in Two Heteroscedastic Error Models. ms., Indian Statistical Institute, Calcutta and U. of New England, Armidale, Australia.

ZELLNER, A. (1975). Bayesian Analysis of Regression Errors. *J. Amer. Statist. Assoc.,* **70**, 138-144.

— (1980). The Current State of Bayesian Econometrics. Invited address presented at the Canadian Conference on Applied Statistics, Montreal, April 29-May 1, 1981. To be published in the Conference Proceedings Volume.

— (1982). On Assessing Prior Distributions and Bayesian Regression Analysis with *g*-Prior Distributions. *Bayesian Inference and Decision Techniques with Applications: Essays in Honor of Bruno de Finetti,* in preparation. P. Goel and A. Zellner (eds.).

— (1983). Applications of Bayesian Analysis in Econometrics, *The Statistician,* **32**, 23-34.

ZELLNER, A. and PALM, F. (1974). Time Series Analysis and Simultaneous Equation Models. *J. of Econometrics,* **2**, 17-54.

ZELLNER, A. and ROSSI, P.E. (1984). Bayesian Analysis of Dichotomous Quantal Response Models. *J. of Econometrics,* **25**, 365-393.

ZELLNER, A. and VANDAELE, W. (1975). Bayes-Stein Estimators for *k*-Means, Regression and Simultaneous Equation Models. *Studies in Bayesian Econometrics and Statistics in Honor of Leonard J. Savage.* Amsterdam: North-Holland, 627-653.

ZELLNER, A. and WILLIAMS, A. (1973). Bayesian Analysis of the Federal Reserve-MIT-Penn Model's Almon Lag Consumption Function. *J. of Econometrics,* **1**, 267-299.

APPENDIX *

ON ASSESSING INFORMATIVE PRIOR DISTRIBUTIONS FOR REGRESSION COEFFICIENTS

In assessing informative prior distributions for regression coefficients, it is often difficult to assign values for prior covariances of regression coefficients. Since such quantities are needed to perform Bayesian regression analyses with informative prior distributions such as the Raiffa-Schlaifer natural conjugate and Dickey prior distributions, it is important to have operational techniques for assigning values to prior covariances of regression coefficients. Below procedures are suggested that may be helpful in solving this problem and also providing checks on the consistency of prior beliefs regarding regression coefficients.

To illustrate the general approach, we first consider the following standard regression equation with two independent variables:

$$y_i = x_{1i}\beta_1 + x_{2i}\beta_2 + u_i \quad , \quad i = 1,2,\ldots,n . \tag{1}$$

* This appendix was written in November, 1972.

Let $\eta_i = x_{1i}\beta_1 + x_{2i}\beta_2$ be the mean of y_i. Our problem is to provide an operational procedure for evaluating the prior covariance between β_1 and β_2, $\text{cov}(\beta_1,\beta_2) = \sigma_{12}$. One procedure that accomplishes this objective is the following one:

a) Assign values to the prior means and variances of β_1 and β_2, that is $\bar{\beta}_1 = E\beta_1$, $\bar{\beta}_2 = E\beta_2$, $\sigma_{11} = \text{var}\,\beta_1$ and $\sigma_{22} = \text{var}\,\beta_2$. It is often possible to assign these values without much difficulty.

b) Given values for x_{1i} and x_{2i}, assign prior values for the mean and variance of η_i, that is $E\eta_i = \bar{\eta}_i$ and $\text{var}\,\eta_i = \sigma^2_{\eta_i}$.

b.1) Check that $\bar{\eta}_i$ is in agreement with $x_{1i}\bar{\beta}_1 + x_{2i}\bar{\beta}_2$.

c) Since $\eta_i - \bar{\eta}_i = x_{1i}(\beta_1 - \bar{\beta}_1) + x_{2i}(\beta_2 - \bar{\beta}_2)$,

$$\sigma^2_{\eta_i} = x^2_{1i}\sigma_{11} + x^2_{2i}\sigma_{22} + 2x_{1i}x_{2i}\sigma_{12} \tag{2}$$

and thus

$$\sigma_{12} = (\sigma^2_{\eta_i} - x^2_{1i}\sigma_{11} - x^2_{2i}\sigma_{22})/2x_{1i}x_{2i} \tag{3}$$

is the implied value for σ_{12}, the prior covariance for β_1 and β_2.

As a check on the consistency of (3), various (x_{1i},x_{2i}), $i = 1,2,\ldots,N$, can be employed in connection with (2). Each pair will generate a value for σ_{12} via (3). If these are reasonably consistent*, we can average (2) as follows

$$\sum_{i=1}^{N} \sigma^2_{\eta_i}/N = \sigma_{11}\Sigma x^2_{1i}/N + \sigma_{22}(\Sigma x^2_{2i}/N) + 2\sigma_{12}\Sigma x_{1i}x_{2i}/N \tag{4a}$$

or

$$\bar{\sigma}^2_{\eta} = \sigma_{11}m_{11} + \sigma_{22}m_{22} + 2\sigma_{12}m_{12}\,. \tag{4b}$$

Then

$$\sigma_{12} = (\bar{\sigma}^2_{\eta} - \sigma_{11}m_{11} - \sigma_{22}m_{22})/2m_{12}\,. \tag{5}$$

The above procedure generalizes to apply to cases in which we have more than two coefficients, e.g. $\beta' = (\beta_1,\beta_2,\ldots,\beta_k)$. In this case we have to assign values to $m = k(k-1)/2$ prior covariances. Let $\eta_i = \mathbf{x}_i'\beta$, where $\mathbf{x}_i'$ is a $1\times k$ given vector and η_i is the mean of y_i given $\mathbf{x}_i$. To evaluate the $m = k(k-1)/2$ prior covariances proceed as follows:

a′) Assign values for the prior means and variances of the elements of β, i.e. $E\beta_i = \bar{\beta}_i$ and $\text{var}\,\beta_i = \sigma_{ii}$, $i = 1,2,\ldots,k$.

b′) Given $\mathbf{x}_i$, assign a prior mean and variance for η_i, i.e. $E\eta_i = \bar{\eta}_i$ and $\text{var}\,\eta_i = \sigma^2_{\eta_i}$.

b.1′) Check that η_i is consistent with $\mathbf{x}_i'\bar{\beta}$, where $\bar{\beta}' = (\bar{\beta}_1,\bar{\beta}_2,\ldots,\bar{\beta}_k)$, the prior mean vector for β.

c′) From $\eta_i - \bar{\eta}_i = \mathbf{x}_i'(\beta - \bar{\beta})$, the prior variance of η_i is given by

$$\sigma^2_{\eta_i} = \sum_{j=1}^{k} x^2_{ij}\sigma_{jj} + 2\sum_{j=1}^{k}\sum_{\substack{\ell=1 \\ j\neq\ell}}^{k} x_{ij}x_{i\ell}\sigma_{j\ell} \tag{6}$$

* If they are found to vary considerably, this may be due to carelessness and/or an incorrect specification of the functional form of (1). Also, as pointed out by Robert C. Blattberg, it is important to check that the following condition is satisfied

$$\begin{vmatrix} \sigma_{11} & \sigma_{12} \\ \sigma_{21} & \sigma_{22} \end{vmatrix} > 0, \text{ i.e. } \sigma_{11}\sigma_{22} - \sigma^2_{12} > 0.$$

where x_{ij} is the j'th element of $\mathbf{x}_i' = (x_{i1}, x_{i2}, \ldots, x_{ij}, \ldots, x_{ik})$.

(6′) If (6) is written for m' choices of the vector $\mathbf{x}_i$, with $m' \geq m$, we can write the m' relations as follows

$$\sigma_\eta = D\sigma_v + 2C\sigma_c \tag{7}$$

where

$\sigma_\eta' = (\sigma^2_{\eta_1}, \sigma^2_{\eta_2}, \ldots, \sigma^2_{\eta_m})$, a $1 \times m'$ vector,

$\sigma_v' = (\sigma_{11}, \sigma_{22}, \ldots, \sigma_{kk})$, a $1 \times k$ vector,

$\sigma_c' = (\sigma_{12}, \sigma_{13}, \ldots, \sigma_{1k}, \sigma_{23}, \ldots, \sigma_{2k}, \ldots, \sigma_{k-1,k})$, a $1 \times m$ vector,

D = an $m' \times k$ matrix with i'th row, $\mathbf{d}_i' = (x^2_{i1}, x^2_{i2}, \ldots, x^2_{ik})$

C = an $m' \times m$ matrix with i'th row $\mathbf{c}_i' = (c_{12i}, c_{13i}, \ldots, c_{1ki}, c_{23i}, \ldots, c_{2ki}, \ldots, c_{k-1,k,i})$ with $c_{ij\ell} = x_{ij}x_{i\ell}$.

Then on multiplying both sides of (7) by C', we can solve for the vector σ_c given that $C'C$ has an inverse, that is, we have

$$\sigma_c = \tfrac{1}{2}(C'C)^{-1}C'(\sigma_n - D\sigma_v). \tag{8}$$

Since the elements of σ_c are the prior covariances, the prior variance-covariance matrix for the elements of β has been completely evaluated*.

In (a′) the prior variances of the regression coefficients have been assigned. Equation (8) provides the prior covariances. Thus values for all distinct elements of the regression coefficient prior covariance matrix have been obtained. In the case of a Raiffa-Schlaifer natural conjugate prior pdf,

$$p(\beta,\sigma|\cdot) = p_N(\beta|\sigma,\cdot)p_{IG}(\sigma|\cdot) \qquad \begin{array}{c} 0 < \sigma < \infty \\ -\infty < \beta_i < \infty \\ i = 1,2,\ldots,k \end{array} \tag{9}$$

with $p_N(\beta|\sigma\cdot) \propto \sigma^{-k}\exp\{-\frac{1}{2\sigma^2}(\beta-\bar{\beta})'A(\beta-\bar{\beta})\}$, A p.d.s., and

$p_{IG}(\sigma|\cdot) \propto \sigma^{-(\nu_0+1)}, \exp\{-\nu_0 s_0^2/2\sigma^2\}$, $\nu_0 s_0 > 0$, the marginal prior for β is

$$p_s(\beta|\cdot) \propto \{\nu_0 s_0^2 + (\beta-\bar{\beta})'A(\beta-\bar{\beta})\}^{-\nu_0+k/2} \tag{10}$$

which is in the form of a multivariate Student t pdf with ν_0 degrees of freedom. The prior covariance matrix for β, $\Sigma_\beta = \{\sigma_{ij}\}$, is given by:

$$\Sigma_\beta = \{\sigma_{ij}\} = A^{-1}\nu_0 s_0^2/(\nu_0 - 2) \qquad \nu_0 > 2. \tag{11}$$

Letting a^{ij} be a typical element of A^{-1}, we have from (11)

$$a^{ij} = \sigma_{ij}(\nu_0 - 2)/\nu_0 s_0^2 \qquad i,j = 1,2,\ldots,k \tag{12}$$

Given that σ_{ij}'s value has been obtained from (8) and that the values of ν_0 and s_0^2, parameters of the prior pdf for σ, are available, equation (12) yields a value for a^{ij}, a typical element of A^{-1}.

* If $\Sigma_\beta = \{\sigma_{ij}\}$ denotes the prior variance-covariance matrix for β, assessed as explained above, it is necessary to check that $|\Sigma_\beta| > 0$ and conditions on the principal minors needed for Σ_β to be pds.

DISCUSSION

M. MOUCHART (*Université Catholique de Louvain, Belgium*)

These comments will handle both the contents and some omissions of the paper.

As far as the content is concerned, no summary is needed; indeed Professor Zellner should be congratulated for his crystal-clear exposition.

A significant change (at least in my opinion) from his textbook of 1971 is that natural-conjugate informative priors are taken as prior distributions of reference while non-informative priors are taken as particular cases (I personally tend to consider Bayesian analysis under non-informative priors as a large sample approximation). One even finds in the paper a (finite) mixture of natural conjugate distributions: these are steps in the right direction. After this meeting should we not expect that those mixtures are going to be considered as a reference form for a prior distribution?

Although I feel in sympathy with most statements in the paper, I nevertheless found that some of the topics presented might raise questions of interest. The questions will be proposed in the order of presentation of the paper.

(i) *Bayesian computer programs*

Our experience at CORE reveals that we were more successful in developing an interest in numerical analysis among statisticians and econometricians than in attracting numerical analysts' interest for (Bayesian) statistical problems. Is this experience particular to local conditions or is it more general? How far should statisticians allocate their time to be personally involved in the development of computer programs and how far should we try to involve numerical analysts? Also, how can we improve the diffusion, in particular the portability, of those programs? How far is a pure market solution efficient?

(ii) *Diagnostic checking of models, priors and data*

No doubt this is an important and difficult area and significant progress is still to be expected. A general question is: given the data, what should a Bayesian consider as "good" model? If we agree that a Bayesian model is the joint specification of the prior distribution and of the sampling process, should a "good" model make the data "astonishing" (i.e. "far" in the tail of the predictive distribution and leading to strong revision of the prior distribution) or "as expected" (i.e. close to the center of the predictive distribution and leading to minor revision of the prior distribution). If we consider more specifically the omitted variables problem, I wonder why the author has not found it necessary to stress the relevance of the following distinction. One problem is to ask whether the parameters of interest are those of the sampling process generating $(y|X_1)$ or those of the sampling process generating $(y|X_1, X_2)$. Another problem is to ask whether the coefficient of X_1 is (exactly or roughly) the same in those two models. Furthermore, if the decision maker is convinced that the parameters of interest are those of the process generating $(y|X_1,X_2)$, although X_2 is not available, why did the author not mention a "natural" solution, viz. to compute $p(\beta_1,\beta_2,\sigma^2\|y,X_1)$ as $\int p(\beta_1,\beta_2,\sigma^2|y,X_1,X_2)$ $p(X_2|y,X_1)dX_2$ and then to discuss the specification of $p(X_2|y,X_1)$ and the evaluation of such an integral?

(iii) *Procedures for assessing prior distributions*

I completely share the author's doubts about the use of non-informative priors, particularly in the case of many parameters. I found it interesting to learn that the appendix had been written in November 1972 and wondered whether a variation-free parametrizaton of the prior distribution would not be suitable i.e. assign successively the parameters of $p(\beta_1)$, $p(\beta_2|\beta_1)$ $p(\beta_3|\beta_1,\beta_2)$ etc. As only one-dimensional distributions would be involved, coherence would be easier to achieve but, at the same time, some possibility of cross-checking could be lost.

(iv) *Analysis of non-standard models*

The (small) sample of examples given in this section indicates an important direction of current methodological progress. Let me suggest the following question: if a posterior distribution is evaluated by numerical integration, how should we *analyze* the "learning-by-observing" process (i.e. the revision prior-to-posterior); for instance, how to analyze residuals, outlying or influential observations?

(v) *Simplification of current models*

I feel much in sympathy with this section but in view of the modern theory of reduction for Bayesian experiments, I would tend to present those problems in terms of approximate sufficiency on the parameter space.

I want to conclude those comments by two remarks somewhat out of the scope of this paper but suggesting how far econometrics may be a provocative field of application for Bayesian statisticians and how Bayesian methods may be used to enrich in the future the statistical analysis of economic facts.

Modelling human behaviour often leads to very complex data generating processes, this being due to the necessarily arbitrary frontiers separating the fields of economics, sociology, psychology, etc. and to the time instability of parameters. Also the aggregation of individual behaviours involves the non-independence of individuals with respect to a group and the recognition that constraints may operate differently on the individual and on the aggregate level. This leads one to introduce a large number of nuisance parameters and, therefore, gives a crucial role to "out-of-sample" information. As a matter of fact, a sample very often adds only a little bit to what was already known on the parameters.

In this framework Bayesian methods offer a most natural way, both to a unified treatment of nuisance parameters (while this topic is often controversial in sampling-theory) and to a systematic blending of "sample" and "non-sample" information (leaving irrelevant the arbitrary character of the distinction between "sample" and "non-sample").

A final remark relies on the observation that all human populations have a finite size. This suggests that finite population models should probably receive more attention in econometric modelling. In this field assumptions of independence and of exchangeability have a role comparable to that of ergodicity in stochastic processes, viz. to make inference from the observation of a finite trajectory possible. It is my hope that such models, thanks to a genuinely probabilistic treatment of aggregation in economics, would lead to more appealing specifications.

S. GEISSER (*University of Minnesota*)

I would suggest that rather than use the ad hoc influence measure namely that of formula (7), Zellner use the measures that appear in Johnson and Geisser (1981, 1983). There we measured estimative and predictive influence with respect to actual observables while Zellner is also interested in the influence of the prior specification as embodied in his pseudo-observations individually or as a set. There would be virtually no problem in adapting the Johnson-Geisser Bayesian approach (whether predictive or estimative) to serve Zellner's needs. Moreover a predictive significance test, as outlined in the aforementioned references, is available to assess the consistency of the prior specification with the actual observations. These methods are reviewed in my paper in this symposium.

D.J. POIRIER (*University of Toronto, Canada*)

While Professor Zellner's aversion to large models is understable, I do not feel that simply counting the number of parameters in the model always serves as an adequate measure of its size. Professor Zellner's discussion of vector autoregressive models is a case in point. After downgrading such models for their large number of parameters, Professor Zellner praises the vector autoregressive model of Litterman for its forecasting accuracy, despite the fact that it too involves hundreds of parameters! The explanation for this apparent contradiction is that Litterman's model is small in a different sense, namely, there is a comparatively small number of free hyperparameters to be specified in the rather ingenuous informative prior distribution that he uses. Indeed it is precisely this informative prior distribution that presumably lies behind the impressive forecasting record of Litterman's model versus a vector autoregression with an "uninformative" prior. The quantification of a model's "size" is a complex task, and I think this example illustrates that simply counting the number of parameters may not always be an adequate measure.

REPLY TO THE DISCUSSION

Thanks to Dr. Mouchart for his kind words about my paper and for his many questions which I hoped my paper would provoke.

On prior distributions, there is a need for diffuse priors and informative priors, both of which were used in my 1971 book. Getting a good representation of the state of our information, whether diffuse or non-diffuse was and is my main objective. Thus, contrary to what Mouchart says under his point (iii), I do not have doubts about the proper use of non-informative priors.

On development and diffusion of computer programs, *good* programs made available at reasonable prices will have no trouble in being diffused. Also, I believe that we can rely on individual researchers to make wise decisions vis à vis their participation in the development of computer programs and/or of Bayesian statistics.

With respect to what is a good model, in general a good model is one that explains past data well and is useful in prediction and policy-making. I have trouble understanding Mouchart's comments on this issue. On the LOV problem, the testing problems that he mentions were explicitly considered in my and Siow's last Valencia Conference paper on Bayesian analysis of regression hypotheses. In addition, Mouchart's last suggestion on this topic should be compared with the extended model approach that I introduced in my discussion of the LOV problem. On new methods for assessing priors and the use of approximate sufficiency, I hope that Mouchart will work out the details of what he has in mind.

In the last part of his discussion, Mouchart states, "Modelling human behaviour often leads [necessarily?] to very complex data generating processes..." As pointed out in my paper, I know of no very complex data generating processes or models in the social or physical sciences that work well in explaining past data and in prediction. As regards a number of very complicated macroeconometric models, Christ (1975, p. 59) has written, "...they [the models] disagree so strongly about the effects of important monetary and fiscal policies that they cannot be considered reliable guides to such policy effects, until it can be determined which of them are wrong in this respect and which (if any) are right''. It was with results like this in mind that in my paper I cited examples and references to sophisticatedly simple models in economics and other areas that work reasonably well in practice. Over the years, I have urged economists and other social scientists to emphasize sophisticated simplicity in model-building in accord with the Jeffreys-Wrinch Simplicity Postulate, Ockham's Razor, the Principle of Parsimony, etc. Indeed, many of the best economists, including the Nobel Prize winners, Jan Tinbergen, Ragnar Frisch, Milton Friedman, Kenneth Arrow, George Stigler, James Tobin and Theodore Schultz, as well as many natural scientists and statisticians, appear to appreciate the importance of sophisticated simplicity in model-building.

I thank Seymour Geisser for calling my attention to some more of his and Johnson's influential work on influence. Their procedures employing information measures, etc., are certainly very valuable. However, as I stated in my paper, ''...measures of influence that reflect not only the sample data's impact on analyses but also the prior distribution's impact will be presented''. The results following equation (5a) of my paper relate to *both* the influence of the prior and of the data on analyses and appear to be new and not contained in the Johnson-Geisser papers.

As regards the generalized Cook distance measure given in (7) of my paper, it is but one useful measure that can be computed. As I mentioned in my paper, various types of residuals should also be carefully studied. Also, our Bayesian Regression Analysis Program (BRAP) permits users to compute readily posteriors pdfs, predictive pdfs, residuals, etc., with complete and partial sets of data. Thus users can examine many computational results in their efforts to determine how features of their priors and data influence their analyses.

I did not say in my paper that the number of parameters in a model ''always serves as an adequate measure of its size'', as Dale Poirier apparently suggests. However, it is obviously important to know the number of parameters in a model, particularly in relation to the number of available observations. The ratio of the number of observations to the number of parameters in a model is just a rough measure that is useful in alerting analysts to possible problems in analyzing models. Of course other features of models, for example properties of design matrices, unit roots in time series models, etc., deserve attention too. Model ''size'' and ''complexity'' are concepts that were not the main concern of my paper. However, in this connection, see Jeffreys (1967, pp. 47-49) for a suggested quantitative measure of model complexity that relates to differential equations, the forms of many laws in science, namely, ''We could define the complexity of a differential equation, cleared of roots and fractions, by the sum of the order, the degree, and the absolute values of the coefficients''. (p. 47). It would be useful to relate this and other measures of complexity which have appeared in the literature to the problem of characterizing the degree of complexity of various types of econometric and statistical models.

REFERENCES IN THE DISCUSSION

CHRIST, C.F. (1975). Judging the Performance of Econometric Models of the U.S. Economy. *Internat. Econ. Rev.* **16**, 54-74.

JEFFREYS, H. (1967). Theory of Probability. Oxford: University Press.

JOHNSON, W. and GEISSER, S. (1981). On the architecture of estimative influence functions. *Tech. Rep.* **23**, University of California.

— (1983). A predictive view of the detection and characterization of influential observations in regression analysis. *J. Amer. Statist. Assoc.* **78, 381**, 137-144.

CONTRIBUTED PAPERS

BAYESIAN STATISTICS 2, pp. 589-602
J.M. Bernardo, M.H. DeGroot, D.V. Lindley, A.F.M. Smith (Eds.)

Bayesian Estimation Methods for Incomplete Two-Way Contingency Tables Using Prior Beliefs of Association

JAMES H. ALBERT
Bowling Green State University

SUMMARY

Suppose that a user possesses vague prior information about the association structure of a 2×2 contingency table and consider the problem of estimating the cell probabilities of the table using this prior information together with sample counts from an incomplete table. The prior distribution of Albert and Gupta (1983) is used to reflect vague prior beliefs about an odds ratio and this distribution is used in the development of approximate posterior means and variances for the vector of cell probabilities. The computation of these Bayesian estimates is illustrated in the special case when the classification variables are believed independent.

Keywords: CROSS-PRODUCT RATIO; INDEPENDENCE; MIXTURES; PARTIALLY CROSS-CLASSIFIED DATA.

1. INTRODUCTION

In this paper, Bayesian estimation methods are proposed for the cell probabilities of a 2×2 contigency table, when both completely and partially cross-classified data are collected. To illustrate the sampling scheme, consider data on 456 premature live births, given in Chen and Fienberg (1974) and presented in Table 1. The classification variables in this example are the infant's health index score (low, high) and their serum bilirium reading (low, high). Of the entire sample, 279 infants are completely classified with respect to both variables; 24 are partially classified with respect to their serum bilirium reading and the remaining 153 are classified only with respect to their health index. It is a trivial problem to estimate the cell probabilities of the 2×2 table using solely the completely classified counts. A nontrivial problem is how to use the counts in the two partially classified tables together with the completely classified counts to estimate the cell probabilities.

Many authors, e.g. Hocking and Oxspring (1974), Chen and Fienberg (1974), and Fuchs (1982), have found maximum likelihood estimates (MLE's) of the cell probabilities. As will be shown in Section 3.1, these estimates allocate the partially classified counts to the 2×2 table using proportions that are obtained from the completely classified table. The manner in which the partially classified counts are allocated to the complete table depends primarily on the association structure in the table. In the MLE procedure, the association structure is "estimated" by the completely classified counts.

TABLE 1

Data of premature infants classified with respect to Health Index and Serum Bilirium level

(from Chen and Fienberg (1974))

		Health Index: Low	Health Index: High		
Serum Bilirium Level	Low	35	75		11
	High	57	112		13
				279	24
		117	36	153	

Consider the situation where only a small portion of the total number of counts are completely classified. In this situation, the completely classified counts provide little information about the manner in which the partially classified counts are allocated to the table. In the extreme case where all of the counts are partially classified, the cell probabilities are not even estimable by the data. However, if prior information exists about the association structure in the table, then this information can be used (together with the completely classified counts) to allocate the partially classified counts to the table and give estimates for the cell probabilities. As Antleman (1972) explains, this Bayesian approach is necessary when all the data collected is partially classified.

To use the Bayesian method, the main task is to find a prior distribution which can reflect the typical vague form of prior information about the association structure of the table. To this end, Albert and Gupta (1983) introduced a class of priors, a mixture of Dirichlet distributions, which is designed to reflect vague prior beliefs about the cross-product ratio α, a common measure of association in a 2×2 table. (The rationale for the use of this class versus the use of the conjugate class is given in Albert and Gupta (1983)). One advantage of this class is that only two parameters are elicited from the user; basically these parameters reflect a guess at the association structure of the table and a statement of the precision of this guess.

Before we proceed, some notation will be given. Suppose that n observations are completely classified with respect to classification variables A and B and n_1 (n_2) observations are partially classified with respect to variable $A(B)$, resulting in the observed counts below.

	B			
A	x_{11}	x_{12}	$x_{1\cdot}$	y_1^A
	x_{21}	x_{22}	$x_{2\cdot}$	y_2^A
	$x_{\cdot 1}$	$x_{\cdot 2}$	n	n_1
	y_1^B	y_2^B	n_2	

(The dot notation represents summation over the appropriate index). It is of interest to estimate $\mathbf{p} = (p_{11}, p_{12}, p_{21}, p_{22})$, where p_{ij} denotes the probability of falling in the (i,j) cell. If observations are classified from an infinite population, then the likelihood is given by

$$\Pi p_{ij}^{x_{ij}} \Pi p_{i.}^{y_i^A} p_{.i}^{y_i^B} . \tag{1.1}$$

Since the prior used is a mixture of Dirichlet distributions, it will be convenient to define

$$f_D(\mathbf{p} \mid K\eta) = \Gamma(K) \prod_{i,j} p_{ij}^{Kf_{ij}-1} / \Gamma(Kf_{ij}) , \tag{1.2}$$

the Dirichlet density with prior mean vector $\eta = (f_{11}, f_{12}, f_{21}, f_{22})$ and precision parameter K.

In the special case where the data is entirely completely classified ($n_1 = n_2 = 0$), Albert and Gupta (1983) show that the posterior distribution is a mixture of Dirichlet densities, and use this representation to develop a point estimator and credible ellipsoid for $\mathbf{p}$. In this paper, we consider the more general situation where at least part of the data is partially classified, and show in Section 2 that the posterior distribution is representable as a mixture of Dirichlet-beta densities. In Section 3, simple approximations are developed for the posterior means which show how the partially classified counts are allocated in the complete table. In Section 4, we conclude our discussion by illustrating the computation of the posterior means and variances for the data in Chen and Fienberg (1974) in the situation where the user believes that the two classification variables are independent.

2. PRIOR TO POSTERIOR ANALYSIS

2.1. *The Prior Distribution*

Albert and Gupta (1983) introduced the following two-stage prior distribution to reflect prior beliefs about association in a 2×2 table.

Stage 1: The vector $\mathbf{p}$ is given the Dirichlet distribution (1.2), where the components of η have row margins η_a, $1-\eta_a$, column margins η_b, $1-\eta_b$, and cross-product ratio α_0. Equivalently, the set of prior means satisfy the configuration

$f_{11}(\eta_a,\eta_b)$	$\eta_a - f_{11}(\eta_a,\eta_b)$	η_a	(2.1)
$\eta_b - f_{11}(\eta_a,\eta_b)$	$1-\eta_a-\eta_b+f_{11}(\eta_a,\eta_b)$	$1-\eta_a$	
η_b	$1-\eta_b$		

where $\alpha_0 = [f_{11}(\cdot,\cdot)(1-\eta_a-\eta_b+f_{11}(\cdot,\cdot)]/[(\eta_a-f_{11}(\cdot,\cdot))(\eta_b-f_{11}(\cdot,\cdot))]$.

Stage II: The vector of hyperparameters (η_a,η_b) is given a uniform distribution on the unit square.

The resulting prior density on $\mathbf{p}$ is given by

$$\pi_m(\mathbf{p}) = \int\int f_D(\mathbf{p} \mid K\eta^*) d\eta_a \, d\eta_b, \tag{2.2}$$

where $\eta^* = (f_{11}, f_{12}, f_{21}, f_{22})$ is the vector of prior means with configuration (2.1) (for ease of notation, we will write f_{11} instead of $f_{11}(\eta_a,\eta_b)$, although it is understood that the prior mean is a function of the parameters η_a and η_b).

The prior distribution (2.2) is design to accept the typical form of vague prior information about the association structure in the table. Two parameters are elicited from the user; the parameter α_0 is a guess at the cross-product ratio α and the parameter K

reflects the sureness of one's guess at α. It is illustrated in Albert and Gupta (1983) that the induced prior on $\ell n\ \alpha$ is approximately bell-shaped and symmetric about $\ell n\ \alpha_0$. Therefore, by the specification of an interval which is thought to contain α with probability .9 and the use of Figure 2 in Albert and Gupta (1983), one can obtain values of α_0 and K.

2.2. *Posterior Analysis*

If **p** is given the prior (2.2), then the posterior density of **p** is proportional to

$$\int\int \Pi\, [p_{ij}^{x_{ij} + Kf_{ij} - 1} / \Gamma(Kf_{ij})] \Pi p_{i.}^{y_i^A}\ p_{.i}^{y_i^B}\ d\eta_a d\eta_b\,. \tag{2.3}$$

This density can be seen to be a mixture of densities with kernel

$$\Pi p_{ij}^{a_{ij}-1}\ \ \Pi p_{i.}^{y_i^A}\ p_{.i}^{y_i^B}\,, \tag{2.4}$$

where $a_{ij} = x_{ij} + Kf_{ij}$. The family of distributions, with kernel (2.4) is called by Antleman (1972) the Dirichlet-Beta ($D\beta$) family. It will be convenient first to summarize some facts about the $D\beta$ distribution in Sections 2.2.1 and 2.2.2, and then apply these results in obtaining expressions for the posterior moments in Sections 2.2.3 and 2.2.4.

2.2.1. *The Simple Dirichlet-Beta Distribution*

Unfortunately, the general $D\beta$ distribution is not very tractable; thus, as in Antleman, we first consider the special case where $y_1^B = y_2^B = 0$. (The results given here can be easily adjusted for the case where $y_1^A = y_2^A = 0$). After some manipulation, it can be shown that this density, called the simple $D\beta$ density, can be represented as

$$\pi_s(\mathbf{p}\,|\,\mathbf{a},\mathbf{y}^A) = \sum_{i=0}^{y_1^A} \sum_{j=0}^{y_2^A} f_{Bb}(i\,|\,y_1^A, a_{11}, a_{12}) f_{Bb}(j\,|\,y_2^A, a_{21}, a_{22})$$

$$\bullet f_D(\mathbf{p}\,|\,a_{11} + i,\ a_{12} + y_1^A - i,\ a_{21} + j,\ a_{22} + y_2^A - j)\,, \tag{2.5}$$

where $\mathbf{a} = (a_{11}, a_{12}, a_{21}, a_{22})$, $\mathbf{y}^A = (y_1^A, y_2^A)$ and f_{Bb} denotes the beta-binomial density given by

$$f_{Bb}(k\,|\,m,b,c) = \binom{m}{k} B(b + k,\ c + m - k)/B(b,c)\,.$$

The representation (2.5) is useful in computing posterior moments. For example, using the rules of conditional expectation,

$$E(p_{11}\,|\,\mathbf{a},\mathbf{y}^A) = E[E(p_{11}\,|\,\mathbf{a},\mathbf{y}^A, i, j)]$$

$$= E\left(\frac{a_{11} + i}{a_{..} + n_1}\right)$$

$$= \frac{a_{11} + y_1^A a_{11}/a_{1.}}{a_{..} + n_1} = \frac{(a_{11}\, a_{1.} + y_1^A)}{a_{1.}(a_{..} + n_1)} = \lambda_{11} \tag{2.6}$$

and

$$\mathrm{Var}(p_{11}\,|\,\mathbf{a},\mathbf{y}^A) = E[\mathrm{Var}(p_{11}\,|\,\mathbf{a},\mathbf{y}^A,i,j)] + \mathrm{Var}[E(p_{11}\,|\,\mathbf{a},\mathbf{y}^A,i,j)]$$

$$= E[(a_{..} + n_1 + 1)^{-1} \frac{a_{11} + i}{a_{..} + n_1} (1 - \frac{a_{11} + i}{a_{..} + n_1})]$$

$$+ \operatorname{Var}\left(\frac{a_{11} + i}{a_{..} + n_1}\right) \tag{2.7}$$

$$= (a_{..} + n_1 + 1)^{-1}[\lambda_{11}(1-\lambda_{11}) + \frac{y_1^A \lambda_{11} a_{12}}{(a_{1\cdot} + 1)a_{1\cdot}}].$$

In general, the mean vector of $\mathbf{p}$ is given by $\lambda = (\lambda_{11},\lambda_{12},\lambda_{21},\lambda_{22})$, where

$$\lambda_{ij} = E(p_{ij} | \mathbf{a},\mathbf{y}^A) = \frac{a_{ij}}{a_{i\cdot}} \frac{(a_{i\cdot} + y_i^A)}{(a_{..} + n_1)}. \tag{2.8}$$

The posterior covariance matrix is given by

$$\operatorname{cov}(\mathbf{p} | \mathbf{a},\mathbf{y}^A) = (a_{..} + n_1 + 1)^{-1} (\operatorname{diag}\{\lambda_{11},\lambda_{12},\lambda_{21},\lambda_{22}\} - \lambda\lambda'$$

$$+ \begin{bmatrix} \mathbf{A}_1 & 0 \\ 0 & \mathbf{A}_2 \end{bmatrix}), \tag{2.9}$$

where

$$\mathbf{A}_i = \frac{y_i^A \lambda_{i1} a_{i2}}{(a_{i\cdot} + 1)a_{i\cdot}} \cdot \begin{bmatrix} 1 & -1 \\ -1 & 1 \end{bmatrix}, \quad i = 1,2.$$

2.2.2. *The General Dirichlet-Beta Distribution*

Antleman (1972) shows that the general $D\beta$ density is given by

$$\pi_G(\mathbf{p} | \mathbf{a},\mathbf{y}) = (I(\mathbf{a},\mathbf{y}))^{-1} \Pi p_{ij}^{a_{ij}-1} \Pi p_{i.}^{y_i^A} p_{.i}^{y_i^B}, \tag{2.10}$$

where

$$I(\mathbf{a},\mathbf{y}) = C \sum_{i=0}^{y_1^A} \sum_{j=0}^{y_2^A} \binom{a_{11} + i - 1}{i} \binom{a_{12} + y_1^A - 1 - i}{y_1^A - i} \binom{a_{21} + j - 1}{j}$$

$$\cdot \binom{a_{22} + y_2^A - 1 - j}{y_2^A - j} \binom{a_{\cdot 2} + n_1 + y_2^B - i - j - 1}{y_2^B} \binom{a_{\cdot 1} + i + j + y_1^B - 1}{y_1^B} \tag{2.11}$$

$\mathbf{y} = (\mathbf{y}^A,\mathbf{y}^B)$ and $C = \Pi\Gamma(a_{ij})\Pi\Gamma(y_i^A + 1)\Gamma(y_i^B + 1)/\Gamma(a_{..} + n_1 + n_2)$.
From this representation, one can show that moments of the distribution are ratios of normalizing constants. For example,

$$E(p_{11} | \mathbf{a},\mathbf{y}) = I((a_{11} + 1, a_{12},a_{21},a_{22}),\mathbf{y})/I(\mathbf{a},\mathbf{y}). \tag{2.12}$$

These expressions are tedious to compute, especially for large values of y_1^A and y_2^A. Thus, in Antleman (1972), we propose to approximate the general $D\beta$ distribution by a simple $D\beta$ distribution as follows. First, write the $D\beta$ kernel as

$$[\Pi p_{ij}^{a_{ij}-1} \Pi p_{i.}^{y_i^A}] \Pi p_{.j}^{y_j^B} = k_1 \cdot k_2, \tag{2.13}$$

a product of a simple $D\beta$ kernel k_1 and the beta kernel k_2. Next, approximate the simple $D\beta$ density with kernel k_1 by a Dirichlet density with kernel $\Pi p_{ij}^{g_{ij}-1}$ such that the first moments and one particular function of the second moments agree. If the first moments agree, then the parameters $\{g_{ij}\}$ must satisfy

$$g_{ij}/g.. = \lambda_{ij}, \quad i = 1,2, \quad j = 1,2,$$

where λ_{ij} is given in (2.8). Our ultimate aim will be the development of credible intervals of the components of $\mathbf{p}$, so the sum of the parameters $g..$ will be chosen so that $\Sigma \operatorname{Var}(p_{ij} | \mathbf{a},\mathbf{y}^A)$ is preserved. A routine calculation shows that, with this restriction,

$$g.. = (n_1 + a.. + 1)\{1 + 2[\Sigma \frac{y_i^A \lambda_{i1} a_{i2}}{(a_{i\cdot} + 1)a_{i\cdot}}] \div [\Sigma \lambda_{ij}(1 - \lambda_{ij})]\} - 1$$

$$= L \tag{2.14}$$

Thus the general $D\beta$ density will be approximated by a simple $D\beta$ density with kernel

$$\Pi p_{ij}^{L\lambda_{ij}-1} \Pi p_{\cdot j}^{y_j^B}. \tag{2.15}$$

Moments of the density (2.15) can be found using expressions in Section 2.2.1.

Note that an alternative way of approximating the general $D\beta$ density is to write the kernel as

$$[\Pi p_{ij}^{a_{ij}-1} \Pi p_{\cdot j}^{y_j^B}] \Pi p_{i\cdot}^{y_i^A} = k_3 \cdot k_4, \tag{2.16}$$

and approximate the simple $D\beta$ density with kernel k_3 by a Dirichlet density with kernel $\Pi p_{ij}^{h_{ij}-1}$ such that the first moments and the sum of the variances match. This results in the general $D\beta$ density being approximated by a simple $D\beta$ density with kernel

$$\Pi p_{ij}^{H\beta_{ij}-1} \Pi p_{i\cdot}^{y_i^A}, \tag{2.17}$$

where

$$\beta_{ij} = \frac{a_{ij}}{a_{\cdot j}} \frac{(a_{\cdot j} + y_j^B)}{(a.. + n_2)},$$

and

$$H = (n_2 + a.. + 1)\{1 + 2[\Sigma \frac{y_j^B \beta_{1j} a_{2j}}{(a_{\cdot j} + 1)a_{\cdot j}}] \div [\Sigma \beta_{ij}(1 - \beta_{ij})] - 1.$$

We will refer to the first approximation (2.15) as a "row approximation" since the "row kernel" $\Pi p_{i\cdot}^{y_i^A}$ is first combined with the "two-way kernel" $\Pi p_{ij}^{a_{ij}-1}$. The second approximation (2.17) first combines the "column kernel" $\Pi p_{\cdot j}^{y_j^B}$ with the two-way kernel; this will be referred to as a "column approximation".

2.2.3. *Posterior Moments in the Case* $y_1^B = y_2^B = 0$

In the special case $y_1^B = y_2^B = 0$, the posterior density (2.3) can be represented by

$$\pi(\mathbf{p} | \mathbf{x},\mathbf{y}^A) = \int\int \pi_1(\eta_a,\eta_b | \mathbf{x},\mathbf{y}^A) \pi_s(\mathbf{p} | \eta,\mathbf{x},\mathbf{y}^A) d\eta_a d\eta_b, \tag{2.18}$$

where

$$\pi_1(\eta_a,\eta_b | \mathbf{x},\mathbf{y}^A) \propto \Pi \frac{\Gamma(Kf_{ij} + x_{ij})}{\Gamma(Kf_{ij})} \frac{\Gamma(K\eta_a + x_{1\cdot} + y_1^A)}{\Gamma(K\eta_a + x_{1\cdot})}$$

$$\cdot \frac{\Gamma(K(1-\eta_a) + x_{2\cdot} + y_2^A)}{\Gamma(K(1-\eta_a) + x_{2\cdot})} \tag{2.19}$$

and $\pi_s(\mathbf{p}\,|\,\eta,\mathbf{x},\mathbf{y}^A)$ is given by (2.5) with $a_{ij} = Kf_{ij} + x_{ij}$. Using expressions in Section 2.2.1 and rules of conditional expectation,

$$E(p_{ij}\,|\,\mathbf{x},\mathbf{y}^A) = E[E(p_{ij}\,|\,\eta,\mathbf{x},\mathbf{y}^A)] = E[\lambda_{ij}\,|\,\mathbf{x},\mathbf{y}^A] \tag{2.20}$$

$$\begin{aligned}\mathrm{Var}(p_{ij}\,|\,\mathbf{x},\mathbf{y}^A) &= E[\mathrm{Var}(p_{ij}\,|\,\eta,\mathbf{x},\mathbf{y}^A)] + \mathrm{Var}[E(p_{ij}\,|\,\eta,\mathbf{x},\mathbf{y}^A)]\\ &= v_{ij}^1 + v_{ij}^2 + v_{ij}^3\end{aligned}$$

where

$$\begin{aligned}v_{1j}^1 &= (n + K + n_1 + 1)^{-1} E[\lambda_{ij}(1-\lambda_{ij})\,|\,\mathbf{x},\mathbf{y}^A]\\ v_{2j}^2 &= (n + K + n_1 + 1)^{-1} E\Big[\frac{y_i^A \lambda_{i1}(x_{i2} + Kf_{i2})}{u_i(u_i+1)}\,\Big|\,\mathbf{x},\mathbf{y}^A\Big]\\ v_{3j}^3 &= \mathrm{Var}[\lambda_{11}\,|\,\mathbf{x},\mathbf{y}^A]\,,\\ \lambda_{ij} &= (x_{ij} + Kf_{ij})(u_i + y_i^A)/[u_i(n + K + n_1)]\,,\end{aligned} \tag{2.21}$$

$u_1 = x_{1\cdot} + K\eta_a$, $u_2 = x_{2\cdot} + K(1-\eta_a)$, and the expectations and variance in (2.21) are taken with respect to the posterior distribution of (η_a,η_b) (2.19). In Section 3.1, we will illustrate the computation of the above moments in the special case when the two classification variables are believed independent.

2.2.4. *Posterior Moments in the General Case*

As in the conditional general $D\beta$ case of Section 2.2.2, one can develop exact expressions for the posterior moments. To illustrate, the posterior mean of p_{11} is given by

$$E(p_{11}\,|\,\mathbf{x},\mathbf{y}) = \int p_{11}\pi(\mathbf{p}\,|\,\mathbf{x},\mathbf{y})d\mathbf{p} \Big/ \int \pi(\mathbf{p}\,|\,\mathbf{x},\mathbf{y})d\mathbf{p}, \tag{2.22}$$

where $\pi(\mathbf{p}\,|\,\mathbf{x},\mathbf{y})$ is proportional to (2.3). Using equation (2.10), one can rewrite (2.22) as

$$E(p_{11}\,|\,\mathbf{x},\mathbf{y}) = \frac{\int\int I((a_{11}+1,a_{12},a_{21},a_{22}),\mathbf{y})/\Pi\Gamma(Kf_{ij})d\eta_a\,d\eta_b}{\int\int I(\mathbf{a},\mathbf{y})/\Pi\Gamma(Kf_{ij})d\eta_a\,d\eta_b}, \tag{2.23}$$

where $I(\mathbf{a},\mathbf{y})$ is given in (2.11) with a_{ij} replaced by $Kf_{ij} + x_{ij}$. These exact expressions are even more tedious than the expression obtained when the parameters $\{a_{ij}\}$ are held fixed. Thus, as in Section 2.2.2, an approximation will proposed for the general posterior distribution. First, note from (2.3) and (2.18) that this distribution can be written as

$$\pi(\mathbf{p}\,|\,\mathbf{x},\mathbf{y}) \propto \int\int q_1(\eta_a,\eta_b)\pi_s(\mathbf{p}\,|\,\eta,\mathbf{x},\mathbf{y}^A)\Pi p_{\cdot j}^{y_j^B}\,d\eta_a\,d\eta_b, \tag{2.24}$$

where

$$q_1(\eta_a,\eta_b) = \Pi\,\frac{\Gamma(Kf_{ij} + x_{ij})}{\Gamma(Kf_{ij})}\;\Pi\,\frac{\Gamma(u_i + y_i^A)}{\Gamma(u_i)} \tag{2.25}$$

Then, as in Section 2.2.2, approximate the conditional (on η) simple $D\beta$ density $\pi_s(\mathbf{p}\,|\,\eta,\mathbf{x},\mathbf{y}^A)$ by a Dirichlet density which preserves the conditional means $\{E(p_{ij}\,|\,\eta,\mathbf{x},\mathbf{y}^A)\}$ and the sum of the conditional variances $\Sigma\,\mathrm{Var}(p_{ij}\,|\,\eta,\mathbf{x},\mathbf{y}^A)$. Thus

$$\pi_s(\mathbf{p}\,|\,\eta,\mathbf{x},\mathbf{y}^A) \cong f_D(\mathbf{p}\,|\,L\,\lambda), \tag{2.26}$$

where λ_{ij} and L are given by (2.8) and (2.14), and a_{ij} is replaced by $x_{ij} + Kf_{ij}$. By replacing π_s by the approximate Dirichlet density f_D in (2.24), the following approximation to the posterior is obtained:

$$\pi(\mathbf{p}|\mathbf{x},\mathbf{y}) \approx \int\int q_1(\eta_a,\eta_b)\Gamma(L) \prod_{i,j} [p_{ij}^{L\lambda_{ij}-1}/\Gamma(L\lambda_{ij})]\prod p_{.j}^{y_j^B}\, d\eta_a\, d\eta_b$$

$$\propto \int\int q_1(\eta_a,\eta_b)q_2(\eta_a,\eta_b)\pi_s(\mathbf{p}|L\lambda,\mathbf{y}^B)d\eta_a\, d\eta_b, \tag{2.27}$$

where

$$q_2(\eta_a,\eta_b) = \frac{\Gamma(L)}{\Gamma(L+n_2)} \prod \frac{\Gamma(L\lambda_{.j} + ^{®}{}^B_j)}{\Gamma(L\lambda_{.j})}. \tag{2.28}$$

As in Section 2.2.2, the approximate posterior will be referred to as a "row approximation" since the row kernel $\prod p_{i.}^{y_i^A}$ is first combined with the kernel $\prod p_{ij}^{x_{ij}+Kf_{ij}}$ inside the integral. Using the approximate density (2.27), it can be shown (using techniques similar to those of Section 2.2.3) that the approximate mean and variance of p_{ij} are given by

$$E(p_{ij}|\mathbf{x},\mathbf{y}) \cong E[\frac{\lambda_{ij}}{\lambda_{.j}} \frac{L\lambda_{.j} + y_j^B}{L + n_2} |\mathbf{x},\mathbf{y}] \tag{2.29}$$

$$\text{Var}(p_{ij}|\mathbf{x}) \cong E[(L + n_2 + 1)^{-1}\lambda_{ij}(1 - \lambda_{ij})|\mathbf{x},\mathbf{y}]$$

$$+ E[(L + n_2 + 1)^{-1} \frac{y_j^B(L\lambda_{.j} + y_j^B)}{(L + n_2)(L\lambda_{.j} + 1)} \frac{\lambda_{1j}\lambda_{2j}}{\lambda_{.j}^2} |\mathbf{x},\mathbf{y}]\,. \tag{2.30}$$

The expectations in (2.29) and (2.30) are taken with respect to the approximate posterior density

$$\pi(\eta_a,\eta_b|\mathbf{x},\mathbf{y}) \propto q_1(\eta_a,\eta_b)\cdot q_2(\eta_a,\eta_b). \tag{2.31}$$

As in Section 2.2.2, an alternative method of approximating the posterior distribution, the "column approximation", first writes the posterior as

$$\pi(\mathbf{p}|\mathbf{x},\mathbf{y}) \propto \int\int q_1^*(\eta_a,\eta_b)\pi_s(\mathbf{p}|\eta,\mathbf{x},\mathbf{y}^B)\prod p_{i.}^{y_i^A}\, d\eta_a\, d\eta_b, \tag{2.32}$$

where

$$q_1^*(\eta_a,\eta_b) = \prod \frac{\Gamma(Kf_{ij} + x_{ij})}{\Gamma(Kf_{ij})} \frac{\Gamma(K\eta_b + x_{.1} + y_1^B)}{\Gamma(K\eta_b + x_{.1})} \frac{\Gamma(K(1-\eta_b) + x_{.2} + y_1^B)}{\Gamma(K(1-\eta_b) + x_{.2})}, \tag{2.33}$$

and approximates the simple $D\beta$ distribution in (2.32) by a Dirichlet distribution. In Section 3.2 we will evaluate the accuracy of the posterior means and variances calculated by both the row and column approximations.

3. NUMERICAL STUDY

3.1. *The case* $y_1^B = y_2^B = 0$

The posterior expressions in (2.20) and (2.21) are not written in closed form and, therefore, it is difficult to see how these moments incorporate the information contained in the prior and the sample counts. In this section, the computation of these expressions is discussed, and simple approximations are proposed in the independence case which illuminate the behavior of the posterior means and variances.

First note that the posterior quantities (2.20) and (2.21) involve expectations using the posterior density of (η_a,η_b) (2.19), which is not expressible in closed form. Thus it is necessary to compute expectations of the form

$$E[g(\eta_a,\eta_b)\,|\,\mathbf{x},\mathbf{y}^A] = \frac{\int_0^1\int_0^1 g(\eta_a,\eta_b)\pi_1(\eta_a,\eta_b)d\eta_a\,d\eta_b}{\int_0^1\int_0^1 \pi_1(\eta_a,\eta_b)d\eta_a\,d\eta_b} \tag{3.1}$$

where g is an arbitrary function of η_a and η_b. One efficient way of computing the integrals in (3.1) uses the notion of importance sampling. The first step of this simulation technique finds a simple approximation for the posterior density π_1. Since it can be shown for $\alpha_0 = 1$ that

$$\begin{aligned}\lim_{K\to\infty} \pi_1(\eta_a,\eta_b\,|\,\mathbf{x},\mathbf{y}^A) &= \pi_L(\eta_a,\eta_b)\,|\,\mathbf{x},\mathbf{y}^A) \\ &= f_\beta\,(\eta_a\,|\,x_{1\cdot} + y_1^A + 1, x_{2\cdot} + y_2^A + 1) \\ &\quad \cdot f_\beta\,(\eta_b\,|\,x_{\cdot 1} + 1, x_{\cdot 2} + 1),\end{aligned} \tag{3.2}$$

the limiting distribution π_L can serve as a rough approximation to π_1 for values of α_0 near one. Next, rewrite the expectation (3.1) as

$$E[g(\eta_a,\eta_b)\,|\,\mathbf{x},\mathbf{y}^A] = \frac{\int\int g(\eta_a,\eta_b)\dfrac{\pi_1(\pi_a,\pi_b)}{\pi_L(\eta_a,\eta_b)}\pi_L(\eta_a,\eta_b)d\eta_a\,d\eta_b}{\int\int \dfrac{\pi_1(\eta_a,\eta_b)}{\pi_L(\eta_a,\eta_b)}\pi_L(\eta_a,\eta_b)d\eta_a\,d\eta_b}\,. \tag{3.3}$$

Finally, to approximate the integrals in (3.3) using simulation, N_0 values of (η_a,η_b) are randomly generated from the beta densities in (3.2). Call the randomly generated values (e_{ai},e_{bi}), $i = 1,\dots,N_0$. Then (3.3) is approximated by

$$\frac{\sum_{i=1}^{N_0} g(e_{ai},e_{bi})\pi_1(e_{ai},e_{bi})/\pi_L(e_{ai},e_{bi})}{\sum_{i=1}^{N_0} \pi_1(e_{ai},e_{bi})/\pi_L(e_{ai},e_{bi})} \tag{3.4}$$

In the example which follows, we will consider the situation where the user believes a priori that the two classification variables are independent. The prior parameter α_0 will be set to one (reflecting a belief in independence) and the value of the parameter K selected will reflect the precision of a user's belief in independence.

Since the posterior means are, in some sense, a compromise between estimates from an unrestricted model and estimates from an independence model, we will first discuss the computation of these "traditional" estimates. Consider the hypothetical sample counts presented in Table 2. Under the unrestricted model, the MLE estimates a cell probability by allocating the partially classified counts according to the counts in the completely classified table. In this example the 30 counts partially classified in category one are allocated into the (1,1) (1,2) cells in the complete table by the proportions 100/(100 + 50), 50/(100 + 50), respectively. In general, the MLE of p_{ij} is given by

$$\hat{p}_{ij} = \frac{x_{ij} + y_i^A x_{ij}/x_{i\cdot}}{n + n_1}\,, \tag{3.5}$$

and the values of these estimates are given in Table 3. To understand the computation of the MLE under an independence model, first note that if the partially classified counts are ignored, then the expected cell count in cell (1,1) is $x_{1\cdot}x_{\cdot 1}$. Then the 30 partially classified counts are allocated into the (1,1), (1,2) cells by the "pooled" proportions 175/300, 125/300, respectively. The independence MLE of p_{ij} is given by

$$\tilde{p}_{ij} = \frac{x_{i\cdot}x_{\cdot j} + y_i^A x_{\cdot j}/n}{n + n_1} . \tag{3.6}$$

TABLE 2

Some Hypothetical Sample Counts

	B			
A	100	50	150	30
	75	75	150	60
	175	125	300	90

The posterior mean (2.20) can be rewritten as

$$E(p_{ij}|\mathbf{x},\mathbf{y}^A) = (n + K + n_1)^{-1}[x_{ij} + KE(f_{ij}|\mathbf{x},\mathbf{y}^A) + y_i^A E((x_{ij} + Kf_{ij})/u_i|\mathbf{x},\mathbf{y}^A)]. \tag{3.7}$$

Using techniques similar to those discussed in Albert and Gupta (1982), the following approximation to (3.7) is proposed:

$$E(p_{ij}|\mathbf{x},\mathbf{y}^A) \cong (n + K + n_1)^{-1}[x_{ij} + K\tilde{p}_{ij} + y_i^A \hat{\gamma}_{ij}] = p_{ij}^* \tag{3.8}$$

where

$$\hat{\gamma}_{ij} = \frac{n}{n + K}\frac{x_{ij}}{x_{i\cdot}} + \frac{K}{n + K}\frac{x_{\cdot j}}{n} .$$

Values of the exact posterior means together with the approximate values are also given in Table 3. To illustrate the computation of (3.7), note that for the (1,1) cell the observed count 100 is first added to the count 26.6, reflecting a shift of the observed count towards an expected count assuming an independence model. Then the 30 partially classified counts are allocated to the (1,1), (1,2) cells by the probabilities .646, .354, respectively. The probability .646 is a compromise between the allocation probabilities assuming an unrestricted model and an independence model. Thus the posterior means allocate the partially classified counts to the complete table in a way which reflects the vague prior beliefs in independence.

In Table 4, values of posterior variances of the cell probabilities (2.21) are calculated for the same example. As in (2.21), each variance is written as a sum of three terms. Note that

$$v_{ij}^1 \approx (n + K + n_1)^{-1} E(p_{ij}|\mathbf{x},\mathbf{y}^A)(1 - E(p_{ij}|\mathbf{x},\mathbf{y}^A))$$

is the usual form of a variance of the estimate $E(p_{ij}|\mathbf{x},\mathbf{y}^A)$ if the sample size is $n' = n + K + n_1$. However, this variance term, by itself, ignores two types of uncertainty:

TABLE 3

Computed Values of MLE's, exact and approximate posterior means for data of Table 2

	Expected cell counts		Probability estimates	
MLE, unrestricted	100 + 30(.667)	50 + 30(.333)	.308	.154
	75 + 60(.500)	75 + 60(.500)	.269	.269
MLE, independence	87.5 + 30(.583)	62.5 + 30(.417)	.269	.192
	87.5 + 60(.583)	62.5 + 60(.417)	.314	.224
Exact posterior means	100 + 100(.266) + 30(.646)	50 + 100(.192) + 30(.354)	.298	.163
	75 + 100(.310) + 60(.520)	75 + 100(.226) + 60(.480)	.280	.258
Approximate posterior means	100 + 100(.269) + 30(.646)	50 + 100(.192) + 30(.354)	.299	.163
	75 + 100(.314) + 60(.521)	75 + 100(.224) + 60(.479)	.281	.257

TABLE 4

Computed values of exact posterior variances for data of Table 2

(one unit $= 10^{-6}$)

426 + 33 + 87	278 + 33 + 87
411 + 80 + 96	390 + 80 + 93

(1) the uncertainty of the location of the n_1 partially classified counts in the complete table, and

(2) the uncertainty or lack of precision of the prior belief in independence.

The variance terms v_{ij}^2 and v_{ij}^3 reflect the degree of uncertainty of types (1) and (2), respectively. In future work, we plan to develop useful approximations for these terms, so that posterior credible intervals for the components of $\mathbf{p}$ can easily be calculated.

3.2. *The General Case*

In Section 2.4, two approximations (the "row" and "column" approximations) were proposed for the general posterior distribution. In this section, an example will be used to briefly evaluate the accuracy of these approximations and, in addition, to compare the values of the posterior means with values of the classical (ML) estimates.

In Tables 5 and 6, for the data set $\mathbf{x} = (8,2,4,6)$, $\mathbf{y}^A = (5,3)$, $\mathbf{y}^B = (6,4)$, values of the posterior means and variances are calculated using the exact posterior (see (2.23)) and using the approximate "row" and "column" posteriors (see (2.29) and (2.30) for the "row" expressions). The posterior moments are computed for four sets of the prior parameters α_0 and K. (In each case $\alpha_0 = 1$, so the user believes a priori in independence). From observing Table 5, it appears that both the "row" and "column" approximations are very accurate in the computation of the posterior means. However, the two types of approximations appear to be less accurate in computing values of the posterior variances, as shown in Table 6. The errors in the computation of the posterior variance of particular components of $\mathbf{p}$ can be as high as 10 %, but the error in the computation of the "total" posterior variance (as measured by the sum of the posterior variances) is only one per cent

TABLE 5

Exact and Approximate Posterior Means

$\mathbf{x} = (8,2,4,6)$, $\mathbf{y} = (5,3,6,4)$, $\alpha_0 = 1$

(one unit $= 10^{-3}$)

	Exact		Approximation: Row		Approximation: Column	
$K = 1$	419	114	419	115	418	113
	187	280	188	279	191	278
$K = 6$	397	134	397	134	396	133
	203	266	204	265	206	265
$K = 30$	359	174	360	174	360	173
	238	229	238	228	238	229
$K = 500$	321	210	323	211	323	211
	276	193	275	191	275	191

MLE, unrestricted	425	109
	184	282
MLE, independence	321	214
	279	186

TABLE 6

Exact and Approximate Posterior Variances

$\mathbf{x} = (8,2,4,6)$, $\mathbf{y} = (5,3,6,4)$, $\alpha_0 = 1$

(one unit $= 10^{-5}$)

	Exact		Approximation Row		Error	Approximation Column		Error
$K = 1$	804	427	836	381	+1 %	833	404	+1 %
	569	636	578	664		543	681	
$K = 6$	741	436	758	405	+1 %	757	424	+1 %
	543	572	558	594		528	601	
$K = 30$	649	424	664	412	+1 %	665	422	+1 %
	509	440	520	451		509	452	
$K = 500$	492	342	523	368	+7 %	523	369	+7 %
	445	299	478	312		478	313	

in the typical situations considered in Table 6 where small values of the prior parameter K are used. Thus both the "row" and "column" approximation techniques appear accurate in summarizing the posterior distribution of $\mathbf{p}$.

In Table 6, values of the classical MLE's have also been given for this example at the bottom of the table. The ML estimates assuming an independence model are given by

$$\tilde{p}_{ij} = \frac{(x_{i\cdot} + y_i^A)(x_{\cdot j} + y_j^B)}{(n + n_1)(n + n_2)} .$$

The MLE's under an unrestricted model are not given in closed form, but can be quickly computed by means of the EM algorithm (Dempster, Laird and Rubin (1977)). Note that as the value of the prior parameter K changes from 1 to 500, the values of the posterior means move from the unrestricted MLE values towards the independence MLE values. Thus, as in the previous example, the posterior appears to incorporate the prior belief of independence in a very natural fashion.

4. AN EXAMPLE

To illustrate the application of the Bayesian estimation procedures proposed in this paper, consider the Chen and Fienberg (1974) data discussed in Section 1. Suppose the user believes a priori that an infant's health index score is unrelated to his/her serum bilirium reading. Equivalently, the odds of a low health index score infant having a high serum bilirium reading are believed to be equal to the odds of a high health index score infant

TABLE 7

Approximate Posterior Means and Variances for Chen and Fienberg data, $\alpha_0 = 1$, $K = 150$

Posterior Means

.186	.211
.291	.313

Posterior Variances (unit = 10^{-6})

395	416
550	510

having a high serum bilirium reading. In addition, suppose that the user is 90 per cent confident that the ratio of the above odds is between .2 and 5. Using the Albert and Gupta (1982) table, this prior belief is translated to the values of the prior parameters $\alpha_0 = 1$, $K = 150$. Using the prior (2.2) with this prior knowledge, Table 7 gives the (approximate) posterior means and variances. Thus moments can be used to construct approximate credible intervals for the components of **p**. For example, by assuming that the marginal posterior distribution of p_{11} is approximately normal, the interval

$$E(p_{11}|\mathbf{x},\mathbf{y}) \pm 2(\mathrm{Var}(p_{11}|\mathbf{x},\mathbf{y}))^{1/2}$$
$$= .186 \pm 2(395\cdot 10^{-6})^{1/2}$$
$$= .186 \pm .040$$

is an approximate 95 per cent credible interval. These procedures are attractive alternatives to the usual classical procedures when vague prior beliefs exist about the association structure in the 2×2 table. For future research, we plan to identify situations where vague prior beliefs exist and suggest ways of eliciting these beliefs so they can be used in the estimation process.

REFERENCES

ALBERT, J.H. and GUPTA, A.K. (1983). Bayesian estimation methods for 2×2 contingency tables using mixtures of Dirichlet distributions. *J. Amer. Statist. Assoc.* **78**, 708-717.

ANTLEMAN, G.R. (1972). Interrelated Bernoulli processes. *J. Amer. Statist. Assoc.* **67**, 831-841.

CHEN, T. and FIENBERG, S.E. (1974). Two-dimensional contingency tables with both completely and partially cross-classified data. *Biometrics,* **30**, 629-642.

DEMPSTER, A.P. LAIRD, N.M. and RUBIN, D.B. (1977). Maximum likelihood from incomplete data via the EM algorithm. *J. Roy. Statist. Soc. B,* **1**, 1-18 (with discussion).

FUCHS, C. (1982). Maximum likelihood estimation and model selection in contingency tables with missing data. *J. Amer. Statist. Assoc.,* **77**, 270-278.

HOCKING, R.R. and OXSPRING, H.H. (1974). The analysis of partially categorized contingency data. *Biometrics,* **30**, 469-483.

BAYESIAN STATISTICS 2, pp. 603-612
J.M. Bernardo, M.H. DeGroot, D.V. Lindley, A.F.M. Smith (Eds.)

Measures of Information in the Predictive Distribution

M.A. AMARAL-TURKMAN and I.R. DUNSMORE
University of Lisboa *University of Sheffield*

SUMMARY

The use of simple measures of the information in the predictive distribution is investigated in problems in which the aim is to gain knowledge about some future event (or observation). Inter-relationships with the work of Geisser (1971) and Aitchison (1975) are noted. Derivations are provided for the gamma model, and illustrations are given in two situations, namely the selection of sample size in fixed sample size and sequential sample procedures.

Keywords: INFORMATION MEASURES; KULLBACK-LEIBLER MEASURES; PREDICTIVE DISTRIBUTION; SAMPLE SIZE; SEQUENTIAL PROCEDURES.

1. INTRODUCTION

For situations in which it is desired to gain some knowledge about the true value of a parameter by means of experimentation, Lindley (1956) suggested a measure of the information given by an experiment. In the case of statistical prediction problems, although the uncertainty about the future may come from uncertainty about the true value of a parameter θ, the final objective is not to make inferences about the parameter but rather to make inferences about the outcome y of some future experiment F in the light of the outcome x of an informative experiment E. The uncertainty about the future outcome is described by the probability density function $p(y|\theta)$ $(y \in Y;\ \theta \in \Theta)$. In the Bayesian approach to prediction problems the main role is played by the predictive density function $p(y|x)$ which expresses the plausibility of y in the light of the results x of an informative experiment described by $p(x|\theta)$ with prior assessment $p(\theta)$ on Θ. Its importance as a surrogate for $p(y|\theta)$ has been stressed, for example, by Geisser (1971), and it is understood that $p(y|x)$ should converge in some sense to the true density function $p(y|\theta)$ as the sample size in E increases. One should expect the information we have about $p(y|\theta)$ in $p(y|x)$ to increase with the number of observations in the informative experiment; i.e. the more data we obtain the more information we have about $p(y|\theta)$, or the less uncertain we are about the true density of y.

It then seems reasonable to expect that the introduction of a suitable measure of information in the predictive distribution will be helpful in some problems in which the aim is to gain knowledge about some future event (or observation).

Geisser (1971) introduced a measure of uncertainty inherent in the predictive distribution in consideration of the convergence of $p(y|x)$ to $p(y|\theta)$. However this measure depends on θ and so does not seem to be 'fully predictive'. Aitchison (1975) discusses a goodness-of-prediction fit problem which used the Kullback and Leibler (1951) directed measure of divergence.

Measures about the information that a random variable X gives about another random variable Y are of course well known in the context of communications engineering following from the work of Shannon (1948); see, for example, Fano (1961), Blachman (1966, 1968), Gallager (1968).

The measure of information in the predictive distribution which we propose in Section 2 is an adaptation of one of these measures in communication and information theory. Also it can be considered as a particular case of the measure defined by Lindley (1956) when the parameter of interest is the future event. It should be noted in passing that Bernardo (1979b) used the same measure as a tool to derive his 'reference predictive distributions'. In Section 3 we describe the relationship with the proposals of Geisser (1971) and Aitchison (1975). The evaluation of the measures for some of the standard parametric models is discussed in Section 4. Finally in Section 5 we provide two applications of the measures, namely to the problem of the selection of sample size both in fixed sample size and in sequential procedures.

2. MEASURES OF INFORMATION IN THE PREDICTIVE DISTRIBUTION

Prior to the informative experiment being performed, the predictive density for y is given by

$$p(y) = \int_{\Theta} p(y|\theta)\, p(\theta)\, d\theta.$$

After the data $x_1,x_2,\ldots,x_n$ are obtained from the informative experiment E_n, the predictive density function is updated to

$$p(y|x_{(n)}) = \int_{\Theta} p(y|\theta)\, p(\theta|x_{(n)})\, d\theta\,,$$

where $x_{(n)}$ may represent either the data $x_1,x_2,\ldots,x_n$ or some sufficient statistic, and where $p(\theta|x_{(n)})$ is the posterior density function.

If we define

$$I_0 = \int_{Y} p(y) \log p(y)\, dy \quad \text{and} \quad I_1(x_{(n)}) = \int_{Y} p(y|x_{(n)}) \log p(y|x_{(n)})\, dy,$$

then the gain in information about Y provided by the observation $x_{(n)}$ in $E_{(n)}$ may be defined by

$$I_Y(x_{(n)}) = I_1(x_{(n)}) - I_0.$$

This measure depends on $x_{(n)}$ and clearly some observations will be more informative than others. We may then define the expected gain in information about Y provided by $X_{(n)}$ to be

$$I(Y;X_{(n)}) = \int_{X_{(n)}} I_Y(x_{(n)})\, p(x_{(n)})\, dx_{(n)}$$

$$= \int_{X_{(n)}} \int_{Y} p(y|x_{(n)})\, p(x_{(n)}) \log \frac{p(y|x_{(n)})}{p(y)}\, dy\, dx_{(n)}\,.$$

where $X_{(n)}$ denotes the range space of $x_{(n)}$. This expected gain in information is a special case of the Kullback-Leibler mean information for discrimination; see Kullback and Leibler (1951) and Kullback (1959). It has the usual properties; for example:

(i) $I(Y;X_{(n)}) \geq 0$;

(ii) $I(Y;X_n|X_{(n-1)}) = I(Y;X_{(n)}) - I(Y;X_{(n-1)}) \geq 0$;

(iii) $I(Y;X_{(n)}) = I(Y;X_{(1)}) + I(Y;X_2|X_{(1)}) + \ldots + I(Y;X_n|X_{(n-1)})$.

For the case of replicates in E_n it might be expected that, in a manner similar to Lindley (1956),

(a) $I(Y;X_n|X_{(n-1)}) \leq I(Y;X_n)$,

and

(b) $I(Y;X_{(n)})$ is a concave function of n.

However Blachman (1966) showed that in general (a) and (b) are not necessarily true. We note that (b) implies (a), and that (b) implies that the average gain in the information obtained by taking an additional observation decreases with the size of sample. It transpires that (a) and (b) are in fact valid for several standard predictive models; see Section 4.

An alternative to $I_Y(x_{(n)})$ is the Kullback-Leibler directed measure of divergence to judge the relative closeness of $p(y)$ to $p(y|x_{(n)})$, namely

$$J_Y(x_{(n)}) = \int_Y p(y|x_{(n)}) \log \frac{p(y|x_{(n)})}{p(y)}\, dy. \tag{1}$$

This is perhaps the more widely accepted measure; see Blachman (1966, p. 185). We see that

$$I_Y(x_{(n)}) = J_Y(x_{(n)}) + \int_Y \{p(y|x_{(n)}) - p(y)\} \log p(y)\, dy,$$

and that

$$\int_{X_{(n)}} J_Y(x_{(n)})\, p(x_{(n)})\, dx_{(n)} = I(Y;X_{(n)}).$$

Blachman (1968) discusses the various properties of $I_Y(x_{(n)})$ and $J_Y(x_{(n)})$. Unfortunately neither possesses all the satisfactory properties of $I(Y;X_{(n)})$. For example:

(iv) $J_Y(x_{(n)}) \geq 0$ whilst $I_Y(x_{(n)})$ can be positive or negative;

(v) $I_Y(x_{(n)})$ has the additivity property, in an obvious notation,

$$I_Y(x_1,\ldots,x_n) = I_Y(x_1) + I_Y(x_2|x_{(1)}) + \ldots + I_Y(x_n|x_{(n-1)}),$$

whereas $J_Y(x_{(n)})$ does not;

(vi) $J_Y(x_{(n)})$ is coordinate independent, whereas $I_Y(x_{(n)})$ is not; that is, if $U(Y)$ is a $1-1$ transformation of Y, then $J_U(x_{(n)}) = J_Y(x_{(n)})$, but $I_U(x_{(n)}) \neq I_Y(x_{(n)})$.

Because of concern over coordinate independence $J_Y(x_{(n)})$ is the more commonly used information measure. In the particular area of application discussed here, where $p(y|x_{(n)})$ is interpreted as the Bayesian predictive density function of y, the derivation of $J_Y(x_{(n)})$ provides severe computational problems. Whilst approximations for $J_Y(x_{(n)})$ could be developed, we concentrate in this paper on the more readily computed $I_Y(x_{(n)})$. The initial derivation of $I_Y(x_{(n)})$ seems to be the more intuitively satisfactory in this predictive context anyway. The lack of coordinate independence in the continuous case implies that if, for example, a sampling scheme is adopted in which we try to obtain a prescribed amount of information about Y, it will in general be different from one in which we want to achieve the same amount of information about $U(Y)$. With the use of $I_Y(x_{(n)})$ we need to take care to avoid criteria based on such absolute measures.

3. RELATIONSHIPS WITH OTHER MEASURES

Both Geisser (1971) and Aitchison (1975) have been concerned with the relationships between $p(y|x_{(n)})$ and $p(y|\theta)$. Geisser (1971) suggests that in order that $p(y|x_{(n)})$ can be considered as a surrogate for the sampling distribution we should ask that

$$\text{(i)} \quad \lim_{n\to\infty} p(y|x_{(n)}) = p(y|\theta)\,,$$

and

$$\begin{aligned} \text{(ii)} \quad & \text{(a)}\ u_n(\theta) \geq u_{n+1}(\theta) \text{ for all } \theta \in \Theta,\ n > 1\,, \\ & \text{(b)}\ \lim_{n\to\infty} u_n(\theta) = u(\theta)\,, \end{aligned} \tag{2}$$

where

$$u_n(\theta) = -\int_{X_{(n)}} I_1(x_{(n)})\,p(x_{(n)}|\theta)\,dx_{(n)}\,,$$

$$u(\theta) = -\int_Y p(y|\theta)\log p(y|\theta)\,dy\,.$$

In a similar vein Aitchison (1975) and Murray (1977) have used a measure of the relative closeness of $p(y|x_{(n)})$ to $p(y|\theta)$ based on the Kullback-Leibler measure of divergence and given by

$$MD(Y;X_{(n)},\theta) = \int_{X_{(n)}}\int_Y p(x_{(n)}|\theta)\,p(y|\theta)\log\frac{p(y|\theta)}{p(y|x_{(n)})}\,dy\,dx_{(n)}\,. \tag{3}$$

In general this measure will depend on θ, and a natural measure of relative closeness in the Bayesian context is then given by

$$AMD(Y;X_{(n)}) = \int_\Theta MD(Y;X_{(n)},\theta)\,p(\theta)\,d\theta\,.$$

The interrelationship between the different approaches is illustrated by the following easily proved result: if Geisser's conditions (i) and (ii) are satisfied and if

$$\begin{aligned} &\int_\Theta u_n(\theta)\,p(\theta)\,d\theta < \infty \text{ for all } n, \\ &\int_\Theta u(\theta)\,p(\theta)\,d\theta < \infty\,, \end{aligned} \tag{4}$$

then

$$I(Y;X_{(\infty)}) = \lim_{n\to\infty} I(Y;X_{(n)}) \text{ is finite}$$

and

$$I(Y;X_{(\infty)}) - I(Y;X_{(n)}) = AMD(Y;X_{(n)})\,.$$

This result supplies a sufficient condition for $I(Y;X_{(\infty)})$ to be finite, and supplements the necessary condition given by Osteyee and Good (1974, p. 32). We find that in all the standard models mentioned later $I(Y;X_{(\infty)})$ is finite provided the prior knowledge is not vague. If vague priors are assumed then $I(Y;X_{(n)})$ and $AMD(Y;X_{(n)})$ are not defined. However as Aitchison (1975) shows there are models (for example, normal and gamma) for which $MD(Y;X_{(n)},\theta)$ is independent of θ, and in such cases $I(Y;X_{(\infty)}) - I(Y;X_{(n)})$ is finite. We also note that if (4) hold we may substitute Geisser's conditions (2) above by the weaker condition

$$\lim_{n\to\infty}\int_\Theta u_n(\theta)\,p(\theta)\,d\theta = \int_\Theta u(\theta)\,p(\theta)\,d\theta\,,$$

since this is sufficient for $AMD(Y;X_{(n)}) \to 0$ as $n \to \infty$ and alone guarantee the requirement that 'the larger n, the less uncertain is $p(y|x_{(n)})$'.

4. EXAMPLES OF STANDARD FAMILIES

4.1. *Gamma model*

Suppose that the future experiment consists of m replicates of an experiment described by a random variable with an exponential distribution with mean θ^{-1}, and that interest rests in the total y of these m observations. Then $p(y|\theta)$ is $Ga(m,\theta)$, that is,

$$p(y|\theta) = \frac{\theta^m y^{m-1} e^{-\theta y}}{\Gamma(m)} \quad (y > 0).$$

The informative experiment E_n provides data $x_1, x_2, \ldots, x_n$ from n replicates of the same experiment, and we may summarize these by the sufficient statistic $x_{(n)} = x_1 + x_2 + \ldots + x_n$, where $p(x_{(n)}|\theta)$ is $Ga(n,\theta)$.

If we assume a conjugate prior for θ of the form $Ga(g,h)$, then $p(\theta|x_{(n)})$ is $Ga(G = g + n, H = h + x_{(n)})$. With the notation that a random variable z is said to have an $InBe(m,g,h)$ distribution if

$$p(z) = \frac{h^g z^{m-1}}{B(m,g)\,(h + z)^{m+g}} \quad (z > 0),$$

then the relevant marginal and predictive distributions are given by

$$p(x_{(n)}) = InBe(n,g,h)\,;\; p(y) = InBe(m,g,h)\,;\; p(y|x_{(n)}) = InBe(m,G,H).$$

The following standard results are useful in the evaluations of the various information measures; namely, if $p(z)$ is $InBe(m,g,h)$, then

$$\int_0^\infty p(z) \log z \, dz = \log h + \psi(m) - \psi(g),$$

$$\int_0^\infty p(z) \log (h + z) \, dz = \log h + \psi(m + g) - \psi(g),$$

where $\psi(s) = d\log\Gamma(s) / ds$ is the *psi* or digamma function; see, for example, Abramowitz and Stegun (1972, p. 258).

The gain in information $I_Y(x_{(n)})$ is then given by

$$I_Y(x_{(n)}) = \log\left\{\frac{Gh}{gH}\right\} + \left(\frac{1}{g} - \frac{1}{G}\right) + \{f(g+m) - f(g+1)\} - \{f(G+m) - f(G+1)\},$$

where $f(s) = s\psi(s) - \log\Gamma(s) - s$, a function considered by Brooks (1982). Further the expected gain in information $I(Y;X_{(n)})$ is

$$I(Y;X_{(n)}) = \{f(g + m) - f(g)\} - \{f(G + m) - f(G)\}.$$

It is straightforward to show that here $I(Y;X_{(n)})$ is a concave function of n. Notice also that $I(Y;X_{(n)})$ is independent of the prior scale hyperparameter h. Further we see that

$$I(Y;X_{(\infty)}) = f(g + m) - f(g),$$

and that the measure of missing information in the predictive distribution is given by

$$I(Y;X_{(\infty)}) - I(Y;X_{(n)}) = f(G + m) - f(G).$$

It is easy to check that this coincides with $MD(Y;X_{(n)},\theta)$ given by (3), which in this case is independent of θ and so is identical to $AMD(Y;X_{(n)})$. Since, for large s, $f(s) \cong \frac{1}{2} \log s$ –

$\frac{1}{2}\log(2\pi e)$, we have as a simple approximation that $I(Y;X_{(\infty)}) - I(Y;X_{(n)}) = \frac{1}{2}\log\{(G+m)/G\}$.

For the particular case of $m = 1$, some simplifications occur. We define

$$\Lambda(s) = f(s+1) - f(s)$$
$$= \psi(s+1) - \log s\,,$$

which, for large s, can be approximated by $(2s)^{-1}$. Then

$$I_Y(x_{(n)}) = \log\left\{\frac{Gh}{gH}\right\} + \left(\frac{1}{g} - \frac{1}{G}\right),$$
$$I(Y;X_{(n)}) = \Lambda(g) - \Lambda(G)\,,$$
$$I(Y;X_{(\infty)}) = \Lambda(g)\,,$$

and

$$I(Y;X_{(\infty)}) - I(Y;X_{(n)}) = \Lambda(G)\,.$$

In Figure 1 we illustrate $I(Y;X_{(n)})$ for the case $m = 1$ with $g = 0.1$, 1.0 and 2.0. The asymptotes represent the value $I(Y;X_{(\infty)})$ in each case.

Also we find that

$$J_Y(x_{(n)}) = I_Y(x_{(n)}) - \frac{g+1}{g} + \frac{g+1}{G}F(1,G,G+1;\frac{x_{(n)}}{H}),$$

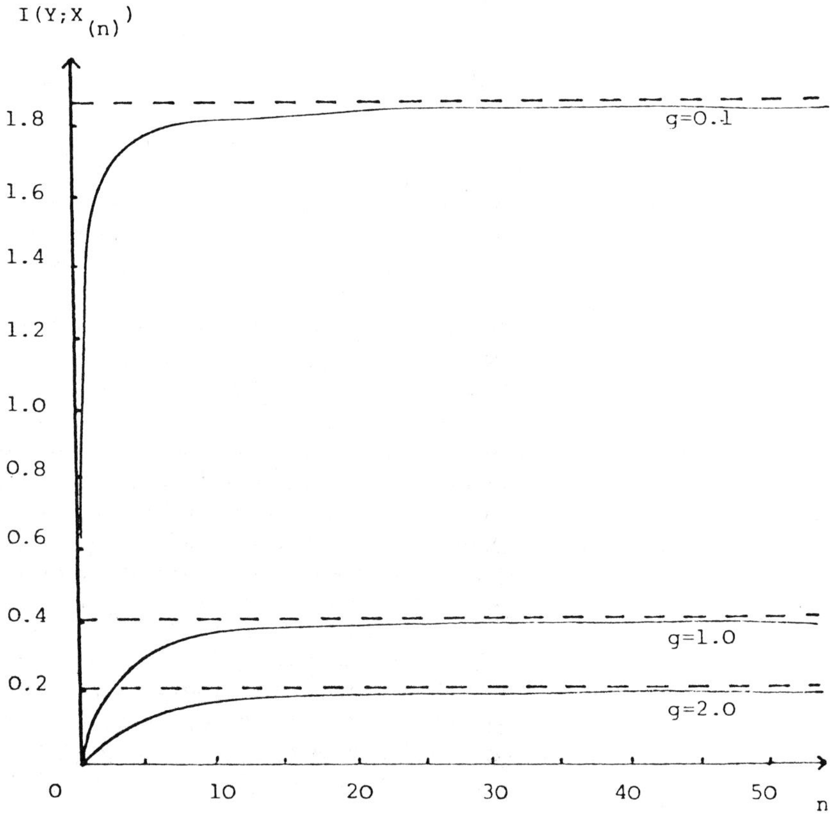

where F is the hypergeometric function; see, for example, Gradshteyn and Ryzhik (1965, p. 1035). We note that we have been unable to obtain an explicit expression for $J_Y(x_{(n)})$ for general m, but approximations are being sought.

4.2. *Other models*

We may evaluate similarly the various measures of information for other underlying families of models. Details are given in Amaral (1980) for the normal, two-parameter exponential and binomial families. Again severe computational problems exist in the evaluation of $J_Y(x_{(n)})$ for the first two families. The concavity property for $I(Y;X_{(n)})$ holds for the first two families, and we conjecture (from numerical investigations) that it also holds for the binomial model.

5. APPLICATIONS

5.1. *Fixed sample size procedures*

The construction of a predictive density function $p(y|x_{(n)})$ requires the performance of an informative experiment E_n to supply $x_1,x_2,\ldots,x_n$. We consider cases in which n is to be fixed, and suggest several criteria based on information measures to find the optimal sample size for situations in which our purpose is to obtain a 'good' predictive distribution for y.

Criterion C1: Choose the smallest n for which the measure of missing information, $I(Y;X_{(\infty)}) - I(Y;X_{(n)})$ is less than some prespecified (small) value, ϵ.

Criterion C2: For cases in which $I(Y;X_{(n)})$ is a concave function of n we can define $G_n(Y) = I(Y;X_{(n+1)}) - I(Y;X_{(n)})$ as the increase in the expected gain of information obtained by taking one more observation after n have been taken. Then $G_n(Y)$ is a decreasing function of n, and we select the smallest n for which $G_n(Y) < \delta$ for some preselected (small) value δ. If we can assign a cost for each observation and if we can relate that cost to the gain in information, δ is in a sense the minimum value of the gain in information which still 'pays' the cost of one more observation.

Criterion C3: That $I(Y;X_{(n)})$ is a suitable measure to use can be illustrated by the arguments of Bernardo (1975, 1979a), who suggests that the expected utility of the experiment E_n is $aI(Y;X_{(n)})$, where a is to be interpreted as the expected utility of one unit of information about Y. If we can assign a cost c_n to a sample of size n, we might select n which maximizes $aI(Y;X_{(n)}) - c_n$. For example we may set $c_n = c_0 + c_1 n$.

Criterion C4: As we have seen the relative size of $I(Y;X_{(n)})$ or of $I(Y;X_{(\infty)}) - I(Y;X_{(n)})$ depends on the situation in hand. For example, as illustrated in Figure 1, it varies with different values of the prior hyperparameters. Thus the specification of ϵ or δ may be difficult, and a more realistic and comparable criterion may be to limit the proportional amount of missing information. For example, since the total expected information available in the system is $I(Y;X_{(\infty)}) + I_0$, we might choose n such that

$$\frac{I(Y;X_{(\infty)}) - I(Y;X_{(n)})}{I(Y;X_{(\infty)}) + I_0} < \alpha ,$$

for some prespecified (small) value α. It must be noted that problems can occur with this criterion in that it is possible for the denominator to become negative, since I_0 may be negative.

5.2. *Illustration for the gamma model*

In the case of the gamma family detailed in Section 4.1, the results in Table 1 are obtained for the case $m = 1$. Optimal values of n can then be found easily from the function $\Lambda(\bullet)$. Use of the approximation for $\Lambda(\bullet)$ provides the further simplifications below:

C1: $$g + n > (2\epsilon)^{-1};$$

C2: $$(g + n)(g + n - 1) > (2\delta)^{-1};$$

C3: $$g + n \cong \left(\frac{a}{2c_1}\right)^{1/2};$$

C4: $$g + n > \{2\alpha(\psi(g) - 1 - \log h)\}^{-1}.$$

Notice that for appropriate values with $\delta = c_1/a$, criteria *C2* and *C3* provide approximately the same answer, thus providing some justification for the earlier interpretation of δ. Notice that the asymptotic equivalence of these two procedures is due to the concavity of $I(Y;X_{(n)})$ and the linearity of c_n, and not to the particular model involved.

TABLE 1
Optimal values n^* for criteria C1 - C4 for the gamma model

Criterion		*Optimal n^**
C1	minimum n s.t.	$\Lambda(g + n) < \epsilon$
C2	minimum n s.t.	$\Lambda(g + n - 1) - \Lambda(g + n) < \delta$
C3	n which maximizes	$-a\,\Lambda(g + n) - c_1 n$
C4	minimum n s.t.	$\dfrac{\Lambda(g + n)}{\Lambda(g) - 1 - \log h} < \alpha$

As an illustration of these criteria Table 2 shows the optimal n^* for $\alpha = 0.01$ in *C4* for the case where $h = 0.1$. Notice that if $g \geq 13$ the prior predictive density provides a sufficient measure of information to require no sampling.

TABLE 2
Optimal n^* for criterion C4 for the gamma model with $h = 0.1$ and $\alpha = 0.01$

g	1	2	3	4	5	6	7	8	9	10	11	12	13
n^*	68	27	20	16	13	11	9	7	6	4	3	2	0

5.3. *Sequential procedures*

Our objective in this section is to investigate the viability of the use of the measures of information $J_Y(x_{(n)})$ and $I_Y(x_{(n)})$ (or equivalently $I_1(x_{(n)})$) in stopping rules in the design of sequential experiments. Similar procedures have been used, for example, by Lindley (1957), DeGroot (1962) and El-Sayyad (1969) based on the measures of information provided by the posterior distribution.

Consider first the measure $J_Y(x_{(n)})$ in (1). An interpretation one would give to a high value of $J_Y(x_{(n)})$ is that the result $x_{(n)}$ of the experiment E_n is highly informative. According to the arguments of Bernardo (1975, 1979a) $J_Y(x_{(n)})$ measures the utility one expects to obtain when E_n is performed and $x_{(n)}$ is observed. So a sensible stopping rule in a sequential setting is 'Stop when $J_Y(x_{(n)}) > \xi$', where ξ is some prescribed level. This stopping rule, contrary to the stopping rules introduced by Lindley (1957), leads to invariant sampling schemes. The problem of its use is a practical one. As we have seen, even in the simplest cases, $J_Y(x_{(n)})$ tends to lack mathematical tractability. For example in the case of exponential sampling as in Section 5.2, the rule is to stop when

$$\log\left(\frac{g+n}{h+x_{(n)}}\right) - \frac{n}{g+n} + \sum_{k=1}^{\infty}\left(\frac{g+1}{g+n+k}\right)\left(\frac{x_{(n)}}{x_{(n)}+k}\right)^k > \xi + \text{constant},$$

where the constant depends only on g and h.

Similarly we may consider the use of $I_Y(x_{(n)})$ or equivalently $I_1(x_{(n)})$. However $I_1(x_{(n)})$, rather than increasing with n, converges, under suitable conditions on $p(y|x_{(n)})$ and $p(y|\theta)$, to the unknown, but usually finite, value

$$I(\theta) = \int_Y p(y|\theta)\log p(y|\theta)\,dy.$$

For this reason a possible stopping rule to investigate is to stop sampling when

$$|I_1(x_{(n)}) - E_{\theta|x_{(n)}}\{I(\theta)\}| < \eta\,,$$

where η is small. Under suitable conditions we have

$$E_{\theta|x_{(n)}}\{I(\theta)\} - I_1(x_{(n)}) = E_{\theta|x_{(n)}}\{\int_Y p(y|\theta)\log\frac{p(y|\theta)}{p(y|x_{(n)})}\,dy\} > 0. \qquad (5)$$

For the gamma, normal and two-parameter exponential models mentioned in Section 4 (5) is independent of the particular $x_{(n)}$ observed, and so we have

$$E_{\theta|x_{(n)}}\{I(\theta)\} - I_1(x_{(n)}) = I(Y;X_{(\infty)}) - I(Y;X_{(n)})\,.$$

In other words this rule leads to the fixed sample size procedure in Criterion *C1* of Section 5.1 with $\epsilon = \eta$.

Application to the binomial model does not lead exactly to a fixed sample size procedure, but as n increases the variability in (5) is so small that they are effectively equivalent.

It is possible that different stopping rules would lead to more satisfactory results, and more work needs to be undertaken here.

ACKNOWLEDGEMENTS

The first author thanks Instituto Nacional de Investigacas Cientifica, Portugal, for financial support during her period of study at the University of Sheffield.

REFERENCES

ABRAMOWITZ, M. & STEGUN, I. (1972). *Handbook of Mathematical Functions.* New York: Dover.

AITCHISON, J. (1975). Goodness of prediction fit. *Biometrika* **62**, 547-554.

AMARAL, M.A. (1980). Some theoretical and practical applications of predictive distributions. Unpublished Ph.D. thesis, University of Sheffield.

BERNARDO, J.M. (1975). Tamaño óptimo de una muestra: Solución Bayesiana. *Trabajos de Estadística* **26**, 83-92.

— (1979a). Expected information as expected utility. *Ann. Statist.* **17**, 686-690.

— (1979b). Reference posterior distributions for Bayesian inference. *J. Roy. Statist. Soc. B* **41**, 113-147 (with discussion).

BLACHMAN, N. (1966). *Noise and its effect on communication.* New York: McGraw-Hill.

— (1968). The amount of information that *y* gives about *X*. *IEEE Trans. Information Theo.* **14**, 27-31.

BROOKS, R.J. (1982). On the loss of information through censoring. *Biometrika* **69**, 137-144.

DEGROOT, M.H. (1962). Uncertainty, information and sequential experiments. *Ann. Math. Statist.* **33**, 404-419.

EL-SAYYAD, G.M. (1969). Information and sampling from the exponential distribution. *Technometrics* **11**, 41-45.

FANO, R.M. (1961). *Transmission of Information.* New York: Wiley.

GALLAGER, R.G. (1968). *Information Theory and Reliable Communication.* New York: Wiley.

GEISSER, S. (1971). The inferential use of predictive distributions. *Foundations of Statistical Inference.* (V.P. Godambe and D.A. Sprott, eds.), pp. 456-469. Toronto: Holt, Rinehart and Winston.

GRADSHTEYN, I.S. & RYZHIK, I.M. (1965). *Tables of Integrals, Series and Products.* New York: Academic Press.

KULLBACK, S. (1959). *Information Theory and Statistics.* New York: Wiley.

KULLBACK, S. & LEIBLER, R.A. (1951). On information and sufficiency. *Ann. Math. Statist.* **22**, 79-86.

LINDLEY, D.V. (1956). On a measure of information provided by an experiment. *Ann. Math. Statist.* **27**, 986-1005.

— (1957). Binomial sampling schemes and the concept of information. *Biometrika* **44**, 179-186.

MURRAY, G.D. (1977). A note on the estimation of probability density functions. *Biometrika* **64**, 150-152.

OTEYEE, D. & GOOD, I.J. (1974). Information, weight of evidence, the singularity between probability measures and signal detection. *Lectures Notes in Mathematics* **376**. Berlin: Springer-Verlag.

SHANNON, C.E. (1948). A mathematical theory of communication. *Bell System Technical J.* **27**, 379-423, 623-656.

BAYESIAN STATISTICS 2, pp. 613-618
J.M. Bernardo, M.H. DeGroot, D.V. Lindley, A.F.M. Smith (Eds.)

Bayesian Analysis of M/M/1/∞/ Fifo Queues

C. ARMERO
University of Valencia, Spain

SUMMARY

In this paper, the posterior distribution of traffic intensity and the posterior predictive distributions of the waiting time and number of customers for a M/M/1/∞/FIFO queue are obtained given two independent samples of arrival and service times. A practical illustration is given.

Keywords: BAYESIAN INFERENCE; INTERARRIVAL TIMES; QUEUING TIMES; SERVICE TIMES; TRAFFIC INTENSITY.

1. INTRODUCTION

This paper approaches the statistical analysis of a queuing system from a practical point of view. It is assumed that observational data are available only for service times and for interarrival times, a practical situation which arises when the installation of a new service is considered. Information on the local demand of this service may be available from the unit which has detected this need, while information on the effectiveness of the service is obtained from another unit that offers this service in other geographical region.

Typically, the utility of a queuing system depends on the costs of installation and maintenance, waiting times and number of customers in the system. This is better understood if we think in terms of a situation where one of several competing systems offering the service must be chosen.

The random variables that are relevant to these situations are: the traffic intensity, ϱ, the number of customers in the system at a fixed point of time, Z, and the queuing time for a customer, t.

The posterior distribution of traffic intensity, the posterior predictive distributions of queuing time and the number of customers in the system at an arbitrary point of time, given $\varrho < 1$, are obtained in Section 2 for a *M/M/1/∞/FIFO* queue.

In Section 3 these results are applied to real data obtained from the Valencia Teaching Hospital.

2. THE INFERENCE PROBLEM

2.1. *Posterior Distribution of Traffic Intensity*

Let $z_1 = \{x_1, x_2, .., x_n\}$ be the random sample of service times from the exponential model with density function $\mathrm{Ex}(x|\theta) = \theta e^{-\theta x}$, $x > 0$, and let $z_2 = \{y_1, y_2, \ldots, y_m\}$ be the random sample of interarrival times from the model $\mathrm{Ex}(y|\lambda)$. The joint likelihood of θ and λ will be:

$$p(z_1,z_2|\theta,\lambda) = \theta^n\lambda^m\exp(-\theta r_1-\lambda r_2) \tag{1}$$

where $r_1=\Sigma x_i$ and $r_2=\Sigma y_i$ are the relevant sufficient statistics.

We shall assume that the initial information about θ and λ may be described by a product of independent Gamma distributions, so that

$$p(\theta,\lambda) = \mathrm{Ga}(\theta|a_0,b_0)\,\mathrm{Ga}(\lambda|\alpha_0,\beta_0) \tag{2}$$

where $\mathrm{Ga}(x|a,b) = (b^a/\Gamma(a))x^{a-1}\exp(-bx)$, $x>0$, is the probability density function of a Gamma random variable with parameters a, b.

Moreover, using Bernardo's (1979a) method of deriving reference prior distributions, which in this case gives the same answer as the Jeffreys' multivariate rule, we get:

$$\pi(\theta,\lambda) \propto |J(\theta,\lambda)|^{1/2} = \theta^{-1}\lambda^{-1} \tag{3}$$

where $J(\theta,\lambda)$ represents the Fisher information matrix for the model (1). Clearly, (3) may be considered as a limiting case of (2) when all its parameters tend to zero.

Applying Bayes' theorem, the posterior distribution for θ and λ will be:

$$p(\theta,\lambda|z_1,z_2) = Ga(\theta|a_n,b_n)\,Ga(\lambda|\alpha_m,\beta_m) \tag{4}$$

where $a_n=a_0+n$, $b_n=b_0+r_1$, $\alpha_m=\alpha_0+m$ and $\beta_m=\beta_0+r_2$.

Since the traffic intensity is λ/θ (Lindley, 1965, p. 188), and noting that $X\sim Ga(a,b)$ implies $2bX\sim\chi^2(2a)$, the posterior distribution of the traffic intensity ϱ turns out to be:

$$\varrho \sim \frac{\alpha_m b_n}{a_n\beta_m} F(2\alpha_m, 2a_n) \tag{5}$$

where $F(a,b)$ represent a random variable having an F-distribution with degrees of freedom a, b.

If no prior information is considered, (5) will lead to Bayesian credible intervals that will coincide numerically with the classical confidence intervals for the traffic intensity ϱ.

2.2. *Posterior predictive distribution of the numbers of customers in the system*

We shall assume that the traffic intensity ϱ is less than one. Indeed, if $\varrho\geq 1$, a steady solution does not exist since too many customers would then arrive for the server to be able to serve them. Thus, we must find the posterior distribution of θ and λ given the constraint $\lambda<\theta$, which can be expressed in the form

$$p(\theta,\lambda|z_1,z_2,\lambda<\theta) = p(\theta,\lambda|z_1,z_2)/P(\varrho<1|z_1,z_2)\,,\ \lambda<\theta \tag{6}$$

where $P(\varrho<1|z_1,z_2)$ is the posterior probability that a steady solution exist for the process, and can be obtained from the tabulated values of the F-distribution.

Now, the posterior predictive distribution of the number of customers in the system, derived from (6) and

$$p(Z=k|\lambda,\theta,\lambda<\theta) = (\lambda/\theta)^k(1-\lambda/\theta) \quad ,\ k=0,1,2,, \tag{7}$$

(see Lindley, 1965, p. 188), turns out to be

$$p(Z=k|z_1,z_2,\lambda<\theta) = \frac{b^a\beta^\alpha\Gamma(a+\alpha)\,{}_2F_1(2,a+\alpha;k+\alpha+2;\beta/(\beta+b))}{\Gamma(a)\Gamma(\alpha)\,(\beta+b)^{(a+\alpha)}\,zp(\varrho<1|z_1,z_2)\,(k+\alpha)\,(k+\alpha+1)} \tag{8}$$

(see Gradshteyn & Ryzhik, 1965, formulae 3.381.1, 6.455.2, 9.100, 9.137.18) where ${}_2F_1(a,b;c;z)$ is the usual notation for hypergeometric functions and, to simplify the notation, we have omitted the obvious subscripts m,n of the parameters a,b,α and β governing the posterior distribution of θ and λ.

Note that $p(Z=0|z_1,z_2,\lambda<\theta)$ is the posterior predictive probability that the system is empty, given $\varrho<1$.

2.3. *Posterior predictive distribution of the queuing time for a customer*

The same initial comment as that made in the preceding subsection applies here. It only makes sense to derive the predictive distribution of queuing time in the case where the process has a steady solution. Thus, our starting point will be the posterior distribution of θ and λ conditioned on this event.

From (6) and

$$p(t|\lambda,\theta,\lambda<\theta) = \begin{cases} 1-\varrho & , \; t=0 \\ \varrho \mathrm{Ex}(t|\theta-\lambda) & , \; t>0 \end{cases} \tag{9}$$

(see Lindley, 1965 p. 191), the predictive distribution turns out to be:

$$p(t|z_1, z_2,\lambda<\theta) = \begin{cases} p(Z=0|z_1,z_2,\lambda<\theta) & , t=0 \\ C\cdot {}_2F_1(2,a+\alpha+1;\alpha+3;\ (\beta-t)/(b+\beta)) & , \; t>0 \end{cases} \tag{10}$$

where $C=b^a\beta^\alpha\,\Gamma(a+\alpha+1)/P(\lambda<\theta|z_1,z_2)\,\Gamma(a)\Gamma(\alpha)(\alpha+1)(\alpha+2)(b+\beta)^{(a+\alpha+1)}$

For $t>0$, the theoretical expression (10) is very difficult to handle because it does not have a convergent series representation, and its integral representation is rather hard to work with.

From a practical point of view, Monte-Carlo numerical integration provides an effective solution, and the computations for the example in the next section have been carried out with a Monte-Carlo algorithm.

3. APPLICATION

Congestion problems are frequent in Hospital queuing systems. In 1982, in the Valencia Teaching Hospital, a statistical analysis of the queuing system of one of its high priority services was undertaken. This high priority service had only one server during the whole day, the queue discipline was *FIFO*, and there were no restrictions on the kind of customers.

Twenty-five observations of service times and twenty observations of interarrival times were made, and it was assumed that their distributions were independent and exponential. The relevant sufficient statistics were found to be $r_1=32.17$ hours for service times and $r_2=34.45$ hours for interarrival times.

Assuming that there was weak prior information, the posterior distribution of the mean arrival rate, λ, and the mean service rate, θ, was found to be:

$$p(\theta,\lambda|r_1,r_2) = Ga(\theta|25, 32.17).\ Ga(\lambda|20, 34.45) \tag{11}$$

From this, the posterior distribution of traffic intensity, ϱ, is found to be $1.34\varrho \sim F(40,50)$ with mean $E(\varrho)=0.78$ and standard deviation $D(\varrho)=0.24$. The posterior probability, $P(\varrho < 1 \mid r_1, r_2)$ is found to be 0.83, a reasonably high probability that the server will be able to cope with the demand satisfactorily.

Figure 1 shows two posterior probability density functions of ϱ. Density (1) correspond to the weak (non-informative) prior, while density (2) correspond to prior opinions of the queue server of the form $p(\theta,\lambda) = Ga(\theta \mid 10,12).Ga(\lambda \mid 8,8)$.

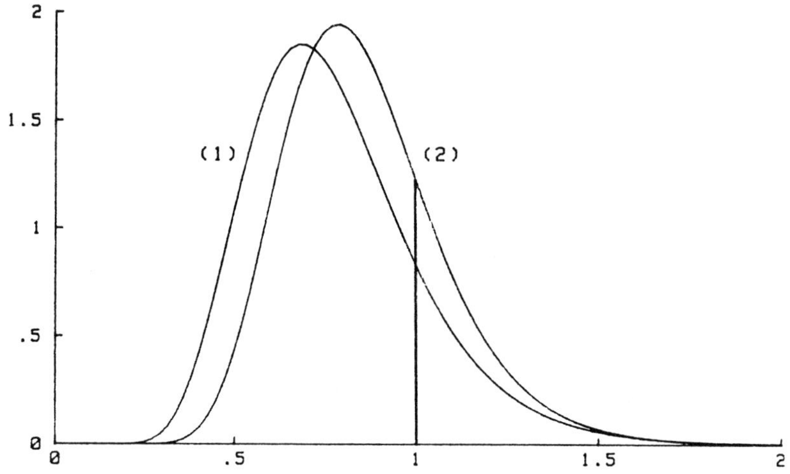

FIGURE 1: *Posterior probability density functions of* ϱ

Table 1 shows the posterior probabilities of the number of customers in the system, Z, given $\varrho < 1$, for the two different states of prior information, and the corresponding classical probabilities obtained from (7) conditioning on $\varrho = \hat{\varrho}$.

TABLE 1

z	$\varrho = \hat{\varrho}$	non-inf. prior	inf. prior
0	0.253	0.301	0.237
1	0.189	0.186	0.162
2	0.141	0.120	0.115
3	0.106	0.081	0.084
4	0.079	0.058	0.063
5	0.059	0.042	0.048
6	0.044	0.031	0.038
7	0.038	0.024	0.030
8	0.025	0.019	0.024
9	0.018	0.015	0.020
10	0.014	0.012	0.017

Note that, given $\varrho < 1$, $P(t=0 \mid \varrho = \hat{\varrho})$ is 0.253, while the posterior predictive probability of $t=0$ is 0.301 using no prior information and 0.237 using the queue server prior opinions about λ and θ.

Figure 2 shows (1) the reference posterior probability density of the queuing time given $\varrho < 1$, (2) its classical approximation obtained from (9) by the substitution of θ and λ by their respective estimators $\hat{\theta} = 0.777$ and $\hat{\lambda} = 0.581$ and (3) the posterior probability density of t given $\varrho < 1$ and the server prior information, $p(\theta,\lambda) = Ga(\theta \mid 10,12).Ga(\lambda \mid 8,8)$.

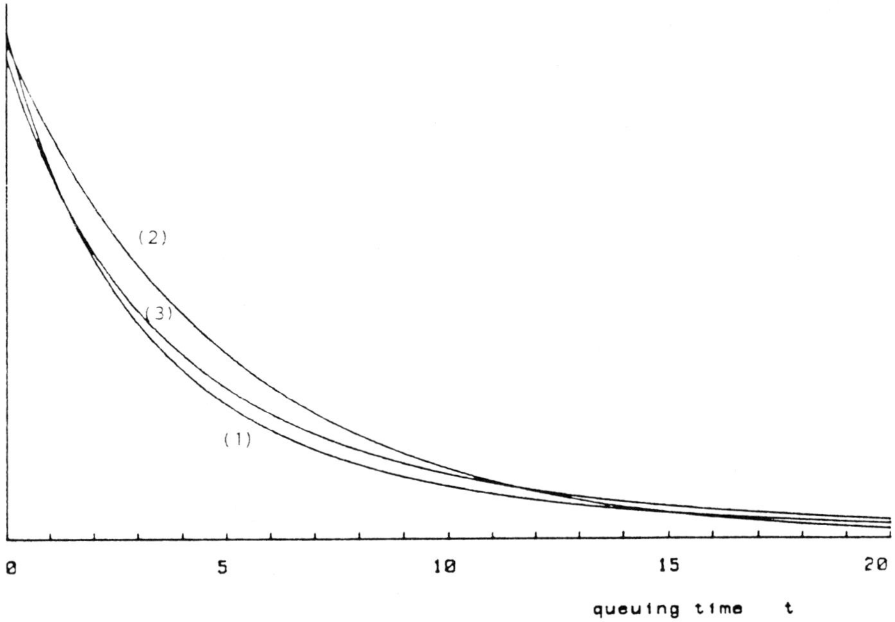

FIGURE 2: *Posterior densities for queuing time*

3. DISCUSSION

From the results obtained in last section, Classical and Bayesian methodologies lead to similar numerical results when non-informative priors are used. However, only the Bayesian approach can handle appropriately the condition $\varrho < 1$ and, indeed, the classical approach completely fails is $\hat{\varrho} > 1$. Moreover, from a practical point of view, the posterior probability of traffic intensity being less than one is a very useful quantity, and has no parallel in clasical methodology.

Bayesian methodology allows the natural incorporation of prior beliefs in the analysis, and this if of vital importance when there are few observations.

REFERENCES

BERNARDO, J.M. (1979a). Reference posterior distributions for Bayesian inference. *J. Roy. Statist. Soc. B,* **41**, 113-147 (with discussion).

GRADSHTEYN, I.S. & RYZHIK, I.M. (1965). *Tables of integrals and products.* New York: Academic Press.

LINDLEY, D.V. (1965). *Introduction to Probability and Statistics from a Bayesian Viewpoint.* Cambridge: University Press.

BAYESIAN STATISTICS 2, pp. 619-628
J.M. Bernardo, M.H. DeGroot, D.V. Lindley, A.F.M. Smith (Eds.)

A Decision Theoretic Structure For Robust Bayesian Analysis with Applications to the Estimation of a Multivariate Normal Mean

L. MARK BERLINER
Ohio State University

SUMMARY

A robust Bayesian viewpoint that subjective prior information can only be quantified in terms of a class Γ of priors is assumed. It is then argued that it may be necessary to consider frequency criteria in choosing a decision rule corresponding to Γ. An example concerning the estimation of a multivariate normal mean is considered. It is shown that under sum of squared errors loss, an estimator composed of several (more than one) independent Stein estimators can be improved upon in Bayes risk for suitably chosen classes of conjugate priors.

Keywords: ADMISSIBILITY; EMPIRICAL BAYES; MULTIVARIATE NORMAL MEANS; ROBUST BAYES ANALYSIS.

1. INTRODUCTION

1.1. *The Robust Bayesian viewpoint*

A major obstacle in a Bayesian analysis of a statistics problem is that, in practice, it is very difficult, if not impossible, to completely quantify subjective prior information by a single prior distribution. In a discussion of the foundational arguments used to equate rational behavior and Bayesian analysis in decision theoretic settings, Rubin (1974) states the case quite strongly:

> "That a procedure corresponds to a particular loss function and prior does not by itself make it appear reasonable unless the loss function and prior are themselves reasonable approximations. Thus the unfortunate position of the rationalist is that non-Bayesian behavior is unreasonable, and that Bayesian behavior is impossible".

Many authors have considered various aspects of the problems associated with the accurate specification of priors. Discussions and references may be found in Berger (1980a), (1983). For brevity, we simply leap to the assumption that subjective prior information can only be summarized in terms of a class Γ of possible prior distributions. The goal of an analysis is then to make decisions which are both (i) "robust" over Γ, yet (ii) able to significantly incorporate prior information. (Such a viewpoint can be traced back at least as far as Good (1950).) To have any mathematical handle on the problem, it is necessary to define robustness. For the Bayesian (fortified by the likelihood principle) facing a decision problem, the most natural definitions of robustness are based on conditional or *a posteriori* measures of loss. A robust decision is relatively insensitive with respect to the "conditional" loss as the prior varies over Γ.

Several possible methods for the implementation of a robust Bayesian analysis have been discussed in the literature. (Again, see Berger (1983) for a review). Perhaps the most natural of these options to the Bayesian is to simply put a "hyperprior" on Γ and proceed with a standard analysis. This approach, of course, corresponds to using a single, fixed prior. However, the construction of Γ should have exhausted the available prior information. Hence, there is no subjective way of choosing the "hyperprior". It is also interesting to note that the prior corresponding to a hyperprior should *ideally* already be an element of Γ. Hence, Γ should be convex.

The above comments should be moderated. Rarely will Γ be sufficiently refined to justify all of the above assertions. More often, Γ will be chosen conveniently to reflect some partial prior information (for example, an appropriate collection of conjugates). In such cases hyperprior analyses often work very well. For example, see Lindley and Smith (1972). Hopefully, the approach discussed below will also be of some use in such cases.

1.2. Γ-*Admissibility*

Rigorous definitions and assumptions will essentially be ignored in this section. Assume X is an observable random variable (or vector) with distribution depending on an unknown parameter θ, an element of a parameter space Θ. A realization of X is denoted by x. Assume that X has a density function $f(\bullet\,|\,\theta)$ with respect to some measure. Let π denote a prior (either density or distribution, as is convenient) for θ. When all quantities exist, let $\pi(d\theta\,|\,x)$ denote the posterior density of θ given $X=x$ and let $m(x\,|\,\pi)$ be the corresponding marginal or predictive density of X; that is,

$$\pi(d\theta\,|\,x) = f(x\,|\,\theta)\pi(d\theta)/m(x\,|\,\pi)$$

where

$$m(x\,|\,\pi) = E^{\pi}[f(x\,|\,\theta)].$$

(In all expectations, subscripts denote constants and superscripts denote either random variables or their distributions with respect to which expectations are taken).

Consider a decision problem in which θ is to be estimated by an action a, contained in a space A, subject to a loss function $L(\theta,a)$. (L: $\Theta \times A \rightarrow \mathbb{R}^1$). A (nonrandomized) decision rule d is a (measurable) function on the sample space of X into A. A Bayes decision rule d^{π} corresponding to π is defined pointwise in x by minimizing the posterior expected loss $E^{\pi}[L(\theta,\bullet)\,|\,X=x]$; that is

$$E^{\pi}[L(\theta,d^{\pi}(x))\,|\,X=x] = \inf_{a\in A} E^{\pi}[L(\theta,a)\,|\,X=x]\,.$$

Thoughout this paper this problem is called the "direct problem".

Another decision problem is now derived from the direct problem. Define a loss function $L^*(\pi,a,x)$ by

$$L^*(\pi,a,x) = E^{\pi}[L(\theta,a)\,|\,X=x] - E^{\pi}[L(\theta,a^{\pi})\,|\,X=x] \tag{1.1}$$

where $a^{\pi} = d^{\pi}(x)$. Define the "risk" function, R, corresponding to L^*, of a decision rule δ to be

$$R(\pi,\delta) = E^{m(\cdot|\pi)}[L^*(\pi,\delta(X),X]. \tag{1.2}$$

(Obviously, under very mild conditions (bounding L from below usually will do) R is the difference in Bayes risks of δ and d^{π} with respect to π for the direct problem). Now define the decision problem based on L^* and R where π is thought of as the unknown parameter, lying in the parameter space Γ, as the "Γ-problem".

Next, consider standard decision theoretic definitions applied to the Γ-problem. (Note the immediate relationship to Γ-minimax or Γ-minimax regret analysis. See Berger (1980a) for references).

Definitions:

1. A rule δ Γ-*dominates* δ' if

   i) $R(\pi,\delta) \leq R(\pi,\delta')$ for all π in Γ,

   and ii) $R(\pi,\delta) < R(\pi,\delta')$ for some π in Γ.

2. A rule δ is Γ-*inadmissible* if there exists some rule δ' that Γ-dominates δ. If no such δ' exists, δ is Γ-*admissible.*

We now partially justify interest in these definitions. First, note that $L^*(\pi,a,x)$ given in (1.1) is precisely the quantity of interest conditionally. That is, for a fixed x, an action a_x, is posterior robust if the range of $L^*(\pi,a_x,x)$ is relatively small as π varies over Γ. Though this approach is foundationally sound, it may be difficult to implement in practice; it may be that for some x no appropriate a_x can be found. Also, it can be argued that the implementation of frequentist criteria can serve as useful checks on Bayes robustness. See Berger (1982, 1983). Such checks are also useful in justifying Bayesian approaches to frequentists. Second, various connections between admissibility, Bayes, and coherence are known. See Stein (1955) and Hill (1974). Finally, as in the role of admissibility studies in classical decision theory, we do not imply that any Γ-admissible rule is necessarily "good" or, in particular, Bayes robust. However, it can be argued that Γ-inadmissible rules should be eliminated from consideration.

1.3. *Applications to Empirical Bayes Analysis*

The sort of analysis discussed here is especially useful in some empirical Bayes settings. In fact, Section 2 presents precisely such an analysis concerning James-Stein estimators viewed as empirical Bayes procedures. By an empirical Bayes approach we mean any method that estimates the prior, or some hyperparameters of the prior, based on the data. See Robbins (1955), Maritz (1970) and Morris (1983a) for discussion and references. An empirical Bayes approach which estimates the prior subject to constraints dictated by Γ may be thought of as possible methodology for the implementation of the robust Bayesian viewpoint. Hence, one reasonable criterion for the robust Bayesian to apply to empirical Bayes procedure is Γ-admissibility.

At least one such analysis has been done. Efron and Morris (1972), (1976) consider the estimation of a matrix of multivariate normal means. In our notation, the associated Γ-problem they derived is that of estimating the inverse of a covariance matrix. (Also, see Haff (1980), (1983)). However, the perspective here is different. (I think). Γ-admissibility should (ideally) only be useful when Γ is meaningful from a subjective viewpoint, not just convenient. In practice, difficulty in analysis will usually force some compromise toward convenience. The degree to which compromises can be made, however, is a delicate matter in that robust rules should still make effective use of prior information.

2. Γ-INADMISSIBLE ESTIMATORS OF A MULTIVARIATE NORMAL MEAN

2.1. *Preliminaries*

Let $\mathbf{X}^{(1)},\ldots,\mathbf{X}^{(q)}$ be independent, multivariate normal random vectors. For $i=1,\ldots,q$ assume

$$\mathbf{X}^{(i)} \sim N(\theta^{(i)},\sigma^2 I_{p_i})$$

where $\theta^{(i)} = (\theta_1^{(i)},\ldots,\theta_{p_i}^{(i)})^t$ is unknown and σ^2 is known and, hence-forth, assumed to be one. (Later, we consider possibly different, but still known, variances $\sigma_1^2,\ldots,\sigma_q^2$). Let X and θ denote the vectors of length $\Sigma_{i=1}^{q} p_i$ obtained by stringing out $(X^{(1)},\ldots,X^{(q)})$ and $(\theta^{(1)},\ldots,\theta^{(q)})$, respectively. The direct problem to be considered is the estimation of θ under the sum of squared error loss given by

$$L(\theta,a) = \sum_{i=1}^{q} \sum_{j=1}^{p_i} (\theta_j^{(i)} - a_j^{(i)})^2$$

$$= \sum_{i=1}^{q} |\theta^{(i)} - \mathbf{a}^{(i)}|^2 = |\theta - a|^2,$$

where a is configured in the same way θ is configured.

To review some results and perform some preliminary computations, it is convenient for now to let $q=1$ and thereby suppress superscripts on $\mathbf{X}$ and θ. Let a decision rule be denoted $\mathbf{d}(\mathbf{x}) = (d_1(\mathbf{x}),\ldots,d_p(\mathbf{x}))^t$. Let the usual frequentist risk, mean squared error, of $\mathbf{d}$ be denoted by M

$$M(\theta,\mathbf{d}) = E_\theta L(\theta,\mathbf{d}(\mathbf{X})).$$

It is by now well known that the standard estimator $\mathbf{d}^o(\mathbf{x}) = \mathbf{x}$ is dominated in mean squared error by

$$\mathbf{d}^{JS}(\mathbf{x}) = (1-(p-2)/|\mathbf{x}|^2)\mathbf{x}$$

if $p \geq 3$ and $\mathbf{d}^o$ is admissible if $p<3$. See James and Stein (1960). Since this result became known, there has been a rapid development of the field of Stein estimation. No attempt at an overview is made here. The reader can find discussion especially related to this paper and references in Berger (1980b), (1982).

Next, consider the following Bayesian analysis of this problem. When the prior π is the conjugate multivariate normal, $N(\mathbf{0}, \tau^2 I_p)$, we have that the posterior distribution of θ, $\pi(\theta\,|\,\mathbf{x})$, is

$$N((\tau^2/(\tau^2+1))\mathbf{x},\ \tau^2/(\tau^2+1)I_p)$$

and the marginal distribution of $\mathbf{X}$, $m(\mathbf{x}\,|\,\pi)$, is $N(\mathbf{0},(\tau^2+1)I_p)$. Furthermore, the Bayes rule (the posterior mean of θ) is given by

$$\mathbf{d}^\pi(\mathbf{x}) = [1-1/(\tau^2+1)]\mathbf{x}$$

(The prior mean has been assumed to be the zero vector without loss of generality, since a simple translation handles a nonzero mean).

A corresponding Γ-problem can now be derived. Suppose Γ is of the form

$$\Gamma = \{N(\mathbf{0},\tau^2 I_p):\ \tau^2 \epsilon \gamma'\}$$

where γ' is a fixed subset of $[0,\infty)$. For an estimator of θ of the form

$$\mathbf{d}(\mathbf{x}) = (1-\delta(|\mathbf{x}|))\mathbf{x}, \tag{2.2}$$

a standard calculation implies that (1.1) is given by

$$\begin{aligned} L^*(\pi,\delta,\mathbf{x}) &= E^\pi[\,|\mathbf{d}-\theta|^2 - |\mathbf{d}^\pi-\theta|^2\,|\,X=x] \\ &= |\mathbf{d}-\mathbf{d}^\pi|^2 \\ &= [\delta(|\mathbf{x}|) - 1/(\tau^2+1)]^2|\mathbf{x}|^2. \end{aligned} \tag{2.3}$$

(The form (2.2) of **d** was chosen because a hyperprior on Γ corresponds to a spherically symmetric (scale mixture of normals) priors for θ. A Bayes rule for any such prior is of the form (2.2). However, in the absence of a complete class theorem for rules of this form in the Γ-problem, the version of the Γ-problem derived here can only be used to infer Γ-inadmissibility). The corresponding risk is then given by

$$R(\pi,\delta) = \int_{\mathbb{R}^p} (\delta(|\mathbf{x}|)-(1+\tau^2)^{-1})^2|\mathbf{x}|^2\,[2\pi(1+\tau^2)]^{-p/2}\exp\{-|\mathbf{x}|^2/[2(1+\tau^2)]\}dx_1...dx_p. \quad (2.4)$$

By change of variables where

$$\eta = (1+\tau^2)^{-1} \text{ and } y = |\mathbf{x}|^2/2, \quad (2.5)$$

it can be shown that

$$R(\pi,\delta) = 2\int_0^\infty (\delta(y)-\eta)^2 y\,\{y^{p/2-1}\eta^{p/2}\,e^{-y\eta}/\Gamma(\tfrac{p}{2})\}\,dy.$$

Therefore, the Γ-problem can be posed as follows:

Let $\alpha = p/2$. *Let* $\gamma = \{\eta: \eta^{-1}\epsilon\gamma'\}$ *Assume Y has a gamma distribution with density*

$$f(y|\eta) = y^{\alpha-1}\eta^{\alpha}e^{-y\eta}/\Gamma(\alpha).$$

Estimate $\eta\,\epsilon\,\gamma$ *subject to the loss function*

$$L(\eta,a,y) = (a-\eta)^2 y.$$

Finally, returning to the original direct problem ($q \geq 1$), assume the $\theta^{(i)}$ have independent, $N(\mathbf{0},\tau_i^2 I_{p_i})$ priors where $\tau_i^2\,\epsilon\,\gamma_i'$. The corresponding Γ problem is:

$$\textit{Let } \alpha_i = p_i/2. \textit{ Let } \gamma_i = \{\eta_i: \eta_i^{-1}\epsilon\gamma_i'\}. \textit{ Assume } \mathbf{Y} = (Y_1,...,Y_q)' \quad (2.6)$$

is a vector of independent, gamma random variables where

$$f_i(y_i|\eta_i) = y_i^{\alpha_i-1}\eta_i^{\alpha_i}e^{-y_i\eta_i}/\Gamma(\alpha_i).$$

Estimate $\eta = (\eta_1,...,\eta_q)'$, *where* $\eta_i\,\epsilon\,\gamma_i$, *subject to*

$$L(\eta,\mathbf{a},\mathbf{y}) = \sum_{i=1}^{q} (a_i-\eta_i)^2 y_i.$$

We close this section with a brief motivation for the analysis considered. In the direct problem, suppose $q=2$ (for simplicity). Consider the following estimators (assume both $p_1, p_2 \geq 3$):

$$d_s(x) = \begin{bmatrix}(1-(p_1-2)/|\mathbf{x}^{(1)}|^2)\mathbf{x}^{(1)}\\(1-(p_2-2/|\mathbf{x}^{(2)}|^2)\mathbf{x}^{(2)}\end{bmatrix}$$

and

$$d_c(x) = [1-(p_1+p_2-2)/(|\mathbf{x}^{(1)}|^2+|\mathbf{x}^{(2)}|^2)]\begin{bmatrix}\mathbf{x}^{(1)}\\\mathbf{x}^{(2)}\end{bmatrix}. \quad (2.7)$$

Both of these estimators dominate the usual rule $d_o(x) = x$.

A possible Bayesian interpretation of the Stein phenomenon is that by shrinking d_o toward one's prior mean of θ (in our case, **0**), it is possible to significantly improve upon d_o when θ is near **0**. However, if even a single component of the vector x is far from zero, the entire Stein rule d_c collapses to the usual estimator d_0, thereby eliminating the possible improvement in other coordinates. Hence, in some cases d_s may be preferable to d_c. For example, if our prior means for $\theta^{(1)}$ and $\theta^{(2)}$ are both **0**, but we are more confident in the case of $\theta^{(1)}$, d_s may be recommended. This paper investigates the Γ-admissibility of d_s. Further discussion of this point may be found in Stein (1973), Efron and Morris (1972), (1973), and Berger and Dey (1980).

2.2. *Main Results*

Again, we begin by assuming $q=1$ and $p \geq 3$. As above, the Γ-problem has risk function given by

$$R(\eta,\delta) = \int_0^\infty (\delta(y)-\eta)^2 \, y^\alpha \eta^\alpha e^{-y\eta}/\Gamma(\alpha)\, dy$$

$$= \alpha \int_0^\infty (\delta(y)-\eta)^2 \eta^{-1} (y^\alpha \eta^{\alpha+1} e^{-y\eta}/\Gamma(\alpha+1))\, dy$$

From this representation, a simple computation implies that, if η is restricted only to be positive, the best invariant estimator of η is $\delta^o(y) = (\alpha-1)/y$. Note that transforming to the original problem, $\delta^o(y)$ corresponds to the Stein estimator

$$d^{JS}(\mathbf{x}) = (1-(p-2)/|\mathbf{x}|^2)\mathbf{x},$$

since

$$\alpha = p/2 \text{ and } y = |\mathbf{x}|^2/2.$$

However, δ^o is Γ-inadmissible (so d^{JS} is Γ-inadmissible), since, at the least, we know $\eta \in (0,1)$. It is clear that $\delta^*(y)$ where $\delta^* = \delta^o$ if $\delta^o \leq 1$ and $\delta^* = 1$ if $\delta^o > 1$ Γ-dominates δ^o. In the direct problem, application of δ^* corresponds to the use of the "positive part, Stein estimator". [This positive part rule is known to dominate d^{JS} in mean squared error.] Furthermore, in applications it may often be assumed that $\eta \in \gamma \subset (0,1]$. If so, δ^o should be truncated to lie in γ.

The main theorem to be given is now stated:

Theorem 2.1. For the decision problem given in (2.6), *assume* $\alpha_i > 1$, $i=1, \ldots, q$ *and the (unrestricted parameter space for* η *is* $\gamma_1 \times \ldots \times \gamma_q$ *where* $\gamma_i = (0,\infty)$, $i=1, \ldots, q$. *Let* $\delta^o(\mathbf{y}) = (\delta_1^o(\mathbf{y}),\ldots,\delta_q^o(\mathbf{y}))'$ where

$$\delta_i^o(\mathbf{y}) = (\alpha_i - 1)/y_i. \qquad (2.8)$$

Define $\delta(\mathbf{y}) = (\delta_1(\mathbf{y}),\ldots,\delta_q(\mathbf{y}))'$, *where for some constants c and b,*

$$\delta_i(\mathbf{y}) = (\alpha_i - 1)/y_i + c/(b + \sum_{i=1}^{q} y_i). \qquad (2.9)$$

If $q \geq 2$, $b \geq 0$, *and*

$$0 < c < 2(q-1)$$

then δ *dominates* δ^o.

Theorem 2.1. implies the following result for the Γ-inadmissibility of separate Stein estimators in the direct problem. The result follows directly from Theorem 2.1 but is stated as a theorem for completeness.

Theorem 2.2. In the notation of Section 2.1, assume $p_i \geq 3$, $i=1,\ldots,q$, *and that* $q \geq 2$. *Let the estimator* d^o *be defined by*

$$d^o = (\mathbf{d}_1^o,\ldots,\mathbf{d}_q^o)$$

where

$$\mathbf{d}_i^o(x) = (1 - \frac{p_i^{-2}}{|\mathbf{x}^{(i)}|^2})\,\mathbf{x}^{(i)} .$$

For the sums of squared error loss, d^o is Γ-inadmissible and is Γ-dominated by estimators of the form

$$d = (\mathbf{d}_1,\ldots,\mathbf{d}_q)$$

where

$$\mathbf{d}_i(x) = (1 - \{\frac{p_i^{-2}}{|\mathbf{x}^{(i)}|^2} + \frac{c}{b + |x|^2}\})\mathbf{x}^{(i)} ,$$

if $b \geq 0$ and $0 < c < 4(q-1)$.

Recall that Theorems 2.1 and 2.2 where derived employing no limitation on η is the Γ-problem. In practice, we would emply the constraints that $\eta_i \in \gamma_i \subset (0,1]$, $i=1,\ldots,q$. However, since the Γ-domination of δ over δ^o (Theorem 2.1) holds for all η, δ Γ-dominates δ^o for restricted η. Also, if δ^* is an estimator obtained by simply truncating each component δ_i to lie in γ_i, then δ^* Γ-dominates δ. These truncations then carry over easily to the direct problem (Theorem 2.2).

2.3. *Proof of Theorem 2.1*

Stein (1973) introduced a powerful technique for proving inadmissibility. Other pertinent references for this paper are Hudson (1978) and Berger (1980c). To prove the inadmissibility of δ^o (in the notation of Theorem 2.1), it is sufficient to find a competitor δ such that

$$\Delta^*(\eta) = R(\eta,\delta) - R(\eta,\delta^o) < 0, \ \forall_\eta .$$

Typically, $\Delta^*(\eta) = E_\eta[H(\eta,\mathbf{Y})]$ for some function H. Stein's method is to use an "integration by parts" -type argument to find a function $\Delta(\mathbf{Y})$ such that

$$\Delta^*(\eta) = E_\eta\Delta(\mathbf{Y}).$$

Once this accomplished, we simply try to show that $\Delta(\mathbf{Y}) < 0$. The relation $\Delta(\mathbf{Y}) < 0$ typically involves various (partial) derivatives of δ^o and δ and is therefore known as a differential inequality.

Our proof relies heavily on the general methods developed in Berger (1980c). Hence, we will be brief.

Lemma (Berger (1980c)). *Assume Y has a gamma distribution with density $f(y|\eta) = = y^{\alpha-1}\eta^\alpha e^{-y\eta}/\Gamma(\alpha)$.*

Let $h(y)$ be a real valued function satisfying

(i) $S(y) = (\alpha-1)y^{\alpha-2}h(y)$ *is absolutely continuous on* $(0,\infty)$.

(ii) $\int_0^\infty |S'(y)|e^{-y\eta}dy < \infty$, *and*

(iii) $\lim y \to {}_0[S(y)e^{-y\eta}] = \lim y \to \infty[S(y)e^{y\eta}] = 0$ *for all* $\eta \in (0,\infty)$.

Then

$$E_\eta[\eta h(y)] = E_\eta[(\alpha-1)h(y)/y + h'(y)].$$

Proof: Integrate by parts.

For a function $h(\mathbf{y})$: $\mathbb{R}^q \to \mathbb{R}^1$, let

$$h^{(i)}(\mathbf{y}) = \frac{\partial}{\partial y_i} h(\mathbf{y}).$$

To prove the theorem, let $\delta_i(\mathbf{y})$ be of the form

$$\delta_i(\mathbf{y}) = \delta_i^o(\mathbf{y}) - \phi_i(\mathbf{y})/y_i$$

for some functions ϕ_i. Then $\Delta(\eta)$ is given by, after some arithmetic,

$$\Delta^*(\eta) = R(\eta,\delta) - R(\eta,\delta^o)$$
$$= E_\eta[\Sigma_{i=1}^q [\phi_i(\mathbf{y})(-2\delta_i^o(\mathbf{y}) + \phi_i^2(\mathbf{y})/y_i)]$$
$$+ 2\Sigma_{i=1}^q \eta_i\phi_i(\mathbf{y})].$$

Assuming the ϕ_i satisfy the requirements of the lemma, Δ^* can be written as

$$\Delta^*(\eta) = E_\eta\{2\Sigma_{i=1}^q [(\alpha_i - 1)\phi_i(\mathbf{y})/y_i + \phi_i^{(i)}(\mathbf{y})]$$
$$+ \Sigma_{i=1}^q \phi_i^2(\mathbf{y})/y_i - 2\sum_{i=1}^q \phi_i(\mathbf{y})\delta_i^o(\mathbf{y})\}.$$

Hence, we need a solution to the differential inequality

$$\Delta(\mathbf{y}) = 2\Sigma_{i=1}^q \phi_i^{(i)}(\mathbf{y}) + \Sigma_{i=1}^q y_i^{-1}\phi_i^2(\mathbf{y}) < 0. \qquad (2.10)$$

Fortunately, Berger (1980c) has done the hard part for us by solving a general version of (2.10). The details are omitted here but Berger's Theorem implies that it is sufficient to find conditions on c and b so that

$$\phi_i(\mathbf{y}) = \frac{-cy_i}{b + \Sigma_{j=1}^q y_i}$$

is a solution to (2.10). Substitution of ϕ_i into (2.10) implies that we require

$$\sum_{i=1}^q \{2(b + \sum_{j=1}^q y_j) - 2y_i - cy_i\} > 0$$

or,

$$b \geq 0 \text{ and } 0 < c < 2(q-1).$$

Finally, recalling the definition of δ, the proof is completed. (The conditions of the lemma are easily checked).

3. GENERALIZATIONS AND COMMENTS

1. As noted by Ghosh and Parsian (1980), it is possible to obtain more general solutions to differential inequalities like (2.10). Applying their results the following extends Theorem 2.1. The proof is a simple modification of that given in Section 2.3 in our case and is omitted.

Theorem 3.1. Assume all of the hypotheses and notation of Theorem 2.1. Further, let $s = \Sigma_{i=1}^q y_i$. *Let* r *be an absolutely continuous function on* $(0,\infty)$ *such that*

$$0 < r(s) < 2(q-1)$$

and

$$r(s)s^{q-1}/\{2(q-1) - r(s)\}$$

is strictly increasing in s. *Then the estimator, defined component-wise, by*

$$\delta_i(y) = (\alpha_i - 1)y_i + r(s)s^{-1}$$

dominates δ^o.

2. An important generalization of Theorem 2.1 is to weighted loss functions of the form

$$L(\eta,\mathbf{a},\mathbf{y}) = \sum_{i=1}^{q} w_i y_i (a_i - \eta_i)^2 \tag{3.1}$$

where the w_i are positive constants. This loss function arises by transformation if, in the direct problem of Section 2.1,

$$\mathbf{X}^{(i)} \sim N(\theta^{(i)}, \sigma_i^2 I p_i), \qquad i=1, \ldots, q,$$

where the σ_i^2 are known.

A method of handling (3.1) given by Berger (1979) (a generalization of Bhattacharya (1966)) considers a decomposition of (3.1) into appropriate subproblems with losses $\Sigma_{i=1}^{j} y_i (a_i - \eta_i)^2$. Assume $w_1 \geq \ldots \geq w_q$. Define for $i, j = 1, \ldots, q$

$$\beta_i^j = \begin{cases} 0 & \text{if} \quad j < i \\ (w_j - w_{j+1})/w_j & \text{if} \quad j \geq i \;, \end{cases}$$

where $w_{q+1} = 0$. Let $\delta_{(j)}(y_1, \ldots, y_j)$ be an estimator which dominates

$$\delta^{\circ}_{(j)} = (\delta^{\circ}_{1(j)}, \ldots, \delta^{\circ}_{j(j)})^t$$

for the jth subproblem ($\delta_{(1)} = \delta^{\circ}_{(1)}$). Then for (3.1) δ' defined componentwise by

$$\delta_i(\mathbf{y}) = \sum_{j=1}^{q} \beta_i^j \delta_{i(j)}(\mathbf{y})$$

dominates δ° it at least one non-trivial $\delta_{(j)}$ was found.

3. Natural generalizations of Section 2 would be for direct problems involving nonnormal distributions. Depending on the loss functions, the key to such extensions is the nature of the marginal distributions corresponding to the priors in Γ. Marginals that are members of the exponential family can often be studied in terms of possible Stein-Berger phenomena. Morris (1983b) may be of some aid in finding such cases.

4. The estimators proposed in Section 2.2 that were obtained via truncation so as to be consistent with the restrictions on hyper-parameters dictated by Γ are themselves anticipated to be Γ-inadmissible. This follows from the observation made in 1.2 that Γ-admissible rules should be (essentially) Bayes rules. However, it is well known that in our setting, Bayes rules satisfy certain smoothness or analyticity conditions. Hence, rules obtained via truncation cannot be Bayes. On the other hand, the truncated rules are simple to compute and it is doubtful that they can be significantly improved upon. (For example, though the positive part Stein estimator ($q=1$) is inadmissible in the direct problem, no one has ever explicitly presented a rule which dominates it).

ACKNOWLEDGEMENT

I wish to thank Jim Berger for several useful discussions on these matters. The comments of an anonymous referee were very helpful and highly appreciated.

REFERENCES

BERGER, J.O. (1979). Multivariate estimation with nonsymmetric loss functions. *Optimizing Methods in Statistics.* (J. Rustagi, ed.) New York: Academic Press.

— (1980a). *Statistical Decision Theory.* New York: Springer Verlag.

— (1980b). A robust generalized Bayes estimator and confidence region of a multivariate normal mean. *Ann. Statist.* **8**, 716-761.

— (1980c). Improving on inadmissible estimators in continuous exponential families with applications to simultaneous estimation of gamma scale parameters. *Ann. Statist.* **8**, 545-571.

— (1982). Bayesian robustness and the Stein effect. *J. Amer. Statist. Assoc.* **77**, 358-368.

— (1983). The robust Bayesian viewpoint. *Tech. Rep.* Statistics Department, Purdue University.

BHATTACHARYA, P.K. (1966). Estimating the mean of a multivariate normal population with general quadratic loss function. *Ann. Math. Statist.* **37**, 1819-1824.

DEY, D. and BERGER, J. (1980). Combining coordinates in simultaneous estimation of normal means. *Tech. Rep.* **80-32**, Statistic Department, Purdue University.

EFRON, B. and MORRIS, C. (1972). Empirical Bayes on vector observations. An extension of Stein's method. *Biometrika* **59**, 335-347.

— (1973). Combining possibly related estimation problems. *J. Roy. Statist. Soc., B,* **35**, 379-421.

— (1976). Multivariate empirical Bayes and estimation of covariance matrices. *Ann. Statist.* **4**, 22-32.

GOOD, I.J. (1950). *Probability and the Weighing of Evidence.* London: Charles Griffin.

HAFF, L. (1979). Estimation of the inverse covariance matrix: Random mixtures of the inverse Wishart matrix and the identity. *Ann. Statist.* **7**, 1264-1276.

— (1982). Solutions of the Euler-LaGrange equations for certain multivariate normal estimation problems. *Tech. Rep.*

HILL, B.M. (1974). On coherence, inadmissibility and inference about many parameters in the theory of least squares. *Studies in Bayesian Econometrics and Statistics.* S. Feinberg and A. Zellner, eds.) Amsterdam: North Holland.

HUDSON, H.M. (1978). A natural identity for exponential families with applications in multiparameter estimation. *Ann. Statist.* **6**, 473-484.

JAMES, W. and STEIN, C. (1960). Estimation with quadratic loss. *Proc. Fourth Berkely Symp. Math. Statist. Prob. I,* 361-379. University of California Press.

LINDLEY, D.V. and SMITH, A.F.M. (1972). Bayes estimates for linear model. *J. Roy. Statist. Soc. B* **34**, 1-41.

MARITZ, J.S. (1970). *Empirical Bayes Method.* London: Methuen.

MORRIS, C.N. (1983a). Parametric empirical Bayes inference: theory and applications (with discussion). *J. Amer. Statist. Assoc.* **78**, 47-65.

— (1983b). Natural exponential families with quadratic variance functions: statistical theory. *Ann. Statist.* **11**, 515-529.

ROBBINS, H. (1955). An empirical Bayes approach to statistics. *Proc. Third Berkeley Symp. Math. Statist. Prob. I,* 157-164.

RUBIN, H. (1974). The application of decision theory to statistical practice. Mimeo Series 399, Statistic Department, Purdue University.

STEIN, C. (1955). A necessary and sufficient condition for admissibility. *Ann. Math. Statist.* **26**, 518-522.

— (1973). Estimation of the mean of a multivariate distribution. *Proc. Prague Symp. Asymp. Statist.* 345-381.

BAYESIAN STATISTICS 2, pp. 629-644
J.M. Bernardo, M.H. DeGroot, D.V. Lindley, A.F.M. Smith (Eds.)

Invariant Normal Bayesian Linear Models and Experimental Designs

G. CONSONNI and A.P. DAWID
(*University College London, England* and *"L. Bocconi" University, Italy*)
(*University College London, England*)

SUMMARY

An experimental design may be characterised by its *symmetries*, i.e. unit permutations preserving the layout and treatment allocation. Assuming normality and unit-treatment additivity, these symmetries lead to a representation of the experiment as an *invariant normal linear model*, with a covariance structure which is invariant, and a mean-structure which is equivariant, under the symmetry group. Similar considerations apply to the prior distribution. Under such invariance, the problem of Bayesian inference decomposes naturally into several independent subproblems. Within any such subproblem, if additional irreducibility conditions hold, the posterior expectation of any parameter will be a fixed scalar multiple of its unique unbiased estimator.

Keywords: BLOCK STRUCTURE: TREATMENT STRUCTURE; GROUP REPRESENTATIONS; ANALYSIS OF VARIANCE; BAYESIAN ANALYSIS.

1. INTRODUCTION AND SUMMARY

Symmetry considerations play an important role in many scientific disciplines, the best known example being perhaps physics.

Statistics has also benefited from the application of the concept of symmetry, a very good example being provided by the notion of *exchangeability* together with de Finetti's celebrated theorem.

The most typical area in which symmetry naturally arises is perhaps that of *experimental design*. Thus, for concreteness, this paper will refer to this branch of statistics, although the underlying argument is more general.

Section 2 discusses what we mean by symmetry and how symmetry arises in experimental designs. The notion of *block structure* proves to be crucial and its statistical implications on the *covariance structure* of a statistical model are examined in section 3. Section 4 deals with *treatment allocation* and its reflections on the *mean structure* of a statistical model. Having assumed that the covariance structure is known, Section 5 is specific to the Bayesian analysis and considers the implications of both the block structure and the treatment allocation on the prior distribution for the "mean" parameters. Section 6 recalls very briefly the theory of invariant normal Bayesian linear models developed in Consonni and Dawid (1983) with particular emphasis on the *decomposition* of Bayesian statistical problems. Section 7 discusses the *analysis* of a single Bayesian subproblem under some special conditions. Section 8 contains a brief critical discussion of the method described in the previous sections.

Throughout this paper we have attempted to illustrate all the main concepts and results by means of four examples, which make their first appearance in section 2 and are then recalled and elaborated upon in each subsequent section.

2. BLOCK STRUCTURE AND TREATMENT STRUCTURE

How does symmetry arise in an experimental design? We start with a *unit set* **U**, which contains all the experimental units under consideration. Furthermore we consider a set **T** of *treatments.* A single treatment $t \in \mathbf{T}$ is allocated to each unit $u \in \mathbf{U}$. We let

$$\tau : \mathbf{U} \to \mathbf{T}$$

denote the *treatment allocation* mapping. We assume that τ is surjective i.e. that every treatment in **T** is allocated to some unit. If this were not the case simply redefine the treatment set accordingly.

The result of an experiment will consist in the observation on each unit u of a measurable quantity denoted by X_u. In agricultural experiments, for example, X_u is often the yield of a certain produce. We call X_u a *basic observable.*

Throughout this paper we shall make the assumption of *unit-treatment additivity.* This says that each basic observable X_u can be written as the sum of two components, i.e.

$$X_u = \Theta_{\tau(u)} + Z_u , \tag{2.1}$$

where the first component depends only on the treatment applied to u, whereas the second is specific to the unit only. We call $\Theta_{\tau(u)}$ the *treatment effect* and Z_u the *unit effect.*

Prior to the experiment the basic observable X_u is clearly uncertain. We turn (2.1) into a statistical model by assigning a joint distribution to (Z_u) and interpreting the quantity $\Theta_{\tau(u)}$ as the sampling mean of X_u (this clearly implies that $E(Z_u) = 0$). Furthermore we assume that (Z_u) is independent of $(\Theta_{\tau(u)})$. From a sampling viewpoint this means that $(\Theta_{\tau(u)})$ does not appear in the sampling distribution of (Z_u); moreover if $(\Theta_{\tau(u)})$ is assigned a (prior) distribution, then the vectors $(\Theta_{\tau(u)})$ and (Z_u) are independent in the usual probabilistic sense. While not always justified, the assumption of unit-treatment additivity is often applicable to experimental designs, see Nelder (1965).

The experimental interpretation of model (2.1) is obtained by distinguishing two basic features of any experimental design, namely a *block structure* and a *treatment structure,* see Nelder (1965), Bailey (1981) and Houtman (1981). We shall first examine the former.

Regardless of which treatment allocation is chosen, the unit set **U** possesses an internal structure of its own. Such a structure is defined by a set of *categories* together with a set of *operations* describing their mutual relations. Two types of category operations which are particularly useful in practice are the *nesting* and *crossing* operations.

Example 1. A nested structure occurs in a randomized block layout (with no treatments). We recognize two categories: i) the "blocks" of the randomized design and ii) the "plots" nested within each block. For example, a block might be a litter and a plot a pig within that litter.

Example 2. A crossed structure occurs in a row-column design. Here we distinguish two block categories: rows and columns (e.g. "machines" and "workers"). Whereas in the nested structure we do not usually recognize any connection between two units having the *same* plot label but being in *different* blocks, such a connection does exist in a crossed structure between two units sharing the same column (or row) but lying in different rows (or columns).

Nested and crossed structure might appear simultaneously in the same block structure.

Example 3. Consider Example 2 and suppose that each worker produces N items on each machine. Thus we have a crossed structure with N replications nested within each combination of the above.

When a block structure can be described in terms of nesting and crossing operations only, then it is called a *simple* block structure. Most of the block structures actually used are simple. An example of a non-simple block structure occurs in the so-called diallele-crosses, see e.g. James (1982).

To each block structure we can associate a permutation group G which *preserves* the block structure itself. Informally, this means that G operates a rearreangement of the unit labels which leaves the structure of the original layout unaltered.

Example 1 (ctd). Let B be the number of blocks and P the number of plots within each block. Consider a group G which, independently for each block, permutes plot labels within the same block and, also, permutes block labels among themselves. Since the block structure is obviously independent of the choice of the plot labels attached to each unit within the same block and similarly for the choice of the block labels, G preserves the block structure, for it simply rearranges labels in a way which is consistent with the original layout. If we denote by S_N the symmetric group of degree N (i.e. the group of *all* permutations of N objects), then G is called the wreath product of S_B and S_P (in this order), written $S_B \, wr \, S_P$.

Example 2 (ctd). If R denotes the number of rows and C the number of columns, then the corresponding group G is given by $S_R \times S_C$, the ordinary *direct product* group. Notice that G permutes whole rows among themselves and, independently, whole columns among themselves, as the actual layout requires.

Example 3 (ctd). Here the relevant group is given by $(S_R \times S_C) \, wr \, S_N$, a natural combination of the group structures described in the previous two examples.

In all the examples mentioned so far the group G associated with the block structure has been obtained by a suitable combination of symmetric groups. The next example shows how, in some circumstances, other types of constituent groups might be more appropriate.

Example 4 (Olkin and Press, 1969). Consider a point source located at the geocenter of a regular polygon of N sides from which a signal is transmitted. Identical signal receiver (with identical noise characteristics) are positioned at the N vertices (or units) denoted sequentially by $u_0,\ldots,u_{N-1}$. A useful criterion of classification of the N units in this case, essentially due to the "circular" arrangements of the units, turns out to be the number of vertices separating any two units. Thus if $N = 2M$ or $N = 2M + 1$, there are $M + 1$ possible levels of classifications. The group G must then permute the labels attached to the vertices in such a way that if units u and u' are l vertices apart ($l = 0,\ldots,M$), then gu and gu' are still l vertices apart for any $g \in G$. It is not difficult to realize that such a group is represented by the cyclic group of order N, written C_N.

A treatment structure may be described in a way very similar to what has been done for the block structure, the only difference being of course that treatments rather than units are being classified. We shall not discuss the treatment structure any further, however, since it will play a subsidiary role in our analysis (see however sect. 5 and sect. 8).

3. BLOCK STRUCTURE AND COVARIANCE STRUCTURE

Section 2 introduced the idea of symmetry in experimental designs without attempting to elaborate on its statistical consequences. This aspect will be pursued in this section.

Consider a particular experimental design and let G be the group associated with its block structure. How does this information influence the statistical model (2.1)?

Since the block structure pertains to the units only, it follows from the nature of model (2.1) that it will only affect the distribution of the vector of unit effects (Z_u). A useful way, from a probabilistic viewpoint, of interpreting the nature of group G is to regard it as a means of describing the type of arbitrariness inherent in the labelling of the units, which is nevertheless consistent with the block structure of the experiment. Thus if we decide that only the knowledge of such a structure (and no other considerations) should be used to describe the stochastic nature of (Z_u), then its distribution should be *invariant* with respect to (*w.r.t.*) the group G. Assuming that the distribution of (Z_u) is normal and using the fact that $E(Z_u) = 0$, invariance of the distribution *w.r.t* G clearly amounts to the following requirement

$$\mathrm{Cov}(Z_u, Z_{u'}) = \mathrm{Cov}(Z_{gu}, Z_{gu'}), \qquad u,u' \in \mathbf{U}\,; g \in G; \tag{3.1}$$

in other words the *covariance matrix* of (Z_u) is *G-invariant*.

Although we did not explicitly mention it, the invariance of the distribution of the unit effects *w.r.t* G is closely related to the idea of randomization in experimental designs. While randomization does not seem to be logically necessary in a subjective framework, it might nevertheless make assumption (3.1) particularly compelling.

Example 1 (ctd). We identify a unit u with the pair (i,j), where $i=1,\ldots,B$, denotes block and $j=1,\ldots,P$, plot within block i. Letting $\gamma_{i_1 j_1, i_2 j_2} = \mathrm{Cov}(Z_{i_1 j_1}, Z_{i_2 j_2})$, the most general type of covariance matrix invariant under $G = S_B\ wr\ S_P$ is given by

$$\gamma_{i_1 j_1, i_2 j_2} = \begin{cases} \gamma_0 & i_1=i_2;\ j_1=j_2 \\ \gamma_1 & i_1=i_2;\ j_1\neq j_2 \\ \gamma_2 & i_1\neq i_2 \end{cases}$$

where γ_0, γ_1 and γ_2 must be such as to ensure that the resulting matrix is non-negative definite.

Example 2 (ctd). Again we identify the unit u with the pair (i,j), where $i=1,\ldots,R$, denotes row and $j=1,\ldots,C$, denotes column. The most general type of covariance matrix invariant under $G = S_R \times S_C$ is given by

$$\gamma_{i_1 j_1, i_2 j_2} = \begin{cases} \gamma_0 & i_1=i_2;\ j_1=j_2 \\ \gamma_1 & i_1=i_2;\ j_1\neq j_2 \\ \gamma_2 & i_1\neq i_2;\ j_i=j_2 \\ \gamma_3 & i_1\neq i_2;\ j_i\neq j_2 \end{cases}$$

where again the constants γ_0, γ_1, γ_2 and γ_3 are such that the resulting covariance matrix is non-negative definite.

Example 3 (ctd). This time we need three suffixes conveniently to identify a unit. Thus if $u = (i, j, k)$, $i=1,\ldots,R$; $j=1,\ldots,C$; $k=1,\ldots,N$, it means that u is in row i, column j and corresponds to replication k. The most general type of covariance matrix invariant under $G = (S_R \times S_C)\ wr\ S_N$ is given by

$$\gamma_{i_1j_1k_1,i_2j_2k_2} = \begin{cases} \gamma_0 & i_1=i_2;\ j_1=j_2;\ k_1=k_2 \\ \gamma_1 & i_1=i_2;\ j_1=j_2;\ k_1\neq k_2 \\ \gamma_2 & i_1\neq i_2;\ j_1=j_2 \\ \gamma_3 & i_1=i_2;\ j_1\neq j_2 \\ \gamma_4 & i_1\neq i_2;\ j_1\neq j_2 \end{cases}$$

with the usual constraints on the constants involved.

Example 4 (ctd). The most general type of covariance matrix invariant under $G = C_N$ is the so-called *circulant* matrix namely, if $N = 2M + 1$

$$\begin{pmatrix} \gamma_0 & \gamma_1 & \gamma_2 & \cdots & \gamma_M & \gamma_M & \cdots & \gamma_1 \\ \gamma_1 & \gamma_0 & \gamma_1 & \cdots & \gamma_{M-1} & \gamma_M & \cdots & \gamma_2 \\ \vdots & & & & & & & \vdots \\ \vdots & & & & & \gamma_0 & & \gamma_1 \\ \gamma_1 & \cdots & \cdots & \cdots & \cdots & \gamma_1 & & \gamma_0 \end{pmatrix}$$

whereas, if $N = 2M$,

$$\begin{pmatrix} \gamma_0 & \gamma_1 & \gamma_2 & \cdots & \gamma_M & \gamma_{M-1} & \cdots & \gamma_1 \\ \gamma_1 & \gamma_0 & \gamma_1 & \cdots & \gamma_{M-1} & \gamma_{M-2} & \cdots & \gamma_2 \\ \vdots & & & & & & & \vdots \\ \vdots & & & & & \gamma_0 & & \gamma_1 \\ \gamma_1 & \cdots & \cdots & \cdots & \cdots & \gamma_1 & & \gamma_0 \end{pmatrix}$$

4. TREATMENT ALLOCATION AND EQUIVARIANT STATISTICAL MODELS

Having examined the block structure in Section 3, we now turn our attention to treatments. Recall the notion of treatment allocation mapping

$$\tau : \mathbf{U} \to \mathbf{T} \quad .$$

Next consider a permutation group H acting on the unit set $\mathbf{U}$. We say that H *preserves* the treatment allocation if

$$\tau(u) = \tau(u') \Rightarrow \tau(hu) = \tau(hu'), \qquad u,u' \in \mathbf{U}\ ;\ h \in H. \tag{4.1}$$

In other words, H rearranges units in such a way that, whenever u and u' are given the same treatment, their permutation images hu and hu' are also given the same treatment (which might of course differ from the one administered to u and u').

Now let G be the group associated with the block structure of an experimental design and define the following group

$$G^* = \{g \in G: \tau(u) = \tau(u') \Rightarrow \tau(gu) = \tau(gu'), \qquad u,u' \in \mathbf{U}\}. \tag{4.2}$$

Clearly G^* is a subgroup of G, possibly coinciding with G itself. By definition, if $g^* \in G^*$ then g^* preserves both the block structure and the treatment allocation. We call G^* the group of *allowable permutations.*

The group G^* naturally induces a permutation group $\bar{G}$ (say) on the treatment set $\mathbf{T}$. Indeed let $g^* \in G^*$, then g^* determines a permutation $\bar{g} \in \bar{G}$ and $t' = \bar{g}\,t$ if, for some u, $t = \tau(u)$ and $t' = \tau(g^*u)$. Notice that $\bar{G}$ is well defined since, for any $t \in \mathbf{T}$, there exists a $u \in \mathbf{U}$ such that $t = \tau(u)$, because of surjectivity of τ. Furthermore, because of (4.1), the image of t under $\bar{g}$ is unique, being independent of the particular choice of the inverse image of t under τ.

Notice that from (2.1) and the last paragraph we have, for any specification θ of $\Theta = (\Theta_t;\, t \in \mathbf{T})$,

$$E(X_u|\theta) = \theta_t \Rightarrow E(X_{g^*u}|\theta) = \theta_{\bar{g}t}\,. \tag{4.3}$$

In other words the statistical model (2.1) is *equivariant under* G^* (or G^*-equivariant).

In practice, one tends to specify the treatment allocation directly in terms of the treatment effects Θ as in (4.3) instead of first specifying the set of treatments $\mathbf{T}$ and then the treatment allocation mapping τ. This is the approach which will be taken in the ensuing examples. Furthermore, whenever a specific index set is attached to $\mathbf{T}$, e.g. $\mathbf{T} = \{t_i: i=1,\ldots,B\}$, then, instead of writing $E(X|\theta) = \theta_{t_i}$ (say), we shall write $E(X_u|\theta) = \theta_i$ for simplicity.

The next examples are intended purely as illustrations of the abstract theory, not as recommended designs.

Example 1 (ctd). Suppose that the same treatment is applied to all the plots within the same block, each block having a distinct treatment. We can thus write

$$E(X_{ij}|\theta) = \theta_i\,. \tag{4.4}$$

It is easy to verify that in this case $G = G^*$. Thus the model is equivariant under the original group G which preserves the block structure. Furthermore $\bar{G} = S_B$.

Notice that if we had set

$$E(X_{ij}|\theta) = \theta_{ij}\,,$$

then the model would still be equivariant (trivially) under G and $\bar{G} = G$.

Example 2 (ctd). There are two simple types of treatment allocation in this case: i) same treatment to all the units belonging to the same row with a distinct treatment for each row, and ii) same treatment to all the units belonging to the same column with a distinct treatment for each column. Under i) we clearly have

$$E(X_{ij}|\theta) = \theta_i\,, \tag{4.5}$$

whereas under ii) we have

$$E(X_{ij}|\theta) = \theta_j\,. \tag{4.6}$$

In both cases a G-invariant model is obtained (i.e. $G^* = G$). Furthermore, under (4.5) $\bar{G} = S_R$, whereas under (4.6) $\bar{G} = S_C$.

Consideration of Example 3 will deferred to section 5. We thus turn directly to Example 4.

Example 4 (ctd). Suppose that $N = 2M$ and that the N basic observables are written $X_0,\ldots,X_{N-1}$. Then the following treatment allocation

$$E(X_{2l}|\theta) = \theta_1, \qquad l=0,\ldots,M\text{-}1$$
$$E(X_{2l+1}|\theta) = \theta_2, \qquad l=0,\ldots,M\text{-}1 \tag{4.7}$$

induces a G-equivariant model (i.e. we have again $G^* = G$) and $\bar{G} = S_2$.

The above example are untypical of real experimental designs: for example, treatment and block effects are completely confounded in Example 1 (ctd). Such confounding will in fact always occur when $G^* = G$. More realistically, consider the layout of Example 1 as subject to the usual randomized block allocation, and relabel plots within blocks to agree with the labelling of the allocated treatment. Then we obtain

Example 1 (ctd). Now we have

$$E(X_{ij}|\theta) = \theta_j\,. \tag{4.8}$$

Then clearly this model is not equivariant *w.r.t.* the original group $G = S_B \text{ wr } S_P$. Indeed from (4.8) we have, for example, that unit (1,2) and unit (2,2) have the same treatment. Now consider the permutation $g \in G$ which permutes plots within block 1 leaving all plots within all other blocks unaltered, and suppose that $g(1,2) = (1,4)$. Then we should have

$$E(X_{14}|\theta) = E(X_{22}|\theta)\,,$$

contrary to (4.8).

From Example 2 in this section, or by direct inspection, it follows that in this case $G^* = S_B \times S_P$. Thus under (4.8) the model is only G^*-equivariant, with G^* being the direct product of S_B and S_P. The corresponding group $\bar{G}$ is given by S_P. Further analysis of this G^*-equivariant model is now carried out exactly as for Example 2 (ctd) above.

5. INVARIANT PRIOR DISTRIBUTIONS

We shall assume that the covariance matrix of the unit effects (Z_u) is known. Thus inference is required only for the treatment effects $\Theta = (\Theta_t\colon t \in \mathbf{T})$.

Following a Bayesian approach we thus have to specify a prior distribution on Θ. If we assume that such a distribution is normal, then all we have to specify is the mean and the covariance structure. The main assumption we shall make is that this distribution is $\bar{G}$-invariant, i.e. for all $t,t' \in \mathbf{T}$ and all $\bar{g} \in \bar{G}$

$$E(\Theta_t) = E(\Theta_{\bar{g}t}) \tag{5.1}$$

$$\mathrm{Cov}(\Theta_t, \Theta_{t'}) = \mathrm{Cov}(\Theta_{\bar{g}t}, \Theta_{\bar{g}t'})\,. \tag{5.2}$$

Notice that conditions (5.1) and (5.2) will be certainly satisfied if the prior distribution on Θ is taken to be exchangeable. However exchangeability might be an unnecessarily strong requirement if the group $\bar{G}$ is appreciably smaller than the corresponding symmetric group.

Another important instance in which (5.1) and (5.2) could be deemed appropriate is when the set $\mathbf{T}$ possesses a treatment structure (e.g. a factorial structure) of its own which is preserved by $\bar{G}$ (recall the discussion on the block structure).

Example 3 (ctd). Suppose that

$$E(X_{ijk}|\theta) = \theta_{ij}\,. \tag{5.3}$$

Then one can easily see that the original group $G = (S_R \times S_C) \text{ wr } S_N$ is allowable, i.e. $G = G^*$. Moreover it can be verified that the group $\bar{G}$ induced on the treatment set T is $S_R \times S_C$.

Thus if **T** is endowed with a factorial (i.e. crossed) treatment structure with two types of treatments, one with R levels and the other with C levels, the induced group G does preserve the treatment structure and thus assumptions (5.1) and (5.2) would be regarded as particularly appropriate.

6. INVARIANT NORMAL BAYESIAN LINEAR MODELS: DECOMPOSITION OF STATISTICAL PROBLEMS

In this section we abstract the essential features discussed so far and recall the main points and results of the theory of invariant normal Bayesian linear models developed in Consonni and Dawid (1983). Andersson (1975) represents the seminal paper for the theory of invariant normal models.

Some preliminary results and definitions are necessary, see for example Ledermann (1977).

Consider a finite dimensional real vector space $\mathbf{V}$. We assume that $\mathbf{V}$ is equipped with a (possibly semi-definite) inner product (i.p.) Δ . Let now H be a finite group and consider a linear representation π of H on $\mathbf{V}$.

$$\pi : H \rightarrow GL(\mathbf{V}) ,$$

where $GL(\mathbf{V})$ is the general linear group on $\mathbf{V}$ and π is a group-homomorphism, i.e.

$$\pi(h_1)\,\pi(h_2) = \pi(h_1 h_2) , \qquad h_1, h_2 \in H.$$

Notice that for any $V \in \mathbf{V}$ and any $h \in H$, $\pi(h)V \in \mathbf{V}$, i.e. $\mathbf{V}$ is H-invariant.

The representation π induces a unique (up to rearrangement of terms) decomposition of $\mathbf{V}$ into the direct sum of isotropic H-invariant subspaces, which we write

$$\mathbf{V} = \underset{l}{\oplus} \mathbf{V}_l . \tag{6.1}$$

We call (6.1) the *canonical decomposition* of $\mathbf{V}$.

Now assume that the i.p. Δ is H-invariant, i.e.

$$\Delta(\pi(h)V, \pi(h)W) = \Delta(V, W), \qquad V, W \in \mathbf{V};\, h \in H. \tag{6.2}$$

Then it can be shown, see for example Consonni (1983), that under (6.2) the subspaces $\mathbf{V}_l$ which appear in (6.1) are pairwise orthogonal *w.r.t.* Δ, written

$$\mathbf{V}_{l_1} \perp_\Delta \mathbf{V}_{l_2} , \qquad l_1 \neq l_2. \tag{6.3}$$

Result (6.3) constitutes the cornerstone of all our subsequent theory.

We let $\mathbf{X}$ be the real space spanned by the basic observables $\{X_u : u \in \mathbf{U}\}$. We call $\mathbf{X}$ the *observable space* and $Y \in \mathbf{X}$ is called an *observable*. Notice that the assumption of normality of the joint sampling distribution of $X = (X_u)$ is equivalent to the assumption that each observable is distributed according to a univariate normal distribution.

Next consider the group of allowable permutations G^*, see (4.2). We define a linear representation ϱ of G^* on $\mathbf{X}$

$$\varrho : G^* \rightarrow GL(\mathbf{X})$$

in the following way:

$$\varrho(g^*)X_u = X_{g^* u}, \tag{6.4}$$

and then extend it by linearity. The representation ϱ as defined in (6.4) is called the *permutation representation* of G^* on $\mathbf{X}$.

Now let Ω be the real space spanned by the treatment effects $\{\Theta_t: t \in \mathbf{T}\}$. We call Ω the *parameter space* and any $\Psi \in \Omega$ is called a *parameter*.

Next we endow $\mathbf{X}$ with an i.p. Γ, defined for $Y = \Sigma\, \alpha_u X_u$ and $Z = \Sigma\, \beta_u X_u$ as

$$\Gamma(Y,Z) = \Sigma\Sigma\, \alpha_u \beta_{u'} \mathrm{Cov}(X_u, X_{u'} \mid \Omega),$$

where $\mathrm{Cov}(\cdot,\cdot \mid \Omega)$ denotes the sampling covariance. We call Γ the *covariance i.p.* on $\mathbf{X}$.

Notice that Γ is G^*-invariant if and only if the covariance matrix of (X_u) is G^*-invariant. Thus from the discussion contained in section 3 it follows that the assumption of G^*-invariance of Γ is justified. (Indeed Γ is invariant *w.r.t.* the larger group G, but we shall omit this extra information in this section; see however section 8).

Next consider the canonical decomposition of $\mathbf{X}$ induced by ϱ, which we write

$$\mathbf{X} = \bigoplus_q \mathbf{X}_q\,. \tag{6.5}$$

Each G^*-invariant subspace is called an *observable stratum*. From (6.3) we derive the first basic result, namely

$$X_{q_1} \perp_\Gamma X_{q_2}, \qquad q_1 \neq q_2, \tag{6.6}$$

i.e. the observable strata are pairwise uncorrelated in the sampling distribution.

The linear model (2.1) together with the definition of the group of allowable permutations imply that the representation ϱ naturally induces a representation of $\bar{G}$ on Ω

$$\nu : \bar{G} \rightarrow GL(\Omega)\,.$$

We thus obtain the canonical decomposition of the parameter space Ω which we write as

$$\Omega = \bigoplus_r \Omega_r\,, \tag{6.7}$$

and each $\bar{G}$-invariant subspace Ω_r is called a *parameter stratum*.

If we now let Λ be the prior covariance i.p on Ω, then, from the discussion of section 5, it follows that we can assume that Λ is $\bar{G}$-invariant. From (6.3) it now follows that

$$\Omega_{r_1} \perp_\Lambda \Omega_{r_2}, \qquad r_1 \neq r_2\,. \tag{6.8}$$

Because of normality we can finally write (6.6) and (6.8) as

$$\mathbf{X}_{q_1} \perp \mathbf{X}_{q_2} \mid \Omega, \qquad q_1 \neq q_2, \tag{6.9}$$

$$\Omega_{r_1} \perp \Omega_{r_2}, \qquad r_1 \neq r_2\,, \tag{6.10}$$

where $\perp$ now indicates the conditional independence, the notation introduced by Dawid (1979).

It is shown by Consonni and Dawid (1983) that each parameter stratum represents the mean space of one, and only one, observable stratum. We can therefore rewrite decomposition (6.5) as

$$\mathbf{X} = (\bigoplus_r \mathbf{X}_r) \oplus (\bigoplus_s X_s)\,, \tag{6.11}$$

where Ω_r is the mean space of $\mathbf{X}_r$ and each observable belonging to any $\mathbf{X}_s$ has mean zero. Henceforth the labelling of the observable strata will be derived from (6.11).

From (6.7), (6.9), (6.10) and (6.11) it follows that the original problem of making inference on the whole parameter space Ω, given the observable space $\mathbf{X}$, can be reduced to a collection of a finite number of *independent subproblems*, a typical subproblem involving only the observable stratum $\mathbf{X}_r$ and the corresponding parameter stratum Ω_r. The individual results can then be pieced together at the end of the analysis. Thus for any

Ψ, $\Phi \in \Omega$, writing the posterior expectation of Ψ as $E(\Psi \mid \mathbf{X})$ and the posterior covariance of Ψ and Φ as $\mathrm{Cov}(\Psi, \Phi \mid \mathbf{X})$ we have

$$E(\Psi \mid \mathbf{X}) = \Sigma\, E(\Psi_r \mid \mathbf{X}_r)$$

and

$$\mathrm{Cov}(\Psi, \Phi \mid \mathbf{X}) = \Sigma\, \mathrm{Cov}(\Psi_r, \Phi_r \mid \mathbf{X}_r).$$

where

$$\Psi = \Sigma\, \Psi_r$$

and

$$\Phi = \Sigma\, \Phi_r$$

are the decompositions corresponding to (6.7).

Example 1 (ctd). It can be shown that the observable space $\mathbf{X}$ decomposes into three observable strata, i.e.

$$\mathbf{X} = \mathbf{X}_0 \oplus \mathbf{X}_1 \oplus \mathbf{X}_2.$$

The corresponding projections of a single basic observable are given by

$$X_{ij} = X_{**} + (X_{i*} - X_{**}) + (X_{ij} - X_{i*}),$$

where "$*$" denotes averaging over the replaced suffix.

Under parametrization (4.4) the group $\overline{G}$ is the symmetric group S_B. Any symmetric group induces a decomposition of the corresponding representation space into two subspaces, which, using a statistical terminology, can be termed the grand mean stratum and the deviation from the grand mean stratum. Thus we obtain

$$\Omega = \Omega_0 \oplus \Omega_1,$$

or, in terms of the corresponding projections of a single treatment effect Θ_i

$$\Theta_i = \Theta_* + (\Theta_i - \Theta_*).$$

Example 2 (ctd). In this case the observable space decomposes into four observable strata, i.e.

$$\mathbf{X} = \mathbf{X}_0 \oplus \mathbf{X}_1 \oplus \mathbf{X}_2 \oplus \mathbf{X}_3.$$

In terms of the corresponding projections of a single basic observable, we have

$$X_{ij} = X_{**} + (X_{i*} - X_{**}) + (X_{*j} - X_{**}) + (X_{ij} - X_{i*} - X_{*j} + X_{**}).$$

Since under both parametrizations (4.5) and (4.6) $\overline{G}$ is a symmetric group, Ω decomposes into the usual grand mean and deviation from the grand mean strata.

Example 3 (ctd). Here there are five observable strata, namely

$$\mathbf{X} = \mathbf{X}_0 \oplus \mathbf{X}_1 \oplus \mathbf{X}_2 \oplus \mathbf{X}_3 \oplus \mathbf{X}_4$$

and the corresponding projections of X_{ijk} are given by

$$X_{ijk} = X_{***} + (X_{i**} - X_{***}) + (X_{*j*} - X_{***}) + (X_{ij*} - X_{i**} - X_{*j*} + X_{***}) + (X_{ijk} - X_{ij*}).$$

Under the parametrization discussed in Section 5 we have $\overline{G} = S_R \times S_C$ and so Ω decomposes into four parameter strata whose description parallels that of the observable space contained in Example 2 in this section.

Example 4 (ctd). Let $N = 2M$. Then one can show, see for example Speed (1981), that $\mathbf{X}$ decomposes into the direct sum of $M+1$ subspaces

$$\mathbf{X} = \mathbf{X}_0 \oplus \mathbf{X}_1 \oplus \ldots \oplus \mathbf{X}_{M-1} \oplus \mathbf{X}_M , \tag{6.12}$$

where $\mathbf{X}_0$ is the grand mean stratum and hence one-dimensional, $\mathbf{X}_1,\ldots,\mathbf{X}_{M-1}$ are two-dimensional spaces and $\mathbf{X}_M$ is again one-dimensional.

The corresponding projections of a basic observable are given by

$$X_u = X_* + \sum_{l=1}^{M-1} \Big(\frac{2}{N} \sum_{k=0}^{N-1} X_k \cos(\frac{2\pi}{N}(u-k)) + \frac{1}{N} \sum_{k=0}^{N-1} X_k \cos(\frac{2\pi}{N}(u-k)M).$$

Thus for example if $N = 4$ we have the following projections

$$\begin{aligned}
X_0 &= X_* + \tfrac{1}{2}(X_0-X_2) + \tfrac{1}{4}(X_0-X_1+X_2-X_3) , \\
X_1 &= X_* + \tfrac{1}{2}(X_1-X_3) + \tfrac{1}{4}(-X_0+X_1-X_2+X_3) , \\
X_2 &= X_* + \tfrac{1}{2}(-X_0+X_2) + \tfrac{1}{4}(X_0-X_1+X_2-X_3) , \\
X_3 &= X_* + \tfrac{1}{2}(-X_1+X_3) + \tfrac{1}{4}(-X_0+X_1-X_2+X_3) .
\end{aligned} \tag{6.13}$$

7. INVARIANT NORMAL BAYESIAN LINEAR MODELS: ANALYSIS OF A SINGLE STATISTICAL PROBLEM

From section 6 it follows that all we need to consider each time is a single parameter stratum together with its corresponding observable stratum.

Thus we let in the sequel Ξ be a parameter stratum and $\mathbf{Y}$ its corresponding observable stratum. One of the characteristics of each stratum is that it is invariant *w.r.t.* its appropriate group. If we take Ξ, for example, then we know that, for any $\Psi \in \Xi$, $\nu(\bar{g})\Psi \in \Xi$ for all $\bar{g} \in \bar{G}$. If Ψ does not admit a proper $\bar{G}$-invariant subspace, then we say that Ξ is *irreducible*. A similar definition applies to $\mathbf{Y}$ with $\bar{G}$ replaced by G^* and ν by ϱ.

Two useful results proved by Consonni and Dawid (1983) are the following. If Ξ is irreducible, then:

(i) the posterior expectation of any $\Psi \in \Xi$ is equal to a scalar multiple of an unbiased estimator of Ψ (if Ξ is the grand mean stratum then a constant term must be added), and (ii) the posterior covariance of any two parameters Ψ and $\Phi \in \Xi$ is a scalar multiple of any fixed symmetric positive definite invariant form on Ξ calculated for (Ψ,Φ). It is important to emphasize that in both cases the scalars involved are *constant* within each stratum.

These two results become particularly useful in applications if the condition of irreducibility is also shared by the observable stratum $\mathbf{Y}$. For in this case the unbiased estimator mentioned in (i) is unique and this allows us to determine directly the scalars referred to in (i) and (ii) by means of Ericson's Theorem (1969), as shown by Dawid (1977).

Thus let Ξ be an irreducible parameter stratum (other than the grand mean stratum) and $\mathbf{Y}$ its corresponding irreducible observable stratum. We can write for any $\Psi \in \Xi$

$$E(\Psi \mid \mathbf{X}) = aY,$$

where Y is the unbiased estimator of Ψ.

Letting

$$u_1^{-1} = E(\mathrm{Var}(Y \mid \Omega)) = \mathrm{Var}(Y \mid \Omega); \qquad u_2^{-1} = \mathrm{Var}\Psi ,$$

we derive from Ericson's Theorem

$$a = u_1/(u_1+u_2) .$$

If Ξ is the grand mean stratum then we have an additional constant b (say) so that

$$E(\Psi|\mathbf{X}) = aY + b,$$

and

$$b = u_2 E(\Psi)/(u_1+u_2)\,.$$

Consider now the posterior covariance. Again we assume that Ξ and $\mathbf{Y}$ are irreducible. Because of the general result previously mentioned we can write for any $\Psi, \Phi \in \Xi$

$$\mathrm{Cov}(\Psi,\Phi|\mathbf{X}) = c\,\mathrm{Cov}(\Psi,\Phi)\,, \qquad c>0\,,$$

where we have chosen the fixed symmetric positive invariant form on Ξ to be the prior covariance i.p.. Thus in order to determine c, we simply need to compute, for example, the variance of any Ψ. Let Y be the unbiased estimator of Ψ and suppose for simplicity that Ξ is not the grand mean stratum. Then $E(\Psi|\mathbf{X}) = aY$ and so we have

$$\begin{aligned}\mathrm{Var}\Psi &= E(\mathrm{Var}(\Psi|\mathbf{X})) + \mathrm{Var}(E(\Psi|\mathbf{X}))\\ &= \mathrm{Var}(\Psi|\mathbf{X}) + a^2\,\mathrm{Var}\,Y\\ &= c\,\mathrm{Var}\Psi + a^2\,\mathrm{Var}(Y|\Omega) + a^2\,\mathrm{Var}\Psi.\end{aligned}$$

Thus

$$u_2^{-1} = cu_2^{-1} + a^2\,(u_1^{-1} + u_2^{-1})\,,$$

whence, recalling that $a = u_1/(u_1+u_2)$,

$$c = u_2/(u_1+u_2)\,.$$

Notice that the same result continues to apply even if Ξ is the grand mean stratum.

We can thus conclude that, under irreducibility of both the parameter and the observable stratum, a typically multivariate problem reduces to a univariate one.

Example 1 (ctd; see also Dawid, 1977). First we observe that the three strata which appear in the decomposition of $\mathbf{X}$ are all irreducible. The same applies to the two strata which appear in the decomposition of Ω (this is always the case when the underlying group is the symmetric group).

Let now $Y_0 = X_{**}$; $Y_i = (X_{i*} - X_{**})$; $\Psi_0 = \Theta_*$; $\Psi_i = (\Theta_i - \Theta_*)$.

Then we have

$$E(\Psi_0|\mathbf{X}) = a_0 Y_0 + b;\; E(\Psi_i|\mathbf{X}) = a_1 Y_i.$$

Setting now

$$v_1^{-1} = \mathrm{Var}(Y_0|\Omega);\; v_2^{-1} = \mathrm{Var}\,\Psi_0;\; w_1^{-1} = \mathrm{Var}(Y_i|\Omega);\; w_2^{-1} = \mathrm{Var}\Psi_i;$$

we derive

$$a_0 = v_1/(v_1+v_2);\; b = v_2 E\,(\Psi_0)/(v_1+v_2);\; a_1 = w_1/(w_1+w_2).$$

Similarly for the posterior covariance structure, let

$$\mathrm{Var}(\Psi_0|\mathbf{X}) = c_0\,\mathrm{Var}(\Psi_0);\; \mathrm{Cov}(\Psi_i,\Psi_j|\mathbf{X}) = c_1\,\mathrm{Cov}(\Psi_i,\Psi_j);$$

then

$$c_0 = v_2/(v_1+v_2)\,;\; c_1 = w_2/(w_1+w_2).$$

Example 2 (ctd). Again all four strata which make up the observable space are irreducible and similarly for the parameter strata under both parametrization (4.5) and (4.6). For the sake of definiteness suppose that parametrization (4.6) is adopted. Then, letting $Y_0 = X_{**}$; $Y_j = (X_{*j} - X_{**})$; $\Psi_j = (\Theta_j - \Theta_*)$, we have

$$E(\Psi_0|\mathbf{X}) = a_0\, Y_0 + b;\, E(\Psi_j|\mathbf{X}) = a_1\, Y_j$$

$$\mathrm{Var}(\Psi_0|\mathbf{X}) = c_0\,\mathrm{var}\Psi_0;\;\; \mathrm{Cov}(\Psi_i,\Psi_{i'}|\mathbf{X}) = c_1\,\mathrm{Cov}(\Psi_i,\Psi_{i'}),$$

and the constants a_0, a_1, b, c_0 and c_1 are derived in the usual way.

Example 3 (ctd; see also Dawid, 1977). Again all five observable strata are irreducible. On the other hand the parametrization described in (5.3) induces a decomposition of Ω into four irreducible parameter strata

$$\Omega = \Omega_0 \oplus \Omega_1 \oplus \Omega_2 \oplus \Omega_3,$$

and the corresponding projections of a single treatment effect are given by

$$\Theta_{ij} = \Theta_{**} + (\Theta_{i*} - \Theta_{**}) + (\Theta_{*j} - \Theta_{**}) + (\Theta_{ij} - \Theta_{i*} - \Theta_{*j} + \Theta_{**}).$$

Thus letting $Y_0 = X_{***}$; $Y_i = (X_{i**} - X_{***})$; $\mathbf{Y}_j' = (X_{*j*} - X_{***})$;

$$Y_{ij} = (X_{ij*} - X_{i**} \text{-} X_{*j*} + Y_{***});\; \Psi_0 = \Theta_{**};\; \Psi_i = (\Theta_{i*} - \Theta_{**});\; \Psi_j' = (\Theta_{*j} - \Theta_{**});$$

$$\Psi_{ij} = (\Theta_{ij} - \Theta_{i*} - \Theta_{*j} + \Theta_{**}), \text{ we have}$$

$$E(\Psi_0|\mathbf{X}) = a_0 Y_0 + b;\, E(\Psi_i|\mathbf{X}) = a_1 Y_i;\, E(\Psi_j'|\mathbf{X}) = a_2 Y_j';\, E(\Psi_{ij}|\mathbf{X}) = a_3 Y_{ij}.$$

We omit the expressions for the posterior covariance structure which can be deduced in the obvious way. The determination of the constants is then straightforward.

Example 4 (ctd). Since all $M+1$ strata which appear in (6.12) are irreducible, we can apply the usual procedure. Suppose for definiteness that $N = 4$, so that we have the projections given in (6.13). If we adopt parametrization (4.7) Ω is two-dimensional. Letting $Y_0 = X_*$; $Y_1 = (X_0 - X_1 + X_2 - X_3)/2$; $\Psi_0 = \Theta_*$ and $\Psi_1 = \Theta_1 - \Theta_2$, we have

$$E(\Psi_0|\mathbf{X}) = a_0 Y_0 + b; \qquad E(\Psi_1|\mathbf{X}) = a_1 Y_1;$$

$$\mathrm{Var}(\Psi_0|\mathbf{X}) = c_0\,\mathrm{Var}\Psi_0; \qquad \mathrm{Var}(\Psi_1|\mathbf{X}) = c_1\,\mathrm{Var}\Psi_1,$$

and we can solve for a_0, a_1, b, c_0 and c_1 in the usual way.

8. DISCUSSION

The theory of invariant normal Bayesian linear models provides a general framework within which many experimental designs can be succesfully analyzed in a simple way. The assumption of normality is clearly inessential if we restrict our attention to linear Bayesian estimates, see Hartigan (1969), Mouchart and Simar (1980), and the results will be the same. One of the practical difficulties in implementing the theory concerns the possibility of identifying the strata which appear in the decomposition of the observable and parameter space. Fortunately we already know how to do this for simple block structures, Nelder (1965), and for some other types of structures which often occur in practice, Speed (1981). James (1982) has shown how character theory can be usefully exploited for more complicated designs, although his method is not straightforward.

We have omitted discussion of the treatment structure and have contented ourselves with the decomposition of the parameter space induced by the group G. The relationship between the block and the treatment structure, however, is an intriguing one and has led to the concept of general balance in experimental designs (Nelder, 1965; Houtman, 1981; Speed, 1983).

Perhaps the main constraints in our analysis have been the assumption of known covariance structure and of irreducibility of the parameter and observable strata. Concerning the former point, we remark that if G^* is a proper subgroup of G (as in Example 1* of Section 4), the additional information relating to the G-invariance of the sampling covariance i.p. on $\mathbf{X}$ (which was *not* used in section 6) can be usefully exploited to make inference on "variance components", as we hope to show in another paper. oncerning irreducibility, we must distinguish two cases: (i) the parameter stratum is irreducible and the corresponding observable stratum is reducible; (ii) both the parameter and the observable strata are reducible. In either case inference becomes inevitably more complicated, although, at least under condition (i), some of the simplifying features which characterize the Bayesian analysis of this paper might still be retained.

ACKNOWLEDGEMENTS

Consiglio Nazionale delle Ricerche, Italy, provided financial support for the first author, whose research was undertaken within the framework of CNR-GNAFA.

REFERENCES

ANDERSSON, S. (1975). Invariant normal models. *Ann. Statist.* **3**, 132-154.

BAILEY, R.A. (1981). A unified approach to design of experiments. *J.R. Statist. Soc. A* **144**, 214-223.

CONSONNI, G. (1983). Decomposition of statistical problems by means of invariance: mathematical tools and results. *Research Report N° 22. Department of Statistical Science, University College London.*

CONSONNI, G. and DAWID, A.P. (1983). Decomposition and Bayesian of invariant normal analysis linear models. Unpublished manuscript.

DAWID, A.P. (1977). Invariant distributions and analysis of variance models. *Biometrika* **64**, 291-297.

— (1979). Conditional independence in statistical theory. *J.R. Statist. Soc. B* **41**, 1-31. (with discussion).

ERICSON, W.A. (1969). A note on the posterior mean of a population mean. *J.R. Statist. Soc. B* **31**, 332-334.

HARTIGAN, J.A. (1969). Linear Bayesian methods. *J.R. Statist. Soc. B* **31**, 446-454.

HOUTMAN, A.M. (1981). *The Analysis of Designed Experiments.* Ph.D. Thesis. Princeton University.

JAMES, A.T. (1982). Analyses of variance determined by symmetry and combinatorial properties of zonal polynomials. In *Statistics and Probability: Essays in Honor of C.R. Rao.* (Kallianpur, G., Krishnaiah, P.R. and Ghosh, J.K. eds.), pp. 329-341. Amsterdam: North-Holland.

LEDERMANN, W. (1977). *Introduction to Group Characters.* Cambridge University Press, Cambridge.

MOUCHART, M. and SIMAR, L. (1980). Least squares approximation in Bayesian analysis. In *Bayesian Statistics.* (Bernardo, J.M.; DeGroot, M.H.; Lindley, D.V. and Smith, A.F.M. eds.), pp. 207-222. Valencia: University Press.

NELDER, J.A. (1965). The analysis of randomized experiments with orthogonal block structure. Part I: Block structure and the null analysis of variance. *Proc. Roy. Soc. London A* **283**, 147-162. Part II: Treatment structure and the general analysis of variance. *Proc. Roy. Soc. London A,* **283**, 163-178.

OLKIN, I. and PRESS, S.J. (1969). Testing and estimation for a circular stationary model. *Ann. Math. Statist.* **40,** 1358-1373.

SPEED, T.P. (1981). The analysis of variance. Unpublished manuscript.

— (1983). General balance. In *Encyclopedia of Statistical Sciences*, Volume 3 (Kotz, S., Johnson, N.L. and Read, C.B., eds.), 320-326. New York: Wiley-Interscience.

BAYESIAN STATISTICS 2, pp. 645-654
J.M. Bernardo, M.H. DeGroot, D.V. Lindley, A.F.M. Smith (Eds.)

Bayesian Inference on Mahalanobis Distance: An Alternative Approach to Bayesian Model Testing

J.R. FERRANDIZ
Universidad de Valencia, Spain

SUMMARY

In Bayesian hypothesis testing, the extreme sensitivity of the posterior probability associated with a sharp null hypothesis has led in practice to the use of posterior odds ratios and highest posterior density regions.

This paper proposes a new Bayesian approach which consists in making inferences on the distance, in the parametric space, between the unknown parameter vector and the subspace defined by the sharp null hypothesis.

The method is illustrated using the Mahalanobis distance to test $\mu=0$ in a multivariate normal model $N(\mu,\sigma^2 I)$

Keywords: BAYESIAN INFERENCE; HYPOTHESES TESTING; MAHALANOBIS DISTANCE; NON-INFORMATIVE PRIORS

1. INTRODUCTION

In Bayesian hypotheses testing, the extreme sensitivity of the posterior probability associated with a sharp null hypothesis has led, in practice, to the use of posterior odds ratios and highest posterior density regions.

This paper proposes a new Bayesian approach which consists in making inferences on the distance, in the parameter space, between the unknown parameter vector and the subspace defined by the sharp null hypothesis.

To be more concise, let $X \epsilon S$ be an observable vector whose random behaviour is governed by the parameter vector $\theta \epsilon \Theta$, and let $p(x|\theta)$ be the corresponding family of probability density functions with respect the appropriate dominant measure in S. Let $\Theta_0 \subset \Theta$ be the subspace related to the null hypothesis H_0, while $\Theta_1 = \Theta - \Theta_0$ corresponds to the alternative H_1.

If Θ_1 has positive Lebesgue measure in the Euclidean space $\mathbb{R}^k$, while Θ_0 is a null set, and we want Θ_0 to have positive prior probability, we must use some kind of mixed probability measure that is not singular nor absolutely continuous with respect to Lebesgue measure in $\mathbb{R}^k$.

Typically, a hierarchical model is specified by means of the probability densities $p(x|H_i,\theta) = p(x|\theta)$, $p(\theta|H_i)$, $p(H_i)$ $i=0,1$ and it is well-known (Bartlett (1975) and Bernardo (1979b)) that, even in the simplest cases, the posterior probabilities $p(H_i|x)$ are very sensitive to the assignment of $p(\theta|H_i)$.

From a decision-theoretic viewpoint, the problem can be stated in the following terms: there are only two decisions, d_0 (accept H_0) and d_1 (reject H_0), and the parameter values $\theta \epsilon \Theta$ represent the states of nature. The utility functions $u(d_i,\theta)$ indicate what information on θ is required to solve the problem. If the utility functions $u(d_i,\theta)$ depend on θ only through some specified function $\psi(\theta)$, then, we will call it the *relevant parametric function* of the decision problem.

For instance, if we took $u(d_i,\theta)=c_i\chi_{\Theta_i}$ $c_i>0$ $i=0$, 1, where χ_{Θ_i} is the characteristic function of the set Θ_i, we would only need the posterior density $p(\chi_{\Theta_0}|x)$ in order to solve the decision problem, and we would be led to the inferential problem of hypotheses testing as it was introduced above.

Two ideas are worth noticing now:

(i) Under the utility structure defined above, we do not need the whole $p(\theta|x)$ to solve the problem, but only $p(\chi_{\Theta_0}|x)$. That is, irrespective od the dimension of θ, the decision-maker is interested, through the utility function, in a one-dimensional function of the parameter θ, that we have called relevant parametric function. The posterior distribution of this function contains all the information on θ that is relevant to the decision-maker.

(ii) The relevant parametric function χ_{Θ_0}, here produces excessive sensitivity of the inferential process because of its sharpness. If we smooth it, making it continuous for instance, we will recover the usual robustness of bayesian inference.

These arguments lead us to consider that a judicious choice of other relevant parametric functions can benefit the statistical analysis in at leas two aspects.

First, the new relevant parametric function can reflect the true interests of the decision-maker in a less crude way than χ_{Θ_0} does, allowing for a more realistic specification of the *deviation* from the null hypothesis which the decision-maker wants to detect. Moreover, because of this realism and the reduction of dimensionality, the posterior distribution of this relevant parametric function becomes a good tool in describing the information on θ that is relevant to the decision maker.

Second, a convenient choice of the relevant parametric function will allow us to avoid the sensitivity problem raised by the use of χ_{Θ_0}, yet respect the practical aspects mentioned before.

In the rest of the paper, these very general ideas will be applied to the case of testing $\mu=0$ against $\mu\neq 0$ in a multivariate normal model $N(\mu,\sigma^2 I)$.

First of all, we must choose an appropriate relevant parametric function. If the decision-maker does not mind the direction of departure of μ, from the origin, but only cares for its magnitude, any euclidean distance between μ and the origin would seem a good candidate.

The Mahalanobis distance $(\mu'\mu/\sigma^2)^{1/2}$ has an special appeal because of its invariance under linear transformations of the observable vector $X \sim N(\mu,\sigma^2 I)$, and, in particular, it will let us to disregard the possibly different scales used to measure its components. Moreover, each component of μ contributes according to the precision of the corresponding component of X, that is, the more the gain in information from X which is expected for a component of μ, the more attention we pay to this component, what seems a sensible behaviour.

Consequently, we shall assume in what follows that the utility functions $u(d_i,\theta)$ depend on μ, σ^2 only through $\psi=(\mu'\mu/\sigma^2)^{1/2}$ and the next step will be to make inferences on ψ.

2. INFERENCE ON THE MAHALANOBIS DISTANCE ψ

Let $\{x_i\}_{i=1}^n$ be a random sample from the model $N(\mu,\sigma^2 I)$, of dimension k, and let $\bar{x}=n^{-1}\Sigma_{i=1}^n x_i$, $S=\Sigma_{i=1}^n x_i'x_i$ be, respectively, the sample vector mean and total sum of squares. To get simpler formulas in what follows, we shall work with the precision $h=\sigma^{-2}$ instead of the variance σ^2.

After the prior $p(\mu,h)$ has been specified, the posterior probability density of μ, h will be:

$$p(\mu,h\,|\,x_1,x_2,\ldots,x_n) \propto p(\mu,h)p(\bar{x},S\,|\,\mu,h) \tag{1}$$

$\bar{x}$, S being sufficient statistics.

To obtain the posterior density of ψ we first transform μ to generalized polar coordinates $\psi>0$, $\omega\epsilon\Omega=(0,\pi)^{k-2}\times(0,2\pi)$ leaving h unchanged. Then:

$$\mu = \psi h^{-1/2}\,u(\omega) \tag{2}$$

$$u_i(\omega) = \cos\omega_i \,\Pi_{j=1}^{i-1}\sin\omega_j \qquad i=1,2,\ldots,k-1 \tag{3}$$

$$u_k(\omega) = \Pi_{i=1}^{k-1}\sin\omega_i \tag{4}$$

$$h = h \tag{5}$$

The Jacobian of this transformation is

$$D = |\partial(\mu,h)/\partial(\psi,\omega,h)| = \psi^{k-1}h^{-k/2}g(\omega) \tag{6}$$

$$g(\omega) = \Pi_{i=2}^{k-1}\sin^{k-i}\omega_{i-1} \tag{7}$$

and, in order to compute the posterior density $p(\psi\,|\,\bar{x},S)$, we must integrate out ω, h from the posterior $p(\psi,\omega,h\,|\,\bar{x},S)=p(\mu,h\,|\,\bar{x},D)\,D$

We undertake now the task of choosing a suitable prior for te problem.

2.1. *Non-informative prior*

As Stein's (1959) example shows for the case σ^2 known, some care must be taken in specifying our prior beliefs on μ, if our final goal is ψ. Indeed, the usual diffuse priors for μ lead to J-shaped priors for ψ, which dominate its marginal likelihood for moderate sample sizes (Cox & Hinkley (1974) pp. 383-86)). Similar results may be expected when σ^2 is unknown.

Thus, we shall resort to Bernardo's (1979a) method of deriving reference prior distributions. In the present case, where the regularity conditions for the asymptotic normality of posterior distributions apply, all we need is to compute the determinants $|J(\psi,\omega,h)_{\omega h,\omega h}|$ and $|(J(\psi,\omega,h)^{-1})_{\psi,\psi}|$, where $J(\psi,\omega,h)$ is the Fisher Information Matrix for the re-parametrized model, and the suffixes indicate the rows and columns defining the submatrix considered.

In our case (see Proposition 1 in the appendix) these turn out to be:

$$|(J(\psi,\omega,h)^{-1})_{\psi,\psi}| = (1+\psi^2/2k) \tag{8}$$

$$|J(\psi,\omega,h)_{\omega h,\omega h}| = (1+\psi^2/2k)\psi^{2k-2}\,g(\omega)^2(k/2)h^{-2} \tag{9}$$

and thus, using Bernardo's method, we shall take:

$$\pi(\psi) \propto (1+\psi^2/2k)^{-1/2} \tag{10}$$

$$\pi(\omega,h) \propto h^{-1}\,g(\omega) \tag{11}$$

Professor Lindley pointed out to me that equation (10) is related to equation (16) in section 5.2 of Jeffreys (1967). It would be interesting to compare our argument with his; this will be done in a near future.

2.2. *Reference posterior distribution*

From (10), (11) and the likelihood function $p(\bar{x},S|\psi,\omega,h)$, the joint posterior distribution of ψ, ω, h is

$$\pi(\psi,\omega,h|\bar{x},S) \propto \pi(\psi)\,\mathrm{Ga}(h|nk/2,S/2)\exp(-n\psi^2/2+nh^{1/2}\psi\bar{x}'u(\omega))\,g(\omega) \tag{12}$$

where $\pi(\psi)$, $u(\omega)$ and $g(\omega)$ are as defined in (10), (3) and (7) respectively, and $\mathrm{Ga}(h|a,b)$ stands for a gamma density of parameters a, b.

Now, using Prop. 2 of the appendix to integrate out ω, we get

$$\pi(\psi,h|\bar{x},X) \propto \pi(\psi)\exp(-n\psi^2/2)\,\mathrm{Ga}(h|nk/2,S/2)\sum_{i=0}^{\infty}\frac{((n\psi/2)^2\bar{x}'\bar{x}h)^i}{i!\Gamma(k/2+i)} \tag{13}$$

and taking expectations with respect to h

$$\pi(\psi|\bar{x},S) \propto \pi(\psi)\exp(-n\psi^2/2)\sum_{i=0}^{\infty}\frac{\Gamma(nk/2+i)(n\psi^2/2)^i(n\bar{x}'\bar{x}/S)^i}{i!\Gamma(k/2+i)} \tag{14}$$

It may be checked that $f=n\bar{x}'\bar{x}/S = F/(1+F)$ where F is the usual classical ratio of two independent Chi-squares to test $\mu=0$ against $\mu\neq 0$ for the problem at hand. Thus, the sampling distribution of f is a non-central Beta with parameters $k/2$, $(n-1)k/2$ and $n\psi^2/2$ as non-centrality parameter, and the right hand side of (14) factorizes into the prior $\pi(\psi)$ times the likelihood directly provided by the statistic f.

This is a remarkable result because it reveals that the prior $\pi(\omega,h)$ of (11) has the merit of avoiding the marginalization paradox of Dawid, Stone & Zidek (1973) because the sampling distribution of f depends on μ, h only through ψ. The derivation of (14) through (12), (13), makes clear that (11) is precisely responsible for this fact. Thus, we are able to take $\pi(\psi)$ of (10) or change it to incorporate prior subjective beliefs without loosing this interesting feature.

2.3 *Behaviour of the reference posterior distribution*

We could approximate (10) by a uniform prior for ψ because the resulting posterior distributions will scarcely differ given the flat shape of (10) and, in so doing, we shall get simpler mathematical expressions. Moreover, it will make easier to understand how a new prior will influence the likelihood function.

If we take $\pi(\psi)\propto 1$, the posterior distribution (14) may be represented as a mixture of gamma densities:

$$\pi(\psi|f) = \sum_{i=0}^{\infty} p(i|f)\,\mathrm{Ga}^{1/2}(\psi|i+\tfrac{1}{2},n/2) \tag{15}$$

$$p(i|f) = C^{-1}\frac{\Gamma(nk/2+i)\,\Gamma(\tfrac{1}{2}+i)}{\Gamma(k/2+i)\,\Gamma(1+i)}f^i \tag{16}$$

$$C = {}_2F_1(\tfrac{1}{2},nk/2;k/2;f)\,\Gamma(nk/2)\pi^{1/2}/\Gamma(k/2) \tag{17}$$

where $\mathrm{Ga}^{1/2}(x|a,b)$ stands for the probability density function of the positive square-root of a Gamma random variable with parameters a, b, and C is expressed in the usual notation for hypergeometric functions.

The main characteristics of the posterior distribution are easily derived from (15) to (17). Here we shall only comment some features that will help in understanding how it works in practice.

(i) The posterior mode will be $\psi = 0$ until f exceeds the threshold $f = 1/n$, which is its expected value under $\psi = 0$. In that case, an approximate value of the mode will be $(k(f - 1/n)/(1-f))^{1/2}$. This approximation improves as f approaches 1.

(ii) The posterior distribution function strictly decreases as f increases, that is, $f_1 < f_2$ implies $F(\psi | f_1) > F(\psi | f_2)$ for all $\psi > 0$. This is used in the derivation of the relationship between utility functions and classical significance levels that will be described in section 3 (s. proposition 3 in the appendix).

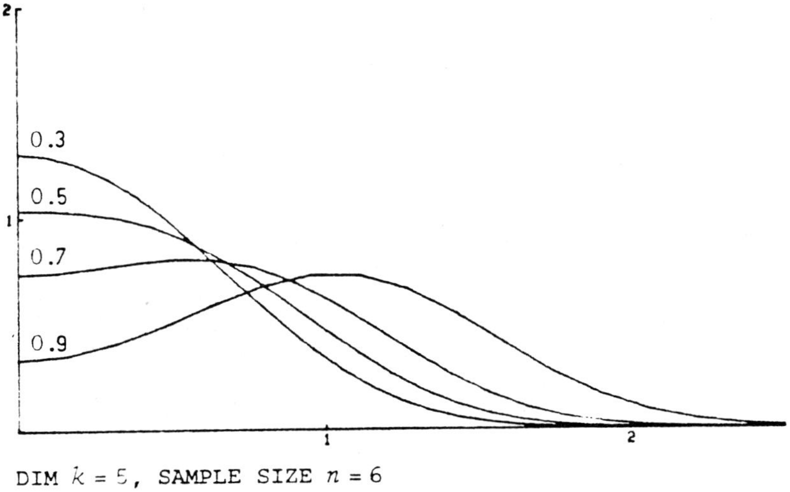

FIGURE 1. *Posterior densities* $\pi(\psi | f)$

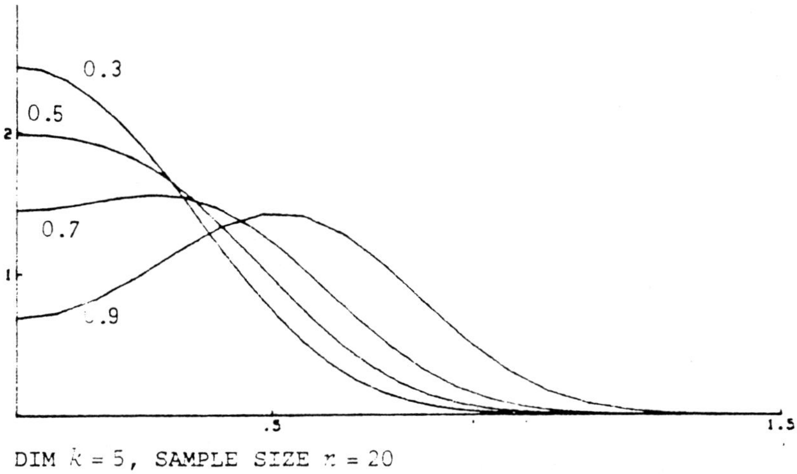

FIGURE 2. *Posterior densities* $\pi(\psi | f)$

Figures 1 and 2 show the posterior densities of ψ for some values of the statistic f. Each density is labelled with the value $F(f|\psi=0)$ of the corresponding f, where $F(f|\psi=0)$ stands for the sampling distribution function of f under the null hypothesis $\psi=0$. These figures demonstrate how the posterior distribution of ψ moves away from the origin as f increases, making less credible the null hypothesis.

3. UTILITY FUNCTIONS AND SIGNIFICANCE LEVELS

To solve the decision problem stated in the introduction of this paper, once the posterior distribution of ψ has been found, it only remains to compute the expected value of the utility functions $u(d_i,\psi)$. The decision-maker must accept the null hypothesis if $E_{\psi|f}(u(d_0,\psi))$ is greater than $E_{\psi|f}(u(d_1,\psi))$. If we define $l(\psi) = u(d_1,\psi)-u(d_0,\psi)$, the preceding procedure is equivalent to accepting the null hypothesis whenever $E_{\psi|f}(l(\psi))$ is negative.

Logically, $u(d_0,\psi)$ must be a decreasing function of ψ, while $u(d_1,\psi)$ is, at least, not decreasing. Thus, it is not unreasonable to think of $l(\psi)$ as a strictly increasing function of ψ. If this condition is met, $E_{\psi|f}\, l(\psi))$ will increase strictly as f increases because the posterior distribution function $F(\psi|f)$ is strictly decreasing, as pointed out in the preceding section. That means that there is a unique threshold value f_i such that $E_{\psi|f}(l(\psi)) \gtreqless 0$ whenever $f \gtreqless f_i$.

Thus, from the point of view of the decision reached, $l(\psi)$ acts as the classical significance level $\alpha = 1-F(f_i|\psi=0)$ in testing $\mu=0$ against $\mu\neq 0$. Moreover, we can construct a monotone family of such $l(\psi)$ functions that will cover all possible $\alpha\epsilon(0,1)$. This is illustrated in Figure 3 using the family $l_\beta(\psi) = -1+\beta\psi \quad \beta\geq 0$. For each significance level α we find its equivalent $\beta=E(\psi|f_\alpha)^{-1}$, where f_α stands for the $(1-\alpha)$-quantile of the sampling distribution of f under $\psi=0$. The value $\beta=0$ implies always accepting H_0, that is, taking $\alpha=0$. In the other side, for β greater than some threshold (that depends on the model dimension and sample size) we always reject H_0, i.e., we reach $\alpha=1$.

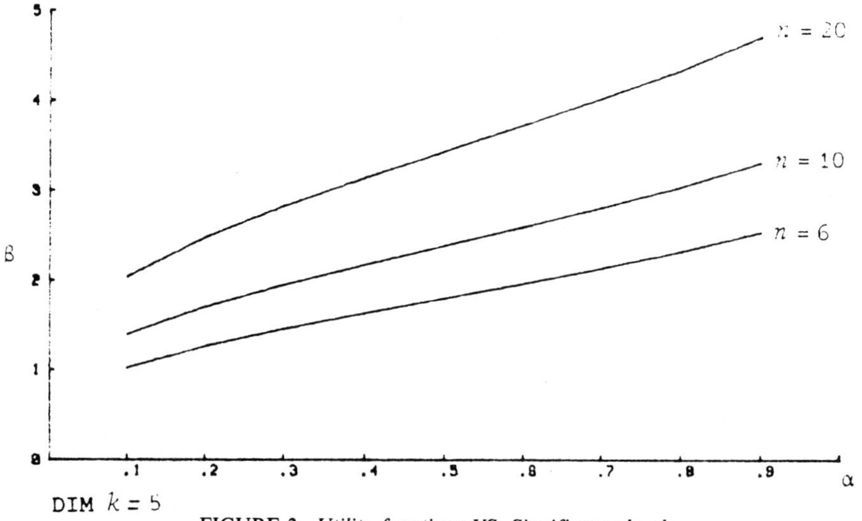

FIGURE 3. *Utility functions VS. Significance levels*

Figure 3 also demonstrates how, as the sample size increases, a greater β is related to each α. Thus, if we keep the α-level fixed while n increases, we are changing our utility functions towards an $l(\psi)$ with greater tendence to reject the null hypothesis. This is the same effect that Lindley pointed out in his (1957) paper on a statistical paradox. Conversely, to keep $l(\psi)$ fixed provides a rule for coherently changing the significance level α as n increases.

4. DISCUSSION

As with many other statistical problems, the decision-theoretical approach has been illuminating. Decision theory provides a general framework which permits a detailed analysis of all the relevant components in the problem. However, in each particular case, there is a need for constructing a bridge between the abstract-theoretical level of this theory and the practical aspects of its application. We suggest that the concept of relevant parametric function provides this bridge in Bayesian hypotheses testing. Indeed, this makes easier the practical assessment of utility functions because it concentrates on describing the type of "deviation from", or "distance to", the null hypothesis to which the interest of the decision maker are related, and it also provides a more informative final solution. Indeed, the relevant parametric function clearly indicates the aspects of the problem on which the inferential process should focus, and its posterior distribution becomes the appropriate tool to describe the useful available knowledge.

The Mahalanobis distance is just an example which has been developed in this paper, but it has raised some interesting questions. Indeed, with the non-informative prior derived in section 2 using Bernardo's method, the classical statistic becomes sufficient and this has led to a paralelism between bayesian and classical results; once again classical results appear as particular cases of bayesian results. Moreover, the bayesian methodology has provided some kind of justification for the use of classical significance levels, but demonstrates the inherent inconsistency of selecting a fixed significance level irrespective of the sample size; conversely, it has provided a practical rule on how to change coherently the significance level as the sample size increases.

The preceding discussion also applies to the usual tests of hypotheses in linear models (Ferrándiz (1981)) and its extension to the normal multivariate model with general dispersion matrix is straightforward.

ACKNOWLEDGEMENTS

I am grateful to the referee and to Professor Lindley for helpful comments to an earlier draft of this paper.

APPENDIX: *Some mathematical results needed in the preceding sections*

Proposition 1: Let $X \sim N(\mu, h^{-1}I)$ of dimension k and let ψ, ω, h, $g(\omega)$ be as defined in (2) to (7). Then

$$|J(\psi,\omega,h)_{\omega h, \omega h}| \propto (1 + \psi^2/2k)\, \psi^{2k-2}\, g(\omega)^2 h^{-2} \tag{19}$$

$$|(J(\psi,\omega,h)^{-1})_{\psi,\psi}| \propto (1 + \psi^2/2k) \tag{20}$$

Proof:

It may be checked that Fisher Information Matrix for the model $N(\mu, h^{-1}I)$ may be expressed as:

$$J(\mu,h) = \left(\begin{array}{c|c} hI_k & 0 \\ \hline 0 & kh^{-2}/2 \end{array}\right) \tag{21}$$

Thus, Fisher Information Matrix for the re-parametrized model will be:

$$J(\psi,\omega,h) = (\partial(\mu,h)/\partial(\psi,\omega,h))' J(\mu,h)\, (\partial(\mu,h)/\partial(\psi,\omega,h)) \tag{22}$$

where $\partial(\mu,h)/\partial(\psi,\omega,h)$ stands for the Jacobian matrix of the transformation of (2) to (5). Then

$$|J(\psi,\omega,h)| = |\partial(\mu,h)/\partial(\psi,\omega,h)|^2\, |J(\mu,h)| \propto h^{-2}(\psi^{k-1}\, g(\omega))^2 \tag{23}$$

from (6), and it may be checked that:

$$\begin{aligned} |(J(\psi,\omega,h)^{-1})_{\psi,\psi}| &= h^{-1}(\partial\psi/\partial\mu)'(\partial\psi/\partial\mu) + 2h^2(\partial\psi/\partial h)^2/k \\ &= 1 + \psi^2/2k \end{aligned} \tag{24}$$

Now, using a well-known expression for determinants of partitioned positive definite matrices:

$$|J(\psi,\omega,h)_{\omega h,\omega h}| = |J(\psi,\omega,h)|\, |(J(\psi,\omega,h)^{-1})_{\omega,\omega}| \tag{25}$$

and (19), (20) are obtained.

Proposition 2: Let $u(\omega)$, $g(\omega)$ be as defined in (3), (7) and $l \epsilon \mathbb{R}^k$. Then

$$\int_\Omega \exp(l'u(\omega))\, g(\omega)\, d\omega = 2\pi^{k/2} \sum_{i=0}^{\infty} (l'l/4)^i/(i!\Gamma(k/2+i)) \tag{26}$$

Proof:
If we transform $X \sim N_k(m,I)$ according to (2) to (7) and define $r=(X'X)^{1/2}$, the marginal probability density function of r is

$$(2\pi)^{-k/2} r^{k-1} \exp(-(m'm+r^2)/2) \int_\Omega \exp(rm'u(\omega))\, g(\omega)\, d\omega \tag{27}$$

and must coincide with the probability density function of the positive square root of a non-central Chi-square with k degrees of freedom and $m'm$ as non-centrality parameter. Comparing both densities and letting $l=rm$, the required result is obtained.

Proposition 3: The distribution function which corresponds to (15), (17) is a strictly decreasing function of f.

Proof:
From (15) to (17), the posterior cumulative distribution function of ψ may be expressed as

$$F(\psi|f) = \sum_{i=0}^{\infty} p(i|f)\, F(\psi|i) \tag{28}$$

where $F(\psi|i)$ stands for the distribution function of the positive square-root of a Gamma random variable with parameters $i+\frac{1}{2}$, $n/2$.

The derivative of (28) with respect to f turns out to be:

$$\partial F(\psi|f)/\partial f = f^{-1} \sum_{i=0}^{\infty} i\, p(f|i) F(\psi|i) - F(\psi|f)(\partial \log C/\partial f) \tag{29}$$

on noting that

$$\partial \log C/\partial f = f^{-1} E(i|f) \tag{30}$$

where $E(i|f)$ stands for the expected value of i with respect to $p(i|f)$

From (29) and (30) we can write (29) as

$$\partial F(\psi|f)\partial f = f^{-1} E((i-E(i|f))F(\psi|i)|f) \tag{31}$$

and the required result is a consequence of $F(\psi \mid i)$ being a strictly decreasing positive function of i.

REFERENCES

BARTLETT, M.S. (1957). A comment on D.V. Lindley's statistical paradox. *Biometrika* **44**, 533-34.

BERNARDO, J.M. (1979a). Reference posterior distributions for Bayesian Inference. *J. R. Statist. Soc. B* **41**, 113-147 (with discussion).

— (1979b). A Bayesian analysis of classical hypotheses testing. *Bayesian Statistics* (Bernardo, J.M. *et al.* eds.). Valencia: University Press.

COX, D.R. and HINKLEY, D.V. (1974). *Theoretical Statistics.* London: Chapman and Hall.

DAWID, A.P., STONE, M., and ZIDEK, I.V. (1973). Marginalization paradoxed in Bayesian and Structural Inference. *J. R. Statist. Soc. B* **35**, 189-233.

FERRANDIZ, J.R. (1981). *A Bayesian Alternative to Hypotheses Testing.* Ph.D. Thesis University of Valencia (in spanish).

JEFFREYS, H. (1939/67). *Theory of Probability.* Oxford: University Press.

LINDLEY, D.V. (1957). A statistical paradox. *Biometrika* **44**, 187-92.

STEIN, C. (1959). An example of the wide discrepancy between fiducial and confident interval. *Ann. Math. Statist.* **30**, 877-80.

BAYESIAN STATISTICS 2, pp. 655-662
J.M. Bernardo, M.H. DeGroot, D.V. Lindley, A.F.M. Smith (Eds.)

Optimal Block Designs under a Hierarchical Linear Model

A. GIOVAGNOLI
Universitá di Perugia, Italy

I. VERDINELLI
Universitá di Roma, Italy

SUMMARY

A three-stage linear model is assumed for an experiment with v treatments and a control, and b blocks. Optimal designs (in particular D-, A- and E-optimal designs) are found under suitable assumptions for the prior distributions of the parameters and hyperparameters.

Keywords: BAYES DESIGNS; BLOCK DESIGNS; HIERARCHICAL MODELS; OPTIMAL DESIGN.

1. INTRODUCTION

In a recent paper (Giovagnoli and Verdinelli, 1983) the problem of designing an experiment for comparing treatments with a control in the presence of blocks of fixed sizes was considered from a Bayesian standpoint. A linear model with no interactions between treatments and blocks was assumed, together with a multivariate normal prior for the treatment and the block parameters. Optimality was defined with respect to a given functional of the inverse posterior covariance matrix for the treatment effects.

Here we continue the investigation when the model has the three-stage hierarchical structure discussed by Lindley and Smith (1972). When the treatments are exchangeable and weak information is assumed at the third stage both for the treatments and for the blocks, the results obtained for the two-stage model can be utilized to find the explicit expression of the optimal designs for a wide class of optimality criteria, including the Schur-criteria as well as the convex decreasing criteria introduced by Kiefer (1975).

Whatever the criterion, the optimal design for the model under consideration is shown —under a further prior assumption on the blocks— to allocate a number of observations proportional to the block size to all the treatments other than the control. We determine the optimal proportions in the case of the ϕ_p—criteria, in particular for A-, D- and E-optimality.

2. THE MODEL

Let $\tau_o,\tau_1,\ldots,\tau_v$ denote the effects of $v+1$ treatments, the first treatment acting as a control. Suppose that the experimental units are grouped into b homogeneous blocks of fixed sizes $k_1,\ldots,k_b$, and let $\beta = (\beta_1,\ldots,\beta_b)^T$ denote the vector of the block effects. We are interested in inference on the treatment differences $\tau_i-\tau_o$, so we take τ_o as "origin of reference" ($\tau_o=0$) and use the symbols τ_i for the differences $\tau_i-\tau_o$ $(i=1,\ldots,v)$. The problem of finding an optimal plan for such an experiment was considered by Owen (1970) and Giovagnoli and Verdinelli (1983) for the case of a two-stage linear model, with multivariate normal prior distributions for β and for the vector $\tau = (\tau_1,\ldots,\tau_v)^T$.

Here we carry on the investigation for a three-stage model, being particularly interested in the case of vague knowledge at the third stage, which is the case considered by Smith and Verdinelli (1980) for a one-way ANOVA model. Our model is

$$y|\tau,\beta \sim N\left[(F:G)\begin{pmatrix}\tau\\ \beta\end{pmatrix}, C_1\right]$$

$$\begin{pmatrix}\tau\\ \beta\end{pmatrix}|\mu_\tau,\mu_\beta \sim N\left[\begin{pmatrix}A_{21} & 0\\ 0 & A_{22}\end{pmatrix}\cdot\begin{pmatrix}\mu_\tau\\ \mu_\beta\end{pmatrix}, \begin{pmatrix}T & 0\\ 0 & B\end{pmatrix}\right] \tag{1}$$

$$\begin{pmatrix}\mu_\tau\\ \mu_\beta\end{pmatrix} \sim N\left[\begin{pmatrix}A_{31} & 0\\ 0 & A_{32}\end{pmatrix}\begin{pmatrix}\theta_1\\ \theta_2\end{pmatrix}, \begin{pmatrix}C_{31} & 0\\ 0 & C_{32}\end{pmatrix}\right]$$

where $\mathbf{y}$ is an $n\times 1$ vector of observations, M_τ and μ_β are vectors and F,G are $(0,1)$ – matrices such that the $v\times b$ matrix:

$$N = \begin{pmatrix}n_{o1} & \dots & n_{ob}\\ & F^TG & \end{pmatrix} = \begin{pmatrix}\mathbf{n}_o^T\\ \bar{N}\end{pmatrix}$$

is the incidence matrix of the design. We assume the matrices C_1, T, B, C_{31}, C_{32} known. Let $C_1 = \sigma^2 I_n$, σ^2 known. It can be shown that finding an optimal design under this assumption also solves the problem for a more general form of C_1, namely

$$C_1 = \sigma^2 \operatorname{diag}(e_1 I_{k_1},\dots,e_b I_{k_b}) \tag{2}$$

i.e. if different error variances are assumed in different blocks ($e_1,e_2,\dots,e_b$ known). It is worthwhile pointing out that correlations $e_{jj'}$, ($j,j'=1,\dots,b$) between observations in (different or coincident) blocks j and j' may be left out of the expression for C_1, since they may be thought of as correlations between different block effects. In Owen's (1970) notation, this corresponds to replacing the matrix $B+\hat{E}$ by a different B.

From (1) it follows that the inverse posterior covariance matrix for $(\tau^T,\beta^T)^T$ has the form

$$\Sigma^{-1} = \sigma^{-2}\begin{pmatrix}\operatorname{diag}(\bar{N}\mathbf{1}_v) & \bar{N}\\ \bar{N}^T & K\end{pmatrix} + \begin{pmatrix}P_T & 0\\ 0 & P_B\end{pmatrix}$$

where $K = \operatorname{diag}(k_1,\dots,k_b)$ and

$$P_T = \sigma^{-2}\{T^{-1} - T^{-1}A_{21}(A_{21}^T T^{-1}A_{21} + C_{31}^{-1})^{-1}A_{21}^T T^{-1}\}$$

$$P_B = \sigma^{-2}\{B^{-1} - B^{-1}A_{22}(A_{22}^T B^{-1}A_{22} + C_{32}^{-1})^{-1}A_{22}^T B^{-1}\} \tag{3}$$

The inverse covariance matrix for the marginal posterior distribution of τ, disregarding σ^{-2}, is

$$D^{-1} = -\bar{N}(K+P_B)^{-1}\bar{N}^T + \operatorname{diag}(\bar{N}\mathbf{1}_v) + P_T \tag{4}$$

The matrix $M=D^{-1}$ is the Bayesian information matrix of the design and can also be written as $M=M_o+H$, where $M_o=\operatorname{diag}(\bar{N}\mathbf{1}_v)-\bar{N}K^{-1}\bar{N}^T$ is the information matrix for the least squares estimator of τ (see Bechhofer and Tamhane, 1981) and

$H = P_T - \bar{N}K^{-1}(K^{-1} + P_B^{-1})^{-1} K^{-1}\bar{N}^T$. It follows that $M \cong M_o$ when vague prior distributions are considered at the second stage in (1) i.e. when $T^{-1} \cong 0$, $B^{-1} \cong 0$.

Following the usual practice, we may further allow the entries n_{ij} $(i=0,\dots,v$ $j=1,\dots,b)$ of N to take on any real non-negative value, provided $\sum_{i=0}^{v} n_{ij} = k_j$ for all values of j. Such "continuous" designs are to be regarded as approximations to the ones with integer entries, and the relative information matrix is defined by (4).

3. EXCHANGEABLE TREATMENTS

Suppose now that the prior knowledge on the treatment parameters is symmetric under permutations of the treatments, which is equivalent to assuming $\tau_i - \tau_o$ $(i=1,\dots,v)$ exchangeable. This assumption corresponds to

$$A_{21} = \mathbf{1}_v, \quad \mu_\tau = \text{a scalar}, \; T = \sigma_T^2 \{(1-\varrho) I_v + \varrho J_v\}, \quad C_{31} = w^2. \tag{5}$$

It follows that P_T which plays the role of T^{-1} in the two-stage model is completely symmetric:

$$P_T = \frac{\gamma}{1-\varrho} \{I_v - \frac{\varrho + \eta}{v\eta + [1 + (v-1)\varrho]} J_v\},$$

where $\gamma = \sigma^2/\sigma_T^2$ and $\eta = w^2/\sigma_T^2$. The eigenvalues of P_T are $\delta_1 = \gamma(1-\varrho)^{-1}$ with multiplicity $v-1$ and $\delta_2 = \gamma[v\eta + \{1+(v-1)\varrho\}]^{-1}$ with multiplicity 1. We want to investigate which designs N are optimal under this assumption.

The possible singularity of P_T, which arises for example if we let $w^2 \to \infty$, does not affect the results.

Since the treatments are assumed a priori exchangeable, it may still be suitable to consider the optimality criteria ϕ defined by Kiefer (1975), namely real functions from the set $\boldsymbol{B}$ of non-negative definite matrices satisfying

a) ϕ is convex;

b) ϕ is matrix-non-increasing, i.e., if $B_1 \geq B_2, B_1, B_2 \in \boldsymbol{B}$, then $\phi(B_1) \leq \phi(B_2)$;

c) ϕ is invariant under the same permutation of rows and columns.

In Kiefer (1975) instead of b) it is assumed only that

b)$_K$ ϕ is non-increasing w.r.t. multiplication by a non-negative scalar.

However, replacement of b)$_K$ by b) in this context seems justified since it has been pointed out (see Sinha and Mukerjee, 1982) that b)$_K$ is unsuitable for matrices of full rank. Criteria satisfying a), b) and c) are the functions ϕ_p $(p=0,1,\dots,\infty)$ introduced by Kiefer, where $\phi_p(M) = (v^{-1} tr\, M^{-p})^{1/p}$. These criteria include D-, A- and E-optimality: for a Bayesian justification see Giovagnoli and Verdinelli (1983). Another such criterion is $\phi(M) = \mathbf{1}_v^T M^{-1} \mathbf{1}_v$ which corresponds to minimizing the posterior variance of the average effect of the treatments (see also Pilz, 1983).

In the recent literature on block designs (Hedayat, 1981; Bondar, 1983; Giovagnoli and Wynn, 1981 and 1983) a different class of criteria has been considered, namely Schur-convex non-increasing functions of the eigenvalues of the information matrix. The definitions and properties of Schur-convexity can be found in Marshall and Olkin (1979). Schur criteria are appealing in so far as they demand both maximum balance (the eigenvalues as equal as possible) and maximum information (the eigenvalues as large as possible). It is not clear however whether there are Schur criteria other than A-, D- and

E-optimality that are susceptible to a Bayesian interpretation.

We can extend Theorem 2 of Giovagnoli and Verdinelli (1983) to a three-stage model, taking into account both types of optimality criteria. Because condition b) is weaker than the b) condition in Giovagnoli and Verdinelli (1983) the optimal designs may no longer be unique.

Theorem

For a three-stage model with exchangeable treatments the design

$$N^* = \begin{pmatrix} n_{o1} & \dots & n_{ob} \\ v^{-1}(k - n_{o1})\mathbf{1}_v \dots & & v^{-1}(k_b - n_{ob})\mathbf{1}_v \end{pmatrix} = \begin{pmatrix} \mathbf{n}_o^T \\ \bar{N}^* \end{pmatrix} \tag{6}$$

is optimal among all designs with fixed observations $(n_{01},\dots,n_{ob})$ on the control in each block.

Proof. Let N be any design with the same k_j's and n_{oj}'s as N^*. Then $\bar{N}^* = \frac{1}{v!}\sum_{\Pi} \Pi\bar{N}$ where the summation is over all the $v \times v$ permutation matrices Π.

Let $M^* = M(\bar{N}^*)$, $M = M(\bar{N})$. From

$$M(\alpha\bar{N}_1 + (1-\alpha)\bar{N}_2) \geq \alpha M(\bar{N}_1) + (1-\alpha)M(\bar{N}_2)$$

for all $\bar{N}_1,\bar{N}_2$ and $0 \leq \alpha \leq 1$ (Owen, 1970, Lemma 1), it follows that

$$M^* \geq \frac{1}{v!}\sum_{\Pi} M(\Pi\bar{N}) = \frac{1}{v!}\sum_{\Pi} \Pi M \Pi^T \tag{7}$$

since
$$\begin{aligned} M(\Pi\bar{N}) &= -\Pi\bar{N}(K + P_B)^{-1}\bar{N}^T\Pi^T + \mathrm{diag}(\Pi\bar{N}\mathbf{1}_v)\mathbf{1} + P_T \\ &= -\Pi(\bar{N}(K + P_B)^{-1}\bar{N}^T)\Pi^T + \Pi(\mathrm{diag}(\bar{N}\mathbf{1}_v)\Pi^T + \Pi P_T\Pi^T. \end{aligned}$$

Observe that K is not affected by row permutations. If a), b), c) hold, (7) implies that $\phi(M^*) \leq \phi(M)$.

From (7) $\quad \lambda(M^*) \geq \lambda(\frac{1}{v!}\sum_{\Pi} \Pi M\, \Pi^T)$

where λ (B) indicates the v-vector of eigen-values of $\boldsymbol{B} \in \boldsymbol{B}$, with entries arranged in decreasing order. By a well-known result (see Marshall and Olkin, 1979, p. 478).

$$\lambda(\frac{1}{v!}\sum_{\Pi} \Pi M\Pi^T) <^* \frac{1}{v!}\sum_{\Pi} \lambda(\Pi M\Pi^T) = \frac{1}{v!}\sum_{\Pi} \lambda(M) = \lambda(M)$$

where $<^*$ denotes majorization. For ϕ non-increasing and Schur-convex on the eigenvalues

$$\begin{aligned} \phi(\lambda(M^*)) &\leq \phi\,\{\lambda(\frac{1}{v!}\sum_{\Pi} \Pi M\Pi^T)\} \\ &\leq \phi\,(\lambda(M))\,; \end{aligned}$$

this gives the ϕ-optimality of N^*.

A similar result in the classical set-up was obtained by Giovagnoli and Wynn (1983).

All the results of §4 of Giovagnoli and Verdinelli (1983) which depend only on Theorem 2 can be applied in the present case with P_T replacing T^{-1} and hold true also for

Schur-convex functions. In particular from (4) and (6) it follows that the Bayesian information matrix of the optimal design (6) is completely symmetric.

Denote by $\mathbf{x} = (x_1,\ldots,x_b)^T$ $(x_j = v^{-1}(k_j - n_{oj}),\ j = 1,\ldots,b)$ the rows of $\bar{N}^*$. Then M^* is a function of $\mathbf{x}$ whose eigenvalues are $\lambda_1(\mathbf{x}) = \mathbf{1}_b^T\,\mathbf{x} + \delta_1$ and $\lambda_2(\mathbf{x}) = \mathbf{1}_b^T\,\mathbf{x} - v\mathbf{x}^T(K+P_B)^{-1}\,\mathbf{x} + \delta_2$, with multiplicities $v-1$ and 1 respectively. Also every function of M^* can be expressed as a function ψ of $\lambda_1(\mathbf{x})$ and $\lambda_2(\mathbf{x})$ alone, which we want to minimize over the set $R = \{\mathbf{x}: \mathbf{0} \le \mathbf{x} \le v^{-1}\mathbf{k}\}$ where $\mathbf{k} = (k_1,\ldots,k_b)^T$.

The same two-step approach suggested by Owen (1970) can be usefully employed here. First we fix $\pi = \mathbf{1}_b^T\,\mathbf{x}$ (i.e. we fix the total number of observations for the control). For a given $\pi\epsilon[0,n/v]$, λ_1 is fixed and χ attains its minimum value at

$$\mathbf{x}_o(\pi) = \frac{\pi}{n+\mathbf{1}_b^T P_B\,\mathbf{1}_b}(\mathbf{k} + P_B\,\mathbf{1}_b)\,,$$

which is the point that maximizes $\lambda_2(\mathbf{x})$. If $\mathbf{x}_o(\pi)\ \epsilon\ R$ for all $\pi\ \epsilon\ [0,n/v]$, then an optimal design is given by $\mathbf{x}_o(\hat{\pi})$, where $\hat{\pi}$ is the value which minimizes $\psi[\lambda_1\{\mathbf{x}_o(\pi)\},\ \lambda_2\{\mathbf{x}_o(\pi)\}]$ over $[0,n/v]$. In particular by Corollary 4.1 of Giovagnoli and Verdinelli (1983) the above condition is satisfied when

$$P_B\mathbf{1}_b = (h-1)\mathbf{k} \qquad \text{for some } h \ge 1 \tag{8}$$

which gives $\mathbf{x}_o(\pi) = n^{-1}\,\pi\mathbf{k}$, $\lambda_1(\pi) = \pi+\delta_1$, $\lambda_2(\pi) = \pi - v(\pi^2/nh) + \delta_2$. In this case the optimal (including Schur-optimal) allocation in each block is proportional to the block size, namely

$$x_j^* = \frac{\hat{\pi}}{n}\,k_j \qquad (j = 1,\ldots,b)\,, \tag{9}$$

with $\hat{\pi}$ dependent on the function ψ, i.e., on the optimality criterion considered. We point out that the row sum j of the matrix $\sigma^{-2}P_B$ can be viewed as a global measure of prior precision for the block parameter β_j $(j = 1,\ldots,b)$. Condition (8) states that the ratio between prior and observation precision equals h-1, i.e., is constant over the blocks.

4. WEAK KNOWLEDGE AT THE THIRD STAGE

Vague knowledge of the vectors of hyperparameters μ_T, μ_B corresponds to the choice $C_{31}^{-1} \cong 0$, $C_{32}^{-1} \cong 0$ in (1). With the same prior assumption for the treatment parameters as in §3, i.e. exchangeability, this implies setting $w^{-2} = 0$ in (5), hence $P_T = \gamma(1-\varrho)^{-1}(I_v - v^{-1}J_v)$ and $\delta_2 = 0$. Let us consider now two different prior specifications for the block parameters, both leading to (8) being satisfied with $h = 1$; the optimal allocation (9) can, in this case, be obtained explicitly.

In the first instance assume that the prior information on the block parameters is such that the β_j's have a common prior mean. In terms of (1):

$$A_{22} = \mathbf{1}_b, \qquad \mu_B \text{ a scalar.}$$

Then

$$P_B = \sigma^2\{B^{-1} - (\mathbf{1}_b^T B^{-1}\mathbf{1}_b)^{-1} B^{-1}J_b\,B^{-1}\}\,.$$

It is straightforward to verify that $P_B\mathbf{1}_b = \mathbf{0}$ and it can be seen that the optimal design (9) does not depend on B.

A similar result holds true whenever the observed units are blocked according to the levels $z_1 < z_2 < \ldots < z_b$ of a given factor; for example age, income, values of a physiological parameter, etc. Then (Smith, 1973) it may be reasonable to assume that the

β_j lie approximately on a polynomial response curve of degree r ($r < b$) but the information about its actual numerical form is vague ($C_{32}^{-1} \cong 0$). Such a prior distribution for the block parameters can be expressed as

$$\beta_j \sim N(\alpha_o + \alpha_1 \xi_1(z_j) + \ldots + \alpha_r \xi_r(z_j), \sigma_B^2)$$

where $\xi_s(z)$ ($s=1,..,r$) are orthogonal polynomials, i.e., $\Sigma_j \xi_\ell(z_j) \xi_s(z_j)=0 \quad \forall \ell \neq s$, and different β_j are independent. This corresponds to taking

$$A_{22} = \begin{pmatrix} 1 & \xi_1(z_1) & \ldots & \xi_r(z_1) \\ 1 & \xi_1(z_2) & \ldots & \xi_r(z_2) \\ \vdots & & & \\ 1 & \xi_1(z_b) & \ldots & \xi_r(z_b) \end{pmatrix}, \mu_B = \begin{pmatrix} \alpha_o \\ \alpha_1 \\ \vdots \\ \alpha_r \end{pmatrix}, B = \sigma_B^2 I_b$$

in (1). $Q = A_{22}^T A_{22}$ is a diagonal matrix with $q_{11}=b$. Substituting into (3) gives

$$P_B = (\sigma^2/\sigma_B^2)(I_b - A_{22} Q^{-1} A_{22}^T)$$

so that

$$P_B \mathbf{1}_b = (\sigma^2/\sigma_B^2)(\mathbf{1}_b - A_{22} Q^{-1} A_{22}^T \mathbf{1}_b) = \mathbf{0}.$$

Again, (8) is satisfied and since $h=1$ the optimal design does not depend on σ_B^2.

More generally, assume that (8) holds, i.e. $P_B \mathbf{1}_b = (h-1)\mathbf{k}$ for a fixed $h \geq 1$. Since $\lambda_1(\pi) = \pi + \delta_1$ and the expression for λ_2 is now

$$\lambda_2(\pi) = \pi - v\,(nh)^{-1}\pi^2\,,$$

both λ_1 and λ_2 are increasing for $0 \leq \pi \leq (2v)^{-1}nh$, whereas for $\pi > (2v)^{-1}nh$, $\lambda_2(\pi)$ is decreasing. Hence if $h \geq 2$, the optimal design (9) for every criterion is $x_j = k_j/v$ for all $j=1,\ldots,b$. If $1 \leq h < 2$, we need to minimize $\psi(\lambda_1(\pi), \lambda_2(\pi))$ over $(2v)\bar{n}h \leq \pi \leq n/v$.

If we take ϕ_p-optimality, this involves solving the following equation (it can be shown that the solution is unique in $(\frac{nh}{2v}, \frac{n}{v})$):

$$(v-1)\,(\pi - \frac{v}{nh}\pi^2)^{p+1} + (1 - \frac{2v}{nh}\pi)\,(\pi + \frac{\gamma}{1-\varrho})^{p+1} = 0. \tag{10}$$

It may be useful to recall what the symbols stand for: σ^2 is the error variance, σ_T^2 and ϱ are the prior variance and correlation respectively for the treatment differences $\tau_i - \tau_o$ ($i=1,\ldots,v$) under exchangeability, h is defined by (8) and $\gamma = \sigma^2/\sigma_T^2$.

From (10) a D-optimal design is $x_j = \frac{\hat{\pi}}{n} k_j$ ($j=1,\ldots,b$) with

$$\hat{\pi} = \frac{nh(1-\varrho) - 2\gamma + (4\gamma^2 + 4(1-\varrho)v^{-1}nh\gamma + n^2h^2(1-\varrho)^2)^{1/2}}{2(1-\varrho)\,(v+1)}$$

The solution for A-optimality is obtained from (10) by letting $p=1$.

Observe that if $h=1$ and $\varrho=0$ these designs are the same as those found by Smith and Verdinelli (1980) for a 3-stage model without blocks.

The solution for E-optimality is $x_j = \frac{1}{2} v^{-1} k_j$ ($j=1,\ldots,b$), the same as the classical one.

We end with a comment on the fact that there are some optimal designs with no observations on the control (in fact this is true if $h \geq 2$). Other optimal designs have fewer observations on the control than on the treatments whereas classical optimal designs always put more observations on the control than on the other treatments. This is due to the assumption $\tau_o = 0$ as a consequence of which (see also Giovagnoli and Verdinelli, 1983) observations on the control in block j give us information on the block parameter β_j only ($j=1,...,b$). Intuitively, the greater the prior precision of the block parameter, the fewer observations we would want on the control. If $h \geq 2$ the prior precision of every block is greater than the observation precision of the block.

ACKNOWLEDGEMENTS

We wish to thank Prof. A.F.M. Smith for useful conversations on this topic and the Italian Science Research Council (C.N.R.) and the Italian Ministery of Education (M.P.I) for financial support given throughout.

REFERENCES

BECHHOFER, R.E. and TAMHANE, A.C. (1981). Incomplete block designs for comparing treatments with a control: General theory. *Technometrics,* **23,** 45-57.

BONDAR, J. (1983). Universal optimality of experimental designs: definitions and a criterion. *Can. J. Statist.,* **11,** 4.

GIOVAGNOLI, A. and VERDINELLI, I. (1983). *D*-optimal and *E*-optimal Bayes block designs. *Biometrika,* **70,** 3, 695-706.

GIOVAGNOLI, A. and WYNN, H.P. (1981). Optimum continuous block designs. *Proc. R. Soc. Lon. A,* **377,** 405-16.

— (1983). Schur-optimal continuous block designs for treatments with a control. *Proc. Berkely Conf. in honor of Jerzy Neyman and Jack Kiefer,* June, 1983. (To appear).

HEDAYAT, A. (1981). Study of optimality criteria in design of experiments. *Statistics and Related Topics.* (Görgö et al. eds.). Amsterdam: North-Holland. 39-56.

KIEFER, J. (1975). Construction and optimality of generalized Youden Designs. *A Survey of Statistical Designs and Linear Models.* (J. Srivastava ed.). Amsterdam: North Holland. 333-353.

LINDLEY, D.V. and SMITH, A.F.M. (1972). Bayes estimates for the linear model. *J.R. Statist. Soc. B,* **34,** 1-42. (With discussion).

MARSHALL, A.W. and OLKIN, I. (1979). *Inequalities: Theory of Majorization and its Applications.* New York: Academic Press.

OWEN, R.J. (1970). The optimum design of a two-factor experiment using prior information. *Ann. Math. Statist.* **41,** 1917-1934.

PILZ, J. (1983). *Bayesian Estimation and Experimental Design in Linear Regression Models.* Leipzig: B.G. Teubner.

SINHA, B.K. and MUKERJEE, R. (1982). A note on the universal optimality criterion for full rank models. *J. Statist. Planning. Inf.* **7,** 101-105.

SMITH, A.F.M. (1973). Bayes estimates in one-way and two-way models. *Biometrika,* **60,** 319-329.

SMITH, A.F.M. and VERDINELLI, I. (1980). A note on Bayes designs for inference using a hierarchical linear model. *Biometrika,* **67,** 613-9.

BAYESIAN STATISTICS 2, pp. 663-672
J.M. Bernardo, M.H. DeGroot, D.V. Lindley, A.F.M. Smith (Eds.)

Bayesian Estimation of Undetected Errors

W.S. JEWELL
University of California, Berkeley

SUMMARY

An unknown number, N, of errors or defects exist in a certain product, and I inspectors with unknown competencies are put to work to find the errors. Given the lists of errors found by each inspector, how can we estimate the number of undetected errors? A similar problem arises in capture-recapture sampling in population biology, where the MLE of N, attributed to Petersen, Chapman, and Darroch, has been known for many years. Our Bayesian model assumes that N is Gamma-mixed-Poisson, that errors are equally difficult to detect, and that inspector error detection probabilities are independent and Beta-distributed, a priori. The predictive density for undetected errors is obtained as a simple, recursive relationship that gives Negative Binomial tails. The predictive mode for undetected errors is given by a generalized Petersen-Chapman-Darroch form involving credibility formulae; as the prior parameter variances increase without limit, this predictive mode approaches the classical estimator.

Keywords: BAYESIAN PREDICTION; CAPTURE-RECAPTURE CENSUS; CREDIBILITY THEORY; ERROR DETECTION; QUALITY CONTROL; SOFTWARE RELIABILITY.

1. INTRODUCTION

A number of estimation problems in reliability can be described as follows: a certain product has an unknown number, N, of defects. A group of I inspectors each allocates a given amount of independent effort to finding and removing the defects. After finding, say, n_T total defects, what is the estimated number, $n_o = N - n_T$, of *undetected defects* still left in the product?

For example, in manufacturing quality control, the product may be a certain production lot for which the inspectors may use visual or machine-aided techniques to inspect a portion or all of the items. Estimation of the number of undiscovered defects in the sample scrutinized is the first step in setting quality assurance levels for the entire lot.

In software reliability, the defects correspond to program errors or bugs that can be detected and removed by programmers using some combination of visual scanning of program code and of experimental running of the program on typical input. The estimation of undetected errors remaining in the program not only helps certify the application-readiness of the software, but also provides an indication of the effort that will be needed for customer support and for the upgrading of future program releases. A similar interpretation arises in the proofreading of manuscripts for misprints.

Superficially, this model is similar to the problem of estimating the ultimate failure rate of a product during the reliability growth (learning curve) phase of product testing and development (see, e.g., Jewell (1984)). However, in that application and unspecified external process of design improvement reduces the stochastic rate of recurrence of product "failures" according to some given law, whose parameters are to be estimated. In this model, on the other hand, an inspector is assumed to actually remove (or at least to identify) one of a finite number of defects or errors, so that, at the end of inspection, there remain only a smaller number of unfound errors. Further, as we shall see below, there is an advantage to having the inspectors in parallel on the same product, rather than in series, as this helps make more precise any uncertainty in the inspection efficiencies of the different examiners, and thus improves the estimate of undetected errors.

2. BASIC MODEL; SERIES AND PARALLEL SEARCH STRATEGIES

Suppose that the error inspection process is such that:

a) Each error present has the same probability of being detected by a given inspector;

b) The probability that inspector i will find any given error is p_i, $(i = 1,2,\ldots,I)$, independent of previous errors found by i or by any other inspector.

The simplest possible strategy for organizing a search by I inspectors is a *serial* one, in which inspector i, examining a product with $N-(n_1+n_2+\ldots+n_{i-1})$ errors, finds and removes n_i errors $(i = 1,2,\ldots,I)$. It follows from the assumptions above that each of the unknown n_i is conditionally Binomially distributed, so that the joint density of the I pieces of data, $(\mathbf{n}_1,n_2,\ldots,\mathbf{n}_I|N;\,\mathbf{p})$ is easily found. More importantly, since each error, if present, is missed by inspector i with probability $q_i = 1-p_i$, the total overlook probability (probability of being undetected by any inspector) for every error is $Q = \prod_{i=1}^{I} q_i$, and thus the conditional density of undetected errors, $(\mathbf{n}_o|N;\mathbf{p})$, is Binomial (Q,N).

A *parallel* search strategy is more complicated, since here we assume, either that the inspectors all work independently on identical copies of the product, or that they work in some sequence on a single product, (secretly) identifying, but not removing, the defects which they find. With this strategy, there will usually be duplication in the defects found by different inspectors, and the lists of defects reported by each will have to be reconciled, classifying and counting the errors in the following mutually exclusive and collectively exhaustive categories:

n_i – the number of defects found only by inspector i;

n_{ij} – the number of defects found jointly only by i and j $(i<j)$;

n_{ijk} – the number of defects found jointly only by i, j, and k $(i<j<k)$

$\vdots$

$n_{123\cdots I}$ – the number of defects found jointly by all inspectors.

Thus, there will be 2^I-1 *separate pieces of observed data:*

$$\nu = \{(n_i);(n_{ij});(n_{ijk});\ldots;n_{123\cdots I}\} \quad .$$

Inspector i finds, in total:

$$n(i) = n_i + \sum_j n_{ij} + \sum_{j<k}\sum n_{ijk} + \ldots + n_{123\cdots I} \tag{2.1}$$

defects, and the total number of distinct defects found by all inspectors is:

$$n_T = \sum_i n_i + \sum\sum_{i<j} n_{ij} + \sum\sum\sum_{i<j<k} n_{ijk} + \ldots + n_{123\ldots I} \tag{2.2}$$

$$= \sum_i n(i) - 1 \sum\sum_{i<j} n_{ij} - 2 \sum\sum\sum_{i<j<k} n_{ijk} - \ldots - (I-1)n_{123\ldots I} .$$

The joint conditional density of ν and $n_o = N - n_T$ is derived in the next Section. Note that, in spite of the additional complexity of parallel search, it again follows from the assumptions that the total overlook probability for each error is Q, and hence $(\mathbf{n}_o | N; \mathbf{p})$ is again Binomial(Q,N). Thus, for fixed N and $\mathbf{p}$, the density of undetected errors is independent of the search strategy.

Why, then, would one be interested in parallel search? The answer lies in the fact that, by permitting duplicate errors to be found, we gain additional information about the detection probabilities (p_i), so that if they are unknown quantities at the beginning of inspection, the increased data set associated with parellel search will provide increased precision in the posterior densities of both $\mathbf{p}$ and $\mathbf{n}_o$. Henceforth, we shall assume that a parallel search for errors has been made.

3. THE PETERSEN-CHAPMAN-DARROCH ESTIMATORS

We begin with some classical point estimators for N. For $I = 2$, we can argue as in Polya (1976) that:

$$\hat{N} = \frac{n(1)n(2)}{n_{12}} = n_T + \frac{n_1 n_2}{n_{12}} \; ; \hat{p}_i = \frac{n(i)}{\hat{N}} \qquad (i = 1,2) \; . \tag{3.1}$$

Note that this argument is symmetric with respect to the two inspectors, and that both singly-found and jointly-found defects are important.

For $I > 2$, a slightly different heuristic argument (see the original paper, Jewell (1983)) gives:

$$\hat{N} = n_T + \hat{N} \prod_{i=1}^{I} \left[1 - \frac{n(i)}{\hat{N}} \right] ; \tag{3.2}$$

$$\hat{p}_i = \frac{n(i)}{\hat{N}} \qquad (i = 1,2,\ldots,I) \, ; \quad \hat{Q} = 1 - \frac{n_T}{\hat{N}} \; . \tag{3.3}$$

For $I = 2$, these formulae reduce to (3.1), while for $I = 3$, they require the solution of a quadratic equation, etc. A variety of approximating and iterative procedures are available for (3.2); see Seber (1982). A good initial approximation in the general case is:

$$\hat{N} = \frac{\sum\sum_{i<j} n(i)n(j)}{\sum\sum_{i<j} n_{ij}} , \tag{3.4}$$

which is reminiscent of (3.1). It is easy to show that, if all $n(i)$ are equal to each other and to n_T, then $\hat{N} = n_T$; otherwise, (3.2) has a unique finite root $\hat{N} > n_T$.

In spite of the appearance of n_{ij} in (3.1) and (3.4), it should be clear from (3.2) that only the $I + 1$ pieces of information in the *reduced data set*, $\nu^* = \{(n(i));n_T\}$, are needed to estimate $\hat{N}$.

(3.1) has a long history in the statistical literature; it was apparently first used by LaPlace in 1783 to estimate the population of France. In population biology, it arises in the capture-recapture sampling of a fixed, but unknown animal population, where it is called the Petersen Method (Seber (1982)). The case $I>2$ corresponds to multiple capture-recapture sampling, in what is called a Schnabel census; the estimator (3.2) was first obtained by Chapman (1952), and analized by Darroch (1958). Since that time, there has been an explosion of generalizations of these formulae in the biometric literature, as well as adaptions to epidemiology and other fields (Seber (1982)). Some numerical examples of the behaviour of $\hat{N}$ are given in Jewell (1983).

There is a limited literature on Bayesian models (Gaskell and George (1972), Carle and Strub (1978), Yang et. al. (1982), Freeman (1973), Casteldine (1981)), all of which, except for Casteldine, are quite limited or specialized in scope. Casteldine is discussed in Jewell (1983).

4. POISSON ERRORS

Using the assumptions of Section 2, it follows that the joint density of $\mathbf{n}_o$ and ν under parallel search is the multinomial distribution:

$$p(n_o,\nu \mid N, \mathbf{p}) = \binom{N}{n_o \;\; \nu} Q^{n_o} \prod_{i=1}^{I} (\omega_i Q)^{n_i} , \tag{4.1}$$

where $\omega_i = p_i/q_i$ is the odds-ratio for inspector i, and of course $n_o = N - n_T$, with n_T given by (2.2). As might have been expected, (5.1) then simplifies to:

$$p(n_o,\nu \mid N, \mathbf{p}) = \binom{N}{n_o \;\; \nu} Q^{N} \prod_{i=1}^{I} \omega_i^{n(i)} , \tag{4.2}$$

In the reliability applications of interest, it seems natural to assume that the total number of defects or errors would be generated by a Poisson process, with parameter λ; in the next Section, additional modelling flexibility will be added by permiting both λ and $\mathbf{p}$ to be random quantities. With the Poisson asssumption, and eliminating N in favor of n_o, the joint density can be rearranged into:

$$p(n_o,\nu \mid \lambda,\mathbf{p}) \propto \frac{(\lambda Q)^{n_o} e^{-\lambda}}{n_o!} \lambda^{n_T} \prod_{i=1}^{I} p_i^{n(i)} q_i^{n_T - n(i)} .$$

It follows that the conditional density of undetected errors is Poisson(λQ):

$$p(n_o \mid \lambda, \mathbf{p}) = \frac{(\lambda Q)^{n_o} e^{-\lambda Q}}{n_o!} , \tag{4.3}$$

(which is expected from first principles), and the data likelihood is:

$$p(\nu^* \mid \lambda,\mathbf{p}) \propto \lambda^{n_T} e^{-\lambda(1-Q)} \prod_{i=1}^{I} p_i^{n(i)} q_i^{n_T - n(i)} . \tag{4.4}$$

In other words, the reduced data set, $\nu^* = \{(n(i));n_T\}$ is sufficient for both λ and $\mathbf{p}$. From this, we find that the MLE of λ also satisfies (3.2), that is, $\hat{\lambda} = \hat{N}$.

5. A BAYESIAN MODEL

The model thus far is unsatisfactory in most applications because it depends upon both λ and $\mathbf{p}$ being known exactly, giving, for example, $E\{\mathbf{n}_o|\lambda,\mathbf{p}\} = \lambda Q$. We will henceforth assume that these are random quantities, with known prior distributions; this assumption raises no special problems in the applications areas of interest. For computational convenience, we shall, in fact, assume independent natural-conjugate priors:

$$p(\lambda;\mathbf{p}\,|\,a,b\,;\alpha,\beta) = \frac{b^a\lambda^{a-1}e^{-b\lambda}}{\Gamma(a)} \prod_{i=1}^{I} \frac{\Gamma(\alpha_i+\beta_i)}{\Gamma(\alpha_i)\Gamma(\beta_i)} (p_i)^{\alpha_i-1}(q_i)^{\beta_i-1} , \tag{5.1}$$

that is, Gamma(a,b) Π Beta$_i(\alpha_i,\beta_i)$. Let $\gamma_i = \alpha_i + \beta_i$.

From (4.4), we note that, *except for the term* $e^{+\lambda Q}$, we would have independent updating of each component of the prior according to:

$$a' = a+n_T;\ b' + b+1;\ \alpha_i'=\alpha_i+n(i);\ \beta_i'=\beta_i+n_T-n(i);\ \gamma_i' = \gamma_i'+n_T \ . \tag{5.2}$$

But the coupling term can be expanded into a power series, so the posterior-to-data joint parameter density becomes:

$$p(\lambda,\mathbf{p}\,|\,\nu^*) = \left[\sum_{k=0}^{\infty} d_k\right]^{-1} \sum_{j=0}^{\infty} d_j\,\text{Gamma}(a'+j,b') \prod_{i=1}^{I} \text{Beta}_i(\alpha_i',\beta_i'+j) ,$$

$$d_j = d_j(\nu^*) = \frac{(b')^{-j}}{j!}\frac{\Gamma(a'+j)}{\Gamma(a')} \prod_{i=1}^{I} \frac{\Gamma(\beta_i'+j)}{\Gamma(\beta_i')}\frac{\Gamma(\gamma_i')}{\Gamma(\gamma_i'+j)} . \tag{5.3}$$

Using (4.3) and (5.1), we find that the marginal density of $\mathbf{n}_o$, prior to inspection, is a rather complex sum of products, similar to (5.3), even though $E\}\mathbf{n}_o\} = (a/b)\,\Pi\,(\alpha_i/\beta_i)$, a priori.

Surprisingly, however, the posterior-to-data predictive density of $\mathbf{n}_o$ is much simpler, as there is a fortuitous cancellation of the nuisance term $e^{+\lambda Q}$ of (4.4) with $e^{-\lambda Q}$ in (4.3)! Perhaps the simplest way of writing the predictive density is in recursive form:

$$\frac{p(n_o+1\,|\,\nu^*)}{p(n_o\,|\,\nu^*)} = \frac{1}{n_o+1}\ \frac{a'+n_o}{b'} \prod_{i=1}^{I} \frac{\beta_i'+n_o}{\gamma_i'+n_o} . \tag{5.4}$$

Numerical computation is very efficient; by setting $p(0\,|\,\nu^*) = 1$, one in fact computes the coefficients d_{n_o} of (5.3), and then gets the predictive density through normalization. Moments of $\mathbf{n}_o$ are thus best found numerically. It is also easy to see that, as $n_o \to \infty$, the predictive density has a Negative Binomial tail.

(5.4) is also remarkable in that the terms in square brackets are *credibility (linear least-squares) estimators* for λ and the $(\mathbf{q}_i)$, respectively; further details are given in Jewell (1983).

6. THE POSTERIOR MODE

Although (5.4) permits the calculation of any moment of $\mathbf{n}_o$, it is difficult to compare these numerical results with the classical point estimators of Section 3.

However, the *mode* of the predictive density is easily found as the smallest integer, $\check{n}_o$, for which $p(\check{n}_o + 1 \mid \nu^*) \leq p(\check{n}_o \mid \nu^*)$.
After some rearranging, we find that $\check{n}_o$ is the smallest integer not less than the solution n_o^* to:

$$n_o^* + 1 = [n_o^* + n_T + \frac{b}{b+1}(E\{\lambda\} - n_o^* - n_T)] \prod_{i=1}^{I} \left[1 - \frac{n(i) + \gamma_i E\{\mathbf{p}_i\}}{n_o^* + n_T + \gamma_i}\right], \quad (6.1)$$

which should be compared with (3.2), rewritten with $\hat{n}_o = \hat{N} - n_T$:

$$\hat{n}_o = [\hat{n}_o + n_T] \prod_{i=1}^{I} \left[1 - \frac{n(i)}{\hat{n}_o + n_T}\right], \quad (6.2)$$

whence we can easily see the effect of adding prior opinion.

If $b \to 0$ and $\gamma_i \to 0$, with *constant* prior means $E\{\lambda\}$ and $E\{\mathbf{p}_i\}$, n_o^* approaches $\hat{n}_o$, so that this would correspond to "diffuse" prior knowledge (although, for the Beta density, $\alpha = \beta = 1$ and $\gamma = 2$ is usually considered the diffuse case). Conversely, letting $b \to \infty$ or $\gamma_i \to \infty$ leads to the special cases where λ or p_i is known with certainty. Thus, (6.1) is a natural generalization of the Petersen-Chapman-Darroch estimator (3.2).

7. OTHER MODELS

A variety of related error-detection models can be developed along the above lines; for example, if, as in model *I* of Casteldine (1981), one assumes that all the p_i are identical and Beta(α,β), then one obtains similar formulae with only the error-detection probability changed. One can also assume that the error-detection or -correction process is defective, that new errors enter randomly during inspection, that detection probabilities vary with different error types, and so on. Point estimators for many of these variations have already been obtained in animal population studies (Otis, et. al. (1978), Seber (1982)). A Bayesian comparison between serial and parallel inspection strategies will be the subject of a forthcoming paper.

8. NUMERICAL BEHAVIOR OF THE BAYESIAN ESTIMATOR

To obtain some idea of the numerical properties of (5.4), simulations were run using various priors, and various values of *I*.

For the detection probabilities, it was assumed that for the Beta priors, $\alpha_i = \beta_i = 1.0$, which gives uniform densities for all *i*. Three cases of error rate prior were examined:

	$E\{\lambda\}$	$V\{\lambda\}$
I	50	1250
II	100	5000
III	200	20000

The shape parameter *a* of the Gamma prior was kept constant at $a = 2$, with *b* adjusted to give the above moments. Since N_{true} was 100, it can be seen that these correspond to low, O.K., and high prior estimates, though of course $N = 100$ could have occurred from any prior.

Then, one sample of data was obtained for $I = 1,2,4$, and 8, with assumed values $p_i = 0.5$ for all i. The data sets obtained were:

$$\begin{array}{lll} I = 1 & n_T = 45 & \mathbf{n} = (45) \\ I = 2 & n_T = 79 & \mathbf{n} = (55,47) \\ I = 4 & n_T = 95 & \mathbf{n} = (48,52,57,45) \\ I = 8 & n_T = 99 & \mathbf{n} = (50,55,42,47,50,44,51,50)\ . \end{array}$$

Of course, the results would have been quite different in another simulation. The classical estimator, $\hat{N}$, does not exist for $I = 1$; but would have given values of 112.39, 101.31, and 99.45, that is, $\hat{n}_o = 33.39, 6.31$, and 0.45 for $I = 2,4,8$ respectively.

Figure 1 shows the density $p(n_o|\nu)$ for $I = 1$, for the three priors given above; the effect of the priors on the predictive mean, though not on the shape can be clearly seen. Figure 2 shows that the predictive density develops an interior mode when $I = 2$, although the difference due to different priors is less perceptible. For $I = 4$ and 8, the effect of the priors is barely perceptible, so that Figure 3 shows just case II above; for $I = 8$, the mode is again at $n_o = 0$.

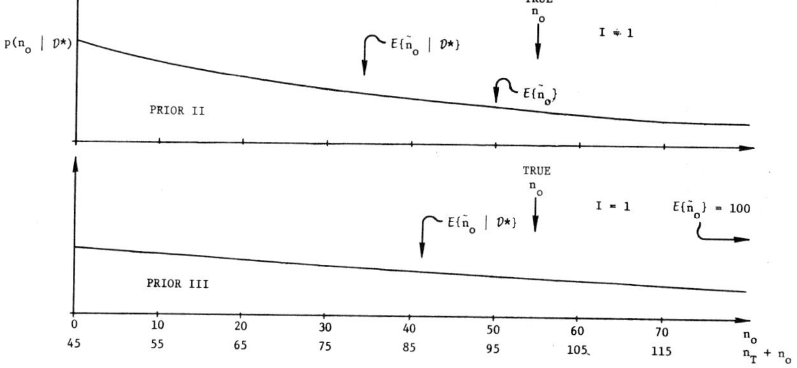

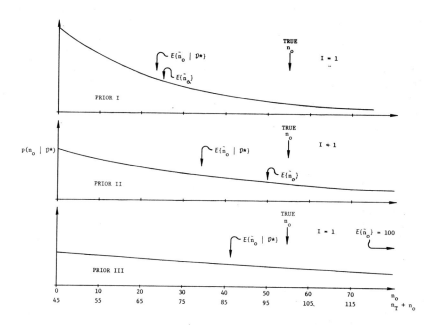

FIGURE 1. *Predictive Density for one sample from one observer, three different priors (Continuous curve for convenience)*

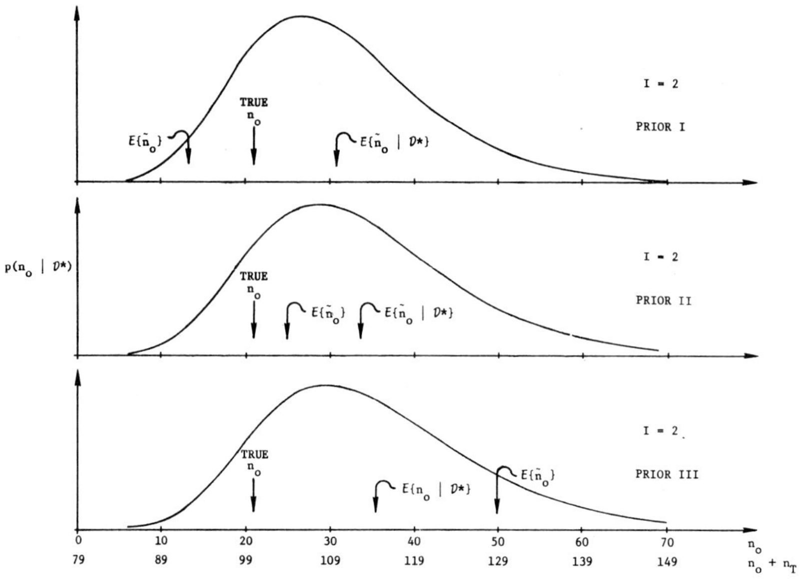

FIGURE 2. *Predictive Density for one sample from two observers, three different priors (Continuous curve for convenience)*

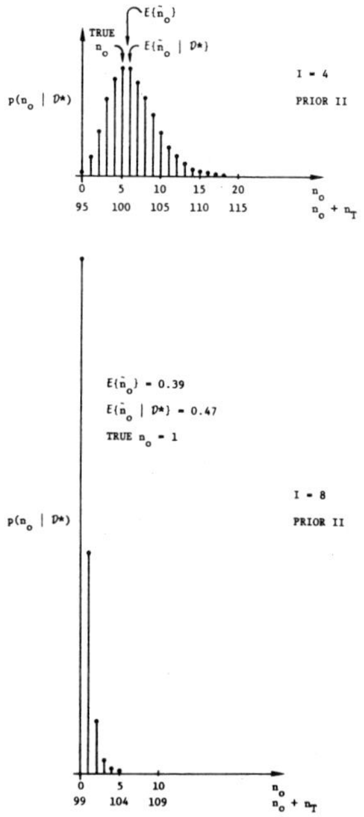

FIGURE 3. *Predictive Densities for one sample from four and eight observers, respectively, using prior density II*

ACKNOWLEDGEMENTS

I would like to express my appreciation to Sheldon Ross, who introduced me to this problem through the paper of Polya (1976), and to Dennis Lindley, who pointed out the connection with capture-recapture census methods. This research was supported by the Office of Scientific Research, USAF, under Grant AFOSR-81-0122.

REFERENCES

CARLE, F.L. and STRUB, M.R. (1978). A New Method for Estimating Population Size From Removal Data. *Biometrics,* **34**, 621-30.

CHAPMAN, D.G. (1952). Inverse Multiple and Sequential Sample Censuses. *Biometrics,* **8**, 286-306.

DARROCH, J.N. (1958). The Multiple-Recapture Census. I: Estimation of a Closed Population. *Biometrika,* **45**, 343-59.

FREEMAN, P.R. (1973). Sequential Recapture. *Biometrika,* **60**, 141-53.

GASKELL, T.J. and GEORGE, B.J. (1972). A Bayesian Modification of the Lincoln Index. *J. Appl. Ecol.,* **41**, 377-84.

JEWELL, W.S. (1984). A General Framework for Learning-Curve Reliability Growth Models, *Operations Research*, **32**, 547-58.

— (1983). Bayesian Estimation of Undetected Errors. ORC 83-11. Operations Research Center, University of California, Berkeley.

OTIS, D.L., BURNHAM, K.P., WHITE, G.C. and ANDERSON, D.R. (1978). Statistical Inference for Capture Data from Closed Populations. *Wildlife Monograph,* **62**, The Wildlife Society, Washington, D.C.

PÓLYA, G. (1976). Probabilities in Proofreading. *Amer. Math. Monthly,* **83**, 42.

SEBER, G.A.F. (1982). *The Estimation of Animal Abundance* (Second Ed.). New York: MacMillan.

YANG, M.C.K. WACKERLY, D.D. and ROSALSKY, A. (1982). Optimal Stopping Rules in Proof-Reading. *J. Appl. Prob.,* **19**, 723-9.

BAYESIAN STATISTICS 2, pp. 673-680
J.M. Bernardo, M.H. DeGroot, D.V. Lindley, A.F.M. Smith (Eds.)

A Bayesian Criterion for the Selection of Binary Features in Classification Problems

G.E. KOKOLAKIS
National Technical University of Athens, Greece

SUMMARY

In order to select, for use in classification, the k "best" binary variables from d available $(1 \leq k < d)$, we present a criterion based on the posterior weights of a mixture of $(d)_k$ priors, each one corresponding to a particular ordering of k variables and, *a priori*, implying that each successive variable is less useful than the previous one. It is shown that this criterion belongs to a class of information measures based on ϕ-entropy functionals, and is related to Lindley's measure of information. The criterion is compared with a number of other information maximizing criteria.

Keywords: BHATTACHARYYA COEFFICIENT; DIRECTED DIVERGENCES; EXPECTED SHANNON INFORMATION; INVARIANT PRIORS.

1. INTRODUCTION

A generalization of the Dirichlet distribution has been derived by Kokolakis (1983) and used for classification problems with binary variables. This generalized Dirichlet has the following form

$$g(\theta) \propto \{ \prod_{i=1}^{n} \theta_i^{\alpha_i^{(d)}-1} \} / D(\theta), \qquad \theta \epsilon S^{(n)}, \tag{1}$$

where

$$D(\theta) = \{ (\sum_{i=1}^{n/2} \theta_i)^{\alpha_1^{(2)} + \alpha_2^{(2)} - \alpha_1^{(1)}} (\sum_{i=\frac{1}{2}n+1}^{n} \theta_i)^{\alpha_3^{(2)} + \alpha_4^{(2)} - \alpha_2^{(1)}} \}$$

$$\times \{ (\sum_{i=1}^{n/4} \theta_i)^{\alpha_1^{(3)} + \alpha_2^{(3)} - \alpha_1^{(2)}} \ldots (\sum_{i=3n/4+1}^{n} \theta_i)^{\alpha_7^{(3)} + \alpha_8^{(3)} - \alpha_4^{(2)}} \} \times \ldots$$

$$\times \{ (\sum_{i=1}^{2} \theta_i)^{\alpha_1^{(d)} + \alpha_2^{(d)} - \alpha_1^{(d-1)}} \ldots (\sum_{i=n-1}^{n} \theta_i)^{\alpha_{n-1}^{(d)} + \alpha_n^{(d)} - \alpha_{n/2}^{(d-1)}} \},$$

$S^{(n)}$ is the n-dimensional simplex, $n = 2^d$, d is the feature dimensionality and the hyperparameters α's are positive constants.

A d-dimensional binary vector $\mathbf{y} \epsilon \{0,1\}^d$ is chosen to correspond to the cell-index $i = n - \Sigma_1^d 2^{k-1} y_k$; that is, the string $(y_1, \ldots, y_d)$ is the binary representation of $(n - i)$. The above distribution of the cell chance vector θ is equivalent to assuming that the conditional chances

$$u_{ij} = Pr(y_i = 1 \mid y_1, \ldots, y_{i-1}) \qquad (i = 1, \ldots, d; j = 1, \ldots, 2^{i-1})$$

with $j = 2^{i-1} - \Sigma_1^{i-1} 2^{k-1} y_k$, have independent $\beta(\alpha_{2j-1}^{(i)}, \alpha_{2j}^{(i)})$ distributions. If

$$\alpha_{2j-1}^{(i)} + \alpha_{2j}^{(i)} = \alpha_j^{(i-1)} \qquad (i=2,\ldots,d; j=1,\ldots,2^{i-1})$$

we obtain from (1) the Dirichlet density with parameter vector $\alpha^{(d)} = (\alpha_1^{(d)},\ldots,\alpha_n^{(d)})^T$, while if

$$\alpha_1^{(1)} = \alpha_2^{(1)} = \alpha \qquad (\alpha > 0)$$

and

$$\alpha_{2j-1}^{(i)} = \alpha_{2j}^{(i)} = \tfrac{1}{2}\,\alpha_j^{(i-1)} \qquad (i=2,\ldots,d; j=1,\ldots,2^{i-1})$$

we get the symmetric Dirichlet density with parameter $2\alpha/n$.

2. A MODEL FOR THE SELECTION VARIABLES

A simple and operational density can be now derived from (1) if we make the following assumptions

$$\alpha_{2j-1}^{(i)} = \alpha_{2j}^{(i)} = \alpha(>0) \qquad (i=1,\ldots,k; j=1,\ldots,2^{i-1})$$

$$\alpha_{2j-1}^{(i)} = \alpha_{2j}^{(i)} = \tfrac{1}{2}\,\alpha_j^{(i-1)} \qquad (i=k+1,\ldots,d; j=1,\ldots,2^{i-1})$$

with the extreme cases, $k=1$ and $k=d$, corresponding to the symmetric Dirichlet and the prior density (4) in Kokolakis (1983), respectively. Under the above assumptions, we have the following density

$$g(\theta) \propto h(\theta) \equiv \{\prod_{i=1}^{n} \theta_i^{2\alpha/\nu - 1}\}/D(\theta), \qquad \theta \epsilon S^{(n)}, \tag{2}$$

where

$$D(\theta) = \{(\sum_{i=1}^{n/2} \theta_i)(\sum_{i=\frac{1}{2}n+1}^{n} \theta_i)\} \times \{(\sum_{i=1}^{n/4} \theta_i)\ldots(\sum_{i=3n/4+1}^{n} \theta_i)\} \times \ldots$$
$$\{(\sum_{i=1}^{\nu} \theta_i)\ldots(\sum_{i=n-\nu+1}^{n} \theta_i)\}^{\alpha}$$

and $\nu = 2^{d-k+1}$.

The posterior density of θ, given the sufficient statistic $\mathbf{r} = (r_1,\ldots,r_n)^T$, will be

$$p(\theta/\mathbf{r}) = c(\alpha;\mathbf{r})\{\prod_{i=1}^{n} \theta_i^{r_i + \alpha/\nu - 1}\}/D(\theta), \qquad \theta \epsilon S^{(n)}, \tag{3}$$

where the normalizing constant is given by

$$c(\alpha;\mathbf{r}) = \frac{\Gamma(\sum_{i=1}^{n} r_i + 2\alpha)}{\prod_{i=1}^{n} \Gamma(r_i + 2\alpha/\nu)} \times \{\frac{\Gamma(\sum_{i=1}^{n/2} r_i + 2\alpha)\Gamma(\sum_{i=\frac{1}{2}n+1}^{n} r_1 + 2\alpha)}{\Gamma(\sum_{i=1}^{n/2} r_i + \alpha)\Gamma(\sum_{i=\frac{1}{2}n+1}^{n} r_i + \alpha)} \cdots$$

$$\frac{\Gamma(\sum_{i=1}^{\nu} r_i + 2\alpha)\ldots\Gamma(\sum_{i=n-\nu+1}^{n} r_i + 2\alpha)}{\Gamma(\sum_{i=1}^{\nu} r_i + \alpha)\ldots\Gamma(\sum_{i=n-\nu+1}^{n} r_i + \alpha)}\} . \tag{4}$$

Furthermore, it can be shown that the chance of y_i being equal to one is *a priori* symmetrically distributed around ½ with variance

$$\mathrm{Var}[\Pr(y_i=1)] = \begin{cases} \dfrac{1}{4(1+2\alpha)} \left\{\dfrac{1+\alpha}{1+2\alpha}\right\}^{i-1} & (i=1,\ldots,k) \\ \dfrac{1}{4(1+2\alpha)} \left\{\dfrac{1+\alpha}{1+2\alpha}\right\}^{k-1} & (i=k+1,\ldots,d) \end{cases}$$

which is a decreasing function of i $(i=1,\ldots,k)$. This means that the $(i+1)th$ test, corresponding to the variable y_{i+1}, is *a priori* expected to be less useful that the *ith* test $(i=1,\ldots,k)$. Thus, the prior (2) is not invariant under permutations of tests. Specifically, using results by Kokolakis (1983, Section 2) we can see that the prior (2) is not invariant under permutations of the first k tests, but is invariant under permutations of the last $(d-k)$ tests.

Invariance can be now achieved by taking the mixture

$$p(\theta) \propto \frac{1}{N} \sum_{i=1}^{N} h\{Q_i(\theta)\}, \qquad \theta \epsilon S^{(n)}, \quad N = \Gamma(d+1)/\Gamma(d-k+1) = (d)_k\,,$$

where $Q_i(\theta)$ is a permutation of the θ's corresponding to the ordering $(i_1,\ldots,i_k)$ of k tests, from the set $\{1,\ldots,d\}$, considered first. The weights w_i occurring in the posterior mixture

$$p(\theta\,|\,\mathbf{r}) = \sum_{i=1}^{N} w_i\, c\{\alpha;Q_i(\mathbf{r})\} h\{Q_i(\theta)\} \prod_{i=1}^{n} \theta_i^{r_i}\,, \qquad \theta \epsilon S^{(n)}$$

which are given by

$$w_i = \frac{1/c\{\alpha;Q_i(\mathbf{r})\}}{\sum_{j=1}^{N} 1/c\{\alpha;Q_j(\mathbf{r})\}} \qquad (i=1,\ldots,N), \tag{5}$$

will reflect the degree to which the data agree with the *a priori* assumption that the $(i_{j+1})th$ test is less useful than the $(i_j)th$ test $(j=1,\ldots,k)$.

To make the argument clearer, let us consider the expression (5) for $\alpha=1$ and $i=1$. In this case, $Q_1(\mathbf{r})=(r_1,\ldots,r_n)$ and from (4) we get

$$w_1 \propto \frac{1}{(\sum_{i=1}^{n/2} r_i+1)(\sum_{i=(n/2)+1}^{n} r_i+1)} \times \frac{1}{(\sum_{i=1}^{n/4} r_i+1)\ldots(\sum_{i=3n/4+1}^{n} r_i+1)} \times \ldots$$

$$\times \frac{1}{(\sum_{i=1}^{\nu} r_i+1)\ldots(\sum_{i=n-\nu+1}^{n} r_i+1)}\,, \qquad (\nu=2^{d-k+1})\,.$$

We realize now that the product of every two successive terms in the denominator of the *jth* ratio above ($j=1,\ldots,k$) is approximately proportional to the sampling variance of the variable y_j given $y_1,y_2,\ldots,y_{j-1}$. Since the larger the sampling variance of a variable y_j, conditional on $y_1,y_2,\ldots,y_{j-1}$, the less useful for classification the variable is expected to be, it follows that the smaller the weight w_1 the less useful the ordering $(y_1,\ldots,y_k)$. Similar conclusions can be drawn for any other positive value of the hyperparameter α.

The usefulness of a subset of k variables $\{y_{s_1},\ldots,y_{s_k}\}$, say, $(s=1,\ldots,\binom{d}{k})$ would be expressed by the product $\prod_{i\in S} w_i$, where S the set of $k!$ elements from the set $\{1,\ldots,N\}$ which correspond to the $k!$ permutations of variables y_{s_j} $(j=1,\ldots,k)$.

3. THE SELECTION CRITERION AND RELATED DIVERGENCE MEASURES

We consider now the two-class classification problem. Let C_1 and C_2 be two classes with prior probabilities p_1 and p_2 respectively, $(p_1+p_2=1)$. Three sets of weights, namely $\{w_i^{(1)}\}$, $\{w_i^{(2)}\}$ and $\{w_i^{(0)}\}$ $(i=1,\ldots,N)$ can be now derived, the first two from the training data from C_1 and C_2, respectively, and the last one from the whole set of the training data considered as if they had been derived from the same class. In accordance with the interpretation of the within class weights, given in the previous section, we can here conclude that the larger the weight $w_i^{(0)}$ the less useful is the ordering $(y_{i_1},\ldots,y_{i_k})$ for classification.

We introduce now the following criterion for selecting the best k $(1\leq k<d)$ variables among the available d. For a subset of k variables $\{y_{s_1},\ldots,y_{s_k}\}$, say, from the set $\{y_1,\ldots,y_1\}$, we consider the ratio

$$R = \prod_{i\in S} \frac{\{w_i^{(1)}\}^{p_1}\{w_i^{(2)}\}^{p_2}}{w_i^{(0)}}, \tag{6}$$

where S is again the set of $k!$ elements from the set $\{1,\ldots,N\}$ which correspond to the $k!$ permutations of the subset $\{y_{s_1},\ldots,y_{s_k}\}$.

To obtain further insight into the criterion R, consider the logarithm of the general term in the product above for $i=1$ and $\alpha=1$. After introducing (4) and (5), we get

$$r = \log \frac{\{w_1^{(1)}\}^{p_1}\{w_1^{(2)}\}^{p_2}}{w_1^{(0)}} = c + \sum_{\ell=1}^{k} 2^{\ell} G_{\ell} \tag{7}$$

with

$$G_{\ell} = \log G(\mathbf{p}_{\ell}^{(0)}) - p_1 \log G(\mathbf{p}_{\ell}^{(1)}) - p^2 \log G(\mathbf{p}_{\ell}^{(2)}), \tag{8}$$

where $\mathbf{p}_{\ell}^{(j)}$ $(\ell=1,\ldots,k; j=0,1,2)$ is a 2^{ℓ}-dimensional probability vector whose elements are the flattened relative frequencies of the ℓ-dimensional training data from class C_j, with flattening constant $\alpha=1$: that is,

$$\mathbf{p}_{\ell}^{(j)} = \left(\frac{R^{(j)}_{\bar{y}_{s_1}\cdots y_{s_\ell}}+1}{m_j+2^{\ell}},\ldots,\frac{R^{(j)}_{\bar{y}_{s_1}\cdots y_{s_\ell}}+1}{m_j+2^{\ell}}\right)^T,$$

with $\bar{y}_i=1-y_i$ $(i=1,\ldots,d)$, and $G(\mathbf{p}) = \{\prod_{i=1}^{n} p_i\}^{1/n}$ the geometric mean of an n-

dimensional probability vector $\mathbf{p} = (p_1,\ldots,p_n)^T$. It can now be seen that, apart from a multiplicative constant, the result (8) is the Jensen difference between $\mathbf{p}_\ell^{(1)}$ and $\mathbf{p}_\ell^{(2)}$, with $\mathbf{p}_\ell^{(0)}$ in place of the mixture $p_1\mathbf{p}_\ell^{(1)} + p_2\mathbf{p}_\ell^{(2)}$, based on the ϕ-entropy functional H_ϕ with $\phi(x) = -\log x$. Specifically

$$2^\ell G_\ell = J_\phi^p(\mathbf{p}_\ell^{(1)},\mathbf{p}_\ell^{(2)}) = H_\phi(\mathbf{p}_\ell^{(0)}) - p_1 H_\phi(\mathbf{p}_\ell^{(1)}) - p_2 H_\phi(\mathbf{p}_\ell^{(2)})$$

with

$$H_\phi(\mathbf{p}) = -\sum_{i=1}^{n} \phi(p_i) = \sum_{i=1}^{n} \log p_i, \qquad \mathbf{p}\epsilon S^{(n)}.$$

We have therefore

$$r = c + \sum_{\ell=1}^{k} J_\phi^p(\mathbf{p}_\ell^{(1)},\mathbf{p}_\ell^{(2)}),$$

with the constant c being the same for all subsets of k variables and their permutations.

Several properties of information measures based on ϕ-entropy functionals can be found in Burbea and Rao (1982a, 1982b). It must be noticed here that when $\phi(x) = x\log x$ the Jensen difference $J_\phi^p(\mathbf{p}^{(1)},\mathbf{p}^{(2)})$ reduces to Lindley's measure of information provided by an experiment (Lindley, 1956),

$$L = H(\mathbf{p}^{(0)}) - p_1 H(\mathbf{p}^{(1)}) - p_2 H(\mathbf{p}^{(2)}), \tag{9}$$

where $H(\mathbf{p})$ is the Shannon entropy of the probability vector $\mathbf{p}$.

If we introduce in (8) the Kullback-Leibler directed divergences between a probability vector $\mathbf{p}\epsilon S^{(n)}$ and the uniform probability vector $\mathbf{u} = (\frac{1}{n},\ldots,\frac{1}{n})^T$, namely

$$I(\mathbf{p},\mathbf{u}) = \sum_{i=1}^{n} p_i \log \frac{p_i}{u_i} = -H(\mathbf{p}) + \log n$$

and

$$I(\mathbf{u},\mathbf{p}) = \sum_{i=1}^{n} u_i \log \frac{u_i}{p_i} = -\log G(\mathbf{p}) - \log n,$$

then (7) can be written

$$r = c + \sum_{\ell=1}^{k} 2^\ell L_\ell + \sum_{\ell=1}^{k} 2^\ell D_\ell \tag{10}$$

with L_ℓ given by (9) and

$$D_\ell = D(\mathbf{p}_\ell^{(0)}) - p_1 D(\mathbf{p}_\ell^{(1)}) - p_2 D(\mathbf{p}_\ell^{(2)}),$$

where

$$D(\mathbf{p}) = I(\mathbf{p},\mathbf{u})\text{-}I(\mathbf{u},\mathbf{p}) = \log G(\mathbf{p}) - H(p) + 2\log n, \qquad \mathbf{p}\epsilon S^{(n)}.$$

Since now $D(\mathbf{p}) = o(\|\mathbf{p}\text{-}\mathbf{u}\|^2)$, we can conclude that apart from extreme probability vectors our proposed criterion R and Lindley's measure of information L will behave similarly.

4. COMPARISON WITH OTHER CRITERIA

Two other familiar measures of divergence between two probability distributions, $\mathbf{p}^{(1)}$ and $\mathbf{p}^{(2)}$ say, used as criteria for the selection of variables are the following. The Jeffreys' divergence

$$J = \sum_{i=1}^{n} (p_i^{(1)} - p_i^{(2)})(\log p_i^{(1)} - \log p_i^{(2)})$$

and the Bhattacharyya coefficient $B = -\log\varrho$, where

$$\varrho = \left\{ \sum_{i=1}^{n} p_i^{(1)} p_i^{(2)} \right\}^{1/2}$$

is a measure of the affinity between two distributions.

TABLE 1

Ordering of pairs according to the criterion R and values of L, J, B and estimated probability of correct classification P

(i)	(ii)	(iii)	(iv)	(v)	(vi)
pairs	R	L	J	B	P
(y_3,y_7)	.957	.100	1.347	.534	.856
(y_7,y_8)	.954	.075	.820	.465	.856
(y_5,y_7)	.719	.080	.954	.480	.856
(y_2,y_7)	.687	.072	.805	.461	.856
(y_6,y_7)	.644	.074	.819	.465	.856
(y_4,y_7)	.643	.075	.845	.468	.856
(y_1,y_7)	.636	.068	1.092	.450	.856
(y_1,y_8)	.515	.059	.735	.451	.797
(y_3,y_8)	.452	.066	.802	.461	.803
(y_2,y_8)	.417	.037	.400	.408	.797
(y_6,y_8)	.411	.043	.468	.418	.797
(y_1,y_3)	.409	.072	.969	.479	.800
(y_4,y_8)	.372	.040	.438	.413	.797
(y_5,y_8)	.371	.041	.466	.416	.797
(y_1,y_5)	.278	.046	.600	.433	.741
(y_1,y_6)	.237	.054	.674	.443	.797
(y_3,y_4)	.202	.054	.626	.440	.782
(y_3,y_6)	.200	.056	.654	.443	.794
(y_1,y_4)	.192	.047	.541	.428	.762
(y_1,y_2)	.187	.044	.518	.425	.765
(y_2,y_3)	.184	.050	.573	.433	.768
(y_3,y_5)	.179	.053	.608	.438	.762
(y_5,y_6)	.122	.028	.312	.398	.753
(y_4,y_5)	.115	.023	.273	.392	.741
(y_2,y_6)	.112	.024	.257	.391	.753
(y_4,y_6)	.111	.026	.279	.394	.753
(y_2,y_5)	.105	.020	.215	.386	.741
(y_2,y_4)	.099	.019	.204	.384	.741

In Table 1 all possible pairs from 8 variables referring to 340 patients, 88 of which had cardiac pain and 252 other forms of chest pain, have been ordered according to the criterion R. The corresponding values of L, J and B, based the predictive distributions with priors of the form (2) with $\alpha=1$ and $k=d=2$, are given in columns (iii), (iv) and (v) respectively. Column (vi) gives the corresponding estimates of the probability of correct classification P using the training data as testing data. It can be seen that all criteria, although they do not agree completely in the ordering of pairs, succeed in picking out the same best pair, namely (x_3,x_7).

ACKNOWLEDGEMENTS

This paper is a part of the work carried out while visiting Nottingham University under an SERC Visiting fellowship. I wish to thank Professor A.F.M. Smith for inviting me and for the many challenging discussions I had with him.

REFERENCES

BURBEA, T. and RAO, C.R. (1982a). On the complexity of some divergence measures based on entropy functions. *IEEE, Trans. Inform. Theory, IT*-**28**, 489-495.

— (1982b). Entropy differential metric, distance and divergence measures in probability spaces: A unified approach. *J. Mult. Analysis,* **12**, 575-596.

KOKOLAKIS, G.E. (1983). A new look at the problem of classification with binary data. *The Statistician,* **32**, 144-152.

LINDLEY, D.V. (1956). On a measure of the information provided by an experiment. *Ann. Math. Statist.,* **27**, 986-1005.

BAYESIAN STATISTICS 2, pp. 681-696
J.M. Bernardo, M.H. DeGroot, D.V. Lindley, A.F.M. Smith (Eds.)

An Application of Non-linear Bayesian Forecasting to Television Advertising

H.S. MIGON and P.J. HARRISON
Warwick University

SUMMARY

This paper describes work which was carried out for a market research and survey company. The objective was to build a descriptive model relating television advertising to consumer awareness. The model has been used succesfully for two years both for retrospective and prospective analysis. Some aspects of its application are discussed in Colman and Brown (1983). The main models are particular members of an important class of models which is introduced. Linear and non-linear formulations are presented, and their performance assessed by application to a number of fast moving consumer goods.

Keywords: BAYESIAN FORECASTING; DYNAMIC LINEAR MODELS; DYNAMIC NON-LINEAR MODELS; KALMAN FILTERING; STOCHASTIC TRANSFER FUNCTIONS.

1. INTRODUCTION

The background to the particular problem of relating consumer awareness and television advertising is described in Section 2 of this paper. Earlier work by Broadbent (1979) is discussed.

Section 3 shows that extra facilities and improvements can be obtained within the class of Dynamic Linear Models, Bayesian Forecasting, (Harrison and Stevens, 1971 and 1976) and Kalman Filtering, (Kalman, 1963). Two linear models are developed. The first is called Broadym, since it introduces a dynamic element to Broadbent's model. The second, called the Local Linear Model, further extends Broadbent's concepts. After discussion and criticism of the linearity assumption, it is concluded that these models are not always satisfactory.

From a descriptive point of view it is clear that the response in consumer awareness to T.V. advertising is non-linear. Section 4 states the propositions on which the non-linear models are founded and lists the facilities required by the users. The effect of advertising and of consumer memory ageing are modelled and combined to produce a basic guide relationship on which the non-linear models are based. The observation model is discussed, as are measures for assessing the effectiveness of advertising and for comparing different advertisements.

Section 5 introduces an important general class of Dynamic Non-linear Models which the authors have used not only in this application but in short and long term forecasting and for the on-line identification of stochastic transfer functions; Migon (1984). An example of how a Binomially distributed observation and regional T.V. variation can be taken into account is illustrated by considering a linear guide relationship based upon the Broadym model. The two non-linear models are then described.

Section 6 shows applications of the models and gives a comparison between the performance of the linear and non-linear models over a set of fast moving consumer goods. The general conclusion is that the non-linear models provide a significantly better description of the relationship and give marked improvements in forecast performance.

2. BACKGROUND TO THE APPLICATION

In any market where a company has one or more brands supported by a substantial amount of advertising at least some of which is on television (T.V.), there is considerable benefit to be derived from setting up a vehicle for continuously monitoring the consumer response to advertising; see Millward Brown (1983). Such data is valuable for reviewing previous advertising; for correcting the tendency of many current pre-testing methods to give a systematically misleading feed-back to the creative team; and for leading to more rational advertising decisions. Measures of brand image, intention to buy and so on are often derived from the data. However, this paper is concerned with the relationship between consumer awareness and television advertising. Such advertising usually takes place in bursts so that the population consumer awareness varies considerably, creating a dynamic situation which merits frequent and perhaps continual study.

T.V. advertising is measured by JICTAR as weekly T.V. ratings (T.V.R.'s) over populations which are usually different from those in the consumer survey. They are taken over the same T.V. areas. A common measure for commercials of different lengths is required. This is usually standardised on 30 seconds, increasing the T.V.R.'s for longer advertisements and decreasing them for shorter ones in proportion to their costs. Further discussion of T.V.R.'s may be found in Broadbent (1979). Awareness may be measured in a variety of ways. For some products a sample of around one hundred people in selected regions of the country is taken by Millward Brown in many weeks. Each person may be shown a list of brands and each respondent asked 'which of these have you seen advertised on T.V. recently?' One awareness measure for a brand may be based upon the sampled number and the number of positive responses for that brand.

Broadbent discusses the relationship between awareness and T.V. advertising. Although he acknowledges the non-linearity, his operational methods use linear relationships and least squares. One of his main concepts is that of adstock which, at any time t, may be thought of as measuring the current worth of all the advertising effort that has been expended. Denoting the adstock at time t by a_t, the actual T.V.R.'s for week t by x_t, and a known discount factor by λ, where $0<\lambda<1$, Broadbent calculates adstock recursively by

$$a_t = \lambda a_{t-1} + x_t \tag{1}$$

Hence, in the absence of further advertising, the adstock decreases exponentially to zero as time increases. In practice x_t may be some function of the actual T.V.R.'s for weeks $t-1$ and t reflecting the sampling period with its timing variation and delayed effects. He then models the expected awareness as a linear function of adstock. For example, letting Y_t be the percentage awareness of the sample for week t, α and η be unknown constants, and ν_t a random error term with zero mean, the statistical model is

$$Y_t = \alpha + \eta a_t + \nu_t \tag{2}$$

Given historical values of the Y_t's as y_t's, λ and the x_t's, the pairs (y_t, a_t) are known. The two constants (α,η) are then calculated using ordinary least squares. This model has given satisfactory results in some cases, presumably those in which the adstock did not vary greatly. However, in many others it is far from satisfactory. The effect of advertising on awareness is essentially non-linear. Although linear models can capture local descriptions, the range over which such a description is adequate is often far too small to be of practical value. This can be partly overcome by making (α,η) dynamic as in the following Dynamic Linear Models. However, there are still major disadvantages in interpreting the linear model and great dangers in using such models to extrapolate to predict the outcome of proposed advertising campaigns. A number of other points also need to be made. Ordinary least squares is questionable in these applications since the random error term does not have a constant variance. Given a fixed sample size and simply arguing on a Binomially distributed response, it is seen that the variance associated with an awareness level of about 50 % is roughly three times as large as the variance associated with a 10 % or 90 % awareness level. Further, the variance obviously depends upon the sample size, varying roughly in proportion to it. This indicates that sequential dynamic regression with an appropriate variance law, should improve performance. The need for λ to be known is a disadvantage. So too is the fact that a change in the televised advertisement can lead to a major sudden change in awareness. The next section considers how some of these disadvantages can be partially or wholly overcome within the framework of Dynamic Linear Models.

3. DYNAMIC LINEAR MODELS

3.1. *General*

This section considers how, by reformulating Broadbent's models in terms of Bayesian forecasting using Dynamic Linear Models (D.L.M.'s), improvements can be obtained.

Proportional awareness varies between 0 and 1. Hence its relationship with adstock a_t can be thought of as non-linear. For example, for a given advertisement, a generally acceptable form of the relationship is as shown in Fig. 1.

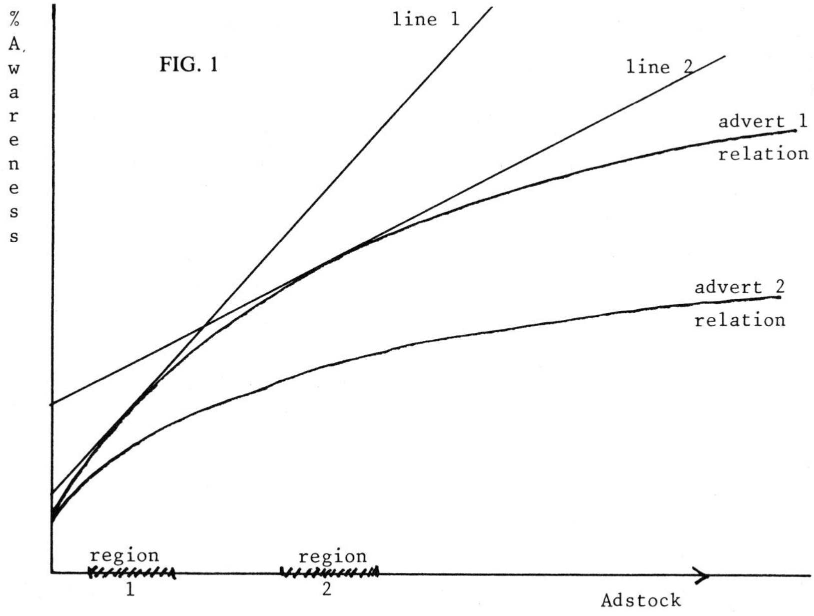

FIG. 1

Linear models approximate the graph by straight lines. In Broadbent's original model a single line is proposed. Its descriptive and predictive powers are obviously limited. With a Bayesian D.L.M. the appropriate line is continually updated so that in different regions of adstock, different lines are proposed. Since adstock varies relatively slowly some improvements can be obtained. When a totally different form of T.V. advertisement is used, this usually has the effect of completely altering the quantitative relationship as can be seen from fig. 1. The Bayesian approach can be very effective in assessing such major changes.

3.2. *The BROADYM Model*

It is a simple matter to extend Broadbent's original model to a dynamic model. Note that the original regression model of Section 2 can be written as

$$Y_t = \alpha + \eta a_t + \nu_t \ , \ \ \nu_t \sim N[0; V]$$

$$\begin{pmatrix} \alpha \\ \eta \\ a_t \end{pmatrix} = \begin{bmatrix} 1 & 0 & 0 \\ 0 & 1 & 0 \\ 0 & 0 & \lambda \end{bmatrix} \begin{pmatrix} \alpha \\ \eta \\ a_{t-1} \end{pmatrix} + \begin{pmatrix} 0 \\ 0 \\ x_t \end{pmatrix} .$$

The dynamic form of the model is then

$$Y_t = \alpha_t + \eta_t a_t + \nu_t \ , \ \ \nu_t \sim N[\bar{\nu}_t; V_t]$$

$$\begin{pmatrix} \alpha \\ \eta \\ a \end{pmatrix}_t = \begin{bmatrix} 1 & 0 & 0 \\ 0 & 1 & 0 \\ 0 & 0 & \lambda \end{bmatrix} \begin{pmatrix} \alpha \\ \eta \\ a \end{pmatrix}_{t-1} + \begin{pmatrix} w_1 \\ w_2 \\ x \end{pmatrix}_t$$

where $\begin{pmatrix} w_1 \\ w_2 \end{pmatrix}_t \sim N[\bar{\mathbf{w}}_t; \mathbf{W}_t]$, and x_t is the known T.V.R. measure for week t.

The variance V_t may more properly reflect the sampling situation by setting $V_t = qf_t(1-f_t)$ where $f_t = E[Y_t | D_{t-1}]$ is the expected awareness for time t based upon the information D_{t-1} up to time $t-1$; q is either specified as a constant, a function of the sample number, or it is estimated on line. Except at times of intervention, $\bar{\nu}_t$ and $\bar{\mathbf{w}}_t$ are zero. One important use of changing $(\bar{\mathbf{w}}_t, \mathbf{W}_t)$ is when a completely new type of advert replaces an existing one. The variance $\mathbf{W}_t$ can be enlarged to reflect the extra uncertainty that then exists concerning the effect of the new advertisement and if required $\bar{\mathbf{w}}_t$ can convey the expected 'improved effect' of the new advertisement as estimated by a marketing department. This is a simple example of stochastic transfer effects and their estimation is dealt with in Migon (1984).

Details of the formulation of this model using the Normal Discount Bayesian Models (N.D.B.M.'s) of Ameen and Harrison (1984) is given in the Appendix and Discount Bayesian Regression Models are discussed in Harrison and Johnston (1983).

3.3. *The Local Linear Model*

A weakness of the previous models is the need to pre-specify the unknown decay factor λ. It is an advantage if λ can be estimated on-line and possibly be allowed to vary with time and type of advertisement. Taking equations (1) and (2) of Broadbents original model, given in Section 2, it is easily seen that

$$Y_t - \lambda Y_{t-1} = \mu + \eta x_t + \nu_t - \lambda \nu_{t-1}$$

where $\mu = (1-\lambda)\alpha$. This leads to the proposal of the D.L.M. $\{\mathbf{F}_t, \mathbf{I}, V_t, \mathbf{W}_t\}$ where $\mathbf{F}_t = (1, x_t, y_{t-1}, e_{t-1})$ and $\theta_t' = (\mu,\eta,\lambda,\gamma)_t$. The model may be written in full as

$$Y_t = \mu_t + \eta_t x_t + \lambda_t y_{t-1} + \gamma_t e_{t-1} + \nu_t$$

$$\begin{pmatrix} \mu \\ \eta \\ \lambda \\ \gamma \end{pmatrix}_t = \begin{pmatrix} \mu \\ \eta \\ \lambda \\ \gamma \end{pmatrix}_{t-1} + \mathbf{w}_t$$

where e_t is the usual one step ahead forecast error, $\nu_t \sim N[\bar{\nu}_t; V_t]$ and $w_t \sim N[\bar{\mathbf{w}}_t; \mathbf{W}_t]$. Again $\bar{\nu}_t$ and $\bar{\mathbf{w}}_t$ are zero apart from at times of intervention and V_t reflects the binomial type variance law. Generally $\mathbf{W}_t$ is a small variance matrix allowing the parameter vector θ to vary through time. However, when exceptional events occur, such as changes in the form of advertisement, the appropriate elements of $\mathbf{W}_t$ are enlarged to provide a rapid means of estimating the effect of the event.

3.4. *Comments*

The D.L.M.'s can adapt well to slow changes in the intercept and slope of the local line but obviously, from fig. 1, their extrapolative validity is unsatisfactory, as is their global description of the true relationship. Nevertheless for short-term description and short term retrospection, the model will be reasonable for the greater part of the time. Further, the models are designed to deal quickly with sudden major changes in awareness, which for example, may occur with a qualitative change in advertising. Referring to fig. 1, if advert 1 is to be replaced by advert 2 whose awareness response form is unknown, a signal of the new advertisement can raise the uncertainty associated with this response and provide a very fast means of adaptation. Even so comparison between adverts is not easy. If linear characterisations are used, the estimated difference may be confounded by differing amounts of expenditure on the two adverts. Missing observations cannot be properly handled by many classical time series models but, as is well known, Bayesian methods have no difficulty with them. Further observed awareness is more properly modelled based on the Binomial distribution. The varying variance V_t can be designed to reflect this and provided the sample is sufficiently large a Normal distributions can be assumed for Y_t. However, for smaller samples and either low or high awareness levels it may be desired that a Binomial Distribution is used. An example of such a linear model is given as a particular case of the non-linear models in section 5.3. Further, that example demonstrates how the extra variation due to sampling in what might often be an unrepresentative T.V. region can be modelled.

4. THE NON-LINEAR MODELS

4.1. *Assumptions*

In this particular T.V. application, the non-linear models have been developed based on the propositions that there is (i) a diminishing return rate from increased advertising; (ii) an exponential decay in awareness; (iii) a threshold awareness in the absence of T.V. advertising; (iv) a maximum (or asymptotic) level of awareness; (v) a sampling variation which is dependent upon the awareness level and also the sample size; (vi) a general slow change in the quantities describing (i) to (v); (vii) in response to a major change in advertisement, a likelihood of a sudden marked change in the awareness response.

4.2. *Required Facilities*

In addition to providing estimates of the model parameters, after discussion, the following facilities were requested by the Market Research Company: (i) a capability for handling weeks when no sample data is collected; (ii) a means of including subjective information so that (a) initially modelling and prediction can begin with no data, (b) at any time the parameter estimates and their associated uncertainties can be changed; (iii) an ability to respond quickly to signalled changes in advertisement and to measure the effect of such changes; (iv) a 'what if' facility for obtaining an estimate of a projected advertising campaign (and also for assessing model adequacy); (v) a filtering facility for estimating the underlying weekly awareness for week t based upon data up to and possibly including week $t+5$. (This filters the random variation and aims to give a smoother retrospective picture of underlying awareness.)

4.3. *Notation*

In what follows: t denotes week t; μ_t the threshold awareness; $\mu_t + d_t$ the asymptotic awareness; λ_t the memory decay rate; x_t the number of T.V.R.'s in week t (possibly delayed); N_t the number of people sampled in week t; u_t the number of people 'aware' in week t; p_t the proportional awareness in the population; $Y_t = 100u_t/N_t$ the observed percentage awareness in week t; E_t the effect of T.V. advertising up to and including time t on the awareness at time t; k_t the advertising response factor.

4.4. *The Assumed Relationship Between Awareness and Advertising*

(i) The Effect of Advertising: Consider a moment of time at which the awareness *effect* E is derived from an adstock a, then the hypothesis of diminishing returns expresses the relationship as

$$E = d\{1-\exp(-k\,a)\}$$

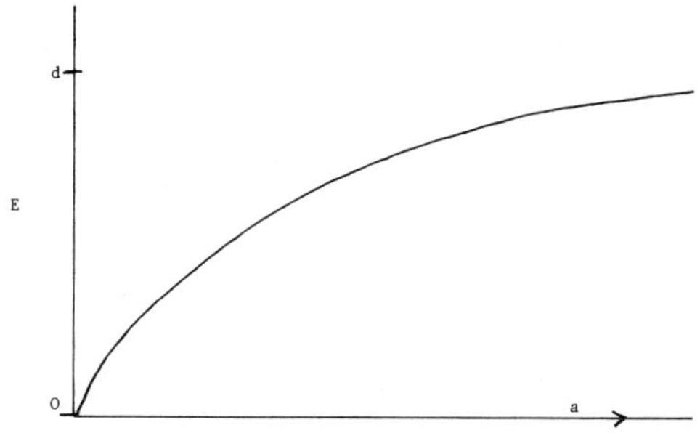

(ii) The Decay of Awareness: At time t denote the time effect of a single advertising campaign in week $t=0$ by E_t. Assuming that after this week, $t=0$, there is no further advertising, it is hypothesised that the awareness (not the adstock) decays exponentially to some threshold value, so that E_t decays to zero. The decay rate is denoted as λ where $0 \leq \lambda < 1$ giving

$E_t = \lambda^t . E_0$ where E_0 is the initial response.

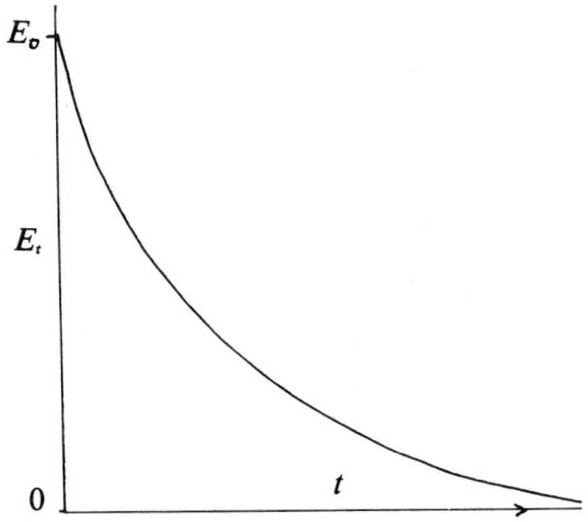

The consequence of (i) and (ii) is that, in the absence of an advertising input in week t, the adstock reduces from a_{t-1} to a_t where the quantities are related according to

$$a_t = \frac{1}{k} \log[1 - \lambda(1 - \exp(-ka_{t-1}))]$$

Hence, in these models adstock does not reduce exponentially with time: it is the memory *effect* of adstock that decays exponentially.

(iii) The Population Awareness: The population awareness p_t is then modelled using a derived guiding relationship

$$p_t \cong \mu_t + d_t - [d_t - \lambda_t E_{t-1}] \exp(-k_t x_t) .$$

Since the weekly awareness sample is taken in selected T.V. regions the relevant regional awareness ψ_t will vary around p_t and is modelled as

$$\psi_t \sim [p_t; \sigma^2] .$$

This can be expressed as $\psi_t = p_t + \delta$ with $\delta \sim [0; \sigma^2]$.

(iv) The Observed Awareness: Letting N_t people be sampled in week t, the number of aware people U_t is modelled as a conditional Binomial

$$(U_t | \psi_t, N_t) \sim \text{Binomial}[N_t, \psi_t] .$$

If N_t is sufficiently large, as is often the case, both U_t and the percentage awareness Y_t can be well approximated by a Normal Distribution with

$$(Y_t | \psi_t, N_t) \sim [100\psi_t / N_t ; 10^4 \psi_t (1 - \psi_t)/N_t]$$

5. THE DYNAMIC NON-LINEAR MODELS

5.1. *General*

In this section a useful class of Dynamic Non-Linear Model is introduced. An illustrative example is given applied to the linear Broadym model and then the applied non-linear models are described.

5.2. *A Class of Dynamic Non-Linear Model*

Let θ be a parameter vector with $(\theta_{t-1}|D_{t-1}) \sim [\mathbf{m}_{t-1}; \mathbf{C}_{t-1}]$. A specified Dynamic Model, which may be non-linear, applied to this posterior distribution gives the prior $(\theta_t|D_{t-1}) \sim [\hat{\theta}_t; \mathbf{R}_t]$. Let ψ be a scalar parameter related to θ through a stated 'guide relationship' such that

$$\left(\begin{matrix}\psi_t \\ \theta_t\end{matrix} \;\middle|\; D_{t-1}\right) \sim \left[\begin{matrix}\hat{\psi}_t \\ \theta_t\end{matrix} ; \begin{matrix}r & \mathbf{r} \\ \mathbf{r}' & \mathbf{R}\end{matrix}\right] \sim \left[\hat{\phi}_t ; \mathbf{Q}_t\right].$$

The guide relationship may be deterministic or stochastic as for example when introducing 'between region variation' in the T.V. survey work.

$$\hat{\psi}_t = E[g(\theta_t)|D_{t-1}];\; r = V[g(\theta_t)|D_{t-1}];\; \mathbf{r}' = \text{cov}[g(\theta_t),\theta_t|D_{t-1})]\,.$$

The marginal distribution of ψ_t is defined by the modeller and hence is not necessarily Normal. Often it will be taken to have a conjugate form to that of the observation r.v. Y_t or be such that a bijective function of ψ_t has the conjugate distribution. However conjugacy is not necessary. In many non-linear models the higher moments of θ are needed and often we take the marginal distribution of θ_t as Normal. The information that Y_t gives on θ_t is conveyed only through ψ and the posterior distribution $(\psi_t|y_t\, D_{t-1})$ and is derived using Bayes Theorem as

$$\text{Log}\, P(\psi_t|y_t\, D_{t-1}) = S(\psi_t|y_t) + \text{Log}\, P(\psi_t|D_{t-1}) + \text{const.}$$

where

$S(\psi_t|y_t)$ is the log likelihood of ψ_t given y_t.

Let the posterior mean and variance of $(\psi_t|y_t\, D_{t-1})$ be

$$E[\psi_t|D_t] = m$$

$$V[\psi_t|D_t] = \sigma_{11}\,.$$

The information is then conveyed back to θ_t so that

$$\left(\begin{matrix}\psi_t \\ \theta_t\end{matrix} \;\middle|\; D_t\right) \sim \left[\begin{matrix}m \\ \mathbf{m}_t\end{matrix} ; \begin{matrix}\sigma_{11} & \cdot \\ \cdot & C_t\end{matrix}\right] \sim \left[\mathbf{m};\; \Sigma\right]$$

where

$$(\theta_t|D_t) \sim [\mathbf{m}_t; \mathbf{C}_t] \text{ and}$$

$$\mathbf{m}_t = \hat{\theta}_t + \mathbf{r}'(m-\hat{\psi}_t)/r$$

$$\mathbf{C}_t = \mathbf{R}_t - \mathbf{r}'\,\mathbf{r}(r-\sigma_{11})/r^2\,.$$

The original basis of this pair $(\mathbf{m}_t; \mathbf{C}_t)$ is given below and an alternative for Generalized Linear Models is to be found in West, Harrison and Migon (1985).

(i) $\mathbf{m}$ is that value of $\alpha = (\alpha_1 \ldots \alpha_{n+1})'$ which minimizes

$$(\hat{\phi}_t - \alpha)' \, \mathbf{Q}_t^{-1} (\hat{\phi}_t - \alpha)$$

subject to the constraint $\alpha_1 = m$

and

(ii) $\Sigma = [\sigma_{ij}]$ is that variance matrix $\mathbf{M}$ which minimizes

$$\left| \ell' \left(\mathbf{M}^{-1} - \left(\mathbf{Q}_t^{-1} + \begin{bmatrix} \sigma_{11}^{-1} - r^{-1} & \mathbf{0} \\ \mathbf{0}' & \mathbf{0}_n \end{bmatrix}\right)\right) \ell \right|$$

$\forall\, \ell \in \mathbb{R}^{n+1}$ subject to the lead element being σ_{11}.

5.3. *Example*

A simple illustration of the procedure is an application to linear models and in particular to Broadym.

(i) Let $(\theta_{t-1} | D_{t-1}) \sim N[\mathbf{m}_{t-1}; \mathbf{C}_{t-1}]$

then $(\theta_t | D_{t-1}) \sim N[\mathbf{G}\, \mathbf{m}_{t-1}; \mathbf{R}_t]$

which is obtained via the relationship $\theta_t = \mathbf{G}\, \theta_{t-1} + \mathbf{w}_t, \mathbf{w}_t \sim N[\mathbf{0}; \mathbf{W}]$.

$p_t \cong \mathbf{L}\, \theta_t$ for a known vector $\mathbf{L}$ and $\psi_t \cong p_t + \delta$ with $\delta \sim [0; \sigma^2]$.

This gives the joint distribution

$$\begin{pmatrix} \psi_t \\ \theta_t \end{pmatrix} \Bigg| D_{t-1} \sim \left[\begin{matrix} \hat{\psi}_t \\ \hat{\theta}_t \end{matrix} ; \begin{bmatrix} r & \mathbf{r} \\ \mathbf{r}' & \mathbf{R}_t \end{bmatrix} \right]$$

with $\hat{\psi}_t = \mathbf{L}\, \hat{\theta}_t = \mathbf{L}\, \mathbf{G}\, \mathbf{m}_{t-1}; r = \mathbf{L}\, \mathbf{R}_t\, \mathbf{L}' + \sigma^2; \mathbf{r} = \mathbf{L}\, \mathbf{R}_t$

(ii) Let the marginal distribution $(\psi_t | D_{t-1}) \sim \text{Beta}(s,t)$

where $s = \hat{\psi}_t[\hat{\psi}_t(1 - \hat{\psi}_t) - r]/r$

$t = s(1 - \hat{\psi}_t)/\hat{\psi}_t$

ensures that the first two moments are $(\hat{\psi}_t; r)$

and $P(\psi_t = \psi | D_{t-1}) \propto \psi^{s-1} (1 - \psi)^{t-1}$.

(iii) Let $(Y_t | \psi_t, N) \sim \text{Binomial}(N, \psi_t)$

so that the likelihood of ψ_t given $Y_t = y$ is

$$L(\psi_t | y, N) \propto \psi_t^y (1 - \psi_t)^{N-y}$$

(iv) The posterior distribution for ψ_t is

$$(\psi_t | D_t) \sim \text{Beta}(s_1, t_1)$$

where $s_1 = s + y$; $t_1 = t + N - y$ so that the mean and variance are

$$m = s_1/(s_1 + t_1) \text{ and } \sigma_{11} = m(1-m)/(s_1 + t_1 - 1).$$

(v) The updating for θ_t is then

$$(\theta_t | D_t) \sim N[\mathbf{m}_t; \mathbf{C}_t],$$

$$\mathbf{m}_t = \hat{\theta}_t + \mathbf{r}'(m - \hat{\psi}_t)/r,$$

$$\mathbf{C}_t = \mathbf{R}_t - \mathbf{r}'\, \mathbf{r}(r - \sigma_{11})/r^2.$$

(vi) With respect to Broadym, as given in 3.2, we may take

$$\theta' = (\alpha,\eta); \qquad \mathbf{L} = \mathbf{F} = (1,a_t); \mathbf{G} = \mathbf{I} \quad ; p_t = \alpha_t + \eta_t\, a_t$$

and a_t as given. The random variable δ_t introduces any required extra variation arising from the sampled T.V. region being unrepresentative of the country as a whole. Referring to the Appendix Discount formulation, **w** is dropped and $\mathbf{R}_t$ is calculated via a discount factor β as C_{t-1}/β.

5.4. *The Applied Models*

Following 4.4 (iii) the guide relationship for the regional proportional awareness ψ_t is to be taken as

$$g(\theta_t) \cong \mu_t + d_t - [d_t - \lambda_t E_{t-1}] \exp(-k_t x_t) + \delta_t$$

where

$$\theta_t' = (\mu_t, d_t, \lambda_t, k_t, E_{t-1}), \delta_t \sim [0; \sigma^2].$$

The corresponding guide relationship for E_t is

$$E_t \cong d_t - [d_t - \lambda_t E_{t-1}] \exp(-k_t x_t).$$

With the Normality assumption for the marginal distribution for θ_t there is no technical problem since the first two moments of $g(\theta_t)$ can be calculated. However it was requested that two models be constructed:

Model 1: here the value of the asymptote $\mu_t + d_t =$ constant, is to be specified.

Model 2: here the value k_t is to be assumed known.

For model 1 the condition is easily satisfied by appropriate settings of the initial covariance matrices and innovation matrices $\mathbf{W}_t$. However, for computational ease, in Model 1, a Taylor Series Expansion is used just in updating, so that for that computer routine we use

$$\exp(-k_t x_t) \cong [1 - x_t(k_t - \hat{k}_t)] \exp[-\hat{k}_t x_t]$$

where $\hat{k}_t$ is the prior expectation for k_t given D_{t-1}.

The guide relationship can be written as

$$g(\theta_t) \cong \mu_t + E_t + \delta_t.$$

In both models the link between $(\theta_t | D_{t-1})$ and $(\theta_{t-1} | D_{t-1})$ is

$$\theta_t = \begin{bmatrix} \mathbf{I}_4 & 0 & 0 \\ 0 & 0 & 1 \end{bmatrix} \begin{pmatrix} \theta_{t-1} \\ E_{t-1} \end{pmatrix} + \mathbf{w}_t.$$

$\left(\begin{matrix} E_t \\ \theta_t \end{matrix} \middle| D_{t-1}\right)$ is then derived using the above guide relationship for E_t and $\psi_t = g(\theta_t)$ provides $\left(\begin{matrix} \psi_t \\ E_t \\ \theta_t \end{matrix} \middle| D_{t-1}\right)$. The observation model may be specified as Binomial as in example 5.3 or as Normal using a variance governed by the Binomial characterisation. Following the principles described in the sections dealing with linear models all the requested facilities as given in 4.2 are provided.

Full details are given in Migon (1984) where further classes of non-linear model are developed. In obtaining the Forecast function which provides future point estimates of consumer awareness the full non-linear form $g(\hat{\theta}_t) = \hat{\mu}_t + d_t - [d_t - \lambda_t E_{t-1}] \exp(-k_t x_t)$ is used with modal (maximum likelihood) estimates of θ_t being used for all future point estimates.

6. EXAMPLES AND COMPARISON

A number of examples of the application of the models are given in Colman and Brown (1983) where they are discussed from a practitioners viewpoint. Estimates of some of the parameters as they evolve throughout time and of the effectiveness of different types of advertisement may be found in that paper. In this paper we concentrate on comparing the relative performance of the linear and non-linear models. It is shown that, over a sample of products, there is little to choose between the two linear models and that there is also little difference between the two non-linear models. However, the predictive performance and the descriptive power of the non-linear models is seen to be far better than that of the linear models.

In fig. 2 the data on product *P5* is given. This is a fast moving consumer good. Both the observed weekly percentage consumer awareness and the weekly T.V.R.'s are plotted for a period of more than three years. Periods of missing observations, which correspond to no awareness samples, are indicated. Plotted also are the one week ahead point predictions derived using model 1. The decay in awareness during periods of no advertising is evident on a number of occasions and the sharp non-linear response to an advertising burst can also be appreciated. For this product, there are five periods of missing observations.

FIG. 2

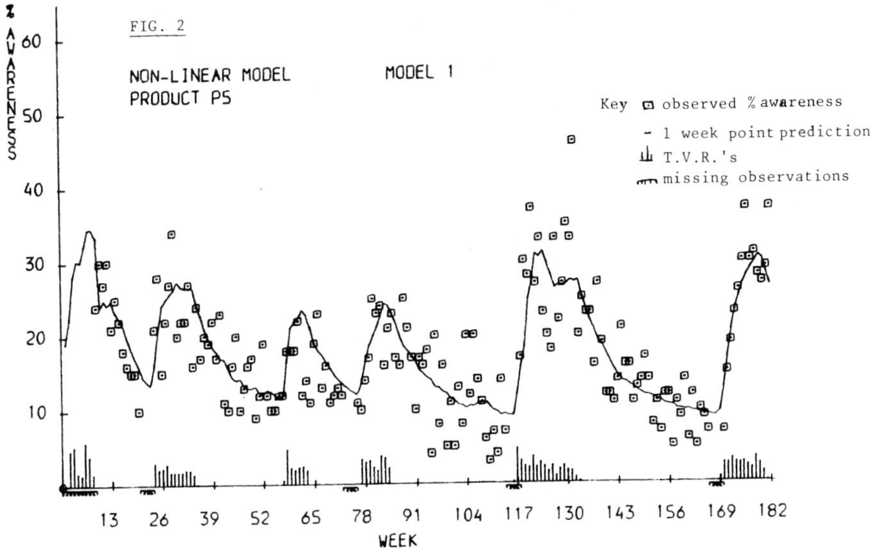

A comparison between the Local Linear model and the non-Linear model 1 is shown in fig. 3 using data on Product *P2*. The point predictions are for four weeks ahead. Fig. 4 shows a similar comparison for the linear model Broadym and the non-linear model 2. The data here relates to a chocolate bar which is also examined in Colman and Brown (1983) where later data is included. The interesting feature of this product is the use of three distinct advertising campaigns. The first advertisement was successful in generating consumer awareness. However in week 96 it was replaced by a qualitatively different advertisement which unfortunately was soon assessed by the models to be relatively poor.

In fact this latter advertisement was estimated to require almost 5 times as many T.V.R.'s as the former advertisement in order to achieve the same increase in awareness response. In turn this advertisement was replaced in week 158 by another qualitatively different advertisement that was estimated to improve the awareness response over its immediate predecessor by a factor of about 3.

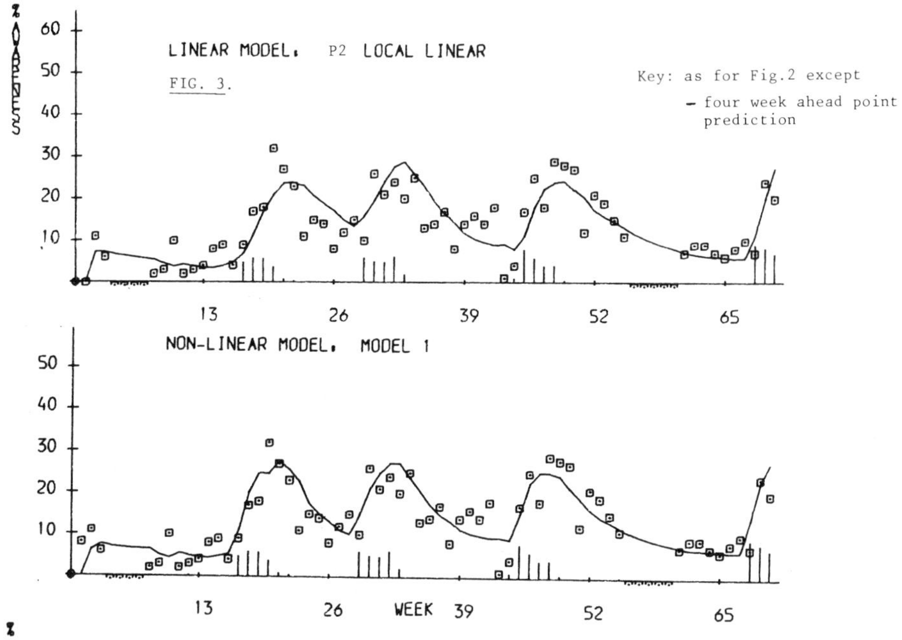

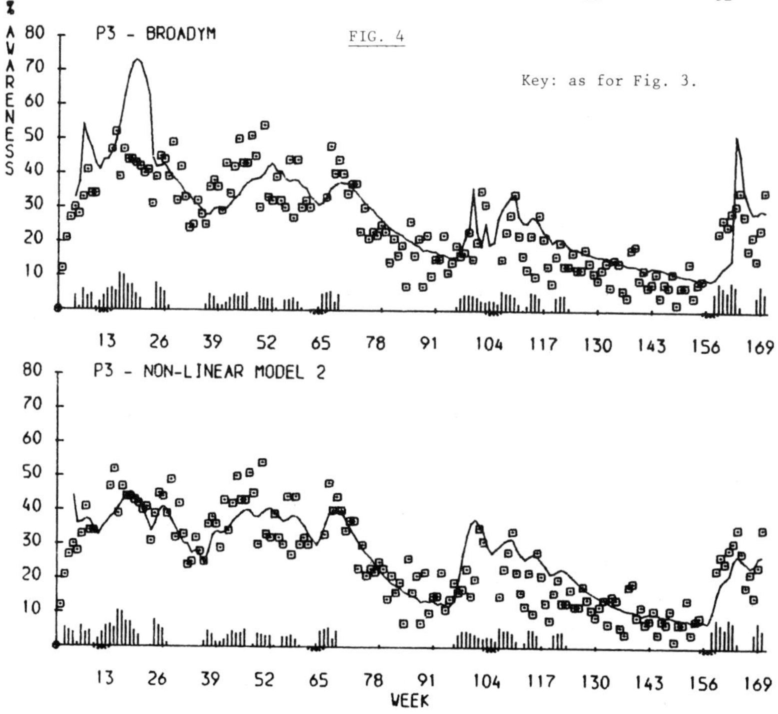

FIG. 5. P2: Point Prediction made at week 39 for the next 31 weeks

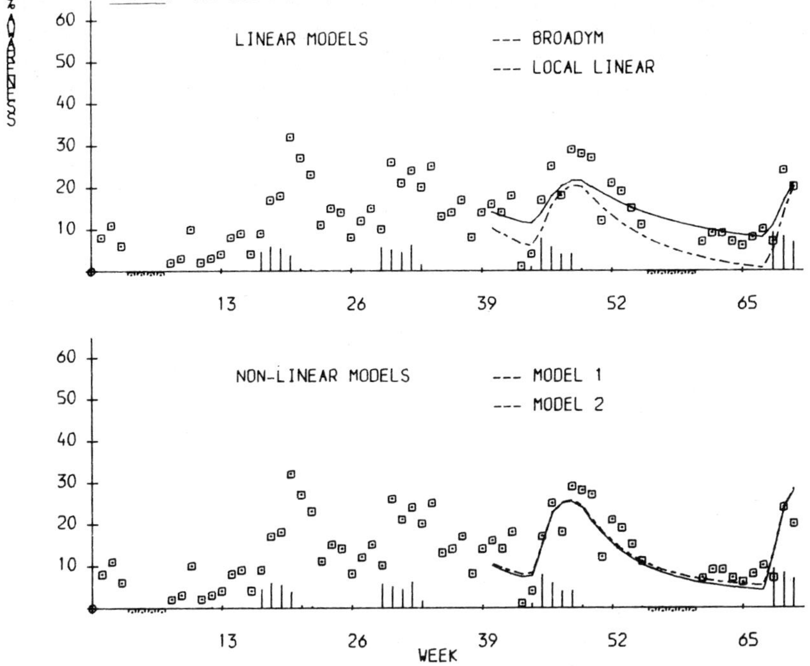

FIG. 6. P5: Point Prediction made at week 129 for the next 53 weeks

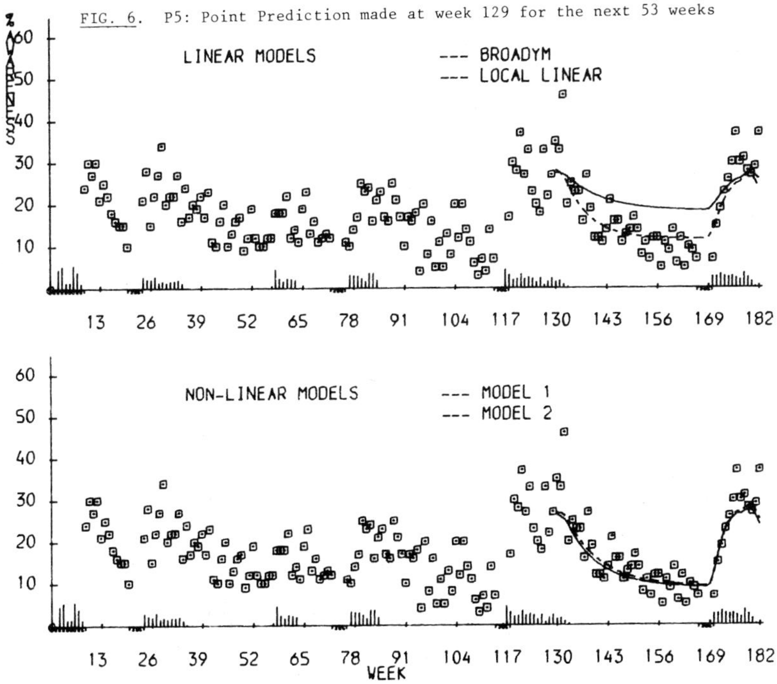

Figures 5 and 6 show the extrapolative inadequacy of the linear models. In Fig. 5 product *P2*, which is not a chocolate product, is shown with the point predictions made at week 39 for the future advertising programme over the next 31 weeks. Clearly compared to the performance of the non-linear model, the linear model fails to give a satisfactory prediction. Similarly in fig. 6 with product *P5*, the point predictions made at week 129 for the future advertising programme over the next year of 53 weeks is given. Again the performance of the linear model is totally unsatisfactory and shows that for 'what if' analyses the linear models are not to be recommended.

Table 1 gives a comparison of the four models relative to the performance of non-linear model 1. The criteria is an inverted Normal loss function which in this comparison gives results, when expressed as integers, identical to that derived using a quadratic loss function.

The performance for four steps ahead using non-linear model 1 was also assessed without signalling the change in the quality of advertisement by increasing the uncertainty of the relevant parameters. The loss was 32 % greater showing the large benefits to be gained by considering the response to different advertisements as stochastic.

TABLE 1. A COMPARISON OF THE FOUR MODELS

The tabulated entries are the percentage loss in excess of that incurred by model 2, using an inverted Normal loss function

One week ahead point forecast performance

Product	Broadym	Local Linear	Model 2	Model 1
P1	35	47	2	0
P2	21	26	5	0
P3	18	9	1	0
P4	2	-1	-5	0
P5	8	20	6	0
MEAN	17	20	2	0

Four week ahead point forecast performance

Product	Broadym	Local Linear	Model 2	Model 1
P1	81	80	0	0
P2	31	41	11	0
P3	56	15	-7	0
P4	4	21	-9	0
P5	31	32	3	0
MEAN	41	38	0	0

The models have been used effectively since their developments in 1981. One of the uses of the awareness estimates has been as an input to models relating sales and advertising. The resulting significant relationships are discussed in Colman and Brown (1983).

APPENDIX

A N.D.B.M. Broadym model may be written as

$$Y_t = \alpha_t + \eta_t a_t + \nu_t \qquad \nu_t \sim N[0; V_t]$$

Let the posterior distribution $\begin{pmatrix}\alpha\\ \eta\end{pmatrix}_{t-1} D_{t-1} \sim N\left[\mathbf{m}_{t-1}; \mathbf{C}_{t-1}\right]$.

Let $0 < \beta \leq 1$, be an appropriate discount factor, then the model is

$$\begin{pmatrix}\alpha\\ \eta\end{pmatrix}_t D_{t-1} \sim N\left[\mathbf{m}_{t-1}; \mathbf{C}_{t-1}/\beta\right].$$

Defining $a_t = \lambda a_{t-1} + x_t$, $\mathbf{F} = (1, a_t)$ and $R = \mathbf{C}_{t-1}/\beta$, the one step ahead forecast distribution is

$$(Y_t | x_t, D_{t-1}) \sim N[\hat{y}_t; \hat{Y}_t]$$

$$\hat{y}_t = \mathbf{F}\,\mathbf{m}_{t-1};\ \hat{Y}_t = \mathbf{F}\,\mathbf{R}\,\mathbf{F}' + V_t.$$

Writing $\mathbf{A}_t = \mathbf{R}\,\mathbf{F}'/\hat{Y}_t$ and $e_t = y_t - \hat{y}_t$, where y_t is the observed percentage awareness at time t, the necessary recurrence relationships are

$$\begin{pmatrix}\alpha\\ \eta\end{pmatrix}_t D_t \sim N[\mathbf{m}_t; \mathbf{C}_t]$$

with

$$\mathbf{m}_t = \mathbf{m}_{t-1} + \mathbf{A}_t e_t$$

$$\mathbf{C}_t = \mathbf{R}\,[\mathbf{I} - \mathbf{F}'\mathbf{A}'].$$

If variance learning is required let N_t be the number sampled at time t, u_t the number aware and $\hat{y}_t$ the expected percentage of people aware. Then a variance law can be formulated as

$$V_t = q_{t-1}\hat{y}_t(100 - \hat{y}_t)/N_t$$

where

$q_t = S_t/n_t$ is recurrently calculated as

$n_t = n_{t-1} + N_t$

$S_t = S_{t-1} + d_t^2.$

$d_t = (y_t - \hat{y}_t)\sqrt{N_t} \,/\, \sqrt{y_t(100 - y_t)}$

and $y_t = 100u_t/N_t$.

Further details of variance learning can be found in West (1982), Ameen and Harrison (1984) and Migon (1984).

ACKNOWLEDGEMENTS

We would to thank the Universidade Federal do Rio de Janeiro and C.N.P.q Brazil for generously supporting the researches of Helio Migon. In addition we wish to acknowledge the Millward Brown company who initially posed and helped to formulate the applied problem. In particular we would thank Gordon Brown, as well as Sue Heenan and Ann Fairburn.

REFERENCES

AMEEN, J.R.M. and HARRISON, P.J. (1984). Normal Discount Bayesian Models. This volume.

BROADBENT, S. (1979). One way T.V. advertisements work. *J. Mkt. Res. Soc. Vol. 21,* **3**, 139-165.

COLMAN, S. and BROWN, G. (1983). Advertising tracking studies and sales effects. *J. Mkt. Res. Soc. Vol. 25,* **2**, 165-183.

HARRISON, P.J. and JOHNSTON, F.R. (1984). A regression method with non stationary parameters. *J. Op. Res.* (To appear).

HARRISON, P.J. and STEVENS, C.F. (1976). Bayesian Forecasting. *J. Roy. Statist. Soc. B* **38**, 205-247 (with discussion).

KALMAN, R.E. (1963). New methods in Wiener filtering. *Proc. first symposium on engineering application of random function theory and probability.* (J.L. Bogdanoff and F. Kozin, eds.). New York: Wiley.

MIGON, H.S. (1984). An approach to non-linear Bayesian Forecasting problems with applications. Ph.D. thesis, University of Warwick.

MILLWARD, B. (1983). Advertising Tracking Studies. Research report, Millward Brown, Ince House, 60, Kenilworth Road, Leamington Spa, CV32 6JY.

WEST, M. (1982). Aspects of recursive Bayesian estimation. Ph.D. thesis, University of Nottingham.

WEST, M. HARRISON, P.J. and MIGON, H.S. (1985). Dynamic Generalized Linear Models and Bayesian Forecasting. *J. Amer. Statist. Assoc.* (To appear).

BAYESIAN STATISTICS 2, pp. 697-710
J.M. Bernardo, M.H. DeGroot, D.V. Lindley, A.F.M. Smith (Eds.)

Shoulders in Hierarchical Models

A. O'HAGAN
University of Warwick

SUMMARY

The paradoxical behaviour of posterior modes, particularly in hierarchical models, has led to disenchantment with them as measures of location or as approximations to posterior means. We argue that to interpret a quantity such as mean or mode as a measure of location presupposes an appreciation of the shape of the posterior density. We examine a particular example of one-way analysis of variance. When the shape of the posterior joint density is analysed, and in particular when we recognise the presence of the feature we call a shoulder, the behaviour of the modes is readily understood.

We see this work as a small contribution to the study of summarisation, which is the process of translating a complicated density function into more comprehensible statements of beliefs.

Keywords: HIERARCHICAL MODELS; ONE-WAY ANALYSIS OF VARIANCE; MODES; SHAPE; SHOULDERS; SUMMARISATION.

1. SUMMARISATION AND SHAPE

A Bayesian analysis of data focuses on the formulation of a posterior distribution usually by applying Bayes' theorem. If required, the next step is to derive formal inferences from the posterior by the use of decision theory. It is well known, for instance, that the optimal estimator under squared-error loss is the mean, therefore a posterior mean is an example of formal inference. The literature is full of both theoretical and practical work in which the computation of posterior means follows directly after the derivation of the posterior distribution. Rarely, however, are they offered in the spirit of formal inference. Instead, their role is informal and descriptive. It is true that the posterior distribution represents a complete statement of what the statistician's beliefs about the parameter would, or should, be after seeing the data, but it expresses those beliefs in an obscure way. The posterior density is typically a highly complicated function of many paramenters, i.e., in many dimensions. In order to understand just what beliefs that function represents it is necessary to translate it into more comprehensible forms. We call this translation exercise *summarisation*.

The posterior mean is a valuable *summary*. It is a measure of location, and as such represents a point "near" which the parameters are now believed to lie. If we add a measure of spread, such as the posterior variance, we know how "near" to that point the parameters are expected to lie.

It may seem grandiose to dignify such a simple and common practice with a title like "summarisation", but we contend that summarisation is much more than the mere computation of a few low-order moments and deserves to be studied seriously in its own right.

We have described summarisation as a translation exercise, and meant this quite literally. It is the process of translating a density function into more comprehensible statements of beliefs. The term also indicates, however, that this translation is never complete. The posterior density always implies infinitely many more particular beliefs than can be captured and expressed in a few summaries. The nature and limitations of summarisation can also be seen in the specification of the prior and likelihood. How is a prior density specified in practice? Not by stating enormous numbers of individual probabilities. The usual procedure is to identify certain features that the prior should have. For instance we may specify that the parameters should have certain prior means and variances, that they should be independent, and that their marginal densities should be unimodal and symmetric. The prior density is then chosen as any convenient (e.g, conjugate) density function which has the required features. In short, the specification of a prior distribution is summarisation in reverse. The features which we specify for the prior are summaries, and because no amount of summaries will serve to specify the prior completely we have a choice of densities at the end, which we can exploit to achieve maximum tractability.

The fact that we choose to express prior beliefs through summaries makes it clear that they are the preferred language of beliefs. We resort to probability distributions because they allow us to use mathematical analysis, in the form of probability theory. But the benefits of that analysis can only be gained by means of a translation exercise at either end. Progress in Bayesian statistics depends on our ability to do this, and this is why we maintain that summarisation deserves serious study.

The aim of summarisation must be to extract the most important features of a density in the minimum number of summaries. Means and variances may seem to do this, but if they are all we know then our knowledge is poor indeed. As Box and Tiao (1973) have pointed out, we cannot interpret these quantities until we appreciate the *shape* of the posterior density.

The quantities and qualities which we would use to study and describe shape are important summaries. They include the number, locations, heights and widths of modes, the locations and heights of antimodes and saddle-points, the directions and degrees of skewness about modes, and measures of tail thickness. Summaries of shape are the most simple and fundamental of all summaries. An understanding of shape is prerequisite to interpreting more sophisticated summaries such as moments. An important practical distinction is that shape summaries are typically based on differentiation of the posterior density, whereas most other summaries —including moments, medians and credible regions— require integration. Because differentiation is so much easier to perform than integration, both analytically and numerically, shape summaries are available to us even when the complexity and high dimensionality of the posterior prevents the computation of integration-based summaries. Furthermore, whenever integration can be performed numerically, an appreciation of shape is still a prerequisite to applying numerical integration techniques effectively. This is true of quadrature but most obvious in the Monte Carlo method, recently enjoying considerable interest stemming from the work of Kloek and van Dijk (1978). The accuracy of Monte Carlo integration depends intimately on how well the sampling density mimics the density to be integrated. Therefore the key to effective use of Monte Carlo is a thorough summarisation of shape. See for instance van Dijk and Kloek's (1983) efforts to mimic a highly skew posterior.

To recap, we have argued that summarisation is an important technique. Although frequently used (generally perfunctorily but sometimes with real effect) it is rarely recognised explicitly and has never been studied in its own right. We believe that such a study is long overdue. Summarisation begins with the analysis of shape, and indeed ends there if the posterior is too intractable for analytical integration and in too many dimensions for numerical integration. Therefore the study of shape summaries is particularly important, especially in complex, high-dimensional cases. The remainder of this paper is a small contribution to this study. We consider a particular example of one-way analysis of variance and begin by demonstrating the (well-known) discrepancies that can arise between joint and marginal modes. We find that these discrepancies are explained by the presence of a shape feature which we call a *shoulder*. We discuss briefly the prevalance and detection of shoulders.

2. THE ONE-WAY ANALYSIS OF VARIANCE MODEL

Lindley and Smith (1972) introduced the following model for data having the one-way analysis of variance structure. It is the simplest hierarchical model of any practical value. The mk observations $\{y_{ij} : 1 \leq i \leq k,\ 1 \leq j \leq m\}$ are distributed independently, given the parameters, as

$$y_{ij} \sim N(\theta_i, \sigma^2). \tag{1}$$

Conditional on σ^2 and two further parameters ("hyperparameters") ξ and τ^2, the k parameters $\theta_1, \theta_2, \ldots, \theta_k$ are independently distributed as

$$\theta_i \sim N(\xi, \tau^2). \tag{2}$$

In the third and final stage of the hierarchy a distribution is assigned to ξ, σ^2 and τ^2. We shall assume for simplicity that they are uniformly distributed, i.e.

$$f(\xi, \sigma^2, \tau^2) \propto 1, \tag{3}$$

$(-\infty < \xi < \infty,\ 0 \leq \sigma^2 < \infty,\ 0 \leq \tau^2 < \infty)$.

As Lindley and Smith pointed out, this model provides a Bayesian representation of one-way analysis of variance in *either* the fixed-effects *or* random-effects sense. To interpret it in the fixed-effects sense the first stage, equation (1), is the conventional model and provides the likelihood function for the parameters $\theta_1, \theta_2, \ldots, \theta_k$ and σ^2. The remaining two stages constitute an hierarchical construction of the prior distribution. The extra parameters ξ and τ^2 can be integrated out immediately to yield an exchangeable prior for the θ_is. In practice, however, they are generally left in as extra nuisance parameters to be eliminated at the posterior analysis stage.

The random-effects model sees ξ, σ^2 and τ^2 as the parameters, and the third stage (equation (3) for instance) of the hierarchy provides their prior distribution. Equations (1) and (2) form an hierarchical construction of the random-effects model, whose likelihood results from integrating out the extra parameters $\theta_1, \theta_2, \ldots, \theta_k$. Again, following Lindley and Smith we would generally leave the θ_i's in the model as nuisance parameters for the posterior analysis. The two versions of the model are united in a common three-stage construction and differ only as to which parameters are regarded as nuisance parameters in the posterior.

In either version, the posterior joint density of all the parameters is

$$f(\theta_1, \theta_2, \ldots, \theta_k, \xi, \sigma^2, \tau^2 \mid y_{11}, \ldots, y_{km})$$

$$\propto (\sigma^2)^{-(mk)/2} \exp\{-\frac{1}{2\sigma^2} \sum_{i=1}^{k} \sum_{j=1}^{m} (y_{ij}-\theta_i)^2\} (\tau^2)^{-(k)/2} \exp\{-\frac{1}{2\tau^2} \sum_{i=1}^{k} (\theta_i-\xi)^2\}. \quad (4)$$

For a fixed-effects analysis of the data, we can integrate out the nuisance parameters ξ and τ^2 to obtain the posterior joint density of the θ_i's and σ^2. In fact, interest generally centres mainly on the θ_i's so we shall also integrate out σ^2. The result is

$$f(\theta_1, \theta_2, \ldots, \theta_k | y_{11}, \ldots, y_{km})$$

$$\propto \{ \sum_{i=1}^{k} \sum_{j=1}^{m} (y_{ij}-\theta_i)^2\}^{-(mk-2)/2} \{ \sum_{i=1}^{k} (\theta_i-\theta_{.})^2\}^{-(k-3)/2}, \quad (5)$$

where $\theta_{.} = k^{-1} \Sigma_i \theta_i$. For a random-effects analysis we integrate out the θ_i's to obtain the posterior joint density of ξ, σ^2 and τ^2. Here also interest usually centres on a subset of the parameters, in this case the variance components σ^2 and τ^2. So we also integrate out ξ to leave

$$f(\sigma^2, \tau^2 | y_{11}, \ldots, y_{km})$$

$$\propto (\sigma^2)^{-(m-1)k/2} (\sigma^2+m\tau^2)^{-(k-1)/2} \exp\{-\tfrac{1}{2} \left(\frac{s_w^2}{\sigma^2} + \frac{s_B^2}{\sigma^2+m\tau^2} \right) \}, \quad (6)$$

where

$$s_W^2 = \sum_{i=1}^{k} \sum_{j=1}^{m} (y_{ij}-y_{i.})^2, \qquad s_B^2 = m \sum_{i=1}^{k} (y_{i.}-y_{..})^2$$

are the usual within-group and between-group sums of squares, respectively.

We shall be studying the shapes of these three densities. In this section we shall find their modes. The simplest of the three densities is the last, equation (6), which is the product of two inverse-chi-square forms. The constraint that $\tau^2 \geq 0$ adds a minor complication. There is a unique mode at $(\hat{\sigma}^2, \hat{\tau}^2)$ where

$$\hat{\sigma}^2 = \frac{s_W^2}{k(m-1)}, \quad \hat{\tau}^2 = \frac{s_B^2}{m(k-1)} - \frac{\hat{\sigma}^2}{m} \quad (7)$$

provided $\hat{\tau}^2 \geq 0$, i.e. provided $F \geq 1$ where F is the conventional F-statistic

$$F = \frac{k(m-1)\, s_B^2}{(k-1)\, s_W^2} . \quad (8)$$

Otherwise, its unique mode is at $(\tilde{\sigma}^2, 0)$, where

$$\tilde{\sigma}^2 = \frac{s_W^2 + s_B^2}{mk-1} . \quad (9)$$

Next consider the mode of (5) and assume that $k>3$. (Unless $k>3$ and $m>1$ the posterior is improper). The second term causes the density to become infinite whenever $\theta_1=\theta_2=\ldots=\theta_k$. All these points trace out a line where the density climbs to an infinitely high ridge that we shall call a *spine*. Therefore the mode is also infinitely high. Seemingly, all points on the spine are equally high and all could be said to be the mode. But there is a sense in which we can identify a unique point on the spine as *the* mode. This is the point given by

$$\theta_1=\theta_2=\ldots=\theta_k=y_{..},$$

where $y_{..}$ is the grand mean of the mk observations. The sense in which this is the mode is that it is the point to which lines of steepest ascent converge from all points in the vicinity of the spine. So we somehow get to infinity "quicker" when approaching this point than any other point on the spine.

In addition to this spinal mode, there may also be an ordinary, finitely-high, local maximum that we shall call a *secondary mode*. This secondary mode, if it exists is located at the point $(\hat{\theta}_1,\ldots,\hat{\theta}_k)$, where

$$\hat{\theta}_i = \lambda y_{..} + (1-\lambda)y_{i.}, \qquad (i=1,2,\ldots,k) \tag{10}$$

and where

$$\lambda = \frac{(mk-2) - [(mk-2)^2 - 4k(k-3)(m-1)(mk+k-5)/\{(k-1)F\}]^{1/2}}{2(mk+k-5)} . \tag{11}$$

The secondary mode exists if and only if the real square root in (11) exists, i.e. if

$$F \geq 4 . \frac{k}{k-1} . \frac{(m-1)(k-3)}{mk-2} . \frac{mk+k-5}{mk-2} . \tag{12}$$

For large m and k this critical value of F tends to 4 from below.

The mode of (4) exhibits very similar behaviour. This density also has a spine. It goes to infinity at all points

$$(\theta_1,\theta_2,\ldots,\theta_k, \xi,\sigma^2,\tau^2) = (\theta,\theta,\ldots,\theta,\theta,v,0)$$

for arbitrary θ and $v \geq 0$. This spine therefore covers a two-dimensional subspace of the full $(k+3)$-dimensional parameter space. In the same sense as before, we can identify the point

$$(\theta_1,\theta_2,\ldots,\theta_k, \xi,\sigma^2,\tau^2) = (y_{..},y_{..},\ldots,y_{..},y_{..}, \tilde{\sigma}^2, 0),$$

where

$$\tilde{\sigma}^2 = (s_W^2 + s_B^2)/(mk),$$

as *the* spinal mode. Again there may be a secondary mode, at

$$(\theta_1,\theta_2,\ldots,\theta_k, \xi,\sigma^2,\tau^2) = (\hat{\theta}_1,\hat{\theta}_2,\ldots,\hat{\theta}_k,\hat{\xi},\hat{\sigma}^2,\hat{\tau}^2)$$

where

$$\begin{aligned} \hat{\theta}_i &= \lambda y_{..} + (1-\lambda)y_{i.}\,, \qquad (i=1,2,\ldots,k) \\ \hat{\xi} &= y_{..} \\ \hat{\sigma}^2 &= (s_W^2 + \lambda^2 s_B^2)/(mk), \\ \hat{\tau}^2 &= (1-\lambda)^2 s_B^2/(mk), \end{aligned} \tag{13}$$

and where

$$\lambda = \frac{m - [m^2 - 4k(m^2-1)/\{(k-1)F\}]^{1/2}}{2(m+1)} . \tag{14}$$

This secondary mode exists if and only if

$$F \geq 4 . \frac{k}{k-1} . \frac{m^2-1}{m^2} , \tag{15}$$

which also approaches 4 for large m and k. It actually exceeds 4 (and therefore exceeds (12)) for most practical cases, i.e. whenever $m^2 > k$.

The existence of spines in (4) and (5), the possible existence of secondary modes, and the simpler form of (6) are all common knowledge. Equally well-known is the fact that the modes of the three different densities can give very conflicting measures of location for the parameters. In the next section we present a numerical example to illustrate this conflict.

3. MODES AS APPROXIMATIONS TO MEANS

Like posterior means, posterior modes have frequently been interpreted as *estimates* of parameters, either informally or in the formal sense of being optimal under zero-one loss. The use of modes as estimates was particularly popular for a short time following the work of Lindley (1971) and Lindley and Smith (1972). Their hierarchical models produced posterior distributions whose means could only be computed by lengthy numerical integration. They suggested using posterior modes as simple approximations to posterior means. Moreover, since numerical integration would also have been required to compute the marginal density of an individual parameter, their suggestion went beyond using the modes of marginal densities. To estimate a parameter θ, the θ-coordinate of the mode of some convenient joint density of θ and some other parameters was used. For instance, if in the one-way analysis of variance model we wished to estimate σ^2 we could use the σ^2-coordinate of the mode of the joint density (6) of σ^2 and τ^2, i.e. $\hat{\sigma}^2$ in (7). Equally, however, we might use the joint density of all the parameters and use the σ^2-coordinate of the mode (13).

Experience with posterior modes in hierarchical models was not encouraging. Modal estimates derived from different joint densities could be very different from one another. Some of them would be very inaccurate approximations to posterior means, but it was never clear which joint density should be used. Other problems, like bimodality, were encountered. Interest in posterior modes as estimates declined.

In this section we illustrate many of the difficulties just described, with the aid of a numerical example. In the following section we see how those "difficulties" can be understood and explained by a more thorough investigation of shape. We hope that, quite apart from demonstrating the importance of shape, this work may help to rehabilatate modes as "estimates".

Suppose we wish to estimate the θ's in a one-way analysis of variance. There are two obvious modal estimates available through our work in the last section. First we could estimate each θ_i by the θ_i-coordinate of the mode of the density (5). The spinal mode is not a very plausible estimate because it does not depend on the data at all. Far more interesting is the secondary mode, particularly because through its use of equation (10) it qualifies as a member of the important class of *shrinkage* estimates. The quantity λ, given in equation (11), represents the degree of shrinkage. Another suggestion is to use the secondary mode of the full joint density (4). The $\hat{\theta}_i$-coordinates of this mode constitute another shrinkage estimator, with degree of shrinkage given by (14). The obvious structural similarity of these two expressions for λ is due to the fact, explained in O'Hagan (1976), that both modes can be obtained by solving very similar sets of equation. However, the actual values they yield can be very different if m and k are not large. Figure 1 plots the two expressions (11) (14) as functions of F in the case $k=6$, $m=5$. The mode of the density (5), which we shall call the "marginal mode", entails only about half as much shrinkage as the mode of the full joint density (4), which we shall call the "joint mode".

The difference is particularly marked, of course, between the two critical values (12) and (15), that is for $2.278 \leq F < 4.608$, where the "marginal" density possesses a secondary mode but the full joint density does not. In fact, a proper definition of our modal estimator is that we use the secondary mode *whenever* it exists, and otherwise use the spinal mode. Thus the "marginal mode" is the shrinkage estimator (10) with degree of

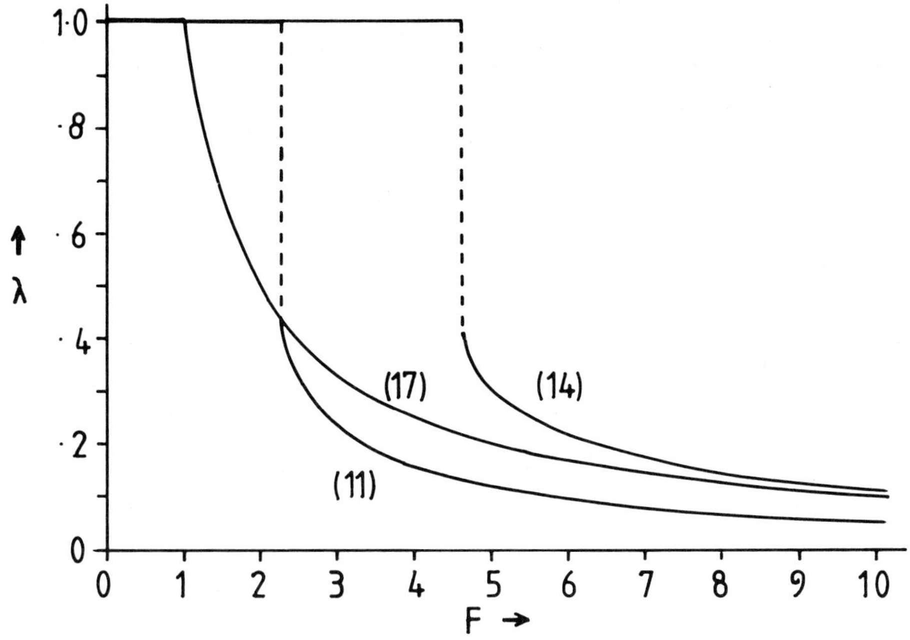

FIGURE 1. *Three expressions for* λ *in the case* $k=6$, $m=5$

shrinkage λ given by (11) provided (12) holds, and otherwise $\lambda=1$. The spinal mode corresponds to complete shrinkage. The form of this estimator is of a *preliminary test estimator*, because the condition (12) acts like a test of the null hypothesis that the θ_is are all equal, rejecting that hypothesis if the conventional test statistic F is sufficiently large. The "joint mode" is defined similarly and also corresponds to a preliminary test estimator, using the condition (15). To the non-Bayesian both conditions seems like unrealistically stringent tests, a point which was made in a similar context by Efron and Morris (1973, p. 417) in their reply to Leonard (1973). This is an interesting side-issue, but lack of space prevents us pursuing it here.

Figure 1 also compares these two functions with a third expression derived from the mode of the density (6). The basis of this comparison is that equation (14) comes essentially from solving the equation

$$\lambda = \sigma^2/(\sigma^2+m\tau^2), \tag{16}$$

where the estimates $\hat{\sigma}^2$ and $\hat{\tau}^2$ given in (13) are substituted for σ^2 and τ^2. Using the methods of O'Hagan (1976) this equation figures in the derivation of the modes of all three densities, with different estimates of the variance ratio τ^2/σ^2 in each case. In particular, O'Hagan (1976) suggests that another shrinkage estimator of the θ_is can be obtained from the mode of (6) by using equations (10) and (16) and substituting instead the modal values $\hat{\sigma}^2$ and $\hat{\tau}^2$ from (7) into (16). This substitution yields the very simple expression

$$\lambda = F^{-1} \tag{17}$$

for the degree of shrinkage, which is the third function plotted in Figure 1. (The non-Bayesian will see in F^{-1} an obvious, if biased, estimator of the ratio (16)).

For the sake of more specific illustration we take the case of some data from Davies and Goldsmith (1972, p. 131) concerning yields of dyestuff in a chemical plant. These data comprise $k=6$ groups each for $m=5$ observations, with sums of squares $s_W{}^2 = 58830$ and $s_B{}^2 = 56357.5$, so that the value of the F statistic is 4.598. We therefore obtain the following three λ values.

Equation (11), "marginal mode",	$\lambda = 0.1308$.
Equation (14), "joint mode",	$\lambda = 1$ (no secondary mode).
Equation (17), "variance mode",	$\lambda = 0.2175$.

To these three widely differing values we could add a fourth; if F had been only a tiny amount larger, at $F = 4.608$, the "joint mode" would have been given by $\lambda = 0.4167$ since then the joint density would have had a secondary mode.

Results such as these led to a rapid loss of confidence in posterior modes as approximations to posterior means. They cannot all be good approximations. Which is best? For these particular data it is relatively easy to resolve the issue by computing the posterior mean. The posterior mean of the θ_is turns out to be a shrinkage estimator whose degree of shrinkage is the posterior mean of the *parameter* λ defined by (16). From (6) we find that the posterior density of λ is

$$f(\lambda \mid y_{11},\ldots,y_{km}) \propto \lambda^{(k-5)/2}(s_W{}^2+\lambda s_B{}^2)^{-(mk-5)/2} \tag{18}$$

for $0 \le \lambda \le 1$. This is (after appropriate scaling) an F-distribution with $k-3$ and $k(m-1)$-2 degrees of freedom, but truncated by the constraint that $\lambda \le 1$. This density is plotted for the dyestuffs data in Figure 2, and is seen to be markedly skew. Its mode is at $\lambda = 0.0435$, suggesting very little shrinkage indeed, but its mean is readily found by numerical integration to be about 0.155. Therefore, for these data at least, the "marginal mode" is best. On the other hand, if we had had data with a lower F value, such that the "marginal" density (5) had no secondary mode, the "variance mode" would almost certainly have been better.

But Lindley and Smith had advocated modes to *avoid* computing means by numerical integration. For their idea to work it is necessary to have rules to determine which joint density will yield the most accurate modal approximation to the mean. Lindley and Smith

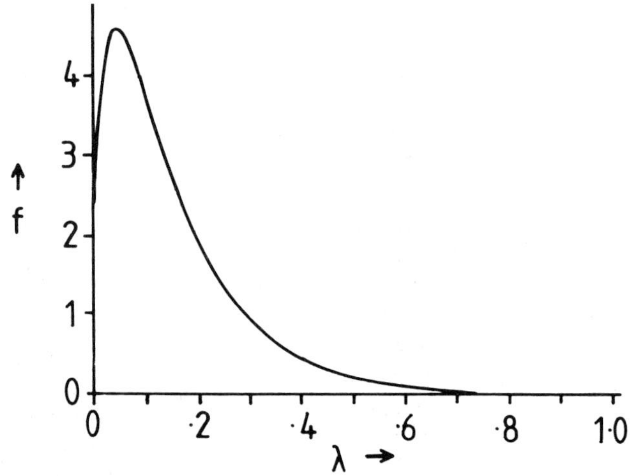

FIGURE 2. *Marginal density of* λ

felt that "marginal" should be better than "joint". O'Hagan (1976), also arguing intuitively, suggested that the "variance mode" should be more reliable. Our example supports neither view fully, and demonstrates the elusiveness of any simple rule of thumb.

Why should modes be such unreliable estimators? The answer lies in asking what we want estimators for. If a formal inference is required, using a well-defined loss function, then the reason is obvious: posterior means and modes are optimal with respect to very different loss functions, and it is hardly surprising that they do not approximate each other well. If, however, an estimate is required in the informal sense of a "measure of location" then neither mode nor mean can be given such an interpretation without qualification. The argument of Box and Tiao (1973) applies equally to mean or mode - in order to interpret them we must first appreciate the underlying shape.

4. SHOULDERS

The large differences between the various modes may be quite insignificant. If the spread of the posterior is sufficiently great then they may all be equally valid as measures of location. Having found modes, it is important to measure their widths in some appropriate way. In the present case we would wish to see whether the width of each of our modes was sufficient to cover such a wide range of degrees of shrinkage, and so cause them to overlap substantially. Going beyond width, the skewness of modes may be an alternative answer to the discrepancies. One or more of our conflicting modes may be found to be misleading measures of location because of extreme skewness. Figure 2 is a nice illustration of how skewness is important in interpreting quantities like modes or mean as measures of location.

There are several useful measures for the widths and skewnesses of modes, and our next step in exploring shape in the present example should be to compute some of these important summaries. That would necessitate introducing and explaining those measures, and so to avoid lengthening this paper unduly we shall cheat a little. The summaries we shall use have been chosen to illuminate the shapes of these particular densities as succinctly as possible. Our presentation will not reflect good technique in exploring the shape of a genuinely unfamiliar density.

The shape of the full joint density (4), using the dyestuffs data, is deceptively simple. There is no secondary mode, and so the density falls away continuously from the spine in all directions. There is no suggestion that there could be any appreciable amount of probability except in some neighbourhood of the spine. If estimation of the θ's is required then complete shrinkage is indicated, i.e. the estimate of each θ_i is the grand mean of the data, $y_{..} = 1527.5$. The inadequacy of such a superficial look at shape becomes clear when we consider the other two densities.

The "marginal" density (5) has a spine, and its spinal mode is also the completely shrunk estimator $\theta_i = 1527.5$ $^*i$. But it also has a secondary mode correponding to the degree of shrinkage $\lambda = 0.1308$, giving estimates of the θ_is ranging from 1477.5 to 1590.5. Since there are two modes it is natural to compare their heights and widths and to see how well separated they are. Figure 3 plots the "marginal" density along the line joining these two modes. That is, it is a cross-section through the density and shows the density at all points corresponding to shrinkage estimators. We can see the density falling away from the infinitely high spine on either side of the point $\lambda = 1$. As λ decreases it continues to fall until about $\lambda = 0.75$ then rises again to the secondary mode at $\lambda = 0.1308$. The exact value of the point where this cross-section gives a minimum may be obtained from (11) by reversing the square root; it is $\lambda = 0.7724$. This point is a saddle-point in the 6-dimensional density,

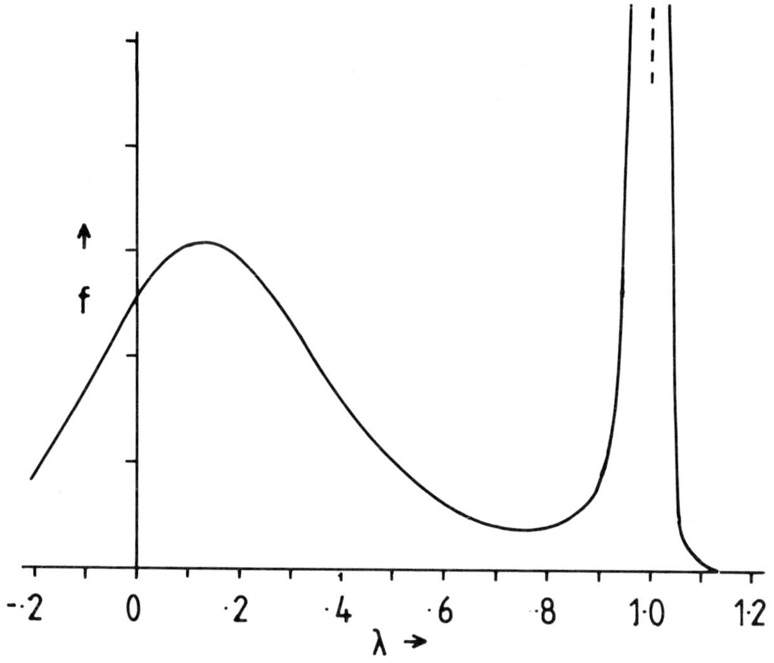

FIGURE 3: *Cross-section through "marginal" density*

and its height is only one-eighth of that of the secondary mode. Clearly the secondary mode is well separated from the spine and accounts for a genuine and appreciable concentration of probability.

The density of the variance components, (6), has a unique mode at $\hat{\sigma}^2 = 2451$, $\hat{\tau}^2 = 1764$, corresponding to the value $\lambda = 0.2175$ already given in Section 3. Figure 4 is a contour map of this density. In view of our findings in the other two densities it is interesting to ask how much support this density gives to $\lambda = 1$, which corresponds to $\tau^2 = 0$. In fact the mode is more than twenty times as high as the highest point on the axis $\tau^2 = 0$. This contour map is a very powerful summary, and shows quite conclusively that the amount of posterior probability in the neighbourhood of $\lambda = 1$ is actually very small.

The evidence of these two densities is that our first attempt to visualise the shape of the joint density was too superficial. The picture we had of the density falling away smoothly and evenly on all sides of the spine was wrong. Figure 5 shows that the fall-off of density is by no means symmetric or even. This diagram is cross-section through the joint density in a very similar way to figure 3, although its interpretation is less clear because it is not a cross-section along a straight line. It is a plot of the joint density at points given by equations (13) as λ varies. As such it covers the spectrum of shrinkage estimators of the θ's. It passes through the spinal mode at $\lambda = 1$ and would also pass through a secondary mode and saddle-point if they existed. We can see how the fact that the expression to be square-rooted in (14) is only just negative results in a mode almost appearing around $\lambda = 0.41$. This feature, which is not quite a mode, is what we call a *shoulder* (although its resemblance to a hunch-back makes the term "quasimode" very tempting!).

Figure 5 indicates that the explanation for the discrepancies between modes lies in skewness - the attachment of the shoulder to the spinal mode in the joint density makes it

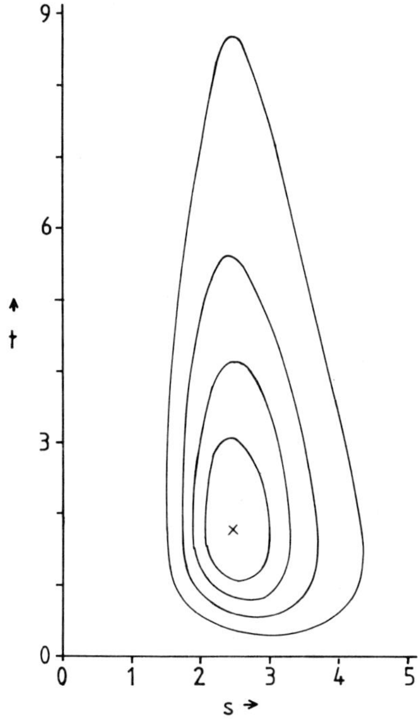

FIGURE 4: *Joint density of* $s = \sigma^2/1000$ and $t = \tau^2/1000$

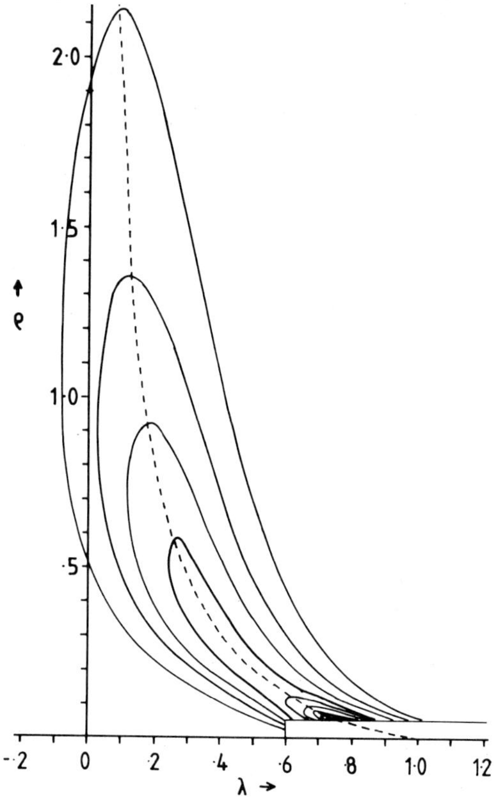

FIGURE 6: *Two-dimensional cross-section through "joint" density*

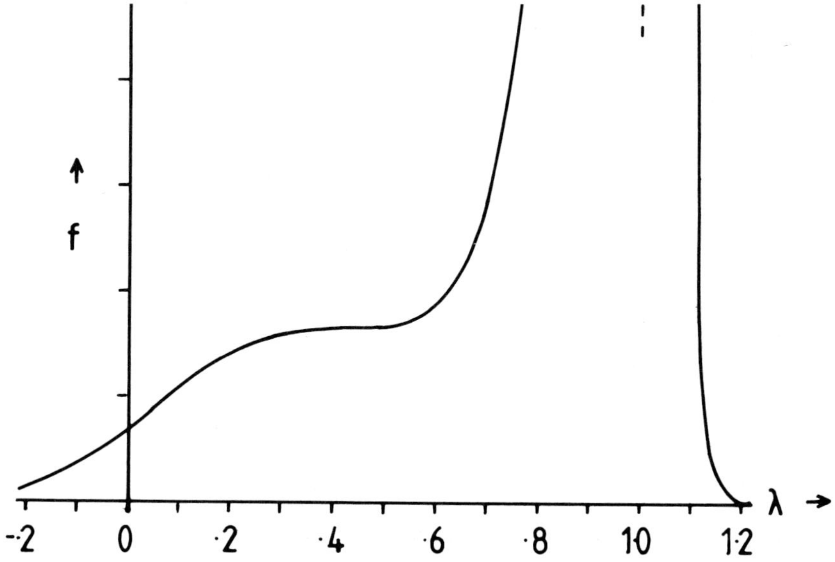

FIGURE 5: *Cross-section through "joint" density*

highly skew. But Figure 5 does not yet offer a complete explanation of those discrepancies. It is a very different picture from Figure 3 and there is nothing in our understanding of the joint density yet to suggest a concentration of probability detached from the spine, nor to suggest much support for λ as low as 0.1 or 0.2. These suggestions only begin to emerge when we look at the shoulder in two or more dimensions.

Figure 6 is a contour map of a two-dimensional cross-section through the joint density. It shows the density at points given by

$$\theta_i = \lambda y_{..} + (1-\lambda) y_{i.}\,, \qquad (i=1,2,\ldots,6)$$

$$\xi = y_{..},$$

$$\tau^2/\sigma^2 = \varrho\,,$$

$$\sigma^2 = \{s_W^2 + \lambda^2 s_B^2 + (1-\lambda)^2 s_B^2/(m\varrho)\}/\{(m+1)k\},$$

plotted as a function of λ and ϱ. Figure 5 is obtained from Figure 6 by maximising over ϱ. In other words, it is a cross-section through Figure 6 along the line

$$\lambda = (1+5\varrho)^{-1}$$

shown as a dotted line in Figure 6. The spinal mode is the point $\lambda = 1$, $\varrho = 0$. Extending out from this point along the dotted line we see a shoulder which becomes broader as we move away from the spine. It is this broadening of the shoulder that explains the shape of the "marginal" dennsity of the θ's. The parameter ϱ is a function of σ^2 and τ^2, and if we imagine integrating over ϱ, instead of maximising, then it is easy to see that we will obtain a diagram much more like Figure 3 than Figure 5, with a maximum around $\lambda = 0.2$.

5. THE IMPORTANCE AND DETECTION OF SHOULDERS

Our example has illustrated the principal reason for studying shoulders. In several dimensions a shoulder may broaden out so much as we move away from the mode to which it is attached that, in terms of the amounts of probability they contain, the shoulder may dominate the mode. Therefore shoulders are potentially important shape features, and detecting and locating them is as important as finding modes.

We have said that summarisation of shape should include summaries of width and skewness of modes, and it may seem that by the use of these measures shoulders could always be detected as mere extensions of the modes to which they are attached. This is not the case, which explains why we regard the shoulder as a shape feature in its own right, and not just a special form of skewness. The density may fall away from the mode to a minuscule fraction of the modal height before developing into a shoulder. However we attempt to define width or skewness, a shoulder which is sufficiently low and sufficiently far from the mode may elude detection by measures of width or skewness. And yet it may broaden out so much (and through so many dimensions) that it still dominates the mode.

Another reason why we consider shoulders to be so important is that they can be expected to arise whenever we construct hierarchical models with unknown variances. Use of an improper prior for the variances in hierarchical models always leads to a spine in the joint density for all the parameters. Whenever we fail to find a secondary mode we must expect and search for a shoulder. The literature contains at least two instances where authors have advanced totally-shrunk, spinal modes as estimates of parameters in such models. We have already mentioned Leonard (1973), who gives the spinal mode in estimating several normal means, and a shoulder is certainly to be found in this case. We have not tackled the much more complicated econometric example of Smith (1973), but strongly suspect a shoulder here too. Both authors used improper priors for their variances, and their posterior densities exhibit spines. The fact that the spinal mode is completely shrunk in such cases makes the resulting estimates stand out so clearly that further investigation in obviously called for. It is most important to realise that shoulders arise almost as readily *even* if we put proper priors on the variances, and *even* if those priors are such as to cancel out the spine. In our example we used the uniform prior (3) for tractability and ease of exposition. Suppose that we use the prior

$$f(\xi,\sigma^2,\tau^2) \propto (\tau^2)^{-2}\exp(-200/\tau^2) \qquad (19)$$

instead of (3). This retains uniform priors for ξ and σ^2 but gives τ^2 a proper inverse-chi-square distribution. Although (19) still expresses weak prior information on the variances, it tends to zero fast enough, as τ^2 tends to zero, to eliminate the spine in the posterior density. The joint posterior density is the product of (4) and (19). It has a unique mode, whose θ_i-coordinates are shrinkage estimates with $\lambda=0.943$. If we fail to detect the fact that it also has a significant shoulder we are in an even worse position than if we had used the uniform prior (3). Not only is the mode still a misleading indicator of location but it is no longer completely shrunk, therefore it does not appear so unusual or signal a need for more investigation.

For the detection of shoulders we can suggest two distinct approaches. The first relies on identifying shoulders as "quasi-modes". Consider for example the density (4). In searching for its modes we can look for intersections of conditional mode surfaces. Specifically, if we plot the (unique) conditional mode of $(\theta_1,\theta_2,\ldots,\theta_k,\xi)$ given (σ^2,τ^2) as a function of σ^2 and τ^2 the result is a 2-dimensional surface in the full $(k+3)$-dimensional parameter space. Similarly, the (unique) conditional mode of (σ^2,τ^2) given $(\theta_1,\theta_2,\ldots,\theta_k,\xi)$ yields a $(k+1)$-dimensional surface in the same space. The points at which these surfaces intersect are alternately modes and saddle-points of the joint density. (They intersect on the spine at the unique point we have called *the* spinal mode). Lindley and Smith's (1972) iterative procedure for finding modes operates by zig-zagging between these surfaces. A shoulder should appear whenever such surfaces come close together, without actually intersecting. To detect these "close encounters" we need first to measure the separation between conditional mode surfaces, and then to search for points of minimal separation. An appropriate separation measure might be the length of zig or zag in Lindley-Smith iteration.

The second approach is to use the curvature, as measured by the hessian matrix, of the surface. The curvature on a shoulder, like the curvature at a mode, is negative definite. Between mode and shoulder there is a transitional period where the hessian becomes non-

definite. Therefore we can search for shoulders by searching for regions of negative curvature. There is a need here for ways of narrowing the search, because to search a high-dimensional space thoroughly is a daunting task. On the other hand, the more significant a shoulder is the larger should be the area of negative curvature which marks it presence.

REFERENCES

BOX, G.E.P. and TIAO, G.C. (1973). *Bayesian Inference in Statistical Analysis,* Reading, Mass.: Addison-Wesley.

DAVIES, O.L. and GOLDSMITH, P.L. (1972). Editors, *Statistical Methods in Research and Production.* Edinburgh: Oliver and Boyd.

EFRON, B. and MORRIS, C. (1973). Combining possibility related estimation problems. *J.R. Statist. Soc.* B**35**, 379-421 (with discussion).

KLOEK, T. and VAN DIJK, H.K. (1978). Bayesian estimates of equation system parameters: an application of integration by Monte Carlo. *Econometrica*, **46**, 1-19.

LEONARD, T. (1973). Discussion of Efron and Morris. *J.R. Statist. Soc.* B**35**, 415-416.

LINDLEY, D.V. (1971). The estimation of many parameters. In *Foundations of Statistical Inference*, V.P. Godambe and D.A. Sprott, editors, 435-455. Toronto: Holt, Rinehart and Winston.

LINDLEY, D.V. and SMITH, A.F.M. (1972). Bayes estimates for the linear model. *J.R. Statist. Soc.* B**34**, 1-41 (with discussion).

O'HAGAN, A. (1976). On posterior joint and marginal modes. *Biometrika* **63**, 329-333.

SMITH, A.F.M. (1973). A general Bayesian linear model *J.R. Statist. Soc.* B**35**, 67-75.

VAN DIJK, H.K. and KLOEK, T. (1983). Monte Carlo analysis of skew posterior distributions: an illustrative econometric example. *The Statistican* **32**, 216-223.

BAYESIAN STATISTICS 2, pp. 711-722
J.M. Bernardo, M.H. DeGroot, D.V. Lindley, A.F.M. Smith (Eds.)

Bayesian Hypothesis Testing in Linear Models with Continuously Induced Conjugate Priors Across Hypotheses

D.J. POIRIER
University of Toronto, Canada

SUMMARY

In the context of competing specifications for the appropriate regressors in a normal linear regression model, a natural conjugate prior is postulated for the unrestricted model and priors for restricted models are chosen to be identical to the conditional distribution induced in the usual way from the unrestricted model. Of particular concern is the behavior of Bayes' factors and posterior odds under a particular form of a "noninformative prior" derived as a limiting case of an informative prior for the unrestricted model.

Keywords: BAYES' FACTORS; MODEL SELECTION; NON- INFORMATIVE PRIORS; POSTERIOR PROBABILITIES.

1. INTRODUCTION

Although statistical hypothesis testing has its critics, it persists as one of the major activities of econometricians.[1] This paper focuses on Bayesian hypothesis testing in the context of competing specifications for the appropriate regressors in a normal linear regression model. The distinguishing feature of the analysis is that a natural conjugate prior distribution is postulated for the unrestricted model, and then the prior distributions for any smaller models are chosen to be identical to the conditional distribution induced in the usual way from the larger unrestricted model. Such analysis is reminiscent of the study of Dickey (1974) which utilized a multivariate t-distribution on the regression coefficients and an independent gamma prior distribution for the precision. The use here of the natural conjugate prior will provide results which are substantially different than Dickey's in some respects. Of particular concern here will be the behavior of Bayes' factors and posterior odds under a particular form of a "non-informative prior" derived as a limiting case of an informative prior for the unrestricted model.

1. The use of fixed levels of significance which are independent of sample size has been criticized since at least Berkson (1938). From a decision theory standpoint, Kadane and Dickey (1980) question the appropriateness of most model selection activities in econometrics. From a forecasting standpoint, hypothesis testing for the purposes of model selection is questionable since weighted forecasts from various models can be expected to dominate single forecasts [see Granger and Newbold (1977, Chapter 8)]. Finally, hypothesis testing under diffuse prior information is plagued by arbitrariness [see Leamer (1978, pp. 111-114)].

2. BASIC MODEL

Let $y = [y_1, y_2, \ldots, y_T]'$, V, W and Z denote $T\times 1$, $T\times K_v$, $T\times K_w$ and $T\times K_z$ matrices respectively. Define $X_0 = V$, $X_1 = [V, W]$, $X_2 = [V, Z]$ and $X_3 = [V, W, Z]$. Let $K_0 = K_v$, $K_1 = K_v + K_w$, $K_2 = K_v + K_z$ and $K_3 = K_v + K_w + K_z$ so that X_j is $T\times K_j$ $(j=0,1,2,3)$. Then consider the standard linear normal regression models

$$H_j\colon y\,|\,V,W,Z \sim N(X_j\beta_j,\ \sigma_j^2 I_T) \qquad (j=0,1,2,3) \tag{2.1}$$

where $N(X_j\beta_j,\ \sigma_j^2 I_T)$ denotes a T-dimensional multivariate normal distribution with mean $X_j\beta_j$ and covariance matrix $\sigma_j^2 I_T$, β_j is a K_j-dimensional column vector of unknown regression coefficients and σ_j^2 is a positive scalar.

It is most important to note that all of the models in (2.1) are defined conditional on all potential regressors, namely, V, W and Z. If, as has unfortunately often been done in the literature, H_j is postulated conditional only on X_j, then $H_j (j=0,1,2,3)$ are *not* in conflict since they refer to different distributions. The importance of this observation lies in the fact that it implies that the parameters in each model are not as unrelated as the usual notation suggests.

To clarify this point, consider the parameters in H_3:

$$\beta_3 \equiv \begin{bmatrix} \beta^{yv.wz} \\ \beta^{yw.vz} \\ \beta^{yz.vw} \end{bmatrix} = \frac{\partial E(y_t\,|\,x_{3t})}{\partial x_{3t}} \equiv \beta \qquad (t=1,2,\ldots,T), \tag{2.2}$$

$$\sigma_3^2 \equiv \operatorname{Var}(y_t\,|\,x_{3t}) \equiv \sigma^2 \qquad (t=1,2,\ldots,T), \tag{2.3}$$

where $x_{3t} = [v_t', w_t', z_t']$ is the t-th row of X_3, and the superscript notation in (2.2) emphasizes the fact that the elements in β_3 are partial regression coefficients. H_j $(j=0,1.2)$ can be written in terms of (2.2) as

$$H_0\colon\ \beta^{yz.vw} = 0_{K_z},\ \beta^{yw.vz} = 0_{K_w}, \tag{2.4}$$

$$H_1\colon\ \beta^{yz.vw} = 0_{K_z}, \tag{2.5}$$

$$H_2\colon\ \beta^{yw.vz} = 0_{K_w}, \tag{2.6}$$

where 0_i denotes an i-dimensional column vector of zeros. Hence, the regression coefficients in H_j $(j=0,1,2)$ are

$$\beta_0 = \beta^{yv.wz}, \quad \beta_1 = \begin{bmatrix} \beta^{yv.wz} \\ \beta^{yw.vz} \end{bmatrix}, \quad \beta_2 = \begin{bmatrix} \beta^{yv.wz} \\ \beta^{yz.vw} \end{bmatrix}. \tag{2.7}$$

Note that $\beta^{yv.wz}$ appears in β_j $(j=0,1,2,3)$. There are no new parameters in H_j $(j=0,1,2)$ that do not appear in H_3. Even the variances σ_j^2 $(j=0,1,2)$ really only refer to the single parameter σ^2 defined in (2.3). The standard subscript notation $j=0,1,2,3$ in (2.1) merely suggests conditioning on H_j for the purposes of estimation and inference concerning the common parameters β and σ^2.

Hypotheses H_j $(j=0,1,2,3)$ cover a wide range of interesting situations that arise when comparing and testing hypotheses in the standard linear normal regression model. While H_j $(j=0,1,2)$ are all nested within H_3 and H_0 is also nested within both H_1 and H_2, H_1 and

H_2 constitute non-nested hypotheses. The fact that H_j ($j=0,1,2$) contain no new parameters and that all are nested within H_3 [see (2.4)-(2.6)], is the motivation for referring to H_3 as the *comprehensive model*, i.e., the data-generating distribution for which all hypotheses in question impose testable competing restrictions. While the comprehensive model is of fundamental importance in both Bayesian and non-Bayesian analyses, it has added importance in Bayesian analysis since once prior beliefs about β and σ^2 are assessed for comprehensive model H_3, prior beliefs about the unconstrained parameters in H_j, conditional on H_j, can be derived in a straightforward manner for $j=0,1,2$. Prior assessments made in such a manner are said to be induced continuously as in Dickey (1974) and such assessments are the topic of the next section.

3. FORMULATION OF CONTINUOUS INDUCED PRIORS

The conjugate prior distribution for the parameters of the normal linear regression model is the normal-gamma distribution. Working in terms of the precision σ^{-2} and in the context of comprehensive model H_3, the probability density function (p.d.f.) of the normal-gamma distribution $NG(\bar{b}_3, \bar{\Omega}_3, \bar{s}_3^2, \bar{\nu}_3)$ is defined by

$$\begin{aligned} p_3(\beta, \sigma^{-2}) &\equiv p_3(\beta, \sigma^{-2} | \bar{b}_3, \bar{\Omega}_3, \bar{s}_3^2, \bar{\nu}_3) \\ &\equiv \phi_{K_3}(\beta | \bar{b}_3, \sigma^2\bar{\Omega}_3)\gamma(\sigma^{-2} | \bar{s}_3^{-2}, \bar{\nu}_3) \end{aligned} \tag{3.1}$$

where $\phi_{K_3}(\bullet)$ denotes a K_3-dimensional multivariate normal p.d.f. and $\gamma(\bullet)$ denotes a univariate gamma p.d.f. In other words, natural conjugate prior (3.1) implies that the distribution of β given σ^{-2} is multivariate normal with mean $\bar{b}_3$ and covariance matrix $\sigma^2\bar{\Omega}_3$, denoted $N(\bar{b}_3, \sigma^2\bar{\Omega}_3)$, and the marginal distribution of σ^{-2} is a gamma distribution with mean $(\bar{s}_3^2)^{-1}$ and variance $2/\bar{\nu}_3\bar{s}_3^4$, denoted $G(\bar{s}_3^2, \bar{\nu}_3)$. The marginal p.d.f. of β is a t-distribution with mean $\bar{b}_3$ (if $\bar{\nu}_3>1$), covariance matrix $\bar{\nu}_3\bar{s}_3^2\bar{\Omega}_3/(\bar{\nu}_3-2)$ (if $\bar{\nu}_3>2$), and degrees of freedom $\bar{\nu}_3$, denoted $t_{K_3}(\bar{b}_3, -\bar{s}_3^2\bar{\Omega}_3, \bar{\nu}_3)$.

It follows from the properties of the normal-gamma distribution [see Raiffa and Schlaifer (1961, p. 344)] that

$$p_0(\beta_0, \sigma^{-2} | \beta^{yz.vw} = 0_{K_z}, \beta^{yw.vz} = 0_{K_w}) = \phi_{K_v}(\beta_0 | \bar{b}_0, \sigma^2\bar{\Omega}_0)\gamma(\sigma^{-2} | \bar{s}_0^2, \bar{\nu}_0), \tag{3.2}$$

$$p_1(\beta_1, \sigma^{-2} | \beta^{yz.vw} = 0_{K_z}) = \phi_{K_1}(\beta_1 | \bar{b}_1, \sigma^2\bar{\Omega}_1)\gamma(\sigma^{-2} | \bar{s}_1^2, \bar{\nu}_1), \tag{3.3}$$

$$p_2(\beta_2, \sigma^{-2} | \beta^{yw.vz} = 0_{K_w}) = \phi_{K_2}(\beta_2 | \bar{b}_2, \sigma^2\bar{\Omega}_2)\gamma(\sigma^{-2} | \bar{s}_2^2, \bar{\nu}_2), \tag{3.4}$$

where

$$\bar{b}_3 \equiv \begin{bmatrix} \bar{b}_3^{yv.wz} \\ \bar{b}_3^{yw.vz} \\ \bar{b}_3^{yz.vw} \end{bmatrix}, \quad \bar{\Omega}_3 \equiv \begin{bmatrix} \bar{\Omega}_{3,vv} & \bar{\Omega}_{3,vw} & \bar{\Omega}_{3,vz} \\ \bar{\Omega}_{3,vw}' & \bar{\Omega}_{3,ww} & \bar{\Omega}_{3,wz} \\ \bar{\Omega}_{3,vz}' & \bar{\Omega}_{3,wz}' & \bar{\Omega}_{3,zz} \end{bmatrix}, \tag{3.5}$$

$$\bar{b}_0 \equiv \bar{b}_0^{yv.wz} = \bar{b}_3^{yv.wz} - [\bar{\Omega}_{3,vw}, \bar{\Omega}_{3,vz}] \begin{bmatrix} \bar{\Omega}_{3,ww} & \bar{\Omega}_{3,wz} \\ \bar{\Omega}_{3,wz}' & \bar{\Omega}_{3,zz} \end{bmatrix}^{-1} \begin{bmatrix} \bar{b}_3^{yw.vz} \\ \bar{b}_3^{yz.vw} \end{bmatrix}, \tag{3.6}$$

$$\bar{\Omega}_0 \equiv \bar{\Omega}_{0,vv} = \bar{\Omega}_{3,vv} - [\bar{\Omega}_{3,vw}, \bar{\Omega}_{3,vz}] \begin{bmatrix} \bar{\Omega}_{3,ww} & \bar{\Omega}_{3,wz} \\ \bar{\Omega}_{3,wz}' & \bar{\Omega}_{3,zz} \end{bmatrix}^{-1} \begin{bmatrix} \bar{\Omega}_{3,vw}' \\ \bar{\Omega}_{3,vz}' \end{bmatrix}, \tag{3.7}$$

$$\bar{\nu}_0 \equiv \bar{\nu}_3 + K_w + K_z \tag{3.8}$$

$$\bar{s}_0^2 \equiv \bar{\nu}_0^{-1} \quad \bar{\nu}_3 \bar{s}_3^2 + [(\bar{b}_3^{yw.vz})',(\bar{b}_3^{yz.vw})']' \begin{bmatrix} \bar{\Omega}_{3,ww} & \bar{\Omega}_{3,wz} \\ \bar{\Omega}_{3,wz}' & \bar{\Omega}_{3,zz} \end{bmatrix}^{-1} \begin{bmatrix} \bar{b}_3^{yw.vz} \\ \bar{b}_3^{yz.vw} \end{bmatrix}, \tag{3.9}$$

$$\bar{b}_1 \equiv \begin{bmatrix} \bar{b}_1^{yv.wz} \\ \bar{b}_1^{yw.vz} \end{bmatrix} = \begin{bmatrix} \bar{b}_3^{yv.wz} \\ \bar{b}_3^{yw.vz} \end{bmatrix} - \begin{bmatrix} \bar{\Omega}_{3,vz} \\ \bar{\Omega}_{3,wz} \end{bmatrix} \bar{\Omega}_{3,zz}^{-1} \bar{b}_3^{yz.vw}, \tag{3.10}$$

$$\bar{\Omega}_1 \equiv \begin{bmatrix} \bar{\Omega}_{1,vv} & \bar{\Omega}_{1,vw} \\ \bar{\Omega}_{1,vw}' & \bar{\Omega}_{1,ww} \end{bmatrix} = \begin{bmatrix} \bar{\Omega}_{3,vv} & \bar{\Omega}_{3,vw} \\ \bar{\Omega}_{3,vw}' & \bar{\Omega}_{3,ww} \end{bmatrix} - \begin{bmatrix} \bar{\Omega}_{3,vz} \\ \bar{\Omega}_{3,wz} \end{bmatrix} \bar{\Omega}_{3,zz}^{-1} \begin{bmatrix} \bar{\Omega}_{3,vz} \\ \bar{\Omega}_{3,wz} \end{bmatrix}, \tag{3.11}$$

$$\bar{\nu}_1 \equiv \bar{\nu}_3 + K_z, \tag{3.12}$$

$$\bar{s}_1^2 \equiv \bar{\nu}_1^{-1} [\bar{\nu}_3 \bar{s}_3^2 + (\bar{b}_3^{yz.vw})' \bar{\Omega}_{3,zz}^{-1} \bar{b}_3^{yz.vw}], \tag{3.13}$$

$$\bar{b}_2 \equiv \begin{bmatrix} \bar{b}_2^{yv.wz} \\ \bar{b}_2^{yz.vw} \end{bmatrix} = \begin{bmatrix} \bar{b}_3^{yv.wz} \\ \bar{b}_3^{yz.vw} \end{bmatrix} - \begin{bmatrix} \bar{\Omega}_{3,vw} \\ \bar{\Omega}_{3,zw} \end{bmatrix} \bar{\Omega}_{3,ww}^{-1} \bar{b}_3^{yw.vz}, \tag{3.14}$$

$$\bar{\Omega}_2 \equiv \begin{bmatrix} \bar{\Omega}_{2,vv} & \bar{\Omega}_{2,vz} \\ \bar{\Omega}_{2,vz} & \bar{\Omega}_{2,zz} \end{bmatrix} = \begin{bmatrix} \bar{\Omega}_{3,vv} & \bar{\Omega}_{3,vz} \\ \bar{\Omega}_{3,vz} & \bar{\Omega}_{3,zz} \end{bmatrix} - \begin{bmatrix} \bar{\Omega}_{3,vw} \\ \bar{\Omega}_{3,zw} \end{bmatrix} \bar{\Omega}_{3,ww}^{-1} \begin{bmatrix} \bar{\Omega}_{3,vw} \\ \bar{\Omega}_{3,zw} \end{bmatrix}, \tag{3.15}$$

$$\bar{\nu}_2 \equiv \bar{\nu}_3 + K_w, \tag{3.16}$$

$$\bar{s}_2^2 \equiv \bar{\nu}_2^{-1} [\bar{\nu}_3 \bar{s}_3^2 + (\bar{b}_3^{yw.vz})' \bar{\Omega}_{3,ww}^{-1} \bar{b}_3^{yw.vz}]. \tag{3.17}$$

Note that (3.2)-(3.4), which are the prior distributions implied by natural conjugate prior (3.1) for H_0, H_1 and H_2, are also of natural conjugate form. Equations (3.6)-(3.9), (3.10)-(3.13) and (3.14)-(3.17) characterize the parameters of the distributions for H_0, H_1 and H_2, respectively.

Since the natural conjugate prior (3.1) is continuous, the exact zero restrictions upon which (3.2)-(3.4) are conditioned are assigned zero probability under (3.1), and hence so are hypotheses H_0, H_1 and H_2. In order to make the choice among H_j $(j=0,1,2,3)$ non-trivial, it is necessary to employ a "mixed" prior which assigns positive probability to the singleton points in the parameter space identified by sharp null hypotheses H_j $(j=0,1,2)$. Given prior probabilities

$$\bar{\pi}_j \equiv \text{Prob}\,(H_j) \qquad (j=0,1,2,3) \tag{3.18}$$

such that $\bar{\pi}_0 + \bar{\pi}_1 + \bar{\pi}_2 + \bar{\pi}_3 = 1$, a mixed prior density/mass function which does this is [2]

$$p(\beta,\sigma^{-2}) = \begin{cases} \bar{\pi}_0 p_0(\beta_0,\sigma^{-2} \mid \beta^{yz.vw} = 0_{K_z}, \beta^{yw.vz} = 0_{K_w}), & \text{if } \beta^{yz.vw} = 0_{K_z} \text{ and } \beta^{yw.vz} = 0_{K_w} \\ \bar{\pi}_1 p_1(\beta_1,\sigma^{-2} \mid \beta^{yz.vw} = 0_{K_z}), & \text{if } \beta^{yz.vw} = 0_{K_z} \\ \bar{\pi}_2 p_2(\beta_2,\sigma^{-2} \mid \beta^{yw.vz} = 0_{K_w}), & \text{if } \beta^{yw.vz} = 0_{K_w} \\ \bar{\pi}_3 p_3(\beta_3,\sigma^{-2}). & \text{otherwise} \end{cases} \tag{3.19}$$

Note that $p_0(\bullet)$, $p_1(\bullet)$ and $p_2(\bullet)$ are not arbitrary, but rather they are connected by their mutual derivation from $p_3(\bullet)$. As a result (3.19) requires assessment of only the four prior probabilities $\bar{\pi}_j$ $(j=0,1,2,3)$, the K elements in $\bar{b}_3$, the $K_3(K_3+1)/2$ distinct elements in $\bar{\Omega}_3$ and the two scalars $\bar{\nu}_3$ and $\bar{s}_3^2$. If the parameters in $p_0(\bullet)$, $p_1(\bullet)$ and $p_2(\bullet)$ are not tied to those in $p_3(\bullet)$, then substantially more quantities need to be assessed, namely,

$$3K_v + K_z + K_w + \tfrac{1}{2}\,[K_v(K_v+1) + (K_v+K_z)(K_v+K_z+1) + (K_v+K_w)(K_v+K_w+1)] + 6 \tag{3.20}$$

in number. Clearly, the saving by use of (3.19) is large when the number of common regressors, K_v, is large.

Prior probabilities $\bar{\pi}_j$ $(j=0,1,2,3)$ in (3.18) are usually treated as unconstrained. Sometimes, however, it may be desirable to also tie them in some way to the prior $p_3(\beta,\sigma^{-2})$ for the comprehensive model. For example, it may seem peculiar to assign, say, a high weight to H_1 if $p_3(\beta,\sigma^{-2})$ is small in the region covered by H_1. One way to overcome such difficulties is to set $\bar{\pi}_j = \tilde{\pi}_j/d$ $(j=0,1,2,3)$, with

$$\tilde{\pi}_j \equiv p_3(\tilde{\beta}_j, \tilde{\sigma}_j^2) \qquad (j=0,1,2,3), \tag{3.21}$$

$$d \equiv \sum_{j=0}^{3} \tilde{\pi}_j \ , \tag{3.22}$$

where $\tilde{\beta}_j$ sets all elements of β equal to those values restricted by H_j and all remaining elements equal to their modal values given in $\bar{b}_3$, and $\tilde{\sigma}_j^2$ is the modal value of σ^{-2} given $\tilde{\beta}_j$:

$$\tilde{\sigma}_j^2 \equiv \frac{K_3 + \bar{\nu}_3 - 2}{\bar{\nu}_3 \bar{s}_3^2 + (\tilde{\beta}_j - \bar{b}_3)' \,\bar{\Omega}_3^{-1}(\tilde{\beta}_j - \bar{b}_3)} \qquad (j=0,1,2,3) \ . \tag{3.23}$$

If this procedure is followed, then four fewer quantities need be assigned in formulating prior (3.19). If in addition,

$$\bar{b}_3^{yw.vz} = 0_{K_w} \ , \tag{3.24}$$

$$\bar{b}_3^{yz.vw} = 0_{K_z} \ , \tag{3.25}$$

2. Dickey (1976) views a mixed prior distribution such as (3.19) as an approximation to a continuous distribution with a high local concentration of probability near the atoms of positive probability. He obtains bounds on the ratios of posterior densities and Bayes' factors for the continuous prior relative to those of the mixed prior. In Dickey's terminology, $p_3(\beta, \sigma^{-2})$ is a "device density" used to insure "conditional continuity". Unlike Dickey, β and σ^{-2} are treated here as a priori dependent.

then $\bar{\pi}_j = .25$ $(j=0,1,2,3)$, i.e., if $H_j(j=0,1,2,3)$ equate $\beta^{yw.vz}$ and $\beta^{yz.vw}$ to their prior modal values, then all hypotheses are equally likely.

Use of (3.19) as opposed to natural conjugate prior (3.1) implies very different prior beliefs. Marginal and conditional distributions under (3.19) are mixtures over H_j $(j=0,1,2,3)$. In the case of $\beta^{yw.vz}$ and $\beta^{yz.vw}$ two of the distributions being mixed are degenerate spikes at zero. In the case of $\beta^{yv.wz}$ and σ^{-2} all four distributions are nondegenerate.

4. POSTERIOR ANALYSIS

Since prior p.d.f.'s (3.1)-(3.4) are conjugate priors given H_3, H_0, H_1 and H_2, respectively, the posterior distribution of β_j and σ^{-2} given H_j is

$$\beta_j,\sigma^{-2}|y,H_j \sim NG(\bar{\bar{b}}_j, \bar{\bar{\Omega}}_j, \bar{\bar{s}}_j^2, \bar{\bar{\nu}}_j) \qquad (j=0,1,2,3)\,, \tag{4.1}$$

where

$$\bar{\bar{\nu}}_j \equiv \bar{\nu}_j + T\,, \tag{4.2}$$

$$\bar{\bar{\Omega}} \equiv (\bar{\Omega}_j^{-1} + X_j'X_j)^{-1}\,. \tag{4.3}$$

$$\bar{\bar{b}}_j \equiv \bar{\bar{\Omega}}[\bar{\Omega}_j^{-1}\bar{b}_j + X_j'X_jb_j] \qquad (j=0,1,2,3)\,, \tag{4.4}$$

$$\bar{\bar{s}}_j^2 \equiv \bar{\bar{\nu}}_j^{-1}[\bar{\nu}_j\bar{s}_j^2 + Q_j] \qquad (j=0,1,2,3)\,, \tag{4.5}$$

with

$$b_j \equiv (X_j'X_j)^{-1}X_j'y \qquad (j=0,1,2,3) \tag{4.6}$$

$$s_j^2 \equiv \nu_j\text{SSE}_j \equiv (T-K_j)^{-1}(y-X_jb_j)'(y-X_jb_j) \quad (j=0,1,2,3) \tag{4.7}$$

denoting the familiar OLS estimators of β_j and σ^2 under H_j, and

$$Q_j = (y-X_j\bar{b}_j)'[I_T - X_j\bar{\bar{\Omega}}_jX_j']^{-1}(y-X_j\bar{b}_j)\quad . \tag{4.8}$$

Comparison of the relative merits of the hypotheses H_i and H_j can be made from the posterior odds ratio

$$\frac{\bar{\bar{\pi}}_i}{\bar{\bar{\pi}}_j} = \frac{\text{Prob}\,(H_i|y)}{\text{Prob}\,(H_j|y)} = B_{ij}\left(\frac{\bar{\pi}_i}{\bar{\pi}_j}\right) \tag{4.9}$$

where B_{ij} is the Bayes' factor (i.e., ratio of marginalized likelihoods) given by

$$B_{ij} = \left[\frac{\Gamma(\bar{\bar{\nu}}_i/2)/\Gamma(\bar{\nu}_i/2)}{\Gamma(\bar{\bar{\nu}}_j/2)/\Gamma(\bar{\nu}_j/2)}\right]\cdot\left[\frac{|\bar{\Omega}_i^{-1}|/|\bar{\bar{\Omega}}_i^{-1}|}{|\bar{\Omega}_j^{-1}|/|\bar{\bar{\Omega}}_j^{-1}|}\right]^{-1/2}\frac{[\bar{\nu}_i\bar{s}_i^2]^{\bar{\nu}_i/2}}{[\bar{\nu}_j\bar{s}_j^2]^{\bar{\nu}_j/2}}$$

$$\cdot\frac{[\bar{\nu}_i\bar{s}_i^2 + \nu_is_i^2 + (b_i-\bar{b}_i)'\{\bar{\Omega}_i + (X_i'X_i)^{-1}\}^{-1}(b_i-\bar{b}_i)]^{-\bar{\bar{\nu}}_i/2}}{[\bar{\nu}_j\bar{s}_j^2 + \nu_js_j^2 + (b_j-\bar{b}_j)'\{\bar{\Omega}_j + (X_j'X_j)^{-1}\}^{-1}(b_j-\bar{b}_j)]^{-\bar{\bar{\nu}}_j/2}}\;. \tag{4.10}$$

While (4.10) is of the standard form [e.g., see Leamer (1978, p. 111)], it is important to remember that unlike in the usual case, prior and posterior parameters in (4.10) are connected across H_j $(j=0,1,2,3)$ because $p_j(\beta_j,\sigma^{-2}|H_j)$ $(j=0,1,2)$ are derived from natural conjugate prior (3.1) for the comprehensive hypothesis H_3.

Such connections are usually ignored when computing posterior odds of competing hypotheses. Their importance can be appreciated by considering the behaviour of Bayes' factor (4.10) as prior information becomes diffuse. This topic has received a great deal of attention in the literature [e.g., see Leamer (1978, pp. 110-114)] and the general conclusion that emerges is that the answer is indeterminate - different answers arise depending on the manner in which prior information becomes diffuse under the two competing hypotheses. Since in the present situation the conditional priors are related, some ambiguity is removed here.

To see this, consider becoming diffuse in conjugate prior (3.1) for comprehensive model H_3. We will do this by first assuming that prior information on β, given σ^{-2}, is a replication of sample information in the sense that

$$\underline{\Omega}^{-1} = gX_3'X_3 . \tag{4.11}$$

for some scalar $g \geq 0$. Condition (4.11) has often been used in Bayesian analyses [e.g., see Zellner (1981)] and it yields a highly simplified expresssion for the posterior mean (4.4) of β_j under H_j:

$$\bar{\bar{b}}_j = (g+1)^{-1}[g\bar{b}_j + b_j] \qquad (j=0,1,2,3) . \tag{4.12}$$

In other words (4.12) implies the posterior means lies on the line segment joining the prior mean $\bar{b}_j$ and the OLS estimator b_j. In terms of Bayes' factor (4.10) assumption (4.11) has the attractive property that it simplifies the second factor in (4.10), the square root of the ratio of relative prior to posterior precision under each hypothesis, to be independent of the sample design:

$$\left[\frac{|\underline{\Omega}_i^{-1}|/|\bar{\bar{\Omega}}_i^{-1}|}{|\underline{\Omega}_j^{-1}|/|\bar{\bar{\Omega}}_j^{-1}|}\right]^{1/2} = \left(\frac{g}{g+1}\right)^{(K_i-K_j)/2} \tag{4.13}$$

Note that in this approach the question of whether the factor g differs across hypotheses [see Leamer (1978, p. 111)] does not arise since it is introduced in (4.11) in terms of the comprehensive model from which the competing hypotheses are derived.

Putting $g=\underline{\nu}_3$ we will consider diffuseness in the comprehensive model to occur as $g = \underline{\nu}_3 \rightarrow 0$. This implies that the effective sample size $\underline{\nu}_3$ of the prior goes to zero at the same rate as the relative precision g of the prior goes to zero. As $g=\underline{\nu}_3 \rightarrow 0$ prior precision (4.11) approaches zero, posterior mean (4.12) approaches the OLS estimator b_j, $Q_j \rightarrow \nu_j s_j^2 \equiv$ SSE$_j$ $(j=0,1,2,3)$ [see (4.8)] and $\bar{\nu}_j \rightarrow \bar{\nu}_j^*(j=0,1,2)$ where

$$\bar{\nu}_j^* \equiv \begin{cases} K_w + K_z , & \text{if } j = 0 \\ K_z \quad , & \text{if } j = 1 \\ K_w \quad , & \text{if } j = 2 \end{cases} . \tag{4.14}$$

Hence, as $g = \underline{\nu}_3 \rightarrow 0$ the posterior distribution $\beta_j,\sigma^{-2}|y,H_j \rightarrow NG(b_j, (X_j'X_j)^{-1}, s_j^2, T)$ mimicking the OLS results except for degrees of freedom.

Before investigating the behaviour of Bayes' factor (4.10) under (4.11) and $g = \underline{\nu}_3 \rightarrow 0$, first note that (3.9), (3.13), (3.17), (4.11) and $g = \underline{\nu}_3$ imply

$$\underline{\nu}_3 \underline{s}_i^2 = \underline{\nu}_3 \delta_i \qquad (i=0,1,2) , \tag{4.15}$$

where

$$\delta_0 \equiv \bar{s}_3^2 + [(\bar{b}_3^{yw.vz})',(\bar{b}_3^{yz.vw})'] \begin{bmatrix} A^{ww} & A^{wz} \\ A^{zw} & A^{zz} \end{bmatrix}^{-1} \begin{bmatrix} \bar{b}_3^{yw.vz} \\ \bar{b}_3^{yz.vw} \end{bmatrix} , \tag{4.16}$$

$$\delta_1 \equiv \bar{s}_3^2 + (\bar{b}_3^{yz.vw})'(A^{zz})^{-1}(\bar{b}_3^{yz.vw}) , \tag{4.17}$$

$$\delta_2 \equiv \bar{s}_3^2 + (\bar{b}_3^{yw.vz})'(A^{ww})^{-1}(\bar{b}_3^{yw.vz}) , \tag{4.18}$$

and

$$A \equiv \begin{bmatrix} A^{vv} & A^{vw} & A^{vz} \\ A^{wv} & A^{ww} & A^{wz} \\ A^{zv} & A^{zw} & A^{zz} \end{bmatrix} \equiv (X_3'X_3)^{-1} . \tag{4.19}$$

The quantities δ_j ($j=0,1,2$) reflect the anticipated fits for $H_j(j=0,1,2)$ under the prior beliefs for the comprehensive model. These fits equal the sum of the location parameter $\bar{s}_3^2$ in the prior for σ^2 and a quadratic form, which under the prior for H_3, is proportional to a chi-square variable measuring the deviation of H_j from the prior in the comprehensive model.

Combining (4.13) and (4.14), it follows that under (4.11) the middle two factors in Bayes' factor (4.10) satisfy (for $i=0,1,2; j=1,2$):

$$\operatorname*{limit}_{g=\bar{\nu}_3\to 0} \left[\frac{|\bar{\Omega}_i^{-1}|/|\bar{\bar{\Omega}}_i^{-1}|}{|\bar{\Omega}_j^{-1}| \quad |\bar{\bar{\Omega}}_j^{-1}|}\right]^{-1/2} \frac{[\bar{\nu}_i \bar{s}_i^2]^{\bar{\nu}_i/2}}{[\bar{\nu}_j \bar{s}_j^2]^{\bar{\nu}_j/2}} = \frac{\delta_i^{\bar{\nu}_i^*}}{\delta_j^{\bar{\nu}_j^*}} . \tag{4.20}$$

Assuming (4.11) and defining

$$B_{ij}^* = \operatorname*{limit}_{g=\bar{\nu}_3\to 0} B_{ij} \qquad (i,j = 0,1,2,3; i < j), \tag{4.21}$$

it follows from (3.9), (3.13), (3.17), (4.10) and (4.20) that

$$B_{01}^* = \left[\frac{\Gamma[(T+K_z+K_w)/2]/\Gamma[(K_z+K_w)/2]}{\Gamma[(T+K_z)/2]/\Gamma(K_z/2)}\right]\left[\frac{SSE_1}{SSE_0}\right]^{T/2} \frac{[SSE_1/\delta_1]^{K_z/2}}{[SSE_0/\delta_0]^{(K_w+K_z)/2}} , \tag{4.22}$$

$$B_{02}^* = \left[\frac{\Gamma[(T+K_z+K_w)/2]/\Gamma[(K_z+K_w)/2]}{\Gamma[(T+K_w)/2]/\Gamma(K_w/2)}\right]\left[\frac{SSE_2}{SSE_0}\right]^{T/2} \frac{[SSE_2/\delta_2]^{K_w/2}}{[SSE_0/\delta_0]^{(K_w+K_z)/2}} , \tag{4.23}$$

$$B_{12}^* = \left[\frac{\Gamma((T+K_z)/2]/\Gamma(K_z/2)}{\Gamma[(T+K_w)/2]/\Gamma(K_w/2)}\right]\left[\frac{SSE_2}{SSE_1}\right]^{T/2} \frac{[SSE_2/\delta_2]^{K_w/2}}{[SSE_1/\delta_1]^{K_z/2}} , \tag{4.24}$$

$$B_{i3}^* = \infty \qquad (i=0,1,2) . \tag{4.25}$$

Limiting expressions (4.22)-(4.25) are most interesting. Firstly, only in cases involving comprehensive model H_3, i.e., (4.25), does a trivial answer emerge, namely, the Bayes' factor equals infinity favouring the smaller models H_j ($j=0,1,2$). This result can be anticipated from the literature [see Leamer (1978, p. 111)]. In all other cases the Bayes' factor does not obviously favour one model or the other.

Secondly, in the other cases [i.e., (4.22)-(4.24)] the Bayes' factor consists of three factors: the first factor favours the smaller of the two models, the second factor compares the *absolute* goodness-of-fit of the two models, and the third factor compares the *relative* fit of the two models (i.e., the fit in terms of the data relative to the anticipated fit based on prior opinions about the comprehensive model). Thus (4.22)-(4.24) all involve nontrivial "parsimony" versus "fit" tradeoffs. Given that these Bayes' factors are all derived under diffuseness, this is a curious result [cf. Leamer (1978, p. 111)], and it warrants careful examination.

The parsimony factor can be substantial even when the competing models differ only slightly in size.[3] As sample size increases the degree to which the small model is favoured also increases. In the case of non-nested models of the same size, the parsimony factor equals unity and (4.24) reduces to

$$B_{12}^{*} = \left[\frac{SSE_2}{SSE_1}\right]^{T/2}\left[\frac{SSE_2/\delta_2}{SSE_1/\delta_1}\right]^{\bar{K}/2}, \tag{4.26}$$

where $\bar{K} = K_w = K_z$. If in addition both models are expected a priori to fit equally well (i.e., $\delta_1 = \delta_2$), then Bayes' factor (4.26) simply favours the model yielding the smaller sum of squared errors.

While the second factors in (4.22)-(4.24) comparing the absolute fits of the two competing models commonly appears in Bayes' factors derived under diffuse prior information, the third factor comparing relative fits does not. Its appearance here is surprising since it involves prior location parameters ($\bar{s}_3^2$, $\bar{b}_3^{yw.vz}$, $\bar{b}_3^{yz.vw}$) for comprehensive model H_3 through δ_j ($j=0,1,2$), and these quantities normally vanish from a problem under a diffuse prior. In the present case, however, as the degrees of freedom $\bar{\nu}_3$ in the prior for σ^{-2} given H_3 approaches zero, the degrees of freedom $\bar{\nu}_j$ ($j=0,1,2$) in the priors for σ^{-2} given H_j ($j=0,1,2$) do *not* approach zero [see (4.14)]. Nonetheless, the modes and the variances of these priors are driven off to infinity suggesting diffuseness. However, the rates at which diffuseness arises in H_j ($j=0,1,2$) are restricted in such a manner that (4.21) converges to a quantity involving location parameters of the prior (3.1).

The existence of δ_j ($j=0,1,2$) in Bayes' factors (4.21)-(4.24) serves two purposes. Firstly, since the sum of squared errors in the third factors are raised to different powers [except in the case of (4.24) and $K_z = K_w$], unless each is divided by a quantity measured in the square of the units of y_t, then these Bayes' factors would not be invariant to changes in the units of measurement. The presence of δ_j ($j=0,1,2$) in (4.22)-(4.24) ensures such invariance. Secondly, the presence of δ_j ($j=0,1,2$) in (4.22)-(4.24) implies that how well each model fits the data should be measured relative to how well a priori the model is expected to fit the data. Somewhat counter-intuitively, $\partial B_{ij}^{*}/\partial\delta_i > 0$ in (4.22)-(4.24) suggesting that if the anticipated fit if H_i increases, then the Bayes' factor B_{ij}^{*} also increases suggesting the data give additional support to H_i. It must be remembered, however, that such a change is *ceteris paribus*, holding fixed the actual observed fits SSE_i and SSE_j of the competing hypotheses. Given the observed fits, increasing the anticipated fit δ_i of H_i implies that the observed fit is, relatively speaking, better than anticipated, and hence the

3. Note that it is possible in the non-nested case of (4.24) that the smaller model may actually fit better absolutely (i.e., yield a smaller sum of squared errors) and yet not be favoured by (4.24). This somewhat perverse result reflects the importance of fit relative to anticipated a priori fit rather than merely just absolute fit.

data provide relatively more support for H_i. Of course if one accepts the argument leading to the assignment of prior odds $\bar{\pi}_i/\bar{\pi}_j$ according to (3.21), then increasing δ_i also decreases $\bar{\pi}_i$ and hence, *ceteribus paribus*, decreases the posterior odds $\bar{\bar{\pi}}_i/\bar{\bar{\pi}}_j$.

To investigate the net effect on the posterior odds of increasing δ_i, suppose that $\bar{\pi}_j$ $(j=0,1,2)$ are determined by (3.21). Then under the same assumptions used to derive (4.22)-(4.24), it follows that

$$\frac{\bar{\pi}_i^*}{\bar{\pi}_j^*} \equiv \lim_{g=\nu_3\to 0} \frac{\bar{\pi}_i}{\bar{\pi}_j} = \left(\frac{\delta_i}{\delta_j}\right)^{(2-K_3)/2} \tag{4.27}$$

for $i,j = 0,1,2$; $i<j$. Combining (4.27) with (4.22)-(4.24) it follows that the limiting posterior odds are

$$\frac{\bar{\bar{\pi}}_i^*}{\bar{\bar{\pi}}_j^*} \equiv \lim_{g=\nu_3\to 0} \frac{\bar{\bar{\pi}}_i}{\bar{\bar{\pi}}_j} = B_{ij}^* \frac{\bar{\pi}_i^*}{\bar{\pi}_j^*}, \tag{4.28}$$

or more specifically,

$$\frac{\bar{\bar{\pi}}_0^*}{\bar{\bar{\pi}}_1^*} = \left[\frac{\Gamma[(T+K_z+K_w)/2]/\Gamma[(K_z+K_w)/2]}{\Gamma[(T+K_z)/2]/\Gamma(K_z/2)}\right] \frac{[SSE_1]^{(T+K_z)/2}}{[SSE_0]^{(T+K_w+K_z)/2}} \frac{\delta_0^{(2-K_v)/2}}{\delta_1^{(2-K_v-K_w)/2}}, \tag{4.29}$$

$$\frac{\bar{\bar{\pi}}_0^*}{\bar{\bar{\pi}}_2^*} = \left[\frac{\Gamma[(T+K_z+K_w)/2]/\Gamma[(K_z+K_w)/2]}{\Gamma[(T+K_w)/2]/\Gamma(K_w/2)}\right] \frac{[SSE_2]^{(T+K_w)/2}}{[SSE_0]^{(T+K_w+K_z)/2}} \frac{\delta_0^{(2-K_v)/2}}{\delta_2^{(2-K_v-K_z)/2}}, \tag{4.30}$$

$$\frac{\bar{\bar{\pi}}_1^*}{\bar{\bar{\pi}}_2^*} = \left[\frac{\Gamma[(T+K_z)/2]/\Gamma(K_z/2)}{\Gamma[(T+K_w)/2]/\Gamma(K_w/2)}\right] \frac{[SSE_2]^{(T+K_w)/2}}{[SSE_1]^{(T+K_z)/2}} \frac{\delta_1^{(2-K_v-K_w)/2}}{\delta_2^{(2-K_v-K_z)/2}}. \tag{4.31}$$

Note that in (4.29)-(4.31), $\partial(\bar{\bar{\pi}}_i^*/\bar{\bar{\pi}}_j^*)/\partial\delta_i < 0$ indicating that the net effect of an increased anticipated fit in H_i is to lower the posterior probability of H_i. This result illustrates the importance of tying both the prior probabilities of hypotheses, as well as the conditional prior distributions for the unrestricted parameters in the hypotheses, to the common beliefs concerning the comprehensive model.

Thirdly and finally, the novelty of the nontrivial forms of Bayes' factors (4.22)-(4.24) and posterior odds (4.29)-(4.31) under diffuse prior information for comprehensive model H_3 must be tempered by the realization that prior parameters δ_j $(j=0,1,2)$ must be assessed. This assessment is the "price" to be paid for obtaining the nontrivial results. Of course if interest lies, say, only in comparing the posterior odds to unity, then the researcher can simply equate $\bar{\bar{\pi}}_i^*/\bar{\bar{\pi}}_j^*$ to unity yielding a curve in the δ_i vs. δ_j plane for which all points on one side favour H_i and all points on the other side favour H_j. If both (3.24) and (3.25) hold, then $\delta_j = \bar{s}_j^2$ $(j=0,1,2)$ and the problem simplifies to deciding on what side of some critical value $\bar{s}_3^2$ lies.

To gain insight into the case where (3.24) and/or (3.25) do not hold, consider the following. Suppose

$$\bar{b}_3^{yw.vz} = \epsilon_w(\bar{s}_3 r_w), \tag{4.32}$$

$$\bar{b}_3^{yz.vw} = \epsilon_z(\bar{s}_3 r_z), \tag{4.33}$$

where ϵ_w and ϵ_z are nonnegative scalars, and r_w and r_z are vectors with elements whose absolute values equal the reciprocal of the sample standard deviations of each regressor in

W and Z, respectively. We will assume signs are attached to the elements of r_w and r_z so that they indicate the direction of the most interesting departures of the elements in $\bar{b}_3^{yw.vz}$ and $\bar{b}_3^{yz.vw}$ from the zero values postulated by (3.24) and (3.25). The parts of (4.32) and (4.33) in parentheses can be interpreted as a vector in which each element represents a change in a standardized unit if the dependent variable [standardized in terms of the standard error of the regression as estimated by the prior hyperparameter ($\bar{s}_3$)] with respect to a change in a standardized unit of a particular regressor (standardized by the respective sample standard deviation). When $\epsilon_w = \epsilon_z = 0$, (3.24) and (3.25) are satisfied. Increasing values of ϵ_w and ϵ_z indicate increased departures in a particular direction of the modal values in the prior for $\beta^{yw.vz}$ and $\beta^{yz.vw}$ from the zero values postulated by H_j ($j=0,1,2$). Note that ϵ_w and ϵ_z are unit free. They measure departures in terms of how many standardized units the mean of y_t will change when a particular regressor is changed one standardized unit. By varying the three scalars $\bar{s}_3^2$, ϵ_w and ϵ_z a simple sensitivity analysis of the posterior odds can be conducted with respect to changes in the location of the prior for the comprehensive model.

5. CONCLUDING COMMENTS

Although the preceding results for a limiting noninformative prior have only been derived under the assumption of a conjugate prior for the comprehensive model, they can be extended to other situations provided the prior dependency between β and σ^2 is maintained. Without such dependency no learning about σ^2 occurs when restriction on β are imposed, and the relative rates at which diffuseness occurs in the prior for σ^2 in competing models will be indeterminate. This rules out use of the Student-gamma prior of Dickey (1974) which reduces to $p(\beta,\sigma) \propto \sigma^{-1}$ in the case of diffuseness. If dependency between β and σ^2 is a priori tolerable, then the natural conjugate prior distribution for H_3, and derived consistent prior (3.19), are attractive tractable priors.

In closing it is worth emphasizing that the noninformative framework discussed here results in a fairly strong tendency toward parsimony. The comprehensive hypothesis H_3 is never preferred to any of the restricted hypotheses, and the leading parsimony factors in (4.22)-(4.24) and (4.29)-(4.31) are important factors in comparing H_0, H_1 and H_2. If prior beliefs are centred over H_0, then only the quantity $\bar{s}_3^2$ needs be assessed in order to compare the latter three hypotheses. To the extent that such comparisons are important, the framework suggested here adds a new criterion to the extensive list of model selection criteria already in the literature. The framework suggested here, however, like these competitors, does not escape the criticism noted in footnote 1.

REFERENCES

BERKSON, J. (1938). Some difficulties of interpretation encountered in the application of the chi-square test. *J. Amer. Statist. Assoc.* **33**, 526-536.

DICKEY, J.M. (1974). Bayesian Alternatives to the *F* Test and the Least Squares Estimate in the Normal Linear Model. *Studies in Bayesian Econometrics and Statistics* (S.E. Fienberg and A. Zellner eds.) 515-554. Amsterdam: North-Holland.

— (1976). Approximate posterior distributions. *J. Amer. Statist. Assoc.* **71**, 680-689.

GRANGER, C.W.J. and NEWBOLD, P. (1977). *Forecasting Economic Time Series.* New York: Academic Press.

KADANE, J.B. and DICKEY, J.M. (1980). Bayesian Decision Theory and the Simplification of Models. *Evaluation of Econometric Models* (J. Kmenta and J.B. Ramsey eds.) 245-268. New York: Academic Press.

LEAMER, E.E. (1978). *Specification Searches: Ad Hoc Inference with Nonexperimental Data.* New York: John Wiley and Sons.

RAIFFA, H. and SCHLAIFER, R. (1961). *Applied Statistical Decision Theory.* Boston: Harvard Business School.

ZELLNER, A. (1981). On assessing prior distributions and Bayesian regression analysis with *g*-prior distributions. Unpublished manuscript (revised), University of Chicago.

BAYESIAN STATISTICS 2, pp. 723-732
J.M. Bernardo, M.H. DeGroot, D.V. Lindley, A.F.M. Smith (Eds.)

Hierarchical Models for Seasonal Time Series

WOLFGANG POLASEK
University of Vienna

SUMMARY

This paper describes two kinds of recently developed Bayesian inference techniques for seasonal distributed lag models. The first kind of inference technique involves the notion of the "likelihood of a Bayesian model" of Akaike (1982) to estimate the smoothness parameters in a univariate distributed lag model. Using a multivariate time series set-up we generalize this concept to the seasonal distributed lag model with smoothness prior. The second type of hierarchical model assumes exchangeability between seasons. The robustness of the Bayesian inference is checked by the extreme bound analysis (Leamer, 1978) for the posterior mean of the distributed lag model. An example is given based on the quarterly import function for Austria.

Keywords: EXCHANGEABILITY; LAG MODEL.

1. INTRODUCTION

In recent time there has been a rapid development in the application of hierarchical models for various purposes. In this paper we discuss some applications to so-called seasonal distributed lag models, which have gained some popularity in time series analysis and econometrics.

In particular we want to contrast two types of hierarchical models which differ only in the type of hierarchy. The hierarchical model with smoothness prior has prior parameters which are the ratios of variances of the first and the second stage. This model was investigated by Shiller (1973) and Akaike (1982) and will be extended for the seasonal case in this paper. It is contrasted with a hierarchical seasonal distributed lag model which is based on the exchangeability assumption suggested by Lindley and Smith (1972). The parameters of the prior distribution are estimated in both models by the data of the first stage which gives the method an empirical Bayes flavour. While the Akaike approach also allows the determination of the optimal differencing parameters and the lag length of the model, the hierarchical exchangeability model can be used for a robustness analysis by hierarchical extreme bounds as in Polasek (1983).

In section 2 we analyse the distributed lag model with smoothness prior and estimate the hyperparameter by the maximisation of the conditional likelihood. Akaike (1982) uses the name "likelihood of a Bayesian model" because hyperparameters are used in the second stage of this Bayesian analysis. This method is extended to seasonal processes with smoothness prior in section 3.

In section 4 we show how a seasonal distributed lag model can be formulated by a simple hierarchical model if we assume exchangeability between seasons. This model assumes a distributed lag model for every season, and for the regression coefficients of every season a common hyper-distribution. The parameters of this hyper-distribution are estimated by generalised least squares (GLS) as in Lindley and Smith (1972). We demonstrate for the example of the Austrian import function how the extreme bound analysis (EBA) can be used for a Bayesian robustness analysis. Besides the ordinary EBA as in Leamer (1982) we discuss the so-called symmetrical extreme bound analysis for hierarchical models as in Polasek (1983). The posterior means are said to be robust against covariance specifications of the hyperdistribution if the upper and lower bounds of the coefficients have the same sign.

In a concluding section we summarize these two approaches of Bayesian modeling for distributed lag models.

2. DISTRIBUTED LAG MODELS WITH SMOOTHNESS PRIORS

The univariate distributed lag model is given by

$$y_t = \sum_{m=0}^{M} \beta_m x_{t-m} + \epsilon_t, \qquad t = 1,\dots,T \tag{2.1}$$

or compactly written as

$$y = X\beta + \epsilon, \qquad \mathrm{Var}(\epsilon) = \sigma^2 I_T \tag{2.2}$$

where the matrices are defined as

$$y = \begin{pmatrix} y_1 \\ \vdots \\ y_T \end{pmatrix}, X = \begin{pmatrix} x_1 & x_0 & \dots & x_{-M+1} \\ \vdots & \vdots & \ddots & \vdots \\ x_T & x_{T-1} & \dots & x_{T-M} \end{pmatrix} \tag{2.3}$$

The variable y is called the output variable and x the input variable, where the first M observations are used for the initial values $x_0, x_{-1},\dots,x_{-M+1}$. The matrix X is of order $(T\times M)$ and is called the lagged input matrix where M denotes the number of lags. All variables are supposed for simplicity to be mean centered.

The smoothness prior assumption was first introduced by Shiller (1973) and takes the following form: Let R^d be the differencing matrix of order d, defined by

$$R^d = (R_1)^d, \quad \text{where} \quad R_1 = \begin{pmatrix} 1 & -1 & 0 & \dots & 0 \\ 0 & 1 & -1 & & 0 \\ \vdots & & & & \vdots \\ 0 & 0 & 0 & \dots & 0\ 1 \end{pmatrix} \tag{2.4}$$

Then a smoothness prior assumes that the d-th difference of the coefficient vector b is small, i.e.

$$R^d\beta \sim N(0, \tau^2 I) \tag{2.5}$$

If τ is parameterised as ratio λ/σ then the prior distribution becomes

$$p(\beta\,|\,\sigma,\lambda) = \left(\frac{1}{2\pi\sigma^2}\right)^{(M+1)/2} |\lambda^2 R^{d\prime} R^d|^{1/2} \exp\{\frac{\lambda^2}{2\sigma^2}\,\beta' R^{d\prime} R^d \beta\} \tag{2.6}$$

The hierarchy in this model concentrates on the variance parameters and we assume no further informative distribution for the variante ratio λ. As was already shown by Shiller (1973) the posterior mean for this model can be expressed as a matrix weighted average of the form

$$b^{**} = (X'X + \lambda^2 R^{d\prime} R^d)^{-1} X'y \tag{2.7}$$

By augmenting the matrices in (2.3) the posterior mean can be calculated by simple OLS-procedures:

$$b^{**} = (X^{*\prime} X^*)^{-1} X^{*\prime} y^* \tag{2.8}$$

where

$$X^* = \begin{pmatrix} X \\ - \\ \lambda R^d \end{pmatrix}, \qquad y^* = \begin{pmatrix} y \\ - \\ 0 \end{pmatrix} \tag{2.9}$$

This estimation procedure is conditional on a pre-chosen lag length M, a fixed order of smoothness d, and the variance ratio λ. Therefore Akaike (1982) suggested using his approach of Bayesian modeling to determine the optimal parameters in a likelihood framework. Assuming a normal distribution for (2.2) and the prior as in (2.6) we can derive a conditional likelihood function by integrating out β and σ:

$$p(y|\lambda) = \int\int p(y|\beta,\sigma) p(\beta|\sigma,\lambda) \sigma^{-1} \, d\beta d\sigma \tag{2.10}$$

The term σ^{-1} stems from a noninformative prior distribution for the variance σ^2. Using our model assumption we can show after some algebra that the conditional likelihood function is proportional to

$$p(y|\lambda) \alpha\, S(\lambda)^{-N/2} |X'X + \lambda^2 R^{d\prime} R^d|^{-1/2} |\lambda^2 R^{d\prime} R^d|^{1/2} \tag{2.11}$$

where the residual variance is given by

$$S(\lambda) = y'y - b^{**\prime}(X'X + \lambda^2 R^{d\prime} R^d) b^{**} . \tag{2.12}$$

Taking logarithms of (2.11) and ignoring constant terms we obtain the so-called Bayesian information criterion ABIC:

$$\text{ABIC} = N \log S(\lambda) + \log|X'X + \lambda^2 R^{d\prime} R^d| - \log|\lambda^2 R^{d\prime} R^d| . \tag{2.13}$$

Minimizing ABIC analytically with respect to λ is difficult; therefore a numerical search procedure has to be applied. Akaike reexamined the example given in Shiller (1973) and found the minimum ABIC for the parameters $d=3$, $M=19$ and $\lambda = 5.2^{**}5$.

3. SEASONAL DISTRIBUTED LAG MODELS WITH SMOOTHNESS PRIOR

The previous univariate distributed lag model can be extended to the seasonal case using a multivariate time series framework. The smoothness concept is now extended to all seasons. We consider the S seasons simultaneously and determine the easiest type of covariance specification in order to facilitate the computational burden. The model can be written as

$$\text{vec}\, Y = \text{diag}\,(X_{(1)},\ldots,X_{(s)})\, \text{vec}\, B + \text{vec}\, U, \quad \text{Varvec}\, U = D_\sigma \oplus I_T \tag{3.1}$$

with smoothness prior of the form

$$R^p B = 0 + E, \qquad \text{Varvec}\, E = D_{\text{vec}\Lambda} \oplus I_K . \tag{3.2}$$

Now Y is a $(T \times S)$ matrix of the seasonal output series and the X_s, $s=1,\ldots,S$ form a block diagonal matrix where every block is created by the seasonal lagged input variable as in (2.3). The $(M \times S)$ coefficient matrix B contains M lags for every season. For the error matrix U we assume the simplest form of a seemingly unrelated regression system which is expressed by the diagonal matrix $D\sigma = (\sigma_1,\ldots,\sigma_s)$.

In the second stage we extend the univariate smoothness assumption (2.5) to the S-dimensional case. The variance ratios have now to be extended to the $(K \times S)$ matrix Λ. Diag (vecΛ) denotes the diagonal matrix formed by all the elements of the matrix Λ.

In this case we can also apply the augmentation procedure of (2.7)-(2.9) to calculate the posterior mean for every season $s = 1,\ldots,S$:

$$b_s^{**} = (X_s^{*\prime} X_s^*)^{-1} X_s^{*\prime} y_s^*, \qquad s = 1,\ldots,S. \tag{3.3}$$

Now the augmented matrices are

$$y_s^* = \begin{pmatrix} y_s \\ \hline 0 \end{pmatrix}, \; X_s^* = \begin{pmatrix} X_{(1)}\ldots X_{(S)} \\ \hline \lambda_s \oplus R^p \end{pmatrix} \tag{3.4}$$

For the optimal choice of Λ we can maximize the conditional likelihood function $p(Y|\Lambda)$ which depends on the matrix Λ, if the other parameters are integrated out:

$$p(Y|\Lambda) = \int\int p(Y|B,D_\sigma).p(B,D_\sigma|\Lambda)dBdD_\sigma \,. \tag{3.5}$$

Here $p(Y|B,D_\sigma)$ denotes the likelihood function and is given by

$$p(Y|B,D_\sigma) \propto \exp(-\sum_{s=1}^{S} (y_s - X_{(s)}b_s)'(y_s - X_{(s)}b_s)/2\sigma_s^2) \tag{3.6}$$

and the prior distribution depending on Λ has the form

$$p(B,D_\sigma|\Lambda) = p(B|D_\sigma,\Lambda)p(D_\sigma)$$

$$\propto \prod_{s=1}^{S} \exp(-b_s'(D_{\lambda_s} \oplus R^{p\prime} R^p)b_s/2\sigma_s^2).1/\sigma_s \tag{3.7}$$

Assuming a normal distribution for both, the first and the second stage of the model (3.1)-(3.2) and a noninformative prior for D_σ we obtain by integrating out D_σ and B

$$p(Y|\Lambda) \propto \prod_{s=1}^{S} |R^{p\prime} R^p|^{1/2} (\prod_{j=1}^{S} \lambda_{js}^2)|X_s'X_s + D_{\lambda_s} \oplus R^{p\prime} R^p| S(\lambda_s)^{-(k-1)/2}, \tag{3.8}$$

where $S(\lambda_s)$ denotes the residual variances for every seasons:

$$S(\lambda_s) = y_s'y_s - b_s^{**\prime}(X_s'X_s + D_{\lambda_s} \oplus R^{p\prime} R^p)^{-1} b_s^{**}, \qquad s = 1,\ldots,S$$

where $S(\lambda_s)$ denotes the residual variances for every seasons:

$$S(\lambda_s) = y_s'y_s - b_s^{**\prime}(X_s'X_s + D_{\lambda_s} \oplus R^{p\prime} R^p)^{-1} b_s^{**}, \qquad s = 1,\ldots,S. \tag{3.9}$$

Here λ_s is the s-th column of Λ and D_{λ_s} is the diagonal matrix formed by this column vector. Again it is possible to transform the conditional likelihood into a ABIC-type information criterion:

$$\text{SBIC} = \sum_{s=1}^{S} [\log S(\lambda_s) + \log|X_s'X_s + D_{\lambda_s} \oplus R^{p\prime} R^p| - \log(|R^{p\prime} R^p| \sum_{j=1}^{S} \lambda_{js}^2)] \tag{3.10}$$

The abbreviation SBIC stands for seasonal Bayesian information criterion. The numerical search procedure now has to be extended to a grid of variance ratios.

4. SEASONAL DISTRIBUTED LAG MODELS AND EXCHANGEABILITY

Seasonal distributed lag models are simultaneous systems of distributed lag models for every season. Trivedi and Lee (1981) derived various linear hierarchical models along the lines of Lindley and Smith (1972) using different exchangeability assumptions for the distributed lag coefficients. In this section we give an example of such a model. We estimate the quarterly import function of Austria by a seasonal non-informative 3-stage hierarchical model and derive the hierarchical extreme bound analysis as in Polasek (1982).

4.1. *The model of the import function*

For the time-span 1964.1 to 1980.4 we have specified an import function in the manner of Leamer (1972). For every season $s=1,\dots,S(=4)$ we specify a distributed lag model with lag length L:

$$IM_{t,s} = \sum_{l=1}^{L} (P^*_{t-\ell,s}\beta_{1,l} + Y_t\beta_{2,\ell}) + \epsilon_{t,s}, \qquad s = 1,\dots,S. \tag{4.1}$$

where all variables are assumed to be mean corrected. In this equation IM is total real imports in billions of Austrian Schillings, Y is the gross national product (GNP) in constant prices (market value 1964), and P^* is the relative import price defined by PM/PIG71, where PM is the price index of imports (base 1971) and PIG71 is the price index of the GNP (base 1971).

For the matrix notation, we stack together the import variables in the $(T\times S)$ matrix Y and the distributed lag coefficient in the $(N\times S)$ matrix $\boldsymbol{B}$. Now $N=L.K$ is the number of columns of X and K is the number of explanatory time series (in our case $K=2$). The non-informative 3-stage model with exchangeability between seasons (EB), also called the SDL*EB model by Trivedi and Lee (1981), is given by

$$\begin{aligned} &Y_s \sim N(X_s\beta_s, \sigma_s^2 I_T) \\ &\beta_s \sim N(\beta^*, \Sigma)\,, \\ &\beta^* \sim N(0,\infty)\,, \end{aligned} \tag{4.2}$$

where $N(0,\infty)$ denotes a constant density over the entire real line. X_s denotes a $(T\times K)$ matrix of independent variables together with their L lags; I_T is a T dimensional identity matrix; and $D_\omega = \text{diag}(\omega_1,\dots,\omega_s)$ is a diagonal covariance matrix in analogy to a seemingly unrelated regression system specification. Thus, 1_s is the S-dimensional vector of ones and β^* the vector of hyperparameters which represent the parameters of the aggregate (yearly) distributed lag model. Σ denotes the second stage $(N\times N)$ covariance matrix, and the last line in (4.2) stands for a non-informative third stage prior distribution. The posterior means of the first stage can be calculated equation by equation by

$$b_s^{**} = (\omega_s^{-2} X_s'X_s + \Sigma^{-1})^{-1} (\omega_s^{-1}X_s'X_s\hat{b}_s + \Sigma^{-1}\hat{\beta}^*), \qquad s = 1,\dots,S, \tag{4.3}$$

where $\hat{b}_s$ is the ordinary least squares (OLS) estimate for every season, given by $\hat{b}_s = (X_s'X_s)^{-1}X_s'y_s, \qquad s = 1,\dots,S.$

Also, $\hat{\beta}^*$ is the estimate of the second stage hyperparameter which can be expressed as a general weighted matrix average of the OLS-estimates of the first stage (see also Smith, 1973):

$$\hat{\beta}^* = \sum_{s=1}^{S} W_s\hat{b}_s\,, \tag{4.4}$$

$$W_s = \left(\sum_{s=1}^{S} (X_s'X_s\omega_s^{-2} + \Sigma^{-1})^{-1} X_s'X_s\omega_s^{-2}\right)^{-1}. (X_s'X_s\omega_s^{-2} + \Sigma^{-1})^{-1} X_s'X_s\omega_s^{-2}. \tag{4.5}$$

In Table 4.1 we list the OLS-estimates of model (4.1) for lag 0. Looking at the sign pattern we see that the first quarter shows a positive coefficient for the relative import prices. The sensitivity analysis of the next section will show how sensitive this result is with respect to prior covariance specifications. Note that the last quarter is the most sensitive for the import prices while the second quarter shows the highest coefficient with respect to income.

In Table 4.2 we list the hyperparameter estimates based on three methods: The column with the label "Mean" contains just the simple average of the first stage OLS-coefficients. The second column ("GLS1") lists the GLS-estimate based on formula (4.4), while the third column ("GLS2") shows a weighted average of OLS-estimates of the first stage, where the weights are the residual variances for every equation. As can be seen, the estimates do not differ too much. Therefore the sensitivity analysis will not depend very much on the method used to estimate the hyperparameters.

TABLE 4.1
*SDL*EB model: OLS-estimates*

variable/season	1	2	3	4
P^*	.0033	−.0904	−.0489	−.1256
Y	.7003	.7311	.7168	.7051

TABLE 4.2.
*SDL*EB model: Hyperparameter estimates*

variable/method	*Mean*	*GLS1*	*GLS2*
P^*	−.0654	−.0630	−.0587
Y	.7133	.7102	.7128

4.2. *Hierarchical extreme bound analysis*

In this section we apply a Bayesian type of robustness analysis with respect to the second stage covariance matrix as it was developed in Polasek (1982). The idea is based on the ellipsoidal sensitivity analysis of Leamer (1978) and (1982) for matrix weighted averages when the prior covariance matrix is not known. It can be shown that the posterior mean (4.3) for unknown covariance matrix lies in the ellipsoid

$$(b_s^{**} - f_s)'X_s'X_s(b_s^{**} - f_s) < c_s, \qquad s = 1,\ldots,s \tag{4.6}$$

with the parameters

$$f_s = (\hat{b}_s - \beta^*)/2 \tag{4.7}$$

$$c_s = (\hat{b}_s - \beta^*)'X_s'X_s(\hat{b}_s - \beta^*)/4 \tag{4.8}$$

If the OLS-estimate and the hyperparameter estimate are fixed, then all possible posterior means which depend now only on the second stage covariance matrix Σ lie in this so-called feasible ellipsoid. In order to report the region of this ellipsoid we project it on each coordinate axis and obtain in this way the extreme bounds for every coefficient. Let ψ be a vector with known coefficients; then the extreme bounds of the ellipsoid are given by

$$E^{s}_{1\cdot 2} = \psi' f \pm (\psi' X_s' X_s \psi)^{1/2} \qquad (4.9)$$

In Table 4.3 these hierarchical extreme bounds are reported for the quarterly import function with lag 0. We see that all bounds for the coefficient of the relative import prices are negative except for the upper bound of the first quarter. This implies a certain robustness of the posterior mean with respect to changes in the prior covariance matrices. Note that the third quarter has the tightest bounds for both variables, while the first quarter has the widest ones. The second and the fourth quarters allow for highest price sensitivities.

TABLE 4.3.
Hierarchical extreme bounds for the 1st. stage parameters

variable/season	1	2	3	4
P^* *lower*	−.1010	−.1409	−.0647	−.1373
upper	.0414	−.0124	−.0472	−.0512
Y *lower*	.6834	.6974	.7102	.6928
upper	.7270	.7439	.7168	.7224

4.3. *Symmetrical Extreme Bound Analysis*

In order to check how strongly the bounds react to symmetrical restrictions in the second stage covariance matrix Σ, we can extend the above analysis to a symmetrically restricted extreme bound analysis.

The basic idea, as in Leamer (1982), is to restrict the prior covariance matrix symmetrically from above and below, i.e., $\Sigma_L < \Sigma < \Sigma_u$, where $<$ means that the difference of the matrices is positive definite. In particular, we take for our example the following sequence of inequalities

$$4^{-m}\hat{\Sigma} \leq \hat{\Sigma} \leq 4^{m}\hat{\Sigma}\,, \qquad m = 0,2,4,\infty. \qquad (4.10)$$

The extreme bounds for the matrix weighted average (4.3) with the inequality restrictions (4.10) are given by

$$E_{1\cdot 2} = \psi'(\beta^* + f) \pm (\psi' A^{-1} \psi c)^{1/2} \qquad (4.11)$$

where the parameters are given by

$$\begin{aligned} f &= (H_u^{-1} + H_L^{-1})\, H(b-\beta^*)/2 \\ A &= H_u^{-1} - H_L^{-1} \\ c &= \beta^{*\prime} H_u^{-1} (4^{m} - 4^{-m})\Sigma^{-1} H_L^{-1} \beta^*/4 \end{aligned} \qquad (4.12)$$

and the H_L and H_u are defined as

$$\begin{aligned} H_L &= \omega_s^{-2} X_s' X_s + 4^{m}\Sigma^{-1} \\ H_U &= \omega_s^{-2} X_s' X_s + 4^{-m}\Sigma^{-1} \end{aligned} \qquad (4.13)$$

The results of this analysis are given in Table 4.4.

TABLE 4.4
Symmetrical Hierarchical Extreme Bounds

coefficient/exponent	$m = 0$	2	4	Inf.
P^* 1.Q *lo.*	−.04	−.0450	−.0695	−.1015
up.	−.04	−.0372	−.0199	.0419
Y 1.Q *lo.*	.7052	.7047	.6953	.6834
up.	.7052	.7109	.7185	.7273
P^* 2.Q *lo.*	−.078	−.1085	−.1277	−.1404
up.	−.078	−.0531	−.3771	−.0129
Y 2.Q *lo.*	.7137	.7109	.7055	.6977
up.	.7137	.7165	.7200	.7438
P^* 3.Q *lo.*	−.063	−.0639	−.0640	−.0644
up.	−.063	−.0612	−.0569	−.0474
Y^* 3.Q *lo.*	.7101	.7102	.7102	.7104
up.	.7101	.7114	.7132	.7168
P^* 4.Q *lo.*	−.072	−.0800	−.0914	−.1370
up.	−.072	−.0640	−.0596	−.0515
Y^* 4.Q *lo.*	.7119	.7119	.7087	.6930
up.	.7119	.7120	.7180	.7224

The first column in Table 4.4 shows the posterior means based on the first and second stage covariance estimates. They correspond to the first step in a Lindley-Smith model iteration procedure. The sequence of bounds is tighter for the income coefficients, while the price coefficients have larger intervals lying closer to the zero point. Note that the price coefficient in the first quarter had a positive upper extreme bound which now becomes negative under the symmetric variance restriction. Therefore we can conclude that the negative sign of the price coefficient is robust against extreme prior covariance specifications in all 4 quarters of the Austrian import function. Also we see that the coefficient of income lies around the value .71 for almost all reasonable prior covariances in the neighbourhood of the empirical second stage covariance matrices.

5. CONCLUSIONS

In this paper we have contrasted two approaches to seasonal distributed lag models which arise in time series analysis and econometrics.

First we described the Akaike (1982) method for the Bayesian modelling of a seasonal distributed lag model with smoothness prior. We showed how the likelihood conditional on the hyperparameter can be used to derive the information criterion ABIC. This method has the advantage of solving the problem of the optimal lag length and the order of differencing "automatically". In the sequel we have extended the method for a seasonal distributed lag model with smoothness prior. The derivation resulted in the proposal of the seasonal Bayesian information criterion SBIC.

The second approach used the concept of an exchangeable hierarchical model for seasonal distributed lag models and allows also a robust Bayesian inference process by an extreme bound analysis. This approach has the disadvantage that it does not take into account an optimal lag length as in the Akaike method. Therefore we suggest some further research in order to incorporate optimal differencing and lag lengths into an exchangeable hierarchical model and for discriminating between different hierarchical models.

REFERENCES

AKAIKE, H., (1980). Likelihood and the Bayes Procedure, *Bayesian Statistics,* (J.M. Bernardo et al. eds.), Valencia: University Press, 143-166.

— (1982). The Selection of Smoothness Priors for Distributed Lag Estimation, *MCR Tech. Rep.* University of Wisconsin.

DEMPSTER, A.P., LAIRD, N.M. and RUBIN, D.B. (1977). Maximum Likelihood from Incomplete Data via the EM Algorithm. *J. Roy. Statist. Soc. B,* **39**, 1-38.

JONES, R.H., and BRELSFORD, W.M. (1967). Time Series with Periodic Structure, *Biometrika,* **54**, 403-408.

LEAMER, E.E. (1972). A Class of Informative Priors and Distributed Lag Analysis, *Econometrica,* **40**, 1059-1081.

— (1978). *Specification Searches,* New York: Wily.

— (1982). Sets of Posterior Means with Bounded Variance Priors, *Econometrica,* **50**, 725-736.

LINDLEY, D. and SMITH, A.F.M. (1972). Bayes Estimates for the Linear Model. *J. Roy. Statist. Soc. B,* **34**, 1-41.

POLASEK, W. (1982). Hierarchical Bounds in Seasonal Distributed Lag Models. *Time Series Analysis* **1**. Amsterdam: North-Holland, 497-514.

— (1983). Multivariate Regression Systems: Estimation an Sensitivity Analysis for Two-Dimensional Data. *Robustness in Bayesian Statistics,* (J. Kadane, ed.) Amsterdam: North-Holland. 229-309.

SHILLER, R.J. (1973). A Distributed Lag Estimator Derived from Smoothness Priors, *Econometrica,* **41**, 775-788.

SMITH, A.F.M. (1973). A General Bayesian Linear Model. *J. Roy. Statist. Soc. B,* **35**, 67-75.

SWAMY, P.A.V.B. (1970). Efficient Inference in a Random Coefficient Regression Model. *Econometrica,* **38**, 311-323.

TRIVEDI, P.K. and LEE, B.M.S. (1981). Seasonal Variability in a Distributed Lag Model. *Rev. Econom. Studies*, **48**, 497-505.

BAYESIAN STATISTICS 2, pp. 733-740
J.M. Bernardo, M.H. DeGroot, D.V. Lindley, A.F.M. Smith (Eds.)

Bayesian Estimation of *k*-Dimensional Distribution Functions Via Neutral to the Right Priors

IRENE POLI
University of Cagliari, Italy

SUMMARY

The problem of nonparametric estimation of a k-dimensional distribution function is studied from a Bayesian point of view. The property of neutrality defined by Doksum (1974) for random probabilities on the real line is extended in this paper to the k-dimensional real space, and the process priors that are neutral to the right are then assumed for the unknown distribution. Using further the special classes of Dirichlet process priors and of homogeneous process priors, the Bayes estimates are derived and comparisons with other approaches are discussed.

Keywords: BAYESIAN NONPARAMETRIC ESTIMATION; DIRICHLET PROCESS; DISTRIBUTION FUNCTION; HOMOGENEOUS PROCESS; NEUTRAL TO THE RIGHT PROCESS.

1. INTRODUCTION

The problem of nonparametric estimation of a distribution function is encountered in many practical situations. Suppose, for instance, we are asked to give an estimate of size distribution of incomes and consumptions that ought to represent an informational basis for the formulation of a monetary stabilization policy. Interest then lies in an estimate of the proportion of people who fall in some given interval of the two mentioned variables which are defined and measured in a meaningful way. If we consider the usual data sets through the empirical distribution function (or some estimate obtained by smoothing out a functional of it) our conclusions might be inadequate and conceal points of interest for the economic policy maker. Other related knowledge (such as taxes, incomes derived from a second undeclared job or the "submerged economy", and any other pertinent element coming from the economic theory involved) should in fact be considered and introduced in the modelling procedures together with the sample evidence. The inductive process, based on all this information, will then lead to a more explicative description of the proposed real situation.

From a Bayesian point of view, considerable attention has been paid in recent years to approximate linear solutions (Fienberg, 1980; Goldstein, 1975; Mouchart and Simar, 1983), to the Bayesian procedures involving smoothing arguments (Leonard, 1978; Wahba, 1983), and to special developments in robust analysis (Box, 1980; Rubin, 1977; Smith, 1983). An important role in this area of research has also, however, been played by nonparametric methods which, assuming prior distributions on the space of all distributions, allow one to study the problem in a very general framework (Antoniak,

1974; Ferguson, 1973; Doksum, 1974). Several applications have, moreover, exhibited the usefulness of these methods (Ferguson and Phadia, 1979; Kalbfleisch, 1978; Susarla and Van Ryzin, 1976).

We shall assume in this paper that our concern is with providing an estimate of a k-dimensional distribution function in a Bayesian nonparametric framework. More specifically we shall derive this estimate under priors which are neutral to the right. These, up until now, seem to offer the most general approach for solving nonparametric problems.

Section 2 contains the Bayes estimate of the distribution under a square error loss function and Section 3 describes the neutral to the right prior for a k-dimensional distribution function. This prior is then specialized to the Dirichlet processes and to the homogeneous neutral to the right processes, the corresponding Bayes estimates being derived in Section 4.

2. THE BAYES ESTIMATE OF A DISTRIBUTION FUNCTION

Let $F(x_1,\ldots,x_k)$ be a distribution function on R^k and let $\mathbf{x}_r = (x_{r1},\ldots,x_{rk})$, $r=1,\ldots,n$, be a sample of size n from F. The form of F is supposed unknown and we would like to assess the uncertainty about it. With $F(x_1,\ldots,x_k)=P\,[(-\infty,\ldots,-\infty;x_1,\ldots,x_k])$ the random distribution function F is regarded as a stochastic process with a separable version, non-decreasing a.s., right continuous a.s., $\lim_{x_j\to-\infty} F(x_1,\ldots,x_k)=0$, $j=1,\ldots,k$, *a.s.*, and

$$\lim_{x_j\to\infty,\ \mathit{all}\ j} F(x_1,\ldots,x_k)=1,\quad j=1,\ldots,k,\ \text{a.s.}$$

For inference in the model we apply the usual Bayesian paradigm. We assign a suitable prior distribution to the random quantity F and, accounting for the sample of observations, we construct the required estimate from the posterior distribution. The Bayes estimate of F, given the data, is then obtained assuming the loss incurred by using $\hat{F}$ as an estimate of F is of the form

$$L(F,\hat{F}) = \int (F(\mathbf{x}) - \hat{F}(\mathbf{x}))^2\, dW(\mathbf{x}) \tag{1}$$

where $F(\mathbf{x})=F(x_1,\ldots,x_k)$, and W is some finite weight (measure) on R^k.

Hence, the final estimate of the distribution F is the minimizer of the Bayes risk $E(L(F,\hat{F}))$ that is achieved by choosing $\hat{F}(\mathbf{x})$ to be $E(F(\mathbf{x}))$, where E is the average taken with respect to the posterior distribution proposed.

In the following development we will assume as a prior for F the neutral to the right process. This very general class of prior distributions has been discovered by Doksum (1974) based on some independence properties within the probabilities defined for a random distribution function F on R. Doksum gave also an important characterization of this prior in terms of independent increment processes and showed that if the prior is neutral to the right then the posterior is in the same class as the prior. Moreover, the well-known Ferguson's Dirichlet prior (1973) was shown to be a special case of this framework. (The reader is referred to Doksum's paper (1974) for the fundamentals of neutral priors. Related work includes that of Ferguson (1974), James and Mosimann (1980), García Pérez and Quesada Paloma (1983), and Rolin (1982)).

3. THE NEUTRALITY TO THE RIGHT OF A K-VARIATE DISTRIBUTION

A random distribution function $F(\mathbf{x}) = F(x_1,\ldots,x_k)$ can be regarded as neutral to the right if for each $p>1$ and each sequence of points in R^k of the form $\mathbf{x}^{(1)} < \mathbf{x}^{(2)} < \ldots < \mathbf{x}^{(p)}$ (entrywise ordered) there exist independent random variables $V^{(i)}$, $i=1,\ldots,p$, such that

$$(F(\mathbf{x}^{(1)}), F(\mathbf{x}^{(2)}),\ldots,F(\mathbf{x}^{(p)})) = (V^{(1)}, 1-(1-V^{(1)})(1-V^{(2)}),\ldots,1-\Pi_{i=1}^{p}(1-V^{(i)})). \qquad (2)$$

Since by definition $F(\mathbf{x}^{(i)}) = 1-\Pi_{j=1}^{i}(1-V^{(j)})$, it follows that

$$V^{(i)} = [F(\mathbf{x}^{(i)}) - F(\mathbf{x}^{(i-1)})]/[1-F(\mathbf{x}^{(i-1)})] \qquad i=1,\ldots,p$$

with $x^{(0)} = (-\infty, x_2,\ldots,x_k) = \ldots = (x_1,\ldots,x_{k-1},-\infty)$ and $F(\mathbf{x}^{(0)}) = 0$.

Neutrality to the right thus asks that the normalized increments

$$F(\mathbf{x}^{(1)}),\ldots,[F(\mathbf{x}^{(i)}) - F(\mathbf{x}^{(i-1)})]/[1-F(\mathbf{x}^{(i-1)})],\ldots,[F(\mathbf{x}^{(p)}) - F(\mathbf{x}^{(p-1)})]/[1-F(\mathbf{x}^{(p-1)})]$$

are independent for all $\mathbf{x}^{(i-1)} < \mathbf{x}^{(i)}$, $i=1,\ldots,p$.

We see that neutrality concerns a particular independence property that involves the partitioning of R^k in nested ordered sets. Suppose, in fact, that $I_{x^{(i)}}$ is the interval $(-\infty,\ldots,-\infty;\ x_1^{(i)},\ldots,x_k^{(i)}\ ]$ in R^k, with $F(\mathbf{x}^{(i)}) = P(I_{x^{(i)}})$ and let $I_{x^{(i-1)}}$ be a second interval, defined in the same way in R^k, with $\mathbf{x}^{(i-1)} < \mathbf{x}^{(i)}$, $i=1,\ldots,p$. The random probability P is then neutral with respect to every sequence of nested ordered partitions of the form $\{\Pi_m : A_{m,1},\ldots,A_{m,pm}\}$, $m=1,2,\ldots$, with $A_{m,i} = (\mathbf{x}_m^{(i-1)}, \mathbf{x}_m^{(i)}]$, $i=1,\ldots,p_m$, and $P(A_{mi}) = \int_{Ami} dF(\mathbf{x})$. The random probabilities

$$P(A_{m,1}), P(A_{m,2} | A_{m,1}^c), P(A_{m,3} | (A_{m,1} \cup A_{m,2})^c),\ldots, P(A_{m,k_m} | A_{m,k_m}) = 1$$

are then assumed to be independent (moving to the right in the ordered partition).

It is worth noticing that from the suggested k-variate neutral to the right distribution we can derive the marginal distributions $F_j(x_j)$, $j=1,\ldots,k$, that are neutral according to the Definition (3.1) given by Doksum.

We now turn to the characterization of neutral to the right processes in terms of independent increment processes stated by Theorem 3.1 of Doksum. We can see that $F(\mathbf{x})$ is neutral to the right with respect to every sequence of the above mentioned partitions of R^k if and only if it has the same distribution function as $1-\exp[-Y(\mathbf{x})]$ where $Y(\mathbf{x})$ is a process with independent increments (the increments are defined by the nested sequences of sets given above), non-decreasing a.s., right continuous a.s., $\lim_{x_j \to -\infty} Y(x_1,\ldots,x_k) = 0$, $j=1,\ldots,k$, a.s., and $\lim_{x_j \to \infty,\ all\ j} Y(x_1,\ldots,x_k) = \infty$, $j=1,\ldots,k$, a.s.

In fact, let $Y(\mathbf{x})$ be the independent increment process described and $F(\mathbf{x}) = 1-\exp[-Y(\mathbf{x})]$. Then $F(\mathbf{x})$ is neutral to the right according to (2), with the independent random variables $V^{(i)}$ in the form: $1-\exp\{-[Y(\mathbf{x}^{(i)}) - Y(\mathbf{x}^{(i-1)})]\}$ for each entrywise ordered sequence of points on R^k. Let now $F(\mathbf{x})$ be neutral to the right. Then $Y(\mathbf{x}) = -\log(1 - F(\mathbf{x}))$ is the independent increment process proposed since $(Y(\mathbf{x}^{(1)}) - Y(\mathbf{x}^{(0)}),\ldots, Y(\mathbf{x}^{(p)}) - Y(\mathbf{x}^{(p-1)})) = (-\log(1-V^{(1)}),\ldots, -\log(1-V^{(p)}))$, where $V^{(1)},\ldots, V^{(p)}$ are the independent random variables introduced in (2).

4. ESTIMATION VIA NEUTRAL TO THE RIGHT PRIORS

In estimating F with loss (1) we know that the Bayes estimate is the expected value of the random distribution function, which under a neutral to the right prior takes the special form

$$E(F(\mathbf{x})) = 1 - E[\exp(-Y(\mathbf{x}))] \tag{3}$$

This estimate requires the evaluation of the characteristic function of the variable $Y(\mathbf{x})$, $\varphi_y(v) = E[\exp ivY]$ at $v = i$, which reduces to the moment generating function at point 1, $M_Y(1)$, where by definition $M_Y(v) = E[\exp - vY]$.

Then supposing F is neutral to the right and observing a sample of data from F, we can write $E(F(\mathbf{x}) \mid \text{data}) = 1 - M_y(1 \mid \text{data})$. To achieve the estimate we need the posterior moment generating function of $Y(\mathbf{x})$. We can derive this quantity specializing the result of Theorem 4 of Ferguson and Phadia (1979) for the censored data problem to the uncensored case. Let $n(\mathbf{x})$ be the number of distinct observations less than or equal to $\mathbf{x}$ and denote those observations as $x_{(1)},\ldots,x_{(n(x))}$. Then, the posterior moment generating function of $Y(\mathbf{x})$ at point 1 takes the form

$$M_Y(1 \mid \text{data}) = \frac{M_Y(n-n(\mathbf{x})+1)}{M_Y(n-n(\mathbf{x}))} \cdot \Pi_{i=1}^{n(x)} \frac{M^{-}_{x_{(i)}}(n-i+2)}{M^{-}_{x_{(i)}}(n-i+1)} \cdot \frac{M_{x_{(i)}}(n-i)}{M_{x_{(i)}}(n-i+1)} \cdot \frac{C_{x_{(i)}}(n-i+1)}{C_{x_{(i)}}(n-i)} \tag{4}$$

where $C_{x_{(i)}}(\alpha) = \int_0^\infty e^{-\alpha s}(1-e^{-s})dG_{x_{(i)}}(s)$ if $x_{(i)}$ is a prior fixed point of discountinuity of $Y(\mathbf{x})$ and G denotes the prior distribution of the jump in $Y(\mathbf{x})$, while $C_{x_{(i)}}(\alpha) = \int_0^\infty e^{-\alpha s} dH_{x_{(i)}}(s)$ if $x_{(i)}$ is not a prior fixed point of discountinuity of $Y(\mathbf{x})$ and H denotes the posterior distribution of the jump in $Y(\mathbf{x})$. Moreover $M^{-}_{x_{(i)}}(\theta) = \lim_{x_h \nearrow x_i} M_{x_{(h)}}(\theta)$.

In order to provide explicit expressions of the Bayes estimate we specify the non-decreasing independent increment process $Y(\mathbf{x})$ and in particular, we will consider the Dirichlet process and some special homogeneous processes.

4.1. *Dirichlet processes as neutral processes*

Let the prior distribution for the random quantity F be the Dirichlet process with parameter α and notation $D(\alpha)$. Observe, further, a sample of n data from F. Then the posterior distribution is known to be in the same class as the prior, with an enriched parameter and notation $D\,(\alpha + \Sigma_{r=1}^{n} \delta x_{r_1},\ldots, x_{r_k})$, where δ_x is the measure assigning mass one to the point x. (See Ferguson, 1973).

Giving now a characterization of the process through neutrality, assume the parameter α, with $\alpha(x_1,\ldots,x_k) = \alpha((-\infty,\ldots,-\infty; x_1,\ldots,x_k])$, to be a non-decreasing right-continuous function on R^K, such that $\lim_{x_j \to -\infty} \alpha(x_1,\ldots,x_k) = 0$, $j = 1,\ldots,k$, $\lim_{x_j \to \infty \, all j} \alpha(x_1,\ldots,x_k) = \alpha(R^k) < \infty^k$.

Then $F \epsilon Be(\alpha(\mathbf{x}), \alpha(R^k) - \alpha(\mathbf{x}))$ and the density of $Y(\mathbf{x})$, denoting $n_o = \alpha(R^k)$, is given by

$$f_Y(y) = \frac{(1-e^{-y})^{\alpha(x)-1}(e^{-y})^{n_o-\alpha(x)}}{B(\alpha(\mathbf{x}), n_o - \alpha(\mathbf{x}))} = \frac{\Gamma(n_o)}{\Gamma(\alpha(\mathbf{x}))\Gamma(n_o-\alpha(\mathbf{x}))}(1-e^{-y})^{\alpha(x)-1}(e^{-y})^{n_o-\alpha(x)}$$

since $B(v,w)$ is the beta function.

The moment generating function of $Y(\mathbf{x})$ then has the form

$$M_Y(\theta) = \frac{\Gamma(n_o)}{\Gamma(\alpha(\mathbf{x}))\Gamma(n_o-\alpha(\mathbf{x}))} \int_0^\infty (1-e^{-y})^{\alpha(x)-1}(e^{-y})^{n_o-\alpha(x)+\theta}\,dy$$

that reduces at $\theta = 1$ to

$$M_Y(1) = \frac{\Gamma(n_o)\Gamma(n_o - \alpha(\mathbf{x}) + 1)}{\Gamma(n_o - \alpha(\mathbf{x}))\Gamma(n_o + 1)} = \frac{n_o - \alpha(\mathbf{x})}{n_o}$$

under the relation $\Gamma(\beta) = (\beta - 1)\Gamma(\beta - 1)$.

Thus, the prior estimate of F, denoted F_o, is

$$F_o(\mathbf{x}) = \frac{\alpha(x)}{n_o}$$

Accounting for the sample of observations we further derive from (4) the posterior moment generating function that, after some algebra, may be written as

$$M_Y(1 \mid \text{data}) = \frac{n_o - \alpha(\mathbf{x}) + n - n(\mathbf{x})}{n_o + n}$$

where $n(\mathbf{x}) = \Sigma_{r=1}^{n} \delta_{x_{r1},\ldots,x_{rk}}((-\infty,\ldots,-\infty; x_1,\ldots,x_k])$.

Therefore, the Bayes estimate of the distribution function exhibits the known form

$$\hat{F}_D(\mathbf{x} \mid \text{data}) = p\, F_o(\mathbf{x}) + (1 - p)\, F_n(\mathbf{x})$$

where $F_n(\mathbf{x}) = \dfrac{n(\mathbf{x})}{n}$ is the empirical distribution function and $p = n_o/(n_o + n)$ is a weight element.

Our choice will be concerned with $F_o(\mathbf{x})$, the prior guess of $F(\mathbf{x})$, and with the real parameter n_o, that reflects, in a sense, the degree of faith in the proposed distribution $F_o(\mathbf{x})$. The role played by n_o as an index of our uncertainty about the shape of $F(\mathbf{x})$ is essential. In fact, when the initial information is poor, n_o will be close to zero and the Bayes estimate will only be the empirical distribution function. On the other hand, if we presume to know the distribution then n_o will tend to infinity and the model reduces to a parametric model.

It is worth remarking that the Dirichlet process prior can also be viewed as a special case of another class of random probabilities, i.e. the mixtures of Dirichlet processes introduced by Antoniak (1974). Let, in fact, the parameter α of the Dirichlet process prior be itself unknown and dependent on a random quantity u whose distribution H is uncertain. Observing a sample of data, it is known that the posterior distribution is still a mixture of Dirichlet processes denoted $\int_U D(\alpha_u + \Sigma_{r=1}^{n} \delta_{x_{r1},\ldots,x_{rk}})\, dH(u \mid \text{data})$. Under this structure of the model and the loss function (1), we then derive the Bayes estimate of F

$$E(F(\mathbf{x}) \mid \text{data}) = \int_U \frac{\alpha_u(\mathbf{x}) + n(\mathbf{x})}{n_o + n}\, dH(u \mid \text{data})$$

with $\alpha_u(R^k) = \alpha(R^k) = n_o$. This estimate can further be expressed as

$$\hat{F}_{mD}(\mathbf{x} \mid \text{data}) = p \int_U F_{o,u}(\mathbf{x})\, dH(u \mid \text{data}) + (1 - p)\, F_n(\mathbf{x});$$

that is the weighted average of the empirical distribution function $F_n(\mathbf{x})$ and the prior distribution $F_{o,u}(\mathbf{x})$ whose parameters are supposed unknown and estimated through the posterior distribution $H(u \mid \text{data})$. Our choice is now concerned with F_{ou} and n_o, as in the simple Dirichlet process, as well as with $H(u)$, which represents the prior guess at the shape of the distribution of the random quantity u. In the same way it is possible derive the Bayes estimate of the distribution under the product of mixtures priors developed by Cifarelli and Regazzini (1978).

Finally, it is important to mention that subsequent developments on the Dirichlet process, such as the estimate of a symmetric distribution proposed by Dalal (1979), the empirical Bayes estimation defined by Korwar and Hollander (1976) and the sequential estimate given by Ferguson (1982), can be considered in this general framework.

4.2. *The homogeneous neutral to the right processes*

We regard now the independent increment process $Y(\mathbf{x})$ as homogeneous, so that its moment generating function at point one, $M_Y(1)$, takes the known form

$$M_Y(1) = e^{\gamma(x)\int_0^{\infty}(e^{-z}-1)dN(z)}$$

where $\gamma(\mathbf{x})$ is continuous non-decreasing, $\lim_{x_j\to-\infty}\gamma(x_1,\ldots,x_k)=0$, $j=1,\ldots,k$, $\lim_{x_j\to\infty \, all\, j}$ $\gamma(x_1,\ldots,x_k)=\infty$, $j=1,\ldots,k$, and N is any measure on $(0,\infty)$ such that $\int_0^{\infty} z(1+z)^{-1}\,dN(z)<\infty$. Then, we specialize the analysis to the simple homogeneous process, proposed by Ferguson and Phadia (1979) for censored data. This subclass of the homogeneous neutral to the right processes has been characterized for having the following moment generation function

$$M_Y(\theta) = e^{\gamma(x)\int_0^{\infty}(e^{-\theta z}-1)(1-e^{-z})^{-1}e^{-\tau z}\,dz} \qquad \tau>0$$

that, solving the integral, reduces for $\theta=1$, to

$$M_Y(1) = e^{-\gamma(x)/\tau}$$

Firstly, let us consider the problem in the absence of data, the so-called no-sample problem. We see that the estimate of the unknown distribution function is then given by

$$F_o(\mathbf{x}) = 1 - e^{-\gamma(x)/\tau} \tag{5}$$

reflecting the initial opinion of the shape of the distribution.

We may now derive the posterior moment generating function after observing a sample of n observations. Under the above notation, we obtain from (4), after some algebra, the following form

$$M_Y(1\,|\,\text{data}) = e^{-\gamma(x)/(n-n(x)+\tau)}\prod_{i=1}^{n(x)}\frac{n-i+\tau}{n-i+\tau+1}\; e^{\gamma(x_{(i)})/\{(n-i+\tau+1)(n-i+\tau)\}}.$$

Writing this expression in terms of the prior guess F_o given by (5), we achieve the Bayes estimate of the unknown distribution function under a simple homogeneous neutral to the right prior:

$$\hat{F}_H(\mathbf{x}\,|\,\text{data}) = 1-[(1-F_o(\mathbf{x}))^{\tau/(n-n(x)+\tau)}\prod_{i=1}^{n(x)}\frac{n-i+\tau}{n-i+\tau+1}\;.$$
$$.(1-F_o(x_{(i)}))^{-\tau/\{(n-i+\tau+1)(n-i+\tau)\}}]$$

The interpretation of this estimate gives the parameter τ the role of a measure of the strength of belief in the prior guess at F. In fact, for $\tau\to 0$ we have $\hat{F}_H(\mathbf{x}\,|\,\text{data})\to$ $\to [1 - \prod_{i=1}^{n(x)}\frac{n-i}{n-i+1}] = \frac{n(\mathbf{x})}{n}$; that is, the Bayes estimate reduces to the empirical distribution function. On the other hand, for large values of τ the Bayes estimate is close to $F_o(\mathbf{x})$ the prior opinion of the shape of $F(\mathbf{x})$. The interpretation of the value of τ is less direct than that for n_o in the Dirichlet case, but its effect may always be computed.

One may achieve a similar but more complicate result from the homogeneous gamma process. This process is a subclass of the neutral to the right process that presents independent increments in $Y(\mathbf{x})$ having the gamma distribution.

ACKNOWLEDGEMENT

I am very grateful to Dr. P.J. Brown for helpful suggestions.

REFERENCES

ANTONIAK, C.E. (1974). Mixtures of Dirichlet processes with applications to Bayesian nonparametric problems. *Ann. Statist.,* **2**, 1152-1174.

BOX, G.E.P. (1980). Sampling and Bayes' inference in scientific modelling and robustness. *J.R. Statist. Soc.* A. **143**, 383-430. (with discussion).

CIFARELLI, D.M. and REGAZZINI, E. (1978). Problemi statistici non parametrici in condizioni di scambiabilità parziale. Impiego di medie associative. *Quad. Ist. Mat. Finanz. Univ. Torino,* **12**, 1-35.

DALAL, S.R. (1979). Dirichlet invariant processes and applications to nonparametric estimation of symmetric distribution functions. *Stoch. Proc. and Appl.,* **9**, 99-107.

DOKSUM, K. (1974). Tailfree and neutral random probabilities and their posterior distributions. *Ann. Probability,* **2**, 183-201.

FERGUSON, T.S. (1973). A Bayesian analysis of some nonparametric problems. *Ann. Statist.,* **1**, 209-230.

— (1974). Prior distributions on spaces of probability measures. *Ann. Statist.,* **2**, 615-629.

— (1982). Sequential estimation with Dirichlet process priors. *Statistical Decision Theory and Related Topics* (S.S. Gupta and J.O. Berger, eds.). New York: Academic Press.

FERGUSON, T.S. and PHADIA, E.G. (1979). Bayesian nonparametric estimation based on censored data. *Ann. Statist.,* **7**, 163-186.

FIENBERG, S.E. (1980). Linear and quasi-linear Bayes estimators. *Bayesian Analysis in Econometrics and Statistics* (A. Zellner ed.). Amsterdam: North Holland.

GARCÍA PÉREZ, A. and QUESADA PALOMA, V. (1983). On nonparametric Bayesian Statistics. *Tech. Rep.* Univ. Madrid.

GOLDSTEIN, M. (1975). Approximate Bayes solutions to some nonparametric problems. *Ann. Statist.,* **3**, 512-517.

JAMES, I.R. and MOSIMANN, J.E. (1980). A new characterization of the Dirichlet distribution through neutrality. *Ann. Statist.,* **8**, 183-189.

KALBFLEISCH, J.D. (1978). Nonparametric Bayesian analysis of survival time data. *J.R. Statist. Soc.* B, **40**, 214-221.

KORWAR, R.M. and HOLLANDER, M. (1976). Empirical Bayes estimation of a distribution function. *Ann. Statist.,* **4**, 581-588.

LEONARD, T. (1978). Density estimation, stochastic processes and prior information. *J.R. Statist. Soc.* B, **40**, 113-146.

MOUCHART, M. and SIMAR, L. (1983). Theory and applications of least squares approximation in Bayesian analysis. *Specifying Statistical Models.* (J.P. Florens, *et. al.* eds.). Berlin: Springer.

ROLIN, J.M. (1982). Nonparametric Bayesian Statistics: a stochastic process approach. *CORE Discussion Paper N°* **8225**, Louvain-la-Neuve.

RUBIN, H. (1977). Robust Bayesian estimation. *Statistical Decision Theory and Related Topics.* (S.S. Gupta and D.S. Moore eds.). New York: Academic Press.

SMITH, A.F.M. (1983). Bayesian approaches to outliers and robustness. *Specifying Statistical Models* (J.P. Florens *et al.* eds.). Berlin: Springer.

SUSARLA, V. and VAN RYZIN, J. (1976). Nonparametric Bayesian estimation of survival curves from incomplete observations. *J. Amer. Statist. Ass.* **71**, 897-902.

WAHBA, G. (1983). Bayesian "confidence intervals" for the cross-validated smoothing spline. *J. R. Statist. Soc.* B, **45**, 133-150.

BAYESIAN STATISTICS 2, pp. 741-742
J.M. Bernardo, M.H. DeGroot, D.V. Lindley, A.F.M. Smith (Eds.)

A Conflict Between Finite Additivity and Avoiding Dutch Book

T. SEIDENFELD and M.J. SCHERVISH
Washington University, St. Louis — *Carnegie-Mellon University, Pittsburgh*

SUMMARY

For Savage, as for de Finetti, the existence of subjective (personal) probability is a consequence of the normative theory of preference. (De Finetti achieves the reduction of belief to desire with his generalized Dutch Book argument for *previsions*.) Both Savage and de Finetti rebel against legislating countable additivity for subjective probability. They require merely that probability be finitely additive. Simultaneously, they insist that their theories of preference are weak, accommodating all but self-defeating desires. In this paper we dispute these claims by showing that the following three cannot hold at once:

(i) Coherent belief is reducible to rational preference, i.e. the generalized Dutch Book argument fixes standards of coherence.

(ii) Finitely additive probability is coherent.

(iii) Admissible preference structures may be free of *consequences*, i.e. they may lack prizes whose values are robust against all contingencies.

We discuss (§2) conditions under which Savages's generalized sure-thing principle (P-7) remains independent of the other six postulates, and we show the independence is *not* a question of additivity. In Section 3 we point out that Savages's theory does not entail a generalized dominance principle, due to non-conglomerability of finitely (but not countably) additive probability. In the final section we examine the limitations these findings place on de Finetti's Dutch Book argument.

Keywords: COHERENCE; FOUNDATIONS OF DECISION THEORY.

This paper has been published in *Philosophy of Science*, **50** (1983), pp. 398-412.

BAYESIAN STATISTICS 2, pp. 743-750
J.M. Bernardo, M.H. DeGroot, D.V. Lindley, A.F.M. Smith (Eds.)

Exact Bayesian inference on the parameters of a Cauchy distribution with vague prior information

D.J. SPIEGELHALTER
MRC Biostatistics Unit, Cambridge

SUMMARY

Exact expressions are given for the posterior distributions of the location and scale parameters of Cauchy distribution assuming vague prior information. The exact posterior means and variances are seen to take on simple weighted average forms.

Keywords: BAYESIAN INFERENCE; CAUCHY DISTRIBUTION; MAXIMAL INVARIANT; VAGUE PRIOR INFORMATION.

1. INTRODUCTION

Suppose a random sample $x = (x_1,\ldots,x_n)$ has been drawn from a Cauchy distribution, where x_k, $k = 1,\ldots,n$, has density

$$p(x_k \mid \theta,\sigma) = (\sigma/\pi)\{\sigma^2 + (x_k - \theta)^2\}^{-1} \qquad (-\infty < x_k < \infty);$$

θ is the median of the density and σ is the scale parameter, specifically the interquartile range. Approximate methods of obtaining interval estimates of θ and σ include those based on order statistics (for example, Cane, 1974), maximum likelihood (Copas, 1975) and the empirical characteristic function (Koutrouvelis, 1982). In this paper we consider exact inference on θ and σ when the vague, or invariant, prior $p(\theta,\sigma) \propto \sigma^{-1}$ is assumed. We place this within the Bayesian framework, although the exact intervals we obtain could also arise from a likelihood approach when conditioning on a maximal invariant statistic (see, for example, Fisher, 1934, Cox & Hinkley, 1974, p. 221).

The integration techniques used to obtain expressions for the posterior densities $p(\theta \mid x)$, $p(\sigma \mid x)$ and their first two moments are an extension of those adopted by Franck (1981) in his derivation of the most powerful invariant test statistic for testing between the Cauchy and Gaussian shapes. We follow his example, in the intermediate stages of the analysis, of using $\lambda = \sigma^{-1}$ as the parameterisation of scale, in which case the density and prior are written $p(x_k \mid \theta,\lambda) = (\lambda/\pi)/\{1 + \lambda^2(x_k - \theta)^2\}$ and $p(\theta,\lambda) \propto \lambda^{-1}$, respectively.

Sections 2 and 3 deal with scale and location parameters, respectively, while Section 4 contains a numerical example in which the unconditional maximum likelihood approach is contrasted. A brief discussion is contained in Section 5.

2. INFERENCE ON THE SCALE PARAMETER

2.1. *Franck's results*

The aim of Franck's paper was an expression for the integrated likelihood $p(x) = \int\int p(x|\theta,\lambda)\lambda^{-1}d\theta d\lambda$, and as a preliminary step he evaluated the integral

$$p(x|\lambda) = \int_{-\infty}^{\infty} (\lambda/\pi)^n \prod_{k=1}^{n} \{1+\lambda^2(x_k-\theta)^2\}^{-1} d\theta$$

by noting that the integrand has simple poles at $\theta = x_k + i/\lambda$, $k=1,\ldots,n$, $i=\sqrt{-1}$. Using the residue theorem and expanding by partial fractions he obtained

$$p(x|\lambda) = \begin{cases} \dfrac{(-1)^{(n+1)/2}}{(2\pi)^{n-1}} \sum\limits_{j\neq k} \dfrac{c_{jk}}{m_{jk}{}^2(\lambda^2+m_{jk}{}^{-2})} & \text{(n odd)} \\[2ex] \dfrac{(-1)^{n/2-1}}{(2\pi)^{n-1}} \sum\limits_{j\neq k} \dfrac{c_{jk}{}^{\lambda}}{m_{jk}(\lambda^2+m_{jk}{}^{-2})} & \text{(n even)} \end{cases} \tag{2.1}$$

where

$$m_{jk} = (x_j - x_k)/2$$

$$c_{jk} = 2^{1-n} m_{jk}{}^{n-3} / \prod_{l\neq j,k} m_{lj} m_{lk}$$

and $\sum\limits_{j\neq k}$ denotes $\sum\limits_{k=1}^{n}\left(\sum\limits_{j=1}^{k-1} + \sum\limits_{j=k+1}^{n}\right)$. Our notation differs slightly from that of Franck: his $I(\lambda)$ is our $\pi^n p(x|\lambda)/\lambda$, and his A_{jk} is our $2m_{jk}c_{jk}$.

Franck goes on to obtain the required expression

$$p(x) = \int p(x|\lambda)\lambda^{-1}d\lambda = \begin{cases} \dfrac{(-1)^{(n-1)/2}}{(2\pi)^{n-1}} \sum\limits_{j\neq k} c_{jk}\log|m_{jk}| & \text{(n odd)} \\[2ex] \dfrac{(-1)^{n/2-1}}{(2\pi)^{n-1}} \sum\limits_{j\neq k} c_{jk}\dfrac{\pi}{2}\,\mathrm{sign}(m_{jk}) & \text{(n even)} \end{cases} \tag{2.2}$$

(although it should be noted that his expression A.15 is twice the correct value).

The remainder of this section is concerned with generalising a technique used by Franck when, for n odd, he finds an apparently unbounded expression for $p(x)$ proportional to

$$\sum_{j\neq k} c_{jk}\{\log\lambda - \tfrac{1}{2}\log(\lambda^2+m_{jk}{}^{-2})\}\Big|_0^{\infty} = \sum_{j\neq k} c_{jk}\ln|m_{jk}| - \sum_{j\neq k} c_{jk}\log(0).$$

However, he shows that $p(x|\lambda)\lambda^{-1}$ is bounded in the neighbourhood of zero and hence, since this implies the lower limit of the integral must exist, he argues that $\sum\limits_{j\neq k} c_{jk}$ must be equal to zero. We shall find it useful in later sections to extend this device to a more general situation. Specifically, let

$$I_{k,m}(\lambda) = \int_{-\infty}^{\infty} \theta^k \lambda^{-m} p(x|\theta,\lambda)d\theta$$

and we wish to know when $I_{k,m}$ is bounded in the neighbourhood of zero. Now, for $\lambda < (\max\limits_{k=1,\ldots,n} |x_k|)^{-1}$, the integrand is bounded by $g(\theta,\lambda)$, where

$$g(\theta,\lambda) = \begin{cases} \theta^k\lambda^{n-m} & (|\theta| < \lambda^{-1}) \;. \\ \theta^k\lambda^{n-m}\{1+(1-\lambda|\theta|)^2\}^{-1} & (|\theta| > \lambda^{-1}) \end{cases}$$

Since $\int_{-\infty}^{\infty} g(\theta,\lambda)d\theta = \lambda^{n-m-k-1}\{2(k+1)^{-1} + 2\int_0^{\infty} (1+u)^k(1+u^2)^{-n}du\}$,

it follows from the dominated convergence theorem that, when $n \geq m+k+1$, $I_{k,m}(\lambda)$ exists and is continuous in a neighbourhood of zero, and hence is bounded. In Franck's result, $k=0$ and $m=1$, and so $p(x)$ exists for $n \geq 2$.

2.2. *Posterior distribution of σ*

Reverting to the original parameterisation of scale, we may write $p(\sigma|x) =$ $= p(x|\lambda = \sigma^{-1})\sigma^{-1}/p(x)$ and hence from (2.1) and (2.2).

$$p(\sigma|x) = \begin{cases} -\sum_{j\neq k} c_{jk}\dfrac{\sigma}{(\sigma^2+m^2_{jk})} \Big/ \sum_{j\neq k} c_{jk}\log|m_{jk}| & (n \text{ odd}) \\ \dfrac{2}{\pi}\sum_{j\neq k} c_{jk}\dfrac{m_{jk}}{(\sigma^2+m^2_{jk})} \Big/ \sum_{j\neq k} c_{jk}\,\mathrm{sign}(m_{jk}) & (n \text{ even}) \end{cases}$$

2.3. *Posterior mean of σ*

The required expression $E(\sigma|x) = \int\sigma p(\sigma|x)d\sigma$ may be written as $\int p(x|\lambda)\lambda^{-2}d\lambda/p(x)$. Straightforward integration using (2.1) and (2.2) leads, for $n \geq 3$, to

$$E(\sigma|x) = \begin{cases} \dfrac{\pi}{2}\sum_{j\neq k} c_{jk}|m_{jk}| \Big/ \sum_{j\neq k} c_{jk}\log|m_{jk}| & (n \text{ odd}) \\ \dfrac{-2}{\pi}\sum_{j\neq k} c_{jk}m_{jk}\log|m_{jk}| \Big/ \sum_{j\neq k} c_{jk}\,\mathrm{sign}(m_{jk}) & (n \text{ even}) \end{cases}$$

The expression when n is even required the fact that $I_{0,2}(\lambda)$ is bounded in the neighbourhood of zero for $n \geq 3$, and proved in passing that $\sum_{j\neq k} c_{jk}m_{jk} = 0$.

2.4. *Posterior variance of σ*

The posterior variance of σ is given by $V(\sigma|x) = E(\sigma^2|x) - E^2(\sigma|x)$, where $E(\sigma^2|x)$ may be written as $\int p(x|\lambda)\lambda^{-3}\,d\lambda/p(x)$. We find that for $n \geq 4$,

$$E(\sigma^2|x) = \begin{cases} -\sum_{j\neq k} c_{jk}\log|m_{jk}|m_{jk}^2 \Big/ \sum_{j\neq k} c_{jk}\log|m_{jk}| & (n \text{ odd}) \\ -\sum_{j\neq k} c_{jk}\,\mathrm{sign}(m_{jk})m_{jk}^2 \Big/ \sum_{j\neq k} c_{jk}\,\mathrm{sign}(m_{jk}) & (n \text{ even}) \end{cases}$$

We note the simple weighted average form of these expressions. The fact that $I_{0,3}(\lambda)$ is bounded near zero for $n \geq 4$ is used in the derivation when n is odd, for which $\sum_{j\neq k} c_{jk}m_j = 0$.

3. INFERENCE ON THE LOCATION PARAMETER

3.1. *Posterior distribution of* θ

The posterior density is $p(\theta|x) = p(x|\theta)p(\theta)/p(x)$, where $p(\theta)$ is locally uniform and

$$p(x|\theta) = \int_0^\infty (\lambda/\pi)^n \prod_{k=1}^n \{1+\lambda^2(x_k-\theta)^2\}^{-1}\lambda^{-1}\,d\lambda \tag{3.1}$$

Expanding the integrand as partial fractions $\sum_{k=1}^n (A_k\lambda+B_k)/\{1+\lambda^2(x_k-\theta)^2\}$ we obtain the identity

$$\lambda^{n-1} = \sum_{k=1}^n (A_k\lambda+B_k) \prod_{j\neq k} \{1+\lambda^2(x_k-\theta)^2\}$$

To evaluate A_k and B_k, we set $\lambda = i/(x_k-\theta)$, $i=\sqrt{-1}$, and find

$$i^{n-1} = \{A_k i/(x_k-\theta)+B_k\}d_k(\theta) \tag{3.2}$$

where $d_k(\theta) = (x_k-\theta)^{n-1} \prod_{j\neq k} \{1 - (x_j-\theta)^2/(x_k-\theta)^2\}$. The result then depends on whether n is odd or even.

For n odd, (3.2) is wholly real so $A_k = 0$, $k = 1,\ldots,n$, $B_k = (-1)^{(n-1)/2}/d_k(\theta)$ and the integral (3.1) is

$$p(x|\theta) = \frac{(-1)^{(n-1)/2}}{\pi^n} \sum_{k=1}^n \frac{1}{d_k(\theta)} \int_0^\infty \frac{1}{1+\lambda^2(x_k-\theta)^2}\,d\lambda \tag{3.3}$$

For n even, (3.2) is wholly imaginary so $B_k = 0$, $k = 1,\ldots,n$, $A_k = (-1)^{(n/2)-1}(x_k-\theta)/d_k(\theta)$ and the integral (3.1) is

$$p(x|\theta) = \frac{(-1)^{n/2-1}}{\pi^n} \sum_{k=1}^n \frac{(x_k-\theta)}{d_k(\theta)} \int_0^\infty \frac{\lambda}{1+\lambda^2(x_k-\theta)^2}\,d\lambda$$

$$= \frac{(-1)^{n/2-1}}{\pi^n} \sum_{k=1}^n \frac{1}{2d_k(\theta)(x_k-\theta)} \log\{\lambda^2+(x_k-\theta)^{-2}\}\Big|_0^\infty \tag{3.4}$$

Now, it is straightforward to show (3.1) exists for $n \geq 1$, and hence the upper limit in (3.4) exists, which must mean that $\sum\{d_k(\theta)(x_k-\theta)\}^{-1} = 0$. Evaluating (3.3) and (3.4) and dividing by $p(x)$ we finally obtain for $n \geq 2$

$$p(\theta|x) = \begin{cases} 2^{n-2} \sum_{k=1}^n d_k^{-1}(\theta)|x_k-\theta|^{-1} \Big/ \sum_{j\neq k} c_{jk}\log|m_{jk}| & (n \text{ odd}) \\[2ex] \dfrac{2^n}{\pi^2} \sum_{k=1}^n d_k^{-1}(\theta)(x_k-\theta)^{-1}\log|x_k-\theta| \Big/ \sum_{j\neq k} c_{jk}\,\mathrm{sign}(m_{jk}) & (n \text{ even}) \end{cases}$$

3.2. *Posterior mean of* θ

To find the posterior mean $E(\theta|x) = \int\theta p(x|\theta)d\theta/p(x)$ it is most convenient to write the numerator as

$$\int\int \theta p(x|\theta,\lambda)\lambda^{-1}d\theta d\lambda = \int I_{1,1}(\lambda)d\lambda$$

using the notation of Section 2.1. By using the residue theorem, expanding by partial fractions and considering real and imaginary parts, we eventually find for $n \geq 3$

$$E(\theta \mid x) = \begin{cases} \sum_{j \neq k} c_{jk} \log|m_{jk}| a_{jk} \Big/ \sum_{j \neq k} c_{jk} \log|m_{jk}| & (n \text{ odd}) \\ \sum_{j \neq k} c_{jk} \,\text{sign}(m_{jk}) a_{jk} \Big/ \sum_{j \neq k} c_{jk} \,\text{sign}(m_{jk}) & (n \text{ even}) \end{cases}$$

where

$$a_{jk} = (x_j + x_k)/2$$

The posterior mean is therefore seen to be a weighted average of all pairwise averages. In deriving $E(\theta \mid x)$ when n is odd we have used the fact that $\int I_{1,1}(\lambda) d\lambda$ exists for $n \geq 3$, and proved in passing that $\sum_{j \neq k} c_{jk} a_{jk} = 0$.

3.3. *Posterior variance of θ*

Writing $E(\theta^2 \mid x)$ as $\int I_{2,1}(\lambda) d\lambda / p(x)$ and following similar steps to the previous examples provides, for $n \geq 4$, the expression

$$E(\theta^2 \mid x) = \begin{cases} \sum_{j \neq k} c_{jk} \log|m_{jk}| a_{jk}^2 \Big/ \sum_{j \neq k} c_{jk} \log|m_{jk}| & (n \text{ odd}) \\ \sum_{j \neq k} c_{jk} \,\text{sign}(m_{jk}) a_{jk}^2 \Big/ \sum_{j \neq k} c_{jk} \,\text{sign}(m_{jk}) & (n \text{ even}) \end{cases}$$

which again has a pleasing weighted average form. Once again, for n odd, we have used the existence of the integral for $n \geq 4$ which has incidentally proved that $\sum_{j \neq k} c_{jk} a_{jk}^2 = 0$.

The posterior variance is given by $V(\theta \mid x) = E(\theta^2 \mid x) - E^2(\theta \mid x)$.

4. A NUMERICAL EXAMPLE

The ever-popular Darwin's data consists of 15 observations (-67, -48, 6, 8, 14, 16, 23, 24, 28, 29, 41, 49, 56, 60, 75) for which, assuming the first two observations are not 'rogues' (Abraham and Box, 1978), a long-tailed distribution might be appropriate. Figure 1 displays the posterior distributions for θ and σ assuming the Cauchy model, together with the posterior means and variances. It is advisable to standardise the data to have mean zero and standard deviation one before carrying out the calculations and care must be taken over rounding errors.

We note that $p(\theta \mid x)$ is insensitive to the outlying observations, and an approximate 95 % highest posterior density interval is given by (12.2, 40.2); a normal approximation to $p(\theta \mid x)$ would provide a very similar interval $E(\theta \mid x) \pm 1.96 V^{1/2}(\theta \mid x) = (12.2, 39.3)$. The unconditional maximum likelihood estimates are $\hat{\theta} = 0.25$ and $\hat{\sigma} = 15.7$, with estimated variances 34.6 and 33.0, respectively, based on the inverse of the second derivatives of the log-likelihood. Hence a simplistic approach using $\hat{\theta} \pm 1.96 \hat{V}^{1/2}(\hat{\theta})$ as an interval for θ would yield (13.5, 36.5), which can be seen to be rather optimistic. Finally, if we were to assume a Gaussian sampling distribution, standard theory leads to a 95 % interval of (0.0, 41.8) for θ, reflecting the influence of the two low observations.

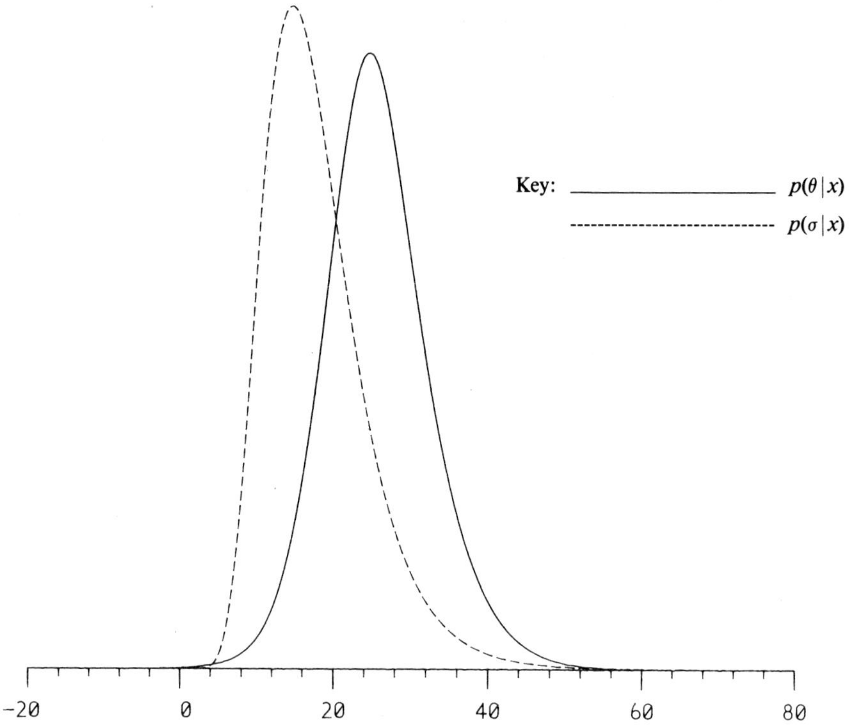

FIGURE 1

Posterior distributions arising from Darwin's data
$\{E(\theta|x) = 25.7,\ V(\theta|x) = 47.9,\ E(\sigma|x) = 17.9,\ V(\sigma|x) = 45.8\}$

5. DISCUSSION

In spite of its rather complex formula the posterior distribution of θ derived in this paper appears to approach a normal shape fairly swiftly. However, for small sample sizes it is possible to obtain a multimodal posterior distribution for θ; for example, when $n = 2$ we find

$$p(\theta|x_1,x_2) = \begin{cases} \dfrac{\operatorname{sign}(x_2-x_1)}{\pi^2(\theta-\bar{x})} \log \dfrac{|x_1-\theta|}{|x_2-\theta|} & (\theta \neq \bar{x}) \\[2ex] \dfrac{4}{\pi^2|x_2-x_1|} & (\theta = \bar{x}) \end{cases}$$

where $\bar{x} = (x_1+x_2)/2$. This function is symmetric with a minimum at $\bar{x}$ but has discontinuities at x_1 and x_2.

It may happen that due to rounding errors two or more x_i's take on the same value. If, say, $x_1 = x_2$ then $c_{jk} = \infty$ if j or k are equal to 1 or 2, making the expressions in this paper inapplicable. Exact, but complicated, expressions may be obtained by letting $m_{12} \to 0$ in the formulae, but, in practice, it is preferable to simply perturb the concordant values by a small quantity in order to distinguish them.

REFERENCES

ABRAHAM, B. & BOX, G.E.P. (1978). Linear models and spurious observations. *Appl. Statist.* **27**, 131-8.

CANE, G. (1974). Linear estimation of the parameters of the Cauchy distribution based on sample quantiles. *J. Amer. Statist. Assoc.,* **65**, 851-9.

COPAS, J.B. (1975). On the unimodality of the likelihood for the Cauchy distribution. *Biometrika* **62**, 701-4.

COX, D.R. & HINKLEY, D.V. (1974). *Theoretical Statistics.* London: Chapman & Hall.

FISHER, R.A. (1934). Two new properties of mathematical likelihood. *Proc. R. Soc. A,* **144**, 285-307.

FRANCK, W.E. (1981). The most powerful invariant test of normal versus Cauchy with applications to stable alternatives. *J. Amer. Statist. Assoc.* **76**, 1002-5.

KOUTROUVELIS, I.A. (1982). Estimation of location and scale in Cauchy distributions using the empirical characteristic function. *Biometrika* **69**, 205-13.

BAYESIAN STATISTICS 2, pp. 751-754
J.M. Bernardo, M.H. DeGroot, D.V. Lindley, A.F.M. Smith (Eds.)

A Bayesian View of Ignorable Designs in Survey Sampling Inference

R.A. SUGDEN
Goldsmiths' College, University of London

SUMMARY

Survey data is often analysed by a different agent than was responsible for the survey design. We investigate the problem of the ignorability of the design in such situations.

Keywords: FACE-VALUE LIKELIHOOD; IGNORABILITY; PARTIAL DESIGN INFORMATION; RESPONSE AND DESIGN VARIABLES; SELECTION MECHANISM.

1. INTRODUCTION

It is a well known result of the Bayesian approach to sampling from finite populations (Ericson, 1969) that the sample design plays no part in the inference. This conclusion can be given formal justification using the approach of Rubin (1976) by treating the unsampled values as 'missing data', provided we assume that the 'design variates' used to construct the design (e.g. stratum/cluster indicators, size measures, etc.) are known for all population units.

However a frequently occurring situation, noted by Scott (1977), is that the survey data are analysed by a research worker or statistician (the *analyst*), whereas the survey itself and the data collection have been conducted by an essentially independent individual or body (the *sampler*). The analyst takes no part in the design state of the survey - indeed the choice of design may well be dictated by administrative costs and constraints rather than considerations of future inference. Consequently, the analyst may well have access to considerably less prior and supplementary information on the population units than was available to the sampler when he chose the design. It is clear now that the design is no longer ignorable by the analyst, although it still is by the sampler. In general, the set of selection probabilities usually reported, and forming the foundation of the classical sampling theoretic approach, will carry design information, the knowledge of which may still allow the analyst to ignore the selection mechanism inherent in his model of the design process.

In this paper we set up a formal framework for these problems and find a condition, satisfied by most commonly used probability sampling designs, for ignorability under given partial design information. The condition corresponds to that of the selection process being 'report-based', studied in more general statistical context by Dawid and Dickey (1977).

2. THE ROLE OF THE DESIGN OR SELECTION MECHANISM

The finite population P is assumed to be identifiable with fixed labels $U = \{1,2,\ldots,N\}$ and represented by a $N \times v$ matrix $\mathbf{x} = (x_1,x_2,\ldots,x_N)'$, where x_i is a column vector of measurements on v variates for the ith unit.

A sample $s \subset U$ is selected according to the discrete probability distribution

$$[s,p(s)] \qquad s \in S \tag{2.1}$$

where S is the set of all feasible samples.

For given s, define the partition $\mathbf{x} \longleftrightarrow (\mathbf{x}_s,\mathbf{x}_{\bar{s}})$ into measurements $\mathbf{x}_s$ on sampled units, which are known, and those on unobserved units $\bar{s}$, generally unknown. Inference is normally required on $\mathbf{x}_{\bar{s}}$ or descriptive statistics such as the finite population mean. The data set is

$$e_s = (s,\mathbf{x}_s) \tag{2.2}$$

so that label information is included. Otherwise see Scott and Smith (1973).

Under the classical or 'fixed population' model, stating simply that the population is a collection of N unknown vectors in R^v, Ericson (1969) and Royall (1968) give the likelihood as

$$p(s,\mathbf{x}_s|\mathbf{X}) = p(s|\mathbf{X}) \prod_{i \in s} \delta(x_i,X_i)$$

where $\delta(a,b) = 1\ (a=b), = 0(a \neq b)$ (2.3)

On the assumption that $S \perp \mathbf{X}$ (in the notation of Dawid, 1979), where S is the sample indicator random variable and $\mathbf{X}$ is the population random matrix with prior distribution $f(\mathbf{x})$, $p(s|\mathbf{X})$ can be replaced by $p(s)$ from (2.1) and posterior inferences are given by

$$f(\mathbf{x}_{\bar{s}}|s,\mathbf{x}_s) \alpha f_{\bar{s}|s}(\mathbf{x}_{\bar{s}}|\mathbf{x}_s)\,, \tag{2.4}$$

which is simply the conditional prior density of unsampled units given the observed values in s, corresponding to a simple 'face-value' analysis.

A strictly correct formulation of the likelihood (2.3) acknowledges the role of selection and the possible informativeness of the sample design modelled in the selection mechanism $p(s|\mathbf{X})$.

The data include observation of the event A_s that sample s has been selected (rather than any other) so that $p(s|\mathbf{X})$ should strictly be replaced by the mathematically equivalent

$$pr(A_s|\mathbf{X};\, \mathbf{X}_s = \mathbf{x}_s) \tag{2.5}$$

for each fixed s. This is a correction factor applied to the face-value likelihood (Dawid and Dickey, 1977).

This is unimportant when the analyst himself designs and implements the survey, but otherwise $S \perp \mathbf{X}$ is unlikely to hold so that a face-value analysis is not valid. Suppose however that (2.5) depends only on $\mathbf{X}$ through the observed $\mathbf{x}_s$, so that

$$A_s \perp \mathbf{X} \mid \mathbf{x}_s \tag{2.6}$$

Once again, inference using (2.4) is valid: (2.6) corresponds to the property of selection mechanisms called *report-based* by Dawid and Dickey (1977). Is it likely to be satisfied for practical surveys which usually involve some randomisation?

Scott (1977) divides the v variates into two classes

response variates Y — whose population characteristics are of interest to the sampler.

design variates Z known and used only by the sampler to select the design. Typically these will include label information such as stratum or cluster indicators as well as concomitant variates such as size measures.

The basic design assumption by the analyst is thus

$$S \perp \mathbf{Y} | \mathbf{Z}, \tag{2.7}$$

so that an analyst who knows $\mathbf{Z} = \mathbf{z}$ can analyse the data at face-value but under partial design information (Sugden and Smith, 1984). the design may or may not be ignorable. Suppose for example, the design employed is stratified random sampling with known stratum sizes and fixed allocation. Then provided the strata to which sampled units belong is known we have ignorability because -

$$S \perp \mathbf{Z} | \mathbf{W}_s \tag{2.8}$$

where $\mathbf{W} = \mathbf{W}(Z)$ is the stratum indicator. The reader is referred to Sugden and Smith (1984) for further examples and a formal framework which deals also with other possible targets of inference, for example the parameters of superpopulations assumed by the analyst to represent his prior knowledge of the finite population.

3. CONCLUSION

The recognition that survey data is often selective and that the selection itself may carry information through the sample design should caution Bayesians who approach such data collected and compiled by others!

Fortunately reasonable conditions can be found which still allow a face-value analysis. In particular, use of randomisation on the design may allow the analyst with some design information to ignore the selection mechanism. This conclusion should also support Bayesian samplers who use randomisation in the design of surveys where the data is intended for use by other analysts.

Other designs may carry information in quite a complex way: in general survey sampling inference *does* depend on the design whenever only partial design information is available.

REFERENCES

DAWID, A.P. (1979). Conditional Independence in statistical theory. *J. Roy. Statist. Soc.* B, **41**, 1-31 (with discussion).

DAWID, A.P. and DICKEY, J.M. (1977). Likelihood and Bayesian Inference from selectively reported data. *J. Amer. Statist. Assoc.,* **72**, 845-850.

ERICSON, W.A. (1969). Subjective Bayesian models in sampling finite populations. *J. Roy. Statist. Soc.* B, **31**, 195-224 (with discussion).

ROYALL, R.M. (1968). An old approach to finite population sampling theory. *J. Amer. Statist. Assoc.* **63**, 1269-1279.

RUBIN, D.B. (1976). Inference and missing data. *Biometrika,* **53**, 581-592.

SCOTT, A.J. (1977). Some comments on the problem of randomisation in surveys. *Sankhyā* C.**39**,1-9.

SCOTT, A.J. and SMITH, T.M.F. (1973). Survey design, symmetry and posterior distributions. *J. Roy. Statist. Soc.* B, **35**, 57-60.

SUGDEN, R.A. and SMITH, T.M.F. (1984). Ignorable and Informative designs in survey sampling inference. To appear in *Biometrika*.

BAYESIAN STATISTICS 2, pp. 755-762
J.M. Bernardo, M.H. DeGroot, D.V. Lindley, A.F.M. Smith (Eds.)

Consistent Prior Distributions for Transformed Models

TREVOR J. SWEETING
(University of Surrey, Guildford, U.K.)

SUMMARY

As recognised by Box and Cox (1964), any transformation of one's data generally changes the level and range of the observations. For consistency, the joint prior distribution of (λ,θ), the transformation and standard model parameters respectively, should reflect this fact. A general approach to the problem of constructing suitable prior distributions in the presence of a transformation parameter is put forward, with particular regard to the case of vague prior knowledge about the location and range of the data. Approximations, based on a given method of analysis for the standard model, to the posterior distributions of λ and $\theta|\lambda$ are derived.

Keywords: BAYESIAN ANALYSIS OF TRANSFORMATIONS; LOCAL NONIDENTIFIABILITY; OUTCOME-DEPENDENT PRIOR.

1. INTRODUCTION

A widely used parametric approach to statistical analysis is to attempt to find a transformation of one's data which will permit the use of some 'standard' developed model. The most common instance is an attempted transformation to a normal linear homoscedastic model, via a (possibly shifted) logarithmic or power transformation. Some other instances are transformation to a Weibull model (by a simple shift of the data, for example), power transformation to a Gamma-based model, or possibly transformation to a nonlinear normal model.

In this paper we assume that some given standard parametric model obtains under an unknown transformation g_λ of the data, where $g_\lambda \epsilon G$, a given family of transformations indexed by the parameter $\lambda \epsilon \Lambda$. Denote the standard model parameters by θ. The assignment of a joint prior distribution for (λ,θ) in a Bayesian approach requires some care, since in a neighbourhood of any value λ' of λ there may well be some lack of identifiability. This normally occurs when the standard model includes unknown location and scale parameters, and G is a family of sufficiently smooth transformations. In this case the transformations g_λ in a small neighbourhood of λ' will be almost linearly related to one another. This lack of local identifiability gives rise to a consistency requirement for the conditional prior distribution of $\theta|\lambda$ at neighbouring values of λ, a fact recognized by Box and Cox (1964), who argued that the changes in the level and range of the data induced by the transformation should be taken into account when forming a prior for (λ,θ). The improper prior distribution proposed by Box and Cox, although operationally very reasonable, depends on the data values. One way of overcoming this data-dependence was suggested by Pericchi (1981), who showed that if an alternative form of improper conditional prior for $\theta|\lambda$ is adopted in the linear model, then the data-dependence does not occur. It is argued in Sweeting (1984) however that this form of prior has a number of drawbacks, since the conditional prior of $\theta|\lambda$ arises as a limiting case of the family of conjugate distributions for θ in the linear model. An alternative family of non-outcome-dependent proper prior distributions for $\theta|\lambda$ was put forward in Sweeting (1984), and the posterior consequences assessed. The corresponding limiting vague prior distribution of (λ,θ) is in fact close in spirit to that proposed by Box and Cox, while removing the data dependence.

In Section 2 we set up the general problem and derive a family of conditional prior distributions for $\theta|\lambda$ which satisfy the consistency requirements when prior knowledge about the level and range of observations is vague. It is shown further how invariance properties of the full (λ,θ) model can help in choosing a particular prior for (λ,θ). Various examples are discussed in Section 3, and in Section 4 approximations, based on a given method of analysing the standard model, to the posterior distributions of λ and $\theta|\lambda$ are derived.

2. CONSTRUCTING THE PRIOR DISTRIBUTION

2.1. *Transformed translation - scale models*

We suppose that observations y_i, $i = 1,\dots,n$ are generated by distributions dominated by some common measure on $\mathbb{R}$, having densities $f_i(\sigma^{-1}(y-\alpha);\beta)$ where $\alpha\in\mathbb{R}$, $\sigma > 0$ are scalar parameters, $\beta\in\mathbb{R}^k$ a (possibly null) vector of parameters, and the f_i are known functions. Write $\theta = (\alpha,\sigma,\beta)$ and let Ω denote the parameter space. We refer to this model as the *standard model*, S, which we assume to be identifiable. Clearly, S is invariant under the group H of translation-scale transformations of the observations; denoting by $\bar{H}$ the group of induced transformations on Ω, $\bar{h} \in \bar{H}$ corresponding to $h(y) = a+by$, $b > 0$, is given by

$$\bar{h}(\alpha,\sigma,\beta) = (a+b\alpha,b\sigma,\beta).$$

Next we introduce a class $G = \{g_\lambda:\lambda\in\Lambda\}$ of strictly increasing transformations on $\mathbb{R}$. Here G need not be a group; indeed, neither the identity nor g_λ^{-1} need be in G. The *full model*, S^*, assumes that the standard model applies to the transformed data $g_\lambda(y_i)$, $i = 1,\dots,n$, for some $\lambda\in\Lambda$. We can therefore write this as

$$g_\lambda(y_i) = \alpha + \sigma\varepsilon_i,\ i = 1,\dots,n \tag{1}$$

where the ε_i are independent random variables with respective densities $f_i(.;\beta)$. We assume that S^* is identifiable, which is readily seen to be equivalent to requiring that (i) S is identifiable, and (ii) S is not invariant under the transformation $g_\lambda g_{\lambda_1}^{-1}$ of the data for any $\lambda,\lambda_1\in\Lambda$, $\lambda \neq \lambda_1$. Although the standard model is not invariant under $g_\lambda g_{\lambda_1}^{-1}$, if $g_\lambda(y)$ is sufficiently smooth in both y and λ then it will be possible to approximate $g_\lambda g_{\lambda_1}^{-1}$ by a linear function of y locally in $\Lambda\times\mathbb{R}$, leading to *local* invariance of S under $g_\lambda g_{\lambda_1}^{-1}$ or, equivalently, local nonidentifiability of S^*.

2.2. *A Family of Proper Prior Distributions*

We assume that in model (1) the parameter α may be interpreted as the 'average level' of the transformed data. There will be some arbitrariness in the choice of parametrization to achieve this, but the final result should hardly be affected by alternative sensible choices here. Often there will be a natural choice; setting $\alpha = E(g_\lambda(y))$ would be appropriate in a normal linear model for example. Some further guidance is given in Section 4, where we consider posterior approximations.

We now assume that $\Lambda = \mathbb{R}^s$, and that if $g_\lambda \epsilon\, G$ then $g_\lambda'(y)$ exists and is continuous in $\Lambda \times \mathbb{R}$. Suppose that the data are generated by the model with parameters (λ_1,θ_1) and let λ be some neighbouring value in Λ. Write $g_{\lambda_1} = g$ for brevity and let $\gamma_\lambda = g_\lambda g^{-1}$. Applying a first-order Taylor expansion to γ_λ about the central value α_1 gives

$$g_\lambda(y) = \gamma_\lambda g(y) = \gamma_\lambda(\alpha_1) + \gamma_\lambda'(u)(g(y)-\alpha_1) \tag{2}$$

where $u = pg(y) + (1-p)\alpha_1$, $0 < p < 1$. The continuity conditions imposed on g_λ' ensure the existence of a neighbourhood $R(\lambda_1)$ of λ_1 in which $\gamma_\lambda'(u)/\gamma_\lambda'(\alpha_1) \cong 1$ to any specified degree of accuracy, and for $\lambda\epsilon R(\lambda_1)$ γ_λ behaves like the transformation $h_\lambda(z) = \gamma_\lambda(\alpha_1) + \gamma_\lambda'(\alpha_1)(z-\alpha_1)$ which, since $\gamma_\lambda'(\alpha) > 0$, is in H and induces the transformation $\bar{h}_\lambda(\alpha_1,\sigma_1,\beta_1) = (\gamma_\lambda(\alpha_1), \gamma_\lambda'(\alpha_1)\sigma_1,\beta_1)$ on Ω. It follows that the transformed parameter $\tau = (g_\lambda^{-1}(\alpha), \{g_\lambda' g_\lambda^{-1}(\alpha)\}^{-1}\sigma,\beta)$ is invariant under $\bar{h}_\lambda$, and a natural local consistency requirement is that the conditional prior distribution of $\tau|\lambda$ should be independent of λ in $R(\lambda_1)$. This conclusion does not depend on the choice of λ_1, and suggests that in the absence of clear prior knowledge we should take λ and τ a priori independent. Note that as the data information increases, the main range R^* of λ dictated by the likelihood function will eventually be contained in $R(\lambda)$ for every $\lambda\epsilon R^*$, and any prior misspecification outside R^* will have virtually no influence on the results.

Let ℓ be some reference value of λ for which we can specify a conditional prior density of θ given $\lambda = \ell$, which we denote by $q(\theta)$. The conditional density of $\tau|\lambda = \ell$ is then $q(g_\ell(\alpha), g_\ell'(\alpha)\sigma,\beta)\{g_\ell'(\alpha)\}^2$ which, if a priori independence of λ and τ is assumed, easily yields the required form of joint prior for (λ,θ). Let $p_o(\lambda)$ be the prior density of λ and write

$$F_\ell(\lambda,\alpha) = \{g_\lambda' g_\lambda^{-1}(\alpha)\}^{-1} g_\ell' g_\lambda^{-1}(\alpha).$$

We find that

$$p(\lambda,\theta) \propto p_o(\lambda) q(g_\ell g_\lambda^{-1}(\alpha), F_\ell(\lambda,\alpha)\sigma,\beta) F_\ell^2(\lambda,\alpha). \tag{3}$$

2.3. *Vague prior knowledge*

We suppose that prior knowledge about (α,σ) on the reference scale $\lambda = \ell$ is vague. One natural method of determining an appropriate form of reference prior for θ is to appeal to invariance properties of the model S. That is, if the model is invariant under the monotone transformation h on $\mathbb{R}$ and $\bar{h}$ is the induced transformation on Ω, we require that the posterior distribution of $\bar{h}(\theta)$ based on the transformed data $h(y)$ be consistent with that of θ based on y. This will be the case iff the prior density q satisfies $q(\bar{h}(\theta))\,|\bar{h}'(\theta)| \propto q(\theta)$. Here S is invariant under the group H of translation-scale transformations, and we require $q(a + b\alpha, b\sigma,\beta) \propto q(\alpha,\sigma,\beta)$ for all a and $b > 0$. If we assume that q is differentiable and nonzero in Ω we are led to solutions of the form

$$q(\alpha,\sigma,\beta) \propto q_1(\beta)\sigma^{r-1} \tag{4}$$

and from (3) we have

$$p(\lambda,\theta) \propto p_o(\lambda) q_1(\beta) \sigma^{r-1} F_\ell^{r+1}(\lambda,\alpha). \tag{5}$$

If r is taken as -1, $p(\lambda,\theta)$ takes on the simple form $p_o(\lambda)q_1(\beta)\sigma^{-2}$. If further $q_1(\beta) = 1$ this effectively coincides with the treatment by Pericchi (1981) of the normal linear homoscedastic power-transformed model. In Sweeting (1984) however it was argued that this value of r is inappropriate in the normal case, as it gives rise to a marginalization paradox regarding σ, arising from the fact that the posterior distribution is not the probability limit of posterior distributions based on a sequence of proper priors. Thus some care needs to be taken in choosing an appropriate value for r. The following discussion refers only to the above normal model. If $q_1(\beta)$ is a proper distribution, then $r = 0$ is the unique value giving a nonparadoxical posterior distribution. (This value arises by proceeding to the limit when α and σ^{-2} have independent normal and Gamma distributions respectively). More difficulties arise if one lets $q_1(\beta) \to 1$, which corresponds to proceeding to the limit in the conjugate family for the standard model. There are two valid ways of dealing with vague prior knowledge about β; either use a proper but relatively diffuse prior $q_1(\beta)$ and take $r = 0$ in (5), or return to (3) and proceed to the limit assuming (μ,β^*) is a priori normally distributed independently of σ^{-2}, which is gamma distributed, where $\beta^* = \sigma\beta$. This last alternative does not lead to any difficulties when one goes to the limit in the posterior distribution, and the limiting prior of (α,σ,β^*) takes the standard form σ^{-1}. This formally corresponds to taking $q_1(\beta) = 1$ and $r = k$ in (5). Furthermore, when $q_1(\beta) = 1$ the choice $r = k$ is the only one which gives a nonparadoxical posterior distribution for $\theta|\lambda = \ell$. In general there may be a choice of r giving a nonparadoxical posterior distribution, but no general procedure for determining such a value is known to the author and we leave r unspecified in (5).

Notice that the scale $\lambda = \ell$ is the only scale on which the prior for S is translation invariant, and this fact may dictate a natural choice for ℓ. In the absence however of any clear prior grounds on which to base this choise, it may be possible to determine ℓ uniquely in order that the posterior distribution of (λ,θ) be invariant under a group $\bar{K}$ of tranformations induced by a group K of transformations on $\mathbb{R}$ which leave the *full* model S^* invariant. Such considerations may also help determine $p_o(\lambda)$. Particular cases will be discussed in the next section.

3. EXAMPLES

3.1. *Box-Cox power transformation*

Here $g_\lambda(y) = \lambda^{-1}(y^\lambda - 1)$, $\lambda \neq 0$, $g_o(y) = \log y$ (Box and Cox (1964)). It is readily seen that

$$F_\ell(\lambda,\alpha) = \begin{cases} (1+\lambda\alpha)^{\ell/\lambda-1} & \lambda \neq 0 \\ e^{\ell\alpha} & \lambda = 0. \end{cases}$$

This differs only notationally from the result in Sweeting (1984) where the case of a normal linear homoscedastic model is discussed.

Vague prior knowledge. We assume that $q(\theta)$ is of the form (4) for an appropriate value of r, as discussed in Section 2.3. (In the normal case, $r = 0$ when $q_1(\beta)$ is proper, $r = k$ when $q_1(\beta) = 1$.) This will also be assumed in succeeding examples. Here the full model is invariant under the group of scale transformations, but it turns out that all priors of the form (5) satisfy the corresponding invariance requirement. However, the problem is also invariant under the group H of power transformations, $h(y) = y^c$, $c > 0$. The induced transformation on Ω is $\bar{h}(\lambda,\theta) = (\lambda/c, c\alpha, c\sigma, \beta)$ and we require

$$p_o(\lambda/c)(1+\lambda\alpha)^{(\ell c/\lambda-1)(r+1)} \propto p_o(\lambda)(1+\lambda\alpha)^{(\ell/\lambda-1)(r+1)}$$

for every $c > 0$. If we demand that p_o be continuous and nonzero at the origin, the only solution is $p_o(\lambda) \propto 1$ and $\ell = 0$, giving the noninformative reference prior

$$p(\lambda,\theta) \propto q_1(\beta)\sigma^{r-1}(1+\lambda\alpha)^{-(r+1)} \,. \tag{8}$$

We have proceeded under the assumption that the g_λ are transformations of $\mathbb{R}$ onto itself, whereas g_λ is from $\mathbb{R}^+$ to $(-\lambda^{-1},\infty)$, the problem arising of course because the standard model is untenable unless $\lambda = 0$. The simplest approach is to ignore the positivity of y initially, and finally check that over the main range of parameter values given by the posterior distribution, the portion of the y_i distributions on $(-\infty,0)$ are negligible. Further discussion in the case of a normal linear homoscedastic model is given in Sweeting (1984).

3.2. *Shifted power transformation*

An extra degree of flexibility is provided by taking $g_\lambda(y) = \lambda_1^{-1}\{(y-\lambda_2)^{\lambda_1}-1\}$, $\lambda_1 \neq 0$, $g_\lambda(y) = \log(y-\lambda_2)$, $\lambda_1 = 0$, where $\lambda = (\lambda_1,\lambda_2)$. Let $\ell = (\ell_1,\ell_2)$; it is readily seen that

$$F_\ell(\lambda,\alpha) = \{(1+\lambda_1\alpha)^{1/\lambda_1} + (\lambda_2-\ell_2)\}^{\ell_1-1}/(1+\lambda_1\alpha)^{1-1/\lambda_1} \,.$$

The inclusion of λ_2 alters the nature of the problem, since now S^* is translation-scale invariant but no longer power invariant. Appealing to the translation-scale invariance gives $p_o(\lambda_1,\lambda_2) \propto p_1(\lambda_1)$ and $\ell_1 = 1$, ℓ_2 arbitrary (details omitted) finally giving $p(\lambda,\theta)$ of the form

$$p(\lambda,\theta) \propto p_1(\lambda_1)q_1(\beta)\sigma^{r-1}(1+\lambda_1\alpha)^{(\lambda_1^{-1}-1)(r+1)} \,. \tag{9}$$

An interesting feature here is that one might also consider λ_2- transformation of a (λ_1,θ) standard model. Since this latter model is scale-power invariant, on transformation to $y' = \log y$ we will have a translation-scale invariant standard model subject to the transformation $g_{\lambda_2}(y') = \log(e^{y'}-\lambda_2)$. Omitting the details, with the natural choice of $\lambda_1^{-1}\log(1+\lambda_1\alpha)$ as the new "α" parameter, it turns out that (9) also satisfies the invariance requirements of the full model $(\lambda_2, (\lambda_1,\theta))$.

3.3. *Shifted logarithmic transformation*

Here $g_\lambda(y) = \log(y-\lambda)$; we find $F_\ell(\lambda,\alpha) = \{e^\alpha+(\lambda-\ell)\}^{-1}e^\alpha$. The problem is translation-scale invariant, but no finite choice of ℓ gives posterior translation invariance here. However, in order that $g_\ell' g_\lambda^{-1}(\alpha)$ be finite we need to choose ℓ such that $\ell < \lambda + e^\alpha$ for all conceivable values of (λ,α), which suggests letting $\ell \to -\infty$. On dividing (5) through by ℓ here, we find the limiting prior is proportional to $p_o(\lambda)q_1(\beta)\sigma^{r-1}e^{(r+1)\alpha}$, which satisfies the translation-scale invariance requirements iff $p_o(\lambda) \propto 1$. (A similar argument could have been used in §3.2 to let $\ell_2 \to -\infty$, yielding the same prior (9).)

3.4. *Folded power transformation and an alternative*

When $0 < y_i < 1$ a possible analogue of §3.1 is $g_\lambda(y) = (y^\lambda-(1-y)^\lambda)/\lambda$, $\lambda \neq 0$, $g_o(y) = \log\{y/(1-y)\}$. Details are omitted, but here the choice of ℓ cannot be made on invariance grounds; in particular, the problem is not power invariant. An alternative class of transformations is $g_\lambda(y) = (\{y/(1-y)\}^\lambda-1)/\lambda$, $\lambda \neq 0$, $g_o(y) = \log\{y/(1-y)\}$ i.e. a Box-Cox transformation of $y/(1-y)$. We could then use the prior density (8), although the power invariance considerations leading to (8) do not have much appeal in terms of the original y variable here. Alternatively, consider the simple group H of transformations generated by $h(y) = 1-y$. If the model is linear and the error distribution *symmetric*, we find that the full model is invariant under H, the induced transformation on Ω being $\bar{h}(\lambda,\theta) = (-\lambda,-\alpha,\sigma,-\beta)$. The requirement $p(\bar{h})|\bar{h}'| \propto p(\lambda,\theta)$ gives $\ell = 0$, and $p_o(\lambda)$ and $q_1(\beta)$ symmetric about the origin. In the case of vague prior knowledge about λ, adoption of (8) would therefore be reasonable on these grounds.

3.5. *A generalized production model*

As a final example, we consider a particular generalized production function discussed in Zellner (1971). Here a certain normal linear homoscedastic model is assumed to apply to $g_\lambda(y) = \log y + \lambda y$, $\lambda > 0$, where y is output. We have $F_\ell(\lambda,\alpha) = (1+\lambda\nu)^{-1}(1+\ell\nu)$ with $\nu = g_\lambda^{-1}(\alpha)$, the unique root of $\alpha = \log\nu + \lambda\nu$. The full model is scale invariant, and it turns out that the invariance requirement is satisfied when $\ell = 0$ or as $\ell \to \infty$, and $p_o(\lambda) \propto \lambda^a$. Taking $\ell = 0$ for example (this value corresponds to constant returns-to-scale), $a = -1$, $q_1(\beta) = 1$ and $r = k = 2$, yields the prior density

$$p(\lambda,\theta) \propto \{\lambda\sigma(1+\lambda\nu)^3\}^{-1}.$$

4. POSTERIOR DISTRIBUTIONS

In this section we derive approximations to posterior distributions based on the prior distributions proposed in Section 2. In most cases explicit expressions are not available, even for the standard model.

We shall assume that there is a 'standard' method of analysis available for the standard model S based on a prior distribution of the form (4), appropriate in the case of vague prior knowledge about (α,σ). By a 'standard' method of analysis we mean that we have a method of analyzing those features of the posterior distribution of interest. The precise method does not concern us; it may be based on classical numerical integration or Monte Carlo integration, theoretical approximations, exact formulae or a mixture of these. Let $\mathbf{Y} = (y_1,\ldots,y_n)$ and denote the standard posterior density by $q(\theta\,|\,\mathbf{Y})$; then

$$q(\theta\,|\,\mathbf{Y}) = T(\mathbf{Y})q_1(\beta)\sigma^{r-1}p(\mathbf{Y}\,|\,\theta)$$

where $p(\mathbf{Y}\,|\,\theta) = \prod_{i=1}^{n} f_i(\sigma^{-1}(y_i-\alpha);\beta)$ and the statistic $T(\mathbf{Y})$ is the constant of integration.

The simplest approach to inference is to choose a value for λ in the light of $p(\lambda\,|\,\mathbf{Y})$, and possibly other practical considerations, and regard this value as 'correct' for subsequent inference about θ, as discussed in Box and Cox (1964). We shall show that the conditional posterior distribution of $\phi = (\sigma,\beta)$ given λ will usually coincide approximately with the 'standard' posterior distribution of ϕ given data $g_\lambda(\mathbf{Y}) = (g_\lambda(y_1),\ldots,g_\lambda(y_n))$. In other words, after transformation the standard method of analysis for ϕ may be applied. Thus the process of selecting λ and analyzing $\phi\,|\,\lambda$ may be carried out by simple modifications to existing computer programs for analyzing S. The transformation will affect the posterior distribution of α slightly; in comparative experiments however, α is usually regarded as a nuisance parameter.

We suppose that the following two approximations are reasonable:

A1: The conditional posterior mean $E(\alpha\,|\,\phi,\mathbf{Y})$ of α in the standard model is approximately independent of ϕ over the main range of $p(\phi\,|\,\mathbf{Y})$.

A2: $F_\ell^{r+1}(\lambda,\alpha)$ is approximately linear in α over the main range of $p(\lambda,\alpha\,|\,\mathbf{Y})$.

Approximation A1 will usually be reasonable, as values of σ and β should hardly affect estimation of the overall level of the data. Indeed, the need for A1 may help in selecting a suitable parametrization for the model. In a normal linear homoscedastic model for which $\mathbf{X}^T\mathbf{e} = \mathbf{0}$, where $\mathbf{X}$ is the matrix of regressor values and $\mathbf{e}^T = (1,1,\ldots,1)$, we have $E(\alpha\,|\,\phi,\mathbf{Y}) = \bar{y}$ exactly. If A1 is reasonable we set $E(\alpha\,|\,\phi,\mathbf{Y}) \cong a(\mathbf{Y})$. Approximation A2 will clearly improve as sample size increases.

The joint posterior density of (λ,α,ϕ) is

$$p(\lambda,\alpha,\phi\,|\,\mathbf{Y}) \propto p_o(\lambda)G(\lambda,\mathbf{Y})T(g_\lambda(\mathbf{Y}))F_\ell^{r+1}(\lambda,\alpha)q(\alpha,\phi\,|\,g_\lambda(\mathbf{Y}))$$

where $G(\lambda,\mathbf{Y}) = \prod_{i=1}^{n} g_\lambda'(y_i)$. Write $q(\theta\,|\,\mathbf{Y}) = q_1(\phi\,|\,\mathbf{Y})q_2(\alpha\,|\,\phi,\mathbf{Y})$; then

$$p(\lambda,\phi\,|\,\mathbf{Y}) \propto p_o(\lambda)G(\lambda,\mathbf{Y})T(g_\lambda(\mathbf{Y}))q_1(\phi\,|\,g_\lambda(\mathbf{Y}))\int F_\ell^{r+1}(\lambda,\alpha)q_2(\alpha\,|\,\phi,g_\lambda(\mathbf{Y}))d\alpha. \tag{10}$$

A first-order Taylor expansion of $F_\ell^{r+1}(\lambda,\alpha)$ about $E(\alpha\,|\,\phi,g_\lambda(\mathbf{Y}))$ gives the desired linear approximation; in fact the integral in (10) is $F_\ell^{r+1}(\lambda,E(\alpha\,|\,\phi,g_\lambda(\mathbf{Y}))+0(n^{-1})$. Applying approximation A2 now gives

$$p(\lambda,\phi\,|\,Y) \propto \{p_o(\lambda)G(\lambda,\mathbf{Y})T(g_\lambda(\mathbf{Y}))F_\ell^{r+1}(\lambda,a(g_\lambda(\mathbf{Y}))\}\{q_1(\phi\,|\,g_\lambda(\mathbf{Y}))\}\,. \tag{11}$$

Since $q_1(\phi\,|\,g_\lambda(\mathbf{Y}))$ is a probability density, the first factor in braces is approximately proportional to $p(\lambda\,|\,\mathbf{Y})$, while $p(\phi\,|\,\lambda,\mathbf{Y}) \cong q_1(\phi\,|\,g_\lambda(\mathbf{Y}))$. From these, the approximate posterior mode of λ may be obtained, or else $p(\lambda\,|\,\mathbf{Y})$ plotted out in order to choose a suitable value for λ, and then the conditional analysis for ϕ is identical to the standard analysis based on $g_\lambda(\mathbf{Y})$.

For an approximation to the posterior distribution $p(\alpha\,|\,\lambda,\mathbf{Y})$ write $q(\theta\,|\,\mathbf{Y}) = q_3(\alpha\,|\,\mathbf{Y})q_4(\phi\,|\,\alpha,\mathbf{Y})$ and note from (10) that $p(\phi\,|\,\lambda,\alpha,\mathbf{Y}) = q_4(\phi\,|\,\alpha,g_\lambda(\mathbf{Y}))$. Then

$$p(\alpha\,|\,\lambda,\mathbf{Y}) = \frac{p(\lambda,\alpha,\phi\,|\,\mathbf{Y})}{p(\phi\,|\,\lambda,\alpha,\mathbf{Y})p(\lambda\,|\,\mathbf{Y})} \propto \left\{\frac{F_\ell(\lambda,\alpha)}{F_\ell(\lambda,a(g_\lambda(\mathbf{Y}))}\right\}^{r+1} q_3(\alpha\,|\,g_\lambda(\mathbf{Y}))$$

$$\cong [1+(r+1)\{\alpha-a(g_\lambda(\mathbf{Y}))\}h_\ell(\lambda,a(g_\lambda(\mathbf{Y})))]q_3(\alpha\,|\,g_\lambda(\mathbf{Y})) \tag{12}$$

where $h_\ell(\lambda,\alpha) = \partial \log F_\ell(\lambda,\alpha)/\partial\alpha$. An extra factor is introduced into the standard posterior distribution of α (unless $\lambda = \ell$), the effect of which will of course diminish as sample size increases. Note that (12) is only intended to be a local approximation of $p(\alpha\,|\,\lambda,\mathbf{Y})$ about $\alpha = a(g_\lambda(\mathbf{Y}))$, and also that the integral of (12) is unity. The approximate posterior mean of α can readily be calculated from the posterior mean and variance of α in the standard model. Denoting the standard posterior variance by $b(\mathbf{Y})$, we see that

$$E(\alpha\,|\,\lambda,\mathbf{Y}) \cong a(g_\lambda(\mathbf{Y})) + (r+1)h_\ell(\lambda,a(g_\lambda(\mathbf{Y})))b(g_\lambda(\mathbf{Y})).$$

The approximate posterior variance of α involves the third moment of the standard posterior; in particular, if $q(\alpha\,|\,\mathbf{Y})$ is approximately symmetric, as will often be the case, then

$$\mathrm{Var}(\alpha\,|\,\lambda,\mathbf{Y}) \cong b(g_\lambda(\mathbf{Y})) - \{E(\alpha\,|\,\lambda,\mathbf{Y}) - a(g_\lambda(\mathbf{Y}))\}^2\,.$$

In the special case of the Box-Cox transformed normal linear homoscedastic model equation (11) yields, on taking $q_1(\beta) = 1$ and $r = k$,

$$p(\lambda\,|\,\mathbf{Y}) \propto \frac{g^{n\lambda}}{m(\lambda)^{k+1}R(\lambda)^{\nu/2}}$$

where $g = (\prod_i y_i)^{1/n}$, $m(\lambda) = n^{-1}\sum_i y_i^\lambda$ and $R(\lambda)$ is the residual sum of squares given data $g_\lambda(\mathbf{Y})$. This result was derived in Sweeting (1984), and a numerical illustration given.

Some further work is required to assess the adequacy of the approximations in (11) in different circumstances. The best practical approach here is to compare the approximations with the 'exact' posterior densities, obtained by Monte Carlo integration for example.

REFERENCES

BOX, G.E.P. and COX, D.R. (1964). An analysis of transformations (with discussion). *J. Roy. Statist. Soc. B* **26**, 211-52.

PERICCHI, L.R. (1981). A Bayesian approach to transformations to normality. *Biometrika,* **68**, 35-43.

SWEETING, T.J. (1984). On the choice of prior distribution for the Box-Cox transformed linear model. *Biometrika,* **71** 127-134.

ZELLNER, A. (1971). *An Introduction to Bayesian Inference in Econometrics*. New York: Wiley.

CONFERENCE PROGRAMME

Tuesday, 6th September

9.45 **Opening Ceremony**

The Dean of the School of Medicine
The Provost of the University of Valencia.
The Governor of the State of Valencia

11.00 **Session 1:** *Foundational Principles*

Chairman: M.H. DeGroot

Invited papers:

Berger J.O.: In defense of the likelihood principle: axiomatics and coherency.

Csiszar, I.: An extended maximum entropy principle and a Bayesian justification.

Invited discussant: **Barnard, G.A.**

17.00 **Session 2:** *Coherence and Prior Specification*

Chairman: A.F.M. Smith

Invited papers:

Goldstein, M.: Temporal coherence.

Dickey, J.M. and **Chen C-H.:** Direct subjective-probability modelling methods using elliptical distributions.

Diaconis, P. and **Ylvisaker, D.:** Quantifying prior opinions.

Invited discussants: **Giron, F.J.** and **Lindley, D.V.**

21.00 **Poster Session I:** *Foundations and General Topics*

Chairman: D.V. Lindley

Wednesday, 7th September

9.00 **Session 3:** *Probability and Evidence*

Chairman: D.V. Lindley

Invited papers:

Good, I.J.: Weight of evidence: a brief survey.

Dempster, A.P.: Probability, evidence and judgment.

Invited discussants: **Rubin, H.** and **Seidenfeld, T.**

11.00 **Session 4:** *Applications*

Chairman: A.F.M. Smith

Invited papers:

Aitchison, J.: Practical Bayesian problems in simplex sample spaces.

Chen, W.C. and **Hill, B.M.** *et al.*: Bayesian analysis of survival curves for cancer patients following treatment. (Read by J.O. Berger).

Invited discussants: **Barlow, R.E.** and **Skene, A.M.**

13.30 **Excursion**

Boat trip to Peñíscola. Reception in the old castle.

Thursday, 8th September

9.00 **Session 5:** *Assessment Problems I*

Chairman: A.F.M. Smith

Invited papers:

French, S.: Group consensus probability distributions: a critical survey.

Lindley, D.V.: Reconciliation of discrete probability distributions.

Invited discussants: **McConway, K.J.** and **O'Hagan, A.**

11.00 **Session 6:** *Assessment Problems II*

Chairman: M.H. DeGroot

Invited papers:

Pratt, J.W. and **Schlaifer, R.:** Repetitive assessment of judgmental probability distributions: a case study.

Press, S.J.: Multivariate group assessment of probabilities of nuclear war.

Invited discussants: **Brown, P.J.** and **DuMouchel, W.H.**

17.00 **Session 7:** *Modelling, Robustness and Diagnostics*

Chairman. D.V. Lindley

Invited papers:

Leonard, T.: Bayesian techniques for statistical modelling. (Read by D.V. Lindley; this paper does not appear in the Proceedings.)

Smith A.F.M. and **Pettit, L.I.:** Bayesian approaches to outlying and influential observations.

West, M.: Generalized linear models: scale parameters, outlier accommodation and prior distributions.

Invited discussants: **Freeman, P.R.** and **Dempster, A.P.**

21.00 **Poster Session II:** *Techniques*

Chairman: M.H. DeGroot

Friday, 9th September

9.00 **Session 8:** *Information and Complexity*

Chairman: M.H. DeGroot

Invited papers:

Kadane, J.B. and **Wasilkowski, G.W.:** Average case epsilon-complexity in computer science, a Bayesian view.

Winkler, R.L.: Information loss in noisy and dependent processes.

Invited discussants: **Smith, J.Q.** and **Viertl, R.**

11.00 **Session 9:** *Time Series and Econometrics*

Chairman: D.V. Lindley

Invited papers:

Harrison, P.J. and **Ameen, J.R.M.:** Normal discount Bayesian models

Zellner, A.: Bayesian statistics in econometrics.

Invited discussants: **Makov, U.E.** (contribution read by A.F.M. Smith) and **Mouchart, M.**

17.00 **Session 10:** Numerical Methods and Optimization

Chairman: M.H. DeGroot

Invited papers:

Stewart, L.: Multiparameter Bayesian inference using Monte Carlo integration: some technique for bivariate analysis.

Van Dijk, H.K. and **Kloek, T.:** Experiments with some alternatives for simple importance sampling in Monte Carlo integration.

Mockus, J.: The Bayesian approach to global optimization (Only the abstract of this paper appears in the Proceedings)

Invited discussants: **Naylor, J.** and **Sweeting, T.J.**

21.00 **Poster Session III:** *Further Techniques and Applications*

Chairman: A.F.M. Smith

Saturday, 10th September

9.00 **Session 11:** *Forecasting and Scoring*

Chairman: D.V. Lindley

Invited papers:

DeGroot, M.H. and **Eriksson, E.A.:** Probability forecasting, stochastic dominance and the Lorenz curve.

Rubin, D.B.: The use of propensity scores in applied Bayesian inference.

Invited discussants: **Goel, P.K.** and **Spiegelhalter, D.J.**

11.00 **Session 12:** *Prediction and Classification*

Chairman: A.F.M. Smith

Invited papers:

Bernardo J.M. and **Bermúdez, J.D.:** The choice of variables in probabilistic classification.

Geisser, S.: On predicting observables: a selective update.

Invited discussants: **Dunsmore, I.R.** and **Piccinato, L.**

17.00 **Session 13:** *Informative Priors*

Chairman: D.V. Lindley

Invited paper:

Jaynes, E.T.: Highly informative priors: the effect of multiplicity on inference.

Invited discussants: **Deely, J.J.** and **DuMouchel, W.H.**

17.45 *Open Forum on Future Bayesian Meetings.*

21.00 *Farewell Dinner*

Guest of Honour: The Minister for Local Government.

CONFERENCE PARTICIPANTS

C.J. ADCOCK
C SQUARED COMPANY LTD.
79-83 Great Portland St.
London W1N 5RA
UNITED KINGDOM

J. AITCHISON
Department of Statistics
University of Hong Kong
Pokfulam Road
HONG KONG

J.H. ALBERT
Dept. of Mathematics and Statistics
Bowling Green State University
Bowling Green, OH 43403
U.S.A.

J.R.M. AMEEN
Department of Statistics
University of Warwick
Coventry CV4 7AL
UNITED KINGDOM

C. ARMERO
Dept. Bioestadística
Fac. Medicina, Univ. Valencia
Ave. Blasco Ibáñez, 17
46010 Valencia
SPAIN

J. BACON-SHONE
Department of Statistics
University of Hong Kong
Pokfulam Road
HONG KONG

B. BARIGELLI
Istituto di Matematica
Fac. Ingegneria, Univ. Ancona
Via della Montagnola, 30
60100 Ancona
ITALY

R.E. BARLOW
Operations Research Center
3115 Etcheverry Hall
University of California
Berkeley, CA 94720
U.S.A.

G.A. BARNARD
Mill House, 54 Hurst Green
Brightlingsea
Colchester CO7 0EH
UNITED KINGDOM

A.P. BASU
Department of Statistics
Univ. of Missouri-Columbia
222 Math. Sciences Bldg.
Columbia, MO 65211
U.S.A.

M.J. BAYARRI
Dept. Bioestadística
Fac. Medicina, Univ. Valencia
Ave. Blasco Ibáñez, 17
46010 Valencia
SPAIN

J. BERGER
Department of Statistics
Purdue University
West Lafayette, IN 47907
U.S.A.

M. BERLINER
Department of Statistics
The Ohio State University
128 Cockins Hall, 1958 Neil Avenue
Columbus, OH 43210
U.S.A.

J.D. BERMUDEZ
Dept. Bioestadística
Fac. Medicina, Univ. Valencia
Ave. Blasco Ibáñez, 17
46010 Valencia
SPAIN

J.M. BERNARDO
Dept. de Bioestadística
Fac. Medicina, Univ. Valencia
Ave. Blasco Ibáñez, 17
46010 Valencia
SPAIN

B. BETRO
CNR - IAMI
Via Cicognara 7
I-20129 Milano
ITALY

G. BOENDER
Erasmus University H2-5
Burg, Oudlaan 50
3062 LN Rotterdam
THE NETHERLANDS

P.J. BROWN
Department of Mathematics
Imperial College
Queen's Gate
London SW7 2BZ
UNITED KINGDOM

H.D. BRUNK
Department of Statistics
Oregon State University,
Corvallis, OR 97331
U.S.A.

R.V. CANFIELD
Applied Statistics, UMC 42
Utah State University
Logan, UT 84322
U.S.A.

D. COCCHI
Dept. di Scienze Statistiche
Universita di Bologna
Via Belle Arti, 41
40126 Bologna
ITALY

G. CONSONNI
IMQ Universita «L. Bocconi»
Via Sarfatti, 25
20136 Milano
ITALY

R. COPPI
Dipartamento di Statistica Probabilita e
Statistiche App.
Univ. di Roma
Pzza. A. Moro 5, 00185 Roma
ITALY

I. CSISZAR
Mathematical Institute
Hungarian Academy of Sciences
H-1395 Budapest, Pf. 428
HUNGARY

D. DAM
Fokker, B.V.
P.O. Box 7600
Schiphol - Oost
THE NETHERLANDS

C.A. DE BRAGANÇA-PEREIRA
Rua Fradique Coutinho 441
Apt. 12
05416 Sao Paulo - SP
BRAZIL

J. DE LA HORRA
Dep. Estadística. Fac. Ciencias
Univ. Autónoma de Madrid
28034 Madrid
SPAIN

J.J. DEELY
Department of Mathematics
University of Canterbury
Christchurch, 1
NEW ZEALAND

A. DEMPSTER
Department of Statistics
Harvard University
1 Oxford Street
Cambridge, MA 02138
U.S.A.

M. DI BACCO
Dipartimento di Scienze Statistiche
Universita di Bologna
Via Belle Arti, 41
40126 Bologna
ITALY

P.W. DIACONIS
Department of Statistics
Stanford University
Stanford, CA 94305
U.S.A.

J.M. DICKEY
Dept. of Mathematics and Statistics
State Univ. of N.Y. at Albany
1400 Washington Ave.,
Albany, NY 12222
U.S.A.

W.H. DU MOUCHEL
M.I.T., Statistics Centre
Rm E40-119, M.I.T.
Cambridge, MA 02139
U.S.A.

I.R. DUNSMORE
Dept. of Probability & Statistics
The University, Sheffield S3 7RH
UNITED KINGDOM

M. DURST
Mathematics & Statistics
Lawrence Livermore Laboratory
Division (1-316), P.O. Box 808
Livermore, CA 94550
U.S.A.

M.H. DeGROOT
Department of Statistics
Carnegie-Mellon University
Schenley Park
Pittsburgh, PA 15213
U.S.A.

B. EDSTAM
TEL.AB. LM Ericsson
S-126 25 Stockholm
SWEDEN

G.M. EL-SAYYAD
Department of Statistics
King Abdul Aziz University
P.O. Box 9028, Jeddah
SAUDI ARABIA

E.A. ERIKSSON
Dept. of Mathematics
Royal Institute of Technology
S-100 44 Stockholm 70
SWEDEN

J. FERRANDIZ
Dept. Bioestadistica
Fac. Medicina, Univ. Valencia
Ave. Blasco Ibáñez, 17
46010 Valencia
SPAIN

P.R. FREEMAN
Department of Mathematics
The University
Leicester, LE 17RH
UNITED KINGDOM

S. FRENCH
Department of Decision Theory
University of Manchester
Manchester, M13 9PL,
UNITED KINGDOM

R. FRIES
CIBA-GEIGY AG
CH-4000 Basel
SWITZERLAND

D. GAMERMAN
Department of Statistics
University of Warwick
Coventry CV4 7AL
UNITED KINGDOM

A. GARCIA PEREZ
Dept. de Estadística Matemática
Facultad de Ciencias
Universidad Autónoma de Madrid
28034 Madrid
SPAIN

P. GARCIA-CARRASCO
Departamento de Estadística
Fac. de Ciencias Matemáticas
Universidad Complutense
28003 Madrid
SPAIN

S. GEISSER
School of Statistics
270 Vincent Hall
206 Church Street S.E., Univ. of Minnessota
Minneapolis, MI 55455
U.S.A.

A.E. GELFAND
Department of Statistics
The College of Liberal Arts
and Sciences, Univ. of Connecticut
Storrs, CT 06268
U.S.A.

A. GIOVAGNOLI
Dipartimento di Matematica
Universita di Perugia
Perugia
ITALY

F.J. GIRON
Departamento de Estadística
Facultad de Ciencias
Universidad de Málaga
Málaga
SPAIN

P.K. GOEL
Dept. of Statistics
Ohio State University
1958 Niel Ave.
Columbus, OH 43210
U.S.A.

M. GOLDSTEIN
Department of Mathematics
The University
Cottingham Rd.,
Hull HU6 7RX
UNITED KINGDOM

M.A. GOMEZ VILLEGAS
Dept. Investigación Operativa y E.
Facultad de Matemáticas
Universidad Complutense
28003 Madrid
SPAIN

I.J. GOOD
Department of Statistics
Hutcheson Hall
Virginia Polythecnic Institute
Blacksburg, VA 24061
U.S.A.

K. GORDON
Dept. of Mathematics.
University Park
Nottingham NG7 2RD
UNITED KINGDOM

S.S. GUPTA
Department of Statistics
Purdue University
West Lafayette, IN 47907
U.S.A.

H.H. HANSEN
Skast 80
6780 Skaerbaek
DENMARK

P.J. HARRISON
Department of Statistics
University of Warwick
Coventry CV4 7AL
UNITED KINGDOM

B.M. HILL
Department of Statistics
The University of Michigan
1447 Mason Hall
Ann Arbor, Michigan 48109
U.S.A.

R.V. HOGG
Dept. Statist. and Actuarial Scien.
University of Iowa
Iowa City, IA 52242
U.S.A.

A. HOULE
2400 Chemin Ste-Foy
Quebec
CANADA G1V 1T2

H.A. HOWLADER
Department of Statistics
The University of Winnipeg
Winnipeg
CANADA R3B 2E9

E.T. JAYNES
Department of Physics
Washington University
St. Louis, MO 63130
U.S.A.

W.S. JEWELL
I.E.O.R., 4173 EH
University of California
Berkeley, CA 94567
U.S.A.

D.G. JONES
H.M. Treasury, Room 89/4
Treasury Chambers
Parliament Street
London SW1
UNITED KINGDOM

V. JULIÁ
Dept. Bioestadística
Fac. Medicina, Univ. Valencia
Ave. Blasco Ibáñez, 17
46010 Valencia
SPAIN

J.B. KADANE
Department of Statistics
Carnegie-Mellon University
Schenley Park
Pittsburgh, PA 15213
U.S.A.

G.K. KANJI
Dept. of Math. Statist. & O.R.
Sheffield City Polytechnic
Sheffield S1 1WB
UNITED KINGDOM

G.E. KOKOLAKIS
National Technical Univ. Athens
13-15, Monis Sykkov, St.
Srythros Stavros
Athens 607
GREECE

R.D. LEITCH
Royal Military College of Science
Shrivenham
Swindon, Wiltshire
UNITED KINGDOM

T. LEONARD
Department of Statistics
Univ. of Wisconsin-Madison
1210 West Dayton Street
Madison, WI 53706
U.S.A.

K.G. LIEBRICH
CIBA-GEIGY AG
CH-4002 Basel
SWITZERLAND

D.V. LINDLEY
2 Periton Lane
Minehead, Somerset, TA24 8AQ
UNITED KINGDOM

M. LUBRANO
G.R.E.Q.E.,
Univ. d'Aix-Marseille II III
41, Rue des Dominicaines
13001 Marseille
FRANCE

U.E. MAKOV
Department of Statistics
University of Haifa
Mount Carmel, Haifa 31999
ISRAEL

A. MARAZZI
Institut Universitaire de Medecine
Sociale et Preventive
Rue Cesar-Roux 29,
1005 Lausanne
SWITZERLAND

J.M. MARTEL
Universite Laval
Faculte de Sciences de l'Administration
Cite Universitaire
Quebec
CANADA G1K 7P4

M.L. MARTINEZ-GARCIA
Dept. de Estadistica
Fac. de Ciencias
Univ. de Málaga
Málaga
SPAIN

F.E. MARTINEZ-ROMERO
Dept. Bioestadística
Fac. Medicina, Univ. Valencia
Ave. Blasco Ibáñez, 17
46010 Valencia
SPAIN

J.B. MEEKER
CIBA GEIGY
Pharmaceutical Division
Research Statistics
Summit, NJ 07901
U.S.A.

M. MENDOZA
Dept. de Matemáticas
Fac. de Ciencias
U.N.A.M., México, D.F.
MEXICO

H.S. MIGON
Instituto de Matemática
Univ. Federal do Rio de Janeiro
Rio de Janeiro-RJ
BRAZIL

V. MOCELLIN
Laboratorio di Statistica
Universita di Venezia
Universita Degli Studi
30100 Venezia
ITALY

J. MOCKUS
Dept. Optimal Decision Theory
Acd. of Sciences of Lithuanian
Institute of Mathematics & Cybernetics
54 K. Pozelos St.,
232600 Vilnius
U.S.S.R.

M. MOUCHART
CORE, Univ. Catholique de Louvain
34 Voie du Roman Pays
B-1348 Louvain-La-Neuve
BELGIUM

K.J. McCONWAY
Faculty of Mathematics
The Open University
Walton Hall,
Milton Keynes, MK7 6AA
UNITED KINGDOM

J.C. NAYLOR
Division of Mathematics
Derbyshire College of Higher Education
Kedleston Road, Derby
UNITED KINGDOM

A. O'HAGAN
Department of Statistics
University of Warwick
Coventry CV4 7AL,
UNITED KINGDOM

J. PASTOR
Inserm-U168
CHR La Grave
31052 Toulouse Cedex
FRANCE

L.R. PERICCHI
Departamento de Matemáticas
Univ. Simón Bolivar
Apartado 80659, Caracas 1080A
VENEZUELA

L.I. PETTIT
Department of Mathematics
Goldsmiths' College
New Cross, London SE14 6NW
UNITED KINGDOM

L. PICCINATO
Dipartimento di Statistica
Probabilita e Statistiche App.
Univ. di Roma
Pzza. A. Moro 5, 00185, Roma
ITALY

D.J. POIRIER
Department of Economics
University of Toronto
150 St. George Street
Toronto, Ontario
CANADA M5S 1A1

W. POLASEK
Institut Für Statistik
University of Wien
A-1090 Wien, Rooseveltplatz 6/2
AUSTRIA

I. POLI
Fac. di Scienze Politiche
Univ. di Cagliari
Via S. Ignazio da Laconi 78
09100 Cagliari (Sardinia)
ITALY

J.W. PRATT
Harvard Business School
325 Morgan Hall
Boston, MA 02163
U.S.A.

M. PRECHT
Datenverarbeitungsstelle
Technical University
Munich-Weihenstephan, 805 Freising 12
WEST GERMANY

S.J. PRESS
Department of Statistics
University of California
Riverside, CA 92521
U.S.A.

M.T. RABENA
Dept. Bioestadistica
Fac. Medicina, Univ. Valencia
Ave. Blasco Ibáñez, 17
46010 Valencia
SPAIN

A. ROPARS
Dept. de Securite des Matier. Nucl.
CEN/FAR, Boite Postale n. 6
92260 Fontenay aux Roses
FRANCE

H. RUBIN
Dept. Statistics
Purdue University
West Lafayette, IN 47907
U.S.A.

D.B. RUBIN
Department of Statistics
The University of Chicago
5734 University Avenue,
Chicago, IL 60637
U.S.A.

T. SEIDENFELD
Dept. of Philosophy
Washington University
St. Louis, MO 63130
U.S.A.

M. SENDRA
Dept. Bioestadistica
Fac. Medicina, Univ. Valencia
Ave. Blasco Ibáñez, 17
46010 Valencia
SPAIN

N. SHARIFF
Medical Comp. & Statist. Unit
University of Edinburgh
Medical School, Teviot Place
Edinburgh
UNITED KINGDOM

L. SIMAR
CORE, Univ. Catholique de Louvain
34 Voie du Roman Pays
B-1348 Louvain-La-Neuve
BELGIUM

A.M. SKENE
Department of Mathematics
University Park
Nottingham NG7 2RD
UNITED KINGDOM

A.F.M. SMITH
Department of Mathematics
University Park
Nottingham NG7 2RD
UNITED KINGDOM

P. SMITH
CIBA-GEIGY AG.
CH-4002 Basel
SWITZERLAND

J.Q. SMITH
Dept. of Statistical Science
University College London
Gower Street, London WC1E 6BT
UNITED KINGDOM

D.J. SPIEGELHALTER
Medical Research Council
MRC Biostatistics Unit
MRC Centre, Hills Road,
Cambridge CB2 2QH
UNITED KINGDOM

G. STEIN
CIBA-GEIGY AG
CH-4002 Basel
SWITZERLAND

G. STEVE
CNR Istituto Tecnologie Biomediche
Via Morgagni 30/E
00161 Roma
ITALY

L. STEWART
Lockheed Palo Alto Research Lab.
3251 Hanover Street, Palo Alto, CA 94304
U.S.A.

R.A. SUGDEN
Department of Mathematics
Goldsmiths' College
New Cross, London SE14 GNW
UNITED KINGDOM

Y. SUZUKI
Fac. of Economics
University of Tokyo
7-3-1 Hongo - Bunkyo-ku
Tokyo
JAPAN

T.J. SWEETING
Dept. Mathematics
Univ. Surrey,
Guilford, GU2 5XH
UNITED KINGDOM

G.B. TRANQUILLI
Dept. Statistica, Probab. & Appl.
Universita di Roma
Pzza. A. Moro 5, 00185 Roma
ITALY

M.P. UPSDELL
Ruakurn Research Station
Private Bag
Hamilton
NEW ZEALAND

G. VALENCIA
Dept. de Matemáticas
Fac. de Ciencias
U.N.A.M., México, D.F.
MEXICO

H.K. VAN DIJK
Econometric Research Center
Erasmus University
Postbus 1738, 3000 DR Rotterdam
THE NETHERLANDS

I. VERDINELLI
Dipartimento di Statistica
Probabilita e Statistiche App.
Universita di Roma
Pzza. A. Moro, 5, 00185 Roma
ITALY

R. VIERTL
Institut fur Statistik und
Wahrscheinlichkeitstheorie
Technische Universitat Wien
Wiedner Hauptstr. 8
A-1040 Wien
AUSTRIA

A.M. WALKER
Dept. of Probability & Statistics
The University, Sheffield, S3 7RH
UNITED KINGDOM

P.P. WALLER
British Telecom
Room 424, Cheapside House, 138 Cheapside
London EC2V 6JH
UNITED KINGDOM

M. WEST
Department of Statistics
University of Warwick
Coventry CV4 7AL
UNITED KINGDOM

G.N. WILKINSON
Biometry Section, Waite Agricul. Research Inst.
University of Adelaide, Glen Osmond
South Australia, 5068
AUSTRALIA

R.L. WINKLER
The Fuqua School of Business
Duke University
Durham, NC 27706
U.S.A.

A. ZELLNER
H.G.B. Alexander Research Found.
Graduate School of Business
University of Chicago
Chicago, IL 60637
U.S.A.

AUTHOR INDEX